Birkhäuser

Birkhäuser Advanced Texts Basler Lehrbücher

This series presents, at an advanced level, introductions to some of the fields of current interest in mathematics. Starting with basic concepts, fundamental results and techniques are covered, and important applications and new developments discussed. The textbooks are suitable as an introduction for students and non–specialists, and they can also be used as background material for advanced courses and seminars.

Sergio Vessella

Unique Continuation Properties for Partial Differential Equations

Introduction to the Stability Estimates for Inverse Problems

Sergio Vessella
DiMaI
Università degli Studi di Firenze
Firenze, Italy

ISSN 1019-6242 ISSN 2296-4894 (electronic)
Birkhäuser Advanced Texts Basler Lehrbücher
ISBN 978-3-031-86368-4 ISBN 978-3-031-86366-0 (eBook)
https://doi.org/10.1007/978-3-031-86366-0

This book is published under the imprint Birkhäuser, www.birkhauser-science.com by the registered company Springer Nature Switzerland AG
The registered company address is: Gewerbestrasse 11, 6330 Cham, Switzerland

Dedicated to my family and especially to my dear wife Luisella and my dear son Luigi

Preface

I must confess that the idea of uniting diverse topics from different courses into a single book stems from two sources: my experiences as a teacher and, alas, my more distant memories as a student. During my advanced mathematics courses, I often had to consolidate the various notations, theorems, and definitions that overlapped across different classes. I would feel uneasy when a professor remarked, "... as you surely learned in previous courses ..." only to discover that the topic had been barely mentioned or treated in a much more elementary way. This led me to revisit and expand those subjects, often beyond the level I had previously studied. While this process proved invaluable in the long run, it also made me appreciate the importance of a resource that could act as a bridge—be it a grand one or a small footpath—connecting the various neighborhoods in the vast domain of mathematics, and in this case, the "city" of partial differential equations. In an era brimming with virtual connections, the real links we establish in our minds remain the only ones that grant us genuine understanding and confidence in our knowledge.

This book is intended for advanced undergraduates, graduates, and researchers seeking an accessible yet rigorous introduction to key topics in Mathematical Analysis and Partial Differential Equations (PDEs), such as the theory of distributions, boundary value problems using Sobolev spaces, unique continuation properties, and inverse problems. The book aims to fill a gap in existing literature: while there are numerous introductory and specialized texts on PDEs, few integrate a comprehensive study of the Cauchy problem, unique continuation properties, and quantitative aspects of inverse problems at a level accessible to students. To address this, the book offers a self-contained introduction to Carleman estimates and their applications—indispensable tools often overlooked in foundational texts. The goal is to familiarize readers with stability estimates in inverse problems and quantitative unique continuation estimates. A solid understanding of Lebesgue measure (briefly reviewed here) and the basics of PDEs and functional analysis is assumed.

Finally, I extend my heartfelt gratitude to all friends, colleagues, and collaborators who directly or indirectly contributed to this book. I am particularly thankful to my colleagues at the "Dipartimento di Matematica e Informatica dell'Università degli Studi di Firenze" (Department of Mathematics and Computer Science at the

University of Florence), my son Luigi for his help in preparing the figures, and my dear wife Luisella for her patience and insightful suggestions. Special thanks also go to Andrea Aspri, Lorenzo Baldassari, and Elisa Francini for the fruitful discussions I had with them and their valuable advice. My sincere gratitude is extended to the referees for their thorough, constructive, and insightful criticism. Of course, any errors, omissions, or oversights are solely my responsibility, and I welcome feedback from readers who wish to bring them to my attention.

Firenze, Italy
December 2024

Sergio Vessella

Declarations

Competing Interests The author has no competing interests to declare that are relevant to the content of this manuscript.

Contents

Chapter 1
Introduction

"Hunc igitur terrorem animi tenebrasque necessest /
non radii solis neque lucida tela diei /
discutiant, sed naturae species ratioque./"
This terror of the mind, then, and these shadows must be dispelled not by the rays of the sun nor by the shining darts of daylight, but by the contemplation and the knowledge of nature.

— Lucretius, De Rerum Natura, III, 91–93

Abstract The main aim of this book is to introduce the study of unique continuation properties and stability estimates for inverse problems in partial differential equations (PDEs). Each topic covered is chosen with inverse problems in mind, though we believe all of them should be included in the training of those pursuing PDEs, especially regarding unique continuation properties. While excellent review articles and books exist on the subject, there remains a lack of a truly introductory text that begins with foundational principles, suitable for graduate or advanced undergraduate students who have some familiarity with Lebesgue measure theory, basic functional analysis, and an introductory PDEs course.

The main aim of this book is to introduce the study of unique continuation properties and stability estimates for inverse problems in partial differential equations (PDEs). Each topic covered is chosen with inverse problems in mind, though we believe all of them should be included in the training of those pursuing PDEs, especially regarding unique continuation properties. While excellent review articles and books exist on the subject, there remains a lack of a truly introductory text that begins with foundational principles, suitable for graduate or advanced undergraduate students who have some familiarity with Lebesgue measure theory, basic functional analysis, and an introductory PDEs course.

To support this goal, we cover additional foundational topics in PDEs theory: Part I addresses existence and regularity for second-order elliptic equations with real coefficients, while Part II covers the classical Cauchy problem for equations

S. Vessella, *Unique Continuation Properties for Partial Differential Equations*,
Birkhäuser Advanced Texts Basler Lehrbücher,
https://doi.org/10.1007/978-3-031-86366-0_1

with analytic coefficients. In Part III, we explore unique continuation properties for equations with non-analytic coefficients.

This Introduction provides a brief overview of the book, with detailed descriptions for each chapter in the respective introductions.

In Chap. 2, which marks the beginning of Part I, the main notations used throughout the book are introduced, along with a review of some elementary formulas. Chapters 3 and 4 connect and complete essential topics in analysis. Chapter 3 includes the fundamental definitions and theorems of measure theory (often without proofs) and presents several significant results in real analysis, such as the extension theorem in $C^{0,\alpha}$, the Lebesgue differentiation Theorem, Rademacher's Theorem, and the divergence Theorem for open sets with Lipschitz boundaries. Chapter 3 also examines distance functions and the Hausdorff distance between compact sets, tools that are valuable for addressing inverse problems with unknown boundaries.

Chapter 4 introduces the basics of distribution theory and the Fourier transform, concluding with a brief section on oscillatory integrals. While not strictly necessary, as the book largely remains within classical analysis, a basic understanding of distribution theory simplifies the presentation of the Sobolev spaces in Chap. 5.

In Chap. 5, we extensively study the Sobolev spaces. After providing the definition and proving the basic properties of such spaces, we present the main density theorems, extension theorems, the definition of the trace of a function in $W^{1,p}$, embedding and compactness theorems, and conclude with the study of Sobolev spaces with non-integer exponents (the case $p = 2$) and the characterization of traces through these spaces.

Chapter 6 discusses the Dirichlet-to-Neumann map, existence, and regularity theory in L^2 for second-order elliptic equations, along with the inverse problem of detecting inclusions, particularly size estimates, closely related to the unique continuation estimates discussed in Part III.

The books most influential for Part I include [69] and [107] (for Chap. 3), [58], [78], [116] (for Chap. 4), and [19], [38], [39] (for Chaps. 5 and 6).

Part II provides, in Chap. 7, a concise discussion of the Cauchy problem for first-order PDEs. Chapter 8 presents the basic properties of real analytic functions needed in Chap. 9, where the Cauchy problem is formulated, and we prove the classical Cauchy–Kovalevskaya, Holmgren, and John theorems for linear PDEs with analytic coefficients. Chapter 10 applies the Holmgren and John theorems and L^2 regularity theory to establish a uniqueness theorem for an inverse problem for the Laplace equation with an unknown boundary. In Chap. 11, the concept of a well-posed problem in the sense of Hadamard is introduced, and we use the Lax–Mizohata Theorem to underscore the relationship between uniqueness, solvability, and continuous dependence in Cauchy problems with C^∞ coefficients. Chapter 12 defines conditional stability (or "well-posed problem in the sense of Tikhonov") and introduces conditional stability estimates, such as Hadamard's three-circle inequality, for derivative calculus and analytic continuation. Chapter 12 serves as a "laboratory" for tools essential to the study of conditional stability in ill-posed problems. Part II concludes with Chap. 13, proving the John stability theorem for the

Cauchy problem and examining its implications. Key references for Part II include [27], [38], [41], [66], and [103] (Chaps. 7, 8, and 9), as well as [56, Ch. V], [58, Vol. II], [95] (Chap. 11), [76] and [129] (Chap. 12).

Part III presents a comprehensive study of Carleman estimates and unique continuation properties for PDEs.

Chapter 14 introduces foundational definitions of unique continuation properties, the general characteristics of Carleman estimates, and their standard applications in establishing uniqueness and stability for the Cauchy problem. To aid understanding of more advanced theory, we present examples proving Carleman estimates and initial unique continuation results for equations from Parts I and II, such as:

$$\Delta U + \sum_{j=1}^{n} a_j(x)\partial_{x_j} U + a_0(x)U = 0, \quad \text{in } \Omega,$$

where Ω is an open subset of $\mathbb{R}^n$ and

$$a_j \in L^\infty(\Omega), \quad j = 0, 1, \cdots, n.$$

In Part III, particularly in Chap. 14, certain themes are reiterated to solidify the foundational aspects of Carleman estimates, preparing the reader to approach specific applications effectively.

In Chap. 15, we consider differential operators of order $m \geq 1$ with general form

$$P(x, D) = \sum_{|\alpha| \leq m} a_\alpha(x) D^\alpha.$$

Here, we concentrate on operators whose leading-order coefficients $a_\alpha(x)$ are in C^∞, while coefficients of lower-order terms are in L^∞. This setup allows us to focus initially on the symbolic properties of these operators, particularly their principal symbol, a homogeneous polynomial in the frequency variable ξ,

$$p_m(x, \xi) = \sum_{|\alpha| = m} a_\alpha(x) \xi^\alpha.$$

This approach offers a foundation in symbols without full pseudodifferential machinery, as often found in more adavanced texts like [58, Vol. IV], [79], and [83]. By using C^∞ regularity on principal coefficients, we prepare the reader to later transition into pseudodifferential calculus, highlighting the advantages of symbol-based techniques particularly for Carleman estimates. In this chapter, we provide the definition and main properties of pseudo-convex and strongly pseudo-convex surfaces, illustrating their role in the study of uniqueness for the Cauchy problem. In a final section, we give a concise overview of pseudodifferential operators and the Sharp Gårding Inequality, illustrating their role in proving Carleman estimates.

Chapter 16 provides Carleman estimates for operators whose principal part is C^1 or Lipschitz continuous, focusing on elliptic operators of order m.

In Chap. 17, we use the Rellich identity for second-order operators with real, Lipschitz continuous coefficients.

In Chap. 18, Carleman estimates with singular weights are used to derive the optimal three-sphere inequality, doubling inequality, and strong unique continuation property for

$$Lu = b(x) \cdot \nabla u + c(x)u,$$

where $b \in L^\infty(\mathbb{R}^n; \mathbb{R}^n)$ and $c \in L^\infty(\mathbb{R}^n)$. For these estimates, we present the transformation proof from [8] and [9], using polar coordinates. While the case where coefficients of lower-order terms are not locally bounded is beyond the scope of this book due to the need for advanced techniques, significant contributions in this area include [64] and [68], where these issues are investigated in depth.

Chapter 19 explores the methods of log-convexity and frequency functions for unique continuation. We also discuss A_p weights in stability and size estimates, and the Runge property for the Laplace operator. The primary references for Part III are [56] and [83].

Part I
The Sobolev Spaces and the Boundary Value Problems

Chapter 2
Main Notations and Basic Formulas

Abstract This chapter introduces the primary notations used throughout the book, reviews essential calculus formulas, and establishes convenient conventions to streamline the presentation.

2.1 Notations

We denote by $\mathbb{N}$ the set of positive integer number and by $\mathbb{N}_0 = \mathbb{N} \cup \{0\}$. We call multi-index any n-uple of elements of $\mathbb{N}_0$

$$\alpha = (\alpha_1, \alpha_2, \cdots, \alpha_n), \quad \alpha_j \in \mathbb{N}_0, \quad j = 1, 2, \cdots, n.$$

For any $\alpha \in \mathbb{N}_0^n$ we denote by

$$|\alpha| = |\alpha_1| + |\alpha_2| \cdots |\alpha_n|, \quad \text{and} \quad \alpha! = \alpha_1!\alpha_2! \cdots \alpha_n!,$$

the length (modulus) and the factorial of α, respectively. For any $x \in \mathbb{R}^n$, $x = (x_1, x_2, \cdots, x_n)$, we set

$$x^\alpha = x_1^{\alpha_1} x_2^{\alpha_2} \cdots x_n^{\alpha_n}.$$

Let $\alpha, \beta \in \mathbb{N}_0^n$ we write $\alpha \leq \beta$ provided $\alpha_j \leq \beta_j$ for $j = 1, 2, \cdots, n$ and we write $\alpha < \beta$ provided $\alpha \leq \beta$ and there exists $j_0 \in \{1, 2, \cdots, n\}$ such that $\alpha_{j_0} < \beta_{j_0}$. For any $x = (x_1, \cdots, x_n) \in \mathbb{R}^n$ we denote, unless otherwise stated, by $x' = (x_1, \cdots, x_{n-1}) \in \mathbb{R}^{n-1}$ and we write $x = (x', x_n)$. Similar convention will be used for the multi-indices.

For any $\alpha, \beta \in \mathbb{N}_0$ and $\alpha \leq \beta$, let us denote by (the binomial "β over α")

$$\binom{\beta}{\alpha} = \frac{\beta!}{(\beta - \alpha)!\alpha!}.$$

S. Vessella, *Unique Continuation Properties for Partial Differential Equations*, Birkhäuser Advanced Texts Basler Lehrbücher,
https://doi.org/10.1007/978-3-031-86366-0_2

Let us denote by ∂_k the operator $\frac{\partial}{\partial x_k}$, $k = 1, 2, \cdots, n$ and by

$$\partial = (\partial_1, \partial_2, \cdots, \partial_n) \quad \text{(the gradient operator)}.$$

Hence, we set

$$\partial^\alpha = \partial_1^{\alpha_1} \partial_2^{\alpha_2} \cdots \partial_n^{\alpha_n} = \frac{\partial^{|\alpha|}}{\partial x_1^{\alpha_1} \partial x_2^{\alpha_2} \cdots \partial x_n^{\alpha_n}}.$$

To denote the gradient operator, we also use the notation ∇; however, **we will not write** ∇^α to denote ∂^α. Of course, we will continue to use the notation u_{x_k} (or other standard symbols) to represent the partial derivative of u with respect to x_k, where $k = 1, 2, \cdots, n$. The Hessian matrix of a smooth function u is denoted by

$$\partial^2 u = \left\{ \partial^2_{jk} u \right\}_{j,k=1}^{n}.$$

We point out that some authors (and also in these notes in some context) reserve the notation D_k to denote the operator $\frac{1}{i} \frac{\partial}{\partial x_k}$, where $i = \sqrt{-1}$, consequently

$$D^\alpha = \left(\frac{1}{i} \right)^{|\alpha|} \partial_1^{\alpha_1} \partial_2^{\alpha_2} \cdots \partial_n^{\alpha_n}.$$

The latter notation is useful especially when an extensive use of the Fourier transform (Sect. 4.8) is done

$$\widehat{u}\,(\xi) = \int_{\mathbb{R}^n} u(x) e^{-ix\cdot\xi} dx.$$

Actually, we have (see Theorem 4.26)

$$\widehat{D^\alpha u}\,(\xi) = \xi^\alpha \widehat{u}\,(\xi).$$

while, using the former notation, we have

$$\widehat{\partial^\alpha u}\,(\xi) = (i\xi)^\alpha \widehat{u}\,(\xi).$$

Let $\ell \in \mathbb{R}^n \setminus \{0\}$. For any $j \in \mathbb{N}_0$ we set

$$\frac{\partial^j}{\partial \ell^j} = \sum_{|\alpha|=j} \ell^\alpha \partial^\alpha,$$

(we mean $\frac{\partial^0 u}{\partial \ell^0} = u$). In particular

$$\frac{\partial}{\partial \ell} = \ell \cdot \partial = \ell \cdot \nabla.$$

As a consequence of the notations introduced above, we denote a polynomial p of degree m in the variables $\xi_1, \xi_2 \cdots, \xi_n$

$$p(\xi) = \sum_{|\alpha| \le m} a_\alpha \xi^\alpha, \tag{2.1}$$

where $a_\alpha \in \mathbb{R}$ (or $a_\alpha \in \mathbb{C}$) for any $|\alpha| \le m$. We say that the homogeneous polynomial

$$p_m(\xi) = \sum_{|\alpha|=m} a_\alpha \xi^\alpha,$$

is the **principal part of the polynomial** p provided that there exists $\alpha_0 \in \mathbb{N}_0^n$ such that $|\alpha_0| = m$ and $a_{\alpha_0} \neq 0$.

If A is a matrix, we will denote its transpose by A^t or A^T, with the choice between these notations depending on the typographical context.

2.2 Some Useful Formulas

In this section we recall some basic and useful formulas.

$$(x+y)^\alpha = \sum_{\beta \le \alpha} \binom{\alpha}{\beta} x^\beta y^{\alpha-\beta}, \quad \forall x, y \in \mathbb{R}^n, \quad \forall \alpha \in \mathbb{N}_0^n. \tag{2.2}$$

$$\partial^\beta x^\alpha = \begin{cases} \frac{\alpha!}{(\alpha-\beta)!} x^{\alpha-\beta}, & \text{for } \alpha \ge \beta \\ \\ 0, & \text{otherwise} \end{cases}. \tag{2.3}$$

$$(x_1 + x_2 + \cdots + x_n)^m = \sum_{|\alpha|=m} \frac{m!}{\alpha!} x^\alpha, \quad \forall m \in \mathbb{N}_0. \tag{2.4}$$

$$\alpha! \le |\alpha|! \le n^{|\alpha|} \alpha!, \quad \forall \alpha \in \mathbb{N}_0^n. \tag{2.5}$$

Let us recall the following **Stirling formula** [28, Vol. I, Ch. 6]

$$\lim_{n\to\infty} \frac{n!}{n^n e^{-n}\sqrt{n}} = \sqrt{2\pi} \ . \tag{2.6}$$

Let f be a smooth function and $m \in \mathbb{N}_0$, we have

$$\frac{d^m}{dt^m} f(x+ty) = \left((\sum_{j=1}^{n} y_j \partial_j)^m f \right)(x+ty) = \sum_{|\alpha|=m} \frac{m!}{\alpha!} y^\alpha \left(\partial^\alpha f\right)(x+ty). \tag{2.7}$$

We recall the **Taylor formula**, centered at $x_0 \in \mathbb{R}^n$, of a polynomial P of degree m

$$P(x) = \sum_{|\alpha|\le m} \frac{1}{\alpha!} \partial^\alpha P(x_0)(x - x_0)^\alpha. \tag{2.8}$$

Let f and g be two smooth functions and $\alpha \in \mathbb{N}_0^n$, we have the **Leibniz formula** for the α-th derivative of the product fg

$$\partial^\alpha (fg) = \sum_{\beta\le\alpha} \binom{\alpha}{\beta} \partial^\beta f \partial^{\alpha-\beta} g. \tag{2.9}$$

Now we **check some formula.**
Formula (2.2) easily follows by the Newton binomial formula. Actually, we have

$$\begin{aligned}
(x+y)^\alpha &= (x_1+y_1)^{\alpha_1} \cdots (x_n+y_n)^{\alpha_n} = \\
&= \sum_{\beta_1\le\alpha_1} \binom{\alpha_1}{\beta_1} x_1^{\beta_1} y_1^{\alpha_1-\beta_1} \cdots \sum_{\beta_n\le\alpha_n} \binom{\alpha_n}{\beta_n} x_n^{\beta_n} y^{\alpha_n-\beta_n} = \\
&= \sum_{\beta\le\alpha} \binom{\alpha}{\beta} x^\beta y^{\alpha-\beta}.
\end{aligned}$$

The proof of (2.3) is immediate. Before checking (2.4) let us notice that if α and β are multi-indices such that $\alpha \le \beta$ and $|\alpha| = |\beta|$, then $\alpha = \beta$. Let us denote by

$$S(x) = \sum_{j=1}^{n} x_j$$

and set

$$P(x) = (S(x))^m \ .$$

Since P is a homogeneous polynomial of degree m we get

$$P(x) = \sum_{|\alpha|=m} c_\alpha x^\alpha .$$

Let us show that $c_\alpha = \frac{m!}{\alpha!}$, for every multi-indices α such that $|\alpha| = m$. Let β be a multi-index satisfying $|\beta| = m$. By what we notice above and by (2.3), we get

$$\partial^\beta P(x) = \sum_{|\alpha|=m} c_\alpha \partial^\beta x^\alpha = c_\beta \beta! \,. \tag{2.10}$$

On the other hand,

$$\begin{aligned}
\partial^\beta P(x) &= \partial^\beta S^m = \partial_1^{\beta_1} \cdots \partial_n^{\beta_n} S^m = \\
&= m \cdots (m - \beta_n + 1) \partial_1^{\beta_1} \cdots \partial_{n-1}^{\beta_{n-1}} S^{m-\beta_n} = \\
&= m \cdots (m - \beta_n - \beta_{n-1} + 1) \partial_1^{\beta_1} \cdots \partial_{n-2}^{\beta_{n-2}} S^{m-\beta_n-\beta_{n-1}} = \cdots = m! \,.
\end{aligned} \tag{2.11}$$

By (2.10) and (2.11) we get

$$c_\beta \beta! = m! \quad \text{for every } \beta \in \mathbb{N}_0^n \text{ such that } |\beta| = m,$$

from which we obtain (2.4). Of course, formula (2.4) can be proved more elementarly. For instance, it can be proved by induction starting from the Newton binomial formula.

Concerning the inequality $\alpha! \le |\alpha|!$ in (2.5), recalling that $h!k! \le (h+k)!$ for every $h, k \in \mathbb{N}_0$ we get

$$\alpha! = \alpha_1! \alpha_2! \cdots \alpha_n! \le (\alpha_1 + \alpha_2)! \alpha_3! \cdots \alpha_n! \le \cdots \le (\alpha_1 + \alpha_2 + \cdots + \alpha_n)! = |\alpha|! \,.$$

Regarding the inequality $|\alpha|! \le n^{|\alpha|} \alpha!$ it suffices to use formula (2.4) and we have

$$n^{|\alpha|} = (\underbrace{1 + 1 + \cdots + 1}_{n})^{|\alpha|} = \sum_{|\beta|=|\alpha|} \frac{|\alpha|!}{\beta!} \ge \frac{|\alpha|!}{\alpha!}.$$

The first equality in (2.7) can be obtained by iterating the formula

$$\frac{d}{dt} f(x + ty) = \left(\sum_{j=1}^n y_j \partial_j f \right) (x + ty).$$

The second equality in (2.7) can be obtained by a formal development of

$$(\sum_{j=1}^{n} y_j \partial_j)^m$$

through (2.4).

Leibniz formula (2.9) can be easily obtained by the namesake formula for the one variable functions

$$\frac{d^k}{dt^k}(fg) = \sum_{h=0}^{k} \binom{k}{h} \frac{d^h f}{dt^h} \frac{d^{k-h} g}{dt^{k-h}},$$

where f and g are two smooth functions in the variable t.

Let $m \in \mathbb{N}_0$. Let a_α, $|\alpha| \leq m$, be some functions defined in an open set $\Omega \subset \mathbb{R}^n$ with values in $\mathbb{R}$ or in $\mathbb{C}$. We say that the operator

$$P(x, \partial) = \sum_{|\alpha| \leq m} a_\alpha(x) \partial^\alpha \tag{2.12}$$

is a **linear differential operator of order** m in Ω, provided that there exists $\alpha \in \mathbb{N}_0^n$, $|\alpha| = m$ such that a_α does not vanish identically in Ω. We say that the functions a_α, $|\alpha| \leq m$, are the **coefficients** of the differential operator (2.12).

We define the **symbol of operator** (2.12) as the following polynomial in the variable ξ

$$p(x, \xi) = \sum_{|\alpha| \leq m} a_\alpha(x)(i\xi)^\alpha. \tag{2.13}$$

Notice that if we write $P(x, \partial)$ as

$$\widetilde{P}(x, D) = \sum_{|\alpha| \leq m} a_\alpha(x)(iD)^\alpha,$$

then the symbol (2.13) can be obtained by formally substituting D to ξ.

We have

$$e^{-ix\cdot\xi} \widetilde{P}(x, D) e^{ix\cdot\xi} = p(x, \xi). \tag{2.14}$$

Note Here and in the sequel we will follow the convention of using a lowercase letter, thus $p(x, \xi)$, to denote the symbol of $P(x, D)$.

It might appear more natural to define the symbol of (2.12) simply as $\sum_{|\alpha| \leq m} a_\alpha(x)\xi^\alpha$. In fact, the contexts in which this symbol is primarily used often align with those in which it is convenient to utilize $D_j = \frac{1}{i}\partial_j$ as the derivative operator. As a result, adhering to the standard definition of the symbol is recommended to ensure consistency with the existing literature. For instance, the symbol of the **Laplace operator**

$$\Delta = \sum_{j=1}^{n} \partial_j^2 = -\sum_{j=1}^{n} D_j^2$$

is given by

$$-\sum_{j=1}^{n} \xi_j^2,$$

the symbol of the **heat operator**

$$\sum_{j=1}^{n} \partial_j^2 - \partial_t = -\sum_{j=1}^{n} D_j^2 - iD_t,$$

is equal to

$$-\sum_{j=1}^{n} \xi_j^2 - i\xi_{n+1}$$

and the symbol of the **wave operator** or **d'Alembertian operator**

$$\square = \Delta - \partial_t^2 = \sum_{j=1}^{n} \partial_j^2 - \partial_t^2 = -\sum_{j=1}^{n} D_j^2 + D_t^2$$

is equal to

$$-\sum_{j=1}^{n} \xi_j^2 + \xi_0^2.$$

We will call the **principal part of operator** (2.12), the differential operator

$$P_m(x, \partial) = \sum_{|\alpha|=m} a_\alpha(x)\partial^\alpha. \tag{2.15}$$

In the sequel, to simplify the notations, we will concentrate our attention to the case in which the coefficients a_α are real-valued functions. However, we warn that what we will establish, in many cases, can easily be extended to the case where the coefficients a_α are complex-valued functions.

If all the coefficients of the operator $P(x, \partial)$ are constants, we will say that $P(x, \partial)$ is an **operator with constant coefficients**. In these cases, to denote the operator $P(x, \partial)$, we will just write $P(\partial)$.

We notice that, by the above definition, the symbol of the principal part of operator (2.12) is the homogeneous polynomial

$$p_m(x, \xi) = i^m \sum_{|\alpha|=m} a_\alpha(x)\xi^\alpha. \tag{2.16}$$

Conventions on the Constants In the sequel, to denote a positive constant we will use very often the letter C. We notice right now that the value of the constants will may change from line to line, but we will generally indicate the dependence of the constants by the various parameters. However, sometimes to be able to better follow the various steps, we will put an index or a sign to C and we will write $C_0, C_1, \overline{C}, \widetilde{C} \ldots$. We will generally omit the dependence of the various constants on the dimension of the space.

Chapter 3
Overview of Measure Theory and Functional Analysis

Abstract This chapter recalls fundamental concepts from real and functional analysis, such as smooth and Hölder spaces, measure theory, and L^p spaces, along with the partition of unity. Key theorems, including the extension theorem, the Lebesgue differentiation theorem, and the divergence theorem for Lipschitz domains, are also presented. The chapter introduces the Hausdorff distance and the distance function, which are crucial for studying inverse problems with unknown boundaries. These foundational results bridge basic analysis and research-level topics. Readers familiar with these concepts may skip this chapter and refer back as needed.

3.1 Introduction

In this chapter we recall the definitions and the main properties of the spaces C^k (Sect. 3.2) and $C^{k,\alpha}$ (Sect. 3.3) and we give the main definitions and theorems (without proof) of measure theory and L^p spaces (Sect. 3.4). In Sect. 3.5 we give the definition and the construction of the partition of unity. In Sects. 3.8 and 3.9 we provide, respectively, a detailed description of the boundary of an open set and the definition of $L^p(\partial\Omega)$.

Moreover we prove some important theorems of real analysis: the extension theorem in $C^{0,\alpha}$ (Sect. 3.3), the Lebesgue differentiation theorem (Sect. 3.6), the Rademacher theorem (Sect. 3.7), and the divergence Theorem over open sets with Lipschitz boundary (Sect. 3.10). These theorems often exist in a sort of limbo for the students, as they are not covered at the basic Analysis level, but rather assumed in courses more geared towards research-oriented topics.

Furthermore, we study the Hausdorff distance between compact sets (Sect. 3.11) and the distance function (Sect. 3.12), which are useful for studying the stability issue of inverse problems with unknown boundaries and unknown domain.

Readers who are already familiar with these topics can skip this chapter and refer to it as needed.

S. Vessella, *Unique Continuation Properties for Partial Differential Equations*, Birkhäuser Advanced Texts Basler Lehrbücher,
https://doi.org/10.1007/978-3-031-86366-0_3

3.2 The Spaces C^k

Let X be a subset of $\mathbb{R}^n$. We will denote by $C^0(X)$ the vector space of continuous functions defined in X with values in $\mathbb{R}$. If $u \in C^0(X)$, we denote the **support of** u by

$$\operatorname{supp} u := \overline{\{x \in X : u(x) \neq 0\}}^X \quad \text{(closure in } X\text{)}.$$

We will denote by $C_0^0(X)$ the space of continuous functions whose support is a compact set of $\mathbb{R}^n$ contained in X. We recall that, if X is an open subset of $\mathbb{R}^n$ then $K \subset X$ is compact w.r.t. the topology induced by X if and only if K is a compact subset of $\mathbb{R}^n$. Therefore, whenever X is open, saying that $u \in C_0^0(X)$ equivalent to saying that $\operatorname{supp} u$ is a compact subset of $\mathbb{R}^n$.

Proposition 3.1 *Let $u \in C^0(X)$. Let $K \subset X$ be a compact set of $\mathbb{R}^n$; for any $r > 0$ we set*

$$K_r = \left\{x \in \mathbb{R}^n : dist(x, K) \leq r\right\},$$

where dist(x, K) denotes the distance of x from K. Suppose that there exists $r_0 > 0$ such that $K_{r_0} \subset X$. Then

$$\lim_{r \to 0} \max_{K_r} u = \max_K u . \tag{3.1}$$

Proof Since K_{r_0} is a compact subset of X and $u \in C^0(X)$, u is uniformly continuous on K_{r_0}. Let $\varepsilon > 0$ and $\delta > 0$ such that

$$|u(x) - u(y)| < \varepsilon, \qquad \text{for all } x, y \in K_{r_0} \text{ such that } |x - y| \leq \delta;$$

we may assume that $\delta < r_0$. Let $r \in (0, \delta]$ and let x be any point of K_r. Hence there exists $y \in K$ such that $|x - y| \leq r$. Therefore

$$u(x) < u(y) + \varepsilon \leq \max_K u + \varepsilon.$$

By the arbitrariness of x in K_r we have

$$\max_K u \leq \max_{K_r} u < \max_K u + \varepsilon,$$

which concludes the proof of (3.1). ∎

Let $u \in C^0(X)$. We define the **modulus of continuity** of u in X

$$\omega(\delta) = \sup\left\{|u(x) - u(y)| : x, y \in X, \ |x - y| \leq \delta\right\}, \qquad \text{for } \delta > 0. \tag{3.2}$$

ω is an increasing function, defined on $[0, +\infty)$ and satisfies $\omega(0) = 0$. Of course, ω may not be finite. It is easy to check that u is uniformly continuous if and only if

$$\lim_{\delta \to 0} \omega(\delta) = 0. \tag{3.3}$$

If the function ω is bounded, it may be convenient to use the **concave modulus of continuity** which is defined as

$$\tilde{\omega}(\delta) = \inf \{ f(\delta) : \ f \text{ concave}, \ f \geq \omega, \ \text{in } [0, +\infty) \}, \quad \text{for } \delta > 0. \tag{3.4}$$

Now we check that

$$\lim_{\delta \to 0} \tilde{\omega}(\delta) = 0. \tag{3.5}$$

Let us denote

$$M = \sup_{\delta \in [0, +\infty)} \omega(\delta) < +\infty.$$

If $M = 0$, (3.5) is trivial. Let us suppose therefore $M > 0$. Let $0 < \varepsilon < M$. From (3.3) it follows that there exists $\delta_0 > 0$ such that

$$0 \leq \omega(\delta) < \frac{\varepsilon}{2}, \quad \forall \delta \in [0, \delta_0].$$

Set

$$g_\varepsilon(\delta) = \begin{cases} \frac{\varepsilon}{2} + \frac{2M - \varepsilon}{2\delta_0} \delta, & \text{for } \delta \in [0, \delta_0] \\ \\ M, & \text{for } \delta \in (\delta_0, +\infty) \end{cases}.$$

It easy to check that g_ε is concave and that $g_\varepsilon \geq \omega$ in $[0, +\infty)$. Furthermore, set

$$\delta_1 = \frac{\varepsilon \delta_0}{2M - \varepsilon},$$

it turns out

$$g_\varepsilon(\delta) < \varepsilon, \quad \forall \delta \in [0, \delta_1).$$

Therefore

$$\tilde{\omega}(\delta) < \varepsilon, \quad \forall \delta \in [0, \delta_1),$$

which gives (3.5).

Remark 3.1 Let $\tilde{\omega}$ be a concave modulus of continuity, then

$$0 < \eta_1 < \eta_2 \Longrightarrow \eta_1 \tilde{\omega}\left(\frac{1}{\eta_1}\right) \leq \eta_2 \tilde{\omega}\left(\frac{1}{\eta_2}\right). \tag{3.6}$$

Let us check (3.6). From the concavity of $\tilde{\omega}$ and recalling that $\tilde{\omega}(0) = 0$ we have, for $0 < \eta_1 < \eta_2$,

$$\eta_1 \tilde{\omega}\left(\frac{1}{\eta_1}\right) = \frac{\tilde{\omega}\left(\frac{1}{\eta_1}\right) - \tilde{\omega}(0)}{\frac{1}{\eta_1} - 0} \leq \frac{\tilde{\omega}\left(\frac{1}{\eta_2}\right) - \tilde{\omega}(0)}{\frac{1}{\eta_2} - 0} = \eta_2 \tilde{\omega}\left(\frac{1}{\eta_2}\right).$$

♦

Let us denote by $C^0_*(X)$ the space of the bounded functions of $C^0(X)$, let us define the norm

$$\|u\|_{C^0_*(X)} = \sup_X |u(x)|, \quad \forall u \in C^0_*(X). \tag{3.7}$$

As it is well-known, the space $C^0_*(X)$, equipped with the norm (3.7), is a **Banach space**.

The following Theorem holds true (see [73, Corollary 1.3, Ch. 3] for a proof)

Theorem 3.2 (Weierstrass Approximation) *Let X be a compact subset of $\mathbb{R}^n$. For every $u \in C^0(X)$ and for every $\varepsilon > 0$ there exists a polynomial P such that*

$$\|u - P\|_{C^0_*(X)} < \varepsilon.$$

We notice that by approximating the polynomial P given in the previous Theorem, by a polynomial with rational coefficients, we derive that $C^0(X)$, with compact X, is a **separable space**. Let us recall that a topological space $\mathcal{S}$ is said to be separable if there exists a countable set $D \subset \mathcal{S}$ such that $\overline{D} = \mathcal{S}$.

The following Proposition holds true

Proposition 3.3 *Let $\mathcal{S}$ be a metric space with distance d. If there exists $\mathcal{Y}$, uncountable subset of $\mathcal{S}$, and $\delta > 0$ such that*

$$d(x, y) > \delta, \qquad \forall x, y \in \mathcal{Y}, \ x \neq y, \tag{3.8}$$

then $\mathcal{S}$ is not a separable space.

Proof We argue by contradiction and we assume that $\mathcal{S}$ is separable. Hence there exists $D = \{u_n\}_{n \in \mathbb{N}}$ such that $\overline{D} = \mathcal{S}$. Consequently, for every $x \in \mathcal{Y}$ there exists u_{n_x} such that

$$d\left(x, u_{n_x}\right) < \frac{\delta}{3}.$$

Therefore, if $x, y \in \mathcal{Y}$, $x \neq y$, the triangle inequality gives

$$d\left(u_{n_x}, u_{n_y}\right) \geq d(x, y) - d\left(x, u_{n_x}\right) - d\left(y, u_{n_y}\right) > \frac{\delta}{3}.$$

In particular, if $x, y \in \mathcal{Y}$, $x \neq y$, then $u_{n_x} \neq u_{n_y}$. Consequently, the map

$$\mathcal{Y} \ni x \to u_{n_x} \in D,$$

is injective, but this fact contradicts that $\mathcal{Y}$ is an uncountable set. Therefore $\mathcal{S}$ is not separable. ■

Remark 3.2 Let us note that the compactness assumption of X cannot be dropped for $C^0_*(X)$ to be separable. We show, for instance, that $C^0_*(\mathbb{R})$ is not separable.

For any $A \in \mathcal{P}(\mathbb{Z}) \setminus \{\emptyset\}$ (where $\mathcal{P}(\mathbb{Z})$ denotes the power set of $\mathbb{Z}$) and any $\varepsilon \in \left(0, \frac{1}{2}\right)$, define

$$u_A = \sum_{g \in A} u_g,$$

where

$$u_g(t) = \begin{cases} 1 - \varepsilon^{-1}|t - g|, & \text{for } t \in [g - \varepsilon, g + \varepsilon] \\ \\ 0, & \text{for } t \in \mathbb{R} \setminus [g - \varepsilon, g + \varepsilon] \end{cases}.$$

We have

$$\|u_A - u_B\|_{C^0_*(\mathbb{R})} = 1, \quad \forall A, B \in \mathcal{P}(\mathbb{Z}) \setminus \{\emptyset\}, \quad A \neq B.$$

Since $\mathcal{P}(\mathbb{Z}) \setminus \{\emptyset\}$ is uncountable, Proposition 3.3 implies that $C^0_*(\mathbb{R})$ is not separable. ♦

Generally we will be interested in the case when X is an open or the closure of an open set Ω of $\mathbb{R}^n$. If Ω is a bounded open set then we may consider

$$C^0\left(\overline{\Omega}\right)$$

as a subspace of $C^0_*(\Omega)$ and we will denote the norm of $C^0\left(\overline{\Omega}\right)$ by

$$\|u\|_{C^0\left(\overline{\Omega}\right)} = \sup_{\overline{\Omega}} |u|, \quad \forall u \in C^0\left(\overline{\Omega}\right). \tag{3.9}$$

In the sequel we will use the following classical theorems on relatively compact sets (see [33]).

Theorem 3.4 *Let (X, d) be a complete metric space and let $Y \subset X$. Then Y is a relatively compact set (i.e., $\overline{Y} = X$) if and only if it is* ***totally bounded*** *that is, for every $\delta > 0$ there exists a finite set $\{x_1, \cdots, x_N\} \subset X$ such that*

$$d\left(y, x_j\right) < \delta, \quad \forall y \in Y, \ j = 1, \cdots N$$

or, equivalently,

$$Y \subset \bigcup_{j=1}^{N} B_\delta(x_j).$$

Theorem 3.5 (Arzelà–Ascoli) *Let Ω a bounded open set of $\mathbb{R}^n$ and let $\{u_k\}$ be a sequence of functions belonging to $C^0\left(\overline{\Omega}\right)$ such that:*

(i) $\{u_k\}$ is equibounded, i.e., there exists $M > 0$ such that

$$\|u_k\|_{C^0(\overline{\Omega})} \leq M, \qquad \forall k \in \mathbb{N};$$

(ii) $\{u_k\}$ is equicontinuous, i.e., for every $\eta > 0$ there exists $\delta > 0$ such that if $|x - y| < \delta$ and $x, y \in \overline{\Omega}$, then

$$|u_k(x) - u_k(y)| \leq \eta, \quad \forall k \in \mathbb{N}.$$

Then there exists a subsequence $\left\{u_{k_j}\right\}$ of $\{u_k\}$ and a function $u \in C^0\left(\overline{\Omega}\right)$ such that

$$\lim_{j \to \infty} \left\|u_{k_j} - u\right\|_{C^0(\overline{\Omega})} = 0.$$

Let $k \in \mathbb{N}$. Let Ω be an open set of $\mathbb{R}^n$, we will denote by $C^k(\Omega)$ the space of functions which satisfy $\partial^\alpha u \in C^0(\Omega)$ for every $\alpha \in \mathbb{N}_0^n$, $|\alpha| \leq k$. Further, we will denote by $C^k\left(\overline{\Omega}\right)$ the space of the functions $u \in C^k(\Omega)$ such that, for every $\alpha \in \mathbb{N}_0^n$, $|\alpha| \leq k$, $\partial^\alpha u$ is extensible to a function $U_\alpha \in C^0\left(\overline{\Omega}\right)$. Of course, if such an extension exists it is unique and we will write $\partial^\alpha u$ instead of U_α. If Ω is a bounded **open set** of $\mathbb{R}^n$, we define the norm on $C^k\left(\overline{\Omega}\right)$ as follows

$$\|u\|_{C^k(\overline{\Omega})} = \sum_{|\alpha| \leq k} \sup_{\overline{\Omega}} |\partial^\alpha u|, \quad \forall u \in C^k\left(\overline{\Omega}\right). \tag{3.10}$$

As it is well-known, the space $C^k\left(\overline{\Omega}\right)$, equipped with the norm (3.10), is a **Banach space**. In some contexts it turns out to be convenient to consider, instead of norm (3.10), an equivalent "dimensionless" norm, e.g.

$$\sum_{|\alpha|\leq k} d_0^{|\alpha|} \sup_{\overline{\Omega}} |\partial^\alpha u|, \quad \forall u \in C^k\left(\overline{\Omega}\right), \tag{3.11}$$

where d_0 is the diameter of Ω.

Proposition 3.6 *Let Ω be a bounded open set of $\mathbb{R}^n$ and $k \in \mathbb{N}_0$, then the space $C^k\left(\overline{\Omega}\right)$, with norm* (3.10) *is a separable space.*

Proof Recall that if a topological space is separable, then every subset of it is a separable space [25, Ch. 3, Sec. 6], [33].

We have already noticed (after Theorem 3.2) that $C^0\left(\overline{\Omega}\right)$ is separable. We consider the case $k = 1$ (the case $k > 1$ can be treated similarly). Let Ψ be the map

$$\Psi : C^1\left(\overline{\Omega}\right) \to \mathcal{X},$$

where

$$\mathcal{X} = \underbrace{C^0\left(\overline{\Omega}\right) \times \cdots \times C^0\left(\overline{\Omega}\right)}_{(n+1)\text{ - times}},$$

$$\Psi(u) = (u, \partial_1 u, \cdots \partial_n u), \quad \forall u \in C^1\left(\overline{\Omega}\right).$$

If we equip $\mathcal{X}$ with the norm

$$\|\mathbf{v}\|_{\mathcal{X}} = \sum_{j=0}^{n} \left\|v_j\right\|_{C^0(\overline{\Omega})}, \quad \forall \mathbf{v} = (v_0, v_1, \cdots, v_n) \in \mathcal{X},$$

Ψ is an isometry.

On the other hand, $\mathcal{X}$ is a separable space as a cartesian product of separable spaces. Thus $\Psi\left(C^1(\overline{\Omega})\right)$ is separable as a subspace of $\mathcal{X}$ and, since Ψ is an isometry, also $C^1\left(\overline{\Omega}\right)$ is separable. ∎

It is evident that if $k, m \in \mathbb{N}_0$ and $k < m$ then $C^m\left(\overline{\Omega}\right) \subset C^k\left(\overline{\Omega}\right)$. We set

$$C_0^k(\Omega) = \left\{u \in C^k(\Omega) : \ \operatorname{supp} u \text{ is a compact set contained in } \Omega\right\},$$

$$C^\infty(\Omega) = \bigcap_{k=0}^{\infty} C^k(\Omega), \quad C^\infty\left(\overline{\Omega}\right) = \bigcap_{k=0}^{\infty} C^k\left(\overline{\Omega}\right), \quad C_0^\infty(\Omega) = \bigcap_{k=0}^{\infty} C_0^k(\Omega).$$

Let K be a compact subset of Ω, we denote by

$$C_K^\infty(\Omega) = \left\{v \in C^\infty(\Omega) : \ \operatorname{supp} v \subset K\right\}.$$

Let $k \in \mathbb{N}$ or $k = \infty$ and $\widetilde{\Omega} \supset \Omega$, we will often adopt the convention of identifying $C_0^k(\Omega)$ with the space of functions u belonging to $C_0^k\left(\widetilde{\Omega}\right)$ and such that $\operatorname{supp} u$ is a compact set contained in Ω.

3.3 The Spaces $C^{k,\alpha}$

Let X be a subset of $\mathbb{R}^n$ and $\alpha \in (0, 1]$, we will denote by $C^{0,\alpha}(X)$ the space of the functions $u \in C^0(X)$ which satisfy

$$[u]_{C^{0,\alpha}(X)} = \sup\left\{\frac{|u(x) - u(y)|}{|x - y|^\alpha} : x, y \in X,\ x \neq y\right\} < +\infty.$$

If u is a function of $C^{0,\alpha}(X)$ it is also said that u is a **Hölder function of order** α. The number α is said the **Hölder exponent** of the space $C^{0,\alpha}(X)$. For any $u \in C^{0,\alpha}(X)$, the number $[u]_{C^{0,\alpha}(X)}$ is called the **Hölder constant** of u. If $\alpha = 1$ we will also say that u is a **Lipschitz continuous function** in X and we call **Lipschitz constant** the number $[u]_{C^{0,1}(X)}$. We observe that if $\alpha > 1$ and X is a connected open set, then the space $C^{0,\alpha}(X)$ consists of only the constant functions (as a matter of fact, if $\alpha > 1$ then any function of $C^{0,\alpha}(X)$ is differentiable with zero gradient in X). It can be easily checked that if X is bounded, and the space $C^{0,\alpha}(X)$ is equipped with the norm

$$\|u\|_{C^{0,\alpha}(X)} = \|u\|_{C^0(X)} + [u]_{C^{0,\alpha}(X)}, \tag{3.12}$$

then $C^{0,\alpha}(X)$ is a **Banach space**. Sometimes it is convenient to consider, instead of the norm (3.12), an equivalent "dimensionless" norm, e.g.

$$\|u\|_{C^0(X)} + d_0^\alpha [u]_{C^{0,\alpha}(X)},$$

where d_0 is the diameter of X.

Let $m \in \mathbb{N}$, we denote by $C^{0,\alpha}(X;\mathbb{R}^m)$ the space of the functions $u \in C^0(X;\mathbb{R}^m)$ satisfying $u_j \in C^{0,\alpha}(X)$, $j = 1, \cdots, m$. We set

$$\|u\|_{C^{0,\alpha}(X;\mathbb{R}^m)} = \|u\|_{C^0(X;\mathbb{R}^n)} + [u]_{C^{0,\alpha}(X;\mathbb{R}^m)},$$

where

$$[u]_{C^{0,\alpha}(X;\mathbb{R}^m)} = \sup\left\{\frac{|u(x) - u(y)|_{\mathbb{R}^m}}{|x - y|^\alpha_{\mathbb{R}^n}} : x, y \in X,\ x \neq y\right\} < +\infty,$$

where $|\cdot|_{\mathbb{R}^m}$ is the Euclidean norm in $\mathbb{R}^m$ (in the sequel we will often omit the subscript $\mathbb{R}^m$ from this norm).

The following Proposition holds true.

Proposition 3.7 *Let X be a bounded set of $\mathbb{R}^n$ and $0 < \beta < \alpha \leq 1$ then*

$$[u]_{C^{0,\beta}(X)} \leq d_0^{\alpha-\beta} [u]_{C^{0,\alpha}(X)}, \tag{3.13}$$

where d_0 is the diameter of X;

$$[u]_{C^{0,\beta}(X)} \leq \left(2 \|u\|_{C^0(X)}\right)^{1-\frac{\beta}{\alpha}} \left([u]_{C^{0,\alpha}(X)}\right)^{\frac{\beta}{\alpha}}. \tag{3.14}$$

In particular we have

$$C^{0,\alpha}(X) \subset C^{0,\beta}(X). \tag{3.15}$$

Proof It suffices to observe that for $x, y \in X$, $x \neq y$, by (3.13), we have

$$\frac{|u(x) - u(y)|}{|x-y|^\beta} = \frac{|u(x) - u(y)|}{|x-y|^\alpha} |x-y|^{\alpha-\beta} \leq [u]_{C^{0,\alpha}(X)} d_0^{\alpha-\beta}.$$

Regarding (3.14), we first note that the case

$$[u]_{C^{0,\alpha}(X)} = 0$$

is trivial. Let us assume, then,

$$[u]_{C^{0,\alpha}(X)} \neq 0$$

and let $r > 0$ be chosen later. Let $x, y \in X$, $x \neq y$. If

$$|x-y| \leq r,$$

then

$$\frac{|u(x) - u(y)|}{|x-y|^\beta} \leq [u]_{C^{0,\alpha}(X)} r^{\alpha-\beta}.$$

If

$$|x-y| > r,$$

then

$$\frac{|u(x) - u(y)|}{|x-y|^\beta} \leq 2r^{-\beta} \|u\|_{C^0(X)}.$$

In any case, we have

$$[u]_{C^{0,\beta}(X)} \leq r^{-\beta} \max\left\{r^{\alpha} [u]_{C^{0,\alpha}(X)}, 2\|u\|_{C^0(X)}\right\}$$

and, choosing

$$r = \left(\frac{2\|u\|_{C^0(X)}}{[u]_{C^{0,\alpha}(X)}}\right)^{1/\alpha},$$

we obtain (3.14). ■

Remark 3.3 By using the Mean Value Theorem, it is easily shown that if Ω is a bounded, convex open set of $\mathbb{R}^n$ then

$$C^1\left(\overline{\Omega}\right) \subset C^{0,\beta}\left(\overline{\Omega}\right) \tag{3.16}$$

and

$$[u]_{0,1,\Omega} \leq \|\nabla u\|_{C^0(\overline{\Omega})},$$

where

$$\|\nabla u\|_{C^0(\overline{\Omega})} = \||\nabla u|\|_{C^0(\overline{\Omega})}.$$

Nevertheless, for a bounded open set Ω it is not necessarily the case that the inclusion (3.16) holds. Let us consider, for instance, the following example. Let

$$\Omega = \left\{(x, y) \in \mathbb{R}^2 : x \leq \sqrt{|y|},\ x^2 + y^2 < 1\right\},$$

$$1 < \beta < 2,$$

$$u(x, y) = \begin{cases} x^{\beta}\operatorname{sgn}(y), & \text{if } x \in \Omega,\ x > 0 \\ \\ 0, & \text{if } x \in \Omega,\ x \leq 0 \end{cases}.$$

We have $u \in C^1\left(\overline{\Omega}\right)$, however if α satisfies $\frac{\beta}{2} < \alpha \leq 1$ then

$$u \notin C^{0,\alpha}\left(\overline{\Omega}\right).$$

As a matter of fact, if $x = \sqrt{|y|}$, we get

$$\frac{|u(x, y) - u(x, -y)|}{(2|y|)^{\alpha}} = 2^{1-\alpha}|y|^{\frac{\beta}{2}-\alpha} \to +\infty, \quad \text{as } y \to 0.$$

♦

Remark 3.4 The space $C^{0,\alpha}\left(\overline{\Omega}\right)$, where $\alpha \in (0, 1]$ and Ω is a bounded open set, is not separable. Let us consider the case $n = 1$ and $\Omega = (0, 1)$ and for any $a \in (0, 1)$, let us define

$$u_a(t) = \begin{cases} 0, & \text{for } t \in [0, a) \\ \\ (t-a)^\alpha, & \text{for } t \in [a, 1] \end{cases}.$$

We have

$$\|u_a - u_b\|_{C^{0,\alpha}([0,1])} \geq 1, \quad \forall a, b \in [0, 1], \ \ a \neq b. \tag{3.17}$$

We check (3.17). Let $a, b \in [0, 1]$, $a < b$ and let us denote

$$v_{a,b} = u_a - u_b.$$

We have

$$v_{a,b}(b) = u_a(b) - u_b(b) = u_a(b) = (b-a)^\alpha$$

and

$$v_{a,b}(a) = u_a(a) - u_b(a) = 0.$$

Therefore

$$[u_a - u_b]_{C^{0,\alpha}([0,1])} = \left[v_{a,b}\right]_{C^{0,\alpha}([0,1])} \geq \frac{\left|v_{a,b}(b) - v_{a,b}(a)\right|}{|b-a|^\alpha} = 1,$$

which implies (3.17). Finally, from the latter and from Proposition 3.3 it follows that $C^{0,\alpha}([0, 1])$ is not separable. ♦

Theorem 3.8 (Extension in $C^{0,\alpha}$) *Let X be a bounded set of $\mathbb{R}^n$ and $u \in C^{0,\alpha}(X)$, $\alpha \in (0, 1]$, then there exists $U \in C^{0,\alpha}(\mathbb{R}^n)$ such that*

$$U(x) = u(x), \quad \forall x \in X, \tag{3.18}$$

$$\|U\|_{C^0(\mathbb{R}^n)} = \|u\|_{C^0(X)}, \tag{3.19}$$

$$[U]_{C^{0,\alpha}(\mathbb{R}^n)} = [u]_{C^{0,\alpha}(X)}. \tag{3.20}$$

Proof Let us denote by

$$M = \|u\|_{C^0(X)}, \qquad m = [u]_{C^{0,\alpha}(X)}$$

and let us define the function

$$v(x) = \sup_{y \in X} \left\{ u(y) - m|x - y|^{\alpha} \right\}, \qquad \text{for } x \in \mathbb{R}^n.$$

We have

$$v(x) = u(x), \quad \forall x \in X. \tag{3.21}$$

We check (3.21). First note that we have trivially

$$u(x) \le v(x), \quad \forall x \in X. \tag{3.22}$$

On the other hand we have

$$u(y) - u(x) \le m|x - y|^{\alpha}, \qquad \forall x, y \in X,$$

hence

$$u(y) - m|x - y|^{\alpha} \le u(x), \qquad \forall x, y \in X.$$

Consequently

$$v(x) \le u(x), \qquad \forall x \in X.$$

By the latter and by (3.22) we get (3.21).
We also notice that

$$v(x) \le M, \quad \forall x \in \mathbb{R}^n. \tag{3.23}$$

Now, for any $x \in \mathbb{R}^n$ let us define

$$U(x) = \begin{cases} v(x), & \text{for } |v(x)| \le M \\ \\ -M, & \text{for } v(x) < -M \end{cases}.$$

Let us note that, by (3.23), U is defined throughout $\mathbb{R}^n$. Let us note also that if $x \in X$, then (3.21) gives (3.18) and

$$\sup_X |U(x)| = M.$$

We also have

$$\sup_{\mathbb{R}^n} |U(x)| = M,$$

Concerning the latter, notice that, if $|v(x)| \leq M$ then $U(x) = v(x)$, hence $|U(x)| \leq M$ and if $v(x) < -M$ then $|U(x)| = M$.

It only remains to prove that $U \in C^{0,\alpha}(\mathbb{R}^n)$ and that (3.20) holds. Let, then, $x, y \in \mathbb{R}^n$ be such that $x \neq y$ and otherwise arbitrary. Let us suppose that $U(x) \neq U(y)$. For instance, let us assume

$$U(x) > U(y). \tag{3.24}$$

Let us check that

$$0 < U(x) - U(y) \leq v(x) - v(y). \tag{3.25}$$

The following cases occur.

(a) $v(x) < -M$ and $|v(y)| \leq M$,
(b) $v(x) < -M$ and $v(y) < -M$,
(c) $|v(x)| \leq M$ and $|v(y)| \leq M$,
(d) $|v(x)| \leq M$ and $v(y) < -M$.

Cases (a) and (b) cannot occur. As a matter of fact, in case (a) we would have

$$U(x) - U(y) = -M - v(y) \leq -M + M = 0,$$

that contradicts (3.24). In case (b) we would have

$$U(x) - U(y) = -M - (-M) = 0,$$

that contradicts (3.24).

In case (c) we have

$$U(x) - U(y) = v(x) - v(y).$$

Finally, in case (d) we have

$$U(x) - U(y) = v(x) - (-M) = v(x) + M < v(x) - v(y).$$

Therefore (3.25) holds true.

We have

$$\begin{aligned} v(x) - v(y) &= \sup_{z \in X} \left\{u(z) - m|x - z|^\alpha\right\} - \sup_{z \in X} \left\{u(z) - m|y - z|^\alpha\right\} \leq \\ &\leq m \sup_{z \in X} \left\{|y - z|^\alpha - |x - z|^\alpha\right\} \leq \\ &\leq m \sup_{\zeta \in X} \left\{(|y - x| + |\zeta|)^\alpha - |\zeta|^\alpha\right\}. \end{aligned} \tag{3.26}$$

Now, let us denote by

$$\omega(t) = t^{\alpha}, \quad \text{if } t \in [0, +\infty).$$

Since ω is concave, we have

$$\omega(t+h) - \omega(t) \leq \omega(h) - \omega(0) = \omega(h), \quad \forall t, h \in [0, +\infty),$$

by the just obtained inequality, by (3.25) and (3.26) we get

$$|U(x) - U(y)| \leq m|y - x|^{\alpha} \tag{3.27}$$

and since we have proved (3.18), we get (3.20). ■

A more comprehensive version of Theorem 3.8, which holds true for uniformly continuous functions, can be found in [107, Chapter 4].

Let us notice that if

$$u : X \to \mathbb{R}^m$$

is a Lipschitz continuous function, then, Theorem 3.8 implies that there exists an extension

$$U : \mathbb{R}^n \to \mathbb{R}^m,$$

such that

$$[U]_{C^{0,1}(\mathbb{R}^n;\mathbb{R}^m)} \leq \sqrt{m}[u]_{C^{0,1}(X;\mathbb{R}^m)}.$$

Actually, this result can be improved. As a matter of fact, the following Theorem of **Kirszbraun** holds true, for the proof of which we refer to [86, cap. 7].

Theorem 3.9 (Kirszbraun) *Let $u : X \to \mathbb{R}^m$ be a Lipschitz continuous function, where $X \subset \mathbb{R}^n$, then there exists $U \in C^{0,1}(\mathbb{R}^n; \mathbb{R}^m)$ such that*

$$U = u, \quad \textit{in } X$$

and

$$[U]_{C^{0,1}(\mathbb{R}^n;\mathbb{R}^m)} = [u]_{C^{0,1}(X;\mathbb{R}^m)}.$$

Let Ω be a bounded open set of $\mathbb{R}^n$, $k \in \mathbb{N}_0$ and $\alpha \in (0, 1]$, we denote by $C^{k,\alpha}(\overline{\Omega})$ the space of the functions $u \in C^k(\overline{\Omega})$, satisfying

$$\left[\partial^{\beta} u\right]_{C^{0,\alpha}(\overline{\Omega})} < +\infty, \quad \forall \beta \in \mathbb{N}_0^n, \ |\beta| = k.$$

It is easily proven that $C^{k,\alpha}\left(\overline{\Omega}\right)$, equipped with the norm

$$\|u\|_{C^{k,\alpha}\left(\overline{\Omega}\right)} = \|u\|_{C^{k}\left(\overline{\Omega}\right)} + [u]_{C^{k,\alpha}\left(\overline{\Omega}\right)}, \quad \forall u \in C^{k}\left(\overline{\Omega}\right), \tag{3.28}$$

is a **Banach space**, where

$$[u]_{C^{k,\alpha}\left(\overline{\Omega}\right)} = \sum_{|\beta|=k} \left[\partial^{\beta} u\right]_{C^{0,\alpha}\left(\overline{\Omega}\right)}.$$

Sometimes, instead of norm (3.28) we will consider the dimensionless norm

$$\|u\|_{C^{k,\alpha}\left(\overline{\Omega}\right)} = \sum_{|\beta|\leq k} d_0^{|\beta|} \sup_{\overline{\Omega}} |\partial^{\beta} u| + d_0^{k+\alpha} [u]_{C^{k,\alpha}\left(\overline{\Omega}\right)}.$$

where d_0 is the diameter of Ω.

We define the space $C_{loc}^{k,\alpha}(\Omega)$, $k \in \mathbb{N}_0$, $0 < \alpha \leq 1$, as the space of functions $u \in C^0(\Omega)$ such that for every bounded open set $\omega \Subset \Omega$, (i.e. $\overline{\omega}$ compact and $\overline{\omega} \subset \Omega$) we have

$$u_{|\omega} \in C^{k,\alpha}\left(\overline{\omega}\right).$$

3.4 Review of Measure Theory and L^p Spaces

In this section, we give, for the convenience of the reader, the main definitions and statements of the main theorems of the Measure Theory and of L^p spaces. Some reference texts are [30], [107], [113] (see also lecture notes [87] and [88]).

3.4.1 Measurable Sets, Measurable Functions, Positive Measures

Definition 3.10 Let X be a nonempty set and $\mathcal{M}$ be a family of subsets of X with the following properties:

(i) $X \in \mathcal{M}$,
(ii) $E \in \mathcal{M} \Longrightarrow CE := X \setminus E \in \mathcal{M}$,
(iii) $E_j \in \mathcal{M}$, $j \in \mathbb{N} \Longrightarrow \bigcup_{j\in\mathbb{N}} E_j \in \mathcal{M}$.

$\mathcal{M}$ is called a σ**-algebra** and the couple $(X, \mathcal{M})$ is called a **measurable space**.

We will be interested almost exclusively in the case where $X = \mathbb{R}^n$ and $\mathcal{M}$ consists of the Lebesgue measurable subsets of $\mathbb{R}^n$.

Definition 3.11 Let $(X, \mathcal{M})$ be a measurable space and let Y be topological space. We say that the function

$$f : X \to Y,$$

is a **measurable function**, provided we have

$$f^{-1}(A) \in \mathcal{M}, \qquad \text{for every open subset } A \text{ of } Y.$$

Let us recall that if $(X, \mathcal{M})$ is a measurable space, Y, Z two topological spaces, $f : X \to Y$ is a measurable function and $g : Y \to Z$ is a continuous function, then $g \circ f : X \to Z$ is a measurable function.

We denote by $\overline{\mathbb{R}} = \mathbb{R} \cup \{-\infty, +\infty\}$, the extended real line equipped with its usual topology.

The following theorems hold true.

Theorem 3.12 *Let $(X, \mathcal{M})$ be a measurable space and $f : X \to \overline{\mathbb{R}}$. Then f is a measurable function if and only if for any $t \in \mathbb{R}$ one of the following level sets*

$$f^{-1}((t, +\infty]), \quad f^{-1}([t, +\infty]), \quad f^{-1}([-\infty, t)), \quad f^{-1}([-\infty, t])),$$

is a measurable set.

Theorem 3.13 *Let $(X, \mathcal{M})$ be a measurable space. Then we have*

(i) If $f, g : X \to \overline{\mathbb{R}}$ are two measurable functions and $f + g$ is well defined, then $f + g$ and λf are measurable functions (we use the convention that $0 \cdot (\pm\infty) = 0$).

(ii) Let $\{f_k\}$ be a sequence of measurable functions, then

$$\sup_{k \in \mathbb{N}} f_k, \quad \inf_{k \in \mathbb{N}} f_k, \quad \liminf_{k \to \infty} f_k, \quad \limsup_{k \to \infty} f_k,$$

are measurable functions.

We define as **simple function** on the measurable space X any function

$$s : X \to \mathbb{R}$$

that assumes a finite set of values.

Theorem 3.14 *Let $(X, \mathcal{M})$ be a measurable space and $f : X \to \overline{\mathbb{R}}$ be a measurable function, then there exists a sequence $\{s_k\}$ of simple functions such that*

$$\lim_{k \to \infty} s_k(x) = f(x), \quad \forall x \in X.$$

If f is bounded then $\{s_k\}$ uniformly converges to f.

Definition 3.15 Let $(X, \mathcal{M})$ a measurable space. We say that

$$\mu : \mathcal{M} \to [0, \infty],$$

is a **positive measure** provided that we have

(i) $\mu(\emptyset) = 0$;
(ii) if $\{E_j\}_{j\in\mathbb{N}}$ is a countable family of measurable sets such that

$$E_i \cap E_j = \emptyset, \qquad \text{for } i \neq j,$$

then we have

$$\mu\left(\bigcup_{j=1}^{\infty} E_j\right) = \sum_{j=1}^{\infty} \mu(E_j).$$

The tern $(X, \mathcal{M}, \mu)$ is called a **measure space.**

If E is a Lebesgue measurable set of $\mathbb{R}^n$, we will denote by $|E|$ its measure. The following theorems hold true.

Theorem 3.16 *Let $(X, \mathcal{M}, \mu)$ be a measure space. The following properties hold true.*

(i) If $\{E_j\}_{1\leq j\leq N}$ is a finite family of measurable sets such that $E_i \cap E_j = \emptyset$, for $i \neq j$, then

$$\mu\left(\bigcup_{j=1}^{N} E_j\right) = \sum_{j=1}^{N} \mu(E_j);$$

(ii) if $E \subset F$ and $E, F \in \mathcal{M}$, then $\mu(E) \leq \mu(F)$;
(iii) if $\{E_j\}_{j\in\mathbb{N}}$ is a countable family of measurable sets such that

$$E_j \subset E_{j+1}, \ \forall j \in \mathbb{N},$$

then

$$\lim_{j\to\infty} \mu(E_j) = \mu\left(\bigcup_{j=1}^{\infty} E_j\right);$$

(iv) if $\{E_j\}_{j\in\mathbb{N}}$ is a countable family of measurable sets such that

$$E_{j+1} \subset E_j, \forall j \in \mathbb{N}$$

and

$$\mu(E_1) < +\infty,$$

then

$$\lim_{j\to\infty} \mu\left(E_j\right) = \mu\left(\bigcap_{j=1}^{\infty} E_j\right).$$

Theorem 3.17 *Let E be a Lebesgue measurable subset of $\mathbb{R}^n$ whose measure is finite. Let $\left\{f_j\right\}$ be a sequence of measurable functions such that there exists the limit*

$$\lim_{j\to\infty} f_j(x)$$

and it is finite almost everywhere. Then, for every $\varepsilon > 0$ there exists a compact set $K \subset E$ which satisfies $|E \setminus K| < \varepsilon$ and

$$f_j \to f, \text{ as } j \to \infty, \quad \textbf{uniformly on} K.$$

Theorem 3.18 (Lusin) *Let E be a Lebesgue measurable subset of $\mathbb{R}^n$ which has finite measure, and let $f : E \to \overline{\mathbb{R}}$ such that*

$$|f(x)| < +\infty, \quad \text{a.e. } x \in E.$$

Then f is a measurable function in E if and only if for each $\varepsilon > 0$ there exists $K \subset E$, K closed, such that $|E \setminus K| < \varepsilon$ and $f_{|K}$ is a continuous function.

Now, let us define **the integral over the measure space** $(X, \mathcal{M}, \mu)$. Let s be a nonnegative simple function

$$s(x) = \sum_{i=1}^{N} c_j \chi_{E_j},$$

where $\left\{E_j\right\}_{1\leq j\leq N}$ is a finite family of measurable set pairwise disjoints and $c_j \geq 0$, $j = 1, \cdots, N$. If $E \in \mathcal{M}$, we set by definition

$$\int_E s(x)d\mu = \sum_{i=j}^{N} c_j \mu\left(E \cap E_j\right),$$

in which the convention $0 \cdot \infty = 0$ occurs. We call $\int_E s(x)d\mu$ "the integral of s over E".

Let us consider the measurable function

$$f : X \to [0, +\infty].$$

We call **the Lebesgue integral of** f *with respect to the measure* μ the following element of $\overline{\mathbb{R}}$

$$\int_E f(x)d\mu := \sup \left\{ \int_E s(x)d\mu : s \text{ simple function}, 0 \leq s \leq f \text{ in } E \right\}.$$

We say that f is a summable function over E if

$$\int_E f(x)d\mu < +\infty.$$

Theorem 3.19 *Let* $(X, \mathcal{M}, \mu)$ *be a measure space. Let*

$$f, g : X \to [0, +\infty]$$

be two measurable functions and let $E, F \in \mathcal{M}$. *The following properties hold true:*

(i) $\int_E f d\mu = \int_X f \chi_E d\mu$;
(ii) if $f \leq g$ *in* E *then* $\int_E f d\mu \leq \int_E g d\mu$;
(iii) if $E \subset F$ *then* $\int_E f d\mu \leq \int_F f d\mu$;
(iv) if $f = 0$ *in* E *then* $\int_E f d\mu = 0$;
(v) if $\mu(E) = 0$ *then* $\int_E f d\mu = 0$.

Theorem 3.20 (Monotone Convergence) *Let* $(X, \mathcal{M}, \mu)$ *be a measure space. Let* $\{f_j\}$ *be a sequence of nonnegative measurable functions which satisfy*

$$f_j(x) \leq f_{j+1}(x), \quad \forall x \in X, \quad \forall j \in \mathbb{N}.$$

Then

$$\int_X \lim_{j \to \infty} f_j(x)d\mu = \lim_{j \to \infty} \int_X f_j(x)d\mu.$$

Theorem 3.21 (Fatou) *Let* $(X, \mathcal{M}, \mu)$ *be a measure space. Let* $\{f_j\}$ *be a sequence of nonnegative measurable functions, then we have*

$$\int_X \liminf_{j \to \infty} f_j(x)d\mu \leq \liminf_{j \to \infty} \int_X f_j(x)d\mu.$$

Theorem 3.22 *Let* $(X, \mathcal{M}, \mu)$ *be a measure space and let*

$$f, g : X \to [0, +\infty]$$

be two measurable functions, then

$$\int_X (f+g)d\mu = \int_X f d\mu + \int_X g d\mu;$$

$$\int_X \lambda f d\mu = \lambda \int_X f d\mu, \quad \forall \lambda \in \mathbb{R},$$

with the convention $0 \cdot \int_X f d\mu = 0$.

Theorem 3.23 *Let* $(X, \mathcal{M}, \mu)$ *be a measure space and let* $\{f_j\}$ *be a sequence of nonnegative measurable functions, then we have*

$$\int_X \sum_{j=1}^{\infty} f_j d\mu = \sum_{j=1}^{\infty} \int_X f_j d\mu.$$

Furthermore, recall that if

$$f : X \to [0, +\infty],$$

is a measurable function and defining

$$\nu(E) = \int_E f d\mu, \quad \forall E \in \mathcal{M},$$

ν turns out to be a measure on X.

Definition 3.24 Let $(X, \mathcal{M}, \mu)$ be a measure space and let

$$f : X \to \mathbb{R}$$

be a measurable function. We say that f is summable over X provided

$$\int_X |f| d\mu < +\infty.$$

In such a case we set

$$\int_X f d\mu = \int_X f_+ d\mu - \int_X f_- d\mu,$$

where $f_+ = \max\{f, 0\}$, $f_+ = -\min\{f, 0\}$. We denote by $\mathcal{L}^1(X)$ the class of summable functions over X.

Theorem 3.25 *Let $(X, \mathcal{M}, \mu)$ be a measure space and let $f \in \mathcal{L}^1(X)$, then*

$$\mu\left(\{x \in X : \quad |f(x)| = +\infty\}\right) = 0.$$

Theorem 3.26 *$\mathcal{L}^1(X)$ is a vector space and the map*

$$\mathcal{L}^1(X) \ni f \to \int_X f d\mu \in \mathbb{R},$$

is linear. Furthermore, if $f, g \in \mathcal{L}^1(X)$ then

$$\max\{f, g\} \in \mathcal{L}^1(X)$$

and

$$\left|\int_X f d\mu\right| \leq \int_X |f| d\mu, \qquad \forall f \in \mathcal{L}^1(X).$$

Theorem 3.27 *If $f \in \mathcal{L}^1(X)$ then*

$$\forall \varepsilon > 0\, \exists \delta > 0 \text{ such that } \forall E \in \mathcal{M} \text{ and } \mu(E) < \delta \text{ we have } \int_E |f| d\mu < \varepsilon.$$

Theorem 3.28 (Dominated Convergence) *Let $\{f_j\}$ be a sequence of functions of $\mathcal{L}^1(X)$. Let us assume*

(i)

$$\lim_{j\to\infty} f_j(x) = f(x), \qquad a.e.\ x \in X,$$

(ii) there exists $g \in \mathcal{L}^1(X)$ such that

$$\left|f_j(x)\right| \leq g(x), \qquad a.e.\ x \in X,\ j \in \mathbb{N}.$$

Then $f \in \mathcal{L}^1(X)$ and

$$\lim_{j\to\infty} \int_X |f_j - f| \, d\mu = 0,$$

$$\lim_{j\to\infty} \int_X f_j d\mu = \int_X f d\mu.$$

The Monotone Convergence Theorem and the Dominated Convergence Theorem give that if $\{f_j\}$ is a sequence of measurable functions in $\mathcal{L}^1(X)$ satisfying

$$\sum_{j=1}^{\infty}\int_X |f_j|\, d\mu < +\infty,$$

then $\sum_{j=1}^{\infty} f_j$ converges almost everywhere to a function of $\mathcal{L}^1(X)$ and

$$\sum_{j=1}^{\infty}\int_X f_j d\mu = \int_X \sum_{j=1}^{\infty} f_j d\mu.$$

Theorem 3.29 (Differentiation Under the Integral Sign) *Let $(X, \mathcal{M}, \mu)$ be a measure space and let A be an open set of $\mathbb{R}^n$. Let*

$$F : A \times X \to \mathbb{R}$$

satisfy

(i) $F(x, \cdot) \in \mathcal{L}^1(X)$ for every $x \in A$,
(ii) $F(\cdot, y) \in C^1(A)$ for almost every $y \in X$.

If, for any $k = 1, \cdots, n$, there exist $g_k \in \mathcal{L}^1(X)$, $g_k \geq 0$ such that

$$\left|\partial_{x_k} F(x, y)\right| \leq g_k(y), \quad \forall y \in X, \quad \forall x \in A,$$

then the function

$$G(x) := \int_X F(x, y) d\mu(y), \quad x \in A$$

belongs to $C^1(A)$ and we have

$$\partial_{x_k} G(x) = \int_X \partial_{x_k} F(x, y) d\mu(y), \quad \forall x \in A,\ k = 1, \cdots, n.$$

Theorem 3.30 (Fubini–Tonelli) *Let*

$$f : \mathbb{R}^n \times \mathbb{R}^m \to \overline{\mathbb{R}}$$

be a measurable function. We have

(i) if $f \geq 0$ then: $f(x, \cdot)$ is measurable for almost every $x \in \mathbb{R}^n$, in addition the function

$$\mathbb{R}^n \ni x \to \int_{\mathbb{R}^m} f(x, y)dy \in [0, +\infty],$$

is measurable over $\mathbb{R}^n$ *and we have*

$$\int_{\mathbb{R}^n} \left(\int_{\mathbb{R}^m} f(x, y)dy \right) dx = \int_{\mathbb{R}^{n+m}} f(x, y)dxdy; \tag{3.29}$$

(ii) if $f \in \mathcal{L}^1(\mathbb{R}^{n+m})$ *then* $f(x, \cdot) \in \mathcal{L}^1(\mathbb{R}^m)$ *for almost every* $x \in \mathbb{R}^n$*, furthermore*

$$\int_{\mathbb{R}^m} f(\cdot, y)dy \in \mathcal{L}^1(\mathbb{R}^n)$$

and (3.29) *holds true.*

3.4.2 The L^p Spaces

Let $p \in [1, +\infty)$. Let $(X, \mathcal{M}, \mu)$ be a measurable space. We say that $f \in \mathcal{L}^p(X)$ if f is measurable and $|f|^p \in \mathcal{L}^1(X)$. $\mathcal{L}^p(X)$ is a vector space. We define $L^p(X)$ as the quotient space $\left(\mathcal{L}^p(X)/ \sim\right)$ where "$\sim$" is the equivalence relation on $\mathcal{L}^p(X)$ defined as follows: $f \sim g$ if and only if $f = g$ almost everywhere. We equip $L^p(X)$ with the norm

$$\|f\|_{L^p(X)} = \left(\int_X |f|^p d\mu \right)^{1/p}.$$

We say that $f \in \mathcal{L}^\infty(X)$ provided

$$\text{ess sup}\,|f| = \inf\{t \in \mathbb{R} : \quad \mu(\{|f(x)| > t\}) = 0\} < +\infty.$$

We define $L^\infty(X)$ similarly as we have previously defined $L^p(X)$, $p < +\infty$. We equip $L^\infty(X)$ with the norm

$$\|f\|_{L^\infty(X)} = \text{ess sup}\,|f|.$$

Minkowski Inequality If $p \in [1, +\infty]$, $f, g \in L^p(X)$ then

$$\|f + g\|_{L^p(X)} \leq \|f\|_{L^p(X)} + \|g\|_{L^p(X)}.$$

Hölder Inequality Let $p \in [1, +\infty]$, let us denote by p' (the conjugate of p) the element of $[1, +\infty]$ satisfying

$$\frac{1}{p} + \frac{1}{p'} = 1.$$

If $f \in L^p(X)$ and $g \in L^{p'}(X)$ then $fg \in L^1(X)$ and

$$\|fg\|_{L^1(X)} \le \|f\|_{L^p(X)} \|g\|_{L^{p'}(X)} .$$

If $\mu(X) < +\infty$, we have

$$p_2 \ge p_1 \Longrightarrow L^{p_2}(X) \subset L^{p_1}(X)$$

and the function

$$p \to \left(\frac{1}{\mu(X)} \int_X |f|^p d\mu\right)^{1/p},$$

turns out to be an increasing function (just apply Hölder inequality).

Moreover

$$\lim_{p\to\infty} \|f\|_{L^p(X)} = \|f\|_{L^\infty(X)} .$$

Theorem 3.31 *Let $(X, \mathcal{M}, \mu)$ be a measure space and let $p \in [1, +\infty]$. Then $L^p(X)$ is a* ***Banach space****. If $p = 2$, $L^2(X)$ is a* ***Hilbert space*** *equipped with the scalar product*

$$(f, g)_{L^2(X)} = \int_X fg d\mu, \quad \forall f, g \in L^2(X).$$

We say that the measure space $(X, \mathcal{M}, \mu)$ is σ-finite, if there exists a countable family $\{X_j\}_{j\in\mathbb{N}} \subset \mathcal{M}$ such that

$$X = \bigcup_{j\in\mathbb{N}} X_j, \quad \text{and} \quad \mu(X_j) < +\infty.$$

Theorem 3.32 *Let $(X, \mathcal{M}, \mu)$ be a σ-finite measure space and let $p \in [1, +\infty)$, then F is a bounded linear functional from $L^p(X)$ to $\mathbb{R}$ if and only if there exists $g \in L^{p'}(X)$ which satisfies*

$$F(f) = \int_X gf d\mu, \quad \forall f \in L^p(X).$$

Density and Separability in L^p. Let $E \subset \mathbb{R}^n$ be a Lebesgue measurable set.

The following theorems hold true.

Theorem 3.33 (Density of Simple Functions in $L^p(E)$, $1 \le p \le \infty$) *If $f \in L^p(E)$ and $p \in [1, +\infty]$, then for every $\varepsilon > 0$ there exists a simple function s such* *that*

$$\|f - s\|_{L^p(E)} < \varepsilon.$$

Theorem 3.34 (Density of $C_0^0(E)$ in $L^p(X)$, $1 \le p < \infty$) *If $f \in L^p(E)$ and $p \in [1, +\infty)$, then for every $\varepsilon > 0$ there exists $g \in C_0^0(E)$, such that*

$$\|f - g\|_{L^p(E)} < \varepsilon.$$

Theorem 3.35 *If $p \in [1, +\infty)$, then $L^p(E)$ is a separable space. If $|E| > 0$ then $L^\infty(E)$ is not a separable space.*

Theorem 3.36 *If $p \in [1, +\infty)$ and $f \in L^p(\mathbb{R}^n)$ then*

$$\lim_{\delta \to 0} \left(\sup_{|h|<\delta} \int_{\mathbb{R}^n} |f(x-h) - f(x)|^p dx \right) = 0.$$

Reflexivity
Let $\mathcal{X}$ be a normed space, let us denote by $\mathcal{X}'$ the space of the bounded linear functionals from $\mathcal{X}$ to $\mathbb{R}$. As it is well-known, $\mathcal{X}'$ is called the **dual** of $\mathcal{X}$ and $\mathcal{X}'$ turns out a Banach space (whether or not $\mathcal{X}$ is a Banach space) equipped with the norm

$$\|f\|_{\mathcal{X}'} = \sup\left\{\langle f, u\rangle : \ \|u\|_{\mathcal{X}} \le 1\right\}, \quad \forall f \in \mathcal{X}',$$

where

$$\langle f, u\rangle = f(u), \quad \forall u \in \mathcal{X},$$

$\langle \cdot, \cdot \rangle$ is said the duality bracket between $\mathcal{X}'$ and $\mathcal{X}$.

Definition 3.37 Let $\mathcal{X}$ be a Banach space. We say that $\mathcal{X}$ is a **reflexive space** provided

$$\forall v \in (\mathcal{X}')' \quad \exists u \in \mathcal{X}, \text{ such that } \langle v, f\rangle = \langle f, u\rangle \quad \forall f \in \mathcal{X}'.$$

Keep in mind that if $u \in \mathcal{X}$ then the map

$$X' \ni f \to j_u(f) := \langle f, u\rangle,$$

is a bounded linear functional, hence $j_u \in (\mathcal{X}')'$ and it can prove (by the Hahn–Banach Theorem [19], see Theorem 3.88) that

$$\|j_u\|_{(\mathcal{X}')'} = \|u\|_{\mathcal{X}}, \quad \forall u \in \mathcal{X}. \tag{3.30}$$

Therefore, it is defined the map j

$$\mathcal{X} \ni u \to j_u \in (\mathcal{X}')'.$$

The map j is injective (applying, again, Hahn–Banach Theorem). Consequently, j is an embedding and, by (3.30) it is an isometry. If $\mathcal{X}$ is a reflexive space then j is also surjective and, by the Open Map Theorem, j^{-1} is continuous too. Ultimately, if $\mathcal{X}$ is a reflexive space, we can identify $(\mathcal{X}')'$ with $\mathcal{X}$ by means of j. Recall that the Hilbert spaces are reflexive.

Definition 3.38 Let $\mathcal{X}$ be a Banach space and let $\{u_k\}$ be a sequence of $\mathcal{X}$. We say that $\{u_k\}$ weakly converges to $u \in \mathcal{X}$ and we write

$$u_k \rightharpoonup u, \quad \text{as } k \to \infty, \text{ (or } \{u_k\} \rightharpoonup u),$$

if

$$\langle f, u_k \rangle \to \langle f, u \rangle, \quad \text{as } k \to \infty, \ \forall f \in \mathcal{X}'.$$

If a weak limit exists, then it is unique (it can be again proved by Hahn–Banach Theorem).

Proposition 3.39 *Let $\mathcal{X}$ be a Banach space and $\{u_k\}$ be a sequence of $\mathcal{X}$.*
We have

(i) if $\{u_k\}$ weakly converges then it is bounded;
(ii) if $\{u_k\}$ weakly converges to u then

$$\|u\|_{\mathcal{X}} \le \liminf_{k\to\infty} \|u_k\|_{\mathcal{X}} .$$

Proof Let us prove (i), since for every $f \in \mathcal{X}'$, $\{\langle f, u_k \rangle\}$ is a converging sequence of $\mathbb{R}$, it is bounded, that is

$$\sup_{k\in\mathbb{N}} |\langle f, u_k \rangle| \le C(f) < +\infty, \quad \forall f \in \mathcal{X}'.$$

Now, applying the Banach–Steinhauss Theorem [19] (see also [88]) to the map

$$\mathcal{X}' \ni f \to T_k(f) := \langle f, u_k \rangle \in \mathbb{R}, \ \ k \in \mathbb{N}$$

we have

$$\sup_{k\in\mathbb{N}} \|u_k\|_{\mathcal{X}} = \sup_{k\in\mathbb{N}} \|T_k\|_{(\mathcal{X}')'} < +\infty.$$

Let us prove (ii). Since $\{u_k\} \rightharpoonup u, k \to \infty$, by (i) we get

$$\sup_{k \in \mathbb{N}} \|u_k\|_X < +\infty.$$

Moreover, for every $f \in X'$, we have

$$\begin{aligned}\langle f, u\rangle &= \lim_{k\to\infty} \langle f, u_k\rangle = \liminf_{k\to\infty} \langle f, u_k\rangle \leq \\ &\leq \liminf_{k\to\infty} \|f\|_{X'} \|u_k\|_X = \\ &= \|f\|_{X'} \liminf_{k\to\infty} \|u_k\|_X .\end{aligned}$$

Therefore

$$\|u\|_X \leq \liminf_{k\to\infty} \|u_k\|_X .$$

■

We recall

Theorem 3.40 (Banach–Alaoglu) *Let X be a reflexive Banach and let $\{u_k\}$ be a bounded sequence of X. Then there exists a subsequence $\{u_{k_j}\}$ which weakly converges.*

Theorem 3.41 *Let E be a measurable subset of $\mathbb{R}^n$. Then, $L^p(E)$ is a reflexive space, for every $p \in (1, +\infty)$.*

Convolution

Let f, g be two measurable functions defined in $\mathbb{R}^n$ with values in $\mathbb{R}$. Let $x \in \mathbb{R}^n$. If the function of the variable y, $f(x-y)g(y)$ is summable, we set

$$(f \star g)(x) = \int_{\mathbb{R}^n} f(x-y)g(y)dy.$$

If the function $(f \star g)(x)$ is defined for almost every $x \in \mathbb{R}^n$, we will call it the **convolution product** (or, simply, the **convolution**) of f and g.

Theorem 3.42 (The Young Inequality) *Let $f \in L^p(\mathbb{R}^n)$ and $g \in L^q(\mathbb{R}^n)$, where*

$$\frac{1}{p} + \frac{1}{q} \geq 1,$$

then $f \star g \in L^r(\mathbb{R}^n)$, where

$$\frac{1}{r} = \frac{1}{p} + \frac{1}{q} - 1, \quad \textit{that is} \quad \frac{1}{r'} = \frac{1}{p'} + \frac{1}{q'}$$

and we have

$$\|f \star g\|_{L^r(\mathbb{R}^n)} \leq \|f\|_{L^p(\mathbb{R}^n)} \|g\|_{L^q(\mathbb{R}^n)} .$$

Theorem 3.43 (The Generalized Young Inequality) *Let K be a measurable function on $\mathbb{R}^n \times \mathbb{R}^n$ (w.r.t. the Lebesgue measure) and let M be a positive number. Assume that*

$$\sup_{x\in\mathbb{R}^n} \int_{\mathbb{R}^n} |K(x, y)|\, dy \leq M, \quad \sup_{y\in\mathbb{R}^n} \int_{\mathbb{R}^n} |K(x, y)|\, dx \leq M. \tag{3.31}$$

Let $1 \leq p \leq +\infty$. For all $f \in L^p(\mathbb{R}^n)$ the following function is defined almost everywhere

$$(Tf)(x) = \int_{\mathbb{R}^n} K(x, y) f(y) dy \tag{3.32}$$

and

$$\|Tf\|_{L^p(\mathbb{R}^n)} \leq M \|f\|_{L^p(\mathbb{R}^n)} . \tag{3.33}$$

Proof If $p = +\infty$, by (3.31) we have

$$\begin{aligned}
|(Tf)(x)| &\leq \int_{\mathbb{R}^n} |K(x, y)|\, |f(y)|\, dy \leq \\
&\leq \|f\|_{L^\infty(\mathbb{R}^n)} \int_{\mathbb{R}^n} |K(x, y)|\, dy \leq \\
&\leq M \|f\|_{L^\infty(\mathbb{R}^n)} .
\end{aligned}$$

If $p = 1$, Theorem 3.30 and (3.31) imply

$$\begin{aligned}
\int_{\mathbb{R}^n} |(Tf)(x)| &\leq \int_{\mathbb{R}^n} \left(\int_{\mathbb{R}^n} |K(x, y)|\, |f(y)|\, dy \right) dx \leq \\
&\leq \int_{\mathbb{R}^n} \left(\int_{\mathbb{R}^n} |K(x, y)|\, dx \right) |f(y)|\, dy \leq \\
&\leq M \|f\|_{L^1(\mathbb{R}^n)} .
\end{aligned}$$

Now let us consider the case $1 < p < \infty$. By the Hölder inequality we have

$$|(Tf)(x)| \leq \left(\int_{\mathbb{R}^n} |K(x,y)|\,dy\right)^{1/p'} \left(\int_{\mathbb{R}^n} |K(x,y)|\,|f(y)|^p\,dy\right)^{1/p} \leq$$
$$\leq M^{1/p'} \left(\int_{\mathbb{R}^n} |K(x,y)|\,|f(y)|^p\,dy\right)^{1/p}.$$

Hence

$$\begin{aligned}
\int_{\mathbb{R}^n} |(Tf)(x)|^p\,dx &\leq M^{p/p'} \int_{\mathbb{R}^n} \left(\int_{\mathbb{R}^n} |K(x,y)|\,|f(y)|^p\,dy\right) dx = \\
&= M^{p/p'} \int_{\mathbb{R}^n} \left(\int_{\mathbb{R}^n} |K(x,y)|\,dx\right) |f(y)|^p\,dy \leq \\
&\leq M^{1+(p/p')} \int_{\mathbb{R}^n} |f(y)|^p\,dy = \\
&= M^p\,\|f\|^p_{L^p(\mathbb{R}^n)}.
\end{aligned}$$

The above inequality concludes the proof of (3.31). ■

Let $\eta \in C_0^\infty(\mathbb{R}^n)$ satisfy

(i) $\operatorname{supp} \eta \subset B_1$,
(ii) $\eta \geq 0$
(iii) $\int_{\mathbb{R}^n} \eta(x)dx = 1$.

η is named a **mollifier**. For instance, let

$$\tilde{\eta}(s) = \begin{cases} c_n \exp\left\{-\frac{1}{1-4s^2}\right\}, & \text{for } s \in [0, 1/2) \\ \\ 0, & \text{otherwise} \end{cases},$$

where

$$c_n = \left(\int_0^{1/2} s^{n-1} \exp\left\{-\frac{1}{1-4s^2}\right\} ds\right)^{-1},$$

then

$$\eta(x) = \tilde{\eta}(|x|),$$

is a mollifier.
Here and in the sequel we set, for $\varepsilon > 0$,

$$\eta_\varepsilon(x) = \varepsilon^{-n} \eta\left(\varepsilon^{-1} x\right).$$

Theorem 3.44

(i) if $f \in L^p(\mathbb{R}^n)$, $p \in [1, +\infty)$, *then*

$$\eta_\varepsilon \star f \to f, \quad \text{as } \varepsilon \to 0, \quad \text{in } L^p(\mathbb{R}^n).$$

(ii) If f *is uniformly continuous and bounded in* $\mathbb{R}^n$, *then*

$$\eta_\varepsilon \star f \to f, \quad \text{as } \varepsilon \to 0, \quad \text{uniformly in } \mathbb{R}^n.$$

Remark Let E be a measurable set of $\mathbb{R}^n$, $f \in L^p(E)$, where $p \in [1, +\infty)$, then

$$\|\eta_\varepsilon \star (f\chi_E) - f\|_{L^p(E)} \to 0, \quad \text{as } \varepsilon \to 0,$$

where

$$\eta_\varepsilon \star f = \eta_\varepsilon \star (f\chi_E) = \int_E \eta_\varepsilon(x-y) f(y) dy.$$

♦

Theorem 3.45 *Let* Ω *be an open set of* $\mathbb{R}^n$, $f \in L^p(\Omega)$, *where* $p \in [1, +\infty)$, *then*

$$\eta_\varepsilon \star f \in C^\infty(\Omega),$$

$$\partial^\alpha (\eta_\varepsilon \star f) = \left(\partial^\alpha \eta_\varepsilon\right) \star f$$

and

$$\eta_\varepsilon \star f \to f, \quad \text{in } L^p(\Omega), \text{ as } \varepsilon \to 0.$$

Hence $C^\infty(\Omega)$ *is dense in* $L^p(\Omega)$.

We also have

Theorem 3.46 (Density of $C_0^\infty(\Omega)$ in $L^p(\Omega)$, $1 \le p < +\infty$) *Let* Ω *be an open set of* $\mathbb{R}^n$, $p \in [1, +\infty)$, *then* $C_0^\infty(\Omega)$ *is dense in* $L^p(\Omega)$.

Let Ω be an open set of $\mathbb{R}^n$ and $p \in [1, +\infty]$. We denote by $L^p_{loc}(\Omega)$ the space of measurable functions defined on Ω such that for every compact set K we have $f_{|K} \in L^p(K)$. Let $\{u_k\}$ be a sequence of $L^p_{loc}(\Omega)$, we write

$$u_k \to u, \quad \text{as } k \to \infty, \text{ in } L^p_{loc}(\Omega),$$

provided $u \in L^p_{loc}(\Omega)$ and for every compact $K \subset \Omega$, we have

$$(u_k)_{|K} \to u_K, \quad \text{as } k \to \infty, \text{ in } L^p(K).$$

Let us define, for any $\varepsilon > 0$,

$$\Omega_\varepsilon = \{x \in \Omega : \operatorname{dist}(x, \partial\Omega) > \varepsilon\}.$$

If $f \in L^1_{loc}(\Omega)$, then $(\eta_\varepsilon \star f)(x)$ is defined for every $x \in \Omega_\varepsilon$, and we may rephrase Theorems 3.44 and 3.45 as follows.

Theorem 3.47 *Let Ω be an open set of $\mathbb{R}^n$ and $f \in L^1_{loc}(\Omega)$. Then*

(i) $\eta_\varepsilon \star f \in C^\infty(\Omega_\varepsilon)$;
(ii) if $f \in C^0(\Omega)$ then $\eta_\varepsilon \star f \to f$, uniformly on the compact sets of Ω as $\varepsilon \to 0$;
(iii) if $p \in [1, +\infty)$ and $f \in L^p_{loc}(\Omega)$ then

$$\eta_\varepsilon \star f \to f, \quad \text{as } \varepsilon \to 0, \text{ in } L^p_{loc}(\Omega).$$

Let f be a measurable function defined on Ω, open set of $\mathbb{R}^n$, and let

$$O = \{A \subset \Omega : A \text{ open and } f = 0 \text{ in } A \text{ a.e.}\},$$

the set

$$\operatorname{supp}_{\text{ess}} f = \Omega \setminus \bigcup_{A \in O} A,$$

is named the **essential support** of f. Hereafter, if there is no ambiguity, instead of "essential support of f" we will simply say "support of f". We recall that then the essential support of a function if $u \in C^0(\mathbb{R}^n)$ is equal to the support defined in Sect. 3.2 (see also Remark 4.11, Sect. 4.4).

Let us conclude this section with the following Proposition that will be useful later.

Proposition 3.48 *Let Ω be an open set of $\mathbb{R}^n$ and let $f \in L^1_{loc}(\Omega)$. We have*

$$\int_\Omega f(x)\phi(x)dx = 0, \quad \forall \phi \in C^\infty_0(\Omega) \Longrightarrow f(x) = 0, \quad a.e.\ x \in \Omega. \tag{3.34}$$

Proof It is not restrictive to assume $f \in L^1(\Omega)$. Let us suppose that

$$\int_\Omega f(x)\phi(x)dx = 0, \quad \forall \phi \in C^\infty_0(\Omega). \tag{3.35}$$

Let η be a mollifier and let us denote by

$$\Omega_k = \left\{x \in \Omega : \operatorname{dist}(x, \partial\Omega) > \frac{1}{k}\right\}.$$

Let $\nu \in \mathbb{N}$ be fixed. If $x \in \Omega_\nu$ and $k \geq \nu$ then

$$\operatorname{supp} \eta_{1/k}(x - \cdot) \subset \Omega_\nu.$$

Hence by (3.35) we have

$$\left(\eta_{1/k} \star f\right)(x) = \int_\Omega f(y)\eta_{1/k}(x-y)dy = 0. \tag{3.36}$$

Let K be a compact subset of Ω_ν. Theorem 3.47 yields

$$\lim_{k\to\infty} \eta_{1/k} \star f = f, \quad \text{in } L^1(K). \tag{3.37}$$

Therefore, by (3.36) and (3.37) we have

$$\int_K |f(x)|\,dx = \lim_{k\to\infty} \int_K \left|\left(\eta_{1/k} \star f\right)(x)\right| dx = 0.$$

Hence, $f = 0$ almost everywhere in K and due to the arbitrariness of K we have

$$f = 0, \quad \text{a.e. in } \Omega.$$

■

3.5 Partition of Unity

Let us start by the following

Lemma 3.49 *Let Ω be an open set of $\mathbb{R}^n$ and let K be a compact set contained in Ω, then there exists $\varphi \in C_0^\infty(\mathbb{R}^n)$ such that* $\operatorname{supp}\varphi \subset \Omega$, $0 \leq \varphi \leq 1$ *and* $\varphi = 1$ *in a neighborhood of K.*

Proof For any $\varepsilon > 0$, let us denote by

$$K^{(\varepsilon)} = \left\{x \in \mathbb{R}^n : \operatorname{dist}(x, K) \leq \varepsilon\right\}.$$

Let ε_0 and ε_1 satisfy

$$0 < \varepsilon_0 < \varepsilon_1 < \varepsilon_0 + \varepsilon_1 < \operatorname{dist}\left(K, \mathbb{R}^n \setminus \Omega\right).$$

Let us define

$$\varphi(x) = \int_{K^{(\varepsilon_1)}} \eta_{\varepsilon_0}(x-y)dy.$$

It can be easily checked that $\varphi \in C_0^\infty(\mathbb{R}^n)$,

$$\operatorname{supp}\varphi \subset K^{(\varepsilon_0+\varepsilon_1)} \subset \Omega$$

and

$$\varphi(x) = 1, \quad \forall x \in K^{(\varepsilon_0)}.$$

■

Lemma 3.50 *Let K be a compact set of $\mathbb{R}^n$ and let $V_1, V_2, \cdots, V_l$ be some open sets of $\mathbb{R}^n$ satisfying*

$$K \subset \bigcup_{j=1}^{l} V_j.$$

Then there exist the functions $\zeta_1, \cdots, \zeta_l \in C_0^\infty(\mathbb{R}^n)$ which satisfy

$$\operatorname{supp}\zeta_j \subset V_j, \quad j = 1, \cdots, l,$$

$$0 \le \zeta_j \le 1, \quad j = 1 \cdots, l; \quad \sum_{j=1}^{l} \zeta_j \le 1, \quad \textit{on } \mathbb{R}^n$$

$$\sum_{j=1}^{l} \zeta_j = 1, \quad \textit{in a neighborhood of } K.$$

Proof Let us denote, for any $\varepsilon > 0$ and $V \subset \mathbb{R}^n$,

$$V_\varepsilon = \{x \in V : \operatorname{dist}(x, \partial V) > \varepsilon\}.$$

We have

$$K \subset \bigcup_{\varepsilon>0} \bigcup_{j=1}^{l} V_{\varepsilon,j}$$

and by the compactness of K it follows that there exists $\varepsilon_0 > 0$ such that

$$K \subset \bigcup_{j=1}^{l} V_{\varepsilon_0,j} \subset \bigcup_{j=1}^{l} \overline{V_{\varepsilon_0,j}} \subset \bigcup_{j=1}^{l} V_j,$$

(because $\overline{V}_{\varepsilon_0,j} \subset V_j$). Hence, denoting by

$$K_j = K \cap \overline{V_{\varepsilon_0 \cdot j}} \quad , j = 1, \cdots, l$$

we have immediately that K_j is a compact set, $K_j \subset V_j$, for any $j = 1, \cdots, l$ and

$$K \subset \bigcup_{j=1}^{l} K_j.$$

By Lemma 3.49, we derive that for every $j \in \{1, \cdots, l\}$ there exist $\varphi_j \in C_0^\infty\left(V_j\right)$ satisfying

$$0 \le \varphi_j \le 1, \quad \varphi_j = 1, \text{ in a neighborhood, } W_j, \text{ of } K_j.$$

Now, defining

$$\zeta_1 = \varphi_1, \quad \zeta_2 = \varphi_2\left(1-\varphi_1\right), \quad \cdots, \quad \zeta_j = \varphi_j\left(1-\varphi_1\right)\cdots\left(1-\varphi_{j-1}\right),$$

we get

$$\begin{aligned}
&\sum_{j=0}^{l} \zeta_j = \varphi_1 + \varphi_2\left(1-\varphi_1\right) + \cdots + \varphi_l\left(1-\varphi_1\right)\cdots\left(1-\varphi_{l-1}\right) = \\
&= 1 - \left(1-\varphi_1\right) + \varphi_2\left(1-\varphi_1\right) + \cdots + \varphi_l\left(1-\varphi_1\right)\cdots\left(1-\varphi_{l-1}\right) = \\
&= 1 - \left(1-\varphi_1\right)\left(1-\varphi_2\right) + \varphi_3\left(1-\varphi_1\right)\left(1-\varphi_2\right) + \\
&\quad + \cdots + \varphi_l\left(1-\varphi_1\right)\cdots\left(1-\varphi_{l-1}\right) = \\
&= \cdots = \\
&= 1 - \left(1-\varphi_1\right)\left(1-\varphi_2\right)\cdots\left(1-\varphi_l\right).
\end{aligned}$$

Therefore, if

$$x \in \bigcup_{j=1}^{l} W_j,$$

there exists $\overline{j} \in \{1, \cdots, l\}$ such that $x \in W_{\overline{j}}$, hence $\varphi_{\overline{j}}(x) = 1$ and

$$\sum_{j=0}^{l} \zeta_j(x) = 1, \quad \forall x \in \bigcup_{j=1}^{l} W_j.$$

Since $\bigcup_{j=1}^{l} W_j$ is a neighborhood of K, the Lemma is proved. ■

In what follows, we will say that the set of functions $\zeta_1, \cdots, \zeta_l$ is a **partition of the unity subordinate to the covering** $\{V_j\}_{1\leq j\leq l}$.

Theorem 3.51 (Partition of Unity) *Let Ω be an open set of $\mathbb{R}^n$ and let $V_1, V_2, \cdots, V_l$ be open sets of $\mathbb{R}^n$ satisfying*

$$\partial\Omega \subset \bigcup_{j=1}^{l} V_j.$$

Then there exist the functions $\zeta_0, \zeta_1, \cdots, \zeta_l \in C^\infty(\mathbb{R}^n)$ such that

$$(\zeta_0)_{|\Omega} \in C_0^\infty(\Omega),$$

$$\operatorname{supp}\zeta_0 \subset \mathbb{R}^n \setminus \partial\Omega; \quad \operatorname{supp}\zeta_j \subset V_j, \quad j = 1, \cdots, l,$$

$$\sum_{j=0}^{l} \zeta_j = 1, \quad on\ \mathbb{R}^n; \quad 0 \leq \zeta_j \leq 1, \quad j = 0, 1 \cdots, l.$$

Proof Let us consider a partition of unity subordinate to the covering, $\{V_j\}_{1\leq j\leq l}$, of $\partial\Omega$. Let us denote by

$$O_j = \operatorname{supp}\varphi_j, \quad j = 1, \cdots, l$$

and set

$$V_0 = \mathbb{R}^n \setminus \bigcup_{j=1}^{l} O_j.$$

Let

$$\zeta_0 = 1 - \sum_{j=1}^{l} \zeta_j.$$

We get trivially $\zeta_0 \in C_0^\infty(\mathbb{R}^n)$ and

$$\sum_{j=0}^{l} \zeta_j(x) = 1, \quad \forall x \in \mathbb{R}^n,$$

$$\operatorname{supp}\zeta_0 \subset \mathbb{R}^n \setminus \partial\Omega.$$

Moreover, by Lemma 3.50 there exists an open neighborhood, $\mathcal{U}$, of $\partial\Omega$ such that

$$\sum_{j=1}^{l} \zeta_j(x) = 1, \quad \forall x \in \mathcal{U}.$$

Hence

$$\zeta_0(x) = 0, \quad \forall x \in \mathcal{U}.$$

Therefore

$$\mathrm{supp}(\zeta_0)_{|\Omega} \subset \overline{\Omega} \setminus \mathcal{U} \subset \Omega,$$

this implies

$$(\zeta_0)_{|\Omega} \in C_0^\infty(\Omega)$$

concluding the proof. ■

Remark By Theorem 3.51 it is evident that $V_0 \cap \Omega, V_1, \cdots, V_l$ is a covering of $\overline{\Omega}$ and $(\zeta_0)_{|\Omega}, \zeta_1, \cdots, \zeta_l$ is a partition of the unity subordinate to that covering. ♦

3.6 The Lebesgue Differentiation Theorem

In this section we prove the following

Theorem 3.52 (Lebesgue Differentiation) *If $f \in L^1_{loc}(\mathbb{R}^n)$ then*

$$\lim_{r\to 0} \fint_{B_r(x)} f(y)dy = f(x), \quad a.e.\ x \in \mathbb{R}^n, \tag{3.38}$$

where

$$\fint_{B_r(x)} f(y)dy = \frac{1}{|B_r(x)|} \int_{B_r(x)} f(y)dy.$$

In order to prove Theorem 3.52 we need some preliminary lemmas and propositions. We start by

Lemma 3.53 (Covering) *Let E be a Lebesgue measurable subset of $\mathbb{R}^n$ and let $\mathcal{B}$ be a family of balls of $\mathbb{R}^n$ satisfying*

$$E \subset \bigcup_{B\in\mathcal{B}} B.$$

and,

$$\sup_{B\in\mathcal{B}} d(B) < +\infty,$$

where $d(B)$ is the diameter of B. Then there exists a countable (or finite) family, $\{B_k\}_{k\in\Lambda} \subset \mathcal{B}$ which satisfies

$$B_j \cap B_k = \emptyset, \quad \textit{for } j \neq k, \ j, k \in \Lambda$$

and

$$\sum_{k\in\Lambda} |B_k| \geq 5^{-n}|E|.$$

(If $|E| = +\infty$ the sum on the left hand side is equal to $+\infty$).

Proof of the Lemma Firstly we construct the family $\mathcal{B}_0 := \{B_k\}_{k\in\Lambda}$. Let $B_1 \in \mathcal{B}$ satisfy

$$d(B_1) \geq \frac{1}{2}\sup\{d(B) : \ B \in \mathcal{B}\}$$

and set

$$d_1 = \sup\{d(B) : \ B \in \mathcal{B}, \ B \cap B_1 = \emptyset\} < +\infty.$$

Let us consider the set

$$\mathcal{F}_1 = \left\{B \in \mathcal{B} : \ B \cap B_1 = \emptyset, \ d(B) \geq \frac{1}{2}d_1\right\}.$$

If $\mathcal{F}_1 = \emptyset$, then we choose $\mathcal{B}_0 := \{B_1\}$; otherwise, if $\mathcal{F}_1 \neq \emptyset$, we choose $B_2 \in \mathcal{F}_1$ and we continue the process. Let us suppose we have chosen $B_1, \cdots, B_i$, let us consider the family

$$\mathcal{F}_i = \left\{B \in \mathcal{B} : \ B \cap \bigcup_{j=1}^{i} B_j = \emptyset, \ d(B) \geq \frac{1}{2}d_i\right\},$$

where

$$d_i = \sup\left\{d(B) : \ B \in \mathcal{B}, \ B \cap \bigcup_{j=1}^{i} B_j = \emptyset\right\}.$$

If $\mathcal{F}_i = \emptyset$, then we choose $\mathcal{B}_0 := \{B_1, \cdots, B_i\}$; otherwise, if $\mathcal{F}_i \neq \emptyset$, then we choose $B_{i+1} \in \mathcal{F}_i$ as above and continue the process. All in all, we construct a finite or infinite, countable family, $\{B_i\}_{i\in\Lambda} \subset \mathcal{B}$ such that

$$B_j \cap B_k = \emptyset, \quad \text{for } j \neq k,\ j,k \in \Lambda$$

and

$$d(B_i) \geq \frac{1}{2} \sup \left\{ d(B) : B \in \mathcal{B},\ B \cap \bigcup_{j=1}^{i-1} B_j = \emptyset \right\}, \quad \text{for } i \geq 2.$$

Let us consider the case in which $\mathcal{B}_0$ is finite. Let

$$\mathcal{B}_0 = \{B_1, \cdots, B_k\}. \tag{3.39}$$

Let us denote by

$$\mathcal{A} = \left\{ B \in \mathcal{B} : B \cap \bigcup_{j=1}^{k} B_j = \emptyset \right\},$$

$$C = \mathcal{B} \setminus \mathcal{A} = \left\{ B \in \mathcal{B} : B \cap \bigcup_{j=1}^{k} B_j \neq \emptyset \right\},$$

$$A = \bigcup_{B \in \mathcal{A}} B,$$

$$C = \bigcup_{B \in C} B.$$

We have

$$E \subset \bigcup_{B \in \mathcal{B}} B = A \cup C. \tag{3.40}$$

Claim We have

$$\mathcal{A} = \emptyset.$$

Proof of Claim We argue by contradiction. Let us assume that $\mathcal{A} \neq \emptyset$. Let us first observe that it cannot occur that

$$d(B) < \frac{1}{2}\sup\left\{d(B) : B \in \mathcal{B},\ B \cap \bigcup_{j=1}^{k} B_j = \emptyset\right\}, \quad \forall B \in \mathcal{A}$$

otherwise we would have

$$\begin{aligned} 0 < \sup\left\{d(B) : B \in \mathcal{B},\ B \cap \bigcup_{j=1}^{k} B_j = \emptyset\right\} \leq \\ \leq \frac{1}{2}\sup\left\{d(B) : B \in \mathcal{B},\ B \cap \bigcup_{j=1}^{k} B_j = \emptyset\right\}, \end{aligned} \tag{3.41}$$

which is evidently absurd.

Therefore, there exists $\tilde{B} \in \mathcal{A}$ such that

$$d\left(\tilde{B}\right) \geq \frac{1}{2}\sup\left\{d(B) : B \in \mathcal{B},\ B \cap \bigcup_{j=1}^{k} B_j = \emptyset\right\}.$$

In particular, we have $\tilde{B} \notin \{B_1, \cdots, B_k\}$ and, consequently, the process of construction of $\mathcal{B}_0$ does not stop, but we had assumed the opposite (i.e., $\mathcal{B}_0$ finite family consisting of k elements) and thus we have a contradiction. **This concludes the proof of the Claim.**

By what is proved in the Claim and by (3.40), we have

$$E \subset C. \tag{3.42}$$

Now, let us prove

$$E \subset \bigcup_{j=1}^{k} B_j^*, \tag{3.43}$$

where B_j^* denotes the ball having the same center as B_j with radius equal to 5 times the radius of B_j. Let $x \in E$, then (3.42) implies that there exists $\hat{B} \in \mathcal{B}$ which satisfies

$$\hat{B} \cap \bigcup_{j=1}^{k} B_j \neq \emptyset.$$

Let $j_0 \in \{1, \cdots, k\}$ be such that

$$x \in \hat{B},$$

$$\hat{B} \cap B_{j_0} \neq \emptyset$$

and

$$\hat{B} \cap \bigcup_{j=1}^{j_0-1} B_j = \emptyset,$$

(if $j_0 = 1$, $\bigcup_{j=1}^{j_0-1} B_j$ is the empty set). By the third relationship we have

$$d\left(B_{j_0}\right) \geq \frac{1}{2} d\left(\hat{B}\right).$$

By the latter and by $\hat{B} \cap B_{j_0} \neq \emptyset$ we easily obtain

$$\hat{B} \subset B_{j_0}^*.$$

Hence

$$x \in \bigcup_{j=1}^{k} B_j^*,$$

and (3.43) is proved. Moreover we have

$$|E| \leq \sum_{j=1}^{k} \left|B_j^*\right| = 5^n \sum_{j=1}^{k} |B_j|.$$

Now let us consider the case where $\mathcal{B}_0$ is infinite. Hence, in such a case we have $\mathcal{B}_0 = \{B_k\}_{k \in \mathbb{N}}$. If

$$\sum_{k=1}^{\infty} |B_k| = +\infty,$$

there is nothing to prove. Let us assume that

$$\sum_{k=1}^{\infty} |B_k| < +\infty,$$

which implies

$$\lim_{k\to\infty} d\left(B_k\right) = 0.$$

If $\tilde{B} \in \mathcal{B}$ and

$$k_0 = \min\left\{j \in \mathbb{N} : \ d\left(B_{j+1}\right) < \frac{1}{2} d\left(\tilde{B}\right)\right\}, \tag{3.44}$$

then

$$\tilde{B} \cap \bigcup_{i=1}^{k_0} B_i \neq \emptyset. \tag{3.45}$$

To prove (3.45) it suffices to notice that if it were

$$\tilde{B} \cap \bigcup_{i=1}^{k_0} B_i = \emptyset,$$

we would have

$$d\left(B_{k_0+1}\right) \geq \frac{1}{2} \sup\left\{d(B) : \ B \in \mathcal{B},\ B \cap \bigcup_{j=1}^{k_0} B_j = \emptyset\right\} \geq \frac{1}{2} d\left(\tilde{B}\right),$$

which contradicts (3.44).

Since (3.45) holds true, we set

$$j_0 = \min\left\{j \in \{1, \cdots, k_0\} : \ B_j \cap \tilde{B} \neq \emptyset\right\} \tag{3.46}$$

obtaining

$$\tilde{B} \cap \bigcup_{j=1}^{j_0-1} B_j = \emptyset.$$

Hence (by (3.46) and by the definition of $\mathcal{B}_0$)

$$\begin{cases} \tilde{B} \cap B_{j_0} \neq \emptyset \\ \\ d\left(B_{j_0}\right) \geq \frac{1}{2} d\left(\tilde{B}\right) \end{cases}.$$

From which it follows that for every $B \in \mathcal{B}$ there exists $B_{j_0} \in \mathcal{B}_0$ such that $B \subset B^*_{j_0}$. Therefore, arguing as in the finite case, we have

$$E \subset \bigcup_{B \in \mathcal{B}} B = \bigcup_{j=1}^{\infty} B^*_j$$

and by the latter the thesis follows. ■

Now we introduce the notion of **maximal function**. Let $f \in L^1(\mathbb{R}^n)$. The following function is called the maximal function associated to f

$$M(f)(x) = \sup_{r>0} \frac{1}{|B_r(x)|} \int_{B_r(x)} |f(y)| dy, \quad \forall x \in \mathbb{R}^n. \tag{3.47}$$

Let us observe that $M(f)$ is a measurable function. More precisely, the following Proposition holds true

Proposition 3.54 *If $f \in L^1(\mathbb{R}^n)$ then $M(f)$ is a lower semicontinuous function.*

Proof If f is identically 0, we have $M(f) \equiv 0$. Let us assume that f is not identically 0. Hence

$$M(f)(x) > 0, \quad \forall x \in \mathbb{R}^n. \tag{3.48}$$

Fix $t \geq 0$ and let us prove that

$$A = \left\{x \in \mathbb{R}^n : M(f)(x) > t\right\}$$

is an open set.

In the case where $t = 0$, we have $A = \mathbb{R}^n$. In the case where $t > 0$, let $x_0 \in A$ and $0 < \varepsilon < M(f)(x_0) - t$. By the definition of $M(f)$, there exists $r_\varepsilon > 0$ such that

$$\frac{1}{\left|B_{r_\varepsilon}(x_0)\right|} \int_{B_{r_\varepsilon}(x_0)} |f(y)| dy > M(f)(x_0) - \varepsilon > t.$$

Now, let $0 < \eta < M(f)(x_0) - t - \varepsilon$. Since $f \in L^1(\mathbb{R}^n)$, there exists $\delta > 0$ such that if $|x_0 - x| < \delta$ then

$$\left| \int_{B_{r_\varepsilon}(x_0)} |f(y)| dy - \int_{B_{r_\varepsilon}(x)} |f(y)| dy \right| < \eta \left|B_{r_\varepsilon}\right|.$$

Hence

$$\begin{aligned}
M(f)(x) &\geq \frac{1}{\left|B_{r_\varepsilon}(x)\right|} \int_{B_{r_\varepsilon}(x)} |f(y)| dy > \\
&> \frac{1}{\left|B_{r_\varepsilon}(x_0)\right|} \int_{B_{r_\varepsilon}(x_0)} |f(y)| dy - \eta > \\
&> M(f)(x_0) - \varepsilon - \eta > t,
\end{aligned}$$

which implies

$$B_\delta(x_0) \subset A.$$

Therefore A is open. ■

Lemma 3.55 *Let $f \in L^1(\mathbb{R}^n)$. Let $M(f)$ be the maximal function of f, then*

$$\left|\left\{x \in \mathbb{R}^n : M(f)(x) > t\right\}\right| \leq \frac{5^n}{t} \int_{\mathbb{R}^n} |f(y)| dy. \tag{3.49}$$

Proof Set

$$E_t = \left\{x \in \mathbb{R}^n : M(f)(x) > t\right\}.$$

If $x \in E_t$, then $M(f)(x) > t$. Hence there exists $r_x > 0$ such that

$$\int_{B_{r_x}(x)} |f(y)| dy > t \left|B_{r_x}(x)\right|.$$

For the sake of brevity, set $B_x = B_{r_x}(x)$ so that we have

$$\frac{1}{t} \int_{B_x} |f(y)| dy > |B_x|. \tag{3.50}$$

Now, we have trivially that $\{B_x\}_{x \in E_t}$ is a covering of E_t; moreover (3.50) and $f \in L^1(\mathbb{R}^n)$ give

$$\sup_{x \in E_t} |B_x| < +\infty.$$

Thus, the assumptions of Lemma 3.53 are satisfied and therefore there exists a finite or countable, pairwise disjoint family of balls, $\{B_k\}_{k \in \Lambda}$, such that

$$\sum_{k \in \Lambda} |B_k| \geq 5^{-n} |E_t|.$$

Now, recalling that $B_k \cap B_j = \emptyset$, for any $j \neq k$, we obtain

$$\begin{aligned}\int_{\mathbb{R}^n} |f(y)|dy &\geq \int_{\bigcup_{k\in\Lambda} B_k} |f(y)|dy = \\ &= \sum_{k\in\Lambda} \int_{B_k} |f(y)|dy \geq \\ &\geq t \sum_{k\in\Lambda} |B_k| \geq \\ &\geq 5^{-n} t |E_t|.\end{aligned}$$

Therefore (3.49) is proved. ■

Remark 3.5 The function $M(f)$ may take the value $+\infty$, however it is almost everywhere finite. As a matter of fact, by Lemma 3.55 we get

$$\begin{aligned}\left|\left\{x \in \mathbb{R}^n : M(f)(x) = +\infty\right\}\right| &\leq \left|\left\{x \in \mathbb{R}^n : M(f)(x) > t\right\}\right| \leq \\ &\leq \frac{5^n}{t} \int_{\mathbb{R}^n} |f(y)|dy, \quad \forall t > 0\end{aligned}$$

hence, passing to the limit as t that goes to $+\infty$, we have

$$\left|\left\{x \in \mathbb{R}^n : M(f)(x) = +\infty\right\}\right| = 0.$$

♦

Remark 3.6 Inequality (3.49), apart from the value of the constant 5^n, cannot be improved. To show this it suffices to consider functions $f \in L^1(\mathbb{R}^n)$ which approximate the Dirac measure concentrated at 0. For instance

$$f_\varepsilon = \frac{\chi_{B_\varepsilon}}{|B_\varepsilon|}.$$

Proceeding formally (the reader takes care of the details), we consider $f = \delta(x)$ (the Dirac delta). For this choice, we have

$$M(f)(x) = \frac{1}{c_n |x|^n},$$

where c_n is the measure of unit ball of $\mathbb{R}^n$. Therefore

$$\left|\left\{x \in \mathbb{R}^n : M(f)(x) > t\right\}\right| = \frac{1}{t} = \frac{1}{t} \int_{\mathbb{R}^n} |f(y)|dy.$$

♦

Remark 3.7 Let us observe that, unless in the trivial case where f is identically equal to 0, we have

$$M(f) \notin L^1\left(\mathbb{R}^n\right).$$

In this respect, we prove

$$M(f)(x) \geq \frac{C}{|x|^n}, \qquad \text{for } |x| \geq 1. \tag{3.51}$$

Indeed, since f does not vanish identically, there exists $t_0 > 0$ such that

$$0 < |E| < +\infty;$$

where

$$E = \left\{x \in \mathbb{R}^n : |f(x)| > t_0\right\}.$$

Let $r_0 > 0$ satisfy

$$\left|E \cap B_{r_0}\right| \geq \frac{1}{2}|E|.$$

For any $x \in \mathbb{R}^n$, we have (since $B_{r_0} \subset B_{r_0+|x|}(x)$)

$$\begin{aligned} M(f)(x) &\geq \frac{1}{\left|B_{r_0+|x|}(x)\right|} \int_{B_{r_0+|x|}(x)} |f(y)|dy \geq \\ &\geq \frac{1}{c_n\left(|x|+r_0\right)^n} \int_{B_{r_0}} |f(y)|dy \geq \\ &\geq \frac{t_0|E|}{2c_n\left(|x|+r_0\right)^n}, \end{aligned}$$

from which we have (3.51) with $C = \frac{t_0|E|}{2c_n(1+r_0)^n}$.

It can be proved that if $f \in L^p\left(\mathbb{R}^n\right)$, where $1 < p \leq +\infty$, then $M(f) \in L^p\left(\mathbb{R}^n\right)$ and

$$\|M(f)\|_{L^p(\mathbb{R}^n)} \leq C\,\|f\|_{L^p(\mathbb{R}^n)}, \quad \forall f \in L^p\left(\mathbb{R}^n\right). \tag{3.52}$$

For more insights into the maximal function, we refer to [123, Ch. 1], [124]. ♦

Proof of Theorem 3.52 Provided that f is replaced by $f\chi_{B_R}$ with arbitrary R, we may assume $f \in L^1(\mathbb{R}^n)$. Let us denote by

$$f_r(x) = \frac{1}{|B_r(x)|}\int_{B_r(x)} f(y)dy, \qquad x \in \mathbb{R}^n$$

and notice that

$$f_r = \varphi_r \star f,$$

where

$$\varphi_r(x) = r^{-n}\varphi_1\left(r^{-1}x\right),$$

$$\varphi_1 = \frac{1}{|B_1|}\chi_{B_1}.$$

Hence

$$\lim_{r\to 0}\|f_r - f\|_{L^1(\mathbb{R}^n)} = 0.$$

Consequently, there exists a sequence $\{r_k\}$ such that

$$\{r_k\} \to 0^+$$

and

$$\lim_{k\to\infty} f_{r_k}(x) = f(x), \qquad \text{a.e. } x \in \mathbb{R}^n. \tag{3.53}$$

Now, let us denote by

$$\Omega f(x) = \limsup_{r\to 0} f_r(x) - \liminf_{r\to 0} f_r(x).$$

By Remark 3.5 we get that $\limsup_{r\to 0^+} f_r(x)$, $\liminf_{r\to 0^+} f_r(x)$ are finite almost everywhere. As a matter of fact, we have

$$\left|\limsup_{r\to 0} f_r(x)\right|, \left|\liminf_{r\to 0} f_r(x)\right| \leq M(f)(x) < +\infty, \qquad \text{a.e. } x \in \mathbb{R}^n. \tag{3.54}$$

Let us now prove that

$$\Omega f(x) = 0, \qquad \text{a.e. } x \in \mathbb{R}^n. \tag{3.55}$$

Claim If $g \in C_0^0(\mathbb{R}^n)$, then

$$g_r \to g, \qquad \text{uniformly as } r \to 0,$$

hence

$$\Omega g(x) = 0, \qquad \forall x \in \mathbb{R}^n. \tag{3.56}$$

Proof of Claim Since g is an uniformly continuous function, for any $\varepsilon > 0$ there exists $\delta > 0$ such that if $|x - y| < \delta$ then

$$|g(x) - g(y)| < \varepsilon.$$

Now

$$g_r(x) - g(x) = \int_{\mathbb{R}^n} \varphi_1(z) \left(g(x - rz) - g(x)\right) dz.$$

Hence, if $0 < r < \delta$, we get

$$|g_r(x) - g(x)| \leq \int_{\mathbb{R}^n} \varphi_1(z) \left|g(x - rz) - g(x)\right| dz \leq \varepsilon, \qquad \forall x \in \mathbb{R}^n.$$

This conclude the proof of the Claim.

Now, let $f \in L^1(\mathbb{R}^n)$. Since $C_0^0(\mathbb{R}^n)$ is dense in $L^1(\mathbb{R}^n)$ (Theorem 3.34) it follows that for any $\sigma > 0$ there exists $g \in C_0^0(\mathbb{R}^n)$ such that

$$\|f - g\|_{L^1(\mathbb{R}^n)} < \sigma.$$

Let $h = f - g$. We have trivially $f = g + h$, $\|h\|_{L^1(\mathbb{R}^n)} < \sigma$ and by (3.56) we get

$$\Omega f(x) \leq \Omega g(x) + \Omega h(x) = \Omega h(x), \qquad \forall x \in \mathbb{R}^n.$$

Hence, for any $\eta > 0$, we have

$$\left|\left\{x \in \mathbb{R}^n : \Omega f(x) > \eta\right\}\right| \leq \left|\left\{x \in \mathbb{R}^n : \Omega h(x) > \eta\right\}\right|. \tag{3.57}$$

On the other hand, we have trivially

$$\Omega h(x) \leq 2M(h)(x), \qquad \forall x \in \mathbb{R}^n,$$

this inequality and (3.57) imply

$$\left|\left\{x \in \mathbb{R}^n : \Omega f(x) > \eta\right\}\right| \leq \left|\left\{x \in \mathbb{R}^n : M(h)(x) > \frac{\eta}{2}\right\}\right|.$$

By the latter, by (3.57) and by Lemma 3.55 we have

$$\left|\left\{x \in \mathbb{R}^n : \Omega f(x) > \eta\right\}\right| \leq \frac{2C}{\eta} \|h\|_{L^1(\mathbb{R}^n)} \leq \frac{2C\sigma}{\eta},$$

where $C = 5^n$. Hence, by choosing

$$\sigma = \eta^2,$$

we obtain

$$\left|\left\{x \in \mathbb{R}^n : \Omega f(x) > \eta\right\}\right| \leq 2C\eta, \tag{3.58}$$

which yields

$$\left|\left\{x \in \mathbb{R}^n : \Omega f(x) > 0\right\}\right| = \lim_{j\to\infty} \left|\left\{x \in \mathbb{R}^n : \Omega f(x) > \frac{1}{j}\right\}\right| = 0$$

hence, (3.55) follows. Taking into account (3.54), we have that the limit

$$\lim_{r\to 0} f_r(x)$$

there exists almost everywhere. Therefore, (3.53) implies

$$\lim_{r\to 0} f_r(x) = \lim_{k\to\infty} f_{r_k}(x) = f(x), \qquad \text{a.e. } x \in \mathbb{R}^n.$$

■

Corollary 3.56 *Let $p \in [1, +\infty)$. If $f \in L^p_{loc}(\mathbb{R}^n)$, then*

$$\lim_{r\to 0} ⨍_{B_r(x)} |f(y) - f(x)|^p dy = 0, \qquad a.e.\ x \in \mathbb{R}^n. \tag{3.59}$$

Proof For any $c \in \mathbb{R}$ let us denote by D_c the subset of $\mathbb{R}^n$ of the points x satisfying

$$\lim_{r\to 0} \left(⨍_{B_r(x)} |f(y) - c|^p dy\right)^{1/p} = |f(x) - c|.$$

Set

$$E_c = \mathbb{R}^n \setminus D_c.$$

Theorem 3.52 implies

$$|E_c| = 0.$$

Consequently, setting

$$E = \bigcup_{q \in \mathbb{Q}} E_q,$$

we obtain

$$|E| = 0.$$

Now, let us prove that

$$\lim_{r \to 0} \left(\fint_{B_r(x)} |f(y) - c|^p dy \right)^{1/p} = |f(x) - c|, \quad \forall x \in \mathbb{R}^n \setminus E, \ \forall c \in \mathbb{R}. \tag{3.60}$$

Let $c \in \mathbb{R}$ and $x \in \mathbb{R}^n \setminus E$. Let $\delta > 0$ and $q \in \mathbb{Q}$ satisfy

$$|q - c| < \delta. \tag{3.61}$$

The triangle inequality gives

$$\begin{aligned} \left(\fint_{B_r(x)} |f(y) - q|^p dy \right)^{1/p} - \delta < \left(\fint_{B_r(x)} |f(y) - c|^p dy \right)^{1/p} < \\ < \left(\fint_{B_r(x)} |f(y) - q|^p dy \right)^{1/p} + \delta. \end{aligned} \tag{3.62}$$

Now, let us denote

$$\Lambda'(x) = \liminf_{r \to 0} \left(\fint_{B_r(x)} |f(y) - c|^p dy \right)^{1/p},$$

$$\Lambda''(x) = \limsup_{r \to 0} \left(\fint_{B_r(x)} |f(y) - c|^p dy \right)^{1/p}.$$

Passing to the limit as $r \to 0$, by (3.62) we obtain

$$|f(x) - q| - \delta \leq \Lambda'(x) \leq \Lambda''(x) \leq |f(x) - q| + \delta. \tag{3.63}$$

Passing again to the limit as $\delta \to 0$ in (3.63) and taking into account (3.61), we obtain

$$\Lambda'(x) = \Lambda''(x) = |f(x) - c|, \quad \forall x \in \mathbb{R}^n \setminus E.$$

Therefore (3.60) holds true. Set therein $c = f(x)$ and we obtain (3.59). ■

3.7 The Rademacher Theorem

Let us recall the definition of **absolutely continuous function** over the interval $[a, b]$, where $a, b \in \mathbb{R}$ and $a < b$.

We say that the function

$$f : [a, b] \to \mathbb{R},$$

is absolutely continuous provided that for every $\varepsilon > 0$ there exists $\delta > 0$ such that, chosen anyway a finite family of pairwise disjoint intervals $(a_j, b_j) \subset [a, b]$, $j = 1, \cdots, N$ satisfying

$$\sum_{j=1}^{N} (b_j - a_j) < \delta,$$

we have

$$\sum_{j=1}^{N} \left| f(b_j) - f(a_j) \right| < \varepsilon.$$

We will denote by $AC([a, b])$ the class of the absolutely continuous functions on $[a, b]$ and we will denote by $AC_{loc}(\mathbb{R})$ the class of functions $f : \mathbb{R} \to \mathbb{R}$ such that for every interval $[a, b]$ we have $f_{|[a,b]} \in AC([a, b])$. Let us recall that if $f \in AC_{loc}(\mathbb{R})$, then f is a differentiable function almost everywhere in $\mathbb{R}$. If f is a Lipschitz continuous function in $\mathbb{R}$, then $f \in AC_{loc}(\mathbb{R})$, hence f is a differentiable almost every everywhere. The main purpose of the present section is to extend this result to several variable Lipschitz continuous functions. Precisely we want to prove

Theorem 3.57 (Rademacher) *If $f \in C^{0,1}(\mathbb{R}^n)$, then f is a differentiable function almost everywhere.*

Proof Let $v \in \mathbb{R}^n$ satisfy $|v| = 1$. Set

$$\partial_v f(x) = \lim_{t \to 0} \frac{f(x + tv) - f(x)}{t}, \qquad \text{provided the limit exists;}$$

notice that, since f is Lipschitz continuous, if the limit above exists it is finite.

Claim 3.1

$$\partial_v f(x), \qquad \text{exists a.e. } x \in \mathbb{R}^n.$$

Proof of Claim 3.1 Let us denote by

$$\overline{\partial}_v f(x) := \limsup_{t\to 0} \frac{f(x+tv)-f(x)}{t},$$

$$\underline{\partial}_v f(x) := \liminf_{t\to 0} \frac{f(x+tv)-f(x)}{t}.$$

Since f is a continuous function, we have

$$\overline{\partial}_v f(x) = \lim_{k\to\infty} \sup_{0<|t|<1/k,\ t\in\mathbb{Q}} \frac{f(x+tv)-f(x)}{t}$$

and

$$\underline{\partial}_v f(x) = \lim_{k\to\infty} \inf_{0<|t|<1/k,\ t\in\mathbb{Q}} \frac{f(x+tv)-f(x)}{t}.$$

hence $\overline{\partial}_v f$ and $\underline{\partial}_v f$ are measurable functions. Consequently the sets

$$A_v := \left\{x \in \mathbb{R}^n : \underline{\partial}_v f(x) < \overline{\partial}_v f(x)\right\},$$

$$\ell_v := \{tv : t \in \mathbb{R}\}$$

are Lebesgue measurable sets.

Let us first consider the case $v = e_n$. The Fubini–Tonelli Theorem gives

$$\begin{aligned} |A_{e_n}| &= \int_{\mathbb{R}^n} \chi_{A_{e_n}}(x)dx = \\ &= \int_{\mathbb{R}^{n-1}} dx' \int_{\mathbb{R}} \chi_{A_{e_n}}\left(x', x_n\right) dx_n = \\ &= \int_{\mathbb{R}^{n-1}} \left|\left(x' + \ell_{e_n}\right) \cap A_{e_n}\right|_1 dx', \end{aligned} \tag{3.64}$$

where $\left|\left(x' + \ell_{e_n}\right) \cap A_{e_n}\right|_1$ is the Lebesgue measure on $\mathbb{R}$ of $\left(x' + \ell_{e_n}\right) \cap A_{e_n}$. For any fixed $x' \in \mathbb{R}^{n-1}$, we have

$$\left(x' + \ell_{e_n}\right) \cap A_{e_n} = \{x'\} \times \left\{t \in \mathbb{R} : \underline{\partial}_n f(x', t) < \overline{\partial}_n f(x', t)\right\}.$$

Now, by denoting

$$\varphi(t) = f\left(x', t\right),$$

since f is Lipschitz continuous, we have $\varphi \in AC_{loc}(\mathbb{R})$. In particular, the function φ is almost everywhere differentiable, hence

$$\left|\left(x' + \ell_{e_n}\right) \cap A_{e_n}\right|_1 = 0.$$

Therefore, by (3.64), we get

$$\left|A_{e_n}\right| = 0.$$

Whenever $v \neq e_n$, let us consider a rotation $\mathcal{R}$ of $\mathbb{R}^n$ such that

$$v = \mathcal{R}(e_n).$$

By setting

$$\tilde{f} = f \circ \mathcal{R},$$

we get (due to the invariance of the Lebesgue measure with respect to rotations)

$$|A_v| = \left|\mathcal{R}^{-1}(A_v)\right|.$$

Moreover, it is easily checked that

$$\mathcal{R}^{-1}(A_v) = \left\{y \in \mathbb{R}^n : \underline{\partial}_n \tilde{f}(y) < \overline{\partial}_n \tilde{f}(y)\right\}.$$

Consequently, since $\tilde{f}$ is Lipschitz continuous, from what we have previously proved, we derive

$$|A_v| = \left|\mathcal{R}^{-1}(A_v)\right| = \left|\left\{y \in \mathbb{R}^n : \underline{\partial}_n \tilde{f}(y) < \overline{\partial}_n \tilde{f}(y)\right\}\right| = 0.$$

This concludes the proof of Claim 3.1.

Set

$$\nabla f(x) = (\partial_1 f(x), \cdots, \partial_n f(x)), \qquad \text{a.e. } x \in \mathbb{R}^n.$$

Claim 3.2 We have

$$\partial_v f(x) = v \cdot \nabla f(x), \qquad \text{a.e. } x \in \mathbb{R}^n \tag{3.65}$$

and

$$|\nabla f(x)| \leq L, \qquad \text{a.e. } x \in \mathbb{R}^n, \tag{3.66}$$

where

$$L = [f]_{0,1,\mathbb{R}^n},$$

Proof of Claim 3.2 Let $\zeta \in C_0^\infty(\mathbb{R}^n)$ be arbitrary. We easily get

$$\int_{\mathbb{R}^n} \frac{f(x+tv)-f(x)}{t}\zeta(x)dx = -\int_{\mathbb{R}^n} f(x)\frac{\zeta(x)-\zeta(x-tv)}{t}dx. \tag{3.67}$$

Now, since

$$\lim_{t\to 0}\frac{f(x+tv)-f(x)}{t} = \partial_v f(x), \qquad \text{a.e. } x \in \mathbb{R}^n$$

and

$$\left|\frac{f(x+tv)-f(x)}{t}\zeta(x)\right| \le L|\zeta(x)|, \quad \forall x \in \mathbb{R}^n, \quad \forall t \in \mathbb{R}\setminus\{0\},$$

we have by Dominated Convergence Theorem and by (3.67),

$$\begin{aligned}
\int_{\mathbb{R}^n} \partial_v f(x)\zeta(x)dx &= \lim_{t\to 0}\int_{\mathbb{R}^n} \frac{f(x+tv)-f(x)}{t}\zeta(x)dx = \\
&= -\lim_{t\to 0}\int_{\mathbb{R}^n} f(x)\frac{\zeta(x)-\zeta(x-tv)}{t}dx = \\
&= -\sum_{j=1}^n v_j\int_{\mathbb{R}^n} f(x)\partial_j\zeta(x)dx = \\
&= \sum_{j=1}^n v_j\int_{\mathbb{R}^n} \partial_j f(x)\zeta(x)dx.
\end{aligned} \tag{3.68}$$

Let us justify the last equality. For any $j \in \{1,\cdots,n\}$ we have that the function

$$x_j \to f(x_1,\cdots,x_{j-1},x_j,x_{j+1},\cdots,x_n),$$

belongs to $AC_{loc}(\mathbb{R})$. Therefore, by considering, for instance, the case $j = n$ (the other cases are similar), we have by Fubini–Tonelli Theorem

$$
\begin{aligned}
&-\int_{\mathbb{R}^n} f(x)\partial_n\zeta(x)dx = -\int_{\mathbb{R}^{n-1}} dx' \int_{\mathbb{R}} f(x',x_n)\partial_n\zeta(x',x_n)dx_n = \\
&= -\int_{\mathbb{R}^{n-1}} dx' \int_{\mathbb{R}} \left(\partial_n\left(f(x',x_n)\zeta(x',x_n)\right) - \partial_n f(x',x_n)\zeta(x',x_n)\right)dx_n = \\
&= \int_{\mathbb{R}^{n-1}} dx' \int_{\mathbb{R}} \partial_n f(x',x_n)\zeta(x',x_n)dx_n = \\
&= \int_{\mathbb{R}^n} \partial_n f(x)\zeta(x)dx.
\end{aligned}
$$

Consequently, (3.68) gives

$$
\int_{\mathbb{R}^n} \partial_v f(x)\zeta(x)dx = \int_{\mathbb{R}^n} \left(v\cdot\nabla f(x)\right)\zeta(x)dx, \quad \forall \zeta \in C_0^{\infty}\left(\mathbb{R}^n\right)
$$

which yields (3.65). Concerning (3.66), it is an immediate consequence of (3.65) and of the Cauchy–Schwarz inequality. **This concludes the proof of Claim 3.2.**

Let $D = \{v_k\}_{k\in\mathbb{N}}$ be such that

$$
|v_k| = 1, \quad \forall k \in \mathbb{N}
$$

and

$$
\overline{D} = \partial B_1.
$$

Moreover, let

$$
x \in \mathbb{R}^n \setminus \bigcup_{k=1}^{\infty} A_{v_k}.
$$

Let us prove that f is differentiable in x whereby we will conclude the proof of the theorem, because

$$
\left|\bigcup_{k=1}^{\infty} A_{v_k}\right| \le \sum_{k=1}^{\infty} |A_{v_k}| = 0.
$$

Let $y \in \mathbb{R}^n \setminus \{x\}$. Set

$$
w = \frac{y-x}{|y-x|}, \qquad t = |y-x|,
$$

we have trivially

$$
y = x + tw.
$$

Moreover, for any $k \in \mathbb{N}$ we have

$$\begin{aligned}
&|f(y) - f(x) - \nabla f(x) \cdot (y - x)| = \\
&= |f(x + tw) - f(x) - t\nabla f(x) \cdot w| \leq \\
&\leq |f(x + tv_k) - f(x) - t\nabla f(x) \cdot w| + \\
&+ |f(x + tw) - f(x + tv_k)| \leq \\
&\leq |f(x + tv_k) - f(x) - t\nabla f(x) \cdot v_k| + \\
&+ t\,|\nabla f(x)|\,|v_k - w| + \\
&+ |f(x + tw) - f(x + tv_k)| \leq \\
&\leq t\left|\frac{f(x + tv_k) - f(x) - t\nabla f(x) \cdot v_k}{t}\right| + \\
&+ 2L\,|w - v_k|\,t.
\end{aligned} \tag{3.69}$$

Let $\varepsilon > 0$. Since D is dense in ∂B_1, there exists k_ε such that

$$\left|w - v_{k_\varepsilon}\right| < \frac{\varepsilon}{2\,(2L + 1)}.$$

Since

$$x \in \mathbb{R}^n \setminus \bigcup_{k=1}^{\infty} A_{v_k},$$

we have

$$\lim_{\tau \to 0} \frac{f(x + \tau v_{k_\varepsilon}) - f(x)}{\tau} = \nabla f(x) \cdot v_{k_\varepsilon}.$$

Therefore, there exists $\delta > 0$ such that, if $0 < |\tau| < \delta$ then

$$\left|\frac{f(x + \tau v_{k_\varepsilon}) - f(x) - \tau\nabla f(x) \cdot v_{k_\varepsilon}}{\tau}\right| < \frac{\varepsilon}{2}.$$

Now taking into account (3.69) (and $t = |x - y|$), we have

$$|f(y) - f(x) - \nabla f(x) \cdot (y - x)| < \varepsilon|x - y|, \quad \forall y \in B_\delta(x) \setminus \{x\},$$

which gives the differentiability of f in x. ■

3.8 Description of the Boundary of an Open Set of $\mathbb{R}^n$

Let $r > 0$, $x \in \mathbb{R}^n$ and $x' \in \mathbb{R}^{n-1}$. We denote by $B_r(x)$ and $B'_r(x')$, the open ball of $\mathbb{R}^n$ centered in x with radius r and the open ball of $\mathbb{R}^{n-1}$ centered in x' with radius $r > 0$ respectively. We will also write B_r (B'_r) instead of $B_r(0)$ ($B'_r(0)$). For any $r, M > 0$ and any $x \in \mathbb{R}^n$, here and in the sequel, we denote by

$$Q_{r,M}(x) = B'_r(x') \times (-Mr + x_n, Mr + x_n).$$

We will write also $Q_{r,M}$ instead of $Q_{r,M}(0)$.

Definition 3.58 Let Ω be an open set of $\mathbb{R}^n$. Let r_0, M_0 be positive numbers and $m \in \mathbb{N}_0$.

(a) We say that Ω has the **boundary of C^m class with constants** r_0, M_0 (or, briefly, Ω is of class C^m with constants r_0, M_0), if for every $P \in \partial\Omega$ there exists an isometry

$$\Phi_P : \mathbb{R}^n \to \mathbb{R}^n,$$

such that

$$\Phi_P(0) = P$$

and

$$\Phi_P^{-1}(\Omega) \cap Q_{r_0,2M_0} = \left\{x \in Q_{r_0,2M_0} : x_n > g_P(x')\right\}, \tag{3.70}$$

where $g_P \in C^m\left(\overline{B'_{r_0}}\right)$,

$$g_P(0) = 0, \quad |\nabla g_P(0)| = 0, \text{ for } m \geq 1$$

and

$$\|g_P\|_{C^m\left(\overline{B'_{r_0}}\right)} \leq M_0 r_0,$$

where

$$\|g_P\|_{C^m\left(\overline{B'_{r_0}}\right)} = \sum_{|\gamma| \leq m} r_0^{|\gamma|} \left\|\partial^\gamma g_P\right\|_{L^\infty\left(B'_{r_0}\right)}.$$

(b) Let $\alpha \in (0, 1]$. We say that Ω has the **boundary of $C^{m,\alpha}$ class with constants** r_0, M_0 (or, briefly, Ω is of class $C^{m,\alpha}$ with constants r_0, M_0) if for every $P \in \partial\Omega$ there exists an isometry

$$\Phi_P : \mathbb{R}^n \to \mathbb{R}^n,$$

such that

$$\Phi_P(0) = P,$$

and

$$\Phi_P^{-1}(\Omega) \cap Q_{r_0,2M_0} = \left\{x \in Q_{r_0,2M_0} : x_n > g_P(x')\right\}, \tag{3.71}$$

where $g_P \in C^m\left(\overline{B'_{r_0}}\right)$,

$$g_P(0) = 0, \quad |\nabla g_P(0)| = 0 \quad (\text{for } m \geq 1),$$

$$\|g_P\|_{C^{m,\alpha}\left(\overline{B'_{r_0}}\right)} \leq M_0 r_0,$$

and

$$\|g_P\|_{C^{m,\alpha}\left(\overline{B'_{r_0}}\right)} = \|g_P\|_{C^m\left(\overline{B'_{r_0}}\right)} + r_0^m \sum_{|\gamma|=m} r_0^{|\alpha|}\left[\partial^\gamma u\right]_{C^{0,\alpha}\left(\overline{B'_{r_0}}\right)}.$$

(c) If there exists $r_0 > 0$ and $M_0 > 0$ such that $\partial\Omega$ has the boundary of class C^m ($C^{m,\alpha}$) with constants $r_0 > 0$ and $M_0 > 0$, then we say that Ω has the boundary of class C^m ($C^{m,\alpha}$).

Exercise Prove that if Ω is an open set of class C^0 then $\overset{\circ}{\overline{\Omega}} = \Omega$. ♠

Let us note that the graph of the function g_P that occurs in the definition above is contained in Q_{r_0,M_0} (see Fig. 3.1).

3.9 The Spaces $L^p(\partial\Omega)$

We now provide a brief review of the definition of $L^p(\partial\Omega)$ spaces, where Ω is a bounded open set of $\mathbb{R}^n$ of class $C^{0,1}$. For any $P \in \partial\Omega$ let us denote by Φ_P an isometry which satisfies (3.71). Let

$$f : \partial\Omega \to \mathbb{R}.$$

We have that $\mathcal{U} = \left\{\Phi_{P_j}\left(Q_{r_0,2M_0}\right)\right\}_{1\leq j\leq l}$ is a finite open covering of $\partial\Omega$. Let us denote

$$g_j = g_{P_j}, \quad j = 1, \cdots, l,$$

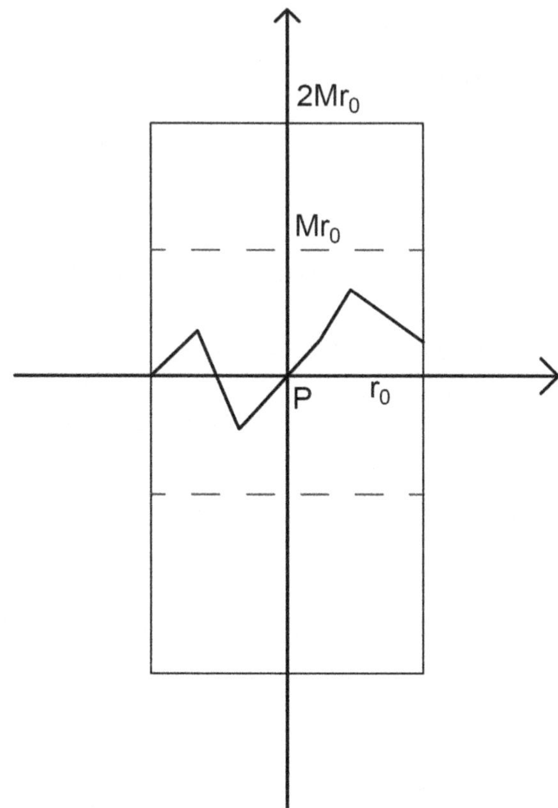

Fig. 3.1 The graph of g_P is shown

and let $\zeta_1, \cdots, \zeta_l$ be a partition of unity subordinate to the covering $\mathcal{U}$. we say that $f \in L^1(\partial\Omega)$ if we have, for any $1 \le j \le l$,

$$f_j \in L^1\left(B'_{r_0}\right),$$

where

$$f_j(x') = \left(f \circ \Phi_{P_j}\right)\left(x', g_j(x')\right), \quad \forall x' \in B'_{r_0}.$$

Let us denote

$$\int_{\partial\Omega} f dS = \sum_{j=1}^{l} \int_{\partial\Omega} f \zeta_j dS, \tag{3.72}$$

where

$$\int_{\partial\Omega} f\zeta_j dS = \int_{B'_{r_0}} f_j\left(x'\right)\tilde{\zeta}_j(x')\sqrt{1+|\nabla g_j(x')|^2}dx', \tag{3.73}$$

$$\tilde{\zeta}_j(x') = \zeta_j\left(\Phi_{P_j}\left(x', g_j(x')\right)\right).$$

Concerning the last integral in (3.73), take into account that it is well defined, because $g \in C^{0,1}\left(\overline{B'_{r_0}}\right)$ is differentiable almost everywhere and its gradient belongs to $L^\infty\left(B'_{r_0}; \mathbb{R}^n\right)$. Let us observe that integral in (3.72) does not depend on the particular partition of unity that we use. As a matter of fact, if $\eta_1, \cdots, \eta_m$ is another partition of unity, then we have

$$\begin{aligned}\sum_{j=1}^{l}\int_{\partial\Omega} f\zeta_j dS &= \sum_{j=1}^{l}\int_{\partial\Omega}\left(\sum_{k=1}^{m}\eta_k f\right)\zeta_j dS = \\ &= \sum_{j=1}^{l}\sum_{k=1}^{m}\int_{\partial\Omega}\zeta_j\eta_k f dS = \\ &= \sum_{k=1}^{m}\int_{\partial\Omega}\left(\sum_{j=1}^{l}\zeta_j f\right)\eta_k dS = \\ &= \sum_{k=1}^{m}\int_{\partial\Omega} f\eta_k dS.\end{aligned}$$

Let us denote by

$$\|u\|_{L^1(\partial\Omega)} = \int_{\partial\Omega}|f|dS.$$

Likewise, we define $L^p(\partial\Omega)$ for $1 \le p < +\infty$ and we set

$$\|u\|_{L^p(\partial\Omega)} = \left(\int_{\partial\Omega}|f|^p dS\right)^{1/p}. \tag{3.74}$$

The space $L^p(\partial\Omega)$ is a separable Banach space, and if $p = 2$, $L^2(\partial\Omega)$ is a Hilbert space.

Let $g \in C^{0,1}\left(\overline{B'_{r_0}}\right)$ be such that

$$\|g\|_{C^{0,1}\left(\overline{B'_{r_0}}\right)} \le M_0 r_0.$$

Let us consider the set

$$W = \left\{ y \in Q_{r_0,2M_0} : \ y_n > g(y') \right\}.$$

We define the field of unit outward normal to the graph of g as follows

$$\nu^{(g)}(y', g(y')) = \frac{\left(\nabla_{y'} g(y'), -1\right)}{\sqrt{1 + \left|\nabla_{y'} g(y')\right|^2}}, \qquad \text{a.e. } y' \in B'_{r_0}. \tag{3.75}$$

Let Ω be a bounded open set of $\mathbb{R}^n$ of $C^{0,1}$ class; let $\mathcal{U} = \left\{ \Phi_{P_j}\left(Q_{r_0,2M_0}\right) \right\}_{1 \le j \le l}$ be the open covering defined above. We define the field of unit outward normal on $\partial\Omega \cap \Phi_{P_j}\left(Q_{r_0,2M_0}\right)$, $j = 1, \cdots, l$, as

$$\nu(x) = \left(\Phi_{P_j} - P_j\right)\left(\nu^{(g_j)}(y', g_j(y'))\right), \qquad x = \Phi_j\left(y', g_j(y')\right). \tag{3.76}$$

If $f \in L^1(\partial\Omega)$, we define, by (3.73),

$$\int_{\partial\Omega} f \nu dS = \sum_{j=1}^{l} \int_{\partial\Omega} f \zeta_j \nu dS. \tag{3.77}$$

3.10 The Divergence Theorem

The purpose of this section is to prove the following

Theorem 3.59 (Divergence) *Let Ω be a bounded open set of $\mathbb{R}^n$ of class $C^{0,1}$ and let f belong to $C^{0,1}\left(\overline{\Omega}\right)$, then we have*

$$\int_{\Omega} \nabla f dx = \int_{\partial\Omega} f \nu dS. \tag{3.78}$$

In order to prove Theorem 3.59 we need some lemmas.

Lemma 3.60 *Let Ω be a bounded open set of $\mathbb{R}^n$ and let $f \in C^{0,1}\left(\overline{\Omega}\right)$. Then there exists a sequence $\{f_m\}$ in $C^{\infty}\left(\mathbb{R}^n\right)$ satisfying*

$$\lim_{m \to \infty} \|f_m - f\|_{C^0(\overline{\Omega})} = 0, \tag{3.79}$$

$$\lim_{m \to \infty} \|\nabla f_m - \nabla f\|_{L^2(\Omega)} = 0. \tag{3.80}$$

Proof Set

$$L = [f]_{0,1,\overline{\Omega}} \cdot$$

Let $\tilde{f} \in C^{0,1}(\mathbb{R}^n)$ be an extension of f which satisfies (see Theorem 3.8)

$$\left[\tilde{f}\right]_{0,1,\mathbb{R}^n} = L.$$

For any $\varepsilon > 0$, let us denote

$$f_\varepsilon(x) = \int_{\mathbb{R}^n} \eta_\varepsilon(x-y)\tilde{f}(y)dy,$$

where η is a mollifier. We have that $f_\varepsilon \in C^\infty(\mathbb{R}^n)$. Moreover, if $x \in \Omega$, we get

$$\begin{aligned}
|f_\varepsilon(x) - f(x)| &= \left|\int_{\mathbb{R}^n} \eta_\varepsilon(y)\left(\tilde{f}(x-y) - \tilde{f}(x)\right)dy\right| \leq \\
&\leq \int_{\mathbb{R}^n} \eta_\varepsilon(y)\left|\tilde{f}(x-y) - \tilde{f}(x)\right|dy \leq \\
&\leq L\int_{\mathbb{R}^n} \eta_\varepsilon(y)|y|dy \leq \\
&\leq L\varepsilon.
\end{aligned}$$

Therefore

$$\lim_{\varepsilon\to 0} \|f_\varepsilon - f\|_{C^0(\overline{\Omega})} = 0,$$

which gives (3.79).

Let us prove (3.80). Let $1 \leq k \leq n$. Theorem 3.57 gives

$$\begin{aligned}
\partial_{x_k} f_\varepsilon(x) &= \int_{\mathbb{R}^n} \partial_{x_k}\left(\eta_\varepsilon(x-y)\right)\tilde{f}(y)dy = \\
&= -\int_{\mathbb{R}^n} \partial_{y_k}\left(\eta_\varepsilon(x-y)\right)\tilde{f}(y)dy = \\
&= \int_{\mathbb{R}^n} \eta_\varepsilon(x-y)\partial_{y_k}\tilde{f}(y)dy.
\end{aligned}$$

Now, for any $x \in \Omega$ we get

$$\left|\partial_{x_k} f_\varepsilon(x) - \partial_{x_k} f(x)\right| \leq \int_{\mathbb{R}^n} \eta_\varepsilon(y) \left|\left(\partial_{x_k} \tilde{f}\right)(x-y) - \partial_{x_k} \tilde{f}(x)\right| dy \leq$$

$$\leq \left(\int_{\mathbb{R}^n} \eta_\varepsilon(y) \left|\left(\partial_{x_k} \tilde{f}\right)(x-y) - \partial_{x_k} \tilde{f}(x)\right|^2 dy\right)^{1/2}.$$

From which we derive

$$\begin{aligned}
&\int_\Omega \left|\partial_{x_k} f_\varepsilon(x) - \partial_{x_k} f(x)\right|^2 dx \leq \\
&\leq \int_\Omega dx \int_{B_\varepsilon} \eta_\varepsilon(y) \left|\left(\partial_{x_k} \tilde{f}\right)(x-y) - \partial_{x_k} \tilde{f}(x)\right|^2 dy = \\
&= \int_{B_\varepsilon} \eta_\varepsilon(y) \int_\Omega \left|\left(\partial_{x_k} \tilde{f}\right)(x-y) - \partial_{x_k} \tilde{f}(x)\right|^2 dx \leq \\
&\leq \sup_{|y|<\varepsilon} \int_\Omega \left|\left(\partial_{x_k} \tilde{f}\right)(x-y) - \partial_{x_k} \tilde{f}(x)\right|^2 dx.
\end{aligned}$$

By the just proved inequality and by Theorem 3.36 we have

$$\lim_{\varepsilon \to 0} \|\nabla f_\varepsilon - \nabla f\|_{L^2(\Omega)} = 0,$$

that gives (3.79). ■

For any $f \in C^{0,1}\left(\overline{\Omega}\right)$ we say that a sequence $\{f_m\}$ which satisfies (3.79) and (3.80) is a **smooth approximating sequence** of f .

Lemma 3.61 *Let r, h be positive numbers. Let $g \in C^\infty\left(\overline{B'_r}\right)$ satisfy*

$$-h < g(x') < h, \quad \forall x' \in B'_r.$$

Let $f \in C_0^\infty\left(Q_{r,h}\right)$, where $Q_{r,h} = B'_r \times (-h, h)$. Denoting by

$$W = \left\{x \in Q_{r,h} : x_n > g(x')\right\},$$

we have

$$\int_W \partial_n f dx = -\int_{B'_r} f\left(x', g(x')\right) dx', \tag{3.81a}$$

$$\int_W \partial_k f dx = \int_{B'_r} f\left(x', g(x')\right) \partial_k g(x') dx', \quad k = 1, \cdots, n-1. \tag{3.81b}$$

Proof Let $1 \leq k \leq n$. If $k = n$, then we have

$$\int_{g(x')}^{h} \partial_n f\left(x', x_n\right) dx_n = -f\left(x', g(x')\right), \quad x' \in B'_r.$$

Hence

$$\begin{aligned}\int_W \partial_n f dx &= \int_{B'_r} dx' \int_{g(x')}^{h} \partial_n f(x', x_n) dx_n = \\ &= -\int_{B'_r} f\left(x', g(x')\right) dx'.\end{aligned} \tag{3.82}$$

If $1 \leq k \leq n-1$, then we have

$$\partial_k \int_{g(x')}^{h} f\left(x', x_n\right) dx_n = \int_{g(x')}^{h} \partial_k f\left(x', x_n\right) dx_n - f\left(x', g(x')\right) \partial_k g(x'),$$

from which we get

$$\begin{aligned}\int_W \partial_k f dx &= \int_{B'_r} dx' \int_{g(x')}^{h} \partial_k f(x', x_n) dx_n = \\ &= \int_{B'_r} dx' \left(\partial_k \int_{g(x')}^{h} f\left(x', x_n\right) dx_n\right) + \\ &+ \int_{B'_r} f\left(x', g(x')\right) \partial_k g(x') dx' = \\ &= \int_{B'_r} f\left(x', g(x')\right) \partial_k g(x') dx,\end{aligned} \tag{3.83}$$

where in the fourth step we used the fact that the function

$$x' \to \int_{g(x')}^{h} f\left(x', x_n\right) dx_n,$$

has the support contained in B'_r. From what was obtained in (3.82) and by (3.83) we derive (3.81). ■

Remark Under the same assumptions of Lemma 3.61, taking into account (3.75), we have

$$\int_W \partial_s f dx = \int_{\Gamma(g)} f \nu_s^{(g)} dS, \quad s = 1, \cdots, n, \tag{3.84}$$

where $\Gamma(g)$ is the graph of g. ◆

Proof Theorem 3.59 Let us begin by considering the case where $f \in C^{\infty}(\mathbb{R}^n)$.

Since $\partial\Omega$ is of class $C^{0,1}$, we may assume that there exist positive numbers r_0, M_0, such that for every $P \in \partial\Omega$ there is an isometry

$$\Phi_P : \mathbb{R}^n \to \mathbb{R}^n,$$

satisfying

$$\Phi_P(0) = P$$

and

$$\Phi_P^{-1}(\Omega) \cap Q_{r_0,2M_0} = W_P,$$

where

$$W_P = \left\{x \in Q_{r_0,2M_0} : x_n > g_P(x')\right\},$$

$$g_P \in C^{0,1}\left(\overline{B'_{r_0}}\right),$$

$$g_P(0) = 0,$$

and

$$\|g_P\|_{C^{0,1}\left(\overline{B'_{r_0}}\right)} \leq M_0 r_0.$$

Since $\partial\Omega$ is compact, there exist $P_1, \cdots, P_l \in \partial\Omega$ such that the family of sets $\left\{\Phi_{P_j}\left(Q_{r_0,2M_0}\right)\right\}_{1\leq j\leq l}$ is a finite covering of $\partial\Omega$. Set

$$V_j = \Phi_{P_j}\left(Q_{r_0,2M_0}\right), \quad j = 1, \cdots, l.$$

By Theorem 3.51 there exist $\zeta_0, \zeta_1, \cdots, \zeta_l \in C_0^{\infty}(\mathbb{R}^n)$ satisfying

$$(\zeta_0)_{|\Omega} \in C_0^{\infty}(\Omega),$$

$$\operatorname{supp}\zeta_0 \subset \mathbb{R}^n \setminus \partial\Omega; \quad \operatorname{supp}\zeta_j \subset V_j, \quad j = 1, \cdots, l,$$

$$\sum_{j=0}^{l} \zeta_j = 1, \quad \text{on } \mathbb{R}^n; \quad 0 \leq \zeta_j \leq 1, \quad j = 0, 1 \cdots, l.$$

We have

$$\int_{\Omega} \nabla (f\zeta_0)\, dx = 0.$$

Hence

$$\int_{\Omega} \nabla f dx = \sum_{j=0}^{l} \int_{\Omega} \nabla (f\zeta_j)\, dx = \sum_{j=1}^{l} \int_{\Omega} \nabla (f\zeta_j)\, dx. \tag{3.85}$$

Now, let us fix $j \in \{1 \cdots, l\}$ and let us denote by $W := W_{P_j}$, $g := g_{P_j}$ and

$$F := f\zeta_j. \tag{3.86}$$

Let $\{g_m\}$ be a sequence of smooth approximating of g. Set

$$\Phi_j = \Phi_{P_j}, \quad j = 1, \cdots l$$

and denoting by

$$U = \Phi_j(W),$$

$$W_m = \left\{x \in Q_{r_0,2M_0} : x_n > g_m(x')\right\},$$

$$U_m = \Phi_j(W_m),$$

we have

$$\int_{\Omega} \nabla (f\zeta_j)\, dx = \int_{U} \nabla F dx = \lim_{m\to\infty} \int_{U_m} \nabla F dx \tag{3.87}$$

(as a matter of fact, we have $|U_m \setminus U| \to 0$ as $m \to \infty$).

Now let us deal with the last integral in (3.87). Since Φ_j is an isometry, there exists a matrix $A = \left\{a_{qs}\right\}_{1\leq q,s\leq n}$ such that

$$\Phi(y) = Ay + P_j, \quad A^T A = I_n,$$

where I_n is the identity matrix $n \times n$. For the sake of brevity, let us set $P := P_j$, $\Phi := \Phi_j$. We have

$$\int_{U_m} \nabla F dx = \int_{W_m} (\nabla_x F)(\Phi(y))\, dy. \tag{3.88}$$

By setting

$$\overline{F} = F \circ \Phi, \tag{3.89}$$

we have

$$\left(\partial_y \Phi(y)\right)^T \left(\nabla_x F\right)(\Phi(y)) = \nabla_y \overline{F}(y)$$

and, recalling that $A = \left(\partial_y \Phi(y)\right)$, $A^T A = I_n$, we get

$$\left(\nabla_x F\right)(\Phi(y)) = A \nabla_y \overline{F}(y), \quad \forall y \in Q_{r_0, 2M_0}. \tag{3.90}$$

Now, since $\overline{F} \in C_0^\infty\left(Q_{r_0,2M_0}\right)$ and, for m large enough, the graph of g_m is contained in $Q_{r_0,2M_0}$ (recall that $\{g_m\}$ is a smooth approximating sequence of g), we obtain, by Lemma 3.61 and by (3.90)

$$\begin{aligned}
\left(\int_{W_m} \left(\nabla_x F\right)(\Phi(y))\, dy\right)_q &= \left(\int_{W_m} A \nabla_y \overline{F}(y) dy\right)_q = \\
&= \sum_{s=1}^{n} \int_{W_m} a_{qs} \partial_s \overline{F}(y) dy = \\
&= -\int_{B'_r} a_{qn} \overline{F}\left(y', g_m(y')\right) dy' + \\
&+ \sum_{s=1}^{n-1} \int_{B'_r} a_{qs} \overline{F}\left(y', g_m(y')\right) \partial_s g_m(y') dy'.
\end{aligned} \tag{3.91}$$

Since $\{g_m\}$ is a smooth approximating sequence of g, we have

$$\overline{F}\left(y', g_m(y')\right) \to \overline{F}\left(y', g(y')\right), \quad \text{as } m \to \infty \text{ in } L^\infty\left(B'_r\right)$$

and, for any $s = 1, \cdots, n-1$,

$$\overline{F}\left(y', g_m(y')\right) \partial_s g_m(y') \to \overline{F}\left(y', g(y')\right) \partial_s g(y'), \text{ as } m \to \infty, \text{ in } L^2\left(B'_r\right).$$

Hence

$$\lim_{m \to \infty} \int_{B'_r} \overline{F}\left(y', g_m(y')\right) dy' = \int_{B'_r} \overline{F}\left(y', g(y')\right) dy' \tag{3.92}$$

and, for any $s = 1, \cdots, n-1$,

$$\lim_{m \to \infty} \int_{B'_r} \overline{F}\left(y', g_m(y')\right) \partial_s g_m(y') dy' = \int_{B'_r} \overline{F}\left(y', g(y')\right) \partial_s g(y') dy'. \tag{3.93}$$

By (3.87), (3.91), (3.92) and (3.93), denoting by $\nu^{(g)}$ the unit outward normal to $\Gamma(g)$ (recall (3.86) and (3.89)) we have

$$\begin{aligned}\int_{\Omega} \nabla\left(f\zeta_j\right) dx &= \lim_{m\to\infty} \int_{U_m} \nabla F dx = \\ &= \lim_{m\to\infty} \int_{W_m} \nabla_x F\left(\Phi(y)\right) dy = \\ &= \int_{\Gamma(g)} \overline{F} A \nu^{(g)} dS = \\ &= \int_{\partial\Omega} f\zeta_j \nu dS.\end{aligned}$$

By what has just been obtained and by (3.85) we derive that if $f \in C^{\infty}\left(\mathbb{R}^n\right)$, then

$$\int_{\Omega} \nabla f dx = \int_{\partial\Omega} f \nu dS.$$

Finally, let us consider the case where $f \in C^{0,1}\left(\overline{\Omega}\right)$. If $\{f_m\}$ is a smooth approximating sequence of f given by Lemma 3.60, then we have

$$\begin{aligned}\int_{\Omega} \nabla f dx &= \lim_{m\to\infty} \int_{\Omega} \nabla f_m dx = \\ &= \lim_{m\to\infty} \int_{\partial\Omega} f_m \nu dS = \\ &= \int_{\partial\Omega} f \nu dS.\end{aligned}$$

■

3.11 The Hausdorff Distance

Let (X, d) be a metric space. Let us recall that the distance of a point $x \in X$ from a subset A of X, $A \neq \emptyset$, is given by

$$d(x, A) = \inf\left\{d(x, y) :\ y \in A\right\}. \tag{3.94}$$

Proposition 3.62 *If $A \subset X$, $A \neq \emptyset$, then we have*

$$|d(x, A) - d(y, A)| \leq d(x, y), \quad \forall x, y \in A. \tag{3.95}$$

In particular, the map

$$X \ni x \to d(x, A) \in \mathbb{R},$$

is Lipschitz continuous.

Proof Let $x, y \in X$. By the triangle inequality we have

$$d(x, A) \leq d(x, z) \leq d(x, y) + d(y, z), \quad \forall z \in A,$$

from which we derive

$$d(x, A) \leq d(x, y) + d(y, A).$$

Hence

$$d(x, A) - d(y, A) \leq d(x, y)$$

and interchanging x and y, we obtain (3.95). ■

We denote by $\mathbf{K}(X)$ the family of nonempty compact sets of X. If $K \in \mathbf{K}(X)$ and $x \in X$ we have

$$d(x, K) = \min \{d(x, y) : \ y \in K\}.$$

Let $K \in \mathbf{K}(X)$. Let us denote by $S(K)$ the set of the points $x \in X$ such that

$$\{y \in K : \ d(x, y) = d(x, K)\}, \text{ has only one point.}$$

Trivially, we have $K \subset S(K)$ and it is well-defined the map

$$p_K : S(K) \to K, \qquad \text{such that } d(x, p_K(x)) = d(x, K). \tag{3.96}$$

If $x \in S(K)$, we call $p_K(x)$ the **point of minimum distance** of x from K or also **the projection** of x on K.

Proposition 3.63 *Let $K \in \mathbf{K}(X)$. Then p_K, defined by* (3.96), *is a continuous map.*

Proof Let us argue by contradiction. Let us assume that p_K is not continuous. Consequently, there exists a point $x_0 \in S(K)$ and a sequence $\{x_n\}$ of $S(K)$ satisfying

$$\{x_n\} \to x_0 \tag{3.97}$$

and

$$\{p_K(x_n)\} \nrightarrow p_K(x_0). \tag{3.98}$$

The latter implies that there exists $\varepsilon > 0$ and a subsequence $\{x_n^*\}$ of $\{x_n\}$ satisfying

$$d\left(p_K\left(x_n^*\right), p_K(x_0)\right) \geq \varepsilon, \quad \forall n \in \mathbb{N}. \tag{3.99}$$

Since for every $n \in \mathbb{N}$ we have $p_K\left(x_n^*\right) \in K$, and K is compact, there exists a subsequence $\{x_n^{**}\}$ of $\{x_n^*\}$ such that $\{p_K\left(x_n^{**}\right)\}$ converges to a point $z \in K$. On the other hand, by (3.97) we get

$$\{x_n^{**}\} \to x_0.$$

Hence

$$d(x_0, K) = \lim_{n\to\infty} d\left(x_n^{**}, K\right) = \lim_{n\to\infty} d\left(x_n^{**}, p_K\left(x_n^{**}\right)\right) = d(x_0, z).$$

Consequently

$$d(x_0, K) = d(x_0, z)$$

and by the definition of $S(K)$ we have

$$z = p_K(x_0).$$

On the other hand, (3.99) gives

$$d\left(p_K(x_0), z\right) = \lim_{n\to\infty} d\left(p_K(x_0), p_k\left(x_n^{**}\right)\right) \geq \varepsilon.$$

We have actually reached a contradiction. Therefore the map $x \to p_K(x)$ is continuous. ■

Definition 3.64 For any $K_1, K_2 \in \mathbf{K}(X)$ we denote by

$$\delta\left(K_1, K_2\right) = \max\left\{d(x, K_2) : x \in K_1\right\}. \tag{3.100}$$

Proposition 3.65 *If $K_1, K_2 \in \mathbf{K}(X)$, then we have*

$$K_1 \subset K_2 \quad \Leftrightarrow \quad \delta\left(K_1, K_2\right) = 0. \tag{3.101}$$

Proof If $K_1 \subset K_2$, then

$$d(x, K_2) = 0, \quad \forall x \in K_1,$$

hence

$$\delta\left(K_1, K_2\right) = 0.$$

Conversely, if $\delta(K_1, K_2) = 0$ then

$$d(x, K_2) = 0, \quad \forall x \in K_1.$$

Since K_1 is a closed set of X, we obtain

$$x \in K_2, \quad \forall x \in K_1,$$

that is $K_1 \subset K_2$. ■

Let us notice that $\delta(\cdot, \cdot)$ **does not** define a distance on $\mathbf{K}(X)$. Actually, by (3.101) we have that $\delta(\cdot, \cdot)$ is not symmetric and

$$\delta(K_1, K_2) = 0 \nRightarrow K_1 = K_2.$$

For any $K_1, K_2 \in \mathbf{K}(X)$, let us denote by

$$d_{\mathcal{H}}(K_1, K_2) = \max\{\delta(K_1, K_2), \delta(K_2, K_1)\}. \tag{3.102}$$

Proposition 3.66 *$d_{\mathcal{H}}(\cdot, \cdot)$, defined by* (3.102)*, is a distance on* $\mathbf{K}(X)$.

Proof It is obvious that $d_{\mathcal{H}}(K_1, K_2) = d_{\mathcal{H}}(K_2, K_1)$ and that $d_{\mathcal{H}}(K_1, K_2) \geq 0$ for every $K_1, K_2 \in \mathbf{K}(X)$. Furthermore, if

$$d_{\mathcal{H}}(K_1, K_2) = 0,$$

then $\delta(K_1, K_2) = 0$ and $\delta(K_2, K_1) = 0$ which imply, respectively, $K_1 \subset K_2$ and $K_2 \subset K_1$, hence $K_1 = K_2$.

It only remains to prove the triangular inequality. We begin by proving that if $K_1, K_2 \in \mathbf{K}(X)$ then for any $L \in \mathbf{K}(X)$ we have

$$\delta(K_1, K_2) \leq \delta(K_1, L) + \delta(L, K_2). \tag{3.103}$$

Let $x \in K_1$. For any $y \in K_2$ and for any $z \in L$ we have

$$d(x, K_2) \leq d(x, y) \leq d(x, z) + d(z, y),$$

from which we have

$$\begin{aligned} d(x, K_2) &\leq d(x, z) + d(z, K_2) \leq \\ &\leq d(x, z) + \delta(L, K_2). \end{aligned}$$

Therefore

$$d(x, K_2) \leq d(x, L) + \delta(L, K_2) \leq$$
$$\leq \delta(K_1, L) + \delta(L, K_2).$$

By the latter we obtain (3.103) and similarly

$$\delta(K_2, K_1) \leq \delta(K_2, L) + \delta(L, K_1). \tag{3.104}$$

Now, let us assume, for instance, that $d_{\mathcal{H}}(K_1, K_2) = \delta(K_1, K_2)$. We have, by (3.103) and (3.104),

$$d_{\mathcal{H}}(K_1, K_2) = \delta(K_1, K_2) \leq$$
$$\leq \delta(K_1, L) + \delta(L, K_2) \leq$$
$$\leq d_{\mathcal{H}}(K_1, L) + d_{\mathcal{H}}(L, K_2).$$

■

Definition 3.67 Let $K_1, K_2 \in \mathbf{K}(X)$, we call $d_{\mathcal{H}}(K_1, K_2)$, defined by (3.102), the **Hausdorff distance** between K_1 and K_2.

In Proposition 3.69 below we give an useful characterization of the Hausdorff distance. For this purpose we introduce the following notation. Let $r \geq 0$ and $K \in \mathbf{K}(X)$. Set

$$[K]_r = \{u \in K : d(x, K) \leq r\},$$

Let us call $[K]_r$ the r-**dilation** of r of the set K.

Proposition 3.68 *If $K \in \mathbf{K}(X)$ and $r \geq 0$ then $[K]_r$ is a closed set of X.*

Proof Let $\{x_n\}$ be any sequence in $[K]_r$ which satisfies

$$\lim_{n\to\infty} x_n = x_0.$$

Since

$$d(x_n, K) \leq r, \quad \forall n \in \mathbb{N},$$

passing to the limit as $n \to \infty$ and taking into account that $x \to d(x, K)$ is continuous, we have

$$d(x_0, K) = \lim_{n\to\infty} d(x_n, K) \leq r.$$

Hence $x_0 \in [K]_r$. Therefore $[K]_r$ is closed set of X.

Remark It generally does not occur that $[K]_r$ is compact. For instance, let X be an infinite dimensional Hilbert space, and let $K = \{a\}$, where $a \in X$. Then we have $[K]_r = \overline{B_r(a)}$ and from Functional Analysis we know that $\overline{B_r(a)}$ is not compact. ♦

Proposition 3.69 *Let $K_1, K_2 \in \mathbf{K}(X)$ and $r \geq 0$. Then we have*

$$d_{\mathcal{H}}(K_1, K_2) \leq r \iff \begin{cases} K_1 \subset [K_2]_r \\ \\ K_2 \subset [K_1]_r \end{cases}, \tag{3.105}$$

$$d_{\mathcal{H}}(K_1, K_2) = \min\{r \geq 0 : \; K_1 \subset [K_2]_r \,, \; K_2 \subset [K_1]_r\} . \tag{3.106}$$

Proof The condition

$$d_{\mathcal{H}}(K_1, K_2) \leq r$$

is equivalent to $\delta(K_1, K_2) \leq r$ and $\delta(K_2, K_1) \leq r$. On the other hand

$$\delta(K_1, K_2) \leq r \iff (d(x, K_2) \leq r, \; \forall x \in K_1) \iff K_1 \subset [K_2]_r \,.$$

Similarly,

$$\delta(K_2, K_1) \leq r \iff K_2 \subset [K_1]_r \,.$$

From what has been obtained above, (3.105) follows.

Now, we prove (3.106). Let us denote by

$$d = d_{\mathcal{H}}(K_1, K_2)$$

and

$$\rho = \inf\{r \geq 0 : \; K_1 \subset [K_2]_r \,, \; K_2 \subset [K_1]_r\} .$$

(3.105) implies (" $\Longrightarrow$")

$$\rho \leq d. \tag{3.107}$$

On the other hand, for any $\varepsilon > 0$, we have

$$K_1 \subset [K_2]_{\rho+\varepsilon} \,, \quad \text{and} \quad K_2 \subset [K_1]_{\rho+\varepsilon}$$

and (3.105) implies (" $\Longleftarrow$")

$$d = d_{\mathcal{H}}(K_1, K_2) \leq \rho + \varepsilon.$$

Therefore, since ε is arbitrary, we obtain

$$d \leq \rho.$$

By the latter and by (3.107) we get

$$d = \rho,$$

from which (3.106) follows. ■

Example Let us consider

$$K_1 = \overline{B_1} \setminus B_\varepsilon, \quad K_2 = \overline{B_1},$$

in $\mathbb{R}^n$, where $\varepsilon \in (0, 1)$.

We have

$$K_1 \subset K_2 \subset [K_2]_\varepsilon$$

and

$$K_2 \subset [K_1]_r, \quad \forall r \geq \varepsilon.$$

Hence

$$d_{\mathcal{H}}(K_1, K_2) = \varepsilon.$$

Let us see what happens with regard to

$$d_{\mathcal{H}}(\partial K_1, \partial K_2).$$

We have

$$\partial K_2 \subset \partial K_1$$

and

$$\partial K_1 \subset [\partial K_2]_r, \quad \forall r \geq 1 - \varepsilon.$$

Hence

$$d_{\mathcal{H}}(\partial K_1, \partial K_2) = 1 - \varepsilon.$$

Therefore, neither of the two relationships holds true

$$d_{\mathcal{H}}(K_1, K_2) \leq d_{\mathcal{H}}(\partial K_1, \partial K_2), \tag{3.108}$$

$$d_{\mathcal{H}}(\partial K_1, \partial K_2) \leq d_{\mathcal{H}}(K_1, K_2). \tag{3.109}$$

As a matter of fact (3.108) is false for $\frac{1}{2} \leq \varepsilon < 1$, and (3.109) is false for $0 < \varepsilon < \frac{1}{2}$. ♣

Proposition 3.70

$$[K_1]_r \cup [K_2]_r = [K_1 \cup K_2]_r, \quad \forall K_1, K_2 \in \mathbf{K}(X), \ \forall r \geq 0.$$

Proof Let $K_1, K_2 \in \mathbf{K}(X)$. Let us begin to prove

$$[K_1]_r \cup [K_2]_r \subset [K_1 \cup K_2]_r. \tag{3.110}$$

Let $x \in [K_1]_r \cup [K_2]_r$ and, for instance, let $x \in [K_1]_r$, then

$$d(x, K_1 \cup K_2) \leq d(x, K_1) \leq r,$$

which implies $x \in [K_1 \cup K_2]_r$. Hence, (3.110) is proved.

Now, let us prove

$$[K_1 \cup K_2]_r \subset [K_1]_r \cup [K_2]_r. \tag{3.111}$$

Let $x \in [K_1 \cup K_2]_r$. We have

$$d(x, K_1 \cup K_2) \leq r.$$

Let $y \in K_1 \cup K_2$ satisfy $d(x, y) = d(x, K_1 \cup K_2)$. Now, if $y \in K_1$, we have

$$d(x, K_1) \leq d(x, y) \leq r,$$

hence $x \in [K_1]_r$. Similarly, if $y \in K_2$ then $x \in [K_2]_r$. In any case $x \in [K_1]_r \cup [K_2]_r$. Hence (3.111) is proved. ■

The following Theorem has been proved in **Kuratowski**, [71, §15, VIII].

Theorem 3.71 *Let $x_0 \in X$. Let us define, for any $K \in \mathbf{K}(X)$, the function*

$$f_K(x) = d(x, K) - d(x, x_0). \tag{3.112}$$

We have:

(i) f_K is bounded,

(ii)

$$d_{\mathcal{H}}(K_1, K_2) = \sup_{x \in X} \left| f_{K_2}(x) - f_{K_1}(x) \right|, \quad \forall K_1, K_2 \in \mathbf{K}(X).$$

Proof Let us prove (i). Let $x \in X$. We have, by the triangle inequality,

$$d(x, K) \leq d(x, y) \leq d(y, x_0) + d(x_0, x), \quad \forall y \in K.$$

Hence

$$d(x, K) \leq d(x_0, K) + d(x, x_0). \tag{3.113}$$

By using again the triangle inequality we have, for any $y, z \in K$

$$d(x, x_0) \leq d(x, y) + d(y, z) + d(z, x_0) \leq d(x, y) + d(z, x_0) + d(K),$$

where $d(K)$ is the diameter of K, by the last inequality we have

$$d(x, x_0) \leq d(x, K) + d(x_0, K) + d(K). \tag{3.114}$$

Now, (3.113) gives

$$f_K(x) = d(x, K) - d(x, x_0) \leq d(x_0, K)$$

and (3.114) gives

$$f_K(x) \geq d(x, K) - (d(x, K) + d(x_0, K) + d(K)) = -d(x_0, K) - d(K).$$

Therefore we have

$$|f_K(x)| \leq d(x_0, K) + d(K), \quad \forall x \in X.$$

Let us now prove (ii). Let $K_1, K_2 \in \mathbf{K}(X)$. It is not restrictive to assume

$$d_{\mathcal{H}}(K_1, K_2) = \delta(K_1, K_2) = \max_{x \in K_1} d(x, K_2).$$

Let $\overline{x} \in K_1$ satisfy

$$d(\overline{x}, K_2) = d_{\mathcal{H}}(K_1, K_2).$$

Since we have trivially $d(\overline{x}, K_1) = 0$, we get

$$d_{\mathcal{H}}(K_1, K_2) = d(\overline{x}, K_2) - d(\overline{x}, K_1) =$$

$$= d(\overline{x}, K_2) - d(\overline{x}, x_0) - (d(\overline{x}, K_1) - d(\overline{x}, x_0)) = \tag{3.115}$$

$$= f_{K_2}(\overline{x}) - f_{K_1}(\overline{x}) \leq \sup_{x \in X} \left| f_{K_1}(x) - f_{K_1}(x) \right| .$$

Now, for any $x \in X$ let $y \in K_1$ satisfy

$$d(x, y) = d(x, K_1) .$$

We have

$$d(x, K_2) \leq d(x, y) + d(y, K_2) = d(x, K_1) + d(y, K_2) .$$

Hence

$$d(x, K_2) - d(x, K_1) \leq d(y, K_2) \leq d_{\mathcal{H}}(K_1, K_2)$$

and, by interchanging K_1 with K_2, we have.

$$d(x, K_1) - d(x, K_2) \leq d_{\mathcal{H}}(K_1, K_2) .$$

Hence

$$\left| f_{K_1}(x) - f_{K_2}(x) \right| = |d(x, K_1) - d(x, K_2)| \leq d_{\mathcal{H}}(K_1, K_2) , \quad \forall x \in X,$$

which implies

$$\sup_{x \in X} \left| f_{K_1}(x) - f_{K_1}(x) \right| \leq d_{\mathcal{H}}(K_1, K_2) . \tag{3.116}$$

Finally, (3.115) and (3.116) imply (ii). ■

3.11.1 Completeness and Compactness of $(\mathbf{K}(X), d_{\mathcal{H}})$

The Main Theorem that we prove in the present section is the following one.

Theorem 3.72 (Completeness) *If (X, d) is a complete metric space, then $(\mathbf{K}(X), d_{\mathcal{H}})$ is complete.*

In order to prove Theorem 3.72 we need the following Lemma.

Lemma 3.73 *Let* (X, d) *be a metric space and let* $\{K_n\}$ *be a Cauchy sequence in* $(\mathbf{K}(X), d_{\mathcal{H}})$. *Let* $\{n_j\}$ *be a strictly increasing sequence in* $\mathbb{N}$ *and let* $\{x_{n_j}\}$ *be a Cauchy sequence in* (X, d) *satisfying*

$$x_{n_j} \in K_{n_j}, \quad \forall j \in \mathbb{N}.$$

Then there exists a Cauchy sequence $\{\overline{x}_n\}$ *in* (X, d) *such that*

$$\overline{x}_{n_j} = x_{n_j}, \quad \forall j \in \mathbb{N}, \quad \textit{and} \quad \overline{x}_n \in K_n, \quad \forall n \in \mathbb{N}. \tag{3.117}$$

Proof Let us define $\{\overline{x}_n\}$ as follows: if $1 \leq n \leq n_1 - 1$, then we choose $\overline{x}_n$ satisfying

$$d\left(x_{n_1}, \overline{x}_n\right) = d\left(x_{n_1}, K_n\right),$$

if $n_j + 1 \leq n \leq n_{j+1} - 1$, $j \in \mathbb{N}$, then we choose $\overline{x}_n$ satisfying

$$d\left(x_{n_j}, \overline{x}_n\right) = d\left(x_{n_j}, K_n\right),$$

finally, if $n = n_j$, $j \in \mathbb{N}$, then we choose

$$\overline{x}_n = x_{n_j}.$$

Notice that (3.117) is satisfied by construction, so we are left to prove that $\{\overline{x}_n\}$ is a Cauchy sequence.

Let us fix any $\varepsilon > 0$ and let $\nu \in \mathbb{N}$ satisfy

$$d\left(x_{n_j}, x_{n_h}\right) < \frac{\varepsilon}{3}, \quad \forall j, h \geq \nu \tag{3.118}$$

and

$$d_{\mathcal{H}}(K_n, K_m) < \frac{\varepsilon}{3}, \quad \forall n, m \geq n_\nu. \tag{3.119}$$

Let $n, m \geq n_\nu$ and let j and h be such that

$$n_j \leq n \leq n_{j+1}, \qquad n_h \leq m \leq n_{h+1}.$$

By the triangle inequality we have

$$d\left(\overline{x}_n, \overline{x}_m\right) \leq d\left(\overline{x}_n, x_{n_j}\right) + d\left(x_{n_j}, x_{n_h}\right) + d\left(\overline{x}_m, x_{n_h}\right). \tag{3.120}$$

Now, (3.119) implies

$$d\left(x_{n_j}, \overline{x}_n\right) = d\left(x_{n_j}, K_n\right) \leq d_{\mathcal{H}}\left(K_{n_j}, K_n\right) < \frac{\varepsilon}{3}.$$

Similarly, we have

$$d\left(x_{n_h}, \overline{x}_m\right) < \frac{\varepsilon}{3}.$$

By these latter inequalities and by (3.118), (3.120) we get

$$d\left(\overline{x}_n, \overline{x}_m\right) < \varepsilon$$

and thereby we have also proved that $\{\overline{x}_n\}$ is a Cauchy sequence. ■

In what follows, for any sequence $\{K_n\}$ in $\mathbf{K}(X)$ and any sequence $\{x_n\}$ in X, we will write simply $\{x_n \in K_n\}$ to denote that

$$x_n \in K_n, \quad \forall n \in \mathbb{N}.$$

Proof of Theorem 3.72 Let $\{K_n\}$ be a Cauchy sequence in $\mathbf{K}(X)$. Let us denote by

$$K = \left\{x \in X : \text{ there exists a sequence } \{x_n \in K_n\} \text{ such that } \lim_{n\to\infty} x_n = x\right\}.$$

Let us prove the following:

(a) $K \neq \emptyset$;
(b) K is closed;
(c)

$$\forall \varepsilon > 0\ \exists n_\varepsilon \in \mathbb{N} \text{ such that } \forall n \geq n_\varepsilon \ \ K \subset [K_n]_\varepsilon\,;$$

(d) K is compact;

and

$$\{K_n\} \to K, \quad \text{in } (\mathbf{K}(X), d_{\mathcal{H}})\,.$$

Proof of (a) Since $\{K_n\}$ is a Cauchy sequence in $(\mathbf{K}(X), d_{\mathcal{H}})$, we have that, for any $\varepsilon > 0$, there exists $n_\varepsilon \in \mathbb{N}$, such that

$$d_{\mathcal{H}}(K_n, K_m) < \varepsilon, \quad \forall n, m \geq n_\varepsilon. \tag{3.121}$$

We may assume $n_\varepsilon \in \mathbb{N}$ be strictly increasing w.r.t. $1/\varepsilon$. For any $j \in \mathbb{N}$ let

$$\varepsilon_j = \frac{1}{2^j}.$$

Set $n_j = n_{\varepsilon_j}$.

Let $x_{n_1} \in K_{n_1}$ be chosen arbitrarily. Since

$$d\left(x_{n_1}, K_{n_2}\right) \leq d_{\mathcal{H}}\left(K_{n_1}, K_{n_2}\right),$$

we may choose $x_{n_2} \in K_{n_2}$ such that

$$d\left(x_{n_1}, x_{n_2}\right) = d\left(x_{n_1}, K_{n_2}\right).$$

Similarly, after choosing $x_{n_1}, \cdots, x_{n_{j-1}}$, we choose $x_{n_j} \in K_{n_j}$. More precisely, let us suppose to have already chosen $x_{n_1}, \cdots, x_{n_{j-1}}$, then we choose $x_{n_j} \in K_{n_j}$ so that

$$d\left(x_{n_{j-1}}, x_{n_j}\right) = d\left(x_{n_{j-1}}, K_{n_j}\right) \leq d_{\mathcal{H}}\left(K_{n_{j-1}}, K_{n_j}\right) < \frac{1}{2^{j-1}}.$$

Now, let us prove that $\left\{x_{n_j}\right\}$ is a Cauchy sequence. For any $h > j$ the triangle inequality gives

$$d\left(x_{n_j}, x_{n_h}\right) \leq \sum_{l=j}^{h-1} d\left(x_{n_l}, x_{n_{l+1}}\right) < \sum_{l=j}^{h-1} \frac{1}{2^l} < \frac{1}{2^{j-1}}.$$

On the other hand

$$x_{n_j} \in K_{n_j}, \quad \forall j \in \mathbb{N}.$$

Hence, Lemma 3.73 implies that there exists a Cauchy sequence in (X, d), $\{\overline{x}_n \in K_n\}$, which satisfies

$$\overline{x}_{n_j} = x_{n_j}, \quad \forall j \in \mathbb{N}.$$

Since (X, d) is a complete metric space, there exists $x \in X$ such that

$$\lim_{n \to \infty} \overline{x}_n = x.$$

Hence, x belongs to K (as we have defined K). Therefore $K \neq \emptyset$.

Proof of (b) We prove that if $\{x_n\}$ is a sequence in K, which converges to x_0, then $x_0 \in K$. By the definition of K, we have that for every $n \in \mathbb{N}$ there exists a sequence

$\left\{y_j^{(n)}\right\}_{j\in\mathbb{N}}$ which satisfies

$$y_j^{(n)} \in K_j, \quad \forall j \in \mathbb{N} \tag{3.122}$$

and

$$y_j^{(n)} \to x_n, \text{ as } j \to \infty, \quad \forall n \in \mathbb{N}.$$

Since $\{x_n\}$ converges to x_0, there exists a strictly increasing sequence in $\mathbb{N}$, $\{n_h\}_{h\in\mathbb{N}}$, which satisfies

$$d\left(x_{n_h}, x_0\right) < \frac{1}{h}, \quad \forall h \in \mathbb{N}. \tag{3.123}$$

In addition, since

$$y_j^{(n_h)} \to x_{n_h}, \text{ as } j \to \infty, \quad \forall h \in \mathbb{N},$$

there exists a strictly increasing sequence in $\mathbb{N}$, $\{m_h\}_{h\in\mathbb{N}}$, which satisfies

$$d\left(y_{m_h}^{(n_h)}, x_{n_h}\right) < \frac{2}{h}, \quad \forall h \in \mathbb{N}. \tag{3.124}$$

By (3.123) and (3.124) we have

$$d\left(y_{m_h}^{(n_h)}, x_0\right) < \frac{1}{h}, \quad \forall h \in \mathbb{N}. \tag{3.125}$$

Now, let us consider the sequence $\left\{y_{m_h}^{(n_h)}\right\}_{h\in\mathbb{N}}$. Since it is convergent, it is a Cauchy sequence and, by (3.122), we have

$$y_{m_h}^{(n_h)} \in K_{m_h}, \quad \forall h \in \mathbb{N}.$$

Therefore by Lemma 3.73, there exists a Cauchy sequence, $\left\{\overline{y}_n\right\}$ which satisfies

$$\overline{y}_{m_h} = y_{m_h}^{(n_h)}, \quad \forall h \in \mathbb{N}$$

and

$$\overline{y}_n \in K_n, \quad \forall n \in \mathbb{N}.$$

Since X is a complete space and $\left\{\overline{y}_n\right\}$ is a Cauchy sequence, it converges. On the other hand, (3.125) implies that the subsequence $\left\{y_{m_h}^{(n_h)}\right\}_{h\in\mathbb{N}}$ converges to x_0.

Therefore the whole sequence $\{\overline{y}_n\}$ converges to x_0 and by the definition of K we have $x_0 \in K$.

Proof of (c) Let $\varepsilon > 0$. Let n_ε satisfy

$$d_{\mathcal{H}}(K_n, K_m) < \varepsilon, \quad \forall n, m \geq n_\varepsilon.$$

Proposition 3.69 gives

$$K_m \subset [K_n]_\varepsilon, \quad \forall n, m \geq n_\varepsilon. \tag{3.126}$$

Let $x \in K$. Let us prove that $x \in [K_n]_\varepsilon$ for every $n \geq n_\varepsilon$. Fix $n \geq n_\varepsilon$. By the definition of K, there exists $\{x_m \in K_m\}$ such that

$$x_m \to x, \quad \text{as } m \to \infty.$$

Now, since $x_m \in K_m \subset [K_n]_\varepsilon$ (by (3.126)), for every $m \geq n_\varepsilon$ and taking into account that $[K_n]_\varepsilon$ is a closed set (Proposition 3.68), we have $x \in [K_n]_\varepsilon$.

Proof of (d) Since X is a complete metric space and K is a closed set (by (b)), by Theorem 3.4, it suffices to prove that K is totally bounded. Let us argue by contradiction. Let us assume that K is not totally bounded. Hence, let us assume that there exists $\delta > 0$ and there exists a sequence $\{x_n\}$ in K so that

$$d(x_n, x_m) \geq \delta.$$

Now, by (c), there exists $\nu \in \mathbb{N}$ such that, if $n \neq m$,

$$K \subset [K_\nu]_{\frac{\delta}{4}}.$$

From which we have that for every $n \in \mathbb{N}$ there exists $y_n \in K_\nu$ such that

$$d(y_n, x_n) < \frac{\delta}{4}.$$

On the other hand, since K_ν is compact, there exists a subsequence of $\{y_n\}$, $\{y_{n_j}\}$, which converges, consequently there exists $\nu' \geq \nu$ such that

$$d\left(y_{n_j}, y_{n_h}\right) < \frac{\delta}{4}, \quad \forall j, h \geq \nu'.$$

Therefore, by the triangle inequality we have, if $j \neq h$,

$$\delta \leq d\left(x_{n_j}, x_{n_h}\right) \leq d\left(x_{n_j}, y_{n_j}\right) + d\left(y_{n_j}, y_{n_h}\right) + d\left(y_{n_h}, x_{n_h}\right) < \frac{3\delta}{4},$$

for $j, h \geq \nu'$, $j \neq h$. This is clearly a contradiction.

Proof of (e) Since we have proved (c), it suffices to prove

$$\forall \varepsilon > 0 \, \exists \nu_\varepsilon \in \mathbb{N} \text{ such that } \forall n \geq \nu_\varepsilon \;\; K_n \subset [K]_\varepsilon . \tag{3.127}$$

Since $\{K_n\}$ is a Cauchy sequence we have that, for any $\varepsilon > 0$ there exists $\nu_\varepsilon \in \mathbb{N}$ such that

$$d_{\mathcal{H}}(K_n, K_m) < \frac{\varepsilon}{2}, \quad \forall n, m \geq \nu_\varepsilon. \tag{3.128}$$

Hence

$$K_m \subset [K_n]_{\frac{\varepsilon}{2}}, \quad \forall n, m \geq \nu_\varepsilon.$$

Let us fix $\overline{n} \geq \nu_\varepsilon$. Inequality (3.128) implies that there exists a strictly increasing sequence $\{n_j\}$ in $\mathbb{N}$ such that $n_j \geq \nu_\varepsilon$, for every $j \in \mathbb{N}$, and

$$d_{\mathcal{H}}\left(K_{n_{j-1}}, K_{n_j}\right) < \frac{\varepsilon}{2^j}.$$

Since $n_1, \overline{n} \geq \nu_\varepsilon$, we get by (3.128)

$$d_{\mathcal{H}}\left(K_{\overline{n}}, K_{n_1}\right) < \frac{\varepsilon}{2}.$$

Hence

$$K_{\overline{n}} \subset \left[K_{n_1}\right]_{\frac{\varepsilon}{2}} . \tag{3.129}$$

Now, let us fix $y \in K_{\overline{n}}$ and let us prove that $y \in [K]_\varepsilon$. By (3.129) we have

$$y \in \left[K_{n_1}\right]_{\frac{\varepsilon}{2}} ,$$

hence there exists $x_{n_1} \in K_{n_1}$ such that

$$d\left(x_{n_1}, y\right) < \frac{\varepsilon}{2}. \tag{3.130}$$

Generally speaking, since

$$K_{n_{j-1}} \subset \left[K_{n_j}\right]_{\frac{\varepsilon}{2^j}}, \quad \forall j \geq 2,$$

there exists a sequence $\{x_{n_j}\}$ which satisfies $x_{n_j} \in K_{n_j}$, for every $j \in \mathbb{N}$ and

$$d\left(x_{n_{j-1}}, x_{n_j}\right) < \frac{\varepsilon}{2^j}, \quad \forall j \in \mathbb{N}.$$

By the latter and by (3.130) we get

$$d\left(y, x_{n_j}\right) \leq d\left(y, x_{n_1}\right) + d\left(x_{n_1}, x_{n_2}\right) + \cdots + d\left(x_{n_{j-1}}, x_{n_j}\right) < \varepsilon \tag{3.131}$$

and

$$d\left(x_{n_j}, x_{n_h}\right) \leq \sum_{l=j}^{h-1} d\left(x_{n_l}, x_{n_{l+1}}\right) < \frac{\varepsilon}{2^j}, \quad \forall h > j \geq \nu_\varepsilon \in \mathbb{N}.$$

In particular, the just obtained inequality implies that for every $\delta > 0$ there exists n_δ such that

$$d\left(x_{n_j}, x_{n_h}\right) < \delta, \quad \forall j, h \geq n_\delta.$$

Hence $\left\{x_{n_j}\right\}$ is a Cauchy sequence and it satisfies

$$x_{n_j} \in K_{n_j}, \quad \forall j \in \mathbb{N}.$$

Now, Lemma 3.73 implies that there exists a Cauchy sequence $\{\overline{x}_n \in K_n\}$ which satisfies

$$\overline{x}_{n_j} = x_{n_j}, \quad \forall j \in \mathbb{N}.$$

Consequently $\{\overline{x}_n \in K_n\}$ converges to a point x and such a point x, by the definition of K, belongs to K. In addition, since $\left\{x_{n_j}\right\}$ is a subsequence of $\{\overline{x}_n\}$, we have

$$\left\{x_{n_j}\right\} \to x.$$

Hence, by (3.131), we have

$$d(x, y) = \lim_{j \to \infty} d\left(y, x_{n_j}\right) \leq \varepsilon.$$

Therefore

$$y \in [K]_\varepsilon.$$

Hence (3.127) is proved.

■

Theorem 3.74 (Compactness) *If* (X, d) *is a compact metric space, then* $(\mathbf{K}(X), d_{\mathcal{H}})$ *is a compact metric space.*

Proof Since (X, d) is a compact space it is complete and totally bounded. On the other hand, by Theorem 3.72, $(\mathbf{K}(X), d_{\mathcal{H}})$ is complete, hence for proving that it is compact, it suffices to prove that $(\mathbf{K}(X), d_{\mathcal{H}})$ is totally bounded.

Let ε be any positive number, since X is totally bounded, there exists a finite set F_ε which satisfies

$$d(x, F_\varepsilon) < \varepsilon, \quad \forall x \in X. \tag{3.132}$$

Let $\mathcal{G}_\varepsilon$ be the family of all subsets of F_ε ($\mathcal{G}_\varepsilon$ is finite because F_ε is finite). Let $K \in \mathbf{K}(X)$. Let us consider the set

$$G = \{p \in F_\varepsilon : d(p, K) < \varepsilon\}.$$

We have $G \neq \emptyset$. As a matter of fact, (3.132) implies that if $x \in K$ then there exists $y \in F_\varepsilon$ such that

$$d(x, y) < \varepsilon,$$

hence

$$d(y, K) \le d(x, y) < \varepsilon,$$

therefore $y \in G$. Notice, that we have trivially $G \in \mathcal{G}_\varepsilon$ and

$$\delta(G, K) = \max_{p \in G} d(p, K) < \varepsilon. \tag{3.133}$$

Now, let us prove

$$\delta(K, G) = \max_{x \in K} d(x, G) < \varepsilon. \tag{3.134}$$

Let $x \in K$. Relationship (3.132) implies that there exists $y \in F_\varepsilon$ such that

$$d(x, y) < \varepsilon,$$

consequently

$$d(y, K) \le d(y, x) < \varepsilon.$$

Therefore $y \in G$, which yields

$$d(x, G) \le d(x, y) < \varepsilon,$$

and (3.134) follows. Hence

$$d_{\mathcal{H}}(K, G) = \max\{\delta(K, G), \delta(G, K)\} < \varepsilon.$$

All in all, we have proved

$$\forall K \in \mathbf{K}(X) \ \exists G \in \mathcal{G}_\varepsilon \text{ such that } d_{\mathcal{H}}(K, G) < \varepsilon,$$

which, since ε is arbitrary, implies that $\mathbf{K}(X)$ is a compact metric space. ■

3.12 The Distance Function

In this section we will give some properties of the function

$$\mathbb{R}^n \ni x \to d_{\partial\Omega}(x) := d(x, \partial\Omega),$$

where Ω is a bounded open set of $\mathbb{R}^n$ whose boundary is of class $C^{1,1}$. When there is no risk of ambiguity, we simply write $d(x)$. In Proposition 3.62 we have proved that $d_{\partial\Omega}(x)$ is a Lipschitz continuous function, because it satisfies the inequality

$$|d_{\partial\Omega}(x) - d_{\partial\Omega}(y)| \leq |x - y|, \quad \forall x, y \in \mathbb{R}^n. \tag{3.135}$$

We say that an open set A of $\mathbb{R}^n$ enjoys the **interior ball property**, if for every point $P \in \partial A$ there exists $P' \in A$ and $r > 0$ such that

$$\overline{B_r(P')} \cap \overline{\Omega} = \{P\}.$$

We say that A enjoys the **exterior ball property** if $\mathbb{R}^n \setminus \overline{\Omega}$ enjoys the property of the interior ball.

The following Proposition holds true.

Proposition 3.75 *Let Ω be a bounded open set of $\mathbb{R}^n$ whose boundary is of class $C^{1,1}$ with constants r_0, M_0. Then Ω enjoys the interior ball property and the exterior ball property. More precisely we have what follows. Denoting by*

$$\mu_0 = \min\left\{\frac{1}{M_0}, M_0\right\}, \tag{3.136}$$

for any $P \in \partial\Omega$ and for any $r \in (0, \mu_0 r_0)$, we have

$$\overline{B_r(P - r\nu(P))} \cap \overline{\Omega} = \{P\} \tag{3.137}$$

and

$$\overline{B_r(P + r\nu(P))} \cap \overline{\mathbb{R}^n \setminus \Omega} = \{P\}, \tag{3.138}$$

where $\nu(P)$ is the unit outward normal to $\partial\Omega$ in P.

Proof Let $P \in \partial\Omega$. Let us consider a local representation of $\partial\Omega$. Hence, let us assume $P = 0$ and let us assume, up to a isometry,

$$\Omega \cap Q_{r_0,2M_0} = \left\{x \in Q_{r_0,2M_0} : x_n > g(x')\right\},$$

where $g \in C^{1,1}\left(\overline{B'}_{r_0}\right)$ satisfies

$$g(0) = |\nabla g(0)| = 0$$

and

$$\|g\|_{C^{1,1}\left(\overline{B'_{r_0}}\right)} \le M_0 r_0.$$

We have

$$\nu(0) = -e_n.$$

Now, notice that

$$g\left(x'\right) = \int_0^1 (1-s)\partial^2 g\left(sx'\right)x' \cdot x' ds.$$

Hence

$$g\left(x'\right) \le \frac{M_0\left|x'\right|^2}{2r_0}.$$

Therefore, in order to satisfy (3.137) it suffices that, besides the condition $r \le r_0$, the following conditions are satisfied

$$\frac{M_0\left|x'\right|^2}{2r_0} < r - \sqrt{r^2 - |x'|^2}, \quad \forall x' \in \overline{B'}_r \setminus \{0\}$$

and

$$r + \sqrt{r^2 - |x'|^2} < 2M_0 r_0, \quad \forall x' \in \overline{B'}_r.$$

It is easy to check that if $r \in (0, \mu_0 r_0)$, the above conditions are satisfied. In a similar way we proceed for the property of the exterior ball. ■

For any $\rho \in (0, \mu_0 r_0)$, set

$$S_\rho = \{x \in \Omega : \, d_{\partial\Omega}(x) < \rho\}. \tag{3.139}$$

We observe that for every $x \in S_\rho$ there exists a unique point $p(x) \in \partial\Omega$ such that

$$|x - p(x)| = d_{\partial\Omega}(x).$$

As a matter of fact, let $\overline{x} \in S_\rho$. Let $p \in \partial\Omega$ be a point which satisfies

$$d_{\partial\Omega}(\overline{x}) = |\overline{x} - p|.$$

We may assume that y belongs to the graph, $\Gamma^{(g)}$, of a function $g \in C^{1,1}\left(\overline{B'}_{r_0}\right)$ such that

$$\Omega \cap Q_{r_0, 2M_0} = \left\{x \in Q_{r_0, 2M_0} : \, x_n > g(x')\right\},$$

and $g(0) = |\nabla g(0)| = 0$. Since p is a minimum point on $\Gamma^{(g)}$ of the function

$$y \to \frac{1}{2}|\overline{x} - y|^2.$$

By The Lagrange Multiplier Theorem, the following conditions need to be fulfilled

$$\begin{cases} p_j - \overline{x}_j + \lambda \partial_j g\left(p'\right) = 0, & 1 \le j \le n-1 \\ p_n - \overline{x}_n - \lambda = 0 \\ g\left(p'\right) - p_n = 0 \end{cases}.$$

Hence

$$p - \overline{x} = \left(p' - \overline{x}', p_n - \overline{x}_n\right) = \lambda\left(-\nabla_{p'} g\left(p'\right), 1\right),$$

which implies

$$p - \overline{x} = |p - \overline{x}|\, \nu(y). \tag{3.140}$$

This relation, in turn, implies that p is the unique point of $\partial\Omega$ that achieves the minimum distance. Indeed, let $B_\rho(z)$ be the interior ball tangent in y to $\partial\Omega$ (such a ball exists by Proposition 3.138), the equality (3.140) ensures us that $\overline{x}$ lies on the segment of extremes z and p. Consequently, by setting $\rho_1 = |\overline{x} - p|$ we have $B_{\rho_1}(\overline{x}) \subset B_\rho(z)$. It is, therefore, evident that the distance of $\overline{x}$ from $\partial B_\rho(z)$ is greater than or equal to ρ_1 and, recalling that $B_\rho(\overline{y})$ is an interior ball to Ω, tangent to $\partial\Omega$ at the unique point y, we obtain that

$$|\overline{x} - p| = \rho_1 \leq \rho < |\overline{x} - \xi|, \quad \forall \xi \in \partial\Omega \setminus \{y\}.$$

Therefore we have proved the following

Proposition 3.76 *If $\rho \in (0, \mu_0 r_0)$, then for any $x \in S_\rho$ there exists a unique point $p(x) \in \partial\Omega$ which attains the minimum of distance from x to $\partial\Omega$. Moreover we have*

$$x = p(x) - d_{\partial\Omega}(x)\nu(p(x)). \tag{3.141}$$

The following Proposition holds true

Proposition 3.77 *If $\rho \in (0, \mu_0 r_0)$, then*

$$S_\rho = \{y - t\nu(y) : \ y \in \partial\Omega, \ 0 \leq t < \rho\}. \tag{3.142}$$

Proof Proposition 3.76 implies

$$S_\rho \subset \{y - t\nu(y) : \ y \in \partial\Omega, \ 0 \leq t < \rho\}.$$

Now, let $x = y - t\nu(y)$, where $y \in \partial\Omega$ and $0 \leq t < \rho$. We have

$$d_{\partial\Omega}(x) \leq |x - y| = t < \rho.$$

Hence $x \in S_\rho$. Therefore

$$\{y - t\nu(y) : \ y \in \partial\Omega, \ 0 \leq t < \rho\} \subset S_\rho.$$

Therefore (3.142) is proved. ■

Let us prove the following

Lemma 3.78 *There exists $\mu_1 \leq \mu_0$ such that if $\rho \in (0, \mu_1 r_0)$ then the maps*

$$S_\rho \ni x \to p(x) \in \partial\Omega, \quad \text{and} \quad S_\rho \ni x \to \nu(p(x)) \in \mathbb{S}^{n-1} \tag{3.143}$$

are Lipschitz continuous, where $p(x)$ is the point that realizes the minimum distance of $x \in S_\rho$ from $\partial\Omega$.

Proof We begin by proving that the map

$$\partial\Omega \ni y \to \nu(y) \in \mathbb{S}^{n-1}, \tag{3.144}$$

is Lipschitz continuous. To prove this, let $y_1, y_2 \in \partial\Omega$ and distinguish two cases

(a) $|y_1 - y_2| \geq r_0$,

(b) $|y_1 - y_2| < r_0$.

In case (a), we have trivially

$$|\nu(y_1) - \nu(y_2)| \leq 2 \leq 2\frac{|y_1 - y_2|}{r_0}. \tag{3.145}$$

In case (b), we may employ a local representation of $\partial\Omega$ assuming that $y_2 = 0$ and $y_1 = g(x')$ where $g \in C^{1,1}\left(\overline{B'_{r_0}}\right)$ and $g(0) = |\nabla g(0)| = 0$. Hence

$$\nu(y_2) = -e_n$$

and

$$\nu(y_1) = \left(\frac{\nabla_{x'} g(x')}{\sqrt{1 + |\nabla_{x'} g(x')|^2}}, \frac{-1}{\sqrt{1 + |\nabla_{x'} g(x')|^2}}\right).$$

Now it is easy to check that

$$|\nu(y_1) - \nu(y_2)| \leq \frac{\sqrt{2}M_0}{r_0}\left|x'\right| = \frac{\sqrt{2}M_0}{r_0}|y_1 - y_2|. \tag{3.146}$$

Therefore, by (3.145) and (3.146) we get

$$|\nu(y_1) - \nu(y_2)| \leq \frac{M_1}{r_0}|y_1 - y_2|, \tag{3.147}$$

where

$$M_1 = \max\left\{2, \sqrt{2}M_0\right\}.$$

Now, let us prove that $x \to p(x)$ is Lipschitz continuous. By (3.141) we have (we omit subscript in $d_{\partial\Omega}$)

$$p(x) = x + d(x)\nu(p(x)).$$

Therefore, recalling (3.135) and (3.147), we get

$$\begin{aligned}|p(x) - p(y)| &\leq |x - y| + d(x)|\nu(p(x)) - \nu(p(y))| + |d(x) - d(y)||\nu(p(y))| \leq \\ &\leq 2|x - y| + \rho\frac{M_1}{r_0}|p(x) - p(y)|.\end{aligned}$$

Hence

$$\left(1 - \rho\frac{M_1}{r_0}\right)|p(x) - p(y)| \leq 2|x - y|.$$

Moreover for any

$$\rho < r_0 \min\left\{\frac{1}{2M_1}, M_0, \frac{1}{M_0}\right\}$$

we have

$$|p(x) - p(y)| \leq 4|x - y|, \quad \forall x, y \in S_\rho. \tag{3.148}$$

The above inequality proves that the map $x \to p(x)$ is Lipschitz continuous provided

$$\mu_1 = \min\left\{\frac{1}{2M_1}, M_0, \frac{1}{M_0}\right\}.$$

Therefore (3.147) and (3.148) imply that $\nu(p(x))$ is Lipschitz continuous. ■

Lemma 3.79 *Let A be on open set of $\mathbb{R}^n$ and let $f \in C^{0,1}_{loc}(A)$. If there exists a function $g \in C^0(A; \mathbb{R}^n)$ which satisfies*

$$\nabla f(x) = g(x), \quad a.e.\ x \in A, \tag{3.149}$$

then $f \in C^1(A)$ and

$$\nabla f(x) = g(x), \quad \forall x \in A. \tag{3.150}$$

Proof Let $x_0 \in A$ and $\delta = \frac{1}{4}d(x, \partial A)$. For any $\varepsilon \in (0, \delta)$ let us consider the function

$$f_\varepsilon(x) = \int_A f(y)\eta_\varepsilon(x - y)dy, \quad \forall x \in B_\delta(x_0),$$

where η is a mollifier. It turns out that $f_\varepsilon \in C^\infty\left(\overline{B_\delta(x_0)}\right)$ and, moreover, the divergence Theorem gives

$$\begin{aligned}
\partial_j f_\varepsilon(x) &= -\int_A f(y)\partial_{y_j}\eta_\varepsilon(x-y)dy = \\
&= -\int_{B_\varepsilon(x)} \left[\partial_{y_j}\left(f(y)\eta_\varepsilon(x-y)\right) - \partial_{y_j} f(y)\eta_\varepsilon(x-y)\right] dy = \\
&= \int_{B_\varepsilon(x)} \partial_{y_j} f(y)\eta_\varepsilon(x-y)dy = \\
&= \int_A \partial_{y_j} f(y)\eta_\varepsilon(x-y)dy, \quad j=1,\cdots n, \quad \forall x \in B_\delta(x_0).
\end{aligned}$$

By what has just been obtained and by (3.149) we have

$$\nabla f_\varepsilon(x) = g_\varepsilon(x) := \int_A g(y)\eta_\varepsilon(x-y)dy, \quad \forall x \in B_\delta(x_0). \tag{3.151}$$

Let v be a unit vector of $\mathbb{R}^n$. By (3.151) we get

$$f_\varepsilon(x_0+tv) - f_\varepsilon(x_0) = \int_0^t g_\varepsilon(x_0+sv)\cdot v ds, \quad \forall t \in [-\delta, \delta]. \tag{3.152}$$

Now, Theorem 3.44 implies that f_ε and g_ε uniformly converges in $B_\delta(x_0)$. Therefore passing to the limit in (3.152) as $\varepsilon \to 0$, we obtain

$$f(x_0+tv) - f(x_0) = \int_0^t g(x_0+sv)\cdot v ds, \quad \forall t \in [-\delta, \delta].$$

On the other hand, since g is continuous, we have

$$\lim_{t\to 0} \frac{f(x_0+tv)-f(x_0)}{t} = \lim_{t\to 0}\frac{1}{t}\int_0^t g(x_0+sv)\cdot v ds = g(x_0)\cdot v.$$

Therefore

$$\frac{\partial f(x_0)}{\partial v} = g(x_0)\cdot v, \ \forall v \in \mathbb{R}^n, \ |v|=1.$$

which implies

$$\nabla f(x_0) = g(x_0),$$

so that, since x_0 is arbitrary in A, we have $\nabla f = g$ in A. That, in turn, by the continuity of g implies $f \in C^1(A)$. ■

Theorem 3.80 *Let μ_1 be the same of Proposition 3.78. Then we have*

$$d_{\partial\Omega} \in C^{1,1}\left(\overline{S_{\mu_1 r_0}}\right)$$

and

$$\nabla d_{\partial\Omega}(x) = -\nu(p(x)), \quad \forall x \in S_{\mu_1 r_0}. \tag{3.153}$$

Proof For the sake of brevity, we omit the subscript in $d_{\partial\Omega}$. Inequality (3.135) implies that d is almost everywhere differentiable, in addition in the points where it is differentiable we have

$$|\nabla d| \leq 1. \tag{3.154}$$

Let $x \in S_{\mu_1 r_0} \setminus \partial\Omega$ be a point in which d is differentiable. Let $t \in (0, d(x))$. Since $x + t\nu(p(x))$ lies on the segment of endpoints x and $p(x)$, we have

$$d(x + t\nu(p(x))) = d(x) - t.$$

Hence

$$\nabla d(x) \cdot \nu(p(x)) = \lim_{t \to 0^+} \frac{d(x + t\nu(p(x))) - d(x)}{t} = -1.$$

Therefore

$$\nabla d(x) \cdot \nu(p(x)) = -1. \tag{3.155}$$

Consequently we get

$$1 = |\nabla d(x) \cdot \nu(p(x))| \leq |\nabla d(x)||\nu(p(x))| \leq 1.$$

Hence, there exists $\lambda \in \mathbb{R}$ such that $\nabla d(x) = \lambda\nu(p(x))$ and by (3.155) we have $\lambda = -1$. This implies

$$\nabla d(x) = -\nu(p(x)), \quad \text{a.e. } x \in S_{\mu_1 r_0} \setminus \partial\Omega. \tag{3.156}$$

Now, by Lemma 3.78 we know that $\nu(p(x))$ is Lipschitz continuous, therefore by Lemma 3.79 we obtain $d \in C^1\left(S_{\mu_1 r_0} \setminus \partial\Omega\right))$. Finally, exploit again (3.156), we have $d \in C^{1,1}\left(\overline{S_{\mu_1 r_0}}\right)$. ■

Corollary 3.81 *For any $\rho \in (0, \mu_1 r_0)$ the boundary of the open set*

$$(\Omega)_\rho = \{x \in \Omega : \ d_{\partial\Omega}(x) > \rho\} \tag{3.157}$$

is of class $C^{1,1}$ and we have

$$\partial(\Omega)_\rho = \Gamma_\rho, \tag{3.158}$$

where

$$\Gamma_\rho = \{x \in \Omega : \ d_{\partial\Omega}(x) = \rho\} . \tag{3.159}$$

Moreover

$$\Gamma_\rho = \{y - \rho\nu(\rho) : \ y \in \partial\Omega\} . \tag{3.160}$$

Proof We first prove (3.158). To prove that $\Gamma_\rho \subset \partial(\Omega)_\rho$ we argue by contradiction. Let us assume that there exists $x \in \partial(\Omega)_\rho$ such that $d_{\partial\Omega}(x) > \rho$. Consequently, x would be an interior point of $(\Omega)_\rho$. If $d_{\partial\Omega}(x) < \rho$ then x would be exterior to $(\Omega)_\rho$. Therefore, $x \in \Gamma_\rho$ and we have $\partial(\Omega)_\rho \subset \Gamma_\rho$. Now, let $x \in \Gamma_\rho$. Since $\rho \in (0, \mu_1 r_0)$ (and $\mu_1 \leq \mu_0$), Proposition 3.76 implies that there is a unique point $p(x) \in \partial\Omega$ which attains the minimum of the distance of x from $\partial\Omega$, moreover

$$x = p(x) - \rho\nu(p(x)).$$

For any $\varepsilon > 0$ small enough, we have

$$x - \varepsilon\nu(p(x) = p(x) - (\rho + \varepsilon)\nu(p(x)) \in (\Omega)_\rho$$

and

$$x + \varepsilon\nu(p(x) = p(x) - (\rho - \varepsilon)\nu(p(x)) \notin (\Omega)_\rho.$$

Hence $x \in \partial(\Omega)_\rho$. Therefore $\Gamma_\rho \subset \partial(\Omega)_\rho$.

In order to prove that $\partial(\Omega)_\rho$ is of class $C^{1,1}$, we exploit (3.158). By Theorem 3.80 we derive $|\nabla d_{\partial\Omega}(x)| = 1$, for every $x \in \Gamma_\rho$, and by applying the Implicit Function Theorem we easily reach the assertion.

Concerning (3.160), let us note that if $x = y - \rho\nu(y)$, taking into account $\rho < \mu_1 r_0 \leq \mu_0 r_0$, then $d_{\partial\Omega}(x) = \rho$. Conversely, if $x \in \Gamma_\rho$ Proposition 3.76 gives

$$x = p(x) - d_{\partial\Omega}(x)\nu(p(x)) = p(x) - \rho\nu(p(x))$$

so that, since $p(x) \in \partial\Omega$ we have $x \in \{y - \rho\nu(\rho) : \ y \in \partial\Omega\}$. ■

We now provide a few words about the map

$$\Phi : \partial\Omega \times (0, \mu_1 r_0) \rightarrow \ \mathbb{R}^n,$$

such that

$$\Phi(y, t) = y - t\nu(y), \ \ \forall (y, t) \in \partial\Omega \times (0, \mu_1 r_0) . \tag{3.161}$$

The following Proposition holds true

Proposition 3.82 *If Ω is a bounded open set of $\mathbb{R}^n$ of class $C^{1,1}$ then we have:*

(a) $\Phi\left(\partial\Omega\times(0,\mu_1 r_0)\right)=S_{\mu_1 r_0}$,
(b) $\Phi\in C^{0,1}\left(\partial\Omega\times[0,\mu_1 r_0]\right)$,
(c) Φ *is injective on* $\partial\Omega\times(0,\mu_1 r_0)$ *and its inverse is a Lipschitz continuous map.*

Proof (a) is a consequence of Proposition 3.77. (b) is a consequence of Lemma 3.78. Now let us prove (c). Let $x\in S_{\mu_1 r_0}$ satisfy

$$x=\Phi(y,t)=y-t\nu(y),\quad (y,t)\in\partial\Omega\times(0,\mu_1 r_0)\,.$$

By Proposition 3.76 and by the interior ball property we get

$$y=p(x),\qquad t=d_{\partial\Omega}(x).$$

Hence

$$\Phi^{-1}(x)=p(x)-d_{\partial\Omega}(x)\nu(p(x)).$$

By the latter and by Lemma 3.78 it follows that Φ^{-1} is Lipschitz continuous. ■

If Ω is of class C^k, $k\geq 2$, other properties of the distance function and the map Φ can be proved. For instance, one can prove that $d_{\partial\Omega}\in C^k$ and $\Phi\in C^{k-1}$. For further details, we refer to [50, Ch. 14, Sect. 6].

We say that a continuous map

$$\gamma:[0,1]\to A,$$

is a **continuous path in a set** $A\subset\mathbb{R}^n$. Let $B\subset A$ and $x,y\in B$. If $\gamma([0,1])\subset B$ and $\gamma(0)=x$, $\gamma(1)=y$, we say that the path γ joins x and y in B. If γ_1 and γ_2 are two continuous paths in A which satisfy $\gamma_2(0)=\gamma_1(1)$, we denote by $\gamma_1\vee\gamma_2$ the following continuous path

$$\left(\gamma_1\vee\gamma_2\right)(t)=\begin{cases}\gamma_1(2t), & \text{for } t\in\left[0,\frac{1}{2}\right)\\ \\ \gamma_2(2t-1), & \text{for } t\in\left[\frac{1}{2},1\right]\end{cases}. \tag{3.162}$$

If $\gamma_1,\cdots\gamma_k$ are $k\geq 2$ continuous paths in A such that $\gamma_j(0)=\gamma_{j-1}(1)$, $j=2,\cdots k$, we set

$$\gamma_1\vee\cdots\vee\gamma_k:=\left(\gamma_1\vee\cdots\vee\gamma_{k-1}\right)\vee\gamma_k.$$

We say that $\gamma_1\vee\cdots\vee\gamma_k$ is the **union of continuous paths** $\gamma_1,\cdots\gamma_k$.

Proposition 3.83 *Let us assume that* Ω *and* $\partial\Omega$ *are connected. If* $\rho \in (0, \mu_1 r_0)$, *then the open set* $(\Omega)_\rho$, *defined by* (3.157), *is connected.*

Proof Let $z, w \in (\Omega)_\rho$. Let $\varepsilon > 0$ be such that

$$\rho + \varepsilon < \min\{\mu_1 r_0, d_{\partial\Omega}(z), d_{\partial\Omega}(w)\}.$$

Then

$$z, w \in (\Omega)_{\rho+\varepsilon},$$

$$\Gamma_{\rho+\varepsilon} \subset (\Omega)_\rho, \tag{3.163}$$

and $\Gamma_{\rho+\varepsilon}$ is connected, as it is the image by Φ, defined in Proposition 3.82, of the connected set $\partial\Omega \times [0, \rho + \varepsilon)$.

Now, since Ω is connected, $\overline{\Omega}$ is also connected (path connected, because $\partial\Omega$ is of class $C^{1,1}$). Be, therefore, $x \in \partial\Omega$ and be

$$\gamma_1 : [0, 1] \to \overline{\Omega} \quad \text{and} \quad \gamma_2 : [0, 1] \to \overline{\Omega},$$

two continuous paths such that

$$\gamma_1(0) = z, \;\; \gamma_1(1) = x, \quad \gamma_2(0) = x, \;\; \gamma_2(1) = w.$$

Let

$$t_1 = \inf\{t \in [0, 1] : \; d(\gamma_1(t), \partial\Omega) < \rho + \varepsilon\}.$$

We have (because $d(\gamma_1(\cdot), \partial\Omega)$ is continuous)

$$y' := \gamma_1(t_1) \in \Gamma_{\rho+\varepsilon}.$$

Similarly, let

$$t_2 = \sup\{t \in [0, 1] : \; d(\gamma_2(t), \partial\Omega) < \rho + \varepsilon\}.$$

We have

$$y'' := \gamma_2(t_2) \in \Gamma_{\rho+\varepsilon}.$$

Since $\Gamma_{\rho+\varepsilon}$ is connected, there exists a continuous path $\widetilde{\gamma} : [0, 1] \to \Gamma_{\rho+\varepsilon}$, such that

$$\widetilde{\gamma}(0) = y', \;\; \widetilde{\gamma}(1) = y''.$$

It is now evident that the path

$$\gamma := \gamma_1 \vee \widetilde{\gamma} \vee \gamma_2,$$

is continuous and it joins z and w in $(\Omega)_\rho$. ■

Remark The assumption that $\partial\Omega$ is connected is not necessary in Proposition 3.83. The proof of this statement can be fashioned similarly to the proof of Proposition 3.83, while considering that the boundedness of Ω and the $C^{1,1}$ nature of $\partial\Omega$ imply a finite number of connected components for $\partial\Omega$. We encourage the reader to explore the details. ♦

3.13 A Brief Collection of Notion on the Topological Vector Spaces

We conclude this chapter with a concise overview of concepts and theorems from Functional Analysis on topological vector spaces. The use of these topics is mainly confined to Sect. 11.3 and, to a small extent, to the Chap. 4. However, we will not provide the proofs of the theorems stated in this section, but we choose W. Rudin's book [114] as the reference text to which we refer for the above-mentioned proofs and for further consideration. We will quote detailed references from of [114] as we go along.

Let us recall the following Theorem of General Topology.

Theorem 3.84 (Baire) *Let $\mathcal{X}$ be a complete metric space. Then for any countable family of closed subset, $\{F_n\}_{n\in\mathbb{N}}$, which satisfies*

$$\mathrm{Int}(F_n) = \emptyset, \quad \forall n \in \mathbb{N}, \tag{3.164}$$

we have

$$\mathrm{Int}\left(\bigcup_{n\in\mathbb{N}} F_n\right) = \emptyset. \tag{3.165}$$

Definition 3.85 (Topological Vector Space) Let $\mathcal{X}$ be a vector space on $\mathbb{C}$ (or $\mathbb{R}$) equipped with a topology τ. We say that $\mathcal{X}$ is a topological vector space

(a) for every $x \in \mathcal{X}$, $\{x\}$ is closed w.r.t. τ
(b) the maps

$$\mathcal{X} \times \mathcal{X} \ni (x, y) \to x + y \in \mathcal{X}$$

and

$$\mathcal{X} \times \mathbb{C} \ni (x, \lambda) \to \lambda x \in \mathcal{X}$$

(or, $\mathcal{X} \times \mathbb{R} \ni (x, \lambda) \to \lambda x \in \mathcal{X}$) are continuous.

Let $\mathcal{X}$ be a vector space, we say that $p : \mathcal{X} \to [0, +\infty)$ is a **seminorm** on $\mathcal{X}$ if we have

$$p(x + y) \leq p(x) + p(y), \quad \forall x, y \in \mathcal{X}$$

and

$$p(\lambda x) \leq |\lambda| p(x), \quad \forall \lambda \in \mathbb{C}, \ \forall x \in \mathcal{X}.$$

Let $\mathcal{X}$ be a topological vector space:

(i) We say that $\mathcal{X}$ is **locally convex** if there exists a local base of neighborhoods of 0 whose members are convex. By the condition (b) of Definition 3.85 it is clear that if $\mathcal{U}$ is a neighborhood of 0 then, for every $x \in \mathcal{X}$, $x + \mathcal{U}$ is a neighborhood of x and conversely. Hence, given a local base of neighborhoods of 0 it turns out defined trivially a local base of neighborhoods of each point of $\mathcal{X}$;
(ii) We say that $\mathcal{X}$ is a **F-space** if its topology is induced by a complete metric d which is invariant w.r.t. translation (i.e. $d(x + z, y + z) = d(x, y)$ for every $x, y, z \in \mathcal{X}$);
(iii) We say that $\mathcal{X}$ is a **Fréchet space** if it is a locally convex F-space.

Let Ω be an open set of $\mathbb{R}^n$. We equip $C^k(\Omega)$, where $k \in \mathbb{N}_0$, or $k = \infty$, with a topology that makes it a Fréchet space.

We start by $C^0(\Omega)$ ($C^0(\Omega, \mathbb{C})$ or $C^0(\Omega, \mathbb{R})$, [114, Ch. 1, Sect. 1.44]). Let $\{K_j\}_{j \in \mathbb{N}}$ be a family of compact sets contained in Ω such that $K_j \neq \emptyset$, $K_j \subset \mathring{K}_{j+1}$ for every $j \in \mathbb{N}$ and

$$\bigcup_{j=1}^{\infty} K_j = \Omega.$$

For any $f \in C^0(\Omega)$, let us denote by

$$p_{0,j}(f) = \max_{K_j} |f|.$$

$\{p_{0,j}\}_{j \in \mathbb{N}}$ is a family of **separating seminorms**, that is, for each $f \in C^0(\Omega)$, which does not vanish identically, there exists $j \in \mathbb{N}$ such that $p_j(f) \neq 0$. The collection of sets [114, Thm. 1.37]

$$\mathcal{V}_j = \left\{ f \in C^0(\Omega) : p_{0,j}(f) < \frac{1}{j} \right\},$$

make up a local base in $C^0(\Omega)$ of convex neighborhoods of 0 which in turn defines a topology induced by the distance

$$d_0(f,g)=\sum_{j=1}^{\infty}\frac{2^{-j}p_{0,j}(f-g)}{1+p_{0,j}(f-g)},\qquad \forall f,g\in C^0(\Omega). \tag{3.166}$$

It is proved that $C^0(\Omega)$ with the metric (3.166) is a Fréchet space (it is quite simple and is left as an exercise).

In a similar way we proceed for $C^k(\Omega)$, k finite. More precisely, we start, rather than from the seminorms $p_{0,j}$, from the seminorms

$$p_{k,j}(f)=\max\left\{\max_{K_j}|\partial^\alpha f| : |\alpha|\le k\right\}$$

and we define the distance d_k on $C^k(\Omega)$ by substituting $p_{0,j}$ by $p_{k,j}$ in (3.166), that is

$$d_k(f,g)=\sum_{j=1}^{k}\frac{2^{-j}p_{k,j}(f-g)}{1+p_{0,j}(f-g)},\qquad \forall f,g\in C^k(\Omega). \tag{3.167}$$

Similarly, it is proved that $C^k(\Omega)$ with the metric (3.167) is a Fréchet space.

Finally, concerning $C^\infty(\Omega)$, the following seminorms (with the corresponding metric), are defined

$$q_N(f)=\max\left\{|\partial^\alpha f(x)| : x\in K_N,\quad |\alpha|\le N\right\}, \tag{3.168}$$

$$d_\infty(f,g)=\sum_{N=1}^{\infty}\frac{2^{-N}q_N(f-g)}{1+q_N(f-g)},\qquad \forall f,g\in C^\infty(\Omega). \tag{3.169}$$

in the sequel, when we are dealing with the convergence of sequences, it will be more convenient to use directly the seminorms instead of the distances d_k, $0\le k\le\infty$. For instance, the sequence $\{f_k\}$ in $C^\infty(\Omega)$ converges to f in the topology induced by the distance d_∞ if and only if

$$\lim_{k\to\infty} q_N(f_k-f)=0,\quad \forall N\in\mathbb{N}.$$

The Fréchet space $C^\infty(\Omega)$ defined above is denoted also by $\mathcal{E}(\Omega)$.

The proof of the following Proposition is a simple exercise that we leave to the reader (hint: use Lemma 3.49).

Proposition 3.86 *For any $\varphi \in C^{\infty}(\Omega)$ there exists a sequence $\{\varphi_k\}$ in $C_0^{\infty}(\Omega)$ such that $\{\varphi_k\} \to \varphi$ in $\mathcal{E}(\Omega)$. Namely*

$$\lim_{k\to\infty} q_N(\varphi_k - \varphi) = 0, \quad \forall N \in \mathbb{N}.$$

We conclude this section with the following important theorems.

Theorem 3.87 (Closed Graph) *Let $\mathcal{X}$, $\mathcal{Y}$ be two F-spaces and*

$$\Lambda : \mathcal{X} \to \mathcal{Y} \tag{3.170}$$

be a linear map. Then Λ is continuous if and only if the graph of Λ

$$\mathcal{G} = \{(x, \Lambda x) : x \in \mathcal{X}\},$$

is closed in $\mathcal{X} \times \mathcal{Y}$.

We refer to [114, Prop. 2.14, Thm 2.15] for a proof. Keep in mind that the most significant implication of Theorem 3.87 consists of

$$\mathcal{G} \text{ closed} \Longrightarrow \Lambda \text{ continuous}.$$

The reverse is true even if Λ is nonlinear, where $\mathcal{X}$ and $\mathcal{Y}$ are topological spaces and $\mathcal{Y}$ is a Hausdorff space.

Theorem 3.88 (Hahn–Banach) *Let $\mathcal{X}$ be a vector space, p a seminorm on $\mathcal{X}$ and let M be a subspace of $\mathcal{X}$. If f is a linear functional on M such that*

$$|f(x)| \leq p(x), \quad \forall x \in M,$$

then there exists a linear functional F on $\mathcal{X}$ such that

$$F_{|M} = f$$

and

$$|F(x)| \leq p(x), \quad \forall x \in M.$$

We refer to [114, Ch. 3] for a proof of Theorem 3.88.

Chapter 4
Notes on the Distribution Theory and Fourier Transform

Abstract This chapter introduces the rudiments of distribution theory and the Fourier transform, with a brief exploration of oscillatory integrals. The treatment is meant as a recap rather than an exhaustive exposition. We highlight the shift in perspective introduced by distributions, developed by Laurent Schwartz, where functions are identified with linear functionals. This approach extends classical notions, such as derivatives, through duality principles, significantly broadening their applicability. We also present key examples, such as the Dirac delta, which illustrates the limitations of classical analysis. This foundational material bridges the gap between classical analysis and the advanced tools used in subsequent chapters. It equips readers with essential concepts for understanding the broader applications of distributions in mathematical analysis and partial differential equations.

4.1 Introduction

In this chapter, we will provide the first rudiments of the theory of distributions and the Fourier transform. In the last section (Sect. 4.10) we give a brief introduction to oscillatory integrals. Let us explicitly state that the entire chapter has more of a recap tone rather than a systematic treatment of the topics. Essentially, the framework we cover in this book is that of classical analysis, although, particularly for Chap. 5, a preliminary understanding of distribution theory can slightly simplify its presentation. Readers seeking further in-depth material can find it in numerous books, among which we mention here [56, Ch. 1], [58, Vol. 1], [114, Ch. 6, 7], [141] and, of course, [119].

The theory of distributions, developed by the mathematician Laurent Schwartz, introduced profound and radical innovations in Mathematical Analysis, particularly in the study of differential equations. These effects arose from a shift in perspective regarding the way functions are considered. Let us start with a particularly simple situation where we have a function f defined and continuous on $\mathbb{R}$. From a classical point of view, we know that assigning such a function is equivalent to assigning the values $f(x)$ for each $x \in \mathbb{R}$. The core of the radical shift introduced by the theory

S. Vessella, *Unique Continuation Properties for Partial Differential Equations*, Birkhäuser Advanced Texts Basler Lehrbücher,
https://doi.org/10.1007/978-3-031-86366-0_4

of distributions lies in identifying the function f with the linear functional, which we will denote as T_f

$$C_0^\infty(\mathbb{R}) \ni \phi \to \langle T_f, \phi \rangle := \int_{-\infty}^{+\infty} f(x)\phi(x)dx \in \mathbb{R}. \tag{4.1}$$

That T_f is well-defined is evident, as it is also evident that if we know f, then we know the mapping T_f, and the converse also holds (this holds, with the appropriate modifications, even when $f \in L^1_{loc}(\mathbb{R})$, see Proposition 3.48). That is, T_f determines f. Therefore, one can assign f either through the values it takes at all points $x \in \mathbb{R}$ or via the linear functional T_f. However, to properly develop the theory, the linearity property of T_f alone is not sufficient; an appropriate notion of continuity is also required, and we will focus on this in the next section.

Without a doubt, one of the most famous distributions is the so-called Dirac delta distribution concentrated at x_0, where x_0 is a fixed point in $\mathbb{R}$ (or more generally in $\mathbb{R}^n$), defined as the functional

$$\langle \delta_{x_0}, \phi \rangle = \phi(x_0), \quad \forall \phi \in C_0^\infty(\mathbb{R}). \tag{4.2}$$

As we will see in the next section, there is no function f, not even one in $L^1_{loc}(\mathbb{R})$, such that $T_f = \delta_{x_0}$.

It is characteristic of distribution theory that many operations already known for functions are defined "by duality" for the distributions. For example, the product of a distribution T by a function $a \in C^\infty(\mathbb{R})$ is defined as the distribution aT, such that

$$\langle aT, \phi \rangle = \langle T, a\phi \rangle, \quad \forall \phi \in C_0^\infty(\mathbb{R}).$$

Similarly, the derivative T' is defined as the distribution such that

$$\langle T', \phi \rangle = -\langle T, \phi' \rangle, \quad \forall \phi \in C_0^\infty(\mathbb{R}). \tag{4.3}$$

This definition of a derivative greatly extends the classical notion. In this Introduction, we will limit ourselves to noting that if $f \in C^1(\mathbb{R})$, definition (4.3) gives us the classical derivative. In fact, integrating by parts, we have

$$\begin{aligned} \langle T_f', \phi \rangle &= -\langle T, \phi' \rangle = \\ &= -\int_{-\infty}^{+\infty} f(x)\phi'(x)dx = \\ &= \int_{-\infty}^{+\infty} f'(x)\phi(x)dx = \\ &= \langle T_{f'}, \phi \rangle. \end{aligned} \tag{4.4}$$

In the following sections, we will show that the notion of a derivative (which we will refer to as the *distributional derivative*) is much broader than the classical notion. An important consequence of this is that differential equation problems can be reformulated in a more relaxed manner, simplifying the proofs of existence theorems, which in the classical context are significantly more difficult.

4.2 Definition of Distributions

The space $\mathcal{D}(\Omega)$

Let Ω be an open set of $\mathbb{R}^n$. In what follows, we will denote by $\mathcal{D}(\Omega)$ the space $C_0^\infty(\Omega)$ with the following notion of convergence. Let $\{\phi_j\} \subset C_0^\infty(\Omega)$ and $\phi \in C_0^\infty(\Omega)$. We say that $\{\phi_j\}$ converges to ϕ in $\mathcal{D}(\Omega)$ if

(a) there exists a compact subset K of Ω such that $\operatorname{supp} \phi_j \subset K$ for every $j \in \mathbb{N}$;
(b)

$$\lim_{j\to\infty} \sup_{\Omega} \left|\partial^\alpha \phi_j - \partial^\alpha \phi\right| = 0, \quad \forall \alpha \in \mathbb{N}_0^n.$$

For brevity we will also write

$$\{\phi_j\} \to \phi, \quad \text{in } \mathcal{D}(\Omega),$$

to indicate that $\{\phi_j\}$ converges to ϕ in $\mathcal{D}(\Omega)$.

Remark 4.8 Let us note that we have not provided an explicit definition of the topology of $\mathcal{D}(\Omega)$. For completeness, we will briefly indicate how such a topology can be introduced. However, we warn that we will not use it later, as the notion of convergence we provided above is sufficient for all our purposes. Let us see how the topology on $C_0^\infty(\Omega)$ can be introduced.

(i) For every compact subset K of Ω, we equip $C_K^\infty(\Omega)$ (see Sect. 3.2 for the definition of $C_K^\infty(\Omega)$) with the topology induced by the family of seminorms

$$p_k(f) = \max\left\{\max_K |\partial^\alpha f| : |\alpha| \le k\right\}, \quad k \in \mathbb{N}_0.$$

We denote by $\mathcal{D}_K(\Omega)$ the space $C_K^\infty(\Omega)$ with this topology. $\mathcal{D}_K(\Omega)$ is a Fréchet space (see [114], [141] and Sect. 3.13).

(ii) We consider the union of the spaces $\mathcal{D}_K(\Omega)$ as K varies over the compact subsets of Ω (more precisely, we consider the inductive limit [141, page 27]) and in this way $\mathcal{D}(\Omega)$ is a topological vector space (more precisely: a locally convex topological vector space). It turns out that the notion of convergence of a sequence $\{\phi_j\}$ in $\mathcal{D}(\Omega)$ is exactly the one we provided above.

♦

The space $\mathcal{D}'(\Omega)$ A linear map

$$T : \mathcal{D}(\Omega) \to \mathbb{R} \text{ (or } \mathbb{C}),$$

is called a **distribution** on Ω if, whenever ($\phi \in C_0^\infty(\Omega)$)

$$\{\phi_j\} \to \phi, \quad \text{in } \mathcal{D}(\Omega),$$

it follows

$$\{T(\phi_j)\} \to T(\phi), \quad \text{in } \mathbb{R} \text{ (or in } \mathbb{C}).$$

Of course, since T is linear, the above definition is equivalent to require that whenever

$$\{\phi_j\} \to 0, \quad \text{in } \mathcal{D}(\Omega),$$

it follows

$$\{T(\phi_j)\} \to 0, \quad \text{in } \mathbb{R} \text{ (or in } \mathbb{C}).$$

We will denote the space of distributions on Ω by $\mathcal{D}'(\Omega)$. In what follows, we will only consider real-valued distributions, since everything can be immediately extended to the case of complex-valued distributions. Instead of $T(\phi)$, we will often write

$$\langle T, \phi \rangle_{\mathcal{D}'(\Omega), \mathcal{D}(\Omega)}$$

or simply

$$\langle T, \phi \rangle,$$

if there is no ambiguity, because this notation makes it easier to follow arguments based on duality.

Example 4.1 Let $x_0 \in \mathbb{R}^n$. We denote by δ_{x_0} the distribution

$$\langle \delta_{x_0}, \phi \rangle = \phi(x_0), \quad \forall \phi \in C_0^\infty(\mathbb{R}^n). \tag{4.5}$$

It is immediately clear that δ_{x_0} is a distribution. Moreover, as mentioned in the introduction, we call it the **Dirac delta distribution concentrated at** x_0. ♣

Example 4.2 Let Ω be an open set of $\mathbb{R}^n$ and $f \in L^1_{loc}(\Omega)$. We define

$$\langle T_f, \phi \rangle := \int_\Omega f(x)\phi(x)dx, \quad \forall \phi \in C_0^\infty(\Omega). \tag{4.6}$$

Also in this case, it is easily verified that $T_f \in \mathcal{D}'(\Omega)$. Due to Proposition 3.48, we can identify the function f with the distribution T_f, and if there is no risk of ambiguity, we will refer to either the function f or the distribution T_f interchangeably. ♣

Remark 4.9 Let us note that, given the distribution δ_{x_0}, there does not exist $f \in L^1_{loc}(\mathbb{R}^n)$ such that $T_f = \delta_{x_0}$. We limit ourselves to the case $n = 1$, $x_0 = 0$, as more general cases can be treated in a completely analogous way.

Let us suppose there exists $f \in L^1_{loc}(\mathbb{R})$ such that

$$\int_{\mathbb{R}} f(x)\phi(x)dx = \phi(0), \quad \forall \phi \in C_0^\infty(\mathbb{R}). \tag{4.7}$$

Let now $\{\phi_k\}_{k\geq 2}$ be the following sequence of functions

$$\phi_k(x) = \begin{cases} e^{k^2 - \frac{k^2}{1-k^2x^2}}, & \text{for } |x| < \frac{1}{k} \\ \\ 0, & \text{for } \frac{1}{k} \leq |x| < 1 \end{cases}.$$

We have $\phi_k \in C_0^\infty(\mathbb{R})$, $\operatorname{supp}\phi_k \subset \left[-\frac{1}{2}, \frac{1}{2}\right]$ for every $k \geq 2$. Moreover

$$\phi_k(0) = 1, \quad \text{and} \quad \lim_{k\to\infty} \phi_k(x) = 0, \quad \text{for } x \neq 0.$$

On the other hand, by (4.7) we have

$$1 = \phi_k(0) = \int_{\mathbb{R}} f(x)\phi_k(x)dx, \qquad \forall k \geq 2, \tag{4.8}$$

but $f \in L^1_{loc}(\mathbb{R})$, hence the Dominated Convergence Theorem implies

$$\lim_{k\to\infty} \int_{-1}^{1} f(x)\phi_k(x)dx = 0.$$

By the latter and by (4.8) we reach a contradiction. ♦

In the following Proposition we give a characterization of a distribution.

Proposition 4.1 *Let*

$$T : \mathcal{D}(\Omega) \to \mathbb{R}$$

be a linear map.

The following sentences are equivalent:

(a) $T \in \mathcal{D}'(\Omega)$,

(b) *for every* K, *nonempty compact subset of* Ω, *there exist* $N \in \mathbb{N}_0$ *and* $C \geq 0$ *such that*

$$|\langle T, \phi \rangle| \leq C \sum_{|\alpha| \leq N} \max_K |\partial^\alpha \phi|, \quad \forall \phi \in C_K^\infty(\Omega). \tag{4.9}$$

Proof It is clear that (b) implies (a). Thus, it remains for us to prove the converse.

Let us argue by contradiction. Let us assume that

$$T \in \mathcal{D}'(\Omega) \tag{4.10}$$

and let us suppose that (b) does not hold. Namely, let us suppose there exists K, nonempty compact subset of Ω, such that for every $j \in \mathbb{N}_0$ and for every $C \geq 0$ there exists $\phi_{j,C} \in C_K^\infty(\Omega)$ such that

$$\left|\langle T, \phi_{j,C} \rangle\right| > C \sum_{|\alpha| \leq N} \max_K \left|\partial^\alpha \phi_{j,C}\right|. \tag{4.11}$$

Let $C = j \in \mathbb{N}$ and set $\phi_j = \phi_{j,j}$, then (4.11) implies

$$\left|\langle T, \phi_j \rangle\right| > j \sum_{|\alpha| \leq N} \max_K \left|\partial^\alpha \phi_j\right|, \quad \forall j \in \mathbb{N}. \tag{4.12}$$

Now, (4.12) implies $\phi_j \neq 0$ for every $j \in \mathbb{N}$ and we can define

$$\psi_j = \frac{\phi_j}{j \sum_{|\alpha| \leq N} \max_K \left|\partial^\alpha \phi_j\right|}, \quad j \in \mathbb{N}.$$

Hence we have

$$\psi_j \in C_K^\infty(\Omega), \quad \forall j \in \mathbb{N}$$

and, for every $\beta \in \mathbb{N}_0^n$,

$$\max_K \left|\partial^\beta \psi_j\right| \leq \frac{1}{j}, \quad \text{for every } j \geq |\beta|.$$

Therefore

$$\left\{\partial^\beta \psi_j\right\} \to 0, \quad \text{uniformly } \forall \beta \in \mathbb{N}_0^n.$$

Thus

$$\{\psi_j\} \to 0, \quad \text{in } \mathcal{D}(\Omega)$$

and by (4.12) we have

$$\left|\langle T, \psi_j\rangle\right| > j, \quad \forall j \in \mathbb{N}.$$

Consequently

$$\lim_{j\to\infty} \left|\langle T, \psi_j\rangle\right| = +\infty,$$

which contradicts (4.10). ■

Let us note that the number N in (4.9) depends on K. We say that the distribution T is of finite order if N does not depend on K. More precisely, $T \in \mathcal{D}'(\Omega)$ is a distribution of **finite order** if there exists $N \in \mathbb{N}_0$ such that for every K, nonempty compact subset of Ω, there exists $C \geq 0$ such that

$$|\langle T, \phi\rangle| \leq C \sum_{|\alpha|\leq N} \max_K \left|\partial^\alpha \phi\right|, \quad \forall \phi \in C^\infty_K(\Omega). \tag{4.13}$$

We denote by $\mathcal{D}'_F(\Omega)$ the class of distribution of finite order. If $T \in \mathcal{D}'_F(\Omega)$ we call the **order of** T the smallest $N \in \mathbb{N}_0$ such that (4.13) holds. For instance, $T_f \in \mathcal{D}'_F(\Omega)$, $f \in L^1_{loc}(\Omega)$ and $\delta_{x_0} \in \mathcal{D}'_F(\mathbb{R}^n)$, for every $x_0 \in \mathbb{R}^n$.

Example 4.3 For any $\phi \in C^\infty_0(\mathbb{R})$ let us consider the limit

$$\lim_{\varepsilon\to 0^+} \int_{|x|\geq\varepsilon} \frac{\phi(x)}{x} dx. \tag{4.14}$$

This limit exists and is finite. In fact we have

$$\begin{aligned}\int_{|x|\geq\varepsilon} \frac{\phi(x)}{x} dx &= \int_\varepsilon^{+\infty} \frac{\phi(x)-\phi(-x)}{x} dx = \\ &= -\left(\phi(\varepsilon)-\phi(-\varepsilon)\right)\log\varepsilon - \int_\varepsilon^{+\infty} \left(\phi'(x)+\phi'(-x)\right)\log x \, dx.\end{aligned}$$

Now we have

$$\lim_{\varepsilon\to 0^+} \left(\phi(\varepsilon)-\phi(-\varepsilon)\right)\log\varepsilon = \frac{\phi(\varepsilon)-\phi(-\varepsilon)}{2\varepsilon}(2\varepsilon\log\varepsilon) = \phi'(0)\cdot 0 = 0$$

and $\log t \in L^1(0,a)$ for every $a > 0$. Therefore limit (4.14) exists and is equal to

$$-\int_0^{+\infty} \left(\phi'(x) + \phi'(-x)\right) \log x \, dx = -\int_{\mathbb{R}} \phi'(x) \log|x| \, dx.$$

We define the **distribution principal value of $\frac{1}{x}$** as

$$\left\langle \mathrm{pv}\, \frac{1}{x}, \phi \right\rangle := \lim_{\varepsilon \to 0^+} \int_{|x| \geq \varepsilon} \frac{\phi(x)}{x} dt = -\int_{\mathbb{R}} \phi'(x) \log|x| \, dx. \tag{4.15}$$

The fact that $\mathrm{pv}\, \frac{1}{x}$ is a distribution follows immediately from Proposition 4.1, noting that, from (4.15), we have

$$\left| \left\langle \mathrm{pv}\, \frac{1}{x}, \phi \right\rangle \right| \leq \left(\int_{\mathrm{supp}\, \phi'} |\log|x|| \, dx \right) \|\phi'\|_{L^\infty(\mathbb{R})}, \quad \forall \phi \in C_0^\infty(\mathbb{R}).$$

♣

Exercise 4.1 Prove that $\mathrm{pv}\, \frac{1}{x}$ is a distribution of order 1. ♠

Exercise 4.2 Let

$$T : \mathcal{D}((0,2)) \to \mathbb{R}$$

be such that

$$\langle T, \phi \rangle = \sum_{j=1}^{\infty} \phi^{(j)}\left(\frac{1}{j}\right), \quad \forall \phi \in \mathcal{D}((0,2)).$$

Prove that:

(i) $T \in \mathcal{D}'((0,2))$,
(ii) T is not of finite order.

♠

Exercise 4.3 Let Ω be a bounded open set of $\mathbb{R}^n$ whose boundary is of class C^1. Set $\Gamma = \partial\Omega$. Let $\varrho, \mu \in C^0(S)$. Prove that the following linear maps, $\mu\delta_\Gamma$ and $-\partial_\nu(\mu\delta_\Gamma)$, are distributions:

$$\langle \varrho\delta_\Gamma, \phi \rangle = \int_\Gamma \varrho(x)\phi(x) dS, \quad \forall \phi \in \mathcal{D}(\mathbb{R}^n), \tag{4.16}$$

$$\langle -\partial_\nu(\mu\delta_\Gamma), \phi \rangle = \int_\Gamma \mu(x)\partial_\nu\phi(x) dS, \quad \forall \phi \in \mathcal{D}(\mathbb{R}^n) \tag{4.17}$$

(ν is the unit outward normal to Γ).

Note: $\varrho\delta_\Gamma$ and $-\partial_\nu(\mu\delta_\Gamma)$ are called **single-layer distribution of density** ϱ and **double-layer distribution of density** μ respectively. ♠

4.3 Differentiation and Other Operations with Distributions

As we mentioned in the Introduction, many operations already known for functions can be extended to distributions. Among these operations, differentiation plays a central role in the entire theory and its applications. For this reason, we will begin this section with the definition of the derivative of a distribution.

We begin with the following

Proposition 4.2 *Let $T \in \mathcal{D}'(\Omega)$. Let $\alpha \in \mathbb{N}_0^n$. Then the linear map*

$$\Lambda_\alpha : C_0^\infty(\Omega) \to \mathbb{R}, \quad \langle \Lambda_\alpha, \phi \rangle = (-1)^{|\alpha|} \left\langle T, \partial^\alpha \phi \right\rangle, \ \forall \phi \in C_0^\infty(\Omega) \tag{4.18}$$

is a distribution.

Proof We apply Proposition 4.1. Let K be a compact subset of Ω. Then there exist $N \in \mathbb{N}_0$ and $C \geq 0$ such that

$$|\langle T, \phi \rangle| \leq C \sum_{|\alpha| \leq N} \max_K \left|\partial^\alpha \phi\right|, \quad \forall \phi \in C_K^\infty(\Omega).$$

Hence, for any $\alpha \in \mathbb{N}_0^n$, we have

$$|\langle \Lambda_\alpha, \phi \rangle| = \left|\langle T, \partial^\alpha \phi \rangle\right| \leq C \sum_{|\alpha| \leq N + |\alpha|} \max_K \left|\partial^\alpha \phi\right|, \quad \forall \phi \in C_K^\infty(\Omega).$$

Therefore by (b) of Proposition 4.1 we have $\Lambda_\alpha \in \mathcal{D}'(\Omega)$. ■

Definition 4.3 Let $T \in \mathcal{D}'(\Omega)$. Let $\alpha \in \mathbb{N}_0^n$. We define the derivative of order α of T as the distribution Λ_α and we will denote it by $\partial^\alpha T$. Thus we have

$$\left\langle \partial^\alpha T, \phi \right\rangle = (-1)^{|\alpha|} \left\langle T, \partial^\alpha \phi \right\rangle, \ \forall \phi \in C_0^\infty(\Omega). \tag{4.19}$$

We also say that $\partial^\alpha T$ is α-th **derivative of T in the sense of distributions**.

Definition 4.3 is justified by the integration by parts formula that, in the case of $u \in C^{|\alpha|}(\Omega)$, gives precisely the derivative $\partial^\alpha f$ in the classical sense. For instance, if $f \in C^1(\Omega)$, we have, for $j = 1, \cdots, n$,

$$\begin{aligned}
\langle \partial_j T_f, \phi \rangle &= \int_\Omega f \partial_j \phi dx = \\
&= \int_\Omega \left[\partial_j (f\phi) - \phi \partial_j f \right] dx = \\
&= - \int_\Omega \phi \partial_j f dx = \\
&= \langle T_{\partial_j f}, \phi \rangle, \quad \forall \phi \in C_0^\infty(\Omega).
\end{aligned}$$

Let us note that by Definition 4.3 we have that, for any $T \in \mathcal{D}'(\Omega)$,

$$\partial^\alpha \partial^\beta T = \partial^{\alpha+\beta} T = \partial^\beta \partial^\alpha T, \quad \forall \alpha, \beta \in \mathbb{N}_0^n. \tag{4.20}$$

In particular the Schwarz Theorem on the equality of mixed partial derivatives holds:

$$\partial_j \partial_k T = \partial_k \partial_j T, \quad j, k = 1, \cdots, n.$$

Example 4.4 Let $\Omega = \mathbb{R}$. Let $f(x) = |x|$. Let us show that

$$f' = \text{sgn}(x) := \begin{cases} 1, & \text{for } x > 0 \\ 0, & \text{for } x = 0 \ , \\ -1, & \text{for } x < 0 \end{cases} \quad \text{in the sense of distributions.}$$

Let us verify the statement. Let $\phi \in C_0^\infty(\mathbb{R})$. Let $a > 0$ be such that $\phi(x) = 0$ for every $x \in (-\infty, a] \cup [a, , +\infty)$. We have

$$\begin{aligned}
\int_{-\infty}^{+\infty} |x|\phi'(x)dx &= \int_0^a x\phi'(x)dx - \int_{-a}^0 x\phi'(x)dx = \\
&= [x\phi(x)]_0^a - \int_0^1 \phi(x)dx - [x\phi(x)]_{-a}^0 + \int_{-a}^0 \phi(x)dx = \\
&= - \int_{-a}^a \text{sgn}(x)\phi(x)dx = \\
&= - \int_{-\infty}^{+\infty} \text{sgn}(x)\phi(x)dx.
\end{aligned}$$

Hence $f' = \text{sgn}(\cdot) \in L^1_{loc}(\mathbb{R})$ is the derivative in the sense of distribution of f. ♣

Example 4.5 Let $\Omega = \mathbb{R}$. Let

$$H(x) = \begin{cases} 1, & \text{for } x \geq 0 \\ \\ 0, & \text{for } x < 0 \end{cases}.$$

H is known as the **Heaviside function**. Let us check that

$$H' = \delta_0, \quad \text{in the sense of distributions.} \tag{4.21}$$

Let $\phi \in C_0^\infty(R)$ be arbitrary. We have

$$\begin{aligned} \left\langle \frac{d}{dx} T_H, \phi \right\rangle &= -\int_{-\infty}^{+\infty} H(x)\phi'(x)dx = \\ &= \int_0^{+\infty} \phi'(x)dx = \\ &= \phi(0) = \langle \delta_0, \phi \rangle . \end{aligned}$$

This gives (4.21). ♣

Exercise 4.4 Let $Q = \{(x, y) \in \mathbb{R}^2 : x > 0 \text{ and } y > 0\}$. Let

$$G(x, y) = \begin{cases} 1, & \text{for } (x, y) \in Q \\ \\ 0, & \text{for } (x, y) \in \mathbb{R}^2 \setminus Q \end{cases}.$$

Prove that

$$\partial^2_{xy} T_G = \delta_0.$$

♠

Exercise 4.5 Let $I = (-1, 1)$. Let $f \in C^1(I \setminus \{0\})$. Let $g \in L^1_{loc}(I)$ be such that $f' = g$ in $I \setminus \{0\}$.

Prove that

(i) The following limits exist and are finite

$$\lim_{x \to 0^\pm} f(x).$$

Set

$$f\left(0^{\pm}\right)=\lim_{x\to 0^{\pm}} f(x).$$

In this way $f\in L^1_{loc}(I)$.

(ii) It follows

$$\frac{d}{dx}T_f=T_g+\left(f\left(0^+\right)-f\left(0^-\right)\right)\delta_0.$$

♠

Exercise 4.6 Prove that

$$\frac{d}{dx}\log|x|=\mathrm{vp}\,\frac{1}{x},\quad \text{in } \mathcal{D}'(\mathbb{R}).$$

♠

Exercise 4.7 Let Ω be a bounded open set of $\mathbb{R}^n$ whose boundary is of class C^∞. Set $\Gamma=\partial\Omega$. Let $\widetilde{\nu}$ be a C^∞-extension of ν to $\mathbb{R}^n$ (ν outward normal to Γ). Prove that

$$\partial_j\chi_\Omega=-\widetilde{\nu}_j\delta_\Gamma,\quad \text{for } j=1,\cdots,n, \text{ in } \mathcal{D}(\mathbb{R}^n),$$

where δ_Γ is the single-layer distribution defined by (4.16). ♠

What unites the definitions of operations on distributions (including the differentiation operation we have just presented) is the method of duality, which finds a clear and practical formulation in the following **Lax extension Theorem**, [78, Appendix B]. To state and prove this theorem, however, we need a few definitions, which we provide below.

1. Let $\{T_j\}$ be a sequence in $\mathcal{D}'(\Omega)$. We say that $\{T_j\}$ **converges to** $T\in\mathcal{D}'(\Omega)$ in $\mathcal{D}'(\Omega)$ whenever

$$\lim_{j\to\infty}\left\langle T_j,\phi\right\rangle=\langle T,\phi\rangle,\quad \forall\phi\in C_0^\infty(\Omega). \tag{4.22}$$

For example the sequence $\{T_{f_j}\}$, where

$$f_j(x)=\sqrt{\frac{j}{2\pi}}e^{-\frac{jx^2}{2}},\quad x\in\mathbb{R},$$

converges in $\mathcal{D}'(\mathbb{R})$ to δ_0. We leave the simple check of this fact to the reader as an exercise.

2. Let Ω_1 and Ω_2 be two open sets of $\mathbb{R}^n$ and let

$$\mathcal{L} : \mathcal{D}(\Omega_1) \to \mathcal{D}(\Omega_2) \tag{4.23}$$

be a linear map. We say that $\mathcal{L}$ is (sequentially) continuous, whenever the following fact holds: for every sequence $\{\phi_j\}$ in $\mathcal{D}(\Omega_1)$ such that

$$\{\phi_j\} \to 0, \quad \text{in } \mathcal{D}(\Omega_1),$$

we have

$$\{\mathcal{L}(\phi_j)\} \to 0, \quad \text{in } \mathcal{D}(\Omega_2).$$

For example, for any $\alpha \in \mathbb{N}_0^n$,

$$\partial^\alpha : \mathcal{D}(\Omega) \to \mathcal{D}(\Omega)$$

is continuous. We invite the reader to confirm this fact as an exercise.

3. Let Ω_1 and Ω_2 be two open sets of $\mathbb{R}^n$ and let

$$\Lambda : \mathcal{D}'(\Omega_1) \to \mathcal{D}'(\Omega_2) \tag{4.24}$$

be a linear map. We say that Λ is (sequentially) continuous, whenever the following fact holds: for every sequence $\{T_j\}$ in $\mathcal{D}'(\Omega_1)$ such that

$$\{T_j\} \to 0, \quad \text{in } \mathcal{D}'(\Omega_1),$$

we have

$$\{\Lambda(T_j)\} \to 0, \quad \text{in } \mathcal{D}'(\Omega_2).$$

4. Let $\mathcal{L}$ be a linear operator like (4.23), we say that the linear map

$$\mathcal{L}^t : \mathcal{D}(\Omega_2) \to \mathcal{D}(\Omega_1) \tag{4.25}$$

is the transpose of $\mathcal{L}$ if we have

$$\begin{aligned} &\int_{\Omega_2} \mathcal{L}(\phi)(y)\psi(y)dy = \\ &= \int_{\Omega_1} \phi(x)\mathcal{L}^t(\psi)(x)dx, \quad \forall\phi \in \mathcal{D}(\Omega_1),\ \forall\psi \in \mathcal{D}(\Omega_2). \end{aligned} \tag{4.26}$$

Theorem 4.4 (Lax Extension) *Let Ω_1 and Ω_2 be two open sets of $\mathbb{R}^n$, let $\mathcal{L}$ be a linear operator as* (4.23) *and let $\mathcal{L}^t$ be its transpose. Assume that $\mathcal{L}$ and $\mathcal{L}^t$ are continuous. Then the operator*

$$\Lambda_{\mathcal{L}} : \mathcal{D}'(\Omega_1) \to \mathcal{D}'(\Omega_2), \tag{4.27}$$

such that for every $T \in \mathcal{D}'(\Omega_1)$ and for every $\psi \in \mathcal{D}(\Omega_2)$

$$\langle \Lambda_{\mathcal{L}}(T), \psi \rangle_{\mathcal{D}'(\Omega_2),\mathcal{D}(\Omega_2)} = \left\langle T, \mathcal{L}^t(\psi) \right\rangle_{\mathcal{D}'(\Omega_1),\mathcal{D}(\Omega_1)} \tag{4.28}$$

is a well-defined linear and continuous operator and it is an extension of $\mathcal{L}$. Namely

$$\Lambda_{\mathcal{L}}(T_\phi) = \mathcal{L}(\phi), \quad \forall \phi \in \mathcal{D}(\Omega_1). \tag{4.29}$$

Moreover, $\Lambda_{\mathcal{L}}$ is the unique linear continuous operator from $\mathcal{D}'(\Omega_1)$ to $\mathcal{D}'(\Omega_2)$ which satisfies (4.29).

Proof First we prove that map (4.27) is well-defined. Let $T \in \mathcal{D}'(\Omega_1)$. Let $\{\psi_j\}$ be a sequence in $\mathcal{D}(\Omega_2)$ that converges to $\psi \in \mathcal{D}(\Omega_2)$ in $\mathcal{D}(\Omega_2)$. By (4.28) and by the continuity of $\mathcal{L}^t$ we have

$$\begin{aligned}\lim_{j\to\infty} \left\langle \Lambda_{\mathcal{L}}(T), \psi_j \right\rangle_{\mathcal{D}'(\Omega_2),\mathcal{D}(\Omega_2)} &= \lim_{j\to\infty} \left\langle T, \mathcal{L}^t(\psi_j) \right\rangle_{\mathcal{D}'(\Omega_1),\mathcal{D}(\Omega_1)} = \\ &= \left\langle T, \mathcal{L}^t(\psi) \right\rangle_{\mathcal{D}'(\Omega_1),\mathcal{D}(\Omega_1)} = \\ &= \langle \Lambda_{\mathcal{L}}(T), \psi \rangle_{\mathcal{D}'(\Omega_2),\mathcal{D}(\Omega_2)} .\end{aligned}$$

Hence $\Lambda_{\mathcal{L}}(T) \in \mathcal{D}'(\Omega_2)$ for every $T \in \mathcal{D}'(\Omega_2)$. Clearly, $\Lambda_{\mathcal{L}}$ is a linear map. Thus $\Lambda_{\mathcal{L}}$ is a well-defined linear operator.

Now we prove that $\Lambda_{\mathcal{L}}$ is continuous. Let $\{T_j\}$ be a sequence in $\mathcal{D}'(\Omega_1)$ that converges to 0 in $\mathcal{D}'(\Omega_1)$. We have

$$\begin{aligned}&\lim_{j\to\infty} \left\langle \Lambda_{\mathcal{L}}(T_j), \psi \right\rangle_{\mathcal{D}'(\Omega_2),\mathcal{D}(\Omega_2)} \\ &\quad = \lim_{j\to\infty} \left\langle T_j, \mathcal{L}^t(\psi) \right\rangle_{\mathcal{D}'(\Omega_1),\mathcal{D}(\Omega_1)} = 0, \quad \forall \psi \in \mathcal{D}(\Omega_2).\end{aligned}$$

Hence

$$\left\{\Lambda_{\mathcal{L}}(T_j)\right\} \to 0, \quad \text{in } \mathcal{D}'(\Omega_2).$$

Thus $\Lambda_{\mathcal{L}}$ is continuous.

Now we prove (4.29). Let $\phi \in \mathcal{D}(\Omega_1)$. We have

$$\begin{aligned}\left\langle \Lambda_{\mathcal{L}}(T_\phi), \psi \right\rangle_{\mathcal{D}'(\Omega_2),\mathcal{D}(\Omega_2)} &= \left\langle T_\phi, \mathcal{L}^t(\psi) \right\rangle_{\mathcal{D}'(\Omega_1),\mathcal{D}(\Omega_1)} = \\ &= \int_{\Omega_1} \phi(x)\mathcal{L}^t(\psi)(x)dx = \\ &= \int_{\Omega_2} \mathcal{L}(\phi)(y)\psi(y)dy, \quad \forall \psi \in \mathcal{D}(\Omega_2).\end{aligned}$$

Hence

$$\Lambda_{\mathcal{L}}(T_\phi) = \phi, \quad \forall \phi \in \mathcal{D}(\Omega_1).$$

To complete the proof, let $\Lambda_{\mathcal{L}}$ and $\widetilde{\Lambda}_{\mathcal{L}}$ be two linear map that satisfy (4.27)–(4.29). We have

$$\begin{aligned}&\left\langle \widetilde{\Lambda}_{\mathcal{L}}(T), \psi \right\rangle_{\mathcal{D}'(\Omega_2),\mathcal{D}(\Omega_2)} \\ &= \left\langle T, \mathcal{L}^t(\psi) \right\rangle_{\mathcal{D}'(\Omega_1),\mathcal{D}(\Omega_1)} = \\ &= \left\langle \Lambda_{\mathcal{L}}(T), \psi \right\rangle_{\mathcal{D}'(\Omega_2),\mathcal{D}(\Omega_2)}, \quad \forall T \in \mathcal{D}'(\Omega_1), \ \forall \psi \in \mathcal{D}(\Omega_2).\end{aligned}$$

Hence

$$\widetilde{\Lambda}_{\mathcal{L}}(T) = \Lambda_{\mathcal{L}}(T), \quad \forall T \in \mathcal{D}'(\Omega_1).$$

That is $\widetilde{\Lambda}_{\mathcal{L}} = \Lambda_{\mathcal{L}}$. ∎

For a given $\mathcal{L}$ that satisfies the hypotheses of Theorem 4.4, we say that $\Lambda_{\mathcal{L}}$ is the **Lax extension** of $\mathcal{L}$.

Remark 4.10 Let us consider $\mathcal{L} = \partial^\alpha$, where $\alpha \in \mathbb{N}_0^n$. We have seen above that $\mathcal{L} : \mathcal{D}(\Omega) \to \mathcal{D}(\Omega)$ is continuous (linear) map. Also we have

$$\mathcal{L}^t = (-1)^{|\alpha|}\partial^\alpha.$$

Therefore, in this case, if $T \in \mathcal{D}(\Omega)$, $\Lambda_{\mathcal{L}}(T)$ gives us exactly the distributional derivative of order α of T. ♦

Multiplication of a Distribution by a C^∞-Function
Let $a \in C^\infty(\Omega)$. Let

$$\mathcal{L} : \mathcal{D}(\Omega) \to \mathcal{D}(\Omega), \quad \mathcal{L}(\phi) = a\phi, \quad \forall \phi \in \mathcal{D}(\Omega).$$

We have trivially

$$\mathcal{L}^t(\psi) = a\psi = \mathcal{L}(\psi), \quad \forall \psi \in \mathcal{D}(\Omega).$$

It is easy to check that the hypotheses of Theorem 4.4 are satisfied. Hence we can define the Lax extension of $\mathcal{L}$. For a given $T \in \mathcal{D}'(\Omega)$ we call $\Lambda_{\mathcal{L}}(T)$ the multiplication of the distribution T by a and we set $aT = \Lambda_{\mathcal{L}}(T)$. In other words we have, for $T \in \mathcal{D}'(\Omega)$ and $a \in \mathcal{D}(\Omega)$,

$$\langle aT, \phi \rangle = \langle T, a\phi \rangle , \quad \forall \phi \in \mathcal{D}(\Omega). \tag{4.30}$$

Example 4.6 Let $a \in C^{\infty}(\Omega)$ and $x_0 \in \Omega$. We have

$$a\delta_{x_0} = a(x_0)\delta_{x_0}.$$

As a matter of fact we have

$$\begin{aligned}\langle a\delta_{x_0}, \phi \rangle &= \langle \delta_{x_0}, a\phi \rangle = \\ &= a(x_0)\phi(x_0) = \\ &= \langle a(x_0)\delta_{x_0}, \phi \rangle , \quad \forall \phi \in \mathcal{D}(\Omega).\end{aligned}$$

♣

Example 4.7 Let $m \in \mathbb{N}$ and $a_\alpha \in C^{\infty}(\Omega)$, for $|\alpha| \leq m$. We define the linear partial differential operator

$$P(x, \partial) : \mathcal{D}'(\Omega) \to \mathcal{D}'(\Omega),$$

such that

$$P(x, \partial)T = \sum_{|\alpha| \leq m} a_\alpha \partial^\alpha T, \quad \forall T \in \mathcal{D}'(\Omega).$$

In other words $P(x, \partial)T$ is defined as

$$\langle P(x, \partial)T, \phi \rangle = \langle T, P^t(x, \partial)\phi \rangle , \quad \forall \phi \in \mathcal{D}(\Omega),$$

where

$$P^t(x, \partial)\phi = \sum_{|\alpha| \leq m} (-1)^{|\alpha|}\partial^\alpha (a_\alpha \phi) .$$

Let $F \in \mathcal{D}'(\Omega)$. When we say that $T \in \mathcal{D}'(\Omega)$ is a solution of the equation

$$P(x, \partial)T = F,$$

in the sense of distributions, we simply mean that

$$\langle T, P^t(x, \partial)\phi\rangle = \langle F, P^t(x, \partial)\phi\rangle, \quad \forall \phi \in \mathcal{D}(\Omega).$$

♣

Exercise 4.8 After proving that

(i) $\{x\phi : \phi \in \mathcal{D}(\mathbb{R})\} = \{\phi \in \mathcal{D}(\mathbb{R}) : \phi(0) = 0\}$,
prove that
(ii) $\{T \in \mathcal{D}'(\mathbb{R}) : xT = 0\} = \{C\delta_0 : C \in \mathbb{R}\}$. ♠

Translation of a Distribution
Let Ω be an open set of $\mathbb{R}^n$ and $h \in \mathbb{R}^n$. The translation operator on $\mathcal{D}(\Omega)$ is defined by

$$\mathcal{T}_h : \mathcal{D}(\Omega) \to \mathcal{D}(h + \Omega), \ \ (\mathcal{T}_h\phi)(x) = \phi(x - h), \ \forall \phi \in \mathcal{D}(\Omega), \ \forall x \in h + \Omega.$$

$\mathcal{T}_h$ is a linear and continuous operator. Moreover

$$\mathcal{T}_h^t = \mathcal{T}_{-h}. \tag{4.31}$$

Indeed, for every $\phi \in \mathcal{D}(\Omega)$ and every $\psi \in \mathcal{D}(h + \Omega)$, we have

$$\begin{aligned}\int_{h+\Omega} (\mathcal{T}_h\phi)(x)\psi(x)dx &= \int_{h+\Omega} \phi(x - h)\psi(x)dx = \\ &= \int_\Omega \phi(x)\psi(x + h)dx = \\ &= \int_\Omega \phi(x)(\mathcal{T}_{-h}\psi)(x)dx.\end{aligned}$$

By Theorem 4.4, if $T \in \mathcal{D}'(\Omega)$, we can define the distribution $\mathcal{T}_hT \in \mathcal{D}'(h+\Omega)$ as

$$\begin{aligned}\langle \mathcal{T}_hT, \phi\rangle_{\mathcal{D}'(h+\Omega),\mathcal{D}(h+\Omega)} &= \langle T, \mathcal{T}_{-h}\phi\rangle_{\mathcal{D}'(\Omega),\mathcal{D}(\Omega)} = \\ &= \langle T, \phi(\cdot + h)\rangle_{\mathcal{D}'(\Omega),\mathcal{D}(\Omega)}, \quad \forall \phi \in \mathcal{D}(h + \Omega).\end{aligned}$$

Example 4.8 We have

$$\mathcal{T}_h\delta_0 = \delta_h, \quad \text{in } \mathcal{D}'(\mathbb{R}^n).$$

Indeed we have

$$\begin{aligned}\langle \mathcal{T}_h\delta_0, \phi\rangle_{\mathcal{D}'(\mathbb{R}^n),\mathcal{D}(\mathbb{R}^n)} &= \langle \delta_0, \phi(\cdot + h)\rangle_{\mathcal{D}'(\mathbb{R}^n),\mathcal{D}(\mathbb{R}^n)} = \\ &= \phi(h) = \\ &= \langle \delta_h, \phi\rangle_{\mathcal{D}'(\mathbb{R}^n),\mathcal{D}(\mathbb{R}^n)}, \quad \forall \phi \in \mathcal{D}(\mathbb{R}^n).\end{aligned}$$

♣

Reflection of a Distribution
Let Ω be an open set of $\mathbb{R}^n$ and let us denote by

$$\widetilde{\Omega} = \{-x : x \in \Omega\}.$$

We define the reflexion operator on $\mathcal{D}(\Omega)$ as

$$\mathcal{R} : \mathcal{D}(\Omega) \to \mathcal{D}\left(\widetilde{\Omega}\right), \tag{4.32}$$

$$(\mathcal{R}\phi)(x) = \check{\phi}(x) := \phi(-x), \ \forall \phi \in \mathcal{D}\left(\widetilde{\Omega}\right), \ \forall x \in \widetilde{\Omega}. \tag{4.33}$$

$\mathcal{R}$ is a linear and continuous operator. Moreover

$$\mathcal{R}^t = \mathcal{R}.$$

For any $T \in \mathcal{D}'(\Omega)$ we define $\mathcal{R}T \in \mathcal{D}'\left(\widetilde{\Omega}\right)$ as (the reader should verify that Theorem 4.4 can be applied)

$$\langle \mathcal{R}T, \psi\rangle_{\mathcal{D}'(\mathcal{D}(\widetilde{\Omega}),\mathcal{D}(\widetilde{\Omega})} = \langle T, \mathcal{R}\psi\rangle_{\mathcal{D}'(\Omega),\mathcal{D}(\Omega)}, \quad \forall \psi \in \mathcal{D}\left(\widetilde{\Omega}\right).$$

Smooth Changes of Variables
Let Ω_1 and Ω_2 be two open sets of $\mathbb{R}^n$ and let

$$\chi : \Omega_1 \to \Omega_2$$

be a C^∞-diffeomorphism. Given $T \in \mathcal{D}'(\Omega_1)$ we want to apply Theorem 4.4 to define the distribution $T \circ \chi^{-1} \in \mathcal{D}'(\Omega_2)$. Let

$$\mathcal{L} : \mathcal{D}(\Omega_1) \to \mathcal{D}(\Omega_2)$$

be defined as

$$\mathcal{L}(\phi) = \phi \circ \chi^{-1}, \quad \forall \phi \in \mathcal{D}(\Omega_1).$$

We have that

$$\mathcal{L}^t : \mathcal{D}(\Omega_2) \to \mathcal{D}(\Omega_1),$$

satisfies

$$\mathcal{L}^t(\psi) = (\psi \circ \chi)\, J_\chi, \quad \forall \psi \in \mathcal{D}(\Omega_2), \tag{4.34}$$

where

$$J_\chi(x) = |\det \partial_x \chi(x)|, \quad x \in \Omega_1.$$

Let us check (4.34). Let $\phi \in \mathcal{D}(\Omega_1)$ and $\psi \in \mathcal{D}(\Omega_2)$. We have

$$\begin{aligned}\int_{\Omega_1} \phi(x)\mathcal{L}^t(\psi)(x)dx &= \int_{\Omega_2} \mathcal{L}(\phi)(y)\psi(y)dy = \\ &= \int_{\Omega_2} \left(\phi \circ \chi^{-1}\right)(y)\psi(y)dy = \\ &= \int_{\Omega_1} \phi(x)\,(\psi \circ \chi)\,(x)J_\chi(x)dx,\end{aligned}$$

from which (4.34) follows. Now, the continuity of $\mathcal{L}$ and $\mathcal{L}^t$ is fairly obvious, so, for any $T \in \mathcal{D}'(\Omega_1)$, we can define through Theorem 4.4,

$$T \circ \chi^{-1} \in \mathcal{D}'(\Omega_2)$$

as follows

$$\left\langle T \circ \chi^{-1}, \psi \right\rangle_{\mathcal{D}'(\Omega_2),\mathcal{D}(\Omega_2)} = \left\langle T, (\psi \circ \chi)\, J_\chi \right\rangle_{\mathcal{D}'(\Omega_1),\mathcal{D}(\Omega_1)}. \tag{4.35}$$

Convolution of Distribution of $\mathcal{D}'(\mathbb{R}^n)$ and a Function of $\mathcal{D}(\mathbb{R}^n)$
Let $\eta \in \mathcal{D}(\mathbb{R}^n)$. Let us define

$$\mathcal{L}_\eta : \mathcal{D}(\mathbb{R}^n) \to \mathcal{D}(\mathbb{R}^n),$$

such that

$$\mathcal{L}_\eta(\phi) = (\eta \star \phi)\,(x) = \int_{\mathbb{R}^n} \eta(x-y)\phi(y)dy, \quad \forall \phi \in \mathcal{D}(\mathbb{R}^n), \ \forall x \in \mathbb{R}^n.$$

$\mathcal{L}_\eta$ is a well-defined and linear operator, now we check that it is continuous.

Let K be a compact set of $\mathbb{R}^n$. Let $\{\phi_j\}$ be a sequence in $\mathcal{D}(\mathbb{R}^n)$ such that (i) $\operatorname{supp}\phi_j \subset K$, for every $j \in \mathbb{N}$, (ii) $\{\partial^\alpha \phi_j\}$ converges uniformly to 0 for every $\alpha \in \mathbb{N}_0^n$. We have

$$\operatorname{supp}(\mathcal{L}_\eta(\phi_j)) \subset \operatorname{supp}\eta + K, \quad j \in \mathbb{N} \tag{4.36}$$

and

$$\partial^\alpha(\mathcal{L}_\eta(\phi_j)) = \int_{\mathbb{R}^n} \eta(x-y)\partial_y^\alpha \phi(y)dy, \quad j \in \mathbb{N}, \quad \forall \alpha \in \mathbb{N}_0^n.$$

Hence for every $\alpha \in \mathbb{N}_0^n$ we have

$$\sup_{\mathbb{R}^n} \left|\partial^\alpha(\mathcal{L}_\eta(\phi_j))\right| \leq C \sup_{\mathbb{R}^n} |\eta| \sup_K \left|\partial^\alpha \phi_j\right| \to 0, \quad \text{as } j \to \infty, \tag{4.37}$$

where $C = |\operatorname{supp}\eta + K|$ (the Lebesgue measure of $\operatorname{supp}\eta + K$). By (4.36) and (4.37) we have

$$\left\{\mathcal{L}_\eta(\phi_j)\right\} \to 0, \quad \text{in } \mathcal{D}(\mathbb{R}^n).$$

Now we check that

$$\mathcal{L}_\eta^t(\psi) = \check{\eta} \star \psi, \quad \forall \psi \in \mathcal{D}(\mathbb{R}^n).$$

As a matter of fact, for any $\phi, \psi \in \mathcal{D}(\mathbb{R}^n)$, by the Fubini–Tonelli Theorem (take into account $\operatorname{supp}(\eta(y-z)\phi(z)\psi(y)) \subset \operatorname{supp}(\phi) \times \operatorname{supp}(\psi)$), we have

$$\begin{aligned}
\int_{\mathbb{R}^n} \mathcal{L}_\eta(\phi)(y)\phi(y)dy &= \int_{\mathbb{R}^n} \left(\int_{\mathbb{R}^n} \eta(y-z)\phi(z)dz\right)\psi(y)dy = \\
&= \int_{\mathbb{R}^n} \left(\int_{\mathbb{R}^n} \eta(y-z)\psi(y)dz\right)\phi(z)dz = \\
&= \int_{\mathbb{R}^n} \left(\check{\eta} \star \psi\right)(z)\phi(z)dz.
\end{aligned}$$

Clearly, $\mathcal{L}_\eta^t$ is continuous.

Therefore, Theorem 4.4 yields that for any $T \in \mathcal{D}'(\mathbb{R}^n)$ we can define $\eta \star T$ as

$$\langle \eta \star T, \psi \rangle = \left\langle T, \check{\eta} \star \psi \right\rangle, \quad \forall \psi \in \mathcal{D}(\mathbb{R}^n). \tag{4.38}$$

Similarly, if instead of starting from $\mathcal{L}_\eta$ we begin with

$$\widetilde{\mathcal{L}}_\eta(\phi) = \phi \star \eta, , \quad \forall \phi \in \mathcal{D}(\mathbb{R}^n),$$

we can define $T \star \eta$, for $T \in \mathcal{D}'(\mathbb{R}^n)$, as

$$\langle T \star \eta, \psi \rangle = \left\langle T, \psi \star \check{\eta} \right\rangle, \quad \forall \psi \in \mathcal{D}(\mathbb{R}^n). \tag{4.39}$$

It is a simple exercise that we leave to the reader to check that

$$\eta \star T = T \star \eta, \quad \forall \eta \in \mathcal{D}(\mathbb{R}^n), \ \forall T \in \mathcal{D}'(\mathbb{R}^n). \tag{4.40}$$

Exercise 4.9 Let $T \in \mathcal{D}'(\mathbb{R}^n)$ $\eta, \zeta \in \mathcal{D}(\mathbb{R}^n)$, $\alpha \in \mathbb{N}_0^n$ and $h \in \mathbb{R}^n$. Check that

(i) $\mathcal{T}_h (\eta \star T) = \eta \star (\mathcal{T}_h T) = (\mathcal{T}_h \eta) \star T$,
(ii) $\partial^\alpha (\eta \star T) = \eta \star (\partial^\alpha T) = (\partial^\alpha \eta) \star T$,
(iii) $(\eta \star \zeta) \star T = \eta \star (\zeta \star T)$,
(iv) $\eta \star (\partial^\alpha \delta_0) = \partial^\alpha \eta$. ♠

4.4 Restriction and Support of a Distribution: Localizzation Principle

Let Ω be an open subset of $\mathbb{R}^n$. Let ω be an open subset of Ω. Here and in the sequel we identify $\mathcal{D}(\omega)$ with the subspace of $\mathcal{D}(\Omega)$

$$\{\phi \in \mathcal{D}(\Omega) : \phi = 0, \text{ in } \Omega \setminus \omega\}.$$

We define the **restriction of a distribution** $T \in \mathcal{D}'(\Omega)$ to ω as

$$T_{|\omega} = T_{|\mathcal{D}(\omega)}.$$

In other words $T_{|\omega}$ is defined by

$$\left\langle T_{|\omega}, \phi \right\rangle_{\mathcal{D}'(\omega), \mathcal{D}(\omega)} = \langle T, \phi \rangle_{\mathcal{D}'(\Omega), \mathcal{D}(\Omega)}, \quad \forall \phi \in \mathcal{D}(\omega). \tag{4.41}$$

It is straightforward to verify that $T_{|\omega} \in \mathcal{D}'(\omega)$.

Theorem 4.5 *Let $\{\omega_i\}_{i \in \mathcal{I}}$ be an open covering of Ω. Then we have:*

(a) if $T \in \mathcal{D}'(\Omega)$ is such that $T_{|\omega_i} = 0$ for every $i \in \mathcal{I}$ then $T = 0$.
(b) if $T_i \in \mathcal{D}'(\omega_i)$, $i \in \mathcal{I}$, satisfies

$$T_{i|\omega_i \cap \omega_j} = T_{j|\omega_i \cap \omega_j}, \quad \textit{for every } i, j \in \mathcal{I}, \tag{4.42}$$

then there exists a unique $T \in \mathcal{D}'(\Omega)$ such that

$$T_{|\omega_i} = T_i, \quad \forall i \in \mathcal{I}. \tag{4.43}$$

Proof

(a) Let $\phi \in \mathcal{D}(\Omega)$. Let $K = \operatorname{supp} \phi$. Since K is a compact subset of Ω, there exists L, finite subset of $\mathcal{I}$, such that $K \subset \bigcup_{j \in L} \omega_j$. Let $\zeta_i \in \mathcal{D}(\Omega)$ be ($i \in L$) a partition of unity subordinate to the covering $\{\omega_i\}_{i\in L}$ (see Sect. 3.5). By $T_{|\omega_i} = 0$, for every $i \in L$, we have

$$\langle T, \phi \rangle = \left\langle T, \sum_{i\in L} \zeta_i \phi \right\rangle = \sum_{i\in L} \langle T, \zeta_i \phi \rangle = 0.$$

Since $\phi \in \mathcal{D}(\Omega)$ is arbitrary, we have $T = 0$.

(b) Let ϕ, K, $\{\omega_i\}_{i\in L}$ and $\{\zeta_i\}_{i\in L}$ be as above. Let us define

$$\langle T, \phi \rangle = \sum_{i\in L} \langle T_i, \zeta_i \phi \rangle . \tag{4.44}$$

First we check (4.43). Let $j \in \mathcal{I}$. Let $\phi \in \mathcal{D}(\omega_j)$. By (4.42) we have

$$\langle T, \phi \rangle = \sum_{i\in L} \langle T_i, \zeta_i \phi \rangle = \sum_{i\in L} \langle T_j, \zeta_i \phi \rangle = \langle T_j, \phi \rangle .$$

To prove that T indeed defines a linear map, we need the following

Claim Let K_0 be a compact subset of Ω. Let M be a finite subset of $\mathcal{I}$ such that $\{\omega_k\}_{k\in M}$ a finite covering of K_0 and let $\eta_k \in \mathcal{D}(\Omega)$ $(k \in M)$ be a partition of unity subordinate to the covering $\{\omega_k\}_{k\in M}$, then

$$\sum_{k\in M} \langle T_k, \eta_k \phi \rangle = \sum_{i\in L} \langle T_i, \zeta_i \phi \rangle . \tag{4.45}$$

Proof of the Claim First of all let us notice that by (4.42) we have

$$\langle T_k, \eta_k \zeta_i \phi \rangle = \langle T_i, \eta_k \zeta_i \phi \rangle , \quad \text{for } i \in L, \; k \in M. \tag{4.46}$$

Hence we get

$$\begin{aligned} \sum_{k\in M} \langle T_k, \eta_k \phi \rangle &= \sum_{k\in M} \sum_{i\in L} \langle T_k, \eta_k \zeta_i \phi \rangle = \\ &= \sum_{k\in M} \sum_{i\in L} \langle T_i, \eta_k \zeta_i \phi \rangle = \\ &= \sum_{i\in L} \sum_{k\in M} \langle T_i, \eta_k \zeta_i \phi \rangle = \\ &= \sum_{i\in L} \langle T_i, \zeta_i \phi \rangle . \end{aligned}$$

This concludes the proof of the Claim.

We prove the linearity of T. By (4.44) we have trivially

$$\langle T, \lambda\phi \rangle = \lambda \langle T, \phi \rangle, \quad \forall \phi \in \mathcal{D}(\Omega), \ \forall \lambda \in \mathbb{R}. \tag{4.47}$$

Now let $\phi, \psi \in \mathcal{D}(\Omega)$. Let K_0 be the support of $\phi + \psi$ and let M be a finite subset of $\mathcal{I}$ such that $\{\omega_k\}_{k \in M}$ is a finite covering of K_0. Let $\eta_k \in \mathcal{D}(\Omega)$ be ($k \in M$) a partition of unity subordinate to the covering $\{\omega_k\}_{k \in M}$. By the proved Claim we have

$$\langle T, \phi \rangle = \sum_{k \in M} \langle T_k, \eta_k \phi \rangle$$

and similarly

$$\langle T, \psi \rangle = \sum_{k \in M} \langle T_k, \eta_k \psi \rangle .$$

Hence by Definition (4.44) we have

$$\begin{aligned} \langle T, (\phi + \psi) \rangle &= \sum_{k \in M} \langle T_k, \eta_k (\phi + \psi) \rangle = \\ &= \sum_{k \in M} \langle T_k, \eta_k \phi \rangle + \sum_{k \in M} \langle T_k, \eta_k \psi \rangle = \\ &= \langle T, \phi \rangle + \langle T, \psi \rangle . \end{aligned} \tag{4.48}$$

Thus, by (4.47) and (4.48) we get the linearity of T. Concerning the continuity of T it is an easy consequence of (4.44) and we leave it to the reader. Hence $T \in \mathcal{D}'(\Omega)$ and it satisfies (4.43). ■

In the following, given $T_1, T_2 \in \mathcal{D}'(\Omega)$ such that

$$T_{1|\omega} = T_{2|\omega},$$

where ω is an open subset of Ω, we will also say that

$$T_1 = T_2, \quad \text{on } \omega.$$

Definition 4.6 Let $T \in \mathcal{D}'(\Omega)$. Let

$$\Lambda = \{x \in \Omega : \ \exists r > 0 \text{ s.t. } T = 0 \text{ on } B_r(x)\} . \tag{4.49}$$

We define the **support** of T, and we denote it with $\operatorname{supp} T$ as the set

$$\operatorname{supp}_{\mathcal{D}} T = \Omega \setminus \Lambda. \tag{4.50}$$

Let us note that Λ is an open set of $\mathbb{R}^n$, so that the support of $T \in \mathcal{D}'(\Omega)$ is a closed set in the topology of Ω. Also we note that if $T \in \mathcal{D}'(\Omega)$ then

$$\begin{aligned} &x_0 \in \Omega \setminus \left(\operatorname{supp}_{\mathcal{D}} T\right) \Longleftrightarrow \\ &\Longleftrightarrow \exists r > 0, \text{ s.t. } \langle T, \phi \rangle = 0, \ \forall \phi \in \mathcal{D}(\Omega), \ \operatorname{supp} \phi \subset B_r(x_0). \end{aligned} \tag{4.51}$$

Remark 4.11 For a function f defined in an open set $\Omega \subset \mathbb{R}^n$, we currently have three definitions of support, which we recall here.

(a) If $f \in C^0(\Omega)$ then

$$\operatorname{supp}_{C^0} f = \overline{\{x \in \Omega : \ f(x) \neq 0\}}^{\Omega}, \quad \text{(closure in } \Omega),$$

see Sect. 3.2.

(b) If $f \in L^1_{loc}(\Omega)$ then

$$\operatorname{supp}_{\text{ess}} f = \Omega \setminus \bigcup_{A \in O} A,$$

where

$$O = \{A \subset \Omega : \ A \text{ open and } f = 0 \text{ in } A \text{ a.e.}\},$$

see Sect. 3.4.2.

(c) If $f \in C^0(\Omega)$ or $f \in L^1(\Omega)$ we can consider the support of T_f that we denote by

$$\operatorname{supp}_{\mathcal{D}} f.$$

Now we show that

(i) If $f \in C^0(\Omega)$ then $\operatorname{supp}_{\mathcal{D}} f = \operatorname{supp}_{C^0} f$,
(ii) If $f \in L^1_{loc}(\Omega)$ then $\operatorname{supp}_{\mathcal{D}} f = \operatorname{supp}_{\text{ess}} f$,
(iii) If $f \in C^0(\Omega)$ then $\operatorname{supp}_{C^0} f = \operatorname{supp}_{\text{ess}} f$.

First notice that if $f \in C^0(\Omega)$ then

$$x_0 \in \Omega \setminus \left(\operatorname{supp}_{C^0} f\right) \Longleftrightarrow \exists r > 0, \text{ s.t. } f(x) = 0, \ \forall x \in B_r(x_0). \tag{4.52}$$

Now we prove (i). Let $f \in C^0(\Omega)$. If $x_0 \in \Omega \setminus \left(\operatorname{supp}_{\mathcal{D}} f\right)$ then by (4.51) there exists $r > 0$ such that

$$\int_{B_r(x_0)} f(x)\phi(x)dx = \langle T, \phi \rangle = 0, \ \forall \phi \in \mathcal{D}(\Omega), \ \operatorname{supp} \phi \subset B_r(x_0) \tag{4.53}$$

($\operatorname{supp}\phi$ is, of course $\operatorname{supp}_{C^0}\phi$) that, as $f \in C^0(\Omega)$, implies $f = 0$ in $B_r(x_0)$. This implies, by (4.52), that $x_0 \in \Omega \setminus (\operatorname{supp}_{\mathcal{D}} f)$. Conversely, if $x_0 \in \Omega \setminus (\operatorname{supp}_{C^0} f)$, then by (4.52) there exists $r > 0$ such that $f(x) = 0$, for every $x \in B_r(x_0)$. Hence $T_f = 0$ in $B_r(x_0)$, that is $x_0 \in \Omega \setminus (\operatorname{supp}_{\mathcal{D}} f)$.

Now we prove (ii). If $x_0 \in \Omega \setminus (\operatorname{supp}_{\mathcal{D}} f)$ then again (4.53) holds, but this time we have, since $f \in L^1_{loc}(\Omega)$ (see Proposition 3.48),

$$f(x) = 0, \quad \text{a.e. } x \in B_r(x_0).$$

Hence $B_r(x_0) \in \mathcal{O}$, therefore $x_0 \in \bigcup_{A \in \mathcal{O}} A = \Omega \setminus (\operatorname{supp}_{\text{ess}} f)$. Conversely, if $x_0 \in \Omega \setminus (\operatorname{supp}_{\text{ess}} f)$ then there exists $r > 0$ such that $B_r(x_0) \in \mathcal{O}$. Thus $f = 0$ almost everywhere in $B_r(x_0)$ hence

$$\langle T, \phi \rangle = \int_{B_r(x_0)} f(x)\phi(x)dx = 0, \ \forall \phi \in \mathcal{D}(\Omega), \ \operatorname{supp}\phi \subset B_r(x_0).$$

Therefore $x_0 \in \Omega \setminus (\operatorname{supp}_{\mathcal{D}} f)$.

Finally (iii) is a trivial consequence of (i) and (ii). ♦

In light of the above Remark, if there is no risk of ambiguity, here and in the sequel we will simply denote the support of a function f (in $C^0(\Omega)$ or $L^1_{loc}(\Omega)$) or distribution T as $\operatorname{supp} f$ or $\operatorname{supp} T$ (without a subscript) respectively.

Exercise 4.10 Check that

(i) $\operatorname{supp}\delta_{x_0} = \{x_0\}$,
(ii) $\operatorname{supp} T_H = [0, +\infty)$, where H is the Heaviside function.

♠

Proposition 4.7 *Let $T \in \mathcal{D}'(\Omega)$ and $\phi \in \mathcal{D}(\Omega)$ satisfy*

$$(\operatorname{supp} T) \cap (\operatorname{supp}\phi) = \emptyset, \tag{4.54}$$

then we have

$$\langle T, \phi \rangle = 0.$$

Proof Set

$$K = \operatorname{supp}\phi.$$

Since $\operatorname{supp} T$ is a closed set in Ω and (4.54) holds, we have

$$\forall y \in K \ \exists r_y > 0 \text{ s.t. } B_{r_y}(y) \subset \Omega \setminus (\operatorname{supp} T)$$

which, from the definition of the support of T, implies that for every $y \in K$

$$\langle T, \varphi \rangle = 0, \ \forall \varphi \in \mathcal{D}'(\Omega), \ \text{s.t. } \operatorname{supp} \varphi \subset B_{r_y}(y). \tag{4.55}$$

Since K is compact and $\left\{B_{r_y}(y)\right\}_{y \in K}$ is an open covering of K, there exists a finite subcovering $\left\{B_{r_{y_j}}(y_j)\right\}_{1 \leq j \leq l}$ of K. Let $\{\zeta_j\}_{1 \leq j \leq l}$ be a partition of unity subordinate to the covering $\left\{B_{r_{y_j}}(y_j)\right\}_{1 \leq j \leq l}$. By (4.55) we have

$$\langle T, \varphi \rangle = \left\langle T, \sum_{j=1}^{l} \zeta_j \varphi \right\rangle = \sum_{j=1}^{l} \langle T, \zeta_j \varphi \rangle = 0.$$

■

4.5 Distributions with Compact Support

The Space $\mathcal{E}'(\Omega)$
In Sect. 3.13, we defined a Fréchet space structure on $C^\infty(\Omega)$ and denoted this space by $\mathcal{E}(\Omega)$. In particular, we recall that a sequence $\{\phi_j\}$ in $\mathcal{E}(\Omega)$ converges to $\phi \in \mathcal{E}(\Omega)$ if and only if, for every compact set $K \subset \Omega$, we have

$$\lim_{j \to \infty} \max_K \left|\partial^\alpha \phi_j - \partial^\alpha \phi\right| = 0 \tag{4.56}$$

and we write

$$\{\phi_j\} \to \phi, \quad \text{in } \mathcal{E}(\Omega).$$

We define $\mathcal{E}'(\Omega)$ as the space of continuous linear functionals

$$L : \mathcal{E}(\Omega) \to \mathbb{R}.$$

We set

$$L(\phi) = \langle L, \phi \rangle_{\mathcal{E}'(\Omega), \mathcal{E}(\Omega)}, \quad \forall \phi \in \mathcal{E}(\Omega).$$

If there is no risk of ambiguity, we will omit the subscripts and simply write $\langle L, \phi \rangle$ instead of $\langle L, \phi \rangle_{\mathcal{E}'(\Omega), \mathcal{E}(\Omega)}$.

In other words, $L \in \mathcal{E}'(\Omega)$ if and only if L is a linear map from $\mathcal{E}(\Omega)$ to $\mathbb{R}$ and for every sequence $\{\phi_j\}$ in $\mathcal{E}(\Omega)$ which satisfies

$$\{\phi_j\} \to 0, \quad \text{in } \mathcal{E}(\Omega),$$

we have

$$\left\{\langle L, \phi_j \rangle\right\} \to 0, \quad \text{in } \mathbb{R}.$$

It is evident that $\mathcal{E}'(\Omega) \subset \mathcal{D}'(\Omega)$ (set-theoretically and topologically). We will prove that the inclusion is proper (see Remark 4.12 below).

As in Sect. 3.13, let $\{K_j\}_{j\in\mathbb{N}}$ be a family of compact sets contained in Ω such that $K_j \neq \emptyset$, $K_j \subset \overset{\circ}{K}_{j+1}$ for every $j \in \mathbb{N}$ and

$$\bigcup_{j=1}^{\infty} K_j = \Omega.$$

Moreover let us define, as (3.168), for $\phi \in \mathcal{E}(\Omega)$

$$q_N(\phi) = \max\left\{|\partial^\alpha \phi(x)| : x \in K_N, \quad |\alpha| \leq N\right\},$$

or, equivalently

$$q_N(\phi) = \sum_{|\alpha|\leq N} \max_{K_N} \left|\partial^\alpha \phi\right|.$$

The following Proposition holds, and its proof follows similarly to that of Proposition 4.1, thus it is left to the reader.

Proposition 4.8 *Let*

$$L : \mathcal{E}(\Omega) \to \mathbb{R}$$

be a linear map.

The following sentence are equivalent:

(a) $L \in \mathcal{E}'(\Omega)$,
(b) there exist $N \in \mathbb{N}$ *and* $C \geq 0$ *such that*

$$|\langle L, \phi \rangle| \leq C q_N(\phi), \quad \forall \phi \in \mathcal{E}(\Omega). \tag{4.57}$$

For the time being we denote by

$$\mathcal{D}'_0(\Omega) = \left\{T \in \mathcal{D}'(\Omega) : \text{ supp}\, T \text{ is a compact of } \Omega\right\}.$$

In Theorems 4.9 and 4.10 below, we prove that $\mathcal{D}'_0(\Omega)$ and $\mathcal{E}'(\Omega)$ can be identified.

Theorem 4.9 *With the notations introduced above, we have*

$$\mathcal{E}'(\Omega) \subset \mathcal{D}_0'(\Omega). \tag{4.58}$$

Proof Let $L \in \mathcal{E}'(\Omega)$. By Proposition 4.8 we have there exist $N \in \mathbb{N}$ and $C \geq 0$ such that

$$\left|\langle L, \phi\rangle_{\mathcal{E}'(\Omega),\mathcal{E}(\Omega)}\right| \leq C q_N(\phi), \quad \forall \phi \in \mathcal{E}(\Omega). \tag{4.59}$$

Let us prove that $\Omega \setminus K_N \subset \Omega \setminus \operatorname{supp} L$. Let $x_0 \in \Omega \setminus K_N$. We have there exists $r > 0$ such that

$$B_r(x_0) \subset \Omega \setminus K_N.$$

Therefore, if $\phi \in \mathcal{D}(\Omega)$ and $\operatorname{supp}\phi \subset B_r(x_0)$ by (4.59) we have $\langle L, \phi\rangle = 0$. Hence $x_0 \in \Omega \setminus \operatorname{supp} L$. Thus, $\operatorname{supp} L \subset K_N$. This concludes the proof of (4.58). ■

Theorem 4.10 *Let $T \in \mathcal{D}_0'(\Omega)$. Then there exists a unique $L \in \mathcal{E}'(\Omega)$ such that*

$$\langle L, \phi\rangle_{\mathcal{E}'(\Omega),\mathcal{E}(\Omega)} = \langle L, \phi\rangle_{\mathcal{D}'(\Omega),\mathcal{D}(\Omega)}, \quad \forall \phi \in C_0^\infty(\Omega). \tag{4.60}$$

Proof First we prove that if there exists $L \in \mathcal{E}'(\Omega)$ which satisfies (4.60) then it is unique. Let us assume that $L_1, L_2 \in \mathcal{E}'(\Omega)$ satisfies (4.60). Set

$$\widetilde{L} = L_1 - L_2 \in \mathcal{E}'(\Omega)$$

and we have

$$\left\langle\widetilde{L}, \phi\right\rangle_{\mathcal{E}'(\Omega),\mathcal{E}(\Omega)} = 0, \quad \forall \phi \in C_0^\infty(\Omega). \tag{4.61}$$

Let $\varphi \in C^\infty(\Omega)$. By Proposition 3.86 there exists a sequence $\{\phi_k\}$ in $C_0^\infty(\Omega)$ such that

$$\{\phi_k\} \to \varphi, \quad \text{in } \mathcal{E}(\Omega).$$

Hence, by (4.61) we have

$$\left\langle\widetilde{L}, \varphi\right\rangle_{\mathcal{E}'(\Omega),\mathcal{E}(\Omega)} = \lim_{k\to\infty} \left\langle\widetilde{L}, \phi_k\right\rangle_{\mathcal{E}'(\Omega),\mathcal{E}(\Omega)} = 0.$$

Therefore $\widetilde{L} = 0$. This concludes the proof of the uniqueness.

Let $T \in \mathcal{D}_0'(\Omega)$. Let us prove the following

Claim There exist K, compact subset of Ω and $C \geq 0$ such that

$$\left|\langle T, \phi\rangle_{\mathcal{D}'(\Omega),\mathcal{D}(\Omega)}\right| \leq C \sum_{|\alpha|\leq N} \max_K \left|\partial^\alpha \phi\right|, \quad \forall \phi \in C_0^\infty(\Omega). \tag{4.62}$$

Proof of the Claim Let

$$\widetilde{K} = \operatorname{supp} T.$$

By Lemma 3.49 there exists $\psi \in C^\infty(\Omega)$ such that $\psi = 1$ in a neighborhood of $\widetilde{K}$. Let us denote by $K = \operatorname{supp} \psi$.

Let $\phi \in C_0^\infty(\Omega)$. Since

$$\widetilde{K} \cap \operatorname{supp}((1-\psi)\phi) = \emptyset,$$

Proposition 4.7 implies (we omit the subscript "$\mathcal{D}'(\Omega)$, $\mathcal{D}(\Omega)$")

$$\langle T, \phi\rangle = \langle T, \psi\phi + (1-\psi)\phi\rangle = \langle T, \psi\phi\rangle\,. \tag{4.63}$$

By Proposition 4.1 there exist $N \in \mathbb{N}_0$ and $C \geq 0$ such that

$$|\langle T, \psi\phi\rangle| \leq C \sum_{|\alpha|\leq N} \max_K \left|\partial^\alpha (\psi\phi)\right|. \tag{4.64}$$

We may assume $K \subset K_N$.

By (4.63) and by (4.64), using by the Leibniz formula we have

$$|\langle T, \phi\rangle| \leq \widetilde{C} \sum_{|\alpha|\leq N} \max_{K_N} \left|\partial^\alpha \phi\right|, \tag{4.65}$$

where $\widetilde{C}$ is a constant that may depend on ψ, while is independent of ϕ. **This concludes the proof of the Claim.**

At this point, it is useful to emphasize that, thanks to (4.65), T, in addition to being a distribution in $\mathcal{D}'(\Omega)$, is continuous in the topology of $\mathcal{E}(\Omega)$. This, by virtue of Proposition 3.86, allows us to extend $\widetilde{T}$ to a map L of $\mathcal{D}'(\Omega)$ as follows. Let $\varphi \in C^\infty(\Omega)$. Let $\{\phi_k\}$ be a sequence in $C_0^\infty(\Omega)$ such that

$$\{\phi_k\} \to \varphi, \quad \text{in } \mathcal{E}(\Omega), \tag{4.66}$$

Let us define

$$\langle L, \varphi\rangle_{\mathcal{E}'(\Omega),\mathcal{E}(\Omega)} = \lim_{k\to\infty} \langle T, \phi_k\rangle_{\mathcal{D}'(\Omega),\mathcal{D}(\Omega)}\,, \tag{4.67}$$

let us note that the limit above exists thanks to (4.62) and (4.66). Moreover, the value of $\langle L, \varphi\rangle_{\mathcal{E}'(\Omega),\mathcal{E}(\Omega)}$ does not depend on the particular sequence that satisfies (4.66). L is linear. Moreover by (4.65), (4.66) and (4.67) we have $L \in \mathcal{E}'(\Omega)$ and satisfies (4.60). ■

Remark 4.12 The two Theorems 4.9, 4.10 allow us to identify $\mathcal{D}'_0(\Omega)$ with $\mathcal{E}'(\Omega)$. Thus, the distributions in $\mathcal{E}'(\Omega)$ are exactly the distributions in $\mathcal{D}'(\Omega)$ with compact support. This tells us, first of all, that the inclusion

$$\mathcal{E}'(\Omega) \subset \mathcal{D}'(\Omega)$$

is **proper**. Moreover, we observe that if $L \in \mathcal{E}'(\Omega)$, then $L \in \mathcal{D}'(\Omega)$ and there exist $N \in \mathbb{N}$ and $C \geq 0$ such that $\operatorname{supp} L \subset K_N$ and

$$\left|\langle L, \phi\rangle_{\mathcal{D}'(\Omega),\mathcal{D}(\Omega)}\right| \leq C \sum_{|\alpha|\leq N} \max_{K_N} \left|\partial^\alpha \phi\right|, \quad \forall \phi \in C_0^\infty(\Omega). \tag{4.68}$$

We also have that L is a distribution of finite order. Indeed, let $\Lambda \subset \Omega$ be a compact set. From (4.68), we have, for every $\phi \in C_\Lambda^\infty(\Omega)$,

$$\begin{aligned} \left|\langle L, \phi\rangle_{\mathcal{D}'(\Omega),\mathcal{D}(\Omega)}\right| &\leq C \sum_{|\alpha|\leq N} \max_{K_N} \left|\partial^\alpha \phi\right| = \\ &= C \sum_{|\alpha|\leq N} \max_{K_N\cap\Lambda} \left|\partial^\alpha \phi\right| \leq \\ &\leq C \sum_{|\alpha|\leq N} \max_{\Lambda} \left|\partial^\alpha \phi\right|. \end{aligned} \tag{4.69}$$

Therefore, there exists $N \in \mathbb{N}$ such that for every compact subset $\Lambda \subset \Omega$, we have

$$\left|\langle L, \phi\rangle_{\mathcal{D}'(\Omega),\mathcal{D}(\Omega)}\right| \leq C \sum_{|\alpha|\leq N} \max_{\Lambda} \left|\partial^\alpha \phi\right|, \quad \forall \phi \in C_\Lambda^\infty(\Omega).$$

Thus $L \in \mathcal{D}'_F(\Omega)$. ♦

Now we give some theorems concerning the **local structure of the distributions**.

Theorem 4.11 *Let Ω be an open subset of $\mathbb{R}^n$, and let ω be an open subset of Ω such that $\omega \Subset \Omega$ (meaning that $\overline{\omega}$ is compact and it is contained in Ω). Suppose $T \in \mathcal{D}'(\Omega)$. Then there exist a function $f \in L^\infty(\omega)$ and $m \in \mathbb{N}_0$ such that*

$$T = \partial_1^m \cdots \partial_n^m f, \quad \textit{in } \mathcal{D}'(\omega), \tag{4.70}$$

namely

$$\langle T, \phi \rangle_{\mathcal{D}'(\Omega),\mathcal{D}(\Omega)} = (-1)^{mn} \int_\Omega f \left(\partial_1^m \cdots \partial_n^m \phi \right), \quad \forall \phi \in C_0^\infty(\omega). \tag{4.71}$$

Proof By Proposition 4.1 there exist $N \in \mathbb{N}_0$ and $C \geq 0$ such that

$$|\langle T, \phi \rangle| \leq C \sum_{|\alpha| \leq N} \sup_\omega \left| \partial^\alpha \phi \right|, \quad \forall \phi \in C_0^\infty(\omega). \tag{4.72}$$

Since $\overline{\omega}$ is compact we can assume that, for some positive number d, we have $\omega \subset [-d, d]^n$. Now, if φ is an arbitrary function of $C_0^\infty(\omega)$ we have, for $x \in \omega$ and $j = 1, \cdots, n$,

$$|\varphi(x)| = \left| \int_{-d}^{x_j} \partial_j \varphi(x_1, \cdots, \underbrace{t}_{j-\text{th}}, \cdots x_n) dt \right| \leq 2d \sup_\omega \left| \partial_j \varphi \right|.$$

Hence

$$\sup_\omega |\varphi| \leq 2d \sup_\omega \left| \partial_j \varphi \right|, \quad \text{for } j = 1, \cdots, n. \tag{4.73}$$

Consequently, by repeatedly applying (4.73), we obtain from (4.72)

$$|\langle T, \phi \rangle| \leq C_1 \sup_\omega \left| \partial_1^N \cdots \partial_n^N \phi \right|, \quad \forall \phi \in C_0^\infty(\omega). \tag{4.74}$$

On the other hand, if φ is an arbitrary function of $C_0^\infty(\omega)$ we have, for every $x \in \omega$,

$$\varphi(x) = \int_{-d}^{x_1} \cdots \int_{-d}^{x_n} \partial_1 \cdots \partial_n \varphi(y) dy.$$

Hence

$$\sup_\omega |\varphi| \leq \int_\omega |\partial_1 \cdots \partial_n \varphi(x)| \, dx. \tag{4.75}$$

By applying (4.75) to the function $\varphi = \partial_1^N \cdots \partial_n^N \phi$, from (4.74) we obtain

$$|\langle T, \phi \rangle| \leq C_1 \int_\omega \left| \partial_1^{N+1} \cdots \partial_n^{N+1} \phi(x) \right| dx, \quad \forall \phi \in C_0^\infty(\omega). \tag{4.76}$$

Now, let us denote by

$$M = \left\{ (-1)^{n(N+1)} \partial_1^{N+1} \cdots \partial_n^{N+1} \phi : \; \phi \in C_0^\infty(\omega) \right\}.$$

By (4.76) we get that the functional

$$M \ni (-1)^{n(N+1)} \partial_1^{N+1} \cdots \partial_n^{N+1} \phi \to \langle T, \phi \rangle$$

(let us note that, by (4.76), this functional is well-defined) is linear and continuous from M, equipped with the norm $L^1(\omega)$, in $\mathbb{R}$. Then, by Hahn–Banach Theorem (Theorem 3.88) and by Theorem 3.32, there exists $f \in L^\infty(\omega)$ such that

$$\langle T, \phi \rangle = (-1)^{n(N+1)} \int_\omega f \left(\partial_1^{N+1} \cdots \partial_n^{N+1} \phi \right), \quad \forall \phi \in C_0^\infty(\omega).$$

Thus (4.71) follows with $m = N + 1$. ■

Corollary 4.12 *Let Ω be an open subset of $\mathbb{R}^n$, and let ω be an open subset of Ω such that $\omega \Subset \Omega$. Suppose $T \in \mathcal{D}'(\Omega)$. Then there exist a function $g \in C^0(\Omega)$ and $m' \in \mathbb{N}_0$ such that*

$$T = \partial_1^{m'} \cdots \partial_n^{m'} g, \quad in\ \mathcal{D}'(\omega), \tag{4.77}$$

Proof By Theorem 4.11 there exists $f \in L^\infty(\omega)$ such that

$$T = \partial_1^m \cdots \partial_n^m f, \quad \text{in } \mathcal{D}'(\omega), \tag{4.78}$$

Let $\widetilde{f}$ be the function

$$\widetilde{f}(x) = \begin{cases} f(x), & \text{in } \omega \\ 0, & \text{in } \Omega \setminus \omega \end{cases}.$$

The following function

$$g(x) = \int_{-\infty}^{x_1} \cdots \int_{-\infty}^{x_n} f(y)dy$$

is continuous in Ω and satisfies

$$\partial_1 \cdots \partial_n g = f, \quad \text{in } \mathcal{D}'(\omega) \tag{4.79}$$

By (4.78) and (4.79) we have

$$T = \partial_1^{m+1} \cdots \partial_n^{m+1} g, \quad \text{in } \mathcal{D}'(\omega).$$

Thus (4.77) follows with $m' = m + 1$. ■

Theorem 4.13 *Let $L \in \mathcal{E}'(\Omega)$. Let $K = \operatorname{supp} L$ (by Theorem 4.9, K is compact). Then, for every open set $\mathcal{U}$ such that $K \subset \mathcal{U} \subset \Omega$, there exist $N \in \mathbb{N}$ and $f_\alpha \in C_0^0(\mathcal{U})$, for $|\alpha| \leq N$ such that*

$$L = \sum_{|\alpha|\leq N} (-1)^{|\alpha|} \partial^\alpha f_\alpha, \quad in\, \mathcal{E}'(\Omega). \tag{4.80}$$

Proof Let ω be an open set such that $K \subset \omega \Subset \mathcal{U}$. Corollary 4.12 implies there exist $m \in \mathbb{N}$ and $g \in C^0(\omega)$ such that

$$L = \partial_1^m \cdots \partial_n^m g, \quad \text{in } \mathcal{D}'(\omega).$$

By Lemma 3.49 there exists $\psi \in C_0^\infty(\Omega)$ such that $\operatorname{supp} \psi \subset \omega$ and $\psi = 1$ in K. Let $\varphi \in C^\infty(\Omega)$ be arbitrary. Then we have

$$\begin{aligned}\langle L, \varphi\rangle &= (-1)^{mn} \int_\Omega g \partial_1^m \cdots \partial_n^m (\psi\varphi) dx = \\ &= (-1)^{mn} \sum_{\alpha_1,\cdots,\alpha_n \leq m} \binom{m}{\alpha_1} \cdots \binom{m}{\alpha_n} \left(g \partial_1^{m-\alpha_1} \cdots \partial_n^{m-\alpha_n} \psi\right) \left(\partial_1^{\alpha_1} \cdots \partial_n^{\alpha_n} \varphi\right) dx.\end{aligned}$$

Hence (4.80) follows with $N = mn$ and

$$f_\alpha = (-1)^{mn} \binom{m}{\alpha_1} \cdots \binom{m}{\alpha_n} \left(g \partial_1^{m-\alpha_1} \cdots \partial_n^{m-\alpha_n} \psi\right).$$

■

Exercise 4.11 Let $L \in \mathcal{E}'(\Omega)$. Then there exists a unique $\widetilde{L} \in \mathcal{E}'(\mathbb{R}^n)$ that satisfies

$$\operatorname{supp} \widetilde{L} \subset \Omega,$$

and

$$\left\langle \widetilde{L}, \varphi \right\rangle_{\mathcal{E}'(\mathbb{R}^n),\mathcal{E}(\mathbb{R}^n)} = \left\langle \widetilde{L}, \varphi \right\rangle_{\mathcal{E}'(\Omega),\mathcal{E}(\Omega)}, \quad \forall \varphi \in C_0^\infty(\mathbb{R}^n), \text{ s.t. } \operatorname{supp} \varphi \subset \Omega.$$

♠

4.6 Further Insights into Convolutions for Distributions

In Sect. 4.3, we provided, among other operations on distributions, the definition of the convolution of a distribution $T \in \mathcal{D}'(\mathbb{R}^n)$ and a function $\eta \in \mathcal{D}(\mathbb{R}^n)$. We denoted such a convolution by $\eta \star T$. In this section, we aim to prove some properties of this operation and give an overview of extensions of the convolution operation. Let us recall that, if T and η are as above, then (see (4.38))

$$\langle \eta \star T, \psi \rangle = \langle T, \check{\eta} \star \psi \rangle, \quad \forall \psi \in \mathcal{D}(\mathbb{R}^n). \tag{4.81}$$

We begin by the following

Lemma 4.14 *Let $\eta \in \mathcal{D}(\mathbb{R}^n)$. Then we have*

$$\lim_{k\to\infty} k^{-n} \sum_{g\in\mathbb{Z}^n} \eta\left(\frac{g}{k}\right) = \int_{\mathbb{R}^n} \eta(x)dx. \tag{4.82}$$

Proof Let $R > 0$ be such that

$$\operatorname{supp} \eta \subset B_R.$$

Let $Q = \left[-\frac{1}{2}, \frac{1}{2}\right]$. Let us note that the cubes $Q_{g,k} = \frac{1}{k}(g + Q)$, $g \in \mathbb{Z}^n$, are pairwise internally disjoint and we have

$$\mathbb{R}^n = \bigcup_{g\in\mathbb{Z}^n} Q_{g,k}.$$

Hence

$$\begin{aligned} &\left| k^{-n} \sum_{g\in\mathbb{Z}^n} \eta\left(\frac{g}{k}\right) - \int_{\mathbb{R}^n} \eta(x)dx \right| = \\ &= \left| \sum_{g\in\mathbb{Z}^n} \int_{Q_{g,k}} \eta\left(\frac{g}{k}\right) dx - \sum_{g\in\mathbb{Z}^n} \int_{Q_{g,k}} \eta(x)dx \right| = \\ &= \left| \sum_{g\in\mathbb{Z}^n} \int_{Q_{g,k}} \left(\eta\left(\frac{g}{k}\right) - \eta(x)\right) dx \right| = \\ &= \sum_{g\in\mathbb{Z}^n} \int_{Q_{g,k}} \left|\eta\left(\frac{g}{k}\right) - \eta(x)\right| dx. \end{aligned} \tag{4.83}$$

Now we have

$$\left|\eta\left(\frac{g}{k}\right) - \eta(x)\right| \leq \frac{M}{k} \chi_{B_R}, \quad \forall x \in Q_{g,k}, \tag{4.84}$$

where

$$M = \sup_{\mathbb{R}^n} |\nabla \eta|.$$

By (4.83) and (4.84) we get

$$\left| k^{-n} \sum_{g \in \mathbb{Z}^n} \eta\left(\frac{g}{k}\right) - \int_{\mathbb{R}^n} \eta(x)dx \right| \leq \frac{M}{k} |B_R|, \quad \forall k \in \mathbb{N},$$

from which (4.82) follows. ■

Theorem 4.15 *Let $T \in \mathcal{D}'(\mathbb{R}^n)$, $\eta \in \mathcal{D}(\mathbb{R}^n)$ and*

$$f : \mathbb{R}^n \to \mathbb{R}$$

be such that

$$f(x) = \langle T, \mathcal{T}_x(\check{\eta}) \rangle, \quad for\, x \in \mathbb{R}^n.$$

Then we have

(a) $f \in C^\infty(\mathbb{R}^n)$,
(b) $T_f = \eta \star T$.

Proof (a) First we prove that f is a continuous function. Let $x_0 \in \mathbb{R}^n$. Let $\{x_k\}$ be a sequence in $\mathbb{R}^n$ that converges to x_0. We have there exists $\delta > 0$ such that

$$\text{supp}\left(\mathcal{T}_{x_k}\check{\eta}\right) \subset [K]_\delta, \quad \forall k \in \mathbb{N} \tag{4.85}$$

(we recall that $[K]_\delta = \{x \in \mathbb{R}^n : \text{dist}(x, K) \leq \delta\}$). Moreover, for every $\alpha \in \mathbb{N}_0^n$, we have

$$\left\{\partial^\alpha \left(\mathcal{T}_{x_k}\check{\eta}\right)\right\} \to \partial^\alpha \left(\mathcal{T}_{x_0}\check{\eta}\right) \quad \text{uniformly.} \tag{4.86}$$

By (4.85)and (4.86) we have

$$\left\{\mathcal{T}_{x_k}\check{\eta}\right\} \to \mathcal{T}_{x_0}\check{\eta}, \quad \text{in } \mathcal{D}(\mathbb{R}^n).$$

Hence

$$\lim_{k\to\infty} f(x_k) = \lim_{k\to\infty} \langle T, \mathcal{T}_{x_k}(\check{\eta}) \rangle = \langle T, \mathcal{T}_{x_0}(\check{\eta}) \rangle = f(x_0).$$

This proves that f is continuous in x_0.

Similarly can be proved that, for $j = 1, \cdots, n$, we have

$$\lim_{h \to 0} \frac{f(x + he_j) - f(x)}{h} = \left\langle T, \mathcal{T}_x \partial_j \left(\check{\eta}\right) \right\rangle, \quad \forall x \in \mathbb{R}^n \tag{4.87}$$

and, consequently, $\partial_j f$ turns out to be continuous in $\mathbb{R}^n$, for $j = 1, \cdots, n$. By induction we have that $f \in C^\infty(\mathbb{R}^n)$ and

$$\partial^\alpha f(x) = \left\langle T, \mathcal{T}_x \partial^\alpha \left(\check{\eta}\right) \right\rangle, \quad \forall \alpha \in \mathbb{N}_0^n.$$

This concludes the proof of (a).

Now we prove (b). Namely we prove

$$\int_{\mathbb{R}^n} \left\langle T, \mathcal{T}_x \check{\eta} \right\rangle \phi(x) dx = \langle \eta \star T, \phi \rangle, \quad \forall \phi \in \mathcal{D}(\mathbb{R}^n). \tag{4.88}$$

Claim Let $\phi \in \mathcal{D}(\mathbb{R}^n)$. Let us denote by

$$\psi_k = k^{-n} \sum_{g \in \mathbb{Z}^n} \phi\left(\frac{g}{k}\right) \mathcal{T}_{\frac{g}{k}}\left(\check{\eta}\right), \quad \text{for } k \in \mathbb{N}.$$

Then we have

$$\{\psi_k\} \to \check{\eta} \star \phi, \quad \text{in } \mathcal{D}(\mathbb{R}^n). \tag{4.89}$$

Proof of the Claim First we note that, if either $\frac{g}{k} \notin \operatorname{supp} \phi$ or $x - \frac{g}{k} \notin \operatorname{supp} \check{\eta}$ (or both) then $x \notin \operatorname{supp} \psi_k$. Hence

$$\operatorname{supp} \psi_k \subset \operatorname{supp} \phi + \operatorname{supp} \check{\eta}, \quad \forall k \in \mathbb{N}. \tag{4.90}$$

On the other hand, proceeding as in Lemma 4.14, we easily have

$$\left\{\partial^\alpha \psi_k\right\} \to \partial^\alpha \left(\check{\eta} \star \phi\right), \quad \text{uniformly.} \tag{4.91}$$

By (4.90) and (4.91) we get (4.89). **This concludes the proof of the Claim.**

Now we prove (4.88). By Lemma 4.14 (recalling (a)) and by (4.89) we have, for every $\phi \in \mathcal{D}(\mathbb{R}^n)$

$$
\begin{aligned}
&\int_{\mathbb{R}^n} \langle T, \mathcal{T}_x \check{\eta} \rangle \phi(x) dx = \\
&= \lim_{k\to\infty} k^{-n} \sum_{g\in\mathbb{Z}^n} \phi\left(\frac{g}{k}\right) \left\langle T, \mathcal{T}_{\frac{g}{k}}\left(\check{\eta}\right)\right\rangle = \\
&= \lim_{k\to\infty} \left\langle T, k^{-n} \sum_{g\in\mathbb{Z}^n} \phi\left(\frac{g}{k}\right) \mathcal{T}_{\frac{g}{k}}\left(\check{\eta}\right)\right\rangle = \\
&= \lim_{k\to\infty} \langle T, \psi_k \rangle = \\
&= \langle T, \check{\eta} \star \phi \rangle = \\
&= \langle \eta \star T, \phi \rangle ,
\end{aligned}
\tag{4.92}
$$

let us note that in the last step we have used (4.81). ■

Remark 4.13 The sentence (b) of the statement of Theorem 4.15 can be made more explicit if we adopt the following conventions, which are also useful in other contexts of distribution theory. If $T \in \mathcal{D}'(\mathbb{R}^n)$, $\phi \in \mathcal{D}(\mathbb{R}^n)$ and $x \in \mathbb{R}^n$, we will denote

$$\langle T, \mathcal{T}_x \phi \rangle$$

by

$$\langle T(y), \phi(y - x) \rangle .$$

Then (b) is nothing but

$$(\eta \star T)(x) = \langle T(y), \eta(x - y) \rangle , \quad \forall x \in \mathbb{R}^n. \tag{4.93}$$

♦

Exercise 4.12 Prove (4.87) and (4.91) in detail. ♠

Exercise 4.13 Let $T \in \mathcal{D}'(\mathbb{R}^n)$. Let $\eta \in C_0^\infty(\mathbb{R}^n)$. Prove that

$$\operatorname{supp}(\eta \star T) \subset \operatorname{supp} \eta + \operatorname{supp} T. \tag{4.94}$$

♠

Exercise 4.14 Let $T \in \mathcal{D}'(\mathbb{R}^n)$. Let $\eta, \zeta \in C_0^\infty(\mathbb{R}^n)$ satisfy

$$\int_{\mathbb{R}^n} \eta(x) dx = 1, \quad \zeta(0) = 1.$$

Define

$$\eta_\varepsilon(x) = \varepsilon^{-n}\eta\left(\varepsilon^{-1}x\right), \quad \zeta_\varepsilon(x) = \zeta\left(\varepsilon x\right), \quad \text{for } \varepsilon > 0.$$

Prove that

(i) $\zeta_\varepsilon T$ and $\eta_\varepsilon \star T$ converge to T in $\mathcal{D}'(\mathbb{R}^n)$ as $\varepsilon \to 0$;
(ii) if, in addition, we have $\zeta(x) = 1$ in a neighborhood of 0 then $\zeta_\varepsilon\left(\eta_\varepsilon \star T\right)$ converges to T in $\mathcal{D}'(\mathbb{R}^n)$ as $\varepsilon \to 0$.

(Note that by Theorem 4.15 we have $\zeta_\varepsilon\left(\eta_\varepsilon \star T\right) \in C_0^\infty(\mathbb{R}^n)$, for every $\varepsilon > 0$, hence (ii) provides the density of $C_0^\infty(\mathbb{R}^n)$ in $\mathcal{D}'(\mathbb{R}^n)$ in the topology of $\mathcal{D}'(\mathbb{R}^n)$). ♠

Exercise 4.15 Let Ω be an open set of $\mathbb{R}^n$ and let $T \in \mathcal{D}'(\Omega)$. Prove that there exists a sequence $\{f_k\}$ in $C_0^\infty(\Omega)$ such that

$$\left\{T_{f_k}\right\} \to T, \quad \text{in } \mathcal{D}'(\Omega).$$

(In other words, $C_0^\infty(\Omega)$ is dense in $\mathcal{D}'(\Omega)$ in the topology of $\mathcal{D}'(\Omega)$).

[Hint: denote by $\Omega_k = B_k \cap \left\{x \in \Omega : \operatorname{dist}(x, \partial\Omega) > \frac{1}{k}\right\}$, for $k \in \mathbb{N}$, note that $\overline{\Omega_k} \subset \Omega_{k+1}$, for every $k \in \mathbb{N}$. Use Lemma 3.49 to construct a sequence $\{\zeta_k\}$ such that $\zeta_k \in C_0^\infty(\Omega)$, $\zeta_k = 1$ on $\overline{\Omega_k}$, $\operatorname{supp}\zeta_k \subset \Omega_{k+1}$, for every $k \in \mathbb{N}$. Let η be a mollifier (see Sect. 3.4.2) and $\eta_\varepsilon(x) = \varepsilon^{-n}\eta\left(\varepsilon^{-1}x\right)$. Let $\{\varepsilon_k\} \downarrow 0$ be such that $\varepsilon_k < \operatorname{dist}\left(\operatorname{supp}\zeta_k, \partial\Omega_{k+1}\right)$, for every $k \in \mathbb{N}$. Set $f_k := \eta_{\varepsilon_k} \star \left(\zeta_k T\right)$. Check that $\operatorname{supp} f_k \subset \Omega_{k+1}$, $f_k \in C_0^\infty(\Omega)$, for every $k \in \mathbb{N}$ and prove that $\left\{T_{f_k}\right\}$ converges to T in $\mathcal{D}'(\Omega)$ as $k \to \infty$.] ♠

Exercise 4.16 Let

$$f_\varepsilon(x) = \frac{x}{x^2 + \varepsilon^2}, \quad \text{for } \varepsilon > 0.$$

Prove that

$$\lim_{\varepsilon \to 0} T_{f_\varepsilon} = \operatorname{vp}\frac{1}{x}, \quad \text{in } \mathcal{D}'(\mathbb{R}).$$

♠

We now wish to explain how to extend the definition of convolution to two distributions, one of which has compact support.

We begin to observe what follows. Let $T \in \mathcal{D}'(\mathbb{R}^n)$. We denote by G the map

$$\mathcal{D}(\mathbb{R}^n) \ni \phi \to G\phi = T \star \phi \in \mathcal{E}(\mathbb{R}^n).$$

It turns out (the reader check) that G is a linear and continuous map. Namely, for every sequence $\left\{\phi_j\right\}$ in $\mathcal{D}(\mathbb{R}^n)$ that converges to 0 then $\left\{T \star \phi_j\right\}$ converges to 0 in

$\mathcal{E}(\mathbb{R}^n)$. Moreover (see Sect. 4.3, Exercise 6-(i)) G commutes with translations, more precisely we have

$$\mathcal{T}_h G = G\mathcal{T}_h, \quad \forall h \in \mathbb{R}^n. \tag{4.95}$$

In the following Theorem, we will prove that the continuity properties of G, along with the property of commuting with translation, characterize the convolution. Specifically, we have

Theorem 4.16 *Let*

$$U : \mathcal{D}(\mathbb{R}^n) \to \mathcal{E}(\mathbb{R}^n) \tag{4.96}$$

be a linear map that satisfies

(i) U is continuous. Namely, for every sequence $\{\phi_j\}$ in $\mathcal{D}(\mathbb{R}^n)$ that converges to 0 then $\{U\phi_j\}$ converges to 0 in $\mathcal{E}(\mathbb{R}^n)$.
(ii) U commutes with translation. Namely

$$\mathcal{T}_h U = U\mathcal{T}_h, \quad \forall h \in \mathbb{R}^n. \tag{4.97}$$

Then there exists a unique $T \in \mathcal{D}'(\mathbb{R}^n)$ such that

$$U\phi = T \star \phi, \quad \forall \phi \in \mathcal{D}(\mathbb{R}^n). \tag{4.98}$$

Proof Let us define the map

$$T : \mathcal{D}(\mathbb{R}^n) \to \mathbb{R} \tag{4.99}$$

such that

$$\langle T, \phi \rangle = \left(U\check{\phi}\right)(0), \quad \forall \phi \in \mathcal{D}(\mathbb{R}^n). \tag{4.100}$$

T is linear and, by (i), it belongs to $\mathcal{D}'(\mathbb{R}^n)$. Now we check that (4.98) is satisfied. We have for every $h \in \mathbb{R}^n$

$$\begin{aligned}(T \star \phi)(-h) &= (\mathcal{T}_h (T \star \phi))(0) = (T \star (\mathcal{T}_h \phi))(0) = \\ &= \langle T(y), \phi(-y-h)\rangle = (U\phi(\cdot - h))(0) = \\ &= (U(\mathcal{T}_h\phi)) \underbrace{=}_{\text{by (ii)}} (\mathcal{T}_h (U\phi))(0) = (U\phi)(-h).\end{aligned} \tag{4.101}$$

Hence

$$(T \star \phi)(-h) = (U\phi)(-h), \quad \forall h \in \mathbb{R}^n. \tag{4.102}$$

This proves equality (4.98).

To show that T is the unique distribution satisfying (4.98), it suffices to retrace the steps taken in (4.101). ■

We note that if $T \in \mathcal{E}'(\mathbb{R}^n)$ then for any $\phi \in \mathcal{E}(\mathbb{R}^n)$ and any $x \in \mathbb{R}^n$ it is well-defined

$$(T \star \phi)(x) = \langle T(y), \phi(x-y) \rangle . \tag{4.103}$$

Moreover it turns out (the reader check) that the above defined function $T \star \phi$ belongs to $\mathcal{E}(\mathbb{R}^n)$ and the linear map

$$\mathcal{E}(\mathbb{R}^n) \ni \phi \to T \star \phi \in \mathcal{E}(\mathbb{R}^n)$$

is continuous in the sense that if $\{\phi_j\}$ is a sequence in $\mathcal{E}(\mathbb{R}^n)$ that converges to 0 then $\{T \star \phi_j\}$ converges to 0 in $\mathcal{E}(\mathbb{R}^n)$. Let us also note, if $\phi \in \mathcal{D}(\mathbb{R}^n)$ then, by (4.94) we have $T \star \phi \in \mathcal{D}(\mathbb{R}^n)$ and the map

$$\mathcal{D}(\mathbb{R}^n) \ni \phi \to T \star \phi \in \mathcal{D}(\mathbb{R}^n)$$

is also linear and continuous (in the topology of $\mathcal{D}(\mathbb{R}^n)$).

We refer to operation (4.103) as the convolution between $T \in \mathcal{E}'(\mathbb{R}^n)$ and $\phi \in \mathcal{E}(\mathbb{R}^n)$ because, for a given $T \in \mathcal{E}'(\mathbb{R}^n)$, it represents the extension of the convolution, $T \star \phi$, already defined for $\phi \in \mathcal{D}(\mathbb{R}^n)$ to $\phi \in \mathcal{E}(\mathbb{R}^n)$.

Now, let $T_1, T_2 \in \mathcal{D}'(\mathbb{R}^n)$, where at least one of them belongs to $\mathcal{E}'(\mathbb{R}^n)$. For instance, let us assume that $T_2 \in \mathcal{E}'(\mathbb{R}^n)$, then, from what we said earlier, we have that the map

$$U : \mathcal{D}(\mathbb{R}^n) \to \mathcal{E}(\mathbb{R}^n)$$

such that

$$U\phi = T_1 \star (T_2 \star \phi) , \quad \text{for } \phi \in \mathcal{D}(\mathbb{R}^n),$$

satisfies (i) and (ii) of Theorem (4.16). Therefore, there exists a unique distribution in $\mathcal{D}'(\mathbb{R}^n)$, which we denote by $T_1 \star T_2$, such that

$$(T_1 \star T_2) \star \phi = U\phi = T_1 \star (T_2 \star \phi) , \quad \forall \phi \in \mathcal{D}(\mathbb{R}^n). \tag{4.104}$$

Similarly we can proceed if $T_1 \in \mathcal{E}'(\mathbb{R}^n)$.

Thus, if $T_1, T_2 \in \mathcal{D}'(\mathbb{R}^n)$, where at least one of them belongs to $\mathcal{E}'(\mathbb{R}^n)$, we can define the convolution between T_1 and T_2 as the unique distribution which satisfies (4.104).

We conclude this section with the following Theorem, for the proof of which we refer the reader to the literature mentioned in the introduction of this chapter, and in particular to [58, Vol. 1, Thm. 4.2.4].

Theorem 4.17 *Let $T_1, T_2 \in \mathcal{D}'(\mathbb{R}^n)$. Suppose that one of them belongs to $\mathcal{E}'(\mathbb{R}^n)$ then we have*

$$T_1 \star T_2 = T_2 \star T_1 \tag{4.105}$$

and

$$\text{supp}\,(T_1 \star T_2) \subset \text{supp}\,T_1 + \text{supp}\,T_2. \tag{4.106}$$

4.7 Distributions with Zero Gradient in a Connected Open Set

We know from elementary analysis that the differentiable functions in a connected open set Ω of $\mathbb{R}^n$ with null gradient are precisely the constant functions in Ω. In this section, we extend this result to distributions. More specifically, we aim to prove the following.

Theorem 4.18 *Let Ω be a connected open set of $\mathbb{R}^n$. Let $T \in \mathcal{D}'(\Omega)$ be such that*

$$\partial_j T = 0, \quad \text{for } j = 1, \cdots, n.$$

Then there exists $C \in \mathbb{R}$ such that

$$\langle T, \phi\rangle = \int_\Omega C\phi dx, \quad \forall\phi \in \mathcal{D}(\Omega)$$

or, in other words, T is constant in Ω.

To prove Theorem 4.18 we make use of the following

Lemma 4.19 *Let Ω be an open set of $\mathbb{R}^n$, $T \in \mathcal{D}'(\Omega)$ and $\phi \in \mathcal{D}(\Omega)$. Let $h \in \mathbb{R}^n$ be such that $\mathcal{T}_{th}\phi \in \mathcal{D}(\Omega)$ for every $t \in [0, 1]$. Then we have*

$$\langle T, \mathcal{T}_h\phi\rangle - \langle T, \phi\rangle = \int_0^1 \sum_{j=1}^n \left\langle \partial_j T, \mathcal{T}_{th}\phi\right\rangle h_j dt. \tag{4.107}$$

Proof of Lemma 4.19 Let $\phi \in \mathcal{D}(\Omega)$. Let us denote by

$$\mathcal{U}_\phi = \left\{y \in \mathbb{R}^n : \mathcal{T}_y\phi \in \mathcal{D}(\Omega)\right\}.$$

As is easy to verify, Ω_ϕ is an open set. Furthermore, proceeding as in the proof of part (a) of Theorem 4.15 we have

$$\mathcal{U}_\phi \ni y \to \left\langle \partial_j T, \mathcal{T}_y \phi \right\rangle \in C^\infty\left(\mathcal{U}_\phi\right).$$

Claim We have

$$\partial_{y_j} \left\langle T, \mathcal{T}_y \phi \right\rangle = \left\langle \partial_j T, \mathcal{T}_y \phi \right\rangle, \quad \text{for } j = 1, \cdots, n. \tag{4.108}$$

Proof of the Claim Let $y \in \mathcal{U}_\phi$. Let $r_0 > 0$ be such that for every $\lambda \in [-r_0, r_0]$ we have $y + \lambda e_j \in \mathcal{U}_\phi$. For every $\lambda \in [-r_0, r_0] \setminus \{0\}$ we have

$$\frac{\left\langle T, \mathcal{T}_{y+\lambda e_j} \phi \right\rangle - \left\langle T, \mathcal{T}_y \phi \right\rangle}{\lambda} = \langle T, \psi_\lambda \rangle, \tag{4.109}$$

where

$$\begin{aligned}\psi_\lambda(x) &= \frac{\left(\mathcal{T}_{y+\lambda e_j} \phi\right)(x) - \left(\mathcal{T}_y \phi\right)(x)}{\lambda} = \\ &= \frac{\phi\left(x - y - \lambda e_j\right) - \phi(x-y)}{\lambda}.\end{aligned}$$

Now we check that

$$\psi_\lambda \to -\partial_{x_j}\left(\mathcal{T}_y \phi\right), \quad \text{as } \lambda \to 0, \quad \text{in } \mathcal{D}(\Omega). \tag{4.110}$$

Indeed we have

$$\psi_\lambda \subset K, \quad \forall \lambda \in [-r_0, r_0] \setminus \{0\}, \tag{4.111}$$

where K is the compact subset of Ω

$$K = \left\{x \in \mathbb{R}^n : \operatorname{dist}(x, \operatorname{supp} \phi(\cdot - y))\right\}$$

and

$$\max_K \left|\partial_x^\alpha \psi_\lambda - \partial_x^\alpha\left(-\partial_{x_j}\left(\mathcal{T}_y \phi\right)\right)\right| \to 0, \quad \text{as } \lambda \to 0. \tag{4.112}$$

Concerning (4.112), it is an immediate consequence of the inequality (obtained by the Mean Value Theorem)

$$\max_K \left|\partial_x^\alpha \psi_\lambda - \partial_x^\alpha\left(-\partial_{x_j}\left(\mathcal{T}_y \phi\right)\right)\right| \le \frac{|\lambda|}{2} \max_K \left|\partial^{\alpha+2e_j} \phi\right|, \ \forall \lambda \in [-r_0, r_0] \setminus \{0\}.$$

Now, by (4.109) and (4.110) we have

$$\begin{aligned}\lim_{\lambda\to 0}\frac{\left\langle T,\mathcal{T}_{y+\lambda e_j}\phi\right\rangle-\left\langle T,\mathcal{T}_y\phi\right\rangle}{\lambda}&=\lim_{\lambda\to 0}\langle T,\psi_\lambda\rangle=\\&=\left\langle T,-\partial_{x_j}\left(\mathcal{T}_y\phi\right)\right\rangle=\\&=\left\langle\partial_j T,\mathcal{T}_y\phi\right\rangle.\end{aligned}$$

This concludes the proof of the Claim.

Let $h\in\mathbb{R}^n$ be such that $\mathcal{T}_{th}\phi\in\mathcal{D}(\Omega)$ for every $t\in[0,1]$. Let us define

$$f:[0,1]\to\mathbb{R}$$

such that

$$f(t)=\langle T,\mathcal{T}_{th}\phi\rangle,\quad \text{for } t\in[0,1].$$

We have $f\in C^\infty([0,1])$. By the Fundamental Theorem of Calculus and by (4.108) we get

$$\begin{aligned}\langle T,\mathcal{T}_h\phi\rangle-\langle T,\phi\rangle&=f(1)-f(0)=\\&=\int_0^1 f'(t)dt=\\&=\int_0^1\sum_{j=1}^n\partial_{y_j}\left\langle T,\mathcal{T}_y\phi\right\rangle_{|y=th}h_j dt=\\&=\int_0^1\sum_{j=1}^n\left\langle\partial_j T,\mathcal{T}_{th}\phi\right\rangle h_j dt.\end{aligned}$$

This concludes the proof of the Lemma. ∎

Proof of Theorem 4.18 Let $\varepsilon_0>0$ be such that for every $\varepsilon\in(0,\varepsilon_0]$, $\Omega_\varepsilon=\{x\in\Omega:\ \text{dist}(x,\partial\Omega)>\varepsilon\}$ is a nonempty open set.

1st Step First we prove that there exists $f\in\mathcal{D}(\Omega)$ such that $T=T_f$. Let $\phi\in\mathcal{D}(\Omega_\varepsilon)$. Let $y\in\mathbb{R}^n$ be such that $|y|<\varepsilon\le\varepsilon_0$. Then we have $\mathcal{T}_{ty}\phi\in\mathcal{D}(\Omega)$, for every $t\in[0,1]$. By Lemma 4.19 we have

$$\left\langle T,\mathcal{T}_y\phi\right\rangle-\langle T,\phi\rangle=\int_0^1\sum_{j=1}^n\left\langle\partial_j T,\mathcal{T}_{ty}\phi\right\rangle y_j dt=0.$$

Hence

$$\left\langle T, \mathcal{T}_y \phi \right\rangle = \langle T, \phi \rangle, \quad \forall y \in B_\varepsilon. \tag{4.113}$$

Let $\psi \in \mathcal{D}(\mathbb{R}^n)$ satisfy $\operatorname{supp} \psi \subset B_\varepsilon$ and $\int_{\mathbb{R}^n} \psi(x) dx = 1$. Since $\phi \in \mathcal{D}(\Omega_\varepsilon)$ we have $\psi \star \phi \in \mathcal{D}(\Omega)$. Multiplying both the sides of (4.113) by ψ and integrating over $\mathbb{R}^n$ we get

$$\int_{\mathbb{R}^n} \psi(y) \left\langle T, \mathcal{T}_y \phi \right\rangle dy = \int_{\mathbb{R}^n} \psi(y) \langle T, \phi \rangle \, dy = \langle T, \phi \rangle$$

and with a slight modification in the proof of the Theorem 4.15, we obtain

$$\begin{aligned}
\langle T, \phi \rangle &= \int_{\mathbb{R}^n} \psi(y) \langle T, \phi \rangle \, dy = \\
&= \left\langle \check{\phi} \star T, \psi \right\rangle = \langle T, \phi \star \psi \rangle = \\
&= \langle T, \psi \star \phi \rangle = \left\langle \check{\psi} \star T, \phi \right\rangle = \\
&= \int_{\mathbb{R}^n} \phi(x) \langle T, \mathcal{T}_x \psi \rangle \, dx.
\end{aligned}$$

Thus we have

$$\int_{\mathbb{R}^n} \phi(x) \langle T, \mathcal{T}_x \psi \rangle \, dx = \langle T, \phi \rangle, \quad \forall \phi \in \mathcal{D}(\Omega), \ \text{ s.t. } \operatorname{supp} \phi \subset \Omega_\varepsilon. \tag{4.114}$$

Now, for any $\varepsilon \in (0, \varepsilon_0]$, let us define

$$f_\varepsilon(x) = \langle T, \mathcal{T}_x \phi \rangle, \quad \text{ for } x \in \Omega_\varepsilon. \tag{4.115}$$

By (a) of Theorem 4.15 we have $f_\varepsilon \in C^\infty(\Omega_\varepsilon)$, for every $\varepsilon \in (0, \varepsilon_0]$. Now we prove that if $\varepsilon_1, \varepsilon_2 \in (0, \varepsilon_0]$, $\varepsilon_1 < \varepsilon_2$, then (let us note $\Omega_{\varepsilon_2} \subset \Omega_{\varepsilon_1}$)

$$f_{\varepsilon_1}(x) = f_{\varepsilon_2}(x), \quad \forall x \in \Omega_{\varepsilon_2}. \tag{4.116}$$

Indeed, let $\phi \in \mathcal{D}(\Omega)$ be such that $\operatorname{supp} \phi \subset \Omega_{\varepsilon_2}$. By (4.114) we have

$$\int_{\Omega_{\varepsilon_2}} \phi f_{\varepsilon_2} dx = \langle T, \phi \rangle = \int_{\Omega_{\varepsilon_1}} \phi f_{\varepsilon_1} dx = \int_{\Omega_{\varepsilon_2}} \phi f_{\varepsilon_1} dx.$$

Hence

$$\int_{\Omega_{\varepsilon_2}} \phi f_{\varepsilon_2} dx = \int_{\Omega_{\varepsilon_2}} \phi f_{\varepsilon_1} dx, \quad \forall \phi \in \mathcal{D}\left(\Omega_{\varepsilon_2}\right),$$

consequently we obtain (4.116). Now we define f as the function

$$f(x) = f_\varepsilon(x), \quad \text{for } x \in \Omega_\varepsilon, \quad \text{for } \varepsilon \in (0, \varepsilon_0]. \tag{4.117}$$

f is well-defined and belongs to $\mathcal{D}\left(\Omega_{\varepsilon_2}\right)$.

2nd Step Now we prove that f is constant. Let $\phi \in \mathcal{D}(\Omega)$. Let $\overline{\varepsilon} \in (0, \varepsilon_0]$ satisfy $\operatorname{supp}\phi \subset \Omega_{\overline{\varepsilon}}$. By (4.114), (4.115) and (4.117) we have

$$\langle T, \phi \rangle = \int_\Omega f_{\overline{\varepsilon}} \phi dx = \int_\Omega f \phi dx = \langle T_f, \phi \rangle. \tag{4.118}$$

Therefore

$$T = T_f,$$

consequently

$$0 = \partial_j T = \partial_j T_f = T_{\partial_j f}, \quad \text{for } j = 1, \cdots, n.$$

Thus we have

$$\partial_j f = 0, \quad \text{for } j = 1, \cdots, n$$

and since Ω is connected, we have $f = C$, where C is a constant. Therefore, by (4.118) we have

$$\langle T, \phi \rangle = \langle T_f, \phi \rangle = \int_\Omega C \phi dx, \quad \phi \in \mathcal{D}(\Omega).$$

This concludes the proof of the Theorem. ∎

Exercise 4.17 Let Ω be a connected open set of $\mathbb{R}^n$. Let $g_j \in C^0(\Omega)$, for $j = 1, \cdots, n$. Suppose that there exists $T \in \mathcal{D}'(\Omega)$ such that

$$\partial_j T = T_{g_j}, \quad \text{for } j = 1, \cdots, n,$$

then there exists $f \in C^1(\Omega)$ such that

$$\partial_j f = g_j, \quad \text{for } j = 1, \cdots, n. \tag{4.119}$$

[Hint: by using the same notation of the proof of Theorem 4.18 prove that, for any $\phi \in \mathcal{D}(\Omega)$, such that $\operatorname{supp}\phi \subset \Omega_\varepsilon$

$$\langle T, \phi \rangle = \langle T, \mathcal{T}_y \phi \rangle - \int_{\Omega_\varepsilon} \left(\int_0^1 \sum_{j=1}^n g_j(x+ty) y_j dt \right) \phi(x) dx.$$

Then prove that

$$\langle T, \phi \rangle = \int_{\Omega_\varepsilon} f_\varepsilon \phi dx,$$

where

$$f_\varepsilon(x) = \langle T, \mathcal{T}_x \psi \rangle - \int_{\mathbb{R}^n} \psi(y) \left(\int_0^1 \sum_{j=1}^n g_j(x+ty) y_j dt \right) dy, \quad \text{for } x \in \Omega_\varepsilon$$

($\psi \in \mathcal{D}(\mathbb{R}^n)$ satisfy $\operatorname{supp} \psi \subset B_\varepsilon$ and $\int_{\mathbb{R}^n} \psi(x) dx = 1$). Then prove that it can be defined a function $f \in C^1(\Omega)$ likewise (4.117) and such a function f satisfies (4.119).] ♠

4.8 The Fourier Transform

In the present section the functions are complex-valued. Let us denote by $D_j = \frac{1}{i} \partial_j$, for $j = 1, \cdots, n$.

4.8.1 *The Fourier Transform in* $L^1(\mathbb{R}^n)$

Definition 4.20 Let $f \in L^1(\mathbb{R}^n)$. We define its Fourier transform by

$$\widehat{f}(\xi) := \mathcal{F}(f)(\xi) := \int_{\mathbb{R}^n} f(x) e^{-ix \cdot \xi} dx, \quad \forall \xi \in \mathbb{R}^n.$$

The following Proposition holds.

Proposition 4.21 *Let $f \in L^1(\mathbb{R}^n)$. Then we have:*

(a)

$$\| \widehat{f} \|_{L^\infty(\mathbb{R}^n)} \leq \| f \|_{L^1(\mathbb{R}^n)}, \quad \forall f \in L^1(\mathbb{R}^n), \tag{4.120}$$

(b)

$$\widehat{f} \in C^0(\mathbb{R}^n), \tag{4.121}$$

(c)

$$\widehat{f}(\xi) \to 0, \quad as\ |\xi| \to +\infty. \tag{4.122}$$

Proof Property (a) is an immediate consequence of Definition 4.20. Property (b) is an immediate consequence of the Dominated Convergence Theorem 3.28. Concerning Property (c) we proceed as follows.

We have

$$\widehat{f}(\xi) = \int_{\mathbb{R}^n} f(x)e^{-ix\cdot\xi}dx = -\int_{\mathbb{R}^n} f(x)e^{-i\left(x+\frac{\pi\xi}{|\xi|^2}\right)\cdot\xi}dx =$$
$$-\int_{\mathbb{R}^n} f\left(x - \frac{\pi\xi}{|\xi|^2}\right)e^{-ix\cdot\xi}dx.$$

Hence

$$2\widehat{f}(\xi) = \int_{\mathbb{R}^n} f(x)e^{-ix\cdot\xi}dx + \int_{\mathbb{R}^n} f(x)e^{-ix\cdot\xi}dx =$$
$$= \int_{\mathbb{R}^n} \left(f(x) - f\left(x - \frac{\pi\xi}{|\xi|^2}\right)\right)e^{-ix\cdot\xi}dx.$$

Consequently, by Theorem 3.36 we have

$$\left|\widehat{f}(\xi)\right| \le \frac{1}{2}\int_{\mathbb{R}^n} \left|f(x) - f\left(x - \frac{\pi\xi}{|\xi|^2}\right)\right| dx \to 0, \quad \text{as } |\xi| \to +\infty.$$

This concludes the proof of (c). ■

Remark 4.14 Property (4.122) is known as **Riemann–Lebesgue Lemma**. ♦

For any orthogonal matrix $\mathbf{U}$ in $n \times n$, we denote by $\mathbf{R_U}$ the operator

$$\mathbf{R_U} : L^1\left(\mathbb{R}^n\right) \to L^1\left(\mathbb{R}^n\right),$$

such that

$$\left(\mathbf{R_U} f\right)(x) = f\left(\mathbf{U}x\right), \quad \forall f \in L^1\left(\mathbb{R}^n\right).$$

For any $\lambda > 0$, we denote by $\mathbf{M}_\lambda$ the dilation operator. Namely

$$\mathbf{M}_\lambda : L^1\left(\mathbb{R}^n\right) \to L^1\left(\mathbb{R}^n\right),$$

such that

$$\left(\mathbf{M}_\lambda f\right)(x) = f_\lambda(x) := f(\lambda x), \quad \forall f \in L^1\left(\mathbb{R}^n\right).$$

Theorem 4.22 *Let $f, g \in L^1(\mathbb{R}^n)$. Then we have:*

$$\widehat{(\mathcal{T}_h f)}(\xi) = e^{-i\xi\cdot h}\widehat{f}(\xi), \quad \forall h \in \mathbb{R}^n, \ \forall \xi \in \mathbb{R}^n \tag{4.123}$$

$$\widehat{\mathbf{R_U} f} = \mathbf{R_U}\left(\widehat{f}\right), \tag{4.124}$$

$$\widehat{\mathbf{M}_\lambda f} = \lambda^{-n}\mathbf{M}_{\lambda^{-1}}\left(\widehat{f}\right), \quad \forall \lambda > 0, \tag{4.125}$$

$$\widehat{f \star g} = \widehat{f}\,\widehat{g}. \tag{4.126}$$

Proof Properties (4.123), (4.124), and (4.125) follow immediately from elementary changes of variables, and we leave the proof to the reader. Now we prove (4.126). Since $f, g \in L^1(\mathbb{R}^n)$, by the Fubini–Tonelli Theorem 3.30 we have

$$\begin{aligned}
\left(\widehat{f \star g}\right)(\xi) &= \int_{\mathbb{R}^n} dx \left(e^{-ix\cdot\xi}\int_{\mathbb{R}^n} f(x-y)g(y)dy\right) = \\
&= \int_{\mathbb{R}^n} dx \left(\int_{\mathbb{R}^n} f(x-y)e^{-i(x-y)\cdot\xi}g(y)e^{-iy\cdot\xi}dy\right) = \\
&= \int_{\mathbb{R}^n} dy \left(g(y)e^{-iy\cdot\xi}\int_{\mathbb{R}^n} f(x-y)e^{-i(x-y)\cdot\xi}dx\right) = \\
&= \int_{\mathbb{R}^n} dy \left(g(y)e^{-iy\cdot\xi}\int_{\mathbb{R}^n} f(x)e^{-ix\cdot\xi}dx\right) = \\
&= \widehat{f}(\xi)\widehat{g}(\xi).
\end{aligned}$$

■

Exercise 4.18 Prove (4.123), (4.124), and (4.125) of Proposition 4.22. ♠

4.8.2 The Fourier Transform in the Space of Schwartz $\mathcal{S}(\mathbb{R}^n)$

Definition 4.23 We denote by $\mathcal{S}(\mathbb{R}^n)$ the linear space of (complex-valued) functions $f \in C^\infty(\mathbb{R}^n)$ which satisfy

$$p_{\alpha,\beta}(f) := \sup_{x\in\mathbb{R}^n}\left|x^\alpha D^\beta f\right| < +\infty, \quad \forall \alpha, \beta \in \mathbb{N}_0^n. \tag{4.127}$$

The topology on $\mathcal{S}(\mathbb{R}^n)$ is defined by the seminorms $p_{\alpha,\beta}(f)$.

According to Definition 4.23, a sequence $\{f_k\}$ in $\mathcal{S}(\mathbb{R}^n)$ converges to $f \in \mathcal{S}(\mathbb{R}^n)$ if and only if

$$\lim_{k\to\infty} p_{\alpha,\beta}(f_k - f) = 0, \quad \forall \alpha, \beta \in \mathbb{N}_0^n.$$

The space $\mathcal{S}(\mathbb{R}^n)$ is known as **the Schwartz space** in $\mathbb{R}^n$ or, also, the space of rapidly decreasing functions; equipped with the family of seminorms $\{p_{\alpha,\beta}\}$, $\mathcal{S}(\mathbb{R}^n)$ is a Frechét space. We have

$$C_0^\infty(\mathbb{R}^n) \subset \mathcal{S}(\mathbb{R}^n) \subset L^p(\mathbb{R}^n), \quad \forall p \in [1, +\infty].$$

It is simple to prove that $C_0^\infty(\mathbb{R}^n)$ is dense in $\mathcal{S}(\mathbb{R}^n)$ (see Exercise 4.23 below). The function $f(x) = e^{-|x|^2}$ is an example of function that does not belong to $C_0^\infty(\mathbb{R}^n)$, but belongs to $\mathcal{S}(\mathbb{R}^n)$. The function $\left(1 + |x|^2\right)^k$, $k \in \mathbb{R}$ belongs to $C^\infty(\mathbb{R}^n)$, but does not belong to $\mathcal{S}(\mathbb{R}^n)$.

Exercise 4.19

(a) Prove that a function $f \in C^\infty(\mathbb{R}^n)$ belongs to $\mathcal{S}(\mathbb{R}^n)$ if and only if

$$\forall \alpha \in \mathbb{N}_0^n, \ \forall k \in \mathbb{N}_0, \ \exists C_{\alpha,k} \text{ s.t. } \left|D^\alpha f\right| \le C_{\alpha,k}\left(1 + |x|^2\right)^{-k}, \ \forall x \in \mathbb{R}^n, \tag{4.128}$$

(b) If $f, g \in \mathcal{S}(\mathbb{R}^n)$ then $fg \in \mathcal{S}(\mathbb{R}^n)$.

♠

Exercise 4.20 Let

$$Q_k(f) = \sum_{|\alpha|,|\beta|\le k} p_{\alpha,\beta}(f), \quad \forall f \in \mathcal{S}(\mathbb{R}^n), \ \forall k \in \mathbb{N}_0.$$

For any $f, g \in \mathcal{S}(\mathbb{R}^n)$ let

$$\mathrm{d}_{\mathcal{S}}(f, g) = \sum_{k=0}^{\infty} \frac{2^{-k} Q_k(f - g)}{1 + Q_k(f - g)}.$$

Prove that $\mathrm{d}_{\mathcal{S}}$ is a metric that induces the topology defined by the family of seminorms $\{p_{\alpha,\beta}\}$ and that $\mathcal{S}(\mathbb{R}^n)$, equipped with the distance $\mathrm{d}_{\mathcal{S}}$, is a complete space. ♠

Exercise 4.21 Prove what follows: (a) if $\{\phi_j\}$ is a sequence in $\mathcal{S}(\mathbb{R}^n)$ such that

$$\{\phi_j\} \to 0, \quad \text{in } \mathcal{S}(\mathbb{R}^n),$$

then

$$\{\phi_j\} \to 0, \quad \text{in } \mathcal{E}(\mathbb{R}^n);$$

(b) the converse does not hold, more precisely there exists a sequence $\{\phi_j\}$ in $\mathcal{E}(\mathbb{R}^n)$ that converges to 0 in $\mathcal{E}(\mathbb{R}^n)$, but not in $\mathcal{S}(\mathbb{R}^n)$. ♠

Exercise 4.22 Prove what follows: (a) if $\{\phi_j\}$ is a sequence in $\mathcal{D}(\mathbb{R}^n)$ such that

$$\{\phi_j\} \to 0, \quad \text{in } \mathcal{D}(\mathbb{R}^n),$$

then

$$\{\phi_j\} \to 0, \quad \text{in } \mathcal{S}(\mathbb{R}^n);$$

(b) the converse does not hold, more precisely there exists a sequence $\{\phi_j\}$ in $\mathcal{D}(\mathbb{R}^n)$ that converges to 0 in $\mathcal{S}(\mathbb{R}^n)$, but not in $\mathcal{D}(\mathbb{R}^n)$. ♠

Exercise 4.23 Prove that $C_0^\infty(\mathbb{R}^n)$ is dense in $\mathcal{S}(\mathbb{R}^n)$ equipped with the topology induced by the family of seminorms $\{p_{\alpha,\beta}\}$.

[Hint: let $\zeta \in C_0^\infty(\mathbb{R}^n)$ satisfy $\zeta(x) = 1$ in B_1 and $\zeta(x) = 0$ in $\mathbb{R}^n \setminus B_2$. Check that, for every $f \in \mathcal{S}(\mathbb{R}^n)$ we have

$$C_0^\infty(\mathbb{R}^n) \ni \zeta\left(k^{-1}\cdot\right) f \to f, \quad \text{as } k \to \infty, \text{ in } \mathcal{S}(\mathbb{R}^n).]$$

♠

The following Lemma will be useful here and in the sequel.

Lemma 4.24 (Peetre Inequality) *If $s \in \mathbb{R}$ then we have*

$$\left(\frac{1+|x|^2}{1+|y|^2}\right)^s \leq 2^{|s|}\left(1+|x-y|^2\right)^{|s|}, \quad \forall x, y \in \mathbb{R}^n. \tag{4.129}$$

Proof Let us consider the case $s \geq 0$. We have, for any $\xi, \eta \in \mathbb{R}^n$,

$$\begin{aligned}1+|x|^2 &\leq 1+(|x-y|+|y|)^2 \leq \\ &\leq 1+2|x-y|^2+2|\eta|^2 \leq 2\left(1+|x-y|^2\right)\left(1+|y|^2\right).\end{aligned}$$

Hence

$$\left(\frac{1+|x|^2}{1+|y|^2}\right)^s \leq 2^s\left(1+|x-y|^2\right)^s. \tag{4.130}$$

If $s < 0$, we have

$$\begin{aligned}\left(\frac{1+|x|^2}{1+|y|^2}\right)^s = \left(\frac{1+|y|^2}{1+|x|^2}\right)^{-s} &\leq \\ &\leq 2^{-s}\left(1+|x-y|^2\right)^{-s}.\end{aligned} \tag{4.131}$$

By (4.130) and (4.131) we obtain (4.129). ∎

Proposition 4.25 *If $f, g \in \mathcal{S}(\mathbb{R}^n)$ then $f \star g \in \mathcal{S}(\mathbb{R}^n)$.*

Proof Let $k \in \mathbb{N}_0$. By Lemma 4.24 we have

$$\begin{aligned}&\left(1+|x|^2\right)^k \left|D^\alpha (f \star g)(x)\right| \leq \\ &\leq \int_{\mathbb{R}^n} |g(y)| \left(1+|x|^2\right)^k \left|(D^\alpha f)(x-y)\right| dy \leq \\ &\leq \int_{\mathbb{R}^n} 2^k \left(1+|y|^2\right)^k |g(y)| \left(1+|x-y|^2\right)^{-k} \left|(D^\alpha f)(x-y)\right| dy \leq \\ &\leq C_k \int_{\mathbb{R}^n} \left(1+|x-y|^2\right)^{-\frac{n+1}{2}} \left(1+|x-y|^2\right)^{-k+\frac{n+1}{2}} \left|(D^\alpha f)(x-y)\right| dy \leq \\ &\leq C_k C'_{k,\alpha} \int_{\mathbb{R}^n} \left(1+|z|^2\right)^{-\frac{n+1}{2}} dz,\end{aligned}$$

where C_k and $C'_{k,\alpha}$ are constants. Hence for every $k \in \mathbb{N}_0$ and every $\alpha \in \mathbb{N}_0^n$ there exists $C''_{k,\alpha} > 0$ such that

$$\left|D^\alpha (f \star g)(x)\right| \leq C''_{k,\alpha}\left(1+|x|^2\right)^{-k}, \quad \forall x \in \mathbb{R}^n.$$

This concludes the proof. ∎

Theorem 4.26 *Let $f \in \mathcal{S}(\mathbb{R}^n)$. Then we have*

(a)

$$\widehat{D_x^\alpha f}(\xi) = \xi^\alpha \widehat{f}(\xi), \quad \forall f \in \mathcal{S}(\mathbb{R}^n),$$

(b) $\widehat{f} \in C^\infty(\mathbb{R}^n)$ and

$$\widehat{(x^\alpha f)}(\xi) = (-1)^{|\alpha|} D_\xi^\alpha \widehat{f}(\xi), \quad \forall f \in \mathcal{S}(\mathbb{R}^n).$$

(c) $\widehat{f} \in \mathcal{S}(\mathbb{R}^n)$. Moreover, the following map

$$\mathcal{F}: \mathcal{S}(\mathbb{R}^n) \to \mathcal{S}(\mathbb{R}^n), \tag{4.132}$$

such that

$$\mathcal{F}(\phi) = \widehat{\phi}, \quad \textit{for } \phi \in \mathcal{S}(\mathbb{R}^n) \tag{4.133}$$

is continuous.

Proof First we prove (a). For any $j = 1, \cdots, n$ integrating by parts we have

$$\begin{aligned}
\widehat{\partial_{x_j} f}(\xi) &= \int_{\mathbb{R}^n} \partial_{x_j} f(x) e^{-ix\cdot\xi} dx = \\
&= \lim_{R\to+\infty} \int_{B_R} \partial_{x_j} f(x) e^{-ix\cdot\xi} dx = \\
&= \lim_{R\to+\infty} \left(\int_{\partial B_R} f(x) e^{-ix\cdot\xi} \nu \cdot e_j dS + i\xi_j \int_{B_R} f(x) e^{-ix\cdot\xi} dx \right) = \\
&= i\xi_j \int_{\mathbb{R}^n} f(x) e^{-ix\cdot\xi} dx.
\end{aligned} \tag{4.134}$$

Let us note that the last step is a consequence of $f \in \mathcal{S}(\mathbb{R}^n)$, which implies

$$\lim_{R\to+\infty} \int_{\partial B_R} f(x) e^{-ix\cdot\xi} \nu \cdot e_j dS = 0.$$

By (4.134)

$$\widehat{D_{x_j} f}(\xi) = \xi_j \widehat{f}(\xi) \tag{4.135}$$

and iterating (4.135) we get (a).

Concerning (b), by Theorem 3.29 we obtain that $\widehat{f} \in C^\infty(\mathbb{R}^n)$ and for any $j = 1, \cdots, n$,

$$D_\xi \widehat{f}(\xi) = -\widehat{(x_j f)}(\xi) \tag{4.136}$$

and iterating (4.136) we get (b).

Now we prove (c). By (a) and (b) we get

$$(-1)^{|\alpha|} \widehat{D_x^\beta (x^\alpha f)}(\xi) = (-1)^{|\alpha|} \xi^\beta \widehat{(x^\alpha f)}(\xi) = \xi^\beta D_\xi^\alpha \widehat{f}(\xi).$$

Hence

$$\xi^\beta D_\xi^\alpha \widehat{f}(\xi) = (-1)^{|\alpha|} \widehat{D_x^\beta (x^\alpha f)}(\xi).$$

Consequently, since $f \in \mathcal{S}(\mathbb{R}^n)$, we have, for every $\alpha, \beta \in \mathbb{N}_0^n$,

$$\begin{aligned} \left|\xi^\beta D_\xi^\alpha \widehat{f}(\xi)\right| &= \left|\int_{\mathbb{R}^n} (-1)^{|\alpha|} D_x^\beta \left(x^\alpha f\right) e^{-ix\cdot\xi} dx\right| \leq \\ &\leq \int_{\mathbb{R}^n} \left|D_x^\beta \left(x^\alpha f\right)\right| dx, \quad \forall \xi \in \mathbb{R}^n. \end{aligned} \tag{4.137}$$

By the Leibniz formula we get

$$\begin{aligned} \left|D_x^\beta \left(x^\alpha f\right)\right| &= \left|\sum_{\gamma\leq\beta} \binom{\beta}{\gamma} \left(D^\gamma x^\alpha\right) D^{\beta-\gamma} f\right| \leq \\ &\leq \frac{C_{\alpha,\beta}}{\left(1+|x|^2\right)^{\frac{n+1}{2}}} Q_N(f), \end{aligned} \tag{4.138}$$

where $C_{\alpha,\beta}$ is a constant (independent of f),

$$Q_N(f) = \sum_{|\gamma|,|\mu|\leq N} p_{\gamma,\mu}(f)$$

and $N = |\alpha| + |\beta| + n + 1$.

Now, by (4.137) and (4.138) we have

$$\sup_{\xi\in\mathbb{R}^n} \left|\xi^\beta D_\xi^\alpha \widehat{f}(\xi)\right| \leq \widetilde{C}_{\alpha,\beta} Q_N(f) < +\infty, \quad \forall \alpha, \beta \in \mathbb{N}_0^n.$$

This proves that $\widehat{f} \in \mathcal{S}(\mathbb{R}^n)$ and that $\mathcal{F}$ is continuous. ■

Lemma 4.27 *For any $\varepsilon > 0$, let*

$$\psi_\varepsilon(x) = \frac{1}{(2\pi)^n} e^{-\frac{\varepsilon|x|^2}{2}}. \tag{4.139}$$

Then we have

$$\widehat{\psi_\varepsilon}(\xi) = \frac{1}{\left(\sqrt{2\pi\varepsilon}\right)^n} e^{-\frac{|\xi|^2}{2\varepsilon}}. \tag{4.140}$$

Proof First we observe that (4.125) implies

$$\widehat{\psi_\varepsilon}(\xi) = \left(\sqrt{\varepsilon}\right)^{-\frac{n}{2}} \widehat{\psi_1}\left(\frac{\xi}{\sqrt{\varepsilon}}\right). \tag{4.141}$$

Now we have

$$
\begin{aligned}
(2\pi)^n \widehat{\psi_1}(\xi) &= \int_{\mathbb{R}^n} e^{-\frac{|x|^2}{2}} e^{-ix\cdot\xi} dx = \\
&= \int_{\mathbb{R}^n} \prod_{j=1}^{n} e^{-\frac{x_j^2}{2}} e^{-ix_j\xi_j} dx = \prod_{j=1}^{n} \int_{\mathbb{R}} e^{-\frac{x_j^2}{2}} e^{-ix_j\xi_j} dx_j.
\end{aligned}
\tag{4.142}
$$

To calculate the last integral in (4.142) we note that by setting

$$g(t) = e^{-\frac{t^2}{2}}, \quad \text{for } t \in \mathbb{R},$$

the following relation holds

$$g'(t) + tg(t) = 0, \quad \forall t \in \mathbb{R}.$$

From which, taking the Fourier transform of both the sides and using (a) and (b) of Theorem 4.26, we have

$$0 = \widehat{g'}(\tau) + \widehat{tg}(\tau) = i\tau\widehat{g}(\tau) + i\widehat{g}'(\tau), \quad \forall \tau \in \mathbb{R}.$$

In other words, $\widehat{g}$ is a solution to differential equation

$$\widehat{g}' + \tau\widehat{g} = 0, \quad \text{in } \mathbb{R}.$$

Hence

$$\widehat{g}(\tau) = \widehat{g}(0)e^{-\frac{\tau^2}{2}}. \tag{4.143}$$

On the other hand we have [28, Vol. II, Ch. 4]

$$\widehat{g}(0) = \int_{\mathbb{R}} e^{-\frac{t^2}{2}} dt = \sqrt{2\pi}. \tag{4.144}$$

By (4.142)–(4.144) we obtain

$$\widehat{\psi_1}(\xi) = \frac{1}{\left(\sqrt{2\pi}\right)^n} \prod_{j=1}^{n} e^{-\frac{\xi_j^2}{2}} = \frac{1}{\left(\sqrt{2\pi}\right)^n} e^{-\frac{|\xi|^2}{2}} \tag{4.145}$$

which, together with (4.141), gives (4.140). ■

Remark 4.15 Let us note that

$$\int_{\mathbb{R}^n} \widehat{\psi_\varepsilon}(\xi)d\xi = \frac{1}{\left(\sqrt{2\pi}\right)^n}\int_{\mathbb{R}^n} e^{-\frac{|\xi|^2}{2}}d\xi = 1. \tag{4.146}$$

♦

Lemma 4.28 *For any $\phi \in \mathcal{S}(\mathbb{R}^n)$ we have*

$$\lim_{\varepsilon\to 0^+}\int_{\mathbb{R}^n} \widehat{\psi_\varepsilon}(x)\phi(x)dx = \phi(0) \tag{4.147}$$

and

$$\lim_{\varepsilon\to 0^+}\int_{\mathbb{R}^n} \widehat{\phi}(\xi)\psi_\varepsilon(\xi)d\xi = \frac{1}{(2\pi)^n}\int_{\mathbb{R}^n} \widehat{\phi}(\xi)d\xi. \tag{4.148}$$

Proof We prove (4.147). Let $\phi \in \mathcal{S}(\mathbb{R}^n)$. Set

$$M = \sup_{x\in\mathbb{R}^n} |\nabla\phi(x)|.$$

$$|\phi(x) - \phi(0)| \leq M|x|. \tag{4.149}$$

Now by (4.146) we have

$$\left|\int_{\mathbb{R}^n} \widehat{\psi_\varepsilon}(x)\phi(x)dx - \phi(0)\right| = \left|\int_{\mathbb{R}^n} \widehat{\psi_\varepsilon}(x)(\phi(x) - \phi(0))dx\right| =$$

$$= \frac{1}{\left(\sqrt{2\pi}\right)^n}\left|\int_{\mathbb{R}^n} e^{-\frac{|x|^2}{2}}(\phi(\varepsilon x) - \phi(0))dx\right| \leq$$

$$\leq \frac{1}{\left(\sqrt{2\pi}\right)^n}\int_{\mathbb{R}^n} e^{-\frac{|x|^2}{2}}|\phi(\varepsilon x) - \phi(0)|dx \leq \frac{M\varepsilon}{\left(\sqrt{2\pi}\right)^n}\int_{\mathbb{R}^n} e^{-\frac{|x|^2}{2}}|x|dx.$$

Hence

$$\lim_{\varepsilon\to 0^+}\left|\int_{\mathbb{R}^n} \widehat{\psi_\varepsilon}(x)\phi(x)dx - \phi(0)\right| = 0.$$

This concludes the proof of (4.147).

We prove (4.148). We have

$$\left|\int_{\mathbb{R}^n} \widehat{\phi}(\xi)\psi_\varepsilon(\xi)d\xi - \frac{1}{(2\pi)^n}\int_{\mathbb{R}^n} \widehat{\phi}(\xi)d\xi\right| = $$
$$= \frac{1}{(2\pi)^n}\left|\int_{\mathbb{R}^n} \widehat{\phi}(\xi)\left(e^{-\frac{\varepsilon|\xi|^2}{2}} - 1\right)d\xi\right| \leq \tag{4.150}$$
$$\leq \frac{1}{(2\pi)^n}\int_{\mathbb{R}^n} \left|\widehat{\phi}(\xi)\right|\left|1 - e^{-\frac{\varepsilon|\xi|^2}{2}}\right|d\xi.$$

Now, set

$$M_1 = \sup_{\xi\in\mathbb{R}^n}\left|\nabla\left(e^{-\frac{|\xi|^2}{2}}\right)\right| < +\infty.$$

We have

$$\left|1 - e^{-\frac{\varepsilon|\xi|^2}{2}}\right| \leq \sqrt{\varepsilon}M_1|\xi|, \quad \forall\xi \in \mathbb{R}^n.$$

By this inequality and by (4.150) we get

$$\left|\int_{\mathbb{R}^n} \widehat{\phi}(\xi)\psi_\varepsilon(\xi)d\xi - \frac{1}{(2\pi)^n}\int_{\mathbb{R}^n} \widehat{\phi}(\xi)d\xi\right| \leq \frac{\sqrt{\varepsilon}M_1}{(2\pi)^n}\int_{\mathbb{R}^n} |\xi|\left|\widehat{\phi}(\xi)\right|d\xi. \tag{4.151}$$

Since $\widehat{\phi} \in \mathcal{S}(\mathbb{R}^n)$ the last integral in (4.151) is finite, therefore (4.148) follows. ■

Proposition 4.29 *If $f, g \in \mathcal{S}(\mathbb{R}^n)$ then we have*

$$\int_{\mathbb{R}^n} f(x)\widehat{g}(x)dx = \int_{\mathbb{R}^n} \widehat{f}(x)g(x)dx, \tag{4.152}$$

Proof Set

$$F(x,\xi) = f(x)g(\xi)e^{-ix\cdot\xi}.$$

By the Fubini–Tonelli Theorem we have

$$\begin{aligned}\int_{\mathbb{R}^n} f(x)\widehat{g}(x)dx &= \int_{\mathbb{R}^n} f(x)dx \int_{\mathbb{R}^n} g(\xi)e^{-ix\cdot\xi}d\xi = \\ &= \int_{\mathbb{R}^n} dx \int_{\mathbb{R}^n} F(x,\xi)d\xi = \\ &= \int_{\mathbb{R}^n} d\xi \int_{\mathbb{R}^n} F(x,\xi)dx = \\ &= \int_{\mathbb{R}^n} g(\xi)d\xi \int_{\mathbb{R}^n} f(x)e^{-ix\cdot\xi}dx = \\ &= \int_{\mathbb{R}^n} \widehat{f}(\xi)g(\xi)d\xi.\end{aligned}$$

■

Theorem 4.30 *For any $f \in \mathcal{S}(\mathbb{R}^n)$ we have*

$$f(x) = \frac{1}{(2\pi)^n}\int_{\mathbb{R}^n} \widehat{f}(\xi)e^{ix\cdot\xi}d\xi, \quad \forall x \in \mathbb{R}^n. \tag{4.153}$$

Proof Let $f \in \mathcal{S}(\mathbb{R}^n)$. First we prove

$$f(0) = \frac{1}{(2\pi)^n}\int_{\mathbb{R}^n} \widehat{f}(\xi)d\xi, \quad \forall x \in \mathbb{R}. \tag{4.154}$$

By Proposition 4.29 we have

$$\int_{\mathbb{R}^n} f(x)\widehat{\psi_\varepsilon}(x)dx = \int_{\mathbb{R}^n} \widehat{f}(\xi)\psi_\varepsilon(\xi)d\xi, \tag{4.155}$$

where ψ_ε is defined by (4.139).

By (4.147) and (4.148) we have respectively

$$\lim_{\varepsilon\to 0^+}\int_{\mathbb{R}^n} f(x)\widehat{\psi_\varepsilon}(x)dx = f(0)$$

and

$$\lim_{\varepsilon\to 0^+}\int_{\mathbb{R}^n} \widehat{f}(\xi)\psi_\varepsilon(\xi)d\xi = \frac{1}{(2\pi)^n}\int_{\mathbb{R}^n} \widehat{f}(\xi)d\xi.$$

Hence, passing to the limit as ε goes to 0^+, (4.155) gives (4.154).

Now we prove (4.153). Let $x \in \mathbb{R}^n$. Set

$$\varphi(t) = f(t+x), \quad \text{for } t \in \mathbb{R}^n.$$

By (4.123) we have

$$\widehat{\varphi}(\xi) = e^{ix\cdot\xi}\widehat{f}(\xi).$$

Hence by (4.154) we get

$$f(x) = \varphi(0) = \frac{1}{(2\pi)^n}\int_{\mathbb{R}^n}\widehat{\varphi}(\xi)d\xi = \frac{1}{(2\pi)^n}\int_{\mathbb{R}^n}\widehat{f}(\xi)e^{ix\cdot\xi}d\xi.$$

This concludes the proof of (4.153). ■

Remark 4.16 By Theorem 4.30 we have immediately that the map

$$\mathcal{F}: \mathcal{S}(\mathbb{R}^n) \to \mathcal{S}(\mathbb{R}^n),$$

defined by (4.132), is injective and surjective. The inverse of $\mathcal{F}$ is given by

$$\mathcal{F}^{-1} = \frac{1}{(2\pi)^n}\mathcal{R}\mathcal{F}, \tag{4.156}$$

where $\mathcal{R}$ is the reflection operator defined by (4.32)–(4.33). In other words we have

$$\mathcal{F}^2 = (2\pi)^n\mathcal{R}. \tag{4.157}$$

or, equivalently,

$$\widehat{\widehat{f}}(x) = (2\pi)^n\check{f}(x), \quad \forall f \in \mathcal{S}(\mathbb{R}^n). \tag{4.158}$$

We had already seen that $\mathcal{F}$ is continuous. As a simple consequence of (4.156) and a trivial modification of the proof of 4.26, we have that map $\mathcal{F}^{-1}$ is also continuous.

Formula (4.153) is known as the **inversion formula for the Fourier transform**. It can be proven that (see, for example, [114, Ch. 7] for a proof) if $f \in L^1(\mathbb{R}^n)$ and $\widehat{f} \in L^1(\mathbb{R}^n)$, we have

$$f(x) = \frac{1}{(2\pi)^n}\int_{\mathbb{R}^n}\widehat{f}(\xi)e^{ix\cdot\xi}d\xi, \quad \text{a.e. in } \mathbb{R}^n. \tag{4.159}$$

♦

Theorem 4.31 *Let $f, g \in \mathcal{S}(\mathbb{R}^n)$. We have*

$$\widehat{(fg)}(\xi) = \frac{1}{(2\pi)^n}(\widehat{f}\star\widehat{g})(\xi), \tag{4.160}$$

$$\|f\|_{L^2(\mathbb{R}^n)} = \frac{1}{(2\pi)^{n/2}}\|\widehat{f}\|_{L^2(\mathbb{R}^n)} \tag{4.161}$$

and

$$\int_{\mathbb{R}^n} f(x)\overline{g(x)}dx = \frac{1}{(2\pi)^n}\int_{\mathbb{R}^n} \widehat{f}(\xi)\overline{\widehat{g}(\xi)}d\xi. \tag{4.162}$$

Proof First we prove (4.160). By (4.126) and (4.157) we have

$$\mathcal{F}\left(\frac{1}{(2\pi)^n}\mathcal{F}(f)\star\mathcal{F}(g)\right) = (2\pi)^n\mathcal{F}^2(f)\mathcal{F}^2(g) = (2\pi)^n\check{f}\check{g}.$$

Hence

$$\frac{1}{(2\pi)^n}\mathcal{F}(f)\star\mathcal{F}(g) = \mathcal{R}\mathcal{F}\left(\check{f}\check{g}\right) = \mathcal{F}(fg),$$

that is (4.160).

Now we prove (4.161). For any $f \in \mathcal{S}(\mathbb{R}^n)$, let

$$\widetilde{f}(x) = \overline{f(-x)}.$$

We have

$$\begin{aligned}\mathcal{F}\left(\widetilde{f}\right)(\xi) &= \int_{\mathbb{R}^n}\overline{f(-x)}e^{-ix\xi}dx = \\ &= \overline{\int_{\mathbb{R}^n} f(-x)e^{ix\xi}dx} = \overline{\int_{\mathbb{R}^n} f(x)e^{-ix\xi}dx} = \overline{\widehat{f}(\xi)}.\end{aligned} \tag{4.163}$$

Let

$$h = f\star\widetilde{f}.$$

By (4.126) we have

$$\widehat{h}(\xi) = \left|\widehat{f}(\xi)\right|^2. \tag{4.164}$$

Also let us note

$$\begin{aligned}h(0) &= \left(\int_{\mathbb{R}^n} f(x-y)\overline{f(-y)}dy\right)_{|x=0} = \\ &= \int_{\mathbb{R}^n}|f(-y)|^2dy = \int_{\mathbb{R}^n}|f(x)|^2dx.\end{aligned} \tag{4.165}$$

Now, by Theorem 4.30 and by (4.163)–(4.165) we have

$$\int_{\mathbb{R}^n} |f(x)|^2 dx = h(0) = \frac{1}{(2\pi)^n} \int_{\mathbb{R}^n} \widehat{h}(\xi) d\xi = \frac{1}{(2\pi)^n} \int_{\mathbb{R}^n} \left|\widehat{f}(\xi)\right|^2 d\xi. \tag{4.166}$$

This gives (4.161).

Concerning (4.162), it is an immediate consequence of (4.166) and of the polarization formula

$$\int_{\mathbb{R}^n} \phi(x)\overline{\psi(x)} dx = \frac{1}{4} \sum_{k=0}^{3} i^k \left\|\phi + i^k \psi\right\|^2_{L^2(\mathbb{R})} \quad \forall \phi, \psi \in \mathcal{S}(\mathbb{R}^n).$$

■

Formulas (4.161) and (4.162) (they are equivalent) are known as the **Parseval formulas**.

Let us notice that the operator $\mathcal{F}$ over $\mathcal{S}(\mathbb{R}^n)$ enjoys the following property

$$\|\mathcal{F}(f)\|_{L^2(\mathbb{R}^n)} = (2\pi)^{n/2} \|f\|_{L^2(\mathbb{R}^n)}, \quad \forall f \in \mathcal{S}(\mathbb{R}^n). \tag{4.167}$$

Let us observe that, since $C_0^\infty(\mathbb{R}^n)$ is dense in $L^2(\mathbb{R}^n)$ and

$$C_0^\infty\left(\mathbb{R}^n\right) \subset \mathcal{S}(\mathbb{R}^n) \subset L^2\left(\mathbb{R}^n\right),$$

then $\mathcal{S}(\mathbb{R}^n)$ is dense in $L^2(\mathbb{R}^n)$. Hence (4.167) implies that the linear operator $\mathcal{F}$ can be extended to a bounded linear operator from $L^2(\mathbb{R}^n)$ to $L^2(\mathbb{R}^n)$. We continue to denote by $\mathcal{F}$ such an extension. Hence it is defined as

$$\widehat{f} := \mathcal{F}(f), \quad \forall f \in L^2\left(\mathbb{R}^n\right).$$

It can be proved that the operator

$$\mathcal{F} : L^2(\mathbb{R}^n) \to L^2(\mathbb{R}^n)$$

is bijective and Theorem 4.31 continue to holds. Moreover,

$$f = \frac{1}{(2\pi)^n} \mathcal{R}\mathcal{F}\left(\widehat{f}\right), \quad \forall f \in L^2\left(\mathbb{R}^n\right). \tag{4.168}$$

If $f \in \mathcal{S}(\mathbb{R}^n)$ or $f \in L^2(\mathbb{R}^n) \cap L^1(\mathbb{R}^n)$ then formula (4.168) is nothing but inversion formula (4.153) or (4.159) respectively. For the proofs and much more details we refer the reader to [38], [58, Vol. I], [114]. Here we limit ourselves to proving the following

Theorem 4.32 *For any $f \in L^2(\mathbb{R}^n)$ we have*

$$\lim_{R\to+\infty}\left\|\widehat{f}-\int_{B_R} f(x)e^{-ix\cdot\xi}dx\right\|_{L^2(\mathbb{R}^n)}=0. \tag{4.169}$$

Proof Let $f \in L^2(\mathbb{R}^n)$. Let us note that

$$\int_{B_R} f(x)e^{-ix\cdot\xi}dx=\widehat{\left(\chi_{B_R}f\right)}(\xi).$$

Therefore (4.169) is equivalent to

$$\lim_{R\to+\infty}\left\|\widehat{f}-\widehat{\left(\chi_{B_R}f\right)}\right\|_{L^2(\mathbb{R}^n)}=0. \tag{4.170}$$

Now, by the Parseval formula and the inversion formula we have

$$\begin{aligned}\left\|\widehat{f}-\widehat{\left(\chi_{B_R}f\right)}\right\|^2_{L^2(\mathbb{R}^n)}&=\frac{1}{(2\pi)^n}\left\|\mathcal{RF}\left(f-\chi_{B_R}f\right)\right\|^2_{L^2(\mathbb{R}^n)}=\\&=(2\pi)^n\left\|f-\chi_{B_R}f\right\|^2_{L^2(\mathbb{R}^n)}.\end{aligned} \tag{4.171}$$

On the other hand,

$$\lim_{R\to+\infty}\left\|f-\chi_{B_R}f\right\|^2_{L^2(\mathbb{R}^n)}=\lim_{R\to+\infty}\int_{\{|x|\geq R\}}|f(x)|^2dx=0. \tag{4.172}$$

Therefore, by (4.171) and (4.172) we get (4.170) and, consequently, (4.169). ■

In the sequel we will use the following (for the definition and first properties of holomorphic function we refer to [24])

Proposition 4.33 *Let $f \in L^1(\mathbb{R}^n)$ be such that* $\operatorname{supp} f$ *is compact. Then the function*

$$\mathbb{C}^n \ni \zeta \to \int_{\mathbb{R}^n} f(x)e^{-ix\cdot\zeta}dx \in \mathbb{C},\quad \left(x\cdot\zeta=\sum_{j=1}^n x_j\zeta_j\right), \tag{4.173}$$

is a holomorphic function.

Proof Let us denote by

$$F(\zeta)=\int_{\mathbb{R}^n} f(x)e^{-ix\cdot\zeta}dx,\quad \text{for } \zeta\in\mathbb{C}^n.$$

Let $\zeta = \xi + i\eta$, $\xi, \eta \in \mathbb{R}^n$. By Theorem 3.29 we have, for $j = 1, \cdots, n$

$$\frac{\partial F(\xi + i\eta)}{\partial \xi_j} = -i \int_{\mathbb{R}^n} x_j f(x) e^{-ix\cdot\zeta} dx$$

and

$$\frac{\partial F(\xi + i\eta)}{\partial \eta_j} = \int_{\mathbb{R}^n} x_j f(x) e^{-ix\cdot\zeta} dx.$$

Hence, $F \in C^1(\mathbb{R}^n)$ and the Cauchy–Riemann equations

$$\frac{\partial F(\xi + i\eta)}{\partial \xi_j} = \frac{1}{i} \frac{\partial F(\xi + i\eta)}{\partial \eta_j}, \quad j = 1, \cdots, n$$

are satisfied. Consequently F is a holomorphic function. ■

Remark 4.17 Proposition 4.173 immediately implies that if $f \in L^1(\mathbb{R}^n)$ is such that $\operatorname{supp} f$ is compact, then the Fourier transform of f is a real analytic function (see Sect. 8.3). As a matter of fact, we have

$$\widehat{f}(\xi) = F(\xi + i\eta)_{|\eta=0}, \quad \forall \xi \in \mathbb{R}^n.$$

♦

4.9 Tempered Distributions

Let us recall that the space $\mathcal{S}(\mathbb{R}^n)$ was defined in Sect. 4.8.2.

The Space $\mathcal{S}'(\mathbb{R}^n)$
A linear map

$$T : \mathcal{S}(\mathbb{R}^n) \to \mathbb{R} \text{ (or } \mathbb{C}),$$

is called a **tempered distribution** if, whenever

$$\{\phi_j\} \to 0, \quad \text{in } \mathcal{S}(\mathbb{R}^n),$$

it follows

$$\{T(\phi_j)\} \to 0, \quad \text{in } \mathbb{R} \text{ (or in } \mathbb{C}).$$

We will denote the space of tempered distributions by $\mathcal{S}'(\mathbb{R}^n)$. In what follows, we will only consider real-valued distributions, since everything can be immediately extended to the case of complex-valued distributions. Instead of $T(\phi)$, we will often write

$$\langle T, \phi\rangle_{\mathcal{S}'(\mathbb{R}^n),\mathcal{S}(\mathbb{R}^n)}$$

or simply

$$\langle T, \phi\rangle ,$$

if there is no ambiguity.

In analogy with Propositions 4.1 and 4.8, it can be proven

Proposition 4.34 *Let*

$$T : \mathcal{S}(\mathbb{R}^n) \to \mathbb{R}$$

be a linear map.

The following sentence are equivalent:

(a) $T \in \mathcal{S}'(\mathbb{R}^n)$,
(b) there exist $N \in \mathbb{N}_0$ *and* $C \geq 0$ *such that*

$$\left|\langle T, \phi\rangle_{\mathcal{S}'(\mathbb{R}^n),\mathcal{S}(\mathbb{R}^n)}\right| \leq C \sum_{|\alpha|,|\beta|\leq N} p_{\alpha,\beta}(\phi), \quad \forall\phi \in \mathcal{S}(\mathbb{R}^n). \tag{4.174}$$

We leave the proof as an exercise for the reader.

Example 4.9 Let $p \in [1, +\infty]$ and $f \in L^p(\mathbb{R}^n)$. Then the linear map

$$T_f : \mathcal{S}(\mathbb{R}^n) \to \mathbb{R},$$

such that

$$\langle T_f, \phi\rangle_{\mathcal{S}'(\mathbb{R}^n),\mathcal{S}(\mathbb{R}^n)} = \int_{\mathbb{R}^n} f(x)\phi(x)dx, \quad \forall\phi \in \mathcal{S}(\mathbb{R}^n) \tag{4.175}$$

is a tempered distribution. To check what claimed above, let us consider first the case $p = 1$. By (4.175) we have

$$\left|\langle T_f, \phi\rangle_{\mathcal{S}'(\mathbb{R}^n),\mathcal{S}(\mathbb{R}^n)}\right| \leq \int_{\mathbb{R}^n} |f(x)||\phi(x)|dx \leq \|f\|_{L^1(\mathbb{R}^n)}\, p_{0,0}(\phi), \quad \forall\phi \in \mathcal{S}(\mathbb{R}^n).$$

Hence, by Proposition 4.34 we get $T_f \in \mathcal{S}'(\mathbb{R}^n)$.

Now, let $p \in (1, +\infty]$. Let $k \in \mathbb{N}$ be such that

$$kp' > n, \tag{4.176}$$

where $p' = \frac{p}{p-1}$. We have there exists C_k such that

$$|\phi(x)| \leq C_k (1 + |x|)^{-k} q_k(\phi), \quad \forall \phi \in \mathcal{S}(\mathbb{R}^n), \tag{4.177}$$

where

$$q_k(\phi) = \sum_{|\alpha| \leq k} p_{\alpha,0}(\phi), \quad \forall \phi \in \mathcal{S}(\mathbb{R}^n).$$

Now by the Hölder inequality and by (4.177) we have

$$\left|\langle T_f, \phi \rangle_{\mathcal{S}'(\mathbb{R}^n),\mathcal{S}(\mathbb{R}^n)}\right| \leq \int_{\mathbb{R}^n} |f(x)||\phi(x)|dx \leq$$

$$\leq C_k q_k(\phi) \int_{\mathbb{R}^n} (1 + |x|)^{-k} |f(x)|dx \leq$$

$$\leq C_k \left(\int_{\mathbb{R}^n} (1 + |x|)^{-kp'} dx \right)^{1/p'} \|f\|_{L^p(\mathbb{R}^n)} q_k(\phi), \quad \forall \phi \in \mathcal{S}(\mathbb{R}^n)$$

and, by (4.176) and Proposition 4.34 we get $T_f \in \mathcal{S}'(\mathbb{R}^n)$. ♣

Remark 4.18 We have

$$\mathcal{E}'(\mathbb{R}^n) \subsetneq \mathcal{S}'(\mathbb{R}^n) \subsetneq \mathcal{D}'(\mathbb{R}^n). \tag{4.178}$$

$\mathcal{E}'(\mathbb{R}^n) \subsetneq \mathcal{S}'(\mathbb{R}^n)$ Actually the inclusion is an immediate consequence of the fact that if $\{\phi_j\}$ is a sequence in $\mathcal{S}(\mathbb{R}^n)$ such that $\{\phi_j\} \to 0$ in $\mathcal{S}(\mathbb{R}^n)$, then $\{\phi_j\} \to 0$, in $\mathcal{E}(\mathbb{R}^n)$. The inclusion is strict: as a matter of fact for any $f \in L^p(\mathbb{R}^n)$, where $p \in [1, +\infty]$, we have $T_f \in \mathcal{S}'(\mathbb{R}^n)$, whereas $T_f \notin \mathcal{E}'(\mathbb{R}^n)$ unless $\operatorname{supp} f$ is compact.

$\mathcal{S}'(\mathbb{R}^n) \subsetneq \mathcal{D}'(\mathbb{R}^n)$ Likewise before, the inclusion is an immediate consequence of the fact that if $\{\phi_j\}$ is a sequence in $\mathcal{D}(\mathbb{R}^n)$ such that $\{\phi_j\} \to 0$ in $\mathcal{S}(\mathbb{R}^n)$, then $\{\phi_j\} \to 0$, in $\mathcal{S}(\mathbb{R}^n)$. We now prove that the inclusion is strict (we limit ourselves to the case $n = 1$) by showing that the distribution $T = T_{e^{x^2}}$ that belongs to $\mathcal{D}'(\mathbb{R})$ does not belong to $\mathcal{S}'(\mathbb{R})$. We argue by contradiction and we assume that $T \in \mathcal{S}'(\mathbb{R})$. Let $\psi \in \mathcal{D}(\mathbb{R})$ be such that, $\psi \geq 0$ in $\mathbb{R}$, $\psi(x) = 1$, for $|x| \leq 1$, and $\psi(x) = 0$, for $|x| \geq 2$. For any $\lambda \geq 1$, let us define

$$\phi_\lambda(x) = e^{-x^2} \psi\left(\frac{x}{\lambda}\right), \quad \text{for } x \in \mathbb{R}.$$

We have $\phi_\lambda \in \mathcal{S}(\mathbb{R})$. Moreover, by

$$\left|\frac{d^m}{dx^m}\psi\left(\frac{x}{\lambda}\right)\right| \leq \left\|\frac{d^m}{dx^m}\psi\right\|_{L^\infty(\mathbb{R})},$$

for every $x \in \mathbb{R}$, $\lambda \geq 1$, $m \in \mathbb{N}_0$ and by the Leibniz formula, we have

$$\left|x_1^m \frac{d^{m_2}}{dx^{m_2}}\phi_\lambda(x)\right| \leq C\left(1+|x|\right)^M e^{-x^2}, \quad \forall x \in \mathbb{R},$$

where C and M are positive numbers which depend on m_1 m_2, but they are independent of λ. Therefore

$$p_{m_1,m_2}(\phi_\lambda) \leq C'_{m_1,m_2} < +\infty,$$

for every $m_1, m_2 \in \mathbb{N}_0$, where C'_{m_1,m_2} is a constant independent of λ. Consequently, by (4.174), there exists C'', independent of λ such that

$$\left|\langle T, \phi_\lambda\rangle_{\mathcal{S}'(\mathbb{R}),\mathcal{S}(\mathbb{R})}\right| \leq C'', \quad \forall\lambda \geq 1. \tag{4.179}$$

On the other hand we have

$$\left|\langle T, \phi_\lambda\rangle_{\mathcal{S}'(\mathbb{R}),\mathcal{S}(\mathbb{R})}\right| = \left|\int_{\mathbb{R}} e^{x^2}\phi_\lambda(x)dx\right| = \int_{\mathbb{R}} \psi\left(\frac{x}{\lambda}\right)dx \geq \int_{-\lambda}^{\lambda} dx = 2\lambda.$$

Therefore

$$\lim_{\lambda\to+\infty}\left|\langle T, \phi_\lambda\rangle_{\mathcal{S}'(\mathbb{R}),\mathcal{S}(\mathbb{R})}\right| = +\infty$$

which contradicts (4.179). Hence $T \notin \mathcal{S}'(\mathbb{R})$. ♦

Exercise 4.24 Prove Proposition 4.34. ♠

Exercise 4.25 Let $s \in \mathbb{R}$. Let $f \in C^0(\mathbb{R}^n)$. Suppose that there exists a constant C such that

$$|f(x)| \leq C(1+|x|)^s, \quad \forall s \in \mathbb{R},$$

then the map

$$\mathcal{S}(\mathbb{R}^n) \ni \phi \to \langle T_f, \phi\rangle_{\mathcal{S}'(\mathbb{R}),\mathcal{S}(\mathbb{R})} = \int_{\mathbb{R}^n} f(x)\phi(x)dx \in \mathbb{R}$$

is a tempered distribution. ♠

4.9.1 Some Operations with Tempered Distributions

In the present subsection, in a similar way to what was done for distributions in $\mathcal{D}'(\Omega)$, we define some operations for tempered distributions. Since the procedure we will follow is entirely analogous to the one in Sect. 4.3, we will limit ourselves here to outlining the main steps of the process.

1. Let $\{T_j\}$ be a sequence in $\mathcal{S}'(\mathbb{R}^n)$. We say that $\{T_j\}$ **converges to** $T \in \mathcal{S}'(\mathbb{R}^n)$ in $\mathcal{S}'(\mathbb{R}^n)$ whenever

$$\lim_{j\to\infty} \langle T_j, \phi \rangle = \langle T, \phi \rangle\,, \quad \forall \phi \in \mathcal{S}(\mathbb{R}^n). \tag{4.180}$$

2. Let

$$\mathcal{L} : \mathcal{S}(\mathbb{R}^n) \to \mathcal{S}(\mathbb{R}^n) \tag{4.181}$$

be a linear map. We say that $\mathcal{L}$ is (sequentially) continuous provided, for every sequence $\{\phi_j\}$ in $\mathcal{S}(\mathbb{R}^n)$ such that

$$\{\phi_j\} \to 0, \quad \text{in } \mathcal{S}(\mathbb{R}^n),$$

we have

$$\{\mathcal{L}(\phi_j)\} \to 0, \quad \text{in } \mathcal{S}(\mathbb{R}^n).$$

3. Let

$$\Lambda : \mathcal{S}'(\mathbb{R}^n) \to \mathcal{S}'(\mathbb{R}^n), \tag{4.182}$$

be a linear map. We say that Λ is (sequentially) continuous whenever the following fact holds: for every sequence $\{T_j\}$ in $\mathcal{S}'(\mathbb{R}^n)$ such that

$$\{T_j\} \to 0, \quad \text{in } \mathcal{S}'(\mathbb{R}^n),$$

we have

$$\{\Lambda(T_j)\} \to 0, \quad \text{in } \mathcal{S}'(\mathbb{R}^n).$$

4. Let $\mathcal{L}$ be a linear operator like (4.181), we say that the linear map

$$\mathcal{L}^t : \mathcal{S}(\mathbb{R}^n) \to \mathcal{S}(\mathbb{R}^n), \tag{4.183}$$

is the transpose of $\mathcal{L}$ if we have

$$\int_{\mathbb{R}^n} \mathcal{L}(\phi)(y)\psi(y)dy = \int_{\mathbb{R}^n} \phi(x)\mathcal{L}^t(\psi)(x)dx, \quad \forall\phi, \psi \in \mathcal{S}(\mathbb{R}^n). \tag{4.184}$$

The proof of the following Theorem is analogous to the proof of the Lax extension Theorem 4.4 and we leave it to the reader.

Theorem 4.35 *Let $\mathcal{L}$ be a linear operator as* (4.181) *and let $\mathcal{L}^t$ be its transpose. Assume that $\mathcal{L}$ and $\mathcal{L}^t$ are continuous. Then the operator*

$$\Lambda_{\mathcal{L}} : \mathcal{S}'(\mathbb{R}^n) \to \mathcal{S}'(\mathbb{R}^n), \tag{4.185}$$

such that for every $T \in \mathcal{S}'(\mathbb{R}^n)$ and for every $\psi \in \mathcal{S}'(\mathbb{R}^n)$

$$\langle \Lambda_{\mathcal{L}}(T), \psi \rangle_{\mathcal{S}'(\mathbb{R}^n),\mathcal{S}(\mathbb{R}^n)} = \left\langle T, \mathcal{L}^t(\psi) \right\rangle_{\mathcal{S}'(\mathbb{R}^n),\mathcal{S}(\mathbb{R}^n)} \tag{4.186}$$

is a well-defined linear and continuous operator and it is an extension of $\mathcal{L}$. That is

$$\Lambda_{\mathcal{L}}(T_\phi) = \mathcal{L}(\phi), \quad \forall\phi \in \mathcal{S}(\mathbb{R}^n). \tag{4.187}$$

Moreover, $\Lambda_{\mathcal{L}}$ is the unique linear continuous operator from $\mathcal{S}(\mathbb{R}^n)$ to $\mathcal{S}(\mathbb{R}^n)$ which satisfies (4.187).

By applying Theorem 4.35, we can define in $\mathcal{S}'(\mathbb{R}^n)$ the operations D^α, $\mathcal{T}_h$, for $h \in \mathbb{R}^n$ and $\mathcal{R}$ as already done in Sect. 4.3. More precisely, if $T \in \mathcal{S}'(\mathbb{R}^n)$ we define the tempered distributions $D^\alpha T, \mathcal{T}_h T, \mathcal{R}T$ as follows:

$$\left\langle D^\alpha T, \phi \right\rangle_{\mathcal{S}'(\mathbb{R}^n),\mathcal{S}(\mathbb{R}^n)} = (-1)^{|\alpha|} \left\langle T, D^\alpha \phi \right\rangle_{\mathcal{S}'(\mathbb{R}^n),\mathcal{S}(\mathbb{R}^n)}, \quad \forall\phi \in \mathcal{S}(\mathbb{R}^n), \tag{4.188}$$

$$\langle \mathcal{T}_h T, \phi \rangle_{\mathcal{S}'(\mathbb{R}^n),\mathcal{S}(\mathbb{R}^n)} = \langle T, \mathcal{T}_{-h}\phi \rangle_{\mathcal{S}'(\mathbb{R}^n),\mathcal{S}(\mathbb{R}^n)}, \quad \forall\phi \in \mathcal{S}(\mathbb{R}^n) \tag{4.189}$$

and

$$\langle \mathcal{R}T, \phi \rangle_{\mathcal{S}'(\mathbb{R}^n),\mathcal{S}(\mathbb{R}^n)} = \langle T, \mathcal{R}\phi \rangle_{\mathcal{S}'(\mathbb{R}^n),\mathcal{S}(\mathbb{R}^n)}, \quad \forall\phi \in \mathcal{S}(\mathbb{R}^n). \tag{4.190}$$

Similarly we can do for the convolution of $T \in \mathcal{S}'(\mathbb{R}^n)$ and a function $\eta \in \mathcal{S}(\mathbb{R}^n)$. Namely we define

$$\langle \eta \star T, \phi \rangle = \left\langle T, \check{\eta} \star \phi \right\rangle, \quad \forall\phi \in \mathcal{S}(\mathbb{R}^n). \tag{4.191}$$

Concerning the multiplication by a smooth function, first we introduce the class $O_M(\mathbb{R}^n)$ of the functions of moderate growth. We say that $f \in C^\infty(\mathbb{R}^n)$ is of moderate growth if the following holds

$$\forall \alpha \in \mathbb{N}_0^n,\ \exists k \in \mathbb{N},\ \exists C > 0 \text{ s.t. } \left|D^\alpha f(x)\right| \leq C(1+|x|)^k,\ \forall x \in \mathbb{R}^n.$$

It is easy to check that if $a \in O_M(\mathbb{R}^n)$ and $\phi \in \mathcal{S}(\mathbb{R}^n)$, then the map

$$\mathcal{S}(\mathbb{R}^n) \ni \phi \to a\phi \in \mathcal{S}(\mathbb{R}^n),$$

is continuous. Thus, arguing as before we define, if $T \in \mathcal{S}'(\mathbb{R}^n)$ and $a \in O_M(\mathbb{R}^n)$, the tempered distribution aT can be defined as

$$\langle aT, \phi\rangle_{\mathcal{S}'(\mathbb{R}^n),\mathcal{S}(\mathbb{R}^n)} = \langle T, a\phi\rangle_{\mathcal{S}'(\mathbb{R}^n),\mathcal{S}(\mathbb{R}^n)}, \quad \forall \phi \in \mathcal{S}(\mathbb{R}^n). \tag{4.192}$$

4.9.2 Fourier Transform of Tempered Distributions

In (c) of Theorem 4.26 we have proved that the map

$$\mathcal{F}: \mathcal{S}(\mathbb{R}^n) \to \mathcal{S}(\mathbb{R}^n),$$

such that

$$\mathcal{F}(\phi) = \widehat{\phi}, \quad \text{for } \phi \in \mathcal{S}(\mathbb{R}^n)$$

is continuous. Moreover, Proposition 4.29 implies that $\mathcal{F}^t = \mathcal{F}$. Therefore, by virtue of Theorem 4.35, we are able to define the Fourier transform $\widehat{T}$ of $T \in \mathcal{S}'(\mathbb{R}^n)$ as follows

$$\left\langle \widehat{T}, \phi\right\rangle_{\mathcal{S}'(\mathbb{R}^n),\mathcal{S}(\mathbb{R}^n)} = \left\langle T, \widehat{\phi}\right\rangle_{\mathcal{S}'(\mathbb{R}^n),\mathcal{S}(\mathbb{R}^n)}, \quad \forall \phi \in \mathcal{S}(\mathbb{R}^n). \tag{4.193}$$

Example 4.10 If $f \in L^1(\mathbb{R}^n)$ then by Proposition 4.29 we have

$$\widehat{T_f} = T_{\widehat{f}}.$$

In other words, the Fourier transform in the sense of distributions of $f \in L^1(\mathbb{R}^n)$ is the same as the one provided by Definition 4.20. ♣

Example 4.11 We have

$$\widehat{\delta_0} = 1. \tag{4.194}$$

As a matter of fact we have

$$\langle\widehat{\delta_0}, \phi\rangle = \langle\delta_0, \widehat{\phi}\rangle = \widehat{\phi}(0) = \int_{\mathbb{R}^n} \phi(x)dx = \langle 1, \phi\rangle, \ \forall\phi \in \mathcal{S}(\mathbb{R}^n). \tag{4.195}$$

♣

Now we prove the Fourier inversion Theorem for tempered distributions.

Theorem 4.36 *If* $T \in \mathcal{S}'(\mathbb{R}^n)$ *then*

$$\widehat{\widehat{T}} = (2\pi)^n \mathcal{R}T. \tag{4.196}$$

Proof By (4.158), (4.190) and (4.193) we have

$$\left\langle\widehat{\widehat{T}}, \phi\right\rangle = \left\langle T, \widehat{\widehat{\phi}}\right\rangle = \langle T, (2\pi)^n \mathcal{R}\phi\rangle = \langle(2\pi)^n \mathcal{R}T, \phi\rangle, \ \forall\phi \in \mathcal{S}(\mathbb{R}^n),$$

which immediately implies (4.196). ■

Remark 4.19 Let us note

$$\widehat{1} = (2\pi)^n \delta_0.$$

As a matter of fact, by (4.194) and (4.196) we have

$$\widehat{1} = \widehat{\widehat{\delta_0}} = (2\pi)^n \mathcal{R}\delta_0 = (2\pi)^n \delta_0.$$

♦

Theorem 4.37 *Let* $T \in \mathcal{S}'(\mathbb{R}^n)$, $\eta \in \mathcal{S}(\mathbb{R}^n)$, $\alpha \in \mathbb{N}_0^n$ *and* $h \in \mathbb{R}^n$. *Then we have:*

$$\widehat{D^\alpha T} = \xi^\alpha \widehat{T} \tag{4.197}$$

$$\widehat{(x^\alpha T)} = (-1)^{|\alpha|} D^\alpha \widehat{T} \tag{4.198}$$

$$\widehat{(\mathcal{T}_h T)} = e^{-i\xi\cdot h}\widehat{T}, \quad \forall h \in \mathbb{R}^n, \tag{4.199}$$

$$\widehat{\eta \star T} = \widehat{\eta}\,\widehat{T}, \tag{4.200}$$

Proof Formulas (4.197)–(4.200) follow almost immediately from (4.193) and Theorems 4.22 and 4.26. By way of example, we will only prove (4.197), leaving the proof of the others as an exercise for the reader.

$$\begin{aligned}\left\langle \widehat{D^\alpha T}, \phi \right\rangle &= \left\langle D^\alpha T, \widehat{\phi} \right\rangle = (-1)^{|\alpha|} \left\langle T, D^\alpha \widehat{\phi} \right\rangle = \\ &= \left\langle T, \widehat{(x^\alpha \phi)} \right\rangle = \left\langle \widehat{T}, x^\alpha \phi \right\rangle = \left\langle x^\alpha \widehat{T}, \phi \right\rangle, \quad \forall \phi \in \mathcal{S}\left(\mathbb{R}^n\right).\end{aligned} \tag{4.201}$$

From which we have (4.197). ■

Exercise 4.26 Prove that the map

$$\mathcal{S}'\left(\mathbb{R}^n\right) \ni T \to \widehat{T} \in \mathcal{S}'\left(\mathbb{R}^n\right),$$

is continuous. ♠

4.10 Brief Introduction to Oscillatory Integrals

In this section, we aim to provide a brief introduction to the concept of oscillatory integrals, with the Fourier transform serving as a prominent example. Although this concept will only be required in Sect. 15.9 of this book, we believe it is beneficial for the reader to gain a preliminary understanding of it. Oscillatory integrals are primarily used to extend properties and calculation rules, which are valid for absolutely integrable functions, to functions that may not be absolutely integrable but exhibit controlled growth (see Theorem 4.41 below). A more in-depth discussion on this topic can be found in [58], [116], [121], [122], [124].

Oscillatory integrals are crucial in harmonic analysis, the theory of pseudodifferential operators, and various other fields, such as number theory. In the literature, a distinction is usually made between first-kind oscillatory integrals, which have the form

$$\int_{\mathbb{R}^n} e^{if(x)} a(x) dx, \tag{4.202}$$

and second-kind oscillatory integrals, which take the form

$$\int_{\mathbb{R}^n} e^{if(x,\theta)} a(x,\theta) dx, \quad \theta \in \Gamma, \tag{4.203}$$

where Γ is an open cone in $\mathbb{R}^N \setminus \{0\}$ (with N not necessarily equal to n). Here, f is a real-valued function (or possibly complex-valued, with $\Im f \geq 0$), and a is a function in C^∞. The functions f and a are referred to as the phase and amplitude of the oscillatory integral, respectively.

In this section, we focus solely on first-kind oscillatory integrals and, following the approach in [116, Sect. 2.2], we will consider only the case where f is a nondegenerate quadratic form, and a belongs to the class $A^m(\mathbb{R}^n)$ (see below for a precise definition). For second-kind integrals of the form (4.203), we refer the reader to [58, Vol. I, Sect. 7.8], [121, Ch. 1], [124, Ch. 9].

We define a to belong to the class $A^m(\mathbb{R}^n)$, where m is a nonnegative real number, if $a \in C^\infty(\mathbb{R}^n)$ and satisfies

$$\gamma_\alpha(a) := \sup_{x\in\mathbb{R}^n} \left|(1+|x|^2)^{-m/2}\partial^\alpha a(x)\right| < +\infty, \quad \forall\alpha \in \mathbb{N}_0^n.$$

For each $j \in \mathbb{N}_0$, we define the norm

$$|||a|||_j = \max_{|\alpha|\le j} \gamma_\alpha(a). \tag{4.204}$$

We begin by the following

Lemma 4.38 *Let q be a nondegenerate real-valued quadratic form on $\mathbb{R}^n$, $\zeta \in C_0^\infty(\mathbb{R}^n)$ such that $\zeta = 0$ in a neighborhood of 0. Let $b \in C^\infty(\mathbb{R}^n)$. Then for every $N \in \mathbb{N}$ there exists $C > 0$ such that, for every $\lambda \ge 1$ we have*

$$\left|\int_{\mathbb{R}^n} e^{i\lambda^2 q(x)} b(\lambda x)\zeta(x)dx\right| \le C\lambda^{-N} \sup_{x\in\operatorname{supp}\zeta, |\alpha|\le N} \left|\left(\partial^\alpha b\right)(\lambda x)\right|. \tag{4.205}$$

Proof First of all, let us note that with a linear change of variables, there exists an integer number r, $0 \le r \le n$ such that we can write,

$$q(x) = \left|x'\right|^2 - \left|x''\right|^2, \tag{4.206}$$

where $x = (x', x'')$, $x' \in \mathbb{R}^r$, $x'' \in \mathbb{R}^{n-r}$, (if $r = 0$ or $r = n$, we set $x' = x$, $x'' = 0$ or $x' = 0$, $x'' = x$, respectively).

Let us define

$$L = \frac{1}{2|x|^2}\left(x'\cdot\partial_{x'} - x''\cdot\partial_{x''}\right).$$

By (4.206) we have

$$L(q) = \frac{1}{2|x|^2}\left(x'\cdot\partial_{x'}q - x''\cdot\partial_{x''}q\right) = \frac{1}{2|x|^2}\left(2\left|x'\right|^2 + 2\left|x''\right|^2\right) = 1.$$

Hence

$$L\left(e^{i\lambda^2 q(x)}\right) = i\lambda^2 e^{i\lambda^2 q(x)}$$

and

$$L^N\left(e^{i\lambda^2 q(x)}\right) = \left(i\lambda^2\right)^N e^{i\lambda^2 q(x)}, \quad \forall N \in \mathbb{N}. \tag{4.207}$$

Let $N \in \mathbb{N}$. By (4.207), integrating by parts and taking into account that $\zeta \in C_0^\infty(\mathbb{R}^n \setminus \{0\})$, we have

$$\begin{aligned}
&\int_{\mathbb{R}^n} e^{i\lambda^2 q(x)} b(\lambda x)\zeta(x)dx = \\
&= \int_{\mathbb{R}^n} \left(i\lambda^2\right)^{-N} L^N\left(e^{i\lambda^2 q(x)}\right) b(\lambda x)\zeta(x)dx = \\
&= \left(i\lambda^2\right)^{-N} \int_{\mathbb{R}^n} e^{i\lambda^2 q(x)} \left(L^t\right)^N \left(b(\lambda x)\zeta(x)\right) dx.
\end{aligned} \tag{4.208}$$

Now, by the Leibniz formula we have

$$\left(L^t\right)^N \left(b(\lambda x)\zeta(x)\right) = \sum_{|\alpha+\beta|\leq N} c_{\alpha\beta}(x)\lambda^{|\alpha|} \left(\partial^\alpha b\right)(\lambda x)\partial^\beta \zeta^\beta(x),$$

where the coefficients $c_{\alpha\beta}(x)$ are bounded in $\operatorname{supp}\zeta$ and they are independent of λ.

Hence

$$\left|\left(L^t\right)^N \left(b(\lambda x)\zeta(x)\right)\right| \leq C\lambda^N \sup_{x\in \operatorname{supp}\zeta, |\alpha|\leq N} \left|\left(\partial^\alpha b\right)(\lambda x)\right|. \tag{4.209}$$

Therefore (recalling that q is a real-valued function), (4.208) and (4.209) give (4.205). ■

Now we give the main theorem of this section, which allows us to define oscillatory integrals.

Theorem 4.39 *Let m be a nonnegative real number Let $a \in A^m(\mathbb{R}^n)$. Let q be a nondegenerate real-valued quadratic form on $\mathbb{R}^n$. Then we have:*

(a) For any $\phi \in \mathcal{S}(\mathbb{R}^n)$ such that $\phi(0) = 1$, the limit

$$\lim_{\varepsilon\to 0} \int_{\mathbb{R}^n} e^{iq(x)} a(x)\phi(\varepsilon x)dx \tag{4.210}$$

exists, is independent of the choice of $\phi \in \mathcal{S}(\mathbb{R}^n)$ (provided that $\phi(0) = 1$), and is equal to the Lebesgue integral of $e^{iq(x)}a(x)$, provided that $a \in L^1(\mathbb{R}^n)$.

(b) If $a \notin L^1(\mathbb{R}^n)$, we continue to denote the limit (4.210) by $\int_{\mathbb{R}^n} e^{iq(x)}a(x)dx$, and we have

$$\left|\int_{\mathbb{R}^n} e^{iq(x)}a(x)dx\right| \leq C|||a|||_{m+n+1}, \tag{4.211}$$

where C depends only on q, m and n only.

Proof The fact that, when $a \in L^1(\mathbb{R}^n)$, the limit (4.210) exists and is equal to the Lebesgue integral follows immediately from the Dominated Convergence Theorem.

To prove (a) and (b) in the general case, let us introduce a real-valued function $\eta \in C_0^\infty(\mathbb{R}^n)$ such that $\eta = 1$ in B_1, $\eta = 0$ in $\mathbb{R}^n \setminus B_2$ and $0 \leq \eta \leq 1$ and let us denote

$$J_k = \int_{\mathbb{R}^n} e^{iq(x)}a(x)\eta\left(2^{-k}x\right)dx, \quad \forall k \in \mathbb{N}_0 \tag{4.212}$$

Claim I The sequence $\{J_k\}$ converges in $\mathbb{C}$ and denoting by

$$J = \lim_{k\to\infty} J_k, \tag{4.213}$$

we have

$$|J| \leq C_{m,n}|||a|||_{m+n+1}, \tag{4.214}$$

where $C_{m,n}$ depends on q, m and n only.

Proof of the Claim I By the simple change of variables $y = 2^k x$ and by the homogeneity of q, we have, for every $k \geq 1$

$$J_k - J_{k-1} = \int_{\mathbb{R}^n} e^{i2^{2k}q(y)}a\left(2^k y\right)(\eta(y) - \eta(2y))\,2^{kn}dy. \tag{4.215}$$

We set

$$\zeta(y) = \eta(y) - \eta(2y), \quad \text{for } y \in \mathbb{R}^n. \tag{4.216}$$

Let us note that

$$\operatorname{supp}\zeta \subset \left\{y \in \mathbb{R}^n : \frac{1}{2} \leq |y| \leq 2\right\}.$$

Hence, recalling that $a \in A^m$, we have, for every $y \in \operatorname{supp}\zeta$ and every $k \in \mathbb{N}$,

$$\left|(\partial^\alpha a)\left(2^k y\right)\right| \leq |||a|||_{|\alpha|}\left(1 + 2^{2k}|y|^2\right)^{m/2} \leq |||a|||_{|\alpha|}2^{m(k+2)}. \tag{4.217}$$

Now we apply Lemma 4.38 when $\lambda = 2^k$ and $b(x) = a(x)$ and by (4.215) and (4.217) we get, for every $N \in \mathbb{N}$,

$$|J_k - J_{k-1}| \leq C_{N,m}|||a|||_N 2^{k(-N+n+m)},$$

where $C_{N,m}$ depends on q, N and m only. Now, choosing $N = m + n + 1$ we have

$$|J_k - J_{k-1}| \leq C_{m,n}|||a|||_{m+n+1}2^{-k} \tag{4.218}$$

where $C_{m,n}$ depends on q, m and n only. By (4.218) we obtain immediately the convergence of $\{J_k\}$ and inequality (4.214). **This concludes the proof of the Claim I.**

Now let $\phi \in \mathcal{S}(\mathbb{R}^n)$ be such that $\phi(0) = 1$. Let us denote by

$$J(\varepsilon) = \int_{\mathbb{R}^n} e^{iq(x)}a(x)\phi(\varepsilon x)dx.$$

We wish to prove that

$$\lim_{\varepsilon \to 0} J(\varepsilon) = J, \tag{4.219}$$

where J is defined by (4.213). Let us note that (4.219) together with (4.214) provide the proof of (a) and (b) of the Theorem. For this purpose, we set

$$J_k(\varepsilon) = \int_{\mathbb{R}^n} e^{iq(x)}a(x)\phi(\varepsilon x)\eta\left(2^{-k}x\right)dx \tag{4.220}$$

and we prove the following

Claim II For every $\varepsilon > 0$ we have

$$\lim_{k \to \infty} J_k(\varepsilon) = J(\varepsilon). \tag{4.221}$$

Moreover, there exists C (independent of ε) such that

$$|J_k - J_k(\varepsilon)| \leq C\varepsilon, \quad \forall k \in \mathbb{N}_0. \tag{4.222}$$

Proof of the Claim II Concerning (4.221), it is an immediate consequence of the Dominated Convergence Theorem.

Now we prove (4.222). Set, for $k \in \mathbb{N}_0$,

$$I_k(\varepsilon) = J_k - J_k(\varepsilon) = \int_{\mathbb{R}^n} e^{iq(x)}a(x)(1 - \phi(\varepsilon x))\eta\left(2^{-k}x\right)dx. \tag{4.223}$$

Recalling that $\phi(0) = 1$ and $\operatorname{supp} \eta \subset B_2$, we have

$$|I_0(\varepsilon)| \leq \int_{\mathbb{R}^n} |a(x)|\,|1 - \phi(\varepsilon x)|\, dx \leq C \, \|a\|_{L^\infty(\mathbb{R}^n)} \, \varepsilon, \tag{4.224}$$

where C depends on n only. By a simple change of variables we have

$$\begin{aligned} &I_k(\varepsilon) - I_{k-1}(\varepsilon) = \\ &= \int_{\mathbb{R}^n} e^{i2^{2k}q(y)} a(2^k y)(1 - \phi(\varepsilon 2^k x))\,(\eta(y) - \eta(2y))\, 2^{kn} dy. \end{aligned} \tag{4.225}$$

In order to estimate $|I_k(\varepsilon) - I_{k-1}(\varepsilon)|$ we proceed in a manner similar to the proof of Claim I, with the only difference being that, in this case, we are interested in an estimate in terms of ε, uniform with respect to k.

We apply Lemma 4.38 when $\lambda = 2^k$ and

$$b(x) = b_\varepsilon(x) = (1 - \phi(\varepsilon x))\, a\,(x)\,.$$

Taking into account that for $0 < \varepsilon \leq 1$ (ζ is defined by (4.216))

$$\left| \left(\partial^\beta (1 - \phi)\right) \left(2^k \varepsilon y\right) \right| \leq C_\beta 2^k \varepsilon, \quad \forall y \in \operatorname{supp} \zeta, \ \forall \beta \in \mathbb{N}_0^n$$

(for $\beta = 0$ we use that $\phi(0) = 1$), where C is independent of ε and k, by Leibniz formula we get

$$\left| \left(\partial^\alpha b_\varepsilon\right) \left(2^k y\right) \right| \leq C_\alpha \varepsilon |||a|||_{|\alpha|} 2^{m(k+2)}, \quad \forall y \in \operatorname{supp} \zeta, \ \forall \alpha \in \mathbb{N}_0^n.$$

Hence, Lemma 4.38 implies

$$|I_k(\varepsilon) - I_{k-1}(\varepsilon)| \leq C\varepsilon 2^{-k}, \quad \forall k \in \mathbb{N}, \ \forall \varepsilon \in (0, 1], \tag{4.226}$$

where C depends on $|||a|||_{m+n+1}$, but is independent of ε and k.

Now, by (4.224) and (4.224) we have

$$\begin{aligned} |J_k - J_k(\varepsilon)| = |I_k(\varepsilon)| &= \\ &= \left| I_0(\varepsilon) + \sum_{j=1}^{n} \left(I_j(\varepsilon) - I_{j-1}(\varepsilon)\right) \right| \leq \\ &\leq |I_0(\varepsilon)| + \sum_{j=1}^{n} \left|(I_j(\varepsilon) - I_{j-1}(\varepsilon)\right| \leq \\ &\leq C\varepsilon, \quad \forall k \in \mathbb{N}_0, \ \forall \varepsilon \in (0, 1], \end{aligned}$$

where C depends on $|||a|||_{m+n+1}$, but is independent of ε and k. **This concludes the proof of Claim II.**

The triangle inequality gives

$$|J(\varepsilon) - J_k| \leq |J(\varepsilon) - J_k(\varepsilon)| + |J_k - J(\varepsilon)|, \quad \forall k \in \mathbb{N}_0, \ \forall \varepsilon > 0.$$

By this inequality and by (4.213), (4.221) and (4.222) we have, for every $\varepsilon \in (0, 1]$

$$|J(\varepsilon) - J| = \lim_{k\to\infty} |J(\varepsilon) - J_k| \leq$$
$$\leq \limsup_{k\to\infty} |J(\varepsilon) - J_k(\varepsilon)| + \limsup_{k\to\infty} |J_k - J(\varepsilon)| \leq C\varepsilon.$$

Hence

$$|J(\varepsilon) - J| \leq C\varepsilon, \tag{4.227}$$

from which the desired relationship (4.219) follows. Therefore, as noted above, the proof of the theorem is complete. ■

Theorem 4.39 leads to the following definition:

Definition 4.40 Let m be a nonnegative real number, and let $a \in A^m$. Let q be a nondegenerate real-valued quadratic form on $\mathbb{R}^n$. Let $\phi \in \mathcal{S}(\mathbb{R}^n)$ satisfy $\phi(0) = 1$. We define the limit

$$\lim_{\varepsilon\to 0} \int_{\mathbb{R}^n} e^{iq(x)} a(x)\phi(\varepsilon x)dx \tag{4.228}$$

as the **oscillatory integral**. If there is no risk of ambiguity, we denote it simply by

$$\int_{\mathbb{R}^n} e^{iq(x)} a(x)dx. \tag{4.229}$$

Example Let $a \in A^m(\mathbb{R}^n)$. Then we have

$$\int_{\mathbb{R}^{2n}} e^{-ix\cdot\xi} a(x)dxd\xi = \int_{\mathbb{R}^{2n}} e^{-ix\cdot\xi} a(\xi)dxd\xi = (2\pi)^n a(0). \tag{4.230}$$

Let $\varphi \in \mathcal{S}(\mathbb{R}^n)$ be such that $\varphi(0) = 1$. Let us apply Definition 4.40 to $\phi(x, \xi) = \varphi(x)\varphi(\xi)$. We have

$$\int_{\mathbb{R}^{2n}} e^{-ix\cdot\xi} a(x)dxd\xi = \lim_{\varepsilon\to 0} \int_{\mathbb{R}^{2n}} e^{-ix\cdot\xi} a(x)\varphi(\varepsilon x)\varphi(\varepsilon\xi)dxd\xi =$$
$$= \lim_{\varepsilon\to 0} \int_{\mathbb{R}^{2n}} e^{-ix\cdot\xi} a(\xi)\varphi(\varepsilon x)\varphi(\varepsilon\xi)dxd\xi = \int_{\mathbb{R}^{2n}} e^{-ix\cdot\xi} a(\xi)dxd\xi.$$

This proves the first equality in (4.230).

Concerning the second equality, we first perform the change of variables

$$y = \varepsilon x, \quad \zeta = \varepsilon^{-1}\xi$$

and we get, by the Fubini–Tonelli Theorem

$$\begin{aligned}
&\int_{\mathbb{R}^{2n}} e^{-ix\cdot\xi} a(\xi)\varphi(\varepsilon x)\varphi(\varepsilon\xi)dxd\xi = \\
&= \int_{\mathbb{R}^{2n}} e^{-iy\cdot\zeta} a(\varepsilon\zeta)\varphi(y)\varphi(\varepsilon^2\zeta)dyd\zeta = \\
&= \int_{\mathbb{R}^n} d\zeta a(\varepsilon\zeta)\varphi(\varepsilon^2\zeta)\int_{\mathbb{R}^n} e^{-iy\cdot\zeta}\varphi(y)dy = \\
&= \int_{\mathbb{R}^n} a(\varepsilon\zeta)\widehat{\varphi}(\zeta)d\zeta.
\end{aligned} \tag{4.231}$$

Now we have, recalling that $\varphi(0) = 1$,

$$\lim_{\varepsilon\to 0}\int_{\mathbb{R}^n} a(\varepsilon\zeta)\widehat{\varphi}(\zeta)d\zeta = a(0)\int_{\mathbb{R}^n}\widehat{\varphi}(\zeta)d\zeta = (2\pi)^n a(0). \tag{4.232}$$

By (4.231) and (4.10) we get the second equality of (4.230). ♣

We conclude this section by presenting the following theorem, which outlines some properties of oscillatory integrals.

Theorem 4.41 *Let q be a non-degenerate real quadratic form. In what follows, we denote by "$\int$" the oscillatory integral (see Definition 4.40). Then we have:*

(a) (Linear change of variables) If M is a nonsingular real matrix and $a \in A^m(\mathbb{R}^n)$, $m \geq 0$, then

$$\int_{\mathbb{R}^n} e^{iq(x)}a(x)dx = \int_{\mathbb{R}^n} e^{iq(My)}a(My)|\det(M)|dy;$$

(b) (Integration by parts) If $a \in A^m(\mathbb{R}^n)$, $b \in A^l(\mathbb{R}^n)$, $m, l \geq 0$, then

$$\int_{\mathbb{R}^n} e^{iq(x)}a(x)\partial^\alpha b(x)dx = (-1)^{|\alpha|}\int_{\mathbb{R}^n}\partial^\alpha\left(e^{iq(x)}a(x)\right)b(x)dx, \quad \forall\alpha \in \mathbb{N}_0;$$

(c) (Differentiation under the integral sign) If $a \in A^m(\mathbb{R}^n \times \mathbb{R}^p)$, then the function

$$\int_{\mathbb{R}^n} e^{iq(x)}a(x, y)dx$$

belongs to $A^m(\mathbb{R}^n)$ *and*

$$\partial_y^\alpha \int_{\mathbb{R}^n} e^{iq(x)} a(x, y) dx = \int_{\mathbb{R}^n} e^{iq(x)} \partial_y^\alpha a(x, y) dx;$$

(d) *(Interversion of the integral sign) If* $a \in A^m(\mathbb{R}^n \times \mathbb{R}^p)$ *and* r *is a non-degenerate real quadratic form, then*

$$\begin{aligned} &\int_{\mathbb{R}^p} e^{ir(y)} \left(\int_{\mathbb{R}^n} e^{iq(x)} a(x, y) dx \right) dy = \\ &= \int_{\mathbb{R}^{n+p}} e^{i(q(x)+r(y))} a(x, y) dx dy = \\ &= \int_{\mathbb{R}^n} e^{iq(x)} \left(\int_{\mathbb{R}^p} e^{ir(y)} a(x, y) dy \right) dx. \end{aligned}$$

We do not provide a proof of the theorem and refer the reader to [116, Sect. 2.2]. However, we encourage the reader to attempt proving it as an exercise, keeping in mind that arguments used in the proof of Theorem 4.39 may be useful.

Chapter 5
The Sobolev Spaces

Abstract This chapter introduces Sobolev spaces, a framework for analyzing functions with weak derivatives and solving Partial Differential Equations in a more flexible way. These spaces allow reformulating classical problems, such as the Dirichlet problem, into a functional analysis setting that ensures well-posedness and unique solutions. In addition to foundational Sobolev spaces, the chapter touches on broader generalizations to provide a connection with existing mathematical literature. While not all material is immediately necessary for the book's objectives, it offers valuable tools for understanding modern applications in analysis and beyond.

5.1 Introduction

This chapter naturally builds upon the previous discussion on distributions, as Sobolev spaces offer a framework to study functions with weak derivatives, which can be understood using the language of distributions. In this sense, the tools developed for distributions will prove helpful in our exploration of these functional spaces. In this context, we present the key definitions and properties that allow us to generalize the concept of solutions to Partial Differential Equations and their corresponding boundary value problems. For example, using the Sobolev space denoted by $H_0^1(\Omega)$, where Ω is an open set in $\mathbb{R}^n$, we can formulate the Dirichlet problem in a broader, more flexible way

$$\begin{cases} -\Delta u = f, & \text{in } \Omega \\ \\ u = 0, & \text{on } \partial\Omega \end{cases}, \tag{5.1}$$

where $f \in L^2(\Omega)$. This problem can be interpreted as a simple problem of Functional Analysis in a Hilbert space, specifically the space $H_0^1(\Omega)$. This interpretation is made possible by a shift in perspective that, while simple, is quite radical

S. Vessella, *Unique Continuation Properties for Partial Differential Equations*,
Birkhäuser Advanced Texts Basler Lehrbücher,
https://doi.org/10.1007/978-3-031-86366-0_5

regarding the concept of a function (as we already observed in the previous chapter). Let us consider a particularly straightforward example where we have a function $w \in L^2(\Omega)$. The essence of this radical shift, introduced by Sobolev spaces, lies in identifying the function w with the linear functional

$$H_0^1(\Omega) \ni \phi \mapsto \int_\Omega w(x)\phi(x)\,dx \in \mathbb{R}.$$

This, in turn, allow to formulate problem (5.1) as follows

$$\begin{cases} \int_\Omega \nabla u \cdot \nabla \phi dx = \int_\Omega f(x)\phi(x), \quad \forall \phi \in H_0^1(\Omega) \\ \\ u \in H_0^1(\Omega) \end{cases} . \tag{5.2}$$

It will then be straightforward to recognize that problem (5.2) admits a unique solution due to the Riesz Representation Theorem, [19], later generalized in the Lax–Milgram Theorem (Sect. 6.2).

The topics covered in this chapter slightly exceed the immediate needs of this book. In fact, here we address not only the spaces $H_0^1(\Omega)$, $H^k(\Omega)$, but also the Sobolev spaces $W^{k,p}(\Omega)$, with $p \geq 1$, which, for $p \neq 2$, are not Hilbert spaces but still Banach spaces. In this way, we believe that we can facilitate the comparison with the extensive existing literature on Sobolev spaces, for instance, see [39, 69, 126].

In the present chapter, except Sect. 5.12, we consider real valued functions,

5.2 Definition of the Sobolev Spaces

In what follows, if $u \in L^1_{loc}(\Omega)$, where Ω is an open set of $\mathbb{R}^n$, we identify the distribution T_u (defined in (4.6)) with u. Additionally, we refer to "α-th derivative in the sense of distributions" as "α-th weak derivative".

Definition 5.1 Let $1 \leq p \leq \infty$, $k \in \mathbb{N}_0$. Let Ω be an open set of $\mathbb{R}^n$, $n \geq 1$. If $k = 0$, set

$$W^{0,p}(\Omega) = L^p(\Omega).$$

If $k \geq 1$, $W^{k,p}(\Omega)$ is the set of $u \in \mathcal{D}'(\Omega)$ satisfying

$$\partial^\alpha u \in L^p(\Omega), \qquad \text{for } |\alpha| \leq k, \tag{5.3}$$

where $\partial^\alpha u$ is the α-th weak derivative of u.

It is easy to check that $W^{k,p}(\Omega)$ is a vector space. Furthermore we define the following norms. If $1 \leq p < +\infty$, we set

$$\|u\|_{W^{k,p}(\Omega)} = \left(\sum_{|\alpha| \leq k} \int_{\Omega} |\partial^{\alpha} u|^p \, dx \right)^{1/p} . \tag{5.4}$$

If $p = +\infty$, we set

$$\|u\|_{W^{\infty,p}(\Omega)} = \sum_{|\alpha| \leq k} \|\partial^{\alpha} u\|_{L^{\infty}(\Omega)} . \tag{5.5}$$

If $p = 2$, we also set

$$H^k(\Omega) = W^{k,2}(\Omega).$$

Let us observe that $H^k(\Omega)$ is a pre-Hilbertian space equipped with the scalar product

$$(u, v)_{H^k(\Omega)} = \int_{\Omega} \sum_{|\alpha| \leq k} \partial^{\alpha} u \partial^{\alpha} v dx, \quad \forall u, v \in H^k(\Omega). \tag{5.6}$$

Whenever we consider complex-valued functions, instead of the scalar product we equip the space $H^k(\Omega)$ of the hermitian form

$$(u, v)_{H^k(\Omega)} = \int_{\Omega} \sum_{|\alpha| \leq k} \partial^{\alpha} u \overline{\partial^{\alpha} v} dx, \quad \forall u, v \in H^k(\Omega). \tag{5.7}$$

Here and in the sequel, for any $k \in \mathbb{N}$, $p \in [1, \infty]$, we denote by $W^{k,p}_{loc}(\Omega)$, $(H^k_{loc}(\Omega))$ the subspace of $L^p_{loc}(\Omega)$ $(L^2_{loc}(\Omega))$ of the functions u such that for every open set $\omega \Subset \Omega$ (i.e. $\overline{\omega} \subset \Omega$) we have $u_{|\omega} \in W^{k,p}(\omega)$ $(u_{|\omega} \in H^k(\omega))$. Let $\{u_m\}$ be a sequence in $W^{k,p}_{loc}(\Omega)$ and $u \in W^{k,p}_{loc}(\Omega)$, we say that

$$u_m \to u, \quad \text{as } m \to \infty, \text{ in } W^{k,p}_{loc}(\Omega)$$

or

$$\{u_m\} \to u \quad \text{in } W^{k,p}_{loc}(\Omega)$$

if

$$(u_m)_{|\omega} \to u_{|\omega}, \quad \text{as } m \to \infty, \text{ in } W^{k,p}(\omega), \quad \forall \omega \Subset \Omega.$$

Exercise 5.27 Check that, if $1 \le p \le \infty$, $k \in \mathbb{N}_0$, then $W^{k,p}(\Omega)$ is a vector subspace of $L^p(\Omega)$ and $\|\cdot\|_{W^{k,p}(\Omega)}$ defines a norm on $W^{k,p}(\Omega)$. ♠

Proposition 5.2 *If $u \in W^{k,p}(\Omega)$, then we have*

(i) $\partial^\alpha u \in W^{k-|\alpha|,p}(\Omega)$
for $|\alpha| \le k$ and $\partial^\beta \partial^\alpha u = \partial^\alpha \partial^\beta u = \partial^{\alpha+\beta} u$ for $|\alpha| + |\beta| \le k$,
(ii) for any $\zeta \in C^\infty\left(\overline{\Omega}\right)$ we have $\zeta u \in W^{k,p}(\Omega)$ and

$$\partial^\alpha(\zeta u) = \sum_{\beta \le \alpha} \binom{\alpha}{\beta} \partial^\beta \zeta \partial^{\alpha-\beta} u.$$

Proof

(i) is a simple consequence of (4.20) and of Definition 5.1.
(ii) Let us consider the case $|\alpha| = 1$. Let $\alpha = e_j$, for $j = 1, \cdots, n$.

We have

$$\partial_j(\zeta u) = (\partial_j u)\zeta + u\partial_j \zeta, \qquad \text{in the weak sense.}$$

Now, we have

$$(\partial_j u)\zeta + u\partial_j \zeta \in L^p(\Omega).$$

Hence $\zeta u \in L^p(\Omega)$.

If $|\alpha| > 1$, one proceeds by applying the Leibniz formula, and we leave the details to the reader. ■

Theorem 5.3 (Completeness of $W^{k,p}(\Omega)$) *The space $W^{k,p}(\Omega)$, $k \in \mathbb{N}_0$, $1 \le p \le \infty$, equipped with the norm* (5.4), (5.5), *is a Banach space. If $p = 2$, $H^k(\Omega)$ is a Hilbert space.*

Proof We limit ourselves to the case $k = 1$. Similarly it can be handle the case $k > 1$. Let $\{u_m\}$ be a Cauchy sequence in $W^{1,p}(\Omega)$. From the definition of norm of $W^{1,p}(\Omega)$ we have that

$$\{u_m\} \quad \text{and} \quad \left\{\partial_j u_m\right\}, \quad j = 1, \cdots, n,$$

are Cauchy sequences in $L^p(\Omega)$. On the other hand, $L^p(\Omega)$ is complete; hence there exist $u, v_1, \cdots, v_n \in L^p(\Omega)$ such that

$$u_m \to u, \qquad \text{as } m \to \infty, \qquad \text{in } L^p(\Omega), \tag{5.8a}$$

$$\partial_j u_m \to v_j, \qquad \text{as } m \to \infty, \qquad \text{in } L^p(\Omega), \; j = 1, \cdots, n. \tag{5.8b}$$

Now, (5.8a) and (5.8b) imply that, for any $\phi \in C_0^\infty(\Omega)$, we have

$$\begin{aligned}\int_\Omega u\partial_j\phi dx &= \lim_{m\to\infty}\int_\Omega u_m\partial_j\phi dx = \\ &= -\lim_{m\to\infty}\int_\Omega \left(\partial_j u_m\right)\phi dx = \\ &= -\int_\Omega v_j\phi dx.\end{aligned}$$

Hence

$$\partial_j u = v_j, \qquad \text{for } j = 1, \cdots, n.$$

Therefore by (5.8a) and (5.8b) we have

$$u_m \to u, \qquad \text{as } m \to \infty, \qquad \text{in } W^{1,p}(\Omega).$$

■

Proposition 5.4 *The space $W^{k,p}(\Omega)$, $k \in \mathbb{N}_0$, $1 \le p < \infty$, equipped with the norm* (5.4) *is a separable space.*

Proof The proof is similar to the one of Proposition 3.6. Let us consider the case $k = 1$. Let

$$\Phi : W^{1,p}(\Omega) \to L^p(\Omega) \times L^p(\Omega; \mathbb{R}^n),$$

$$\Phi(u) = (u, \nabla u), \qquad \forall u \in W^{1,p}(\Omega).$$

Φ turns out to be an isometry, provided that we equip $L^p(\Omega) \times L^p(\Omega; \mathbb{R}^n)$ by the norm

$$\left(\int_\Omega |v_0|^p dx + \sum_{j=1}^n \int_\Omega \left|v_j\right|^p dx\right)^{1/p},$$

for every $v = (v_0, v_1, \cdots, v_n) \in L^p(\Omega) \times L^p(\Omega; \mathbb{R}^n)$. Now, since $p < +\infty$, $L^p(\Omega) \times L^p(\Omega; \mathbb{R}^n)$ is separable because it is the cartesian product of separable spaces. Hence $\Phi\left(W^{1,p}(\Omega)\right)$ is separable as a subspace of $L^p(\Omega) \times L^p(\Omega; \mathbb{R}^n)$ and, since Φ is an isometry, $W^{1,p}(\Omega)$ is separable too. ■

It can be proved that if $1 < p < \infty$, then $W^{1,p}(\Omega)$ is a *reflexive space*. In the sequel we will not make explicitly use this property, however for a proof we refer to [19, Prop. IX.1].

Example 5.12 Let $\Omega = \mathbb{R}$. Let $f(x) = |x|$. In Example 5.12 of Sect. 4.3 We have proven that $f' = \text{sgn}(\cdot)$ in the sense of distributions. Hence $f \in W^{1,\infty}(\mathbb{R})$. ♣

Example 5.13 Let $\Omega = \mathbb{R}$. Let H be the Heaviside function. We have proven in Example 4.2 of Sect. 4.3

$$H' = \delta_0, \quad \text{in the sense of distributions.}$$

On the other hand, in Remark 4.9 of Sect. 4.2 we have proven that does not exists $f \in L^1_{loc}(\mathbb{R})$ such that $\delta_0 = T_f$. Hence, for every p, $1 \leq p \leq \infty$, we $H \notin W^{1,p}_{loc}(\mathbb{R})$. ♣

Example 5.14 Let $\alpha > 0$. Let us consider

$$u(x) = \frac{1}{|x|^\alpha}.$$

We prove that $u \in W^{1,p}(B_1)$ if and only if $p < n$ and $\alpha < \frac{n}{p} - 1$.

We begin by assuming that $u \in W^{1,p}(B_1)$. Then $u \in L^p(B_1)$ and consequently $\alpha p < n$, therefore $p < \infty$. Moreover, for any $j = 1, \cdots, n$ there exists $v_j \in L^p(B_1)$ such that

$$\int_{B_1} \frac{1}{|x|^\alpha} \partial_j \phi dx = -\int_{B_1} v_j \phi dx, \quad \forall \phi \in C_0^\infty(B_1).$$

In particular we have, for any $\phi \in C_0^\infty(B_1 \setminus \{0\})$,

$$-\int_{B_1} v_j \phi dx = \int_{B_1} \frac{1}{|x|^\alpha} \partial_j \phi dx = \int_{B_1} \frac{\alpha x_j}{|x|^{\alpha+2}} \phi dx.$$

Hence, for any $j = 1, \cdots, n$,

$$v_j(x) = -\frac{\alpha x_j}{|x|^{\alpha+2}}, \quad \text{a.e. in } B_1.$$

Now, $v_j \in L^p(B_1)$, therefore

$$\sum_{j=1}^n \int_{B_1} \left| \frac{\alpha x_j}{|x|^{\alpha+2}} \right|^p dx < \infty. \tag{5.9}$$

Let us observe that if $a = (a_1, \cdots, a_n) \in \mathbb{R}^n$, then

$$\frac{1}{n^{p-1}} |a|^p \leq \sum_{j=1}^n |a_j|^p \leq n |a|^p, \tag{5.10}$$

(the first inequality from the left is a simple consequence of the convexity of the function t^p, for $p \geq 1$, the second is trivial). Hence (5.9) is satisfied if and only if

$$\int_{B_1} \frac{dx}{|x|^{(\alpha+1)p}} = \int_{B_1} \left| \frac{x}{|x|^{\alpha+2}} \right|^p dx < \infty,$$

from which we derive

$$\alpha < \frac{n}{p} - 1. \tag{5.11}$$

Conversely, let us assume that (5.11) holds true and that $p < n$. Let us show that $u \in W^{1,p}(B_1)$. Inequality (5.11) implies $\alpha < \frac{n}{p}$ that, in turn implies $u \in L^p(B_1)$. Now, let $\phi \in C_0^\infty(B_1)$ be arbitrary. We have, for any $j = 1, \cdots, n$,

$$\begin{aligned} \int_{B_1} u \partial_j \phi dx &= \lim_{\varepsilon \to 0} \int_{B_1 \setminus B_\varepsilon} u \partial_j \phi dx = \\ &= \lim_{\varepsilon \to 0} \int_{B_1 \setminus B_\varepsilon} \left(\partial_j (u\phi) - \phi \partial_j u \right) dx = \\ &= \lim_{\varepsilon \to 0} \left\{ \int_{\partial B_\varepsilon} u \phi \nu_j dS - \int_{B_1 \setminus B_\varepsilon} \phi \partial_j u dx \right\}. \end{aligned} \tag{5.12}$$

On the other hand we have

$$\left| \int_{\partial B_\varepsilon} u \phi \nu_j dS \right| \leq \omega_n \, \|\phi\|_{L^\infty(B_1)} \, \varepsilon^{-\alpha+n-1},$$

where ω_n is the measure of $|\partial B_1|$. Now, by (5.11) and $p \geq 1$ we have $\alpha < n - 1$. Hence

$$\lim_{\varepsilon \to 0} \left| \int_{\partial B_\varepsilon} u \phi \nu_j dS \right| = 0;$$

coming back to (5.12) and keeping in mind that (by (5.11))

$$\frac{\alpha x_j}{|x|^{\alpha+2}} \in L^p(B_1) \subset L^1(B_1),$$

we get

$$\int_{B_1} u \partial_j \phi dx = \lim_{\varepsilon \to 0} \int_{B_1 \setminus B_\varepsilon} \frac{\alpha x_j}{|x|^{\alpha+2}} \phi dx = \int_{B_1} \frac{\alpha x_j}{|x|^{\alpha+2}} \phi dx.$$

All in all we have

$$\partial_j u = -\frac{\alpha x_j}{|x|^{\alpha+2}} \in L^p(B_1),$$

therefore $u \in W^{1,p}(B_1)$. ♣

Exercise 5.28 Prove that if

$$u \in C^0\left(\overline{B_1}\right) \cap C^1\left(\overline{B_1} \setminus \{0\}\right)$$

then

$$u \in W^{1,p}(B_1) \Longleftrightarrow \nabla u \in L^p(B_1),$$

here ∇u is the gradient of u in $B_1 \setminus \{0\}$ in the classic sense. [Hint: Proceed in a manner similar to Example 5.14.] ♠

5.2.1 *The Spaces* $W_0^{k,p}(\Omega)$

We give the following

Definition 5.5 Let $1 \leq p \leq \infty$, $k \in \mathbb{N}_0$. Let Ω be an open set of $\mathbb{R}^n$, $n \geq 1$. Let us denote by

$$W_0^{k,p}(\Omega),$$

the closure of $C_0^\infty(\Omega)$ in $W^{k,p}(\Omega)$. We write

$$H_0^k(\Omega) = W_0^{k,2}(\Omega).$$

Let us notice that $W_0^{0,p}(\Omega) = L^p(\Omega)$. Moreover $W_0^{k,p}(\Omega)$, equipped with the norm $\|\cdot\|_{W^{k,p}(\Omega)}$, as it is a closed subspace of $W^{k,p}(\Omega)$, is a Banach space.

5.3 Approximation and Density Theorems

Let η be a mollifier, namely $\eta \in C_0^\infty(\mathbb{R}^n)$ satisfies (i) $\operatorname{supp}\eta \subset B_1$, (ii) $\eta \geq 0$, (iii) $\int_{\mathbb{R}^n} \eta(x)dx = 1$. Set, for any $\varepsilon > 0$,

$$\eta_\varepsilon(x) = \varepsilon^{-n}\eta\left(\varepsilon^{-1}x\right).$$

Let Ω be an open set of $\mathbb{R}^n$. Set

$$\Omega_\varepsilon = \{x \in \Omega : \operatorname{dist}(x, \partial\Omega) > \varepsilon\}.$$

The following Theorem holds true.

Theorem 5.6 (Local Approximation by C^∞ Functions) *Let $k \in \mathbb{N}_0$, $p \in [1, +\infty)$. Assume that $u \in W^{k,p}(\Omega)$. Denote by*

$$u^\varepsilon = \eta_\varepsilon \star u, \qquad in\ \Omega_\varepsilon.$$

Then we have

$$u^\varepsilon \in C^\infty(\Omega_\varepsilon) \cap W^{k,p}_{loc}(\Omega_\varepsilon), \quad \forall \varepsilon > 0$$

and

$$u^\varepsilon \to u, \qquad as\ \varepsilon \to 0,\ in\ W^{k,p}_{loc}(\Omega).$$

Proof The fact that $u^\varepsilon \in C^\infty(\Omega_\varepsilon)$ is an immediate consequence of Theorem 3.47. Concerning $u^\varepsilon \in W^{k,p}_{loc}(\Omega_\varepsilon)$, we have by Theorem 3.45

$$\partial^\alpha u^\varepsilon = \left(\partial^\alpha \eta_\varepsilon\right) \star u, \quad \forall \alpha \in \mathbb{N}^n_0.$$

Now, if $|\alpha| \leq k$, we have, for any $x \in \Omega_\varepsilon$,

$$\begin{aligned}
\left(\partial^\alpha \eta_\varepsilon\right) \star u &= \int_\Omega \left(\partial^\alpha_x \eta_\varepsilon\right)(x-y)u(y)dy = \\
&= (-1)^{|\alpha|} \int_\Omega \partial^\alpha_y \left(\eta_\varepsilon(x-y)\right) u(y)dy = \\
&= \int_\Omega \eta_\varepsilon(x-y)\partial^\alpha_y u(y)dy = \\
&= \left(\eta_\varepsilon \star \partial^\alpha u\right)(x).
\end{aligned}$$

Let $\omega \Subset \Omega$. Since $\partial^\alpha u \in L^p(\omega)$ we have, for any $|\alpha| \leq k$,

$$\eta_\varepsilon \star \partial^\alpha u \to \partial^\alpha u, \qquad \text{as } \varepsilon \to 0,\ \text{in } L^p(\omega).$$

Therefore

$$u^\varepsilon \to u, \qquad \text{as } \varepsilon \to 0,\ \text{in } W^{k,p}(\omega).$$

■

Theorem 5.7 (Meyers–Serrin) *Let Ω be a bounded open set of $\mathbb{R}^n$. Let $k \in \mathbb{N}_0$, $p \in [1, +\infty)$. If $u \in W^{k,p}(\Omega)$ then there exists a sequence $\{u_m\}$ in $W^{k,p}(\Omega) \cap C^\infty(\Omega)$ which satisfies*

$$u_m \to u, \qquad \text{as } m \to \infty, \text{ in } W^{k,p}(\Omega).$$

Proof Let

$$\Omega_j = \left\{ x \in \Omega : \operatorname{dist}(x, \partial\Omega) > \frac{1}{j} \right\}, \qquad j \in \mathbb{N}.$$

It is not restrictive to assume $\Omega_j \neq \emptyset$, for every $j \in \mathbb{N}$. We have

$$\overline{\Omega}_j \subset \Omega_{j+1}, \quad \forall j \in \mathbb{N}, \quad \bigcup_{j=1}^{\infty} \Omega_j = \Omega. \tag{5.13}$$

Let $\phi_j \in C^\infty(\mathbb{R}^n)$, $j \in \mathbb{N}$, satisfy $\operatorname{supp} \phi_j \subset \Omega_{j+1}$; $\phi_j(x) = 1$ for every $x \in \Omega_j$; $0 \leq \phi_j(x) \leq 1$, for every $x \in \mathbb{R}^n$.

For any $j \in \mathbb{N}$ we get

$$x \in \mathbb{R}^n \setminus \Omega_{j+1} \implies \phi_j(x) = 0 \leq \phi_{j+1}(x)$$

and

$$x \in \Omega_{j+1} \implies \phi_j(x) \leq 1 = \phi_{j+1}(x).$$

Hence

$$\phi_j \leq \phi_{j+1}, \quad \forall j \in \mathbb{N}, \quad \text{in } \mathbb{R}^n.$$

Set

$$\zeta_0 = \phi_2, \quad \zeta_j = \phi_{j+1} - \phi_j, \quad \forall j \in \mathbb{N}$$

and

$$V_0 = \Omega_2, \quad V_j = \Omega_{j+3} \setminus \overline{\Omega}_j, \quad \forall j \in \mathbb{N}.$$

We have

$$\zeta_j \in C_0^\infty\left(\mathbb{R}^n\right), \quad \operatorname{supp} \zeta_j \subset V_j, \quad \forall j \in \mathbb{N}_0.$$

Moreover

$$0 \leq \zeta_j \leq 1, \quad \forall x \in \mathbb{R}^n, \quad \forall j \in \mathbb{N}_0$$

and

$$\sum_{j=0}^{\infty} \zeta_j(x) = 1, \quad \forall x \in \Omega. \tag{5.14}$$

Let us check (5.14). Let $x \in \Omega$. By (5.13) we have that there exists $\overline{m} \in \mathbb{N}$ such that $x \in \Omega_j$, for every $j \geq \overline{m}$. Let $m \geq \overline{m}$. We have $x \in \Omega_m$, hence $\phi_{m+1}(x) = 1$. Consequently, we have

$$\begin{aligned}\sum_{j=0}^{m} \zeta_j(x) &= \zeta_0(x) + \zeta_1(x) + \cdots + \zeta_m(x) = \\ &= \phi_2(x) + (\phi_3(x) - \phi_2(x)) + \cdots + (\phi_{m+1}(x) - \phi_m(x)) = \\ &= \phi_{m+1}(x) = 1.\end{aligned}$$

Therefore we have checked (5.14).

Now, let $u \in W^{k,p}(\Omega)$ and let us consider the functions $\zeta_j u$, $j \in \mathbb{N}_0$. Proposition 5.2 gives $\zeta_j u \in W^{k,p}(\Omega)$, in addition

$$\zeta_j u = 0, \quad \text{in } \Omega \setminus \overline{V}_j.$$

Let us denote by $W_0 = \Omega_4$, $W_1 = \Omega_5$, $W_j = \Omega_{j+4} \setminus \overline{\Omega}_{j-1}$, $j \geq 2$. Let $\delta > 0$ be fixed and let $0 < \varepsilon_j < \frac{1}{j+4} - \frac{1}{j+3}$ satisfy

$$u^j = \eta_{\varepsilon_j} \star (\zeta_j u) \in C^{\infty}(\Omega) \cap W^{k,p}(\Omega),$$

we have

$$u^j = 0, \quad \text{in } \Omega \setminus \overline{W}_j.$$

Theorem 5.6 implies that for every $j \in \mathbb{N}_0$ there exists $\varepsilon_j > 0$ such that

$$\left\| u^j - \zeta_j u \right\|_{W^{k,p}(\Omega)} = \left\| u^j - \zeta_j u \right\|_{W^{k,p}(W_j)} \leq \frac{\delta}{2^{j+1}}, \quad j \in \mathbb{N}_0. \tag{5.15}$$

We now set

$$v(x) = \sum_{j=0}^{\infty} u^j(x). \tag{5.16}$$

Let us notice that, for any $x \in \Omega$, only a finite number of terms of series (5.16) is different from 0. Moreover, as $u^j \in C^\infty(\Omega) \cap W^{k,p}(\Omega)$, for every $j \in \mathbb{N}_0$, we have $v \in C^\infty(\Omega) \cap W^{k,p}(\Omega)$.

Now, taking into account that

$$u = \sum_{j=0}^{\infty} \zeta_j u,$$

for any $h \in \mathbb{N}$, (5.15) and (5.16) give

$$\begin{aligned}
\|v - u\|_{W^{k,p}(\Omega_h)} &= \left\| \sum_{j=0}^{\infty} u^j - \sum_{j=0}^{\infty} \zeta_j u \right\|_{W^{k,p}(\Omega_h)} \leq \\
&\leq \sum_{j=0}^{\infty} \left\| u^j - \zeta_j u \right\|_{W^{k,p}(\Omega_h)} \leq \\
&\leq \sum_{j=0}^{\infty} \left\| u^j - \zeta_j u \right\|_{W^{k,p}(\Omega)} \leq \\
&\leq \sum_{j=0}^{\infty} \frac{\delta}{2^{j+1}} = \delta.
\end{aligned}$$

All in all, we have

$$\|v - u\|_{W^{k,p}(\Omega_h)} \leq \delta, \quad \forall h \in \mathbb{N}.$$

Hence

$$\|v - u\|_{W^{k,p}(\Omega)} = \lim_{h \to \infty} \|v - u\|_{W^{k,p}(\Omega_h)} \leq \delta.$$

Therefore, the sequence

$$u_m = \sum_{j=0}^{m} u^j(x), \quad m \in \mathbb{N},$$

satisfies the thesis. ■

Exercise 5.29 Prove Theorem 5.7 without the assumption that Ω is bounded. [Hint: consider $\Omega_j \cap B_j(x_0)$, x_0 fixed point, instead of Ω_j.] ♠

The following Theorem holds.

Theorem 5.8 (**C^∞ Approximation to the Boundary**) *Let Ω be a bounded open set of $\mathbb{R}^n$ whose boundary is of class $C^{0,1}$ with constants r_0, M_0. Let $u \in W^{k,p}(\Omega)$, $1 \leq p < +\infty$. Then there exists a sequence of functions $\{u_j\} \subset C^\infty\left(\overline{\Omega}\right)$ such that*

$$u_j \to u, \quad \text{as } j \to \infty, \text{ in } W^{k,p}(\Omega).$$

To prepare the proof of Theorem 5.8, we introduce some notations and we prove a Proposition,

Let Ω be a bounded open set of $\mathbb{R}^n$ whose boundary is of class $C^{0,1}$ with constants r_0, M_0. Let $x_0 \in \partial\Omega$. We may assume (up to isometry) that $x_0 = 0$ and

$$\Omega \cap Q_{r_0,2M_0} = \left\{x \in Q_{r_0,2M_0} : x_n > \varphi(x')\right\},$$

where $\varphi \in C^{0,1}\left(B'_{r_0}\right)$ satisfies

$$\varphi(0) = 0$$

and

$$\|\varphi\|_{C^{0,1}\left(\overline{B'_{r_0}}\right)} \leq M_0 r_0.$$

Set

$$V = \Omega \cap Q_{\frac{r_0}{2},\frac{M_0}{2}}.$$

Let y be any point of V. We look for what $\lambda > 0$ and $\varepsilon > 0$ we have (see Fig. 5.1)

$$B_\varepsilon\left(y^\varepsilon\right) \subset \Omega \cap Q_{r_0,M_0}, \tag{5.17}$$

where

$$y^\varepsilon = y + \varepsilon\lambda e_n,$$

1. Let us check that if ε and λ satisfy

$$\varepsilon < \frac{r_0}{2}, \quad \varepsilon(1+\lambda) < \frac{M_0 r_0}{2}, \tag{5.18}$$

then we have

$$B_\varepsilon\left(y^\varepsilon\right) \subset Q_{r_0,M_0}. \tag{5.19}$$

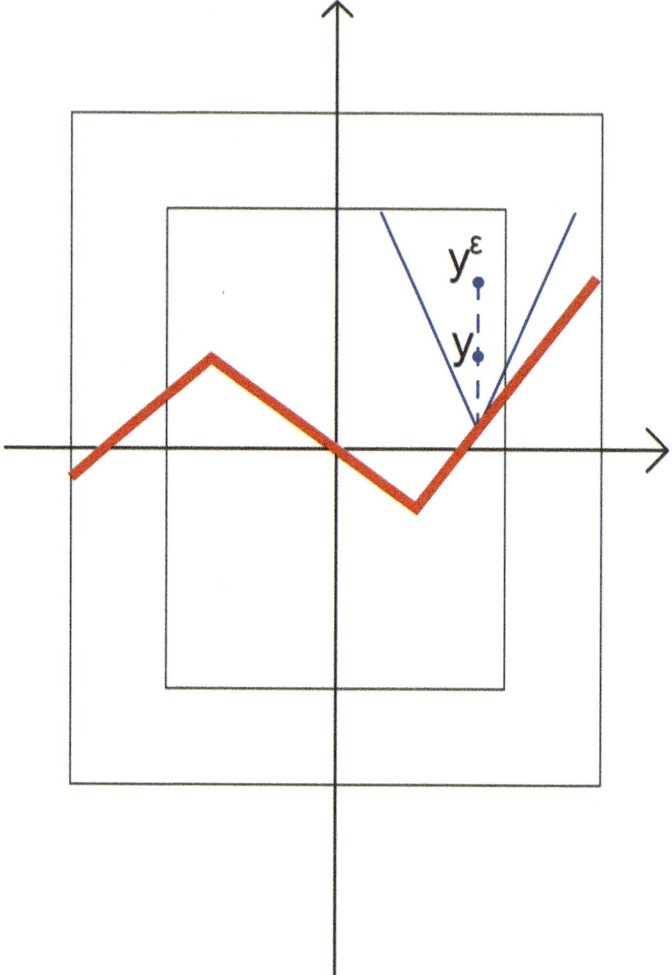

Fig. 5.1 The figure shows the position of y^ε in the cone with vertex $y = (y', \varphi(y'))$

Since $B_\varepsilon(y^\varepsilon) \subset B'_\varepsilon((y^\varepsilon)') \times [y_n^\varepsilon - \varepsilon, y_n^\varepsilon + \varepsilon]$, we have that the first condition of (5.18) implies

$$B'_\varepsilon((y^\varepsilon)') \subset B'_{r_0} \tag{5.20}$$

and second condition of (5.18) implies

$$y_n^\varepsilon + \varepsilon \leq \frac{M_0 r_0}{2} + \varepsilon\lambda + \varepsilon < M_0 r_0$$

and, similarly,

$$y_n^\varepsilon - \varepsilon \geq -\frac{M_0 r_0}{2} + \varepsilon\lambda - \varepsilon > -M_0 r_0.$$

Hence

$$\left[y_n^\varepsilon - \varepsilon, y_n^\varepsilon + \varepsilon\right] \subset [-M_0 r_0, M_0 r_0],$$

which gives (5.19).

2. In order that $B_\varepsilon(y^\varepsilon) \subset \Omega \cap Q_{r_0,M_0}$, it is suffices that, besides conditions (5.18), y^ε have a distance greater or equal to ε from the cone

$$x_n = M_0 \left|x' - y'\right| + \varphi(y').$$

Now, denoting by d_ε this distance, we have

$$\begin{aligned} d_\varepsilon &= \frac{\left|y_n + \varepsilon\lambda - \varphi(y')\right|}{\sqrt{1+M_0^2}} = \\ &= \frac{y_n + \varepsilon\lambda - \varphi(y')}{\sqrt{1+M_0^2}} \geq \\ &\geq \frac{\varepsilon\lambda}{\sqrt{1+M_0^2}}. \end{aligned}$$

Hence, in order that $d_\varepsilon > \varepsilon$ it suffices that $\lambda > \sqrt{1+M_0^2}$. Therefore, by choosing

$$\lambda = \lambda_0 := 2\sqrt{1+M_0^2}$$

and by requiring that

$$\varepsilon < \varepsilon_0 := \min\left\{\frac{r_0}{2}, \frac{M_0 r_0}{2}, \frac{M_0 r_0}{2\sqrt{1+M_0^2}}\right\},$$

we obtain (5.17).

For any $u \in L^p(\Omega)$ and $\varepsilon < \varepsilon_0$ we set

$$u_\varepsilon(x) = u\left(x^\varepsilon\right) = u(x + \lambda_0 \varepsilon e_n), \quad \forall x \in V \tag{5.21}$$

and

$$v^{\varepsilon}(x) = \int_{B_{\varepsilon}(x^{\varepsilon})\cap\Omega} \eta_{\varepsilon}(x + \lambda_0\varepsilon e_n - y)u(y)dy, \quad \forall x \in V, \tag{5.22}$$

where, we recall, $\eta_{\varepsilon}(x) = \varepsilon^{-n}\eta\left(\varepsilon^{-1}x\right)$. Now, since $B_{\varepsilon}\left(x^{\varepsilon}\right) \subset \Omega \cap Q_{r_0,M_0}$, we have

$$\begin{aligned} v^{\varepsilon}(x) &= \int_{B_{\varepsilon}(x^{\varepsilon})} \eta_{\varepsilon}(x + \lambda_0\varepsilon e_n - y)u(y)dy = \\ &= \int_{B_{\varepsilon}} \eta_{\varepsilon}(y)u(x + \lambda_0\varepsilon e_n - y)dy, \quad \forall x \in V. \end{aligned} \tag{5.23}$$

The first equality in (5.23) gives

$$v^{\varepsilon}(x) = \int_{\Omega} \eta_{\varepsilon}(x + \lambda_0\varepsilon e_n - y)u(y)dy.$$

Hence

$$\partial^{\alpha} v^{\varepsilon}(x) = \int_{\Omega} \partial_x^{\alpha}\eta_{\varepsilon}(x + \lambda_0\varepsilon e_n - y)u(y)dy, \quad \forall x \in V \tag{5.24}$$

so that

$$\partial^{\alpha} v^{\varepsilon} \in C^{\infty}\left(\overline{V}\right).$$

Moreover, for any $u \in W^{k,p}(\Omega)$, we have

$$\begin{aligned} \partial^{\alpha} v^{\varepsilon}(x) &= \int_{\Omega} \eta_{\varepsilon}(x + \lambda_0\varepsilon e_n - y)\partial^{\alpha}u(y)dy = \\ &= \int_{B_{\varepsilon}(x^{\varepsilon})} \eta_{\varepsilon}(x + \lambda_0\varepsilon e_n - y)\partial^{\alpha}u(y)dy, \quad \forall x \in V, \end{aligned} \tag{5.25}$$

for every $|\alpha| \leq k$.

We have the following

Proposition 5.9 *If $u \in W^{k,p}(\Omega)$ and $p \in [1, +\infty)$ then*

$$v^{\varepsilon} \to u, \quad \textit{as } \varepsilon \to 0, \textit{ in } W^{k,p}(V).$$

Proof First of all we prove

$$v^{\varepsilon} \to u, \qquad \text{as } \varepsilon \to 0, \text{ in } L^p(V).$$

The triangle inequality gives

$$\left\| v^{\varepsilon} - u \right\|_{L^p(V)} \leq \left\| u_{\varepsilon} - u \right\|_{L^p(V)} + \left\| v^{\varepsilon} - u_{\varepsilon} \right\|_{L^p(V)}. \tag{5.26}$$

Now

$$\left\| u_{\varepsilon} - u \right\|^p_{L^p(V)} = \int_V |u(x + \lambda_0 \varepsilon e_n) - u(x)|^p \, dx$$

and, by Theorem 3.36, we have

$$\lim_{\varepsilon \to 0} \left\| u_{\varepsilon} - u \right\|_{L^p(V)} = 0. \tag{5.27}$$

Moreover, by the second equality in (5.23) we have, for any $x \in V$,

$$v^{\varepsilon}(x) - u_{\varepsilon}(x) = \int_{B_{\varepsilon}} \eta_{\varepsilon}(y) \left(u(x + \lambda_0 \varepsilon e_n - y) - u(x + \lambda_0 \varepsilon e_n) \right) dy.$$

In order to prove that the second term on the right hand side in (5.26) goes to 0 it suffices to repeat the same steps which provide the proof of Theorem 3.44. For completeness, let us repeat these steps.

$$\int_V \left| v^{\varepsilon} - u_{\varepsilon} \right|^p dx \leq \int_V \left(\int_{B_{\varepsilon}} \eta_{\varepsilon}(y) \left| u(x + \lambda_0 \varepsilon e_n - y) - u(x) \right| dy \right)^p dx =$$

$$= \int_V \left(\int_{B_{\varepsilon}} \eta_{\varepsilon}^{1/p'}(y) \eta_{\varepsilon}^{1/p}(y) \left| u(x + \lambda_0 \varepsilon e_n - y) - u(x) \right| dy \right)^p dx \leq$$

$$\leq \int_V dx \left(\int_{B_{\varepsilon}} \eta_{\varepsilon}(y) dy \right)^{p/p'} \int_{B_{\varepsilon}} \eta_{\varepsilon}(y) \left| u(x + \lambda_0 \varepsilon e_n - y) - u(x) \right|^p dy =$$

$$= \int_V dx \int_{B_{\varepsilon}} \eta_{\varepsilon}(y) \left| u(x + \lambda_0 \varepsilon e_n - y) - u(x) \right|^p =$$

$$= \int_{B_{\varepsilon}} \left(\eta_{\varepsilon}(y) \int_V \left| u(x + \lambda_0 \varepsilon e_n - y) - u(x) \right|^p dx \right) dy \leq$$

$$\leq \sup_{|y| \leq \varepsilon} \int_V \left| u(x + \lambda_0 \varepsilon e_n - y) - u(x) \right|^p dx.$$

All in all, we have

$$\int_V \left|v^\varepsilon - u_\varepsilon\right|^p dx \leq \sup_{|y|\leq\varepsilon} \int_V |u(x + \lambda_0 \varepsilon e_n - y) - u(x)|^p dx.$$

Theorem 3.36 now yields

$$\lim_{\varepsilon\to 0} \left\|v^\varepsilon - u_\varepsilon\right\|_{L^p(V)} = 0.$$

By the latter, by (5.26) and by (5.27) we have

$$\lim_{\varepsilon\to 0} \left\|v^\varepsilon - u\right\|_{L^p(V)} = 0.$$

If $u \in W^{k,p}(\Omega)$, we obtain from what has been proven above and from (5.25)

$$\partial^\alpha v^\varepsilon \to \partial^\alpha u, \quad \text{as } \varepsilon \to 0, \text{ in } L^p(V), \text{ for } |\alpha| \leq k.$$

Hence

$$v^\varepsilon \to u, \quad \text{as } \varepsilon \to 0, \text{ in } W^{k,p}(V).$$

■

Proof of Theorem 5.8 Let $x_0 \in \partial\Omega$. Let us denote by $\widetilde{Q}_{r_0,2M_0}(x_0)$ the cylinder isometric to $Q_{r_0,2M_0}$ such that (3.71) holds. As a consequence, $\left\{\widetilde{Q}_{\frac{r_0}{2},\frac{M_0}{2}}(x_0)\right\}_{x_0\subset\partial\Omega}$ is an open covering of the compact set $\partial\Omega$. Let

$$\left\{\widetilde{Q}_{\frac{r_0}{2},\frac{M_0}{2}}(x_i)\right\}_{1\leq i\leq N}$$

be a finite subcovering of $\partial\Omega$. For any $1 \leq i \leq N$ let us denote by

$$V_i = \Omega \cap \widetilde{Q}_{\frac{r_0}{2},\frac{M_0}{2}}(x_i).$$

For any fixed $\delta > 0$ let $v_i \in C^\infty\left(\overline{V_i}\right)$ be the function constructed in (5.22) which satisfies

$$\|v_i - u\|_{W^{k,p}(V_i)} \leq \delta. \tag{5.28}$$

Moreover, let $V_0 \subset \Omega$ be such that

$$\Omega \subset \bigcup_{i=0}^N V_i$$

and let $\{\zeta_i\}_{0\le i\le N}$ be a partition of unity (compare Theorem 3.51) which satisfies $\zeta_i \in C_0^\infty(\mathbb{R}^n)$, $\operatorname{supp}\zeta_i \subset V_i$ for $1 \le i \le N$ and

$$\sum_{i=0}^{N} \zeta_i(x) = 1, \quad \forall x \in \Omega.$$

Let us denote by $v_0 = \zeta_0 u$ and

$$v = \sum_{i=0}^{N} \zeta_i v_i.$$

We have $v \in C^\infty(\overline{\Omega})$ and, taking into account (5.28),

$$\begin{aligned}\left\|\partial^\alpha v - \partial^\alpha u\right\|_{L^p(\Omega)} &= \left\|\sum_{i=0}^{N} \partial^\alpha(\zeta_i v_i) - \sum_{i=0}^{N} \partial^\alpha(\zeta_i u)\right\|_{L^p(\Omega)} \le \\ &\le \sum_{i=0}^{N} \left\|\partial^\alpha(\zeta_i v_i) - \partial^\alpha(\zeta_i u)\right\|_{L^p(V_i)} \le \\ &\le C\sum_{i=0}^{N} \|v_i - u\|_{W^{k,p}(V_i)} \le \\ &\le CN\delta,\end{aligned}$$

for every $|\alpha| \le k$.

Therefore, if $u \in W^{k,p}(\Omega)$ then for every $\eta > 0$ there exists $v \in C^\infty\left(\overline{\Omega}\right)$ such that

$$\|v - u\|_{W^{k,p}(\Omega)} < \eta.$$

The Theorem is proved. ■

We conclude this section with some propositions and exercises.

Proposition 5.10 *Let Ω be a connected open set of $\mathbb{R}^n$ and let $u \in W^{1,1}_{loc}(\Omega)$ satisfy*

$$\nabla u = 0, \quad in\ \Omega,$$

then u is almost everywhere equal to a constant.

Proof The proof follows as a straightforward consequence of Theorem 4.18. We leave the details as an exercise to the reader. ■

Proposition 5.11 *Let $F \in C^1(\mathbb{R})$ be such that F' is bounded. Let Ω be a bounded open set of $\mathbb{R}^n$ and let $u \in W^{1,p}(\Omega)$, $p \in [1, +\infty)$. Let us denote*

$$v := F(u), \quad a.\ e. \ \in \Omega$$

we have $v \in W^{1,p}(\Omega)$ and

$$\partial_j v = F'(u)\partial_j u, \quad a.\ e. \ \in \Omega, \qquad j = 1, \cdots, n.$$

Proof Since F' is a bounded function and Ω is a bounded set, we have $v \in L^p(\Omega)$. As a matter of fact

$$|v| \leq |F(u) - F(0)| + |F(0)| \leq \left\|F'\right\|_{L^\infty(\mathbb{R})} |u| + |F(0)| \in L^p(\Omega).$$

Now, we apply Theorem 5.8 and let $\{u_m\} \subset C^\infty(\Omega) \cap W^{1,p}(\Omega)$ be a sequence such that

$$u_m \to u, \quad \text{as } m \to \infty, \text{ in } W^{1,p}(\Omega). \tag{5.29}$$

We have

$$F(u_m) \to F(u), \quad \text{as } m \to \infty, \text{ in } L^p(\Omega). \tag{5.30}$$

Concerning the latter we have

$$\lim_{m\to\infty} \int_\Omega |F(u_m) - F(u)|^p \, dx \leq \left\|F'\right\|^p_{L^\infty(\mathbb{R})} \lim_{m\to\infty} \int_\Omega |u_m - u|^p \, dx = 0.$$

Now let us check that

$$F'(u_m)\partial_j u_m \to F'(u)\partial_j u, \quad \text{as } m \to \infty, \text{ in } L^p(\Omega), \tag{5.31}$$

for $j = 1, \cdots, n$.

We have

$$\begin{aligned}
\left\|F'(u_m)\partial_j u_m - F'(u)\partial_j u\right\|_{L^p(\Omega)} &\leq \left\|F'(u_m)\left(\partial_j u_m - \partial_j u\right)\right\|_{L^p(\Omega)} + \\
&\quad + \left\|\left(F'(u_m) - F'(u)\right)\partial_j u\right\|_{L^p(\Omega)} \leq \\
&\leq \left\|F'\right\|_{L^\infty(\mathbb{R})} \left\|\partial_j u_m - \partial_j u\right\|_{L^p(\Omega)} + \\
&\quad + \left\|\left(F'(u_m) - F'(u)\right)\partial_j u\right\|_{L^p(\Omega)}.
\end{aligned}$$

Since (5.29) holds, the second-to-last term on the right goes to zero as $m \to \infty$, concerning the last term, it goes to zero by the Dominated Convergence Theorem. Thus, we have checked (5.31).

Now, by (5.30) and (5.31) we have

$$\begin{aligned}\int_\Omega v\partial_j\phi dx &= \lim_{m\to\infty}\int_\Omega F(u_m)\partial_j\phi dx = \\ &= -\lim_{m\to\infty}\int_\Omega \partial_j\left(F(u_m)\right)\phi dx = \\ &= -\lim_{m\to\infty}\int_\Omega F'(u_m)\partial_j u_m\phi dx = \\ &= -\int_\Omega F'(u)\partial_j u\phi dx,\end{aligned}$$

for every $\phi \in C_0^\infty(\Omega)$ and every $j = 1, \cdots, n$. Hence

$$\partial_j v = F'(u)\partial_j u \in L^p(\Omega), \quad j = 1, \cdots, n,$$

so that, taking into account that $v \in L^p(\Omega)$, we have $v \in W^{1,p}(\Omega)$. ■

Proposition 5.12 *Let Ω be a bounded open set of $\mathbb{R}^n$, let $u \in W^{1,p}(\Omega)$ and $p \in [1, +\infty)$. Let us denote by $u_+ = \max\{u, 0\}$, $u_- = \min\{u, 0\}$. We have $u_+, u_- \in W^{1,p}(\Omega)$ and (up to a set of null measure)*

$$\nabla u_+ = \begin{cases}\nabla u, & \text{for } u > 0 \\ \\ 0, & \text{for } u \le 0\end{cases}, \tag{5.32}$$

$$\nabla u_- = \begin{cases}0, & \text{for } u \ge 0 \\ \\ -\nabla u, & \text{for } u < 0\end{cases}, \tag{5.33}$$

$$\nabla |u| = \begin{cases}\nabla u, & \text{for } u > 0 \\ 0, & \text{for } u = 0, \\ -\nabla u, & \text{for } u < 0\end{cases}. \tag{5.34}$$

Proof Let us prove (5.32). For any $\varepsilon > 0$ let us define

$$f_\varepsilon(t) = \begin{cases}\sqrt{t^2+\varepsilon^2} - \varepsilon, & \text{for } t > 0 \\ \\ 0, & \text{for } t \ge 0\end{cases}.$$

Recalling that $u \in W^{1,p}(\Omega)$, by the Dominated Convergence Theorem we get

$$\lim_{\varepsilon\to 0}\int_{\Omega}|f_{\varepsilon}(u)-u_{+}|^{p}\,dx=0,$$

As a matter of fact we have

$$\lim_{\varepsilon\to 0}|f_{\varepsilon}(u)-u_{+}|^{p}=0,\quad \text{in } \Omega$$

and

$$|f_{\varepsilon}(u)-u_{+}|^{p}\leq 2^{p}\left(|f_{\varepsilon}(u)|^{p}+|u_{+}|^{p}\right)\leq 2^{p+1}|u|^{p}\in L^{1}(\Omega).$$

Now, we have

$$f_{\varepsilon}'(t)=\begin{cases}\frac{t}{\sqrt{t^{2}+\varepsilon^{2}}}, & \text{for } t>0\\ \\ 0, & \text{for } t\geq 0\end{cases}$$

and

$$\left|f_{\varepsilon}'(t)\right|\leq 1.$$

Hence, Proposition 5.11 implies

$$f_{\varepsilon}(u)\in W^{1,p}(\Omega).$$

Now, up to a set of null measure, we have

$$\partial_{j}f_{\varepsilon}(u)=\begin{cases}\frac{u\partial_{j}u}{\sqrt{u^{2}+\varepsilon^{2}}}, & \text{for } u>0\\ \\ 0, & \text{for } u\geq 0\end{cases},$$

for $j=1,\cdots,n$. Hence, for any $\phi\in C_{0}^{\infty}(\Omega)$,

$$\begin{aligned}\int_{\Omega}u_{+}\partial_{j}\phi dx&=\lim_{\varepsilon\to 0}\int_{\Omega}f_{\varepsilon}(u)\partial_{j}\phi dx=\\ &=-\lim_{\varepsilon\to 0}\int_{u>0}\frac{u\partial_{j}u}{\sqrt{u^{2}+\varepsilon^{2}}}\phi dx= \qquad (5.35)\\ &=-\int_{u>0}\partial_{j}u\phi dx.\end{aligned}$$

In the last step we have applied the Dominated Convergence Theorem. Therefore

$$\int_\Omega u_+\partial_j\phi dx = -\int_\Omega \partial_j u\chi_{\{u>0\}}\phi dx, \quad \forall\phi \in C_0^\infty(\Omega),$$

from which we get (5.32). Concerning (5.33), it suffices to notice that $u_- = (-u)_+$. All in all, (5.34) follows by (5.32) and (5.33) (recall that $|u| = u_+ + u_-$). ■

Exercise 5.30 Let Ω be an open set of $\mathbb{R}^n$ and let $u, v \in W^{1,p}(\Omega)\cap L^\infty(\Omega)$. Prove that

$$uv \in W^{1,p}(\Omega) \cap L^\infty(\Omega)$$

and

$$\nabla(uv) = v\nabla u + u\nabla v. \tag{5.36}$$

♠

5.4 The Extension Theorems

Let us start by some propositions about the space $W_0^{k,p}(\Omega)$.

Proposition 5.13 *If $k \in \mathbb{N}$ and $p \in [1, +\infty)$ then*

$$W^{k,p}(\mathbb{R}^n) = W_0^{k,p}(\mathbb{R}^n).$$

Proof We limit ourselves to the case $k = 1$, the case $k > 1$ can be proved in a similar way and is left to the reader.

Let $R > 1$. Let $\zeta_R \in C_0^\infty(\mathbb{R}^n)$ satisfy

$$0 \le \zeta_R \le 1, \quad \text{in } \mathbb{R}^n,$$

$$\zeta_R(x) = 1, \quad \forall x \in B_R; \quad \zeta_R(x) = 0, \quad \forall x \in \mathbb{R}^n \setminus B_{2R},$$

$$|\nabla\zeta_R| \le C, \quad \text{in } \mathbb{R}^n,$$

where C is independent of R.

If $u \in W^{1,p}(\mathbb{R}^n)$, we have

$$\|u - \zeta_R u\|_{W^{1,p}(\mathbb{R}^n)} \to 0, \quad \text{as } R \to \infty. \tag{5.37}$$

Let us check (5.37).

$$\|u - \zeta_R u\|_{L^p(\mathbb{R}^n)} \leq \|u\|_{L^p(\mathbb{R}^n \setminus B_R)} \to 0, \quad \text{as } R \to \infty$$

and, for any $j = 1, \cdots, n$,

$$\left\|\partial_j u - \partial_j(\zeta_R u)\right\|_{L^p(\mathbb{R}^n)} = \left\|(1-\zeta_R)\partial_j u - u\partial_j \zeta_R\right\|_{L^p(\mathbb{R}^n)} \leq$$
$$\leq \left\|\partial_j u\right\|_{L^p(\mathbb{R}^n \setminus B_R)} + C\,\|u\|_{L^p(\mathbb{R}^n \setminus B_R)} \to 0, \quad \text{as } R \to \infty.$$

In order to complete the proof, firstly we observe (by Theorem 3.45)

$$(\zeta_R u) \star \eta_\varepsilon \to \zeta_R u, \quad \text{as } \varepsilon \to 0, \text{ in } W^{1,p}(\mathbb{R}^n), \tag{5.38}$$

where η_ε a mollifier. Moreover, let δ be any positive number, let $R_0 > 1$ be such that

$$\left\|u - \zeta_{R_0} u\right\|_{W^{1,p}(\mathbb{R}^n)} < \frac{\delta}{2}$$

and let $\varepsilon_0 > 0$ be such that

$$\left\|\left(\zeta_{R_0} u\right) \star \eta_{\varepsilon_0} - \zeta_{R_0} u\right\|_{W^{1,p}(\mathbb{R}^n)} < \frac{\delta}{2}.$$

From the last two inequalities and the triangle inequality we get

$$\left\|\left(\zeta_{R_0} u\right) \star \eta_{\varepsilon_0} - u\right\|_{W^{1,p}(\mathbb{R}^n)} < \delta.$$

Since $\left(\zeta_{R_0} u\right) \star \eta_{\varepsilon_0} \in C_0^\infty(\mathbb{R}^n)$, the Proposition is proved. ■

Theorem 5.14 (First Poincaré Inequality) *Let Ω be a bounded open set of $\mathbb{R}^n$. Let $p \in [1, +\infty]$, $k \in \mathbb{N}$. The following inequality holds true, for any $\alpha \in \mathbb{N}_0^n$, $|\alpha| \leq k-1$,*

$$\left\|\partial^\alpha u\right\|_{L^p(\Omega)} \leq C d^{k-|\alpha|} \sum_{|\beta|=k} \left\|\partial^\beta u\right\|_{L^p(\Omega)}, \quad \forall u \in W_0^{k,p}(\Omega), \tag{5.39}$$

where d is the diameter of Ω and C depends on n and k only.

Proof We restrict ourselves to the case $k = 1$, actually starting from this case (5.39) can easily be deduced by induction. It is not restrictive to assume $0 \in \Omega$ and

$$\Omega \subset [-d, d]^n.$$

Let $u \in W_0^{1,p}(\Omega)$. Let $\{u_j\}$ be a sequence in $C_0^\infty(\Omega)$ such that

$$\{u_j\} \to u, \qquad \text{in } W^{1,p}(\Omega).$$

For any $p \in [1, +\infty)$ and any $j \in \mathbb{N}$, we have

$$\begin{aligned}
\left|u_j(x)\right| &= \left|u_j(x) - u_j(x', -d)\right| = \\
&= \left|\int_{-d}^{x_n} \partial_y u_j(x', y)dy\right| \leq \\
&\leq \int_{-d}^{d} \left|\partial_y u_j(x', y)\right| dy \leq \\
&\leq (2d)^{1/p'} \left(\int_{-d}^{d} \left|\partial_y u_j(x', y)\right|^p dy\right)^{1/p}.
\end{aligned}$$

Hence

$$\left|u_j(x)\right|^p \leq (2d)^{p-1} \int_{-d}^{d} \left|\partial_y u_j(x', y)\right|^p dy. \tag{5.40}$$

Let us integrate both the sides of (5.40) over $[-d, d]$ w.r.t. x_n. We get

$$\int_{-d}^{d} \left|u_j(x', x_n)\right|^p dx_n \leq (2d)^p \int_{-d}^{d} \left|\partial_y u_j(x', y)\right|^p dy.$$

Now, let us integrate both the sides of the last inequality over $[-d, d]^{n-1}$. We get

$$\left(\int_\Omega \left|u_j(x)\right|^p dx\right)^{1/p} \leq 2d \left(\int_{-d}^{d} \left|\partial_{x_n} u_j(x)\right|^p dx\right)^{1/p}.$$

Passing to the limit as $j \to \infty$, we obtain

$$\|u\|_{L^p(\Omega)} \leq Cd \, \|\nabla u\|_{L^p(\Omega)} \, .$$

If $p = +\infty$, then we have

$$\left|u_j(x)\right| = \left|u_j(x) - u_j(x', -d)\right| \leq 2d \left\|\partial_{x_n} u_j\right\|_{L^\infty(\Omega)},$$

from which, passing to the limit as $j \to \infty$, we have

$$\|u\|_{L^\infty(\Omega)} \leq Cd \, \|\nabla u\|_{L^\infty(\Omega)} \, .$$

∎

Remark 5.20 From the proof of Proposition 5.14 it is evident that inequality (5.39) also holds if Ω is contained in a strip of $\mathbb{R}^n$ of the type $\mathbb{R}^{n-1} \times [-d, d]$ or isometric to it. ♦

Remark 5.21 Proposition 5.14 implies that the following seminorms are actually norms on $W_0^{k,p}(\Omega)$

$$\sum_{|\beta|=k} \left\|\partial^{\beta} u\right\|_{L^p(\Omega)}, \quad \left(\sum_{|\beta|=k} \left\|\partial^{\beta} u\right\|_{L^p(\Omega)}^p\right)^{1/p}.$$

Moreover such norms are equivalent to the norm

$$\|u\|_{W^{k,p}(\Omega)}.$$

♦

Proposition 5.15 *Let $p \in [1, +\infty]$, $k \in \mathbb{N}$. Let Ω and $\widetilde{\Omega}$ be open sets of $\mathbb{R}^n$ such that $\Omega \subset \widetilde{\Omega}$. Let $u \in W_0^{k,p}(\Omega)$.*

Denoting

$$\widetilde{u} = \begin{cases} u, & \text{in } \Omega \\ \\ 0, & \text{in } \widetilde{\Omega} \setminus \Omega \end{cases}, \tag{5.41}$$

we have $\widetilde{u} \in W_0^{k,p}\left(\widetilde{\Omega}\right)$.

Proof Let $u \in W_0^{k,p}(\Omega)$. Let $\{u_j\} \subset C_0^{\infty}(\Omega)$ be a sequence such that

$$u_j \to u, \quad \text{as } j \to \infty, \text{ in } W^{k,p}(\Omega).$$

Hence, by denoting

$$\widetilde{u}_j = \begin{cases} u_j, & \text{in } \Omega \\ \\ 0, & \text{in } \widetilde{\Omega} \setminus \Omega \end{cases}, \tag{5.42}$$

we have $\{\widetilde{u}_j\} \subset C_0^{\infty}\left(\widetilde{\Omega}\right)$ and

$$\widetilde{u}_j \to \widetilde{u}, \quad \text{as } j \to \infty, \text{ in } W^{k,p}\left(\widetilde{\Omega}\right).$$

Therefore $\widetilde{u} \in W_0^{k,p}\left(\widetilde{\Omega}\right)$. ■

The Main Theorem of the present section is the following one.

Theorem 5.16 (Extension in $W^{1,p}$) *Let Ω be a bounded open set of $\mathbb{R}^n$ whose boundary is of class $C^{0,1}$ with constants r_0, M_0. Let d_0 be the diameter of Ω. Let $\widetilde{\Omega}$ be an open set of $\mathbb{R}^n$ such that $\Omega \Subset \widetilde{\Omega}$ and let $p \in [1, +\infty)$.*

Then there exists a linear bounded operator

$$E : W^{1,p}(\Omega) \to W^{1,p}(\mathbb{R}^n), \tag{5.43}$$

which satisfies, for any $u \in W^{1,p}(\Omega)$,

$$Eu = u, \quad in\ \Omega, \tag{5.44}$$

$$\text{supp}(Eu) \subset \widetilde{\Omega}. \tag{5.45}$$

Moreover, there exists a constant C, depending on r_0, M_0, d_0, n and p only, such that

$$\|Eu\|_{W^{1,p}(\mathbb{R}^n)} \leq C \|u\|_{W^{1,p}(\Omega)}, \quad \forall u \in W^{1,p}(\Omega). \tag{5.46}$$

Proof Let $x_0 \in \partial\Omega$. We may assume (up to an isometry) that $x_0 = 0$ and

$$\Omega \cap Q_{r_0,2M_0} = \left\{x \in Q_{r_0,2M_0} : x_n > \varphi(x')\right\},$$

where $\varphi \in C^1\left(\overline{B'}_{r_0}\right)$ satisfies

$$\varphi(0) = 0$$

and

$$\|\varphi\|_{C^1\left(\overline{B'}_{r_0}\right)} = \|\varphi\|_{C^0\left(\overline{B'}_{r_0}\right)} + r_0[\varphi]_{0,1,B'_{r_0}} \leq M_0 r_0.$$

$$V^+ = Q_{\frac{r_0}{4},\frac{M_0}{4}} \cap \Omega, \quad V^- = Q_{\frac{r_0}{4},\frac{M_0}{4}} \setminus \overline{V^+}.$$

Let us notice that, for every $x' \in \overline{B'}_{r_0/4}$ we have

$$\left|\varphi(x')\right| = \left|\varphi(x') - \varphi(0)\right| \leq [\varphi]_{0,1,B'_{r_0}} \left|x'\right| \leq \frac{M_0 r_0}{4}. \tag{5.47}$$

First, we assume that $u \in C^{\infty}\left(\overline{\Omega}\right)$ and we define

$$\nu(x') = \frac{\left(\nabla\varphi(x'), -1\right)}{\sqrt{|\nabla_{x'}\varphi|^2 + 1}}, \quad x' \in B'_{r_0},$$

$$\overline{u}(x) = \begin{cases} u(x), & \text{in } V^+ \\ \\ v(x), & \text{in } V^- \end{cases},$$

where (Fig. 5.2)

$$v(x) = u(x', 2\varphi(x') - x_n).$$

Claim $\overline{u} \in W^{1,p}\left(Q_{\frac{r_0}{4}, \frac{M_0}{4}}\right)$ and

$$\|\overline{u}\|_{W^{1,p}\left(Q_{\frac{r_0}{4}, \frac{M_0}{4}}\right)} \leq C\, \|u\|_{W^{1,p}(\Omega)}, \tag{5.48}$$

where C depends on M_0 only.

Proof of the Claim Let $\Phi \in C_0^{\infty}\left(Q_{\frac{r_0}{4}, \frac{M_0}{4}}\right)$ and $1 \leq i \leq n$. Denoting by Γ the graph of $\varphi_{|B'_{r_0/4}}$, by the divergence Theorem we get

$$\begin{aligned} \int_{Q_{\frac{r_0}{4}, \frac{M_0}{4}}} \overline{u}\partial_i\Phi dx &= \int_{V^+} u\partial_i\Phi dx + \int_{V^-} v\partial_i\Phi dx = \\ &= -\int_{V^+} \partial_i u\Phi dx + \int_{\Gamma} u\Phi(\nu \cdot e_i)dS - \\ &- \int_{V^-} \partial_i v\Phi dx - \int_{\Gamma} v\Phi(\nu \cdot e_i)dS = \\ &= -\int_{Q_{\frac{r_0}{4}, \frac{M_0}{4}}} w_i\Phi dx + \int_{\Gamma} (u - v)\Phi(\nu \cdot e_i)dS, \end{aligned}$$

where

$$w_i(x) = \begin{cases} \partial_i u(x), & \text{in } V^+ \\ \\ \partial_i v(x), & \text{in } V^- \end{cases}. \tag{5.49}$$

On the other hand,

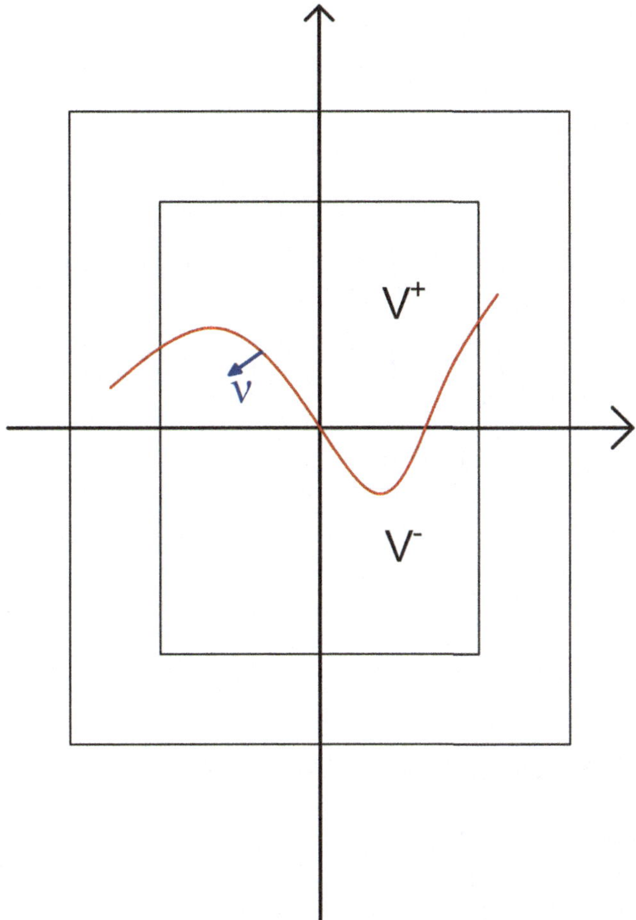

Fig. 5.2 Here, $V^+ = Q_{\frac{r_0}{4},\frac{M_0}{4}} \cap \Omega$ and $V^- = Q_{\frac{r_0}{4},\frac{M_0}{4}} \setminus \overline{V^+}$ are shown, and the red graph represents a portion of $\partial\Omega$

$$(u - v)(x', \varphi(x')) = 0, \quad \forall x' \in B'_{r_0/4},$$

hence

$$\int_{Q_{\frac{r_0}{4},\frac{M_0}{4}}} \overline{u}\partial_i \Phi dx = -\int_{Q_{\frac{r_0}{4},\frac{M_0}{4}}} w_i \Phi dx, \quad \forall \Phi \in C_0^\infty\left(Q_{\frac{r_0}{2},\frac{M_0}{4}}\right). \tag{5.50}$$

Therefore

$$\partial_i \overline{u} = w_i(x), \quad \forall x \in Q_{\frac{r_0}{4},\frac{M_0}{4}}. \tag{5.51}$$

Now, let us notice that

$$\int_{Q_{\frac{r_0}{4},\frac{M_0}{4}}} |\overline{u}(x)|^p\,dx = \int_{V^+} |u(x)|^p\,dx + \int_{V^-} |v(x)|^p\,dx \tag{5.52}$$

and

$$\begin{aligned}\int_{V^-} |v(x)|^p\,dx &= \int_{B'_{r_0/4}} dx' \int_{-\frac{M_0 r_0}{4}}^{\varphi(x')} \left|u(x', 2\varphi(x') - x_n)\right|^p dx_n = \\ &= \int_{B'_{r_0/4}} dx' \int_{\varphi(x')}^{2\varphi(x')+\frac{M_0 r_0}{4}} \left|u(x', \xi_n)\right|^p d\xi_n \le \\ &\le \int_{\Omega} |u(x)|^p\,dx.\end{aligned} \tag{5.53}$$

In the last inequality we have used (5.47). By (5.52) and (5.53), we have,

$$\int_{Q_{\frac{r_0}{4},\frac{M_0}{4}}} |\overline{u}(x)|^p\,dx \le 2\int_{\Omega} |u(x)|^p\,dx. \tag{5.54}$$

Now, (5.49) gives

$$|\nabla\overline{u}(x)| \le C\left|(\nabla u)(x', 2\varphi(x') - x_n)\right|, \quad \forall x \in V^-,$$

where C depends on M_0 only. Hence

$$\begin{aligned}\int_{Q_{\frac{r_0}{4},\frac{M_0}{4}}} |\nabla\overline{u}(x)|^p\,dx &\le \int_{V^+} |\nabla u(x)|^p\,dx + C\int_{V^-} \left|(\nabla u)(x', 2\varphi(x') - x_n)\right|^p dx \le \\ &\le C\int_{\Omega} |\nabla u(x)|^p\,dx.\end{aligned}$$

From the just obtained inequality and from (5.54) we obtain (5.48). **This concludes the proof of the Claim.**

Since $\partial\Omega$ is a compact set, there exist $x_{0,1}, \cdots, x_{0,N} \in \partial\Omega$ such that

$$\partial\Omega \subset \bigcup_{j=1}^{N} \widetilde{Q}_{\frac{r_0}{4},\frac{M_0}{4}}(x_{0,j}),$$

where, for any $j = 1, \cdots, N$, $\widetilde{Q}_{\frac{r_0}{4},\frac{M_0}{4}}(x_{0,j})$ are suitable cylinders which are isometric to $Q_{\frac{r_0}{4},\frac{M_0}{4}}$. Moreover, let us denote by $\overline{u}_j$ the extensions of u on

$\widetilde{Q}_{\frac{r_0}{4},\frac{M_0}{4}}(x_{0,j})$. Let us employ the partition of unity (Lemma 3.51). Set $V_j = \widetilde{Q}_{\frac{r_0}{4},\frac{M_0}{4}}(x_{0,j})$, $j = 1, \cdots, N$, we have that there exist $\zeta_0, \zeta_1, \cdots, \zeta_N \in C_0^\infty(\mathbb{R}^n)$ such that

$$0 \le \zeta_j(x) \le 1, \quad j = 1, \cdots, N, \ \forall x \in \mathbb{R}^n$$

$$\operatorname{supp} \zeta_j \subset V_j, \quad j = 1, \cdots, N, \quad \operatorname{supp} \zeta_0 \subset \mathbb{R}^n \setminus \partial\Omega,$$

$$\sum_{j=0}^{N} \zeta_j(x) = 1, \quad \forall x \in \mathbb{R}^n,$$

and

$$\sum_{j=1}^{N} \zeta_j(x) = 1, \quad \text{for every } x \text{ in a neighborhood of } \partial\Omega.$$

Also, let $\eta \in C_0^\infty\left(\widetilde{\Omega}\right)$ be such that $0 \le \eta(x) \le 1$, $\eta(x) = 1$, for every $x \in \Omega$. Set

$$\widetilde{u} = \eta\left(\zeta_0 u + \sum_{j=1}^{N} \zeta_j \overline{u}_j\right).$$

By (5.48) and by the triangle inequality, we have

$$\|\widetilde{u}\|_{W^{1,p}(\mathbb{R}^n)} \le C\, \|u\|_{W^{1,p}(\Omega)}, \tag{5.55}$$

where C depends on r_0, n and M_0 only. Moreover, we have

$$\widetilde{u}(x) = u(x), \quad \forall x \in \Omega, \tag{5.56}$$

$$\operatorname{supp} \widetilde{u} \subset \widetilde{\Omega}. \tag{5.57}$$

Now, let us denote by

$$Eu := \widetilde{u}, \quad \forall u \in C^\infty\left(\overline{\Omega}\right).$$

E is a linear operator and it satisfies the inequality

$$\|Eu\|_{W^{1,p}(\mathbb{R}^n)} \le C\, \|u\|_{W^{1,p}(\Omega)}. \tag{5.58}$$

Now, let $u \in W^{1,p}(\Omega)$. Let us apply Theorem 5.8. Therefore, let $\{u_m\} \subset C^\infty\left(\overline{\Omega}\right)$ be such that

$$\{u_m\} \to u, \quad \text{in } W^{1,p}(\Omega).$$

We have, by (5.58),

$$\|Eu_m - Eu_{m'}\|_{W^{1,p}(\mathbb{R}^n)} \le C\,\|u_m - u_{m'}\|_{W^{1,p}(\Omega)}\,.$$

Hence $\{Eu_m\}$ is a Cauchy sequence in $W^{1,p}(\mathbb{R}^n)$ consequently it converges to a function which we continue to denote by $\widetilde{u} \in W^{1,p}(\mathbb{R}^n)$ which satisfies trivially (5.55)–(5.57). ■

We merely state, with some comments, the Theorem of extension for $W^{k,p}(\Omega)$, where $k \ge 1$. We refer to [50] for a proof.

Theorem 5.17 (Extension in $W^{k,p}$) *Let $k \ge 1$ and $p \in [1, +\infty)$. Let Ω be a bounded open set of $\mathbb{R}^n$ whose boundary is of class $C^{k-1,1}$ with constants r_0, M_0. Let d_0 be the diameter of Ω. Let $\widetilde{\Omega}$ be an open set of $\mathbb{R}^n$ such that $\Omega \Subset \widetilde{\Omega}$.*

Then there exists a bounded linear operator

$$E : W^{k,p}(\Omega) \to W^{k,p}(\mathbb{R}^n),$$

such that, for any $u \in W^{k,p}(\Omega)$ we have

$$Eu = u, \quad \text{on } \Omega,$$

$$\operatorname{supp}(Eu) \subset \widetilde{\Omega}.$$

Moreover, there exists a constant C, depending on r_0, M_0, d_0, n k and p only, such that

$$\|Eu\|_{W^{k,p}(\mathbb{R}^n)} \le C\,\|u\|_{W^{k,p}(\Omega)}\,, \quad \forall u \in W^{k,p}(\Omega). \tag{5.59}$$

Exercise 5.31

(i) Let $u \in C^\infty\left(\overline{B_r^+}\right)$, where $B_r^\pm = \{x \in \mathbb{R}^n : |x| < r, x_n \gtrless 0\}$. Let us define

$$\overline{u}(x) = \begin{cases} u(x), & \text{in } B_r^+ \\ \\ v(x), & \text{in } B_r^- \end{cases},$$

where

$$v(x) = -3u(x', -x_n) + 4u\left(x', -\frac{x_n}{2}\right).$$

Prove that, if $p \in [1, +\infty)$ then $\overline{u}(x) \in W^{2,p}(B_r)$ and the following inequality holds true

$$\|\overline{u}\|_{W^{2,p}(B_r)} \leq C \|u\|_{W^{2,p}(\Omega)}, \quad \forall u \in W^{2,p}(B_r^+).$$

(ii) Let us define

$$C^\infty\left(\overline{B_r^+}\right) \ni u \to Eu = \overline{u} \in W^{2,p}(B_r).$$

Prove that the operator E can be extended to $W^{2,p}\left(B_r^+\right)$ and that it satisfies $Eu = u$ in B_r^+.

(iii) Let $k \in \mathbb{N}$. Let $c_1, \cdots, c_k$ be such that

$$\sum_{j=1}^{k} c_j \left(-\frac{1}{j}\right)^m = 1, \quad m = 0, 1, \cdots, k-1,$$

(check that such $c_1, \cdots, c_k$ exist); let us define for any $u \in C^\infty\left(\overline{B_r^+}\right)$

$$\overline{u}(x) = \begin{cases} u(x), & \text{in } B_r^+ \\ \\ w(x), & \text{in } B_r^- \end{cases},$$

where

$$w(x) = \sum_{j=1}^{k} c_j u\left(x', -\frac{x_n}{j}\right).$$

Prove that if $p \in [1, +\infty)$ then $\overline{u}(x) \in W^{k,p}(B_r)$ and the following inequality holds true

$$\|\overline{u}\|_{W^{k,p}(B_r)} \leq C \|u\|_{W^{k,p}(B_r^+)}, \quad \forall u \in W^{k,p}(B_r^+).$$

Moreover, deduce that the following operator

$$C^\infty\left(\overline{B_r^+}\right) \ni u \to E_k u = \overline{u} \in W^{k,p}(B_r)$$

can be extended to $W^{k,p}(B_r)$and $E_k u = u$ in B_r^+. ♠

5.5 Traces in $W^{1,p}(\Omega)$

It is well-known that if $u \in C^0\left(\overline{\Omega}\right)$, then we can define its trace on $\partial\Omega$, namely $u_{|\partial\Omega}$. If, on the other hand, $u \in L^p(\Omega)$, generally, it does not make sense to consider its trace on $\partial\Omega$. In the present section we will see that we can define a notion of trace that extends the known one for the functions of $W^{1,p}(\Omega) \cap C^0\left(\overline{\Omega}\right)$.

More precisely we have

Theorem 5.18 (Trace Theorem) *Let Ω be a bounded open set of class $C^{0,1}$ with constants r_0 and M_0. Let $p \in [1, +\infty)$. Let d_0 be the diameter of Ω. Then there exists an unique bounded linear operator*

$$T : W^{1,p}(\Omega) \to L^p(\partial\Omega),$$

which satisfies:

(i) $T(u) = u_{|\partial\Omega}$ *for every* $u \in C^0\left(\overline{\Omega}\right) \cap W^{1,p}(\Omega)$;
(ii)

$$\|T(u)\|_{L^p(\partial\Omega)} \leq \|u\|_{W^{1,p}(\Omega)}, \quad \forall u \in W^{1,p}(\Omega),$$

where C depends by r_0, M_0, d_0 and p;
(iii)

$$\int_\Omega u \operatorname{div} \Phi dx = -\int_\Omega \nabla u \cdot \Phi dx + \int_{\partial\Omega} (\Phi \cdot \nu) T u dS,$$

for every $u \in W^{1,p}(\Omega)$ and for every $\Phi \in C^1\left(\overline{\Omega}, \mathbb{R}^n\right)$.

*The function Tu is called the **trace** of u on $\partial\Omega$.*

Proof We first notice that Φ is arbitrary in $C^1\left(\overline{\Omega}, \mathbb{R}^n\right)$, if T there exists, then it is unique.

Let us prove the existence of T. First, let us consider the case where $u \in C^\infty\left(\overline{\Omega}\right)$. Since $\partial\Omega$ is compact, we may consider a partition of unity subordinate to a finite covering $\left\{V_j\right\}_{1 \leq j \leq N}$, where $V_j = \widetilde{Q}_{\frac{r_0}{2}, M_0}\left(x_{0,j}\right)$, $j = 1, \cdots, N$ ($\widetilde{Q}_{\frac{r_0}{2}, M_0}\left(x_{0,j}\right)$ is a cylinder isometric to $Q_{\frac{r_0}{2}, M_0}$)

$$\zeta_j \in C_0^\infty(\mathbb{R}^n), \quad 0 \leq \zeta_j \leq 1, \quad \operatorname{supp} \zeta_j \subset V_j, \quad \sum_{j=1}^n \zeta_j = 1 \text{ on } \partial\Omega.$$

Let $j \in \{1, \cdots, N\}$ be fixed. Up to isometries we may assume $V_j = Q_{\frac{r_0}{2}, M_0}$ and

$$Q_{r_0, M_0} \cap \Omega = \left\{(x', x_n) \in Q_{r_0, M_0} : x_n > \varphi(x')\right\},$$

where $\varphi \in C^1\left(\overline{B'}_{r_0}\right)$ satisfies $\varphi(0)=0$ and

$$\|\varphi\|_{L^\infty(B'_{r_0})} + r_0\|\nabla\varphi\|_{L^\infty(B'_{r_0})} \leq M_0 r_0.$$

Let $v=\zeta_j u$. For any $t \in \left[0, \frac{M_0 r_0}{2}\right]$ and any $x' \in B'_{r_0/2}$ we have

$$v\left(x', \varphi(x')\right) = v\left(x', \varphi(x')+t\right) - \int_{\varphi(x')}^{\varphi(x')+t} \partial_{x_n} v(x', x_n) dx_n.$$

Hölder inequality gives

$$\left|v\left(x', \varphi(x')\right)\right|^p \leq 2^{p-1}\left|v\left(x', \varphi(x')+t\right)\right|^p + 2^{p-1} t^{p-1} \int_{\varphi(x')}^{M_0 r_0} \left|\partial_{x_n} v\left(x', x_n\right)\right|^p dx_n$$

and if we integrate with respect to x_n over $\left[0, \frac{M_0 r_0}{2}\right]$ both the sides of the last inequality we get

$$\begin{aligned} \frac{M_0 r_0}{2}\left|v\left(x', \varphi(x')\right)\right|^p \leq 2^{p-1} \int_{\varphi(x')}^{\varphi(x')+\frac{M_0 r_0}{2}} \left|v\left(x', x_n\right)\right|^p dx_n + \\ + \frac{2^{p-1}}{p}\left(\frac{M_0 r_0}{2}\right)^p \int_{\varphi(x')}^{M_0 r_0} \left|u \partial_{x_n}\zeta_j + \zeta_j \partial_{x_n} u\right|^p dx_n. \end{aligned}$$

Now, we multiply both the sides of the last inequality by $\sqrt{1+|\nabla_{x'}\varphi(x')|^2}$, and we integrate over $B'_{r_0/2}$ obtaining

$$\int_{\partial\Omega} \left|u\zeta_j\right|^p dS \leq C\int_\Omega |u|^p dx + C\int_\Omega |\nabla u|^p dx,$$

where C depends by M_0 and r_0. Therefore

$$\begin{aligned} \|u\|_{L^p(\partial\Omega)} &= \left\|\sum_{j=1}^N \zeta_j u\right\|_{L^p(\partial\Omega)} \leq \\ &\leq \sum_{j=1}^N \left\|\zeta_j u\right\|_{L^p(\partial\Omega)} \leq \\ &\leq C\|u\|_{W^{1,p}(\Omega)}, \quad \forall u \in C^\infty\left(\overline{\Omega}\right). \end{aligned} \tag{5.60}$$

Set

$$Tu = u_{|\partial\Omega}, \quad \forall u \in C^{\infty}\left(\overline{\Omega}\right).$$

Inequality (5.60) implies

$$\|Tu\|_{L^p(\partial\Omega)} \leq C\,\|u\|_{W^{1,p}(\Omega)}, \quad \forall u \in C^{\infty}\left(\overline{\Omega}\right). \tag{5.61}$$

Now let $u \in W^{1,p}(\Omega)$. From Theorem 5.8 we have that there exists a sequence $\{u_m\}$ in $C^{\infty}\left(\overline{\Omega}\right)$ such that

$$\{u_m\} \to u, \quad \text{in } W^{1,p}(\Omega).$$

In particular, (5.61) implies that $\{Tu_m\}$ is a Cauchy sequence in $L^p(\partial\Omega)$. Set

$$Tu = \lim_{m\to\infty} Tu_m.$$

Now, we observe that if $u \in C^0\left(\overline{\Omega}\right)\cap W^{1,p}(\Omega)$, then the sequence $\{u_m\}$ constructed in the proof of Proposition 5.9 (with $\varepsilon = 1/m$) uniformly converges to u. Hence, for any $u \in C^0\left(\overline{\Omega}\right) \cap W^{1,p}(\Omega)$, we have

$$Tu = \lim_{m\to\infty} Tu_m = \lim_{m\to\infty} u_m = u, \quad \text{in } L^p(\partial\Omega).$$

Therefore, we have proved (i) and (ii), now let us prove (iii). Let $u \in W^{1,p}(\Omega)$. Let $\{u_m\}$ be a sequence in $C^{\infty}\left(\overline{\Omega}\right)$ which converges to u in $W^{1,p}(\Omega)$. We have, for any $\Phi \in C^1\left(\overline{\Omega}, \mathbb{R}^n\right)$,

$$\begin{aligned}\int_{\Omega} u \operatorname{div} \Phi dx &= \lim_{m\to\infty}\int_{\Omega} u_m \operatorname{div} \Phi dx = \\ &= \lim_{m\to\infty}\left(-\int_{\Omega}\nabla u_m\cdot\Phi dx + \int_{\partial\Omega}(\Phi\cdot\nu)u_m dS\right) = \\ &= -\int_{\Omega}\nabla u\cdot\Phi dx + \int_{\partial\Omega}(\Phi\cdot\nu)Tu dS.\end{aligned}$$

■

Remark If $u \in W_0^{1,p}(\Omega)$ then $Tu = 0$. Actually, under the same assumption of Theorem 5.18 the conversely is also valid, but here we omit the proof and refer to Theorem 2 of Ch. 5 of [38]. ♦

The issue of traces will be taken up in Sect. 5.12.

5.6 The Sobolev Spaces of Function of One Variable

In this section, we will briefly explore Sobolev spaces when the spatial dimension is equal to 1. It is important to note that if $I \subset \mathbb{R}$ is a bounded open interval, the theorems established in the preceding sections remain applicable. It is indeed a valuable exercise (which we leave to the reader) to adapt the proofs of these theorems and recognize their simplification in this specific case. Particularly, Theorem 5.8 implies that $C^\infty\left(\overline{I}\right)$ is dense in $W^{k,p}(I)$ for every $p \in [1, +\infty)$, and through the extension Theorem 5.16, it is revealed that if $\widetilde{I} \supset I$, where $\widetilde{I}$ is an open interval in $\mathbb{R}$, there exists a bounded linear operator

$$E : W^{1,p}(I) \to W^{k,p}\left(\widetilde{I}\right),$$

which satisfies

$$\|Eu\|_{W^{1,p}(\widetilde{I})} \leq C\, \|u\|_{W^{1,p}(I)}, \quad \forall u \in W^{1,p}(I),$$

$$Eu_{|I} = u, \quad \text{supp}(Eu) \subset I.$$

Now we investigate the basic relations between the absolutely continuous functions and the functions of $W^{1,p}(I)$. By the above mentioned extension Theorem, we may consider the space $W^{1,p}(\mathbb{R})$, instead of $W^{1,p}(I)$.

Let us recall that if $u \in L^1_{loc}(\mathbb{R})$ then

$$\lim_{r \to 0} \frac{1}{2r} \int_{x-r}^{x+r} |u(t) - u(x)|\, dt = 0, \quad \text{a.e. } x \in \mathbb{R} \tag{5.62}$$

and

$$\lim_{r \to 0} \frac{1}{2r} \int_{x-r}^{x+r} u(t)dt = u(x), \quad \text{a.e. } x \in \mathbb{R}. \tag{5.63}$$

When (5.63) holds true in x, we say that x is a **Lebesgue point** of u.
For any $c \in \mathbb{R}$ we have

$$\mathbb{R} \ni x \to \int_c^x u(t)dt \in AC_{loc}(\mathbb{R}) \tag{5.64}$$

and

$$\left(\int_c^x u(t)dt\right)' = u(x), \quad \text{a.e. } x \in \mathbb{R}. \tag{5.65}$$

When $u \in L^1_{loc}(\mathbb{R})$, we set

$$u^*(x) = \begin{cases} \lim_{r\to 0} \frac{1}{2r} \int_{x-r}^{x+r} u(t)dt, & \text{provided the limit exists} \\ \\ 0, & \text{otherwise} \end{cases}.$$

By (5.63) we have

$$u^*(x) = u(x), \qquad \text{a.e. } x \in \mathbb{R}.$$

The function u^* is called the **precise representative** of u. In the sequel of this section, if $f \in AC_{loc}(\mathbb{R})$ we will denote by f' its derivative, and if $g \in W^{1,p}_{loc}(\mathbb{R})$, we will denote by $\frac{d}{dx}g$ its weak derivative.

Theorem 5.19 *Let $p \in [1, +\infty)$. We have*

(i) if $u \in W^{1,p}_{loc}(\mathbb{R})$, then $u^* \in AC_{loc}(\mathbb{R})$; moreover

$$\left(u^*\right)' \in L^p_{loc}(\mathbb{R}), \quad \textit{and} \quad \frac{d}{dx}u = \left(u^*\right)'.$$

(ii) Let $u \in L^p_{loc}(\mathbb{R})$. If $v \in AC_{loc}(\mathbb{R})$ satisfies

$$u = v, \qquad \textit{a.e. in } \mathbb{R}$$

and $v' \in L^p_{loc}(\mathbb{R})$ then $u \in W^{1,p}_{loc}(\mathbb{R})$ and $\frac{d}{dx}u = v$.

In order to prove the Theorem above we need the following

Lemma 5.20 *Let $u \in L^1_{loc}(\mathbb{R})$. Let η_ε be a mollifier. If x is a Lebesgue point of u we have*

$$u^\varepsilon(x) := (\eta_\varepsilon \star u)(x) \to u^*(x), \qquad \textit{as } \varepsilon \to 0^+.$$

Hence

$$u^\varepsilon(x) \to u(x), \quad \textit{as } \varepsilon \to 0^+, \quad \textit{a.e. } x \in \mathbb{R}.$$

Proof of Lemma Let us recall

$$\eta_\varepsilon = \varepsilon^{-1}\eta\left(\varepsilon^{-1}x\right),$$

where $\operatorname{supp} \eta \subset (-1, 1)$, $\eta \in C_0^\infty(\mathbb{R})$, $\eta \geq 0$ and

$$\int_{\mathbb{R}} \eta(x)dx = 1.$$

If $x \in \mathbb{R}$ is a Lebesgue point of u, then we have

$$\left|u^\varepsilon(x) - u(x)\right| = \left|\frac{1}{\varepsilon}\int_{\mathbb{R}} \eta\left(\frac{x-t}{\varepsilon}\right)(u(t) - u(x))\, dt\right| \leq$$

$$\leq \|\eta\|_{L^\infty(\mathbb{R})} \frac{1}{\varepsilon}\int_{x-\varepsilon}^{x+\varepsilon} |u(t) - u(x)|\, dt \to 0, \qquad \text{as } \varepsilon \to 0^+.$$

■

Proof of Theorem 5.19

1. Let $u \in W^{1,p}_{loc}(\mathbb{R})$. We have $u^\varepsilon \in C^\infty(\mathbb{R})$ and

$$u^\varepsilon(y) = u^\varepsilon(x) + \int_x^y \left(u^\varepsilon\right)'(t)dt, \quad \forall x, y \in \mathbb{R}. \tag{5.66}$$

Let $x_0 \in \mathbb{R}$ be a Lebesgue point of u (hence $u^*(x_0) = u(x_0)$). By (5.66) we have

$$u^\varepsilon(x) = u^\varepsilon(x_0) + \int_x^y \left(u^\varepsilon\right)'(t)dt \tag{5.67}$$

and, for any $\varepsilon, \delta > 0$,

$$\left|u^\varepsilon(x) - u^\delta(x)\right| \leq \left|u^\varepsilon(x_0) - u^\delta(x_0)\right| + \left|\int_{x_0}^x \left|\left(u^\varepsilon\right)' - \left(u^\delta\right)'\right| dt\right|. \tag{5.68}$$

Now, Lemma 5.20 yields

$$u^\varepsilon(x_0) \to u(x_0), \qquad \text{as } \varepsilon \to 0. \tag{5.69}$$

Moreover

$$\left(u^\varepsilon\right)'(t) = \int_{\mathbb{R}} -\partial_y\left(\eta_\varepsilon(t-y)\right)u(y)dy = \int_{\mathbb{R}} \eta_\varepsilon(t-y)\frac{du(y)}{dy}dy,$$

which, in turn, implies

$$\left(u^\varepsilon\right)' \to \frac{du}{dt}, \qquad \text{as } \varepsilon \to 0, \qquad \text{in } L^p_{loc}(\mathbb{R}). \tag{5.70}$$

Now, by (5.68)–(5.70) we have that $\{u^\varepsilon\}$ satisfies the Cauchy property on every compact set of $\mathbb{R}$. Therefore $\{u^\varepsilon\}$ uniformly converges to a continuous function v on every compact set of $\mathbb{R}$. Since we have

$$u^\varepsilon \to u, \quad \text{as } \varepsilon \to 0, \quad \text{in } L^p_{loc}(\mathbb{R}),$$

we get

$$u = v, \quad \text{a.e. in } \mathbb{R}.$$

On the other hand, by (5.67) and by (5.70) (taking into account that $\{u^\varepsilon\} \to v$ on any compact), we have

$$v(x) = v(x_0) + \int_{x_0}^{x} \frac{du(t)}{dt} dt, \quad \forall x \in \mathbb{R},$$

which implies that $v \in AC_{loc}(\mathbb{R})$ and

$$v'(x) = \frac{du(x)}{dx}, \quad \text{a.e. in } \mathbb{R}. \tag{5.71}$$

Moreover,

$$\frac{1}{2r}\int_{x-r}^{x+r} u(t)dt = \frac{1}{2r}\int_{x-r}^{x+r} v(t)dt, \quad \forall x \in \mathbb{R}.$$

Hence, passing to the limit as $r \to 0$ we have

$$u^*(x) = v(x), \quad \forall x \in \mathbb{R}.$$

By the latter and by (5.71) we get

$$\left(u^*\right)'(x) = \frac{du(x)}{dx}, \quad \text{a.e. in } \mathbb{R}.$$

2. Let $u \in L^p_{loc}(\mathbb{R})$ satisfy

$$u = v, \quad \text{a.e. in } \mathbb{R},$$

where $v \in AC_{loc}(\mathbb{R})$ and $v' \in L^p_{loc}(\mathbb{R})$. We have

$$\int_{\mathbb{R}} u\Phi' dx = \int_{\mathbb{R}} v\Phi' dx = -\int_{\mathbb{R}} v'\Phi dx, \quad \forall \Phi \in C_0^\infty(\mathbb{R}).$$

Hence, as $v' \in L^p_{loc}(\mathbb{R})$, we get

$$\frac{du}{dx} = v', \quad \text{and} \quad u \in W^{1,p}_{loc}(\mathbb{R}).$$

■

Remark 5.22 The extension Theorem implies that if $1 \leq p < +\infty$, $a, b \in \mathbb{R}$, $a < b$, denoting by E the extension operator, then we have

(i') if $u \in W^{1,p}(a,b)$ then $(Eu)^* \in AC([a,b])$; in particular u is almost everywhere equal to an absolutely continuous function in $[a,b]$ and the weak derivative of u is equal to the classic derivative of $(Eu)^*$ in (a,b);

(ii') if $u \in L^p(a,b)$ and $v \in AC([a,b])$ satisfies

$$u = v, \quad \text{a.e. in } [a,b]$$

and $v' \in L^p(a,b)$, then $u \in W^{1,p}(a,b)$ and the weak derivative of u is equal to the classic derivative of v. ♦

Theorem 5.19 can be accomplished by the following

Proposition 5.21 *Let $p > 1$ and $u \in W^{1,p}_{loc}(a,b)$. We have that u is almost everywhere equal to a function $C^{0,\alpha}_{loc}(\mathbb{R})$, where $\alpha = 1 - 1/p$. Here $C^{0,\alpha}_{loc}(\mathbb{R})$ denotes the space of the functions u satisfying $u_{|I} \in C^{0,\alpha}(I)$ for every I compact interval of $\mathbb{R}$.*

Proof Let I be a bounded interval. By (5.66), (by using Hölder inequality), we have, for any $x, y \in I$,

$$\left|u^\varepsilon(x) - u^\varepsilon(y)\right| = \left|\int_x^y \left(u^\varepsilon\right)'(t)dt\right| \leq$$

$$\leq |x-y|^{1-1/p}\left(\int_I \left|\left(u^\varepsilon\right)'(t)\right|^p dt\right)^{1/p}.$$

Hence, passing to the limit as $\varepsilon \to 0$, taking into account that $u^\varepsilon \to u^*$ and $u^*(x) = u(x)$ almost everywhere, we obtain

$$|u(x) - u(y)| \leq |x-y|^{1-1/p}\left(\int_I \left|\frac{du}{dt}\right|^p dt\right)^{1/p}, \quad \text{a.e. } x, y \in I.$$

■

5.7 The Embedding Theorems

In this section, our focus lies in proving non-trivial embedding theorems involving $W^{k,p}(\Omega)$ and other function spaces. Specifically, we aim to establish conditions under which $W^{k,p}(\Omega) \subset L^q(\Omega)$ for $q \neq p$, as well as $W^{k,p}(\Omega) \subset C^{m,\gamma}(\Omega)$ for suitable values of $m \in \mathbb{N}_0$ and $0 < \gamma \leq 1$.

First we consider the space $W^{1,p}(\Omega)$ and we distinguish the following three cases

$$\text{(a)} \quad 1 \leq p < n, \qquad \text{(b)} \quad n < p \leq +\infty, \qquad \text{(c)} \quad p = n.$$

About case (c) we will just give brief hints.

5.7.1 *Case $1 \leq p < n$: The Gagliardo–Nirenberg Inequality*

Let us assume

$$1 \leq p < n \tag{5.72}$$

and let us ask ourselves for what $q \in [1, +\infty]$ can be true an estimate like

$$\|u\|_{L^q(\mathbb{R}^n)} \leq C \|\nabla u\|_{L^p(\mathbb{R}^n)}, \quad \forall u \in C_0^\infty(\mathbb{R}^n), \tag{5.73}$$

where C and q do not depend on u.

Let us assume that (5.73) is true and let us prove that, *necessarily*

$$q = \frac{np}{n-p}.$$

First, we examine the case $q \in [1, +\infty)$. Let $u \in C_0^\infty(\mathbb{R}^n)$ be **not identically equal to** 0, and, for any $\lambda > 0$, let

$$u_\lambda(x) = u(\lambda x), \quad \forall x \in \mathbb{R}^n.$$

Of course if (5.73) holds true, then

$$\|u_\lambda\|_{L^q(\mathbb{R}^n)} \leq C \|\nabla u_\lambda\|_{L^p(\mathbb{R}^n)}, \quad \forall u \in C_0^\infty(\mathbb{R}^n), \ \forall \lambda > 0. \tag{5.74}$$

Now

$$\|u_\lambda\|^q_{L^q(\mathbb{R}^n)} = \int_{\mathbb{R}^n} |u(\lambda x)|^q \, dx = \lambda^{-n} \int_{\mathbb{R}^n} |u(x)|^q \, dx;$$

hence

$$\|u_\lambda\|_{L^q(\mathbb{R}^n)} = \lambda^{-\frac{n}{q}} \|u\|_{L^q(\mathbb{R}^n)} \tag{5.75}$$

and

$$\|\nabla u_\lambda\|^p_{L^p(\mathbb{R}^n)} = \int_{\mathbb{R}^n} \lambda^p |(\nabla u)(\lambda x)|^p \, dx = \lambda^{p-n} \int_{\mathbb{R}^n} |\nabla u(x)|^p \, dx;$$

therefore

$$\|\nabla u_\lambda\|_{L^p(\mathbb{R}^n)} = \lambda^{1-\frac{n}{p}} \|\nabla u\|_{L^p(\mathbb{R}^n)} . \tag{5.76}$$

Since (5.75) and (5.76) hold true, we may write (5.74) as follows

$$\|u\|_{L^q(\mathbb{R}^n)} \leq C\lambda^{1+\frac{n}{q}-\frac{n}{p}} \|\nabla u\|_{L^p(\mathbb{R}^n)} , \quad \forall \lambda > 0. \tag{5.77}$$

Now, if

$$q > \frac{np}{n-p},$$

we get

$$1 + \frac{n}{q} - \frac{n}{p} < 1 + n\left(\frac{n-p}{np}\right) - \frac{n}{p} = 0.$$

Consequently

$$\|u\|_{L^q(\mathbb{R}^n)} \leq \lim_{\lambda\to+\infty} C\lambda^{1+\frac{n}{q}-\frac{n}{p}} \|\nabla u\|_{L^p(\mathbb{R}^n)} = 0,$$

this is a contradiction because u does not vanish identically.

On the other hand, if

$$q < \frac{np}{n-p},$$

we have

$$1 + \frac{n}{q} - \frac{n}{p} > 0,$$

hence

$$\|u\|_{L^q(\mathbb{R}^n)} \leq \lim_{\lambda\to 0^+} C\lambda^{1+\frac{n}{q}-\frac{n}{p}} \|\nabla u\|_{L^p(\mathbb{R}^n)} = 0,$$

this is again a contradiction.

Finally, if $q = +\infty$, instead of (5.77) we have

$$\|u\|_{L^\infty(\mathbb{R}^n)} \leq C\lambda^{1-\frac{n}{p}} \|\nabla u\|_{L^p(\mathbb{R}^n)}, \quad \forall\lambda > 0.$$

and, by (5.72), passing to the limit as $\lambda \to +\infty$ we have a contradiction.

Here and in the sequel, if $1 \leq p < n$, we denote by $p^\star$ the number

$$\frac{1}{p^\star} = \frac{1}{p} - \frac{1}{n}$$

and we call $p^\star$ the **Sobolev exponent** or the **Sobolev conjugate** of p . Let us notice

$$p^\star = \frac{pn}{n-p} > p.$$

The Main Theorem of the present subsection is the following one

Theorem 5.22 (Gagliardo–Nirenberg Inequality) *Let*

$$1 \leq p < n.$$

Then there exists C depending on p and n only such that

$$\|u\|_{L^{p^\star}(\mathbb{R}^n)} \leq C \|\nabla u\|_{L^p(\mathbb{R}^n)}, \quad \forall u \in C_0^1\left(\mathbb{R}^n\right). \tag{5.78}$$

The most challenging part of the proof of Theorem 5.78 concerns the case $p = 1$ and this, in turn, is based on the following

Lemma 5.23 *Let $n \geq 2$ and*

$$g_j : \mathbb{R}^{n-1} \to [0, +\infty), \quad j = 1, \cdots, n,$$

be measurable functions. Then

$$\begin{aligned} &\int_{\mathbb{R}^n} \prod_{j=1}^n g_j(x_1, \cdots, x_{j-1}, x_{j+1}, \cdots, x_n) dx_1 \cdots dx_n \leq \\ &\leq \prod_{j=1}^n \left(\int_{\mathbb{R}^{n-1}} g_j^{n-1}(y) dy \right)^{\frac{1}{n-1}}. \end{aligned} \tag{5.79}$$

Proof of Lemma 5.23 Let us proceed by induction on n. If $n = 2$, we have

$$\int_{\mathbb{R}^2} g_1(x_2) g_2(x_1) dx_1 dx_2 = \int_{\mathbb{R}} g_1(x_2) dx_2 \int_{\mathbb{R}} g_2(x_1) dx_1.$$

Therefore, if $n = 2$, (5.79) holds. Now, let us assume that (5.79) holds for n and let us prove it for $n + 1$. Hence, let us assume that **for any** nonnegative measurable functions $g_1, g_2 \cdots$, we have

$$\int_{\mathbb{R}^n} \prod_{j=1}^{n} g_j dx_1 \cdots dx_n \leq \prod_{j=1}^{n} \left(\int_{\mathbb{R}^{n-1}} g_j^{n-1} dy \right)^{\frac{1}{n-1}},$$

let us notice that, to shorten the formula, we have omitted the variables. However, it is important to recall that g_j does not depend on x_j. In what follows we always denote $dx_1 dx_2 \cdots dx_n$ by dx.

By the Hölder inequality we get

$$\begin{aligned} \int_{\mathbb{R}^{n+1}} \prod_{j=1}^{n+1} g_j dx dx_{n+1} &= \int_{\mathbb{R}} dx_{n+1} \int_{\mathbb{R}^n} g_{n+1} \prod_{j=1}^{n} g_j dx \leq \\ &\leq \int_{\mathbb{R}} dx_{n+1} \left(\int_{\mathbb{R}^n} g_{n+1}^n dx \right)^{\frac{1}{n}} \left(\int_{\mathbb{R}^n} \prod_{j=1}^{n} g_j^{\frac{n}{n-1}} dx \right)^{\frac{n-1}{n}} = \\ &= \left(\int_{\mathbb{R}^n} g_{n+1}^n dx \right)^{\frac{1}{n}} \int_{\mathbb{R}} dx_{n+1} \left(\int_{\mathbb{R}^n} \prod_{j=1}^{n} g_j^{\frac{n}{n-1}} dx \right)^{\frac{n-1}{n}}. \end{aligned} \tag{5.80}$$

Now let us apply the inductive assumption to the functions $g_j^{\frac{n}{n-1}}(\cdot, x_{n+1})$, $j = 1, \cdots, n$. We get

$$\int_{\mathbb{R}^n} \prod_{j=1}^{n} g_j^{\frac{n}{n-1}} dx \leq \prod_{j=1}^{n} \left(\int_{\mathbb{R}^{n-1}} g_j^n(y, x_{n+1}) dy \right)^{\frac{1}{n-1}}.$$

Hence, we have trivially

$$\left(\int_{\mathbb{R}^n} \prod_{j=1}^{n} g_j^{\frac{n}{n-1}} dx \right)^{\frac{n-1}{n}} \leq \prod_{j=1}^{n} \left(\int_{\mathbb{R}^{n-1}} g_j^n(y, x_{n+1}) dy \right)^{\frac{1}{n}}.$$

The last inequality and (5.80) imply

$$\begin{aligned}&\int_{\mathbb{R}^{n+1}}\prod_{j=1}^{n+1}g_j dx dx_{n+1}\leq\\&\leq\left(\int_{\mathbb{R}^n}g_{n+1}^n dx\right)^{\frac{1}{n}}\int_{\mathbb{R}}dx_{n+1}\prod_{j=1}^{n}\left(\int_{\mathbb{R}^{n-1}}g_j^n(y,x_{n+1})dy\right)^{\frac{1}{n}}.\end{aligned}\tag{5.81}$$

Now, we set

$$h_j\left(x_{n+1}\right)=\left(\int_{\mathbb{R}^{n-1}}g_j^n(y,x_{n+1})dy\right)^{\frac{1}{n}}$$

and we use the extended Hölder inequality:

$$\begin{aligned}\int_{\mathbb{R}}\prod_{j=1}^{n}\left(\int_{\mathbb{R}^{n-1}}g_j^n(y,x_{n+1})dy\right)^{\frac{1}{n}}dx_{n+1}&=\int_{\mathbb{R}}\prod_{j=1}^{n}h_j\left(x_{n+1}\right)dx_{n+1}\leq\\&\leq\prod_{j=1}^{n}\left(\int_{\mathbb{R}}h_j^n\left(x_{n+1}\right)dx_{n+1}\right)^{\frac{1}{n}}=\\&=\prod_{j=1}^{n}\left(\int_{\mathbb{R}}dx_{n+1}\int_{\mathbb{R}^{n-1}}g_j^n(y,x_{n+1})dy\right)^{\frac{1}{n}}=\\&=\prod_{j=1}^{n}\left(\int_{\mathbb{R}^n}g_j^n(y)dy\right)^{\frac{1}{n}}.\end{aligned}$$

The just obtained inequality and (5.81) yield

$$\int_{\mathbb{R}^{n+1}}\prod_{j=1}^{n+1}g_j dx_1\cdots dx_{n+1}\leq\prod_{j=1}^{n+1}\left(\int_{\mathbb{R}^n}g_j^n dy\right)^{\frac{1}{n}}.$$

The proof of Lemma is concluded. ■

Proof of Theorem 5.22 Let $u\in C_0^1\left(\mathbb{R}^n\right)$. For any $j=1,\cdots,n$, we set

$$\begin{aligned}&f_j\left(x_1,\cdots,x_{j-1},x_{j+1},\cdots,x_n\right)=\\&=\int_{\mathbb{R}}\left|\nabla u\left(x_1,\cdots,x_{j-1},y_j,x_{j+1},\cdots,x_n\right)\right|dy_j.\end{aligned}\tag{5.82}$$

We obtain

$$|u(x)| \leq f_j\left(x_1, \cdots, x_{j-1}, x_{j+1}, \cdots, x_n\right), \quad j = 1, \cdots, n.$$

As a matter of fact, we have

$$u(x) = \int_{-\infty}^{x_j} \partial_{x_j} u\left(x_1, \cdots, x_{j-1}, y_j, x_{j+1}, \cdots, x_n\right) dy_j, \quad j = 1, \cdots, n,$$

from which, for any $j = 1, \cdots, n$ we have

$$\begin{aligned} |u(x)| \leq \int_{\mathbb{R}} \left|\nabla u\left(x_1, \cdots, x_{j-1}, y_j, x_{j+1}, \cdots, x_n\right)\right| dy_j = \\ = f_j\left(x_1, \cdots, x_{j-1}, x_{j+1}, \cdots, x_n\right). \end{aligned} \tag{5.83}$$

Now by multiplying all (5.83) we get

$$|u(x)|^n \leq \prod_{j=1}^{n} f_j,$$

hence

$$|u(x)|^{\frac{n}{n-1}} \leq \prod_{j=1}^{n} f_j^{\frac{1}{n-1}}$$

and, by integrating over $\mathbb{R}^n$ we have

$$\int_{\mathbb{R}^n} |u(x)|^{\frac{n}{n-1}} dx \leq \int_{\mathbb{R}^n} \prod_{j=1}^{n} f_j^{\frac{1}{n-1}} dx.$$

At this stage let us exploit Lemma 5.23. Set

$$g_j = f_j^{\frac{1}{n-1}}, \quad j = 1, \cdots, n$$

and we obtain

$$\int_{\mathbb{R}^n} |u(x)|^{\frac{n}{n-1}} dx \leq \prod_{j=1}^{n} \left(\int_{\mathbb{R}^{n-1}} f_j(\eta) d\eta\right)^{\frac{1}{n-1}}. \tag{5.84}$$

Now, we notice that

$$\int_{\mathbb{R}^{n-1}} f_j(\eta)d\eta =$$
$$= \int_{\mathbb{R}^{n-1}} d\eta \int_{\mathbb{R}} \left|\nabla u\left(\eta_1, \cdots, \eta_{j-1}, y_j, \eta_{j+1}, \cdots, \eta_n\right)\right| dy_j =$$
$$= \int_{\mathbb{R}^n} |\nabla u(x)|\, dx, \quad j = 1, \cdots, n.$$

By the just obtained equality and by (5.84) we get

$$\int_{\mathbb{R}^n} |u(x)|^{\frac{n}{n-1}} dx \leq \left(\int_{\mathbb{R}^n} |\nabla u(x)|\, dx\right)^{\frac{n}{n-1}},$$

which implies

$$\|u\|_{L^{1^\star}(\mathbb{R}^n)} \leq \|\nabla u\|_{L^1(\mathbb{R}^n)}. \tag{5.85}$$

Therefore (5.78) is proved for $p = 1$. Now, let $1 < p < n$, and let $\alpha > 1$ be chosen. By applying (5.85) to $|u|^\alpha$ we have

$$\left(\int_{\mathbb{R}^n} |u(x)|^{\frac{\alpha n}{n-1}} dx\right)^{\frac{n-1}{n}} \leq \int_{\mathbb{R}^n} \left|\nabla\left(|u(x)|^\alpha\right)\right| dx =$$
$$= \int_{\mathbb{R}^n} \alpha\, |u(x)|^{\alpha-1}\, |\nabla u(x)| dx \leq$$
$$\leq \alpha \left(\int_{\mathbb{R}^n} |u(x)|^{\frac{(\alpha-1)p}{p-1}} dx\right)^{\frac{p}{p-1}} \left(\int_{\mathbb{R}^n} |\nabla u(x)|^p\, dx\right)^{\frac{1}{p}}.$$

Now, let us choose α satisfying

$$\frac{\alpha n}{n-1} = (\alpha - 1)\frac{p}{p-1}$$

that is

$$\alpha = \frac{p(n-1)}{n-p},$$

notice that $\alpha > 1$, as $n > p > 1$. The above choice of α gives

$$\frac{\alpha n}{n-1} = (\alpha - 1)\frac{p}{p-1} = \left(\frac{p(n-1)}{n-p} - 1\right)\frac{p}{p-1} = \frac{pn}{n-p}.$$

Hence

$$\left(\int_{\mathbb{R}^n} |u(x)|^{\frac{pn}{n-p}} dx\right)^{\frac{n-1}{n}}$$

$$\leq \frac{p(n-1)}{n-p}\left(\int_{\mathbb{R}^n} |u(x)|^{\frac{pn}{n-p}} dx\right)^{\frac{p-1}{p}}\left(\int_{\mathbb{R}^n} |\nabla u(x)|^p dx\right)^{\frac{1}{p}},$$

that is

$$\left(\int_{\mathbb{R}^n} |u(x)|^{p^\star} dx\right)^{\frac{n-1}{n}-\frac{p-1}{p}} \leq \frac{p(n-1)}{n-p}\left(\int_{\mathbb{R}^n} |\nabla u(x)|^p dx\right)^{\frac{1}{p}}.$$

On the other hand

$$\frac{n-1}{n}-\frac{p-1}{p}=\frac{1}{p}-\frac{1}{n}=\frac{1}{p^\star}.$$

Therefore

$$\left(\int_{\mathbb{R}^n} |u(x)|^{p^\star} dx\right)^{\frac{1}{p^\star}} \leq \frac{p(n-1)}{n-p}\left(\int_{\mathbb{R}^n} |\nabla u(x)|^p dx\right)^{\frac{1}{p}}.$$

■

Theorem 5.24 (Sobolev Inequality) *Let Ω be a bounded open set of $\mathbb{R}^n$ whose boundary is of class $C^{0,1}$ with constants M_0 and r_0. Let d_0 be the diameter of Ω. Assume $1 \leq p < n$.*

Then there exists C depending on M_0, r_0, d_0, p and n only such that

$$\|u\|_{L^{p^\star}(\Omega)} \leq C\,\|u\|_{W^{1,p}(\Omega)}, \quad \forall u \in W^{1,p}(\Omega). \tag{5.86}$$

Proof Since Ω is a bounded open set of $\mathbb{R}^n$ of class $C^{0,1}$, we can apply the extension Theorem 5.16. Hence, there exists $\widetilde{u} \in W^{1,p}(\mathbb{R}^n)$ such that

$$\widetilde{u} = u, \text{ in } \Omega, \qquad \operatorname{supp}\widetilde{u} \text{ compact of } \mathbb{R}^n. \tag{5.87}$$

Moreover, there exists C depending on M_0, r_0, d_0, p and n only such that

$$\|\widetilde{u}\|_{W^{1,p}(\mathbb{R}^n)} \leq C\,\|u\|_{W^{1,p}(\Omega)}, \quad \forall u \in W^{1,p}(\Omega). \tag{5.88}$$

Proposition 5.13 implies that there exists a sequence $\{v_j\} \subset C_0^\infty(\mathbb{R}^n)$ such that

$$\{v_j\} \to \widetilde{u}, \quad \text{in } W^{1,p}(\mathbb{R}^n).$$

Now, by Theorem 5.22 we have

$$\left\|v_j - v_m\right\|_{L^{p^\star}(\mathbb{R}^n)} \leq C\left\|\nabla v_j - \nabla v_m\right\|_{W^{1,p}(\mathbb{R}^n)}, \quad \forall j, m \in \mathbb{N},$$

from which it follows that $\{v_j\}$ is a Cauchy sequence in $L^{p^\star}(\mathbb{R}^n)$, hence there exists $v \in L^{p^\star}(\Omega)$ such that

$$\{v_j\} \to v, \quad \text{in } L^{p^\star}(\mathbb{R}^n).$$

Hence $\widetilde{u} = v$; as a matter of fact, for any $R > 0$ we have $L^{p^\star}(B_R) \subset L^p(B_R)$, consequently

$$\|v - \widetilde{u}\|_{L^p(B_R)} \leq \left\|v - v_j\right\|_{L^p(B_R)} + \left\|v_j - \widetilde{u}\right\|_{L^p(B_R)} \to 0, \quad \text{as } j \to \infty$$

and, as R is arbitrary, we get $\widetilde{u} = v$.

Therefore $\widetilde{u} \in L^{p^\star}(\mathbb{R}^n)$ and passing to the limit in the following inequality

$$\left\|v_j\right\|_{L^{p^\star}(\mathbb{R}^n)} \leq C\left\|\nabla v_j\right\|_{W^{1,p}(\mathbb{R}^n)}, \quad \forall j \in \mathbb{N},$$

we obtain

$$\|\widetilde{u}\|_{L^{p^\star}(\mathbb{R}^n)} \leq C\,\|\nabla\widetilde{u}\|_{W^{1,p}(\mathbb{R}^n)}. \tag{5.89}$$

On the other hand (5.87) implies

$$\|u\|_{L^{p^\star}(\Omega)} \leq \|\widetilde{u}\|_{L^{p^\star}(\mathbb{R}^n)} \tag{5.90}$$

and (5.88) yields

$$\|\nabla\widetilde{u}\|_{L^p(\mathbb{R}^n)} \leq C\,\|u\|_{W^{1,p}(\Omega)}, \quad \forall u \in W^{1,p}(\Omega).$$

Thus, by the latter, by (5.89) and by (5.90) we get

$$\|u\|_{L^{p^\star}(\Omega)} \leq C\,\|u\|_{W^{1,p}(\Omega)}.$$

■

Corollary 5.25 *Let Ω be a bounded open set of $\mathbb{R}^n$ and let d_0 be its diameter. Assume $1 \leq p < n$. We have that, if $u \in W_0^{1,p}(\Omega)$, then $u \in L^{p^\star}(\Omega)$ and there exists C depending on p, n and d_0 only, such that*

$$\|u\|_{L^{p^\star}(\Omega)} \leq C\,\|u\|_{W^{1,p}(\Omega)}.$$

Proof Let $x_0 \in \Omega$ and $R = 2d_0$. Since $u \in W_0^{1,p}(\Omega)$ we have that the function

$$\widetilde{u}(x) = \begin{cases} u(x), & \text{for } x \in \Omega \\ \\ 0, & \text{for } x \in B_R \setminus \Omega \end{cases},$$

belongs to $W_0^{1,p}(B_R(x_0))$. By the first Poincaré inequality (Theorem 5.14) and by Theorem 5.24 we obtain

$$\begin{aligned} \|u\|_{L^{p^\star}(\Omega)} &= \|\widetilde{u}\|_{L^{p^\star}(B_R(x_0))} \leq \\ &\leq C\, \|\widetilde{u}\|_{W^{1,p}(B_R(x_0))} \leq \\ &\leq C\widetilde{C}\, \|\nabla\widetilde{u}\|_{L^p(B_R(x_0))} = \\ &= C\widetilde{C}\, \|\nabla u\|_{L^p(\Omega)}, \end{aligned}$$

where C is the same constant that occurs in the inequality (5.86), hence, it depends on p and n only, and $\widetilde{C}$ is the same constant that occurs in the Poincaré inequality, hence, it depends on R, that is, on d_0, only. ■

Counterexample If $n > 1$ it **does not** happen that $W^{1,n}(\Omega) \subset L^\infty(\Omega)$. Let us prove what is claimed.

Let

$$u(x) = \log\log\left(1 + \frac{1}{|x|}\right), \quad \text{in } B_1.$$

We have $u \notin L^\infty(B_1)$ and

$$\nabla u(x) = \frac{x}{\left(|x|^3 + |x|^2\right)\log\left(1 + \frac{1}{|x|}\right)}.$$

Let us check that $u \in W^{1,n}(B_1)$. We have

$$\begin{aligned} \int_{B_1} |u|^n dx &= \int_{B_1} \left|\log\log\left(1 + \frac{1}{|x|}\right)\right|^n dx = \\ &= \omega_n \int_0^1 \rho^{n-1} \left|\log\log\left(1 + \frac{1}{\rho}\right)\right|^n d\rho < +\infty \end{aligned}$$

and (as $n > 1$)

$$\int_{B_1} |\nabla u|^n dx = \omega_n \int_0^1 \frac{\rho^{n-1}}{(\rho^2+\rho)^n \left(\log\left(1+\frac{1}{\rho}\right)\right)^n} d\rho =$$

$$= \omega_n \int_0^1 \frac{d\rho}{(\rho+1)^n \rho \left(\log\left(1+\frac{1}{\rho}\right)\right)^n} < +\infty.$$

Therefore $u \in W^{1,n}(B_1)$, but $u \notin L^\infty(B_1)$.

Let us observe that if $n = 1$, for what was proved in Sect. 5.6, we have $W^{1,1}(I) \subset L^\infty(I)$, where I is a bounded open interval of $\mathbb{R}$. ♦

5.7.2 The Sobolev–Poincaré Inequality

We begin by the following

Lemma 5.26 *If $p \in [1, +\infty)$, then there exists a constant C depending on p and by n only, such that*

$$\int_{B_r(x)} |u(y) - u(z)|^p \, dy \le C r^{n+p-1} \int_{B_r(x)} |\nabla u(y)|^p \, ||y-z|^{1-n} dy, \tag{5.91}$$

for every $u \in C^1\left(\overline{B_r(x)}\right)$, $r > 0$, $x \in \mathbb{R}^n$ and for every $z \in \overline{B_r(x)}$.

Proof It is not restrictive to assume $x = 0$. By the Fundamental Theorem of Calculus we have, for any $y, z \in B_r$,

$$u(y) - u(z) = \int_0^1 \frac{d}{dt} u(z + t(y-z)) dt = \int_0^1 \nabla u(z + t(y-z)) dt \cdot (y-z);$$

which implies

$$|u(y) - u(z)|^p \le |y-z|^p \int_0^1 |\nabla u(z + t(y-z))|^p \, dt.$$

Let $\rho > 0$. Let us integrate over $B_r \cap \partial B_\rho(z)$ both the sides of the last inequality. We get

$$\begin{aligned}
&\int_{B_r\cap\partial B_\rho(z)} |u(y)-u(z)|^p\,dS_y \\
&\le \int_{B_r\cap\partial B_\rho(z)} dS_y\,|y-z|^p \int_0^1 |\nabla u(z+t(y-z))|^p\,dt = \\
&= \rho^p \int_0^1 dt \int_{B_r\cap\partial B_\rho(z)} |\nabla u(z+t(y-z))|^p\,dS_y = \\
&= \rho^p \int_0^1 dt \int_{B_r\cap\partial B_{t\rho}(z)} |\nabla u(\xi)|^p\,t^{1-n} dS_\xi := (\bigstar),
\end{aligned}$$

where in the last inequality we set $\xi = z + t(y-z)$, so that $t^{n-1}dS_y = dS_\xi$. Now, if $\xi \in \partial B_{t\rho}(z)$, we have $\frac{|\xi-z|}{\rho} = t$; hence

$$\begin{aligned}
(\bigstar) &= \rho^p \int_0^1 dt \int_{B_r\cap\partial B_{t\rho}(z)} |\nabla u(\xi)|^p\,\frac{|\xi-z|^{1-n}}{\rho^{1-n}} dS_\xi = \\
&= \rho^{n+p-1} \int_0^1 dt \int_{B_r\cap\partial B_{t\rho}(z)} |\nabla u(\xi)|^p\,|\xi-z|^{1-n} dS_\xi = \\
&= \rho^{n+p-2} \int_0^\rho d\tau \int_{B_r\cap\partial B_\tau(z)} |\nabla u(\xi)|^p\,|\xi-z|^{1-n} dS_\xi = \\
&= \rho^{n+p-2} \int_{B_r\cap B_\rho(z)} |\nabla u(\xi)|^p\,|\xi-z|^{1-n} d\xi \le \\
&\le \rho^{n+p-2} \int_{B_r} |\nabla u(\xi)|^p\,|\xi-z|^{1-n} d\xi.
\end{aligned}$$

Therefore

$$\int_{B_r\cap\partial B_\rho(z)} |u(y)-u(z)|^p\,dS_y \le \rho^{n+p-2} \int_{B_r} |\nabla u(\xi)|^p\,|\xi-z|^{1-n} d\xi.$$

Now, let us integrate w.r.t. ρ over $[0, r]$ both the sides of the previous inequality. We have

$$\begin{aligned}
&\int_0^r d\rho \int_{B_r\cap\partial B_\rho(z)} |u(y)-u(z)|^p\,dS_y \\
&\le \int_0^r \rho^{n+p-2} d\rho \int_{B_r} |\nabla u(\xi)|^p\,|\xi-z|^{1-n} d\xi = \\
&= \frac{r^{n+p-1}}{n+p-1} \int_{B_r} |\nabla u(\xi)|^p\,|\xi-z|^{1-n} d\xi,
\end{aligned}$$

on the other hand, by using the polar coordinates, we have

$$\int_0^r d\rho \int_{B_r \cap \partial B_\rho(z)} |u(y) - u(z)|^p \, dS_y = \int_{B_r} |u(y) - u(z)|^p \, dy.$$

Therefore, for any $z \in B_r$, we have

$$\int_{B_r} |u(y) - u(z)|^p \, dy \le \frac{r^{n+p-1}}{n+p-1} \int_{B_r} |\nabla u(\xi)|^p \, |\xi - z|^{1-n} d\xi$$

from which we obtain (5.91). ■

In what follows, for any $g \in L^1(B_r(x))$, we set

$$(g)_{x,r} = \fint_{B_r(x)} g(y)dy = \frac{1}{|B_r(x)|} \int_{B_r(x)} g(y)dy.$$

Theorem 5.27 (Sobolev–Poincaré Inequality) *Let $1 \le p < n$. Then there exists a constant C depending on p and n only such that*

$$\left(\fint_{B_r(x)} \left|u(y) - (u)_{x,r}\right|^{p^\star} dy \right)^{\frac{1}{p^\star}} \le Cr \left(\fint_{B_r(x)} |\nabla u(y)|^p \, dy \right)^{\frac{1}{p}}, \tag{5.92}$$

for every $u \in W^{1,p}(B_r(x))$.

Proof First, we assume $x = 0$ and $r = 1$. Let $u \in C^1\left(\overline{B_1}\right)$. Lemma 5.26 and Hölder inequality give

$$\begin{aligned}
\fint_{B_1} \left|u(y) - \fint_{B_1} u(z)dz\right|^p dy &= \fint_{B_1} \left|\fint_{B_1} (u(y) - u(z))dz\right|^p dy \le \\
&\le \fint_{B_1} dy \fint_{B_1} |u(y) - u(z)|^p \, dz \le \\
&\le C \fint_{B_1} dy \fint_{B_1} |\nabla u(z)|^p \, |y - z|^{1-n} \, dz = \\
&= C \fint_{B_1} dz \, |\nabla u(z)|^p \fint_{B_1} |y - z|^{1-n} \, dy \le \\
&\le C_1 \fint_{B_1} dz \, |\nabla u(z)|^p \int_{B_2(z)} |y - z|^{1-n} \, dy = \\
&= C_2 \fint_{B_1} |\nabla u(z)|^p \, dz,
\end{aligned}$$

where $C_1 = \frac{\omega_n}{n} C$, $C_2 = 2\omega_n C_1$. Hence, we have obtained the inequality

$$\fint_{B_1} \left|u(y) - (u)_{0,1}\right|^p dy \leq C_2 \fint_{B_1} |\nabla u(z)|^p dz.$$

Now, let us apply the Sobolev inequality (Theorem 5.24) to $u - (u)_{0,1}$. We get

$$\begin{aligned}\left(\fint_{B_1} \left|u(y) - (u)_{0,1}\right|^{p^\star} dy\right)^{\frac{1}{p^\star}} &\leq C_3 \left[\fint_{B_1} \left|u(y) - (u)_{0,1}\right|^{p^\star} dy\right. \\ &\quad \left. + \fint_{B_1} |\nabla u(y)|^p dy\right]^{\frac{1}{p}} \leq \\ &\leq C_4 \left(\fint_{B_1} |\nabla u(y)|^p dy\right)^{\frac{1}{p}}.\end{aligned}$$

Thus, we have obtained

$$\left(\fint_{B_1} \left|u(y) - (u)_{0,1}\right|^{p^\star} dy\right)^{\frac{1}{p^\star}} \leq C_4 \left(\fint_{B_1} |\nabla u(y)|^p dy\right)^{\frac{1}{p}}, \tag{5.93}$$

where C_4 depend on n and p only.

Let now $u \in C^1\left(\overline{B_r}(0)\right)$. Set

$$v(x) = u(rx), \quad x \in B_1$$

and apply (5.93) to function v. We have

$$(v)_{0,1} = (u)_{0,r}, \tag{5.94}$$

$$\fint_{B_1} |\nabla v(y)|^p dy = r^p \fint_{B_r} |\nabla u(x)|^p dx \tag{5.95}$$

and

$$\fint_{B_1} \left|v(y) - (v)_{0,1}\right|^{p^\star} dy = \fint_{B_r} \left|u(x) - (u)_{0,r}\right|^{p^\star} dx. \tag{5.96}$$

By applying inequality (5.93) to v and taking into account (5.94)–(5.96) we obtain (5.92) by density. ■

5.7.3 The Morrey Inequality

Let E be any measurable of $\mathbb{R}^n$, here and in the sequel we say that $u^* : E \to \mathbb{R}$ is a *version* of a given function $u : E \to \mathbb{R}$ if

$$u = u^*, \qquad \text{a.e. in } E.$$

Lemma 5.28 *If $n < p \leq +\infty$, then there exists a constant C depending on p and n only, such that we have*

$$|u(y) - u(z)| \leq C r^{1-\frac{n}{p}} \int_{B_r(x)} |\nabla u(\xi)|^p \, |d\xi, \tag{5.97}$$

for every $u \in C^1\left(\overline{B_r(x)}\right)$, $r > 0$, $x \in \mathbb{R}^n$ and for every $y, z \in \overline{B_r(x)}$.

Proof Let $u \in C^1\left(\overline{B_r(x)}\right)$. Let us apply Lemma 5.26 for $p = 1$. We have, for any $y, z \in \overline{B_r(x)}$,

$$\begin{aligned} |u(y) - u(z)| &= \fint_{B_r} |u(y) - u(z)| \, d\xi \leq \\ &\leq \fint_{B_r} |u(y) - u(\xi)| \, d\xi + \fint_{B_r} |u(z) - u(\xi)| \, d\xi \leq \\ &\leq C \int_{B_r(x)} \left(|y - \xi|^{1-n} + |z - \xi|^{1-n}\right) |\nabla u(\xi)| \, d\xi, \end{aligned} \tag{5.98}$$

where C depends on n only. Let $n < p < +\infty$. By the Hölder inequality we have

$$\begin{aligned} &\int_{B_r(x)} \left(|y - \xi|^{1-n} + |z - \xi|^{1-n}\right) |\nabla u(\xi)| \, d\xi \leq \\ &\leq \left[\int_{B_r(x)} \left(|y - \xi|^{1-n} + |z - \xi|^{1-n}\right)^{\frac{p}{p-1}} d\xi\right]^{\frac{p-1}{p}} \left(\int_{B_r(x)} |\nabla u(\xi)|^p \, d\xi\right)^{\frac{1}{p}} \leq \\ &\leq 2\Bigg[\underbrace{\int_{B_r(x)} \left(|y - \xi|^{-\frac{(n-1)p}{p-1}} + |z - \xi|^{-\frac{(n-1)p}{p-1}}\right) d\xi}_{I}\Bigg]^{\frac{p-1}{p}} \left(\int_{B_r(x)} |\nabla u(\xi)|^p \, d\xi\right)^{\frac{1}{p}}. \end{aligned} \tag{5.99}$$

Now, let us check that

$$I \leq C 2^{\frac{p-n}{p-1}} \left(\frac{p-1}{p-n}\right) r^{\frac{p-n}{p-1}}, \tag{5.100}$$

where C depends on n only. We have $B_r(x) \subset B_{2r}(y)$, for any $y \in B_r(x)$. Hence (taking into account that $p > n$ implies $\frac{(n-1)p}{p-1} < n$)

$$\begin{aligned}\int_{B_r(x)} |y-\xi|^{-\frac{(n-1)p}{p-1}} d\xi &\leq \int_{B_{2r}(y)} |y-\xi|^{-\frac{(n-1)p}{p-1}} d\xi = \\ &= \int_{\partial B_1} dS \int_0^{2r} \rho^{-\frac{(n-1)p}{p-1}} \rho^{n-1} d\rho = \\ &= \omega_n 2^{\frac{p-n}{p-1}} \left(\frac{p-1}{p-n}\right) r^{\frac{p-n}{p-1}}.\end{aligned}$$

Since a similar estimate holds for the integral

$$\int_{B_r(x)} |z-\xi|^{-\frac{(n-1)p}{p-1}} d\xi,$$

provided $z \in B_r(x)$, we get (5.100). By estimate (5.100) and by (5.98) we obtain

$$|u(y)-u(z)| \leq C_n 2^{2-\frac{n}{p}} r^{1-\frac{n}{p}} \left(\int_{B_r(x)} |\nabla u(\xi)|^p \, d\xi\right)^{\frac{1}{p}}. \tag{5.101}$$

The last obtained estimate implies (5.97) for every $n < p < +\infty$. If $p = +\infty$, we may pass to the limit as $p \to +\infty$ in (5.101). ■

Lemma 5.29 *If $n < p \leq +\infty$, then there exists a constant C depending on p and n such that*

$$|u(x)| \leq C \, \|\nabla u\|_{W^{1,p}(B_1(x))}, \tag{5.102}$$

for every $u \in C^1\left(\overline{B_1(x)}\right)$ and for every $x \in \mathbb{R}^n$.

Proof By (5.97) ($y = x$, $r = 1$ and $z \in B_1(x)$) we get

$$|u(x)| \leq C \, \|\nabla u\|_{L^p(B_1(x))} + |u(z)| \,.$$

Now, by integrating both the sides w.r.t. z on $B_1(x)$ and by applying the Hölder inequality, we have

$$\begin{aligned}|B_1(x)| \, |u(x)| &\leq C \, \|\nabla u\|_{L^p(B_1(x))} + \int_{B_1(x)} |u(z)| dz \leq \\ &\leq C' \left(\|\nabla u\|_{L^p(B_1(x))} + \|u\|_{L^p(B_1(x))}\right) \leq \\ &\leq C'' \, \|u\|_{W^{1,p}(B_1(x))} \,.\end{aligned}$$

Which gives (5.102). ■

Theorem 5.30 (Morrey Inequality) *Let $n < p < +\infty$ and let Ω be a bounded open set of $\mathbb{R}^n$ whose boundary is of class $C^{0,1}$ with constants M_0, r_0. Then there exists a constant C, depending on p, n, M_0 and r_0 only, such that for every $u \in W^{1,p}(\Omega)$ there exists a version of u, $u^* \in C^{0,\gamma}\left(\overline{\Omega}\right)$, where*

$$\gamma = 1 - \frac{n}{p}.$$

Moreover

$$\left\| u^* \right\|_{C^{0,\gamma}\left(\overline{\Omega}\right)} \leq C \left\| u \right\|_{W^{1,p}(\Omega)} . \tag{5.103}$$

Proof Let us begin by proving that

$$\left\| v \right\|_{C^{0,\gamma}(\mathbb{R}^n)} \leq C \left\| v \right\|_{W^{1,p}(\mathbb{R}^n)} , \quad \forall v \in C_0^\infty(\mathbb{R}^n), \tag{5.104}$$

where C depends on p and n only. Indeed, by (5.102) we get trivially

$$\left\| v \right\|_{L^\infty(\mathbb{R}^n)} \leq C \left\| v \right\|_{W^{1,p}(\mathbb{R}^n)} , \quad \forall v \in C_0^\infty(\mathbb{R}^n). \tag{5.105}$$

Now, we set

$$x = \frac{y+z}{2}, \quad \text{and} \quad r = |y-z|,$$

and by (5.97) we get, for any $y, z \in \mathbb{R}^n$,

$$\begin{aligned} |v(y) - v(z)| &\leq C\,|y-z|^\gamma \left(\int_{B_r(x)} |\nabla v(\xi)|^p d\xi \right)^{\frac{1}{p}} \leq \\ &\leq C\,|y-z|^\gamma \left\| v \right\|_{W^{1,p}(\mathbb{R}^n)} . \end{aligned} \tag{5.106}$$

Hence (5.105) and (5.106) give (5.104).

Now, as $\partial\Omega$ is of class $C^{0,1}$ with constants M_0, r_0, by extension Theorem 5.16 we have that if $u \in W^{1,p}(\Omega)$ there exists $\widetilde{u} \in W^{1,p}(\mathbb{R}^n)$ such that

$$\begin{cases} \widetilde{u}(x) = u(x), \ \text{for } x \in \Omega \\ \operatorname{supp} \widetilde{u}, \ \text{is a compact set of } \mathbb{R}^n \\ \left\| \widetilde{u} \right\|_{W^{1,p}(\mathbb{R}^n)} \leq C \left\| u \right\|_{W^{1,p}(\Omega)} \end{cases} . \tag{5.107}$$

By Proposition 5.13 we derive that there exists a sequence $\left\{ v_j \right\}$ in $C_0^\infty(\mathbb{R}^n)$ such that

$$\left\{ v_j \right\} \to \widetilde{u}, \quad \text{in } W^{1,p}\left(\mathbb{R}^n\right) \tag{5.108}$$

and (5.104) implies

$$\left\| v_j - v_k \right\|_{C^{0,\gamma}(\mathbb{R}^n)} \leq C \left\| v_j - v_k \right\|_{W^{1,p}(\mathbb{R}^n)}$$

for every $j, k \in \mathbb{N}$. Hence $\{v_j\}$ is a Cauchy sequence in $C^{0,\gamma}(\mathbb{R}^n)$ and therefore there exists $u^* \in C^{0,\gamma}(\mathbb{R}^n)$ such that

$$v_j \to u^*, \qquad \text{as } j \to \infty, \text{ in } C^{0,\gamma}\left(\mathbb{R}^n\right).$$

By the latter and by (5.107), (5.108) we obtain

$$u^*_{|\Omega} = \widetilde{u}_{|\Omega} = u, \qquad \text{a.e. in } \Omega.$$

Since (5.104) yields

$$\left\| v_j \right\|_{C^{0,\gamma}(\mathbb{R}^n)} \leq C \left\| v_j \right\|_{W^{1,p}(\mathbb{R}^n)}, \quad \forall j \in \mathbb{N},$$

passing to the limit, we have

$$\left\| u^* \right\|_{C^{0,\gamma}(\overline{\Omega})} \leq C \left\| u^* \right\|_{W^{1,p}(\mathbb{R}^n)} \leq C \|\widetilde{u}\|_{W^{1,p}(\mathbb{R}^n)} \leq C'C \|u\|_{W^{1,p}(\Omega)},$$

which concludes the proof. ■

5.7.4 The General Sobolev Inequalities

By Theorems 5.24 and 5.30, proceeding by iteration we obtain the following general theorem, whose proof we leave to the reader.

Theorem 5.31 (Sobolev Embedding) *Let Ω be a bounded open set of class $C^{0,1}$ with constants M_0, r_0, with diameter d_0 and let $u \in W^{k,p}(\Omega)$.*

(i) If

$$k < \frac{n}{p}, \tag{5.109}$$

then $u \in L^q(\Omega)$, where

$$\frac{1}{q} = \frac{1}{p} - \frac{k}{n}. \tag{5.110}$$

Moreover

$$\|u\|_{L^q(\Omega)} \leq C \|u\|_{W^{k,p}(\Omega)}, \tag{5.111}$$

where C depends on M_0, r_0, d_0, k and n only.

(ii) If

$$k > \frac{n}{p}, \tag{5.112}$$

then $u \in C^{m,\alpha}\left(\overline{\Omega}\right)$, where $m = k - [\frac{n}{p}] - 1$ and

$$\alpha = \begin{cases} [\frac{n}{p}] + 1 - \frac{n}{p}, & \text{if } \frac{n}{p} \text{ is not an integer number} \\ \\ \text{any positive number, if } \alpha < 1 \text{ and } \frac{n}{p} \text{ is an integer number} \end{cases} \tag{5.113}$$

and

$$\|u\|_{C^{m,\alpha}(\overline{\Omega})} \leq C \, \|u\|_{W^{k,p}(\Omega)}, \tag{5.114}$$

where C depends on M_0, r_0, d_0, k and n only.

Example If $n = 1$ and $u \in H^1(0,1)$, then $u \in C^{0,1/2}([0,1])$. If $n = 2$ and $u \in H^1(\Omega)$, then $u \in L^q(\Omega)$ for every $1 \leq q < \infty$ and, if $u \in H^2(\Omega)$ then $u \in C^{0,\alpha}(\overline{\Omega})$ for every $\alpha < 1$. Finally, if $u \in H^k(\Omega)$ for every $k \in \mathbb{N}$, then $u \in C^\infty(\overline{\Omega})$. ♣

5.8 The Compactness Theorems

In the previous section we have proved that if Ω is a bounded open set of class $C^{0,1}$ and $1 \leq p < n$ then $W^{1,p}(\Omega) \subset L^{p^\star}(\Omega)$. Moreover the embedding

$$W^{1,p}(\Omega) \hookrightarrow L^{p^\star}(\Omega),$$

is continuous, as inequality (5.86) holds true. Similarly, (Theorem 5.30), for $n < p < +\infty$, the embedding

$$W^{1,p}(\Omega) \hookrightarrow C^{0,1-\frac{n}{p}}\left(\overline{\Omega}\right),$$

is continuous. In this section we will prove compact embedding theorems, in particular, the Rellich–Kondrachov Theorem, which gives the compactness of the embedding

$$W^{1,p}(\Omega) \hookrightarrow L^q(\Omega),$$

for $1 \leq p < n$ and $q < p^\star$. This means that any bounded subset Y of $W^{1,p}(\Omega)$ is relatively compact in $L^q(\Omega)$ (namely, $\overline{Y}$ is compact in $L^q(\Omega)$).

Theorem 5.32 (Rellich–Kondrachov) *Let Ω be a bounded open set of $\mathbb{R}^n$ with boundary of class $C^{0,1}$ and let $1 \leq p < n$, $1 \leq q < p^\star = \frac{np}{n-p}$. Then the embedding of $W^{1,p}(\Omega)$ in $L^q(\Omega)$ is compact.*

In order to prove Theorem 5.32 we need the following [51].

Lemma 5.33 *Let $1 \leq q < +\infty$. Let Ω be a bounded open set of $\mathbb{R}^n$ and let Λ be the subset of $L^q(\Omega)$ defined as follows*

$$\Lambda = \left\{u \in L^q(\Omega) : \ \|u\|_{L^q(\Omega)} \leq 1\right\}.$$

Assume

$$\lim_{\varepsilon \to 0} \left(\sup_{u \in \Lambda} \|u_\varepsilon - u\|_{L^q(\Omega)} \right) = 0, \tag{5.115}$$

where

$$u_\varepsilon(x) = \int_\Omega \eta_\varepsilon(x-y)u(y)dy$$

and $\eta_\varepsilon = \varepsilon^{-n}\eta\left(\varepsilon^{-1}x\right)$ where $\eta \in C_0^\infty(\mathbb{R}^n)$, $\operatorname{supp}\eta \subset B_1$, $\int_{\mathbb{R}^n} \eta(x)dx = 1$.
Then Λ is relatively compact in $L^q(\Omega)$.

Proof of Lemma 5.33 We prove that Λ is a totally bounded set in $L^q(\Omega)$.
Let $\delta > 0$. By (5.115) we have that there exists $\varepsilon_0 > 0$ so that

$$\left\|u_{\varepsilon_0} - u\right\|_{L^q(\Omega)} < \frac{\delta}{2}, \quad \forall u \in \Lambda. \tag{5.116}$$

Let

$$\Lambda_0 = \left\{u_{\varepsilon_0} : \ u \in \Lambda\right\}.$$

Now we prove that Λ_0 is relatively compact in $C^0\left(\overline{\Omega}\right)$.
Let us denote

$$M_0 = \sup_{\mathbb{R}^n} \left|\eta_{\varepsilon_0}\right|, \quad M_1 = \sup_{\mathbb{R}^n} \left|\nabla \eta_{\varepsilon_0}\right|.$$

We have, for any $u \in \Lambda$,

$$\begin{aligned} \left|u_{\varepsilon_0}(x)\right| &= \left|\int_\Omega \eta_{\varepsilon_0}(x-y)u(y)dy\right| \leq \\ &\leq M_0|\Omega|^{1-\frac{1}{q}} \|u\|_{L^q(\Omega)} \leq \\ &\leq M_0|\Omega|^{1-\frac{1}{q}} \end{aligned}$$

and, similarly,

$$\left|\nabla u_{\varepsilon_0}(x)\right| = \left|\int_\Omega \nabla \eta_{\varepsilon_0}(x-y)u(y)dy\right| \leq M_1|\Omega|^{1-\frac{1}{q}}.$$

Therefore Λ_0 is equibounded and equicontinuous. Hence, the Arzelà–Ascoli Theorem implies that Λ_0 is relatively compact in $C^0\left(\overline{\Omega}\right)$.

Now we prove that Λ_0 is relatively compact in $L^q(\Omega)$. The inequality

$$\|w\|_{L^q(\Omega)} \leq |\Omega|^{\frac{1}{q}} \|w\|_{C^0\left(\overline{\Omega}\right)}, \quad \forall w \in C^0\left(\overline{\Omega}\right),$$

implies that, for any $w \in C^0\left(\overline{\Omega}\right) \subset L^q(\Omega)$ and for any $r > 0$ we have

$$B_{r'}^{C^0}(w) \subset B_r^{L^q}(w),$$

where $r' = r|\Omega|^{-\frac{1}{q}}$, $B_{r'}^{C^0}(w)$ is the open ball of $C^0\left(\overline{\Omega}\right)$ centered at w with radius r' and $B_r^{L^q}(w)$ is the open ball of $L^q(\Omega)$ centered at w with radius r. Since Λ_0 is relatively compact in $C^0\left(\overline{\Omega}\right)$, there exist $w_1, \cdots, w_{N_r} \in C^0\left(\overline{\Omega}\right)$ such that

$$\Lambda_0 \subset \bigcup_{j=1}^{N_r} B_{r'}^{C^0}\left(w_j\right) \subset \bigcup_{j=1}^{N_r} B_r^{L^q}\left(w_j\right).$$

All in all, Λ_0 is totally bounded set of $L^q(\Omega)$. Hence

$$\Lambda_0 \subset \bigcup_{j=1}^{N} B_{\delta/2}^{L^q}\left(w_j\right),$$

where N depends by $\delta > 0$. Consequently, if $u \in \Lambda$, then there exists $j_u \in \{1, \cdots, N\}$ so that

$$\left\|u_{\varepsilon_0} - w_{j_u}\right\|_{L^q(\Omega)} < \frac{\delta}{2}.$$

By this inequality and by (5.116) we derive that, if $u \in \Lambda$ then

$$\left\|u - w_{j_u}\right\|_{L^q(\Omega)} < \delta.$$

Hence

$$\Lambda \subset \bigcup_{j=1}^{N} B_\delta^{L^q}\left(w_j\right),$$

which implies the compactness of Λ. ■

Proof of Theorem 5.32 Let us apply Lemma 5.33. Set

$$\Lambda = \left\{u \in W^{1,p}(\Omega) : \ \|u\|_{W^{1,p}(\Omega)} \leq 1\right\}. \tag{5.117}$$

We begin by proving the Theorem for $q = 1$. Hence, let us prove that

$$\lim_{\varepsilon \to 0} \left(\sup_{u \in \Lambda} \|u_\varepsilon - u\|_{L^1(\Omega)} \right) = 0, \tag{5.118}$$

where

$$u_\varepsilon = (\overline{u} \star \eta_\varepsilon),$$

being $\overline{u}$ the extension of u to 0 in $\mathbb{R}^n \setminus \Omega$.

Let $\delta > 0$. Let $\widetilde{\Omega} \Subset \Omega$ satisfy

$$\left|\Omega \setminus \widetilde{\Omega}\right| < \delta^{\frac{p^\star}{p^\star - 1}}. \tag{5.119}$$

We have

$$\int_\Omega |u_\varepsilon(x) - u(x)|\, dx = \int_{\Omega \setminus \widetilde{\Omega}} |u_\varepsilon(x) - u(x)|\, dx + \int_{\widetilde{\Omega}} |u_\varepsilon(x) - u(x)|\, dx. \tag{5.120}$$

Now, by the Hölder inequality we derive

$$\begin{aligned} \int_{\Omega \setminus \widetilde{\Omega}} |u_\varepsilon(x) - u(x)|\, dx &\leq \left|\Omega \setminus \widetilde{\Omega}\right|^{1 - \frac{1}{p^\star}} \|u_\varepsilon - u\|_{L^{p^\star}(\Omega)} \leq \\ &\leq \delta \left(\|u_\varepsilon\|_{L^{p^\star}(\Omega)} + \|u\|_{L^{p^\star}(\Omega)} \right). \end{aligned} \tag{5.121}$$

On the other hand by the Young inequality for convolutions, we have

$$\|u_\varepsilon\|_{L^{p^\star}(\Omega)} \leq \|u\|_{L^{p^\star}(\Omega)}. \tag{5.122}$$

Now, Theorem 5.24 gives

$$\|u\|_{L^{p^\star}(\Omega)} \leq C_1 \|u\|_{W^{1,p}(\Omega)} \leq C_1, \quad \forall u \in \Lambda. \tag{5.123}$$

Therefore by (5.121)–(5.123) we obtain

$$\int_{\Omega \setminus \widetilde{\Omega}} |u_\varepsilon(x) - u(x)|\, dx \leq 2C_1\delta, \quad \forall u \in \Lambda. \tag{5.124}$$

Now, let us consider the second addend on the right-hand side of (5.120). Let $\overline{\varepsilon}$ =dist $\left(\widetilde{\Omega}, \partial\Omega\right)$. For any $x \in \widetilde{\Omega}$ and for any $\varepsilon \in (0, \overline{\varepsilon})$, we have

$$u_\varepsilon(x) = \int_\Omega \eta_\varepsilon(x-y)u(y)dy = \int_\Omega \eta(\xi)u(x-\varepsilon\xi)d\xi.$$

Hence

$$\begin{aligned}
\int_{\widetilde{\Omega}} |u_\varepsilon(x) - u(x)|\, dx &= \int_{\widetilde{\Omega}} dx \left| \int_\Omega \eta(\xi)(u(x-\varepsilon\xi) - u(x))d\xi \right| \leq \\
&\leq \int_\Omega d\xi \int_{\widetilde{\Omega}} \eta(\xi) \left|u(x-\varepsilon\xi) - u(x)\right| dx = \\
&= \int_\Omega \eta(\xi)d\xi \int_{\widetilde{\Omega}} |u(x-\varepsilon\xi) - u(x)|\, dx.
\end{aligned} \tag{5.125}$$

Let now $y \in B_1$. Let $\varepsilon < \overline{\varepsilon}$. By applying Theorem 5.8 we have, for almost every $x \in \widetilde{\Omega}$ and for every $\xi \in B_1$,

$$|u(x-\varepsilon\xi) - u(x)| = \left| \int_0^1 \nabla u(x - t\varepsilon\xi) \cdot \varepsilon\xi dt \right| \leq \varepsilon \int_0^1 |\nabla u(x - t\varepsilon\xi)|\, dt.$$

Hence, for any $u \in \Lambda$, $\xi \in B_1$ and $\varepsilon \in (0, \overline{\varepsilon})$ we have

$$\begin{aligned}
\int_{\widetilde{\Omega}} |u_\varepsilon(x) - u(x)|\, dx &\leq \varepsilon \int_0^1 dt \int_{\widetilde{\Omega}} |\nabla u(x - t\varepsilon\xi)|\, dx = \\
&= \varepsilon \int_0^1 dt \int_{\widetilde{\Omega} - t\varepsilon\xi} |\nabla u(z)|\, dz \leq \\
&\leq \varepsilon \int_\Omega |\nabla u(z)|\, dz \leq \varepsilon |\Omega|^{1-\frac{1}{p}} \|u\|_{W^{1,p}(\Omega)} \leq \\
&\leq \varepsilon |\Omega|^{1-\frac{1}{p}}.
\end{aligned} \tag{5.126}$$

From what we obtained in (5.125) and by (5.126) we get

$$\int_{\widetilde{\Omega}} |u_\varepsilon(x) - u(x)|\, dx \leq \varepsilon |\Omega|^{1-\frac{1}{p}}.$$

By the latter, by (5.120) and by (5.124) we have

$$\int_\Omega |u_\varepsilon(x) - u(x)|\, dx \leq 2C_1\delta + \varepsilon|\Omega|^{1-\frac{1}{p}}, \quad \forall u \in \Lambda,\ \forall \varepsilon \in (0, \overline{\varepsilon}), \tag{5.127}$$

which implies

$$\limsup_{\varepsilon\to 0}\left(\sup_{u\in\Lambda}\|u_\varepsilon-u\|_{L^1(\Omega)}\right)\le 2C_1\delta$$

and, as δ is arbitrary, we get

$$\lim_{\varepsilon\to 0}\left(\sup_{u\in\Lambda}\|u_\varepsilon-u\|_{L^1(\Omega)}\right)=0.$$

Hence, by Lemma 5.33, Λ is relatively compact in $L^1(\Omega)$.

Now we consider the case $1<q<p^\star$. Denoting by $\theta=\frac{q-1}{p^\star-1}$, we have $0<\theta<1$, $q=1-\theta+\theta p^\star$, and

$$\begin{aligned}\int_\Omega |u_\varepsilon(x)-u(x)|^q\,dx&=\int_\Omega |u_\varepsilon(x)-u(x)|^{1-\theta+\theta p^\star}\,dx\le\\&\le\|u_\varepsilon-u\|^{1-\theta}_{L^1(\Omega)}\,\|u_\varepsilon-u\|^{\theta}_{L^{p^\star}(\Omega)}.\end{aligned}$$

Hence

$$\|u_\varepsilon-u\|_{L^q(\Omega)}\le\|u_\varepsilon-u\|^{\frac{1-\theta}{q}}_{L^1(\Omega)}\,\|u_\varepsilon-u\|^{\frac{\theta}{q}}_{L^{p^\star}(\Omega)}.\tag{5.128}$$

On the other hand, by (5.123) we have

$$\|u\|_{L^{p^\star}(\Omega)}\le C_1,\quad \forall u\in\Lambda$$

and by the Young inequality we get

$$\|u_\varepsilon\|_{L^{p^\star}(\Omega)}\le\|u\|_{L^{p^\star}(\Omega)}\le C_1,\quad\forall u\in\Lambda.$$

Hence, (5.127) and (5.128) give

$$\|u_\varepsilon-u\|_{L^q(\Omega)}\le(2C_1)^{\frac{\theta}{q}}\left(2C_1\delta+\varepsilon|\Omega|^{1-\frac{1}{p}}\right)^{\frac{1-\theta}{q}},\quad\forall u\in\Lambda.$$

Consequently

$$\lim_{\varepsilon\to 0}\left(\sup_{u\in\Lambda}\|u_\varepsilon-u\|_{L^q(\Omega)}\right)=0$$

and by Lemma 5.33 we have that Λ is relatively compact in $L^q(\Omega)$. ■

Now we state and prove a compactness theorem in the case $p>n$.

Theorem 5.34 *Let Ω be a bounded open set of $\mathbb{R}^n$ with boundary of class $C^{0,1}$ and let $p > n$, $\alpha \in (0, \gamma)$, where $\gamma = 1 - \frac{n}{p}$. Then the embedding*

$$W^{1,p}(\Omega) \hookrightarrow C^{0,\alpha}\left(\overline{\Omega}\right),$$

is compact.

Proof Let $\{u_j\}$ be a sequence in $W^{1,p}(\Omega)$ satisfying

$$\left\|u_j\right\|_{W^{1,p}(\Omega)} \leq 1, \quad \forall j \in \mathbb{N}.$$

By Theorem 5.30 we have

$$\left\|u_j\right\|_{C^{0,\alpha}\left(\overline{\Omega}\right)} \leq C_1, \quad \forall j \in \mathbb{N}, \tag{5.129}$$

where C_1 depends on p, n and Ω. The Arzelà–Ascoli Theorem yields that there exists a subsequence $\{u_{k_j}\}$ and $u \in C^0\left(\overline{\Omega}\right)$ which satisfy

$$\{u_{k_j}\} \to u, \quad \text{uniformly}. \tag{5.130}$$

By (5.129) and (5.130) we have, for any $x, y \in \overline{\Omega}$, $x \neq y$,

$$\frac{|u(x) - u(y)|}{|x - y|^\gamma} = \lim_{j\to\infty} \frac{\left|u_{k_j}(x) - u_{k_j}(y)\right|}{|x - y|^\gamma} \leq C_1. \tag{5.131}$$

Hence $u \in C^{0,\gamma}\left(\overline{\Omega}\right)$. Therefore $u \in C^{0,\alpha}\left(\overline{\Omega}\right)$ for $0 < \alpha < \gamma$. Now, let us recall the following inequality (see Proposition 3.7):

$$\|f\|_{C^{0,\alpha}\left(\overline{\Omega}\right)} \leq C \|f\|^{\frac{\alpha}{\gamma}}_{C^{0,\gamma}\left(\overline{\Omega}\right)} \|f\|^{1-\frac{\alpha}{\gamma}}_{C^0\left(\overline{\Omega}\right)},$$

For every $f \in C^{0,\gamma}\left(\overline{\Omega}\right)$, where C depends by α, γ and Ω only. By applying such an inequality to $u_{k_j} - u$, taking into account (5.129)–(5.131), we easily obtain

$$\{u_{k_j}\} \to u, \quad \text{in } C^{0,\alpha}\left(\overline{\Omega}\right).$$

■

5.8.1 Counterexamples

1. The Rellich–Kondrachov Theorem does not hold for $q = p^\star$. Indeed, we have the following counterexample. Let $u \in C_0^\infty(B_1 \setminus \{0\})$ be a **not identically**

vanishing function and let

$$u_j(x) = j^{\frac{n}{p^\star}} u(jx), \quad \forall j \in \mathbb{N}, \forall x \in B_1.$$

We have (see beginning of Sect. 5.7.1)

$$\left\|u_j\right\|_{L^{p^\star}(B_1)} = \|u\|_{L^{p^\star}(B_1)}, \quad \forall j \in \mathbb{N}, \tag{5.132}$$

$$\left\|u_j\right\|_{L^p(B_1)} = j^{-1} \|u\|_{L^p(B_1)}, \quad \forall j \in \mathbb{N}, \tag{5.133}$$

$$\left\|\nabla u_j\right\|_{L^p(B_1)} = \|\nabla u\|_{L^p(B_1)}, \quad \forall j \in \mathbb{N}. \tag{5.134}$$

Hence, by (5.133) and (5.134) we have

$$\left\|u_j\right\|_{W^{1,p}(B_1)} = C \|u\|_{W^{1,p}(B_1)} < +\infty, \quad \forall j \in \mathbb{N}. \tag{5.135}$$

Moreover

$$\lim_{j\to\infty} u_j(x) = 0, \quad \forall x \in B_1. \tag{5.136}$$

Now, if the embedding

$$W^{1,p}(B_1) \hookrightarrow L^{p^\star}(B_1)$$

were compact, by (5.135) there should exist a subsequence $\left\{u_{k_j}\right\}$ and $v \in L^{p^\star}(B_1)$ such that

$$\left\{u_{k_j}\right\} \to v, \quad \text{in } L^{p^\star}(B_1). \tag{5.137}$$

Hence, by (5.132) we should have

$$\|v\|_{L^{p^\star}(B_1)} = \|u\|_{L^{p^\star}(B_1)}. \tag{5.138}$$

On the other hand, passing eventually to another subsequence, by (5.137) we should have

$$v(x) = \lim_{j\to\infty} u_{k_j}(x), \quad \text{a.e. } x \in B_1,$$

from the latter and from (5.136) we should have

$$v(x) = 0, \quad \text{a.e. } x \in B_1,$$

that would contradict (5.138).

2. Now, let us consider the case where $\Omega = \mathbb{R}^n$ and let us show that if $1 \leq p < n$ and $q \leq p^\star$, then the embedding

$$W^{1,p}\left(\mathbb{R}^n\right) \hookrightarrow L^q\left(\mathbb{R}^n\right),$$

is not compact.

Let $u \in C_0^\infty\left(\mathbb{R}^n\right)$ be a not identically vanishing function such that $\operatorname{supp} u \subset B_1$ and let

$$u_j(x) = u(x - 2je_1), \quad \forall j \in \mathbb{N}.$$

We obtain

$$\left\|u_j\right\|_{W^{1,p}(\mathbb{R}^n)} = \|u\|_{W^{1,p}(\mathbb{R}^n)}, \quad \forall j \in \mathbb{N}$$

and

$$\left\|u_j - u_k\right\|_{L^q(\mathbb{R}^n)} = 2\,\|u\|_{L^q(\mathbb{R}^n)} > 0, \quad \forall j, k \in \mathbb{N},\ j \neq k. \tag{5.139}$$

Hence $\left\{u_j\right\}$ is a bounded sequence in $W^{1,p}\left(\mathbb{R}^n\right)$ but, as (5.139) holds true, we cannot extract any subsequence that converges in $L^q\left(\mathbb{R}^n\right)$.

3. Let us prove that if Ω is a bounded open set and $p > n$, $\gamma = 1 - \frac{n}{p}$, then the embedding

$$W^{1,p}(\Omega) \hookrightarrow C^{0,\gamma}\left(\overline{\Omega}\right)$$

is not compact.

Let $u \in C_0^\infty(B_1)$ be not identically equal to 0. Let us denote by $\overline{u}$ the extension of u to 0 in $\mathbb{R}^n \setminus B_1$. Let us denote

$$u_j(x) = \frac{1}{j^\gamma}\overline{u}(jx), \quad \forall j \in \mathbb{N}, \forall x \in \overline{B_1}.$$

Now, let us notice (the reader check as an exercise)

$$[\overline{u}]_{0,\gamma,\mathbb{R}^n} = [u]_{0,\gamma,B_1}$$

and

$$\begin{aligned}
\left[u_j\right]_{0,\gamma,B_1} &= \sup_{x,y \in B_1, x \neq y} \frac{\left|u_j(x) - u_j(y)\right|}{|x-y|^\gamma} = \\
&= \sup_{x,y \in B_1, x \neq y} \frac{|\overline{u}(jx) - \overline{u}(jy)|}{|jx - jy|^\gamma} = \\
&= [\overline{u}]_{0,\gamma,\mathbb{R}^n}\,.
\end{aligned}$$

In addition we have

$$\left\|u_j\right\|_{L^p(B_1)} = \frac{1}{j^\gamma}\left(\int_{B_1}|u(jx)|^p dx\right)^{\frac{1}{p}} =$$

$$= \frac{1}{j^{\gamma+\frac{n}{p}}}\left(\int_{B_{1/j}}|u(x)|^p dx\right)^{\frac{1}{p}} =$$

$$= \frac{1}{j}\left(\int_{B_{1/j}}|u(x)|^p dx\right)^{\frac{1}{p}}$$

and

$$\left\|\nabla u_j\right\|_{L^p(B_1)} = \frac{1}{j^\gamma}\left(\int_{B_1}|(\nabla u)(jx)|^p j^p dx\right)^{\frac{1}{p}} =$$

$$= \frac{j^{1-\frac{n}{p}}}{j^\gamma}\left(\int_{B_{1/j}}|\nabla u(x)|^p dx\right)^{\frac{1}{p}} =$$

$$= \left(\int_{B_{1/j}}|\nabla u(x)|^p dx\right)^{\frac{1}{p}}.$$

Hence

$$\left\{u_j\right\} \to 0, \quad \text{in } W^{1,p}(B_1),$$

in particular, $\left\{u_j\right\}$ is a bounded sequence in $W^{1,p}(B_1)$. On the other hand, if there was a subsequence $\left\{u_{k_j}\right\}$ of $\left\{u_j\right\}$ and $v \in C^{0,\gamma}\left(\overline{B_1}\right)$ such that

$$\left\{u_{k_j}\right\} \to v, \quad \text{in } C^{0,\gamma}\left(\overline{B_1}\right),$$

we should necessarily have $v \equiv 0$ and

$$\left[u_j - v\right]_{0,\gamma,B_1} = \left[u_j\right]_{0,\gamma,B_1} = [\overline{u}]_{0,\gamma,\mathbb{R}^n} > 0,$$

which is a contradiction.

4. The case $p > n$, $\Omega = \mathbb{R}^n$, can be handle similarly to the case $p < n$. Let $u \in C_0^\infty(\mathbb{R}^n)$, $\operatorname{supp} u \subset B_1$, u be not identically vanishing function. Let $u_j(x) = u(x - 2je_1)$. We have

$$\left\|u_j\right\|_{W^{1,p}(\mathbb{R}^n)} = \|u\|_{W^{1,p}(\mathbb{R}^n)}, \quad \forall j \in \mathbb{N}$$

and

$$\left[u_j - u_k\right]_{0,\alpha,\mathbb{R}^n} \geq \left[u_j\right]_{0,\alpha,\mathbb{R}^n} = [u]_{0,\alpha,\mathbb{R}^n} > 0, \qquad \text{for } j \neq k,$$

where $\alpha \leq 1 - \frac{n}{p}$. Hence, no extracted sequence of $\{u_j\}$ can be a Cauchy sequence in $C^{0,\alpha}(\mathbb{R}^n)$.

5.9 The Second Poincaré Inequality

In Theorem 5.27 we have proved the Sobolev–Poincaré inequality that, in particular, holds in the following form (see the proof of the above mentioned Theorem)

$$\left(⨍_{B_r(x)} \left|u(y) - (u)_{x,r}\right|^{p\star} dy\right)^{\frac{1}{p^\star}} \leq Cr\left(⨍_{B_r(x)} |\nabla u(y)|^p \, dy\right)^{\frac{1}{p}}, \tag{5.140}$$

for every $u \in W^{1,p}(B_r(x))$, where $p \in [1, +\infty)$ (actually it holds true for $p = +\infty$). We will now prove a more general version of (5.140).

Theorem 5.35 (Second Poincaré Inequality) *Let Ω be a bounded connected open set of $\mathbb{R}^n$ with $\partial\Omega$ of class $C^{0,1}$. Let $p \in [1, +\infty)$ and*

$$u_\Omega = \frac{1}{|\Omega|}\int_\Omega u dx.$$

Then there exists a constant C depending on p, n and Ω only, such that

$$\|u - u_\Omega\|_{L^p(\Omega)} \leq C \|\nabla u\|_{L^p(\Omega)}, \quad \forall u \in W^{1,p}(\Omega). \tag{5.141}$$

Proof We argue by contradiction. Let us assume that (5.141) does not hold. Consequently for any $k \in \mathbb{N}$ there exists $u_k \in W^{1,p}(\Omega)$ such that

$$\|u_k - (u_k)_\Omega\|_{L^p(\Omega)} > k \|\nabla u_k\|_{L^p(\Omega)}.$$

Let us denote by

$$v_k = \frac{u_k - (u_k)_\Omega}{\|u_k - (u_k)_\Omega\|_{L^p(\Omega)}}, \quad \forall k \in \mathbb{N}.$$

We have

$$(v_k)_\Omega = 0,$$

$$\|v_k\|_{L^p(\Omega)} = 1$$

and

$$k \|\nabla v_k\|_{L^p(\Omega)} < 1. \tag{5.142}$$

Hence, there exists $M < +\infty$ such that

$$\|v_k\|_{W^{1,p}(\Omega)} \le M.$$

Therefore, by the Rellich–Kondrachov Theorem we have that there exists a subsequence $\left\{v_{k_j}\right\}$ of $\{v_k\}$, and $v \in L^p(\Omega)$ which satisfy

$$\left\{v_{k_j}\right\} \to v, \qquad \text{in } L^p(\Omega).$$

Hence

$$v_\Omega = 0 \tag{5.143}$$

and

$$\|v\|_{L^p(\Omega)} = \lim_{j\to\infty} \left\|v_{k_j}\right\|_{L^p(\Omega)} = 1. \tag{5.144}$$

On the other hand by (5.142) we have

$$\int_\Omega v \partial_l \phi dx = \lim_{j\to\infty} \int_\Omega v_{k_j} \partial_l \phi dx = -\lim_{j\to\infty} \int_\Omega \partial_l v_{k_j} \phi dx = 0, \quad \forall \phi \in C_0^\infty(\Omega),$$

for $l = 1, \cdots, n$. Consequently

$$\int_\Omega v \partial_l \phi dx = 0, \quad \forall \phi \in C_0^\infty(\Omega), \ l = 1, \cdots, n.$$

Therefore $\nabla v = 0$ in Ω (and, trivially, $v \in W^{1,p}(\Omega)$). Since Ω is a connected open set, Proposition 5.10 yields that there exists a constant $c_0 \in \mathbb{R}$ such that

$$v \equiv c_0,$$

and by (5.143) we have $c_0 = 0$ that contradicts (5.144). Therefore (5.141) holds true. ■

Remark The proof of (5.141) that we have given before is not constructive and this does not allow us to further specify the dependence of the constant C on Ω. To fill this gap we refer the reader to [5]. ♦

5.10 The Difference Quotients

In this section we provide the definition and the main properties of the difference quotients. These topics will turn out to be useful in the study of the regularity of the solutions of second-order elliptic equations.

Definition 5.36 Let V and Ω be open sets of $\mathbb{R}^n$ such that $V \Subset \Omega$. Let $j \in \{1, \cdots, n\}$. Let $u \in L^1_{loc}(\Omega)$. The following function

$$\delta^h_j u(x) = \frac{u(x+he_j) - u(x)}{h}, \quad \forall x \in V. \tag{5.145}$$

is called j-th partial quotient of u with increment $h \in \mathbb{R} \setminus \{0\}$,
$|h| < \text{dist}\,(V, \partial\Omega)$. We denote by

$$\delta^h u(x) = \left(\delta^h_1 u(x), \cdots, \delta^h_n u(x)\right), \quad \forall x \in V. \tag{5.146}$$

We have the following

Theorem 5.37 *Let Ω be an open set of $\mathbb{R}^n$.*

(i) If $p \in [1, +\infty)$, $u \in W^{1,p}(\Omega)$, then

$$\left\|\delta^h u\right\|_{L^p(V)} \leq C \left\|\nabla u\right\|_{L^p(\Omega)}, \quad \textit{for } |h| < \frac{1}{2}\text{dist}(V, \partial\Omega),\ h \neq 0, \tag{5.147}$$

where C depends on n only.

(ii) Let us assume $p \in (1, +\infty)$, $u \in L^p(\Omega)$ and let us assume that there exists $C > 0$ satisfying

$$\left\|\delta^h u\right\|_{L^p(V)} \leq C, \quad \textit{for } |h| < \frac{1}{2}\text{dist}(V, \partial\Omega),\ h \neq 0, \tag{5.148}$$

then

$$u \in W^{1,p}(V) \quad \textit{and} \quad \|\nabla u\|_{L^p(V)} \leq C.$$

Proof In order to prove (i) it suffices to assume $u \in C^\infty(\Omega) \cap W^{1,p}(\Omega)$ and to apply Theorem 5.7.

If $0 < |h| < \text{dist}\,(V, \partial\Omega)$, $j = 1, \cdots, n$ and $x \in V$, we have

$$u(x+he_j) - u(x) = \int_0^1 \nabla u(x+the_j) \cdot (he_j)dt,$$

from which we have

$$\left|u(x+he_j)-u(x)\right| \leq |h| \int_0^1 \left|\nabla u(x+the_j)\right| dt.$$

By using Hölder inequality and by integrating both the sides of the last inequality over V, we get

$$\begin{aligned}\int_V \left|\delta_j^h u\right|^p dx &\leq \int_V dx \int_0^1 \left|\nabla u(x+the_j)\right|^p dt = \\ &= \int_0^1 dt \int_V \left|\nabla u(x+the_j)\right|^p dx \leq \\ &\leq \int_\Omega |\nabla u|^p dx.\end{aligned}$$

Now, let us prove (ii). Let us assume that for some $C>0$ we have

$$\left\|\delta^h u\right\|_{L^p(V)} \leq C, \quad \text{for } 0<|h|<\frac{1}{2}\text{dist}\,(V,\partial\Omega). \tag{5.149}$$

Claim Let $\phi \in C_0^\infty(V)$. Let us denote $K=\operatorname{supp}\phi$, then for any $j\in\{1,\cdots,n\}$ we have

$$\int_V u\delta_j^h\phi dx = -\int_V \delta_j^{-h} u\phi dx, \quad \text{for } 0<|h|<\text{dist}\,(K,\partial V). \tag{5.150}$$

Proof of the Claim Let us notice that

$$K-he_j \subset V \quad \text{for } 0<|h|<\text{dist}\,(K,\partial V),$$

for $j=1,\cdots,n$. Hence we have

$$\begin{aligned}\int_V u\delta_j^h\phi dx &= \frac{1}{h}\left\{\int_V u(x)\phi(x+he_j)dx - \int_V u(x)\phi(x)dx\right\} = \\ &= \frac{1}{h}\left\{\int_{K-he_j} u(x)\phi(x+he_j)dx - \int_V u(x)\phi(x)dx\right\} = \\ &= \frac{1}{h}\left\{\int_K u(x-he_j)\phi(x)dx - \int_V u(x)\phi(x)dx\right\} = \\ &= \frac{1}{h}\left\{\int_V u(x-he_j)\phi(x)dx - \int_V u(x)\phi(x)dx\right\} = \\ &= -\int_V \delta_j^{-h} u\phi dx.\end{aligned}$$

This conclude the proof of the Claim.

Let us fix $j \in \{1, \cdots, n\}$. Since $L^p(V)$ is a reflexive Banach space for $1 < p < +\infty$, by

$$\sup \left\| \delta_j^{-h} u \right\|_{L^p(V)} \le C$$

(recalling Theorems 3.40 and 3.41) there exists a sequence $\{h_k\}$ which goes to 0 and $v_j \in L^p(V)$, such that

$$\left\{ \delta_j^{-h_k} u \right\} \rightharpoonup v_j, \qquad \text{weakly in } L^p(V). \tag{5.151}$$

On the other hand, by the Dominated Convergence Theorem we have, for any $\phi \in C_0^\infty(\Omega)$ such that $\operatorname{supp} \phi \subset V$,

$$\int_\Omega u \partial_j \phi dx = \lim_{k\to\infty} \int_\Omega u \delta_j^{h_k} \phi dx.$$

As a matter of fact

$$u(x)\delta_j^{h_k} \phi(x) \to u(x)\partial_j \phi(x), \quad \forall x \in \Omega, \ \text{ as } k \to \infty$$

and

$$\left| u \delta_j^{h_k} \phi \right| \le |u| \, \|\nabla \phi\|_{L^\infty(\Omega)} \, \chi_{\widetilde{V}}, \quad \forall k \in \mathbb{N},$$

where

$$\widetilde{V} = \left\{ x \in \Omega : \ \operatorname{dist}(x, V) \le \frac{1}{2} \operatorname{dist}(V, \partial\Omega) \right\}.$$

Therefore

$$\begin{aligned}
\int_V u \partial_j \phi dx &= \int_\Omega u \partial_j \phi dx = \\
&= \lim_{k\to\infty} \int_\Omega u \left(\delta_j^{h_k} \phi \right) dx = \\
&= - \lim_{k\to\infty} \int_\Omega \left(\delta_j^{-h_k} u \right) \phi dx = \\
&= - \lim_{k\to\infty} \int_V \left(\delta_j^{-h_k} u \right) \phi dx = \\
&= - \int_V v_j \phi dx.
\end{aligned}$$

Consequently

$$\partial_j u = v_j, \quad \text{in the weak sense for } j = 1, \cdots, n.$$

Hence $\nabla u \in L^p(V, \mathbb{R}^n)$, but $u \in L^p(V)$. Therefore $u \in W^{1,p}(V)$.

Finally, by (5.151) we have

$$\|\nabla u\|_{L^p(V)} \leq \liminf_{k\to\infty} \left\|\delta^{-h_k} u\right\|_{L^p(V)} \leq C,$$

(C is the same constant that occurs in (5.149)). ■

Remark If $p = 1$, then (ii) of Theorem 5.37 does not hold. As a matter of fact, let $\Omega = (-2, 2)$ and

$$u(t) = \chi_{(-1,1)}.$$

We have $u \in L^1(-2, 2)$. Let $V = \left(-\frac{3}{2}, \frac{3}{2}\right)$. Now, dist $(V, \partial\Omega) = \frac{1}{2}$ and for $0 < |h| < \frac{1}{4}$ we have (for $h > 0$)

$$\begin{aligned}
\delta^h u(t) &= \frac{\chi_{(-1,1)}(t+h) - \chi_{(-1,1)}(t)}{h} = \\
&= \frac{\chi_{(-1-h,1-h)}(t) - \chi_{(-1,1)}(t)}{h} = \\
&= \frac{1}{h}\chi_{(-1-h,-1)\cup(1-h,1)}.
\end{aligned}$$

Hence

$$\int_V \left|\delta^h u(t)\right| dt = 2, \quad \text{for } 0 < |h| < \frac{1}{4},$$

but

$$u \notin W^{1,1}(V).$$

♦

In the sequel we will use the following variant of Theorem 5.37.

Theorem 5.38 *Let $r > 0$. We have*

(i) If $p \in [1, +\infty)$ and $u \in W^{1,p}\left(B_r^+\right)$, then for any $k \in \{1, \cdots, n-1\}$ we have, for $0 < |h| < \frac{r}{2}$,

$$\left\|\delta_k^h u\right\|_{L^p\left(B_{r/2}^+\right)} \leq C \left\|\partial_k u\right\|_{L^p\left(B_r^+\right)}, \tag{5.152}$$

where C depends on n only.

(ii) Let $k \in \{1, \cdots, n-1\}$. Let $p \in (1, +\infty)$, $u \in L^p\left(B_r^+\right)$. Suppose that there exists $C > 0$ such that

$$\left\|\delta^h u\right\|_{L^p\left(B_{r/2}^+\right)} \leq C, \quad \text{for } 0 < |h| < \frac{r}{2}, \tag{5.153}$$

then

$$\partial_k u \in L^p\left(B_r^+\right) \quad \text{and} \quad \left\|\partial_k u\right\|_{L^p\left(B_r^+\right)} \leq C.$$

The proof of the above Theorem is completely analogous to the one of Theorem 5.37 and it is left to the reader as an exercise.

5.11 The Dual Space of $H_0^1(\Omega)$

Let Ω be an open set of $\mathbb{R}^n$. We denote by $H^{-1}(\Omega)$ the dual space of $H_0^1(\Omega)$ (i.e. the space of the linear bounded form from $H_0^1(\Omega)$ to $\mathbb{R}$). If $F \in H^{-1}(\Omega)$, we write

$$\langle F, v\rangle := F(v), \quad v \in H_0^1(\Omega)$$

and

$$\|F\|_{H^{-1}(\Omega)} = \sup\left\{\langle F, v\rangle : v \in H_0^1(\Omega), \|u\|_{H_0^1(\Omega)} \leq 1\right\}.$$

The following Theorem holds.

Theorem 5.39 (Characterization of $H^{-1}(\Omega)$) *Let Ω be an open set of $\mathbb{R}^n$.*

(i) $F \in H^{-1}(\Omega)$ if and only if there exist $f_0, f_1, \cdots, f_n \in L^2(\Omega)$ satisfying

$$\langle F, v\rangle = \int_\Omega f_0 v dx + \sum_{j=1}^n \int_\Omega f_j v_{x_j} dx, \quad \forall v \in H_0^1(\Omega). \tag{5.154}$$

(ii)

$$\|F\|_{H^{-1}(\Omega)} = \inf\left\{\left(\sum_{j=0}^n \int_\Omega |f_j|^2 dx\right)^{1/2} : F \text{ satisfies (5.154) for } f_0, f_1, \cdots, f_n \in L^2(\Omega)\right\}.$$

We also write

$$F = f_0 - \sum_{j=1}^{n} \partial_j f_j.$$

If

$$f_1, \cdots, f_n = 0,$$

we will identify the functional

$$\langle F, v\rangle = \int_\Omega f_0 v dx, \qquad \forall v \in H_0^1(\Omega)$$

with f_0 and we will write $F \in L^2(\Omega)$. Similarly, if $f_0 \in H^k(\Omega)$, we will write $F \in H^k(\Omega)$.

Let us note that $f_0, f_1, \cdots, f_n$ are *not* uniquely determined. For instance, if Ω is bounded, then the functional

$$\langle F, v\rangle = \int_\Omega f_0 v dx,$$

where $f_0 \in L^2(\Omega)$, can also be represented by

$$\langle F, v\rangle = \int_\Omega (f_0 + 2x_1) v dx + \int_\Omega x_1^2 \partial_1 v dx, \quad \forall v \in H_0^1(\Omega)$$

and in infinite other ways.

Proof of Theorem 5.39 Let us equip $H_0^1(\Omega)$ with the scalar product

$$(u, v)_{H_0^1(\Omega)} = \int_\Omega (uv + \nabla u \cdot \nabla v) dx, \quad \forall u, v \in H_0^1(\Omega).$$

It is clear that if F is like (5.154), then $F \in H^{-1}(\Omega)$, as a matter of fact, by applying the Cauchy–Schwarz inequality we get

$$\begin{aligned} |\langle F, v\rangle| &= \left| \int_\Omega f_0 v dx + \sum_{j=1}^{n} \int_\Omega f_j \partial_j v dx \right| \leq \\ &\leq \left(\sum_{j=0}^{n} \int_\Omega |f_j|^2 dx \right)^{1/2} \|v\|_{H_0^1(\Omega)}, \quad \forall v \in H_0^1(\Omega). \end{aligned} \tag{5.155}$$

Conversely, let us assume $F \in H^{-1}(\Omega)$. By the Riesz representation Theorem we have there exists a unique $u \in H_0^1(\Omega)$ such that

$$\langle F, v\rangle = (u, v)_{H_0^1(\Omega)} = \int_\Omega (uv + \nabla u \cdot \nabla v)dx, \quad \forall v \in H_0^1(\Omega). \tag{5.156}$$

Hence, denoting

$$f_0 = u, \quad f_j = \partial_j u, \quad j = 1, \cdots, n, \tag{5.157}$$

we have

$$\langle F, v\rangle = \int_\Omega f_0 v dx + \sum_{j=1}^n \int_\Omega f_j \partial_j v dx, \quad \forall v \in H_0^1(\Omega). \tag{5.158}$$

The proof of (i) is concluded.

Now, let us prove (ii). Let $u \in H_0^1(\Omega)$ and $f_j \in L^2(\Omega)$, $j = 0, 1, \cdots, n$, be like in (5.157). Let $g_j \in L^2(\Omega)$, $j = 0, 1, \cdots, n$ satisfy

$$\langle F, v\rangle = \int_\Omega \left(g_0 v + \sum_{j=1}^n g_j \partial_j v\right) dx, \quad \forall v \in H_0^1(\Omega).$$

Let us check that

$$\int_\Omega \sum_{j=0}^n |f_j|^2 dx \le \int_\Omega \sum_{j=0}^n |g_j|^2 dx. \tag{5.159}$$

We have

$$\int_\Omega \left(|u|^2 + |\nabla u|^2\right) dx = \langle F, u\rangle = \int_\Omega \left(g_0 u + \sum_{j=1}^n g_j \partial_j u\right) dx \le$$

$$\le \left(\int_\Omega \sum_{j=0}^n |g_j|^2 dx\right)^{1/2} \left(\int_\Omega \left(|u|^2 + |\nabla u|^2\right) dx\right)^{1/2}.$$

Hence

$$\begin{aligned}\int_\Omega \sum_{j=0}^n \left|f_j\right|^2 dx &= \int_\Omega \left(|u|^2 + |\nabla u|^2\right) dx \leq \\ &\leq \int_\Omega \sum_{j=0}^n \left|g_j\right|^2 dx,\end{aligned} \tag{5.160}$$

which proves (5.159).

In order to complete the proof, let us notice that by (5.155) we get

$$\|F\|_{H^{-1}(\Omega)} \leq \left(\sum_{j=0}^n \int_\Omega |f_j|^2 dx\right)^{1/2}. \tag{5.161}$$

On the other hand, setting

$$\widetilde{v} = \frac{u}{\|u\|_{H_0^1(\Omega)}},$$

we obtain, by (5.158) (recall that u satisfies (5.157)),

$$\begin{aligned}\langle F, \widetilde{v}\rangle &= \int_\Omega (u\widetilde{v} + \nabla u \cdot \nabla \widetilde{v}) dx = \\ &= \|u\|_{H_0^1(\Omega)} = \\ &= \left(\sum_{j=0}^n \int_\Omega |f_j|^2 dx\right)^{1/2}.\end{aligned} \tag{5.162}$$

Hence, by (5.161) and (5.162) we have

$$\|F\|_{H^{-1}(\Omega)} = \left(\sum_{j=0}^n \int_\Omega |f_j|^2 dx\right)^{1/2}.$$

By the just obtained equality and by (5.160) we obtain (ii). ■

Remark 5.23 Let us note that the greatest lower bound in (ii) is actually the minimum. ♦

Exercise 5.32 Let us denote by $H^{-m}(\Omega)$, $m \in \mathbb{N}$, the dual space of $H_0^m(\Omega)$. Prove that $F \in H^{-m}(\Omega)$ if and only if there exist $f_\alpha \in L^2(\Omega)$, $|\alpha| \leq m$ such that

$$\langle F, \widetilde{v} \rangle = \int_\Omega \sum_{|\alpha| \leq m} f_\alpha \partial^\alpha v dx, \quad \forall v \in H_0^m(\Omega)$$

and prove the analogue of the part (ii) of Theorem 5.39. ♠

5.12 The Sobolev Spaces with Noninteger Exponents and Traces

In the present section we will provide a characterization of the traces of $H^k(\Omega)$ function, $k \in \mathbb{N}$. For this purpose we need to extend the notion of the Sobolev space that we have studied so far to the spaces with non integer exponents.

5.12.1 Fourier Transform and $H^k(\mathbb{R}^n)$ Spaces, $k \in \mathbb{N}_0$

Let us state and prove

Theorem 5.40 *Let $k \in \mathbb{N}_0$. The following properties hold*

(i) Let $u \in L^2(\mathbb{R}^n)$. We have that $u \in H^k(\mathbb{R}^n)$ if and only if

$$\left(1 + |\xi|^2\right)^{k/2} \widehat{u}(\xi) \in L^2\left(\mathbb{R}^n\right). \tag{5.163}$$

(ii) There exists a constant $C \geq 1$ depending on k and n only, such that

$$C^{-1} \|u\|_{H^k(\mathbb{R}^n)} \leq \left\| \left(1 + |\xi|^2\right)^{k/2} \widehat{u}(\xi) \right\|_{L^2(\mathbb{R}^n)} \leq C \|u\|_{H^k(\mathbb{R}^n)}, \tag{5.164}$$

for every $u \in H^k(\mathbb{R}^n)$.

Proof If $k = 0$, then (i) and (ii) are obvious. Let us assume $k \geq 1$ and let us begin to prove (i). Since $\mathcal{S}(\mathbb{R}^n) \subset H^k(\mathbb{R}^n)$, we have that if $u \in H^k(\mathbb{R}^n)$ then

$$\int_{\mathbb{R}^n} \partial^\alpha u \varphi dx = (-1)^{|\alpha|} \int_{\mathbb{R}^n} u \partial^\alpha \varphi dx, \quad \text{for } |\alpha| \leq k, \ \forall \varphi \in \mathcal{S}\left(\mathbb{R}^n\right). \tag{5.165}$$

Claim I

$$u \in H^k\left(\mathbb{R}^n\right) \iff (i\xi)^\alpha \widehat{u}(\xi) \in L^2\left(\mathbb{R}^n\right), \text{ for } |\alpha| \leq k. \tag{5.166}$$

Proof of Claim I Let us prove " $\Longrightarrow$". Let $u \in H^k(\mathbb{R}^n)$ and let R be an arbitrary positive number and $\psi \in C_0^\infty(B_R)$. Let us define

$$\varphi(x) = \frac{1}{(2\pi)^n} \int_{\mathbb{R}^n} \overline{\psi(\xi)} e^{ix\cdot\xi} d\xi, \quad \forall x \in \mathbb{R}^n.$$

We have

$$\varphi \in \mathcal{S}(\mathbb{R}^n), \quad \text{and} \quad \overline{\widehat{\varphi}(\xi)} = \psi(\xi), \ \forall \xi \in \mathbb{R}^n.$$

In addition, as $u \in L^2(\mathbb{R}^n)$,

$$(i\xi)^\alpha \widehat{u}(\xi)_{|B_R} \in L^2(B_R), \ \text{for } |\alpha| \leq k,$$

hence $(i\xi)^\alpha \widehat{u}(\xi)\psi(\xi) \in L^2(\mathbb{R}^n)$ and by the Parseval identity we have, for $|\alpha| \leq k$,

$$\begin{aligned}
\int_{\mathbb{R}^n} (i\xi)^\alpha \widehat{u}(\xi)\psi(\xi) d\xi &= \int_{\mathbb{R}^n} (i\xi)^\alpha \widehat{u}(\xi) \overline{\widehat{\varphi}(\xi)} d\xi = \\
&= (-1)^{|\alpha|} \int_{\mathbb{R}^n} \widehat{u}(\xi) \overline{(i\xi)^\alpha \widehat{\varphi}(\xi)} d\xi = \\
&= (-1)^{|\alpha|} (2\pi)^n \int_{\mathbb{R}^n} u(x) \overline{\partial^\alpha \varphi(x)} dx = \\
&= (2\pi)^n \int_{\mathbb{R}^n} \partial^\alpha u(x) \overline{\varphi(x)} dx = \\
&= \int_{\mathbb{R}^n} \widehat{\partial^\alpha u}(\xi) \overline{\widehat{\varphi}(\xi)} d\xi = \\
&= \int_{\mathbb{R}^n} \widehat{\partial^\alpha u}(\xi) \psi(\xi) d\xi.
\end{aligned}$$

Hence

$$\int_{\mathbb{R}^n} \left((i\xi)^\alpha \widehat{u} - \widehat{\partial^\alpha u} \right) \psi d\xi = 0, \ \forall \psi \in C_0^\infty(B_R),$$

but $\widehat{\partial^\alpha u}_{|B_R} \in L^2(B_R)$ for $|\alpha| \leq k$, (because $u \in H^k(\mathbb{R}^n)$), therefore

$$(i\xi)^\alpha \widehat{u} = \widehat{\partial^\alpha u}, \ \text{in } B_R$$

and since R is arbitrary,

$$(i\xi)^\alpha \widehat{u} = \widehat{\partial^\alpha u}, \ \text{in } \mathbb{R}^n.$$

In particular

$$(i\xi)^\alpha \widehat{u} \in L^2\left(\mathbb{R}^n\right),$$

hence " $\Longrightarrow$" is proved.

Now we prove" $\Longleftarrow$". Let us assume that

$$(i\xi)^\alpha \widehat{u}(\xi) \in L^2\left(\mathbb{R}^n\right), \text{ for } |\alpha| \leq k.$$

Let $u_\alpha \in L^2(\mathbb{R}^n)$, $|\alpha| \leq k$ be defined by

$$u_\alpha(x) = \frac{1}{(2\pi)^n} \int_{\mathbb{R}^n} (i\xi)^\alpha \widehat{u}(\xi) e^{ix\cdot\xi} d\xi.$$

For any $\phi \in C_0^\infty(\mathbb{R}^n)$ and any $|\alpha| \leq k$, we have (recall (4.162))

$$\begin{aligned}
\int_{\mathbb{R}^n} \partial^\alpha \phi(x) \overline{u(x)} dx &= \frac{1}{(2\pi)^n} \int_{\mathbb{R}^n} \widehat{\partial^\alpha \phi}(\xi) \overline{\widehat{u}(\xi)} d\xi = \\
&= \frac{1}{(2\pi)^n} \int_{\mathbb{R}^n} (i\xi)^\alpha \widehat{\phi}(\xi) \overline{\widehat{u}(\xi)} d\xi = \\
&= \frac{(-1)^{|\alpha|}}{(2\pi)^n} \int_{\mathbb{R}^n} \widehat{\phi}(\xi) \overline{(i\xi)^\alpha \widehat{u}(\xi)} d\xi = \\
&= \frac{(-1)^{|\alpha|}}{(2\pi)^n} \int_{\mathbb{R}^n} \widehat{\phi}(\xi) \overline{\widehat{u_\alpha}(\xi)} d\xi = \\
&= (-1)^{|\alpha|} \int_{\mathbb{R}^n} \phi(x) \overline{u_\alpha(x)} dx.
\end{aligned}$$

Hence

$$\partial^\alpha u = u_\alpha \in L^2\left(\mathbb{R}^n\right), \qquad \text{for } |\alpha| \leq k.$$

Therefore $u \in H^k(\mathbb{R}^n)$. **This concludes the proof of Claim I.**

Claim II The following conditions are equivalent

(a) $(i\xi)^\alpha \widehat{u} \in L^2(\mathbb{R}^n)$ for $|\alpha| \leq k$,
(b) $\left(1 + |\xi|^2\right)^{k/2} \widehat{u}(\xi) \in L^2(\mathbb{R}^n)$.

Proof of Claim II First, let us note that (a) is equivalent to

$$\int_{\mathbb{R}^n} \sum_{|\alpha| \leq k} |\xi^\alpha|^2 \, |\widehat{u}(\xi)|^2 \, d\xi < +\infty,$$

hence, in order to prove that (a) and (b) are equivalent it suffices to prove that there exists $C \geq 1$ such that

$$C^{-1} \leq \frac{\sum_{|\alpha|\leq k} |\xi^\alpha|^2}{\left(1+|\xi|^2\right)^k} \leq C, \quad \forall \xi \in \mathbb{R}^n. \tag{5.167}$$

To this purpose we notice that the function

$$g(\xi, \tau) = \frac{\sum_{|\alpha|\leq k} \tau^{2(k-|\alpha|)} |\xi^\alpha|^2}{\left(\tau^2+|\xi|^2\right)^k},$$

is homogeneous of degree 0, it is continuous in $\mathbb{R}^{n+1} \setminus \{(0,0)\}$, and

$$g(\xi, \tau) > 0, \quad \forall (\xi, \tau) \in \mathbb{R}^{n+1} \setminus \{(0,0)\}.$$

Hence there exists $C \geq 1$ such that

$$C^{-1} \leq g(\xi, \tau) \leq C$$

so that, if $\tau = 1$ we get (5.167), which, in turn implies the equivalence of (a) and (b). **This concludes the proof of the Claim II.**

By Claim I and Claim II we obtain (i).

Concerning (ii), it is enough to observe that by (4.168) we have

$$\partial^\alpha u(x) = \frac{1}{(2\pi)^n} \int_{\mathbb{R}^n} (i\xi)^\alpha \widehat{u}(\xi) e^{ix\cdot\xi} d\xi, \text{ for } |\alpha| \leq k$$

and by the Parseval identity we have

$$\|u\|_{H^k(\mathbb{R}^n)} = \frac{1}{(2\pi)^n} \int_{\mathbb{R}^n} \sum_{|\alpha|\leq k} |\xi^\alpha|^2 \, |\widehat{u}(\xi)|^2 \, d\xi.$$

Hence by (5.167) we derive (5.164). ∎

5.12.2 The Sobolev Spaces with Noninteger Exponents

Theorem 5.40 justifies the following definition (instead of $\mathbb{R}^n$ we will consider $\mathbb{R}^m$, with $m \in \mathbb{N}$ to avoid confusion later on, when we will need to set $m = n - 1$).

Definition 5.41 Let $m \in \mathbb{N}$ and let s be a real number, we say that $u \in \mathcal{S}'(\mathbb{R}^m)$ belongs to $H^s(\mathbb{R}^m)$, if

$$\left(1+|\xi|^2\right)^{s/2} \widehat{u}(\xi) \in L^2\left(\mathbb{R}^m\right),$$

in this case we denote

$$\|u\|_{H^s(\mathbb{R}^m)} = \left(\frac{1}{(2\pi)^m}\int_{\mathbb{R}^m}\left(1+|\xi|^2\right)^s |\widehat{u}(\xi)|^2\,d\xi\right)^{1/2}. \tag{5.168}$$

It is clear that if $s \in \mathbb{N}$, we once again recover the Sobolev spaces with integer exponents that we have previously examined. However, if $s \notin \mathbb{N}$, we encounter a new set of spaces, referred to as the **Sobolev spaces with noninteger exponents**, or commonly known as 'the Sobolev spaces with fractional exponents.' It is straightforward to verify that the norm $\|\cdot\|_{H^s(\mathbb{R}^m)}$ is induced by the scalar product

$$(u, v)_{H^s(\mathbb{R}^m)} = \frac{1}{(2\pi)^m}\int_{\mathbb{R}^m}\left(1+|\xi|^2\right)^s \widehat{u}(\xi)\overline{\widehat{v}(\xi)}d\xi.$$

We leave the reader to verify that $H^s(\mathbb{R}^m)$ is a Hilbert space.

For example, if $x_0 \in \mathbb{R}^m$, as $\widehat{\delta_{x_0}} = e^{-ix_0\cdot\xi}$, then

$$\delta_{x_0} \in H^s\left(\mathbb{R}^m\right), \quad \text{for every } s < -\frac{m}{2}. \tag{5.169}$$

However, in what follows we consider only the case of

$$s \geq 0,$$

in such cases we have $H^s(\mathbb{R}^m) \subset L^2(\mathbb{R}^m)$ (if $s > 0$, the inclusion is proper).

Theorem 5.42 *If* $0 < s < 1$ *then the norm* $\|u\|_{H^s(\mathbb{R}^m)}$ *is equivalent to the norm*

$$\|u\| = \left(\|u\|^2_{L^2(\mathbb{R}^m)} + |u|^2_{s,\mathbb{R}^m}\right)^{1/2}, \tag{5.170}$$

where

$$|u|^2_{s,\mathbb{R}^m} = \int_{\mathbb{R}^m} dx \int_{\mathbb{R}^m} \frac{|u(x)-u(y)|^2}{|x-y|^{m+2s}}dy. \tag{5.171}$$

Proof We need to prove that there exists $C \geq 1$ such that, for every $u \in H^s(\mathbb{R}^m)$, we have

$$C^{-1}\int_{\mathbb{R}^m}|\xi|^{2s}|\widehat{u}(\xi)|^2\,d\xi \leq |u|^2_{s,\mathbb{R}^m} \leq C\int_{\mathbb{R}^m}|\xi|^{2s}|\widehat{u}(\xi)|^2\,d\xi. \tag{5.172}$$

Let us begin by observing that

$$
\begin{aligned}
|u|^2_{s,\mathbb{R}^m} &= \int_{\mathbb{R}^m} dx \int_{\mathbb{R}^m} \frac{|u(x+z)-u(x)|^2}{|z|^{m+2s}} dz = \\
&= \frac{1}{(2\pi)^m} \int_{\mathbb{R}^m} \frac{1}{|z|^{m+2s}} dz \int_{\mathbb{R}^m} \left| \widehat{u(\cdot+z)} - \widehat{u(\cdot)} \right|^2 d\xi = \\
&= \frac{1}{(2\pi)^m} \int_{\mathbb{R}^m} \frac{1}{|z|^{m+2s}} dz \int_{\mathbb{R}^m} \left| e^{iz\cdot\xi} - 1 \right|^2 |\widehat{u}(\xi)|^2 \, d\xi = \\
&= \frac{1}{(2\pi)^m} \int_{\mathbb{R}^m} \phi(\xi) \, |\widehat{u}(\xi)|^2 \, d\xi,
\end{aligned}
\tag{5.173}
$$

where

$$
\phi(\xi) = \int_{\mathbb{R}^m} \frac{\left| e^{iz\cdot\xi} - 1 \right|^2}{|z|^{m+2s}} dz.
$$

Let us notice that ϕ is a homogeneous function with degree $2s$. As a matter of fact, for any $t > 0$, we have

$$
\begin{aligned}
\phi(t\xi) &= \int_{\mathbb{R}^m} \frac{\left| e^{itz\cdot\xi} - 1 \right|^2}{|z|^{m+2s}} dz = \\
&= \int_{\mathbb{R}^m} \frac{\left| e^{iy\cdot\xi} - 1 \right|^2}{|t^{-1}y|^{m+2s}} \frac{dy}{t^m} = t^{2s} \int_{\mathbb{R}^m} \frac{\left| e^{iy\cdot\xi} - 1 \right|^2}{|y|^{m+2s}} dy = \\
&= t^{2s} \phi(t\xi), \quad \forall \xi \in \mathbb{R}^m.
\end{aligned}
$$

Moreover, ϕ is a continuous function in $\mathbb{R}^m$. In order to prove this, let $\xi_0 \in \mathbb{R}^m$ and let us check that

$$
\lim_{\xi \to \xi_0} \int_{\mathbb{R}^m} \frac{\left| e^{iz\cdot\xi} - 1 \right|^2}{|z|^{m+2s}} dz = \int_{\mathbb{R}^m} \frac{\left| e^{iz\cdot\xi_0} - 1 \right|^2}{|z|^{m+2s}} dz.
\tag{5.174}
$$

We have

$$
\lim_{\xi \to \xi_0} \frac{\left| e^{iz\cdot\xi} - 1 \right|^2}{|z|^{m+2s}} = \frac{\left| e^{iz\cdot\xi_0} - 1 \right|^2}{|z|^{m+2s}}, \quad \forall z \in \mathbb{R}^m
$$

and, if $|\xi - \xi_0| < 1$, we have

$$\frac{\left|e^{iz\cdot\xi} - 1\right|^2}{|z|^{m+2s}} = \frac{\left|e^{iz\cdot\xi} - 1\right|^2}{|z|^{m+2s}}\chi_{B_1}(z) + \frac{\left|e^{iz\cdot\xi} - 1\right|^2}{|z|^{m+2s}}\chi_{\mathbb{R}^m\setminus B_1}(z) \leq$$

$$\leq \frac{C(1+|\xi_0|)^2}{|z|^{m-2(1-s)}}\chi_{B_1}(z) + \frac{4}{|z|^{m+2s}}\chi_{\mathbb{R}^m\setminus B_1}(z) \in L^1\left(\mathbb{R}^m\right).$$

Therefore by the Dominated Convergence Theorem we get (5.174).

Now, since ϕ is continuous in $\mathbb{R}^m$ and $\phi(\xi) > 0$, for every $|\xi| = 1$, we have that there exists $C \geq 1$ such that

$$C^{-1}|\xi|^{2s} \leq \phi(\xi) \leq C|\xi|^{2s}, \quad \forall \xi \in \mathbb{R}^m.$$

By the last inequality and by (5.173) we obtain (5.172). ■

Similarly to the previous Theorem the following one can be proved

Theorem 5.43 *If $s > 0$, $s \notin \mathbb{N}$, then the norm $\|u\|_{H^s(\mathbb{R}^m)}$ is equivalent to the norm*

$$\|u\| = \left(\|u\|^2_{H^{[s]}(\mathbb{R}^m)} + |u|^2_{s,\mathbb{R}^m}\right)^{1/2}, \tag{5.175}$$

where

$$|u|^2_{s,\mathbb{R}^m} = \sum_{|\alpha|=[s]} \int_{\mathbb{R}^m} dx \int_{\mathbb{R}^m} \frac{|\partial^\alpha u(x) - \partial^\alpha u(y)|^2}{|x-y|^{m+2(s-[s])}} dy. \tag{5.176}$$

Theorems 5.42, 5.43 justify the following definition.

Definition 5.44 Let Θ be a bounded open set of $\mathbb{R}^m$ of class $C^{0,1}$. Let $s \notin \mathbb{N}$ be a positive real number. We define $H^s(\Theta)$ as the space of functions $u \in H^{[s]}(\Theta)$ such that

$$|u|^2_{s,\Theta} = \sum_{|\alpha|=[s]} \int_\Theta dx \int_\Theta \frac{|\partial^\alpha u(x) - \partial^\alpha u(y)|^2}{|x-y|^{m+2(s-[s])}} dy < +\infty,$$

equipped with the norm

$$\|u\|_{H^s(\Theta)} = \left(\|u\|_{H^{[s]}(\Theta)} + |u|^2_{s,\Theta}\right)^{1/2}.$$

It is not difficult to prove that the space $H^s(\Theta)$ is complete.

Now let us define $H^s(\partial\Omega)$, where Ω is a bounded open set. If $s \in \mathbb{N}$, we assume that $\partial\Omega$ is of class C^s. If $s \notin \mathbb{N}$ we assume that $\partial\Omega$ is of class $C^{[s],1}$.

We proceed basically as we did in Sect. 3.8 to define $L^p(\partial\Omega)$ (we will use the same notations as Sect. 3.8).

Let us begin by the case $s := k$, positive integer number. Thus, let us assume $\partial\Omega$ of class C^k with constants r_0, M_0 and let us cover $\partial\Omega$ by a finite number, N, of cylinders $\widetilde{Q}_{r_0,2M_0}(X_i)$, $i = 1\cdots, N$, where $X_i \in \partial\Omega$ isometric to $Q_{r_0,2M_0}$. Moreover let us assume that: $\Sigma_i := \widetilde{Q}_{r_0,2M_0}(X_i) \cap \partial\Omega$, for any $i = 1\cdots, N$, up to isometry for which X_i is mapped in 0, is the graph of a function $\varphi_i \in C^k(\overline{B'}_{r_0})$ likewise Definition 3.58. We say that $f \in H^k(\partial\Omega)$ provided that the functions $f(x', \varphi_i(x'))$, $i = 1, \cdots, N$, belong to $H^k(B'_{r_0})$ and we denote

$$\|f\|_{H^k(\partial\Omega)} = \left(\sum_{i=1}^N \|f\|^2_{H^k(\Sigma_i)}\right)^{1/2},$$

where

$$\|f\|_{H^k(\Sigma_i)} = \|f(\cdot, \varphi_i(\cdot))\|_{H^k(B_{r_0})}.$$

Of course, the norm $\|\cdot\|_{H^k(\partial\Omega)}$ depends on the particular family of cylinders that we use as a covering, but they are all equivalent norms. Moreover, $H^k(\partial\Omega)$ is a separable Hilbert space.

If $s \notin \mathbb{N}$, we say that $f \in H^s(\partial\Omega)$ provided that the functions $f(x', \varphi_i(x'))$ belong to $H^s(B'_r)$, for any $i = 1, \cdots, N$ and we denote by

$$\|f\|_{H^s(\partial\Omega)} = \left(\sum_{i=1}^N \|f\|^2_{H^s(\Sigma_i)}\right)^{1/2}.$$

The space $H^s(\partial\Omega)$ is complete and can be equipped of a Hilbert structure.

For an extended discussion of Sobolev spaces with noninteger exponent we refer the reader to [41, Cap. 6], [69, Cap. 6], [100, Cap. 2]. Here we limit ourselves to prove a Theorem that will be useful in the next section.

Theorem 5.45 (Density of $C^\infty(\mathbb{R}^m)$ in $H^s(\mathbb{R}^m)$) *If s is a positive real number then $C_0^\infty(\mathbb{R}^m)$ is dense in $H^s(\mathbb{R}^m)$.*

We begin by stating the following

Lemma 5.46 *Let $\eta \in C_0^\infty(\mathbb{R}^m)$ satisfy*

(i) $\operatorname{supp}\eta \subset B_1$,
(ii) $\eta \geq 0$,
(iii) $\int_{\mathbb{R}^n} \eta(x)dx = 1$.

Let us denote, for any $\varepsilon > 0$, $v \in H^s(\mathbb{R}^m)$, $s > 0$,

$$\eta_\varepsilon(x) = \varepsilon^{-m}\eta\left(\varepsilon^{-1}x\right)$$

and

$$v^{\varepsilon} = \eta_{\varepsilon} \star v, \qquad \text{in } \mathbb{R}^m.$$

Then we have

$$\lim_{\varepsilon \to 0} \|\eta_{\varepsilon} \star v - v\|_{H^s(\mathbb{R}^m)} = 0. \tag{5.177}$$

Proof of the Lemma We have

$$\widehat{\eta_{\varepsilon}}(\xi) = \int_{\mathbb{R}^m} \varepsilon^{-m} \eta\left(\varepsilon^{-1} x\right) e^{-ix\cdot\xi} dx = \int_{\mathbb{R}^m} \eta(y) e^{-i\varepsilon y\cdot\xi} dy = \widehat{\eta}(\varepsilon\xi).$$

Moreover

$$\lim_{\varepsilon \to 0} \widehat{\eta}(\varepsilon\xi) = \widehat{\eta}(0) = \int_{\mathbb{R}^m} \eta(x) dx = 1 \tag{5.178}$$

and

$$|\widehat{\eta}(\varepsilon\xi)| \le \int_{\mathbb{R}^m} \eta(x) dx = 1, \quad \forall \xi \in \mathbb{R}^m \ \forall \varepsilon > 0. \tag{5.179}$$

Hence, for any $v \in H^s(\mathbb{R}^m)$, we have

$$\|\eta_{\varepsilon} \star v - v\|^2_{H^s(\mathbb{R}^m)} = \frac{1}{(2\pi)^m} \int_{\mathbb{R}^m} \left(1 + |\xi|^2\right)^s |\widehat{\eta}(\varepsilon\xi) - 1|^2 |\widehat{v}(\xi)|^2 d\xi.$$

By the last equality, by the Dominated Convergence Theorem (take into account (5.178) and (5.179)) we obtain (5.177). ■

Proof of Theorem 5.45 The case where $s \in \mathbb{N}$ has been proved in Proposition 5.13. Let us consider the case $0 < s < 1$ (if $s > 1$, the proof proceeds in a similar way, and we leave the details to the reader).

Claim Let us denote by $\mathcal{H}_s$ the subspace of the functions of $H^s(\mathbb{R}^m)$ with compact support. Then $\mathcal{H}_s$ is dense in $H^s(\mathbb{R}^m)$.

Proof of the Claim Let $\zeta \in C_0^{\infty}(\mathbb{R}^m)$ satisfy

$$0 \le \zeta \le 1, \quad \text{in } \mathbb{R}^m,$$

$$\zeta(x) = 1, \quad \forall x \in B_1, \quad \zeta(x) = 0, \quad \forall x \in \mathbb{R}^m \setminus B_2$$

and

$$|\nabla\zeta| \le C_0, \quad \text{in } \mathbb{R}^m,$$

where C_0 is a constant. Let $R > 1$ and

$$\zeta_R(x) = \zeta\left(R^{-1}x\right).$$

We have

$$\|u - \zeta_R u\|_{L^2(\mathbb{R}^m)} \leq \|u\|_{L^2(\mathbb{R}^m \setminus B_R)} \to 0, \quad \text{as } R \to \infty.$$

Now, we prove

$$\lim_{R\to\infty} |u - u\zeta_R|_{s,\mathbb{R}^m} = 0. \tag{5.180}$$

In proving the latter, we will obtain as a by-product of the performed calculations that $u\zeta_R$ belongs to $H^s(\mathbb{R}^m)$.

We apply Theorem 5.42 and we write

$$|u - u\zeta_R|^2_{s,\mathbb{R}^m} = \int_{\mathbb{R}^m} dx \int_{\mathbb{R}^m} \frac{\Phi_R^2(x,y)}{|x-y|^{m+2s}} dy,$$

where

$$\Phi_R(x,y) = |(1-\zeta_R(x))\,u(x) - (1-\zeta_R(y))\,u(y)|.$$

We have

$$\begin{aligned}\Phi_R(x,y) &\leq |1-\zeta_R(y)|\,|u(x)-u(y)| + |\zeta_R(x)-\zeta_R(y)|\,|u(x)| \leq \\ &\leq \chi_{\mathbb{R}^m\setminus B_R}(y)\,|u(x)-u(y)| + |\zeta_R(x)-\zeta_R(y)|\,|u(x)|.\end{aligned}$$

Hence

$$\begin{aligned}|u - u\zeta_R|^2_{s,\mathbb{R}^m} \leq\ & 2\int_{\mathbb{R}^m} dx \int_{\mathbb{R}^m\setminus B_R} \frac{|u(x)-u(y)|^2}{|x-y|^{m+2s}} dy + \\ & + 2\int_{\mathbb{R}^m} dx \int_{\mathbb{R}^m} \frac{|\zeta_R(x)-\zeta_R(y)|^2\,|u(x)|^2}{|x-y|^{m+2s}} dy.\end{aligned} \tag{5.181}$$

Set

$$I := \int_{\mathbb{R}^m} dx \int_{\mathbb{R}^m} \frac{|\zeta_R(x)-\zeta_R(y)|^2\,|u(x)|^2}{|x-y|^{m+2s}} dy.$$

Let us notice that

$$|\zeta_R(x) - \zeta_R(y)| = 0, \quad \text{for } |x| \geq 2R \text{ and } |y| \geq 2R.$$

Moreover,

$$|\zeta_R(x) - \zeta_R(y)| \le \min\left\{2, \frac{C_0|x-y|}{R}\right\}, \quad \forall x \in \mathbb{R}^m, \ \forall y \in \mathbb{R}^m.$$

Hence

$$\begin{aligned} I \le & \int_{|x|\le 2R} dx \int_{\mathbb{R}^m} \frac{|\zeta_R(x)-\zeta_R(y)|^2 \, |u(x)|^2}{|x-y|^{m+2s}} dy + \\ & + \int_{\mathbb{R}^m} dx \int_{|y|\le 2R} \frac{|\zeta_R(x)-\zeta_R(y)|^2 \, |u(x)|^2}{|x-y|^{m+2s}} dy := I_1 + I_2. \end{aligned} \tag{5.182}$$

We have

$$\begin{aligned} I_1 \le & \frac{C_0^2}{R^2} \int_{|x|\le 2R} |u(x)|^2 \, dx \int_{|x-y|\le \frac{2R}{C_0}} \frac{dy}{|x-y|^{m-2+2s}} + \\ & + 4 \int_{|x|\le 2R} |u(x)|^2 \, dx \int_{|x-y|> \frac{2R}{C_0}} \frac{dy}{|x-y|^{m+2s}} \le \\ & \le \frac{\omega_m}{2-2s} \frac{C_0^2}{R^2} \left(\frac{2R}{C_0}\right)^{2-2s} \int_{|x|\le 2R} |u(x)|^2 \, dx + \\ & + \frac{\omega_m}{2s} \left(\frac{2R}{C_0}\right)^{-2s} \int_{|x|\le 2R} |u(x)|^2 \, dx \le \\ & \le C R^{-2s} \, \|u\|^2_{L^2(\mathbb{R}^m)}, \end{aligned} \tag{5.183}$$

where C depends on m and s only.

Concerning I_2, we have

$$\begin{aligned} I_2 = & \int_{|y|\le 2R} dy \int_{\mathbb{R}^m} \frac{|\zeta_R(x)-\zeta_R(y)|^2 \, |u(x)|^2}{|x-y|^{m+2s}} dx = \\ = & \int_{|y|\le 2R} dy \int_{|x-y|\le \frac{2R}{C_0}} \frac{|\zeta_R(x)-\zeta_R(y)|^2 \, |u(x)|^2}{|x-y|^{m+2s}} dx + \\ & + \int_{|y|\le 2R} dy \int_{|x-y|> \frac{2R}{C_0}} \frac{|\zeta_R(x)-\zeta_R(y)|^2 \, |u(x)|^2}{|x-y|^{m+2s}} dx \le \\ \le & \frac{C_0^2}{R^2} \int_{|y|\le 2R} dy \int_{|x|\le 2R(1+1/C_0)} \frac{|u(x)|^2}{|x-y|^{m-2+2s}} dx + \\ & + 4 \left(\frac{2R}{C_0}\right)^{-m-2s} \int_{|y|\le 2R} dy \int_{|x-y|> \frac{2R}{C_0}} |u(x)|^2 \, dx. \end{aligned}$$

By interchanging the order of integration in the second-to-last integral and trivially estimating from above the last integral, we obtain

$$\begin{aligned}
I_2 &\leq \frac{C_0^2}{R^2} \int_{|x| \leq 2R(1+1/C_0)} |u(x)|^2 \, dx \int_{|y| \leq 2R} \frac{dy}{|x-y|^{m-2+2s}} + \\
&+ 4\frac{\omega_m}{m} \left(\frac{2R}{C_0} \right)^{-m-2s} (2R)^m \, \|u\|^2_{L^2(\mathbb{R}^m)} \leq \\
&\leq \frac{C_0^2}{R^2} \int_{|x| \leq 2R(1+1/C_0)} |u(x)|^2 \, dx \int_{|y-x| \leq 2R(2+1/C_0)} \frac{dy}{|x-y|^{m-2+2s}} + \\
&+ C R^{-2s} \, \|u\|^2_{L^2(\mathbb{R}^m)} \leq \\
&\leq C' R^{-2s} \, \|u\|^2_{L^2(\mathbb{R}^m)} ,
\end{aligned} \tag{5.184}$$

where C and C' depend on m and s only.

By (5.182), (5.183) and (5.184) we get

$$I = I_1 + I_2 \leq C R^{-2s} \, \|u\|^2_{L^2(\mathbb{R}^m)} ,$$

where C depends on m and s only.

By the just obtained inequality and by (5.181) we have

$$|u - u\zeta_R|^2_{s,\mathbb{R}^m} \leq 2 \int_{\mathbb{R}^m} dx \int_{\mathbb{R}^m \setminus B_R} \frac{|u(x) - u(y)|^2}{|x-y|^{m+2s}} dy + C R^{-2s} \, \|u\|^2_{L^2(\mathbb{R}^m)} .$$

Since the following function belongs to $L^2(\mathbb{R}^m \times \mathbb{R}^m)$ (as $u \in H^s(\mathbb{R}^m)$)

$$\mathbb{R}^m \times \mathbb{R}^m \ni (x, y) \to \frac{|u(x) - u(y)|^2}{|x-y|^{m+2s}} ,$$

we obtain (5.180). **This concludes the proof of the Claim.**

Now, let $\delta > 0$ and let R_0 be (recall (5.180)) such that

$$\left\| u - u\zeta_{R_0} \right\|_{s,\mathbb{R}^m} < \frac{\delta}{2} .$$

Lemma 5.46 implies that there exists $\varepsilon_0 > 0$ such that

$$\left\| u\zeta_{R_0} - \left(u\zeta_{R_0} \right) \star \eta_{\varepsilon_0} \right\|_{s,\mathbb{R}^m} < \frac{\delta}{2} .$$

Hence

$$\left\| u - \left(u\zeta_{R_0} \right) \star \eta_{\varepsilon_0} \right\|_{s,\mathbb{R}^m} < \delta.$$

Since $\left(u\zeta_{R_0} \right) \star \eta_{\varepsilon_0} \in C_0^{\infty}\left(\mathbb{R}^m\right)$, the last inequality concludes the proof. ■

5.12.3 The Theorem of Characterization of the Traces

We preliminarily examine the extension of the notion of trace of a function belonging to $H^1\left(\mathbb{R}^n_+\right)$, where $\mathbb{R}^n_+ = \left\{ x = (x', x_n) \in \mathbb{R}^n : x_n > 0 \right\}$.

If $u \in H^1\left(\mathbb{R}^n_+\right)$, then the function

$$\widetilde{u}\left(x', x_n\right) := u\left(x', |x_n|\right),$$

belongs to $H^1\left(\mathbb{R}^n\right)$ and, as $H^1\left(\mathbb{R}^n\right) = H_0^1\left(\mathbb{R}^n\right)$, there exists a sequence $\left\{v_j\right\}$ in $C_0^{\infty}\left(\mathbb{R}^n\right)$ such that

$$\left\{v_j\right\} \to \widetilde{u}, \quad \text{in } H^1\left(\mathbb{R}^n\right).$$

Hence

$$\left\{ \left(v_j\right)_{|\mathbb{R}^n_+} \right\} \to u, \quad \text{in } H^1\left(\mathbb{R}^n_+\right).$$

Denoting by

$$w = v_j - v_k, \quad j, k \in \mathbb{N},$$

we have

$$w\left(x', 0\right) = w\left(x', x_n\right) - \int_0^{x_n} \partial_{x_n} w\left(x', y\right) dy, \quad \forall x_n > 0,$$

from which we obtain

$$\left|w\left(x', 0\right)\right|^2 \leq 2\left|w\left(x', x_n\right)\right|^2 + 2x_n \int_0^{x_n} \left|\partial_{x_n} w\left(x', y\right)\right|^2 dy.$$

Now, integrating both the sides of the last inequality over $(0, \delta)$, $\delta > 0$, w.r.t. x_n, we have

$$\delta \left|w\left(x', 0\right)\right|^2 \leq 2 \int_0^\delta \left|w\left(x', x_n\right)\right|^2 dx_n + 2\delta \int_0^\delta dx_n \int_0^{x_n} \left|\partial_{x_n} w\left(x', y\right)\right|^2 dy \leq$$

$$\leq 2 \int_0^\delta \left|w\left(x', x_n\right)\right|^2 dx_n + 2\delta^2 \int_0^\delta \left|\partial_{x_n} w\left(x', y\right)\right|^2 dy.$$

Integrating both the sides of the last inequality over $\mathbb{R}^{n-1}$ w.r.t. x', we have

$$\int_{\mathbb{R}^{n-1}} \left|w\left(x', 0\right)\right|^2 dx' \leq \frac{2}{\delta} \int_{\mathbb{R}^n} |w(x)|^2 dx + 2\delta \int_{\mathbb{R}^n} |\nabla w(x)|^2 dx.$$

Starting from this inequality we proceed as in the proof of Theorem 5.18 (inequality (5.61)) and we obtain the extension of the trace operator from the space

$$C_*^\infty\left(\mathbb{R}_+^n\right) := \left\{\widetilde{u}_{|\mathbb{R}_+^n} : \widetilde{u} \in C_0^\infty\left(\mathbb{R}^n\right)\right\}$$

to the space $H^1\left(\mathbb{R}_+^n\right)$. In particular, we have

$$Tu \in L^2\left(\mathbb{R}^{n-1}\right), \quad \forall u \in H^1\left(\mathbb{R}_+^n\right)$$

and

$$\|Tu\|_{L^2(\mathbb{R}^{n-1})} \leq C \|u\|_{H^1(\mathbb{R}^n)}, \quad \forall u \in H^1\left(\mathbb{R}_+^n\right), \tag{5.185}$$

where C depends on n only.

The following Theorem provides a characterization of the image of $H^1(\Omega)$ by means the trace operator.

Theorem 5.47 (Characterization of the Trace) *Let Ω be either a bounded open set of class $C^{0,1}$ with constants M_0, r_0, or $\Omega = \mathbb{R}_+^n$. Let*

$$T : H^1(\Omega) \to L^2(\partial\Omega)$$

be the trace operator defined in Theorem 5.18 (in the case where $\Omega = \mathbb{R}_+^n$ the definition is given at the beginning of this section).

Then we have

$$T\left(H^1(\Omega)\right) = H^{1/2}(\partial\Omega).$$

Moreover

(i)

$$\|T(u)\|_{H^{1/2}(\partial\Omega)} \leq C\,\|u\|_{H^1(\Omega)}\,, \quad \forall u \in H^1(\Omega),$$

where C depends on M_0, r_0 and n only.

(ii) There exists a bounded, linear map

$$\mathcal{T}\colon H^{1/2}(\partial\Omega) \to H^1(\Omega),$$

such that

$$T(\mathcal{T}(h)) = h, \quad \forall h \in H^{1/2}(\partial\Omega).$$

In particular, denoting by $u = \mathcal{T}(h)$, we have

$$\|u\|_{H^1(\Omega)} \leq C\,\|h\|_{H^{1/2}(\partial\Omega)}\,, \quad \forall h \in H^{1/2}(\Omega),$$

where C depends on M_0, r_0 and n only.

In the general case the proof of Theorem 5.47, is quite technical. Here we limit ourselves to the case $\Omega = \mathbb{R}^n_+$. A complete treatment (including the traces of the functions belonging to $W^{k,p}(\Omega)$, $k \in \mathbb{N}$) can be found in [69, Ch. 6] and in [100, Ch. 2, Sects. 2.3–2.5].

Proof of Theorem 5.47 in the case $\Omega = \mathbb{R}^n_+$ We have proved, in (5.185), that $T \in L^2\left(\mathbb{R}^{n-1}\right)$. Now we prove

$$Tu \in H^{1/2}\left(\mathbb{R}^{n-1}\right).$$

Claim Let $v \in C^\infty_*\left(\mathbb{R}^n_+\right)$. Let us denote by

$$h\left(x'\right) = v\left(x', 0\right), \quad \forall x' \in \mathbb{R}^{n-1}.$$

We have

$$\widehat{h}\left(\xi'\right) = \frac{1}{2\pi}\int_{\mathbb{R}} \widehat{v}\left(\xi', \xi_n\right) d\xi_n. \tag{5.186}$$

Proof of the Claim Since

$$v\left(x', x_n\right) = \frac{1}{(2\pi)^n}\int_{\mathbb{R}^n} \widehat{v}\left(\xi\right) e^{ix\cdot\xi} d\xi,$$

we have

$$\begin{aligned}
h\left(x'\right) &= v\left(x', 0\right) = \\
&= \frac{1}{(2\pi)^n} \int_{\mathbb{R}^n} \widehat{v}\,(\xi)\, e^{ix'\cdot\xi'} d\xi = \\
&= \frac{1}{(2\pi)^{n-1}} \int_{\mathbb{R}^{n-1}} \left(\frac{1}{2\pi} \int_{\mathbb{R}} \widehat{v}\,(\xi', \xi_n)\, d\xi_n \right) e^{ix'\cdot\xi'} d\xi'
\end{aligned}$$

and by proposition (a) of Theorem 4.26 we obtain (5.186). **This concludes the proof of the Claim.**

Now, by (5.186) we have

$$\begin{aligned}
&\int_{\mathbb{R}^{n-1}} \left|\widehat{h}(\xi')\right|^2 \left(1 + |\xi'|^2\right)^{1/2} d\xi' \\
&\leq \frac{1}{4\pi^2} \int_{\mathbb{R}^{n-1}} \left(\int_{\mathbb{R}} \left|\widehat{v}(\xi', \xi_n)\right| d\xi_n \right)^2 \left(1 + |\xi'|^2\right)^{1/2} d\xi' = \\
&= \frac{1}{4\pi^2} \int_{\mathbb{R}^{n-1}} \left(1 + |\xi'|^2\right)^{1/2} d\xi' \\
&\quad \times \left(\int_{\mathbb{R}} \left(1 + |\xi|^2\right)^{1/2} \left|\widehat{v}(\xi', \xi_n)\right| \left(1 + |\xi|^2\right)^{-1/2} d\xi_n \right)^2 \leq \\
&\leq \frac{1}{4\pi^2} \int_{\mathbb{R}^{n-1}} \left(1 + |\xi'|^2\right)^{1/2} d\xi' \\
&\quad \times \left(\int_{\mathbb{R}} \left(1 + |\xi|^2\right) \left|\widehat{v}(\xi', \xi_n)\right|^2 d\xi_n \int_{\mathbb{R}} \left(1 + |\xi|^2\right)^{-1} d\xi_n \right).
\end{aligned}$$

Moreover

$$\int_{\mathbb{R}} \left(1 + |\xi|^2\right)^{-1} d\xi_n = \int_{\mathbb{R}} \frac{d\xi_n}{1 + |\xi'|^2 + \xi_n^2} = \frac{\pi}{\left(1 + |\xi'|^2\right)^{1/2}}.$$

Therefore

$$\int_{\mathbb{R}^{n-1}} \left|\widehat{h}(\xi')\right|^2 \left(1 + |\xi'|^2\right)^{1/2} d\xi' \leq \frac{1}{4\pi} \int_{\mathbb{R}^n} \left(1 + |\xi|^2\right) |\widehat{v}(\xi)|^2 d\xi,$$

that is

$$\|T(v)\|_{H^{1/2}(\mathbb{R}^{n-1})} \leq \frac{1}{2\sqrt{\pi}} \|v\|_{H^1(\mathbb{R}^n_+)}, \quad \forall v \in C_*^\infty\left(\mathbb{R}^n_+\right),$$

from which, by density we have

$$\|T(u)\|_{H^{1/2}(\mathbb{R}^{n-1})} \leq \frac{1}{2\sqrt{\pi}} \|u\|_{H^1(\mathbb{R}^n)}, \quad \forall u \in H^1\left(\mathbb{R}^n_+\right). \tag{5.187}$$

Now we prove (ii).

Let $h \in C_0^\infty\left(\mathbb{R}^{n-1}\right)$. For any $\varepsilon > 0$, let

$$u_\varepsilon\left(x', x_n\right) = \frac{1}{(2\pi)^{n-1}} \int_{\mathbb{R}^{n-1}} e^{-(1+|\xi'|)(x_n+\varepsilon)} \widehat{h}\left(\xi'\right) e^{ix'\cdot\xi'} d\xi', \quad \forall x \in \overline{\mathbb{R}^n_+}.$$

By applying Theorem 4.26 and by performing the derivative under the integral sign, it can be easily checked that

$$u_\varepsilon \in C^\infty\left(\overline{\mathbb{R}^n_+}\right)$$

and

$$u_\varepsilon\left(x', 0\right) = \frac{1}{(2\pi)^{n-1}} \int_{\mathbb{R}^{n-1}} e^{-(1+|\xi'|)\varepsilon} \widehat{h}\left(\xi'\right) e^{ix'\cdot\xi'} d\xi', \quad \forall x' \in \mathbb{R}^{n-1}.$$

Now we prove what follows

(a) $u_\varepsilon \in H^1\left(\mathbb{R}^n_+\right)$ and, denoting by

$$u\left(x', x_n\right) = \frac{1}{(2\pi)^{n-1}} \int_{\mathbb{R}^{n-1}} e^{-(1+|\xi'|)x_n} \widehat{h}\left(\xi'\right) e^{ix'\cdot\xi'} d\xi', \quad \forall x \in \mathbb{R}^n,$$

(let us notice that $u(\cdot, 0) = h$) we have,

$$u_\varepsilon \in H^1\left(\mathbb{R}^n_+\right) \quad \text{and} \quad u \in H^1\left(\mathbb{R}^n_+\right).$$

Moreover

$$\|u_\varepsilon\|_{H^1(\mathbb{R}^n)} \leq C \|h\|_{H^{1/2}(\mathbb{R}^{n-1})}, \tag{5.188}$$

and

$$\|u\|_{H^1(\mathbb{R}^n)} \leq C \|h\|_{H^{1/2}(\mathbb{R}^{n-1})}, \tag{5.189}$$

where C depends on n only.

(b)

$$u_\varepsilon \to u, \quad \text{as } \varepsilon \to 0 \text{ in } H^1\left(\mathbb{R}^n_+\right).$$

Proof of (a)
The Parseval identity implies

$$\int_{\mathbb{R}^{n-1}} \left|u_\varepsilon\left(x', x_n\right)\right|^2 dx' = c_n \int_{\mathbb{R}^{n-1}} e^{-2(1+|\xi'|)(x_n+\varepsilon)} \left|\widehat{h}\left(\xi'\right)\right|^2 d\xi',$$

where c_n depends on n only. Hence

$$\begin{aligned}\int_{\mathbb{R}^n_+} \left|u_\varepsilon\left(x', x_n\right)\right|^2 dx'dx_n &= c_n \int_{\mathbb{R}^{n-1}} \left|\widehat{h}\left(\xi'\right)\right|^2 d\xi' \left(\int_0^{+\infty} e^{-2(1+|\xi'|)(x_n+\varepsilon)} dx_n\right) = \\ &= c_n \int_{\mathbb{R}^{n-1}} \left|\widehat{h}\left(\xi'\right)\right|^2 \frac{e^{-2\varepsilon(1+|\xi'|)}}{2\left(1+|\xi'|\right)} d\xi' \leq \\ &\leq \frac{(2\pi)^{n-1} c_n}{2} \|h\|^2_{L^2(\mathbb{R}^{n-1})}.\end{aligned}$$

Therefore

$$\|u_\varepsilon\|_{L^2(\mathbb{R}^n)} \leq C \|h\|_{L^2(\mathbb{R}^{n-1})}, \tag{5.190}$$

where C depends on n only.

Now we estimate from above $\|\nabla u_\varepsilon\|_{L^2(\mathbb{R}^n)}$. We have

$$\partial_{x_n} u_\varepsilon\left(x', x_n\right) = -\frac{1}{(2\pi)^n} \int_{\mathbb{R}^{n-1}} \left(1+|\xi'|\right) e^{-(1+|\xi'|)(x_n+\varepsilon)} \widehat{h}\left(\xi'\right) e^{ix'\cdot\xi'} d\xi'.$$

Hence, arguing as above, we get

$$\begin{aligned}&\int_{\mathbb{R}^n_+} \left|\partial_{x_n} u_\varepsilon\left(x', x_n\right)\right|^2 dx'dx_n = \\ &= c_n \int_{\mathbb{R}^{n-1}} \left|\widehat{h}\left(\xi'\right)\right|^2 d\xi' \left(\int_0^{+\infty} \left(1+|\xi'|\right)^2 e^{-2(1+|\xi'|)(x_n+\varepsilon)} dx_n\right) = \\ &= \frac{c_n}{2} \int_{\mathbb{R}^{n-1}} \left(1+|\xi'|\right) \left|\widehat{h}\left(\xi'\right)\right|^2 e^{-2\varepsilon(1+|\xi'|)} d\xi' \leq \\ &\leq \frac{(2\pi)^{n-1} c_n}{2} \|h\|^2_{H^{1/2}(\mathbb{R}^{n-1})}.\end{aligned} \tag{5.191}$$

Now, if $1 \leq j \leq n-1$, we have

$$\partial_{x_j} u_\varepsilon\left(x', x_n\right) = \frac{1}{(2\pi)^n} \int_{\mathbb{R}^{n-1}} i\xi_j e^{-2(1+|\xi'|)(x_n+\varepsilon)} \widehat{h}\left(\xi'\right) e^{ix'\cdot\xi'} d\xi'$$

and arguing as in (5.191), we get

$$\int_{\mathbb{R}^n_+} \left|\partial_{x_j} u_\varepsilon \left(x', x_n\right)\right|^2 dx' dx_n \leq \frac{(2\pi)^{n-1} c_n}{2} \|h\|^2_{H^{1/2}(\mathbb{R}^{n-1})}. \tag{5.192}$$

By (5.191) and (5.192) we get

$$\|\nabla u_\varepsilon\|^2_{L^2(\mathbb{R}^n)} \leq C \|h\|^2_{H^{1/2}(\mathbb{R}^{n-1})},$$

where C depends on n only. By the just obtained inequality and by (5.190) we derive

$$\|u_\varepsilon\|_{H^1(\mathbb{R}^n)} \leq C \|h\|_{H^{1/2}(\mathbb{R}^{n-1})},$$

where C depends on n only.

As can be easily observed, the calculations performed above also apply to $\varepsilon = 0$ and similarly yield

$$\|u\|_{H^1(\mathbb{R}^n)} \leq C \|h\|_{H^{1/2}(\mathbb{R}^{n-1})},$$

where C depends on n only. Hence we have proved (5.188) and (5.189).**This concludes the proof of (a).**

Proof of (b)

Since

$$(u - u_\varepsilon)\left(x', x_n\right) = \frac{1}{(2\pi)^{n-1}} \int_{\mathbb{R}^{n-1}} e^{-(1+|\xi'|)x_n} \left(1 - e^{-\varepsilon(1+|\xi'|)}\right) \widehat{h}\left(\xi'\right) e^{ix'\cdot\xi'} d\xi,$$

we easily obtain

$$\|u - u_\varepsilon\|_{L^2(\mathbb{R}^n)} \leq C \int_{\mathbb{R}^{n-1}} \frac{\left(1 - e^{-\varepsilon(1+|\xi'|)}\right)^2}{1 + |\xi'|} \left|\widehat{h}\left(\xi'\right)\right|^2 d\xi'$$

and

$$\|\nabla (u - u_\varepsilon)\|_{L^2(\mathbb{R}^n)} \leq C \int_{\mathbb{R}^{n-1}} \left(1 + |\xi'|\right) \left(1 - e^{-\varepsilon(1+|\xi'|)}\right)^2 \left|\widehat{h}\left(\xi'\right)\right|^2 d\xi'.$$

Therefore, by the Dominated Convergence Theorem we get

$$\lim_{\varepsilon\to 0} \|u - u_\varepsilon\|^2_{H^1(\mathbb{R}^n)} = 0.$$

By the previous limit and by the trace Theorem (that is, by (i)) we have

$$T(u) = \lim_{\varepsilon\to 0} T\left(u_\varepsilon\right), \quad \text{in } L^2\left(\mathbb{R}^{n-1}\right). \tag{5.193}$$

On the other hand

$$\begin{aligned} &\lim_{\varepsilon\to 0} \|T(u_\varepsilon) - h\|_{L^2(\mathbb{R}^{n-1})} = \\ &= \lim_{\varepsilon\to 0} c_n \int_{\mathbb{R}^{n-1}} \left(1 - e^{-\varepsilon(1+|\xi'|)}\right)^2 \left|\widehat{h}\left(\xi'\right)\right|^2 d\xi' = 0. \end{aligned} \tag{5.194}$$

Therefore, by (5.193) and (5.194) we get

$$T(u) = h, \quad \forall h \in C_0^\infty\left(\mathbb{R}^{n-1}\right). \tag{5.195}$$

Now, we set

$$\mathcal{T}(h) = u.$$

By (5.189) we have

$$\|\mathcal{T}(h)\|_{H^1(\mathbb{R}^n)} \le C \|h\|_{H^{1/2}(\mathbb{R}^{n-1})}, \quad \forall h \in C_0^\infty\left(\mathbb{R}^{n-1}\right) \tag{5.196}$$

and by (5.195) we have

$$T(\mathcal{T}(h)) = h, \quad \forall h \in C_0^\infty\left(\mathbb{R}^{n-1}\right). \tag{5.197}$$

Now, (5.187) and (5.196) give

$$\|T(\mathcal{T}(h))\|_{L^2(\mathbb{R}^n)} \le C \|h\|_{H^{1/2}(\mathbb{R}^{n-1})}, \quad \forall h \in C^\infty\left(\mathbb{R}^{n-1}\right). \tag{5.198}$$

Finally, by density Theorem 5.45, by (5.196), (5.197) and by (5.198) the thesis follows.

■

5.13 Final Comments and Supplements

In this section we will state, sometimes without proof, some theorems concerning the traces and Lipschitz continuous functions.

Concerning the traces, if $k \ge 1$ and if Ω is a bounded open set of $\mathbb{R}^n$ of class $C^{k-1,1}$, it can be proved that (see [69, 100])

$$H^{k-1/2}(\partial\Omega) = T\left(H^k(\Omega)\right).$$

In the sequel we will be mainly interested in the cases $k = 1, 2$. In particular, if $u \in H^2(\Omega)$, then $\partial_k u \in H^1(\Omega)$ for $k = 1, \cdots, n$, hence $\partial_k u_{|\partial\Omega} \in H^{1/2}(\partial\Omega)$. Moreover, we have

$$H^2(\partial\Omega) \subset H^{3/2}(\partial\Omega) \subset H^1(\partial\Omega).$$

If $u \in H^2(\Omega)$, we can define $\frac{\partial u}{\partial \nu} = \nabla u \cdot \nu$ on $\partial\Omega$ (ν unit outward normal vector) and we have

$$\left\| \frac{\partial u}{\partial \nu} \right\|_{H^{1/2}(\partial\Omega)} \leq C \|u\|_{H^2(\Omega)}, \quad \forall u \in H^2(\Omega), \tag{5.199}$$

where C depends on Ω only.

Similarly to what we saw in Sect. 5.5, the following Theorem can be proved

Theorem 5.48 *Let Ω be a bounded open set of $\mathbb{R}^n$ of class $C^{1,1}$. If $u \in H^2(\Omega)$ then*

$$u \in H_0^2(\Omega) \quad \textit{if and only if} \quad u_{|\partial\Omega} = 0 \textit{ and } \frac{\partial u}{\partial \nu} = 0 \textit{ on } \partial\Omega. \tag{5.200}$$

Theorem 5.47 can be generalized as follows

Theorem 5.49 *Let Ω be a bounded open set of $\mathbb{R}^n$ of class $C^{1,1}$. Then, for every*

$$(\psi_0, \psi_1) \in H^{3/2}(\partial\Omega) \times H^{1/2}(\partial\Omega),$$

there exists $u \in H^2(\Omega)$ such that

$$u_{|\partial\Omega} = \psi_0 \quad \textit{and} \quad \frac{\partial u}{\partial \nu} = \psi \quad \textit{on} \quad \partial\Omega$$

and

$$\|u\|_{H^2(\Omega)} \leq C \left(\|\psi_0\|_{H^{3/2}(\partial\Omega)} + \|\psi_1\|_{H^{1/2}(\partial\Omega)} \right),$$

where C depends on Ω only.

5.13.1 The Space $H^{-1/2}(\partial\Omega)$

Let Ω be a bounded open set of $\mathbb{R}^n$ of class $C^{0,1}$. We denote by $H^{-1/2}(\partial\Omega)$ the dual space of $H^{1/2}(\partial\Omega)$. Thus, $H^{-1/2}(\partial\Omega)$ is the space of the linear functionals

$$\Phi : H^{1/2}(\partial\Omega) \to \mathbb{R},$$

such that for a constant C we have

$$|\Phi(\varphi)| \leq C \, \|\varphi\|_{H^{1/2}(\partial\Omega)}, \quad \forall \varphi \in H^{1/2}(\partial\Omega). \tag{5.201}$$

We define the norm, $\|\Phi\|_{H^{-1/2}(\partial\Omega)}$, of Φ in $H^{-1/2}(\partial\Omega)$ as the greatest lower bound of C satisfying (5.201).

5.13.2 *The Space* $W^{1,\infty}_{loc}(\mathbb{R}^n)$ *and* $C^{0,1}_{loc}(\mathbb{R}^n)$

We say that $u \in C^{0,1}_{loc}(\mathbb{R}^n)$ provided that for any $x_0 \in \mathbb{R}^n$ there exists $r > 0$ such that

$$u_{|\overline{B_r(x_0)}} \in C^{0,1}\left(\overline{B_r(x_0)}\right),$$

Theorem 5.50 *Let* $u : \mathbb{R}^n \to \mathbb{R}$. *We have that* $u \in W^{1,\infty}_{loc}(\mathbb{R}^n)$ *if and only if* $u \in C^{0,1}_{loc}(\mathbb{R}^n)$.

Proof Let $u \in C^{0,1}_{loc}(\mathbb{R}^n)$. Theorem 3.57 implies that u is differentiable almost everywhere and $\nabla u = (\partial_1 u, \cdots \partial_n u) \in L^{\infty}_{loc}(\mathbb{R}^n)$. Hence, for any $1 \leq k \leq n$, we have

$$\int_{\mathbb{R}^n} u \partial_k \varphi dx = \int_{\mathbb{R}^n} (\partial_k (u\varphi) - \partial_k u \varphi) \, dx = -\int_{\mathbb{R}^n} \partial_k u \varphi dx, \quad \forall \varphi \in C^{\infty}_0\left(\mathbb{R}^n\right).$$

Therefore $\partial_k u$ is the weak derivative of u, hence $u \in W^{1,\infty}_{loc}(\mathbb{R}^n)$.

Conversely, let $u \in W^{1,\infty}_{loc}(\mathbb{R}^n)$ and let $B_r(x_0)$ be a ball of $\mathbb{R}^n$. Let us denote

$$u_{\varepsilon}(x) = \int_{\mathbb{R}^n} \eta_{\varepsilon}(x-y) u(y) dy.$$

We have $u_{\varepsilon} \in C^{\infty}(\mathbb{R}^n)$ and

$$\nabla u_{\varepsilon}(x) = \int_{\mathbb{R}^n} \eta_{\varepsilon}(x-y) \nabla u(y) dy.$$

Consequently, for every $\varepsilon \in (0, r]$, we get

$$|\nabla u_{\varepsilon}(x)| \leq \|\nabla u\|_{L^{\infty}(B_{2r}(x_0))} < +\infty, \quad \forall x \in B_r(x_0).$$

Hence

$$|u_{\varepsilon}(x) - u_{\varepsilon}(y)| \leq \|\nabla u\|_{L^{\infty}(B_{2r}(x_0))} |x-y|, \quad \forall x, y \in B_r(x_0). \tag{5.202}$$

Moreover, Theorem 3.47 implies that

$$u_\varepsilon \to u, \quad \text{as } \varepsilon \to 0 \ \text{(uniformly)}.$$

By this and by (5.202), we have

$$|u(x) - u(y)| \le \|\nabla u\|_{L^\infty(B_{2r}(x_0))} |x - y|, \quad \forall x, y \in B_r(x_0). \tag{5.203}$$

Therefore $u_{|\overline{B_r(x_0)}} \in C^{0,1}\left(\overline{B_r(x_0)}\right)$. Since $B_r(x_0)$ is arbitrary, the proof is complete. ■

5.13.3 Almost Everywhere Differentiability of Function Belonging to $W^{1,p}_{loc}(\mathbb{R}^n)$ with $p > n$

Theorem 5.51 *Let $n < p \le +\infty$. If $u \in W^{1,p}_{loc}(\mathbb{R}^n)$. Then u is almost everywhere differentiable in $\mathbb{R}^n$.*

Proof Since $W^{1,\infty}_{loc}(\mathbb{R}^n) \subset W^{1,p}_{loc}(\mathbb{R}^n)$ for $p < +\infty$, we may assume that $n < p < +\infty$. Let $x \in \mathbb{R}^n$ be such that (Corollary 3.56)

$$\lim_{r\to 0} \fint_{B_r(x)} |\nabla u(\xi) - \nabla u(x)|^p d\xi = 0. \tag{5.204}$$

Let us denote by

$$v(\xi) = u(\xi) - u(x) - \nabla u(x) \cdot (\xi - x), \quad \xi \in \mathbb{R}^n,$$

let $y \in \mathbb{R}^n$, $y \ne x$ and $r = |x - y|$. Theorem 5.8 and Lemma 5.28 give

$$\begin{aligned}
&|u(y) - u(x) - \nabla u(x) \cdot (y - x)| = |v(y) - v(x)| \le \\
&\le Cr^{1-\frac{n}{p}} \left(\int_{B_r(x)} |\nabla v(\xi)|^p d\xi\right)^{1/p} = \\
&= C'|x - y| \left(\fint_{B_r(x)} |\nabla u(\xi) - \nabla u(x)|^p d\xi\right)^{1/p}.
\end{aligned}$$

Hence

$$u(y) - u(x) - \nabla u(x) \cdot (y - x) = o(|y - x|), \quad \text{as } y \to x.$$

Therefore u is differentiable in every point x which satisfies (5.204) and by Corollary 3.56 we conclude the proof. ■

Remark By utilizing Theorems 5.50 and 5.51, the Rademacher Theorem can be readily deduced through an alternative proof, distinct from the one presented in Sect. 3.7. We encourage the reader to confirm that adopting this new approach avoids any potential "vicious circles". ♦

Chapter 6
The Boundary Value Problems for Second-order Elliptic Equations and the Dirichlet to Neumann Map

Abstract This chapter examines the Dirichlet problem for second order elliptic operators in bounded domains. We explore the variational formulation of the problem and establish existence and uniqueness of solutions in Sobolev spaces. The chapter also addresses the regularity of solutions, showing that higher regularity of the domain, coefficients, and source terms leads to a corresponding improvement in the solution's regularity. Finally, the discussion extends to a more general class of second-order elliptic operators, incorporating additional terms that allow a broader scope of applications. We will conclude the chapter by defining the Dirichlet to Neumann map and briefly discussing some application to an inverse problem.

6.1 Introduction

Let Ω be a bounded open set of $\mathbb{R}^n$. Let us denote by $\mathbb{M}(n)$ the vector space of the matrices $n \times n$ whose entries are real numbers and let $A \in L^{\infty}(\Omega; \mathbb{M}(n))$, i.e. $A = \left\{a^{jk}\right\}_{j,k=1}^{n}$ is a matrix whose entries a^{jk} belong to $L^{\infty}(\Omega)$, for $j, k = 1, \cdots, n$.

Throughout this chapter we will assume that A satisfies the following condition of **uniform ellipticity**: There exists $\lambda \geq 1$ such that for almost every $x \in \Omega$ we have

$$\lambda^{-1}|\xi|^2 \leq A(x)\xi \cdot \xi \leq \lambda|\xi|^2, \quad \forall \xi \in \mathbb{R}^n. \tag{6.1}$$

We define

$$|A(x)|_{\mathbb{M}(n)} = \left(\sum_{j,k=1}^{n} \left(a^{jk}(x)\right)^2\right)^{\frac{1}{2}}, \quad \text{a.e. } x \in \Omega$$

and

$$\|A\|_{L^{\infty}(\Omega;\mathbb{M}(n))} = \left\| |A(\cdot)|_{\mathbb{M}(n)} \right\|_{L^{\infty}(\Omega)}.$$

S. Vessella, *Unique Continuation Properties for Partial Differential Equations*, Birkhäuser Advanced Texts Basler Lehrbücher,
https://doi.org/10.1007/978-3-031-86366-0_6

Let us notice that if A is symmetric, then the second inequality of (6.1) implies also

$$\sup_{|\xi|=1,|\eta|=1} |A(x)\xi \cdot \eta| \leq \lambda, \quad \forall x \in \Omega \tag{6.2}$$

and

$$\left\| a^{jk} \right\|_{L^\infty(\Omega)} \leq \lambda, \quad \text{for } j, k = 1, \cdots, n. \tag{6.3}$$

Let us check (6.2) and (6.2). Let $\xi, \eta \in \mathbb{R}^n$ be such that $|\xi| = 1$ and $|\eta| = 1$. By the symmetry of $A(x)$ we have

$$\begin{aligned} |A(x)\xi \cdot \eta| &= \frac{1}{4} |A(x)(\xi + \eta) \cdot (\xi + \eta) - A(x)(\xi - \eta) \cdot (\xi - \eta)| \leq \\ &\leq \frac{1}{4} \left(\lambda|\xi + \eta|^2 + \lambda|\xi - \eta|^2\right) = \\ &= \frac{\lambda}{4} \left(2|\xi|^2 + 2|\eta|^2\right) = \\ &= \lambda. \end{aligned}$$

Hence, (6.2) follows. Concerning (6.3), we have, for $j, k = 1, \cdots, n$

$$\left|a^{jk}(x)\right| = |A(x)e_j \cdot e_k| \leq \lambda, \quad \text{a.e. } x \in \Omega.$$

In this chapter we will tackle the Dirichlet problem for the operator $-\text{div}(A\nabla u)$, which formally consists in determining $u \in H^1(\Omega)$ such that

$$\begin{cases} -\text{div}(A\nabla u) = F, & \text{in } \Omega \\ \\ u = \varphi, & \text{on } \partial\Omega \end{cases} . \tag{6.4}$$

where $\varphi \in H^{1/2}(\partial\Omega)$ and $F \in H^{-1}(\Omega)$.

We will address the variational formulation of problem (6.4). Subsequently, we will examine the existence and the uniqueness of the solutions in $H^1(\Omega)$, followed by the proof of various regularity results for the same problem. Concerning the latter, we will establish, roughly speaking, that if Ω, A, and F possess higher regularity, then the solution u also attains a higher degree of regularity.

The investigation on problem (6.4) will guide us to deal with the more general case in which, instead of $-\text{div}(\nabla u)$, we will have the operator

$$-\sum_{j,k=1}^{n} \partial_j \left(a^{jk}\partial_k u + d^j u\right) + \sum_{j=1}^{n} b^j \partial_j u + cu, \tag{6.5}$$

where $b^j, d^j, c \in L^\infty(\Omega)$, for $j = 1, \cdots, n$.

6.2 The Lax–Milgram Theorem and the Fredholm Theorem

Let H be a real Hilbert space, let us denote by $\|\cdot\|$ and $(\cdot, \cdot)$ the scalar product on H and the induced norm induced respectively. As usual we denote by H' the dual space of H.

If $A : H \to H$ is a linear operator we denote by $\mathcal{R}(A)$ the range of A, that is

$$\mathcal{R}(A) := \{Au : u \in H\}$$

and by $\mathcal{N}(A)$ the kernel of A

$$\mathcal{N}(A) := \{u \in H : Au = 0\}.$$

Let

$$a : H \times H \to \mathbb{R} \tag{6.6}$$

be a bilinear form. We say that a is **continuous** if there exists $C > 0$ such that

$$|a(u, v)| \le C \|u\| \|v\|, \quad \forall u, v \in H. \tag{6.7}$$

We say that the bilinear form a is **coercive** if there exists $\alpha > 0$ such that

$$\alpha \|u\|^2 \le a(u, u), \quad \forall u \in H. \tag{6.8}$$

The following Theorem holds true.

Theorem 6.1 (Lax–Milgram) *Let a be a coercive bilinear form and let $F \in H'$. Then there exists a unique $u \in H$ such that*

$$a(u, v) = F(v), \quad \forall v \in H. \tag{6.9}$$

Moreover

$$\|u\| \le \frac{1}{\alpha} \|F\|_{H'}, \tag{6.10}$$

where $\|\cdot\|_{H'}$ is the norm of H'.

Proof The Riesz Representation Theorem implies that there exists a unique $f \in H$ such that

$$F(v) = (f, v), \quad \forall v \in H. \tag{6.11}$$

Moreover

$$\|F\|_{H'} = \|f\|_H \, .$$

Let $u \in H$ be fixed and observe that, as (6.7) holds, the map

$$H \ni v \to a(u, v) \in \mathbb{R},$$

is linear and bounded. The Riesz Representation Theorem implies that there exists a unique $Au \in H$ such that

$$a(u, v) = (Au, v), \quad \forall v \in H.$$

Hence, we have defined the map

$$A : H \to H,$$

such that

$$a(u, v) = (Au, v), \quad \forall u, v \in H. \tag{6.12}$$

By (6.11) and (6.12) we have that (6.9) is equivalent to

$$Au = f, \quad u \in H. \tag{6.13}$$

Now, we prove that the map A is

(i) linear, (ii) bounded, and (iii) bijective

(i) Let $u_1, u_2 \in H, \lambda_1, \lambda_2 \in \mathbb{R}$. We have

$$\begin{aligned}
(A(\lambda_1 u_1 + \lambda_2 u_2), v) &= a(\lambda_1 u_1 + \lambda_2 u_2, v) \\
&= \lambda_1 a(u_1, v) + \lambda_2 a(u_2, v) = \\
&= \lambda_1 (Au_1, v) + \lambda_2 (Au_2, v) = \\
&= (\lambda_1 Au_1 + \lambda_2 Au_2, v), \quad \forall v \in H.
\end{aligned}$$

Hence

$$A(\lambda_1 u_1 + \lambda_2 u_2) = \lambda_1 Au_1 + \lambda_2 Au_2, \quad \forall u_1, u_2 \in H.$$

(ii) By (6.7) we get

$$|(Au, v)| = |a(u, v)| \leq C \|u\| \|v\| , \quad \forall u, v \in H.$$

By the Cauchy–Schwarz inequality we get

$$\|Au\| \leq C \|u\| , \quad \forall u \in H.$$

Therefore A is bounded and

$$\|A\|_{\mathcal{L}(H)} \leq C,$$

where $\mathcal{L}(H)$ is the space of bounded linear map from H in itself.

(iii) Condition (6.8) implies

$$\alpha \|u\|^2 \leq a(u, u) = (Au, u) \leq \|Au\| \|u\| , \quad \forall u \in H.$$

Hence

$$\alpha \|u\| \leq \|Au\| , \quad \forall u \in H. \tag{6.14}$$

Since $\alpha > 0$, A is injective.

In order to prove that A is onto, we first prove that $\mathcal{R}(A)$ is closed. Let $\{w_k\}$ be a sequence in $\mathcal{R}(A)$ such that

$$\{w_k\} \to w, \tag{6.15}$$

and let us check that $w \in \mathcal{R}(A)$. Let $u_k \in H, k \in \mathbb{N}$ satisfy

$$Au_k = w_k, \quad \forall k \in \mathbb{N}.$$

By (6.14) we have

$$\left\|u_k - u_j\right\| \leq \frac{1}{\alpha} \left\|Au_k - Au_j\right\| = \frac{1}{\alpha} \left\|w_k - w_j\right\| , \quad \forall k, j \in \mathbb{N}.$$

Since $\{w_k\}$ converges, it is a Cauchy sequence. Consequently, $\{u_k\}$ is a Cauchy sequence too. Therefore there exists $u \in H$ such that

$$\{u_k\} \to u.$$

Now, by (6.15), as A is continuous, we obtain

$$w = \lim_{k\to\infty} w_k = \lim_{k\to\infty} Au_k = Au.$$

Therefore $w \in \mathcal{R}(A)$.

Now we prove

$$\mathcal{R}(A) = H. \tag{6.16}$$

We argue by contradiction. Let us assume that $\mathcal{R}(A) \subsetneq H$. Since $\mathcal{R}(A)$ is closed, there exists $w \in H \setminus \{0\}$ such that $w \perp \mathcal{R}(A)$, (by this we mean $(w, h) = 0$ for every $h \in \mathcal{R}(A)$). Hence

$$\alpha \|w\|^2 \leq a(w, w) = (Aw, w) = 0.$$

Consequently, we should have $w = 0$ that contradicts $w \neq 0$. Thus (6.16) is proved.

Now, since (6.9) and (6.13) are equivalent, there exists one and only one solution $u \in H$ of the problem

$$a(u, v) = F(v), \quad \forall v \in H. \tag{6.17}$$

Concerning estimate (6.10), it follows immediately by

$$\alpha \|u\|^2 \leq a(u, u) = F(u) \leq \|F\|_{H'} \|u\|. \tag{6.18}$$

■

Remark We observe that, as a is a bilinear form, by (6.10) we obtain that if $F_j \in H'$, $j = 1, 2$ and $u_j \in H$ are solutions to

$$a(u_j, v) = F_j(v) \quad \forall v \in H,$$

then

$$\|u_1 - u_2\| \leq \frac{1}{\alpha} \|F_1 - F_2\|_{H'}, \tag{6.19}$$

from which, in particular, we get again the uniqueness. ♦

We now recall the Fredholm Alternative Theorem [19, 38](see also [88]). Meanwhile, we recall that a linear operator $\mathcal{K}$ from H in itself is said compact, provided that for every bounded set $M \subset H$, $\mathcal{K}(M)$ is relatively compact in H. We recall that a compact operator is necessarily bounded, nevertheless if H does not have finite dimension, the identity on H is a bounded operator, but it is not compact.

Theorem 6.2 (Fredholm Alternative) *Let*

$$\mathcal{K}: H \to H$$

be a linear compact operator. Then we have:

(i) *the dimension of* $\mathcal{N}(I - \mathcal{K})$ *is finite;*
(ii) $\mathcal{R}(I - \mathcal{K})$ *is a closed subspace of* H*;*
(iii) $\mathcal{R}(I - \mathcal{K}) = \mathcal{N}(I - \mathcal{K}^{\star})^{\perp}$*;*
(iv) $\mathcal{N}(I - \mathcal{K}) = \{0\}$ *if and only if* $\mathcal{R}(I - \mathcal{K}) = H$*;*
(v) $\dim \mathcal{N}(I - \mathcal{K}) = \dim \mathcal{N}(I - \mathcal{K}^{\star})$.

Recall that $\mathcal{K}^{\star}$ is the **adjoint operator** of $\mathcal{K}$ defined by

$$(\mathcal{K}u, v) = \left(u, \mathcal{K}^{\star}v\right), \quad \forall u, v \in H.$$

6.3 The Variational Formulation of the Dirichlet Problem: Existence Theorems

Let us begin by clarifying what we mean by **the variational formulation** of Dirichlet problem (6.4). We begin by the case in which the **condition at the boundary is homogeneous**. In this case (6.4) can be written (formally):

$$\begin{cases} -\text{div}(A\nabla u) = F, & \text{in } \Omega \\ \\ u = 0, & \text{on } \partial\Omega \end{cases} \tag{6.20}$$

and the variational formulation of problem above is the following:
Determine u such that

$$\begin{cases} \int_{\Omega} A\nabla u \cdot \nabla v dx = F(v), & \forall v \in H_0^1(\Omega) \\ \\ u \in H_0^1(\Omega) \end{cases} . \tag{6.21}$$

Let us observe that if Ω is an open set of class $C^{0,1}$, $a^{jk} \in C^1(\overline{\Omega})$, $j, k = 1, \cdots, n$ and $u \in C^2\left(\overline{\Omega}\right)$ with $u_{|\partial\Omega} = 0$ and $F \in C^0(\overline{\Omega})$, then (6.20) is equivalent to (6.21). Indeed, under such assumptions, the divergence Theorem implies

$$\int_{\Omega} \text{div}(A\nabla u)v dx = -\int_{\Omega} A\nabla u \cdot \nabla v dx, \qquad \forall v \in C_0^{\infty}(\Omega). \tag{6.22}$$

Therefore, (6.20) implies (6.21) (taking into account the Remark in Sect. 5.5 and that $C_0^{\infty}(\Omega)$ is dense in $H_0^1(\Omega)$). Conversely, if (6.21) holds true, then Theorem 5.18

implies $u_{|\partial\Omega} = 0$ and (6.21) gives

$$\int_{\Omega} (\operatorname{div}(A\nabla u) + F)\, v dx = 0, \quad \forall v \in C_0^{\infty}(\Omega), \tag{6.23}$$

that gives (6.20). Of course, under the general assumptions on Ω, A and F, the formulation (6.20) makes no sense, while the formulation (6.21) makes perfectly sense, and the Lax–Milgram Theorem will tell us easily that it is a well-posed problem. Actually, let $H = H_0^1(\Omega)$ and

$$a(u, v) = \int_{\Omega} A\nabla u \cdot \nabla v dx, \quad \forall u, v \in H_0^1(\Omega). \tag{6.24}$$

The form (6.24) is bilinear. Moreover by the Cauchy–Schwarz inequality we have, for any $u, v \in H_0^1(\Omega)$,

$$\begin{aligned} |a(u, v)| &\leq \|A\|_{L^{\infty}(\Omega;\mathbb{M}(n))} \int_{\Omega} |\nabla u||\nabla v| dx \leq \\ &\leq \|A\|_{L^{\infty}(\Omega;\mathbb{M}(n)} \|\nabla u\|_{L^2(\Omega)} \|\nabla v\|_{L^2(\Omega)} \end{aligned}$$

and by (6.1) we have

$$\lambda^{-1} \|\nabla u\|_{L^2(\Omega)}^2 \leq \int_{\Omega} A\nabla u \cdot \nabla u dx = a(u, u), \quad \forall u \in H_0^1(\Omega).$$

Now, recalling that $\|\nabla u\|_{L^2(\Omega)}$ and $\|u\|_{H^1(\Omega)}$ are two equivalent norms of $H_0^1(\Omega)$, we have only to apply the Lax–Milgram Theorem to conclude that problem (6.21) has a unique solution in $H_0^1(\Omega)$. Moreover the following inequality holds true

$$\|\nabla u\|_{L^2(\Omega)} \leq \lambda \|F\|_{H^{-1}(\Omega)} \cdot \tag{6.25}$$

Inequality (6.25), together with the already proven existence and uniqueness results, implies that problem (6.21) is well-posed in $H_0^1(\Omega)$.

Now we consider the case where the boundary condition is **not homogeneous**, but the equation is still homogeneous.

Let Ω be an open set of $\mathbb{R}^n$ of class $C^{0,1}$, let $\varphi \in H^{1/2}(\partial\Omega)$ and let us assume that (6.1) is satisfied. Formally, the Dirichlet problem can be written as

$$\begin{cases} -\operatorname{div}(A\nabla u) = 0, & \text{in } \Omega \\ \\ u = \varphi, & \text{on } \partial\Omega \end{cases} . \tag{6.26}$$

We wish to give the variational formulation of problem (6.26) and to prove the existence of the solutions in $H^1(\Omega)$ to this problem.

The variational formulation of (6.26) is

$$\begin{cases} \int_\Omega A\nabla u \cdot \nabla v dx = 0, & \forall v \in H_0^1(\Omega) \\ \\ u = \varphi, \quad \text{on } \partial\Omega & \text{(in the sense of the traces)} \end{cases} . \tag{6.27}$$

Let us notice that, in the case where A, u and φ are sufficiently regular, the first equation in (6.27) is equivalent to the first equation of (6.26).

In order to solve (6.26) ((6.27)), we proceed in the following way. Recalling Theorem 5.47, there exists $\Phi \in H^1(\Omega)$ such that

$$\Phi_{|\partial\Omega} = \varphi, \quad \text{(in the sense of the traces)}$$

which in turn (by (ii) of Theorem 5.47) implies

$$\|\Phi\|_{H^1(\Omega)} \leq C \|\varphi\|_{H^{1/2}(\partial\Omega)}, \tag{6.28}$$

where C is a constant depending on Ω only. Set $w = u - \Phi$. Since u and Φ have the same trace on $\partial\Omega$ we have $w_{|\partial\Omega} = 0$, so that problem (6.26) can be written (formally),

$$\begin{cases} -\text{div}(A\nabla w) = \text{div}(A\nabla\Phi), & \text{in } \Omega \\ \\ w = 0, & \text{on } \partial\Omega \end{cases} , \tag{6.29}$$

whose variational formulation is

$$\begin{cases} \int_\Omega A\nabla w \cdot \nabla v dx = -\int_\Omega A\nabla\Phi \cdot \nabla v dx, & \forall v \in H_0^1(\Omega) \\ \\ w \in H_0^1(\Omega) \end{cases} . \tag{6.30}$$

Let us note that the bilinear form is still given by (6.24). The solution of problem (6.26) ((6.27)) is given by

$$u = w + \Phi \in H^1(\Omega). \tag{6.31}$$

Moreover, denoting

$$F(v) = -\int_\Omega A\nabla\Phi \cdot \nabla v dx, \quad \forall v \in H_0^1(\Omega),$$

it turns out that $F \in H^{-1}(\Omega)$. As a matter of fact, by the Cauchy–Schwarz inequality we have

$$|F(v)| \leq \lambda \left\|\nabla\Phi\right\|_{L^2(\Omega)} \left\|\nabla v\right\|_{L^2(\Omega)}, \quad \forall v \in H_0^1(\Omega).$$

By (6.28) and (6.29) we have

$$|F(v)| \leq C\lambda \left\|\varphi\right\|_{H^{1/2}(\partial\Omega)} \left\|v\right\|_{H^1(\Omega)}, \quad \forall v \in H_0^1(\Omega).$$

Therefore $F \in H^{-1}(\Omega)$ and the following inequality holds

$$\left\|F\right\|_{H^{-1}(\Omega)} \leq C' \left\|\varphi\right\|_{H^{1/2}(\partial\Omega)}, \tag{6.32}$$

where C' depends on Ω and λ only .

The Lax–Milgram Theorem implies that problem (6.30) has a unique solution $w \in H_0^1(\Omega)$, moreover by (6.10) we have

$$\left\|\nabla w\right\|_{L^2(\Omega)} \leq \lambda \left\|F\right\|_{H^{-1}(\Omega)} \leq \lambda C' \left\|\varphi\right\|_{H^{1/2}(\partial\Omega)}. \tag{6.33}$$

Now, by using (6.28), (6.31) and (6.33) we get

$$\left\|u\right\|_{H^1(\Omega)} \leq \left\|w\right\|_{H^1(\Omega)} + \left\|\Phi\right\|_{H^1(\Omega)} \leq C'' \left\|\varphi\right\|_{H^{1/2}(\partial\Omega)}, \tag{6.34}$$

where C'' depends by λ and Ω only. Notice that (6.34) implies, in particular, the uniqueness of solution of problem (6.26).

From what we have proved so far, we have the following

Theorem 6.3 *Let Ω be an open set of class $C^{0,1}$. Assume that $A \in L^\infty(\Omega; \mathbb{M}(n))$ and that A satisfies* (6.1). *Let $F \in H^{-1}(\Omega)$ and $\varphi \in H^{1/2}(\partial\Omega)$.*

Then the following problem

$$\begin{cases} -div(A\nabla u) = F, & in\ \Omega \\ \\ u = \varphi, & on\ \partial\Omega \end{cases}, \tag{6.35}$$

whose variational formulation is

$$\begin{cases} \int_\Omega A\nabla u \cdot \nabla v dx = F(v), \quad \forall v \in H_0^1(\Omega) \\ \\ u = \varphi, \quad on\ \partial\Omega \quad (in\ the\ sense\ of\ traces) \end{cases}, \tag{6.36}$$

has a unique solution $u \in H^1(\Omega)$ and we have

$$\|u\|_{H^1(\Omega)} \leq C\left(\|F\|_{H^{-1}(\Omega)} + \|\varphi\|_{H^{1/2}(\partial\Omega)}\right), \tag{6.37}$$

where C depends on λ and Ω only.

Now let L be the following operator

$$Lu = -\sum_{j,k=1}^{n} \partial_j\left(a^{jk}\partial_k u + d^j u\right) + \sum_{j=1}^{n} b^j \partial_j u + cu, \tag{6.38}$$

where $A \in L^\infty(\Omega; \mathbb{M}(n))$, $A = \left\{a^{jk}\right\}_{j,k=1}^{n}$, satisfies (6.1) and

$$b^j,\ d^j,\ c \in L^\infty(\Omega),$$

for $j = 1, \cdots, n$.

Let us consider the Dirichlet problem

$$\begin{cases} Lu = f, & \text{in } \Omega \\ u = 0, & \text{on } \partial\Omega \end{cases}, \tag{6.39}$$

where $f \in L^2(\Omega)$. The variational formulation of above problem is: Determine u such that

$$\begin{cases} a(u, v) = (f, v), & \forall v \in H_0^1(\Omega) \\ u \in H_0^1(\Omega) \end{cases}, \tag{6.40}$$

where

$$a(u, v) = \int_\Omega (A\nabla u \cdot \nabla v + ud \cdot v - b \cdot \nabla uv - cuv)\, dx, \tag{6.41}$$

$d = \left(d^1, \cdots, d^n\right)$ and $b = \left(b^1, \cdots, b^n\right)$.

Problem (6.39) does not always have existence and uniqueness. To show this fact, let us consider the following simple example

$$\begin{cases} -u'' - u = f, & \text{in } (0, \pi) \\ u(0) = u(\pi) = 0 \end{cases}, \tag{6.42}$$

where $f \in L^2(0, \pi)$. The solutions to (6.42) have to be found among the functions of the type

$$C_1 \sin x + C_2 \cos x - \int_0^x \sin(x - t) f(t) dt.$$

By the boundary conditions we have $C_2 = 0$ and

$$\int_0^{\pi} f(t) \sin t dt = 0. \tag{6.43}$$

Therefore, if (6.43) is satisfied, then (6.42) has infinite solutions, given by

$$C \sin x - \int_0^x \sin(x - t) f(t) dt, \quad C \in \mathbb{R}.$$

Whereas if (6.43) is not satisfied, then (6.42) has no solutions. So we cannot expect that the bilinear form (6.41) is always coercive. We can, however, prove the following Theorem that will be useful for establish some conditions of existence and uniqueness to problem (6.40)

Theorem 6.4 *Let Ω be a bounded open set of $\mathbb{R}^n$. Assume that $A \in L^\infty(\Omega; \mathbb{M}(n))$ and that A satisfies condition* (6.1) *and $b, d \in L^\infty(\Omega; \mathbb{R}^n)$, $c \in L^\infty(\Omega)$, for $j = 1, \cdots, n$. Moreover, let*

$$a(u, v) = \int_\Omega (A\nabla u \cdot \nabla v + ud \cdot \nabla v - b \cdot \nabla uv - cuv) \, dx.$$

Then a is a continuous bilinear form and there exist $\gamma_0 \geq 0$ and $\alpha_0 > 0$ such that for any $\gamma \geq \gamma_0$ we have

$$\alpha_0 \|u\|^2_{H^1(\Omega)} \leq a(u, u) + \gamma \|u\|^2_{L^2(\Omega)}, \quad \forall u \in H^1_0(\Omega). \tag{6.44}$$

In particular,

$$a_\gamma(u, v) := a(u, u) + \gamma (u, v)_{L^2(\Omega)}, \tag{6.45}$$

is a continuous and coercive bilinear form for every $\gamma \geq \gamma_0$.

Proof By the Cauchy–Schwarz inequality we have, for any $u, v \in H^1_0(\Omega)$

$$
\begin{aligned}
|a(u,v)| &\leq \int_\Omega \|A\|_{L^\infty(\Omega;\mathbb{M}(n))} |\nabla u|\,|\nabla v|\,dx+ \\
&+ \int_\Omega \left(\|d\|_{L^\infty(\Omega;\mathbb{R}^n)} |u|\,|\nabla v| + \|b\|_{L^\infty(\Omega;\mathbb{R}^n)} |\nabla u|\,|v| + \|c\|_{L^\infty(\Omega)} |u|\,|v|\right) dx \leq \\
&\leq \|A\|_{L^\infty(\Omega;\mathbb{M}(n))} \|\nabla u\|_{L^2(\Omega)} \|\nabla v\|_{L^2(\Omega)} + \|d\|_{L^\infty(\Omega;\mathbb{R}^n)} \|u\|_{L^2(\Omega)} \|\nabla v\|_{L^2(\Omega)} + \\
&+ \|b\|_{L^\infty(\Omega;\mathbb{R}^n)} \|\nabla u\|_{L^2(\Omega)} \|v\|_{L^2(\Omega)} + \|c\|_{L^\infty(\Omega)} \|u\|_{L^2(\Omega)} \|v\|_{L^2(\Omega)} \leq \\
&\leq C\, \|u\|_{H^1(\Omega)} \|v\|_{H^1(\Omega)}\,,
\end{aligned}
$$

where

$$C = \left(\|A\|_{L^\infty(\Omega;\mathbb{M}(n))} + \|d\|_{L^\infty(\Omega;\mathbb{R}^n)} + \|b\|_{L^\infty(\Omega;\mathbb{R}^n)} + \|c\|_{L^\infty(\Omega)}\right).$$

Therefore

$$|a_\gamma(u,v)| \leq (C + |\gamma|)\, \|v\|_{H^1(\Omega)} \|u\|_{H^1(\Omega)}\,, \qquad \forall u, v \in H_0^1(\Omega)$$

which implies the continuity of a_γ for any $\gamma \in \mathbb{R}$.

Concerning coercivity of a_γ, we notice firstly that (6.1) gives

$$\int_\Omega A\nabla u \cdot \nabla u dx \geq \lambda^{-1} \|\nabla u\|^2_{L^2(\Omega)}\,, \qquad \forall u \in H_0^1(\Omega). \tag{6.46}$$

Moreover, let ε be a positive number which we will choose later on. We get

$$
\begin{aligned}
\left|\int_\Omega ud \cdot \nabla u dx\right| &\leq \|d\|_{L^\infty(\Omega;\mathbb{R}^n)} \|u\|_{L^2(\Omega)} \|\nabla u\|_{L^2(\Omega)} \leq \\
&\leq \frac{\varepsilon}{2} \|\nabla u\|^2_{L^2(\Omega)} + \frac{1}{2\varepsilon} \|d\|^2_{L^\infty(\Omega;\mathbb{R}^n)} \|u\|^2_{L^2(\Omega)}\,.
\end{aligned} \tag{6.47}
$$

Similarly,

$$\left|\int_\Omega b \cdot \nabla u u dx\right| \leq \frac{\varepsilon}{2} \|\nabla u\|^2_{L^2(\Omega)} + \frac{1}{2\varepsilon} \|b\|^2_{L^\infty(\Omega;\mathbb{R}^n)} \|u\|^2_{L^2(\Omega)}\,. \tag{6.48}$$

Furthermore

$$\left|\int_\Omega cu^2 dx\right| \leq \|c\|_{L^\infty(\Omega)} \|u\|^2_{L^2(\Omega)}\,. \tag{6.49}$$

Hence, by (6.46)–(6.49) we obtain

$$a_\gamma(u,u) \geq \left(\lambda^{-1} - \varepsilon\right) \|\nabla u\|^2_{L^2(\Omega)} +$$
$$+ \left[\gamma - \left(\frac{1}{2\varepsilon} \|d\|^2_{L^\infty(\Omega;\mathbb{R}^n)} + \frac{1}{2\varepsilon} \|b\|^2_{L^\infty(\Omega;\mathbb{R}^n)} + \|c\|_{L^\infty(\Omega)}\right)\right] \|u\|^2_{L^2(\Omega)} .$$

Now, by choosing

$$\varepsilon = \frac{\lambda^{-1}}{2}$$

and denoting by

$$\gamma_0 = \lambda \|d\|^2_{L^\infty(\Omega;\mathbb{R}^n)} + \lambda \|b\|^2_{L^\infty(\Omega;\mathbb{R}^n)} + \|c\|_{L^\infty(\Omega)} ,$$

we have, for any $\gamma \geq \gamma_0$

$$a_\gamma(u,u) \geq \frac{\lambda^{-1}}{2} \|\nabla u\|^2_{L^2(\Omega)} , \quad \forall u \in H^1_0(\Omega).$$

Finally, the first Poincaré inequality (Theorem 5.14) implies that there exists α_0, depending on the diameter of Ω, such that (6.44) is satisfied. ■

Exercise Prove there that there exists $\delta > 0$ such that if $\mathrm{diam}(\Omega) \leq \delta$ ($\mathrm{diam}(\Omega)$ is the diameter of Ω) then problem (6.39) has a unique solution in $H^1_0(\Omega)$, provide $f \in H^{-1}(\Omega)$. [Hint: use the first Poincaré inequality]. ♠

The following operator is called **the (formal) adjoint** of the operator L

$$L^\star v = -\sum_{j,k=1}^{n} \partial_k \left(a^{jk}\partial_j v + b^k v\right) + \sum_{j=1}^{n} d^j \partial_j v + cv. \tag{6.50}$$

To the operator $L^\star$ corresponds the bilinear form

$$a^\star(v,u) = \int_\Omega \left(A^T \nabla v \cdot \nabla u - vb \cdot \nabla u + d \cdot \nabla vu - cvu\right) dx, \quad \forall u, v \in H^1_0(\Omega),$$

where A^T is the transposed of the matrix A. Let us notice that

$$a^\star(v,u) = a(u,v), \quad \forall u, v \in H^1_0(\Omega).$$

Finally, let $f \in L^2(\Omega)$. We say that $v \in H^1_0(\Omega)$ is a weak solution of the adjoint problem

$$\begin{cases} L^\star v = f, & \text{in } \Omega \\ \\ u = 0, & \text{on } \partial\Omega \end{cases}, \tag{6.51}$$

provided that we have

$$\begin{cases} a^\star(v, u) = (f, u)_{L^2(\Omega)}, & \forall u \in H_0^1(\Omega) \\ \\ v \in H_0^1(\Omega) \end{cases}. \tag{6.52}$$

The following Theorem holds true

Theorem 6.5 *Let L be operator* (6.38) *and $L^\star$ its formal adjoint.*

(i) The following alternative holds true.
either

(a) for any $f \in L^2(\Omega)$ there exists a unique $u \in H_0^1(\Omega)$ such that

$$\begin{cases} Lu = f, & \text{in } \Omega \\ \\ u = 0, & \text{on } \partial\Omega \end{cases} \tag{6.53}$$

or

(b) there exists at least one ***not identically vanishing solution*** *$u \in H_0^1(\Omega)$ to the homogeneous problem*

$$\begin{cases} Lu = 0, & \text{in } \Omega \\ \\ u = 0, & \text{on } \partial\Omega \end{cases}. \tag{6.54}$$

(ii) If (b) holds, then, denoting by N the subspace $H_0^1(\Omega)$ of the solutions to (6.54) *and by $N^\star$ the subspace of $H_0^1(\Omega)$ of the solutions to*

$$\begin{cases} L^\star v = 0, & \text{in } \Omega \\ \\ v = 0, & \text{on } \partial\Omega \end{cases}, \tag{6.55}$$

we have that N and $N^\star$ have finite ***dimension****, moreover*

$$\text{dimension of } N = \text{dimension of } N^\star.$$

(iii) Finally, problem (6.53) *admits a solution in* $H_0^1(\Omega)$ *if and only if*

$$(f, v)_{L^2(\Omega)} = 0, \quad \forall v \in N^\star. \tag{6.56}$$

Proof Let γ_0 be the same of Theorem 6.4. Let us fix $\gamma \geq \gamma_0$. Since a_γ, defined by (6.45), is a continuous and coercive bilinear form, we have that for any $g \in L^2(\Omega)$ there exists a unique $u \in H_0^1(\Omega)$ such that

$$L_\gamma u = g.$$

Set

$$L_\gamma^{-1} g = u.$$

Now, we notice that $u \in H_0^1(\Omega)$ solves the boundary value problem

$$\begin{cases} Lu = f, & \text{in } \Omega \\ u = 0, & \text{on } \partial\Omega \end{cases}, \tag{6.57}$$

if and only if

$$a_\gamma(u, v) = (\gamma u + f, v)_{L^2(\Omega)}, \quad \forall v \in H_0^1(\Omega),$$

which, in turn, is equivalent to

$$u = L_\gamma^{-1}(\gamma u + f). \tag{6.58}$$

Let us denote

$$\mathcal{K}u = \gamma L_\gamma^{-1} u \tag{6.59}$$

and

$$h = L_\gamma^{-1} f. \tag{6.60}$$

Let us notice that $\mathcal{K}$ is linear and it satisfies

$$a_\gamma(\mathcal{K}g, v) = \gamma(g, v)_{L^2(\Omega)}, \quad \forall g \in L^2(\Omega), \ \forall v \in H_0^1(\Omega).$$

Moreover, as (6.58) and (6.59) hold, (6.57) is equivalent to

$$u - \mathcal{K}u = h. \tag{6.61}$$

Also, we notice that we have $h \in H_0^1(\Omega)$ and by the definition of $\mathcal{K}$, we have $\mathcal{K}g \in H_0^1(\Omega)$, for any $g \in L^2(\Omega)$. Therefore, every function of $L^2(\Omega)$ which is a solution to (6.61) belongs to $H_0^1(\Omega)$.

Now we examine the solvability of (6.61) in $L^2(\Omega)$.

Let us begin to check that

$$\mathcal{K} : L^2(\Omega) \to L^2(\Omega),$$

is a well-defined compact operator. Let $g \in L^2(\Omega)$. Set

$$w = \mathcal{K}g.$$

By (6.44) we have

$$\alpha_0 \|w\|_{H^1(\Omega)}^2 \leq a_\gamma(w, w) = \gamma(g, w)_{L^2(\Omega)} \leq \gamma \|g\|_{L^2(\Omega)} \|w\|_{L^2(\Omega)} .$$

Hence

$$\|\mathcal{K}g\|_{H^1(\Omega)} = \|w\|_{H^1(\Omega)} \leq \frac{\gamma}{\alpha_0} \|g\|_{L^2(\Omega)} . \tag{6.62}$$

Therefore the operator $\mathcal{K}$ is well-defined from $L^2(\Omega)$ in itself. In addition $\mathcal{K}$ is compact. As a matter of fact, let M be a bounded set of $L^2(\Omega)$, by (6.62), we have that $\mathcal{K}(M)$ is bounded in $H^1(\Omega)$ so that the Rellich–Kondrachov Theorem implies that $\mathcal{K}(M)$ is relatively compact in $L^2(\Omega)$.

Let us apply Theorem 6.2. By sentence (iv) of such a Theorem we have the following alternative:
either

(j) the equation

$$u - \mathcal{K}u = \widetilde{h}, \tag{6.63}$$

has a unique solution in $L^2(\Omega)$, for every $\widetilde{h} \in L^2(\Omega)$
or

(jj) there exists at least a not identically vanishing solution (in $L^2(\Omega)$) to the equation

$$u - \mathcal{K}u = 0. \tag{6.64}$$

If sentence (j) holds, then, as (6.57) and (6.61) are equivalent, we have that there exists a unique solution to problem (6.53). Whereas, if sentence (jj) holds, then, by (i) and (v) of Theorem 6.2, we have that the subspace $N \neq \{0\}$ of solutions of (6.64) (hence, of the solutions to (6.54)) has finite dimension. Let us observe that, in the latter case we have

$$\gamma \neq 0. \tag{6.65}$$

otherwise we should have $\mathcal{K} = 0$ and, by (6.64), $N = \{0\}$. Moreover the dimension of N is equal to the dimension of $N^\star$, where $N^\star$ is the subspace of the solutions to

$$u - \mathcal{K}^\star u = 0. \tag{6.66}$$

At this point, to conclude (b), let us examine what relationship holds true between $\mathcal{K}^\star$ and $L^\star$.

Claim Let us set

$$L_\gamma^\star = L^\star + \gamma, \tag{6.67}$$

we have

$$\mathcal{K}^\star = \gamma \left(L_\gamma^\star\right)^{-1}. \tag{6.68}$$

Proof of the Claim Firstly, let us recall that

$$\begin{aligned} &\forall g \in L^2(\Omega) \quad \exists u \in H_0^1(\Omega) \text{ (unique) such that} \\ &a_\gamma(u, v) = (g, v)_{L^2(\Omega)}, \quad \forall v \in H_0^1(\Omega), \end{aligned}$$

that is

$$u = L_\gamma^{-1} g. \tag{6.69}$$

By the Lax–Milgram Theorem we have

$$\begin{aligned} &\forall \widetilde{g} \in L^2(\Omega) \quad \exists \widetilde{u} \in H_0^1(\Omega) \text{ (unique) such that} \\ &a_\gamma^\star(\widetilde{u}, \widetilde{v}) = (\widetilde{g}, \widetilde{v})_{L^2(\Omega)}, \quad \forall \widetilde{v} \in H_0^1(\Omega), \end{aligned}$$

that is

$$\widetilde{u} = \left(L_\gamma^\star\right)^{-1} \widetilde{g}. \tag{6.70}$$

Now, let us recall that

$$a_\gamma^\star(\widetilde{u}, \widetilde{v}) = a_\gamma(\widetilde{v}, \widetilde{u}), \quad \forall \widetilde{v} \in H_0^1(\Omega),$$

and let us choose $\widetilde{v} = u$, where u is given by (6.69). We get

$$(\widetilde{g}, u)_{L^2(\Omega)} = a^\star_\gamma (\widetilde{u}, u) = a_\gamma (u, \widetilde{u}) = (g, \widetilde{u})_{L^2(\Omega)} .$$

Hence, (6.69) and (6.70) give

$$\left(\widetilde{g}, L_\gamma^{-1} g\right)_{L^2(\Omega)} = \left(\left(L^\star_\gamma\right)^{-1} \widetilde{g}, g\right)_{L^2(\Omega)},$$

for every $\widetilde{g} \in L^2(\Omega)$ and for every $g \in L^2(\Omega)$. Now, as $\mathcal{K}g = \gamma L_\gamma^{-1} g$ we obtain (6.68). **This conclude the proof of the Claim.**

By (6.67) and (6.68) (taking into account that $\mathcal{K}^\star$ assumes its values in $H^1_0(\Omega)$) we have the equivalences

$$v - \mathcal{K}^\star v = 0 \Longleftrightarrow L^\star_\gamma v - \gamma v = 0 \Longleftrightarrow L^\star v = 0.$$

Therefore, the solutions to (6.66) are all and only the solutions to (6.55) and by this also (b) is proved

Now, we prove (iii). Sentence (iii) of Theorem 6.2 implies that the boundary value problem (6.57) (which, we recall, is equivalent to Eq. (6.61)) admits a solution if and only if

$$(h, v)_{L^2(\Omega)} = 0, \quad \forall v \in N^\star.$$

On the other hand, by $v = \mathcal{K}^\star v$, (6.59) and by (6.60), we have

$$(f, v)_{L^2(\Omega)} = \left(f, \mathcal{K}^\star v\right)_{L^2(\Omega)} = (\mathcal{K}f, v)_{L^2(\Omega)} = \gamma\, (h, v)_{L^2(\Omega)}$$

and, taking into account (6.65), we have that problem (6.53) has a solution in $H^1_0(\Omega)$ if and only if

$$(f, v)_{L^2(\Omega)} = 0, \quad \forall v \in N^\star.$$

■

6.4 The Neumann Problem

Let Ω be a connected open set of $\mathbb{R}^n$ whose boundary is of class $C^{0,1}$ and let A be a matrix whose entries are functions of $L^\infty(\Omega)$. Let us assume that A satisfies (6.1). Formally the Neumann problem for the equation

$$-\operatorname{div}(A\nabla u) = F,$$

may be written as follows

$$\begin{cases} -\operatorname{div}(A\nabla u) = F, & \text{in } \Omega \\ \\ A\nabla u \cdot \nu = g, & \text{on } \partial\Omega \end{cases}. \tag{6.71}$$

Concerning the variational formulation we first have to specify that

$$F \in \left(H^1(\Omega)\right)' \quad \text{and} \quad g \in H^{-1/2}(\partial\Omega),$$

where $\left(H^1(\Omega)\right)'$ is the dual space of $H^1(\Omega)$. Having done this, arguing similarly to the Dirichlet problem, we formulate the Neumann problem as follows:

Determine $u \in H^1(\Omega)$ such that

$$\begin{cases} \int_\Omega A\nabla u \cdot \nabla v dx = F(v) + \langle g, v\rangle_{H^{-1/2},H^{1/2}}, & \forall v \in H^1(\Omega) \\ \\ u \in H^1(\Omega) \end{cases}, \tag{6.72}$$

where we mean

$$\langle g, v\rangle_{H^{-1/2},H^{1/2}} = \langle g, T(v)\rangle_{H^{-1/2},H^{1/2}},$$

here $T(v)$ is the trace of v on $\partial\Omega$.

Let us notice at once that, by setting $v = 1$, in (6.72) we have that a necessary condition (and, as we will see in Theorem 6.6, also sufficient) to ensure that problem (6.72) admits solutions, is

$$F(1) + \langle g, 1\rangle_{H^{-1/2},H^{1/2}} = 0. \tag{6.73}$$

Also, we notice that if $u_0 \in H^1(\Omega)$ is a solution to problem (6.72), then all the solutions to (6.72) are given by

$$u_0 + C,$$

where C is any constant. Indeed, it is immediately checked that $u_0 + C$ is a solution to (6.72). Conversely, if $u \in H^1(\Omega)$ is a solution to (6.72), then

$$\int_\Omega A\nabla(u - u_0) \cdot \nabla v dx = 0, \quad \forall v \in H^1(\Omega).$$

Now, we choose $v = u - u_0$ and (6.1) gives

$$\lambda^{-1}\int_\Omega |\nabla(u-u_0)|^2 \leq \int_\Omega A\nabla(u-u_0)\cdot\nabla(u-u_0)dx = 0.$$

Since Ω is connected, we obtain that $u - u_0$ is a constant in Ω.

Thus, to ensure the uniqueness to the Neumann problem we may formulate it as follows

$$\begin{cases} \int_\Omega A\nabla u\cdot\nabla v dx = F(v) + \langle g, v\rangle_{H^{-1/2},H^{1/2}}, & \forall v\in H^1(\Omega) \\ \\ u\in\left\{w\in H^1(\Omega): \quad \int_\Omega w dx = 0\right\} \end{cases}. \tag{6.74}$$

Concerning the existence, we have

Theorem 6.6 *Let Ω be a connected open set of $\mathbb{R}^n$ whose boundary is of class $C^{0,1}$. Let us assume that $A \in L^\infty(\Omega; \mathbb{M}(n))$ and that A satisfies* (6.1). *Let us assume that $F \in \left(H^1(\Omega)\right)'$ and $g \in H^{-1/2}(\partial\Omega)$ satisfy* (6.73).

Then problem (6.74) *has a unique solution and the following inequality holds true*

$$\|u\|_{H^1(\Omega)} \leq C\left(\|F\|_{(H^1(\Omega))'} + \|g\|_{H^{-1/2}(\partial\Omega)}\right), \tag{6.75}$$

where C depends on λ and Ω only.

Proof Set

$$\widetilde{H} := \left\{w\in H^1(\Omega): \quad \int_\Omega w dx = 0\right\}.$$

Theorem 5.35 implies that $\widetilde{H}$ is a Hilbert space equipped with the norm

$$\|w\|_{\widetilde{H}} = \left(\int_\Omega |\nabla w|^2 dx\right)^{1/2}.$$

Moreover, the bilinear form on $\widetilde{H}$

$$a(u,v) = \int_\Omega A\nabla u\cdot\nabla v dx, \tag{6.76}$$

is coercive and continuous. Now we check that the linear functional

$$\widetilde{H} \ni v \to \widetilde{F}(v) := F(v) + \langle g, v\rangle_{H^{-1/2},H^{1/2}} \in \mathbb{R},$$

is well-defined and continuous on $\widetilde{H}$. As a matter of fact, by Theorem 5.35 we have

$$|F(v)| \leq \|F\|_{(H^1(\Omega))'} \|v\|_{H^1(\Omega)} \leq C \|F\|_{(H^1(\Omega))'} \|v\|_{\widetilde{H}}, \quad \forall v \in \widetilde{H}, \tag{6.77}$$

where C depends on Ω only. Moreover, recalling that $H^{-1/2}(\Omega)$ is the dual space of $H^{1/2}(\Omega)$ (see Sect. 5.13.1), inequality (ii) of Theorem 5.18 gives

$$\begin{aligned}\left|\langle g, v\rangle_{H^{-1/2},H^{1/2}}\right| &\leq \|g\|_{H^{-1/2}(\partial\Omega)} \|v\|_{H^{1/2}(\partial\Omega)} \leq \\ &\leq C \|g\|_{H^{-1/2}(\partial\Omega)} \|v\|_{H^1(\Omega)} \leq \\ &\leq C \|g\|_{H^{-1/2}(\partial\Omega)} \|v\|_{\widetilde{H}},\end{aligned} \tag{6.78}$$

where C depends on Ω only. Therefore, (6.77) and (6.78) give

$$|\widetilde{F}(v)| \leq C \left(\|F\|_{(H^1(\Omega))'} + \|g\|_{H^{-1/2}(\partial\Omega)}\right) \|v\|_{\widetilde{H}}, \quad \forall v \in \widetilde{H}. \tag{6.79}$$

Now, since bilinear form (6.76) is continuous and coercive on $\widetilde{H}$ and since (6.79) holds, $\widetilde{F}$ is a bounded linear functional on $\widetilde{H}$. Therefore by the Lax–Milgram Theorem we have that there exists a unique $u \in \widetilde{H}$ which satisfies

$$a(u, v) = \widetilde{F}(v), \quad \forall v \in \widetilde{H}. \tag{6.80}$$

Moreover

$$\|u\|_{H^1(\Omega)} \leq C \|\nabla u\|_{L^2(\Omega)} \leq C \left(\|F\|_{(H^1(\Omega))'} + \|g\|_{H^{-1/2}(\partial\Omega)}\right), \tag{6.81}$$

where C depends on λ and Ω only.

Now, let v be any function of $H^1(\Omega)$ and let us denote

$$v_\Omega = \frac{1}{|\Omega|}\int_\Omega v dx, \quad \widetilde{v} = v - v_\Omega.$$

Since $\widetilde{v} \in \widetilde{H}$, by (6.73) and (6.80) we obtain

$$\begin{aligned}&\int_\Omega A\nabla u \cdot \nabla v dx = \int_\Omega A\nabla u \cdot \nabla \widetilde{v} dx = \\ &= a(u, \widetilde{v}) = F(\widetilde{v}) + \langle g, \widetilde{v}\rangle_{H^{-1/2},H^{1/2}} = \\ &= F(v) + \langle g, v\rangle_{H^{-1/2},H^{1/2}} - v_\Omega \left(F(1) + \langle g, 1\rangle_{H^{-1/2},H^{1/2}}\right) = \\ &= F(v) + \langle g, v\rangle_{H^{-1/2},H^{1/2}}.\end{aligned} \tag{6.82}$$

Therefore u is a solution to problem (6.74). Estimate (6.74) follows by (6.81). ■

6.5 The Caccioppoli Inequality

Theorem 6.7 (Caccioppoli Inequality) *Let $x_0 \in \mathbb{R}^n$ and $R > 0$. Let A be a symmetric matrix whose entries are measurable functions on $B_R(x_0)$. Assume that A satisfies (6.1) (with $\Omega = B_R(x_0)$).*

Let $b \in L^\infty(B_R(x_0); \mathbb{R}^n)$ and $c \in L^\infty(B_R(x_0))$. Let $u \in H^1_{loc}(B_R(x_0))$ satisfy

$$\int_{B_R(x_0)} A\nabla u \cdot \nabla v dx = \int_{B_R(x_0)} (b\nabla u + cu)\, v dx, \quad \forall v \in H^1_0(B_R(x_0)). \tag{6.83}$$

If $0 < r < \rho < R$, then we have

$$\int_{B_r(x_0)} |\nabla u|^2 dx \leq \frac{C}{(\rho - r)^2} \int_{B_\rho(x_0)} u^2 dx, \tag{6.84}$$

where C depends on λ, $R\, \|b\|_{L^\infty(B_R(x_0);\mathbb{R}^n)}$ and $R^2\, \|c\|_{L^\infty(B_R(x_0);\mathbb{R}^n)}$ only.

Proof It is not restrictive to assume $x_0 = 0$. Let $\eta \in C_0^\infty(B_\rho)$ satisfy

$$0 \leq \eta \leq 1; \qquad \eta = 1, \quad \text{in } B_r \tag{6.85}$$

and

$$|\nabla \eta| \leq \frac{K}{\rho - r}, \tag{6.86}$$

where K is a positive constant. We choose in (6.83)

$$v = \eta^2 u$$

and we have

$$\int_{B_R} A\nabla u \cdot \nabla \left(\eta^2 u\right) dx = \int_{B_R} (b\nabla u + cu)\, \eta^2 u dx. \tag{6.87}$$

Hence

$$\int_{B_R} (A\nabla u \cdot \nabla u)\eta^2 dx = \int_{B_R} (b\nabla u + cu)\,\eta^2 u dx -$$
$$-2\int_{B_R} (A\nabla u \cdot \nabla \eta)\eta u dx \leq$$
$$\leq \int_{B_R} \left(|b||\nabla u||u|\eta^2 + |c|u^2\eta^2\right) dx +$$
$$+2\int_{B_R} (A\nabla u \cdot \nabla u)^{1/2}(A\nabla \eta \cdot \nabla \eta)^{1/2}|u|\eta dx \leq$$
$$\leq \int_{B_R} \left(|b||\nabla u||u|\eta^2 + |c|u^2\eta^2\right) dx +$$
$$+\frac{1}{2}\int_{B_R} (A\nabla u \cdot \nabla u)\eta^2 dx + 2\int_{B_R} (A\nabla \eta \cdot \nabla \eta)u^2 dx.$$

By moving to the left-hand side the second-to-last integral and, by estimating from above the last integral, we have

$$\begin{aligned} &\frac{1}{2}\int_{B_R} (A\nabla u \cdot \nabla u)\eta^2 dx \leq \\ &\leq \int_{B_R} \left(|b||\nabla u||u|\eta^2 + |c|u^2\eta^2\right) dx + \frac{2K^2\lambda}{(\rho - r)^2}\int_{B_\rho} u^2 dx. \end{aligned} \tag{6.88}$$

Now let us estimate from above the first integral on the right hand side of (6.88). We obtain, for $\varepsilon > 0$ to be choosen,

$$\begin{aligned} \int_{B_R} \left(|b||\nabla u||u|\eta^2 + |c|u^2\eta^2\right) dx &\leq \frac{\varepsilon}{2}\int_{B_R} |\nabla u|^2\eta^2 dx + \\ &+\frac{1}{2\varepsilon}\,\|b\|^2_{L^\infty(B_R;\mathbb{R}^n)}\int_{B_R} u^2\eta^2 dx + \\ &+\|c\|_{L^\infty(B_R)}\int_{B_R} u^2\eta^2 dx \leq \\ &\leq \frac{\varepsilon\lambda}{2}\int_{B_R} (A\nabla u \cdot \nabla u)\eta^2 dx + \\ &+C_\varepsilon \int_{B_R} u^2\eta^2 dx, \end{aligned} \tag{6.89}$$

where

$$C_\varepsilon = \frac{1}{2\varepsilon}\,\|b\|^2_{L^\infty(B_R,\mathbb{R}^n)} + \|c\|_{L^\infty(B_R)}\,.$$

Using inequality (6.89) in (6.88), after a few easy calculations, we have

$$\begin{aligned}\frac{1}{2}(1-\varepsilon\lambda)\int_{B_R}(A\nabla u\cdot\nabla u)\eta dx &\leq \\ \leq C_\varepsilon\int_{B_R}u^2\eta^2dx+\frac{2K^2\lambda}{(\rho-r)^2}\int_{B_\rho}u^2dx.\end{aligned} \tag{6.90}$$

Now we choose

$$\varepsilon=\varepsilon_0:=\frac{1}{2\lambda},$$

and we get

$$C_{\varepsilon_0}=\lambda\,\|b\|^2_{L^\infty(B_R;\mathbb{R}^n)}+\|c\|_{L^\infty(B_R)}$$

and by (6.90) we have

$$\begin{aligned}\frac{\lambda^{-1}}{4}\int_{B_r}|\nabla u|^2dx \leq \frac{1}{4}\int_{B_R}(A\nabla u\cdot\nabla u)\eta^2dx &\leq \\ \leq \frac{K_1}{(\rho-r)^2}\int_{B_\rho}u^2,\end{aligned} \tag{6.91}$$

where

$$K_1=2K^2\lambda+R^2C_{\varepsilon_0}.$$

By (6.91) we obtain immediately

$$\int_{B_r}|\nabla u|^2dx\leq\frac{4K_1\lambda}{(\rho-r)^2}\int_{B_\rho}u^2dx,$$

so that (6.84) follows. ■

Exercise 6.33 Let $x_0\in\mathbb{R}^n$ and $R>0$. Let L be the operator

$$Lu=-\sum_{j,k=1}^n\partial_j\left(a^{jk}\partial_k u+d^ju\right)+\sum_{j=1}^n b^j\partial_j u+cu, \tag{6.92}$$

where $A\in L^\infty(B_R(x_0);\mathbb{M}(n))$, $A=\left\{a^{jk}\right\}^n_{j,k=1}$, satisfies (6.1) and $b^j,d^j,c\in L^\infty(B_R(x_0))$, for $j=1,\cdots,n$. Let $f\in L^2(B_R(x_0))$ and let us assume that $u\in$

$H^1(B_R(x_0))$ is a weak solution to

$$Lu = f, \quad \text{in } B_R(x_0).$$

Prove that, if $0 < r < \rho < R$ then we have

$$\int_{B_r(x_0)} |\nabla u|^2 dx \leq \frac{C}{(\rho - r)^2} \int_{B_\rho(x_0)} u^2 dx + C\rho^2 \int_{B_\rho(x_0)} f^2 dx, \tag{6.93}$$

where C depends on λ, $\|A\|_{L^\infty(B_R(x_0);\mathbb{M}(n))}$, $R\|d\|_{L^\infty(B_R(x_0);\mathbb{R}^n)}$, $R\|b\|_{L^\infty(B_R(x_0);\mathbb{R}^n)}$ and $R^2\|c\|_{L^\infty(B_R(x_0))}$ only. ♠

Exercise 6.34 Let $R > 0$ and $x_0 \in \{(x', 0) : x' \in \mathbb{R}^{n-1}\}$. Let L be operator (6.92) and let $u \in H^1(B_R^+(x_0))$ satisfy

$$\begin{cases} Lu = f, & \text{in } B_R^+(x_0), \text{ in the weak sense} \\ \\ u(x', 0) = 0, & x' \in B'_R(x_0) \text{ in the traces sense} \end{cases},$$

then

$$\int_{B_r^+(x_0)} |\nabla u|^2 dx \leq \frac{C}{(\rho - r)^2} \int_{B_\rho^+(x_0)} u^2 dx + C\rho^2 \int_{B_\rho^+(x_0)} f^2 dx, \tag{6.94}$$

where C depends on λ, $\|A\|_{L^\infty(B_R^+(x_0);\mathbb{M}(n))}$, $R\|d\|_{L^\infty(B_R^+(x_0);\mathbb{R}^n)}$, $R\|b\|_{L^\infty(B_R^+(x_0);\mathbb{R}^n)}$ and $R^2\|c\|_{L^\infty(B_R^+(x_0))}$ only. ♠

Exercise 6.35 Let x_0, R and A be like Exercise 2.

(a) Give the variational formulation of problem

$$\begin{cases} \text{div}(A\nabla u) = 0, & \text{in } B_R^+(x_0) \\ \\ (A\nabla u)(x', 0) \cdot e_n = 0, & \text{for } x' \in B'_R(x_0) \end{cases}. \tag{6.95}$$

(b) Prove that

$$\int_{B_r^+(x_0)} |\nabla u|^2 dx \leq \frac{C}{(\rho - r)^2} \int_{B_\rho^+(x_0)} u^2 dx, \tag{6.96}$$

where C depends on λ and $\|A\|_{L^\infty(B_R^+(x_0);\mathbb{M}(n))}$ only. ♠

6.6 The Regularity Theorems

In Sect. 6 (Theorem 6.3) we have proved that if Ω is a bounded open set of $\mathbb{R}^n$, $A \in L^\infty(\Omega; \mathbb{M}(n))$ satisfies (6.1) and $F \in H^{-1}(\Omega)$, then the Dirichlet problem

$$\begin{cases} -\operatorname{div}(A\nabla u) = F, & \text{in } \Omega \\ \\ u = 0, & \text{on } \partial\Omega \end{cases}, \tag{6.97}$$

is well-posed in $H_0^1(\Omega)$.

It is natural to ask whether with more restrictive assumptions on the data Ω, A and F, u is more regular. More precisely, we ask whether there exists $k > 1$ such that $u \in H^k(\Omega)$. In carrying out this investigation it is convenient to distinguish between the **regularity in the interior** and the **regularity at the boundary**. In the investigation of the regularity in the interior we are interested in whether for some $k > 1$ we have $u \in H^k_{loc}(\Omega)$, while in the investigation of regularity at the boundary we are interested in knowing whether for some $k > 1$ it happens that for every $x_0 \in \partial\Omega$ there exists a neighborhood of x_0, $\mathcal{U}$, such that $u_{|\mathcal{U}} \in H^k(\Omega \cap \mathcal{U})$. As we should expect, in the study of regularity in the interior, the regularity of $\partial\Omega$ plays no role. In contrast, in the study of regularity at the boundary, the regularity of $\partial\Omega$ plays a crucial role.

Before going on to the rigorous treatment, let us illustrate in a rough manner the main idea that drives the study of the regularity in the interior; similar arguments can be made for the regularity at the boundary.

Let us consider the equation

$$-\Delta u = f, \qquad \text{in } \Omega, \tag{6.98}$$

where $f \in L^2(\Omega)$. Let $u \in H^1(\Omega)$ be a solution to (6.98); that is.

$$\int_\Omega \nabla u \cdot \nabla v dx = \int_\Omega f v dx, \quad \forall v \in H_0^1(\Omega). \tag{6.99}$$

Let us suppose that we know u be sufficiently regular (say $u \in H^3_{loc}(\Omega)$) so that the operations that we will make are allowed.

Let $x_0 \in \Omega$ and $R > 0$ satisfy $B_{2R}(x_0) \subset \Omega$. Let $\eta \in C_0^\infty(B_{2R}(x_0)$ be such that

$$0 \le \eta \le 1; \qquad \eta = 1, \quad \text{in } B_R(x_0) \tag{6.100}$$

and

$$|\nabla \eta| \le \frac{K}{R}, \tag{6.101}$$

where K is a positive constant. Let $k \in \{1, \cdots, n\}$. Multiply both the sides of (6.98) by $\partial_k \left(\eta^2 \partial_k u\right)$ and integrate over Ω or, equivalently, choose in (6.99))

$$v = \partial_k \left(\eta^2 \partial_k u\right), \tag{6.102}$$

obtaining

$$\int_\Omega \Delta u \partial_k \left(\eta^2 \partial_k u\right) dx = \int_\Omega f \partial_k \left(\eta^2 \partial_k u\right) dx. \tag{6.103}$$

Let us consider the left-hand side of (6.103), integration by parts yields

$$\begin{aligned}
\int_\Omega \Delta u \partial_k \left(\eta^2 \partial_k u\right) dx &= \int_\Omega \sum_{j=1}^n \partial_j^2 u \partial_k \left(\eta^2 \partial_k u\right) dx = \\
&= -\int_\Omega \sum_{j=1}^n \partial_j \left(\partial_{jk}^2 u\right) \eta^2 \partial_k u dx = \\
&= \int_\Omega \sum_{j=1}^n \partial_{jk}^2 u \partial_j \left(\eta^2 \partial_k u\right) dx = \\
&= \int_\Omega \sum_{j=1}^n \left|\partial_{jk}^2 u\right|^2 \eta^2 dx + \\
&+ 2\int_\Omega \sum_{j=1}^n \left(\partial_{jk}^2 u\right) \eta \partial_j \eta \partial_k u dx.
\end{aligned}$$

Concerning the right-hand side of (6.103), we get

$$\int_\Omega f \partial_k \left(\eta^2 \partial_k u\right) dx = \int_\Omega f \eta^2 \partial_k^2 u dx + 2\int_\Omega f \eta \partial_k \eta \partial_k u dx.$$

Using in (6.103) the last two obtained equalities and summing up over k, we have

$$\begin{aligned}
\int_\Omega \sum_{j,k=1}^n \left|\partial_{jk}^2 u\right|^2 \eta^2 dx &= -2\int_\Omega \sum_{j,k=1}^n \left(\partial_{jk}^2 u\right) \eta \partial_j \eta \partial_k u dx + \\
&+ \int_\Omega f \eta^2 \sum_{k=1}^n \partial_k^2 u dx + 2\int_\Omega f \eta \sum_{k=1}^n \partial_k \eta \partial_k u dx := I.
\end{aligned} \tag{6.104}$$

Let $\varepsilon > 0$ be chosen later, let us denote by $\partial^2 u$ the Hessian matrix $\left\{\partial^2_{jk} u\right\}^n_{j,k=1}$. We have

$$I \leq \varepsilon \int_\Omega \left|\partial^2 u\right|^2 \eta^2 dx + \frac{1}{\varepsilon} \int_\Omega |\nabla u|^2 |\nabla \eta|^2 dx +$$
$$+ \frac{\varepsilon}{2} \int_\Omega \left|\partial^2 u\right|^2 \eta^2 dx + \frac{1}{2\varepsilon} \int_\Omega |f|^2 \eta^2 dx +$$
$$+ \int_\Omega |f|^2 \eta^2 dx + \int_\Omega |\nabla \eta|^2 |\nabla u|^2 dx.$$

Now, in (6.104), we move to the left-hand side the terms that contain the second derivatives, and we get

$$\left(1 - \frac{3\varepsilon}{2}\right) \int_\Omega \left|\partial^2 u\right|^2 \eta^2 dx \leq \left(1 + \frac{1}{2\varepsilon}\right) \int_\Omega |\nabla u|^2 |\nabla \eta|^2 dx +$$
$$+ \left(1 + \frac{1}{2\varepsilon}\right) \int_\Omega |f|^2 \eta^2 dx.$$

At this point, we choose $\varepsilon = \frac{1}{3}$ and by (6.100), (6.101) we get

$$\int_{B_R(x_0)} \left|\partial^2 u\right|^2 dx \leq \frac{5K^2}{R^2} \int_{B_{2R}(x_0)} |\nabla u|^2 dx + 5 \int_{B_{2R}(x_0)} |f|^2 dx. \tag{6.105}$$

We observe that (6.105) enables us to estimate the second derivatives of u in $L^2(B_R(x_0))$ using the finite quantity on the right-hand side. **However, it is evident that this estimate alone does not constitute a proof** that $u \in H^2(B_R(x_0))$ for $j, k = 1, \cdots, n$, as the derivation of the estimate relied on a higher regularity of u than what was actually established (!). Nonetheless, in the forthcoming rigorous proofs within this chapter, we will 'retrace' the preceding steps in a sense, employing as the test function—instead of (6.102)—the function

$$v = -\delta_k^{-h} \left(\eta^2 \delta_k^h u\right),$$

where δ_k^{-h} and δ_k^h, are the difference quotients studied in Sect. 5.10.

6.6.1 The Regularity Theorems in the Interior

The Main Theorem of the present subsection is the following.

Theorem 6.8 (Regularity in the Interior) *Let Ω be a bounded open set of $\mathbb{R}^n$. Let*

$$f \in L^2(\Omega). \tag{6.106}$$

Let A be a symmetric matrix. Assume that A satisfies (6.1), $A \in C^{0,1}(\Omega; \mathbb{M}(n))$ and it satisfies

$$|A(x) - A(y)| \leq E|x-y|, \quad \forall x, y \in \Omega, \tag{6.107}$$

where E is a positive number. Let us assume $u \in H^1(\Omega)$ is a solution to

$$-\operatorname{div}(A\nabla u) = f, \quad \text{in } \Omega. \tag{6.108}$$

Then we have

$$u \in H^2_{loc}(\Omega), \tag{6.109}$$

and, for any $B_{2R}(x_0) \subset \Omega$, the following estimate holds true

$$\begin{aligned} \sum_{|\alpha|\leq 2} R^{2|\alpha|} \int_{B_R(x_0)} |\partial^\alpha u|^2 dx \leq C\left(1 + E^2R^2\right) \int_{B_{2R}(x_0)} u^2 dx + \\ + CR^4 \int_{B_{2R}(x_0)} f^2 dx, \end{aligned} \tag{6.110}$$

where C depends on λ only.

Proof It is not restrictive to assume $0 \in \Omega$ and $R < \frac{1}{2}\text{dist}\,(0, \partial\Omega)$. Let $\eta \in C_0^\infty(B_{3R/2})$ satisfy

$$0 \leq \eta \leq 1; \qquad \eta = 1, \quad \text{in } B_R \tag{6.111}$$

and

$$|\nabla \eta| \leq \frac{K}{R}, \tag{6.112}$$

where K is a positive constant.

Since $u \in H^1(\Omega)$ satisfies (6.108), we have

$$\int_\Omega A\nabla u \cdot \nabla v dx = \int_\Omega f v dx, \quad \forall v \in H_0^1(\Omega). \tag{6.113}$$

Let $h \in \left(-\frac{R}{8}, \frac{R}{8}\right) \setminus \{0\}$. Let us note that, if $w_1, w_2 \in H^1(\Omega)$ and $\operatorname{supp} w_1 \subset B_{3R/2}$ (or $\operatorname{supp} w_2 \subset B_{3R/2}$), then, for any $k \in \{1, \cdots, n\}$, we have

$$\int_{\Omega} w_1 \delta_k^{-h} w_2 dx = -\int_{\Omega} w_2 \delta_k^h w_1 dx \tag{6.114}$$

and

$$\delta_k^h (w_1 w_2) = w_1^h \delta_k^h w_2 + w_2 \delta_k^h w_1, \tag{6.115}$$

where $w_1^h(x) = w_1(x + he_k)$. Concerning (6.114), just argue like in the Claim of the proof of Theorem 5.37. While equality (6.114) follows easily by

$$\begin{aligned} h\delta_k^h (w_1 w_2) &= w_1(x + he_k) w_2(x + he_k) - w_1(x) w_2(x) = \\ &= w_1(x + he_k) w_2(x + he_k) - w_1(x + he_k) w_2(x) + \\ &+ w_1(x + he_k) w_2(x) - w_1(x) w_2(x) = \\ &= h \left(w_1^h \delta_k^h w_2 + w_2 \delta_k^h w_1 \right). \end{aligned}$$

Now, let us choose as test function in (6.113)

$$v = -\delta_k^{-h} \left(\eta^2 \delta_k^h u \right). \tag{6.116}$$

We get

$$\begin{aligned} \int_{\Omega} A\nabla u \cdot \nabla v dx &= -\int_{\Omega} A\nabla u \cdot \left[\delta_k^{-h} \nabla \left(\eta^2 \delta_k^h u \right) \right] dx = \\ &= \int_{\Omega} \delta_k^h (A\nabla u) \cdot \nabla \left(\eta^2 \delta_k^h u \right) dx = \\ &= \int_{\Omega} A^h \left(\delta_k^h \nabla u \right) \cdot \nabla \left(\eta^2 \delta_k^h u \right) dx + \\ &+ \int_{\Omega} \left(\delta_k^h A \right) \nabla u \cdot \nabla \left(\eta^2 \delta_k^h u \right) dx = \\ &= \int_{\Omega} \left[A^h \left(\delta_k^h \nabla u \right) \cdot \left(\delta_k^h \nabla u \right) \right] \eta^2 dx + \mathcal{R}, \end{aligned} \tag{6.117}$$

where

$$\begin{aligned} \mathcal{R} &= \int_{\Omega} A^h \left(\delta_k^h \nabla u \right) \cdot \left(2\eta \nabla \eta \delta_k^h u \right) dx + \\ &+ \int_{\Omega} \left(\delta_k^h A \right) \nabla u \cdot \left(\delta_k^h \nabla u \right) \eta^2 dx + \\ &+ \int_{\Omega} \left(\delta_k^h A \right) \nabla u \cdot \left(2\eta \nabla \eta \delta_k^h u \right) dx. \end{aligned}$$

Now, by (6.1) we get

$$\int_{\Omega}\left[A^h\left(\delta_k^h\nabla u\right)\cdot\left(\delta_k^h\nabla u\right)\right]\eta^2dx \geq \lambda^{-1}\int_{\Omega}\left|\delta_k^h\nabla u\right|^2\eta^2dx. \tag{6.118}$$

Concerning $\mathcal{R}$, let ε be a positive number which will choose later on. By (6.107) we have

$$\begin{aligned}|\mathcal{R}| \leq &\frac{c_n\lambda}{R}\int_{\Omega}\left|\delta_k^h\nabla u\right|\left|\delta_k^h u\right|\eta dx + E\int_{\Omega}\left|\delta_k^h\nabla u\right|\left|\nabla u\right|\eta^2dx+\\ &+\frac{E}{R}\int_{\Omega}\left|\nabla u\right|\left|\delta_k^h u\right|\eta^2dx \leq\\ &\leq \varepsilon\int_{\Omega}\left|\delta_k^h\nabla u\right|^2\eta^2dx+\\ &+\frac{C}{\varepsilon}\left(R^{-2}+E^2\right)\int_{B_{3R/2}}\left(\left|\delta_k^h u\right|^2+|\nabla u|^2\right)dx,\end{aligned} \tag{6.119}$$

where c_n depends on n only and C depends on λ and n only. Now let us choose

$$\varepsilon = \frac{\lambda^{-1}}{2}$$

so that, by (6.118) and (6.119), we get

$$\begin{aligned}\int_{\Omega}\left[A^h\left(\delta_k^h\nabla u\right)\cdot\left(\delta_k^h\nabla u\right)\right]\eta^2dx + \mathcal{R} \geq \frac{\lambda^{-1}}{2}\int_{\Omega}\left|\delta_k^h\nabla u\right|^2\eta^2dx-\\ -2C\lambda\left(R^{-2}+E^2\right)\int_{B_{3R/2}}\left(\left|\delta_k^h u\right|^2+|\nabla u|^2\right)dx.\end{aligned} \tag{6.120}$$

Hence, (6.113), (6.116), (6.117) and (6.120) yield

$$\begin{aligned}\frac{\lambda^{-1}}{2}\int_{\Omega}\left|\delta_k^h\nabla u\right|^2\eta^2dx \leq -\int_{\Omega}f\delta_k^{-h}\left(\eta^2\delta_k^h u\right)dx+\\ +C\left(R^{-2}+E^2\right)\int_{B_{3R/2}}\left(\left|\delta_k^h u\right|^2+|\nabla u|^2\right)dx,\end{aligned} \tag{6.121}$$

where C depends on λ and n only.

Now, by Theorem 5.37-(i) (with $V = B_{3R/2}$ and $\Omega = B_{7R/4}$) we have

$$\int_{B_{3R/2}}\left|\delta_k^h u\right|^2dx \leq \int_{B_{7R/4}}|\nabla u|^2dx. \tag{6.122}$$

Moreover

$$\begin{aligned}\int_\Omega \left|\delta_k^{-h}\left(\eta^2\delta_k^h u\right)\right|^2 dx &\le C\int_\Omega \left|\nabla\left(\eta^2\delta_k^h u\right)\right|^2 dx \le \\ &\le C'\int_\Omega \left|\eta\nabla\eta\left(\eta^2\delta_k^h u\right)\right|^2 dx + C'\int_\Omega \eta^2\left|\nabla\delta_k^h u\right|^2 dx \le \\ &\le \frac{C''}{R^2}\int_{B_{3R/2}} \left|\delta_k^h u\right|^2 dx + C'\int_\Omega \eta^2\left|\delta_k^h\nabla u\right|^2 dx.\end{aligned} \tag{6.123}$$

Let now $\sigma > 0$ be chosen, (6.123) implies

$$\begin{aligned}\left|\int_\Omega f\delta_k^{-h}\left(\eta^2\delta_k^h u\right)dx\right| &\le \frac{1}{2\sigma}\int_{B_{2R}} f^2 dx + \frac{\sigma}{2}\int_\Omega \left|\delta_k^{-h}\left(\eta^2\delta_k^h u\right)\right|^2 dx \le \\ &\le \frac{1}{2\sigma}\int_{B_{2R}} f^2 dx + \\ &+ C\sigma\left(R^{-2}\int_{B_{3R/2}} \left|\delta_k^h u\right|^2 dx + \int_\Omega \eta^2\left|\delta_k^h\nabla u\right|^2 dx\right).\end{aligned} \tag{6.124}$$

Now we apply Theorem 5.37-(i) and inequality, (6.93), so that we have

$$R^{-2}\int_{B_{3R/2}} \left|\delta_k^h u\right|^2 dx \le R^{-2}\int_{B_{7R/4}} |\nabla u|^2 dx \le CR^{-4}\int_{B_{2R}} u^2 dx + C\int_{B_{2R}} f^2 dx.$$

By the last obtained estimate and by (6.124) we have

$$\begin{aligned}\left|\int_\Omega f\delta_k^{-h}\left(\eta^2\delta_k^h u\right)dx\right| &\le \frac{1}{2\sigma}\int_{B_{2R}} f^2 dx + \\ &+ C_*\sigma\left(R^{-4}\int_{B_{2R}} u^2 dx + \int_\Omega \eta^2\left|\delta_k^h\nabla u\right|^2 dx\right),\end{aligned}$$

where C_* is a constant depending on λ and n only. Inserting what we have just obtained into (6.121) we get

$$\begin{aligned}\left(\frac{\lambda^{-1}}{2} - C_*\sigma\right)\int_\Omega \left|\delta_k^h\nabla u\right|^2 \eta^2 dx &\le \frac{1}{2\sigma}\int_{B_{2R}} f^2 dx + \\ &+ \frac{C\left(1+\sigma+E^2R^2\right)}{R^4}\int_{B_{2R}} u^2 dx.\end{aligned}$$

Now, choosing

$$\sigma = \frac{\lambda^{-1}}{4C_*},$$

we obtain, for $k = 1, \cdots, n$,

$$\int_{B_R} \left|\delta_k^h \nabla u\right|^2 \eta^2 dx \leq C \int_{B_{2R}} f^2 dx + \frac{C\left(1+E^2R^2\right)}{R^4} \int_{B_{2R}} u^2 dx.$$

By the last obtained inequality and by Theorem 5.37-(ii) we get

$$\partial_k u \in H^1\left(B_R\right),$$

for $k = 1, \cdots, n$. Hence $u \in H^2\left(B_R\right)$ and

$$\sum_{|\alpha|=2} R^{2|\alpha|} \int_{B_R} \left|\partial^\alpha u\right|^2 dx \leq C\left(1+E^2R^2\right) \int_{B_{2R}} u^2 dx + CR^4 \int_{B_{2R}} f^2 dx.$$

Finally, by the latter and by (6.93) we get (6.110). ■

Exercise 6.36 Under the same assumptions of Theorem 6.8, prove that, if $0 < r < \rho$ and $B_\rho(x_0) \subset \Omega$ then

$$\sum_{|\alpha|\leq 2} (\rho - r)^{2|\alpha|} \int_{B_r(x_0)} \left|\partial^\alpha u\right|^2 dx \leq C\left(1+E^2\rho^2\right) \int_{B_\rho(x_0)} u^2 dx + + C(\rho - r)^4 \int_{B_\rho(x_0)} f^2 dx, \tag{6.125}$$

where C depends on λ only. [Hint: consider a finite covering $\overline{B_r(x_0)}$ consisting of balls of the type $B_{\frac{\rho-r}{2}}(x)$, $x \in B_r(x_0)$, and apply (6.110)]. ♠

Exercise 6.37 Under the same assumption of Theorem 6.8, prove that, if

$$\Omega' \Subset \Omega$$

then

$$\sum_{|\alpha|\leq 2} \delta_0^{2|\alpha|} \int_{\Omega'} \left|\partial^\alpha u\right|^2 dx \leq C\left(1+E^2 d_0^2\right) \int_\Omega u^2 dx + + Cd_0^4 \int_\Omega f^2 dx, \tag{6.126}$$

where d_0 is the diameter of Ω, δ_0 =dist $\left(\Omega', \partial\Omega\right)$ and C depends on λ and $d_0\delta_0^{-1}$ only. [Hint: use Exercise 6.36 and a partition of unity]. ♠

Exercise 6.38

(a) Generalize Theorem 6.8 to the equation

$$-\sum_{j,k=1}^{n} \partial_j \left(a^{jk}\partial_k u + d^j u\right) + \sum_{j=1}^{n} b^j \partial_j u + cu = f,$$

where $A = \left\{a^{jk}\right\}_{j,k=1}^{n}$ and f satisfy the same assumptions of Theorem 6.8, $d \in C^{0,1}\left(\overline{\Omega}; \mathbb{R}^n\right)$, $b \in L^\infty(\Omega; \mathbb{R}^n)$, $c \in L^\infty(\Omega)$.

(b) Generalize (a) to the case where A is a nonsymmetric matrix. [Hint to (b): write the operator $\operatorname{div}(A\nabla u)$ like

$$\operatorname{div}(A^s\nabla u) + \text{terms of order less than 2},$$

where A^s is the symmetric part of A]. ♠

Theorem 6.9 (Improved Regularity in the Interior) *Let Ω be a bounded open set of $\mathbb{R}^n$ with diameter d_0. Let*

$$f \in H^m(\Omega). \tag{6.127}$$

Let A be a symmetric matrix. Assume that A satisfies (6.1). *Assume that $A \in C^{m,1}(\overline{\Omega}; \mathbb{M}(n))$ and it satisfies*

$$\|A\|_{C^{m,1}(\overline{\Omega};\mathbb{M}(n))} \le E_m, \tag{6.128}$$

where, E_m is a positive number and, recall,

$$\|A\|_{C^{m,1}(\overline{\Omega};\mathbb{M}(n)))} = \sum_{|\alpha|\le m} d_0^{|\alpha|} \left\|\partial^\alpha A\right\|_{L^\infty(\Omega;\mathbb{M}(n)))} + d_0^{m+1} \sum_{|\alpha|=m} [\partial^\alpha A]_{1,\Omega}.$$

Assume that $u \in H^1(\Omega)$ is a solution to

$$-\operatorname{div}(A\nabla u) = f, \quad \text{in } \Omega. \tag{6.129}$$

Then we have

$$u \in H^{m+2}_{loc}(\Omega), \tag{6.130}$$

moreover, if $B_{2R}(x_0) \subset \Omega$, then the following inequality holds true

$$\sum_{|\alpha|\leq m+2} R^{2|\alpha|} \int_{B_R(x_0)} \left|\partial^\alpha u\right|^2 dx \leq C\left(1+E_m^2\right) \int_{B_{2R}(x_0)} u^2 dx +$$
$$+ C \sum_{|\alpha|\leq m} R^{2(|\alpha|+4)} \int_{B_{2R}(x_0)} \left|\partial^\alpha f\right|^2 dx, \tag{6.131}$$

where C depends on λ only.

Proof We simply consider the case $m = 1$, leaving the reader to complete the proof by induction. Let $l \in \{1, \cdots, n\}$. Let $\widetilde{v}$ be any function belonging to $C_0^\infty\left(B_{3R/2}(x_0)\right)$. Let us choose, as a test function,

$$v = -\partial_l \widetilde{v}.$$

We get

$$-\int_\Omega A\nabla u \cdot \nabla \partial_l \widetilde{v} dx = -\int_\Omega f \partial_l \widetilde{v} dx. \tag{6.132}$$

Since $f \in H^1(\Omega)$, we have

$$-\int_\Omega f \partial_l \widetilde{v} dx = \int_\Omega \partial_l f \widetilde{v} dx. \tag{6.133}$$

Moreover, since $u \in H^2_{loc}(\Omega)$ (by Theorem 6.8), we have

$$-\int_\Omega A\nabla u \cdot \nabla \partial_l \widetilde{v} dx = -\int_\Omega \sum_{j,k=1}^n a^{jk} \partial_k u \partial_l \left(\partial_j \widetilde{v}\right) dx =$$
$$= \int_\Omega \sum_{j,k=1}^n \partial_l \left(a^{jk} \partial_k u\right) \partial_j \widetilde{v} dx =$$
$$= \int_\Omega \sum_{j,k=1}^n a^{jk} \partial_k \left(\partial_l u\right) \partial_j \widetilde{v} dx +$$
$$+ \int_\Omega \sum_{j,k=1}^n \partial_l a^{jk} \partial_k u \partial_j \widetilde{v} dx =$$
$$= \int_\Omega \sum_{j,k=1}^n a^{jk} \partial_k \left(\partial_l u\right) \partial_j \widetilde{v} dx -$$
$$- \int_\Omega \sum_{j,k=1}^n \partial_j \left[\left(\partial_l a^{jk}\right) \partial_k u\right] \widetilde{v} dx.$$

By the equality obtained above, by (6.132) and by (6.133) we have (recall that $C_0^\infty\left(B_{3R/2}(x_0)\right)$ is dense in $H_0^1\left(B_{3R/2}(x_0)\right)$

$$\int_{B_{3R/2}(x_0)} A\nabla\left(\partial_l u\right)\cdot\nabla w dx=\int_{B_R(x_0)}\widetilde{f} w dx,\quad \forall w\in H_0^1\left(B_{3R/2}(x_0)\right), \tag{6.134}$$

where

$$\widetilde{f}=\partial_l f+\sum_{j,k=1}^{n}\partial_j\left[\left(\partial_l a^{jk}\right)\partial_k u\right].$$

Now we have $\widetilde{f}\in L^2(B_{3R/2}(x_0))$. As a matter of fact

$$\begin{aligned}&\int_{B_{3R/2}(x_0)}\left|\widetilde{f}\right|^2 dx\leq\\ &\leq 2\int_{B_{3R/2}(x_0)}\left|\partial_l f\right|^2 dx+cE_1^2 d_0^{-4}\int_{B_{3R/2}(x_0)}\left|\partial^2 u\right|^2 dx+\\ &+cE_1^2 d_0^{-2}\int_{B_{3R/2}(x_0)}\left|\nabla u\right|^2 dx.\end{aligned} \tag{6.135}$$

By the latter, by Theorem 6.8 (more precisely, by (6.125)) and by (6.134), we obtain

$$\begin{aligned}R^4\sum_{|\alpha|=3}\int_{B_R(x_0)}\left|\partial^\alpha u\right|^2 dx\leq C\left(1+E_1^2\right)\sum_{|\alpha|=2}\int_{B_{3R/2}(x_0)}\left|\partial^\alpha u\right|^2 dx+\\ +CR^4\int_{B_{3R/2}(x_0)}\left|\widetilde{f}\right|^2 dx.\end{aligned} \tag{6.136}$$

At this point, we again apply (6.125) to estimate from above the derivatives of order less than or equal to 2 and we obtain (6.131). ■

Exercise 6.39 Generalize Theorem 6.9 to the equation

$$-\sum_{j,k=1}^{n}\partial_j\left(a^{jk}\partial_k u+d^j u\right)+\sum_{j,k=1}^{n}b^j\partial_j u+cu=f$$

where A and f satisfy the same assumption of Theorem 6.9, $d\in C^{m,1}\left(\overline{\Omega};\mathbb{R}^n\right)$, $b\in C^{m-1,1}\left(\overline{\Omega};\mathbb{R}^n\right)$, $c\in C^{m-1,1}\left(\overline{\Omega}\right)$, $m\geq 1$. ♠

Corollary 6.10 (**C^∞ Regularity in the Interior**) *Let Ω be an open set of $\mathbb{R}^n$. Let*

$$f\in C^\infty(\Omega).$$

Let $A \in C^{\infty}(\Omega; \mathbb{M}(n))$. Assume that A satisfies (6.1).

Let $u \in H^1(\Omega)$ be a solution to

$$-\operatorname{div}(A\nabla u) = f, \quad in\ \Omega.$$

Then we have

$$u \in C^{\infty}(\Omega).$$

Proof Let $B_R(x_0) \Subset \Omega$. For any $m \geq 0$, we have $f \in H^m(B_R(x_0))$ and $A \in C^{m,1}(\overline{B_R(x_0)}; \mathbb{M}(n))$. Therefore Theorem 6.9 implies

$$u \in \bigcap_{m=0}^{\infty} H^m(B_R(x_0)) = C^{\infty}(B_R(x_0)),$$

Where the last equality is due to Theorem 5.31. Since $B_R(x_0)$ is arbitrary, the thesis follows. ■

Exercise 6.40 Prove Corollary 6.10 for the equation

$$-\sum_{j,k=1}^{n} \partial_j \left(a^{jk}\partial_k u + d^j u\right) + \sum_{j=1}^{n} b^j \partial_j u + cu = f$$

where A satisfies the same assumption of Corollary 6.10 and $d, b \in C^{\infty}(\Omega; \mathbb{R}^n)$, $c \in C^{\infty}(\Omega)$. ♠

6.6.2 Regularity Theorems at the Boundary-global Regularity

In this section we will study the regularity at the boundary. The following Lemma is a crucial step in the proof of the forthcoming theorems

Lemma 6.11 (Local Regularity at the Boundary) *Let $R > 0$ and*

$$f \in L^2\left(B^+_{2R}\right). \tag{6.137}$$

Let A be a symmetric matrix. Assume that A satisfies (6.1), *$A \in C^{0,1}(\Omega; \mathbb{M}(n))$ and it satisfies*

$$|A(x) - A(\overline{x})| \leq E|x - \overline{x}|, \quad \forall x, \overline{x} \in B^+_{2R}, \tag{6.138}$$

where E is a positive number. Assume that $u \in H^1\left(B^+_{2R}\right)$ satisfies

$$-\operatorname{div}(A\nabla u) = f, \quad in\ B^+_{2R} \tag{6.139}$$

and

$$u(\cdot, 0) = 0, \qquad \textit{in the sense of the traces in } B'_{2R}. \tag{6.140}$$

Then we have

$$u \in H^2\left(B_R^+\right) \tag{6.141}$$

and the following estimate holds true

$$\sum_{|\alpha|\leq 2} R^{2|\alpha|} \int_{B_R^+} \left|\partial^\alpha u\right|^2 dx \leq C\left(1 + E^2 R^2\right) \int_{B_{2R}^+} u^2 dx + \\ + CR^4 \int_{B_{2R}^+} f^2 dx, \tag{6.142}$$

where C depends on λ only.

Proof Let $\eta \in C_0^\infty(B_{3R/2})$ satisfy

$$0 \leq \eta \leq 1; \qquad \eta = 1, \quad \text{in } B_R \tag{6.143}$$

and

$$|\nabla \eta| \leq \frac{K}{R}, \tag{6.144}$$

where K is a positive constant. Let $h \in \left(-\frac{R}{8}, \frac{R}{8}\right) \setminus \{0\}$. Let

$$k \in \{1, \cdots, n-1\}.$$

Let us denote by

$$v = -\delta_k^{-h}\left(\eta^2 \delta_k^h u\right). \tag{6.145}$$

By (6.140), taking into account that $\eta \in C_0^\infty\left(B_{3R/2}\right)$, we have

$$v \in H_0^1\left(B_{2R}^+\right).$$

Therefore, by $u \in H^1\left(B_{2R}^+\right)$, by (6.137), (6.139) and by (6.140), we get

$$\int_{B_{2R}^+} A\nabla u \cdot \nabla v dx = \int_{B_{2R}^+} f v dx. \tag{6.146}$$

At this point we may argue likewise in the proof of Theorem 6.8. Actually, by using Theorem 5.38 instead of 5.37 we have

$$\begin{aligned}\frac{\lambda^{-1}}{2}\int_{B_{2R}^+}\left|\delta_k^h\nabla u\right|^2\eta^2dx \leq &-\int_{B_{2R}^+} f\delta_k^{-h}\left(\eta^2\delta_k^h u\right)dx+\\ &+C\left(R^{-2}+E^2\right)\int_{B_{3R/2}^+}\left(\left|\delta_k^h u\right|^2+|\nabla u|^2\right)dx,\end{aligned} \tag{6.147}$$

for any $k\in\{1,\cdots,n-1\}$, where C depends on λ and n only. Let $\sigma>0$ be chosen, we get (see (6.124))

$$\begin{aligned}\left|\int_{B_{2R}^+} f\delta_k^{-h}\left(\eta^2\delta_k^h u\right)dx\right| \leq &\frac{1}{2\sigma}\int_{B_{2R}^+} f^2dx+\\ &+C\sigma\left(R^{-2}\int_{B_{3R/2}^+}\left|\delta_k^h u\right|^2dx+\right.\\ &\left.+\int_{B_{2R}^+}\eta^2\left|\nabla\delta_k^h u\right|^2dx\right).\end{aligned} \tag{6.148}$$

By applying Theorem 5.38 and inequality (6.94), we get

$$\begin{aligned}R^{-2}\int_{B_{3R/2}^+}\left|\delta_k^h u\right|^2dx &\leq R^{-2}\int_{B_{7R/4}^+}|\nabla u|^2dx\leq\\ &\leq CR^{-4}\int_{B_{2R}^+}u^2dx+C\int_{B_{2R}^+}f^2dx.\end{aligned}$$

By the latter and by (6.148) we have

$$\begin{aligned}\left|\int_\Omega f\delta_k^{-h}\left(\eta^2\delta_k^h u\right)dx\right| \leq &\frac{1}{2\sigma}\int_{B_{2R}} f^2dx+\\ &+C_\star\sigma\left(R^{-4}\int_{B_{2R}}u^2dx+\int_\Omega\eta^2\left|\nabla\delta_k^h u\right|^2dx\right),\end{aligned}$$

where $C_\star$ depends on λ and and n only. By using in (6.147) the just obtained inequality, we obtain

$$\begin{aligned}\left(\frac{\lambda^{-1}}{2}-C_\star\sigma\right)\int_{B_{2R}^+}\left|\delta_k^h\nabla u\right|^2\eta^2dx \leq &\frac{1}{2\sigma}\int_{B_{2R}^+} f^2dx+\\ &+\frac{C\left(1+\sigma+E^2R^2\right)}{R^4}\int_{B_{2R}^+}u^2dx.\end{aligned}$$

Now let us choose

$$\sigma = \frac{\lambda^{-1}}{4C_*}$$

and we get, for $k = 1, \cdots, n-1$,

$$\int_{B_R^+} \left|\delta_k^h \nabla u\right|^2 \eta^2 dx \leq C \int_{B_{2R}^+} f^2 dx + \frac{C\left(1 + E^2 R^2\right)}{R^4} \int_{B_{2R}^+} u^2 dx.$$

By this inequality and by Theorem 5.38-(ii) we get

$$\partial_k u \in H^1\left(B_R^+\right), \quad \text{for } k = 1, \cdots, n-1$$

and applying again (6.94) we have

$$\sum_{\substack{k,j=1 \\ k+j<2n}}^{n} \int_{B_R^+} \left|\partial_{jk}^2 u\right|^2 dx \leq \frac{C\left(1 + E^2 R^2\right)}{R^4} \int_{B_{2R}^+} u^2 dx + C R^4 \int_{B_{2R}^+} f^2 dx, \tag{6.149}$$

where C depends on λ n only.

Now, Theorem 6.8 implies that $u \in H^2_{loc}\left(B_R^+\right)$, this allows us to write the Eq. (6.138) in the form

$$\sum_{k,j=1}^{n} a^{jk} \partial_{jk}^2 u + \sum_{k,j=1}^{n} \partial_j a^{jk} \partial_k u = -f, \quad \text{a.e. in } B_R^+.$$

By the last equality we can find (taking into account that $a^{nn} \geq \lambda^{-1} > 0$)

$$\partial_n^2 u = -\frac{1}{a^{nn}} \left(\sum_{\substack{k,j=1 \\ k+j<2n}}^{n} a^{jk} \partial_{jk}^2 u + \sum_{k,j=1}^{n} \partial_j a^{jk} \partial_k u + f \right). \tag{6.150}$$

By (6.150), (6.149), (6.138) we have $\partial_n^2 u \in H^2\left(B_R^+\right)$. Finally, by (6.94) we obtain (6.142). ∎

Theorem 6.12 (Global Regularity) *Let Ω be a bounded open set of $\mathbb{R}^n$ whose boundary is of class $C^{1,1}$ with constants M_0, r_0. Let*

$$f \in L^2(\Omega) \tag{6.151}$$

and let A be a symmetric matrix. Assume that A satisfies (6.1), $A \in C^{0,1}(\Omega; \mathbb{M}(n))$ *and it satisfies*

$$|A(x) - A(\overline{x})| \leq E|x - \overline{x}|, \quad \forall x, \overline{x} \in \Omega, \tag{6.152}$$

where E is a positive number. Suppose that $u \in H_0^1(\Omega)$ is the solution to

$$- \operatorname{div}(A\nabla u) = f, \quad \text{in } \Omega. \tag{6.153}$$

Then we have

$$u \in H^2(\Omega), \tag{6.154}$$

and the following estimate holds true

$$\sum_{|\alpha|\leq 2} r_0^{2|\alpha|} \int_\Omega |\partial^\alpha u|^2 \, dx \leq C d_0^4 \int_\Omega f^2 dx, \tag{6.155}$$

where d_0 is the diameter of Ω and C depends by λ, E, M_0 and $\frac{d_0}{r_0}$ only.

Proof Let $P \in \partial\Omega$. There exists a rigid transformation of coordinates under which we have $P = 0$ and

$$\Omega \cap Q_{r_0,2M_0} = \left\{x \in Q_{r_0,2M_0} \, : \, x_n > g(x')\right\},$$

where $g \in C^{1,1}\left(\overline{B'_{r_0}}\right)$,

$$g(0) = 0, \quad |\nabla_{x'} g(0)| = 0$$

and

$$\|g\|_{C^{1,1}\left(\overline{B'_{r_0}}\right)} \leq M_0 r_0.$$

Let us consider the change of coordinates

$$\Phi : Q_{r_0,2M_0} \to \mathbb{R}^n, \quad \Phi(x) = \left(x', x_n - g(x')\right).$$

Let us note that Φ "flattens the boundary" i.e.

$$Q_{r_0,2M_0} \cap \Phi\left(\partial\Omega\right) = \left\{(y', 0) \, : \, y' \in B'_{r_0}\right\}.$$

$\Phi \in C^{1,1}\left(Q_{r_0,2M_0} \cap \Omega\right)$ is injective and it is a local diffeomorphism. Let

$$W = Q_{r_0,2M_0} \cap \Phi\left(\Omega\right)$$

and

$$\Psi : W \to Q_{r_0,2M_0} \cap \Omega, \quad \Psi = \Phi^{-1}.$$

Let J be the jacobian matrix of Ψ. Then

$$\det J(y) = 1, \quad \forall y \in W.$$

Let $u \in H_0^1(\Omega)$ be the weak solutions to (6.153). Let $\widetilde{v}$ be any function belonging to $H_0^1(W)$ and set

$$v(x) = \widetilde{v}(\Phi(x)).$$

We have

$$\int_{Q_{r_0,2M_0}\cap\Omega} A\nabla u \cdot \nabla v dx = \int_{Q_{r_0,2M_0}\cap\Omega} f v dx. \tag{6.156}$$

By the change of variables $x = \Psi(y)$ Eq. (6.156) becomes

$$\int_W \widetilde{A}(y)\nabla w(y) \cdot \nabla\widetilde{v}(y)dy = \int_W \widetilde{f}(y)\widetilde{v}(y)dy, \tag{6.157}$$

where

$$w(y) = u(\Psi(y)), \quad \forall y \in W,$$

$$\widetilde{A}(y) = (J(y))^{-1}A(\Psi(y))\left((J(y))^{-1}\right)^T, \quad \forall y \in W$$

and

$$\widetilde{f}(y) = f(\Psi(y)), \quad \forall y \in W.$$

It is easy to check that

$$\widetilde{\lambda}^{-1}|\xi|^2 \leq \widetilde{A}(y)\xi \cdot \xi \leq \widetilde{\lambda}|\xi|^2, \quad \forall y \in W, \ \forall \xi \in \mathbb{R}^n,$$

where $\widetilde{\lambda} \geq 1$ depends on λ and M_0 only. Moreover

$$w\left(y', 0\right) = 0, \quad y' \in B'_{r_0} \text{ in the sense of the traces.}$$

Also we have $\widetilde{A} \in C^{0,1}(W)$ and

$$\left|\widetilde{A}(y) - \widetilde{A}(\overline{y})\right| \leq \widetilde{E}\,|y - \overline{y}|, \quad \forall y, \overline{y} \in W,$$

where

$$\widetilde{E} = C\left(E + M_0 r_0^{-1}\right),$$

and C depends on λ and M_0 only.

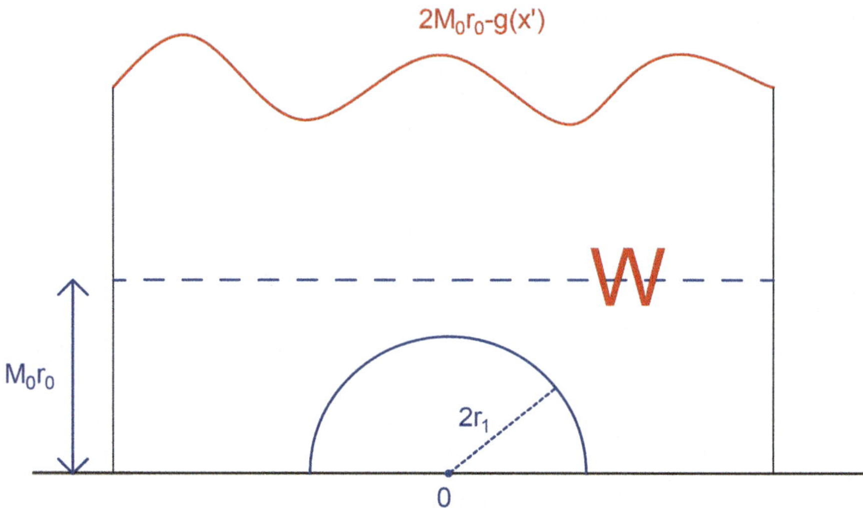

Fig. 6.1 Here, $W = Q_{r_0,2M_0} \cap \Phi(\Omega)$, the black horizontal line represents a flattened portion of the boundary; note that $\partial B^+_{2r_1}$ shares with ∂W only a portion of the black horizontal line

At this stage (Fig. 6.1) we introduce the quantity

$$r_1 = \frac{r_0}{2} \min\{1, M_0\},$$

in such a way that we have

$$B^+_{2r_1} \subset W.$$

Applying Lemma 6.11 we get

$$\begin{aligned} \sum_{|\alpha|\leq 2} r_1^{2|\alpha|} \int_{B^+_{r_1}} |\partial^\alpha w|^2 \, dy \leq C\left(1+\widetilde{E}^2 r_1^2\right) \int_{B^+_{2r_1}} w^2 dy + \\ + C r_1^4 \int_{B^+_{2r_1}} \widetilde{f}^2 dy, \end{aligned} \tag{6.158}$$

where C depends on $\widetilde{\lambda}$ only. Coming back to the original variables, after some calculation and simple estimates, we have

$$\sum_{|\alpha|\leq 2} r_0^{2|\alpha|} \int_{\Psi\left(B^+_{r_1}\right)} |\partial^\alpha u|^2 \, dx \leq C\left(\int_\Omega u^2 dx + r_0^4 \int_\Omega f^2 dx\right), \tag{6.159}$$

where C depends on λ, M_0 and Er_0 only. On the other hand, as is easily checked, there is $\overline{C} \geq 1$, C depending on λ and M_0 only, such that, if $r_2 = \frac{r_0}{\overline{C}}$, we have

$$\Omega \cap B_{r_2}(P) \subset \Psi\left(B^+_{r_1}\right).$$

Therefore by (6.159) we get trivially

$$\sum_{|\alpha|\leq 2} r_0^{2|\alpha|} \int_{\Omega\cap B_{r_2}(P)} \left|\partial^\alpha u\right|^2 dx \leq C\left(\int_\Omega u^2 dx + r_0^4 \int_\Omega f^2 dx\right). \tag{6.160}$$

Now, by the compactness of $\partial\Omega$, we can extract a finite covering by $\left\{B_{r_2}(P)\right\}_{P\in\partial\Omega}$. Let $\left\{B_{r_2}(P_j)\right\}_{1\leq j\leq N}$ be such a finite covering, where $P_j \in \partial\Omega$, and let

$$\Lambda = \Omega \cap \bigcup_{j=1}^N B_{r_2}(P_j) \quad \text{and} \quad \Omega' = \Omega \setminus \Lambda.$$

We can make $\text{dist}(\Omega', \partial\Omega) \geq r_0/C$, where $C \geq 1$ depends on M_0 only, furthermore N depends on M_0 and $\frac{d_0}{r_0}$ only. Inequality (6.160) implies that there is a constant C depending on λ, E, M_0 and $\frac{d_0}{r_0}$ so that

$$\sum_{|\alpha|\leq 2} r_0^{2|\alpha|} \int_\Lambda \left|\partial^\alpha u\right|^2 dx \leq C\left(\int_\Omega u^2 dx + r_0^4 \int_\Omega f^2 dx\right).$$

By the just obtained inequality and by (6.126) we have

$$\sum_{|\alpha|\leq 2} r_0^{2|\alpha|} \int_\Omega \left|\partial^\alpha u\right|^2 dx \leq C\left(\int_\Omega u^2 dx + d_0^4 \int_\Omega f^2 dx\right). \tag{6.161}$$

By the first Poincaré inequality (Proposition (5.14)) and by inequality (6.25), we find

$$\int_\Omega u^2 dx \leq C d_0^4 \int_\Omega f^2 dx,$$

where C depends on λ only. By the last obtained inequality and by (6.161) we get (6.154). ■

Exercise 6.41

(a) Generalize Theorem 6.12 to the boundary value problem

$$\begin{cases} -\sum_{j,k=1}^{n} \partial_j \left(a^{jk}\partial_k u + d^j u\right) + \sum_{j=1}^{n} b^j \partial_j u + cu = f, & \text{in } \Omega \\ u = 0, \quad \text{on } \partial\Omega \end{cases}.$$

where $A = \left\{a^{jk}\right\}_{j,k=1}^{n}$ and f satisfy the same assumptions of Theorem 6.8 and $d \in C^{0,1}\left(\overline{\Omega}; \mathbb{R}^n\right)$, $b \in L^{\infty}(\Omega; \mathbb{R}^n)$, $c \in L^{\infty}(\Omega)$.

(b) Generalize the result obtained in (a) to the case where A is not a symmetric matrix. ♠

Theorem 6.13 (Improved Global Regularity) *Let Ω be a bounded open set of $\mathbb{R}^n$ of class $C^{m+1,1}$ with constants M_0, r_0, let d_0 be the diameter of Ω. Let*

$$f \in H^m(\Omega) \tag{6.162}$$

and let A be a symmetric matrix. Assume that A satisfies (6.1), $A \in C^{m,1}(\overline{\Omega}; \mathbb{M}(n))$ *and it satisfies*

$$\|A\|_{C^{m,1}(\overline{\Omega};\mathbb{M}(n))} \leq E_m,$$

where E_m is a positive number.

Suppose that $u \in H_0^1(\Omega)$ is the solution to

$$-\operatorname{div}(A\nabla u) = f, \quad \text{in } \Omega. \tag{6.163}$$

Then we have

$$u \in H^{m+2}(\Omega), \tag{6.164}$$

and the following estimate holds true

$$\sum_{|\alpha|\leq m+2} r_0^{2|\alpha|} \int_{\Omega} \left|\partial^{\alpha} u\right|^2 dx \leq C \sum_{|\alpha|\leq m} d_0^{2(|\alpha|+4)} \int_{\Omega} \left|\partial^{\alpha} f\right|^2 dx, \tag{6.165}$$

where C depends on λ, E_m, M_0 and $\frac{d_0}{r_0}$ only.

Proof The proof is mostly similar to that of Theorem 6.9, so we focus here by considering, in the case $m = 1$, the steps in which the new proof differs from the proof of Theorem 6.9, leaving the details to the care of the reader. First of all, in analogy to Lemma 6.11, let us consider the following special situation. Let $R > 0$

and

$$f \in H^1\left(B_{2R}^+\right).$$

Let A be a symmetric matrix whose entries are measurable functions in B_{2R}^+. Let us suppose that A satisfies (6.1) and that $A \in C^{1,1}\left(B_{2R}, \mathbb{M}(n)\right)$. Let us suppose that $u \in H^1\left(B_{2R}^+\right)$ is a solution to

$$-\operatorname{div}(A\nabla u) = f, \quad \text{in } B_{2R}^+,$$

and

$$u(\cdot, 0) = 0, \quad \text{in } B'_{2R}, \text{ (in the sense of the traces)}. \tag{6.166}$$

Let us prove that

$$u \in H^3\left(B_r^+\right), \quad \forall r < R. \tag{6.167}$$

To this purpose, let us prove

Claim Let $l = 1, \cdots, n-1$. We have, for every $r \in (0, 2R)$,

$$\partial_l u(\cdot, 0) = 0, \quad \text{in } B'_r, \text{ (in the sense of the traces)}. \tag{6.168}$$

Proof of the Claim First let us note that, since $u \in H^2\left(B_r^+\right)$, for every $r < R$, $\partial_l u(x', 0)$ is well-defined in the sense of traces. Now, let

$$v = \partial_l u$$

and let us denote by $T(v)$ the trace of v on $\{x_n = 0\}$. As a consequence of Theorem 5.18, $T(v)$ is characterised by the identity.

$$-\int_{B'_R} \Phi_n(x', 0) T(v) dx' = \int_{B_R^+} v \operatorname{div} \Phi dx + \int_{B_R^+} \nabla v \cdot \Phi dx, \tag{6.169}$$

for every $\Phi \in C_0^\infty\left(B_R; \mathbb{R}^n\right)$.

Let now Φ be any function belonging to $C_0^\infty\left(B_R; \mathbb{R}^n\right)$. By (6.167) and by Theorem 5.18 (applied to u), we get

$$\begin{aligned}\int_{B_R^+} v(\text{div}\,\Phi)dx &= \int_{B_R^+} \partial_l u(\text{div}\,\Phi)dx = -\int_{B_R^+} u(\partial_l \,\text{div}\,\Phi)dx = \\ &= -\int_{B_R^+} u(\text{div}\,\partial_l \Phi)dx = \\ &= \int_{B_R^+} \nabla u \cdot \partial_l \Phi dx + \int_{B_R'} \partial_l \Phi_n(x',0)T(u)dx' = \\ &= \int_{B_R^+} \nabla u \cdot \partial_l \Phi dx = -\int_{B_R^+} \partial_l \nabla u \cdot \Phi dx = \\ &= -\int_{B_R^+} \nabla v \cdot \Phi.\end{aligned}$$

Hence, (6.169) implies

$$T(v) = 0, \quad \text{in } B_r', \quad \forall r \in (0, 2R).$$

This concludes the proof of the Claim.

We now briefly and only formally show the most significant steps to complete the proof; we encourage the reader to treat the steps in a rigorous manner using appropriately the variational formulation in a similar way as we did in the proof of Theorem (6.9).

By calculating the derivative w.r.t. x_l, $l \in \{1, \cdots, n-1\}$, of both the sides of the equation

$$-\sum_{j,k=1}^{n} \partial_j \left(a^{jk}\partial_k u\right) = f, \tag{6.170}$$

we obtain

$$-\sum_{j,k=1}^{n} \partial_j \left(a^{jk}\partial_k v\right) = \widetilde{f}, \tag{6.171}$$

where $v = \partial_l u$ and

$$\widetilde{f} = \partial_l f + \sum_{j,k=1}^{n} \partial_j \left((\partial_l a^{jk})\partial_k u\right).$$

Now, as $f \in H^1\left(B_{2R}^+\right)$, $a^{jk} \in C^{1,1}\left(B_{2R}^+\right)$, $u \in H^2\left(B_r^+\right)$ for every $r \in (0, 2R)$, we get

$$\widetilde{f} \in L^2\left(B_r^+\right), \quad \forall r \in (0, 2R).$$

on the other hand, for every $r \in (0, 2R)$, we have $v \in H^1\left(B_r^+\right)$ and

$$v(\cdot, 0) = 0, \quad \text{in } x' \in B_r', \text{ (in the sense of the traces).}$$

Therefore by Lemma 6.11 we have

$$\partial_l u = v \in H^2\left(B_r^+\right), \quad \forall r \in (0, 2R), \tag{6.172}$$

for $l \in \{1, \cdots, n-1\}$. Moreover by (6.171) we get (likewise to (6.150))

$$\partial_n^3 u = \partial_n \left[-\frac{1}{a^{nn}} \left(\sum_{\substack{k,j=1 \\ k+j<2n}}^{n} a^{jk} \partial_{jk}^2 u + \sum_{k,j}^{n} \partial_j a^{jk} \partial_k u + f \right) \right], \tag{6.173}$$

from which, taking into account (6.172), we get

$$u \in H^3\left(B_r^+\right), \quad \forall r \in (0, 2R).$$

Finally, inequality (6.165) (for $m = 1$) is obtained applying inequality (6.142) to (6.171). In addition, inequality (6.142) gives the estimates

$$\begin{aligned} R^6 \sum_{\substack{k,j,l=1 \\ k+j+l<3n}}^{n} \left\| \partial_{jkl}^3 u \right\|_{L^2(B_R^+)}^2 &\leq C\left(1 + E_1^2\right) \int_{B_{2R}} u^2 dx + \\ &+ C \sum_{|\alpha| \leq 1} R^{2|\alpha|+4} \int_{B_{2R}^+} \left|\partial^\alpha u\right|^2 dx. \end{aligned}$$

By the last inequality, by means of (6.173) (applying again inequality (6.142) to Eq. (6.170)), we obtain (6.165) (for $m = 1$).

In order to complete the proof, simply follow the proof of Theorem 6.12 taking into account that diffeomorphisms Φ and Ψ are, in this case, of class $C^{2,1}$ ($C^{m+1,1}$ in the general case). ■

Corollary 6.14 (**C^∞ Global Regularity**) *Let Ω be a bounded open set of $\mathbb{R}^n$ with boundary of class C^∞. Let*

$$f \in C^\infty(\overline{\Omega}).$$

Let $A \in C^\infty(\overline{\Omega}; \mathbb{M}(n))$. Assume that A satisfies (6.1).

Assume that $u \in H_0^1(\Omega)$ is a solution to

$$-\operatorname{div}(A\nabla u) = f, \quad \text{in } \Omega$$

then we have

$$u \in C^{\infty}(\overline{\Omega}).$$

Proof By Theorem 6.13 and by the Embedding Theorem 5.31 we have

$$u \in \bigcap_{m=0}^{\infty} H^m(\Omega) = C^{\infty}(\overline{\Omega}).$$

■

6.7 The Dirichlet to Neumann Map

Denote by $\mathbb{M}^S(n)$ the vector space of the symmetric matrices $n\times$ with real entries. Let Ω be a bounded open set of $\mathbb{R}^n$ of class $C^{0,1}$. Let $A \in L^{\infty}(\Omega; \mathbb{M}^S(n))$. Let us suppose that (6.1) holds. Formally, the Dirichlet to Neumann Map can be constructed in the following way. Let $\varphi \in H^{1/2}(\partial\Omega)$. Let $u \in H^1(\Omega)$ be the solution to the problem

$$\begin{cases} \operatorname{div}(A\nabla u) = 0, & \text{in } \Omega \\ \\ u = \varphi, & \text{on } \partial\Omega \end{cases}. \tag{6.174}$$

We have seen that u is uniquely determined by φ hence, *if it would make sense*, we could define the map

$$\varphi \to A\nabla u \cdot \nu, \qquad \text{(conormal derivative of } u \text{ on } \partial\Omega). \tag{6.175}$$

Let us observe that if $u \in C^2(\overline{\Omega})$ and $A \in C^1\left(\overline{\Omega}; \mathbb{M}^S(n)\right)$, by (6.174) we have

$$\int_{\partial\Omega} (A\nabla u \cdot \nu) v dS = \int_{\Omega} A\nabla u \cdot \nabla v dx, \quad \forall v \in C^1(\overline{\Omega}). \tag{6.176}$$

As a matter of fact, if $v \in C^1(\overline{\Omega})$, by the divergence Theorem and by (6.174) we get

$$\int_{\Omega} A\nabla u \cdot \nabla v dx = \int_{\Omega} (\operatorname{div}(vA\nabla u) - \operatorname{div}(A\nabla u)v)\, dx = \int_{\partial\Omega} (A\nabla u \cdot \nu) v dS.$$

Equality (6.176) allows us to "read" $A\nabla u \cdot \nu$ by means the integral on the right-hand side. Now, if $A \in L^{\infty}(\Omega; \mathbb{M}^S(n))$ and $u \in H^1(\Omega)$ then the integral on the right-hand side of (6.176) makes perfectly sense. Based on these insights we will define below $A\nabla u \cdot \nu$ as an element of $H^{-1/2}(\partial\Omega)$ (the dual space of $H^{1/2}(\partial\Omega)$).

First, notice that

$$\int_\Omega A\nabla u \cdot \nabla v dx = 0, \qquad \forall v \in H_0^1(\Omega). \tag{6.177}$$

In fact, recalling (6.30) and (6.31), we have, for each $v \in H_0^1(\Omega)$,

$$\int_\Omega A\nabla u \cdot \nabla v dx = 0.$$

By (6.177), recalling Theorem 5.18, we have that if $\varphi \in H^1(\Omega)$ then the integral

$$\int_\Omega A\nabla u \cdot \nabla v dx, \tag{6.178}$$

depends only by the trace of v on $\partial\Omega$. Indeed, if $v_1, v_2 \in H^1(\Omega)$ have the same trace on $\partial\Omega$ then $v_1 - v_2 \in H_0^1(\Omega)$. Hence (6.177) implies

$$\int_\Omega A\nabla u \cdot \nabla v_1 dx = \int_\Omega A\nabla u \cdot \nabla v_2 dx.$$

Therefore, for every $\varphi \in H^{1/2}(\partial\Omega)$ it turns out to be well-defined the functional

$$L_\varphi : H^{1/2}(\partial\Omega) \to \mathbb{R},$$

which maps $\phi \in H^{1/2}(\partial\Omega)$ in the real number

$$L_\varphi(\phi) = \int_\Omega A\nabla u \cdot \nabla v dx,$$

where $v_{|\partial\Omega} = \phi$ (in the sense of traces).

We prove that L_φ Is a Linear and Bounded Functional
The linearity of L_φ is trivial. Concerning the boundedness, let

$$\phi \in H^{1/2}(\partial\Omega).$$

By Theorem 5.47 we know that there exists $v \in H^1(\Omega)$ so that

$$v_{|\partial\Omega} = \phi \quad \text{(in the sense of traces)},$$

which satisfies

$$\|v\|_{H^1(\Omega)} \leq C \|\phi\|_{H^{1/2}(\partial\Omega)}, \tag{6.179}$$

where C is a constant depending on Ω only. Proceeding in a similar way to what we did to obtain (6.32) and taking into account (6.34), we get

$$\begin{aligned} |L_\varphi(\phi)| = \left|\int_\Omega A\nabla u\cdot\nabla v dx\right| \leq \lambda \|\nabla u\|_{L^2(\Omega)} \|\nabla v\|_{L^2(\Omega)} \leq \\ \leq \overline{C} \|\varphi\|_{H^{1/2}(\partial\Omega)} \|\phi\|_{H^{1/2}(\partial\Omega)}, \end{aligned} \tag{6.180}$$

where $\overline{C}$ is a constant depending on Ω and λ only. Therefore, the functional L_φ is bounded and it satisfies

$$\|L_\varphi\|_{H^{-1/2}(\partial\Omega)} \leq \overline{C} \|\varphi\|_{H^{1/2}(\partial\Omega)}, \quad \forall\varphi\in H^{1/2}(\partial\Omega). \tag{6.181}$$

Moreover, notice that (6.181) implies that the linear operator

$$H^{1/2}(\partial\Omega) \ni \varphi \to L_\varphi \in H^{-1/2}(\partial\Omega),$$

is bounded.

Now we set

$$A\nabla u\cdot\nu := L_\varphi,$$

consequently, we write, for any $v\in H^1(\Omega)$ such that $v_{|\partial\Omega} = \phi$,

$$\langle A\nabla u\cdot\nu, \phi\rangle_{H^{-1/2},H^{1/2}} = L_\varphi(\phi) = \int_\Omega A\nabla u\cdot\nabla v dx, \tag{6.182}$$

where, by $\langle\cdot,\cdot\rangle_{H^{-1/2},H^{1/2}}$ we denote the bracket of duality between $H^{-1/2}(\partial\Omega)$ and $H^{1/2}(\partial\Omega)$. With the notations used so far we should have written $(A\nabla u\cdot\nu)(\phi)$ instead of $\langle A\nabla u\cdot\nu, \phi\rangle_{H^{-1/2},H^{1/2}}$, but the latter notation it is certainly more handleable in the present context.

Finally, we define the **Dirichlet to Neumann Map** as

$$\Lambda_A : H^{1/2}(\partial\Omega) \to H^{-1/2}(\partial\Omega), \quad \Lambda_A(\varphi) = A\nabla u\cdot\nu. \tag{6.183}$$

By (6.181) it follows that $\Lambda_A \in \mathcal{L}\left(H^{1/2}(\partial\Omega), H^{-1/2}(\partial\Omega)\right)$, where, we recall, $\mathcal{L}\left(H^{1/2}(\partial\Omega), H^{-1/2}(\partial\Omega)\right)$ denotes the space of the linear and bounded operators from $H^{1/2}(\partial\Omega)$ to $H^{-1/2}(\partial\Omega)$. From the construction that we have performed so far we get

$$\langle\Lambda_A(\varphi), \phi\rangle_{H^{-1/2},H^{1/2}} = \int_\Omega A\nabla u\cdot\nabla v dx, \quad \forall\varphi,\phi\in H^{1/2}(\partial\Omega), \tag{6.184}$$

where $u \in H^1(\Omega)$ is the solution to (6.174) and v is any function of $H^1(\Omega)$ which satisfies $v_{|\partial\Omega} = \phi$. It is simple to check that

$$\|\Lambda_A\|_{\mathcal{L}(H^{1/2},H^{-1/2})} =$$
$$= \sup\left\{\langle\Lambda_A(\varphi),\phi\rangle_{H^{-1/2},H^{1/2}} : \|\varphi\|_{H^{1/2}(\partial\Omega)} \leq 1, \|\phi\|_{H^{1/2}(\partial\Omega)} \leq 1\right\}.$$

In what follows we prove other simple but important properties of Λ_A.

We first observe that since the right-hand integral in (6.184) is independent of the choice of v (as long as it is a trace of ϕ) we can choose $v = w$ where $w \in H^1(\Omega)$ is the solution of the Dirichlet problem

$$\begin{cases} \operatorname{div}(A\nabla w) = 0, & \text{in } \Omega \\ \\ w = \phi, & \text{on } \partial\Omega \end{cases}. \tag{6.185}$$

From this it follows that the bilinear form

$$H^{1/2}(\partial\Omega) \times H^{1/2}(\partial\Omega) \ni (\varphi,\phi) \to \langle\Lambda_A(\varphi),\phi\rangle_{H^{-1/2},H^{1/2}} \in \mathbb{R}$$

is **symmetric** that is

$$\langle\Lambda_A(\varphi),\phi\rangle_{H^{-1/2},H^{1/2}} = \langle\Lambda_A(\phi),\varphi\rangle_{H^{-1/2},H^{1/2}}, \quad \forall\varphi,\phi \in H^{1/2}(\partial\Omega). \tag{6.186}$$

Indeed we have

$$\begin{aligned} \langle\Lambda_A(\varphi),\phi\rangle_{H^{-1/2},H^{1/2}} &= \int_\Omega A\nabla u \cdot \nabla w dx = \\ &= \int_\Omega A\nabla w \cdot \nabla u dx = \\ &= \langle\Lambda_A(\phi),\varphi\rangle_{H^{-1/2},H^{1/2}}. \end{aligned}$$

An important consequence of (6.186) is the following identity

Theorem 6.15 (Alessandrini Identity) *Let $A_1, A_2 \in L^\infty(\Omega; \mathbb{M}^S(n))$. Assume that A_1, A_2 satisfy (6.1). Let $\varphi, \phi \in H^{1/2}(\partial\Omega)$. Let $u_1, u_2 \in H^1(\Omega)$ be the solutions to*

$$\begin{cases} \operatorname{div}(A\nabla u_1) = 0, & \text{in } \Omega \\ \\ u_1 = \varphi, & \text{on } \partial\Omega \end{cases} \tag{6.187}$$

and

$$\begin{cases} \operatorname{div}(A\nabla u_2) = 0, & \textit{in } \Omega \\ u_2 = \phi, & \textit{on } \partial\Omega \end{cases}, \tag{6.188}$$

then

$$\langle (\Lambda_{A_1} - \Lambda_{A_2})(\varphi), \phi \rangle_{H^{-1/2}, H^{1/2}} = \int_\Omega (A_1 - A_2)\nabla u_1 \cdot \nabla u_2 dx. \tag{6.189}$$

Proof By (6.186) we have (by setting for sake of brevity, $\langle \cdot, \cdot \rangle = \langle \cdot, \cdot \rangle_{H^{-1/2}, H^{1/2}}$)

$$\begin{aligned} \langle (\Lambda_{A_1} - \Lambda_{A_2})(\varphi), \phi \rangle &= \langle \Lambda_{A_1}(\varphi), \phi \rangle - \langle \Lambda_{A_2}(\varphi), \phi \rangle = \\ &= \langle \Lambda_{A_1}(\varphi), \phi \rangle - \langle \Lambda_{A_2}(\phi), \varphi \rangle. \end{aligned} \tag{6.190}$$

Now, by (6.187) and (6.188) we have, respectively,

$$\langle \Lambda_{A_1}(\varphi), \phi \rangle = \int_\Omega A_1 \nabla u_1 \cdot \nabla u_2 dx$$

and

$$\langle \Lambda_{A_2}(\phi), \varphi \rangle = \int_\Omega A_2 \nabla u_2 \cdot \nabla u_1 dx = \int_\Omega A_2 \nabla u_1 \cdot \nabla u_2.$$

By inserting the latter in (6.190) we get

$$\langle (\Lambda_{A_1} - \Lambda_{A_2})(\varphi), \phi \rangle = \int_\Omega (A_1 - A_2)\nabla u_1 \cdot \nabla u_2 dx.$$

■

A simple consequence of the Alessandrini identity is the continuity of the map $A \to \Lambda_A$. Precisely we have the following

Proposition 6.16 *There exists a constant C depending on λ and Ω only so that, if $A_1, A_2 \in L^\infty(\Omega; \mathbb{M}^S(n))$ satisfy* (6.1), *then*

$$\left\| \Lambda_{A_1} - \Lambda_{A_2} \right\|_{\mathcal{L}(H^{1/2}, H^{-1/2})} \leq C \left\| A_1 - A_2 \right\|_{L^\infty(\Omega, \mathbb{M}^S(n))}. \tag{6.191}$$

Proof By the Alessandrini identity and by the Cauchy–Schwarz inequality we have

$$\begin{aligned}\left|\langle(\Lambda_{A_1}-\Lambda_{A_2})(\varphi),\phi\rangle_{H^{-1/2},H^{1/2}}\right| &= \left|\int_\Omega (A_1-A_2)\nabla u_1\cdot\nabla u_2 dx\right| \leq \\ &\leq \|A_1-A_2\|_{L^\infty(\Omega;\mathbb{M}^S(n)))}\|\nabla u_1\|_{L^2(\Omega)}\|\nabla u_2\|_{L^2(\Omega)}\,.\end{aligned} \tag{6.192}$$

Now, (6.34) gives

$$\|\nabla u_1\|_{L^2(\Omega)} \leq C\,\|\varphi\|_{H^{1/2}(\partial\Omega)} \text{ and } \|\nabla u_2\|_{L^2(\Omega)} \leq C\,\|\phi\|_{H^{1/2}(\partial\Omega)}\,,$$

where C depends on λ and Ω only. Inserting the last obtained inequalities in (6.192) we have

$$\begin{aligned}&\left|\langle(\Lambda_{A_1}-\Lambda_{A_2})(\varphi),\phi\rangle_{H^{-1/2},H^{1/2}}\right| \leq \\ &\leq C\,\|A_1-A_2\|_{L^\infty(\Omega;\mathbb{M}^S(n))}\|\varphi\|_{H^{1/2}(\partial\Omega)}\|\phi\|_{H^{1/2}(\partial\Omega)}\,.\end{aligned}$$

From which (6.191) follows. ■

Similarly, one can also define the **Neumann to Dirichlet Map**.

Let $g\in H^{-1/2}(\partial\Omega)$ satisfy

$$\langle g,1\rangle=0.$$

Let us consider the solution $u\in H^1(\Omega)$ to the Neumann problem

$$\begin{cases}\int_\Omega A\nabla u\cdot\nabla v dx=\langle g,\varphi\rangle_{H^{-1/2},H^{1/2}}, & \forall v\in H^1(\Omega)\\ \\ u\in\left\{w\in H^1(\Omega): \quad \int_\Omega w dx=0\right\}\end{cases}, \tag{6.193}$$

we define the Neumann to Dirichlet Map as follows

$$\begin{aligned}&\mathcal{N}_A: H^{-1/2}(\partial\Omega)\to H^{1/2}(\partial\Omega),\\ &\mathcal{N}_A(g)=u_{|\partial\Omega}, \quad \text{(in the sense of the traces).}\end{aligned} \tag{6.194}$$

6.8 The Inclusion Inverse Problem

Let $n=2$ or $n=3$. Let us assume that Ω represents an electrically conductor of conductivity γ, which we assume to be a real-valued function that belongs to $C^{0,1}(\Omega)$ and satisfies the ellipticity condition

$$\lambda^{-1}\leq\gamma(x)\leq\lambda, \quad \forall x\in\Omega, \tag{6.195}$$

where $\lambda\geq 1$ is a given number. We suppose that Ω contains an *inclusion* D of different conductivity, say $k\gamma(x)$, where k is a positive real number such that $k\neq 1$.

We consider the problem of determining D from the knowledge of a density of a prescribed current on $\partial\Omega$ and of the corresponding voltage u measured on $\partial\Omega$.

We provide a mathematical formulation of the problem. Let us assume that Ω is a connected bounded open set of $\mathbb{R}^n$ $(n = 2, 3)$ whose boundary is of class $C^{1,1}$. Let $\phi \in H^{-1/2}(\partial\Omega)$ satisfy

$$\int_{\partial\Omega} \phi dS = 0, \tag{6.196}$$

where we have denoted by

$$\int_{\partial\Omega} \phi dS = \langle \phi, 1 \rangle_{H^{-1/2}, H^{1/2}},$$

ϕ represents the density of a prescribed current on $\partial\Omega$. If the inclusion D is present, the electrostatic potential u is determined, up to an additive constant, as a solution of Neumann problem

$$\begin{cases} \operatorname{div}\left(\gamma(x)\left(1 + (k-1)\chi_D\right)\nabla u\right) = 0, & \text{in } \Omega \\ \\ \gamma \frac{\partial u}{\partial \nu} = \phi, & \text{on } \partial\Omega \end{cases}, \tag{6.197}$$

where D is a measurable subset of Ω. In what follows we assume

$$\int_{\Omega} u(x)dx = 0, \tag{6.198}$$

which yields with the uniqueness of the boundary value problem (6.197).

The **inverse problem** entails determining D by providing a nontrivial input ϕ and measuring the corresponding trace $u_{|\partial\Omega}$. The question of the uniqueness of D remains an open question (even if $\gamma(x) = 1$). It is worth noting that if we have access to the complete Neumann-to-Dirichlet Map (or the Dirichlet-to-Neumann Map), and ∂D possesses sufficient regularity, uniqueness can be established (see [63]). It is important to keep in mind that possessing the entire Dirichlet-to-Neumann Map is equivalent to having the capability to make infinite measurements on $\partial\Omega$. However, the ideas developed in detail in [4], which we will present here, enable us to derive **size estimates (of volume or area)** of the inclusion from specific integrals of the data, ϕ and $u_{|\partial\Omega}$. In this section, we will outline some of the key aspects of the ideas developed in [4], highlighting several interesting connections with the unique continuation properties of the solutions to the equation

$$\operatorname{div}(\gamma \nabla u) = 0, \quad \text{in } \Omega \tag{6.199}$$

that will be studied in a more general context in Part III of this book.

Let us consider the quantity

$$W = \int_{\partial\Omega} \phi u dS, \tag{6.200}$$

where u is the solution to (6.197)–(6.198) and compare W with

$$W_0 = \int_{\partial\Omega} \phi u_0 dS, \tag{6.201}$$

where u_0 represents the solution to the Neumann problem

$$\begin{cases} \operatorname{div}(\gamma(x)\nabla u_0) = 0, & \text{in } \Omega \\ \gamma \frac{\partial u_0}{\partial \nu} = \phi, & \text{on } \partial\Omega \end{cases}, \tag{6.202}$$

with

$$\int_{\Omega} u_0 dx = 0. \tag{6.203}$$

Also we assume that

$$\phi \neq 0. \tag{6.204}$$

W and W_0 represent the power required to sustain the current ϕ in the presence and absence of the inclusion D, respectively.

Now, we anticipate two unique continuation properties for solutions $u \in H^2_{loc}(\Omega)$ of Eq. (6.199) that we will prove in Part III.

(A) Weak unique continuation property: If ω is a non-empty open subset of Ω and $u = 0$ in ω, then $u = 0$ in Ω.

(B) Strong unique continuation property: If $x_0 \in \Omega$ and

$$\int_{B_r(x_0)} u^2 dx = O\left(r^k\right), \quad \text{as } r \to 0, \ \forall k \in \mathbb{N}, \tag{6.205}$$

then $u = 0$ in Ω.

In the subsequent part of this section, we will prove that if $k \neq 1$ and $k > 0$, the following inequality holds:

$$C_1 \int_D |\nabla u_0|^2 \leq |W_0 - W| \leq C_2 \int_D |\nabla u_0|^2, \tag{6.206}$$

where C_1 and C_2 are positive constants depending on k and λ only.

Let us assume, for instance, that D is a connected open set. Inequalities (6.206) implies that if $\phi \neq 0$ then $|D|$ is 0 if and only if $W_0 - W = 0$. Let us prove this sentence. If $|D| = 0$ then by the second inequality we immediately have $W_0 - W = 0$. Conversely, if $W_0 - W = 0$, then by the first inequality we have that if $D \neq \emptyset$ then u_0 is a constant in D, let c be the value of u_0 in D. Then $u_0 - c$ is a solution in $H^2_{loc}(\Omega)$ to Eq. (6.199). Thus

$$u_0 - c = 0, \quad \text{in } D$$

and by (A), we have

$$u_0 = c, \quad \text{in } \Omega,$$

consequently $\phi = 0$, but we have assumed in (6.204) that $\phi \neq 0$, therefore $D = \emptyset$.

Inequalities (6.206) can be proved as a consequence of general properties of the continuous symmetric coercive bilinear forms on a Hilbert space (see [32]).

Let us introduce some notation. Let H be a real Hilbert space and H' its dual space.

Let $\lambda_0, \lambda_1 \in [1, +\infty)$. Let $a_0(\cdot, \cdot)$ and $a_1(\cdot, \cdot)$ be two symmetric bilinear forms on H which satisfy the conditions

$$\lambda_0^{-1} \|u\|^2 \leq a_0(u, u) \leq \lambda_0 \|u\|^2, \qquad \forall u \in H, \tag{6.207a}$$

$$\lambda_1^{-1} \|u\|^2 \leq a_1(u, u) \leq \lambda_1 \|u\|^2, \qquad \forall u \in H. \tag{6.207b}$$

Let us note that (6.207a) and (6.207b) imply, respectively, the continuity of a_0 and of a_1. Just check this for $a_0(\cdot, \cdot)$. We have that (6.207a) implies

$$|a_0(u, v)| \leq \sqrt{a_0(u, u)}\sqrt{a_0(v, v)} \leq \lambda_0 \|u\| \|v\|, \quad \forall u, v \in H.$$

Moreover, let

$$\alpha(u, v) = a_1(u, v) - a_0(u, v), \quad \forall u, v \in H. \tag{6.208}$$

Let $F \in H'$. By the Lax–Milgram Theorem, there exist $u_1, u_0 \in H$ such that

$$a_j(u_j, v) = < F, v > \qquad \forall v \in H, \quad j = 0, 1. \tag{6.209}$$

Let us define

$$W_0 = < F, u_0 >, \quad W_1 = < F, u_1 >, \quad \delta W = W_0 - W_1. \tag{6.210}$$

Now we prove two simple lemmas.

Lemma 6.17 *The following equalities hold true.*

$$a_1(u_1 - u_0, u_1 - u_0) - \alpha(u_0, u_0) = -\delta W, \tag{6.211a}$$

$$a_0(u_0 - u_1, u_0 - u_1) + \alpha(u_1, u_1) = \delta W, \tag{6.211b}$$

$$\alpha(u_1, u_0) = \delta W. \tag{6.211c}$$

Proof Let us check (6.211a).

$$\begin{aligned}
&a_1(u_1 - u_0, u_1 - u_0) - \alpha(u_0, u_0) = \\
&= a_1(u_1 - u_0, u_1 - u_0) - [a_1(u_0, u_0) - a_0(u_0, u_0)] = \\
&= a_1(u_1, u_1) - 2a_1(u_1, u_0) + a_1(u_0, u_0) - a_1(u_0, u_0) + a_0(u_0, u_0) = \\
&= a_1(u_1, u_1) - 2a_1(u_1, u_0) + a_0(u_0, u_0) = \\
&=< F, u_1 > -2 < F, u_0 > + < F, u_0 >= \\
&=< F, u_1 - u_0 >= -\delta W.
\end{aligned}$$

Equality (6.211b) can be obtained similarly and (6.211c) is an immediate consequence of (6.208). ■

Lemma 6.18 *If one of the following conditions is satisfied*

$$\alpha(u, u) \geq 0, \quad \forall u \in H,$$

or

$$\alpha(u, u) \leq 0, \quad \forall u \in H$$

then we have

$$|\alpha(u, v)| \leq |\alpha(u, u)|^{1/2} |\alpha(v, v)|^{1/2}, \quad \forall u, v \in H. \tag{6.212}$$

Proof Let $u, v \in H$. If $\alpha(u, u) = 0$ and $\alpha(v, v) = 0$. Then, assuming $\alpha(w, w) \geq 0$, for every $w \in H$, we have

$$0 \leq \alpha(u + tv, u + tv) = 2t\alpha(u, v), \qquad \forall t \in \mathbb{R},$$

which implies $\alpha(u, v) = 0$ and (6.212) is proved.

If either $\alpha(u, u) \neq 0$ or $\alpha(v, v) \neq 0$, then, assuming, for instance, $\alpha(v, v) > 0$, we have

$$0 \leq \alpha(u + tv, u + tv) = t^2\alpha(v, v) + 2t\alpha(u, v) + \alpha(u, u), \qquad \forall t \in \mathbb{R},$$

hence

$$(\alpha(u, v))^2 - \alpha(u, u)\alpha(v, v) \leq 0,$$

which gives (6.212).

If $\alpha(w, w) \leq 0$, for every $w \in H$, the thesis follows easily by applying the previous procedure to $-\alpha(\cdot, \cdot)$. ■

Now we prove

Theorem 6.19 *Let $F \in H'$. Assume that the bilinear forms $a_0(\cdot, \cdot)$ and $a_1(\cdot, \cdot)$ satisfy conditions* (6.207) *and suppose that u_0, u_1 satisfy* (6.209). *If $\alpha(\cdot, \cdot)$ (defined by* (6.208)*), satisfies*

$$0 \leq \alpha(u, u) \leq C_0 a_0(u, u), \quad \forall u \in H, \tag{6.213}$$

where C_0 is a positive constant, then

$$\delta W \geq 0$$

and

$$\delta W \leq \alpha(u_0, u_0) \leq (1 + C_0)\delta W. \tag{6.214}$$

If $\alpha(\cdot, \cdot)$ satisfies the condition

$$\alpha(u, u) \leq 0, \quad \forall u \in H, \tag{6.215}$$

then

$$\delta W \leq 0$$

and

$$-C\delta W \leq -\alpha(u_0, u_0) \leq -\delta W, \tag{6.216}$$

where C is a positive constant depending on λ_0 and λ_1 only.

Proof First we consider the case in which (6.213) holds. By (6.211b) we have $\delta W \geq 0$ and by (6.211a) we have $-\alpha(u_0, u_0) \leq -\delta W$. Therefore

$$\delta W \leq \alpha(u_0, u_0). \tag{6.217}$$

Now, let us estimate $\alpha(u_0, u_0)$ from above. Lemmas 6.17-b and 6.18 give

$$\alpha(u_0, u_0) = \alpha(u_0 - u_1, u_0 - u_1) + \alpha(u_1, u_1) + 2\alpha(u_0 - u_1, u_1) \leq$$

$$\leq \alpha(u_0 - u_1, u_0 - u_1) + \alpha(u_1, u_1) + 2|\alpha(u_0 - u_1, u_0 - u_1)|^{1/2}|\alpha(u_1, u_1)|^{1/2} \leq$$

$$\leq \alpha(u_0 - u_1, u_0 - u_1) + \alpha(u_1, u_1) + \frac{1}{C_0}\alpha(u_0 - u_1, u_0 - u_1) + C_0\alpha(u_1, u_1) =$$

$$= (1 + C_0)[\frac{1}{C_0}\alpha(u_0 - u_1, u_0 - u_1) + \alpha(u_1, u_1)] \leq$$

$$\leq (1 + C_0) \max\left\{1, \frac{1}{C_0}\right\}[a_0(u_0 - u_1, u_0 - u_1) + \alpha(u_1, u_1)] \leq$$

$$\leq C_1\delta W,$$

where, in the last inequality we have set

$$C_1 = (1 + C_0) \max\left\{1, \frac{1}{C_0}\right\}.$$

Hence we have

$$\alpha(u_0, u_0) \leq C_1\delta W.$$

By the just obtained inequality and by (6.217) we get (6.214).

Now we consider the case in which (6.215) holds. By (6.211a) we get $\delta W \leq 0$ and also

$$|\alpha(u_0, u_0)| \leq -\delta W. \tag{6.218}$$

Now we estimate $|\alpha(u_0, u_0)|$ from below. By (6.211c) we obtain, for $\varepsilon > 0$ to be chosen,

$$-\delta W = -\alpha(u_0, u_1) \leq (-\alpha(u_0, u_0))^{1/2}(-\alpha(u_1, u_1))^{1/2} \leq$$

$$\leq \frac{\varepsilon}{2}(-\alpha(u_1, u_1)) + \frac{1}{2\varepsilon}(-\alpha(u_0, u_0)). \tag{6.219}$$

By (6.211b) we have

$$-\alpha(u_1, u_1) = a_0(u_1 - u_0, u_1 - u_0) - \delta W. \tag{6.220}$$

Moreover, (6.207) gives

$$a_0(u_1 - u_0, u_1 - u_0) \leq \lambda_0 \|u_1 - u_0\|^2 \leq \lambda_0\lambda_1 a_1(u_1 - u_0, u_1 - u_0).$$

By the just obtained inequality and by (6.220) we get

$$-\alpha(u_1, u_1) \leq \lambda_0\lambda_1 a_1(u_1 - u_0, u_1 - u_0) - \delta W.$$

The last inequality together with (6.219) and (6.211a), give (we denote $A = \lambda_0\lambda_1$)

$$\begin{aligned} -\delta W &\leq \frac{\varepsilon}{2}\left[A a_1(u_1 - u_0, u_1 - u_0) - \delta W\right] + \frac{1}{2\varepsilon}(-\alpha(u_0, u_0)) = \\ &= \frac{\varepsilon}{2}\left[A\left(a_1(u_1 - u_0, u_1 - u_0) - \alpha(u_0, u_0)\right) + A\alpha(u_0, u_0) - \delta W\right] + \\ &+ \frac{1}{2\varepsilon}(-\alpha(u_0, u_0)) = \\ &= -\frac{\varepsilon}{2}(1 + A)\delta W + \left(\frac{1}{2\varepsilon} - A\frac{\varepsilon}{2}\right)(-\alpha(u_0, u_0)). \end{aligned}$$

Therefore

$$\left(1 - \frac{\varepsilon}{2}(1 + A)\right)|\delta W| \leq \frac{1 - A\varepsilon^2}{2\varepsilon}|\alpha(u_0, u_0)|.$$

If

$$\varepsilon = \min\left\{\frac{1}{\sqrt{2A}}, \frac{1}{1 + A}\right\}$$

then we have

$$\frac{2\varepsilon\left(1 - \frac{\varepsilon}{2}(1 + A)\right)}{1 - A\varepsilon^2}|\delta W| \leq |\alpha(u_0, u_0)|.$$

Ultimately, we have

$$C|\delta W| \leq |\alpha(u_0, u_0)|,$$

where C depends on λ_0, λ_1 only. ■

Remark 6.24 If (6.213) holds, condition (6.207) can be weakened by assuming that $a_0(\cdot, \cdot)$, $a_1(\cdot, \cdot)$ are semidefinite positive. In turn, if case (6.215) occurs, it suffices to assume that $a_0(\cdot, \cdot)$, $a_1(\cdot, \cdot)$ are semidefinite positive and satisfy

$$a_0(u, u) \leq C_1 a_1(u, u), \qquad \forall u \in H,$$

where C_1 is a positive constant. ♦

Now we apply Theorem 6.19 to the inclusion inverse problem. Let

$$H = \left\{ v \in H^1(\Omega) : \quad \int_\Omega v dx = 0 \right\},$$

$$a_1(u, v) = \int_\Omega \gamma \left(1 + (k - 1)\chi_D\right) \nabla u \cdot \nabla v, \quad u, v \in H,$$

$$a_0(u, v) = \int_\Omega \gamma \nabla u \cdot \nabla v, \quad u, v \in H.$$

Let us assume that $\phi \in H^{-1/2}(\partial\Omega)$ satisfies (6.196). Then Neumann problems (6.197), (6.202) can be formulated (see Sect. 6.4 and the proof of Theorem 6.6) as follows

$$a_1(u, v) = \langle g, v \rangle_{H^{-1/2}, H^{1/2}}, \quad \forall v \in H, \tag{6.221}$$

$$a_0(u, v) = \langle g, v \rangle_{H^{-1/2}, H^{1/2}}, \quad \forall v \in H. \tag{6.222}$$

Hence, in our case we have

$$H \ni v \to \langle F, v \rangle = \langle g, v \rangle_{H^{-1/2}, H^{1/2}} \in \mathbb{R},$$

$$\alpha(u, v) = (k - 1) \int_D \gamma \nabla u \cdot \nabla v, \quad u, v \in H,$$

$$\delta W = \int_{\partial\Omega} \phi(u_0 - u_1),$$

where u_0 and u_1 are solutions to (6.222) and (6.221) respectively.

Case $k > 1$ In this case, if $C_0 = k - 1$ then (6.213) is satisfied. Hence by (6.214) we have

$$\frac{\delta W}{1 - k} \leq \int_D \gamma |\nabla u_0|^2 \leq \frac{k \delta W}{1 - k}. \tag{6.223}$$

Case $k < 1$ In this case (6.215) holds true. Hence, by (6.216) we get

$$-\frac{C \delta W}{1 - k} \leq \int_D \gamma |\nabla u_0|^2 \leq -\frac{\delta W}{1 - k}, \tag{6.224}$$

where C depends on k only. Inequalities (6.223) and (6.224) imply (6.206).

Let us note that, in certain specific cases, estimates (6.206) can be used to straightforwardly obtain a size estimate for D. For example, if $\gamma(x) = 1$ and

$u_0 = x_1 + c$ in Ω, then $\nabla u_0 = e_1$, and from (6.206) we have

$$C_2^{-1}|W_0 - W| \leq |D| \leq C_1^{-1}|W_0 - W|. \tag{6.225}$$

Assigning a particular value to u_0 in Ω effectively imposes specific constraints on the "input" current density. In this case, $\phi = e_1 \cdot \nu$, although it may not always be feasible to implement this choice in practice. Thus, it becomes valuable to derive both upper and lower estimates for the measure of D when considering a more general conductivity γ and a nontrivial ϕ.

In order to delve further into this matter, let us begin by noting that formulating the Neumann problem (6.197) does not necessitate D to be open; it is sufficient for **D to be a Lebesgue measurable set** in $\mathbb{R}^n$. For instance, if we know that

$$\operatorname{dist}(D, \partial\Omega) \geq d > 0,$$

it is not challenging to derive a lower estimate for $|D|$ by exploiting the second inequality in (6.206), i.e., the inequality

$$|\delta W| \leq C_2 \int_D |\nabla u_0|^2, \tag{6.226}$$

where we have used (6.195),

Let us examine in which manner we can find an estimate from below of $|D|$. Let us denote by E the lipschitz constant of γ. Let Ω' be of $C^{0,1}$ class and such that

$$D \subset \Omega' \Subset \Omega_{d/2},$$

(for instance $\Omega' = \Omega_d$) where $\Omega_{d/2} = \left\{x \in \Omega : \operatorname{dist}(x, \partial\Omega) > \frac{d}{2}\right\}$. Since we will applied Theorem 5.24, for the sake of clarity, we fix $n \geq 3$.

By Hölder inequality and by Theorem 5.24 we have

$$\begin{aligned} \int_D |\nabla u_0|^2 \leq |D|^{\frac{2}{n}} \left(\int_D |\nabla u_0|^{\frac{2n}{n-2}}\right)^{1-\frac{2}{n}} &\leq \\ \leq C|D|^{\frac{2}{n}} \|u_0\|^2_{H^2(\Omega')}, \end{aligned} \tag{6.227}$$

where C depends on the $C^{0,1}$ character of Ω' and

$$\|u_0\|^2_{H^2(\Omega')} = \sum_{|\alpha|\leq 2} d_0^{2|\alpha|} \left\|\partial^\alpha u_0\right\|^2_{L^2(\Omega')},$$

here d_0 denotes the diameter of Ω. Now by (5.35) (recalling (6.203)), (6.126) and by (6.75), we have

$$\|u_0\|^2_{H^2(\Omega')} \leq C\,\|u_0\|^2_{L^2(\Omega)} \leq C'\,\|\nabla u_0\|^2_{L^2(\Omega)} \leq C''\,\|\phi\|^2_{H^{-1/2}(\partial\Omega)}, \tag{6.228}$$

where C, C', C'' are constants.

Hence by (6.226), (6.227) and (6.228) we get

$$\left(\frac{|\delta W|}{C_2 C''\,\|\phi\|^2_{H^{-1/2}(\partial\Omega)}}\right)^{\frac{n}{2}} \leq |D|. \tag{6.229}$$

Finding an upper estimate for $|D|$ (expressed in terms of δW) is certainly more demanding. As we have previously indicated, in the case where D is an open set, such an upper estimate is inherently connected to the **unique continuation property** of solutions to Eq. (6.199).

When D is only a Lebesgue measurable set, in order to prove that

$$\delta W = 0 \quad \Longrightarrow |D| = 0 \tag{6.230}$$

is not trivial. The proof of this implication is related to the following one

$$\nabla u_0 = 0, \ \text{ in } D, \ \text{ and } |D| > 0 \Longrightarrow \nabla u_0 = 0, \ \text{ in } \Omega. \tag{6.231}$$

Indeed, arguing by contradiction, let us suppose that $\delta W = 0$ and let us assume that $|D| > 0$. Then the first inequality on the left in (6.206) would imply $\nabla u_0 = 0$ in D, and with the help of (6.231), we would obtain, in particular, $\phi = 0$, which would contradict (6.204), thus yielding (6.230).

Here, we will prove neither implication (6.230) nor (6.231), referring instead to [4] for a treatment of the topic from a quantitative perspective as well. Rather, we will restrict ourselves to proving that if u is a solution of Eq. (6.199), then the following implication holds:

$$u = 0, \ \text{ in } D \text{ and } |D| > 0 \Longrightarrow u = 0, \ \text{ in } \Omega. \tag{6.232}$$

The proof of this implication will nonetheless allow us to highlight an interesting issue concerning the unique continuation properties related to the inverse problem we are considering. Additional brief notes on this issue will be provided in Sect. 19.4.

To present here a proof of (6.232) we need the differentiation Lebesgue Theorem 3.52. In particular such a Theorem 3.52 implies that if D is a Lebesgue measurable set, then

$$\lim_{r\to 0} \frac{|D \cap B_r(x)|}{|B_r(x)|} = 1, \qquad \text{a.e. } x \in D. \tag{6.233}$$

Let us set

$$\widetilde{D} = \left\{ x \in D : \quad \lim_{r \to 0} \frac{|D \cap B_r(x)|}{|B_r(x)|} = 1 \right\}. \tag{6.234}$$

The following Proposition holds ([28, 29]).

Proposition 6.20 *Let Ω be an open set of $\mathbb{R}^n$ and let D be a Lebesgue measurable set such that $\overline{D} \subset \Omega$ and $|D| > 0$. Let $u \in H^1_{loc}(\Omega)$. Assume that u satisfies the condition*

$$u(x) = 0, \quad \forall x \in D. \tag{6.235}$$

Moreover, assume that, in a given point $x_0 \in \widetilde{D}$ we have

$$\int_{B_r(x_0)} |\nabla u|^2 dx \leq \frac{C}{r^2} \int_{B_{2r}(x_0)} u^2 dx, \tag{6.236}$$

for every $r > 0$ such that $\overline{B_{2r}(x_0)} \subset \Omega$, where C is independent of r.

Then we have

$$\int_{B_r(x_0)} u^2 dx = O\left(r^k\right), \quad as\ r \to 0, \quad \forall k \in \mathbb{N}. \tag{6.237}$$

To prove Proposition 6.20 we need the following

Lemma 6.21 *If $R > 0$ and $u \in H^1(B_R)$ then*

$$\left(\int_{B_R} |u|^q dx \right)^{\frac{1}{q}} \leq \frac{C_{n,q}}{|B_R|^{\frac{1}{2} - \frac{1}{q}}} \left(\int_{B_R} \left[R^2 |\nabla u|^2 + u^2 \right] dx \right)^{\frac{1}{2}}, \tag{6.238}$$

where q is an arbitrary number of $(1, +\infty)$ for $n = 2$, and it is equal to $\frac{2n}{n-2}$ for $n \geq 3$. Moreover, $C_{n,q}$ depends on q and n only.

Proof of Lemma 6.21. Set

$$v(y) = u(Ry), \quad \forall y \in B_1,$$

it turns out that $v \in H^1(B_1)$. Now, by the Embedding Sobolev Theorem (Theorem 5.31) and performing the change of variables $y = Rx$, we get

$$\left(\int_{B_R}|u|^q dx\right)^{\frac{1}{q}}=\left(R^n\int_{B_1}|u(Ry)|^q dy\right)^{\frac{1}{q}}=R^{\frac{n}{q}}\left(\int_{B_1}|v(y)|^q dy\right)^{\frac{1}{q}}\leq$$

$$\leq CR^{\frac{n}{q}}\left(\int_{B_1}\left[|\nabla v|^2+|v|^2\right]dy\right)^{\frac{1}{2}}=$$

$$=CR^{\frac{n}{q}}\left(\int_{B_1}\left[R^2|(\nabla u)(Ry)|^2+|u(Ry)|^2\right]dy\right)^{\frac{1}{2}}=$$

$$=CR^{\frac{n}{q}}\left(R^{-n}\int_{B_R}\left[R^2|\nabla u|^2+|u|^2\right]dx\right)^{\frac{1}{2}}=$$

$$=\frac{C\omega_n^{\frac{1}{2}-\frac{1}{q}}}{|B_R|^{\frac{1}{2}-\frac{1}{q}}}\left(\int_{B_R}\left[R^2|\nabla u|^2+|u|^2\right]dx\right)^{\frac{1}{2}}.$$

■

Proof of Proposition 6.20 Let us denote by $\rho=\frac{1}{2}\text{dist}(x_0,\partial\Omega)$. Since $x_0\in\widetilde{D}$, we have that for any $\varepsilon>0$ there exists $r_\varepsilon<2\rho$ such that

$$\frac{|B_r(x_0)\setminus D|}{|B_r(x_0)|}<\varepsilon,\quad \forall r\in(0,r_\varepsilon].$$

By the above inequality and by Lemma 6.21 we have

$$\int_{B_r(x_0)}u^2dx=\int_{B_r(x_0)\setminus D}u^2dx\leq|B_r(x_0)\setminus D|^{1-\frac{2}{q}}\left(\int_{B_r(x_0)\setminus D}|u|^q dx\right)^{\frac{2}{q}}\leq$$

$$\leq|B_r(x_0)\setminus D|^{1-\frac{2}{q}}\left(\int_{B_r(x_0)}|u|^q dx\right)^{\frac{2}{q}}\leq$$

$$\leq C\left(\frac{|B_r(x_0)\setminus D|}{|B_r(x_0)|}\right)^{1-\frac{2}{q}}\int_{B_r(x_0)}\left[r^2|\nabla u|^2+u^2\right]dx\leq$$

$$\leq C\varepsilon^{1-\frac{2}{q}}\int_{B_r(x_0)}\left[r^2|\nabla u|^2+u^2\right]dx,$$

where q is an arbitrary number of $(1,+\infty)$ for $n=2$, and it is equal to $\frac{2n}{n-2}$ for $n\geq 3$. In addition, C depends on q and n only.

By the just obtained inequality and by (6.236) we get

$$\int_{B_r(x_0)}u^2dx\leq C\varepsilon^{1-\frac{2}{q}}\int_{B_{2r}(x_0)}u^2dx,\quad \forall r\in(0,r_\varepsilon], \tag{6.239}$$

where C is independent on r and ε.

Let $k \in \mathbb{N}$ be arbitrary and let $\varepsilon > 0$ satisfy

$$C\varepsilon^{1-\frac{2}{q}} = 2^{-k}.$$

Further, let us denote by r_k the value of r_ε that corresponds to this choice of ε. Let us introduce the following function

$$f(r) = \int_{B_r(x_0)} u^2 dx, \quad r \in (0, 2r_k].$$

Then (6.239) can be written as

$$f(r) \le 2^{-k} f(2r), \quad r \in (0, r_k]. \tag{6.240}$$

Now, for any $0 < r < r_k$, let $m \in \mathbb{N}$ satisfy

$$2^{-m} r_k \le r < 2^{1-m} r_k. \tag{6.241}$$

Iteration of (6.240) gives

$$f(r) \le 2^{-k} f(2r) \le \cdots \le 2^{-km} f\left(2^m r\right) \le 2^{-km} f\left(2r_k\right).$$

On the other hand (6.241) implies

$$2^{-m} \le \frac{r}{r_k},$$

hence

$$f(r) \le \left(\frac{r}{r_k}\right)^k f\left(2r_k\right),$$

which gives (6.237). ■

Remark 6.25 By the proof of Proposition 6.20 it is clear that assumption (6.236) can be replaced by the assumption that there is $p > 2$ such that

$$\left(⨍_{B_r} |u|^p dx\right)^{1/p} \le C \left(⨍_{B_{2r}} |u|^2 dx\right)^{1/2}. \tag{6.242}$$

See also Sect. 19.4. ♦

Theorem 6.22 *Let Ω be a connected bounded open set of $\mathbb{R}^n$ and let*

$$D \subset \Omega$$

be a Lebesgue measurable set of positive measure. Let $\gamma \in C^{0,1}(\Omega)$ *be a real-valued function which satisfies ellipticity condition* (6.195)*. Let* u *be a solution to the equation*

$$\operatorname{div}(\gamma \nabla u) = 0, \quad \text{in } \Omega. \tag{6.243}$$

Assume that

$$u = 0, \quad \text{in } D.$$

Then

$$u = 0, \quad \text{in } \Omega. \tag{6.244}$$

Remark 6.26 As will become clear from the proof, the boundedness assumption of Ω is not essential: we have introduced it solely for the purpose of easing the proof. We leave the simple extension to the reader. ♦

Proof of Theorem 6.22 Since

$$D = \bigcup_{j=1}^{\infty} \left(D \cap \Omega_j\right), \tag{6.245}$$

where

$$\Omega_j = \{x \in \Omega : \quad \operatorname{dist}(x, \partial\Omega) > 1/j\},$$

we have

$$0 < |D| = \lim_{j\to\infty} |D \cap \Omega_j|,$$

hence $|D \cap \Omega_j| > 0$ for j large enough, in addition we have $\overline{D \cap \Omega_j} \subset \Omega$. Hence, provided to replace $D \cap \Omega_j$, to D, we may always assume $\overline{D} \subset \Omega$.

Let us apply Proposition 6.20. Inequality (6.236) is nothing more than the Caccioppoli inequality proved in Theorem 6.7. Hence we have, for $x_0 \in \widetilde{D}$ ($\widetilde{D}$ is defined by (6.234)),

$$\int_{B_r(x_0)} u^2 dx = O\left(r^k\right), \quad \text{as } r \to 0, \quad \forall k \in \mathbb{N} \tag{6.246}$$

and by the strong unique continuation property stated above the thesis follows. ■

Remark 6.27 To conclude this section, we must mention recent advances in quantitative aspects related to implication (6.232). For example, one might ask: if

$|D| > 0$ and we know that u, a solution of (6.199), is "small" on D, how can we control the error in $u(x)$ at other points x in Ω? More precisely, if

$$|u| \leq \varepsilon, \quad \text{in } D, \tag{6.247}$$

where $\varepsilon > 0$, how can we estimate $|u(x)|$ in terms of ε for $x \in \Omega$? This is the problem of smallness propagation estimates. As we shall see in Parts II and III of this book, condition (6.247) alone does not ensure the propagation of the error to points in $\Omega \setminus D$; additional information on u is required, such as boundedness over Ω. Even with this restriction, however, the matter is far from straightforward. In this regard, we recall the results obtained in [85, 89, 99, 109, 136], and, in cases where u is analytic, in [98, 135]. ♦

Part II
Cauchy Problem for PDEs and Stability Estimates

Chapter 7
The Cauchy Problem for the First-Order PDEs

Abstract This chapter introduces the Cauchy problem for partial differential equations, starting with first-order equations in a real context. It serves two purposes: to familiarize readers with the Cauchy problem and to prepare key concepts for later chapters. The first-order Cauchy problem acts as an intermediate case between ordinary and more general partial differential equations. A brief review of ordinary differential equations is provided for convenience. We focus on the method of characteristic lines for linear equations and give a brief overview of quasilinear equations and fully nonlinear equations.

7.1 Introduction

In this chapter, starting with first-order equations in a real setting, we begin our study of the Cauchy problem for partial differential equations. We will work within a classical framework, as this chapter serves a dual purpose. First, we aim to help readers start familiarizing themselves with the Cauchy problem for partial differential equations, and second, to provide definitions and results that will be applied in later chapters of this book. It should be noted that, in a certain sense, the Cauchy problem for first–order partial differential equations stands as an intermediate case between ordinary differential equations and more general partial differential equations, thus serving as an effective preparation for studying the latter. In Sect. 7.2, we provide a brief review of ordinary differential equations. While these facts should be well known to the reader, we include them here for convenience. Regarding the first–order partial differential equations, we will focus primarily on the classical method of characteristic lines for linear equations and provide a brief look at quasilinear equations. Section 7.6, which deals with fully non–linear equations, can be skipped on a first reading. The results from Sect. 7.7 will be partially applied in Sect. 18.6.

S. Vessella, *Unique Continuation Properties for Partial Differential Equations*, Birkhäuser Advanced Texts Basler Lehrbücher,
https://doi.org/10.1007/978-3-031-86366-0_7

7.2 Review of Ordinary Differential Equations

In this section we give, without proof, some results on ordinary differential equations that we will need later on. For further discussion we refer to [104].

Let $t_0 \in \mathbb{R}$, $x_0 \in \mathbb{R}^n$, $x_0 = \left(x_{0,1}, \cdots, x_{0,n}\right)$, $a > 0$, $b > 0$. Set

$$Q = \left\{(t, x) \in \mathbb{R}^{n+1} : \quad |t - t_0| \leq a, \quad \left|x_i - x_{0,i}\right| \leq b\right\}$$

and let

$$f : Q \to \mathbb{R}^n$$

be a **continuous function** in Q which is **Lipschitz continuous with respect to the variable** x, that is

$$|f(t, x) - f(t, y)| \leq L\,|x - y|\,, \qquad \forall (t, x), (t, y) \in Q. \tag{7.1}$$

Let us consider the following Cauchy problem: determine the function $x(t)$ differentiable in a neighborhood of t_0 and satisfying

$$\begin{cases} \dot{x}(t) = f\,(t, x(t)) \\ \\ x(t_0) = x_0 \end{cases}, \tag{7.2}$$

here $\dot{x}$ is the derivative of x w.r.t. t . The following Theorem holds.

Theorem 7.1 *Let $f \in C^0(Q)$ satisfy* (7.1). *Then there exists $\delta > 0$ and there exists a unique solution $x \in C^1\left([t_0 - \delta, t_0 + \delta]; \mathbb{R}^n\right)$ to problem* (7.2). *Moreover, setting*

$$M_i = \max_Q |f_i|\,, \qquad M = \max_{1 \leq i \leq n} M_i,$$

we can choose $\delta = \min\left\{a, \frac{b}{M}\right\}$.

The following Lemma will be very useful

Lemma 7.2 (Gronwall) *Let I be an interval of $\mathbb{R}$, $\alpha \in I$ and $c \geq 0$. Moreover, let $u, v \in C^0(I, \mathbb{R})$ where $v(t) \geq 0$ and $u(t) \geq 0$, for every $t \in I$.*

What follows holds true.

(i) If

$$v(t) \leq c + \int_\alpha^t u(s) v(s) ds, \quad \forall t \geq \alpha,$$

then

$$v(t) \leq c e^{\int_\alpha^t u(s)ds}, \quad \forall t \geq \alpha.$$

(ii) If

$$v(t) \leq c + \int_t^\alpha u(s)v(s)ds, \quad \forall t \leq \alpha,$$

then

$$v(t) \leq c e^{\int_t^\alpha u(s)ds}, \quad \forall t \leq \alpha.$$

The Gronwall Lemma makes it simple to prove the continuous dependence result of the solution to (7.2) by the data t_0, x_0, f. More precisely, we have

Theorem 7.3 (Continuous Dependence by the Data) *Let $f, \widetilde{f} \in C^0(Q)$ satisfy* (7.1)*. Let $\sigma_1, \sigma_2, \varepsilon$ be positive numbers. Let us suppose that*

$$\left|t_0 - \widetilde{t_0}\right| \leq \sigma_1, \quad \left|x_0 - \widetilde{x}_0\right| \leq \sigma_2, \quad \max_Q \left|f - \widetilde{f}\right| \leq \varepsilon.$$

Set

$$M = \max_Q |f|, \quad \widetilde{M} = \max_Q \left|\widetilde{f}\right|.$$

Then the following fact occurs:

There exists $\sigma_0 > 0$ depending on $a, b, M, \widetilde{M}$ such that if $\sigma_1, \sigma_2 < \sigma_0$, then there is $\delta > 0$ and $x, \widetilde{x} \in C^1([t_0 - \delta, t_0 + \delta], \mathbb{R}^n)$ that satisfy what follows:

$$\begin{cases} \dot{x}(t) = f(t, x(t)) \\ \\ x(t_0) = x_0 \end{cases},$$

$$\begin{cases} \dot{\widetilde{x}}(t) = f(t, \widetilde{x}(t)) \\ \\ \widetilde{x}\left(\widetilde{t_0}\right) = \widetilde{x}_0 \end{cases}$$

and

$$|x(t) - \widetilde{x}(t)| \leq C(\sigma_1 + \sigma_2 + \varepsilon), \qquad \forall t \in [t_0 - \delta, t_0 + \delta],$$

where C is a constant that depends on $a, b, L, M, \widetilde{M}$ only.

Theorem 7.4 (Regularity) *Let $k \in \mathbb{N}$ and $f \in C^k(Q)$. Then the solution to Cauchy problem (7.2) belongs to $C^{k+1}([t_0 - \delta, t_0 + \delta]; \mathbb{R}^n)$, where $\delta = \min\left\{a, \frac{b}{M}\right\}$.*

We now describe the Theorem of **differentiability of the solution** of the Cauchy problem with respect to a parameter and with respect to the inital values. Let $\lambda_1, \lambda_2 \in \mathbb{R}$ be such that $\lambda_1 < \lambda_2$ and let

$$\widetilde{Q} = \left\{(t, x; \lambda) \in \mathbb{R}^{n+2} : \quad |t - t_0| \leq a, \quad \left|x_i - x_{0,i}\right| \leq b, \quad \lambda_1 \leq \lambda \leq \lambda_2\right\}.$$

Moreover, let $f \in C^1\left(\widetilde{Q}, \mathbb{R}^n\right)$. If $\left(t_0, x_0, \overline{\lambda}\right) \in \widetilde{Q}$ then there exists a unique solution to the Cauchy problem

$$\begin{cases} \dot{x}(t) = f\left(t, x(t); \overline{\lambda}\right) \\ \\ x(t_0) = x_0 \end{cases} . \tag{7.3}$$

Let us denote by

$$\overline{x}\left(t, t_0, x_0; \overline{\lambda}\right) \tag{7.4}$$

this solution. The following Theorem holds true (for a proof we refer to [104, Ch. 1])

Theorem 7.5 *Function $\overline{x}$ defined in (7.4) is differentiable with respect to all the variables and the derivatives are continuous functions.*

Here we give short outline of the proof of Theorem 7.5. In order to calculate (formally) the derivatives

$$\frac{\partial \overline{x}}{\partial t_0}, \qquad \frac{\partial \overline{x}}{\partial x_{0,j}}, \qquad \frac{\partial \overline{x}}{\partial \overline{\lambda}}, \quad j = 1, \cdots, n,$$

we proceed in the following way: we write the system (7.3) in the form

$$\begin{cases} \frac{\partial}{\partial t}\overline{x}\left(t, t_0, x_0; \overline{\lambda}\right) = f\left(t, \overline{x}\left(t, t_0, x_0; \overline{\lambda}\right); \overline{\lambda}\right) \\ \\ \overline{x}\left(t_0, t_0, x_0; \overline{\lambda}\right) = x_0 \end{cases} , \tag{7.5}$$

next, we make the derivatives of both the sides of (7.5) obtaining a linear first-order system with Cauchy conditions in the "new unknowns"

$$\frac{\partial \overline{x}}{\partial t_0}, \ \frac{\partial \overline{x}}{\partial x_{0,k}}, \ \frac{\partial \overline{x}}{\partial \lambda}.$$

For instance, in the case $n = 1$, whether we are interested in calculating $\frac{\partial \overline{x}}{\partial t_0}$, we make the derivatives of both the sides of Eq. (7.5) with respect to t_0 and we get

$$\frac{\partial}{\partial t_0}\frac{\partial}{\partial t}\overline{x}\left(t, t_0, x_0; \overline{\lambda}\right) = \frac{\partial f}{\partial x}\left(t, \overline{x}\left(t, t_0, x_0; \overline{\lambda}\right); \overline{\lambda}\right)\frac{\partial}{\partial t_0}\overline{x}\left(t, t_0, x_0; \overline{\lambda}\right)$$

and by the initial datum, making the derivative with respect to t_0, we have

$$\left(\frac{\partial}{\partial t_0}\overline{x}\left(t, t_0, x_0; \overline{\lambda}\right) + \frac{\partial}{\partial t}\overline{x}\left(t, t_0, x_0; \overline{\lambda}\right)\right)_{|t=t_0} = 0.$$

From which, taking into account (7.5), we have

$$\frac{\partial}{\partial t_0}\overline{x}\left(t_0, t_0, x_0; \overline{\lambda}\right) = -f\left(t_0, x_0; \overline{\lambda}\right).$$

Now, set

$$U\left(t, t_0, x_0; \overline{\lambda}\right) = \frac{\partial}{\partial t_0}\overline{x}\left(t, t_0, x_0; \overline{\lambda}\right), \tag{7.6}$$

$$A\left(t, t_0, x_0; \overline{\lambda}\right) = \frac{\partial f}{\partial x}\left(t, \overline{x}\left(t, t_0, x_0; \overline{\lambda}\right); \overline{\lambda}\right)$$

and interchanging the order of derivatives $\frac{\partial}{\partial t_0}\frac{\partial}{\partial t}\overline{x}$ (in the rigorous proof it is proved that this step is admissible, compare [104, Chap. 1]) we have

$$\begin{cases} \frac{\partial U}{\partial t} = A\left(t, t_0, x_0; \overline{\lambda}\right) U \\ \\ U\left(t_0, t_0, x_0; \overline{\lambda}\right) = -f\left(t_0, x_0; \overline{\lambda}\right) \end{cases},$$

that is a Cauchy problem for an ordinary differential equation in the new unknown U. When U is determined, also $\frac{\partial}{\partial t_0}\overline{x}$ turns out determined by (7.6). Similarly we proceed when $n > 1$ and for the others derivatives.

Likewise, if $f \in C^k\left(\widetilde{Q}, \mathbb{R}^n\right)$, it can be proved that $\overline{x}\left(\cdot, t_0, x_0; \overline{\lambda}\right)$ has continuous derivatives w.r.t. t_0, $x_{0,j}$ and $\overline{\lambda}$ up to the order k.

7.3 First-Order Linear PDEs

We begin by giving some definitions. Let Ω be a connected open set of $\mathbb{R}^n$ and let

$$a : \Omega \to \mathbb{R}^n, \qquad a(x) = (a_1(x), \cdots, a_n(x))$$

be a vector field on Ω. Let us denote by $P(x, \partial)$ the following linear differential operator

$$P(x, \partial) = a \cdot \nabla = \sum_{j=1}^{n} a_j(x)\partial_j, \quad x \in \Omega. \tag{7.7}$$

Throughout this chapter we will call the **symbol of the operator** $P(x, \partial)$ the following homogeneous polynomial of first degree w.r.t. the variables $\xi_1, \cdots, \xi_n$

$$p(x, \xi) = a(x) \cdot \xi = \sum_{j=1}^{n} a_j(x)\xi_j, \quad x \in \Omega, \quad \xi \in \mathbb{R}^n. \tag{7.8}$$

We say that $\xi \in \mathbb{R}^n \setminus \{0\}$ is a **characteristic direction for** $P(x, \partial)$ **at the point** $x_0 \in \Omega$ if

$$p(x_0, \xi) = 0. \tag{7.9}$$

A surface

$$\Gamma = \{x \in \Omega : \quad \phi(x) = \phi(x_0)\},$$

where $\phi \in C^1(\Omega)$ and

$$\nabla\phi(x_0) \neq 0,$$

is said a **characteristic surface at the point** $x_0 \in \Omega$ **for** $P(x, \partial)$ if $\nabla\phi(x_0)$ is a characteristic direction for $P(x, \partial)$ at the point x_0. That is

$$p(x_0, \nabla\phi(x_0)) = a(x_0) \cdot \nabla\phi(x_0) = \sum_{j=1}^{n} a_j(x_0)\partial_j\phi(x_0) = 0, \tag{7.10}$$

let us note

$$p(x, \nabla\phi) = P(x, \partial)\phi.$$

We say that Γ is a characteristic surface for $P(x, \partial)$ if

$$p(x, \nabla\phi(x)) = 0, \quad \forall x \in \Gamma. \tag{7.11}$$

Let us consider the following unit vectors (Fig. 7.1)

$$\nu(x_0) = -\frac{\nabla\phi(x_0)}{|\nabla\phi(x_0)|} \quad \text{and} \quad -\nu(x_0) = \frac{\nabla\phi(x_0)}{|\nabla\phi(x_0)|}.$$

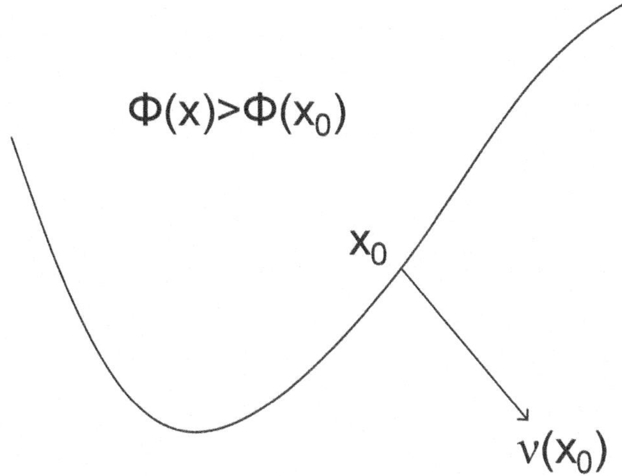

Fig. 7.1 $\nu(x_0) = -\frac{\nabla\phi(x_0)}{|\nabla\phi(x_0)|}$ is the unit outward normal to $\Gamma = \{\phi(x) = \phi(x_0)\}$ in x_0

Let us note that the unit vector $\nu(x_0)$ is directed toward the region $\{x \in \Omega : \phi(x) < \phi(x_0)\}$. We agree to say that $\nu(x_0)$ is the *unit outward normal* to Γ in x_0 and $-\nu(x_0)$ is the *inner outward normal* to Γ in x_0, respectively.

Notice that Γ is a characteristic surface at x_0 for $P(x, \partial)$ if and only if

$$p(x_0, \nu(x_0)) = a(x_0) \cdot \nu(x_0) = 0, \tag{7.12}$$

In other words, Γ is a characteristic surface at x_0 if and only if $a(x_0)$ is a tangent vector to Γ at x_0.

Of course, when $n = 2$, instead of the characteristic surfaces we will simply speak of the characteristic lines (or curves).

In this section we will study the following **Cauchy problem** for the **first–order linear differential equation**. Given the vector field $a(\cdot)$, and the function $h : \Gamma \to \mathbb{R}$, determine u such that

$$\begin{cases} P(x, \partial)u = c(x)u + f(x) \\ \\ u_{|\Gamma} = h \end{cases} . \tag{7.13}$$

We will specify the assumptions on a, c, f, h in a while. Let us now premise some simple example to the investigation of problem (7.13). We call Γ the **initial surface** and h the **initial datum** of Cauchy problem (7.13).

Example 7.1 Let $\Omega = \mathbb{R}^2$. Let

$$P(x, y, \partial) = \partial_x + \partial_y.$$

Let us consider the equation

$$\partial_x u + \partial_y u = 0, \qquad \text{in } \mathbb{R}^2. \tag{7.14}$$

We can easily determine all the $C^1\left(\mathbb{R}^2\right)$ solutions to Eq. (7.14). Actually, setting

$$\mu = (1, 1),$$

we can write (7.14)

$$\frac{\partial u}{\partial \mu} = 0, \qquad \text{in } \mathbb{R}^2, \tag{7.15}$$

where $\frac{\partial u}{\partial \mu}$ denotes the derivatives of u w.r.t. direction μ (defined in Sect. 2.1).

It is clear that the functions $u \in C^1\left(\mathbb{R}^2\right)$ which satisfy (7.15) are all and only the functions which are constant on the lines parallel to the vector μ. Hence, for any fixed x_0, we have

$$u(x_0 + t, t) = F(x_0), \quad t \in \mathbb{R}$$

from which by eliminating t, we have

$$u(x, y) = F(x - y). \tag{7.16}$$

Therefore it suffices to assume that $F \in C^1(\mathbb{R})$ for obtaining, by (7.16), all the solutions to (7.14).

Having (7.16) available, the study of the Cauchy problem for Eq. (7.14) is quite simple, and here we take the opportunity to highlight some important facts.

Let us consider the following Cauchy problem

$$\begin{cases} u_x + u_y = 0 \\ \\ u(x, 0) = h(x) \end{cases}, \tag{7.17}$$

where $h \in C^1(\mathbb{R})$. By (7.16), taking into account of the initial datum in (7.17), we have

$$h(x) = u(x, 0) = F(x), \quad x \in \mathbb{R}.$$

Therefore, the unique solution of problem (7.17) is given by

$$u(x, y) = h(x - y), \quad (x, y) \in \mathbb{R}^2. \tag{7.18}$$

Let us now consider a somewhat more general situation and let us assume that the Cauchy datum is assigned on a regular curve Γ of parametric equations

$$x = \overline{x}(\tau), \qquad y = \overline{y}(\tau), \quad \tau \in I, \tag{7.19}$$

where I is an interval. Let us examine what happens when Γ is a **characteristic line**. We therefore consider the problem

$$\begin{cases} u_x + u_y = 0 \\ \\ u(\tau, \tau) = h(\tau), \quad \tau \in \mathbb{R} \end{cases}. \tag{7.20}$$

We immediately realize that if h is not constant, the problem (7.20) has no solutions: we had, indeed, already observed that every $C^1\left(\mathbb{R}^2\right)$ which is a solution of the equation $u_x + u_y = 0$ must be constant on the lines parallel to the vector $\mu = (1, 1)$ and therefore, in particular, they must be constant on the line

$$\{(\tau, \tau) : \tau \in \mathbb{R}\}.$$

Moreover, if h is constant then Cauchy problem (7.20) has **infinite solutions**. As a matter of fact if $h \equiv 0$ then every solutions to (7.20) is given by

$$u(x, y) = F(x - y), \quad \text{with } F(0) = 0.$$

More generally, if we have to face the Cauchy problem

$$\begin{cases} u_x + u_y = 0 \\ \\ u\left(\overline{x}(\tau), \overline{y}(\tau)\right) = h(\tau), \quad \tau \in \mathbb{R} \end{cases} \tag{7.21}$$

and if a characteristic line intersects Γ at two distinct, say $P_0 = (\overline{x}(\tau_0), \overline{y}(\tau_0))$ and $P_1 = (\overline{x}(\tau_1), \overline{y}(\tau_1))$ where $\tau_0 \neq \tau_1$, then in order that problem (7.21) has solution, it is necessary that h satisfies the condition

$$h(\tau_0) = h(\tau_1)$$

and this imposes, in turn, some restrictions on the datum h itself. In other words, **not every initial value is suitable for the Cauchy problem** (7.21). This limitation arises because the values of u on Γ are determined by the values of u on a

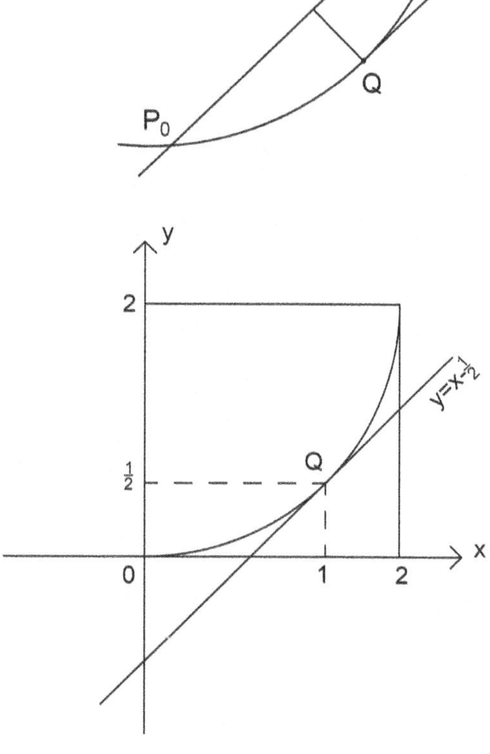

Fig. 7.2 A characteristic line $x - y = c$ intersects the initial line Γ at two distinct points, P_0 and P_1, which imposes the condition $h(P_0) = h(P_1)$ on the initial datum h

Fig. 7.3 The characteristic line $y = x - \frac{1}{2}$ is tangent to $\Gamma = \left\{\left(\tau, \frac{\tau^2}{2}\right) : 0 < \tau < 2\right\}$ at the point Q

smaller portion of Γ itself. It is important to note that in the specific situation under consideration, there must exist a point Q on Γ between P_0 and P_1 where the direction of the characteristic is tangential to Γ. In other words, Γ is a characteristic line with respect to the operator $\partial_x + \partial_y$ at the point Q (see Fig. 7.2).

Let us further illustrate what has just been said. Let it be, then, (Fig. 7.3)

$$\Gamma = \left\{\left(\tau, \frac{\tau^2}{2}\right) : \quad 0 < \tau < 2\right\}$$

and let us consider the following Cauchy problem

$$\begin{cases} u_x + u_y = 0 \\ \\ u\left(\tau, \frac{\tau^2}{2}\right) = h(\tau), \quad \tau \in (0, 2) \end{cases} . \tag{7.22}$$

Notice that Γ is a characteristic line w.r.t. the operator $\partial_x + \partial_y$ at the point $Q = \left(1, \frac{1}{2}\right)$. Moreover, we check what follows.

(i) There exists a solution $u \in C^1\left(\mathbb{R}^2\right)$ to problem (7.22) if and only if $h \in C^1((0,2))$ and

$$h(x) = h(2-x), \quad \forall x \in (0,2). \tag{7.23}$$

(ii) If $h \in C^1(0,2)$ satisfies (7.23), then for any $0 < r < 1$, there exist infinite solutions to the Cauchy problem

$$\begin{cases} u_x + u_y = 0, & \text{in } B_r(Q) \\ \\ u_{|\Gamma \cap B_r(Q)} = h \end{cases}. \tag{7.24}$$

Let us check (i) By (7.16) we have $u(x,y) = F(x-y)$. Consequently, in order to

$$u\left(x, \frac{x^2}{2}\right) = h(x),$$

we need to have

$$h(x) = F\left(x - \frac{x^2}{2}\right) = F\left(\frac{1}{2} - \frac{1}{2}(1-x)^2\right), \qquad \forall x \in (0,2),$$

from which (7.23) follows.

Let us check (ii) By the linearity of problem (7.24), we can choose $h \equiv 0$. Moreover, let $g \in C^1((0,2))$ be an arbitrary function which satisfies

$$g\left(\frac{1}{2}\right) = g'\left(\frac{1}{2}\right) = 0.$$

Then it is easily checked that the functions

$$u(x,y) = \begin{cases} 0, & \text{in } B_r(Q) \cap \left\{y \geq x - \frac{1}{2}\right\} \\ \\ g(x-y), & \text{in } B_r(Q) \cap \left\{y < x - \frac{1}{2}\right\} \end{cases}, \tag{7.25}$$

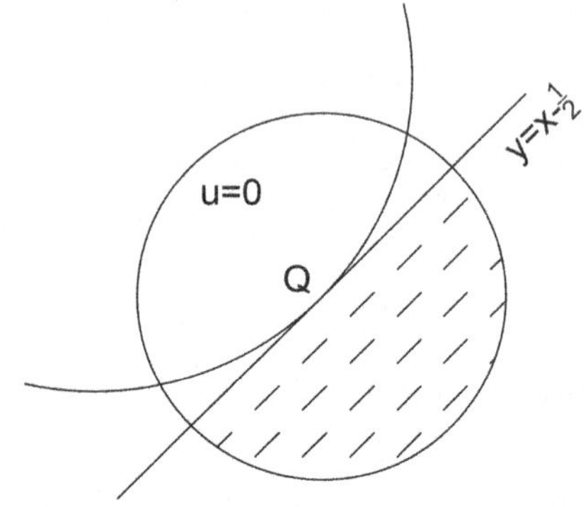

Fig. 7.4 In the shaded region, u can take values different from zero; therefore, the Cauchy problem (7.26) does not have uniqueness

are all solutions of Cauchy problem (Fig. 7.4)

$$\begin{cases} u_x + u_y = 0, & \text{in } B_r(Q) \\ \\ u_{|\Gamma \cap B_r(Q)} = 0 \end{cases} . \tag{7.26}$$

7.4 The Method of Characteristics: The Linear Case

Let Ω be a connected open set of $\mathbb{R}^n$ and let $a \in C^1(\Omega, \mathbb{R}^n)$,

$$a(x) = (a_1(x), \cdots, a_n(x)).$$

Let us consider the operator

$$P(x, \partial) = \sum_{j=1}^{n} a_j(x)\partial_j, \quad x \in \Omega. \tag{7.27}$$

We will call **characteristic line** of $P(x, \partial)$ any solution to the system of ordinary differential equations—**characteristic equations**

$$\frac{dX(t)}{dt} = a(X(t)). \tag{7.28}$$

Let $u \in C^1(\Omega)$. Let us observe that, if we set

$$z(t) = u(X(t)),$$

then we have

$$\frac{dz(t)}{dt} = \frac{dX(t)}{dt} \cdot (\nabla u)(X(t)) = a(X(t)) \cdot (\nabla u)(X(t)). \tag{7.29}$$

This simple relationship is the starting point of the method of characteristics by which we will tackle and we will solve the Cauchy problem

$$\begin{cases} P(x, \partial)u = c(x)u + f(x) \\ \\ u_{|\Gamma} = h \end{cases}, \tag{7.30}$$

where Γ is a portion of regular surface of parametric equations

$$x = \overline{x}(y), \qquad y \in B_1', \tag{7.31}$$

$\overline{x} \in C^1(B_1')$ and

$$\text{Rank}\left(\frac{\partial \overline{x}}{\partial y}(y)\right) = \text{Rank}\begin{pmatrix} \partial_{y_1}\overline{x}_1 & \cdots & \partial_{y_{n-1}}\overline{x}_1 \\ \vdots & \cdots & \vdots \\ \partial_{y_1}\overline{x}_n & \cdots & \partial_{y_{n-1}}\overline{x}_n \end{pmatrix} = n - 1. \tag{7.32}$$

Moreover, we assume

$$c \in C^1(\Omega), \qquad f \in C^1(\Omega). \tag{7.33}$$

The initial condition $u_\Gamma = h$ is expressed by

$$u(\overline{x}(y)) = h(\overline{x}(y)) := \overline{h}(y), \quad \forall y \in B_1', \tag{7.34}$$

where

$$\overline{h} \in C^1(B_1'). \tag{7.35}$$

The **method of characteristics** consists of constructing a local change of coordinates of class C^1,

$$(-\delta, \delta) \times B_r' \ni (t, y) \to X(t, y) \in \mathbb{R}^n, \tag{7.36}$$

for suitable $\delta > 0$ and $r \in (0, 1)$. Where X has the following properties

$$X(0, y) = \overline{x}(y), \quad \forall y \in B_r' \tag{7.37}$$

and

$$\partial_t X(t, y) = a(X(t, y)), \quad \forall (t, y) \in (-\delta, \delta) \times B'_r. \tag{7.38}$$

In this way, setting

$$z(t, y) = u(X(t, y)), \quad C(t, y) = c(X(t, y)), \quad F(t, y) = f(X(t, y))$$

and taking into account (7.29), Cauchy problem (7.30) becomes

$$\begin{cases} \partial_t z(t, y) = C(t, y) z(t, y) + F(t, y) \\ \\ z(0, y) = \overline{h}(y) \end{cases} . \tag{7.39}$$

By the assumptions made on c, f, and h, see Sect. 7.2, it turns out that $z \in C^1\left((-\delta, \delta) \times B'_r\right)$ and, as we will see, the function (under appropriate conditions)

$$u(x) = z\left(X^{-1}(x)\right), \tag{7.40}$$

is a solution to (7.30).

Let us begin to see under what conditions, the transformation defined by (7.36)–(7.38) is a diffeomorphism in a neighborhood of $0 \in \mathbb{R}^n$. Since, $a \in C^1(\Omega)$ and $\overline{x} \in C^1\left(B'_1\right)$ we have (for the differentiability w.r.t. the parameters, see Sect. 7.2)

$$X \in C^1\left(J \times B'_1\right),$$

where J is a suitable neighborhood of $0 \in \mathbb{R}$.

Moreover

$$\frac{\partial X}{\partial(t, y)}(0, 0) = \Big(\underbrace{\partial_t X(0, 0), \partial_{y_1} X(0, 0), \cdots, \partial_{y_{n-1}} X(0, 0)}_{\text{column vectors}}\Big) =$$

$$= \left(a(x_0), \partial_{y_1}\overline{x}(0), \cdots, \partial_{y_{n-1}}\overline{x}(0)\right)$$

and, since

$$\partial_t X(0, 0) = a(X(0, 0)) = a(x_0),$$

$$\text{Rank}\left(a(x_0), \partial_{y_1}\overline{x}(0), \cdots, \partial_{y_{n-1}}\overline{x}(0)\right) \geq n - 1,$$

we have that the following conditions are equivalent

$$\text{Rank}\frac{\partial X}{\partial(t,y)}(0,0)=n \tag{7.41}$$

and

$$a(x_0)\notin\left\langle\partial_{y_1}\overline{x}(0),\cdots,\partial_{y_{n-1}}\overline{x}(0)\right\rangle, \tag{7.42}$$

where, for any $v_1, v_2, \cdots, v_{n-1} \in \mathbb{R}^n$, $\langle v_1, v_2, \cdots, v_{n-1}\rangle$ is the vector space generated by $v_1, v_2, \cdots, v_{n-1}$. Condition (7.42) is, in turn, equivalent to the condition that $a(x_0)$ **is not tangent to** Γ **in** x_0.

All in all, **if** Γ **is noncharacteristic in** x_0 **for operator** (7.27) then there exists $\delta > 0$ and $r \in (0, 1)$ such that X is a diffeomorphism in $(-\delta, \delta) \times B'_r$.

Now, we denote by $\mathcal{U}_{x_0} = X\left((-\delta,\delta)\times B'_r\right)$ and by

$$\Psi(x)=X^{-1}(x), \quad x\in\mathcal{U}_{x_0}. \tag{7.43}$$

Let us check that

$$u(x)=z(\Psi(x)),$$

solves Cauchy problem (7.30).

Regarding the **initial condition**, by (7.37) we have immediately

$$\overline{x}(y)=X(0,y), \qquad \forall y\in B'_r.$$

Hence, by (7.43) and recalling (7.39), we get

$$u\left(\overline{x}(y)\right)=z\left(\Psi(X(0,y))\right)=z(0,y)=\overline{h}(y), \qquad \forall y\in B'_r. \tag{7.44}$$

Concerning **the equation**

$$\sum_{j=1}^{n}a_j(x)\partial_j u=c(x)u+f(x),$$

recall that by (7.43) we have

$$\left(\frac{\partial\Psi(x)}{\partial x}\right)\left(\frac{\partial X\left(\Psi(x)\right)}{\partial(t,y)}\right)=I_n, \tag{7.45}$$

where I_n is the identity matrix $n \times n$. In particular, considering the first column on the right–hand side and the first column on the left–hand side of (7.45), we have

$$\begin{cases} \sum_{j=1}^{n} \partial_{x_j} \Psi_1 \partial_t X_j = 1 \\ \\ \sum_{j=1}^{n} \partial_{x_j} \Psi_k \partial_t X_j = 0, \quad k = 2, \cdots, n \end{cases} . \tag{7.46}$$

Now

$$\partial_{x_j} u(x) = \partial_t z(\Psi(x)) \partial_{x_j} \Psi_1(x) + \partial_{y_1} z(\Psi(x)) \partial_{x_j} \Psi_2(x) + \\ \cdots + \partial_{y_{n-1}} z(\Psi(x)) \partial_{x_j} \Psi_n(x).$$

Hence (multiplying by $a_j(x)$ and summing up on j)

$$\begin{aligned} \sum_{j=1}^{n} a_j(x) \partial_{x_j} u &= \partial_t z(\Psi(x)) \sum_{j=1}^{n} \partial_{x_j} \Psi_1(x) a_j(x) + \\ &+ \partial_{y_1} z(\Psi(x)) \sum_{j=1}^{n} \partial_{x_j} \Psi_2(x) a_j(x) + \cdots \\ &+ \partial_{y_{n-1}} z(\Psi(x)) \sum_{j=1}^{n} \partial_{x_j} \Psi_n(x) a_j(x). \end{aligned} \tag{7.47}$$

On the other hand by (7.38) we know

$$a_j(x) = \left(\partial_t X_j\right)(\Psi(x)).$$

By this equality, by (7.46) and by the equation in (7.39) we have

$$\sum_{j=1}^{n} a_j(x) \partial_{x_j} u(x) = \partial_t z(\Psi(x)) = c(x) u(x) + f(x).$$

Hence, u is a solution to Cauchy problem (7.30).

Finally, we observe that, by the hypotheses (7.33) (actually, it suffices $c \in C^0(\Omega)$), u **is the unique solution** of class C^1 to problem (7.30) in the neighborhood $\mathcal{U}_{x_0}$. Indeed, if $u_1, u_2 \in C^1\left(\mathcal{U}_{x_0}\right)$ are two solutions then, setting

$$w = u_1 - u_2,$$

we have

$$\begin{cases} P(x, \partial)w = c(x)w \\ \\ w_{|\Gamma} = 0 \end{cases}$$

and setting

$$\widetilde{z}(t, y) = w(X(t, y)),$$

by (7.39) we have

$$\begin{cases} \partial_t \widetilde{z}(t, y) = C(t, y)\widetilde{z}(t, y) \\ \\ \widetilde{z}(0, y) = 0 \end{cases}.$$

From which we have $\widetilde{z} = 0$ in $(-\delta, \delta) \times B'_r$, therefore $w = 0$ in $\mathcal{U}_{x_0}$.

The construction we have illustrated and the local uniqueness hold for any point of Γ.

Hence we have proved

Theorem 7.6 *Let $a \in C^1(\Omega, \mathbb{R}^n)$, $c \in C^1(\Omega)$ and $f \in C^1(\Omega)$. Let Γ be a non characteristic surface of parametric equations $x = \overline{x}(y)$, where $\overline{x} \in C^1\left(B'_1\right)$ and satisfying* (7.32). *Let h be a function C^1 on Γ (i.e. $h \circ \overline{x} \in C^1\left(B'_1\right)$).*

Then there exists a neighborhood $\mathcal{U}$ of Γ such that there exists a unique solution u in $C^1(\mathcal{U})$ to the Cauchy problem

$$\begin{cases} \sum_{j=1}^n a_j(x)\partial_j u = c(x)u + f(x), \quad x \in \mathcal{U} \\ \\ u_{|\Gamma} = h \end{cases}. \tag{7.48}$$

Given a surface Γ in $\mathbb{R}^n$ we call **domain of dependence** of Γ with respect to the equation

$$\sum_{j=1}^n a_j(x)\partial_j u = c(x)u,$$

the largest closed set D_Γ for which we have

$$\begin{cases} \sum_{j=1}^n a_j(x)\partial_j u = c(x)u, \quad x \in D_\Gamma \\ \\ u_{|\Gamma} = 0 \end{cases} \implies u = 0 \quad \text{in } D_\Gamma.$$

Exercise 7.1 Let $0 < r < 1$ and

$$\Gamma = \left\{ \left(x', -\sqrt{1-|x'|^2} \right) : \quad |x'| < r \right\}.$$

Construct a vector field $a \in C^1\left(\overline{B_1'}\right)$ such that the domain of dependence of Γ with respect to the equation

$$a(x) \cdot \nabla u = 0,$$

contains $\overline{B_1}$. ♠

Exercise 7.2 Apply the characteristic method to prove that the functions $u \in C^1\left(\mathbb{R}^n \setminus \{0\}\right)$ which satisfy

$$\sum_{j=1}^{n} x_j \partial_j u = \alpha u,$$

are the homogeneous function of degree α. ♠

Exercise 7.3 Let b be a vector of $\mathbb{R}^n$ and let $f \in C^0\left(\mathbb{R}^{n+1}\right)$. Apply the characteristic method to solve the following Cauchy problem

$$\begin{cases} \partial_t u + b \cdot \nabla u = f(x,t) \\ \\ u(x,0) = 0 \end{cases}.$$

The equation $\partial_t u + b \cdot \nabla u = f(x,t)$ is known as the **transport equation**. ♠

7.5 The Method of Characteristics: Quasilinear Case

Let J be an open interval of $\mathbb{R}$, let Ω be a connected open set of $\mathbb{R}^n$. Let $a \in C^1\left(J \times \Omega, \mathbb{R}^n\right)$ and $c \in C^1\left(J \times \Omega\right)$. The following equation

$$a(x,u) \cdot \nabla u = c(x,u), \tag{7.49}$$

is called a **first-order quasilinear equation**. Of course, linear equations are special case of the quasilinear equations.

With minor modifications, the characteristics method studied in Sect. 7.4 can be adapted to handle Eq. (7.49) and the related Cauchy problem. In the case of Eq. (7.49), the characteristic equations are the following one

$$\begin{cases} \frac{dX}{dt}(t) = a(X(t), z(t)) \\ \\ \frac{dz}{dt}(t) = c(X(t), z(t)) \end{cases} . \tag{7.50}$$

Let us note that in the linear case, the equation

$$\frac{dz}{dt}(t) = c(X(t), z(t)),$$

is precisely the one satisfied by $z(t) = u(X(t))$ when u is a solution of the linear equation

$$a(x) \cdot \nabla u = c(x)u + f(x).$$

We continue to call **characteristic line**, the curve of parametric equations

$$(X, z) = (X(t), z(t)), \tag{7.51}$$

where $X(t), z(t)$ is a solution of system (7.50). When there is no risk of ambiguity, we will call "characteristic line" also the projection on $\mathbb{R}^n$ of the line (7.51).

Let us consider the Cauchy problem

$$\begin{cases} a(x, u) \cdot \nabla u = c(x, u) \\ \\ u_{|\Gamma} = h \end{cases} , \tag{7.52}$$

where Γ is a portion of regular surface of parametric equations

$$x = \overline{x}(y), \qquad \forall y \in B_1'.$$

To solve (7.52), we proceed similarly to what we did in the linear case. Namely, we consider $X(t, y)$ and $z(t, y)$ such that

$$\begin{cases} \partial_t X(t, y) = a(X(t, y), z(t, y)) \\ \partial_t z(t, y) = c(X(t, y), z(t, y)) \\ X(0, y) = \overline{x}(y) \\ z(0, y) = h(y) \end{cases} \tag{7.53}$$

and it can be checked, exactly as in Sect. 7.4 that if the map

$$(-\delta, \delta) \times B_r' \ni (t, y) \to X(t, y) \in \mathbb{R}^n,$$

for some $\delta > 0$ and $r \in (0, 1)$, is local change of coordinates of $\mathbb{R}^n$, then the function

$$u(x) := z\left(X^{-1}(x)\right), \tag{7.54}$$

is a solution to Cauchy problem (7.52). More precisely: setting $x_0 = \overline{x}(0)$, there exists a neighborhood $\mathcal{U}_{x_0}$ such that the function u defined by (7.54) satisfies

$$\begin{cases} a(x,u) \cdot \nabla u = c(x,u), & \text{in } \mathcal{U}_{x_0} \\ \\ u_{|\Gamma \cap \mathcal{U}_{x_0}} = h \end{cases}. \tag{7.55}$$

In order to the map X be a local diffeomorphism, it suffices to have

$$\text{Rank} \frac{\partial X}{\partial(t,y)}(0,0) = n \tag{7.56}$$

and since

$$\begin{aligned} \frac{\partial X}{\partial(t,y)}(0,0) &= \left(\partial_t X(0,0), \partial_{y_1} X(0,0), \cdots, \partial_{y_{n-1}} X(0,0)\right) = \\ &= \left(a(x_0, h(x_0)), \partial_{y_1}\overline{x}(0), \cdots, \partial_{y_{n-1}}\overline{x}(0)\right) \end{aligned}$$

and

$$\text{Rank}\left(\frac{\partial \overline{x}}{\partial y}(0)\right) = n - 1,$$

condition (7.56) is equivalent to

$$a(x_0, h(x_0)) \notin \left\langle \partial_{y_1}\overline{x}(0), \cdots, \partial_{y_{n-1}}\overline{x}(0) \right\rangle,$$

(compare this condition with (7.42)).

Now we briefly consider the issue of continuous dependence by the initial data in problem (7.52).

Let u_k, $k = 1, 2$ satisfy

$$\begin{cases} a(x,u_k) \cdot \nabla u_k = c(x,u_k) \\ \\ u_{k|\Gamma} = h_k \end{cases}. \tag{7.57}$$

We have

$$c(x, u_1) - c(x, u_2) = (u_1 - u_2) \int_0^1 \partial_u c\left(x, u_2(x) + t\left(u_1(x) - u_2(x)\right)\right) dt$$

and, similarly

$$a(x, u_1) \cdot \nabla u_1 - a(x, u_2) \cdot \nabla u_2 = \overline{a}(x) \cdot \nabla\left(u_1 - u_2\right) + b(x),$$

where

$$\overline{a}(x) = a(x, u_1(x))$$

and

$$b(x) = -\nabla u_2(x) \cdot \int_0^1 \partial_u a\left(x, u_2(x) + t\left(u_1(x) - u_2(x)\right)\right) dt.$$

Set

$$\overline{c}(x) = b(x) + \int_0^1 \partial_u c\left(x, u_2(x) + t\left(u_1(x) - u_2(x)\right)\right) dt,$$

$$w = u_1 - u_2,$$

$$\overline{h} = h_1 - h_2.$$

By (7.57) we have

$$\begin{cases} \overline{a}(x) \cdot \nabla w = \overline{c}(x) w \\ \\ w_{|\Gamma} = \overline{h} \end{cases} . \tag{7.58}$$

Now, if $\overline{h} \equiv 0$ and if $a(x_0, h_1(x_0))$ is not tangent to Γ at x_0, then there exists a neighborhood $\mathcal{U}_{x_0}$ of x_0 such that $w \equiv 0$ in $\mathcal{U}_{x_0}$, that is

$$u_1 \equiv u_2, \quad \text{in } \mathcal{U}_{x_0}.$$

As a matter of fact, the equation

$$\overline{a}(x) \cdot \nabla w = \overline{c}(x) w,$$

is linear and Theorem 7.6 applies.

We leave as an exercise to the reader to prove that if $a(x_0, h_1(x_0))$ is not tangent to Γ at x_0 then there exists a neighborhood $\mathcal{V}_{x_0}$ such that

$$\|w\|_{L^\infty(\mathcal{V}_{x_0})} \leq K \left\|\overline{h}\right\|_{L^\infty(\Gamma\cap\mathcal{V}_{x_0})}, \tag{7.59}$$

that is

$$\|u_1 - u_2\|_{L^\infty(\mathcal{V}_{x_0})} \leq K \|h_1 - h_2\|_{L^\infty(\Gamma\cap\mathcal{V}_{x_0})},$$

where K is a constant which depends on the C^1 norms of a and c and on the (convex) angle between the vector $a(x_0, h_1(x_0))$ and the unit outward normal to Γ in x_0.

We conclude this section by the following

Example Let us consider the Cauchy problem

$$\begin{cases} u_y + uu_x = 0 \\ \\ u(x,0) = h(x), \quad x \in \mathbb{R} \end{cases}, \tag{7.60}$$

where $h \in C^1(\mathbb{R})$.

The characteristic equations are given by

$$\begin{cases} \frac{\partial x(t,s)}{\partial t} = z \\ \\ \frac{\partial y(t,s)}{\partial t} = 1 \\ \\ \frac{\partial z(t,s)}{\partial t} = 0 \end{cases}, \tag{7.61}$$

the initial conditions are

$$x(0,s) = s, \qquad y(0,s) = 0 \qquad z(0,s) = h(s). \tag{7.62}$$

By (7.61) and (7.62) we have easily

$$\begin{cases} x(t,s) = s + th(s) \\ y(t,s) = t \\ z(t,s) = h(s) \end{cases}. \tag{7.63}$$

By the method shown in this section, the solution to (7.60) is given by a function u such that

$$u\left(x(t,s),y(t,s)\right)=z(t,s)=h(s). \tag{7.64}$$

To express u in the variables x and y we eliminate s and t from the first two equations of (7.63). We have

$$\begin{cases} x=s+yh(s) \\ \\ t=y \end{cases}.$$

To obtain s from the first equation it is necessary that $s\to s+yh(s)$ be injective, that is, it is necessary that

$$0\neq\frac{d}{ds}\left(s+yh(s)\right)=1+yh'(s).$$

For instance, if $h'>0$, let us assume

$$Y_0:=\sup_{s\in\mathbb{R}}-\frac{1}{h'(s)}<0,$$

then we have that the solution to (7.60) is defined for all y such that

$$y>Y_0.$$

If $h'(s)<0$, we assume

$$Y_1:=\inf_{s\in\mathbb{R}}-\frac{1}{h'(s)}>0,$$

then the solution to (7.60) is defined for all y such that

$$y<Y_1.$$

Let us dwell for a while on the latter case and examine what happens above the line $y=Y_1$. Let us come back to (7.64); this relation tells us that u is constant on the projection of the characteristic passing through the point $(s,0)$ and there it is equal to $h(s)$. Let now $s_1,s_2\in\mathbb{R}$ satisfy $s_1<s_2$, then the straight lines whose equations are given by

$$x=s_1+yh(s_1)$$

and

$$x=s_2+yh(s_2),$$

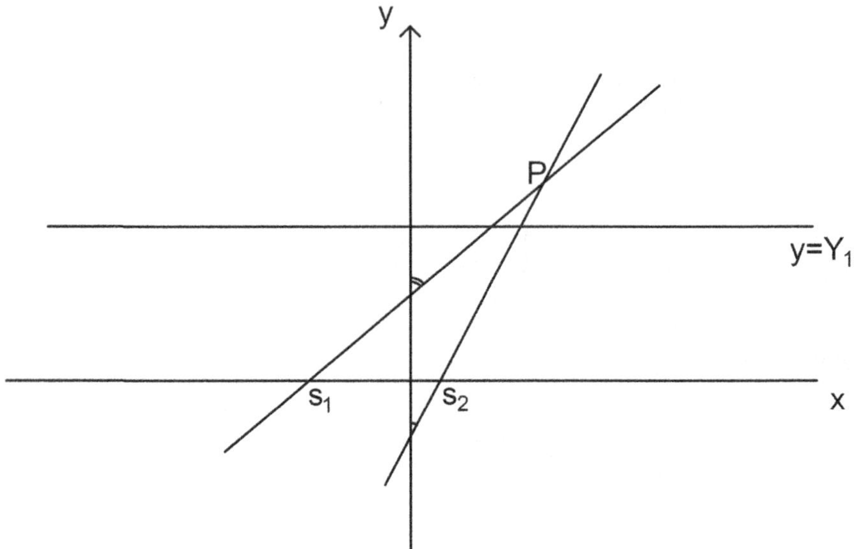

Fig. 7.5 The characteristic lines $x = s_1 + yh(s_1)$ and $x = s_2 + yh(s_2)$ intersect at the point P, therefore the solution u to problem (7.60) cannot be continuous in P

intersect at the point

$$P = \left(-\frac{s_2 h(s_1)}{h(s_2) - h(s_1)}, -\frac{s_2 - s_1}{h(s_2) - h(s_1)}\right),$$

that implies that the function u *cannot be continuous* in P. Let us observe that the point P is situated either on the line $y = Y_1$ or above such a line $y = Y_1$ as, for an appropriate $\overline{s} \in (s_1, s_2)$, we have (Fig. 7.5)

$$-\frac{s_2 - s_1}{h(s_2) - h(s_1)} = -\frac{1}{h'(\overline{s})} \geq Y_1.$$

It is, actually, of some physical interest to include (in an appropriate sense) the discontinuous solution among the solutions to problem (7.60) since they correspond to "shock waves". We refer for insights to [38, Ch. 3, Sec. 4].

7.6 Brief Review on the Fully Non–linear Case

In this section we wish briefly consider the method of the characteristics to solve the Cauchy problem for the fully non–linear equation

$$F\,(x, u(x), \nabla u(x)) = 0.$$

Namely, the Cauchy problem

$$\begin{cases} F\left(x, u(x), \nabla u(x)\right) = 0, & \text{in } \Omega \\ \\ u_{|\Gamma} = g \end{cases}, \tag{7.65}$$

where Ω is an open set $\mathbb{R}^n$, Γ is a regular surface of $\mathbb{R}^n$ contained Ω,

$$F : \Omega \times \mathbb{R} \times \mathbb{R}^n \to \mathbb{R}, \tag{7.66}$$

is a function of class $C^2(\Omega \times \mathbb{R} \times \mathbb{R}^n)$. The variables of F are $x \in \Omega$, $z \in \mathbb{R}$ and $p \in \mathbb{R}^n$. Moreover $g : \Gamma \to \mathbb{R}$ is a function defined on Γ. The assumptions on Γ, F and g will be specified in more detail later on. We notice that in the linear and quasilinear cases investigated, respectively, in Sects. 7.3 and 7.5, we have

$$F(x, z, p) = a(x) \cdot p - c(x)z - f(x), \quad F(x, z, p) = a(x, z) \cdot p - c(x, z).$$

Let us notice that in the non–linear case, generally, we cannot expect the uniqueness of the solutions to Cauchy (7.65). The following simple example will help us to understand this fact. Let us consider the Cauchy problem

$$\begin{cases} u_{x_1}^2 + u_{x_2}^2 = 1, & \text{in } \mathbb{R}^2 \\ \\ u(x_1, 0) = g(x_1), & \text{for } x_1 \in \mathbb{R} \end{cases}. \tag{7.67}$$

We note that all we can say about $u_{x_2}(x_1, 0)$ is that it satisfies the condition

$$g'^2(x_1) + u_{x_2}^2(x_1, 0) = 1,$$

which leaves undetermined the sign of $u_{x_2}(x_1, 0)$. If, for instance, $g = 0$, then $u = x_2$ and $u = -x_2$ are both solutions of Cauchy problem (7.67).

We have already explored the method of characteristics for both linear and quasi-linear equations. In the non–linear case, we adopt a procedure similar to the previous two cases, but identifying the characteristic equations is less straightforward. To tackle this, certain geometric insights could prove valuable. However, we won't delve into those geometric considerations here, directing interested readers to [66, Ch. 1] for further exploration. Let us begin by considering the following

Definition 7.7 Let F be the function (7.66). Assume that F is of class $C^2(\Omega \times \mathbb{R} \times \mathbb{R}^n)$. We call the characteristic equations related to the partial differential equation

$$F\left(x, u(x), \nabla u(x)\right) = 0, \text{ in } \Omega, \tag{7.68}$$

the following system of ordinary differential equations

$$\begin{cases} \frac{dX(t)}{dt} = \nabla_p F(X(t), z(t), p(t)) \\ \\ \frac{dz(t)}{dt} = \nabla_p F(X(t), z(t), p(t)) \cdot p(t) \\ \\ \frac{dp(t)}{dt} = -\partial_z F(X(t), z(t), p(t)) p(t) - \nabla_x F(X(t), z(t), p(t)) \end{cases} . \tag{7.69}$$

The function $X(\cdot), z(\cdot), p(\cdot)$ are called the **characteristic lines of the equation** (7.68). $X(\cdot)$ is called the **ray** or the **projected characteristic lines** on $\mathbb{R}^n$.

If F does not depend on z, system (7.69) is decoupled in z, while the first and the third equations constitute the **Hamilton–Jacobi system**:

$$\begin{cases} \frac{dX(t)}{dt} = \nabla_p F(X(t), p(t)) \\ \\ \frac{dp(t)}{dt} = -\nabla_x F(X(t), p(t)) \end{cases} . \tag{7.70}$$

Remarks

1. Let us note that if $(X(\cdot), z(\cdot), p(\cdot))$ is a characteristic line for Eq. (7.68) then

$$F(X(t), z(t), p(t)) = \text{constant}. \tag{7.71}$$

As a matter of fact, exploiting (7.69), we have

$$\begin{aligned} \frac{d}{dt} F(X(t), z(t), p(t)) &= \nabla_x F(X(t), z(t), p(t)) \cdot \frac{dX}{dt} + \\ &+ F_z(X(t), z(t), p(t)) \frac{dz}{dt} + \nabla_p F(X(t), z(t), p(t)) \cdot \frac{dp}{dt} = \\ &= \nabla_x F \cdot \nabla_p F + F_z(\nabla_p F \cdot p(t)) + \\ &+ \nabla_p F \cdot (-F_z p(t) - \nabla_x F) = 0, \end{aligned}$$

In the last step, for the sake of brevity, we have omitted the arguments $X(t), z(t), p(t)$ in F.

2. Let us suppose that u is a solution of class C^2 to the equation

$$F(x, u(x), \nabla u(x)) = 0, \tag{7.72}$$

we wish to look for an equation for $X(t)$ *such that, setting*

$$z(t) = u(X(t)), \qquad p(t) = \nabla u(X(t)), \tag{7.73}$$

it happens that $(X(t), z(t), p(t))$ *solves system* (7.69).

We have

$$\frac{dz(t)}{dt} = p(t) \cdot \frac{dX(t)}{dt}, \tag{7.74}$$

$$\frac{dp_i(t)}{dt} = \sum_{j=1}^{n} \partial_{ij}^2 u(X(t)) \frac{dX_j(t)}{dt}, \quad i = 1, \cdots, n. \tag{7.75}$$

Now, calculating the derivatives of both the sides of Eq. (7.72) w.r.t. x_i, $i = 1, \cdots n$, we have

$$\sum_{j=1}^{n} \partial_{p_j} F \partial_{ij}^2 u(x) = -\partial_{x_i} F - \partial_z F \partial_{x_i} u(x), \tag{7.76}$$

where the argument of F in (7.76) is $(x, u(x), \nabla u(x))$. Now, let us observe what follows: if

$$\frac{dX_j(t)}{dt} = \partial_{p_j} F(X(t), z(t), p(t)), \quad j = 1, \cdots, n, \tag{7.77}$$

then by (7.75) and by (7.76), calculated for $x = X(t)$, we have, for $i = 1, \cdots, n$

$$\frac{dp_i(t)}{dt} = -\partial_{x_i} F(X(t), z(t), p(t)) - \partial_z F(X(t), z(t), p(t)) p_i(t) \tag{7.78}$$

(recall that, by the second of (7.73), $\partial_{x_i} u(X(t)) = p_i(t)$) and by (7.74) we have

$$\frac{dz(t)}{dt} = \sum_{j=1}^{n} \partial_{p_j} F(X(t), z(t), p(t)) p_j(t). \tag{7.79}$$

Equations (7.77), (7.78) and (7.79) are just the equations of system (7.69). ♦

To solve the Cauchy problem (7.65), we will adopt an approach similar to the one used in the linear (and quasilinear) cases. This involves utilizing the projected characteristic lines $X(t)$, which play a role akin to the characteristic lines in the linear case.

In what follows we will consider the case

$$\Gamma = \{x \in \Omega : x_n = 0\}. \tag{7.80}$$

We observe that we can always lead back to this situation, at least locally, even if Γ is given by

$$\Gamma = \{x \in \Omega : \phi(x) = 0\},$$

where $\phi \in C^3(\Omega)$ and $\phi(x_0) = 0$ for a given $x_0 \in \Omega$ and

$$\nabla\phi(x_0) \neq 0. \tag{7.81}$$

Indeed, thanks to (7.81), a neighborhood $\mathcal{U}$ of x_0 exists such that $\Gamma \cap \mathcal{U}$ forms a graph of a function of $n-1$ variables. For instance, let us consider the case where $\phi_{x_n}(x_0) \neq 0$. Up to a translation that relocates x_0 to the origin, we can assume that, for a suitable $\delta > 0$, we have

$$\Gamma \cap \mathcal{U} = \left\{(x', \varphi(x')) : x' \in B'_\delta\right\}, \tag{7.82}$$

where $\varphi \in C^3(B'_\delta)$, $\varphi(0) = |\nabla_{x'}\varphi(0)| = 0$. Now, let

$$\Lambda : B_\delta \subset \mathbb{R}^n_x \to \mathbb{R}^n_y, \quad \Lambda(x) = \left(x', x_n - \varphi(x')\right),$$

$$\Lambda(\Gamma) = \left\{(y', 0) : y' \in B'_\delta\right\} = \{y \in B_\delta : -y_n = 0\}$$

and, setting

$$v(y) = u\left(\Lambda^{-1}(y)\right),$$

we easily obtain that problem (7.65) takes the form

$$\begin{cases} \widetilde{F}\left(y, v(y), \nabla_y v(y)\right) = 0, & \text{in } \mathcal{V} \\ \\ v(y) = \widetilde{g}(y), & \text{for } y \in \Lambda(\Gamma) \cap \mathcal{V} \end{cases}, \tag{7.83}$$

where $\mathcal{V}$ is a neighborhood of 0 and $\widetilde{F} : \mathcal{V} \times \mathbb{R} \times \mathbb{R}^n \to \mathbb{R}$ is a function of class $C^2(\mathcal{V} \times \mathbb{R} \times \mathbb{R}^n)$.

Theorem 7.8 *Let $R > 0$ and $F \in C^2(B_R \times \mathbb{R} \times \mathbb{R}^n)$. Let $g \in C^2(B'_R)$. Let $\eta \in \mathbb{R}$ satisfy*

$$F\left(0, g(0), \nabla_{x'} g(0), \eta\right) = 0 \tag{7.84}$$

and

$$F_{p_n}\left(0, g(0), \nabla_{x'} g(0), \eta\right) \neq 0, \tag{7.85}$$

then for some $r \in (0, R)$ there exists a unique solution $u \in C^2(B_r)$ to the initial–value problems

$$\begin{cases} F\left(x, u(x), \nabla u(x)\right) = 0, & in\ B_r \\ u(x', 0) = g(x'), & for\ x' \in B'_r \\ u_{x_n}(0,0) = \eta \end{cases} . \tag{7.86}$$

Proof Let us begin by proving **the uniqueness**. It will suffice to prove that if u is a solution to

$$\begin{cases} F\left(x, u(x), \nabla u(x)\right) = 0 \\ u(x', 0) = g(x') \\ u_{x_n}(0,0) = \eta \end{cases} , \tag{7.87}$$

then there exists a neighborhood of 0 in which u is uniquely determined.

Set

$$z_0 = g(0), \qquad p'_0 = \nabla_{x'} g(0).$$

By (7.84) and (7.85) we have

$$F\left(0, z_0, p'_0, \eta\right) = 0, \qquad F_{p_n}\left(0, z_0, p'_0, \eta\right) \neq 0.$$

By applying the Implicit Function Theorem, we have that there exists $\delta > 0$ such that, setting

$$\mathcal{U} = B_\delta \times (z_0 - \delta, z_0 + \delta) \times B'_\delta(p'_0) \times (\eta - \delta, \eta + \delta)\,,$$

the set

$$\{(x, z, p) \in \mathcal{U} :\ F(x, z, p) = 0\}\,,$$

is equal to the graph of the function

$$\psi : B_\delta \times (z_0 - \delta, z_0 + \delta) \times B'_\delta(p'_0) \to (\eta - \delta, \eta + \delta)\,, \tag{7.88}$$

where ψ is of class C^2 and

$$\psi(0, z_0, p'_0) = \eta.$$

Now, since u satisfies (7.87), we have

$$F\left(x', 0, g(x'), \nabla_{x'} g(x'), u_{x_n}(x', 0)\right) = 0,$$

$$u_{x_n}(0,0) = \eta,$$

and we have

$$u_{x_n}(x', 0) = \psi\left(x', 0, g(x'), \nabla_{x'}g(x')\right), \quad \forall x' \in B'_\delta.$$

Set

$$p_n(x') = \psi\left(x', 0, g(x'), \nabla_{x'}g(x')\right)$$

and

$$p^{(0)}(x') = \left(\nabla_{x'}g(x'), p_n(x')\right), \tag{7.89}$$

we have

$$\begin{cases} F\left(x, u(x), \nabla u(x)\right) = 0, & \text{in } B_\delta \\ u(x', 0) = g(x'), & \text{for } x' \in B'_\delta \\ \nabla u(x', 0) = p^{(0)}(x'), & \text{for } x' \in B'_\delta \end{cases}. \tag{7.90}$$

We denote by y an arbitrary point of B'_δ and recalling that, by Remark 2 of the present section, the function

$$t \to (X(t), z(t), p(t)) := (X(t, y), u(X(t, y)), \nabla u(X(t, y))$$

is a solution to the characteristic equations (for each $y \in B'_\delta$)

$$\begin{cases} \frac{dX}{dt} = \nabla_p F(X, z, p) \\ \\ \frac{dz}{dt} = \nabla_p F(X, z, p) \cdot p \\ \\ \frac{dp}{dt} = -\partial_z F(X, z, p)p - \nabla_x F(X, z, p) \end{cases} \tag{7.91}$$

and

$$\begin{cases} X(0, y) = (y, 0) \\ z(0, y) = u(X(0, y)) = g(y) \\ p(0, y) = \nabla u(X(0, y)) = p^{(0)}(y) \end{cases}. \tag{7.92}$$

Therefore, due to the uniqueness of the solution to Cauchy problem (7.91) and (7.92) it turns out that $u(X(t, y))$ is determined for every $y \in B'_\delta$ and for every t in a neighborhood of 0 (this neighborhood depends on y). To conclude the proof, it suffices, therefore, to prove that the map

$$(t, y) \to X(t, y), \tag{7.93}$$

is a diffeomorphism in a neighborhood of 0. From what we said in Sect. 7.2 (final part), map (7.93) is of class C^2. To establish that it is a local diffeomorphism, it suffices to check that the Jacobian matrix of

$$(t, y) \to X(t, y),$$

is nonsingular in 0. Now, from (7.92) we have

$$\frac{\partial X(0,0)}{\partial(t,y)} = \begin{pmatrix} \partial_t X_1(0,0) \; \partial_{y_1} X_1(0,0) & \cdots & \partial_{y_{n-1}} X_1(0,0) \\ \vdots \; \vdots & \cdots & \vdots \\ \partial_t X_n(0,0) \; \partial_{y_1} X_n(0,0) & \cdots & \partial_{y_{n-1}} X_n(0,0) \end{pmatrix}. \tag{7.94}$$

On the other hand

$$\partial_{y_i} X_j(0,0) = \delta_{ij}, \text{ for } 1 \le i \le n-1, \;\; 1 \le j \le n-1,$$

$$\partial_{y_i} X_n(0,0) = 0, \text{ for } 1 \le i \le n-1$$

and, for $1 \le j \le n-1$,

$$\partial_t X_j(0,0) = \partial_{p_j} F(X(0,0), z(0,0), p(0,0)) = \partial_{p_j} F\left(0, g(0), \nabla_{x'} g(0), \eta\right).$$

Hence

$$\det\left(\frac{\partial X(0,0)}{\partial(t,y)}\right) = (-1)^n \partial_{p_n} F\left(0, g(0), \nabla_{x'} g(0), \eta\right) \neq 0, \tag{7.95}$$

from which it follows that map (7.93) is a local diffeomorphism. The proof of uniqueness is complete.

Now, let us prove **the existence** of the solution to problem (7.86). Let $p^{(0)}(x')$ be defined by (7.89) and let $(X(t, y), z(t, y), p(t, y))$ be the solution of the Cauchy problem comprising the system (7.91) and the initial conditions

$$\begin{cases} X(0, y) = (y, 0) \\ z(0, y) = g(y) \\ p(0, y) = p^{(0)}(y) \end{cases}. \tag{7.96}$$

Set

$$f(t, y) = F\left(X(t, y), z(t, y), p(t, y)\right),$$

we have

$$f(0, y) = F\left((y, 0), g(y), p^{(0)}(y)\right) =$$
$$= F\left((y, 0), g(y), \nabla_y g(y), \psi\left((y, 0), g(y), \nabla_y g(y)\right)\right) = 0,$$

where ψ is given by (7.88). Hence, by (7.71), we have

$$f(t, y) = F\left(X(t, y), z(t, y), p(t, y)\right) = 0. \tag{7.97}$$

Moreover, in a completely similar way to what has been done above for the uniqueness, we have that there exists $\delta_1 > 0$ and a neighbourhood of 0, $\mathcal{U}_0$, such that

$$B_{\delta_1} \ni (t, y) \to X(t, y) \in \mathcal{U}_0,$$

is a diffeomorphism of class $C^2\left(B_{\delta_1}(0)\right)$. We denote the inverse of $X(\cdot, \cdot)$ by

$$X^{-1}(x) = (t(x), y(x))$$

and set

$$u(x) = z(t(x), y(x)), \quad p(x) = p(t(x), y(x)).$$

The remaining part of the proof consists of proving that u satisfies (7.86). First of all, we check that

$$u(x', 0) = g(x'). \tag{7.98}$$

To this purpose we note that

$$t(x', 0) = 0, \quad y(x', 0) = x'.$$

Hence

$$u(x', 0) = z\left(t(x', 0), y(x', 0)\right) = z\left(0, x'\right) = g\left(x'\right).$$

Therefore, we have (7.98). We will check the condition $\partial_{x_n} u(0, 0) = \eta$ later on, now we check that u satisfies the equation

$$F\left(x, u(x), \nabla u(x)\right) = 0. \tag{7.99}$$

Firstly, we observe that from (7.97) we have

$$F\left(x, u(x), p(x)\right) = f(t(x), y(x)) = 0, \quad \forall x \in \mathcal{U}_0. \tag{7.100}$$

Therefore, to prove (7.99) it suffices to prove that

$$p(x) = \nabla u(x), \quad \forall x \in \mathcal{U}_0. \tag{7.101}$$

To this purpose we prove the following claims:
Claim I

$$\partial_t z(t, y) = \sum_{j=1}^{n} p_j(t, y)\partial_t X_j(t, y), \quad \forall (t, y) \in B_{\delta_1}. \tag{7.102}$$

Claim II

$$\partial_{y_i} z(t, y) = \sum_{j=1}^{n} p_j(t, y)\partial_{y_i} X_j(t, y), \quad \forall (t, y) \in B_{\delta_1}. \tag{7.103}$$

Claim I follows by the first and the second equation of (7.91). As a matter of fact, we have

$$\partial_t z(t, y) = \nabla_p F\left(X(t, y), z(t, y), p(t, y)\right) \cdot p = \partial_t X(t, y) \cdot p(t, y).$$

The proof of *Claim II* is less immediate than Claim I. Set

$$h_i(t, y) = \partial_{y_i} z(t, y) - \sum_{j=1}^{n} p_j(t, y)\partial_{y_i} X_j(t, y). \tag{7.104}$$

By (7.89) and recalling that

$$\begin{aligned} &\partial_{y_i} X_j(0, y) = \delta_{ij}, \quad \text{for } 1 \leq i, j \leq n-1, \\ &\partial_{y_i} X_n(0, y) = 0, \quad \text{for } 1 \leq i \leq n-1, \end{aligned} \tag{7.105}$$

we have, for $i = 1, \cdots, n-1$,

$$h_i(0, y) = \partial_{y_i} z(0, y) - \sum_{j=1}^{n} p_j(0, y)\partial_{y_i} X_j(0, y) = \partial_{y_i} g(y) - p_i^{(0)}(y) = 0. \tag{7.106}$$

Let us prove that $h_i(\cdot, y)$ satisfies

$$\partial_t h_i(t, y) = -\partial_z F\left(X(t, y), z(t, y), p(t, y)\right) h_i(t, y). \tag{7.107}$$

By (7.102) we have

$$\partial^2_{ty_i} z = \sum_{j=1}^{n} \left(\partial_{y_i} p_j \partial_t X_j + p_j \partial^2_{ty_i} X_j \right). \tag{7.108}$$

Now making the derivative w.r.t. t of both the sides of (7.104) we have

$$\partial_t h_i = \partial^2_{ty_i} z - \sum_{j=1}^{n} \left(\partial_t p_j \partial_{y_i} X_j + p_j \partial^2_{ty_i} X_j \right).$$

By this equality, by (7.108) and by (7.91) we have

$$\begin{aligned} \partial_t h_i &= \sum_{j=1}^{n} (\partial_{y_i} p_j \partial_t X_j - \partial_t p_j \partial_{y_i} X_j) = \\ &= \sum_{j=1}^{n} (\partial_{y_i} p_j \partial_{p_j} F - (-\partial_{x_j} F - \partial_z F p_j) \partial_{y_i} X_j) = \\ &= \sum_{j=1}^{n} (\partial_{y_i} p_j \partial_{p_j} F + \partial_{x_j} F \partial_{y_i} X_j + \partial_z F p_j \partial_{y_i} X_j). \end{aligned} \tag{7.109}$$

Now by (7.97), making the derivative w.r.t. y_i of both the sides, we get

$$\sum_{j=1}^{n} (\partial_{y_i} p_j \partial_{p_j} F + \partial_{x_j} F \partial_{y_i} X_j) = -\partial_z F \partial_{y_i} z$$

and inserting the latter into (7.109) we get

$$\begin{aligned} \partial_t h_i &= -\partial_z F \partial_{y_j} z + \sum_{j=1}^{n} \partial_z F p_j \partial_{y_i} X_j = \\ &= -\partial_z F \left(\partial_{y_i} z - \sum_{j=1}^{n} p_j \partial_{y_i} X_j \right) = \\ &= -\partial_z F h_i. \end{aligned}$$

All in all, by the latter and by (7.106) we get, for $i = 1, \cdot, n-1$,

$$\begin{cases} \partial_t h_i = -\partial_z F h_i \\ \\ h_i(0, y) = 0 \end{cases},$$

from which we have

$$h_i(t, y) = 0, \quad \text{for } i = 1, \cdot, n-1.$$

This conclude the proof of the Claim II.

Now, let us prove (7.101). First, we recall that

$$u(x) = z\left(X^{-1}(x)\right) = z(t(x), y(x)).$$

We have by (7.102) and (7.103),

$$\begin{aligned}
\partial_{x_i} u &= \partial_t z \partial_{x_i} t + \sum_{j=1}^{n-1} \partial_{y_j} z \partial_{x_i} y_j = \\
&= \left(\sum_{k=1}^{n} p_k \partial_t X_k\right) \partial_{x_i} t + \sum_{j=1}^{n-1} \sum_{k=1}^{n} p_k \partial_{y_j} X_k \partial_{x_i} y_j = \\
&= \sum_{k=1}^{n} p_k \left(\partial_t X_k \partial_{x_i} t + \sum_{j=1}^{n-1} \partial_{y_j} X_k \partial_{x_i} y_j\right) = \\
&= \sum_{k=1}^{n} p_k \partial_{x_i} \left(X_k \left(X^{-1}(x)\right)\right) = \\
&= \sum_{k=1}^{n} p_k \delta_{ik} = p_i,
\end{aligned}$$

for $i = 1, \cdots, n$. From which we have (7.101) and, in particular,

$$\partial_{x_n} u(0) = p_n(0) = \eta.$$

Which concludes the proof. ■

7.7 Geodesics and Hamilton–Jacobi Equations

We would like to note that throughout this section, we will employ the convention of repeated indices. Additionally, we will rigorously adhere to the notation of indices (upper or lower) for the components of a tensor.

We start by outlining the essential concepts of Hamilton-Jacobi equations. It is worth mentioning that these topics can be explored in a broader context. For more in-depth information, we recommend referring to [38].

Let Ω be an open set of $\mathbb{R}^n$, we say that a real–valued function,

$$L \in C^{\infty}\left(\Omega \times \mathbb{R}^n\right),$$

is a **Lagrangian** on Ω. An example of Lagrangian that we are interested in is given by

$$L(x,q) = \frac{1}{2} g_{ij}(x) q^i q^j, \quad \forall x \in \Omega, \ \forall q \in \mathbb{R}^n, \tag{7.110}$$

where $\left\{g_{ij}(x)\right\}_{i,j=1}^n$ is a real symmetric nonsingular matrix, $n \times n$, whose entries belong to $C^{\infty}(\Omega)$.

Given a Lagrangian L we will call **equation of Euler–Lagrange** the differential equation in the unknown $x = x(t)$

$$\frac{d}{dt}\left(\nabla_q L\left(x(t), \dot{x}(t)\right)\right) - \nabla_x L\left(x(t), \dot{x}(t)\right) = 0. \tag{7.111}$$

Here and in the sequel we will indistinctly denote by

$$\frac{df}{dt} \text{ or by } \dot{f}(t),$$

the derivative with respect to t of a differentiable function f. The solutions $x : [t_0, t_1] \to \mathbb{R}^n$ of (7.111) are also called the **extremal** of the functional

$$\int_{t_0}^{t_1} L\left(x(t), \dot{x}(t)\right) dt. \tag{7.112}$$

Assumption I In what follows we suppose that, for every $p \in \mathbb{R}^n$, the equation

$$\nabla_q L(x,q) = p, \tag{7.113}$$

has a unique solution of class $C^{\infty}(\Omega \times \mathbb{R}^n)$. We denote such a solution by $q(x,p)$.

The function

$$H(x, p) = p \cdot q(x, p) - L\left(x, q(x, p)\right). \tag{7.114}$$

is called the **Hamiltonian** associated to L

We have

Theorem 7.9 *Let $x = x(t)$ be a solution to Euler–Lagrange equation*

$$\frac{d}{dt}\left(\nabla_q L\left(x(t), \dot{x}(t)\right)\right) - \nabla_x L\left(x(t), \dot{x}(t)\right) = 0. \tag{7.115}$$

Then, setting

$$p(t) = \nabla_q L\left(x(t), \dot{x}(t)\right),$$

it turns out that $(x(t), p(t))$ is a solution of the Hamilton–Jacobi system

$$\begin{cases} \frac{dx}{dt} = \nabla_p H(x(t), p(t)) \\ \\ \frac{dp}{dt} = -\nabla_x H(x(t), p(t)) \end{cases}. \tag{7.116}$$

Moreover

$$H(x(t), p(t)) = constant. \tag{7.117}$$

Proof Let $x(t)$ be a solution to Eq. (7.115). Then, since Eq. (7.113) has a unique solution, $q(x, p)$, and since

$$p(t) = \nabla_q L\left(x(t), \dot{x}(t)\right),$$

we have

$$\dot{x}(t) = q(x(t), p(t)). \tag{7.118}$$

Now we have, for $i = 1, \cdots, n$,

$$\begin{aligned} &\partial_{p_i} H(x, p) = \partial_{p_i}\left(p \cdot q(x, p) - L\left(x, q(x, p)\right)\right) = \\ &= p_k \partial_{p_i} q^k(x, p) + q^i(x, p) - \partial_{q^k} L(x, q(x, p)) \partial_{p_i} q^k(x, p) = \\ &= \partial_{p_i} q^k(x, p)\left(p_k - \partial_{q^k} L(x, q(x, p))\right) + q^i(x, p) = \\ &= q^i(x, p). \end{aligned}$$

Hence, recalling (7.118), we have, for $i = 1, \cdots, n$,

$$\frac{dx^i(t)}{dt} = \partial_{p_i} H(x(t), p(t)),$$

which is the first equation of system (7.116). Concerning the second equation, for $i = 1, \cdots, n$, we have

$$\begin{aligned}
&\partial_{x^i} H(x, p) = \partial_{x^i} (p \cdot q(x, p) - L(x, q(x, p))) = \\
&= p_k \partial_{x^i} q^k(x, p) - \partial_{x^i} L(x, q(x, p)) - \partial_{q^k} L(x, q(x, p)) \partial_{x^i} q^k(x, p) = \\
&= \partial_{x^i} q^k(x, p) \left(p_k - \partial_{q^k} L(x, q(x, p))\right) - \partial_{x^i} L(x, q(x, p)) = \\
&= -\partial_{x^i} L(x, q(x, p)).
\end{aligned} \tag{7.119}$$

On the other hand, by (7.115) and (7.118), we have, for $i = 1, \cdots, n$,

$$\begin{aligned}
\partial_{x^i} L(x(t), q(x(t), p(t)) &= \partial_{x^i} L(x(t), \dot{x}(t)) = \\
&= \frac{d}{dt} \partial_{q^i} L(x(t), \dot{x}(t)) = \\
&= \frac{dp_i(t)}{dt}.
\end{aligned}$$

By the just obtained equality and by (7.119) we get

$$\frac{dp_i(t)}{dt} = -\partial_{x^i} H(x(t), p(t)), \qquad \text{for } i = 1, \cdots, n. \tag{7.120}$$

which is the second equation of system (7.116).

Finally, (7.117) follows by (7.116) and by

$$\begin{aligned}
&\frac{d}{dt} H(x(t), p(t)) = \partial_{x^i} H(x(t), p(t)) \frac{dx^i(t)}{dt} + \partial_{p_i} H(x(t), p(t)) \frac{dp_i(t)}{dt} = \\
&= \partial_{x^i} H(x(t), p(t)) \partial_{p_i} H(x(t), p(t)) - \partial_{p_i} H(x(t), p(t)) \partial_{x^i} H(x(t), p(t)) = 0.
\end{aligned}$$

■

Assumption II Let H be the Hamiltonian associated to the Lagrangian L which satisfies Assumption I. Let us suppose that, for every $q \in \mathbb{R}^n$, the equation

$$\nabla_p H(x, p) = q,$$

has a unique solution of class $C^\infty(\Omega \times \mathbb{R}^n)$. We denote by $p(x, q)$ such a solution. Let us notice that (7.114) implies trivially

$$L(x, q) = p(x, q) \cdot q - H(x, p(x, q)). \tag{7.121}$$

We now prove the converse of Theorem 7.9.

Theorem 7.10 *Let H be the Hamiltonian associated to L and suppose that Assumption I and II hold true. Moreover, suppose that $(x(t), p(t))$ is a solution to the Hamilton–Jacobi equations*

$$\begin{cases} \frac{dx(t)}{dt} = \nabla_p H(x(t), p(t)) \\ \\ \frac{dp(t)}{dt} = -\nabla_x H(x(t), p(t)) \end{cases}. \tag{7.122}$$

Then $x(t)$ is a solution to Euler–Lagrange equation

$$\frac{d}{dt}\left(\nabla_q L\left(x(t), \dot{x}(t)\right)\right) - \nabla_x L\left(x(t), \dot{x}(t)\right) = 0. \tag{7.123}$$

Proof By Assumption II and by (7.122), in particular by

$$\nabla_p H(x(t), p(t)) = \dot{x}(t),$$

we have

$$p(t) = p\left(x(t), \dot{x}(t)\right).$$

Now, by (7.121) we have

$$L\left(x, \dot{x}(t)\right) = p(t) \cdot \dot{x}(t) - H\left(x(t), p(t)\right). \tag{7.124}$$

On the other hand

$$\nabla_q L(x, q) = p(x, q),$$

hence

$$\nabla_q L\left(x(t), \dot{x}(t)\right) = p\left(x(t), \dot{x}(t)\right) = p(t).$$

Therefore, taking into account (7.122), we have

$$\frac{d}{dt}\nabla_q L\left(x(t), \dot{x}(t)\right) = \dot{p}(t) = -\nabla_x H(x(t), p(t)). \tag{7.125}$$

Now, let us make the derivatives w.r.t. x^i of both the sides of (7.121)

$$\begin{aligned}
&\partial_{x^i} L(x,q) = q^k \partial_{x^i} p_k(x,q) - \partial_{x^i} H(x, p(x,q)) - \\
&- \partial_{p_k} H(x, p(x,q)) \partial_{x^i} p_k(x,q) = \\
&= \left(q^k - \partial_{p_k} H(x, p(x,q))\right) \partial_{x^i} p_k(x,q) - \partial_{x^i} H(x, p(x,q)),
\end{aligned}$$

from which we have

$$\begin{aligned}
&\partial_{x^i} L\left(x(t), \dot{x}(t)\right) = \\
&= \left(\frac{dx^k}{dt} - \partial_{p_k} H(x(t), p(t))\right) \partial_{x^i} p_k(x(t), p(t)) - \\
&- \partial_{x^i} H(x(t), p(t)) = \\
&= -\partial_{x^i} H(x(t), p(t)).
\end{aligned}$$

By the just obtained inequality and by (7.125) we have

$$\frac{d}{dt} \nabla_q L\left(x(t), \dot{x}(t)\right) = \nabla_x L\left(x(t), \dot{x}(t)\right).$$

■

Now let us consider the **geodesic lines** with respect to the Riemannian metric

$$g_{ij}(x) dx^i \otimes dx^j,$$

where $\left\{g_{ij}(x)\right\}_{i,j=1}^n$ is a symmetric real matrix $n \times n$ whose entries belong to $C^\infty(\Omega)$. Let us suppose

$$\lambda^{-1} |\xi|^2 \leq g_{ij}(x) \xi^i \xi^j \leq \lambda |\xi|^2, \quad \forall \xi \in \mathbb{R}^n, \forall x \in \Omega, \tag{7.126}$$

where $\lambda \geq 1$. Let us denote by $\left\{g^{ij}(x)\right\}_{i,j=1}^n$ the inverse matrix of $\left\{g_{ij}(x)\right\}_{i,j=1}^n$.

Definition 7.11 We say that the path

$$\gamma : [t_0, t_1] \to \Omega,$$

Is a **geodesic line** with respect to the Riemannian metric $g_{ij}(x) dx^i \otimes dx^j$, if $\gamma \in C^\infty([t_0, t_1], \Omega)$ and it solves the equations

$$\frac{d^2 \gamma^h(t)}{dt^2} + \Gamma_{ij}^h(\gamma(t)) \frac{d\gamma^i(t)}{dt} \frac{d\gamma^j(t)}{dt} = 0, \quad h = 1, \cdots, n, \tag{7.127}$$

where (Christoffel symbols), for $i, j, h = 1, \cdots, n$,

$$\Gamma^h_{ij}(x) = \frac{1}{2} g^{hk}(x) \left[\partial_i g_{kj}(x) + \partial_j g_{ki}(x) - \partial_k g_{ij}(x) \right]. \tag{7.128}$$

The following Proposition holds true.

Proposition 7.12 *The path $\gamma : [t_0, t_1] \to \Omega$ is a geodesic line w.r.t. the Riemannian metric $g_{ij}(x) dx^i \otimes dx^j$ if and only if γ is an extremal of the Lagrangian*

$$L(x, q) = g_{ij}(x) q^i q^j, \quad x \in \Omega, \ q \in \mathbb{R}. \tag{7.129}$$

Proof Let us write the Euler–Lagrange equation

$$\frac{d}{dt}\left(\nabla_q L\left(x(t), \dot{x}(t) \right) \right) - \nabla_x L\left(x(t), \dot{x}(t) \right) = 0. \tag{7.130}$$

We have

$$\partial_{q_k} L\left(x(t), \dot{x}(t) \right) = 2 g_{kj}(x(t)) \frac{dx^j(t)}{dt},$$

$$\frac{d}{dt}\left(\partial_{q_k} L\left(x(t), \dot{x}(t) \right) \right) = 2 g_{kj}(x(t)) \frac{d^2 x^j(t)}{dt^2} + \\ + 2 \partial_i g_{kj}(x(t)) \frac{dx^i(t)}{dt} \frac{dx^j(t)}{dt}$$

and

$$\partial_{x_k} L\left(x(t), \dot{x}(t) \right) = \partial_k g_{ij}(x(t)) \frac{dx^i(t)}{dt} \frac{dx^j(t)}{dt}.$$

Therefore the Euler–Lagrange equation can be written (we omit the variables, for the sake of brevity)

$$g_{kj} \frac{d^2 x^j}{dt^2} = \left(\frac{1}{2} \partial_k g_{ij} - \partial_i g_{kj} \right) \frac{dx^i}{dt} \frac{dx^j}{dt}.$$

From which we have

$$\begin{aligned} \frac{d^2 x^h}{dt^2} &= g^{hk} \left(\frac{1}{2} \partial_k g_{ij} - \partial_i g_{kj} \right) \frac{dx^i}{dt} \frac{dx^j}{dt} = \\ &= \frac{1}{2} g^{hk} \left(\partial_k g_{ij} - \partial_i g_{kj} - \partial_j g_{ki} \right) \frac{dx^i}{dt} \frac{dx^j}{dt} = \\ &= -\Gamma^h_{ij} \frac{dx^i}{dt} \frac{dx^j}{dt}. \end{aligned}$$

Hence the equation

$$\frac{d^2x^h}{dt^2} + \Gamma^h_{ij}\frac{dx^i}{dt}\frac{dx^j}{dt} = 0,$$

is equivalent to the Euler–Lagrange equation related to L and by (7.127) the thesis follows. ■

Remark 7.2 Let $\{g_{ij}(x)\}_{i,j=1}^n$ be a matrix like in Proposition 7.12, set

$$L(x, q) = \frac{1}{2}g_{ij}(x)q^iq^j, \quad \forall x \in \Omega, \ \forall q \in \mathbb{R}^n.$$

Notice that, by (7.110), we can write (7.113) as

$$g_{ij}q^j = p_i, \qquad \text{for } i = 1, \cdots, n.$$

Hence

$$H(x, p) = g^{ij}p_jp_i - \frac{1}{2}g_{ij}g^{ih}p_hg^{ik}p_k = \frac{1}{2}g^{ij}p_jp_i.$$

♦

The following Theorem holds true

Theorem 7.13 *Let* $\{g_{ij}(x)\}_{i,j=1}^n$ *be a real symmetric matrix* $n \times n$ *whose entries belong to* $C^\infty(\Omega)$ *and assume that it satisfies* (7.126)*. Let* $u \in C^\infty(\Omega)$ *be a solution to the eikonal equation*

$$g^{ij}(x)\,\partial_{x^i}u\partial_{x^j}u = 1 \tag{7.131}$$

and let $x = \gamma(t)$ *be a solution to the system*

$$\frac{dx^i(t)}{dt} = g^{ij}(x(t))\,\partial_{x^j}u(x(t)), \quad i = 1, \cdots, n.$$

Then $x = \gamma(t)$ *is a geodesic line w.r.t. the Riemannian metric*

$$g_{ij}(x)dx^i \otimes dx^j.$$

Proof Let

$$L(x, q) = \frac{1}{2}g_{ij}(x)q^iq^j, \quad \forall x \in \Omega, \ \forall q \in \mathbb{R}^n.$$

Let H be the Hamiltonian of L, that is

$$H(x, p) = \frac{1}{2} g^{ij}(x) p_j p_i \quad \forall x \in \Omega, \ \forall p \in \mathbb{R}^n.$$

Set

$$p(t) = \nabla u(\gamma(t)), \tag{7.132}$$

where u is a solution to Eq. (7.131) and γ is a solution to the equations

$$\frac{d\gamma^i(t)}{dt} = g^{ij}(\gamma(t))\, \partial_{x^j} u(\gamma(t)) \left(= \partial_{p_i} H(\gamma(t), p(t))\right), \tag{7.133}$$

for $i = 1, \cdots, n$.

Now, we make the derivative w.r.t. x^k of both the sides of Eq. (7.131) and we get

$$2\left(g^{ij}(x)\, \partial^2_{x^i x^k} u\right) \partial_{x^j} u + \left(\partial_{x^k} g^{ij}(x)\right) \partial_{x^i} u \partial_{x^j} u = 0. \tag{7.134}$$

By (7.132), (7.133) and (7.134) we have, for $i = 1, \cdots, n$,

$$\begin{aligned}
\frac{dp_i(t)}{dt} &= \partial^2_{x^i x^k} u(\gamma(t)) \frac{d\gamma^k(t)}{dt} = \\
&= \partial^2_{x^i x^k} u(\gamma(t)) g^{kj}(\gamma(t))\, \partial_{x_j} u(\gamma(t)) = \\
&= \left(g^{kj}(\gamma(t))\, \partial^2_{x^i x^k} u(\gamma(t))\right) \partial_{x^j} u(\gamma(t)) = \\
&= -\frac{1}{2} \partial_{x^i} g^{jk}(\gamma(t))\, \partial_{x^j} u(\gamma(t)) \partial_{x^k} u(\gamma(t)) = \\
&= -\partial_{x^i} H(\gamma(t), p(t)).
\end{aligned}$$

The just obtained equality and (7.133) implies that $(\gamma(t), p(t))$ is a solution to the system

$$\begin{cases}
\frac{d\gamma(t)}{dt} = \nabla_p H(x(t), \gamma(t)) \\
\\
\frac{dp(t)}{dt} = -\nabla_x H(\gamma(t), p(t))
\end{cases}. \tag{7.135}$$

Therefore, by Theorem 7.10 and by Proposition (7.12) the thesis follows. ■

Remark 7.3 Let us observe that if u is a solution to the equation

$$g^{ij}(x)\, \partial_{x^i} u \partial_{x^j} u = 1$$

and $\gamma(t)$ is a solution to the system

$$\frac{d\gamma^i(t)}{dt} = g^{ij}(\gamma(t))\,\partial_{x^j}u(\gamma(t)), \quad i = 1, \cdots, n,$$

then t is the natural parameter (in the Riemannian metric) of the path $x = \gamma(t)$. As a matter of fact we have

$$\begin{aligned} g_{ij}(\gamma(t))\frac{d\gamma^i(t)}{dt}\frac{d\gamma^j(t)}{dt} &= g_{ij}(\gamma(t))\left(g^{ik}(\gamma(t))\,\partial_{x^k}u(\gamma(t))\right) \\ &\times \left(g^{jl}(\gamma(t))\,\partial_{x^l}u(\gamma(t))\right) = \\ &= \delta^k_j\partial_{x^k}u(\gamma(t))g^{jl}(\gamma(t))\,\partial_{x^l}u(\gamma(t)) = \\ &= g^{jl}(\gamma(t))\,\partial_{x^l}u(\gamma(t))\partial_{x^j}u(\gamma(t)) = 1. \end{aligned}$$

Moreover, for fixed $\overline{x} \in \mathbb{R}^n$, set

$$u(\overline{x}) = R_0.$$

If γ is the solution to the Cauchy problem

$$\begin{cases} \frac{d\gamma^i(t)}{dt} = g^{ij}(\gamma(t))\,\partial_{x^j}u(\gamma(t)), & i = 1, \cdots, n \\ \\ \gamma(R_0) = \overline{x} \end{cases}.$$

then

$$u(\gamma(t)) = t, \quad \forall t \in I \tag{7.136}$$

(I is the maximal interval of the solution γ). As a matter of fact we have

$$\begin{aligned} \frac{d}{dt}u(\gamma(t)) &= \partial_{x^i}u(\gamma(t))\frac{d\gamma^i(t)}{dt} = \\ &= \partial_{x^i}u(\gamma(t))g^{ij}(\gamma(t))\,\partial_{x^j}u(\gamma(t)) = 1, \end{aligned}$$

hence

$$u(\gamma(t)) = t + C,$$

where C is a constant which can be determined easily in the following way

$$R_0 = u(\overline{x}) = u(\gamma(R_0)) = R_0 + C,$$

hence $C = 0$ and (7.136) is proved. These comments will be used in Chap. 18. ♦

Let us conclude this section by some propositions on the extremal and other comments on the geodetics lines.

Proposition 7.14 *Let $L \in C^\infty(\Omega \times \mathbb{R}^n)$. We have what follows.*

(i) if $\overline{x} \in C^\infty([t_0, t_1], \mathbb{R}^n)$ is an extremal of the functional

$$\int_{t_0}^{t_1} L\left(x(t), \frac{dx(t)}{dt}\right) dt, \tag{7.137}$$

we have

$$\nabla_q L\left(\overline{x}(t), \frac{d\overline{x}(t)}{dt}\right) \cdot \frac{d\overline{x}(t)}{dt} - L\left(\overline{x}(t), \frac{d\overline{x}(t)}{dt}\right) = constant. \tag{7.138}$$

(ii) If $L(x, q)$ is an homogeneous function w.r.t. q of degree $\alpha \neq 1$ and $\overline{x}(t)$ is an extremal of functional (7.137) *then*

$$L\left(\overline{x}(t), \frac{d\overline{x}(t)}{dt}\right) = constant. \tag{7.139}$$

Proof

(i) Set

$$F(t) = \nabla_q L\left(\overline{x}(t), \frac{d\overline{x}(t)}{dt}\right) \cdot \frac{d\overline{x}(t)}{dt} - L\left(\overline{x}(t), \frac{d\overline{x}(t)}{dt}\right).$$

We have

$$\begin{aligned}\frac{dF(t)}{dt} &= \frac{d}{dt}\left(\nabla_q L\left(\overline{x}(t), \frac{d\overline{x}(t)}{dt}\right)\right) \cdot \frac{d\overline{x}(t)}{dt} \\ &+ \nabla_q L\left(\overline{x}(t), \frac{d\overline{x}(t)}{dt}\right) \cdot \frac{d^2\overline{x}(t)}{dt^2} - \\ &- \nabla_x L\left(\overline{x}(t), \frac{d\overline{x}(t)}{dt}\right) \cdot \frac{d\overline{x}(t)}{dt} - \nabla_q L\left(\overline{x}(t), \frac{d\overline{x}(t)}{dt}\right) \cdot \frac{d^2\overline{x}(t)}{dt^2} = \\ &= \left[\frac{d}{dt}\left(\nabla_q L\left(\overline{x}(t), \frac{d\overline{x}(t)}{dt}\right)\right) - \nabla_x L\left(\overline{x}(t), \frac{d\overline{x}(t)}{dt}\right)\right] \cdot \frac{d\overline{x}(t)}{dt} = 0.\end{aligned}$$

From which the thesis follows.

(ii) By point (i) we have

$$\nabla_q L\left(\overline{x}(t), \frac{d\overline{x}(t)}{dt}\right) \cdot \frac{d\overline{x}(t)}{dt} - L\left(\overline{x}(t), \frac{d\overline{x}(t)}{dt}\right) = \text{constant}.$$

On the other hand by the homogeneity of $L(x, \cdot)$ we get

$$\nabla_q L\left(\overline{x}(t), \frac{d\overline{x}(t)}{dt}\right) \cdot \frac{d\overline{x}(t)}{dt} = \alpha L\left(\overline{x}(t), \frac{d\overline{x}(t)}{dt}\right).$$

Therefore

$$(\alpha - 1)L\left(\overline{x}(t), \frac{d\overline{x}(t)}{dt}\right) = \text{constant}$$

and recalling that $\alpha \neq 1$, the thesis follows.

■

Proposition 7.15 *Let $L \in C^\infty(\Omega \times \mathbb{R}^n)$ be an homogeneous function w.r.t. q of degree 2. If $\overline{x} \in C^\infty([t_0, t_1], \mathbb{R}^n)$ is an extremal of the functional*

$$\int_{t_0}^{t_1} L\left(x(t), \frac{dx(t)}{dt}\right) dt, \tag{7.140}$$

and

$$L\left(\overline{x}(t), \frac{d\overline{x}(t)}{dt}\right) > 0, \quad \forall t \in [t_0, t_1], \tag{7.141}$$

then $\overline{x}$ is an extremal of the functional

$$\int_{t_0}^{t_1} \sqrt{L\left(x(t), \frac{dx(t)}{dt}\right)}\, dt. \tag{7.142}$$

Proof By Proposition (7.14) and by (7.141) we may set

$$c_0^2 = L\left(\overline{x}(t), \frac{d\overline{x}(t)}{dt}\right) > 0,$$

where c_0 is a positive constant. Since $\overline{x}$ is an extremal of the functional (7.140), we have

$$\begin{aligned}
\frac{d}{dt}\left(\nabla_q \sqrt{L\left(\overline{x}(t), \frac{d\overline{x}(t)}{dt}\right)}\right) &= \frac{1}{2c_0}\frac{d}{dt}\left(\nabla_q L\left(\overline{x}(t), \frac{d\overline{x}(t)}{dt}\right)\right) = \\
&= \frac{1}{2c_0}\nabla_x L\left(\overline{x}(t), \frac{d\overline{x}(t)}{dt}\right) = \\
&= \nabla_x \sqrt{L\left(\overline{x}(t), \frac{d\overline{x}(t)}{dt}\right)}.
\end{aligned}$$

Hence $\overline{x}$ is an extremal of functional (7.142). ■

Proposition 7.16 *Suppose that L satisfies the same assumptions of Proposition 7.15. Let φ be an extremal of functional*

$$\int_{t_0}^{t_1} \sqrt{L\left(x(t), \frac{dx(t)}{dt}\right)}\, dt. \tag{7.143}$$

Suppose

$$L\left(\varphi(t), \frac{d\varphi(t)}{dt}\right) > 0, \quad \forall t \in [t_0, t_1],$$

then there exists a unique parametrization $t(\tau)$, $t'(\tau) > 0$ in $[\tau_0, \tau_1]$ ($t(\tau_0) = t_0$ and $t(\tau_1) = t_1$), such that, setting $\psi(\tau) = \varphi(t(\tau))$, we have

$$L\left(\psi(\tau), \frac{d\psi(\tau)}{d\tau}\right) = constant. \tag{7.144}$$

Moreover the path $x = \psi(\tau)$ is an extremal of the functional

$$\int_{\tau_0}^{\tau_1} L\left(x(\tau), \frac{dx(\tau)}{d\tau}\right) d\tau. \tag{7.145}$$

Proof Let

$$f(t) = \sqrt{L\left(\varphi(t), \frac{d\varphi(t)}{dt}\right)}, \quad \forall t \in [t_0, t_1].$$

Let $c_0 > 0$ be a constant and let $t(\tau)$ satisfy

$$\int_{t_0}^{t(\tau)} f(t)dt = c_0\tau, \ \ \forall \tau \in [\tau_0, \tau_1].$$

Set $\psi(\tau) = \varphi(t(\tau))$; we get, by the homogeneity of $L(x, \cdot)$,

$$\sqrt{L\left(\psi(\tau), \frac{d\psi(\tau)}{d\tau}\right)} = \sqrt{(t'(\tau))^2 L\left(\varphi(t(\tau)), \frac{d\varphi}{dt}(t(\tau))\right)} =$$

$$= t'(\tau)\sqrt{L\left(\varphi(t(\tau)), \frac{d\varphi}{dt}(t(\tau))\right)} = c_0.$$

Hence (7.144) is proved.

Now, let us prove that $x = \psi(\tau)$ is an extremal of functional (7.145). Set

$$F(t) = \frac{1}{f(t)} \nabla_q L \left(\varphi(t), \frac{d\varphi(t)}{dt} \right)$$

and

$$G(t) = \frac{1}{f(t)} \nabla_x L \left(\varphi(t), \frac{d\varphi(t)}{dt} \right).$$

Since φ is an extremal of functional (7.143), we have

$$\frac{dF(t)}{dt} = G(t). \tag{7.146}$$

Now, recalling

$$\sqrt{L \left(\psi(\tau), \frac{d\psi(\tau)}{d\tau} \right)} = c_0,$$

we have (by the homogeneity of L w.r.t. q)

$$\begin{aligned}
\nabla_q L \left(\psi(\tau), \frac{d\psi(\tau)}{d\tau} \right) &= t'(\tau) \nabla_q L \left(\varphi(t(\tau)), \frac{d\varphi}{dt}(t(\tau)) \right) = \\
&= t'(\tau) \sqrt{L \left(\varphi(t(\tau)), \frac{d\varphi}{dt}(t(\tau)) \right)} F(t(\tau)) = \\
&= c_0 F(t(\tau)).
\end{aligned}$$

Hence, recalling (7.146) (and the homogeneity of L w.r.t. q), we have

$$\begin{aligned}
\frac{d}{d\tau} \left(\nabla_q L \left(\psi(\tau), \frac{d\psi(\tau)}{d\tau} \right) \right) &= c_0 \frac{d}{d\tau} (F(t(\tau))) = \\
&= c_0 t'(\tau) \frac{dF}{dt}(t(\tau)) = c_0 t'(\tau) G(t(\tau)) = \\
&= \frac{c_0 t'(\tau)}{f(t(\tau))} \nabla_x L \left(\varphi(t(\tau)), \frac{d\varphi}{dt}(t(\tau)) \right) = \\
&= \frac{c_0}{t'(\tau) f(t(\tau))} \nabla_x L \left(\psi(\tau), \frac{d\psi(\tau)}{d\tau} \right) = \\
&= \nabla_x L \left(\psi(\tau), \frac{d\psi(\tau)}{d\tau} \right).
\end{aligned}$$

Hence ψ is an extremal of functional (7.145). ■

Remark 7.4 Let $\left\{g_{ij}(x)\right\}_{i,j=1}^{n}$ be a real symmetric matrix $n \times n$ whose entries belong to $C^{\infty}(\Omega)$ and let us assume that it satisfies (7.126). Let

$$L(x,q) = g_{ij}(x)\, q^i q^j, \quad \forall x \in \Omega,\ \forall q \in \mathbb{R}^n.$$

By Proposition 7.15 we have that, if $x = \gamma(t)$ is a geodesic line, i.e. it is an extremal of the functional

$$\int_{t_0}^{t_1} L\left(x(t), \frac{dx(t)}{dt}\right) dt = \int_{t_0}^{t_1} g_{ij}(x(t))\, \frac{dx^i}{dt}\frac{dx^j}{dt} dt, \tag{7.147}$$

then $x = \gamma(t)$ is also an extremal of the functional

$$\int_{t_0}^{t_1} L\left(x(t), \frac{dx(t)}{dt}\right) dt = \int_{t_0}^{t_1} \sqrt{g_{ij}(x(t))\, \frac{dx^i}{dt}\frac{dx^j}{dt}} dt. \tag{7.148}$$

On the other hand, by Proposition 7.16, we have that if $x = \gamma(t)$ is an extremal of functional (7.148) and if $t(\tau)$ is strictly increasing and it satisfies

$$t'(\tau)\sqrt{g_{ij}(\gamma(t(\tau)))\, \frac{d\gamma^i}{dt}(t(\tau)\frac{dx^j}{dt}(t(\tau)} = c, \tag{7.149}$$

where $c > 0$ is a positive constant, then $x = \gamma(t(\tau)$ is an extremal of

$$\int_{t_0}^{t_1} L\left(x(t), \frac{dx(t)}{dt}\right) dt = \int_{\tau_0}^{\tau_1} g_{ij}(x(\tau))\, \frac{dx^i}{d\tau}\frac{dx^j}{d\tau} d\tau, \tag{7.150}$$

where τ_0 and τ_1 satisfy $t(\tau_0) = t_0$ and $t(\tau_1) = t_1$. Let us notice that if $c = 1$, then condition (7.149) means that τ is the natural parameter of the path $x = \gamma(t)$ (extremal of (7.148)) in the Riemanniann metric $g_{ij}(x) dx^i \otimes dx^j$. ♦

Chapter 8
Real Analytic Functions

Abstract In this brief chapter, we cover the foundational concepts of real analytic functions needed for Part II, especially for proving the Cauchy–Kovalevskaya and Holmgren theorems for the Cauchy problem with analytic coefficients, as well as the John Theorem on continuous dependence estimates. The key idea is the power series in multiple variables, an extension of the single-variable case. We also introduce majorant functions and series, essential for the Cauchy–Kovalevskaya Theorem and useful for showing that analytic functions are closed under sum, multiplication, and composition.

8.1 Introduction

In this brief chapter, we will provide the foundational concepts of real analytic functions necessary throughout Part II, particularly in proving the Cauchy–Kovalevskaya and Holmgren existence and uniqueness theorems for the Cauchy problem with differential equations that have analytic coefficients (Chap. 9), as well as the John Theorem on continuous dependence estimates for the Cauchy problem, which will be presented in Chap. 13. The fundamental element in studying real analytic functions will be the notion of power series in multiple variables, a natural extension of the single-variable power series with which we assume familiarity. In this chapter, we will also introduce the concepts of majorant functions and majorant series (Sect. 8.4). This notion, essential to the proof of the Cauchy–Kovalevskaya Theorem given in this book, also serves as a useful tool for easily proving that the class of analytic functions is closed under the operations of sum, multiplication, and composition.

S. Vessella, *Unique Continuation Properties for Partial Differential Equations*,
Birkhäuser Advanced Texts Basler Lehrbücher,
https://doi.org/10.1007/978-3-031-86366-0_8

8.2 Power Series

In this chapter we will consider the multiple series

$$\sum_{\alpha\in\mathbb{N}_0^n} c_\alpha, \tag{8.1}$$

where $c_\alpha \in \mathbb{R}$ (or $c_\alpha \in \mathbb{C}$).

When we say that the series (8.1) converges, we will mean *always* that it is **absolutely convergent**. That is

$$\sum_{\alpha\in\mathbb{N}_0^n} |c_\alpha| < +\infty.$$

Therefore, if the series (8.1) converges, the value of the sum in (8.1) does not depend on the order of the terms c_α. If $c_\alpha(x)$ are functions, we will naturally extend the notions of uniform, total convergence, $C^k(\overline{\Omega})$ convergence and so on. For instance, we will say that

$$\sum_{\alpha\in\mathbb{N}_0^n} c_\alpha(x), \tag{8.2}$$

uniformly converges to a function f in a set $K \subset \mathbb{R}^n$ provided that:

(i) for every $x \in K$, $\sum_{\alpha\in\mathbb{N}_0^n} |c_\alpha(x)|$ converges,
(ii) we have

$$f(x) = \sum_{\alpha\in\mathbb{N}_0^n} c_\alpha(x), \quad \forall x \in K$$

and
(iii)

$$\lim_{N\to+\infty} \sup_{x\in K} \left| f(x) - \sum_{|\alpha|\leq N} c_\alpha(x) \right| = 0.$$

Let $c_\alpha \in \mathbb{R}$ (or $c_\alpha \in \mathbb{C}$). We call **power series** a series like

$$\sum_{\alpha\in\mathbb{N}_0^n} c_\alpha x^\alpha. \tag{8.3}$$

For any $y \in \mathbb{R}^n$ set

$$Q_y = \{x \in \mathbb{R}^n : |x_j| \leq |y_j|, j = 1, \cdots, n\}.$$

We have

Proposition 8.1 *If series* (8.3) *converges at a point* $y \in \mathbb{R}^n$ *then the series uniformly converges in* Q_y.

Proof The convergence at y of (8.3) is equivalent to

$$\sum_{\alpha\in\mathbb{N}_0^n} |c_\alpha||y^\alpha| < +\infty.$$

Hence, we have

$$\sum_{\alpha\in\mathbb{N}_0^n} \sup_{Q_y} |c_\alpha x^\alpha| \leq \sum_{\alpha\in\mathbb{N}_0^n} |c_\alpha||y^\alpha| < +\infty.$$

From which we get the total convergence and, consequently, the uniform convergence, of series (8.3). ■

Proposition 8.1 implies that the sum of series (8.3) is continuous in Q_y.

The differentiability will be proved in Proposition 8.3 to prove such a Proposition we need

Lemma 8.2 *Denote by* $\upsilon = (1, 1, \cdots, 1)$. *If* $|x_j| < 1$ $j = 1, \cdots, n$, *we have*

$$\sum_{\alpha\in\mathbb{N}_0^n} x^\alpha = \frac{1}{(\upsilon - x)^\upsilon}, \tag{8.4}$$

and

$$\sum_{\alpha\geq\beta} \frac{\alpha!}{(\alpha-\beta)!} x^{\alpha-\beta} = \partial^\beta \left(\frac{1}{(\upsilon - x)^\upsilon}\right) = \frac{\beta!}{(\upsilon - x)^{\upsilon+\beta}}. \tag{8.5}$$

Proof Concerning the convergence of series (8.4), we have, for $|x_j| < 1$, $j = 1, \cdots, n$,

$$\begin{aligned}
\sum_{\alpha\leq\upsilon N} |x^\alpha| &= \sum_{\alpha\leq\upsilon N} \left|x_j^{\alpha_j}\right| = \\
&= \prod_{j=1}^{n} \sum_{\alpha_j\leq N} \left|x_j^{\alpha_j}\right| = \\
&= \prod_{j=1}^{n} \frac{1-|x_j|^{N+1}}{1-|x_j|} \to \prod_{j=1}^{n} \frac{1}{1-|x_j|}, \qquad \text{as } N \to \infty.
\end{aligned}$$

Concerning the sum of the series we have, similarly,

$$\sum_{\alpha\leq \upsilon N} x^\alpha = \lim_{N\to\infty}\prod_{j=1}^{n}\sum_{\alpha_j\leq N} x_j^{\alpha_j} = \frac{1}{(1-x_1)\cdots(1-x_n)} = \frac{1}{(\upsilon - x)^\upsilon}.$$

Now, let us prove (8.5). Recalling (2.3)

$$\partial^\beta x^\alpha = \begin{cases} \frac{\alpha!}{(\alpha-\beta)!}x^{\alpha-\beta}, & \text{for } \alpha\geq\beta \\ \\ 0, & \text{otherwise} \end{cases},$$

we have

$$\begin{aligned}\sum_{\alpha\geq\beta}\frac{\alpha!}{(\alpha-\beta)!}x^{\alpha-\beta} &= \sum_{\alpha\in\mathbb{N}_0^n}\partial^\beta x^\alpha = \\ &= \prod_{j=1}^{n}\sum_{\alpha_j\in\mathbb{N}_0}\partial^{\beta_j}x_j^{\alpha_j} = \\ &= \prod_{j=1}^{n}\partial_j^{\beta_j}\frac{1}{1-x_j} = \partial^\beta\left(\frac{1}{(\upsilon-x)^\upsilon}\right) = \\ &= \frac{\beta!}{(\upsilon-x)^{\upsilon+\beta}}.\end{aligned}$$

■

Proposition 8.3 *If series* (8.3) *converges at the point* $y\in\mathbb{R}^n$ *and* $y_j>0$ *for every* $j=1,\cdots,n$ *then, denoted by* f *the sum of such a series, we have* $f\in C^\infty\left(Int(Q_y)\right)$, *where* $Int(Q_y)$ *is the interior part of* Q_y.

Moreover

$$\partial^\alpha f(0) = \alpha! c_\alpha. \tag{8.6}$$

Proof In order to prove that $f\in C^\infty\left(Int(Q_y)\right)$ it suffices to prove that for every $q\in(0,1)$ and for every $\beta\in\mathbb{N}_0^n$ we have

$$\sum_{\alpha\in\mathbb{N}_0^n}\sup_{Q_{qy}}\left|\partial^\beta\left(c_\alpha x^\alpha\right)\right| < +\infty.$$

By (2.3) we have

$$\partial^\beta\left(c_\alpha x^\alpha\right)=\begin{cases}\frac{\alpha!}{(\alpha-\beta)!}c_\alpha x^{\alpha-\beta}, & \text{for } \alpha\geq\beta\\ \\ 0, & \text{otherwise}\end{cases}.$$

Hence

$$\begin{aligned}\sum_{\alpha\in\mathbb{N}_0^n}\sup_{Q_{qy}}\left|\partial^\beta\left(c_\alpha x^\alpha\right)\right| &\leq \sum_{\alpha\geq\beta}\frac{\alpha!}{(\alpha-\beta)!}|c_\alpha|\left|(qy)^{\alpha-\beta}\right| = \\ &= \frac{1}{\left|y^\beta\right|}\sum_{\alpha\geq\beta}\frac{\alpha!|c_\alpha y^\alpha|}{(\alpha-\beta)!}q^{|\alpha-\beta|}.\end{aligned} \tag{8.7}$$

Now, since (8.3) converges in $y\in\mathbb{R}^n$, we get

$$\left|c_\alpha y^\alpha\right|\leq\mu_y:=\sum_{\alpha\in\mathbb{N}_0^n}\left|c_\alpha y^\alpha\right|<+\infty. \tag{8.8}$$

By the above obtained inequality and by (8.7) we have

$$\sum_{\alpha\in\mathbb{N}_0^n}\sup_{Q_{qy}}\left|\partial^\beta\left(c_\alpha x^\alpha\right)\right|\leq\frac{\mu_y}{\left|y^\beta\right|}\sum_{\alpha\geq\beta}\frac{\alpha!}{(\alpha-\beta)!}q^{|\alpha-\beta|}. \tag{8.9}$$

Applying (8.5) with $x=qv=q(1,1\cdots,1)$, we have

$$\begin{aligned}\sum_{\alpha\geq\beta}\frac{\alpha!}{(\alpha-\beta)!}q^{|\alpha-\beta|} &= \sum_{\alpha\geq\beta}\frac{\alpha!}{(\alpha-\beta)!}(qv)^{\alpha-\beta} = \\ &= \frac{\beta!}{(v-vq)^{v+\beta}} = \\ &= \frac{\beta!}{(1-q)^{n+|\beta|}}.\end{aligned} \tag{8.10}$$

By the just obtained equality and by (8.9) we obtain

$$\sum_{\alpha\in\mathbb{N}_0^n}\sup_{Q_{qy}}\left|\partial^\beta\left(c_\alpha x^\alpha\right)\right|\leq\frac{\mu_y}{\left|y^\beta\right|}\frac{\beta!}{(1-q)^{n+|\beta|}}<+\infty. \tag{8.11}$$

We have proved so far that for each $q\in(0,1)$ we have $f\in C^\infty\left(Q_{yq}\right)$. Therefore $f\in C^\infty\left(\mathrm{Int}(Q_y)\right)$.

Concerning (8.6) we have

$$\partial^{\beta} f(x) = \sum_{\alpha\in\mathbb{N}_0^n} \partial^{\beta}\left(c_{\alpha}x^{\alpha}\right) = \sum_{\alpha\geq\beta} \frac{\alpha!}{(\alpha-\beta)!} c_{\alpha}x^{\alpha-\beta}.$$

Therefore

$$\partial^{\beta} f(0) = \beta! c_{\beta}.$$

■

Exercise 8.1 Let $f(x)$ be the sum of the series

$$\sum_{\alpha\in\mathbb{N}_0^n} c_{\alpha}x^{\alpha},$$

in Q_{ϱ} where $\varrho = (\varrho_1, \cdots, \varrho_n)$, with $\varrho_j > 0$, $j = 1, \cdots, n$. Let $q \in (0, 1)$. Set

$$r = (1-q) \min_{1\leq j\leq n} \varrho_j, \quad \text{and} \quad \mu_{\varrho} = \sum_{\alpha\in\mathbb{N}_0^n} \left|c_{\alpha}\varrho^{\alpha}\right|. \tag{8.12}$$

Then we have

$$\sup_{Q_{q\varrho}} \left|\partial^{\beta} f\right| \leq (1-q)^{-n} \mu_{\varrho} r^{-|\beta|} \beta!. \tag{8.13}$$

Solving Exercise 8.1
Is an immediate consequence of (8.11). ♠

Exercise 8.2 Prove that

$$\sum_{\alpha\in\mathbb{N}_0^n} \frac{|\alpha|!}{\alpha!} x^{\alpha} = \frac{1}{1-\sum_{j=1}^n x_j}, \quad \text{for} \quad \sum_{j=1}^n |x_j| < 1, \tag{8.14}$$

$$\sum_{\alpha\geq\beta} \frac{|\alpha|!}{(\alpha-\beta)!} x^{\alpha-\beta} = \frac{|\beta|!}{\left(1-\sum_{j=1}^n x_j\right)^{1+|\beta|}}, \quad \text{for} \quad \sum_{j=1}^n |x_j| < 1. \tag{8.15}$$

Solving Exercise 8.2
By (2.4) we have

$$(x_1 + x_2 + \cdots + x_n)^m = \sum_{|\alpha|=m} \frac{m!}{\alpha!} x^{\alpha}, \quad m \in \mathbb{N}_0.$$

Hence, for $\sum_{j=1}^{n} |x_j| < 1$, we have

$$\sum_{\alpha \in \mathbb{N}_0^n} \frac{|\alpha|!}{\alpha!} x^\alpha = \sum_{m=0}^{\infty} \sum_{|\alpha|=m} \frac{|\alpha|!}{\alpha!} x^\alpha =$$

$$= \sum_{m=0}^{\infty} (x_1 + x_2 + \cdots + x_n)^m =$$

$$= \frac{1}{1 - (x_1 + x_2 + \cdots + x_n)},$$

from which we get (8.14).

Concerning (8.15), it suffices to note that by (8.14), we have, for $\sum_{j=1}^{n} |x_j| < 1$,

$$\sum_{\alpha \geq \beta} \frac{|\alpha|!}{(\alpha - \beta)!} x^{\alpha - \beta} = \sum_{\alpha \in \mathbb{N}_0^n} \frac{|\alpha|!}{\alpha!} \partial^\beta (x^\alpha) =$$

$$= \partial^\beta \left(\frac{1}{1 - (x_1 + x_2 + \cdots + x_n)} \right) =$$

$$= \frac{|\beta|!}{(1 - (x_1 + x_2 + \cdots + x_n))^{1+|\beta|}}.$$

♠

8.3 Analytic Functions in an Open Set of $\mathbb{R}^n$

Definition 8.4 Let Ω be an open set of $\mathbb{R}^n$ and $f : \Omega \to \mathbb{R}$ (or $\mathbb{C}$) a function. We say that f is an **analytic function of real variables** in $x_0 \in \Omega$ if there exist a neighborhood $\mathcal{U}_{x_0}$ of x_0, $c_\alpha \in \mathbb{R}$ ($\mathbb{C}$), $\alpha \in \mathbb{N}_0^n$, such that

$$f(x) = \sum_{\alpha \in \mathbb{N}_0^n} c_\alpha (x - x_0)^\alpha, \quad \forall x \in \mathcal{U}_{x_0}.$$

We say that f is an **analytic function of real variables** in Ω if it is an analytic function of real variables in every $x_0 \in \Omega$.

We say that $f : \Omega \to \mathbb{R}^m$ (or $\mathbb{C}^m$) where $m \in \mathbb{N}$, $f = (f_1, \cdots, f_m)$ is an **analytic function of real variables** in $x_0 \in \Omega$ (or in Ω) provided f_j are analytic functions of real variables in $x_0 \in \Omega$ (or in Ω).

In what follows, if there is no ambiguity, we will simply say **"analytic functions"**, omitting the expression "of real variables".

We will denote by $C^\omega(\Omega; \mathbb{R}^m)$ ($C^\omega(\Omega; \mathbb{C}^m)$) the class of analytic function defined on Ω with values in $\mathbb{R}^m$ ($\mathbb{C}^m$), if $m = 1$ we write, if there is no ambiguity, simply $C^\omega(\Omega)$ to denote $C^\omega(\Omega; \mathbb{R})$ ($C^\omega(\Omega; \mathbb{C})$).

Proposition 8.3 immediately gives:

$$C^\omega(\Omega) \subset C^\infty(\Omega). \tag{8.16}$$

Moreover, for every $x_0 \in \Omega$ there exists a neighborhood $\mathcal{U}_{x_0}$ of x_0 such that

$$f(x) = \sum_{\alpha \in \mathbb{N}_0^n} \frac{1}{\alpha!} \partial^\alpha f(x_0)(x - x_0)^\alpha, \quad \forall x \in \mathcal{U}_{x_0} \tag{8.17}$$

and there exist $M > 0$, $r > 0$ (depending on x_0) and $\widetilde{\mathcal{U}}_{x_0}$, neighborhood of x_0, with $\widetilde{\mathcal{U}}_{x_0} \Subset \mathcal{U}_{x_0}$, such that

$$\left|\partial^\alpha f(x)\right| \leq M|\alpha|! r^{-|\alpha|}, \quad \forall \alpha \in \mathbb{N}_0^n, \quad \forall x \in \mathcal{U}_{x_0}, \tag{8.18}$$

((8.18) follows by (8.13)).

We recall that inclusion (8.16) is proper. As a matter of fact, the function

$$f(t) = \begin{cases} e^{-1/t^2}, & \text{for } t \neq 0 \\ \\ 0, & \text{for } t = 0 \end{cases}, \tag{8.19}$$

belongs to $C^\infty(\mathbb{R})$, but it is not analytic. Indeed, since we have $f^{(k)}(0) = 0$, for every $k \in \mathbb{N}_0$, we have

$$\sum_{k=0}^{\infty} \frac{1}{k!} f^{(k)}(0) t^k = 0 \neq f(t), \quad \forall t \in \mathbb{R} \setminus \{0\}.$$

Therefore, for every $t \neq 0$, $f(t)$ is different from the sum of its Taylor series.

The analytic functions enjoy the **unique continuation property.** Indeed, the following holds.

Theorem 8.5 *Let Ω be a connected open set of $\mathbb{R}^n$ and $f \in C^\omega(\Omega)$. Let $x_0 \in \Omega$. Then we have*

$$\partial^\alpha f(x_0) = 0, \quad \forall \alpha \in \mathbb{N}_0^n \quad \Longrightarrow \quad f \equiv 0, \ in\ \Omega.$$

In particular, if f vanishes identically in an open set (not empty) of Ω then f vanishes identically in Ω.

Proof Let

$$\widetilde{\Omega} = \left\{x \in \Omega : \partial^\alpha f(x) = 0, \quad \forall \alpha \in \mathbb{N}_0^n\right\}.$$

It is clear that $\widetilde{\Omega} \neq \emptyset$ because $x_0 \in \widetilde{\Omega}$. Therefore, whether we prove that $\widetilde{\Omega}$ is at the same time an open and a closed set in Ω (in the topology induced by $\mathbb{R}^n$), as Ω is connected, we have $\Omega = \widetilde{\Omega}$ and the thesis follows.

Since f is continuous, the set $\widetilde{\Omega}$ is closed in Ω. Now we prove that $\widetilde{\Omega}$ is an open set of Ω.

Let $\widetilde{x} \in \widetilde{\Omega}$. By the analyticity of f we have that there exists a neighborhood $\mathcal{U}_{\widetilde{x}}$ of $\widetilde{x}$ such that

$$f(x) = \sum_{\alpha \in \mathbb{N}_0^n} \frac{1}{\alpha!} \partial^\alpha f(\widetilde{x})(x - \widetilde{x})^\alpha, \quad \forall x \in \mathcal{U}_{\widetilde{x}}.$$

On the other hand, $\partial^\alpha f(\widetilde{x}) = 0$ for every $\alpha \in \mathbb{N}_0^n$, hence $f(x) = 0$ for every $x \in \mathcal{U}_{\widetilde{x}}$. Hence

$$\partial^\alpha f(x) = 0, \quad \forall \alpha \in \mathbb{N}_0^n, \quad \forall x \in \mathcal{U}_{\widetilde{x}},$$

from which $\mathcal{U}_{\widetilde{x}} \subset \widetilde{\Omega}$. Therefore $\Omega = \widetilde{\Omega}$. ■

In Theorem 8.7 we will prove that the analytic functions can be characterized by the growth of their derivatives. We premise the following

Definition 8.6 Let $m \in \mathbb{N}$ and let Ω be an open set of $\mathbb{R}^n$. Let $f : \Omega \to \mathbb{R}^m$ (or $\mathbb{C}^m$), $f = (f_1, \cdots, f_m)$. Let $x_0 \in \Omega$ and $M, r > 0$, $j = 1, \cdots, m$. We write

$$f \in C_{M,r}(x_0),$$

provided that $f \in C^\infty$ in a neighborhood of x_0 and we have

$$\left|\partial^\alpha f_j(x_0)\right| \leq M|\alpha|!r^{-|\alpha|}, \quad \forall \alpha \in \mathbb{N}_0^n, \quad j = 1, \cdots, m.$$

Theorem 8.7 *Let Ω be an open set of $\mathbb{R}^n$ and $f : \Omega \to \mathbb{R}^m$ (or $\mathbb{C}^m$). The following conditions are equivalent:*

(i) f is analytic in Ω,
(ii) for any compact K, $K \subset \Omega$, there exist $M, r > 0$ (depending on K) such that

$$f \in C_{M,r}(x_0), \quad \forall x_0 \in K.$$

Proof It suffices to consider the case $m = 1$.

We prove that (i) $\Rightarrow$(ii).

By (8.18) we know that if $y \in \Omega$, there exist $M_y, r_y > 0$ and a neighborhood $\mathcal{U}_y$ of y such that

$$\sup_{\mathcal{U}_y} \left|\partial^\alpha f\right| \leq M_y |\alpha|! r_y^{-|\alpha|}, \quad \forall \alpha \in \mathbb{N}_0^n. \tag{8.20}$$

Let $K \subset \Omega$ be a compact, then $\left\{\mathcal{U}_y\right\}_{y \in K}$ is an open covering of K, hence there exist $\mathcal{U}_{y_1}, \cdots, \mathcal{U}_{y_N}$ such that

$$K \subset \bigcup_{j=1}^{N} \mathcal{U}_{y_j}.$$

Set

$$M = \max_{1 \leq j \leq N} M_{y_j}, \quad r = \min_{1 \leq j \leq N} r_{y_j}.$$

By (8.20) we have

$$\sup_K \left|\partial^\alpha f\right| \leq \sup_{\bigcup_{j=1}^{N} \mathcal{U}_{y_j}} \left|\partial^\alpha f\right| \leq M |\alpha|! r^{-|\alpha|}, \quad \forall \alpha \in \mathbb{N}_0^n. \tag{8.21}$$

We prove that (ii) $\Rightarrow$(i).

Let us suppose that (ii) holds. Let $x_0 \in \Omega$ and $\rho > 0$ satisfy $\overline{B_\rho(x_0)} \subset \Omega$. It is not restrictive to assume that $x_0 = 0 \in \Omega$. Let us choose $K = \overline{B_\rho}$ and let $M, r > 0$ satisfy

$$\left|\partial^\alpha f(x)\right| \leq M |\alpha|! r^{-|\alpha|}, \quad \forall x \in \overline{B_\rho}, \quad \forall \alpha \in \mathbb{N}_0^n. \tag{8.22}$$

Let us fix d such that $0 < d < \min\{r, \rho\}$. We now prove

$$f(x) = \sum_{\alpha \in \mathbb{N}_0^n} \frac{1}{\alpha!} \partial^\alpha f(0) x^\alpha, \quad \text{for every } x \text{ such that } \sum_{j=1}^{n} \left|x_j\right| \leq d.$$

Let x satisfy $\sum_{j=1}^{n} \left|x_j\right| \leq d$ and set

$$\phi(t) = f(tx), \quad t \in [0, 1].$$

We have, for any $m \in \mathbb{N}$,

$$f(x) = \phi(1) = \sum_{k=0}^{m-1} \frac{1}{k!} \phi^{(k)}(0) + \mathcal{R}_m = \sum_{|\alpha| \leq m-1} \frac{1}{\alpha!} \partial^\alpha f(0) x^\alpha + \mathcal{R}_m, \tag{8.23}$$

where

$$\mathcal{R}_m = \frac{1}{(m-1)!}\int_0^1 (1-t)^{(m-1)}\phi^{(m)}(t)dt.$$

By (2.4), (8.22) and by $\sum_{j=1}^n |x_j| \leq d$ we have

$$\begin{aligned}
\left|\frac{1}{m!}\phi^{(m)}(t)\right| &= \left|\sum_{|\alpha|=m}\frac{1}{\alpha!}\partial^\alpha f(0)x^\alpha\right| \leq \\
&\leq \sum_{|\alpha|=m}\frac{|\alpha|!}{\alpha!}Mr^{-|\alpha|}\left|x^\alpha\right| = \\
&= Mr^{-m}\sum_{|\alpha|=m}\frac{|\alpha|!}{\alpha!}|x_1|^{\alpha_1}\cdots|x_n|^{\alpha_n} = \\
&= Mr^{-m}\left(\sum_{j=1}^n |x_j|\right)^m \leq \\
&\leq M\left(\frac{d}{r}\right)^m.
\end{aligned}$$

Therefore

$$\begin{aligned}
|\mathcal{R}_m| &\leq \frac{1}{(m-1)!}\int_0^1 (1-t)^{(m-1)}\left|\phi^{(m)}(t)\right| dt \leq \\
&\leq \frac{1}{(m-1)!}\int_0^1 (1-t)^{(m-1)} m! M\left(\frac{d}{r}\right)^m dt = \\
&= M\left(\frac{d}{r}\right)^m.
\end{aligned} \tag{8.24}$$

All in all, by (8.23) and (8.24) we get, if $\sum_{j=1}^n |x_j| \leq d$ (recall $d < r$),

$$\left|f(x) - \sum_{|\alpha|\leq m}\frac{1}{\alpha!}\partial^\alpha f(0)x^\alpha\right| \leq M\left(\frac{d}{r}\right)^m \to 0, \quad \text{as } m \to \infty.$$

■

8.4 Majorant Functions

In the proof of the Cauchy–Kowalevskaya Theorem (in Sect. 9.5) we will make use the **method of majorant series**. We give the following

Definition 8.8 Let $m \in \mathbb{N}$. Let $\mathcal{U}_{x_0}$ be a neighborhood of $x_0 \in \mathbb{R}^n$ and let $f : \mathcal{U}_{x_0} \to \mathbb{R}^m$ (or $\mathbb{C}^m$), $F : \mathcal{U}_{x_0} \to \mathbb{R}^m$, $f = (f_1, \cdots, f_m)$, $F = (F_1, \cdots, F_m)$. Suppose that $f_j, F_j \in C^\infty(\mathcal{U}_{x_0})$, $j = 1, \cdots, m$. We say that F is **a majorant** of f or F **majorazes** f **in** x_0 and we write

$$f \preccurlyeq F, \qquad \text{in } x_0,$$

provided

$$\left|\partial^\alpha f_j(x_0)\right| \leq \partial^\alpha F_j(x_0), \quad \forall \alpha \in \mathbb{N}_0^n, \quad j = 1, \cdots, m.$$

Remark Let us notice that if $f \preccurlyeq F$ in x_0, then, we have $f \preccurlyeq \widetilde{F}$ in x_0, where

$$\widetilde{F}(x) = F(a_1^{-1}x_1, \cdots, a_n^{-1}x_n),$$

for every $a_j \in (0, 1]$, $j = 1, \cdots, n$. ♦

In what follows we will assume, without any restriction that $x_0 = 0$ and we will write simply (if there is no ambiguity) "$f \preccurlyeq F$" instead of "$f \preccurlyeq F$ in 0".

Proposition 8.9 *Let $f : \mathcal{U}_0 \to \mathbb{C}^m$, where $\mathcal{U}_0$ is a neighborhood of $0 \in \mathbb{R}^n$ and, for given $M, r > 0$, let*

$$\phi_{M,r}(x) = \frac{Mr}{r - (x_1 + \cdots + x_n)}.$$

Then we have

(i) $f \in C_{M,r}(0)$ if and only if $f \preccurlyeq \upsilon_m \phi_{M,r}$ (here $\upsilon_m = \underbrace{(1 \cdots, 1)}_{m}$),

(ii) $f \in C_{M,r}(0)$ and $f(0) = 0$ if and only if $f \preccurlyeq \upsilon_m \left(\phi_{M,r} - M\right)$.

Proof (i) is an immediate consequence of (8.14). We have, indeed,

$$\phi_{M,r}(x) = \frac{M}{1 - (\frac{x_1 + \cdots + x_n}{r})} = \sum_{\alpha \in \mathbb{N}_0^n} \frac{M|\alpha|!r^{-|\alpha|}}{\alpha!} x^\alpha.$$

Hence $\partial^\alpha \phi_{M,r}(0) = M|\alpha|!r^{-|\alpha|}$, from which we get (i). (ii) is a trivial consequence of (i). ■

From the differentiation rules we have

Proposition 8.10 *Let $f, g : \mathcal{U}_0 \to \mathbb{C}$; $F, G : \mathcal{U}_0 \to \mathbb{R}$ where $\mathcal{U}_0$ is a neighborhood of $0 \in \mathbb{R}^n$. If*

$$f \preccurlyeq F, \quad \text{and} \quad g \preccurlyeq G.$$

Then

$$f + g \preccurlyeq F + G, \quad fg \preccurlyeq FG.$$

Proof The proof is left as an exercise to the reader. ■

Proposition 8.11 *Let $n, m, p \in \mathbb{N}$ and*

$$f, F : \mathcal{U}_0 \subset \mathbb{R}^n \to \mathbb{R}^m,$$

where $\mathcal{U}_0$ is a neighborhood of 0. Let us suppose

$$f(0) = F(0) = 0,$$

and let $\mathcal{V}$ be an open set of $\mathbb{R}^m$ such that $\mathcal{U}_0 \subset \mathcal{V}$. Let

$$g, G : \mathcal{V} \to \mathbb{R}^p.$$

Let us assume that f, F, g, G be of class C^∞ and

$$f \preccurlyeq F, \quad g \preccurlyeq G.$$

Then we have

$$g \circ f \preccurlyeq G \circ F.$$

Proof Set $h = g \circ f$ and $H = G \circ F$. By the chain rule we have that for every $\alpha \in \mathbb{N}_0^n$ there exists a polynomial P_α with positive coefficients *indipendent* of f, g, F, G such that, for $j = 1, \cdots, p$, we have

$$\partial^\alpha h_j(0) = P_\alpha \left(\partial^\beta g_l(0), \cdots, \partial^\gamma f_k(0)\right),$$

$$\partial^\alpha H_j(0) = P_\alpha \left(\partial^\beta G_l(0), \cdots, \partial^\gamma F_k(0)\right).$$

Since

$$\left|\partial^\gamma f_k(0)\right| \leq \partial^\gamma F_k(0), \quad \forall \beta \in \mathbb{N}_0^n, \quad k = 1, \cdots, m,$$

$$\left|\partial^\beta g_l(0)\right| \leq \partial^\beta G_l(0), \quad \forall \beta \in \mathbb{N}_0^m, \quad l = 1, \cdots, p,$$

and taking into account that the coefficients of P_α are positive, we get

$$\begin{aligned}\left|\partial^\alpha h_j(0)\right| &\leq P_\alpha\left(\left|\partial^\beta g_l(0)\right|,\cdots,\left|\partial^\gamma f_k(0)\right|\right) \leq \\ &\leq P_\alpha\left(\left|\partial^\beta G_l(0)\right|,\cdots,\left|\partial^\gamma F_k(0)\right|\right) = \\ &= \partial^\alpha H_j(0).\end{aligned}$$

■

Proposition 8.12 *Let $n, m, p \in \mathbb{N}$. Let M, r, μ, ρ be positive numbers and $\widetilde{x} \in \mathbb{R}^n$, $\widetilde{y} \in \mathbb{R}^m$. Let $f \in C_{M,r}(\widetilde{x})$ be a function with values in $\mathbb{R}^m$, let $g \in C_{\mu,\rho}(\widetilde{y})$ be a function with values in $\mathbb{R}^p$, $\widetilde{y} = f(\widetilde{x})$. Then*

$$h := g \circ f \in C_{\mu, \frac{\rho r}{Mm+\rho}}(\widetilde{x}). \tag{8.25}$$

Proof Set

$$g^*(y) = g(y + \widetilde{y}), \quad f^*(x) = f(x + \widetilde{x}) - f(\widetilde{x}).$$

We have

$$h(x + \widetilde{x}) = g\left(\widetilde{y} + f(x + \widetilde{x}) - f(\widetilde{x})\right) = g^*\left(f^*(x)\right)$$

and

$$f^* \in C_{M,r}(0), \quad g^* \in C_{\mu,\rho}(0).$$

Proposition 8.9 implies

$$f^* \preccurlyeq \upsilon_m\left(\phi_{M,r} - M\right). \tag{8.26}$$

$$g^* \preccurlyeq \upsilon_p \phi_{\mu,\rho}. \tag{8.27}$$

Set

$$\chi(x) = \phi_{\mu,\rho}\left(\left(\phi_{M,r}(x) - M\right)\right).$$

By Proposition 8.11, by (8.26) and (8.27) we have

$$h(x + \widetilde{x}) = \left(g^* \circ f^*\right)(0) \preccurlyeq \upsilon_p \chi(x). \tag{8.28}$$

Moreover it is simple to obtain

$$\begin{aligned}\chi(x) &= \phi_{\mu,\rho}\left(\phi_{M,r}(x) - M\right) = \\ &= \frac{\mu\rho}{\rho - m\left(\phi_{M,r}(x) - M\right)} = \\ &= \frac{\mu\rho\left(r - (x_1 + \cdots + x_n)\right)}{\rho r - (\rho + mM)(x_1 + \cdots + x_n)}.\end{aligned}$$

Now we check that

$$\chi(x) \preccurlyeq \frac{\mu\rho r}{\rho r - (\rho + mM)(x_1 + \cdots + x_n)}. \tag{8.29}$$

To this purpose, set $A = \rho r$, $B = \rho + mM$, $t = x_1 + \cdots + x_n$,

$$\widetilde{\chi}(t) = \mu\rho\frac{r - t}{A - Bt}.$$

We have

$$\chi(x) = \widetilde{\chi}(x_1 + \cdots + x_n). \tag{8.30}$$

and

$$\begin{aligned}\frac{r - t}{A - Bt} &= \frac{r}{A}\frac{1}{1 - \frac{B}{A}t} - \frac{t}{A}\frac{1}{\left(1 - \frac{B}{A}t\right)} = \\ &= \frac{r}{A}\sum_{k=0}^{\infty}\left(\frac{Bt}{A}\right)^k - \frac{t}{A}\sum_{k=0}^{\infty}\left(\frac{Bt}{A}\right)^k = \\ &= \frac{r}{A} + \frac{1}{A}\sum_{k=1}^{\infty}\left[\left(\frac{B}{A}\right)^k r - \left(\frac{B}{A}\right)^{k-1}\right]t^k \preccurlyeq \\ &\preccurlyeq \frac{r}{A} + \frac{r}{A}\sum_{k=1}^{\infty}\left(\frac{B}{A}\right)^k t^k = \frac{r}{A - Bt}.\end{aligned} \tag{8.31}$$

By (8.30) and (8.31) we obtain (8.29). Finally, by (8.28), (8.29) and by Proposition 8.9 we get (8.25). ∎

Theorem 8.13 *Let $n, m, p \in \mathbb{N}$. Let Ω_1 be an open set of $\mathbb{R}^n$, and Ω_2 be an open set of $\mathbb{R}^m$. Let $f \in C^\omega(\Omega_1, \mathbb{R}^m)$ satisfy $f(\Omega_1) \subset \Omega_2$. Let $g \in C^\omega(\Omega_2, \mathbb{R}^p)$. Then $g \circ f \in C^\omega(\Omega_1, \mathbb{R}^p)$.*

Proof It is an immediate consequence of Proposition 8.12 and of Theorem 8.7. ∎

In what follows we will use the **Inverse Function Theorem** and the **Implicit Function Theorem for analytic functions** .

For instance, we will exploit the following fact. If $\phi : \Omega \to \mathbb{R}$ is an analytic function in Ω and

$$\nabla\phi(x_0) \neq 0,$$

$x_0 \in \Omega$, then there exist $r, \delta > 0$ and an isometry

$$\Psi : \mathbb{R}^n \to \mathbb{R}^n,$$

such that

$$\Psi(0) = x_0,$$

$$\Psi^{-1}\left(\{x \in \Omega : \phi(x) = \phi(x_0)\}\right) \cap Q_{r,2M} = \left\{(x', \varphi(x')) : x' \in B'_r\right\},$$

where $\varphi \in C^\omega\left(B'_r; \mathbb{R}\right)$ and it satisfies

$$\varphi(0) = 0, \quad |\nabla\varphi(0)| = 0,$$

and

$$\|\varphi\|_{C^1\left(\overline{B'_{r_0}}\right)} \leq Mr.$$

We will not prove this Theorem which can be proved by the method of the majorant functions that we will learn to use in the next chapter.

Chapter 9
The Cauchy Problem for PDEs with Analytic Coefficients

Abstract In this chapter, we study the Cauchy problem for partial differential equations with analytic coefficients, focusing on the linear case but also considering results for non–linear equations. The formulation of the problem and the concept of a characteristic surface do not require analyticity assumptions and are essential for both this chapter and those that follow. Our main result is the Cauchy–Kovalevskaya Theorem, which guarantees the existence and uniqueness of solutions within the class of analytic functions, assuming all data are analytic. We provide a classical proof using majorant functions and discuss alternative approaches. This theorem, though significant, has limitations, prompting further investigation into uniqueness, particularly in the Holmgren Theorem, which relaxes regularity assumptions.

9.1 Introduction

In this chapter, we will study the Cauchy problem for partial differential equations with analytic coefficients, focusing primarily on the linear case. We will, however, also consider some general results for non–linear equations. The formulation of the Cauchy problem and the concept of a characteristic surface—extending the one provided in Sect. 7.3 for first–order equations—do not require analyticity assumptions on the coefficients and will be essential in both this chapter and those that follow.

Our central result is the Cauchy–Kovalevskaya Theorem, which establishes the existence of a solution within the class of analytic functions under the assumption that all data (such as equation coefficients, initial surface, initial conditions) are analytic. Uniqueness follows as an immediate consequence of the principle of analytic continuation. We provide a classical proof based on the method of majorant functions, introduced in Chap. 8. Although this proof is standard, it is not the only approach; we also reference alternative proofs in [56, Chapter 5] and [133, Chapter 17].

While the Cauchy–Kovalevskaya Theorem is an important starting point for the study of the Cauchy problem for partial differential equations, it is far from

S. Vessella, *Unique Continuation Properties for Partial Differential Equations*,
Birkhäuser Advanced Texts Basler Lehrbücher,
https://doi.org/10.1007/978-3-031-86366-0_9

exhaustive due to its restriction to the analytic setting. As we will observe here and in later chapters, especially Chap. 11, this limitation invites further investigation. Sect. 9.6 of this chapter provides various examples illustrating the range of possible situations for the existence and uniqueness of solutions to a Cauchy problem, highlighting the need for deeper study.

In this regard, we will concentrate here and subsequently on uniqueness questions, examining the Holmgren Theorem in Sect. 9.7. This theorem significantly relaxes the regularity assumptions on the initial surface (though not on the equation coefficients) and broadens the functional class of solution within which uniqueness for the Cauchy problem is guaranteed.

9.2 Formulation of the Cauchy Problem

In this introductory section we will give a fairly general formulation of the Cauchy problem. Although we are mainly interested in the linear operators, the formulation that we will give also applies to the fully non–linear operators.

Let Ω be a connected open set of $\mathbb{R}^n$. Let $x_0 \in \Omega$, $\phi \in C^m(\Omega; \mathbb{R})$, where $m \in \mathbb{N}$. Let us suppose that

$$\phi(x_0) = 0. \tag{9.1}$$

Set

$$\Gamma = \{x \in \Omega : \phi(x) = 0\} \tag{9.2}$$

and let us assume

$$\nabla\phi(x) \neq 0, \quad \forall x \in \Gamma. \tag{9.3}$$

Let us denote by

$$\nu(x) = -\frac{\nabla\phi(x)}{|\nabla\phi(x)|}, \quad \forall x \in \Gamma. \tag{9.4}$$

Let be given the function $g_0, g_1, \cdots, g_{m-1}$, defined on Γ, and let $F(x, (p_\alpha)_{|\alpha|\leq m})$ be a function defined on $\Omega \times \mathbb{R}^{N_m}$, where $N_m \in \mathbb{N}$ depends on m only. The **Cauchy problem** is formulated as follows.

Determine u of class C^m in a neighborhood $\mathcal{U}$ of x_0 such that

$$\begin{cases} F\left(x, (\partial^\alpha u)_{|\alpha|\leq m}\right) = 0, & \forall x \in \mathcal{U} \\ \frac{\partial^j u(x)}{\partial \nu^j} = g_j(x), \quad j = 0, 1, \cdots, m-1, & \forall x \in \Gamma \cap \mathcal{U} \end{cases}. \tag{9.5}$$

The functions $g_0, g_1, \cdots, g_{m-1}$ and Γ are called, respectively, the **initial data** or the **initial values** and the **initial surface** of Cauchy problem (9.5). The equations

$$\frac{\partial^j u(x)}{\partial \nu^j} = g_j(x), \quad j = 0, 1, \cdots, m-1, \ \forall x \in \Gamma \cap \mathcal{U}, \tag{9.6}$$

are called the **initial conditions of the Cauchy problem**. Of course, it makes sense and interest to set more general initial conditions. For instance, instead of the vector field $\nu(x)$, we may consider a vector field $\ell(x)$ in a neighborhood of Γ, of class C^{m-1}, requiring that

$$\frac{\partial^j u(x)}{\partial \ell^j} = g_j(x), \quad j = 0, 1, \cdots, m-1, \ \forall x \in \Gamma \cap \mathcal{U}. \tag{9.7}$$

We can easly check that if the vector field $\ell(x)$ and the functions g_j are smooth enough and if $\ell(x) \cdot \nu(x) \neq 0$, for any $x \in \Gamma$, then conditions (9.6) and (9.7) are equivalent. To realize this, let us consider the simple case where $m = 2$, $\Omega = \mathbb{R}^n$, $\phi(x) = x_n$, hence $\nu(x) = -e_n$ for every $x \in \Gamma$ and $\ell = (\ell', \ell_n)$, where $\ell' = (\ell_1, \cdots, \ell_{n-1})$, is a vector field such that $\ell_n(x') \neq 0$ for every $x' \in \mathbb{R}^{n-1}$. Let us assume $u \in C^2(\mathbb{R}^n)$ and

$$u(x', 0) = g_0(x'), \quad \partial_n u(x', 0) = -g_1(x'), \quad \forall x' \in \mathbb{R}^{n-1}, \tag{9.8}$$

where $g_0 \in C^1(\mathbb{R}^{n-1})$ and $g_1 \in C^0(\mathbb{R}^{n-1})$. By the first equation in (9.8) we have

$$\nabla_{x'} u(x', 0) = \nabla_{x'} g_0(x'), \quad \forall x' \in \mathbb{R}^{n-1}$$

that, together with $\partial_n u(x', 0) = -g_1(x')$, gives

$$\frac{\partial u}{\partial \ell}(x', 0) = \ell' \cdot \nabla_{x'} g_0(x') - \ell_n g_1(x'), \quad \forall x' \in \mathbb{R}^{n-1}. \tag{9.9}$$

Conversely, let us suppose that

$$u(x', 0) = \widetilde{g}_0(x'), \quad \frac{\partial}{\partial \ell} u(x', 0) = \widetilde{g}_1(x'), \quad \forall x' \in \mathbb{R}^{n-1}, \tag{9.10}$$

where $\widetilde{g}_0 \in C^1(\mathbb{R}^{n-1})$ and $\widetilde{g}_1 \in C^0(\mathbb{R}^{n-1})$. We have, by the first equation in (9.10) and taking into account that $\ell_n \neq 0$,

$$\partial_n u(x', 0) = \frac{1}{\ell_n(x')} \left(-\ell'(x') \cdot \nabla_{x'} \widetilde{g}_0(x') + \widetilde{g}_1(x') \right).$$

We also notice that if $\ell_n = 0$ at some point $x'_0 \in \mathbb{R}^{n-1}$ then between (9.8) and (9.10) there is no equivalence. Actually, if (9.8) holds, then we can equally get (9.9), but by (9.10) we see that between $\widetilde{g}_0$ and $\widetilde{g}_1$ the following condition of compatibility needs to be fulfilled

$$-\ell'(x_0')\cdot\nabla_{x'}\widetilde{g}_0(x_0')+\widetilde{g}_1(x_0')=0.$$

On the other hand, if this condition is satisfied, the value of $\partial_n u(x_0',0)$ is indeterminate.

9.3 The Characteristic Surfaces

The notion of the **characteristic surface** has a fundamental importance in the investigation of the Cauchy problem. Roughly speaking, we say that the **surface** Γ is noncharacteristic if, assuming $\phi, u, F, g_j\in C^\infty$, for $j=0,1,\cdots,m-1$, all the derivatives of u on Γ can be determined from (9.5).

Of course, we need to specify this notion and arrive at a formal definition, however in what follows we will not tackle problem (9.5) in its full generality, but we will limit ourselves to the linear case, i.e.

$$F\left(x,(\partial^\alpha u)_{|\alpha|\le m}\right)=P(x,\partial)u-f(x)=\sum_{|\alpha|\le m}a_\alpha(x)\partial^\alpha u-f(x),$$

Where we will assume, unless explicitly otherwise stated,

$$\phi, a_\alpha, f\in C^\infty(\Omega),\quad |\alpha|\le m \tag{9.11}$$

and that (9.3) holds.

To motivate the definition of a characteristic surface that we will give, we begin by considering the following Cauchy problem

$$\begin{cases}P(x,\partial)u=f(x), & \forall x\in B_R\\ \\ \partial_n^j u(x',0)=g_j(x'), & j=0,1,\cdots,m-1,\ \forall x'\in B_R'\end{cases}. \tag{9.12}$$

Here $\phi(x)=-x_n$, $\Gamma=\{x\in B_R: x_n=0\}$.

Now, if g_j, $j=0,1,\cdots,m-1$, are of class C^∞, and if there exists a solution $u\in C^\infty(B_R)$ to problem (9.12), we have, by the initial conditions,

$$\partial^{\alpha'}\partial_n^j u(x',0)=\partial^{\alpha'}g_j(x'),\quad \forall\alpha'\in\mathbb{N}_0^{n-1},\quad j=0,1,\cdots,m-1. \tag{9.13}$$

Let us notice that by (9.13) we *cannot* determine the derivatives

$$\partial_n^j u(x',0),\quad j\ge m \tag{9.14}$$

and a fortiori, we cannot determine the derivatives $\partial^{\alpha'}\partial_n^j u(x',0)$ for $j \geq m$, $\alpha' \in \mathbb{N}_0^{n-1}$. To gain such derivatives we should exploit the equation

$$P(x,\partial)u = f(x),$$

which we write in the form

$$a_{(0',m)}(x',0)\partial_n^m u(x',0) = -\sum_{|\alpha'|\leq m,\ \alpha_n<m} a_\alpha(x',0)\partial^\alpha u(x'0) + f(x',0). \tag{9.15}$$

Let us note that to the right–hand side of (9.15), by (9.13), all the derivatives that appear can be expressed in terms of the initial data and their derivatives. Therefore, if

$$a_{(0',m)}(x',0) \neq 0, \quad \forall x' \in B'_R, \tag{9.16}$$

by (9.15) we determine $\partial_n^m u(x',0)$, hence we can calculate

$$\partial^{\alpha'}\partial_n^m u(x',0), \quad \forall \alpha' \in \mathbb{N}_0^{n-1},\ \forall x' \in B'_R. \tag{9.17}$$

Actually, condition (9.16) allows us to deduce all the derivatives of $u(x',0)$. As a matter of fact, by calculating the derivatives w.r.t. x_n of both the sides of (9.15), we have

$$\begin{aligned} a_{(0',m)}(x)\partial_n^{m+1}u(x) = & \underbrace{-\partial_n a_{(0',m)}(x)\partial_n^m u(x)}_{\text{known for } x_n=0 \text{ (by (9.17))}} - \\ & \underbrace{-\partial_n\Big(\sum_{|\alpha'|\leq m,\ )\alpha_n<m} a_\alpha(x)\partial^\alpha u(x)\Big)+}_{\text{known for } x_n=0 \text{ (by (9.13))}} \\ & + \underbrace{\partial_n f(x)}_{\text{known for } x_n=0}. \end{aligned} \tag{9.18}$$

Again, condition (9.16) allows us to derive from (9.18) the derivative

$$\partial_n^{m+1}u(x',0), \quad \forall x' \in B'_R.$$

Of course, we can further make the derivatives of both the sides of (9.18) and we determine $\partial_n^{m+2}u(x', 0)$. Iterating the procedure we obtain the derivatives

$$\partial_n^k u(x', 0), \quad \forall k \in \mathbb{N}_0, \ \forall x' \in B_R'$$

and then calculating the derivatives w.r.t. $x_1, x_2 \cdots, x_{n-1}$ we can calculate all the derivatives of u at the points $(x', 0) \in B_R'$.

At this point we note that condition (9.16) can be written

$$p_m((x', 0), \nu) = i^m \sum_{|\alpha|=m} a_\alpha(x', 0)\nu^\alpha \neq 0, \quad \forall \alpha' \in \mathbb{N}_0^{n-1}, \ \forall x' \in B_R' \tag{9.19}$$

being, in this case, $\nu = e_n$ and recalling that $p_m(x, \xi) = i^m \sum_{|\alpha|=m} a_\alpha(x)\xi^\alpha$.

Now we give the following general definition

Definition 9.1 Let Ω be an open set of $\mathbb{R}^n$ and $x_0 \in \Omega$. Let $a_\alpha \in C^0(\Omega)$, for any $|\alpha| \leq m$. Let

$$P(x, \partial) = \sum_{|\alpha|\leq m} a_\alpha(x)\partial^\alpha$$

be a linear differential operator of order m.

We say that $\ell \in \mathbb{R}^n$ is a **characteristic direction** for the operator $P(x, \partial)$ in x_0 if

$$p_m(x_0, \ell) = 0. \tag{9.20}$$

Let $\phi \in C^1(\Omega)$ satisfy

$$\nabla\phi(x_0) \neq 0.$$

We say that $\Gamma = \{x \in \Omega : \phi(x) = \phi(x_0)\}$ is a **characteristic surface in** x_0 for the operator $P(x, \partial)$ provided $\nu(x_0) = -\frac{\nabla\phi(x_0)}{|\nabla\phi(x_0)|}$ is a characteristic direction for $P(x, \partial)$ in x_0 that is, if

$$p_m(x_0, \nu(x_0)) = i^m \sum_{|\alpha|=m} a_\alpha(x_0)\nu^\alpha(x_0) = 0,$$

or, equivalently,

$$p_m(x_0, \nabla\phi(x_0) = 0.$$

In the sequel, we will say that Γ is a **noncharacteristic surface in** x_0 for the operator $P(x, \partial)$, provided that

$$p_m(x_0, \nabla\phi(x_0) \neq 0.$$

We will say that Γ is a **characteristic surface** for the operator $P(x, \partial)$ if it is characteristic **at each point of** Γ. Finally, we say that Γ is a **noncharacteristic surface** for the operator $P(x, \partial)$ as long as it is noncharacteristic **at every point of** Γ. Let us notice that being a "non characteristic" is more restrictive than the negation of "characteristic surface." This little abuse will simplify the form of expression later on.

We notice that the previous definition **involves only the principal part of the operator** $P(x, \partial)$.

For completeness, we also give a definition of a noncharacteristic surface for the quasilinear operator of order m

$$\mathcal{P}(u) = \sum_{|\alpha|=m} a_\alpha \left(x, (\partial^\beta u)_{|\beta|\leq m-1}\right) \partial^\alpha u + a_0 \left(x, (\partial^\beta u)_{|\beta|\leq m-1}\right), \tag{9.21}$$

where $a_\alpha \left(x, (p^\beta)_{|\beta|\leq m-1}\right)$, a_0 are given functions.

Definition 9.2 Let Ω, x_0, ϕ, Γ be like in Definition 9.1 and let $\mathcal{P}$ be like in (9.21). We say that Γ is a **noncharacteristic surface in** x_0 for the operator $\mathcal{P}$ if Γ is a noncharacteristic surface in x_0 for the operator

$$\sum_{|\alpha|=m} a_\alpha \left(x, (p^\beta)_{|\beta|\leq m-1}\right) \partial^\alpha$$

for each value of p^β, for $|\beta| \leq m-1$. We say that Γ is a **noncharacteristic surface** for the operator $\mathcal{P}$ if it is a noncharacteristic in each point of Γ.

Remark Definition 9.2 is actually more restrictive than that would be needed to determine $\partial^\gamma u(x', 0)$ for all $\gamma \in \mathbb{N}_0^n$. Let us consider, for instance, the Cauchy problem

$$\begin{cases} \mathcal{P}(u) = 0, & \forall x \in B_R \\ \\ \partial_n^j u(x', 0) = g_j(x'), & j = 0, 1, \cdots, m-1, \ \forall x' \in B'_R \end{cases} . \tag{9.22}$$

Let us suppose that a_α, a_0, g_j are functions C^∞. We have proved before that the derivatives $\partial^\gamma u(x', 0)$, for $|\gamma| \leq m-1$, depend on the initial data only. Hence the value of $a_\alpha \left((x', 0), (\partial^\beta u(x', 0))_{|\beta|\leq m-1}\right)$ are determined by the Cauchy data only. Therefore, it should be more natural to say that Γ is a noncharacteristic surface for the operator $\mathcal{P}$ provided that Γ is a noncharacteristic surface in x_0 for the operator

$$\sum_{|\alpha|=m} a_\alpha \left((x', 0), (\partial^\beta u(x', 0))_{|\beta|\leq m-1}\right) \partial^\alpha .$$

♦

9.4 Transformation of a Linear Differential Operator

We wish to examine the transformation of the principal part of the linear differential operator

$$P(x, \partial) = \sum_{|\alpha| \leq m} a_\alpha(x) \partial^\alpha,$$

under the action of

$$\Lambda \in C^m(\Omega, \mathbb{R}^n),$$

where Ω is an open set of $\mathbb{R}^n$ and $\Lambda = (\Lambda_1, \cdots, \Lambda_n)$ is a diffeomorphism of class C^m. By this we mean that Λ is injective and it satisfies

$$\det(\partial_x \Lambda(x)) \neq 0, \quad \forall x \in \Omega,$$

where $\partial_x \Lambda(x)$ is the jacobian matrix of Λ. Let $u \in C^m(\Omega)$. Set

$$v(y) = u\left(\Lambda^{-1}(y)\right), \quad \forall y \in \widetilde{\Omega} := \Lambda(\Omega),$$

we have $v \in C^m(\widetilde{\Omega})$.

Moreover by

$$u(x) = v(\Lambda(x)), \quad \forall x \in \Omega,$$

we have

$$\partial_{x_j} u(x) = \sum_{k=1}^{n} (\partial_{y_k} v)(\Lambda(x))\, \partial_{x_j} \Lambda_k(x), \quad j = 1, \cdots, n, \quad \forall x \in \Omega,$$

$$\begin{aligned}
\partial^2_{x_j x_i} u(x) = & \sum_{h,k=1}^{n} (\partial_{y_h y_k} v)(\Lambda(x))\, \partial_{x_i} \Lambda_h(x) \partial_{x_j} \Lambda_k(x) + \\
& + \sum_{k=1}^{n} (\partial_{y_k} v)(\Lambda(x))\, \partial_{x_j x_i} \Lambda_k(x) = \\
& = \left(\left((\partial_x \Lambda(x))^t \partial_y \right)_i \left((\partial_x \Lambda(x))^t \partial_y \right)_j v \right)(\Lambda(x)) + \\
& + (\text{first–order terms}).
\end{aligned}$$

In general we have

$$\partial_x^\alpha u(x) = \left(\left((\partial_x \Lambda(x))^t \partial_y\right)^\alpha v\right)(\Lambda(x)) + (\text{terms of order less than } |\alpha|). \tag{9.23}$$

Now, let us denote by $P^\bullet(y, \partial_y)$ the transformed operator of $P(x, \partial_x)$ through Λ, that is the operator satisfying

$$\left(P^\bullet(y, \partial_y)v(y)\right)_{|y=\Lambda(x)} = P(x, \partial_x)u(x). \tag{9.24}$$

By (9.24) we have that the principal part of $P^\bullet(y, \partial_y)$, $P_m^\bullet(y, \partial_y)$, is given by

$$\begin{aligned} P_m^\bullet(y, \partial_y) &= \sum_{|\alpha|=m} a_\alpha(\Lambda^{-1}(y)) \left((\partial_x \Lambda(x))^t \partial_y\right)^\alpha_{|x=\Lambda^{-1}(y)} = \\ &= P(x, \partial_x(\Lambda(x))^t \partial_y)_{|x=\Lambda^{-1}(y)} \end{aligned} \tag{9.25}$$

and its symbol $p_m^\bullet(y, \eta)$ is given (up to the multiplicative constant i^m) by

$$p_m^\bullet(y, \eta) = \sum_{|\alpha|=m} a_\alpha(\Lambda^{-1}(y)) \left((\partial_x \Lambda(x))^t \eta\right)^\alpha_{|x=\Lambda^{-1}(y)} . \tag{9.26}$$

From what we have so far established, we easily obtain

Theorem 9.3 (Invariant Property of the Characteristic Surfaces) *Let Ω be an open set of $\mathbb{R}^n$. Let $x_0 \in \Omega$, $\phi \in C^m(\Omega)$,*

$$\Gamma = \{x \in \Omega : \phi(x) = \phi(x_0)\}.$$

Suppose that

$$\nabla\phi(x) \neq 0, \quad \forall x \in \Gamma.$$

Moreover, let $P(x, \partial)$ be a linear differential operator of order m and $\Lambda \in C^m(\Omega, \mathbb{R}^n)$ be a diffeomorphism of class C^m.

Then Γ is a noncharacteristic surface for $P(x, \partial)$ if and only if $\Lambda(\Gamma)$ is a noncharacteristic surface for the operator $P^\bullet(y, \partial_y)$.

Proof It is not restrictive to assume $x_0 = 0$ and $\Lambda(x_0) = 0$. Now, since $\Gamma = \{x \in \Omega : \phi(x) = 0\}$, we have

$$\Lambda(\Gamma) = \left\{y \in \Lambda(\Omega) : \widetilde{\phi}(y) = 0\right\},$$

where

$$\widetilde{\phi} = \phi \circ \Lambda^{-1}.$$

On the other hand, Γ is a noncharacteristic surface for $P(x, \partial)$ if and only if

$$p_m(x, \nabla\phi(x)) \neq 0, \quad \forall x \in \Gamma,$$

but we have

$$\nabla_y \widetilde{\phi}(y) = \left((\partial_x \Lambda(x))^t\right)^{-1}_{|x=\Lambda^{-1}(y)} (\nabla_x \phi) \left(\Lambda^{-1}(y)\right).$$

Therefore by (9.26) we have

$$p^{\bullet}_m(y, \nabla_y \widetilde{\phi}(y)) = i^m \sum_{|\alpha|=m} a_\alpha(x) \left(\nabla_x \phi(x)\right)^\alpha = p_m(x, \nabla_x \phi(x)),$$

from which the thesis follows. ■

The following Proposition holds true.

Proposition 9.4 *Let Ω be an open set of $\mathbb{R}^n$ and let $x_0 \in \Omega$ and $\phi \in C^\infty(\Omega)$ satisfy*

$$\nabla\phi(x_0) \neq 0.$$

Let $f, a_\alpha \in C^\infty(\Omega)$ for $|\alpha| \leq m$. Assume that

$$\Gamma = \{x \in \Omega : \phi(x) = \phi(x_0)\},$$

*is a **noncharacteristic surface in** x_0 for the operator $P(x, \partial)$. Let $g_j \in C^\infty(\Omega)$, for $j = 0, 1, \cdots, m-1$.*

If u is a C^∞ solution in a neighborhood $\mathcal{U}$ of x_0 of the Cauchy problem

$$\begin{cases} P(x, \partial)u = f(x), & \forall x \in \mathcal{U}, \\ \\ \frac{\partial^j u(x)}{\partial \nu^j} = g_j(x), & j = 0, 1, \cdots, m-1, \ \forall x \in \Gamma \cap \mathcal{U} \end{cases} \tag{9.27}$$

then the derivatives $\partial^\alpha u(x_0)$ are uniquely determined for every $\alpha \in \mathbb{N}^n_0$.

Proof Although the proof follows the line indicated in the particular case $\phi(x) = -x_n$ we want to dwell on some details that may be useful later.

We split the proof into two steps. In **Step I** we determine the derivatives of u of order less than m on Γ in, **Step II** we determine the higher order derivatives of u.

Step I We begin by proving that if $\phi \in C^k(\Omega)$, $k \geq 1$, $\nabla\phi(x_0) \neq 0$ and v is any C^k function in a neighborhood of x_0 then there exists a neighborhood $\mathcal{V}$ of x_0 such that the derivatives $\frac{\partial^j v(x)}{\partial \nu^j}$ on $\Gamma \cap \mathcal{V}$, for $j = 0, 1, \cdots, k$ determine the derivatives $\partial^\alpha v$ on $\Gamma \cap \mathcal{V}$ for each $|\alpha| \leq k$.

Since $\nabla\phi(x_0) \neq 0$, we may limit ourselves, up to isometries, to consider the case where, we have, for a suitable $\delta > 0$,

$$\Gamma = \left\{(x', \varphi(x')) : x' \in B'_\delta\right\}, \tag{9.28}$$

where $\varphi \in C^k(\overline{B'_\delta})$,

$$\varphi(0) = |\nabla\varphi(0)| = 0.$$

We have

$$\nu((x', \varphi(x'))) = \frac{(\nabla_{x'}\varphi(x'), -1)}{\sqrt{1 + |\nabla_{x'}\varphi|^2}}. \tag{9.29}$$

Also set

$$\mu(x') := \nu((x', \varphi(x')))\sqrt{1 + |\nabla_{x'}\varphi|^2} = (\nabla_{x'}\varphi(x'), -1). \tag{9.30}$$

We proceed by induction on the order s of derivatives. If $s = 0$, then v is known on Γ, but to better understand the procedure, we also consider the case $k = 1$. In such a case we have, for $x' \in B'_\delta$,

$$v(x', \varphi(x')) = g_0(x'), \tag{9.31}$$

$$\begin{aligned} &\sum_{j=1}^{n-1}(\partial_j v)(x', \varphi(x'))\partial_j\varphi(x') - (\partial_n v)(x', \varphi(x')) = \\ &= g_1(x')\sqrt{1 + |\nabla_{x'}\varphi(x')|^2}. \end{aligned} \tag{9.32}$$

Computing the derivatives w.r.t. x_i of both the sides of (9.31), for $i = 1, \cdots, n-1$, we get, (for the sake of brevity, omit the variables)

$$\partial_i v + \partial_n v \partial_i \varphi = \partial_i g_0.$$

Hence

$$\partial_i v = \partial_i g_0 - \partial_n v \partial_i \varphi$$

and, inserting these derivatives in (9.32), we have

$$\sum_{j=1}^{n-1} \partial_j g_0 \partial_j \varphi - \left(1 + |\nabla_{x'}\varphi|^2\right)\partial_n v = g_1\sqrt{1 + |\nabla_{x'}\varphi|^2},$$

from which we get

$$\partial_n v = \frac{1}{1 + |\nabla_{x'}\varphi|^2}\left(\nabla_{x'} g_0 \cdot \nabla_{x'}\varphi - g_1\sqrt{1 + |\nabla_{x'}\varphi|^2}\right).$$

Now let us prove that if $\partial^\alpha v$ are determined by $|\alpha| \le s$ on Γ (with $s \le k-1$) then $\partial^\alpha v$ are determined on Γ for $|\alpha| \le s+1$.

Let

$$\partial^\alpha v(x', \varphi(x')) = h_\alpha(x'), \quad \text{for } |\alpha| = s, \quad \text{on } \Gamma. \tag{9.33}$$

Set, for $j = 0, 1, \cdots, k$,

$$\widetilde{g}_j(x') = \left(1 + |\nabla_{x'}\varphi|^2\right)^{j/2} \frac{\partial^j v}{\partial \nu^j} = \sum_{|\alpha|=j} \mu^\alpha(x')\partial^\alpha v(x', \varphi(x')).$$

Then, besides (9.33), we know that

$$\sum_{|\alpha|=s+1} \mu^\alpha \partial^\alpha v = \widetilde{g}_{s+1}, \quad \text{on } \Gamma, \tag{9.34}$$

which we write

$$\sum_{j=0}^{s+1} \sum_{|\alpha'|=s+1-j} \mu'^{\alpha'} \mu^j \partial_n^j \partial^{\alpha'} v = \widetilde{g}_{s+1} \quad \text{on } \Gamma, \tag{9.35}$$

where $\alpha' = (\alpha_1, \cdots, \alpha_{n-1})$ and $\mu' = (\mu_1, \cdots, \mu_{n-1})$.

At this point we express $\partial_n^j \partial^{\alpha'} v$, for $|\alpha'| = s+1-j$, through the functions $\partial_n^{s+1} v$ and h_γ for $|\gamma| \le s$.

If $j < s+1$, there exists $i \in \{1, \cdots, n-1\}$ such that $\alpha' - e_i \ge 0$. Set $\beta' = \alpha' - e_i$ and let us recall that by (9.33) we have

$$\partial_n^j \partial^{\beta'} v = h_{(\beta', j)},$$

from which, computing the derivative w.r.t. x_i we have

$$\underbrace{\partial_n^j \partial_i \partial^{\beta'} v}_{\partial^{\alpha'} v} + \mu_i \partial_n^{j+1} \partial^{\beta'} v = \partial_i h_{(\beta', j)},$$

hence

$$\partial_n^j \partial^{\alpha'} v = \partial_i h_{(\beta', j)} - \mu_i \partial_n^{j+1} \partial^{\beta'} v.$$

Now, if $\beta' \neq 0$, we proceed in a similar manner for $\partial_n^j \partial^{\alpha'} v$ and then we iterate. Denoting by $H_{(\alpha',j)}$ the functions of the type

$$\sum_{k=0}^{j} \sum_{|\beta'|=s-j} c_{(\beta',k)} h_{(\beta',k)},$$

where $c_{(\beta',k)}$ are known functions expressable by means of μ_i, $i = 1, \cdots, n-1$, we obtain

$$\partial_n^j \partial^{\alpha'} v = H_{(\alpha',j)} + (-1)^{s+1-j} \mu'^{\alpha'} \partial_n^{s+1} v, \qquad \text{for } |\alpha| = s+1-j. \tag{9.36}$$

Inserting (9.36) in (9.35) we get (recall $\mu_n = -1$)

$$\widetilde{g}_{s+1} = \sum_{j=0}^{s+1} \sum_{|\alpha'|=s+1-j} (-1)^j \mu'^{\alpha'} \left(H_{(\alpha',j)} + (-1)^{s+1-j} \mu'^{\alpha'} \partial_n^{s+1} v \right) =$$

$$= \sum_{j=0}^{s+1} \sum_{|\alpha'|=s+1-j} (-1)^j \mu'^{\alpha'} H_{(\alpha',j)} + \left(\sum_{j=0}^{s+1} \sum_{|\alpha'|=s+1-j} (-1)^j \mu'^{2\alpha'} \right) \partial_n^{s+1} v.$$

From which we have

$$\begin{aligned} &(-1)^{s+1} \left(1 + \sum_{1 \leq |\alpha'| \leq s+1} \mu'^{2\alpha'} \right) \partial_n^{s+1} v = \\ &= \widetilde{g}_{s+1} - \sum_{j=0}^{s+1} \sum_{|\alpha'|=s+1-j} (-1)^j \mu'^{\alpha'} H_{(\alpha',j)}. \end{aligned} \tag{9.37}$$

Since

$$\sum_{1 \leq |\alpha'| \leq s+1} \mu'^{2\alpha'} \geq 0,$$

by (9.37) we determine $\partial_n^{s+1} v$.

Step II Let us consider the Cauchy problem

$$\begin{cases} P(x, \partial) u = f(x), & \forall x \in \mathcal{U} \\ \\ \dfrac{\partial^j u(x)}{\partial \nu^j} = g_j(x), & j = 0, 1, \cdots, m-1, \ \forall x \in \Gamma \cap \mathcal{U} \end{cases} . \tag{9.38}$$

Like Step I we assume that Γ is the graph (9.28), where $\varphi \in C^{\infty}(\overline{B'_{\delta}})$, $\varphi(0) = |\nabla_{x'}\varphi(0)| = 0$. By what was proved in Step I we can determine, from the functions $g_0, g_1, \cdots, g_{m-1}$ only, the derivatives

$$\partial^{\alpha}u, \quad \text{for } |\alpha| \leq m-1, \quad \text{on } \Gamma. \tag{9.39}$$

Now let us show in which a way we determine the other derivatives of u on Γ. Let

$$\Lambda : B_{\delta} \subset \mathbb{R}^n_x \to \mathbb{R}^n_y, \quad \Lambda(x) = \left(x', x_n - \varphi(x')\right).$$

We have (Fig. 9.1)

$$\Lambda(\Gamma) = \left\{(y', 0) : y' \in B'_{\delta}\right\} = \{y \in B_{\delta} : -y_n = 0\}$$

and, set

$$v(y) = u\left(\Lambda^{-1}(y)\right).$$

Equality (9.23) implies that the derivatives

$$\partial^{\alpha}v(y', 0), \quad \text{for } |\alpha| \leq m-1, \quad \forall y' \in B'_{\delta},$$

are all uniquely determined. In particular, the following derivatives are determined

$$\partial_n^j v(y', 0), \quad \text{for } j = 0, 1, \cdots, m-1, \quad \forall y' \in B'_{\delta}. \tag{9.40}$$

On the other hand, by (9.24) we have that v, solves the following equation in a neighborhood of 0

$$P^{\bullet}(y, \partial_y)v(y) := \sum_{|\alpha| \leq m} b_{\alpha}(y)\partial_y^{\alpha}v(y) = \widetilde{f}(y),$$

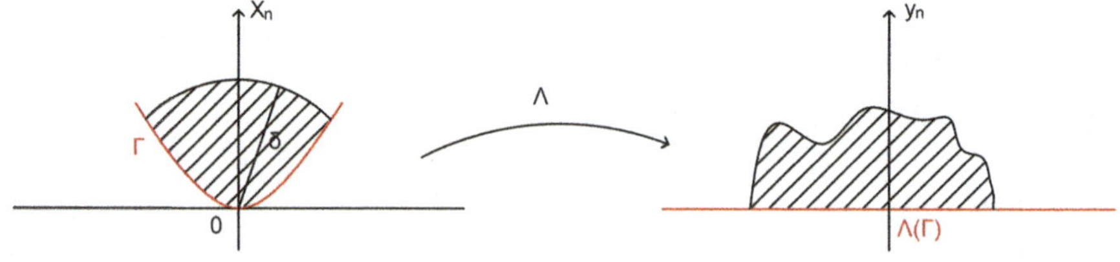

Fig. 9.1 The map Λ flattens the boundary Γ

where $\widetilde{f}(y) = f\left(\Lambda^{-1}(y)\right)$. Now, by Proposition 9.3 we have that $\Lambda(\Gamma)$ is a noncharacteristic surface for $\widetilde{P}(y, \partial_y)$. Since $\Lambda(\Gamma) = \{y \in B_\delta : -y_n = 0\}$ we have

$$b_{(0,m)}(y', 0) = \sum_{|\alpha|=m} b_\alpha(y', 0)e_n^\alpha \neq 0, \quad \forall y' \in B'_\delta.$$

We are therefore reduced to the same situation examined at the beginning of this section and we then calculate the derivatives $\partial_y^\alpha v(y', 0)$ for $|\alpha| \geq m$ from the derivatives of g_j, $j = 0, 1, \cdots, m-1$ and the coefficients (and their derivatives) of $P^\bullet(y, \partial_y)$. Finally, by exploiting formula (9.23) we obtain the derivatives $\partial_x^\alpha u$ on Γ for $|\alpha| \geq m$. ■

We conclude this section with some examples and remarks.

Remarks Let $P(x, \partial)$, $x \in \Omega$ be a linear differential operator whose principal part is $P_m(x, \partial)$.

1. We call **characteristic equation** the non–linear first–order equation in the unknown ϕ

$$p_m(x, \nabla\phi(x)) = 0. \tag{9.41}$$

2. We say that $P(x, \partial)$ is **elliptic in the point** $x_0 \in \Omega$ if

$$p_m(x_0, \xi) \neq 0, \quad \forall \xi \in \mathbb{R}^n \setminus \{0\}.$$

It is evident that the elliptic operators have not characteristic surface. If $P(x, \partial)$ is elliptic in each point of Ω we say that $P(x, \partial)$ is **elliptic in** Ω. Each linear differential operators of one variable are elliptic

$$P(t, \frac{d}{dt}) = a_m(t)\frac{d^m}{dt^m} + \cdots + a_0(t), \quad t \in I,$$

where $a_m(t) \neq 0$ for $t \in I$, where I is an interval of $\mathbb{R}$. As a matter of fact

$$p_m(t, \xi) = i^m a_m(t)\xi^m \neq 0, \quad t \in I, \quad \forall \xi \in \mathbb{R} \setminus \{0\}.$$

A remarkable example of an elliptic operator is the operator of Cauchy-Riemann

$$P((x, y), \partial_x, \partial_y) = \partial_x + i\partial_y.$$

If $m = 2$ and a_{jk}, where $a_{jk}(x) = a_{kj}(x)$ for $x \in \Omega$, $j, k = 1, \cdots, n$ are real–valued functions in $\mathbb{R}$, we define a **uniformly elliptic** operator with bounded coefficients an operator of the type

$$P(x, \partial) = \sum_{j,k=1}^{n} a_{jk}(x)\partial_{jk}^2 + \sum_{j=1}^{n} b_j(x)\partial_j + c(x),$$

such that there exists a constant $\lambda \geq 1$ satisfying

$$\lambda^{-1}|\xi|^2 \leq \sum_{j,k=1}^{n} a_{jk}(x)\xi_j\xi_k \leq \lambda|\xi|^2, \quad \forall x \in \Omega, \quad \forall \xi \in \mathbb{R}^n.$$

Exercise Let us consider the following operator with constant coefficients

$$P(\partial) = \sum_{j,k=1}^{n} a_{jk}\partial_{jk}^2, \tag{9.42}$$

where $\{a_{jk}\}$ is a real symmetric matrix. If $\det\{a_{jk}\} \neq 0$ then there exists a nonsingular matrix C such that, setting $y = Cx$, operator (9.42), is transformed in

$$P(\partial_y) = \sum_{j=1}^{n} \varkappa_j \partial_{y_j}^2,$$

where $\varkappa_j$, $j = 1, \cdots, n$, is equal either to 1 or to -1. ♠

9.5 The Cauchy–Kovalevskaya Theorem

In what follows we denote by Ω an open set of $\mathbb{R}^n$, $x_0 \in \Omega$ and by $\phi \in C^\omega(\Omega, \mathbb{R})$ a function such that

$$\nabla\phi(x) \neq 0, \quad \forall x \in \Gamma := \{x \in \Omega : \phi(x) = \phi(x_0)\}. \tag{9.43}$$

By (9.43) we have that for every $\widehat{x} \in \Gamma$ there exist $r, M > 0$ and an isometry Ψ under which we have $\Psi(0) = \widehat{x}$, and

$$\Psi^{-1}(\Gamma) \cap Q_{r,2M} = \left\{(x', \varphi(x')) : x' \in B_r'\right\},$$

where $\varphi \in C^\omega\left(B_r'; \mathbb{R}\right)$ and it satisfies

$$\varphi(0) = 0, \quad |\nabla\varphi(0)| = 0,$$

$$\|\varphi\|_{C^1\left(\overline{B_r'}\right)} \leq Mr.$$

We say that a function

$$h : \Gamma \to \mathbb{C},$$

is analytic on Γ, provided

$$(h \circ \Psi)(\cdot, \varphi(\cdot)) \in C^{\omega}\left(B'_r\right).$$

In this section we will prove

Theorem 9.5 (Cauchy–Kovalevskaya) *Let $m \in \mathbb{N}$ and let Ω be an open set of $\mathbb{R}^n$, $x_0 \in \Omega$, $\phi \in C^{\omega}(\Omega, \mathbb{R})$ which satisfies (9.43). Moreover, let $P(x, \partial)$ be the linear differential operator*

$$P(x, \partial) = \sum_{|\alpha| \le m} a_{\alpha}(x)\partial^{\alpha}, \tag{9.44}$$

where $a_{\alpha} \in C^{\omega}(\Omega)$, for $|\alpha| \le m$. Let $g_0, g_1, \cdots, g_{m-1}$ be ***analytic functions on*** *Γ. Assume that Γ is a noncharacteristic surface for the operator $P(x, \partial)$. Let $f \in C^{\omega}(\Omega)$.*

Then for every $\widetilde{x} \in \Gamma$ there exists a neighborhood $\mathcal{U}_{\widetilde{x}}$ of $\widetilde{x}$ such that the Cauchy problem

$$\begin{cases} P(x, \partial)u(x) = f(x), & \forall x \in \mathcal{U}_{\widetilde{x}} \\ \frac{\partial^j u(x)}{\partial \nu^j} = g_j(x), & j = 0, 1, \cdots, m-1, \quad \forall x \in \Gamma \cap \mathcal{U}_{\widetilde{x}} \end{cases} \tag{9.45}$$

has a unique analytic solution in $\mathcal{U}_{\widetilde{x}}$.

In order to prove Cauchy-Kovalevskaya Theorem we need two preliminary steps

(i) local flatness of the initial surface;
(ii) transformation of problem (9.45) to a Cauchy problem for a first–order system.

We have already considered **point (i)** in the context of of the proof of Step II of Proposition 9.4. Here it suffices to add that, referring to the notations used in the above proof, the function φ introduced there is, not only C^{∞}, but also analytic in B'_{δ} and, consequently, the map

$$\Lambda : B_{\delta} \subset \mathbb{R}^n_x \to \mathbb{R}^n_y, \quad \Lambda(x) = \left(x', x_n - \varphi(x')\right), \tag{9.46}$$

which, we recall, flatten Γ in the sense that

$$\Lambda(\Gamma) = \left\{(y', 0) : y' \in B'_{\delta}\right\} = \{y \in B_{\delta} : -y_n = 0\}.$$

Moreover, setting

$$v(y) = u\left(\Lambda^{-1}(y)\right), \tag{9.47}$$

the operator $P^{\bullet}(y, \partial_y)$, defined by

$$\left(P^{\bullet}(y, \partial_y)v(y)\right)_{|y=\Lambda(x)} = P(x, \partial_x)u(x), \tag{9.48}$$

has its principal part $P_m^{\bullet}(y, \partial_y)$, given by

$$P_m^{\bullet}(y, \partial_y) = \sum_{|\alpha|=m} b_\alpha(y)\partial^\alpha, \tag{9.49}$$

where b_α are analytic functions on B_δ.

Again by Proposition 9.4 we know that the new initial data $\widetilde{g}_j(y') = \partial_n^j v(y', 0)$, $j = 0, 1, \cdots, m-1$, are determined by the initial data g_j and that $\widetilde{g}_j$ are analytic in a neighborhood of $0 \in \mathbb{R}^{n-1}$. Moreover, we know that the surface

$$\{y \in B_\delta : -y_n = 0\},$$

is noncharacteristic. Therefore

$$b_{(0,m)}(y', 0) \neq 0, \quad \forall y' \in B'_\delta(0).$$

Hence, setting

$$\widetilde{b}_\alpha(y) = -\frac{b_\alpha(y)}{b_{(0,m)}(y)}, \quad \widetilde{f}(y) = \frac{f\left(\Lambda^{-1}(y)\right)}{b_{(0,m)}(y)},$$

we write problem (9.45) as

$$\begin{cases} \partial_n^m v = \sum_{|\alpha|\leq m, \alpha_n \leq m-1} \widetilde{b}_\alpha(y)\partial^\alpha v + \widetilde{f}(y) \\ \\ \partial_n^j v(y', 0) = \widetilde{g}_j(y'), \quad j = 0, 1, \cdots, m-1 \end{cases}. \tag{9.50}$$

We may easily transform problem (9.50) into a Cauchy problem with homogeneous initial conditions. To this purpose it suffices to define the function

$$H(y) = \sum_{j=0}^{m-1} \frac{y_n^j}{j!}\widetilde{g}_j(y')$$

and set

$$w = v - H,$$

obtaining

$$\begin{cases} \partial_n^m w = \sum_{|\alpha| \le m, \alpha_n \le m-1} \widetilde{b}_\alpha(y) \partial^\alpha w + F(y) \\ \\ \partial_n^j w(y', 0) = 0, \quad j = 0, 1, \cdots, m-1 \end{cases} . \tag{9.51}$$

where

$$F(y) = \widetilde{f}(y) - \partial_n^m H(y) + \sum_{|\alpha| \le m, \alpha_n \le m-1} \widetilde{b}_\alpha(y) \partial^\alpha H(y). \tag{9.52}$$

(ii). The concept of this transformation is straightforward and somewhat parallels the process commonly used to reduce a Cauchy problem for ordinary differential equations of order m to a first-order system's Cauchy problem. However, in our situation, the unknown is dependent on $n > 1$ variables, necessitating additional adaptations.

To emphasize the key stages, we will exemplify the procedure in the scenario where $n = 2$ and the operator equals its principal part exclusively. Subsequently, we will provide an overview of the approach for the general case.

Let us consider the Cauchy problem

$$\begin{cases} \partial_t^m u = \sum_{j \le m-1} \sum_{i+j=m} a_{i,j}(x,t) \partial_x^i \partial_t^j u + f(x,t) \\ \\ \partial_t^j u(x, 0) = 0, \quad j = 0, 1, \cdots, m-1, \ \ \forall x \in \mathbb{R} \end{cases} . \tag{9.53}$$

Let us assume that $m \ge 2$ and let $u(x,t)$ be a C^∞ solution of (9.53). We set

$$V_{i,j} = \partial_x^i \partial_t^j u, \qquad \text{for } i + j \le m - 1.$$

It is simple to check what follows

$$\partial_t V_{i,j} = V_{i,j+1}, \qquad \text{for } i + j < m - 1, \tag{9.54a}$$

$$\partial_t V_{i,j} = \partial_x V_{i-1,j+1}, \qquad \text{for } i + j = m - 1, \quad i > 0, \tag{9.54b}$$

$$\partial_t V_{0,m-1} = \sum_{i+j=m-1} a_{i+1,j}(x,t) \partial_x V_{i,j} + f(x,t), \tag{9.54c}$$

$$V_{i,j}(x, 0) = 0 \quad \text{for } i + j \le m - 1. \tag{9.54d}$$

Hence, if $V_{0,0}$ is a solution to Cauchy problem (9.53), then it is a solution to Cauchy problem (9.54a)–(9.54d). Thus, in order to prove the equivalence of problem (9.53) and problem (9.54a)–(9.54d) it suffices to prove the converse. Let us suppose, therefore, that $V = \left(V_{i,j}\right)_{i+j\leq m-1}$ is a C^∞ a solution to problem (9.54a)–(9.54d) and let us prove that $V_{0,0}$ is a solution to problem (9.53).

1. We prove that

$$\text{If} \quad i+j=m-1,\ i>0 \quad \text{then} \quad V_{i,j}=\partial_x V_{i-1,j}. \tag{9.55}$$

Proof Let $i+j\leq m-1$ and $i>0$. By (9.54b). We have

$$\partial_t V_{i,j}=\partial_x V_{i-1,j+1}, \qquad \text{for } i+j=m-1, \quad i>0. \tag{9.56}$$

Now, since $(i-1)+j=m-2$, by (9.54a) we have

$$V_{i-1,j+1}=\partial_t V_{i-1,j}. \tag{9.57}$$

Hence, by (9.56) and (9.57) we have

$$\partial_t\left(V_{i,j}-\partial_x V_{i-1,j}\right)=0. \tag{9.58}$$

On the other hand, by (9.54d) we have

$$\left(V_{i,j}-\partial_x V_{i-1,j}\right)(x,0)=0,$$

this equality and (9.58) implies

$$V_{i,j}=\partial_x V_{i-1,j}.$$

2. We prove that

$$\text{If} \quad i+j\leq m-1,\ i>0 \quad \text{then} \quad V_{i,j}=\partial_x V_{i-1,j}. \tag{9.59}$$

Proof Set $l=(m-1)-(i+j)$ and let us proceed by induction on l. If $l=0$, then (9.59) holds true, because it is nothing but (9.55). Now let us suppose that (9.59) holds true for l and let us prove it for $l+1$. Hence, let us suppose that

$$\text{if} \quad i+j=(m-1)-l,\ i>0 \quad \text{then} \quad V_{i,j}=\partial_x V_{i-1,j}. \tag{9.60}$$

Let i, j satisfy $i+j=(m-1)-(l+1)$ and $i>0$. Since $i+j<m-1$, by (9.54a) we have

$$\partial_t V_{i,j}=V_{i,j+1}. \tag{9.61}$$

Since we have $i + (j + 1) = (m - 1) - l$, (9.60) gives

$$V_{i,j+1} = \partial_x V_{i-1,j+1}. \tag{9.62}$$

Since $(i - 1) + j = (m - 1) - l - 2 < m - 1$, (9.54a) gives

$$V_{i-1,j+1} = \partial_t V_{i-1,j}, \qquad \text{for } i + j < m - 1. \tag{9.63}$$

Hence, by (9.62) and (9.63) we have

$$V_{i,j+1} = \partial_x V_{i-1,j+1} = \partial_t \partial_x V_{i-1,j},$$

by the latter and by (9.61) we have

$$\partial_t \left(V_{i,j} - \partial_x V_{i-1,j}\right) = 0 \tag{9.64}$$

so, by (9.54d) we have that, if $i + j = (m - 1) - (l + 1)$, $i > 0$, then

$$V_{i,j} = \partial_x V_{i-1,j}.$$

(9.59) is proved.

Conclusions Iteration of (9.59) implies what follows:

$$\text{if} \quad i + j \leq m - 1,\ i > 0 \quad \text{then} \quad V_{i,j} = \partial_x^i V_{0,j}. \tag{9.65}$$

On the other hand, (9.54a) gives

$$V_{0,j} = \partial_t V_{0,j-1} = \cdots = \partial_t^j V_{0,0}. \tag{9.66}$$

All in all, by (9.65) and (9.66) we get

$$V_{i,j} = \partial_x^i \partial_t^j V_{0,0}$$

and using this equality into (9.54c)–(9.54d) we have that $V_{0,0}$ solves (9.53).

We may rewrite problem (9.54a)–(9.54d) in a more concentrated form as follows

$$\begin{cases} \partial_t V(x,t) = B(x,t)\partial_x V + F(x,t) \\ \\ V(x,0) = 0 \end{cases}, \tag{9.67}$$

where, for an appropriate $N \in \mathbb{N}$, V is a function with values in $\mathbb{R}^N$, B is an $N \times N$ matrix whose entries are analytic and $F = f e_N$.

In the case $n > 2$ one may similarly reduce Cauchy problem (9.45) to a Cauchy problem for a first–order system. We outline the procedure (the details of which we leave to the reader). First, it is convenient to introduce the following notation. If $\alpha \in \mathbb{N}_0^{n-1} \setminus \{0\}$ is a multi-index, we set

$$i(\alpha) = \min\{i : \alpha_i > 0\}.$$

In addition, we set $t = x_n$ and

$$V_{\alpha,j} = \partial_t^j \partial_{x'}^\alpha w, \qquad \text{for } |\alpha| + j \le m - 1.$$

By (9.51) we have

$$\partial_t V_{\alpha,j} = V_{\alpha,j+1}, \qquad \text{for } |\alpha| + j < m - 1, \tag{9.68a}$$

$$\partial_t V_{\alpha,j} = \partial_{x_{i(\alpha)}} V_{\alpha - e_{i(\alpha)},j+1}, \qquad \text{for } |\alpha| + j = m - 1, |\alpha| > 0, \tag{9.68b}$$

$$\partial_t V_{0,m-1} = \sum_{|\alpha|+j=m, j<m} c_{\alpha,j} \partial_{x_{i(\alpha)}} V_{\alpha - e_{i(\alpha)},j} + \sum_{|\alpha|+j \le m} d_{\alpha,j} V_{\beta,j} + F, \tag{9.68c}$$

$$V_{\alpha,j}(x', 0) = 0 \quad \text{for } |\alpha| + j \le m - 1, \tag{9.68d}$$

where $c_{\alpha,j}$, $d_{\alpha,j}$, F are analytic functions in the variables x' and t.

Proof of the Cauchy–Kovalevskaya Theorem Taking into account what has been done in (i) and (ii) and changing the notations a little, we may reformulate Cauchy problem (9.51) as follows.

$$\begin{cases} \partial_t U_k(x,t) = \sum_{j=1}^{n-1} \sum_{l=1}^{N} B_j^{lk} \partial_{x_j} U_l + \sum_{l=1}^{N} C^{lk} U_l + F_k \\ \\ U_k(x, 0) = 0 \end{cases}, \tag{9.69}$$

for $k = 1, \cdots, N$, where B_j^{lk}, C^{lk}, F_k, $j = 1, \cdots, n-1$, $l, k = 1, \cdots, N$ are analytic functions in a neighborhood of 0.

In order to prove the existence and the uniqueness for Cauchy problem (9.69) we proceed as follows:

Step I For any function $U \in C^\infty$ that *we suppose* to satisfy (9.69), we will calculate the derivatives

$$\partial^\alpha U_k(0, 0) := U_{k,\alpha}, \quad \forall \alpha \in \mathbb{N}_0^n, \quad k = 1, \cdots, N.$$

Setting, $\partial^\alpha = \partial_x^{\alpha'}\partial_t^{\alpha_n}$, for $\alpha = (\alpha', \alpha_n)$, we will have

$$U_{k,(\alpha',0)} = 0, \quad \forall \alpha' \in \mathbb{N}_0^{n-1}, \quad k = 1, \cdots, N. \tag{9.70}$$

Step II We will show what follows. Let us assume that the functions $\widetilde{B}_j^{lk}$, $\widetilde{C}^{lk}$, $\widetilde{F}$ and $\widetilde{\varphi}$ (the latter is independent of t) satisfy the following conditions

(a) $B_j^{lk} \preccurlyeq \widetilde{B}_j^{lk}$, $C^{lk} \preccurlyeq \widetilde{C}^{lk}$, for $j = 1, \cdots, n-1, l, k = 1, \cdots, N$, $F \preccurlyeq \widetilde{F}, 0 \preccurlyeq \widetilde{\varphi}$
and let us assume that
(b) it occurs that for any $\widetilde{U} \in C^\infty$ solution to the Cauchy problem

$$\begin{cases} \partial_t \widetilde{U}_k = \sum_{j=1}^{n-1}\sum_{l=1}^{N} \widetilde{B}_j^{lk}\partial_{x_j}\widetilde{U}_l + \sum_{l=1}^{N}\widetilde{C}^{lk}\widetilde{U}_l + \widetilde{F}_k \\ \\ \widetilde{U}_k(x,0) = \widetilde{\varphi}_k(x) \end{cases}, \tag{9.71}$$

we will have

$$\left|U_{k,\alpha}\right| \le \partial^\alpha \widetilde{U}_k(0,0), \quad \forall \alpha \in \mathbb{N}_0^n, \quad k = 1, \cdots, N. \tag{9.72}$$

Step III We will construct majorants $\widetilde{B}_j^{lk}$, $\widetilde{C}^{lk}$, $\widetilde{F}$, and $\widetilde{\varphi}$ such that Cauchy problem (9.71) **indeed possesses an analytic solution**. Let us denote this solution once again as $\widetilde{U}$. Through Step II and, specifically, using (9.72), it will be deduced that the power series

$$\sum_{\alpha \in \mathbb{N}_0^n} \frac{1}{\alpha!} U_\alpha x^{\alpha'} t^{\alpha_n},$$

will converge in a neighborhood of 0, and its sum, which we denote as V, will satisfy the system:

$$\partial_t V_k(x,t) = \sum_{j=1}^{n-1}\sum_{l=1}^{N} B_j^{lk}(x,t)\partial_{x_j} V_l + \sum_{l=1}^{N} C^{lk}(x,t)V_l + F_k(x,t). \tag{9.73}$$

For $k = 1, \cdots, N$, the analytic functions

$$\partial_t V_k$$

and

$$\sum_{l=1}^{N} B_j^{lk}(x,t)\partial_{x_j} V_l + \sum_{l=1}^{N} C^{lk}(x,t)V_l + F_k(x,t),$$

will have (due to construction) all their derivatives equal at 0. Moreover, in accordance with (9.70), we will have

$$V_k(x,0)=0, \quad k=1,\cdots,N. \tag{9.74}$$

This completes the proof of the existence of a solution for Cauchy problem (9.69).

Step IV The uniqueness in the class of analytic functions in a connected neighborhood of 0 for Cauchy problem (9.69) will be a consequence of **Step I** and of the unique continuation property for the analytic functions (Theorem 8.5).

Step I By the initial conditions $U(x,0)=0$ we have

$$\partial_x^{\alpha'} U_k(x,0)=0, \quad k=1,\cdots,N, \tag{9.75}$$

which implies (9.70). Now, for every $\alpha\in\mathbb{N}_0^n$, where $\alpha_n>0$, we have, for $k=1,\cdots,N$,

$$\begin{aligned}\partial^\alpha U_k(0,0)=P_{k,\alpha}\left(\partial^\gamma B_j^{kl},\cdots,\partial^\delta C^{kl},\cdots\partial^\beta U_h\right)_{x=0,t=0}+\\+\partial_x^{\alpha'}\partial_t^{\alpha_n-1}F_k(0,0),\end{aligned} \tag{9.76}$$

where $P_{k,\alpha}$ **is a polynomial with positive coefficients** and the multi-indices β in the derivatives $\partial^\beta U_h$ satisfy

$$|\beta|\le|\alpha| \quad \text{and} \quad \beta_n\le\alpha_n-1.$$

In particular, (9.76) is a recursive relation on the derivatives of U. To prove (9.76) it suffices to make the derivatives of both the sides of the equations in (9.69). Concerning the positivity of the coefficients of $P_{k,\alpha}$ it suffices to keep in mind that we only use the rules of differentiation of a product and a sum of functions.

For instance, if $i=1,\cdots,n-1$, we have

$$\begin{aligned}\partial_t\partial_{x_i}U_k=\sum_{j=1}^{n-1}\sum_{l=1}^{N}\left(B_j^{lk}\partial^2_{x_ix_j}U_l+\partial_{x_i}B_j^{lk}\partial_{x_j}U_l\right)+\\+\sum_{l=1}^{N}\left(C^{lk}\partial_{x_i}U_l+\partial_{x_i}C^{lk}U_l\right)+\partial_{x_i}F_k\end{aligned} \tag{9.77}$$

and, taking into account (9.75), we have

$$\partial_t\partial_{x_i}U_k(0,0)=\partial_{x_i}F_k(0,0).$$

Analogously,

$$\partial_t \partial_x^{\alpha'} U_k(0,0) = \partial_x^{\alpha'} F_k(0,0)$$

and

$$\begin{aligned} \partial_t^2 U_k = &\sum_{j=1}^{n-1}\sum_{l=1}^{N}\left(B_j^{lk}\partial_t\partial_{x_j}U_l + \partial_t B_j^{lk}\partial_{x_j}U_l\right) + \\ &+\sum_{l=1}^{N}\left(C^{lk}\partial_t U_l + \partial_t C^{lk}U_l\right) + \partial_t F_k. \end{aligned} \tag{9.78}$$

Let us observe that all the derivatives of U in $(0,0)$ that occur in (9.78) can be obtained by (9.77) and by (9.75). A similar argument applies to $\partial_t^2\partial^{\alpha'}U(0,0), \cdots, \partial_t^j\partial^{\alpha'}U(0,0)$, $j = 1,\cdots,\alpha' \in \mathbb{N}_0^{n-1}$.

Step II Let $\widetilde{U}$ be a solution to problem (9.71). In a similar way to what we have done in the **Step I** we obtain, for $\alpha \in \mathbb{N}_0^n$ with $\alpha_n > 0$, that for each $k = 1,\cdots,N$,

$$\begin{aligned} \partial^\alpha \widetilde{U}_k(0,0) = &P_{k,\alpha}\left(\partial^\gamma \widetilde{B}_j^{kl},\cdots,\partial^\delta \widetilde{C}^{kl},\cdots \partial^\beta \widetilde{U}_h\right)_{x=0,t=0} + \\ &+ \partial_x^{\alpha'}\partial_t^{\alpha_n-1}\widetilde{F}_k(0,0), \end{aligned} \tag{9.79}$$

where $P_{k,\alpha}$ is the same polynomial with positive coefficients that occurs in (9.76) and (as in (9.76)) the multi-indices β in the derivatives $\partial^\beta U_h$ satisfy $|\beta| \le |\alpha|$ and $\beta_n \le \alpha_n - 1$. Furthermore, we have

$$\partial_x^{\alpha'}\widetilde{U}_k(0,0) = \partial_x^{\alpha'}\widetilde{\varphi}_k(0) \tag{9.80}$$

and by, $0 \preccurlyeq \widetilde{\varphi}$, we have

$$0 \le \partial_x^{\alpha'}\widetilde{\varphi}_k(0). \tag{9.81}$$

In order to prove (9.72) one can proceed by induction on the order α_n of the derivative with respect to t. If $\alpha_n = 0$ then we have

$$\left|U_{k,(\alpha',0)}\right| \le \partial_x^{\alpha'}\widetilde{\varphi}_k(0) = \partial_x^{\alpha'}\widetilde{U}_k(0,0), \quad \forall\alpha \in \mathbb{N}_0^{n-1}, \quad k = 1,\cdots,N. \tag{9.82}$$

Now, let us suppose that, for a given α_n, the following **inductive hypothesis** is satisfied

$$\left|U_{k,(\alpha',\alpha_n)}\right| \le \partial_t^{\alpha_n}\partial_x^{\alpha'}\widetilde{U}_k(0,0), \quad \forall\alpha \in \mathbb{N}_0^{n-1} \quad k = 1,\cdots,N. \tag{9.83}$$

We have, for $k = 1, \cdots, N$,

$$U_{k,(\alpha',\alpha_n+1)} = \partial_t^{\alpha_n+1}\partial_x^{\alpha'} U_k(0,0) = \partial_x^{\alpha'}\partial_t^{\alpha_n} F_k(0,0)+$$
$$+ P_{k,(\alpha',\alpha_n+1)}\left(\partial^\gamma B_j^{kl}, \cdots, \partial^\delta C^{kl}, \cdots \partial^\beta U_h\right)_{x=0,t=0},$$

where

$$|\beta| \leq |\alpha|, \quad \text{and} \quad \beta_n \leq \alpha_n.$$

Then, since the coefficients of $P_{k,(\alpha_n+1)}$ are positive, using **(a)** of the **Step II** and the inductive hypothesis (9.83), we have

$$\begin{aligned}
&\left|U_{k,(\alpha',\alpha_n+1)}\right| \leq \left|\partial_x^{\alpha'}\partial_t^{\alpha_n} F_k(0,0)\right| + \\
&+ P_{k,(\alpha',\alpha_n+1)}\left(\left|\partial^\gamma B_j^{kl}\right|, \cdots, \left|\partial^\delta C^{kl}\right|, \cdots \left|U_{h,\beta}\right|\right)_{x=0,t=0} \leq \\
&\leq \partial_x^{\alpha'}\partial_t^{\alpha_n} \widetilde{F}_k(0,0)+ \qquad\qquad (9.84)\\
&+ P_{k,(\alpha',\alpha_n+1)}\left(\partial^\gamma \widetilde{B}_j^{kl}, \cdots, \partial^\delta \widetilde{C}^{kl}, \cdots \partial^\beta \widetilde{U}_h\right)_{x=0,t=0} = \\
&= \partial_t^{\alpha_n+1}\partial_x^{\alpha'} \widetilde{U}_k(0,0).
\end{aligned}$$

Step III We may assume that for appropriate positive numbers M_1, M_2 and ρ_1, ρ_2 with $M_1 \geq 1$, we have

$$B_j^{lk}, C^{lk} \in C_{M_1,\rho_1}(0), \quad j = 1, \cdots, n-1, \quad l, k = 1, \cdots, N, \tag{9.85}$$

$$F_k \in C_{M_2,\rho_2} \quad k = 1, \cdots, N. \tag{9.86}$$

We set

$$M = \frac{M_2}{M_1}, \qquad \rho = \min\{\rho_1, \rho_2\}.$$

By Proposition 8.9, we may choose, for $j = 1, \cdots, n-1, l, k = 1, \cdots, N$,

$$\widetilde{B}_j^{lk} = \widetilde{C}^{lk} = \frac{M_1\rho}{\rho - (\sigma^{-1}t + x_1 + \cdots + x_{n-1})}$$

and

$$\widetilde{F}_k = \frac{M_2\rho}{\rho - (\sigma^{-1}t + x_1 + \cdots + x_{n-1})},$$

where $\sigma \in (0, 1]$ is to be chosen. System (9.71) becomes, for $k = 1, \cdots, N$,

$$\partial_t \widetilde{U}_k(x,t) =$$

$$= \frac{M_1 \rho}{\rho - (\sigma^{-1} t + x_1 + \cdots + x_{n-1})} \left(\sum_{j=1}^{n-1} \sum_{l=1}^{N} \partial_{x_j} \widetilde{U}_l + \sum_{l=1}^{N} \widetilde{U}_l + M \right). \tag{9.87}$$

At this point we look for a solution to Eq. (9.87) of the form

$$\widetilde{U}_k(x,t) = w\left(\sigma^{-1} t + x_1 + \cdots + x_{n-1}\right). \tag{9.88}$$

Set $s = \sigma^{-1} t + x_1 + \cdots + x_{n-1}$ and

$$\phi(s) = \frac{M_1}{1 - \rho^{-1} s}, \tag{9.89}$$

we have

$$\left(\sigma^{-1} - \phi(s) N(n-1)\right) \frac{dw}{ds} = N\phi(s) w + M\phi(s). \tag{9.90}$$

Now we choose $\sigma > 0$ so that we have, in a neighborhood of 0,

$$\sigma^{-1} - \phi(s) N(n-1) > 0.$$

For instance, we choose

$$\sigma = \sigma_0 := \frac{1}{2NM_1(n-1)}. \tag{9.91}$$

We get

$$\sigma_0^{-1} - \phi(s) N(n-1) = NM_1(n-1)\left(1 - 2s/\rho\right) > 0, \qquad \text{for } |s| < \frac{\rho}{2}.$$

Setting

$$h(s) = \frac{\phi}{\sigma_0^{-1} - N(n-1)\phi} = \frac{1}{N(n-1)} \frac{\rho}{\rho - 2s},$$

Equation (9.90) becomes

$$\frac{dw}{ds} = Nh(s) w + Mh(s). \tag{9.92}$$

Let w_0 be the solution to (9.92) such that

$$w_0(0) = 0. \tag{9.93}$$

We have

$$w_0 = \frac{M}{N}\left[\exp\left(N\int_0^s h(\eta)d\eta\right) - 1\right] = \frac{M}{N}\left[\left(\frac{\rho}{\rho - 2s}\right)^{\frac{\rho}{2(n-1)}} - 1\right]. \tag{9.94}$$

In particular, w_0 is analytic in $\left(-\frac{\rho}{2}, \frac{\rho}{2}\right)$ and

$$0 \preccurlyeq w_0. \tag{9.95}$$

The latter relationship can be easily checked by using formula (9.94) or can be also easily derived from (9.92) and (9.94), by expressing the derivatives of w_0 in 0 by means of those of lower order and noticing that they are all nonnegative. Now, for $k = 1, \cdots, N$, let us consider the following functions

$$\widetilde{U}_k(x,t) = w_0\left(\sigma_0^{-1}t + x_1 + \cdots + x_{n-1}\right), \tag{9.96}$$

we have that $\widetilde{U}_k$ are solutions to Eqs. (9.87) (when $\sigma = \sigma_0$) and by Proposition 8.11, they are analytic. Moreover by (2.4) we have

$$\begin{aligned}\widetilde{\varphi}_k(x) = \widetilde{U}_k(x,0) = w_0\left(x_1 + \cdots + x_{n-1}\right) = \\ = \sum_{m=0}^{\infty} w_0^{(m)}(0) \sum_{|\alpha'|=m} \frac{1}{\alpha'!}x^{\alpha'}.\end{aligned} \tag{9.97}$$

From which, taking into account (9.95), it is obvious that

$$0 \preccurlyeq \widetilde{\varphi}_k, \qquad \text{for } k = 1, \cdots, N. \tag{9.98}$$

All in all, $\widetilde{U}$ is an analytic solution in a neighborhood of $(0,0)$, of the Cauchy problem

$$\begin{cases}\partial_t \widetilde{U}_k = H(x,t)\left(\sum_{j=1}^{n-1}\sum_{l=1}^{N} \partial_{x_j}\widetilde{U}_l + \sum_{l=1}^{N}\widetilde{U}_l + M\right) \\ \\ \widetilde{U}_k(x,0) = \widetilde{\varphi}_k(x)\end{cases}, \tag{9.99}$$

where

$$H(x,t)=\frac{M_1\rho}{\rho-(\sigma_0^{-1}t+x_1+\cdots+x_{n-1})}.$$

Since $\widetilde{U}$ is analytic, (9.84) implies that the following power series converges in a neighborhood $\mathcal{U}$ of $(0,0)$

$$\sum_{\alpha\in\mathbb{N}_0^n}\frac{1}{\alpha!}U_\alpha x^{\alpha'}t^{\alpha_n},$$

in addition its sum, U, solves Cauchy problem (9.69) in $\mathcal{U}$.

Step IV The uniqueness to Cauchy problem (9.69) in the class of analytic functions in a connected neighborhood of 0 is a consequence of **Step I** and of the unique continuation property for the analytic functions. As a matter of fact, if V' V'' are analytic solutions of (9.69), then

$$\partial^\alpha V_k'(0,0)=U_{k,\alpha}=\partial^\alpha V_k''(0,0),\quad k=1,\cdots,N,$$

so that, by Theorem 8.5, we have $V'=V''$ in a neighborhood of 0.

■

Remark on the Neighborhood in Which There Exist Solutions of the Cauchy Problem

In the upcoming discussions, our focus will be on acquiring detailed insights into how the neighborhood $\mathcal{U}$ is influenced by the given term f and, consequently, by the initial data of Cauchy problem (9.45). Through Step III of the previous proof, we can assert that the neighborhood $\mathcal{U}$ remains independent of the constant M_2. To elucidate this assertion, we need only apply Proposition 8.12 to the composite function.

$$w_0\left(\sigma_0^{-1}t+x_1+\cdots+x_{n-1}\right).$$

Let us observe that using (9.94) and setting $\kappa=\frac{\rho}{2(n-1)}$, we can establish the existence of a constant $c_\kappa\geq 1$, which depends on κ only, such that (recall that $M=\frac{M_2}{M_1}$)

$$0\leq w_0^{(m)}(0)\leq\frac{c_\kappa M}{N}\left(c_\kappa\rho^{-1}\right)^m m!,\quad\forall m\in\mathbb{N}_0.$$

Hence

$$w_0\in C_{\frac{c_\kappa M}{N},\frac{\rho}{c_\kappa}}.$$

On the other hand we get trivially

$$\left(\sigma_0^{-1}t + x_1 + \cdots + x_{n-1}\right) \in C_{(\sigma_0^{-1}+n-1),1}.$$

Hence

$$w_0\left(\sigma_0^{-1}t + x_1 + \cdots + x_{n-1}\right) \in C_{\frac{c_K M}{N},R},$$

where

$$R = \frac{\rho}{c_K\sigma_0^{-1} + n - 1}$$

and by (9.91) it turns out that σ_0 does not depend by M_2. Therefore R depends on M_1, ρ, n and N only. Moreover, we can choose

$$\mathcal{U} = \{(x,t) \in \mathbb{R}^n : |t| + |x_1| + \cdots + |x_{n-1}| < R\},$$

as the neighborhood in which the Cauchy problem (9.69) admits a solution.

In preparation for what we will do later, let us go back to consider the case of a linear differential operator of order m given by

$$P(x,\partial) = \sum_{|\alpha|\le m} a_\alpha(x)\partial^\alpha \tag{9.100}$$

and let us consider the following Cauchy problem

$$\begin{cases} P(x,\partial)u(x) = f(x) \\ \\ \partial_n^j u(x',0) = 0, \quad j = 0,1,\cdots,m-1, \ \forall x' \in B_1' \end{cases} . \tag{9.101}$$

Let us suppose that, for given M_0, ρ_0, we have

$$P_m((x',0),e_n) \neq 0, \quad \forall x' \in \overline{B_1'}, \tag{9.102}$$

$$a_\alpha \in C_{M_0,\rho_0}(z), \quad |\alpha| \le m, \quad \forall z \in \overline{B_1'} \times [-\delta_0,\delta_0], \tag{9.103}$$

$$f \text{ be a polynomial.} \tag{9.104}$$

Then the solution to problem (9.101) exists in $\overline{B_1'} \times [-\delta,\delta]$ (actually, in a neighborhood of $\overline{B_1'} \times \{0\}$) where $\delta > 0$ depends on M_0, ρ_0, δ_0 and on

$$\min\{|P_m((x',0),e_n)| : x' \in \overline{B_1'}\},$$

but does not depend by the polynomial f. In order to check this assertion, let h be the degree of the polynomial f. Set

$$K = 1 + \sum_{|\beta| \leq h} \max_{\overline{B'_1} \times [-\delta_0, \delta_0]} \left| \partial^\beta f \right|.$$

It is evident that, setting.

$$\widetilde{u} = \frac{u}{K}, \qquad \widetilde{f} = \frac{f}{K},$$

u solves Cauchy problem (9.101) if and only if v is solves the following Cauchy problem

$$\begin{cases} P(x, \partial)\widetilde{u}(x) = \widetilde{f}(x) \\ \\ \partial_n^j \widetilde{u}(x', 0) = 0, \quad j = 0, 1, \cdots, m-1, \ \forall x' \in B'_1 \end{cases} . \tag{9.105}$$

On the other hand, because of the way we defined $\widetilde{f}$ we can certainly state that

$$\widetilde{f} \in C_{1,1}(z), \quad \forall z \in \overline{B'_1} \times [-\delta_0, \delta_0]. \tag{9.106}$$

We can then return to problem (9.69). Hence by applying the Cauchy–Kovalevskaya Theorem and taking into account that a_α, $|\alpha| \leq m$, are analytic functions in a neighborhood of $\overline{B'}_1(0) \times \{0\}$, we conclude that the solution of Cauchy problem (9.101) exists and it is analytic in $\overline{B'_1} \times [-\delta, \delta]$, where $\delta > 0$ depends on M_0, ρ_0, δ_0 and on

$$\min\{|P_m((x', 0), e_n)| : x' \in \overline{B'_1}\},$$

but does not depend on f.

Obviously, if (9.102), (9.103), (9.104) hold, similar conclusions are valid to the Cauchy problem

$$\begin{cases} P(x, \partial)u(x) = f(x), & \text{for } x \in \mathcal{U}_{\widetilde{x}} \\ \\ \partial_n^j u(x', 0) = g_j, & j = 0, 1, \cdots, m-1, \ \forall x' \in B'_1 \end{cases} . \tag{9.107}$$

provided that g_j are polynomials for $j = 0, 1, \cdots, m-1$. ♦

9.6 Further Comments on the Cauchy–Kovalevskaya Theorem: Examples

9.6.1 A Few Brief Note on the Qualisinear and the Fully Non–linear Case

It is not difficult to adapt the proof of the Cauchy–Kovalevskaya Theorem to the case of a quasilinear operator

$$\sum_{|\alpha|=m} a_\alpha\left(x, (\partial^\beta u)_{||\beta|\leq m-1}\right)\partial^\alpha u + a_0\left(x, (\partial^\beta u)_{||\beta|\leq m-1}\right).$$

In this case, we recall, the Cauchy problem is

$$\begin{cases} \mathcal{P}(u) = 0 \\ \\ \partial_n^j u(x) = g_j(x), \quad j = 0, 1, \cdots, m-1, \ \forall x \in \Gamma \end{cases} . \tag{9.108}$$

One can proves that if a_α, a_0, g_j are analytic functions, Γ is analytic and noncharacteristic (Definition 9.2) then for every $\widetilde{x} \in \Gamma$ there exists a neighborhood $\mathcal{U}_{\widetilde{x}}$ in which Cauchy problem (9.108) has an analytic solution. For the proof we refer to [38].

In (9.5) we have formulated the Cauchy problem for the fully non–linear equation

$$\begin{cases} F\left(x, (\partial^\alpha u)_{|\alpha|\leq m}\right) = 0 \\ \\ \frac{\partial^j u(x)}{\partial \nu^j} = g_j(x), \quad j = 0, 1, \cdots, m-1, \quad \forall x \in \Gamma \end{cases} . \tag{9.109}$$

Here we only outline the proof of the existence of the solutions to problem (9.109) referring for more details to [27, Chapter 1], [41, Chapter 1].

Let us consider the case where $\Gamma = \{x_n = 0\}$. We know that we may always reduce to this case by "flattening" Γ (by map (9.46)). So, instead of the conditions $\frac{\partial^j u(x)}{\partial \nu^j} = g_j(x)$, $j = 0, 1, \cdots, m-1$, for $x \in \Gamma$, we may consider

$$\partial_n^j u(x', 0) = g_j(x'), \quad j = 0, 1, \cdots, m-1, \quad \forall x' \in B'_r, \tag{9.110}$$

where $r > 0$. In the first part of the proof of Proposition 9.4 we have already seen that conditions (9.110) allow us to determine the derivatives

$$\partial_{x'}^{\alpha'} \partial_n^j u(x', 0) = \partial_{x'}^{\alpha'} g_j(x'), \quad j = 0, 1, \cdots, m-1, \ \alpha' \in \mathbb{N}_0^{n-1}, \ x' \in B'_r \tag{9.111}$$

without involving the equation $F\left(x, (\partial^\alpha u)_{|\alpha|\le m}\right) = 0$. Let us recall that in order to calculate the derivative $\partial_n^m u(x', 0)$ we need to use the equation. More precisely, we have

$$F\left(x', 0, (\partial_{x'}^{\alpha'} g_j)_{|\alpha'|+j\le m, j\le m-1}(x'), \partial_n^m u(x', 0)\right) = 0. \tag{9.112}$$

To find $z = \partial_n^m u(x', 0)$ from Eq. (9.112) we need that the equation

$$F\left(x', 0, (\partial_{x'}^{\alpha'} g_j)_{|\alpha'|+j\le m, j\le m-1}(x'), z\right) = 0, \tag{9.113}$$

admits a solution. If, for instance, we require

$$(\partial_z F)\left(x', 0, (\partial_{x'}^{\alpha'} g_j)_{|\alpha'|+j\le m, j\le m-1}(x'), z\right) \ne 0, \tag{9.114}$$

then we may express z as a function of variable x' of class C^1 (provided F is of class C^1) or as an analytic function, provided F is analytic. Let us observe that condition (9.114) makes it possible to write, in a neighborhood of Γ, the equation

$$F\left(x, (\partial^\alpha u)_{|\alpha|\le m}\right) = 0,$$

like

$$\partial_n^m u = G\left(x, (\partial^\alpha u)_{|\alpha|\le m, \alpha_n < m}\right), \tag{9.115}$$

where G is analytic (provided F is analytic). Hence, in the fully non–linear case, condition (9.114) may replace the condition that Γ is noncharacteristic surface for a linear (or quasilinear) operator. Furthermore, since we can find $\partial_n^m u(x', 0)$ from (9.112), we set $g_m(x') = \partial_n^m u(x', 0)$, by making the derivatives of both the sides of (9.115) w.r.t. x_n we have

$$\begin{cases} \partial_n^{m+1} u = \sum_{|\alpha|\le m, j<m} a_{\alpha',j} \partial_{x'}^{\alpha'} \partial_n^{j+1} u + (\partial_n G)\left(x, (\partial^\alpha u)_{||\alpha|\le m-1, \alpha_n}\right) \\ \\ \frac{\partial^j u(x',0)}{\partial \nu^j} = g_j(x'), \quad j = 0, 1, \cdots, m, \quad \forall x \in \Gamma \end{cases},$$

where

$$a_{\alpha',j} = \left(\partial_{p_{\alpha',j}} G\right)\left(x, (\partial_{x'}^{\alpha'} \partial_n^j u)_{||\alpha'|+j\le m, j<m}\right),$$

for $|\alpha'| + j \le m$, $j < m$.

In simpler terms, if conditions (9.113) and (9.114) are satisfied, it becomes possible to reframe Cauchy problem (9.109) as a Cauchy problem for a quasilinear equation of order $m + 1$. This allows us to leverage the existence results obtained in the quasilinear case.

9.6.2 *Comments About the Existence and the Uniqueness of Solutions: Examples and Counterexamples*

In this subsection we return to the **linear case.** In general, if Γ is a characteristic surface for the operator $P(x, \partial)$, neither existence nor uniqueness for Cauchy problem (9.45) can be expected. Let us look at some example.

(a) Let $p_m(\xi)$ be a homogeneous polynomial of degree m and let us assume

$$p_m(N) = 0,$$

where $N \in \mathbb{R}^n \setminus \{0\}$. Let $P_m(\partial)$ be the differential operator whose symbol is equal to $p_m(\xi)$. Then the hyperplane

$$\pi = \{x \cdot N = 0\}$$

is a characteristic surface for the operator $P_m(\partial)$. Now, let us consider the functions

$$u^{(t)}(x) = e^{tx \cdot N} - \sum_{k=0}^{m-1} \frac{t^k (x \cdot N)^k}{k!}, \quad t \in \mathbb{R},$$

where $t \neq 0$. It easy to check that, for every $t \in \mathbb{R}$, $u^{(t)}$ solves the Cauchy problem

$$\begin{cases} P_m(\partial)u(x) = 0 \\ \\ \frac{\partial^j u}{\partial N^j} = 0, \quad j = 0, 1, \cdots, m-1, \text{ on } \pi \end{cases} . \tag{9.116}$$

Hence, does not hold the uniqueness for problem (9.116).

(b) The Cauchy–Kovalevskaya Theorem guarantees the existence and uniqueness of solutions to the Cauchy problem when initial data is provided on a non-characteristic surface. This holds *provided that all the data of the problem are analytic*. If we aim to maintain the same level of generality as the Theorem, we cannot relax the requirement of analyticity. To prove this, consider the following Cauchy problem

$$\begin{cases} \partial_y^2 u + \partial_x^2 u = 0 \\ u(x, 0) = g_0(x), \quad \text{for } x \in (-r, r) \\ \partial_y u(x, 0) = g_1(x), \quad \text{for } x \in (-r, r) \end{cases} . \tag{9.117}$$

Since u satisfies the Laplace equation $\partial_y^2 u + \partial_x^2 u = 0$ in a neighborhood of $(0, 0)$ (see, for instance [38, Ch. 2]) it is analytic in such a neighborhood, therefore the initial data, $u(x, 0) = g_0(x)$ and $\partial_y u(x, 0) = g_1(x)$ also need to be analytic.

Incidentally, even though we consider the "one–sided" Cauchy problem the situation do not change in a significant way. Let us consider, indeed, the problem

$$\begin{cases} \partial_y^2 u + \partial_x^2 u = 0, & \text{for } (x, y) \in B_r^+ \\ u(x, 0) = g_0(x), & \text{for } x \in (-r, r) \\ \partial_y u(x, 0) = g_1(x), & \text{for } x \in (-r, r) \end{cases},$$

where $u \in C^2(B_r^+) \cap C^1(\overline{B_r^+})$. Let us suppose for simplicity that $g_1 \equiv 0$ and let us consider the even reflection w.r.t. x–axis of u

$$v(x, y) = u(x, |y|)$$

then, by the Schwarz reflection principle, we have

$$\Delta v = 0, \text{ in } B_r$$

and again we have that v is an analytic function and therefore g_0 is an analytic function.

Exercise 9.1 Let us suppose $g_0, g_1 \in C^\omega(-r, r)$ in (9.117). Prove that the solution to Cauchy problem (9.117) is given by

$$u(x, y) = \sum_{n=0}^{\infty} (-1)^n \left(\frac{g_0^{(2n)}(x) y^{2n}}{(2n)!} + \frac{g_1^{(2n+1)}(x) y^{2n+1}}{(2n+1)!} \right). \tag{9.118}$$

♠

(c) Let us consider the Cauchy problem

$$\begin{cases} \partial_t u - \partial_x^2 u = 0, & \text{for } (x, t) \in \mathbb{R}^2 \\ u(x, 0) = g_0(x), & \text{for } x \in \mathbb{R} \\ \partial_t u(x, 0) = g_1(x), & \text{for } x \in \mathbb{R} \end{cases}. \tag{9.119}$$

In this case the straight line $\{t = 0\}$ is a characteristic line for the operator $\partial_t - \partial_x^2$. It is evident that if we do not require

$$g_1(x) = g_0''(x), \ \forall x \in \mathbb{R}, \tag{9.120}$$

then Cauchy problem (9.119) has no solutions. As a matter of fact, if problem (9.119) admits solutions, even just of class C^2, then

$$g_0''(x) = \partial_x^2 u(x,0) = \partial_t u(x,0) = g_1(x), \quad \text{for } |x| < 1$$

and therefore (9.120) hold.

We now check that even condition (9.120) is satisfied we can exhibit $g_0 \in C^\omega$ such that problem (9.119) has no solutions. First, it is evident that if (9.120) is satisfied, then problem (9.119) can be be formulated as

$$\begin{cases} \partial_t u - \partial_x^2 u = 0 \\ \\ u(x,0) = g_0(x), \ \text{for } x \in \mathbb{R} \end{cases} . \tag{9.121}$$

Let

$$g_0(x) = \frac{1}{1+x^2} .$$

We have that $g_0 \in C^\omega(\mathbb{R})$ and

$$g_0^{(2k)}(0) = (-1)^k (2k)! , \qquad \text{for } k \in \mathbb{N}_0.$$

Now, if there exists a solution, u, of problem (9.121) which is **analytic** in a neighborhood of $(0,0)$, then $u(0,t)$ should be expanded in Taylor series in $t = 0$. Now we have

$$\partial_t^j u(0,0) = \partial_x^{2j} u(0,0) = g_0^{(2j)}(0) = (-1)^j (2j)!,$$

therefore the power series

$$\sum_{j=0}^{\infty} \frac{1}{j!} \partial_t^j u(0,t) t^j = \sum_{j=0}^{\infty} \frac{(-1)^j (2j)!}{j!} t^j$$

has the radius of convergence equal to zero. Hence $u(0,t)$ is not analytic.

Also, we note that the Cauchy problem

$$\begin{cases} \partial_t u - \partial_x^2 u = 0 \\ \\ u(x,0) = 0, \ \text{for } x \in \mathbb{R} \end{cases} , \tag{9.122}$$

admits only one analytic solution, namely the null solution. As a matter of fact, the null function is a trivial solution of problem (9.122) and if u is an analytic solution of (9.122) then we have, for each $i, j \in \mathbb{N}_0$

$$\partial_x^i \partial_t^j u(0,0) = \partial_x^{i+2j} u(0,0) = 0,$$

hence $u \equiv 0$.

Of course the one-sided Cauchy problems are also of interest

$$\begin{cases} \partial_t u - \partial_x^2 u = 0, & \text{for } x \in \mathbb{R}, \quad t > 0 \\ \\ u(x,0) = g_0(x), & \text{for } x \in \mathbb{R} \end{cases} . \tag{9.123}$$

For this problem it turns out that (if g_0 is regular enough)

$$u(x,t) = \frac{1}{\sqrt{4\pi t}} \int_{-\infty}^{+\infty} g_0(\xi) e^{-\frac{(x-\xi)^2}{4t}} d\xi$$

is a solution to (9.123). In particular if $g_0 = \frac{1}{1+x^2}$, then Cauchy problem (9.123) admits solutions (of course, nonanalytic w.r.t. t).

Keep in mind that the problems that we have considered in the point (c) are not written in the form (9.53). As a matter of fact, the term on the right–hand side of equation $\partial_t u = \partial_x^2 u$ has order 2 greater than the order of derivative $\partial_t u$, on the left–hand side.

(d) In stark contrast to the example explored in (b), we now present a different example for the vibrating string equation. Let us examine the Cauchy problem

$$\begin{cases} \partial_t^2 u - \partial_x^2 u = 0 \\ u(x,0) = g_0(x), & \text{for } x \in (-1,1) \\ \partial_t u(x,0) = g_1(x), & \text{for } x \in (-1,1) \end{cases} . \tag{9.124}$$

Let us first prove that problem (9.124) has at most one solution $u \in C^2(\overline{Q})$ where $Q = \{(x,t) \in \mathbb{R}^2 : |x| + |t| < 1\}$ (Fig. 9.2).

To this purpose we observe that

$$0 = \left(\partial_t^2 u - \partial_x^2 u\right) \partial_t u = \frac{1}{2} \left(\partial_t (\partial_t u)^2 - 2\partial_x (\partial_x u \partial_t u) + \partial_t (\partial_x u)^2\right). \tag{9.125}$$

Moreover, let us suppose that $g_0 = 0$ and $g_1 = 0$ in (9.124). We integrate both the sides of (9.125) over

$$\mathcal{T}_\delta = Q \cap \{(x,t) \in \mathbb{R}^2 : 0 < t < 1 - \delta\},$$

where $\delta \in (0,1)$ is arbitrary. We obtain, by the divergence Theorem,

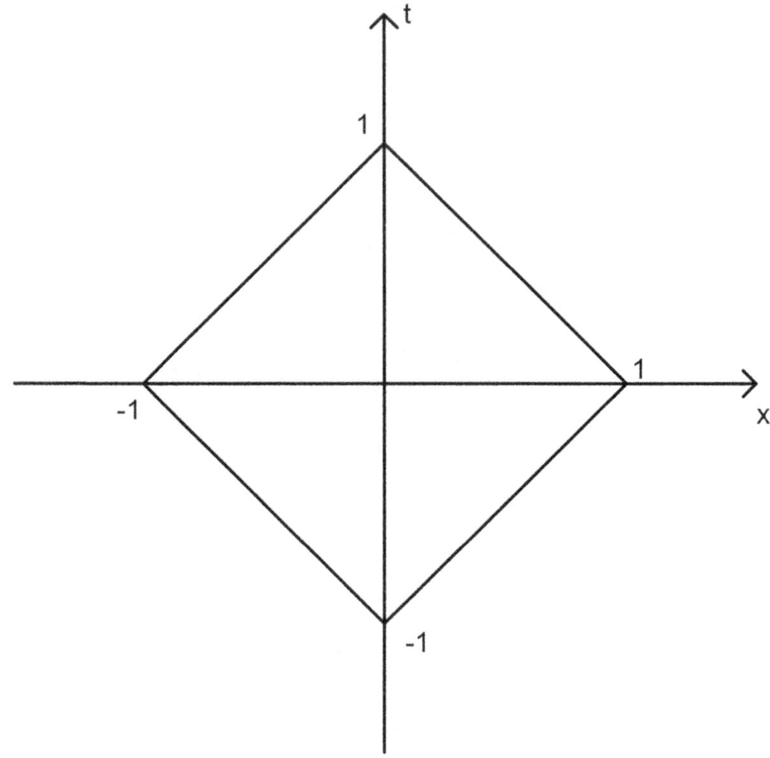

Fig. 9.2 The square $Q = \{(x, t) \in \mathbb{R}^2 : |x| + |t| \le 1\}$

$$
\begin{aligned}
0 = &\iint_{\mathcal{T}_\delta} \left(\partial_t^2 u - \partial_x^2 u\right) \partial_t u dx dt = \\
&= \frac{1}{2} \int_{\partial \mathcal{T}_\delta} \left((\partial_t u)^2 \nu_t - 2(\partial_x u \partial_t u)\nu_x + (\partial_x u)^2 \nu_x\right) dS = \\
&= -\frac{1}{2} \int_{-1}^{1} \left((\partial_t u)^2(x, 0) + (\partial_x u)^2(x, 0)\right) dx + \\
&+ \frac{1}{2\sqrt{2}} \int_{1-\delta}^{1} (\partial_t u(x, 1 - x) - \partial_x u(x, 1 - x))^2 \, dx + \\
&+ \frac{1}{2} \int_{-1+\delta}^{1-\delta} \left((\partial_t u)^2(x, \delta) + (\partial_x u)^2(x, \delta)\right) dx + \\
&+ \frac{1}{2\sqrt{2}} \int_{-1}^{-1+\delta} (\partial_t u(x, 1 + x) + \partial_x u(x, 1 + x))^2 \, dx \ge \\
&\ge \frac{1}{2} \int_{-1+\delta}^{1-\delta} \left((\partial_t u)^2(x, \delta) + (\partial_x u)^2(x, \delta)\right) dx.
\end{aligned}
\tag{9.126}
$$

Hence

$$\int_{-1+\delta}^{1-\delta} \left((\partial_t u)^2(x,\delta) + (\partial_x u)^2(x,\delta) \right) dx = 0,$$

from which we have $(\partial_t u)^2(x,\delta) + (\partial_x u)^2(x,\delta) = 0$ and, since δ is arbitrary, we have $\partial_x u = \partial_t u = 0$ in $Q \cap \{(x,t) \in \mathbb{R}^2 : 0 \leq t\}$. Finally, since $u(x,0) = \partial_x u(x,0) = 0$, we have $u = 0$ in $Q \cap \{(x,t) \in \mathbb{R}^2 : 0 \leq t\}$. Similarly, we obtain $u = 0$ in $Q \cap \{(x,t) \in \mathbb{R}^2 : t \leq 0\}$. Therefore $u = 0$ in Q.

The existence of solutions also does not require that the Cauchy data g_0 and g_1 are analytic. As a matter of fact it is checked straightforwardly that, if $g_0 \in C^2([-1,1])$ and $g_1 \in C^1([-1,1])$, then the solution to Cauchy problem (9.124) is given by

$$u(x,t) = \frac{g_0(x+t) + g_0(x-t)}{2} + \frac{1}{2}\int_{x-t}^{x+t} g_1(\eta)d\eta. \tag{9.127}$$

Conclusion Based on the brief discussion in this section, we can conclude that, within the realm of partial differential equations, the Cauchy–Kovalevskaya Theorem serves more as a starting point than a final destination. As we move on to the next section, our focus will shift towards the matter of uniqueness, exploring assumptions that are less stringent than requiring the analyticity of all the data.

9.7 The Holmgren Theorem

Let us start by recalling the **divergence Theorem**. Let D a bounded open of $\mathbb{R}^n$ such that its boundary ∂D is of class $C^{0,1}$. Then we have

$$\int_D \partial_j u dx = \int_{\partial D} u\nu_j dS, \quad j = 1, \cdots, n \quad \forall u \in C^1\left(\overline{D}\right), \tag{9.128}$$

where $\nu = (\nu_1, \cdots, \nu_n)$ is the unit outward normal to ∂D and dS is the $(n-1)$–element of surface.

Let $m \in \mathbb{N}$, $a_\alpha \in C^m\left(\overline{D}\right)$, $|\alpha| \leq m$, and

$$P(x,\partial) = \sum_{|\alpha| \leq m} a_\alpha(x)\partial^\alpha. \tag{9.129}$$

We will assume, solely for the purpose of simplifying the notation, that the coefficients of $P(x,\partial)$ are real–valued functions.

We call the **(formal) adjoint operator** of $P(x, \partial)$ the following operator

$$C^m\left(\overline{D}\right) \ni u \to P^\star(x,\partial)u = \sum_{|\alpha|\leq m} (-1)^{|\alpha|}\partial^\alpha\left(a_\alpha(x)u\right). \tag{9.130}$$

Let us note that, up to the sign, the principal part of $P^\star(x, \partial)$ is equal to the principal part of $P(x, \partial)$.

The following **Green identity** holds true, for any $u, v \in C^m\left(\overline{D}\right)$,

$$\int_D \left(vP(x,\partial)u - uP^\star(x,\partial)v\right)dx = \int_{\partial D} \mathcal{M}(u,v;\nu)dS, \tag{9.131}$$

where $\mathcal{M}(u, v; \nu)$ is linear w.r.t. u, v and ν. Moreover

$$\mathcal{M}(u,v;\nu) = \sum_{|\beta|+|\gamma|\leq m-1} c_{\beta\gamma}(x)\partial^\beta u\partial^\gamma v, \tag{9.132}$$

where $c_{\beta\gamma}$, $|\beta| + |\gamma| \leq m - 1$, belong (at least) to $C^0(\partial D)$.

Identity (9.131) can be obtained by applying repeatedly the following simple identity

$$v(x)a(x)\partial_j u(x) = \partial_j\left(v(x)a(x)u(x)\right) - \partial_j\left(v(x)a(x)\right)u(x),$$

obtaining

$$\begin{aligned} v(x)a_\alpha(x)\partial^\alpha u(x) &= v(x)a_\alpha(x)\partial_j\partial^{\alpha-e_j}u(x) = \\ &= \partial_j\left(v(x)a_\alpha(x)\partial^{\alpha-e_j}u(x)\right) - \partial_j\left(v(x)a_\alpha(x)\right)\partial^{\alpha-e_j}u(x) = \\ &= \cdots = \\ &= \operatorname{div}\left(F_\alpha\right) + (-1)^{|\alpha|}\partial^\alpha\left(a_\alpha(x)v(x)\right)u(x), \end{aligned} \tag{9.133}$$

where F_α is a suitable vector field. Next we add up the identities obtained in (9.133), we integrate over D the obtained new identity, and by the divergence Theorem we get (9.131).

Before stating the Holmgren Theorem, let us introduce some notation.

Let Ω be an open set of $\mathbb{R}^n$, $x_0 \in \Omega$ and let $\phi \in C^2(\Omega)$. Let us denote by

$$\Gamma = \{x \in \Omega : \phi(x) = \phi(x_0)\}.$$

We will assume

$$\nabla\phi(x) \neq 0, \quad \forall x \in \Gamma. \tag{9.134}$$

If $\mathcal{U}$ is a neighborhood of x_0 we denote by $\mathcal{U}_+$ the set

$$\mathcal{U}_+ = \mathcal{U} \cap \{x \in \Omega : \phi(x) \geq \phi(x_0)\}. \tag{9.135}$$

Theorem 9.6 (Holmgren) *Let $a_\alpha \in C^\omega(\Omega)$, $|\alpha| \leq m$. Suppose that Γ is a noncharacteristic surface in x_0 for the operator*

$$P(x, \partial) = \sum_{|\alpha| \leq m} a_\alpha(x)\partial^\alpha. \tag{9.136}$$

Then there exists a neighborhood $\mathcal{U}$ of x_0 such that we have:
if $u \in C^m\left(\overline{\mathcal{U}_+}\right)$ satisfies

$$\begin{cases} P(x, \partial)u = 0, & \text{in} \quad \mathcal{U}_+ \\ \\ \partial^\alpha u(x) = 0, & \text{for } |\alpha| \leq m-1, \quad x \in \Gamma \cap \mathcal{U}_+ \end{cases} . \tag{9.137}$$

Then we have

$$u \equiv 0 \quad \text{in} \quad \mathcal{U}_+. \tag{9.138}$$

Remarks Before starting with the proof of Theorem 9.6 we observe what follows.

(i) In (9.137), u is required to be a solution to the equation $P(x, \partial)u = 0$ in $\mathcal{U}_+$, unlike the Cauchy-Kovalevskaya Theorem in which u is required to be a solution in a full neighborhood of x_0. Furthermore, it is only required that $u \in C^m\left(\overline{\mathcal{U}_+}\right)$ (Fig. 9.3).
(ii) The initial surface Γ is assumed to be of class C^2, thus, not analytic like in the Cauchy-Kovalevskaya Theorem. Also, let us note that in (9.137) we require $\partial^\alpha u = 0$ for $|\alpha| \leq m-1$, on $\Gamma \cap \mathcal{U}_+$ and not just that

$$\frac{\partial^j u}{\partial \nu^j} = 0, \quad \text{for } j = 0, 1, \cdots, m-1.$$

Of course, if we want to assume the latter conditions we should require $\phi \in C^{m-1}(\Omega)$ (see the proof of the first part of Proposition 9.4).

♦

Proof of the Holmgren Theorem We may assume Γ be a graph of a function. More precisely we may assume that

$$\Gamma = \{(x', \psi(x')) : x' \in B'_{r_0}\}, \tag{9.139}$$

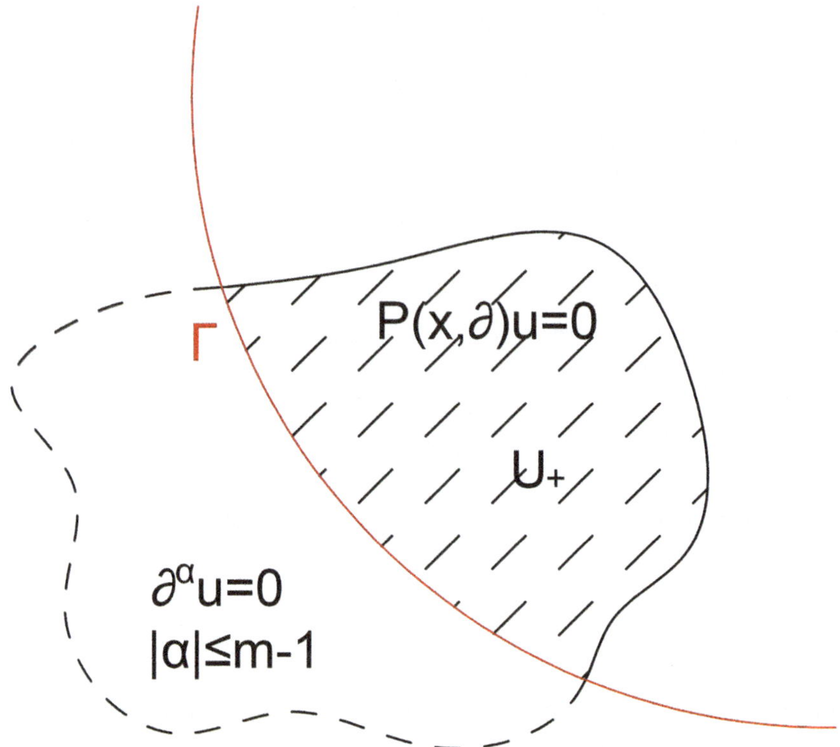

Fig. 9.3 $u \in C^m(\mathcal{U})$ is a solution of $P(x, \partial)u = 0$ only in the shaded region $\mathcal{U}_+$, and all derivatives of u of order less than or equal to $m-1$ vanish on Γ

where $\psi \in C^2\left(\overline{B'_{r_0}}\right)$ satisfies

$$\psi(0) = |\nabla_{x'}\psi(0)| = 0. \tag{9.140}$$

We will divide the proof of the Theorem into two steps. In the first step we will assume that ψ is a **strictly convex function**. In the second step we will reduce to the first part by means of the so–called Holmgren transformation.

Step I Let ψ strictly convex (Fig. 9.4), let R_0 satisfy

$$\Gamma \subset Q_{R_0} := B'_{R_0} \times (-R_0, R_0) \tag{9.141}$$

and, by the assumptions,

$$a_\alpha \in C^\omega\left(Q_{2R_0}\right). \tag{9.142}$$

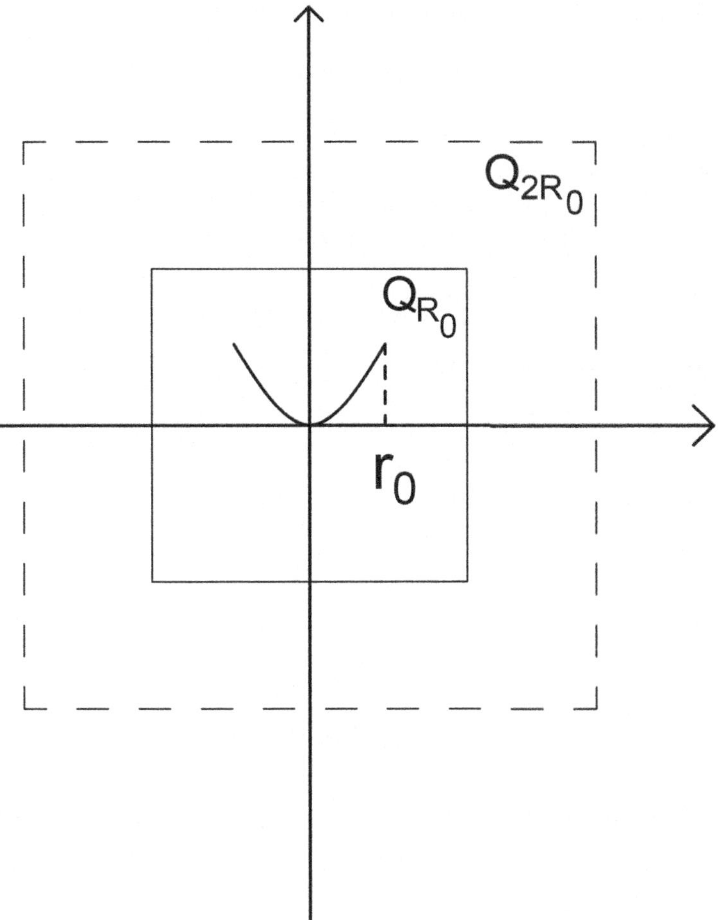

Fig. 9.4 Here, the position of the graph of ψ (strictly convex) within the cylinder Q_{2R_0} is depicted

Since Γ is a noncharacteristic surface in 0 we may assume, recalling (9.140),

$$|p_m(0, e_n)| = c_0 > 0. \tag{9.143}$$

By the continuity of the coefficients of $P_m(x, \partial)$ and by (9.143) we have that there exists $\rho_1 > 0$ such that

$$\left|p_m\left((x', h), e_n\right)\right| \geq \frac{c_0}{2}, \quad \forall x' \in B'_{\rho_1}(0), \ |h| \leq \rho_1. \tag{9.144}$$

This implies that for every $h \in [-\rho_1, \rho_1]$ the flat surface

$$\{(x', h) : x' \in B'_{r_0}\},$$

is noncharacteristic.

Now, let f be a polynomial. By the Cauchy–Kovalevskaya Theorem and by the Remark which follows such a Theorem, there exists ρ_2, $0 < \rho_2 < \rho_1$, ρ_2 *independent of* f such that, there exists an analytic solution in

$$\overline{B'_{\rho_2}(0)} \times [h - \rho_2, h + \rho_2]$$

to the following Cauchy problem

$$\begin{cases} P^\star(x, \partial)w = f, \\ \\ \partial^\alpha w(x', h) = 0, \quad |\alpha| \leq m - 1 \end{cases} . \tag{9.145}$$

Let

$$h_0 = \min_{\partial B'_{\rho_2}(0)} \psi,$$

$$h_1 = \min\left\{h_0, \frac{\rho_2}{2}\right\}.$$

Let us notice that by the strict convexity of ψ, h_1 is positive. Let us choose in (9.145)

$$h = h_1. \tag{9.146}$$

Let us consider the set

$$\widetilde{\Omega} = \left\{(x', x_n) : x' \in B'_{\rho_2}(0), \quad \psi(x') < x_n < h_1\right\}.$$

It is worth mentioning that $\widetilde{\Omega}$ exhibits a "lens" shape, specifically lacking any vertical segments on its boundary (Fig. 9.5). Let us take note that, due to our selection of h_1, we are able to establish $w \in C^\omega\left(\overline{\widetilde{\Omega}}\right)$

Now, by the assumptions, u belongs to $C^m\left(\overline{\widetilde{\Omega}}\right)$ and it is a solution to the Cauchy problem

$$\begin{cases} P(x, \partial)u = 0, \quad \text{in} \quad \widetilde{\Omega} \\ \\ \partial^\alpha u = 0, \qquad \text{for } |\alpha| \leq m - 1, \quad x \in \Gamma \cap \overline{\widetilde{\Omega}} \end{cases} . \tag{9.147}$$

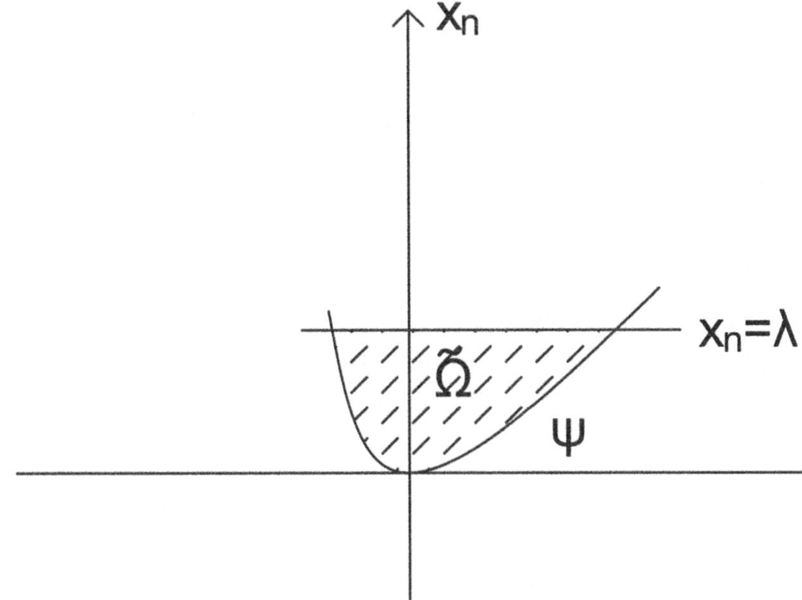

Fig. 9.5 In the figure, $0 < \lambda \leq h_1$, and it is important to note the lens-shaped form of the shaded region $\widetilde{\Omega}$, which plays a crucial role in the proof of Holmgren's Theorem

By the Green identity (9.131), we have

$$\begin{aligned}\int_{\widetilde{\Omega}} f u dx &= \int_{\widetilde{\Omega}} u P^{\star}(x, \partial) w dx = \\ &= \int_{\widetilde{\Omega}} \big(u P^{\star}(x, \partial) w - w P(x, \partial) u\big)\, dx = \\ &= \int_{\partial \widetilde{\Omega}} \mathcal{M}(u, w; \nu) dS = 0.\end{aligned} \tag{9.148}$$

To prove that

$$\int_{\partial \widetilde{\Omega}} \mathcal{M}(u, w; \nu) dS = 0,$$

it suffices to write the integral over $\partial \widetilde{\Omega}$ as a sum of two integrals, say I_1 and I_2, with the same integrand $\mathcal{M}(u, w; \nu)$, where I_1 is the integral over a portion of the graph of ψ, on which $\partial^{\alpha} u = 0$, for $|\alpha| \leq m - 1$, and I_2 is the integral over a portion of the hyperplane $\{x_n = h_1\}$ on which, by (9.145) and (9.146), we have $\partial^{\alpha} w = 0$, for $|\alpha| \leq m - 1$. Hence, by (9.132), both I_1 and I_2 are equal to zero. Therefore, by

(9.148) we get

$$\int_{\widetilde{\Omega}} fudx = 0, \qquad \text{for every polynomial } f$$

and since the set of polynomials is dense in $C^0\left(\overline{\widetilde{\Omega}}\right)$ (Theorem 3.2) we have

$$u \equiv 0, \quad \text{in } \widetilde{\Omega}.$$

The first part of proof is concluded.

Step II Now, we suppose that Γ satisfies (9.140) and (9.139), but we *do not suppose* that ψ is strictly convex. We may reduce to the case discussed in Step I using the following **Holmgren transformation**

$$\Lambda : \mathbb{R}^n_x \to \mathbb{R}^n_y, \quad x \to y = \Lambda(x', x_n) = \left(x', x_n + \frac{A}{2}|x'|^2\right), \tag{9.149}$$

where $A > 0$ is to be chosen. Let us note that Λ is a diffeomorphism. Furthermore, let

$$\widetilde{\Gamma} := \Lambda(\Gamma) = \left\{(x', \widetilde{\psi}(x')) : x' \in B'_{r_0}\right\}, \tag{9.150}$$

where

$$\widetilde{\psi} := \psi(x') + \frac{A}{2}|x'|^2. \tag{9.151}$$

Let us choose A in such a way that $\widetilde{\psi}$ is strictly convex. For this purpose it suffices to have

$$A > \left\|\partial^2\psi\right\|_{L^\infty(B'_{r_0})},$$

where $\partial^2\psi$ is the Hessian matrix of ψ. Fix such a number A. Let us denote by $P^\bullet(y, \partial_y)$ the transformed operator by means of Λ

$$P^\bullet(y, \partial_y)v(y)_{|y=\Lambda(x)} = P(x, \partial_x)u(x),$$

where v is defined by

$$v(\Lambda(x)) = u(x).$$

Let us notice that the coefficients of $\widetilde{P}$ are analytic functions. Setting

$$\widetilde{\mathcal{U}}_+ = \Lambda(\mathcal{U}_+),$$

we have that $v \in C^m\left(\widetilde{\mathcal{U}_+}\right)$ is a solution to the Cauchy problem

$$\begin{cases} P^\bullet(y, \partial_y)v = 0, & \text{in} \quad \mathcal{U}_+ \\ \\ \partial^\alpha v(y) = 0, & \text{for } |\alpha| \leq m-1, \quad y \in \widetilde{\Gamma} \cap \mathcal{U}_+ \end{cases}.$$

Let us recall (see (9.26)) that the symbol of the principal part of $p_m^\bullet(y, \eta)$ is given by

$$p_m^\bullet(y, \eta) = p_m(x, \partial_x(\Lambda(x))^t \eta)_{|x=\Lambda^{-1}(y)}.$$

Since

$$(\Lambda(0))^t e_n = e_n,$$

we have

$$p_m^\bullet(0, e_n) = p_m(0, e_n) \neq 0.$$

In short, we come back to the case already treated in Step I. Therefore, for a suitable neighborhood $\mathcal{U}$ of 0, we have $v \equiv 0$ in $\mathcal{U}_+$ which implies $u \equiv 0$ in $\mathcal{U}_+$. ■

Remarks About the Holmgren Theorem

1. If Γ is a noncharacteristic surface, Theorem 9.6 allows us to say that there exists an open set S such that $\Gamma \subset S$ and such that, denoting by $S_+ = S \cap \{x \in \Omega : \phi(x) \geq \phi(x_0)\}$, it occurs that if $v \in C^m\left(\overline{S_+}\right)$ is a solution of the Cauchy problem

$$\begin{cases} P(x, \partial)u = 0, & \text{in} \quad S_+ \\ \\ \partial^\alpha u(x) = 0, & \text{for } |\alpha| \leq m-1, \quad x \in \Gamma \end{cases}, \tag{9.152}$$

then $u \equiv 0$ in S_+.

However, the statement of Holmgren's Theorem doesn't provide sufficient clarity about the potential size of the set S (or S_+). In reality, one might anticipate that $\partial S \setminus \Gamma$ should primarily consist of characteristic surfaces. In simpler terms, the uniqueness for the Cauchy problem should prevail until a characteristic surface is encountered.

When dealing with an elliptic operator $P(x, \partial)$ possessing analytic coefficients in an open connected set Ω of $\mathbb{R}^n$ (as discussed in Section 9.4), and when Γ constitutes a portion of a regular surface, the ellipticity of $P(x, \partial)$ assures us that Γ isn't characteristic. Under this condition, we would naturally expect that the same ellipticity of $P(x, \partial)$ would imply a solution of $P(x, \partial)u = 0$ in Ω, with vanishing Cauchy data on Γ, is identically null over Ω. For instance, this holds true for $P(x, \partial) = \Delta$, as the solutions of $\Delta u = 0$ are analytic in Ω, implying $u \equiv 0$.

If we have the vibrating string operator $\partial_t^2 - \partial_x^2$ we know that if u is a solution of $\partial_t^2 u - \partial_x^2 u = 0$ with zero initial conditions on $\Gamma = (-R, R)$ then u vanishes in the square $\{|x|+|t| < R\}$ that is, u vanishes in a region bounded by characteristic lines parallel to $\{x + t = 0\}$, $\{x - t = 0\}$.

Neither the scenario described for the Laplace equation nor the one depicted for the vibrating string operator directly follows from the statement of Theorem 9.6. A comprehensive resolution to these issues is offered by the Global Uniqueness Theorem established by F. John. In the subsequent sections, we will present the theorem's statement and delve into some of its implications.

2. The assumptions of Theorem 9.6 can be weakened. Here we merely give a few hints and refer to [56, Theorem 5.3.1] the interested reading in learning more about the topic. We point out, in particular, that:

(i) one may assume $\phi \in C^1(\Omega)$,
(ii) one may give a distributional formulation of Cauchy problem (9.137) and in this framework prove the uniqueness of the solution.

3. It is important to note that while extensions of the Holmgren uniqueness Theorem to first–order non–linear equations have been proven, as shown by Métivier in [91], the same author, in [92], proven the non–extendibility of Holmgren uniqueness Theorem for higher–order non–linear equations. For further information we refer the reader to [48]. ♦

9.7.1 *Statement of the Holmgren–John Theorem: Examples*

Here, we will solely present the Holmgren–John Global Uniqueness Theorem, referring readers to [66] for the proof. Alternatively, within this book, the proof can be found in Chap. 13, where we establish F. John's Stability Theorem, from which uniqueness can be trivially inferred.

In order to state the Global Uniqueness Theorem, we need the following definition (in it we follow the terminology introduced in [66]).

Definition 9.7 Let

$$F : B_1' \times (0, 1) \to \mathbb{R}^n \tag{9.153}$$

be a function satisfying the following properties (let us denote by $y' \in B_1'$ and $\lambda \in (0, 1)$ the independent variables):

(i) F is injective,
(ii) F is analytic in $B_1' \times (0, 1)$,
(iii) for every $(y', \lambda) \in B_1' \times (0, 1)$ the jacobian matrix $\partial_{y',\lambda} F(y', \lambda)$ is nonsingular.

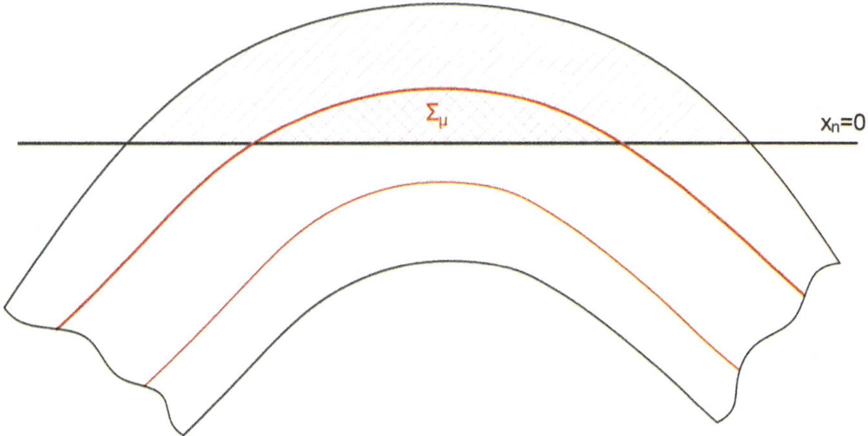

Fig. 9.6 Here, the sets Σ and Σ_μ are depicted for $0 < \mu < 1$. Note that the intersections of these sets with $\{(x', x_n) \in \mathbb{R}^n : x_n > 0\}$ (shaded region) have a lens-shaped form

We call **analytic field** in $\mathbb{R}^n$ the family of sets $\{S_\lambda\}_{\lambda\in(0,1)}$, where

$$S_\lambda = \left\{F(y', \lambda) : y' \in B_1'\right\}, \qquad \text{for } \lambda \in (0, 1). \tag{9.154}$$

We call **support of the analytic field** the open set

$$\Sigma := \bigcup_{\lambda\in(0,1)} S_\lambda.$$

We denote, for any $\mu \in (0, 1)$, (Fig. 9.6)

$$\Sigma_\mu := \bigcup_{\lambda\in(0,\mu]} S_\lambda.$$

Theorem 9.8 (Holmgren–John) *Let S_λ be an analytic field in $\mathbb{R}^n$ and let Σ be its support. Let*

$$P(x, \partial) = \sum_{|\alpha|\leq m} a_\alpha(x)\partial^\alpha$$

be a linear differential operator of order m, where $a_\alpha \in C^\omega(\Sigma)$. Define the sets (see Fig. 9.6)

$$\mathcal{R} = \left\{x \in \mathbb{R}^n : x \in \Sigma, \quad x_n \geq 0\right\}, \tag{9.155}$$

$$\mathcal{Z} = \left\{(x', 0) : (x', 0) \in \Sigma\right\}. \tag{9.156}$$

Suppose

(a) $\mathcal{Z}$ *and* S_λ, $\lambda \in (0, 1)$, *are noncharacteristic for* $P(x, \partial)$,
(b) for every $\mu \in (0, 1)$, $\mathcal{R} \cap \Sigma_\mu$, *is a* ***closed set*** *of* $\mathbb{R}^n$.

Then, if $u \in C^m(\mathcal{R})$ *is a solution to the Cauchy problem*

$$\begin{cases} P(x, \partial)u = 0, & in \quad \mathcal{R} \\ \\ \partial^\alpha u = 0, & for \ |\alpha| \leq m-1, \quad x \in \mathcal{Z} \end{cases}, \tag{9.157}$$

we have

$$u \equiv 0.$$

Let us provide some clarification on the assumptions of Theorem 9.8. Notably, $\mathcal{Z}$ is a portion of the hyperplane $\{x_n = 0\}$—this choice is purely for the sake of exposition. Actually, $\mathcal{Z}$ could represent any C^2 noncharacteristic surface for $P(x, \partial)$. Moreover, the condition outlined in (b) ensures the absence of vertical segments along the boundary of $\mathcal{R} \cap \Sigma_\mu$, where $\mu \in (0, 1)$. This property implies that $\mathcal{R} \cap \Sigma_\mu$ assumes a "lens" shape—a shape encountered earlier in the proof of Theorem 9.6.

Example 1: Elliptic Equations with Analytic Coefficients
Let

$$P(x, \partial)$$

be a **linear elliptic operator** of order m whose coefficients are analytic functions. Let us begin by proving the following unique continuation property.

Proposition 9.9 *Let* ρ, R *be such that* $0 < \rho < R$. *Let* $u \in C^m(B_R)$ *and suppose that*

$$\begin{cases} P(x, \partial)u = 0, & in \quad B_R \\ \\ u = 0, & in \ B_\rho \end{cases}, \tag{9.158}$$

then

$$u \equiv 0, \quad in \ B_R.$$

Proof Let $x_0 \in B_R \setminus \overline{B_\rho}$. Let us prove that $u(x_0) = 0$. We may assume that x_0 lies on the x_n–axis, because, by means of a rotation of $\mathbb{R}^n$, we may always reduce to this

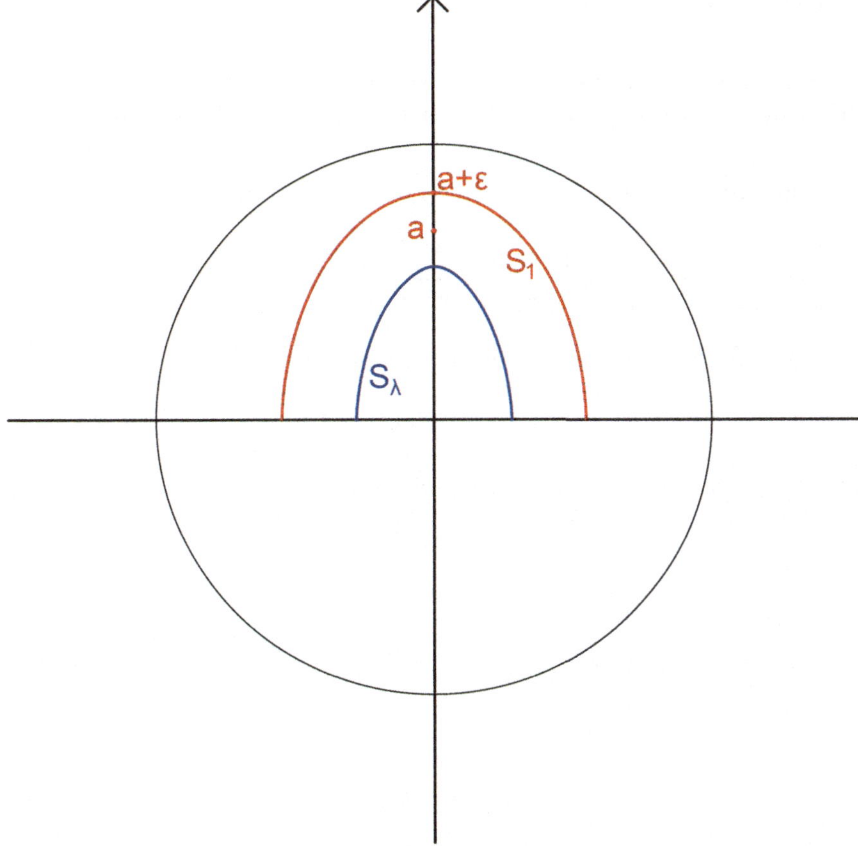

Fig. 9.7 The figure shows the construction of some of the surfaces S_λ involved in the proof of Proposition 9.9

case (Fig. 9.7). Hence let us suppose that

$$x_0 = ae_n,$$

where $\rho < a < R$.

Trivially we have

$$\partial^\alpha u(x', 0) = 0, \quad |\alpha| \le m, \quad x' \in B'_\rho. \tag{9.159}$$

Let ε be such that $0 < \varepsilon < \min\{R - a, \rho\}$ and let S_λ be defined by

$$x_n = (a + \varepsilon)^2 \left[\lambda - \frac{|x'|^2}{(\rho - \varepsilon)^2}\right], \quad x' \in B'_{\rho-\varepsilon}, \quad 0 < \lambda < 1.$$

Recalling that an elliptic operator has no characteristics, by Theorem 9.8 we have $u(ae_n) = 0$. ■

It is evident from the proof of the above Proposition that if $u \in C^m(B_R)$, is a solution to $P(x, \partial)u = 0$ in B_R and it is null in $B_\rho(\widetilde{x}) \subset B_R$ we have that $u \equiv 0$ in B_R.

Theorem 9.10 *Let Ω be a connected open set of $\mathbb{R}^n$ and let*

$$P(x, \partial) = \sum_{|\alpha| \leq m} a_\alpha(x)\partial^\alpha,$$

be a linear differential operator of order m, elliptic in $\overline{\Omega}$, where

$$a_\alpha \in C^\omega(\Omega \cup \Gamma),$$

and $\Gamma \subset \partial\Omega$. Assume that Γ is a local graph of class C^2.
If $u \in C^m(\Omega \cup \Gamma)$ is a solution to the Cauchy problem

$$\begin{cases} P(x, \partial)u = 0, & in \quad \Omega \\ \\ \partial^\alpha u = 0, & |\alpha| \leq m - 1 \; on \; \Gamma \end{cases}, \tag{9.160}$$

then

$$u = 0, \quad in \; \Omega.$$

Proof By Theorem 9.6 (or 9.8) we that have there exists an open set (in the topology induced on $\Omega \cup \Gamma$) $\mathcal{U}_+ \subset \Omega$ such that $\partial\mathcal{U}_+ \cap \Gamma \neq \emptyset$ and such that

$$u = 0, \quad \text{in } \mathcal{U}_+. \tag{9.161}$$

Set

$$A = \left\{x \in \Omega : \exists \rho_x > 0 \text{ such that } u = 0, \text{ in } B_{\rho_x}(x)\right\}. \tag{9.162}$$

By (9.161) we have $A \neq \emptyset$ and, trivially, we have that A is an open set in Ω. In order to prove the assertion, it suffices to prove that A is also closed in Ω and, since Ω is connected, we have $A = \Omega$ from which we will obtain the thesis. In order to prove that A is closed in Ω it suffices to prove that if $\left\{x_j\right\}_{j \in \mathbb{N}}$ is a sequence of A such that

$$\lim_{j \to \infty} x_j = x_0, \tag{9.163}$$

where $x_0 \in \Omega$, then $x_0 \in A$.

Let $\varepsilon > 0$ satisfy $B_\varepsilon(x_0) \subset \Omega$. By (9.163), there exists j_0 such that

$$\left|x_0 - x_{j_0}\right| < \frac{\varepsilon}{4}.$$

Since $x_{j_0} \in A$, there exists $\rho_{x_{j_0}} > 0$ such that

$$u = 0, \quad \text{in } B_{\rho_{x_{j_0}}}(x_{j_0}) \subset \Omega. \tag{9.164}$$

Set $\overline{\rho} = \min\left\{\frac{\varepsilon}{4}, \rho_{x_{j_0}}\right\}$, we have $B_{\overline{\rho}}(x_{j_0}) \subset B_\varepsilon(x_0)$ and, by (9.164), we have $u = 0$ in $B_{\overline{\rho}}(x_{j_0})$, then, by Proposition 9.9 and subsequent remarks, we have $u = 0$ in $B_\varepsilon(x_0)$. Therefore $x_0 \in A$ and the thesis follows. ■

Remarks on Theorem 9.10

1. At this juncture, we provide a side note for informational purposes. If $P(x, \partial)$ is an elliptic operator in Ω, $f \in C^\omega(\Omega)$, and $u \in C^m(\Omega)$ is a solution to the equation $P(x, \partial)u = f$ in Ω, then it follows that $u \in C^\omega(\Omega)$—a proof of this can be found in [96]. Notably, it is worth mentioning that, although this regularity property of solutions exists, we did not rely on it to establish Theorem 9.10. Although Theorem 9.10 can be deduced from the unique continuation property for analytic functions (and the Holmgren Theorem) through the utilization of this regularity result, we were able to prove Theorem 9.10 without requiring this particular level of regularity.

2. Let us consider some interesting consequences of Theorem 9.10. Let Ω, Γ and $P(x, \partial)$ be as in Theorem 9.10, further let us suppose that Γ is of class C^m. We denote by $\mathcal{X}_\Gamma$ the class of functions

$$g : \Gamma \to \mathbb{R}^m, \;\; g(x) = (g_0(x), g_1(x), \cdots, g_{m-1}(x)), \;\; \forall x \in \Gamma,$$

such that there exists $u \in C^m(\Omega \cup \Gamma)$ solution to the Cauchy problem

$$\begin{cases} P(x, \partial)u = 0, & \text{in } \;\Omega \\ \\ \frac{\partial^j u}{\partial \nu^j} = g_j, & j = 0, 1, \cdots, m-1 \text{ on } \Gamma \end{cases} . \tag{9.165}$$

The class $\mathcal{X}_\Gamma$ enjoys the following property.

For every $\Gamma_0 \subset \Gamma$, Γ_0 open in Γ in the induced topology, it occurs that

$$g \in \mathcal{X}_\Gamma, \text{ and } g = 0 \text{ on } \Gamma_0 \quad \Longrightarrow \quad g = 0 \text{ on } \Gamma \tag{9.166}$$

and, therefore, by linearity, if $g, \widetilde{g} \in \mathcal{X}_\Gamma$ and $g = \widetilde{g}$ on Γ_0 then $g = \widetilde{g}$ on Γ. In other words, the class $\mathcal{X}_\Gamma$ must enjoy the unique continuation property (9.166). The proof of the assertion above is very simple. As a matter of fact, let $g \in \mathcal{X}_\Gamma$

and let $u \in C^k(\Omega \cup \Gamma)$, $k \geq m$ be a solution of problem (9.166). Let us suppose

$$g = 0, \quad \text{on} \quad \Gamma_0.$$

Then applying Theorem 9.10 (with Γ_0 in the place of Γ) we have that $u = 0$ in Ω and being $u \in C^k(\Omega \cup \Gamma)$, we have $g = u_{|\Gamma} = 0$.

By this fact, when the initial data of a Cauchy problem for $P(x, \partial)$ belong to the class of functions $C^k(\Gamma, \mathbb{R}^m)$, for any $k \geq m$, such a Cauchy problem generally cannot have solutions. The reason is that the class $C^k(\Gamma, \mathbb{R}^m)$ does not satisfy property (9.166). ♦

Example 2: Wave Equation
Let us denote by $x \in \mathbb{R}^n$ and $t \in \mathbb{R}$ the independent variables, let

$$K = \left\{(x, t) \in \mathbb{R}^{n+1} : |x| < 1 - |t|\right\}. \tag{9.167}$$

Let us prove that if $u \in C^2(K)$ is a solution to the Cauchy problem

$$\begin{cases} \partial_t^2 u - \Delta_x u = 0, & \text{in} \quad K \\ u(x, 0) = 0, & \text{for } |x| < 1 \,, \\ \partial_t u(x, 0) = 0, & \text{for } |x| < 1 \end{cases} \tag{9.168}$$

then $u = 0$ in K.

We apply Theorem 9.8
In this case we have

$$P(\partial_t, \partial_x) = \partial_t^2 u - \Delta_x.$$

Let $\varepsilon \in (0, 1)$ be fixed and for $\lambda \in (0, 1)$ let

$$S_\lambda^\varepsilon = \left\{(x, t) \in \mathbb{R}^{n+1} : t = 1 - \sqrt{(1 - \lambda + \varepsilon)^2 + |x|^2}, \quad |x| < \sqrt{1 - \varepsilon^2}\right\}.$$

Let

$$F : B_{\sqrt{1-\varepsilon^2}}(0) \times (0, 1) \to \mathbb{R}^{n+1},$$

$$F(y, \lambda) = \left(y, 1 - \sqrt{(1 - \lambda + \varepsilon)^2 + |y|^2}\right).$$

Moreover, set

$$\Sigma^{\varepsilon} = \bigcup_{\lambda\in(0,1)} S^{\varepsilon}_{\lambda} =$$

$$= \left\{1-\sqrt{(1+\varepsilon)^2+|x|^2} < t < 1-\sqrt{\varepsilon^2+|x|^2}, \quad |x| < \sqrt{1-\varepsilon^2}\right\},$$

$$\mathcal{R}^{\varepsilon} = \left\{(x,t)\in\mathbb{R}^{n+1} : 0 \le t < 1-\sqrt{\varepsilon^2+|x|^2}\right\},$$

$$\mathcal{Z}^{\varepsilon} = \left\{(x,0) : |x| < \sqrt{1-\varepsilon^2}\right\}$$

and

$$\Sigma^{\varepsilon}_{\mu} = \left\{(x,t)\in\mathbb{R}^{n+1} : 0 \le t \le 1-\sqrt{(1-\mu+\varepsilon)^2+|x|^2}\right\}.$$

For each $0 < \mu < \varepsilon$ we have $\mathcal{R}^{\varepsilon} \cap \Sigma^{\varepsilon}_{\mu} = \emptyset$ and, for each $\varepsilon \le \mu < 1$, we have

$$\Sigma^{\varepsilon}_{\mu} = \mathcal{R}^{\varepsilon} \cap \Sigma^{\varepsilon}_{\mu}.$$

Finally, let us check that $\mathcal{Z}$ and $S^{\varepsilon}_{\lambda}$ are noncharacteristic. We have trivially that $\mathcal{Z}$ is a noncharacteristic surface. Concerning $S^{\varepsilon}_{\lambda}$, we notice that

$$S^{\varepsilon}_{\lambda} = \left\{(x,t)\in\mathbb{R}^{n+1} : \phi(x,t) = (1-\lambda+\varepsilon)^2, \quad t < 1\right\},$$

where

$$\phi(x,t) = (1-t)^2 - |x|^2.$$

Therefore we have

$$\nabla_{x,t}\phi(x,t) = (-2x, 2(t-1)), \quad p(\nabla_{x,t}\phi(x,t)) = 4\left((1-t)^2 - |x|^2\right).$$

Hence, if $(x,t)\in S^{\varepsilon}_{\lambda}$, then

$$p(\nabla_{x,t}\phi(x,t)) = 4(1-\lambda+\varepsilon)^2 > 0.$$

Therefore $S^{\varepsilon}_{\lambda}$ is noncharacteristic.

By Theorem 9.8 we get

$$u = 0, \quad \text{in } \mathcal{R}^{\varepsilon}$$

and, since ε is arbitrary, we have

$$u = 0, \qquad \text{for } 0 \le t < 1 - |x|.$$

Similarly we can check that $u = 0$ for $0 \le |x| - 1 < t \le 0$. Therefore $u = 0$ in K. ♣

Example 3: One-Dimensional Heat Equation
We consider the following Cauchy problem

$$\begin{cases} \partial_x^2 u - \partial_t u = 0, & \text{in} \quad D := (0,1) \times (0,1) \\ u(0,t) = 0, & \text{for } t \in (a,b) \\ \partial_x u(0,t) = 0, & \text{for } t \in (a,b) \end{cases}, \tag{9.169}$$

where a, b are given numbers and such that $0 < a < b < 1$. Let us notice that problem (9.169) is a **noncharacteristic Cauchy problem**, since the initial line is $\{x = 0\}$.

By using the same arguments exploited in **Example 1** (Proposition 9.9) and in **Example 2** it can be proved easily that if $u \in C^2(\overline{D})$ then $u = 0$ in $[0,1] \times [a,b]$. The details are left to the reader (it is useful to keep in mind that the only characteristics of the operator $\partial_x^2 - \partial_t$ are the straight lines $t = t_0$, for any $t_0 \in \mathbb{R}$).

It is quite natural to wonder whether a solution of (9.169) is null in D. The answer to this question is negative as proved in an example due to **Tychonoff**, [130], which we will discuss below.

First, let us consider the following Cauchy problem

$$\begin{cases} \partial_x^2 u - \partial_t u = 0, & \\ u(0,t) = \varphi(t), & \text{for } t \in \mathbb{R} \\ \partial_x u(0,t) = 0, & \text{for } t \in \mathbb{R} \end{cases}. \tag{9.170}$$

For the time being, let us just assume that $\varphi \in C^\infty(\mathbb{R})$ and let us search, at first just formally, a solution of the type

$$u(x,t) = \sum_{j=0}^{\infty} a_j(t) x^j, \tag{9.171}$$

where a_j are functions to be found.

Obviously we have to require that

$$a_0(t) = \varphi(t) \quad \text{and} \quad a_1(t) = 0$$

and, by requiring that (9.171) is a solution to the equation $\partial_x^2 u - \partial_t u = 0$, we need to require

$$\sum_{j=0}^{\infty} a_j'(t)x^j - \sum_{j=2}^{\infty} j(j-1)a_j(t)x^{j-2} = 0,$$

from which we have

$$a_{j+2}(t) = \frac{1}{(j+2)(j+1)}a_j'(t), \quad \forall j \in \mathbb{N}_0.$$

Hence

$$a_{2k}(t) = \frac{\varphi^{(k)}(t)}{(2k)!}, \quad a_{2k+1}(t) = 0, \quad \forall k \in \mathbb{N}_0.$$

Therefore

$$u(x,t) = \sum_{k=0}^{\infty} \frac{\varphi^{(k)}(t)}{(2k)!}x^{2k}. \tag{9.172}$$

In order that (9.172) is actually a solution to Cauchy problem (9.170) (in a neighborhood of $\{0\} \times \mathbb{R}$) it suffices to require that there exist two positive numbers c and M such that

$$\left|\varphi^{(k)}(t)\right| \leq cM^k(2k)!, \quad \forall k \in \mathbb{N}_0. \tag{9.173}$$

Now, let

$$\varphi(t) = \begin{cases} e^{-\frac{1}{t^2}}, & \text{for } t > 0 \\ \\ 0, & \text{for } t \leq 0 \end{cases}. \tag{9.174}$$

We show now that (9.173) is satisfied. We will prove, indeed,

$$\left|\varphi^{(k)}(t)\right| \leq \left(\frac{9k}{2e}\right)^{\frac{k}{2}} k!, \quad \forall k \in \mathbb{N}_0. \tag{9.175}$$

which (by the Stirling formula, (2.6)) implies (9.173).

To prove (9.175) we use the Cauchy formula for the holomorphic functions. Therefore let $t > 0$ and let S be the circumference centered at $t + i0$ and with radius $\frac{t}{2}$ in the complex plane, i.e.

$$S = \left\{ t\left(1 + \frac{1}{2}e^{i\vartheta}\right) : \vartheta \in [0, 2\pi) \right\}.$$

We have, for any $k \in \mathbb{N}_0$,

$$\varphi^{(k)}(t) = \frac{k!}{2\pi i} \int_S \frac{e^{-\frac{1}{z^2}}}{(z-t)^{k+1}} dz.$$

Now, when $z = t\left(1 + \frac{1}{2}e^{i\vartheta}\right) \in S$, it is easily checked that

$$\frac{1}{z} = \frac{4}{3t} + \frac{2}{3t}e^{-i\vartheta}.$$

Therefore

$$\Re\left(\frac{1}{z^2}\right) = \left(\frac{4}{3t}\right)^2 \left[\left(1 + \frac{1}{2}\cos\vartheta\right)^2 - \left(\frac{1}{2}\sin\vartheta\right)^2\right] =$$
$$= \left(\frac{4}{3t}\right)^2 \left[\frac{1}{4} + \frac{1}{2}(1 + \cos\vartheta)^2\right] \geq \frac{4}{9t^2}.$$

Therefore we have

$$\begin{aligned} \left|\varphi^{(k)}(t)\right| &\leq \frac{k!}{2\pi} \int_{|z-t|=\frac{t}{2}} \left| \frac{e^{-\frac{1}{z^2}}}{(z-t)^{k+1}} \right| ds \leq \\ &\leq \frac{k!}{2\pi} \int_{|z-t|=\frac{t}{2}} \frac{e^{-\frac{4}{9t^2}}}{|z-t|^{k+1}} ds = \\ &= \frac{k!}{2\pi} \left(2\pi \frac{t}{2}\right) \frac{1}{(t/2)^{k+1}} e^{-\frac{4}{9t^2}} = \\ &= \frac{2^k k!}{t^k} e^{-\frac{4}{9t^2}}. \end{aligned} \tag{9.176}$$

Since

$$\sup_{(0,+\infty)} \frac{1}{t^k} e^{-\frac{4}{9t^2}} = \left(\frac{9k}{8e}\right)^{k/2},$$

by (9.176) we get (9.175) which in turn implies that the series in (9.172) converges for every $x \in \mathbb{R}$ and its sum, u, is indeed the solution to Cauchy problem (9.170).

From what we have just established, it turns out that, denoting by ψ the following function

$$\psi(t) = \begin{cases} e^{-\frac{1}{(t-b)^2}}, & \text{for } t > b \\ 0, & \text{for } a \leq t \leq b \\ e^{-\frac{1}{(a-t)^2}}, & \text{for } t < a \end{cases},$$

we have that

$$\widetilde{u} = \sum_{k=0}^{\infty} \frac{\psi^{(k)}(t)}{(2k)!} x^{2k},$$

is the solution to the Cauchy problem

$$\begin{cases} \partial_x^2 \widetilde{u} - \partial_t \widetilde{u} = 0, & \text{in } \mathbb{R}^2 \\ \widetilde{u}(0,t) = \psi(t), & \text{for } t \in \mathbb{R} \\ \partial_x \widetilde{u}(0,t) = 0, & \text{for } t \in \mathbb{R} \end{cases}$$

and $\widetilde{u} = 0$ in $\mathbb{R} \times [a, b]$, but if $t_0 \notin [a, b]$ then $\widetilde{u}(\cdot, t_0)$ does not identically vanish, more precisely $\widetilde{u}(\cdot, t_0)$ does not vanish in any open set of $\mathbb{R}$ because, as $\widetilde{u}(\cdot, t_0)$ is analytic (as it is the sum of a series of powers in the variable x) one would have that $\psi^{(k)}(t_0) = 0$ for every $k \in \mathbb{N}_0$ that is false.

We conclude this discussion about the heat equation by observing that does not hold the uniqueness in $C^2(\mathbb{R} \times [0, +\infty))$ to the following **characteristic** Cauchy problem

$$\begin{cases} \partial_x^2 u - \partial_t u = 0, & \text{in } \mathbb{R}^2 \\ \\ u(x,0) = 0, & \text{for } x \in \mathbb{R} \end{cases}. \tag{9.177}$$

It suffices to consider the function u defined by (9.172) where φ defined by (9.174) and we have that u is the solution of problem (9.177), but it does not vanish identically. ♣

Exercise Let $T, r > 0$. Denote by

$$K = \left\{(x,t) \in \mathbb{R}^{n+1} : |x| + |t| < T + r\right\}.$$

Prove that if $u \in C^2(K)$ satisfies

$$\begin{cases} \partial_t^2 u - \Delta_x u = 0, & \text{in } K \\ \\ u = 0, & \text{in } B_r \times (-T, T) \end{cases},$$

then

$$u = 0, \quad \text{in } K.$$

♠

Chapter 10
Uniqueness for an Inverse Problem

Abstract In this chapter, we present an inverse problem for the Laplace equation. The direct problem is the Dirichlet problem, where we seek the solution to the Laplace equation in a domain with specified boundary values. The inverse problem arises when a portion of the boundary, called the "inaccessible" part, is unknown, and we are given the normal derivative of the solution on "accessible" part of the boundary. The goal is to determine the inaccessible part based on the given information. This problem is overdetermined, but we can reformulate it as finding a solution to a related equation involving the inaccessible part. The inverse problem then becomes the task of determining the inaccessible part such that a certain condition holds on the boundary, and, in particular on an accessible part.

10.1 Introduction

In this short chapter we present an inverse problem for the Laplace equation. The direct problem is nothing but the Dirichlet problem

$$\begin{cases} \Delta u = 0, & \text{in } \Omega \\ \\ u = \varphi, & \text{on } \partial\Omega \end{cases} . \tag{10.1}$$

For simplicity, we assume that Ω is a bounded and connected open of $\mathbb{R}^n$ whose boundary is of class C^∞ and $\varphi \in C^\infty(\partial\Omega)$. In Chap. 6 we saw (see in particular, Corollary 6.14) that, under these assumptions, there exists a unique $u \in C^\infty\left(\overline{\Omega}\right)$, which is the solution to (10.1).

Now, let us suppose that a portion of $\partial\Omega$, which we will call $\Gamma^{(i)}$, is unknown and that we have

$$\varphi(x) = 0, \quad \forall x \in \Gamma^{(i)} \tag{10.2}$$

S. Vessella, *Unique Continuation Properties for Partial Differential Equations*, Birkhäuser Advanced Texts Basler Lehrbücher,
https://doi.org/10.1007/978-3-031-86366-0_10

and let us suppose that we know

$$\frac{\partial u}{\partial \nu} = \psi, \quad \text{on} \quad \Sigma, \tag{10.3}$$

where $\Sigma \subset \partial\Omega \setminus \Gamma^{(i)}$. We are interested in determining $\Gamma^{(i)}$. This is our **inverse problem**. Let us note that if we consider all the data of the problem we should write

$$\begin{cases} \Delta u = 0, & \text{in } \Omega \\ u = \varphi, & \text{on } \partial\Omega \ . \\ \frac{\partial u}{\partial \nu} = \psi, & \text{on } \Sigma \end{cases} \tag{10.4}$$

Clearly, this is an overdetermined problem, from which we can anticipate a compatibility relationship among φ, ψ, and $\partial\Omega$. Thus, if we assume that φ, ψ, and $\partial\Omega \setminus \Gamma^{(i)}$ are known, we can reasonably expect to extract information about $\Gamma^{(i)}$ itself. Notably, it is evident that if $\varphi \equiv 0$ on $\partial\Omega$, we are left without any informative input about $\Gamma^{(i)}$. It will become apparent that this straightforward scenario is the only case (under precise conditions) where $\Gamma^{(i)}$ remains non-uniquely determined.

Rather than approaching the inverse problem as an overdetermined one, it proves more effective to consider the problem from the perspective we are about to outline. Given that the direct problem boasts a unique solution $u \in C^{\infty}\left(\overline{\Omega}\right)$ for any bounded open set of class C^{∞}, and for any $\varphi \in C^{\infty}(\partial\Omega)$ that satisfies (10.2), we find that the derivative

$$\frac{\partial u}{\partial \nu}_{|\Sigma},$$

becomes a function of $\Gamma^{(i)}$. We shall denote this function as

$$\mathbf{U}\left(\Gamma^{(i)}\right),$$

leading us to formulate the inverse problem of our interest as follows:

$$\text{Determine the solution to the equation } \mathbf{U}\left(\Gamma^{(i)}\right) = \psi$$

In the next section we will specify the assumptions and we will formulate the uniqueness theorem for the inverse problem above.

10.2 Statement of the Uniqueness Theorem for the Inverse Problem

Let us suppose that Ω is as above and let us suppose that $\partial\Omega$ is the union of two internally disjoint portions, $\Gamma^{(a)}$ ("accessible" portion) and $\Gamma^{(i)}$ ("inaccessible"

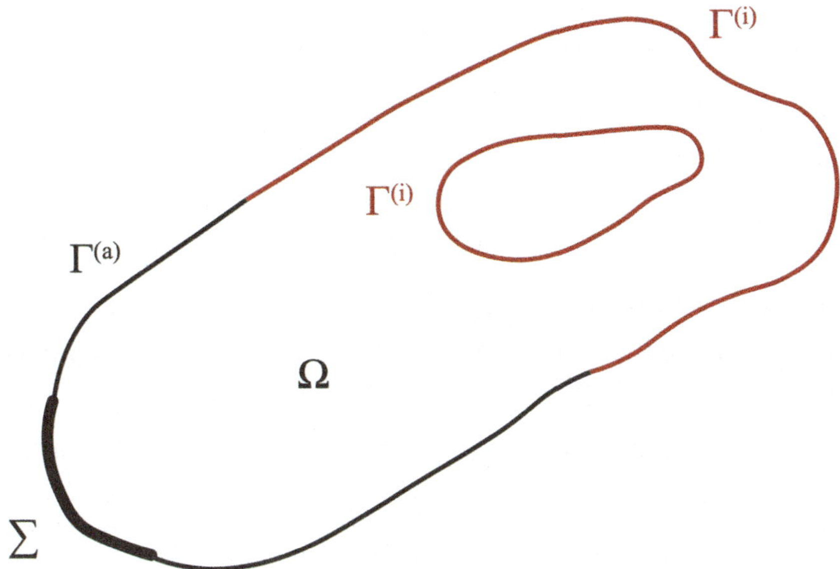

Fig. 10.1 $\Gamma^{(a)}$ (black) and $\Gamma^{(i)}$ (red), representing the accessible and inaccessible portions, respectively, together form $\partial\Omega$

portion) (Fig. 10.1). More precisely, let us suppose that:

$$\partial\Omega = \Gamma^{(a)} \cup \Gamma^{(i)}, \tag{10.5}$$

$$\Gamma^{(a)} \text{ and } \Gamma^{(i)} \text{ closed in } \partial\Omega \quad \text{and} \quad \overline{\overset{\circ}{\Gamma^{(a)}}} = \Gamma^{(a)}, \quad \overline{\overset{\circ}{\Gamma^{(i)}}} = \Gamma^{(i)}, \tag{10.6}$$

we equip $\partial\Omega$ with the topology induced by the Euclidean topology of $\mathbb{R}^n$,

$$\overset{\circ}{\Gamma^{(a)}} \cap \overset{\circ}{\Gamma^{(i)}} = \emptyset. \tag{10.7}$$

Hence

$$\Gamma^{(i)} = \partial\Omega \setminus \overset{\circ}{\Gamma^{(a)}} \quad \text{and} \quad \Gamma^{(a)} = \partial\Omega \setminus \overset{\circ}{\Gamma^{(i)}} \tag{10.8}$$

and

$$\overset{\circ}{\Gamma^{(a)}} \ \textbf{is connected}, \tag{10.9}$$

therefore, (10.6) implies $\Gamma^{(a)}$ is connected.

Let Σ be a compact subset of $\partial\Omega$ such that

$$\Sigma \Subset \overset{\circ}{\Gamma^{(a)}}, \qquad \overset{\circ}{\Sigma} \neq \emptyset. \tag{10.10}$$

Let $\varphi \in C^\infty(\partial\Omega)$ satisfy

$$\operatorname{supp}\varphi \subset \overset{\circ}{\Gamma^{(a)}}. \tag{10.11}$$

We now state the Theorem

Theorem 10.1 (Uniqueness) *Let Ω_k, $k = 1, 2$ be two bounded connected open sets of $\mathbb{R}^n$ whose boundary is of class C^∞. Assume that*

$$\partial\Omega_1 = \Gamma^{(a)} \cup \Gamma_1^{(i)} \quad \text{and} \quad \partial\Omega_2 = \Gamma^{(a)} \cup \Gamma_2^{(i)}, \tag{10.12}$$

where $\Gamma^{(a)}, \Gamma_k^{(i)}$, $k = 1, 2$, satisfies (10.6), (10.7) *and* (10.9). *Assume that $\varphi_k \in C^\infty(\partial\Omega_k)$ do not vanish identically and satisfying* (10.11). *Moreover, let us assume*

$$\varphi_1 = \varphi_2, \quad \text{on} \quad \Gamma^{(a)}. \tag{10.13}$$

Let $u_k \in C^\infty\left(\overline{\Omega_k}\right)$, $k = 1, 2$ be the solutions to

$$\begin{cases} \Delta u_k = 0, & \text{in } \Omega_k \\ \\ u_k = \varphi, & \text{on } \partial\Omega_k \end{cases} . \tag{10.14}$$

Assume that

$$\frac{\partial u_1}{\partial \nu} = \frac{\partial u_2}{\partial \nu}, \quad \text{on } \Sigma, \tag{10.15}$$

where Σ is a compact subset of $\partial\Omega$ which satisfies (10.10).
Then we have

$$\Omega_1 = \Omega_2, \tag{10.16}$$

and, consequently,

$$\Gamma_1^{(i)} = \Gamma_2^{(i)}.$$

The proof will be given in the next section.

10.3 Proof of the Uniqueness

The idea of the proof of the uniqueness Theorem is quite simple, but it requires some propositions of general topology that we will prove separately for the purpose of not breaking the main argument of the proof.

We argue by contradiction. We assume that one of the two sets $\Omega_1 \setminus \overline{\Omega}_2$, $\Omega_2 \setminus \overline{\Omega}_1$, is not empty. For instance, let us assume

$$\Omega_1 \setminus \overline{\Omega}_2 \neq \emptyset. \tag{10.17}$$

Define G as

$$G = \bigcup_{A \in \mathcal{A}} A, \tag{10.18}$$

where

$$\mathcal{A} = \left\{ A \text{ open set of } \mathbb{R}^n : \ A \subset \Omega_1 \cap \Omega_2, \ \Gamma^{(a)} \subset \overline{A}, \ A \text{ connected} \right\}. \tag{10.19}$$

As we will prove in Proposition 10.2 (and as can be expected), G is a connected open set, moreover $\Gamma^{(a)} \subset \overline{G}$ and $G \subset \Omega_1 \cap \Omega_2$. From the latter and from (10.17) we have

$$\Omega_1 \setminus \overline{G} \neq \emptyset. \tag{10.20}$$

Set

$$u := u_1 - u_2, \quad \text{in} \quad G.$$

We have

$$\begin{cases} \Delta u = 0, & \text{in } G \\ u = 0, & \text{on } \Sigma \\ \frac{\partial u}{\partial \nu} = 0, & \text{on } \Sigma \end{cases} .$$

Theorem 9.10 implies

$$u \equiv 0, \quad \text{in} \quad G. \tag{10.21}$$

Claim

$$u_1 = 0, \quad \text{on} \quad \partial \left(\Omega_1 \setminus \overline{G} \right). \tag{10.22}$$

Proof of Claim We exploit the following relationship (proved in Proposition 10.4)

$$\partial\left(\Omega_1 \setminus \overline{G}\right) \subset \Gamma_1^{(i)} \cup \left(\Gamma_2^{(i)} \cap \partial G\right). \tag{10.23}$$

Let $x \in \partial\left(\Omega_1 \setminus \overline{G}\right)$. Then by (10.23) we distinguish two cases:

(a) $x \in \Gamma_1^{(i)}$,
(b) $x \in \Gamma_2^{(i)} \cap \partial G$.

In case (a) $u_1(x) = 0$ because $u = \varphi_1$ on $\partial\Omega_1$ and $\varphi_1 = 0$ on $\Gamma_1^{(i)}$.

In case (b), we have $u_2(x) = 0$ because $u = \varphi_2$ on $\partial\Omega_2$ and $\varphi_2 = 0$ on $\Gamma_2^{(i)}$. On the other hand, by (10.21) and taking into account that u is continuous in $\overline{G}$, we have

$$u_1(x) = u_1(x) - u_2(x) = u(x) = 0.$$

This concludes the proof of the Claim.

Therefore u_1 solves the Dirichlet problem

$$\begin{cases} \Delta u_1 = 0, & \text{in } \Omega_1 \setminus \overline{G} \\ \\ u_1 = 0, & \text{on } \partial\left(\Omega_1 \setminus \overline{G}\right) \end{cases},$$

and the maximum principle implies

$$u_1 = 0, \quad \text{in} \quad \Omega_1 \setminus \overline{G}.$$

Now, taking into account (10.20), the unique continuation property gives

$$u_1 = 0, \quad \text{in} \quad \Omega_1.$$

From which we have

$$\varphi_1 = 0, \quad \text{on} \quad \partial\Omega_1,$$

this contradicts the assumption that φ_1 does not vanish identically. Therefore $\Omega_1 \setminus \overline{\Omega_2} = \emptyset$. Hence $\Omega_1 \subset \overline{\Omega_2}$. Similarly we have $\Omega_2 \subset \overline{\Omega_1}$. Therefore, $\overline{\Omega_1} = \overline{\Omega_2}$. On the other hand (see Exercise of Sect. 3.8)

$$\Omega_1 = \overset{\circ}{\overline{\Omega}}_1 = \overset{\circ}{\overline{\Omega}}_2 = \Omega_2.$$

■

Proposition 10.2 *G, defined by* (10.18), *is an open nonempty set and it enjoys the following properties*

(a) $G \subset \Omega_1 \cap \Omega_2$,
(b) $\Gamma^{(a)} \subset \overline{G}$,
(c) G *is connected.*

Proof Since Ω_k, $k = 1, 2$, are of class C^∞, there exist r_0, M_0, positive numbers, such that Ω_k, $k = 1, 2$, are of class $C^{1,1}$ with constant r_0, M_0. Proposition 3.82 implies that there exists $\mu_1 > 0$ such that, the following map is continuous

$$\Phi : \partial\Omega \times (0, \mu_1 r_0) \to \mathbb{R}^n,$$

$$\Phi(y, t) = y - t\nu(y), \quad \forall (y, t) \in \partial\Omega \times (0, \mu_1 r_0)\,.$$

For any $\delta \in (0, \mu_1 r_0)$ we denote

$$\Lambda_\delta = \Phi\left(\overset{\circ}{\Gamma^{(a)}} \times (0, \delta)\right).$$

Let us check

$$\Lambda_\delta \in \mathcal{A}, \tag{10.24}$$

where $\mathcal{A}$ is defined in (10.19). It evident that $\Lambda_\delta \neq \emptyset$. Moreover, by the continuity of Φ^{-1} (see (c) of Proposition 3.82), Λ_δ is an open set (see (a) of Proposition 3.82)

$$\Lambda_\delta \subset \Omega_1, \quad \text{and} \quad \Lambda_\delta \subset \Omega_2.$$

Also, Λ_δ is connected, as the image of the connected set $\Gamma^{(a)} \times (0, \delta)$ by means of the continuous map Φ. To complete the proof of (10.24), it suffices to check

$$\Gamma^{(a)} \subset \overline{\Lambda}_\delta. \tag{10.25}$$

Let $x \in \Gamma^{(a)}$ and let $r > 0$ be arbitrary. Since, by (10.6),

$$\overline{\overset{\circ}{\Gamma^{(a)}}} = \Gamma^{(a)},$$

we have $B_r(x) \cap \overset{\circ}{\Gamma^{(a)}} \neq \emptyset$. Let $y \in B_r(x) \cap \overset{\circ}{\Gamma^{(a)}}$. If t is a positive number small enough, we have

$$y - t\nu(y) \in B_r(x) \cap \Lambda_\delta.$$

Therefore

$$B_r(x) \cap \Lambda_\delta \neq \emptyset, \quad \forall r > 0.$$

Hence $x \in \overline{\Lambda}_\delta$ and (10.25) is proved. Now, let us notice that

$$\Lambda_\delta \cap A \neq \emptyset, \quad \forall A \in \mathcal{A}. \tag{10.26}$$

Let us fix $A \in \mathcal{A}$ and let $x \in \overset{\circ}{\Gamma^{(a)}}$. Since x is an interior point of $\Gamma^{(a)}$, $\Gamma^{(a)} \subset \partial\Omega_1 \cap \partial\Omega_2$ and since Ω_1 and Ω_2 are of class $C^{1,1}$, there exists $\overline{r} > 0$ such that

$$B_{\overline{r}}(x) \cap \Omega_1 = B_{\overline{r}}(x) \cap \Omega_2 \subset \Lambda_\delta.$$

Hence

$$B_{\overline{r}}(x) \cap A \subset \Lambda_\delta.$$

On the other hand, since $\overset{\circ}{\Gamma^{(a)}} \subset \overline{A}$, we have $x \in \overline{A}$. Hence

$$\emptyset \neq B_{\overline{r}}(x) \cap A \subset \Lambda_\delta \cap A,$$

which implies (10.26).

Now, since

$$G = \bigcup_{A \in \mathcal{A}} A \tag{10.27}$$

and since $\Lambda_\delta \in \mathcal{A}$, we have that $G \neq \emptyset$ and (trivially)

$$G \subset \Omega_1 \cap \Omega_2, \quad \Gamma^{(a)} \subset \overline{\Lambda}_\delta \subset \overline{G}.$$

Hence (a) and (b) are proved. It remains to prove (c). Let $x, y \in G$ and let $A, B \in \mathcal{A}$ satisfy $x \in A$ and $y \in B$. By (10.26), we have $\Lambda_\delta \cap A \neq \emptyset$ and $\Lambda_\delta \cap B \neq \emptyset$. Let

$$z \in \Lambda_\delta \cap A \quad \text{and} \quad w \in \Lambda_\delta \cap B$$

and let γ_1 be a continuous path that joins x and z in A, γ_2 be a continuous path that joins z and w in Λ_δ and γ_3 be a continuous path that joins w and y in B. Set

$$\gamma = \gamma_1 \vee \gamma_2 \vee \gamma_3.$$

γ is a continuous path that joins x and y in $A \cup \Lambda_\delta \cup B$. Hence $A \cup \Lambda_\delta \cup B \in \mathcal{A}$ and we have $A \cup \Lambda_\delta \cup B \subset G$. Consequently γ joins x and y in G. All in all, G is connected. ■

We have

Proposition 10.3 *Let C be a nonempty set of $\mathbb{R}^n$ and A an open set of $\mathbb{R}^n$. then*

$$A \cap \overline{C} \neq \emptyset \Longleftrightarrow A \cap C \neq \emptyset \tag{10.28}$$

Proof The implication " $\Longleftarrow$" is trivial. Concerning the implication " $\Longrightarrow$", let $z \in A \cap \overline{C}$ and $B_r(z) \subset A$. Since $z \in \overline{C}$, we have $\emptyset \subsetneq B_r(z) \cap C \subset A \cap C$. Therefore $A \cap C \neq \emptyset$. ∎

Proposition 10.4 *Let G be defined by* (10.18). *Let us suppose*

$$\Omega_1 \setminus \overline{\Omega_2} \neq \emptyset. \tag{10.29}$$

Then we have

$$\partial\left(\Omega_1 \setminus \overline{G}\right) \subset \Gamma_1^{(i)} \cup \left(\Gamma_2^{(i)} \cap \partial G\right). \tag{10.30}$$

Proof **Step I** First, let us notice that, since $\Omega_1 \setminus G \supset \Omega_1 \setminus \left(\overline{\Omega_1 \cap \Omega_2}\right) \supsetneq \emptyset$ and $\Omega_1 \setminus \overline{G} \neq \mathbb{R}^n$, we have $\partial\left(\Omega_1 \setminus \overline{G}\right) \neq \emptyset$.

Now we prove that

$$\partial\left(\Omega_1 \setminus \overline{G}\right) \subset \partial\Omega_1 \cup \partial\Omega_2. \tag{10.31}$$

We argue by contradiction. Let

$$x_0 \in \partial\left(\Omega_1 \setminus \overline{G}\right)$$

and let us suppose that

$$x_0 \notin \partial\Omega_1 \cup \partial\Omega_2. \tag{10.32}$$

It cannot be the case that $x_0 \notin \overline{\Omega_1}$ because if it were, it would exist $\rho > 0$ such that $B_\rho(x_0) \cap \left(\Omega_1 \setminus \overline{G}\right) \subset B_\rho(x_0) \cap \Omega_1 = \emptyset$, that contradicts $x_0 \in \partial\left(\Omega_1 \setminus \overline{G}\right)$. Now, let us examine the following two cases:

(a) $x_0 \in \Omega_1 \cap \Omega_2$,
(b) $x_0 \in \Omega_1 \cap \left(\mathbb{R}^n \setminus \overline{\Omega_2}\right)$.

Case (a) Let $\rho > 0$ be such that

$$B_\rho(x_0) \subset \Omega_1 \cap \Omega_2. \tag{10.33}$$

On the other hand, $x_0 \in \partial\left(\Omega_1 \setminus \overline{G}\right)$, hence

$$\begin{cases} B_\rho(x_0) \cap \left(\Omega_1 \setminus \overline{G}\right) \neq \emptyset \\ \\ B_\rho(x_0) \cap \left(\mathbb{R}^n \setminus \left(\Omega_1 \setminus \overline{G}\right)\right) \neq \emptyset \end{cases} . \tag{10.34}$$

By (10.33) and by the second of (10.34), we have

$$\begin{aligned} \emptyset \subsetneq B_\rho(x_0) \cap \left(\mathbb{R}^n \setminus \left(\Omega_1 \setminus \overline{G}\right)\right) = B_\rho(x_0) \cap \left[\left(\mathbb{R}^n \setminus \Omega_1\right) \cup \left(\overline{G} \cap \Omega_1\right)\right] = \\ = B_\rho(x_0) \cap \left(\overline{G} \cap \Omega_1\right) \subset \\ \subset B_\rho(x_0) \cap \overline{G}. \end{aligned}$$

All in all, $B_\rho(x_0) \cap \overline{G} \neq \emptyset$ and, by Proposition 10.3 we get

$$B_\rho(x_0) \cap G \neq \emptyset. \tag{10.35}$$

Moreover, by the first relationship of (10.34), we obtain

$$B_\rho(x_0) \cap \left(\mathbb{R}^n \setminus \overline{G}\right) \supset B_\rho(x_0) \cap \left(\Omega_1 \setminus \overline{G}\right) \supsetneq \emptyset. \tag{10.36}$$

Therefore (recalling (10.33))

$$G \subsetneq B_\rho(x_0) \cup G \subset \Omega_1 \cap \Omega_2. \tag{10.37}$$

■

Claim $B_\rho(x_0) \cup G$ is a connected set.

Proof of the Claim Let $x, y \in B_\rho(x_0) \cup G$. Let us prove that there exists a continuous path γ that joins x and y in $B_\rho(x_0) \cup G$. If $x, y \in G$ we have nothing to prove because (by Proposition 10.2) G is connected. If $x, y \in B_\rho(x_0)$, of course we have nothing to prove. Hence, let us suppose that $x \in G$ and $y \in B_\rho(x_0)$. By (10.35) there exists $z \in B_\rho(x_0) \cap G$. Let γ_1 be a continuous path that joins x and z in G and γ_2 be a continuous path that joins z and y in $B_\rho(x_0)$. Let

$$\gamma = \gamma_1 \vee \gamma_2.$$

γ is a continuous path that joins x and y in $B_\rho(x_0) \cup G$. **This concludes the proof of the Claim.**

Now, from (b) of Proposition 10.2 we have $\overline{B_\rho(x_0) \cup G} \supset \Gamma^{(a)}$ and recalling the definition of G we would have

$$G = B_\rho(x_0) \cup G,$$

which contradicts (10.37).

Case (b) Let $\overline{\rho} > 0$ satisfy

$$B_{\overline{\rho}}(x_0) \subset \Omega_1 \cap \left(\mathbb{R}^n \setminus \overline{\Omega}_2\right). \tag{10.38}$$

Since $x_0 \in \partial\left(\Omega_1 \setminus G\right)$, we again obtain (10.35). Therefore

$$\emptyset \subsetneq B_{\overline{\rho}}(x_0) \cap G \subset \Omega_1 \cap \Omega_2 \subset \Omega_2,$$

hence $B_{\overline{\rho}}(x_0) \cap \Omega_2 \neq \emptyset$, that contradicts (10.38). Hence (10.31) is proved.

Step II Now we prove

$$\partial\left(\Omega_1 \setminus G\right) \subset \Gamma_1^{(i)} \cup \Gamma_2^{(i)}. \tag{10.39}$$

To this aim, let us prove

$$\overset{\circ}{\Gamma}{}^{(a)} \cap \partial\left(\Omega_1 \setminus \overline{G}\right) = \emptyset. \tag{10.40}$$

Let us suppose that (10.40) does not hold. Hence, there exists $\overline{x}$ such that

$$\overline{x} \in \overset{\circ}{\Gamma}{}^{(a)} \cap \partial\left(\Omega_1 \setminus \overline{G}\right). \tag{10.41}$$

Since $\overline{x} \in \overset{\circ}{\Gamma}{}^{(a)}$ and $\Gamma^{(a)} \subset (\partial\Omega_1) \cap (\partial\Omega_2)$, taking into account that $\partial\Omega_k$, $k = 1, 2$, are of class $C^{1,1}$, there exist $\overline{r}$, $\overline{M}$ positive numbers, such that

(i) $Q_{\overline{r},2\overline{M}}\left(\overline{x}\right) \cap \partial\Omega_1 = Q_{\overline{r},2\overline{M}}\left(\overline{x}\right) \cap \partial\Omega_2 \subset \overset{\circ}{\Gamma}{}^{(a)}$,
(ii) $Q_{\overline{r},2\overline{M}}\left(\overline{x}\right) \cap \Omega_1 = Q_{\overline{r},2\overline{M}}\left(\overline{x}\right) \cap \Omega_2$ and they are connected.

Now we prove that

$$G \subsetneq G \cup \left(Q_{\overline{r},2\overline{M}}\left(\overline{x}\right) \cap \Omega_1\right) \subset \Omega_1 \cap \Omega_2 \tag{10.42}$$

and

$$G \cup \left(Q_{\overline{r},2\overline{M}}\left(\overline{x}\right) \cap \Omega_1\right) \quad \text{is connected.} \tag{10.43}$$

From these we will arrive to a contradiction.

First, by $G \subset \Omega_1 \cap \Omega_2$ we have

$$G \cap \left(Q_{\overline{r},2\overline{M}}\left(\overline{x}\right) \cap \Omega_1\right) \subset \Omega_1 \cap \Omega_2. \tag{10.44}$$

On the other hand, since, by (10.41) and $\overline{x} \in \partial\left(\Omega_1 \setminus \overline{G}\right)$, we obtain

$$\emptyset \subsetneq Q_{\overline{r},2\overline{M}}\left(\overline{x}\right) \cap \left(\Omega_1 \setminus \overline{G}\right) \subset \left(Q_{\overline{r},2\overline{M}}\left(\overline{x}\right) \cap \Omega_1\right) \setminus G.$$

Hence

$$G \subsetneq G \cup \left(Q_{\overline{r},2\overline{M}}\left(\overline{x}\right) \cap \Omega_1\right)$$

and (10.42) is proved.

In order to prove that the set $G \cup \left(Q_{\overline{r},2\overline{M}}\left(\overline{x}\right) \cap \Omega_1\right)$ is connected, we first of all check that

$$\left(Q_{\overline{r},2\overline{M}}\left(\overline{x}\right) \cap \Omega_1\right) \cap G \neq \emptyset, \tag{10.45}$$

As a matter of fact, by $\overline{x} \in \partial\left(\Omega_1 \setminus \overline{G}\right)$ we have

$$\begin{aligned} \emptyset \subsetneq \left(Q_{\overline{r},2\overline{M}}\left(\overline{x}\right) \cap \Omega_1\right) \cap \left[\mathbb{R}^n \setminus \left(\Omega_1 \setminus \overline{G}\right)\right] = \\ = \left(Q_{\overline{r},2\overline{M}}\left(\overline{x}\right) \cap \Omega_1\right) \cap \left[\left(\mathbb{R}^n \setminus \Omega_1\right) \cup \left(\Omega_1 \cap \overline{G}\right)\right] = \\ = \left(Q_{\overline{r},2\overline{M}}\left(\overline{x}\right) \cap \Omega_1\right) \cap \overline{G}. \end{aligned}$$

Hence

$$\left(Q_{\overline{r},2\overline{M}}\left(\overline{x}\right) \cap \Omega_1\right) \cap \overline{G} \neq \emptyset$$

and by Proposition 10.3 we get (10.45).

At this point, in order to prove that $\left(Q_{\overline{r},2\overline{M}}\left(\overline{x}\right) \cap \Omega_1\right) \cup G$ is connected, we need only recall (ii) and to proceed as in the proof of the Claim in Step I. Therefore (10.43) is proved. Now, since

$$\overline{G \cup \left(Q_{\overline{r},2\overline{M}}\left(\overline{x}\right) \cap \Omega_1\right)} \supset \Gamma^{(a)},$$

by (10.42), (10.43) and by the definition of G, we arrive to a contradiction. Hence (10.40) holds, which combined with (10.31) implies

$$\begin{aligned}\partial\left(\Omega_1\setminus G\right)\subset\left(\partial\Omega_1\cup\partial\Omega_2\right)\setminus\overset{\circ}{\Gamma^{(a)}}&=\\=\left(\partial\Omega_1\setminus\overset{\circ}{\Gamma^{(a)}}\right)\cup\left(\partial\Omega_2\setminus\overset{\circ}{\Gamma^{(a)}}\right)&=\\=\Gamma_1^{(i)}\cup\Gamma_2^{(i)},&\end{aligned}$$

which gives (10.39).

Step III We conclude the proof of (10.30). Let

$$x\in\partial\left(\Omega_1\setminus\overline{G}\right).$$

Let us distinguish two cases:

(j) $x\in\mathbb{R}^n\setminus\overline{\Omega}_1$,
(jj) $x\in\overline{\Omega}_1$.

Case (j) cannot occur, because if $x\in\mathbb{R}^n\setminus\overline{\Omega}_1$ then there exists $r>0$ such that $B_r(x)\subset\mathbb{R}^n\setminus\overline{\Omega}_1$, hence

$$B_r(x)\cap\left(\Omega_1\setminus\overline{G}\right)\subset B_r(x)\cap\Omega_1=\emptyset,$$

but this cannot hold because $x\in\partial\left(\Omega_1\setminus\overline{G}\right)$.

Let us consider case (jj). If $x\in\Gamma_1^{(i)}$ then trivially $x\in\Gamma_1^{(i)}\cup\left(\Gamma_2^{(i)}\cap\partial G\right)$. Instead, if $x\notin\Gamma_1^{(i)}$ then, by (10.39), we have

$$x\in\Gamma_2^{(i)}. \tag{10.46}$$

Moreover, since, by (10.40), we have $x\notin\overset{\circ}{\Gamma^{(a)}}$ and since, by (10.8), we have $\partial\Omega_1=\Gamma_1^{(i)}\cup\overset{\circ}{\Gamma^{(a)}}$, we get $x\notin\partial\Omega_1$ and, taking into account that $x\in\overline{\Omega}_1$, we have $x\in\Omega_1$. Hence there exists $r>0$ such that

$$B_r(x)\subset\Omega_1. \tag{10.47}$$

Now, let $s\in(0,r]$ be arbitrary. Since $x\in\partial\left(\Omega_1\setminus\overline{G}\right)$ we have

$$\begin{aligned}\emptyset\subsetneq B_s(x)\cap\left[\mathbb{R}^n\setminus\left(\Omega_1\setminus\overline{G}\right)\right]&=\\=B_s(x)\cap\left(\Omega_1\cap\overline{G}\right).&\end{aligned}$$

Hence

$$B_s(x)\cap\overline{G}\neq\emptyset,\quad\forall s\in(0,r], \tag{10.48}$$

therefore

$$x \in \overline{G}. \tag{10.49}$$

On the other hand, since $x \in \partial \left(\Omega_1 \setminus \overline{G}\right)$, we have for every $s > 0$

$$\emptyset \subsetneq B_s(x) \cap \left(\Omega_1 \setminus \overline{G}\right) \subset B_s(x) \cap \left(\mathbb{R}^n \setminus G\right).$$

Therefore

$$x \in \overline{\mathbb{R}^n \setminus G},$$

that combined with (10.49) gives $x \in \partial G$, which in turn combined with (10.46) concludes proof. ■

Chapter 11
The Hadamard Example: Solvability of the Cauchy Problem and Continuous Dependence by the Data

Abstract In this chapter, we explore continuous dependence on initial data for the Cauchy problem. We start with Hadamard's example, which shows that small errors in initial data for the Laplace equation can lead to large effects on the solution. We then prove the Lax-Mizohata Theorem, establishing the connection between solvability and continuous dependence on initial data.

11.1 Introduction

In this brief chapter, we initiate the study of continuous dependence on initial data for partial differential equations for the Cauchy problem. We first introduce the Hadamard famous example, [54], which heuristically shows that for the Laplace equation, even small errors in the initial data of a Cauchy problem can lead to uncontrollable effects on the solution. Following this, we provide a proof of the Lax-Mizohata Theorem, which highlights the deeper connection between the solvability of the Cauchy problem and continuous dependence on the initial data.

11.2 The Hadamard Example

We present the Hadamard example relating to the Cauchy problem.

Let us consider the following Cauchy problem

$$\begin{cases} \partial_x^2 u + \partial_y^2 u = 0 \\ u(x,0) = \varphi(x), & \text{for } x \in (0,1) \\ \partial_y u(x,0) = \psi(x), & \text{for } x \in (0,1) \end{cases} . \tag{11.1}$$

If φ and ψ are analytic, by the Cauchy–Kovalevskaya Theorem, there exists a unique analytic solution to problem (11.1). Such a solution, by the Holmgren Theorem, is

S. Vessella, *Unique Continuation Properties for Partial Differential Equations*, Birkhäuser Advanced Texts Basler Lehrbücher,
https://doi.org/10.1007/978-3-031-86366-0_11

the unique solution of class C^2 in a neighborhood of $(0, 1) \times \{0\}$. Now, let

$$\varphi_\nu(x) = e^{-\sqrt{\nu}} \sin \nu x, \quad \text{and} \quad \psi_\nu(x) = 0, \quad \nu \in \mathbb{N}.$$

It is immediately checked that

$$u_\nu(x, y) = e^{-\sqrt{\nu}} \sin \nu x \sinh \nu y, \quad \nu \in \mathbb{N},$$

is the solution to the Cauchy problem

$$\begin{cases} \partial_x^2 u_\nu + \partial_y^2 u_\nu = 0 \\ u_\nu(x, 0) = \varphi_\nu(x), & \text{for } x \in (0, 1) \\ \partial_y u_\nu(x, 0) = \psi_\nu(x), & \text{for } x \in (0, 1) \end{cases} . \tag{11.2}$$

Let us note that for every $k \in \mathbb{N}_0$

$$\sup_{(0,1)} \left| \frac{d^k \varphi_\nu}{dx^k} \right| \to 0, \quad \text{as } \nu \to \infty,$$

on the other hand, for every $a, b \in (0, 1)$, $a < b$ and for every $\delta > 0$ we have

$$\sup_{[a,b]\times[-\delta,\delta]} |u_\nu| \to +\infty, \quad \text{as } \nu \to \infty.$$

In essence, **"small errors" in the data of Cauchy problem** (11.1) **result in unmanageable errors in the solution**. This phenomenon renders problem (11.1) highly challenging for practical applications. In applied problems, the data—here, the initial data—are typically acquired through measurements, which inherently involve approximations. Consequently, in order to make the mathematical solution practically applicable, its dependence on the data must be continuous.

In a broad way, we may present the notion of **well-posed problem in the sense of Hadamard** as follows. Let X and Y be two metric spaces and be

$$A : X \to Y,$$

a map from X to Y. Let us consider the problem of determining $x \in X$ such that

$$A(x) = f, \tag{11.3}$$

where $f \in Y$.

We say that problem (11.3) is well-posed in the sense of Hadamard provided that we have:

1. **(Existence)** for any $f \in Y$, there exists at least one $x \in X$ such that $A(x) = f$.

2. **(Uniqueness)** for any $x_1, x_2 \in X$ which satisfy $A(x_1) = A(x_2)$ we have $x_1 = x_2$;
3. **(Continuous dependence by the data)** let us suppose that condition 2 is satisfied, then the map

$$A^{-1} : A(X) \to X,$$

is continuous ($A(X)$ with the topology induced by Y).

In the next section we will see that in the Cauchy problem there is an interesting relationship between the first two points above (existence and uniqueness) and the third point (continuous dependence by the data).

11.3 Solvability of the Cauchy Problem and Its Relations to the Continuous Dependence on the Data

Let

$$P(x, \partial) = \sum_{|\alpha| \leq m} a_\alpha(x) \partial^\alpha \tag{11.4}$$

be a linear differential operator of order m and $a_\alpha \in C^\infty(\mathbb{R}^n, \mathbb{C})$, for $|\alpha| \leq m$, we say that the Cauchy problem with initial surface $\{x_n = 0\}$ for $P(x, \partial)$ enjoys the **local uniqueness property (in** 0) provided there exists $\delta > 0$ such that we have: if $u \in C^m\left(\overline{B_\delta}\right)$ satisfies

$$\begin{cases} P(x, \partial)u = 0, & \text{in} \quad B_\delta \\ \\ \partial_n^j u(x', 0) = 0, & \text{for } j = 0, 1, \cdots, m-1, \quad \forall x' \in B'_\delta \end{cases}, \tag{11.5}$$

then

$$u \equiv 0 \text{ in } B_\delta.$$

For instance, if the coefficients of $P(x, \partial)$ are analytic (in a neighborhood of 0) and $p_m(0, e_n) \neq 0$, the Holmgren Theorem implies that the **local uniqueness property** is satisfied.

We say that the following Cauchy problem

$$\begin{cases} P(x, \partial)u = 0 \\ \\ \partial_n^j u(x', 0) = g_j(x'), \quad \text{for } j = 0, 1, \cdots, m-1 \end{cases}, \tag{11.6}$$

is **locally solvable (in the origin)** in C^∞ if the following occurs:
for every open neighborhood $\mathcal{U}$ of 0 in $\mathbb{R}^{n-1}$ and for every

$$g = (g_0, g_1, \cdots, g_{m-1}) \in C^\infty(\mathcal{U}, \mathbb{C}^m),$$

there exists $\mathcal{V}$, open neighborhood of 0 in $\mathbb{R}^n$ and there exists $u \in C^\infty(\mathcal{V})$ such that

$$\begin{cases} P(x, \partial)u = 0, & \text{in} \quad \mathcal{V} \\ \\ \partial_n^j u = g_j, & \text{for } j = 0, 1, \cdots, m-1, \quad \text{in } \mathcal{V} \cap \{x_n = 0\} \end{cases} . \tag{11.7}$$

The definitions of the **local uniqueness property** and of the **local solvability in the origin** of the Cauchy problem with initial surface the hyperplane

$$\left\{x \in \mathbb{R}^n : N \cdot x = 0\right\},$$

where N is a unit vector of $\mathbb{R}^n$, is formulated in obvious way or, simply, by reconducting them to the case $N = e_n$ by means of an isometry of $\mathbb{R}^n$.

The subsequent pair of theorems, along with their immediate implications, are well-recognized in the literature as the Lax–Mizohata Theorem. Here, we present them with a slight modification. Specifically, Theorem 11.1 is attributed to Lax, while Theorem 11.2 is credited to Mizohata. These theorems are extensively discussed in works such as [95].

Theorem 11.1 *Let $P(x, \partial)$ be operator* (11.4) *whose coefficients belong to $C^\infty(\mathbb{R}^n, \mathbb{C})$. Suppose that $P(x, \partial)$ enjoys the* ***local uniqueness property*** *and that Cauchy problem* (11.6) *is* ***locally solvable*** *in C^∞. Then for every $\mathcal{U}$, neighborhood of 0 in $\mathbb{R}^{n-1}$, there exists $r > 0$ such that for every*

$$g = (g_0, g_1, \cdots, g_{m-1}) \in C^\infty\left(\mathcal{U}, \mathbb{C}^m\right),$$

there exists a unique $u \in C^m\left(\overline{B_r}\right)$ such that

$$\begin{cases} P(x, \partial)u = 0, & \text{in } B_r \\ \\ \partial_n^j u = g_j, & \text{for } j = 0, 1, \cdots, m-1, \quad \text{in } B'_r \end{cases} . \tag{11.8}$$

Remark 11.1 Let us observe that in Theorem 11.1, unlike in the definition of local solvability, r, hence the neighborhood of 0, B_r, **does not** depend on g. Of course r depends on $\mathcal{U}$. ♦

Proof Let $\mathcal{U} \in \mathbb{R}^{n-1}$ be a neighborhood of 0 and let $\sigma \in (0, 1)$ be fixed. Since Cauchy problem (11.6) is locally solvable in 0, by local uniqueness property we have that, for every

$$g = (g_0, g_1, \cdots, g_{m-1}) \in C^\infty\left(\mathcal{U}, \mathbb{C}^m\right),$$

there exists $\delta > 0$ such that $B'_\delta \subset \mathcal{U}$ and such that for every $\rho \in (0, \delta]$ there exists a unique solution $u \in C^{m,\sigma}\left(\overline{B_\rho}\right)$ of problem $\mathcal{P}_{g,\rho}$:

$$(\mathcal{P}_{g,\rho}) \qquad \begin{cases} P(x, \partial)u = 0, & \text{in } B_\rho \\ \\ \partial_n^j u = g_j, & \text{for } j = 0, 1, \cdots, m-1, \quad \text{in } B'_\rho \end{cases}.$$

Let $\{\rho_k\}$ be a strictly decreasing sequence such that

$$\lim_{k\to\infty} \rho_k = 0.$$

For any $k, M \in \mathbb{N}$, let us consider the sets

$$\mathcal{A}_{k,M} =$$
$$= \left\{g \in C^\infty(\mathcal{U}, \mathbb{C}^m) : \text{ there exists } u \text{ solution to } \mathcal{P}_{g,\rho_k} \text{ and } \|u\|_{C^{m,\sigma}\left(\overline{B_{\rho_k}}\right)} \le M\right\}.$$

It is evident that $\mathcal{A}_{k,M}$ is symmetric, for any $k, M \in \mathbb{N}$ ($g \in \mathcal{A}_{k,M} \Rightarrow -g \in \mathcal{A}_{k,M}$) and convex for every $k, M \in \mathbb{N}$.

Step 1 Let us check that

$$C^\infty\left(\mathcal{U}, \mathbb{C}^m\right) = \bigcup_{k,M\in\mathbb{N}} \mathcal{A}_{k,M}. \tag{11.9}$$

Of course, it suffices to check " $\subset$". Let $g \in C^\infty(\mathcal{U}, \mathbb{C}^m)$. By the local solvability there exists $\mathcal{V} \subset \mathbb{R}^n$, neighborhood of 0, such that Cauchy problem (11.7) admits a solution $u \in C^\infty(\mathcal{V})$. It is enough then to choose $k \in \mathbb{N}$ such that $\overline{B_{\rho_k}} \subset \mathcal{V}$ and M such that $M \ge \|u\|_{C^{m,\sigma}\left(\overline{B_{\rho_k}}\right)}$ and we have $g \in \mathcal{A}_{k,M}$.

Step 2 Now we prove that for every $k, M \in \mathbb{N}$, $\mathcal{A}_{k,M}$ is closed in $C^\infty\left(\mathcal{U}, \mathbb{C}^m\right)$ equipped with the topology induced by the metric d_∞.

We fix $k, M \in \mathbb{N}$ and let $\{g_\nu\}_{\nu\in\mathbb{N}}$ be a sequence in $\mathcal{A}_{k,M}$ such that

$$\{g_\nu\} \to \widetilde{g}, \qquad \text{in } C^\infty(\mathcal{U}, \mathbb{C}^m). \tag{11.10}$$

let us check that $\widetilde{g} \in \mathcal{A}_{k,M}$.

First of all, we have

$$\widetilde{g} \in C^\infty(\mathcal{U}, \mathbb{C}^m). \tag{11.11}$$

Moreover, denoting by u_ν the solution of problem $\mathcal{P}_{g_\nu,\rho_k}$ we have, by the definition of $\mathcal{A}_{k,M}$,

$$\|u_\nu\|_{C^{m,\sigma}\left(\overline{B_{\rho_k}}\right)} \leq M, \quad \forall \nu \in \mathbb{N}.$$

Hence, by the Arzelà–Ascoli Theorem there exists a subsequence $\left\{u_{\nu_q}\right\}$ of $\{u_\nu\}$ and $\widetilde{u} \in C^{m,\sigma}\left(\overline{B_{\rho_k}}\right)$ such that

$$u_{\nu_q} \to \widetilde{u}, \quad \text{as } q \to \infty, \text{ in } C^m\left(\overline{B_{\rho_k}}\right). \tag{11.12}$$

Moreover we have

$$\|\widetilde{u}\|_{C^{m,\sigma}\left(\overline{B_{\rho_k}}\right)} \leq M. \tag{11.13}$$

Concerning the justification of the latter inequality, we notice that the sequence $\left\{u_{\nu_q}\right\}$ converges to $\widetilde{u}$ in $C^m\left(\overline{B_{\rho_k}}\right)$, but not necessarily it converges in $C^{m,\sigma}\left(\overline{B_{\rho_k}}\right)$, nevertheless (11.13) holds, since for $x \neq y \in \overline{B_{\rho_k}}$ we have, for any $q \in \mathbb{N}$,

$$\sum_{|\alpha|=m} \frac{\left|\partial^\alpha u_{\nu_q}(x) - \partial^\alpha u_{\nu_q}(y)\right|}{|x-y|^\sigma} \leq M - \left\|u_{\nu_q}\right\|_{C^m\left(\overline{B_{\rho_k}}\right)};$$

hence, from the punctual convergence of $\left\{\partial^\alpha u_{\nu_q}\right\}$ for $|\alpha| = m$ and by

$$\left\|u_{\nu_q}\right\|_{C^m\left(\overline{B_{\rho_k}}\right)} \to \|u\|_{C^m\left(\overline{B_{\rho_k}}\right)}, \quad \text{as } q \to \infty,$$

we obtain (11.13).

By (11.12) we have easily

$$0 = P(x, \partial)u_{\nu_q} \to P(x, \partial)\widetilde{u}, \quad \text{as } q \to \infty, \text{ in } C^0\left(\overline{B_{\rho_k}}\right).$$

Therefore

$$P(x, \partial)\widetilde{u} = 0, \quad \text{in } \overline{B_{\rho_k}}. \tag{11.14}$$

By (11.10) and (11.12) we have, for $j = 0, 1, \cdots, m-1$,

$$\partial_n^j u(x', 0) = \lim_{q\to\infty} \partial_n^j u_{\nu_q} u(x', 0) = \lim_{q\to\infty} g_{j,\nu_q}(x') = \widetilde{g}, \quad \forall x' \in \overline{B'_{\rho_k}}. \tag{11.15}$$

By (11.11), (11.13)–(11.15) we get $\widetilde{g} \in \mathcal{A}_{k,M}$.

Step 3 Now, recalling that $C^\infty(\mathcal{U}, \mathbb{C}^m)$ is a Frechét space, by Theorem 3.84 we have that there exist $k_0, M_0 \in \mathbb{N}$ such that

$$\operatorname{Int}\left(\mathcal{A}_{k_0,M_0}\right) \neq \emptyset. \tag{11.16}$$

To prove (11.16) we argue by contradiction. Let us assume (11.16) does not hold. Consequently we have

$$\mathrm{Int}\left(\mathcal{A}_{k,M}\right) = \emptyset, \quad \forall k, M \in \mathbb{N},$$

and recalling that $\mathcal{A}_{k,M}$ is closed for every $k, M \in \mathbb{N}$, (11.9) and Theorem 3.84 imply

$$C^{\infty}\left(\mathcal{U}, \mathbb{C}^m\right) = \mathrm{Int}\left(C^{\infty}\left(\mathcal{U}, \mathbb{C}^m\right)\right) = \mathrm{Int}\left(\bigcup_{k,M\in\mathbb{N}} \mathcal{A}_{k,M}\right) = \emptyset.$$

This is clearly a contradiction, then (11.16) needs to hold.

On the other hand, (11.16) implies that there exists $\psi \in \mathrm{Int}\left(\mathcal{A}_{k_0,M_0}\right)$, and since $\mathcal{A}_{k_0,M_0}$ is symmetric and convex, we have $-\psi \in \mathrm{Int}\left(\mathcal{A}_{k_0,M_0}\right)$ therefore

$$0 = \frac{1}{2}(\psi - \psi) \in \mathrm{Int}\left(\mathcal{A}_{k_0,M_0}\right).$$

All in all, we have

$$0 \in \mathrm{Int}\left(\mathcal{A}_{k_0,M_0}\right).$$

Consequently there exists $\mathcal{W}_0 \subset \mathrm{Int}\left(\mathcal{A}_{k_0,M_0}\right)$, where $\mathcal{W}_0$ is an element of local base of neighborhood of 0 in $C^{\infty}(\mathcal{U}, \mathbb{C}^m)$, hence it is of the type

$$\mathcal{W}_0 = \bigcap_{|\alpha|\leq h, 1\leq j\leq h'} \left\{ f \in C^{\infty}(\mathcal{U}, \mathbb{C}^m) : \quad \max_{K_j} |\partial^{\alpha} f| < \varepsilon_{\alpha,j} \right\},$$

where $h, h', \varepsilon_{\alpha,j}$ are suitable positive numbers and $\{K_j\}_{j\in\mathbb{N}}$ is a family of compact subset of $\mathcal{U}$ which satisfies $K_j \neq \emptyset$, $K_j \subset \overset{\circ}{K}_{j+1}$ for every $j \in \mathbb{N}$ and

$$\bigcup_{j=1}^{\infty} K_j = \mathcal{U}.$$

We recall that the local uniqueness property implies that there exists $r > 0$, that we may assume less or equal to ρ_{k_0}, such that if $w \in C^m\left(\overline{B_r}\right)$ satisfies

$$\begin{cases} P(x, \partial)w = 0, & \text{in } B_r \\ \\ \partial_n^j w = 0, & \text{for } j = 0, 1, \cdots, m-1, \quad \text{in } B_r' \end{cases},$$

then we have

$$w \equiv 0 \text{ in } B_r.$$

Now, let $g \in C^{\infty}(\mathcal{U})$. Then there exists trivially $\lambda > 0$ such that

$$\lambda^{-1} g \in \mathcal{W}_0 \subset \operatorname{Int}\left(\mathcal{A}_{k_0, M_0}\right)$$

and by the definition of $\mathcal{A}_{k_0,M_0}$ we have there exists $v \in C^m\left(\overline{B_{\rho_{k_0}}}\right)$ (actually, $v \in C^{m,\sigma}\left(\overline{B_{\rho_{k_0}}}\right)$) solution to $\mathcal{P}_{\lambda^{-1}g,\rho_{k_0}}$. Hence $v_{|B_r}$ is the unique solution in $C^m\left(\overline{B_r}\right)$ to the Cauchy problem

$$\begin{cases} P(x,\partial)v = 0, & \text{in } B_r \\ \\ \partial_n^j v = \lambda^{-1} g_j, & \text{for } j = 0, 1, \cdots, m-1, \quad \text{in } B_r' \end{cases}.$$

Therefore, $u = \lambda v$ is the unique solution in $C^m\left(\overline{B_r}\right)$ to the Cauchy problem

$$\begin{cases} P(x,\partial)u = 0, & \text{in } B_r \\ \\ \partial_n^j u = g_j & \text{for } j = 0, 1, \cdots, m-1, \quad \text{in } B_r' \end{cases}.$$

■

Remark 11.2 It is evident from the proof of Theorem 11.1 that if in the definition of local solvability we require the existence of u in $C^{m'}(\mathcal{V})$ with $m' > m$, then we reach the same conclusions. ♦

Theorem 11.2 (Lax–Mizohata) *Suppose that $P(x, \partial)$ satisfies the same assumption of Theorem 11.1. Let $\mathcal{U} \subset \mathbb{R}^{n-1}$ be a neighborhood of 0 and let $r > 0$ be defined in Theorem 11.1. Let Λ be the map*

$$\Lambda : C^{\infty}\left(\mathcal{U}, \mathbb{C}^m\right) \to C^m\left(\overline{B_r}\right),$$

$$C^{\infty}\left(\mathcal{U}, \mathbb{C}^m\right) \ni g \to \Lambda(g) = u \textit{ solution to the Cauchy problem}:$$

$$\begin{cases} P(x,\partial)u = 0, & \textit{in } B_r \\ \\ \partial_n^j u = g_j, & \textit{for } j = 0, 1, \cdots, m-1, \quad \textit{in } B_r' \end{cases}. \tag{11.17}$$

Then Λ is continuous.

Proof Let us prove that the graph of Λ is closed and then we apply Theorem 3.87. Let $\{(g_\nu, u_\nu)\}_{\nu\in\mathbb{N}}$ be a sequence in $C^\infty(\mathcal{U}, \mathbb{C}^m) \times C^m\left(\overline{B_r}\right)$ which satisfies

$$\begin{cases} P(x, \partial)u_\nu = 0, & \text{in } B_r \\ \\ \partial_n^j u_\nu = g_{\nu,j}, & \text{for } j = 0, 1, \cdots, m-1, \quad \text{in } B_r' \end{cases} \tag{11.18}$$

and

$$\{(g_\nu, u_\nu)\} \to (g, v), \quad \text{in } C^\infty\left(\mathcal{U}, \mathbb{C}^m\right) \times C^m\left(\overline{B_r}\right). \tag{11.19}$$

Then

$$P(x, \partial)v = \lim_{\nu\to\infty} P(x, \partial)u_\nu = 0$$

and, for $0 \leq j \leq m-1$,

$$\partial_n^j v(x', 0) = \lim_{\nu\to\infty} \partial_n^j u_\nu(x', 0) = \lim_{\nu\to\infty} g_{\nu,j}(x') = g_j(x').$$

Hence v solves Cauchy problem (11.17) and by Theorem 11.1, we have

$$v = \Lambda g.$$

Therefore the graph of Λ is closed, hence Theorem 3.87 implies that Λ is continuous. ■

From Theorem 11.2 and the Hadamard example, it becomes evident (once again) that the Cauchy problem for the Laplace equation, denoted as (11.1), lacks the local solvability in C^∞. To clarify further, the Holmgren Theorem establishes that this Cauchy problem for the Laplace equation possesses the property of local uniqueness in C^2, yet, as the Hadamard example proves, the map Λ defined for this Cauchy problem is not continuous.

Similarly, Theorem 11.2 can be employed to derive necessary conditions for the local solvability of the Cauchy problem. In this discussion, we will focus solely on the case of operators with constant coefficients that are equivalent to their principal parts. For the broader context of operators with constant coefficients, including general cases, we refer interested readers to sources such as [56, Ch.5, Sect.4] and [58, Ch.12, Sect.3] and to [58, Ch.23, Sect.3] for the operators with C^∞ coefficients, warning the reader, however, that (especially in [58, Ch.23, Sect. 3]) quite advanced tools are used.

Let

$$p_m(\xi) = \sum_{|\alpha|=m} a_\alpha \xi^\alpha \tag{11.20}$$

be a **homogeneous polynomial** of degree m with coefficients $a_\alpha \in \mathbb{C}$, for $|\alpha| = m$ and let N be a unit vector of $\mathbb{R}^n$. We say that p_m is hyperbolic with respect to the direction N provided we have

(a) $p_m(N) \neq 0$
(b) for every $\xi \in \mathbb{R}^n$ the algebraic equation in z

$$p_m(\xi + zN) = 0,$$

has real roots only.

Theorem 11.3 *Let N be a unit vector of $\mathbb{R}^n$ and let $P_m(\partial)$ be the differential operator with constant coefficients*

$$P_m(\partial) = \sum_{|\alpha|=m} a_\alpha \partial^\alpha. \tag{11.21}$$

Suppose that the Cauchy problem for operator (11.21) *with initial surface*

$$\left\{x \in \mathbb{R}^n : N \cdot x = 0\right\}, \tag{11.22}$$

enjoys the local solvability and the local uniqueness property.

Then the polynomial $p_m(\xi)$ is hyperbolic w.r.t. the direction N.

Proof We have already seen, in Sect. 9.6.2 that by the local uniqueness for the Cauchy problem for operator (11.21) with initial surface (11.22), we have

$$p_m(N) \neq 0. \tag{11.23}$$

We argue by contradiction to prove that, for every $\xi \in \mathbb{R}^n$, there exist real roots only to the equation

$$p_m(\xi + zN) = 0. \tag{11.24}$$

Let us suppose that there exist $\xi_0 \in \mathbb{R}^n$ and $z_0 = \Re z_0 + i\Im z_0$, such that $\Im z_0 \neq 0$ and

$$p_m(\xi_0 + z_0 N) = 0. \tag{11.25}$$

Let us denote by

$$\eta = \xi_0 + \Re z_0 N, \quad \text{and} \quad \tau = \Im z_0$$

and by

$$u_\nu(x) = \exp\left\{-|\nu|^{1/2} + \nu\left(i\eta \cdot x - \tau N \cdot x\right)\right\}, \quad \nu \in \mathbb{Z}.$$

We have, by (11.25) (recalling that $\xi_0+z_0N=\eta+i\tau N$ and $P_m(\xi)$ is a homogeneous polynomial)

$$P_m(\partial)u_\nu(x)=u_\nu(x)p_m\left(\nu\left(i\eta-\tau N\right)\right)= \\ =u_\nu(x)(i\nu)^m p_m\left(\eta+i\tau N\right)=0. \tag{11.26}$$

Set

$$g_{\nu,j}=\frac{\partial^j}{\partial N^j}u_\nu, \quad \text{for } \nu\in\mathbb{Z} \quad \text{and } j=0,1\cdots,m-1,$$

we have trivially that u_ν solves the Cauchy problem

$$\begin{cases} P_m(x,\partial)u_\nu=0, & \text{in } \mathbb{R}^n \\ \\ \frac{\partial^j}{\partial N^j}u_\nu=g_{\nu,j}, & \text{for } j=0,1,\cdots,m-1, \quad N\cdot x=0 \end{cases}.$$

On the other hand it is easy to check that

$$g_\nu\rightarrow 0, \quad \text{as } |\nu|\rightarrow+\infty, \text{ in } C^\infty.$$

Moreover, if $\tau N\cdot x>0$ then we have

$$|u_\nu(x)|\rightarrow+\infty, \quad \text{as } \nu\rightarrow-\infty$$

and if $\tau N\cdot x<0$ then we have

$$|u_\nu(x)|\rightarrow+\infty, \quad \text{as } \nu\rightarrow+\infty.$$

Hence, the map Λ defined in Theorem 11.2 is not continuous. Since $P_m(\partial)$ enjoys the local uniqueness property, $P_m(\partial)$ cannot enjoy the local solvability. Thus we have a contradiction. Therefore equation (11.24) has real roots only. ■

If the polynomial $p(\xi)$ is not homogeneous, one can prove (with more efforts) a theorem which is similar to Theorem 11.3 that we merely state here (see [56, Ch.5, Sect. 4] , [58, Ch.12, Sect. 3]).

Theorem 11.4 *Let $p(\xi)$ be a polynomial of degree m, N a unit vector of $\mathbb{R}^n$ and $P(\partial)$ the following differential operator with constant coefficient*

$$P(\partial)=\sum_{|\alpha|\leq m}a_\alpha\partial^\alpha, \tag{11.27}$$

where $a_\alpha \in \mathbb{C}$ *for* $|\alpha| \leq m$*. Let us suppose that the Cauchy problem for operator* (11.27) *with initial surface*

$$\left\{x \in \mathbb{R}^n : N \cdot x = 0\right\},$$

enjoys local solvability property and let us suppose

$$p_m(N) \neq 0, \tag{11.28}$$

where p_m *is the principal part of* p*.*

Then

$$\exists \tau_0 \in \mathbb{R} \text{ such that } \forall \xi \in \mathbb{R}^n \text{ and } \forall \tau < \tau_0 \quad p(i(\xi + i\tau N)) \neq 0. \tag{11.29}$$

A polynomial that enjoys properties (11.28) and (11.29) is called **hyperbolic polynomial w.r.t. the direction** N. It is not difficult to prove (see the literature quoted above) that in the case where $p(\xi)$ is a homogeneous polynomial the two definitions of hyperbolicity coincide. Moreover, if $p(\xi)$ is a hyperbolic polynomial (not necessarily homogeneous), denoted by $p_m(\xi)$ its principal part, then we have

(a) $p_m(N) \neq 0$,
(b) for every $\xi \in \mathbb{R}^n$ the equation in z

$$p_m(\xi + zN) = 0,$$

has real roots only.

However, if the principal part of $p(\xi)$ satisfies (a) and (b) it does not imply that $P(\partial)$ is hyperbolic w.r.t. N. It can be proved that if the equation $p_m(\xi + zN) = 0$, has **simple real roots** only, then the hyperbolicity condition with respect to a direction N is also sufficient for the local solvability of the Cauchy problem with initial surface $\{N \cdot x = 0\}$.

11.4 Concluding Remarks

1. In this chapter we have considered the properties of the local uniqueness and the local solvability for the Cauchy problem, but with simple and natural changes one could consider the same properties for the one-sided Cauchy problem. Actually, it suffices to replace each neighborhood of $0 \in \mathbb{R}^n$, say $\mathcal{V}$, by $\mathcal{V}_+ = \mathcal{V} \cap \mathbb{R}^{n-1} \times [0, +\infty)$, and in defining the seminorms of $C^\infty(\mathcal{V}_+)$ we consider a family of compacts $K_j \subset \text{Int}_{\mathcal{V}_+}(K_{j+1})$ for every $j \in \mathbb{N}$ and $\bigcup_{j=1}^\infty K_j = \mathcal{V}_+$. In this way provided that the appropriate modifications are introduced, Theorems 11.2 and 11.1 preserve their validity, in particular, the ball B_r should be replaced with the half-ball $B_r \cap \left(\mathbb{R}^{n-1} \times [0, +\infty)\right)$.

2. In Theorem 11.2 we have seen that if the properties of the local uniqueness and of the local solvability in C^∞ hold true then the map Λ, defined in Theorem 11.2, is continuous. In what follows, we check that if the operator $P(x, \partial)$ has analytic coefficients, the converse also holds (in a sense).

Let us suppose that

$$p_m(0, e_n) \neq 0,$$

and let us suppose that there exists $r > 0$ such that only the null function solves (in $C^\infty(B_r)$) the Cauchy problem

$$\begin{cases} P(x, \partial)u = 0, & \text{in } B_r \\ \partial_n^j u = 0, & \text{for } j = 0, 1, \cdots, m-1, \quad \text{in } B'_r \end{cases}. \tag{11.30}$$

Then by the Cauchy-Kovalevskaya Theorem, there exists $\widetilde{r} \leq r$ such that the following map $\widetilde{\Lambda}$ is well-defined

$$C^\omega\left(B'_r, \mathbb{C}^m\right) \ni g \to \widetilde{\Lambda}(g) = u \in C^\omega(B_{\widetilde{r}}) \text{ solution to the Cauchy problem:}$$

$$\begin{cases} P(x, \partial)u = 0, & \text{in } B_{\widetilde{r}} \\ \partial_n^j u = g_j, & \text{for } j = 0, 1, \cdots, m-1, \quad \text{in } B'_r \end{cases}. \tag{11.31}$$

We prove what follows: let us assume that $\widetilde{\Lambda}$ is a continuous map provided that we equip the spaces $C^\omega\left(B'_r, \mathbb{C}^m\right)$ and $C^\omega(B_{\widetilde{r}})$ with the metric d_∞, we have that Cauchy problem (11.31) satisfies the local solvability property in C^∞ (i.e. if the initial data of the Cauchy belong to C^∞ there exist solutions to the Cauchy problem).

Let us suppose that $\widetilde{\Lambda}$ is continuous and let $g \in C^\infty\left(B'_r, \mathbb{C}^m\right)$. Let $\left\{g_{,\nu}\right\}$ be the sequence in $C^\omega\left(B'_r, \mathbb{C}^m\right)$ defined as follows

$$g_{,\nu}(x') = \left(\frac{\nu}{2\pi}\right)^{(n-1)/2} \int_{B'_r} e^{-\frac{\nu|x'-y'|^2}{2}} g(x')dx'.$$

The sequence $\left\{g_{,\nu}\right\}$ converges to g in $C^\infty\left(B'_r, \mathbb{C}^m\right)$ and since $g_{,\nu} \in C^\omega\left(B'_r, \mathbb{C}^m\right)$, for every $\nu \in \mathbb{N}$, the Cauchy-Kovalevskaya Theorem yields the existence of a solution u_ν which is unique in $C^\omega(B_{\widetilde{r}})$. By the continuity of $\widetilde{\Lambda}$ and since $\{g_\nu\}_{\nu\in\mathbb{N}}$ is a Cauchy sequence in $C^\infty\left(B'_r, \mathbb{C}^m\right)$ we derive that $\{u_\nu\}$ is a Cauchy sequence in $C^\infty(B_{\widetilde{r}}(0))$. The completeness of $C^\infty(B_{\widetilde{r}})$ implies that $\{u_\nu\}_{\nu\in\mathbb{N}}$ converges to a function $u \in C^\infty(B_{\widetilde{r}})$.

Moreover, by

$$P(x, \partial)u_\nu \to P(x, \partial)u, \quad \text{as } \nu \to \infty$$

and

$$\partial_n^j u_\nu(x', 0) = g_{j,\nu}(x') \to g_j(x'), \qquad \partial_n^j u_\nu(x', 0) \to \partial_n^j u(x', 0), \text{ as } \nu \to \infty,$$

for $j = 0, 1, \cdots, m-1$, we derive that $u \in C^\infty(B_{\widetilde{r}})$ solves the Cauchy problem

$$\begin{cases} P(x, \partial)u = 0, & \text{in } B_{\widetilde{r}} \\ \partial_n^j u = g_j, & \text{for } j = 0, 1, \cdots, m-1, \quad \text{in } B'_{\widetilde{r}} \end{cases}.$$

3. Keep in mind that the definition of hyperbolicity that we provided above, involves not only the operator (or, more precisely, its symbol), but also the direction N. Let us consider, for instance, the wave operator in space dimension two

$$P(\partial_t, \partial_{x_1}, \partial_{x_2}) = \partial_t^2 - \left(\partial_{x_1}^2 + \partial_{x_2}^2\right).$$

The symbol of $P(\partial_t, \partial_{x_1}, \partial_{x_2})$ is the polynomial

$$p(\eta, \xi_1, \xi_2) = -\eta^2 + \left(\xi_1^2 + \xi_2^2\right),$$

which is hyperbolic w.r.t. the direction $(1, 0, 0)$, but **is not hyperbolic w.r.t. the directions** $(0, 1, 0)$ **and** $(0, 0, 1)$.

Elliptic operators (with constant coefficients), as is checked easily, are not hyperbolic with respect to any direction. Hence, the Cauchy problem for the elliptic operators does not enjoy the property of local solvability in C^∞, nor, at the light of what was shown in Sect. 11.3, it may happen that there is a continuous dependence in C^∞ (or from C^∞ in C^m).

Chapter 12
Ill-Posed Problems: Conditional Stability

Abstract In the preceding chapter, we introduced the concept of a well-posed problem in the sense of Hadamard. We also observed that certain Cauchy problems do not fall under the category of well-posed problems. We realized that these problems might lack continuous dependence of solutions on the initial data. This phenomenon poses a significant challenge in the analysis of problems arising from real-world applications. In practical situations, data measurements are inherently subject to approximation errors. Consequently, the impact of these errors must always be considered to ensure that theoretical results remain applicable. Importantly, non-continuous dependence can also arise in simpler cases, such as approximating derivatives or estimating velocity from a time function. A small error in the time function can cause a larger error in the velocity, illustrating that a well-approximated time function may not yield a good approximation of the velocity. This example highlights the issue that the operator from time functions to their derivatives may not be continuous, as small errors in the input can result in large discrepancies in the output. Throughout the discussion, when referring to a problem as ill-posed in the sense of Hadamard, we imply that while solutions may exist and be unique, the problem lacks continuous dependence on the initial data, and this should be understood in the context of the specific applied problem.

12.1 Introduction

In the preceding chapter, we introduced the concept of a well-posed problem in the sense of Hadamard. We also observed that certain Cauchy problems do not fall under the category of well-posed problems. Notably, we realized that these problems might lack a form of continuous dependence of solutions on the initial data. This phenomenon poses a significant challenge in the analysis of problems arising from real-world applications. In practical situations, data measurements are inherently subject to approximation errors. Consequently, the impact of these errors must always be taken into consideration to ensure that the theoretical results obtained retain reasonable applicability.

S. Vessella, *Unique Continuation Properties for Partial Differential Equations*,
Birkhäuser Advanced Texts Basler Lehrbücher,
https://doi.org/10.1007/978-3-031-86366-0_12

Furthermore, it is essential not to believe that the occurrence of non-continuous dependence of solutions on the data is restricted solely to intricately situations, such as the case of Cauchy problems. These phenomena can manifest even in more commonplace contexts, such as the approximate computation of a derivative or—taking an "applied" perspective—the estimation of velocity from a given time function. Consider an object undergoing rectilinear motion with a time function given by $x = s(t)$. We aim to ascertain its velocity $v(t)$. As we know very well

$$v(t) = s'(t).$$

However, now, let us suppose that we only have an approximation of the time law. Let $\varepsilon > 0$ and let s_ε be an approximate measure of the time law of our object. For instance let us suppose that

$$\sup_{t\in[0,T]} |s(t) - s_\varepsilon(t)| \leq \varepsilon, \tag{12.1}$$

where $T > 0$ is the time in which the motion is performed (0 is the initial time). It would be desired that $v_\varepsilon = s'_\varepsilon$ be itself an approximation of v. Nevertheless, this does not happen. As a matter of fact, let

$$s_\varepsilon(t) = s(t) + \varepsilon \sin\left(\varepsilon^{-2} t\right), \quad t \in [0, T].$$

Then (12.1) is satisfied, hence

$$\lim_{\varepsilon\to 0} \sup_{t\in[0,T]} |s(t) - s_\varepsilon(t)| = 0, \tag{12.2}$$

on the other side

$$v(t) - v_\varepsilon(t) = s'(t) - s'_\varepsilon(t) = \varepsilon^{-1} \cos\left(\varepsilon^{-2} t\right) \nrightarrow 0, \quad \text{as } \varepsilon \to 0.$$

Consequently, lacking supplementary information about the object's motion, deriving an approximation of its velocity solely from an approximation of its time law is unattainable.

The example we have just discussed can be succinctly summarized by stating that the operator

$$\frac{d}{dt} : X \to Y,$$

where

$$X = C^1([0, T]), \quad \text{equipped with norm } \|\cdot\|_{C^0([0,T])}$$

and

$$Y = C^0([0,T]), \quad \text{equipped with norm } \|\cdot\|_{C^0([0,T])}$$

is not continuous.

Throughout the upcoming discussion, when we refer to a problem as not being well-posed or ill-posed in the sense of Hadamard, we will consistently imply (unless there is potential for confusion) that despite the existence and uniqueness of solutions, the problem lacks continuous dependence on the initial data. The concept of continuous dependence should be interpreted within the context of the specific topology dictated by the nature of the particular applied problem under scrutiny.

12.2 Conditional Well-Posed Problems

Functional analysis offers a wealth of examples that illustrate the concept of not being well-posed in the sense specified above. For instance, it is known that if X is a Banach space and $A \in \mathcal{L}(X)$ (where $\mathcal{L}(X)$ denotes the space of linear and continuous operators from X into itself) is injective then

$$A^{-1} : \mathcal{R}(A) \to X,$$

where

$$\mathcal{R}(A) = \{Au : u \in X\},$$

is continuous if and only if $\mathcal{R}(A)$ is closed in X. In particular, if $\mathcal{R}(A)$ is not closed, then the problem

$$A(u) = f, \tag{12.3}$$

is not well-posed in the sense of Hadamard. This is the case of the integral operator

$$L^2(0,1) \ni u \to (Au)(t) = \int_0^t u(s)ds \in L^2(0,1). \tag{12.4}$$

Similarly, if X has not a finite dimension and A is a compact and injective operator from X into itself, then again A^{-1} is not continuous and therefore (12.3) is an ill-posed problem.

As it is well known, operator (12.4) is compact. More generally, if $k \in L^2([0,1] \times [0,1])$ then the operator

$$L^2(0,1) \ni u \to (Ku)(t) = \int_0^1 k(t,s)u(s)ds \in L^2(0,1), \tag{12.5}$$

is a compact operator. Hence the integral equation

$$\int_0^1 k(t,s)u(s)ds = f(t), \quad t \in [0,1], \tag{12.6}$$

is certainly an ill-posed problem in the sense of Hadamard since, even if it admits solutions they do not depend continuously (in $L^2(0,1)$) by f.

As previously noted, a problem lacking continuous dependence on the data, without supplementary information, remains impractical to address. Addressing such a problem necessitates additional insights. This concept leads us to the notion of a **conditionally well-posed problem**, which can be formally defined as follows.

Definition 12.1 Let (X, d_1) and (Y, d_2) be two metric spaces and let

$$A : X \to Y$$

be a map from X to Y. We say that the problem of determining $u \in X$ such that

$$A(u) = f, \tag{12.7}$$

where $f \in Y$, is a **conditional well-posed problem** or likewise **well-posed problem in the sense of Tikhonov** [129] with respect to $\mathcal{K} \subset X$ provided that we have

(i) $A_{|\mathcal{K}} : \mathcal{K} \to Y$ is injective,
(ii) $\left(A_{|\mathcal{K}}\right)^{-1} : A(\mathcal{K}) \to \mathcal{K}$ is continuous.

Remark Notice that the above definition does not impose a prerequisite for the existence of solutions to the problem $A(u) = f$. Essentially, the problem under consideration can be summarized as follows.

Given $f \in A(\mathcal{K})$, determine u such that

$$\begin{cases} A(u) = f \\ u \in \mathcal{K} \end{cases} . \tag{12.8}$$

Incorporating the set $\mathcal{K}$ into the definition of a well-posed problem in the Tikhonov sense is tantamount to introducing **additional information** or **a priori information** (as commonly referred in the literature), [106], to the problem at hand. To encapsulate these conditions succinctly, we can state that

$$\begin{aligned} &\text{Determine } u \in X \text{ such that } A(u) = f \\ &\textit{with the a priori information } u \in \mathcal{K}. \end{aligned} \tag{12.9}$$

The a priori information is tailored to the distinct nature of the problem at hand, guided by the practical context in which the problem arises. Seeking a **stability estimate for problem** (12.8) or (12.9) involves determining a suitable estimate for the modulus of continuity ω of the inverse $\left(A_{|\mathcal{K}}\right)^{-1}$, which, as a reminder, is defined by

$$\omega(\delta) = \sup\left\{d_1\left(A^{-1}(f_1), A^{-1}(f_2)\right) : f_1, f_2 \in A(\mathcal{K}), \quad d_2(f_1, f_2) \le \delta\right\}.$$

Certainly, determining the exact value of ω is an ideal goal, but often achieving a precise asymptotic estimate of $\omega(\delta)$ as δ approaches 0 can be deemed satisfactory and highly useful for numerous practical applications. ♦

If the problem

$$A(u) = f,$$

is ill-posed in the sense of Hadamard, and if A is a continuous and injective mapping, then there always exist sets $\mathcal{K}$ for which problem (12.7) becomes conditionally well-posed. This theorem can be stated as follows:

Theorem 12.2 (Tikhonov) *Let $\mathcal{K}$ and Y be two metric spaces. Let us assume $\mathcal{K}$ is a compact. Moreover, let*

$$F : \mathcal{K} \to Y,$$

satisfy

(i) F is injective,
(ii) F is continuous.

Then

$$F^{-1} : F(\mathcal{K}) \to \mathcal{K},$$

is continuous.

Proof Let C be a closed subset of $\mathcal{K}$, let us prove that

$$F(C) = \left(F^{-1}\right)^{-1}(C),$$

is closed in Y. We have

$$\begin{aligned} C \subset \mathcal{K}, C \text{ closed } \mathcal{K} \Rightarrow F(C) \text{ compact in } Y \Rightarrow \\ \Rightarrow F(C) \text{ closed in } Y. \end{aligned}$$

Therefore F is continuous. ■

Example 12.1 To illustrate what we have said so far, let us return to the problem of calculating the derivative of a function and we consider a simplified version of it, i.e. the following one: calculate the derivative of $f \in C^1([0,1])$ which satisfies

$$f'(0) = f'(1) = 0. \tag{12.10}$$

Let us note that also this problem is ill-posed in $L^2(0,1)$. As a matter of fact, let

$$f_n(t) = \frac{1}{\sqrt{n}} \cos \pi n t, \quad t \in [0,1], \quad n \in \mathbb{N}.$$

Then (12.10) is satisfied and

$$\|f_n\|_{L^2(0,1)} = \frac{1}{\sqrt{2n}} \to 0, \quad \text{as} \quad n \to \infty$$

and

$$\|f_n'\|_{L^2(0,1)} = \frac{\pi\sqrt{n}}{\sqrt{2}} \to +\infty, \quad \text{as } n \to \infty.$$

We reformulate the problem as a functional equation. Let

$$X = \left\{u \in C^0([0,1]) : u(0) = u(1) = 0\right\},$$

$$Y = \left\{f \in C^1([0,1]) : f'(0) = f'(1) = 0\right\}$$

and

$$T : X \to Y, \quad (T(u))(t) := \int_0^t u(s)ds.$$

Equip X and Y with the $L^2(0,1)$ norm.

The above problem becomes: given $f \in Y$ determine $u \in X$ such that

$$T(u) = f, \tag{12.11}$$

which has, trivially, a unique solution given by

$$u(t) = f'(t), \quad t \in [0,1].$$

We have seen above that there is no continuous dependence of the solutions by the datum. On the other hand T is continuous and, setting

$$\mathcal{K} = \left\{ u \in X : u \in C^1([0,1]), \quad \|u'\|_{L^2(0,1)} \leq E \right\},$$

where E is a positive number, since $\mathcal{K}$ is a compact of X, the problem

$$\begin{cases} T(u) = f \\ \\ u \in \mathcal{K} \end{cases}, \tag{12.12}$$

is well-posed in the sense of Tikhonov. Hence, denoting by ω_E the modulus of continuity of $\left(T_{|\mathcal{K}}\right)^{-1}$ we have

$$\begin{aligned} \left\| T^{-1}(f_1) - T^{-1}(f_2) \right\|_{L^2(0,1)} &= \|f_1' - f_2'\|_{L^2(0,1)} \leq \\ &\leq \omega_E \left(\|f_1 - f_2\|_{L^2(0,1)} \right), \end{aligned} \tag{12.13}$$

for every $f_1, f_2 \in T(\mathcal{K})$.

An estimate of ω_E can be easily proved as follows. If $f \in C^2([0,1])$ and $f'(0) = f'(1) = 0$ then, integrating by parts and applying the Cauchy–Schwarz inequality, we get

$$\begin{aligned} \int_0^1 f'^2(t)dt &= \int_0^1 f'(t)f'(t)dt = -\int_0^1 f''(t)f(t)dt \leq \\ &\leq \left(\int_0^1 f''^2(t)dt \right)^{1/2} \left(\int_0^1 f^2(t)dt \right)^{1/2}. \end{aligned}$$

Hence

$$\int_0^1 f'^2(t)dt \leq \left(\int_0^1 f''^2(t)dt \right)^{1/2} \left(\int_0^1 f^2(t)dt \right)^{1/2}. \tag{12.14}$$

Now, let $f_1, f_2 \in T(\mathcal{K})$ satisfy

$$\|f_1 - f_2\|_{L^2(0,1)} \leq \varepsilon,$$

then by (12.14) we have

$$\left\| T^{-1}(f_1) - T^{-1}(f_2) \right\|_{L^2(0,1)} \leq (2E\varepsilon)^{1/2}. \tag{12.15}$$

Therefore

$$\omega_E(\varepsilon) \leq (2E\varepsilon)^{1/2}. \tag{12.16}$$

In particular, for fixed E, we have

$$\omega_E(\varepsilon) = O\left((\varepsilon)^{1/2}\right) \quad \text{as } \varepsilon \to 0, \tag{12.17}$$

the reader is invited to check that the exponent $1/2$ in the estimate (12.17) cannot be improved, however on this kind of issue we will return to further in this chapter.

12.3 Interpolation Estimates for the Derivatives of a Function

In this section we will prove some estimates among functions and their derivatives. These estimates can be considered as conditional stability estimates of some not well-posed problem. We start by the following.

Proposition 12.3 *If $f \in C^2([a,b])$, where $a, b \in \mathbb{R}$, $a < b$, then we have*

$$\begin{aligned} &\left\|f'\right\|_{L^\infty(a,b)} \leq \\ &\leq c_0 \left(((b-a)^{-2} \left\|f\right\|_{L^\infty(a,b)} + \left\|f''\right\|_{L^\infty(a,b)}\right)^{1/2} \left\|f\right\|^{1/2}_{L^\infty(a,b)}, \end{aligned} \tag{12.18}$$

where $c_0 \leq 8$ is a positive constant.

Proof The proof of (12.18) can be reduced to the case where $[a,b] = [0,1]$. To this aim it suffices to consider, instead of f, the function

$$[0,1] \ni x \to f(a + (b-a)x) \in \mathbb{R}.$$

Let us continue to denote by f this function. Let us fix $x \in \left[0, \frac{1}{2}\right]$ and let $h \in \left(0, \frac{1}{2}\right]$. We have

$$f'(x) = \left(f'(x) - \frac{f(x+h) - f(x)}{h}\right) + \frac{f(x+h) - f(x)}{h}. \tag{12.19}$$

By the Taylor formula we have that there exist ξ such that

$$x < \xi < x + h$$

and

$$f'(x) - \frac{f(x+h) - f(x)}{h} = -\frac{h}{2} f''(\xi).$$

Hence

$$\left| f'(x) - \frac{f(x+h) - f(x)}{h} \right| \leq \frac{h}{2} \left\| f'' \right\|_{L^\infty(0,1)}. \tag{12.20}$$

By the just obtained inequality and by (12.19) we have, for every $x \in \left[0, \frac{1}{2}\right]$ and for every $h \in \left(0, \frac{1}{2}\right]$,

$$\begin{aligned} \left| f'(x) \right| &\leq \frac{h}{2} \left\| f'' \right\|_{L^\infty(0,1)} + h^{-1} \left(\left| f(x+h) \right| + \left| f(x) \right| \right) \leq \\ &\leq \frac{h}{2} \left\| f'' \right\|_{L^\infty(0,1)} + 2h^{-1} \left\| f \right\|_{L^\infty(0,1)}. \end{aligned} \tag{12.21}$$

Instead, if $x \in \left[\frac{1}{2}, 1\right]$, then it suffices to replace (12.19) by

$$f'(x) = \left(f'(x) - \frac{f(x-h) - f(x)}{-h} \right) + \frac{f(x-h) - f(x)}{-h},$$

for every $h \in \left(0, \frac{1}{2}\right]$ and we obtain

$$\left| f'(x) \right| \leq \frac{h}{2} \left\| f'' \right\|_{L^\infty(0,1)} + 2h^{-1} \left\| f \right\|_{L^\infty(0,1)}, \tag{12.22}$$

for every $h \in \left(0, \frac{1}{2}\right]$. By (12.21) and (12.22) we get

$$\left\| f' \right\|_{L^\infty(0,1)} \leq \frac{h}{2} \left\| f'' \right\|_{L^\infty(0,1)} + 2h^{-1} \left\| f \right\|_{L^\infty(0,1)}, \quad \forall h \in \left(0, \frac{1}{2}\right]. \tag{12.23}$$

Now, set

$$E = \left\| f'' \right\|_{L^\infty(0,1)} \quad \text{and} \quad \varepsilon = \left\| f \right\|_{L^\infty(0,1)} \tag{12.24}$$

and let us determine the minimum of the function (in h variable) on the right-hand side of (12.23), i.e.

$$\left(0, \frac{1}{2}\right] \ni h \to \Phi(h) = \frac{h}{2} E + 2h^{-1} \varepsilon.$$

It turns out that if $\left(\frac{4\varepsilon}{E}\right)^{1/2} \leq \frac{1}{2}$ then, for $h = h_0 := \left(\frac{4\varepsilon}{E}\right)^{1/2}$,

$$\min_{(0,1/2]} \Phi = \Phi(h_0) = 2\sqrt{E\varepsilon}, \tag{12.25}$$

while, if $\left(\frac{4\varepsilon}{E}\right)^{1/2} \geq \frac{1}{2}$ then

$$\min_{(0,1/2]} \Phi = \Phi(1/2) = \frac{1}{4}E + 4\varepsilon,$$

but, since in this case $16\varepsilon \geq E$, we get

$$\min_{(0,1/2]} \Phi \leq 8\varepsilon. \tag{12.26}$$

By (12.25), (12.26) and (12.23) we have

$$\|f'\|_{L^\infty(0,1)} \leq 2\sqrt{E\varepsilon} + 8\varepsilon \leq 8(E+\varepsilon)^{1/2}\varepsilon^{1/2}$$

and recalling (12.24) we obtain (12.18). ■

Remarks

1. By Proposition 12.3 it follows the equivalence of the norms

$$\|f\|_{C^2([a,b])} = \|f\|_{L^\infty(a,b)} + (b-a)\|f'\|_{L^\infty(a,b)} + (b-a)^2\|f''\|_{L^\infty(a,b)},$$

and

$$\|f\| = \|f\|_{L^\infty(a,b)} + (b-a)^2\|f''\|_{L^\infty(a,b)},$$

(reader chek: use the inequality $2AB \leq A^2 + B^2$).

2. Inequality (12.18) is a **stability estimate for the calculation of the first derivative** provided we have the a priori information

$$(b-a)^2\|f''\|_{L^\infty(a,b)} \leq E. \tag{12.27}$$

As a matter of fact, if

$$\|f\|_{L^\infty(a,b)} \leq \varepsilon, \tag{12.28}$$

then

$$\|f'\|_{L^\infty(a,b)} \leq \frac{c_0}{b-a}(E+\varepsilon)^{1/2}\varepsilon^{1/2}.$$

More precisely, setting

$$\mathcal{K}_E = \left\{f \in C^2([a,b]) : (b-a)^2\|f''\|_{L^\infty(a,b)} \leq E\right\}$$

and denoting by ω the modulus of continuity of

$$\mathcal{K}_E \ni f \to f' \in C^1([a,b]),$$

we have

$$\omega(\varepsilon) \le \frac{c_0}{b-a}(E+\varepsilon)^{1/2}\varepsilon^{1/2}, \quad \forall \varepsilon > 0. \tag{12.29}$$

Also, we observe

$$\omega(\varepsilon) \ge \frac{1}{b-a}(E+\varepsilon)^{1/2}\varepsilon^{1/2}, \quad \forall \varepsilon > 0. \tag{12.30}$$

In order to check (12.30), for the sake of brevity, let us consider the case $[a,b] = [0,1]$ and let us denote by

$$f_\varepsilon(x) = \frac{E\varepsilon}{E+\varepsilon} \sin \frac{x}{\sqrt{\varepsilon/(\varepsilon+E)}}. \tag{12.31}$$

We have

$$\|f_\varepsilon\|_{L^\infty(0,1)} \le \varepsilon, \quad \left\|f''_\varepsilon\right\|_{L^\infty(0,1)} \le E.$$

Hence $f_\varepsilon \in \mathcal{K}_E$ and we have

$$\omega(\varepsilon) \ge \left\|f'_\varepsilon\right\|_{L^\infty(0,1)} = (E+\varepsilon)^{1/2}\varepsilon^{1/2},$$

from which we get (12.30).

Inequality (12.30) implies that the exponent $\frac{1}{2}$ in stability estimate (12.18) is **optimal**, that is it cannot be improved by a bigger exponent. ♦

In what follows we extend Proposition 12.3 to higher order derivatives. To this aim we need some notations. Let $a \in \mathbb{R}$. We recall that the translation operator, τ_a, is defined as

$$(\tau_a f)(x) = f(x-a), \tag{12.32}$$

where f is any one real variable function. Also we denote

$$\tau^*_a = \tau_{-a}. \tag{12.33}$$

Let us denote by I the identity operator. Moreover, if $h \in \mathbb{R}$ let us denote by $\Delta_h f$ the difference operator

$$(\Delta_h f)(x) = \left(\left(\tau^*_h - I\right) f\right)(x) = f(x+h) - f(x). \tag{12.34}$$

Let $\mathfrak{e}_0, \mathfrak{e}_1, \cdots \mathfrak{e}_j$, $j \in \mathbb{N}_0$, be the polynomials

$$\mathfrak{e}_0(x) = 1,$$

$$\mathfrak{e}_1(x) = x,$$

$$\mathfrak{e}_2(x) = x(x-h),$$

$$\cdots$$

$$\mathfrak{e}_j(x) = x(x-h)\cdots(x-(j-1)h)\,.$$

Let us note that $\mathfrak{e}_j$, has degree j for every $j \in \mathbb{N}_0$. Let $h \neq 0$ and let us denote by P_{n-1}, $n \in \mathbb{N}$, the **Newton interpolation polynomial centered at x_0 and with degree $n-1$**, that is

$$P_{n-1}(x) = \sum_{j=0}^{n-1} \frac{\left(\Delta_h^j f\right)(x_0)}{j!h^j} \mathfrak{e}_j(x-x_0). \tag{12.35}$$

Let us notice that if $1 \leq j \leq n-1$ and $s \in \{0, 1, \cdots, n-1\}$ then we have

$$\mathfrak{e}_j(sh) = \begin{cases} 0, & \text{for } 0 \leq s \leq j-1 \\ \\ \binom{s}{j} j!h^j, & \text{for } j \leq s \leq n-1 \end{cases}. \tag{12.36}$$

We check (12.36). If $0 \leq s \leq j-1$ then one of the factors of $\mathfrak{e}_j(sh)$ vanishes. While, if $j \leq s \leq n-1$ (where $1 \leq j \leq n-1$) we have

$$\begin{aligned} \mathfrak{e}_j(sh) &= (sh)(sh-h)\cdots(sh-(sh-(j-1)h)) = \\ &= h^j s(s-1)\cdots(h-(s-(j-1))) = \\ &= h^j \binom{s}{j} j!\,. \end{aligned}$$

Also, let us note that if $0 \leq s \leq n-1$, we get

$$f(x_0+sh) = \left(\left(\tau_h^*\right)^s f\right)(x_0) \tag{12.37}$$

and

$$\left(\tau_h^*\right)^s = (\Delta_h + I)^s = \sum_{j=0}^{n-1} \binom{s}{j} \Delta_h^j. \tag{12.38}$$

Hence (12.37) and (12.38) yield

$$
\begin{aligned}
f(x_0 + sh) &= \sum_{j=0}^{n-1} \binom{s}{j} \left(\Delta_h^j f\right)(x_0) = \\
&= \sum_{j=0}^{n-1} \frac{1}{j!h^j} \left(\Delta_h^j f\right)(x_0)\mathfrak{e}_j(x_0 + sh - x_0) = \\
&= P_{n-1}(x_0 + sh).
\end{aligned}
$$

All in all, we have

$$f(x_0 + sh) = P_{n-1}(x_0 + sh), \quad \text{for } s = 0, 1, \cdots, n-1. \tag{12.39}$$

We can now state and prove the following

Proposition 12.4 *Let $n \geq 2$ and $f \in C^n([a, b])$, where $a, b \in \mathbb{R}$, $a < b$. For $1 \leq k \leq n-1$ we have*

$$
\begin{aligned}
&\left\| f^{(k)} \right\|_{L^\infty(a,b)} \leq \\
&\leq c_{k,n} \left(((b-a)^{-n} \|f\|_{L^\infty(a,b)} + \left\| f^{(n)} \right\|_{L^\infty(a,b)} \right)^{\frac{k}{n}} \|f\|_{L^\infty(a,b)}^{1-\frac{k}{n}},
\end{aligned}
\tag{12.40}
$$

where $c_{k,n}$ is a positive constant which depends on k and n only.

Proof As in the proof of Proposition 12.3 we may reduce to the case $[a, b] = [0, 1]$. We begin to prove (12.40) when $k = n - 1$. Let $f \in C^n([0, 1])$. Fix $x_0 \in \left[0, \frac{1}{2}\right]$ and let $h \in \left(0, \frac{1}{2(n-1)}\right]$. Set

$$R(x) = f(x) - P_{n-1}(x), \quad x \in [0, 1].$$

By (12.39) we have

$$R(x_0) = R(x_0 + h) = \cdots = R(x_0 + (n-1)h) = 0,$$

from which, by repeatedly applying the Rolle Theorem, we have that there exists $\xi \in (x_0, x_0 + (n-1)h)$ such that

$$f^{(n-1)}(\xi) - P_{n-1}^{(n-1)}(\xi) = R^{(n-1)}(\xi) = 0.$$

Therefore there exists $\xi \in (x_0, x_0 + (n-1)h)$ such that

$$f^{(n-1)}(\xi) = P_{n-1}^{(n-1)}(\xi). \tag{12.41}$$

On the other hand we have

$$P_{n-1}^{(n-1)}(x) = \frac{\mathfrak{e}_{n-1}^{(n-1)}(x - x_0)}{(n-1)!} \frac{\left(\Delta_h^{n-1} f\right)(x_0)}{h^{n-1}},$$

and

$$\mathfrak{e}_{n-1}^{(n-1)}(x) = (n-1)!\,.$$

By the latter and by (12.41) we have

$$f^{(n-1)}(\xi) = \frac{\left(\Delta_h^{n-1} f\right)(x_0)}{h^{n-1}}. \tag{12.42}$$

Therefore

$$f^{(n-1)}(x_0) = \left(f^{(n-1)}(x_0) - f^{(n-1)}(\xi)\right) + \frac{\left(\Delta_h^{n-1} f\right)(x_0)}{h^{n-1}} \tag{12.43}$$

and by the Lagrange Theorem, we get

$$\begin{aligned} \left|f^{(n-1)}(x_0) - f^{(n-1)}(\xi)\right| &\le \left\|f^{(n)}\right\|_{L^\infty(0,1)} |x_0 - \xi| \le \\ &\le \left\|f^{(n)}\right\|_{L^\infty(0,1)} (n-1)h. \end{aligned} \tag{12.44}$$

Moreover

$$\begin{aligned} \left(\Delta_h^{n-1} f\right)(x_0) &= \left(\left(\tau_h^* - I\right)^{n-1} f\right)(x_0) = \\ &= \sum_{j=0}^{n-1} \binom{n-1}{j} (-1)^{n-1-j} f(x_0 + jh), \end{aligned} \tag{12.45}$$

therefore

$$\left|\left(\Delta_h^{n-1} f\right)(x_0)\right| \le \sum_{j=0}^{n-1} \binom{n-1}{j} |f(x_0 + jh)| \le 2^{n-1} \|f\|_{L^\infty(0,1)}\,. \tag{12.46}$$

By (12.43), (12.44) and (12.46) we have

$$\left|f^{(n-1)}(x_0)\right| \le \left\|f^{(n)}\right\|_{L^\infty(0,1)} (n-1)h + \left(\frac{2}{h}\right)^{n-1} \|f\|_{L^\infty(0,1)}\,. \tag{12.47}$$

Applying (12.47) to $f(1-x)$ we also obtain the estimate for $x_0 \in \left[\frac{1}{2}, 1\right]$. All in all, we have for $0 < h \leq \frac{1}{2(n-1)}$,

$$\left\| f^{(n-1)} \right\|_{L^\infty(0,1)} \leq E(n-1)h + \left(\frac{2}{h}\right)^{n-1} \varepsilon, \tag{12.48}$$

here we set

$$E := \left\| f^{(n)} \right\|_{L^\infty(0,1)}, \quad \varepsilon := \|f\|_{L^\infty(0,1)}.$$

Now we find the minimum of the function

$$\left(0, \frac{1}{2(n-1)}\right] \ni \to \Phi(h) = E(n-1)h + \left(\frac{2}{h}\right)^{n-1} \varepsilon.$$

By elementary calculation we have

$$\min_{\left(0,\frac{1}{2(n-1)}\right]} \Phi \leq c_n (E+\varepsilon)^{1-\frac{1}{n}} \varepsilon^{\frac{1}{n}},$$

where c_n depends on n only. Hence

$$\left\| f^{(n-1)} \right\|_{L^\infty(0,1)} \leq c_n \left(\left\| f^{(n)} \right\|_{L^\infty(0,1)} + \|f\|_{L^\infty(0,1)} \right)^{1-\frac{1}{n}} \|f\|^{\frac{1}{n}}_{L^\infty(0,1)}. \tag{12.49}$$

Now, let $1 \leq k \leq n-1$. By iteration of (12.49) we obtain

$$\begin{aligned}
\left\| f^{(k)} \right\|_{L^\infty(0,1)} &\leq c_{k+1} \left(\left\| f^{(k+1)} \right\|_{L^\infty(0,1)} + \|f\|_{L^\infty(0,1)} \right)^{\frac{k}{k+1}} \|f\|^{\frac{1}{k+1}}_{L^\infty(0,1)} \leq \\
&\leq c_{k+1} c_{k+2}^{\frac{1}{k+2}} \left(\left\| f^{(k+2)} \right\|_{L^\infty(0,1)} + \|f\|_{L^\infty(0,1)} \right)^{\frac{k}{k+2}} \|f\|^{\frac{2}{k+2}}_{L^\infty(0,1)} \leq \\
&\leq \cdots \leq \\
&\leq c_{k,n} \left(\left\| f^{(n)} \right\|_{L^\infty(0,1)} + \|f\|_{L^\infty(0,1)} \right)^{\frac{k}{n}} \|f\|^{1-\frac{k}{n}}_{L^\infty(0,1)}.
\end{aligned}$$

where $c_{k,n}$ depends on k and n only. By the above obtained inequality, coming back $[a, b]$ we get (12.40). ∎

Estimates like (12.40) can be easily derived for L^p norm, $1 \le p \le \infty$. Let

$$\|f\|_{L^p(a,b)} = \left(\int_a^b |f(x)|^p\right)^{1/p}. \tag{12.50}$$

We have

Proposition 12.5 *Let $n \geq 2$ and $f \in C^n([a,b])$, where $a, b \in \mathbb{R}$, $a < b$. For $1 \le k \le n-1$ we have*

$$\left\|f^{(k)}\right\|_{L^p(a,b)} \le$$
$$\le c_{k,n}\left(((b-a)^{-n}\|f\|_{L^p(a,b)} + \left\|f^{(n)}\right\|_{L^p(a,b)}\right)^{\frac{k}{n}} \|f\|_{L^p(a,b)}^{1-\frac{k}{n}}, \tag{12.51}$$

where $c_{k,n}$ is a positive constant which depends on k and n only.

Proof Similarly to the proof of the previous Proposition, we may reduce to the case $[a,b] = [0,1]$. Let us prove (12.51) when $k = n-1$. Let $t \in \left[0, \frac{1}{2}\right]$ $h \in \left(0, \frac{1}{2(n-1)}\right]$. By (12.43) (with t replacing x_0) we have

$$f^{(n-1)}(t) = \left(f^{(n-1)}(t) - f^{(n-1)}(\xi)\right) + \frac{\left(\Delta_h^{n-1} f\right)(t)}{h^{n-1}}. \tag{12.52}$$

Set

$$\widetilde{f^{(n)}}(\tau) = \begin{cases} f^{(n)}(\tau), & \text{for } \tau \in [0,1] \\ \\ 0, & \text{for } \tau \notin [0,1] \end{cases}.$$

We have

$$\left|f^{(n-1)}(t) - f^{(n-1)}(\xi)\right| = \left|\int_\xi^t f^{(n)}(\tau)d\tau\right| \le$$
$$\le \int_t^{t+(n-1)h} \left|f^{(n)}(\tau)\right| d\tau = \int_{\mathbb{R}} \left|\widetilde{f^{(n)}}(\tau)\right| \chi_{(0,(n-1)h)}(\tau - t)d\tau,$$

where $\chi_{(0,(n-1)h)}$ is the characteristic function of the interval $(0, (n-1)h)$. Hence, by (12.52), for $t \in \left[0, \frac{1}{2}\right]$ and $h \in \left(0, \frac{1}{2(n-1)}\right]$, we have

$$\left|f^{(n-1)}(t)\right| \le \int_{\mathbb{R}} \left|\widetilde{f^{(n)}}(\tau)\right| \chi_{(0,(n-1)h)}(\tau - t)d\tau + \frac{\left(\Delta_h^{n-1} f\right)(t)}{h^{n-1}}. \tag{12.53}$$

At this point we use the triangle inequality in L^p and the Young inequality for convolutions:

$$\|F \star G\|_{L^p(\mathbb{R})} \leq \|F\|_{L^p(\mathbb{R})} \|G\|_{L^1(\mathbb{R})},$$

where $F = \widetilde{f^{(n)}}$, $G = \chi_{(0,(n-1)h)}$ and we get

$$\left(\int_0^{1/2} \left|f^{(n-1)}(t)\right|^p dt\right)^{1/p} \leq (n-1)h \left\|f^{(n)}\right\|_{L^p(0,1)} + \left(\frac{2}{h}\right)^{n-1} \|f\|_{L^p(0,1)}.$$

Similarly, we have

$$\left(\int_{1/2}^{1} \left|f^{(n-1)}(t)\right|^p dt\right)^{1/p} \leq (n-1)h \left\|f^{(n)}\right\|_{L^p(0,1)} + \left(\frac{2}{h}\right)^{n-1} \|f\|_{L^p(0,1)}.$$

Hence, by $h \in \left(0, \frac{1}{2(n-1)}\right]$, we have

$$\left\|f^{(n-1)}\right\|_{L^p(0,1)} \leq (n-1)h \left\|f^{(n)}\right\|_{L^p(0,1)} + \left(\frac{2}{h}\right)^{n-1} \|f\|_{L^p(0,1)}.$$

From now on we proceed as in the proof of Proposition 12.4. ■

Remark 12.2 It can be proved (see exercise below) that the exponent $1 - \frac{k}{n}$ of the estimate (12.40) is optimal. Regarding the constant we report here, without a proof (we refer to [52]), the following sharp estimate ($k, m \in \mathbb{N}_0$, $m > 0$)

$$\left\|f^{(k)}\right\|_{L^\infty(a,b)} \leq 4e^{2k} m^k \|f\|_{L^\infty(a,b)}^{1-\frac{1}{m}} M_{km}^{\frac{1}{m}},$$

where

$$M_{km} = \max\left\{\frac{(km)!}{(b-a)^{nm}} \|f\|_{L^\infty(a,b)}^{1-\frac{1}{m}}, \left\|f^{(km)}\right\|_{L^\infty(a,b)}\right\}.$$

♦

Exercise 12.1 Prove the optimality of the exponent $1-\frac{k}{n}$ in inequalities (12.40) and (12.51). (hint: note that inequalities (12.40) and (12.51) hold for complex-valued functions. After that, instead of trigonometric functions like (12.31) use complex exponential). ♠

Exercise 12.2

(i) Let $0 < \alpha < \beta \leq 1$. Prove that for every $f \in C^{0,\beta}([a, b])$ we have

$$|f|_{\alpha,[a,b]} \leq C(b-a)^{-\alpha}\left[(b-a)^{\beta}|f|_{\beta,[a,b]} + \|f\|_{L^{\infty}(a,b)}\right]^{\frac{\alpha}{\beta}} \|f\|_{L^{\infty}(a,b)}^{1-\frac{\alpha}{\beta}},$$

where C depends on α and β only.

(ii) Let $0 < \alpha \leq 1$. Prove that for every $f \in C^{1,\alpha}([a,b])$ we have

$$\|f'\|_{L^{\infty}(a,b)} \leq C(b-a)^{-1}\left[(b-a)^{1+\alpha}\left|f'\right|_{\alpha,[a,b]} + \|f\|_{L^{\infty}(a,b)}\right]^{\frac{1}{1+\alpha}} \|f\|_{L^{\infty}(a,b)}^{\frac{\alpha}{1+\alpha}},$$

where C depends on α only.

Hint to (ii) note that instead of (12.20) we have

$$\left|f'(x) - \frac{f(x+h)-f(x)}{h}\right| \leq |f|_{\alpha,[a,b]}h^{\alpha}.$$

♠

We conclude this section with two estimates for the derivatives of several variables functions.

Proposition 12.6 *Let $f \in C^2\left(\overline{B_1}\right)$. We have*

$$\|\nabla f\|_{L^{\infty}(B_1)} \leq c\left(\left\|\partial^2 f\right\|_{L^{\infty}(B_1)} + \|f\|_{L^{\infty}(B_1)}\right)^{\frac{1}{2}} \|f\|_{L^{\infty}(B_1)}^{\frac{1}{2}}, \tag{12.54}$$

where c is a positive constant which depends on n only.

Proof Let $h \in (0,1]$ be to choose later and let $x \in B_1$. Set

$$\Omega_h(x) = B_h(x) \cap B_1.$$

For $j = 1, \cdots, n$ we have

$$\begin{aligned} f_{x_j}(x) &= f_{x_j}(x) - \frac{1}{|\Omega_h(x)|}\int_{\Omega_h(x)} f_{x_j}(y)dy + \\ &+ \frac{1}{|\Omega_h(x)|}\int_{\Omega_h(x)} f_{x_j}(y)dy = \\ &= \frac{1}{|\Omega_h(x)|}\int_{\Omega_h(x)}\left(f_{x_j}(x) - f_{x_j}(y)\right)dy + \\ &+ \frac{1}{|\Omega_h(x)|}\int_{\partial\Omega_h(x)} f(y)\nu_j dS. \end{aligned} \tag{12.55}$$

We have, for a suitable $\overline{x}$ on the segment of extremes x and y,

$$f_{x_j}(x) - f_{x_j}(y) = \nabla f_{x_j}(\overline{x}) \cdot (x - y), \tag{12.56}$$

in addition we have

$$|\Omega_h(x)| \geq C_1 h^n, \quad |\partial\Omega_h(x)| \geq C_2 h^{n-1}, \tag{12.57}$$

where C_1 and C_2 depend on n only. From what was obtained in (12.55), (12.56) and (12.57) we get

$$|\nabla f(x)| \leq C\left(h\left\|\partial^2 f\right\|_{L^\infty(B_1)} + h^{-1}\|f\|_{L^\infty(B_1)}\right),$$

where C depends on n only. Now we minimize the function on the right-hand side of (12.57) and we obtain (12.54). ■

Proposition 12.7 *Let $f \in C^1\left(\overline{B_1}\right)$. We have*

$$\|f\|_{L^\infty(B_1)} \leq c\left(\|\nabla f\|_{L^\infty(B_1)} + \|f\|_{L^2(B_1)}\right)^{\frac{n}{n+2}} \|f\|_{L^2(B_1)}^{\frac{n}{n+2}}, \tag{12.58}$$

where c depends on n only.

Proof Let $h \in (0, 1]$ be to choose, $x \in B_1$ and $\Omega_h(x)$ like in the previous proof. We have

$$f(x) = \frac{1}{|\Omega_h(x)|}\int_{\Omega_h(x)} (f(x) - f(y))\, dy + \frac{1}{|\Omega_h(x)|}\int_{\Omega_h(x)} f(y)dy. \tag{12.59}$$

By

$$|f(x) - f(y)| \leq \|\nabla f\|_{L^\infty(B_1)} |x - y|$$

we have

$$\frac{1}{|\Omega_h(x)|}\int_{\Omega_h(x)} |f(x) - f(y)|\, dy \leq h\, \|\nabla f\|_{L^\infty(B_1)}. \tag{12.60}$$

On the other hand, by the Cauchy–Schwarz inequality we have

$$\begin{aligned}\left|\frac{1}{|\Omega_h(x)|}\int_{\Omega_h(x)} |f(y)|\, dy\right| &\leq \frac{1}{|\Omega_h(x)|^{1/2}} \|f\|_{L^2(B_1)} \leq \\ &\leq \frac{1}{(c_1 h^n)^{1/2}} \|f\|_{L^2(B_1)}.\end{aligned} \tag{12.61}$$

Hence

$$|f(x)| \leq C\left(h^{-n/2}\,\|f\|_{L^2(B_1)} + h\,\|\nabla f\|_{L^\infty(B_1)}\right), \tag{12.62}$$

where C depends on n only. Now we minimize the function on the right-hand side of (12.62) and we get (12.58). ■

12.4 Stability Estimates for the Continuation of Holomorphic Functions

In what follows we will identify $\mathbb{C}$ with $\mathbb{R}^2$. Let us recall very quickly the definition and some properties of the holomorphic functions.

1. Let Ω be an open set of $\mathbb{C}$, $f : \Omega \to \mathbb{C}$ be a complex-valued function defined in Ω and $z_0 \in \Omega$. We say that f is **holomorphic in** z_0 if the following limit there exists (in $\mathbb{C}$)

$$\lim_{z \to z_0} \frac{f(z) - f(z_0)}{z - z_0}. \tag{12.63}$$

   In such a case we denote by $f'(z_0)$ the value of limit (12.63) and we say that it is the **derivative of** f **in** z_0. We say that f is **holomorphic in** Ω provided it is holomorphic in each point of Ω. For instance, z, z^n are holomorphic functions in $\mathbb{C}$ while $\Re z$, $\Im z$, $\overline{z}$ are not.
2. From what we say in **1** it follows that if f is holomorphic in $z_0 = x_0 + iy_0$ then f, as a function of the real variables x and y, is differentiable in (x_0, y_0) and

$$\frac{\partial f}{\partial x} + i\frac{\partial f}{\partial y} = 0, \quad \text{in } (x_0, y_0). \tag{12.64}$$

   Denoting by $P = \Re f$, $Q = \Im f$, (12.64) we may write

$$\frac{\partial P}{\partial x} = \frac{\partial Q}{\partial y}, \quad \frac{\partial P}{\partial y} = -\frac{\partial Q}{\partial x}, \quad \text{in } (x_0, y_0). \tag{12.65}$$

   Equations (12.64) and (12.65) are known as the **Cauchy-Riemann equations (or condition)**. By introducing the notations

$$\frac{\partial}{\partial z} = \frac{1}{2}\left(\frac{\partial}{\partial x} - i\frac{\partial}{\partial y}\right), \quad \frac{\partial}{\partial \overline{z}} = \frac{1}{2}\left(\frac{\partial}{\partial x} + i\frac{\partial}{\partial y}\right), \tag{12.66}$$

the Cauchy-Riemann conditions can be written as

$$\frac{\partial f}{\partial \overline{z}} = 0, \quad \text{in } (x_0, y_0). \tag{12.67}$$

Also we have, setting $dz = dx + idy$, $d\overline{z} = dx - idy$ and by considering f as a function of z and $\overline{z}$

$$df = \frac{\partial f}{\partial x}dx + \frac{\partial f}{\partial y}dy = \frac{\partial f}{\partial z}dz + \frac{\partial f}{\partial \overline{z}}d\overline{z}. \tag{12.68}$$

3. **The Cauchy Theorem** If $f : \Omega \to \mathbb{C}$ is holomorphic in Ω then the differential form

$$f(z)dz = f(x + iy)dx + if(x + iy)dy$$

is locally exact. That is, for every $(x_0, y_0) \in \Omega$ there exist $\delta > 0$ and

$$F : B_\delta(x_0, y_0) \to \mathbb{C},$$

F differentiable in $B_\delta(x_0, y_0)$ such that

$$\frac{\partial F}{\partial x} = f, \quad \frac{\partial F}{\partial y} = if.$$

4. It can be proved that if $f \in C^0(\Omega)$ is holomorphic in $\Omega \setminus L$, where L is a straight line then $f(z)dz$ is holomorphic in Ω. In particular, if $f \in C^0(\Omega)$ and f is holomorphic in $\Omega \setminus \{a\}$ where $a \in \Omega$, then $f(z)dz$ is locally exact. This in turn enables the proof of the
Cauchy Integral Formula. Let f be holomorphic in Ω. Let $a \in \Omega$ and $r > 0$ satisfy $\overline{B_r(a)} \subset \Omega$. Setting $\gamma(t) = a + re^{it}$, $t \in [0, 2\pi)$, we have

$$f(a) = \frac{1}{2\pi i}\int_\gamma \frac{f(z)}{z - a}dz. \tag{12.69}$$

5. The Cauchy formula implies that if f is holomorphic in B_ρ then f can be expanded in a power series in B_ρ, that is there exists $\{a_n\}_{n\geq 0}$, sequence of $\mathbb{C}$, such that

$$f(z) = \sum_{n=0}^{\infty} a_n z^n, \quad \forall z \in B_\rho. \tag{12.70}$$

Since a holomorphic function can be expanded in a power series in each point of an open set Ω, we have $f : \Omega \to \mathbb{C}$ is holomorphic in Ω if and only if f is a complex analytic function in Ω, i.e. if and only if for every $a \in \Omega$ there exists δ such that

$$f(z) = \sum_{n=0}^{\infty} \frac{f^{(n)}(a)}{n!}(z-a)^n, \quad \forall z \in B_\delta(a). \tag{12.71}$$

The "if $\cdots$ then" part of the equivalence follows immediately by the properties of differentiability of power series and by (12.63). Keep in mind, however, that the expression "analytic complex function" should not be confused with the expression "analytic complex-valued function". For instance $f(z, \overline{z}) = z^2 - \overline{z}^2$ is complex-valued analytic function, but not analytic complex function, as it is not holomorphic.

6. From what we have said in **4**, we get the converse of the Cauchy Theorem. That is to say: if $f \in C^0(\Omega)$ and $f(z)dz$ is locally exact in Ω then f is holomorphic in Ω. As a matter of fact, if $f(z)dz$ is locally exact in Ω then it has locally a primitive hence, there exists locally, $F \in C^1$ such that $\frac{\partial F}{\partial x} = f$, $\frac{\partial F}{\partial y} = if$, from which we have that F satisfies the Cauchy-Riemann conditions, hence F is holomorphic and $f = F'$, on the other hand since the derivative of a complex analytic function is still a complex analytic function, hence holomorphic, $f = F'$ is holomorphic.

We also have that if $f \in C^0(\Omega)$ and $\overline{B_r(a)} \subset \Omega$ then

$$f^{(n)}(a) = \frac{n!}{2\pi i} \int_\gamma \frac{f(z)}{(z-a)^{n+1}} dz, \quad \forall n \in \mathbb{N}_0, \tag{12.72}$$

where $\gamma(t) = a + re^{it}$, $t \in [0, 2\pi)$. Moreover, again by (12.69), we obtain the **mean property**

$$f(a) = \frac{n!}{2\pi} \int_0^{2\pi} f(a + re^{it})dt, \tag{12.73}$$

as soon as $\overline{B_r(a)} \subset \Omega$. From the mean property it follows the **Maximum Modulus Principle.** Let $\Omega \subset \mathbb{C}$ be a bounded open set and $f \in C^0(\overline{\Omega})$ be a holomorphic function in Ω, then

$$\max_{\overline{\Omega}} |f| = \max_{\partial\Omega} |f|.$$

Moreover, if Ω is connected and there exists $a \in \Omega$ such that

$$|f(a)| = \max_{\overline{\Omega}} |f|$$

then f is constant in Ω.

7. Let us now return our attention to the analyticity of holomorphic functions and let us recall what follows. If

$$f : \Omega \to \mathbb{C},$$

is holomorphic in Ω, connected open set of $\mathbb{C}$, then if $a \in \Omega$ we have

$$f^{(n)}(a) = 0, \quad \forall n \in \mathbb{N}_0 \quad \Longrightarrow \quad f \equiv 0 \quad \text{in } \Omega. \tag{12.74}$$

From which we have that, if D is nonempty open set contained in Ω, then

$$f = 0, \qquad \text{in } D \quad \Longrightarrow \quad f \equiv 0 \quad \text{in } \Omega. \tag{12.75}$$

Moreover, if f does not vanish identically in Ω then **the set of zeros of f has no accumulation points in** Ω. As a matter of fact, if f does not vanish identically in Ω then (12.74) implies that for every $a \in \Omega$ there exists $k \in \mathbb{N}_0$ such that

$$f^{(k)}(a) \neq 0.$$

Denoting by $k_0 \in \mathbb{N}_0$ the minimum of such k, (12.71) gives

$$f(z) = (z-a)^{k_0} \sum_{n=k_0}^{\infty} \frac{f^{(n)}(a)}{n!}(z-a)^{n-k_0} := (z-a)^{k_0}\varphi(z), \quad \forall z \in B_\delta(a),$$

where $\varphi(a) \neq 0$, hence $f(z) \neq 0$ in $B_{\delta_1}(a) \setminus \{a\}$ for some $\delta_1 > 0$. The above point can also be expressed in the following way: let

$$\{z_n : n \in \mathbb{N}\}$$

be a **infinite set which has at least an accumulation point in** Ω, then

$$f(z_n) = 0, \quad \forall n \in \mathbb{N} \quad \Longrightarrow \quad f \equiv 0 \quad \text{in } \Omega \tag{12.76}$$

or also

$$f(z) = O\left(|z-a|^N\right), \quad \forall N \in \mathbb{N} \quad \Longrightarrow \quad f \equiv 0 \quad \text{in } \Omega. \tag{12.77}$$

Let us notice that (12.76) does not hold if $\{z_n : n \in \mathbb{N}\}$ has accumulations points on $\partial\Omega$ only. Let us consider

$$\Omega = \{x + iy : x > 0, \quad |y| < x\}$$

and let

$$f(z) = e^{-\frac{1}{z}}.$$

We have $f(z) = O\left(|z|^N\right)$, for every $N \in \mathbb{N}$, but $f \not\equiv 0$.

8. In this concluding part of this summary, we prove

Proposition 12.8 *Let Ω be a connected open set of $\mathbb{C}$. Let us suppose that*

$$\overset{o}{\overline{\Omega}} = \Omega. \tag{12.78}$$

Let $z_0 \in \partial\Omega$ and $\Gamma = \partial\Omega \cap B_R(z_0) \neq \emptyset$, where $R > 0$. In addition, let $f \in C^0(\Omega \cup \overline{\Gamma})$, f be holomorphic in Ω which satisfies

$$f = 0 \quad on\ \Gamma, \tag{12.79}$$

then $f \equiv 0$ in Ω.

Proof In what follows we will need some simple topological relationships that we will prove (for the convenience of the reader) in the concluding part of the main proof. By (12.78) we have immediately

$$\partial\Omega = \overline{\Omega} \cap \overline{\left(\mathbb{C} \setminus \overline{\Omega}\right)}. \tag{12.80}$$

Now, let us fix $\delta \in \left(0, \frac{R}{4}\right)$. Since $z_0 \in \partial\Omega$, we have by (12.80)

$$B_\delta(z_0) \cap \Omega \neq \emptyset, \quad B_\delta(z_0) \cap \left(\mathbb{C} \setminus \overline{\Omega}\right) \neq \emptyset. \tag{12.81}$$

Now, let a and b be such that

$$b \in B_\delta(z_0) \cap \Omega, \quad a \in B_\delta(z_0) \cap \left(\mathbb{C} \setminus \overline{\Omega}\right).$$

Since

$$|b - a| \leq |b - z_0| + |z_0 - a| < 2\delta < R - 2\delta,$$

we get

$$b \in B_{R-2\delta}(a) \cap \Omega \subset B_R(z_0) \cap \Omega,$$

in particular $B_{R-2\delta}(a) \cap \Omega \neq \emptyset$, (the second inclusion relationship follows from the triangle inequality). Let us denote by (Fig. 12.1)

$$r = R - 2\delta.$$

We have also (see below the "concluding part of the proof")

$$\partial\left(B_r(a) \cap \Omega\right) \setminus \Gamma \subset \left(\partial B_r(a)\right) \cap \overline{\Omega}. \tag{12.82}$$

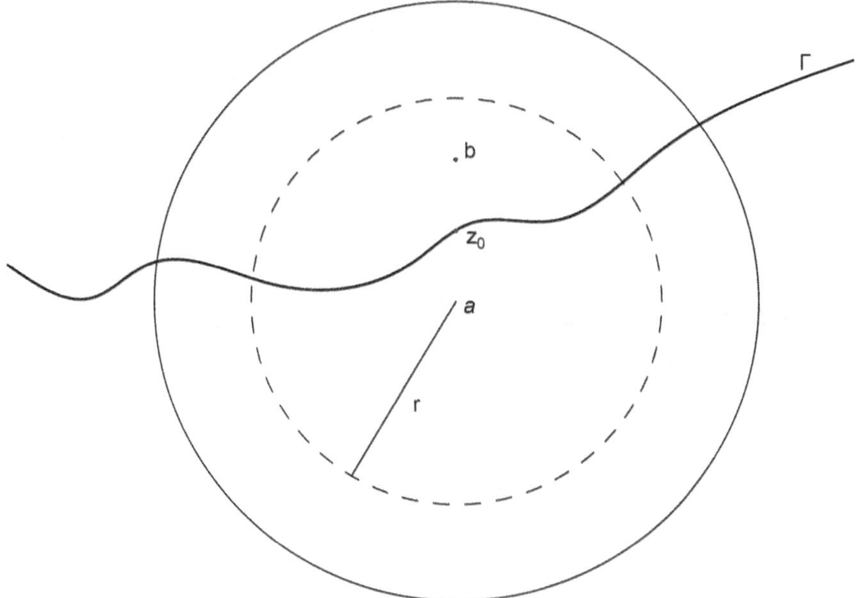

Fig. 12.1 The figure shows that $B_r(a)\cap\Omega\neq\emptyset$ (since $b\in B_r(a)\cap\Omega$) and that $\partial\left(B_r(a)\cap\Omega\right)\setminus\Gamma\subset(\partial B_r(a))\cap\overline{\Omega}$

Set now

$$\varphi(z)=(z-a)^{-n}f(z),\quad n\in\mathbb{N}. \tag{12.83}$$

It turns out that φ is holomorphic in $B_r(a)\cap\Omega$ and continuous in $\overline{B_r(a)\cap\Omega}$. By the maximum modulus principle we obtain

$$|\varphi(z)|\leq\max_{\partial(B_r(a)\cap\Omega)}|\varphi|,\quad\forall z\in B_r(a)\cap\Omega \tag{12.84}$$

and by (12.82) we get, recalling that $f=0$ on Γ,

$$\begin{aligned}\max_{\partial(B_r(a)\cap\Omega)}|\varphi|&\leq\max_{\partial(B_r(a)\cap\Omega)\setminus\Gamma}|\varphi|+\max_{\Gamma}|\varphi|\leq\\&\leq\max_{\partial(B_r(a))\cap\overline{\Omega}}|\varphi|=\max_{\partial(B_r(a))\cap\overline{\Omega}}\left(|z-a|^{-n}|f|\right)=\\&=r^{-n}\max_{\partial(B_r(a))\cap\overline{\Omega}}|f|.\end{aligned} \tag{12.85}$$

Now let $z\in B_r(a)\cap\Omega$. By (12.83), (12.84) and (12.85) we derive

$$|f(z)| \leq \left(\frac{|z-a|}{r}\right)^n \max_{\partial(B_r(a))\cap\overline{\Omega}} |f|, \quad \forall n \in \mathbb{N} \tag{12.86}$$

and passing to the limit as n goes to infinity we deduce

$$f(z) = 0, \quad \forall z \in B_r(a) \cap \Omega.$$

Since $B_r(a) \cap \Omega$ is a nonempty open set and Ω is a connected open set, by (12.75) we have $f \equiv 0$ in Ω.

Concluding Part of the Proof We prove (12.82). Let us recall

$$\partial(A \cap B) \subset \partial A \cup \partial B. \tag{12.87}$$

Now, let

$$x \in \partial\left(B_r(a) \cap \Omega\right) \setminus \Gamma.$$

We wish to prove that

$$x \in (\partial B_r(a)) \cap \overline{\Omega}. \tag{12.88}$$

First, we have trivially

$$\partial\left(B_r(a) \cap \Omega\right) \setminus \Gamma \subset \overline{\Omega}. \tag{12.89}$$

By (12.87) we have

$$x \in (\partial B_r(a)) \cup \partial\Omega \quad \text{and} \quad x \notin \Gamma. \tag{12.90}$$

Now, by (12.90) we have that if $x \notin \partial B_r(a)$ then $x \in \partial\Omega$. Moreover we have $x \in B_r(a)$. Because, if $x \notin B_r(a)$, as $x \notin \partial B_r(a)$, we would have $x \notin \overline{B_r(a)}$, hence it would exist $\rho > 0$ such that $B_\rho(x) \cap B_r(a) = \emptyset$. Consequently, for such ρ we would have $B_\rho(x) \cap (B_r(a) \cap \Omega) = \emptyset$ which contradicts $x \in \partial\left(B_r(a) \cap \Omega\right)$.

All in all, if $x \notin \partial B_r(a)$ then

$$x \in \partial\Omega \cap B_r(a) \subset \partial\Omega \cap B_R(z_0) = \Gamma,$$

but this cannot occur because, by (12.90), $x \notin \Gamma$. Therefore, (12.90) implies that

$$x \in \partial B_r(a).$$

Finally, since (12.89) holds we get (12.88), hence (12.82) is proved. ∎

Remark Assumption $\overset{o}{\overline{\Omega}} = \Omega$ excludes, for instance, that $\Omega = B_1 \setminus \{z_1, \cdots, z_n\}$ where $z_j \in B_R$ for $j = 1, \cdots, n$, where $R < 1$. In this case, Proposition 12.8 does not hold for $R < 1$ and $\Gamma = \{z_1, \cdots, z_n\}$. ♦

12.5 The Hadamard Three Circle Inequality and Other Examples of Stability Estimates

In the previous section we focused on the unique continuation property for the holomorphic functions. As we have seen it takes on several facets corresponding to (12.74)–(12.77) and to Proposition 12.8. In particular, we have that if a holomorphic function $f : \Omega \to \mathbb{C}$ is known in a set $D \subset \Omega$ which admits at least one accumulation point in Ω and if Ω is connected, then f is uniquely determined in Ω. The problem of determining effectively the values of f on Ω from $f_{|D}$ has an interest in applications, [93]. However this problem is not well posed in the sense of Hadamard as can be inferred from the following simple example.

Example 12.3 Let $\Omega = B_1$ and $D = B_r$ where $r \in (0, 1)$. Then any holomorphic function on B_1 is uniquely determined by $f_{|B_r}$. Nevertheless, small errors in the evaluation of $f_{|B_r}$ may produce uncontrollable errors on f. Let indeed

$$f_n(z) = \frac{1}{n}\left(\frac{z}{r}\right)^n, \quad n \in \mathbb{N}.$$

We have

$$\max_{\overline{B_r}} |f_n| = \frac{1}{n} \to 0, \quad \text{as } n \to \infty,$$

on the other hand, if $|z| > r$, then we have

$$|f_n(z)| = \frac{1}{n}\left(\frac{|z|}{r}\right)^n \to \infty, \quad \text{as } n \to \infty.$$

♣

The conditional stability question for the analytic extension problem may be formulated as follows

Let Ω be a connected open set of $\mathbb{C}$ and $D \subset \Omega$ which has at least one accumulation point in Ω. Let f be any holomorphic function in Ω, continuous on $\overline{\Omega}$, which satisfies

$$\max_{\overline{\Omega}} |f| \leq E \tag{12.91}$$

and

$$\max_{\overline{D}} |f| \leq \varepsilon. \tag{12.92}$$

We are interested in finding a stability estimate like the following one

$$|f(z)| \leq E\eta\left(\frac{\varepsilon}{E}; z\right), \quad \forall z \in \Omega, \tag{12.93}$$

where

$$\eta(s; z) \to 0 \quad \text{as } s \to 0, \ \forall z \in \Omega.$$

There is a fairly general treatment of estimates of stability (12.93), but here we will examine only a few examples that are particularly significant.

Example 12.4 (The Hadamard Three Circle Inequality) Let $0 < r \leq \rho < R$ (Fig. 12.2). Let f be a holomorphic function in B_R and continuous in $\overline{B_R}$. Let us denote by

$$M(s) := \max_{\overline{B_s}} |f|, \quad \text{for } 0 < s \leq R. \tag{12.94}$$

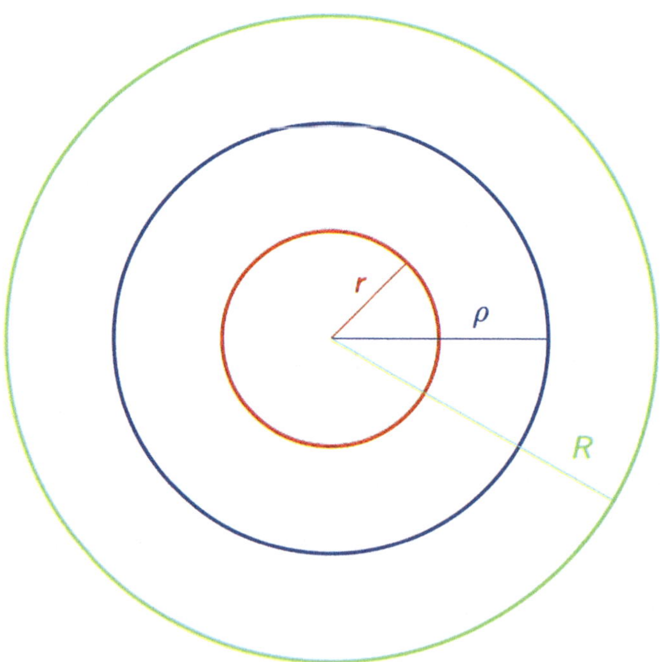

Fig. 12.2 The figure shows three concentric circles with radii $0 < r < \rho < R$, as featured in Hadamard's three circle inequality

We have

$$M(\rho) \leq (M(r))^{\theta_0} (M(R))^{1-\theta_0}, \tag{12.95}$$

where

$$\theta_0 = \frac{\log \frac{R}{\rho}}{\log \frac{R}{r}}. \tag{12.96}$$

Proof of (12.95) Let n and m be two integer numbers, $m > 0$. Let us consider the function

$$F(z) = z^{-n} (f(z))^m, \quad \text{for } z \in B_R \setminus \{0\}. \tag{12.97}$$

F is holomorphic in $B_R \setminus \{0\}$ and it is continuous in $\overline{B_R} \setminus B_r$. We can apply the maximum modulus principle. Set

$$\widetilde{M}(s) := \max_{\partial B_s} |f|, \quad \text{for } 0 < s \leq R.$$

We have, for any $\rho \in (r, R)$,

$$\begin{aligned}
\rho^{-n} \left(\widetilde{M}(\rho)\right)^m &= \max_{\partial B_\rho} |F| \leq \\
&\leq \max\left\{\max_{\partial B_r} |F|, \max_{\partial B_R} |F|\right\} = \\
&= \max\left\{r^{-n} \left(\widetilde{M}(r)\right)^m, R^{-n} \left(\widetilde{M}(R)\right)^m\right\},
\end{aligned}$$

which gives

$$\widetilde{M}(\rho) \leq \max\left\{\left(\frac{\rho}{r}\right)^{\frac{n}{m}} \widetilde{M}(r), \left(\frac{\rho}{R}\right)^{\frac{n}{m}} \widetilde{M}(R)\right\}. \tag{12.98}$$

Since $\mathbb{Q}$ is dense in $\mathbb{R}$, by (12.98) we have

$$\widetilde{M}(\rho) \leq \max\left\{\left(\frac{\rho}{r}\right)^{\alpha} \widetilde{M}(r), \left(\frac{\rho}{R}\right)^{\alpha} \widetilde{M}(R)\right\}, \quad \forall \alpha \in \mathbb{R}. \tag{12.99}$$

Now, let us choose α in such a way that

$$\left(\frac{\rho}{r}\right)^{\alpha} \widetilde{M}(r) = \left(\frac{\rho}{R}\right)^{\alpha} \widetilde{M}(R),$$

that is, let

$$\alpha = \frac{\log\left(\frac{\widetilde{M}(R)}{\widetilde{M}(r)}\right)}{\log\frac{R}{\rho}},$$

and (12.99) implies

$$\widetilde{M}(\rho) \le \left(\widetilde{M}(r)\right)^{\theta_0}\left(\widetilde{M}(R)\right)^{1-\theta_0}, \tag{12.100}$$

where θ_0 is given by (12.96). Finally, by the maximum modulus principle, we get (12.95). ■

Remarks

1. If f is holomorphic in $B_R\backslash\overline{B_r}$ and it is continuous in $\overline{B_R}\backslash B_r$, inequality (12.100) still applies.
2. It is evident that (12.95) is a stability estimate for the problem:
   Determine $f \in C^0\left(\overline{B_R}\right)$, f holomorphic in B_R which satisfies

$$\max_{\overline{B_r}} |f| \le \varepsilon$$

   and

$$\max_{\overline{B_R}} |f| \le E.$$

3. Let us notice that inequality (12.95) is equivalent to the convexity of the function

$$t \to \log M(e^t).$$

4. The inequality (12.95) cannot be improved. More precisely, the following facts apply.
   For every $C > 0$ independent of f we have

$$\theta_0 = \sup\left\{\theta : M(\rho) \le C\left(M(r)\right)^{\theta}, \quad M(R) = 1\right\}. \tag{12.101}$$

   In other words, the exponent θ_0 in (12.95) is the best exponent. Moreover

$$\inf\left\{C > 0 : M(\rho) \le C\left(M(r)\right)^{\theta_0}, \quad M(R) = 1\right\} = 1, \tag{12.102}$$

   that is the constant 1 in (12.95) is the best constant.
   Proof of (12.101)
   It suffices to prove that if $\theta \in \mathbb{R}$ satisfies

$$M(\rho) \le C\left(M(r)\right)^{\theta}, \tag{12.103}$$

for every $f \in C^0\left(\overline{B_R}\right)$, f holomorphic in B_R and such that $M(R) = 1$ then

$$\theta \leq \theta_0. \tag{12.104}$$

Now, let

$$f_n(z) = \left(\frac{z}{R}\right)^n, \quad n \in \mathbb{N}.$$

We have

$$M(\rho) = \left(\frac{\rho}{R}\right)^n, \quad M(r) = \left(\frac{r}{R}\right)^n$$

and by (12.103) we have

$$n \log \frac{\rho}{R} \leq \log C + \theta n \log \frac{r}{R}, \quad \forall n \in \mathbb{N},$$

from which (recalling that $r < \rho < R$) we have

$$\frac{\log(\rho/R)}{\log(r/R)} \geq \theta + \frac{\log C}{n \log(r/R)}, \quad n \in \mathbb{N}$$

and passing to the limit as $n \to \infty$ we obtain (12.104).
Proof of (12.102)
It suffices to prove that if $C > 0$ satisfies

$$M(\rho) \leq C\left(M(r)\right)^{\theta_9},$$

for every $f \in C^0\left(\overline{B_R}\right)$, f holomorphic in B_R and $M(R) = 1$ then

$$C \geq 1. \tag{12.105}$$

It suffices to choose

$$f(z) = \frac{z}{R}$$

and we have trivially

$$\left(M(r)\right)^{\theta_0} = \left(\frac{r}{R}\right)^{\theta_0} = \frac{\rho}{R} = M(\rho),$$

from which (12.105) follows.

5. It is interesting to note that the mere inequality (12.95) implies the following unique continuation property

$$f(z) = O\left(|z|^N\right), \text{ as } z \to 0, \ \forall N \in \mathbb{N} \Longrightarrow f \equiv 0 \text{ in } B_R. \tag{12.106}$$

indeed, let us assume that

$$f(z) = O\left(|z|^N\right), \text{ as } z \to 0, \ \forall N \in \mathbb{N} \tag{12.107}$$

and, arguing by contradiction, let us suppose that

$$f \not\equiv 0 \quad \text{in } B_R. \tag{12.108}$$

Then there exists $\rho \in (0, R)$ such that

$$M(\rho) > 0, \tag{12.109}$$

hence it not restrictive to assume

$$M(R) = 1.$$

Moreover, (12.107) implies

$$M(r) \leq C_N \left(\frac{r}{R}\right)^N, \quad \forall N \in \mathbb{N}$$

for some constant C_N (independent of r). By this inequality and by (12.95) we have

$$\begin{aligned} M(\rho) &\leq (M(r))^{\theta_0} \leq \\ &\leq \left(C_N \left(\frac{r}{R}\right)^N\right)^{\theta_0} = \\ &= \exp\left\{\frac{\log R/\rho}{\log R/r}\left[-N\log\frac{R}{r} + \log C_N\right]\right\}, \quad \forall r \in (0, \rho), \forall N \in \mathbb{N} \end{aligned}$$

and passing to the limit as $r \to 0$, we have

$$M(\rho) \leq \exp\left[-N\log\frac{R}{\rho}\right], \quad \forall N \in \mathbb{N} \tag{12.110}$$

now, again passing to the limit as $N \to \infty$ we have

$$M(\rho) = 0$$

which contradicts (12.109). Hence $f \equiv 0$ in B_R.

6. We can easily prove an inequality in L^2 similar to (12.95). More precisely the following inequalities ($0 < r < \rho < R$) hold true

$$\begin{aligned}&\int_0^{2\pi} \left|f\left(\rho e^{i\phi}\right)\right|^2 d\phi \leq \\ &\leq \left(\int_0^{2\pi} \left|f\left(re^{i\phi}\right)\right|^2 d\phi\right)^{\theta_0} \left(\int_0^{2\pi} \left|f\left(Re^{i\phi}\right)\right|^2 d\phi\right)^{1-\theta_0},\end{aligned} \tag{12.111}$$

$$\int_{B_\rho} |f|^2 \, dxdy \leq \left(\int_{B_r} |f|^2 \, dxdy\right)^{\theta_0} \left(\int_{B_R} |f|^2 \, dxdy\right)^{1-\theta_0}. \tag{12.112}$$

Proof of (12.111)
First, let us observe that

$$\rho = r^{\theta_0} R^{1-\theta_0}. \tag{12.113}$$

By the assumption on f we have

$$f(z) = \sum_{n=0}^{\infty} a_n z^n, \quad |z| \leq R.$$

Hence, by (12.113) and by the Hölder inequality we have

$$\begin{aligned}&\int_0^{2\pi} \left|f\left(\rho e^{i\phi}\right)\right|^2 d\phi = \sum_{n=0}^{\infty} \rho^{2n} |a_n|^2 = \\ &= \sum_{n=0}^{\infty} \left(r^{2n}|a_n|^2\right)^{\theta_0} \left(R^{2n}|a_n|^2\right)^{1-\theta_0} \leq \\ &\leq \left(\sum_{n=0}^{\infty} r^{2n}|a_n|^2\right)^{\theta_0} \left(\sum_{n=0}^{\infty} R^{2n}|a_n|^2\right)^{1-\theta_0} = \\ &= \left(\int_0^{2\pi} \left|f\left(re^{i\phi}\right)\right|^2 d\phi\right)^{\theta_0} \left(\int_0^{2\pi} \left|f\left(Re^{i\phi}\right)\right|^2 d\phi\right)^{1-\theta_0}.\end{aligned}$$

Now, **let us prove** (12.112). We have

$$\begin{aligned}\int_{B_\rho} |f|^2 \, dxdy &= \int_0^{\rho} s \left(\int_0^{2\pi} \left|f\left(se^{i\phi}\right)\right|^2 d\phi\right) ds = \\ &= \int_0^1 t\rho \left(\int_0^{2\pi} \left|f\left(t\rho e^{i\phi}\right)\right|^2 d\phi\right) dt.\end{aligned}$$

From which, by using (12.111), (12.113) and by the Hölder inequality, we get

$$\int_{B_\rho} |f|^2\,dxdy \le$$

$$\le \rho\left\{\int_0^1\left[\int_0^{2\pi} t\left|f\left(tre^{i\phi}\right)\right|^2 d\phi\right]^{\theta_0}\left[\int_0^{2\pi} t\left|f\left(tRe^{i\phi}\right)\right|^2 d\phi\right]^{1-\theta_0} dt\right\} \le$$

$$\le \rho\left[\int_0^1\int_0^{2\pi} t\left|f\left(tre^{i\phi}\right)\right|^2 d\phi dt\right]^{\theta_0}\left[\int_0^1\int_0^{2\pi} t\left|f\left(tRe^{i\phi}\right)\right|^2 d\phi dt\right]^{1-\theta_0} =$$

$$= \left[\int_0^1\int_0^{2\pi} tr\left|f\left(tre^{i\phi}\right)\right|^2 d\phi dt\right]^{\theta_0}\left[\int_0^1\int_0^{2\pi} tR\left|f\left(tRe^{i\phi}\right)\right|^2 d\phi dt\right]^{1-\theta_0} =$$

$$= \left(\int_{B_r} |f|^2\,dxdy\right)^{\theta_0}\left(\int_{B_R} |f|^2\,dxdy\right)^{1-\theta_0}.$$

One can also prove L^p versions of the inequalities (12.111) and (12.112), for these we refer the interested reader to [34, Ch. 1]. ♦

Exercise Let u be a harmonic function in $B_R \subset \mathbb{R}^2$ such that $u \in C^0\left(\overline{B}_R\right)$. Prove that if $0 < r < \rho < R$ then the following inequality holds true

$$\int_{\partial B_\rho} u^2 dS \le \left(\int_{\partial B_r} u^2 dS\right)^{\theta_0}\left(\int_{\partial B_R} u^2 dS\right)^{1-\theta_0}, \tag{12.114}$$

$$\int_{B_\rho} u^2 dxdy \le \left(\int_{B_r} u^2 dxdy\right)^{\theta_0}\left(\int_{B_R} u^2 dxdy\right)^{1-\theta_0}, \tag{12.115}$$

where θ_0 is given by (12.96). [Hint: recall the solution formula for Dirichlet problem in polar coordinates

$$u(\varrho,\phi) = \frac{a_0}{2} + \sum_{n=1}^{\infty} \varrho^n\left(a_n\cos n\phi + b_n\sin n\phi\right)$$

and apply it to obtain (12.115)]. ♠

Example 12.5 (Stability Estimate on the Bisector of an Angle ([21])) Let S be a sector (open set) of $\mathbb{C}$ whose boundary is made up of two segments, OA e OB such that $\widehat{AOB} = \pi\alpha$, $0 < \alpha < 2$, and by a Jordan curve Γ of extremes A and B (Fig. 12.3). Let $z_0 \in S$ and let us assume that z_0 belongs to the bisector of the angle $\widehat{AOB}$. Let $f \in C^0\left(\overline{S}\right)$ be holomorphic in S. Let us denote by

$$E = \max_{\overline{S}} |f|, \quad \varepsilon = \max_{\overline{\Gamma}} |f|.$$

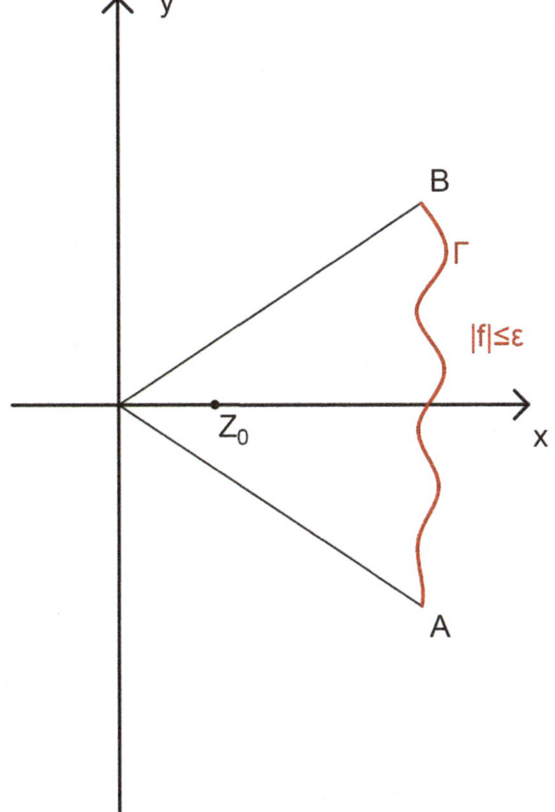

Fig. 12.3 The figure illustrates the sector S. On the Jordan curve Γ (red line), the modulus of the function f does not exceed ε. The point z_0, where f is evaluated, lies on the angle bisector, which in this case coincides with the real axis

Then

$$|f(z_0)| \leq E^{1-\left(\frac{|z_0|}{R}\right)^{1/\alpha}} \varepsilon^{\left(\frac{|z_0|}{R}\right)^{1/\alpha}}, \tag{12.116}$$

where R denotes the diameter of S.

Proof of (12.116) Let $\sigma > 0$ be chosen and

$$F(z) = f(z) \exp \sigma \left(\frac{z}{z_0}\right)^{1/\alpha}.$$

Set $|z_0| = r$. We have, for $z = \rho e^{\pm\frac{i\alpha\pi}{2}}$,

$$\left|F\left(\rho e^{\pm\frac{i\alpha\pi}{2}}\right)\right| = \left|f\left(\rho e^{\pm\frac{i\alpha\pi}{2}}\right)\right|.$$

Hence

$$|F(z)| \leq E, \quad \text{on } OA \text{ and } OB.$$

Moreover

$$|F(z)| \leq \varepsilon \exp\sigma\left(\frac{R}{r}\right)^{1/\alpha}, \quad \text{on } \Gamma.$$

Hence, by the maximum modulus principle, we have

$$|f(z_0)|e^{\sigma} = |F(z_0)| \leq \max\left\{E, \varepsilon \exp\sigma\left(\frac{R}{r}\right)^{1/\alpha}\right\},$$

from which we have

$$|f(z_0)| = e^{-\sigma}|F(z_0)| \leq \max\left\{Ee^{-\sigma}, \varepsilon \exp\sigma\left[\left(\frac{R}{r}\right)^{1/\alpha} - 1\right]\right\}.$$

Now, we choose σ such that

$$e^{\sigma} = \left(\frac{E}{\varepsilon}\right)^{\left(\frac{r}{R}\right)^{1/\alpha}}$$

and we obtain (12.116). ■

Stability estimates for the analytic continuation can be proved even for more general sets than those considered in examples 1 and 2. We present the following result without providing its proof (see [61]), [76, cap. III].

Let $\Omega \subset \mathbb{C}$ be a bounded simply connected open set whose boundary is of C^1 class. Let $\Gamma = \partial\Omega$ and let us assume that $\Gamma = \Gamma_1 \cup \Gamma_2$ where Γ_1 and Γ_2 is a regular path such that $\Gamma_1 \cap \Gamma_2 = \emptyset$. Let us assume that $f \in C^0\left(\overline{\Omega}\right)$, f holomorphic in Ω satisfying

$$|f(z)| \leq \varepsilon, \quad \forall z \in \Gamma_1, \qquad |f(z)| \leq E, \quad \forall z \in \Gamma_2,$$

then

$$|f(z)| \leq E^{1-\omega(z)}\varepsilon^{\omega(z)}, \quad \forall z \in \Omega, \tag{12.117}$$

where $\omega(z)$ is the harmonic function in Ω such that

$$\omega(z) = 1, \quad \forall z \in \Gamma_1, \qquad \omega(z) = 0, \quad \forall z \in \Gamma_2.$$

ω is called *harmonic measure associated to* Γ_1 *in* Ω.

Example 12.6 (Stability Estimate for the Continuation of Real Analytic Functions) In Theorem 8.5 we have seen that if $f : \Omega \to \mathbb{R}$ (or $\mathbb{C}$), where Ω is a connected open set of $\mathbb{R}^n$, is an analytic function and $D \subset \Omega$ is a (nonempty) open set then

$$f = 0 \quad \text{in } D \quad \Longrightarrow \quad f = 0 \quad \text{in } \Omega. \tag{12.118}$$

That is, a real analytic function on a connected open Ω is determined by its values on any nonempty open set $D \subset \Omega$. Nevertheless, small errors on $f_{|D}$ can have uncontrollable effects on $f(z)$ for $z \in \Omega \setminus \overline{D}$.

In the present Example 12.6, as application of Example 12.5, we will find an error estimate for the analytic continuation problem.

Let us consider the following particular situation: let Ω be a star shaped open set of $\mathbb{R}^n$ w.r.t. $x_0 \in \Omega$ (i.e. for every $x \in \Omega$ we have $x_0 + t\,(x - x_0) \in \Omega$, for every $x \in \Omega$). Let us assume that $\overline{B_r(x_0)} \subset \Omega$ for some $r > 0$. Let $f : \Omega \to \mathbb{R}$ satisfy $(E, \varepsilon > 0)$

$$f \in C_{E,\rho}(x), \quad \forall x \in \Omega, \tag{12.119}$$

that is (see Definition 8.6)

$$\left|\partial^\alpha f(x)\right|, \leq E\rho^{-|\alpha|}|\alpha|! \quad \forall \alpha \in \mathbb{N}_0^n, \ \forall x \in \Omega, \tag{12.120}$$

and

$$|f(x)| \leq \varepsilon, \quad \forall x \in B_r. \tag{12.121}$$

We want to prove the following stability estimate

$$|f(x)| \leq (2E)^{1-\theta}\varepsilon^\theta, \quad \forall x \in \Omega, \tag{12.122}$$

where $\theta \in (0, 1)$ and θ depends on n, $\frac{\rho}{r}$ and $\frac{\rho}{d}$ only, where d is the diameter of Ω.

Proof of (12.121) It is not restrictive to assume $x_0 = 0$. The idea of the proof is as follows: let us fix $x \in \Omega \setminus \overline{B_r}$ and let us consider the function

$$\varphi(t) = f(tx), \quad t \in [0, 1]; \tag{12.123}$$

we extend such a function holomorphically to a function φ in a neighborhood (in $\mathbb{C}$) of $\{t + i0 : t \in [0, 1]\}$ and by (12.120), (12.121) and the result of Example 12.5, we reach (12.122).

By formula (2.7) we have, for $t_0 \in [0, 1]$ and $k \in \mathbb{N}_0$,

$$\varphi^{(k)}(t_0) = \sum_{|\alpha|=k} \frac{k!}{\alpha!} x^\alpha \left(\partial^\alpha f\right)(t_0 x).$$

By (12.120) we have

$$\begin{aligned}\left|\varphi^{(k)}(t_0)\right| &\leq \sum_{|\alpha|=k} \frac{k!}{\alpha!}|x|^{|\alpha|}\left|\left(\partial^\alpha f\right)(t_0 x)\right| \leq \\ &\leq Ek!\left(\frac{|x|}{\rho}\right)^k \sum_{|\alpha|=k}\frac{k!}{\alpha!} = \\ &= Ek!\left(\frac{n|x|}{\rho}\right)^k.\end{aligned} \tag{12.124}$$

Hence the power series

$$\sum_{k=0}^{\infty}\frac{1}{k!}\varphi^{(k)}(t_0)\,(z-t_0)^k, \quad z=t+i\tau\in\mathbb{C}, \tag{12.125}$$

converges in

$$B_h(t_0)=\{z\in\mathbb{C}:|z-t_0|<h\},$$

where $h=\frac{\rho}{2n|x|}$. Moreover the sum of power series (12.125) is holomorphic in $B_h(t_0)$. The above extension can be performed for every $t_0\in[0,1]$ therefore the function φ can be holomorphically extended in (Fig. 12.4)

$$K=\{z\in\mathbb{C}: \quad \operatorname{dist}(z,I)<h\},$$

where $I:=\{t+i0:t\in[0,1]\}$. The extension φ to K is formally written as $\varphi(t+i\tau)$ and by (12.124) we have

$$|\varphi(t+i\tau)|\leq 2E, \quad \text{for every } t+i\tau\in K. \tag{12.126}$$

On the other hand, by (12.121) and by (12.123) we have

$$|\varphi(t+i0)|\leq\varepsilon, \quad \text{for} \quad |t|\leq\frac{r}{|x|}. \tag{12.127}$$

At this point it suffices to prove an estimate from above of $|\varphi(t+i0)|=|f(x)|$. This estimate can be obtained by applying twice estimate (12.116). First we apply estimate (12.116) in the triangles

$$S_+=\left\{t+i\tau: \quad |t|\leq s, 0\leq\tau\leq h\left(1-\frac{|t|}{s}\right)\right\},$$

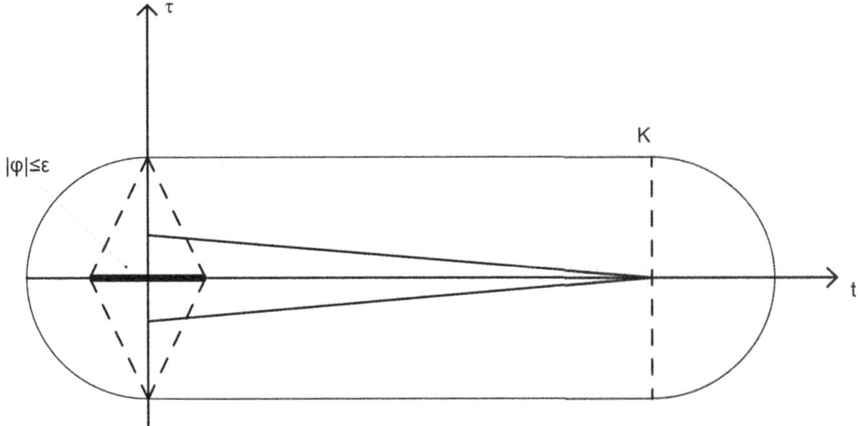

Fig. 12.4 The function φ has modulus less than or equal to ε along the black-marked segment on the t-axis (the interval $\left(-\frac{r}{|x|}, \frac{r}{|x|}\right)$). By applying inequality (12.116) twice, φ can first be estimated along the bisectors of the triangles with dashed sides and then along the bisector of the triangle with solid sides

$$S_- = \left\{t + i\tau : \quad |t| \le s, \ -h\left(1 - \frac{|t|}{s}\right) \le \tau \le 0\right\},$$

where

$$s := \min\{\frac{r}{|x|}, h\}.$$

We get

$$|\varphi(i\tau)| \le (2E)^{1-\vartheta}\varepsilon^{\vartheta}, \quad \text{for } |\tau| \le \frac{h}{2}, \tag{12.128}$$

where

$$\vartheta = \left(\frac{h}{2R_0}\right)^{1/\alpha_0} = \frac{1}{\left(2\sqrt{\frac{s^2}{h^2} + 1}\right)^{1/\alpha_0}}, \tag{12.129}$$

$$\alpha_0 = \frac{2}{\pi} \arctan \frac{s}{h},$$

$$R_0 = \sqrt{s^2 + h^2},$$

(hence, ϑ depends on $\frac{r}{\rho}$ and n only).

Now, we use (12.128) and (12.126) to apply (12.116) in the triangle

$$T = \left\{ t + i\tau : \quad |\tau| \leq \frac{h}{2}, 0 \leq t \leq (h+1)\left(1 - \frac{2|\tau|}{h}\right) \right\}.$$

To this aim, set

$$\alpha_1 = \frac{2}{\pi} \arctan \frac{h}{h+1},$$

$$R_1 = \sqrt{\left(\frac{h}{2}\right)^2 + (1+h)^2}$$

and we have

$$|\varphi(1 + i0)| \leq (2E)^{1-\vartheta\widetilde{\vartheta}} \varepsilon^{\vartheta\widetilde{\vartheta}}, \tag{12.130}$$

where

$$\widetilde{\vartheta} = \left(\frac{1}{R_1}\right)^{1/\alpha_1}.$$

Therefore we have proved (12.122) with $\theta = \vartheta\widetilde{\vartheta}$. ■

12.6 Concluding Remarks

Come back to the holomorphic functions. The Hadamard three circle inequality allows us to estimate $|f(z)|$ for $z \in B_R$ provided we know that f is holomorphic in B_R and, in addition, we know

$$\sup_{|z| \leq r} |f(z)| \leq \varepsilon \text{ (error)}, \tag{12.131}$$

and

$$\sup_{|z| \leq R} |f(z)| \leq E \text{ (a priori information)}. \tag{12.132}$$

As a matter of fact, we have

$$|f(z)| \leq \varepsilon^{\theta_{|z|}} E^{1-\theta_{|z|}}, \tag{12.133}$$

where

$$\theta_{|z|} = \frac{\log (R/|z|)}{\log (R/r)}. \tag{12.134}$$

It is immediately checked that the a priori information

$$\sup_{|z|\leq R} |f(z)| \leq E,$$

is not sufficient to control the error on $\{|z| = R\}$. It is enough to consider $f_n(z) = \left(\frac{z}{R}\right)^n$, obtaining

$$\sup_{|z|\leq r} |f_n(z)| \to 0 \text{ as } n \to \infty, \quad \text{and} \quad |f_n(z)| = 1, \text{ for } |z| = R.$$

We will now show that by "strengthening" the a priori information (12.132) we can find a stability estimate for $|f(z)|$ when $|z| = R$. Let us consider, for instance, the following a priori information: $f \in C^{0,\alpha}\left(\overline{B_R}\right)$, $0 < \alpha \leq 1$

$$\sup_{|z|\leq R} |f(z)| + [f]_{0,\alpha} \leq E_\alpha, \tag{12.135}$$

where

$$[f]_{0,\alpha} = \sup_{z,w\in B_R, z\neq w} \frac{|f(z) - f(w)|}{|z - w|^\alpha}. \tag{12.136}$$

Let us assume, for brevity, that $R = 1$. Let $z_0 \in \partial B_1$. Set $z_t = z_0(1 - t)$, with $t \in [0, 1)$ to be chosen, we have

$$\begin{aligned} |f(z_0)| &\leq |f(z_t) - f(z_0)| + |f(z_t)| \leq \\ &\leq E_\alpha |z_t - z_0|^\alpha + |f(z_t)| \leq \\ &\leq E_\alpha t^\alpha + |f(z_t)|. \end{aligned}$$

On the other hand by (12.133) we have

$$|f(z)| \leq \varepsilon^{\widetilde{\theta}_t} E_\alpha^{1-\widetilde{\theta}_t},$$

where

$$\widetilde{\theta}_t = \frac{\log \frac{1}{1-t}}{\log (1/r)}.$$

Hence

$$|f(z_0)| \leq E_\alpha \left(t^\alpha + \varepsilon_1^{\log \frac{1}{1-t}}\right), \quad \forall t \in [0,1),$$

where

$$\varepsilon_1 = \left(\frac{\varepsilon}{E_\alpha}\right)^{\frac{1}{|\log r|}}. \tag{12.137}$$

Now we have

$$t^\alpha + \varepsilon_1^{\log \frac{1}{1-t}} = t^\alpha + \exp\left(\log(1-t)\,|\log \varepsilon_1|\right).$$

On the other hand, we have

$$\log(1-t) \leq -t.$$

Hence

$$|f(z_0)| \leq E_\alpha \left(t^\alpha + \exp\left(-t\,|\log \varepsilon_1|\right)\right), \quad \forall t \in [0,1). \tag{12.138}$$

Now we note that, if $\varepsilon_1 < 1$ then

$$0 < |\log \varepsilon_1|^{-1} \left(\log |\log \varepsilon_1|\right) \leq e^{-1}$$

and we can choose

$$t = |\log \varepsilon_1|^{-1} \left(\log |\log \varepsilon_1|\right) \in [0,1).$$

We obtain

$$\begin{aligned} |f(z_0)| &\leq E_\alpha \left(|\log \varepsilon_1|^{-\alpha} \left(\log |\log \varepsilon_1|\right) + |\log \varepsilon_1|^{-1}\right) \leq \\ &\leq C E_\alpha\, |\log \varepsilon_1|^{-\alpha} \left(\log |\log \varepsilon_1|\right), \end{aligned} \tag{12.139}$$

where C depends on α only. If $\varepsilon_1 \geq 1$ then we have trivially

$$|f(z_0)| \leq E_\alpha \leq E_\alpha \varepsilon_1. \tag{12.140}$$

Thus, by (12.139) and (12.140) we have the following **stability estimate**, for every $z_0 \in \partial B_1$

$$|f(z_0)| \leq \widetilde{C} E_\alpha\, |\log \varepsilon_1|^{-\alpha} \left(\log |\log \varepsilon_1|\right). \tag{12.141}$$

where $\widetilde{C}$ depends on α and ε_1 is given by (12.137). ♦

Chapter 13
The John Stability Theorem for the Cauchy Problem for PDEs with Analytic Coefficients

Abstract In this chapter we present the classical stability proof of John Theorem, which was partly introduced regarding uniqueness. The theorem provides a logarithmic stability estimate for the Cauchy problem with a non-characteristic initial surface in linear PDEs with analytic coefficients. While the continuous dependence here is modest due to the strong regularity assumptions, John showed that for certain equations, such as the wave equation in two or more spatial dimensions, the logarithmic stability cannot be improved.

13.1 Introduction

This chapter is closely related to Sect. 9.7, as it presents the classical stability proof of the F. John Theorem, [65], partially introduced in Sect. 9.7.1 regarding uniqueness (an immediate consequence of the stability). The John Theorem provides a conditional logarithmic stability estimate for the Cauchy problem with a non-characteristic initial surface in linear partial differential equations with analytic coefficients. While the continuous dependence shown here is modest given the strong regularity assumptions on the coefficients, John proved that for certain equations, such as the wave equation in spatial dimensions of two or higher, the logarithmic nature of stability cannot be improved (see Remarks in Sect. 13.3).

The proof's essential components are:

1. The Green identity and the construction of a suitable solution for the adjoint operator (introduced in Sect. 9.7).
2. Stability estimates for the analytic continuation problem, discussed in Chap. 12.

S. Vessella, *Unique Continuation Properties for Partial Differential Equations*,
Birkhäuser Advanced Texts Basler Lehrbücher,
https://doi.org/10.1007/978-3-031-86366-0_13

13.2 Statement of the Theorem

In what follows we will consider the following linear system

$$u_t(x,t) = \sum_{j=1}^{n} A_j(x,t)u_{x_j}(x,t) + A_0(x,t)u(x,t), \tag{13.1}$$

where $u := \left(u^1, \cdots, u^N\right)^T$, $x \in \mathbb{R}^n$, $t \in \mathbb{R}$, $A_j(x,t)$, $j = 0, 1, \cdots, n$ are $N \times N$ matrices. Moreover, let us introduce the following notations

$$\gamma(x) = \left(1 - |x|^2\right)^{n+1}, \tag{13.2}$$

For any $\lambda \in \mathbb{R}$ let us denote by S_λ the surface

$$S_\lambda = \{(x, \lambda\gamma(x)) : x \in B_1\} \tag{13.3}$$

and, for any $\lambda_1 < \lambda_2$, set

$$\mathcal{R}_{\lambda_1,\lambda_2} = \left\{(x,t) \in \mathbb{R}^{n+1} : x \in B_1 \quad \lambda_1\gamma(x) < t < \lambda_2\gamma(x)\right\}. \tag{13.4}$$

Theorem 13.1 (John Stability Estimate) *Let $c_0, L, M, \rho, E, \varepsilon$ be positive numbers. Let A_j, $j = 1, \cdots, n$ and B be $N \times N$ matrices whose entries are analytic in $\overline{\mathcal{R}_{0,L}}$ and satisfy*

$$A_j \in C_{M,\rho}\left(\overline{x}, \overline{t}\right), \quad j = 0, 1, \cdots, n, \quad \forall \left(\overline{x}, \overline{t}\right) \in \mathcal{R}_{0,L}. \tag{13.5}$$

Set

$$A(x,\lambda) = I + \lambda \sum_{j=1}^{n} A_j(x, \lambda\gamma(x))\gamma_{x_j}(x). \tag{13.6}$$

Assume that

$$|\det A(x,\lambda)| \geq c_0, \quad \forall x \in \overline{\mathcal{R}_{0,L}}, \quad \forall \lambda \in [0, L], \tag{13.7}$$

(that is S_λ is a noncharacteristic surface for every $\lambda \in [0, L]$).

Let $u \in C^{n+1}\left(\overline{\mathcal{R}_{0,L}}\right)$ satisfy

$$u_t(x,t) = \sum_{j=1}^{n} A_j(x,t)u_{x_j}(x,t) + A_0(x,t)u(x,t), \quad \forall (x,t) \in \mathcal{R}_{0,L} \tag{13.8a}$$

$$\|u(\cdot,0)\|_{L^\infty(B_1)} \le \varepsilon, \tag{13.8b}$$

$$\|u\|_{C^{n+1}(\overline{\mathcal{R}_{0,L}})} \le E. \tag{13.8c}$$

Then, for every $r \in (0,1)$, we have

$$|u(x,t)| \le \frac{C(E+2\varepsilon)}{(1-r)^{n+1}} \left|\log \frac{\varepsilon}{E+2\varepsilon}\right|^{-1}, \quad \forall x \in \overline{\mathcal{R}_{0,L}} \cap \left(\overline{B_{1-r}} \times \mathbb{R}\right), \tag{13.9}$$

where C depends on M, L, ρ, c_0 and n only.

13.3 Proof of the Theorem

We begin by stating the following

Lemma 13.2 *Let A_j be as in Theorem 13.1. Suppose that*

$$A_j, \in C_{M_0,\rho_0}\left(\overline{x},\overline{t}\right), \quad \forall \left(\overline{x},\overline{t}\right) \in \overline{B_1} \times [-\delta_0,\delta_0]. \tag{13.10}$$

Let $W \in \mathbb{R}^N$ be such that $|W| = 1$ and let $\xi \in \mathbb{R}^n$ be arbitrary.
Let U be the solution to the Cauchy problem

$$\begin{cases} U_t = \sum_{j=1}^n A_j(x,t)U_{x_j} + A_0(x,t)U \\ \\ U(x,0) = e^{-ix\cdot\xi}W, \quad \forall x \in B_1 \end{cases} . \tag{13.11}$$

Then there exist M, ρ, δ positive numbers depending by M_0, ρ_0, δ_0, but independent of ξ such that

$$U \in C_{Me^{|\xi|},\rho}\left(\overline{x},\overline{t}\right), \quad \forall \left(\overline{x},\overline{t}\right) \in \overline{B_1} \times [-\delta,\delta], \tag{13.12}$$

that is

$$\left|\partial^\alpha U\right| \le Me^{|\xi|}\rho^{-|\alpha|}|\alpha|!, \quad \forall \left(\overline{x},\overline{t}\right) \in \overline{B_1} \times [-\delta,\delta], \ \forall \alpha \in \mathbb{N}_0^{n+1}. \tag{13.13}$$

Proof of Lemma 13.2 Let

$$\psi(x) = e^{-|\xi|}e^{-i\xi\cdot x}W.$$

Let us consider the Cauchy problem

$$\begin{cases} V_t = \sum_{j=1}^{n} A_j(x,t)V_{x_j} + A_0(x,t)V \\ \\ V(x,0) = \psi(x), \quad \forall x \in B_1 \end{cases} . \tag{13.14}$$

We have, trivially,

$$U(x,t) = e^{|\xi|}V(x,t). \tag{13.15}$$

On the other hand

$$\begin{aligned} \left|\partial^{\alpha'}\psi(x)\right| = \left|(i\xi)^{\alpha'}e^{-|\xi|}e^{-\xi\cdot x}W\right| = |\xi|^{|\alpha'|}e^{-|\xi|} \leq \\ \leq |\alpha'|^{|\alpha'|}e^{-|\alpha'|} \leq |\alpha'|!, \quad \forall\alpha' \in \mathbb{N}_0^n, \ \forall x \in \overline{B_1}, \end{aligned}$$

hence

$$\psi \in C_{1,1}\left(\overline{x}\right), \quad \forall \overline{x} \in \overline{B_1}.$$

Therefore there exist M, ρ, δ which depend on M_0, ρ_0, δ_0, but independent of ξ such that

$$V \in C_{1,1}\left(\overline{x},\overline{t}\right), \quad \forall\left(\overline{x},\overline{t}\right) \in \overline{B_1} \times [-\delta,\delta]$$

and by (13.15) we have (13.12). ■

Proof of Theorem 13.1

Step 1. The Green Identity
We have

$$\begin{aligned} v^T\left(u_t - \sum_{j=1}^{n} A_j u_{x_j} - A_0 u\right) = \\ = \partial_t(v^T u) - \partial_{x_j}\left(v^T\sum_{j=1}^{n} A_j u\right) - \\ - \left(v_t^T - \sum_{j=1}^{n}\left(v^T A_j\right)_{x_j} + v^T A_0\right)u. \end{aligned} \tag{13.16}$$

Now let $v \in C^1\left(\overline{\mathcal{R}_{\lambda_1,\lambda_2}}\right)$ be a solution to the adjoint system

$$v_t - \sum_{j=1}^{n} \left(A_j^T v\right)_{x_j} + A_0^T v = 0, \quad \text{in } \mathcal{R}_{\lambda_1,\lambda_2}. \tag{13.17}$$

Since u is a solution to system (13.8a), integrating both the sides of (13.16) over $\mathcal{R}_{\lambda_1,\lambda_2}$ we have

$$\begin{aligned}
0 &= \int_{\mathcal{R}_{\lambda_1,\lambda_2}} \left[\partial_t(v^T u) - \partial_{x_j}\left(v^T \sum_{j=1}^{n} A_j u\right)\right] dxdt = \\
&= \int_{\partial\mathcal{R}_{\lambda_1,\lambda_2}} \left[(v^T u)(\nu \cdot e_{n+1}) - \left(v^T \sum_{j=1}^{n} A_j u(\nu \cdot e_j)\right)\right] dS = \\
&= \int_{B_1} \left(v^T u + \lambda_2 v^T \sum_{j=1}^{n} \gamma_{x_j} A_j u\right)(x, \lambda_2\gamma(x))dx - \\
&- \int_{B_1} \left(v^T u + \lambda_1 v^T \sum_{j=1}^{n} \gamma_{x_j} A_j u\right)(x, \lambda_1\gamma(x))dx.
\end{aligned}$$

Hence, recalling (13.6) we get

$$\begin{aligned}
&\int_{B_1} v^T(x, \lambda_2\gamma(x)) A(x, \lambda_2) u(x, \lambda_2\gamma(x)) dx = \\
&= \int_{B_1} v^T(x, \lambda_1\gamma(x)) A(x, \lambda_1) u(x, \lambda_1\gamma(x)) dx.
\end{aligned} \tag{13.18}$$

Step 2. Construction of an Appropriate Solution to (13.17)
Let $W \in \mathbb{R}^N$ be such that $|W| = 1$ and let $\xi \in \mathbb{R}^n$ be arbitrary. Let us denote by w the function

$$w(x) = e^{-ix\cdot\xi} W. \tag{13.19}$$

Let $\lambda \in [0, L]$. Let us consider the following Cauchy problem

$$\begin{cases}
v_t = \sum_{j=1}^{n} \left(A_j^T(x,t)v\right)_{x_j} - A_0^T(x,t)v \\
\\
v(x, \lambda\gamma(x)) = \gamma(x)\left(A^T(x,\lambda)\right)^{-1} w(x), \quad \forall x \in B_1
\end{cases} . \tag{13.20}$$

Let us prove that there exists $\delta > 0$, depending on M, ρ, c_0 only, such that there exists the solution $v(x, t; \lambda)$ of (13.20), and it is analytic in $\overline{\mathcal{R}_{\lambda-\delta,\lambda+\delta}}$. To this purpose we perform some change of variables. First, we set

$$s = \frac{t}{\gamma(x)}, \quad v(x, t; \lambda) = \gamma(x) V\left(x, \frac{t}{\gamma(x)}; \lambda\right)$$

and we have

$$v_t(x, t; \lambda) = V_s\left(x, \frac{t}{\gamma(x)}; \lambda\right),$$

$$\begin{aligned} v_{x_j} &= \gamma V_{x_j} + \gamma_{x_j} V - \frac{t\gamma_{x_j}}{\gamma} V_s = \\ &= \gamma V_{x_j} + \gamma_{x_j} V - s\gamma_{x_j} V_s. \end{aligned}$$

Inserting what obtained above in system (13.20), we get

$$\begin{aligned} V_s = v_t &= \sum_{j=1}^{n} A_j^T v_{x_j} + \left(\sum_{j=1}^{n} A_{j,x_j}^T - A_0^T\right) v = \\ &= \sum_{j=1}^{n} A_j^T \left(\gamma V_{x_j} - s\gamma_{x_j} V_s\right) + \left[\sum_{j=1}^{n} \left(A_j^T \gamma_{x_j} + A_{j,x_j}^T \gamma\right) - A_0^T \gamma\right] V. \end{aligned}$$

From which (recalling (13.6)), we get

$$\begin{aligned} A^T(x, s) V_s &= \gamma \sum_{j=1}^{n} A_j^T V_{x_j} + \left(\sum_{j=1}^{n} A_{j,x_j}^T - A_0^T\right) V = \\ &= \sum_{j=1}^{n} A_j^T \left(\gamma V_{x_j} - s\gamma_{x_j} V_s\right) + \left[\sum_{j=1}^{n} \left(A_j^T \gamma_{x_j} + A_{j,x_j}^T \gamma\right) - A_0^T \gamma\right] V. \end{aligned}$$

Now, set

$$\overline{A}_j(x, s) = \gamma(x) \left(A^T(x, s)\right)^{-1} A_j^T(x, s\gamma(x)), \quad j = 1, \cdots, n,$$

$$\overline{A}_0(x,s) = \left(A^T(x,s)\right)^{-1}\left[\sum_{j=1}^{n} A_j^T(x,s\gamma(x))\gamma_{x_j} + \right.$$

$$\left. +\gamma\left(\sum_{j=1}^{n} A_{j,x_j}^T(x,s\gamma(x)) - A_0^T(x,s\gamma(x))\right)\right].$$

Therefore problem (13.20) can be written as

$$\begin{cases} V_s(x,s;\lambda) = \sum_{j=1}^{n} \overline{A}_j(x,s)V_{x_j}(x,s;\lambda) + \overline{A}_0(x,s)V(x,s;\lambda) \\ \\ V(x,s;\lambda)_{|s=\lambda} = \left(A^T(x,\lambda)\right)^{-1} w(x), \quad \forall x \in B_1 \end{cases}. \tag{13.21}$$

Now we denote

$$Z(x,s;\lambda) = \left(A^T(x,\lambda)\right)^{-1} V(x,s+\lambda;\lambda) \tag{13.22}$$

and by (13.19), (13.22) we have

$$\begin{cases} Z_s = \sum_{j=1}^{n} \widetilde{A}_j(x,s;\lambda)Z_{x_j} + \widetilde{A}_0(x,s;\lambda)Z \\ \\ Z(x,0;\lambda) = e^{-ix\cdot\xi}W, \quad \forall x \in B_1 \end{cases}, \tag{13.23}$$

where

$$\widetilde{A}_j(x,s;\lambda) = A^T(x,\lambda)\overline{A}_j(x,s+\lambda)\left(A^T(x,\lambda)\right)^{-1}, \quad j = 1,\cdots,n,$$

and

$$\widetilde{A}_0(x,s;\lambda) =$$

$$= A^T(x,\lambda)\left[\overline{B}(x,s+\lambda)\left(A^T(x,\lambda)\right)^{-1} + \sum_{j=1}^{n}\overline{A}_j(x,s+\lambda)\partial_{x_j}\left(A^T(x,\lambda)\right)^{-1}\right].$$

Now $\widetilde{A}_j(x,s;\lambda)$, $j = 1,\cdots,n$ and $\widetilde{A}_0(x,s;\lambda)$ are analytic functions in (x,s,λ). In addition, for every $\left(\overline{x},\overline{s},\overline{\lambda}\right) \in \overline{B_1} \times [0,L] \times [0,L]$, we have

$$\widetilde{A}_j \in C_{M',\rho'}\left(\overline{x},\overline{s},\overline{\lambda}\right), \quad j = 0,1,\cdots,n,$$

where M' and ρ' depend on M, ρ, c_0 (and n, which we will omit in the sequel) only.

By Lemma 13.2, there exist M'', ρ'', δ depending on M, ρ, c_0 and L only such that there is Z which is the solution to (13.23), it is analytic in $\overline{B_1} \times [-\delta, \delta] \times [0, L]$ and satisfies

$$Z \in C_{M'' e^{|\xi|}, \rho''}\left(\overline{x}, \overline{s}, \overline{\lambda}\right), \quad \forall \left(\overline{x}, \overline{s}, \overline{\lambda}\right) \in \overline{B_1} \times [-\delta, \delta] \times [0, L]. \tag{13.24}$$

Coming back to problem (13.20), we have that there exists v, solution to (13.20) in $\mathcal{R}_{\lambda-\delta,\lambda+\delta}$, such that

$$\left|\partial^{\alpha}_{x,t,\lambda} v(x, t; \lambda)\right| \leq M_0 e^{|\xi|} \rho^{-|\alpha|} |\alpha|!, \ \forall \alpha \in \mathbb{N}_0^{n+2}, \tag{13.25}$$

for all $(x, t; \lambda) \in \overline{\mathcal{R}_{\lambda-\delta,\lambda+\delta}} \times [0, L]$, where M_0, δ_0 and δ depend on M, ρ, c_0 and L only. Let us note that to obtain (13.25) for every $\lambda \in [0, L]$ it suffices to consider (13.24) and the similar relationships on the coefficients corresponding to $\lambda = 0$.

Step 3. Planning the Concluding Part of the Proof

We employ (13.18), where v is the solution to problem (13.20) for some $\lambda \in [0, L]$. Let $\lambda_0 \in [0, L]$ be fixed and let λ satisfy $|\lambda - \lambda_0| < \delta$. By (13.18) we have

$$\begin{aligned} g(\lambda) &:= \int_{B_1} \gamma(x) w^T(x) u(x, \lambda\gamma(x)) dx = \\ &= \int_{B_1} v^T(x, \lambda_0\gamma(x); \lambda) A(x, \lambda_0) u(x, \lambda_0\gamma(x)) dx. \end{aligned} \tag{13.26}$$

The function g is **analytic** because the integrand in the second integral of (13.26) depends analytically by λ.

We are first interested in proving an estimates from above for $g(\lambda)$ from which, subsequently, we will derive the estimates from above for u. Setting $\lambda_0 = 0$ in (13.26) we have

$$g(\lambda) = \int_{B_1} v^T(x, 0; \lambda) u(x, 0) dx, \quad \forall \lambda \in [0, \delta). \tag{13.27}$$

By (13.8b) and by (13.25), we have, for $\alpha = 0$,

$$|g(\lambda)| \leq \int_{B_1} \left|v^T(x, 0; \lambda)\right| |u(x, 0)| dx \leq c M_0 e^{|\xi|} \varepsilon, \ \forall \lambda \in [0, \delta), \tag{13.28}$$

where $c \geq 1$ depends on n only.

Let λ_0 be an arbitrary point of $[0, L]$, by (13.25) and (13.26) we get

$$\begin{aligned}\left|g^{(k)}(\lambda)\right| &= \left|\int_{B_1} \partial_\lambda^k v^T(x, \lambda_0\gamma(x); \lambda) A(x, \lambda_0) u(x, \lambda_0\gamma(x)) dx\right| \leq \\ &\leq \int_{B_1} \left|\partial_\lambda^k v^T(x, \lambda_0\gamma(x); \lambda)\right| |A(x, \lambda_0)| |u(x, \lambda_0\gamma(x))| dx \leq \\ &\leq C_1 E M_0 e^{|\xi|} \rho_0^{-k} k!,\end{aligned}$$

where $C_1 \geq 1$ depends on M, ρ, c_0, L only.

Summarizing we have

$$|g(\lambda)| \leq c M_0 e^{|\xi|} \varepsilon, \quad \forall \lambda \in [0, \delta) \tag{13.29}$$

and

$$\left|g^{(k)}(\lambda)\right| \leq C_1 M_0 e^{|\xi|} \rho_0^{-k} k!, \quad \forall \lambda \in [0, L], \ \forall k \in \mathbb{N}_0. \tag{13.30}$$

Inequality (13.30) implies that g can be extended analytically in a neighborhood of $[0, L] \times \{0\} \subset \mathbb{C}$. In addition, for any $\lambda_\star \in [0, L]$, we have that the power series

$$\sum_{k=0}^{\infty} \frac{g^{(k)}(\lambda_\star)}{k!} (z - \lambda_\star)^k,$$

has the radius of convergence equal to ρ_0 and

$$\left|\sum_{k=0}^{\infty} \frac{g^{(k)}(\lambda_\star)}{k!} (z - \lambda_\star)^k\right| \leq C_1 M_0 e^{|\xi|} \frac{\rho_0}{\rho_0 - |z - \lambda_\star|}, \tag{13.31}$$

for $|z - \lambda_\star| < \rho_0$. Therefore the sum of the power series (12.125) is holomorphic in $B_{\frac{\rho_0}{2}}(\lambda_\star)$ and the function g can be extended holomorphically in (Fig. 13.1)

$$J = \{z \in \mathbb{C} : \text{dist}(z, [0, L] \times \{0\}) < \rho_0/2\}.$$

The extension of g to J is formally written as $g(\lambda + i\tau)$ and by (13.31) we have

$$|g(\lambda + i\tau)| \leq 2 C_1 E M_0 e^{|\xi|}, \quad \text{for } t + i\tau \in J,$$

$$|g(\lambda + i0)| \leq c M_0 e^{|\xi|} \varepsilon, \quad \forall \lambda \in [0, \delta).$$

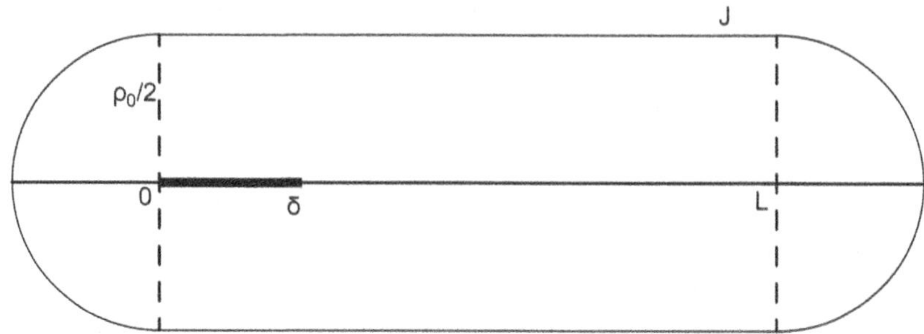

Fig. 13.1 Here, J denotes the set $\{z \in \mathbb{C} : \text{dist}(z, [0, L] \times \{0\}) < \rho_0/2\}$, and on the segment marked in black, an estimate of $|g|$ in terms of ε is at our disposal

Now, proceeding in a similar way to what we did to prove (12.130), we get

$$|g(\lambda + i0)| \leq \left(2C_1 E M_0 e^{|\xi|}\right)^{1-\vartheta} \left(c E M_0 e^{|\xi|} \varepsilon\right)^{\vartheta} \leq$$
$$\leq C_2 e^{|\xi|} E^{1-\vartheta} \varepsilon^{\vartheta}, \qquad \forall \lambda \in [0, L],$$

where $\vartheta \in (0, 1)$ depends on $L\rho_0^{-1}$ only and $C_2 = 2cC_1 M_0$. By the definition of g given in (13.26) and recalling that $w(x) = e^{-|\xi|} e^{-i\xi \cdot x} W$, we have

$$\left| \int_{B_1} \gamma(x) W^T u(x, \lambda\gamma(x)) e^{-i\xi \cdot x} dx \right| \leq C_2 e^{|\xi|} E^{1-\vartheta} \varepsilon^{\vartheta}, \qquad \forall \lambda \in [0, L]. \tag{13.32}$$

Step 4. Conclusion of the Proof
Let us fix $\lambda \in [0, L]$. Let $W = e_j$, for $j = 1, \cdots, N$. Set

$$f_j(x) = \begin{cases} \gamma(x) u_j(x, \lambda\gamma(x)), & \text{for } |x| \leq 1 \\ \\ 0, & \text{for } |x| \geq 1 \end{cases} . \tag{13.33}$$

Let us fix $j = 1, \cdots, N$ and, in the sequel, let us omit the index j by f_j. By (13.32) and (13.33) we get

$$|\widehat{f}(\xi)| = \left| \int_{\mathbb{R}^n} f(x) e^{-i\xi \cdot x} dx \right| \leq C_2 e^{|\xi|} E^{1-\vartheta} \varepsilon^{\vartheta}, \quad \forall \xi \in \mathbb{R}^n, \tag{13.34}$$

where $\widehat{f}$ is the Fourier transform of f (see Sect. 4.8).

The proof will be completed as soon as we estimate $|f(x)|$ by means of (13.34). Let us recall that

$$f(x) = (2\pi)^{-n} \int_{\mathbb{R}^n} \widehat{f}(\xi) e^{i\xi\cdot x} d\xi, \quad \forall x \in \mathbb{R}^n. \tag{13.35}$$

Let s be a positive number which we will choose later. By (13.34) and (13.35) we have, for every $x \in \mathbb{R}^n$

$$\begin{aligned} |f(x)| &\leq (2\pi)^{-n} \int_{\mathbb{R}^n} \left|\widehat{f}(\xi)\right| d\xi = \\ &= (2\pi)^{-n} \left(\int_{|\xi|\leq s} \left|\widehat{f}(\xi)\right| d\xi + \int_{|\xi|>s} \left|\widehat{f}(\xi)\right| d\xi \right) \leq \\ &\leq (2\pi)^{-n} \left(C_3 s^n e^s E^{1-\vartheta} \varepsilon^{\vartheta} + \int_{|\xi|>s} \left|\widehat{f}(\xi)\right| d\xi \right), \end{aligned} \tag{13.36}$$

where $C_3 = \frac{\omega_n}{n} C_2$ (ω_n is the measure of unit ball of $\mathbb{R}^n$).

To estimate from above the last integral in (13.36) we proceed as follows. First of all we note that by the definition of γ and f we have $\partial^\alpha f(x) = 0$ for every $x \in \partial B_1$ and for every $\alpha \in \mathbb{N}_0^n$ such that $|\alpha| \leq n$ from which we have, for $k = 1, \cdots, n$, by using integration by parts

$$\begin{aligned} (-i\xi_k)^{n+1} \widehat{f}(\xi) &= \int_{\mathbb{R}^n} (-i\xi_k)^{n+1} e^{-i\xi\cdot x} f(x) dx = \\ &= \int_{B_1} \partial_k^{n+1} \left(e^{-i\xi\cdot x}\right) f(x) dx = (-1)^{n+1} \int_{B_1} e^{-i\xi\cdot x} \partial_k^{n+1} f(x) dx. \end{aligned}$$

Hence (13.8c) implies

$$|\xi_k|^{n+1} \left|\widehat{f}(\xi)\right| \leq \int_{B_1} \left|\partial_k^{n+1} f(x)\right| dx \leq C_4 E, \tag{13.37}$$

where C_4 depends on L (and on n) only. So that, we have trivially

$$|\xi|^{n+1} \left|\widehat{f}(\xi)\right| \leq C_5 E, \tag{13.38}$$

where $C_5 = n^{\frac{n+1}{2}} C_4$. Now, by (13.37) we have

$$\begin{aligned} \int_{|\xi|>s} \left|\widehat{f}(\xi)\right| d\xi &= \int_{|\xi|>s} |\xi|^{-(n+1)} |\xi|^{n+1} \left|\widehat{f}(\xi)\right| d\xi \leq \\ &\leq C_5 E \int_{|\xi|>s} |\xi|^{-(n+1)} d\xi = \omega_n C_5 \frac{E}{s}. \end{aligned} \tag{13.39}$$

Now we use in (13.39) what we obtained in (13.39) and by the trivial inequality $E < E + 2\varepsilon$ we get

$$|f(x)| \leq C_6(E + 2\varepsilon)\left[\left(\frac{\varepsilon}{E + 2\varepsilon}\right)^{\vartheta} s^n e^s + \frac{1}{s}\right], \quad \forall x \in \mathbb{R}^n, \ \forall s > 0, \tag{13.40}$$

where C_6 depends on M, L, ρ, c_0 only.

In order to choose s we proceed as follows. Set

$$\sigma = \left(\frac{\varepsilon}{E + 2\varepsilon}\right)^{\vartheta}$$

and rewrite the term on the right-hand side of (13.40) as

$$\phi(s) := \exp\left(s + \log s - |\log \sigma|\right) + \frac{1}{s}.$$

Now we choose

$$s : s_0 = \frac{1}{2}|\log \sigma|, \tag{13.41}$$

taking into account that $0 < \sigma \leq 2^{-\vartheta} < 1$, we have

$$\phi(s_0) = \sqrt{\sigma}\left|\log \sqrt{\sigma}\right| + 2|\log \sigma|^{-1} \leq c_{\vartheta}|\log \sigma|^{-1},$$

where c_{ϑ} depends on ϑ. Hence, by (13.41) and (13.40) we have

$$|f(x)| \leq C_7(E + 2\varepsilon)\left|\log\left(\frac{\varepsilon}{E + 2\varepsilon}\right)\right|^{-1}, \tag{13.42}$$

where $C_7 = C_6(c_{\vartheta} + 2)\vartheta^{-1}$. Finally, recalling that $f(x) = \gamma(x)u_j(x, \lambda\gamma(x))$ for $|x| \leq 1$, we have, for any $r \in (0, 1)$

$$|u(x, \lambda\gamma(x))| \leq \frac{C_7 N^{1/2}}{(1 - r)^{n+1}}(E + 2\varepsilon)\left|\log\left(\frac{\varepsilon}{E + 2\varepsilon}\right)\right|^{-1}, \tag{13.43}$$

for $|x| \leq 1 - r$, from which the thesis follows.

■

Remark 13.1 Let us consider the case in which instead of (13.8) we have, for $(x, t) \in \mathcal{R}_{0,L}$,

$$u_t(x,t) = \sum_{j=1}^{n} A_j(x,t)u_{x_j}(x,t) + A_0(x,t)u(x,t) + F(x,t), \tag{13.44a}$$

$$\|u(\cdot,0)\|_{L^\infty(B_1)} \le \varepsilon, \tag{13.44b}$$

$$\|u\|_{C^{n+1}(\overline{\mathcal{R}_{0,L}})} \le E, \tag{13.44c}$$

$$\|F\|_{L^\infty(\mathcal{R}_{0,L}} \le \varepsilon_1, \tag{13.44d}$$

where F *is not necessarily analytic*. Let us continue to denote (even though $\lambda_1 > \lambda_2$) by $\mathcal{R}_{\lambda_1,\lambda_2}$ the subset of $\mathbb{R}^{n+1}$ enclosed by S_{λ_1} and S_{λ_2}. In such a way (13.18) becomes

$$\begin{aligned}
&\int_{B_1} v^T(x,\lambda_2\gamma(x))A(x,\lambda_2)u(x,\lambda_2\gamma(x))dx+ \\
&+\operatorname{sgn}(\lambda_1-\lambda_2)\int_{\mathcal{R}_{\lambda_1,\lambda_2}} v^T(x,t)F(x,t)dxdt = \\
&=\int_{B_1} v^T(x,\lambda_1\gamma(x))A(x,\lambda_1)u(x,\lambda_1\gamma(x))dx.
\end{aligned} \tag{13.45}$$

We construct $v(x,t;\lambda)$ likewise the Step 2 of Theorem 13.2. Consequently, instead of (13.26), for a fixed λ_0 in $[0,L]$ and setting

$$\begin{aligned}
\widetilde{g}(\lambda) = &\int_{B_1} w^T(x)u(x,\lambda\gamma(x))dx+ \\
&+\operatorname{sgn}(\lambda_0-\lambda)\int_{\mathcal{R}_{\lambda,\lambda_0}} v^T(x,t;\lambda)F(x,t)dxdt,
\end{aligned}$$

we have

$$\widetilde{g}(\lambda) = \int_{B_1} v^T(x,\lambda_0\gamma(x);\lambda)A(x,\lambda_0)u(x,\lambda_0\gamma(x))dx, \tag{13.46}$$

for $|\lambda-\lambda_0| < \delta$ and $\lambda \in [0,L]$. Exactly like the Step 3 of the proof of Theorem 13.2 we get the estimate

$$|\widetilde{g}(\lambda+i0)| \le C_2 M_0 e^{|\xi|} E^{1-\vartheta}\varepsilon^{\vartheta}, \quad \forall\lambda\in[0,L]. \tag{13.47}$$

Now by (13.25) and by (13.44d) we have

$$\left|\int_{\mathcal{R}_{\lambda,\lambda_0}} v^T(x,t;\lambda)F(x,t)dxdt\right| \leq \varepsilon_1 \int_{\mathcal{R}_{\lambda,\lambda_0}} \left|v^T(x,t;\lambda)\right| dxdt \leq \tag{13.48}$$
$$\leq cLM_0 e^{|\xi|}\varepsilon_1,$$

where c depends on n only. By (13.47) and (13.48) we have

$$\left|\int_{B_1} \gamma(x)w^T(x)u(x,\lambda\gamma(x))dx\right| \leq C_2 M_0 e^{|\xi|} E^{1-\vartheta}\varepsilon^{\vartheta} + cLM_0 e^{|\xi|}\varepsilon_1.$$

From now on we may argue like Step 4 of the proof of Theorem 13.1 and we find

$$|u(x,t)| \leq \frac{C(E+2\varepsilon_2)}{(1-r)^{n+1}}\left|\log\frac{\varepsilon_2}{E+2\varepsilon_2}\right|^{-1}, \tag{13.49}$$

for all x in $\overline{\mathcal{R}_{0,L}} \cap \left(\overline{B_{1-r}} \times \mathbb{R}\right)$, where C depends on M, L, ρ, c_0 and n only and

$$\varepsilon_2 = E^{1-\vartheta}\varepsilon^{\vartheta} + \varepsilon_1.$$

Remark 13.2 The a priori information (13.8c) occurs in the proof of stability Theorem especially to obtain (13.39), while in the other parts of the proof what is needed to know about u is only that

$$\|u\|_{L^\infty(\mathcal{R}_{0,L})} \leq E.$$

It is not difficult, with some care, to weaken the a priori information (13.8c) and arrive at an estimate essentially analogous to that of the Theorem. We encourage the reader to work out the details when, instead of assuming (13.8c), we assume that

$$\|u\|_{C^1\left(\overline{\mathcal{R}_{0,L}}\right)} \leq E. \tag{13.50}$$

Another useful (though simple) investigation that we propose for the reader's consideration involves examining, from the perspective of stability estimates, the case where the initial surface in the Cauchy problem is not a portion of a hyperplane $\{t = 0\}$. In this case, one may consider the following Cauchy problem

$$\begin{cases} u_t = \sum_{j=1}^n A_j(x,t)u_{x_j} + A_0(x,t)u + F(x,t) \\ \\ u(x,\varphi(x)) = g(x), \quad x \in B_1 \end{cases}, \tag{13.51}$$

where $\varphi \in C^2\left(\overline{B_1}\right)$ and it satisfies $\varphi(0) = |\nabla\varphi(0)| = 0$. Moreover, Where A_j, $j = 0, 1, \cdots, n$ are analytic functions. We can require that the surface $\{t = \varphi(x) | x \in$

$B_1\}$ is noncharacteristic. This is equivalent to require that the "algebraic" system

$$\begin{cases} u_t(x,\varphi(x)) - \sum_{j=1}^n A_j(x,\varphi(x))u_{x_j}(x,\varphi(x)) = \widetilde{f}(x) \\ \\ u_t(x,\varphi(x))\varphi_{x_i}(x) + u_{x_i}(x,\varphi(x)) = g_{x_i}(x), \quad i = 1,\cdots,n \end{cases}, \tag{13.52}$$

has a unique solution. ♦

Remark 13.3 Stability estimate (13.9) is a *logarithmic estimate* and, while it is still a stability estimate, it is a rather modest estimate. John, in [65], called "well-behaved" the problems for which a Hölder conditional stability holds and "not well-behaved" the problems for which the conditional stability is at best of logarithmic type. This terminology is still in use today. Of course, in order to be able to say that a class of problems is "well-behaved" or "not well-behaved" with respect to certain a priori information, it needs to be shown that the estimate in question is optimal in that class of problems with those certain a priori information. Concerning Theorem 13.1, the class of problems is the class of the Cauchy problems for partial differential equations with analytic coefficients and the a priori bounds concern a finite numbers of derivatives of the solutions. Now, with respect to the class of problems and of the a priori information that we have considered above, John himself, in [65], proved that the Cauchy problem is "not well-behaved." The example constructed by John concerns the following Cauchy problem for the wave equation

$$\begin{cases} u_{xx} + u_{yy} - u_{tt} = 0, & x^2 + y^2 < 1, \ t \in \mathbb{R} \\ u = g_0, & x \in \partial B_\rho \times \mathbb{R} \\ \frac{\partial u}{\partial \nu} = g_1, & x \in \partial B_\rho \times \mathbb{R} \end{cases}. \tag{13.53}$$

where $\rho < 1$ and ν is the unit outward normal to $\partial B_\rho \times \mathbb{R}$.

More precisely, set

$$\varepsilon = \|g_0\|_{L^\infty(\partial B_\rho \times \mathbb{R})} + \|g_1\|_{L^\infty(\partial B_\rho \times \mathbb{R})}, \tag{13.54}$$

John has proved that for every $m \in \mathbb{N}$ there exists $u \in C^m\left(\overline{B_1} \times \mathbb{R}\right)$ solution to (13.53), where g_0, g_1 satisfy (13.54), such that

$$\|u\|_{C^m\left(\overline{B_1} \times \mathbb{R}\right)} = 1$$

and such that

$$\|u\|_{L^\infty(B_r \times \mathbb{R})} \geq C\,|\log \varepsilon|^{-\alpha},$$

where $r \in (\rho, 1)$, $C > 0$ and $\alpha > 0$ depend on r. ♦

Concluding Remark At the conclusion of this chapter, we should certainly note that Theorem 13.1 is primarily of historical interest, as the strong regularity assumptions on the coefficients have been substantially relaxed. For certain classes of equations, such as elliptic equations (see [65]) or other equations where Carleman estimates are available Hölder-type estimates can be established, as proved in [10] and [118]. However, the most suitable framework for generalizations of the Holmgren–John Theorem is that of operators with non-analytic coefficients, particularly that of operators with partially analytic coefficients. Starting from Chap. 14, which marks the beginning of Part III of this book, we will investigate the issues of uniqueness and stability for the first type of operators, specifically those with entirely non-analytic coefficients. We will not address the second type, which lies in an intermediate setting between the cases of analytic and completely non-analytic coefficients. In this regard, we should cite [74], which establishes the aforementioned stability estimates. This article follows a line of research that qualitatively began with [111] and was subsequently advanced in works such as [59, 112], and [127], which we will briefly revisit later. ♦

Part III
Carleman Estimates and Unique Continuation Properties

Chapter 14
Carleman Estimates: A First Look with Simple Examples and Basic Applications

Abstract In this chapter, we begin by examining unique continuation properties for operators with non-analytic coefficients. We define the weak and strong unique continuation properties for linear differential equations. The weak property implies that if a solution is zero in some subset of the domain, it must be zero in the entire domain. The strong property extends this by stating that if a solution decays rapidly near a point, it must be zero in the entire domain. The strong property implies the weak one. The main goal of this chapter is to lay the groundwork for Carleman estimates, a technique introduced by Carleman in the 1930s. These weighted integral estimates depend on a parameter and have become a powerful tool in studying unique continuation properties, particularly for equations with non-analytic coefficients. This chapter also includes simple examples of Carleman estimates, preparing the reader for more general theory in later chapters. We will briefly discuss operators with constant coefficients, which, while now primarily of historical and pedagogical interest, help clarify the concept of Carleman estimates. Finally, we will present a theorem by Hörmander on the necessary conditions for Carleman estimates to hold for a differential operator.

14.1 Introduction

In this chapter we begin to study the unique continuation properties for **operators with non analytic coefficients**. First we give some definitions. Let Ω be an open connected set of $\mathbb{R}^n$, we say that the linear differential equation

$$Lu = 0 \quad \text{in } \Omega, \tag{14.1}$$

enjoys the **weak unique continuation property**, if for any open subset ω of Ω, we have

$$P(x, \partial)u = 0 \quad \text{in } \Omega \text{ and } u = 0 \text{ in } \omega \quad \Longrightarrow u = 0 \text{ in } \Omega.$$

S. Vessella, *Unique Continuation Properties for Partial Differential Equations*,
Birkhäuser Advanced Texts Basler Lehrbücher,
https://doi.org/10.1007/978-3-031-86366-0_14

We say that Eq. (14.1) enjoys the **strong unique continuation property**, if for any point $x_0 \in \Omega$ and for any solution u which satisfies

$$\lim_{r\to 0} r^{-k} \int_{B_r(x_0)} u^2 = 0, \quad \forall k \in \mathbb{N},$$

it follows that

$$u = 0, \quad \text{in } \Omega.$$

It is obvious that the strong unique continuation property implies the weak unique continuation property.

As we will see later, the weak unique continuation property is strictly related to the uniqueness of the Cauchy problem for Eq. (14.1).

The primary objective of this chapter is to establish the foundation for the Carleman estimates, which will be thoroughly explored in the subsequent chapters. These types of estimates were introduced by Carleman in [22] and [23] (in 1933 and 1939 respectively). Carleman estimates are specific types of weighted integral estimates that depend on a parameter (see Sect. 14.2 for further details). Through these estimates, a significant qualitative advancement is achieved in the study of unique continuation properties for partial differential equations, especially in the context of the Cauchy problem. Utilizing the Carleman estimates, it becomes feasible to establish unique continuation properties for differential equations possessing nonanalytic coefficients. While the estimates originally presented in [22] and [23] pertain to partial differential equations with two variables, Carleman's conceptual framework has proven highly fruitful, giving rise to a technique that stands as one of the most comprehensive and potent tools—although not the sole one—for tackling unique continuation issues.

The study of Carleman estimates has seen extensive and diverse development over the last 70 years [56, 58, 83], with significant applications in the study of inverse problems [63, 76] and control theory [46]. Thus, the bibliographic references we have provided here, and those we will present throughout this work, can scarcely do justice to the breadth of literature on the subject. We will, however, limit ourselves to citing the books and articles that have most inspired us in the preparation of this Part III.

To introduce the topic, in this chapter we will also provide several simple examples (see Sect. 14.3) of Carleman estimates that prepare the reader for the more general theory presented mainly in Chaps. 15 and 16, as well as for understanding how they can be applied to prove unique continuation properties for differential equations and the corresponding stability estimates (see Sect. 14.4). In particular, in this chapter we will focus (see Sect. 14.5) a bit on the case of operators whose principal part has constant coefficients. Nowadays, operators with constant coefficients in the principal part, within the context of the study of unique continuation properties and Carleman estimates, primarily hold historical and pedagogical importance (see [58, Vol. II], [131]). However, precisely for these reasons, we believe that a brief treatment like the one we propose in this section can provide additional clarity on the topic of Carleman estimates. This is especially true as the proof techniques we present here are particularly straightforward and accessible. In the final part of this chapter (see Sect. 14.6), we will make some remarks and state a theorem due to Hörmander on the necessary conditions for a Carleman estimate to hold for a differential operator of order m.

14.2 Basic Properties

Let Ω be an open set of $\mathbb{R}^n$. Let us consider the operator

$$P(x, D) = \sum_{|\alpha| \leq m} a_\alpha(x) D^\alpha, \qquad x \in \Omega, \tag{14.2}$$

where $D_j = \frac{1}{i}\partial_j$, $j = 1, \cdots, n$ and $a_\alpha \in L^\infty(\Omega; \mathbb{C})$, for every $\alpha \in \mathbb{N}_0^n$, $|\alpha| \leq m$.

We denote by $P_m(x, D)$ the principal part of $P(x, D)$, namely

$$P_m(x, D) = \sum_{|\alpha| = m} a_\alpha(x) D^\alpha \tag{14.3}$$

(of course we assume that a_α is not identically zero for at least one α such that $|\alpha| = m$). In the sequel, if there is no risk of ambiguity, we will also denote $L^\infty(\Omega; \mathbb{C})$ by $L^\infty(\Omega)$, we will do the same with other function spaces.

Let φ be a sufficiently regular real-valued function.

We call **Carleman estimate** for the operator $P(x, D)$ with **exponential weight** φ, an inequality of the following type: there exists $\mu > 0$ such that for every K, compact subset of Ω, there exist $C > 0$ and $\tau_0 > 0$ such that

$$\tau^\mu \sum_{|\alpha| \leq m-1} \int_\Omega \left|D^\alpha u\right|^2 e^{2\tau\varphi} dx \leq C \int_\Omega |P(x, D)u|^2 e^{2\tau\varphi} dx, \tag{14.4}$$

for every $u \in C_K^\infty(\Omega)$ and for every $\tau \geq \tau_0$. Of course, we will continue to call Carleman estimates the inequalities of type (14.4) that hold for every $u \in C_0^\infty(\Omega)$.

Proposition 14.1 *Let $P(x, D)$ be operator* (14.2) *and $\mu > 0$. Carleman estimate* (14.4) *is equivalent to the following Carleman estimate: for every K, compact subset of Ω, there exist $C > 0$ and $\tau_1 > 0$ such that*

$$\tau^\mu \sum_{|\alpha| \leq m-1} \int_\Omega \left|D^\alpha u\right|^2 e^{2\tau\varphi} dx \leq C \int_\Omega |P_m(x, D)u|^2 e^{2\tau\varphi} dx, \tag{14.5}$$

for every $u \in C_K^\infty(\Omega)$ and for every $\tau \geq \tau_1$.

Proof Let us suppose that estimate (14.4) holds and let us denote by

$$R(x, D) = P_m(x, D) - P(x, D),$$

we have

$$|R(x, D)u| = \left| \sum_{|\alpha| \le m-1} a_\alpha(x) D^\alpha u \right| \le M \sum_{|\alpha| \le m-1} \left| D^\alpha u \right|, \tag{14.6}$$

where

$$M = \max_{|\alpha| \le m-1} \|a_\alpha\|_{L^\infty(\Omega)} .$$

Let K be a compact subset of Ω, by (14.4) there exist $C > 0$ and $\tau_0 > 0$ such that

$$\tau^\mu \sum_{|\alpha| \le m-1} \int_\Omega \left| D^\alpha u \right|^2 e^{2\tau\varphi} dx \le C \int_\Omega |P(x, D)u|^2 e^{2\tau\varphi} dx \le$$

$$\le 2C \int_\Omega |P_m(x, D)u|^2 e^{2\tau\varphi} dx + 2C \int_\Omega |R(x, D)u|^2 e^{2\tau\varphi} dx \le$$

$$\le 2C \int_\Omega |P_m(x, D)u|^2 e^{2\tau\varphi} dx + C_* M^2 \sum_{|\alpha| \le m-1} \int_\Omega \left| D^\alpha u \right|^2 e^{2\tau\varphi} dx,$$

for every $u \in C^\infty_K(\Omega)$ and for every $\tau \ge \tau_0$. Moving the last sum to the left-hand side, we have

$$\begin{aligned} \left(\tau^\mu - C_* M^2\right) \sum_{|\alpha| \le m-1} \int_\Omega \left| D^\alpha u \right|^2 e^{2\tau\varphi} dx \le \\ \le 2C \int_\Omega |P_m(x, D)u|^2 e^{2\tau\varphi} dx, \end{aligned} \tag{14.7}$$

for every $u \in C^\infty_K(\Omega)$ and for every $\tau \ge \tau_0$. Now, set

$$\tau_1 = \max \left\{ \tau_0, \left(2C_* M^2\right)^{1/\mu} \right\},$$

by (14.7) we obtain

$$\tau^\mu \sum_{|\alpha| \le m-1} \int_\Omega \left| D^\alpha u \right|^2 e^{2\tau\varphi} dx \le 4C \int_\Omega |P_m(x, D)u|^2 e^{2\tau\varphi} dx,$$

for every $u \in C^\infty_K(\Omega)$ and for every $\tau \ge \tau_1$. Hence, if (14.4) holds then (14.5) holds. The converse can be similarly proved. ■

It should be observed at once that estimates of type (14.4) have a local character in the sense specified in the following

Lemma 14.2 (Local Character of the Carleman Estimates) *Let Ω be an open set of $\mathbb{R}^n$, let $\varphi \in C^0(\Omega)$ be a real-valued function. Let $P(x, D)$ be operator* (14.2) *and $\mu > 0$. Let K be a compact subset of Ω. Assume that for each $y \in K$ there exist $\delta_y > 0$, $C_y > 0$ and $\tau_y \in \mathbb{R}$ such that $B_{\delta_y}(y) \subset \Omega$ and*

$$\begin{aligned} \tau^{\mu} \sum_{|\alpha|\leq m-1} \int_{B_{\delta_y}(y)} \left|D^{\alpha}u\right|^2 e^{2\tau\varphi}dx \leq \\ \leq C_y \int_{B_{\delta_y}(y)} |P(x, D)u|^2 e^{2\tau\varphi}dx, \end{aligned} \tag{14.8}$$

for every $u \in C_0^{\infty}\left(B_{\delta_y}(y)\right)$ and for every $\tau \geq \tau_y$. Then there exist $C > 0$ and $\tau_0 \in \mathbb{R}$ such that

$$\tau^{\mu} \sum_{|\alpha|\leq m-1} \int_{\Omega} \left|D^{\alpha}u\right|^2 e^{2\tau\varphi}dx \leq C \int_{\Omega} |P(x, D)u|^2 e^{2\tau\varphi}dx, \tag{14.9}$$

for every $u \in C_K^{\infty}(\Omega)$ and for every $\tau \geq \tau_0$.

Proof Since K is compact, there exists a finite set of points, $y_1, \cdots, y_N$, such that

$$K \subset \bigcup_{j=1}^{N} B_{\delta_{y_j}}(y_j).$$

Let $\left\{\eta_j\right\}_{1\leq j\leq N}$ be a partition of unity (Lemma 3.50) such that

$$\eta_j \in C_0^{\infty}\left(B_{\delta_{y_j}}(y_j)\right); \quad 0 \leq \eta_j \leq 1; \quad \sum_{j=1}^{N} \eta_j = 1, \quad \text{in } K.$$

Let $u \in C_K^{\infty}(\Omega)$. Let us denote by

$$u_j = u\eta_j, \qquad \text{for } j = 1, \cdots, N.$$

For every $\alpha \in \mathbb{N}_0^n$, by the Cauchy–Schwarz inequality we have

$$\left|D^{\alpha}u\right|^2 = \left|\sum_{j=1}^{N} D^{\alpha}u_j\right|^2 \leq N \sum_{j=1}^{N} \left|D^{\alpha}u_j\right|^2. \tag{14.10}$$

Now, we notice that

$$P(x, D)u_j = \eta_j P(x, D)u + R_{m-1}(x, D)u,$$

where $R_{m-1}(x, D)$ is an operator of order $m - 1$ whose coefficients depend on the coefficients of $P(x, D)$ (but not on their derivatives) and on the functions η_j, $j = 1, \cdots N$ and their derivatives of order less or equal to m. We have

$$\left|P(x, D)u_j\right| \leq \eta_j \left|P(x, D)u\right| + M \sum_{|\alpha|\leq m-1} \left|D^\alpha u\right|, \tag{14.11}$$

where M depends on the L^∞-norms of the coefficients of $P(x, D)$ and on the L^∞-norms of the derivatives of η_j of order less or equal to m.

By (14.10) and (14.11) we get

$$\tau^\mu \sum_{|\alpha|\leq m-1} \int_\Omega \left|D^\alpha u\right|^2 e^{2\tau\varphi} dx \leq$$

$$\leq CN\tau^\mu \sum_{j=1}^{N} \sum_{|\alpha|\leq m-1} \int_{B_{\delta_{y_j}}(y_j)} \left|D^\alpha u_j\right|^2 e^{2\tau\varphi} dx \leq$$

$$\leq N \sum_{j=1}^{N} \int_{B_{\delta_{y_j}}(y_j)} \left|P(x, D)u_j\right|^2 e^{2\tau\varphi} dx \leq$$

$$\leq \widetilde{C} \int_\Omega |P(x, D)u|^2 e^{2\tau\varphi} dx + \widetilde{C} \sum_{|\alpha|\leq m-1} \int_\Omega \left|D^\alpha u\right|^2 e^{2\tau\varphi} dx,$$

for every $\tau \geq \tau_0$, where

$$\tau_0 = \max_{1\leq j\leq N} \tau_{y_j}$$

and $\widetilde{C}$ is a constant. Now, we move the last integral to the left hand side and we proceed exactly as did above to prove Proposition 14.1 and we obtain (14.9). ■

14.3 Some Straightforward Examples of Carleman Estimates and Applications

In this section, we will provide some simple and straightforward Carleman estimates for some operators and we apply such estimates to prove various form of the unique continuation properties.

1. The Cauchy–Riemann Operator Let P be the Cauchy–Riemann operator

$$P = \partial_{\bar{z}} = \frac{1}{2}\left(\partial_x + i\partial_y\right). \tag{14.12}$$

We are interested in proving a suitable Carleman estimate for operator P and we will apply it to prove the strong unique continuation property for the equation

$$\partial_{\bar{z}} U + a(x, y) U = 0, \tag{14.13}$$

where a is a bounded complex-valued function.

We begin by the following

Proposition 14.3 *Let P be operator* (14.12) *and let $\varphi \in C^2\left(\mathbb{R}^2\right)$ be a real-valued function. Then there exists a constant C such that the following inequality holds*

$$\tau \int_{\mathbb{R}^2} \Delta\varphi \, |u|^2 e^{2\tau\varphi} dxdy \leq C \int_{\mathbb{R}^2} |Pu|^2 e^{2\tau\varphi} dxdy, \tag{14.14}$$

for every $u \in C_0^\infty\left(\mathbb{R}^2\right)$ and for every $\tau > 0$.

Proof For any $u \in C_0^\infty\left(\mathbb{R}^2\right)$ let us set

$$v = e^{\tau\varphi} u.$$

We have

$$P_\tau v := e^{\tau\varphi} P\left(e^{-\tau\varphi} v\right) = \frac{1}{2}\left(\partial_{\bar{z}} v - \tau v \partial_{\bar{z}} \varphi\right).$$

Hence, (14.14) is equivalent to (we omit the domain of integration)

$$\tau \int \Delta\varphi \, |v|^2 dxdy \leq C \int |P_\tau v|^2 dxdy, \tag{14.15}$$

for every $v \in C_0^\infty\left(\mathbb{R}^2\right)$ and for every $\tau > 0$.

We have

$$2P_\tau v = \left(i\partial_y v - \tau v \partial_x \varphi\right) + \left(\partial_x v - i\tau v \partial_y \varphi\right),$$

$$\begin{aligned} 4\,|P_\tau v|^2 &= \left|i\partial_y v - \tau v \partial_x \varphi\right|^2 + \left|\partial_x v - i\tau v \partial_y \varphi\right|^2 + \\ &+ 2\Re\left(\left(i\partial_y v - \tau v \partial_x \varphi\right)\overline{\left(\partial_x v - i\tau v \partial_y \varphi\right)}\right). \end{aligned} \tag{14.16}$$

Now we have

$$\begin{aligned}
&2\Re\left(\left(i\partial_y v - \tau v\partial_x\varphi\right)\overline{\left(\partial_x v - i\tau v\partial_y\varphi\right)}\right) = \\
&= 2\Re\left(i\partial_y v\overline{\partial_x v}\right) - \tau\partial_x\varphi\partial_x\left(|v|^2\right) - \tau\partial_y\varphi\partial_y\left(|v|^2\right) = \\
&= 2\Re\left(i\partial_y v\overline{\partial_x v}\right) + \tau|v|^2\Delta\varphi - \tau\,\mathrm{div}\left(|v|^2\nabla\varphi\right).
\end{aligned} \tag{14.17}$$

Concerning the term

$$2\Re\left(i\partial_y v\overline{\partial_x v}\right),$$

we set

$$v_1 = \Re(v), \quad v_2 = \Im(v)$$

and we get

$$\begin{aligned}
\Re\left(i\partial_y v\overline{\partial_x v}\right) &= \partial_y v_1\partial_x v_2 - \partial_y v_2\partial_x v_1 = \\
&= \partial_x\left(v_2\partial_y v_1\right) - v_2\partial^2_{xy}v_1 - \partial_y\left(v_2\partial_x v_1\right) + v_2\partial^2_{xy}v_1 = \\
&= \partial_x\left(v_2\partial_y v_1\right) - \partial_y\left(v_2\partial_x v_1\right).
\end{aligned} \tag{14.18}$$

By (14.17) and (14.18), taking into account that $v \in C_0^\infty\left(\mathbb{R}^2\right)$, we get

$$\int 2\Re\left(\left(i\partial_y v - \tau v\partial_x\varphi\right)\overline{\left(\partial_x v - i\tau v\partial_y\varphi\right)}\right)dxdy = \tau\int \Delta\varphi|v|^2 dxdy.$$

This identity and (14.16) trivially imply

$$\begin{aligned}
4\int |P_\tau v|^2\,dxdy &\geq \int 2\Re\left(\left(i\partial_y v - \tau v\partial_x\varphi\right)\overline{\left(\partial_x v - i\tau v\partial_y\varphi\right)}\right)dxdy = \\
&= \tau\int \Delta\varphi|v|^2 dxdy.
\end{aligned}$$

By this inequality and by (14.15) we obtain (14.14). ■

In the following proposition we prove a Carleman estimate for operator (14.12).

Proposition 14.4 *Let P be operator* (14.12). *Define*

$$\varphi(x, y) = \psi\left(\log\frac{1}{\sqrt{x^2+y^2}}\right), \tag{14.19}$$

where

$$\psi(s) = s + e^{-s}, \quad s \geq 0. \tag{14.20}$$

Then there exists a constant C such that following inequality holds

$$\tau \int_{\mathbb{R}^2} |z|^{-1} |u|^2 e^{2\tau\varphi} dxdy \leq C \int_{\mathbb{R}^2} |Pu|^2 e^{2\tau\varphi} dxdy, \tag{14.21}$$

for every $u \in C_0^\infty\left(\mathbb{R}^2 \setminus \{(0,0)\}\right)$ *and for every* $\tau > 0$.

Proof Set

$$h(x,y) = \log\left(\frac{1}{\sqrt{x^2+y^2}}\right)$$

and let us choose in (14.14)

$$\varphi = \psi \circ h.$$

Taking into account that

$$\Delta h = 0, \quad \text{in } \mathbb{R}^2 \setminus \{(0,0)\},$$

we have

$$\Delta\varphi = \psi''(h)|\nabla h|^2 + \psi'(h)\Delta h = \frac{1}{\sqrt{x^2+y^2}}, \quad \text{in } \mathbb{R}^2 \setminus \{(0,0)\},$$

and by (14.14) we get (14.21). ■

Remark 14.1 Note that by (14.19) we have

$$e^{\varphi(x,y)} \sim \frac{1}{\sqrt{x^2+y^2}}, \quad \text{as } (x,y) \to (0,0). \tag{14.22}$$

♦

In the following theorem we prove a strong unique continuation property for Eq. (14.13) and a three circles inequality.

Theorem 14.5 *Let* $U \in H^1(B_1)$ *be a solution to*

$$\partial_{\bar{z}} U + a(x,y) U = 0, \tag{14.23}$$

where

$$a \in L^{\infty}(B_1). \tag{14.24}$$

(a) Assume that

$$\|U\|_{L^2(B_r)} = O(r^N), \quad as\; r \to 0,\ \forall N \in \mathbb{N}_0. \tag{14.25}$$

Then

$$U \equiv 0, \quad in\ B_1. \tag{14.26}$$

(b) Let $r_0 \in (0, 1/2)$ and $R_1 \in (1/2, 1)$. Then for every $\rho \in (2r_0, 1-2\delta)$ the following inequality holds

$$\int_{B_\rho} |U|^2\, dxdy \leq$$
$$\leq C\left(\rho r_0^{-2}\int_{B_{r_0}} |U|^2\, dxdy\right)^{\theta}\left((1-R_1)^{-2}\int_{B_1} |U|^2\, dxdy\right)^{1-\theta}, \tag{14.27}$$

where C is a constant that depends on ρ, δ and $\|a\|_{L^\infty(B_1)}$ only and

$$\theta = \frac{\phi(R_1) - \phi(\rho)}{\phi(R_1) - \phi(r_0/2)}, \tag{14.28}$$

where

$$\phi(t) = t + \log\frac{1}{t}, \quad t \in (0, 1). \tag{14.29}$$

Proof Let us denote by

$$\|a\|_{L^\infty(B_1)} = M.$$

Let $U \in H^1(B_1)$ be a solution to (14.23). By density we can apply estimate (14.21) to $u = \eta U$, where $\eta \in C_0^1(B_1) \setminus \{(0,0)\}$.

Let $\delta \in (0, 1/4)$, $\rho \in (0, 1-2\delta)$. Let us choose η as follows. Let $r \in (0, \rho/2)$. Let $\eta \in C_0^1(B_1 \setminus \{(0,0)\})$ satisfy

$$0 \leq \eta(x, y) \leq 1, \quad \forall (x, y) \in B_1,$$

$$\eta(x, y) = 0, \quad \forall (x, y) \in B_r \cup (B_1 \setminus B_{1-\delta})$$

and

$$\eta(x,y)=1, \quad \forall (x,y)\in B_{1-2\delta}\setminus B_{2r}.$$

Moreover, let us assume that

$$\sup_{B_{2r}\setminus B_r}|\nabla\eta|\leq Cr^{-1} \text{ and } \sup_{B_{1-\delta}\setminus B_{1-2\delta}}|\nabla\eta|\leq C\delta^{-1},$$

where C is a constant.

By (14.23) we have

$$\partial_{\overline{z}}(\eta U)=-\eta\partial_{\overline{z}}U-U\partial_{\overline{z}}\eta=-\eta aU-U\partial_{\overline{z}}\eta,$$

from which we derive

$$|\partial_{\overline{z}}(\eta U)|\leq M\,|\eta U|+C\left(r^{-1}\chi_{B_{2r}\setminus B_r}+\delta^{-1}\chi_{B_{1-\delta}\setminus B_{1-2\delta}}\right)|U|. \tag{14.30}$$

By inequality (14.21) we get

$$\begin{aligned}&\tau\int_{B_1}|z|^{-1}\,|\eta U|^2\,e^{2\tau\varphi}dxdy\leq C\int_{B_1}|\partial_{\overline{z}}(\eta U)|^2\,e^{2\tau\varphi}dxdy\leq\\ &\leq C_1M^2\int_{B_1}|\eta U|^2\,e^{2\tau\varphi}dxdy+C_1r^{-2}\int_{B_{2r}\setminus B_r}|U|^2\,e^{2\tau\varphi}dxdy+\\ &+C_1\delta^{-2}\int_{B_{1-\delta}\setminus B_{1-2\delta}}|U|^2\,e^{2\tau\varphi}dxdy,\end{aligned}$$

for every $\tau>0$, where C_1 is a constant.

Therefore

$$\begin{aligned}&\tau\int_{B_1}\left(|z|^{-1}-C_1M^2\tau^{-1}\right)|\eta U|^2\,e^{2\tau\varphi}dxdy\leq\\ &\leq C_1r^{-2}\int_{B_{2r}\setminus B_r}|U|^2\,e^{2\tau\varphi}dxdy+C_1\delta^{-2}\int_{B_{1-\delta}\setminus B_{1-2\delta}}|U|^2\,e^{2\tau\varphi}dxdy,\end{aligned} \tag{14.31}$$

for every $\tau>0$.

Now, denoting by $\tau_0=2C_1M^2$, we have trivially

$$|z|^{-1}-C_1M^2\tau^{-1}\geq\frac{1}{2}|z|^{-1}, \quad \forall\tau\geq\tau_0, \quad 0<|z|\leq 1.$$

By this inequality and by (14.31) we have

$$\begin{aligned}
&\tau \int_{B_1} |z|^{-1} |\eta U|^2 e^{2\tau\varphi} dxdy \leq \\
&\leq Cr^{-2} \int_{B_{2r}\setminus B_r} |U|^2 e^{2\tau\varphi} dxdy + \\
&+ C\delta^{-2} \int_{B_{1-\delta}\setminus B_{1-2\delta}} |U|^2 e^{2\tau\varphi} dxdy,
\end{aligned} \tag{14.32}$$

for every $\tau \geq \tau_0$, where C is a constant.

Now, by (14.29) we have

$$\varphi(x, y) = \phi\left(\sqrt{x^2 + y^2}\right)$$

and since ϕ is decreasing in $(0, 1)$, by (14.32) we have, for any $\rho \in (2r, 1 - 2\delta)$,

$$\begin{aligned}
&\tau\rho^{-1} e^{2\tau\phi(\rho)} \int_{B_\rho} |\eta U|^2 dxdy \leq \tau \int_{B_1} |z|^{-1} |\eta U|^2 e^{2\tau\varphi} dxdy \leq \\
&\leq Cr^{-2} e^{2\tau\phi(r)} \int_{B_{2r}} |U|^2 dxdy + C\delta^{-2} e^{2\tau\phi(1-2\delta)} \int_{B_1} |U|^2 dxdy,
\end{aligned} \tag{14.33}$$

for every $\tau \geq \tau_0$.

By the above obtained inequality and by (14.25) we derive

$$\begin{aligned}
&\rho^{-1} e^{2\tau\phi(\rho)} \int_{B_\rho} |\eta U|^2 dxdy \leq \\
&\leq C_N r^{2N-2} e^{2\tau\phi(r)} + C\delta^{-2} e^{2\tau\phi(1-2\delta)} \int_{B_1} |U|^2 dxdy,
\end{aligned} \tag{14.34}$$

for every $N \in \mathbb{N}_0$ and for every $\tau \geq \tau_0$, where C_N is a sequence of positive numbers (independent of r). For $N > \max\{\tau_0, 2\}$ let us choose in (14.34)

$$\tau = \frac{N}{2}$$

and we get

$$\begin{aligned}
&\rho^{-1} e^{N\phi(\rho)} \int_{B_\rho} |\eta U|^2 dxdy \leq \\
&\leq C_N e^{Nr} r^{N-2} + C\delta^{-2} e^{N\phi(1-2\delta)} \int_{B_1} |U|^2 dxdy,
\end{aligned} \tag{14.35}$$

for every $N > \max\{\tau_0, 2\}$.

Now, passing to the limit as $r \to 0$ in (14.35) we get

$$\int_{B_\rho} |U|^2 \, dxdy \leq C\rho\delta^{-2} e^{N(\phi(1-2\delta)-\phi(\rho))} \int_{B_1} |U|^2 \, dxdy. \tag{14.36}$$

Passing to the limit as $N \to \infty$, taking into account that $\phi(1-2\delta) - \phi(\rho) < 0$ we have

$$\int_{B_\rho} |U|^2 \, dxdy = 0.$$

Hence $U = 0$, in B_ρ. Finally, since $\rho \in (0, 1-2\delta)$ and δ is arbitrary in $(0, 1)$, we get

$$U = 0, \quad \text{in } B_1.$$

This concludes the proof of (a).

In order to prove (b), we use (14.33) and, setting

$$R_1 = 1 - 2\delta,$$

we obtain

$$\int_{B_\rho \setminus B_{2r}} |U|^2 \, dxdy \leq$$

$$\leq C\rho r^{-2} e^{2\tau(\phi(r)-\phi(\rho))} \int_{B_{2r}} |U|^2 \, dxdy + C\delta^{-2} e^{2\tau(\phi(R_1)-\phi(\rho))} \int_{B_1} |U|^2 \, dxdy,$$

for every $\tau \geq \tau_0$. Adding to both sides of the above inequality the integral

$$\int_{B_{2r}} |U|^2 \, dxdy,$$

we have, with perhaps a different constant C,

$$\int_{B_\rho} |U|^2 \, dxdy \leq C e^{2\tau(\phi(r)-\phi(\rho))} A + e^{2\tau(\phi(R_1)-\phi(\rho))} B,$$

for every $\tau \geq \tau_0$, where

$$A = \rho r^{-2} \int_{B_{2r}} |U|^2 \, dxdy \quad \text{and} \quad B = \delta^{-2} \int_{B_1} |U|^2 \, dxdy. \tag{14.37}$$

Now let us denote by

$$\overline{\tau}=\frac{\log\left(BA^{-1}\right)}{2\left(\phi(r)-\phi(R_1)\right)}.$$

Note that

$$e^{2\overline{\tau}(\phi(r)-\phi(\rho))}A=e^{2\overline{\tau}(\phi(R_1)-\phi(\rho))}B.$$

Hence, if

$$\overline{\tau}\geq\tau_0,$$

then we choose $\tau=\overline{\tau}$ in (14.37) and we get

$$\int_{B_\rho}|U|^2\,dxdy\leq 2CA^\theta B^{1-\theta}, \tag{14.38}$$

where θ is given by (14.28)

On the other hand, if

$$\overline{\tau}<\tau_0,$$

then we have trivially

$$\int_{B_\rho}|U|^2\,dxdy\leq B=B^\theta B^{1-\theta}\leq e^{2\tau_0(\phi(\rho)-\phi(R_1))}A^\theta B^{1-\theta}. \tag{14.39}$$

Finally, by (14.38) and (14.39) we get (14.27). ■

Remark 14.2 Let us observe that, keeping ρ and R_1 fixed, we have by (14.28)

$$\theta\sim\frac{\log\left(R_1/\rho\right)}{\log\left(R_1/r_0\right)},\quad \text{as } r_0\to 0, \tag{14.40}$$

Due to this asymptotic behavior, proceeding as in Sect. 12.5 we can obtain again part (a) of Theorem 14.5 as a consequence of inequality (14.27). ♦

Remark 14.3 By Theorem 14.5 it is easy to prove the uniqueness for the following Cauchy problem.

Let $U\in H^1_{loc}\left(\mathbb{R}^2\right)$ be a solution to

$$\begin{cases}\partial_{\overline{z}}U+a\left(x,y\right)U=0\\ \\ U(x,y)=0, \quad \text{for } y\leq g(x)\end{cases}, \tag{14.41}$$

where, $a \in L^\infty(\mathbb{R}^2)$ and $g \in C^2(\mathbb{R})$ (g real-valued) satisfies

$$g(0) = g'(0) = 0.$$

In order to prove that

$$U = 0, \quad \text{in } \mathbb{R}^2, \tag{14.42}$$

let us recall that by Proposition 3.75 we know that there exists $r_0 \in (0, 1)$ such that

$$\overline{B_{r_0}(-r_0 e_n)} \cap \left\{(x, y) \in \mathbb{R}^2 : \ y \geq g(x)\right\} = \{(0, 0)\}.$$

Hence, applying Theorem 14.5 we get

$$U = 0, \quad \text{in } B_1(-r_0 e_n),$$

on the other hand, $B_1(-r_0 e_n) \cap \left\{(x, y) \in \mathbb{R}^2 : \ y > g(x)\right\} \neq \emptyset$ and, using iteratively Theorem 14.5 we obtain (14.42). ♦

2. The Laplace Operator in $\mathbb{R}^2$ Now we exploit the well-known identity

$$\Delta = 4\partial_z \partial_{\overline{z}}, \quad \text{in } \mathbb{R}^2, \tag{14.43}$$

to obtain in a very simple way a Carleman estimate for the Laplace operator in $\mathbb{R}^2$. Actually, the following proposition holds.

Proposition 14.6 *Let φ be the function defined by* (14.19)*. Then there exists a constant C such that*

$$\begin{aligned} \tau^2 \int_{B_1} |z|^{-2}|u|^2 e^{2\tau\varphi} dxdy + \tau \int_{B_1} |z|^{-1}|\nabla u|^2 e^{2\tau\varphi} dxdy \leq \\ \leq C \int_{B_1} |\Delta u|^2 e^{2\tau\varphi}, \end{aligned} \tag{14.44}$$

for every $u \in C^\infty(B_1 \setminus \{(0, 0)\})$ and for every $\tau > 0$.

Proof Proposition 14.4 gives

$$\tau \int_{\mathbb{R}^2} |z|^{-1} |u|^2 e^{2\tau\varphi} dxdy \leq C \int_{\mathbb{R}^2} |\partial_{\overline{z}} u|^2 e^{2\tau\varphi} dxdy, \tag{14.45}$$

for every $u \in C_0^\infty(B_1 \setminus \{(0, 0)\})$ and for every $\tau > 0$. Similarly we have

$$\tau \int_{\mathbb{R}^2} |z|^{-1} |u|^2 e^{2\tau\varphi} dxdy \leq C \int_{\mathbb{R}^2} |\partial_z u|^2 e^{2\tau\varphi} dxdy, \tag{14.46}$$

for every $u \in C_0^\infty(B_1 \setminus \{(0, 0)\})$ and for every $\tau > 0$.

By applying (14.45) to $\partial_z u$ and applying (14.46) to $\partial_{\bar{z}} u$ we get, respectively

$$\tau \int_{B_1} |z|^{-1} |\partial_z u|^2 e^{2\tau\varphi} dxdy \leq C \int_{B_1} |\partial_{\bar{z}} (\partial_z u)|^2 e^{2\tau\varphi} dxdy \leq$$
$$\leq \frac{C}{4} \int_{B_1} |\Delta u|^2 e^{2\tau\varphi} dxdy,$$

and

$$\tau \int_{B_1} |z|^{-1} |\partial_{\bar{z}} u|^2 e^{2\tau\varphi} dxdy \leq \frac{C}{4} \int_{B_1} |\Delta u|^2 e^{2\tau\varphi} dxdy, \tag{14.47}$$

for every $u \in C_0^\infty (B_1 \setminus \{(0,0)\})$ and for every $\tau > 0$. Summing up the inequalities above we have

$$\tau \int_{B_1} |z|^{-1} |\nabla u|^2 e^{2\tau\varphi} dxdy \leq C \int_{B_1} |\Delta u|^2 e^{2\tau\varphi} dxdy, \tag{14.48}$$

for every $u \in C_0^\infty (B_1 \setminus \{(0,0)\})$ and for every $\tau > 0$.

Let us note that by (14.19) we have

$$|z|^{-1} \leq e^{\varphi} \leq e|z|^{-1}, \quad \forall z \in B_1 \setminus \{(0,0)\}.$$

By this inequality, by (14.45) and by (14.47) we have

$$\tau \int_{B_1} |z|^{-2} |u|^2 e^{2\tau\varphi} dxdy \leq \left(\tau + \frac{1}{2}\right) \int_{B_1} |z|^{-1} |u|^2 e^{2\left(\tau+\frac{1}{2}\right)\varphi} dxdy \leq$$
$$\leq 4 \int_{B_1} |\partial_{\bar{z}} u|^2 e^{2\left(\tau+\frac{1}{2}\right)\varphi} dxdy \leq$$
$$\leq 4e \int_{B_1} |z|^{-1} |\partial_{\bar{z}} u|^2 e^{2\tau\varphi} dxdy \leq$$
$$\leq \frac{16e}{\tau} \int_{B_1} |\Delta u|^2 e^{2\tau\varphi} dxdy.$$

Hence

$$\tau^2 \int_{B_1} |z|^{-2} |u|^2 e^{2\tau\varphi} dxdy \leq 16e \int_{B_1} |\Delta u|^2 e^{2\tau\varphi} dxdy.$$

for every $u \in C_0^\infty (B_1 \setminus \{(0,0)\})$ and for every $\tau > 0$. This inequality and (14.48) yield (14.44). ■

Remark 14.4 We will see later (see Chap. 18), with a more elaborate procedure, that estimate (14.44) can be proved and (improved) for the n-dimensional Laplacian.

However, we can observe right now that (14.44) allows, with a procedure similar to the one followed in Theorem 14.5, that if $U \in H^2(B_1)$ is a solution to

$$\Delta U + a(x, y)\,\partial_x U + b(x, y)\partial_y U + c(x, y)U = 0, \tag{14.49}$$

where

$$a, b, c \in L^\infty(B_1) \tag{14.50}$$

and

$$\|U\|_{L^2(B_r)} = O(r^N), \quad \text{as } r \to 0,\ \forall N \in \mathbb{N}_0,$$

then

$$U \equiv 0, \quad \text{in } B_1.$$

Moreover, similarly to Remark 14.3, it is easy to prove that if $U \in H^2_{loc}(\mathbb{R}^2)$ is a solution of the following Cauchy problem

$$\begin{cases} \Delta U + a(x, y)\,\partial_x U + b(x, y)\partial_y U + c(x, y)U = 0 \\ \\ U(x, y) = 0, \quad \text{for } y \le g(x) \end{cases},$$

where $g \in C^2(\mathbb{R})$ satisfies $g(0) = g'(0) = 0$, then

$$U = 0, \quad \text{in } \mathbb{R}^2.$$

♦

3. The Laplace Operator in $\mathbb{R}^n$ In the two cases studied previously, we have proved relatively simple Carleman estimates with singular weight, from which we have derived the strong unique continuation property for the considered operators. Subsequently, from the strong unique continuation property, we have derived uniqueness for the Cauchy problem. If one is only interested in uniqueness for the Cauchy problem, this is not, as one might easily imagine, the easiest path because it requires Carleman estimates with singular weight, which are more complicated to obtain compared to those with non-singular weights and which do not hold for important classes of operators. However, the theory that we will develop in this chapter and in Chaps. 15–17 focuses specifically on Carleman estimates with non-singular weights, while in Chap. 18 we will address the case with singular weights in greater detail. In this third example, we will show how, by a Carleman estimate with non-singular weight and simple property of the Laplace operator, one can achieve uniqueness for the Cauchy problem for the equation

$$\Delta U+\sum_{j=1}^{n}a_j(x)\partial_{x_j}U+a_0(x)U=0,\quad \text{in } \Omega, \tag{14.51}$$

where Ω is an open set of $\mathbb{R}^n$ and

$$a_j\in L^{\infty}(\Omega),\quad j=0,1,\cdots,n. \tag{14.52}$$

We set

$$M=\max_{0\le j\le n}\left\|a_j\right\|_{L^{\infty}(\Omega.\mathbb{C})}, \tag{14.53}$$

We begin by the following ([110])

Lemma 14.7 (The Rellich Identity) *Let $\beta\in C^{0,1}(\Omega,\mathbb{R}^n)$, $\beta=\left(\beta^1,\cdots,\beta^n\right)$ and let $v\in C^2(\Omega)$, v be a real-valued function, then*

$$\begin{aligned}2(\beta\cdot\nabla v)\Delta v&=\operatorname{div}\left(2(\beta\cdot\nabla v)\nabla v-\beta|\nabla v|^2\right)+\\&+(\operatorname{div}\beta)|\nabla v|^2-2\partial_k\beta^j\partial_j v\partial_k v,\quad a.e.\ x\in\Omega,\end{aligned} \tag{14.54}$$

(in (14.54) *we have used the Einstein notation of repeated indices).*

Proof We have

$$\begin{aligned}2(\beta\cdot\nabla v)\Delta v&=2\left(\beta^j\partial_j v\right)\Delta v=2\left(\partial_k\left(\beta^j\partial_j v\partial_k v\right)-\partial_k\left(\beta^j\partial_j v\right)\partial_k v\right)=\\&=2\left(\partial_k\left(\beta^j\partial_j v\partial_k v-\left(\partial_k\beta^j\right)\partial_j v\partial_k v-\beta^j\partial^2_{jk}v\partial_k v\right)\right)=\\&=2\operatorname{div}\left[(\beta\cdot\nabla v)\nabla v\right]-2\left(\partial_k\beta^j\right)\partial_j v\partial_k v-\beta^j\partial_j\left(|\nabla v|^2\right)=\\&=2\operatorname{div}\left[(\beta\cdot\nabla v)\nabla v\right]-2\left(\partial_k\beta^j\right)\partial_j v\partial_k v-\partial_j(\beta^j|\nabla v|^2)+(\operatorname{div}\beta)|\nabla v|^2=\\&=\operatorname{div}\left[2(\beta\cdot\nabla v)\nabla v-\beta|\nabla v|^2\right]-2(\partial_k\beta^j)\partial_j v\partial_k v+(\operatorname{div}\beta)|\nabla v|^2,\end{aligned}$$

almost everywhere in Ω. ■

Remark 14.5 By (14.54) we have immediately

$$\int_{\Omega}2(\beta\cdot\nabla v)\Delta v dx=\int_{\Omega}\left((\operatorname{div}\beta)|\nabla v|^2-2\partial_k\beta^j\partial_j v\partial_k v\right)dx, \tag{14.55}$$

for every $v\in C_0^{\infty}(\Omega)$. ♦

Let $\varphi \in C^{\infty}(\Omega)$ be a real-valued function and let us denote by L the operator Δ. Let $u \in C_0^{\infty}(\Omega)$ be a real-valued function. Set $v = e^{\tau\varphi}u$. Let us calculate

$$\partial_j u = e^{-\tau\varphi}\left(\partial_j v - \tau\partial_j\varphi v\right),$$

$$\partial_j^2 u = e^{-\tau\varphi}\left(\partial_j^2 v - 2\tau\partial_j\varphi\partial_j v - \tau\partial_j^2\varphi v + \tau^2 v(\partial_j\varphi)^2\right).$$

We obtain

$$L_\tau v = e^{\tau\varphi}L(e^{-\tau\varphi}v) = \Delta v - \tau(\Delta\varphi)v - 2\tau\nabla\varphi\cdot\nabla v + \tau^2|\nabla\varphi|^2 v. \tag{14.56}$$

Now we focus on the operator

$$\widetilde{L}_\tau v := L_\tau v + \tau(\Delta\varphi)v = \Delta v - 2\tau\nabla\varphi\cdot\nabla v + \tau^2|\nabla\varphi|^2 v. \tag{14.57}$$

Let us notice that, since

$$L_\tau v = \widetilde{L}_\tau v - \tau(\Delta\varphi)v,$$

we have

$$|L_\tau v|^2 \geq \frac{1}{2}\left|\widetilde{L}_\tau v\right|^2 - \tau^2\left|\Delta\varphi\right|^2|v|^2. \tag{14.58}$$

Let us denote

$$S_\tau v = \Delta v + \tau^2|\nabla\varphi|^2 v,$$

$$A_\tau v = -2\tau\nabla\varphi\cdot\nabla v.$$

Of course,

$$\widetilde{L}_\tau v = S_\tau v + A_\tau v.$$

The following proposition holds true.

Proposition 14.8 *With the notation introduced above, we have*

$$\begin{aligned}\int_\Omega \left|\widetilde{L}_\tau v\right|^2 dx &= \int_\Omega |S_\tau v|^2 dx + \int_\Omega |A_\tau v|^2 dx + \\ &+ 4\tau\int_\Omega\left(\partial_{jk}^2\varphi\partial_j v\partial_k v + \tau^2(\partial_{jk}^2\varphi\partial_j\varphi\partial_k\varphi)v^2\right)dx - \\ &- 2\tau\int_\Omega \Delta\varphi(|\nabla v|^2 - \tau^2|\nabla\varphi|^2 v^2)dx,\end{aligned} \tag{14.59}$$

for every $v \in C_0^{\infty}(\Omega)$ and for every $\tau \in \mathbb{R}$.

Proof By (14.57) we have

$$\int \left|\widetilde{L}_\tau v\right|^2 dx = \int |S_\tau v|^2 dx + \int |A_\tau v|^2 dx + 2\int S_\tau v A_\tau v dx, \tag{14.60}$$

for brevity, we omit the domain of integration. Now we use identity (14.55), where

$$\beta = \nabla\varphi.$$

We have

$$2\int (\nabla\varphi\cdot\nabla v)\Delta v\, dx = \int [\Delta\varphi|\nabla v|^2 - 2\partial^2_{jk}\varphi\partial_j v\partial_k v]\, dx.$$

We get

$$2\int S_\tau v A_\tau v dx = -2\tau \int \left(2(\nabla\varphi\cdot\nabla v)\Delta v + 2\tau^2|\nabla\varphi|^2(\nabla\varphi\cdot\nabla v)v\right) dx =$$

$$= -2\tau \int \left(\Delta\varphi|\nabla v|^2 - 2\partial^2_{jk}\varphi\partial_j v\partial_k v + \tau^2|\nabla\varphi|^2\nabla\varphi\cdot\nabla(v^2)\right) dx =$$

$$= -2\tau \int \Big(\Delta\varphi|\nabla v|^2 - 2\partial^2_{jk}\varphi\partial_j v\partial_k v$$

$$- 2\tau^2\left(\partial^2_{jk}\varphi\partial_j\varphi\partial_k\varphi\right)v^2 - \tau^2\left(|\nabla\varphi|^2\Delta\varphi\right)v^2\Big)dx =$$

$$= -2\tau \int \left(\Delta\varphi(|\nabla v|^2 - \tau^2|\nabla\varphi|^2 v^2) - 2\partial^2_{jk}\varphi\partial_j v\partial_k v - 2\tau^2(\partial^2_{jk}\varphi\partial_j\varphi\partial_k\varphi)v^2\right) dx.$$

Therefore

$$2\int S_\tau v A_\tau v dx = 4\tau \int \left(\partial^2_{jk}\varphi\partial_j v\partial_k v + \tau^2(\partial^2_{jk}\varphi\partial_j\varphi\partial_k\varphi)v^2\right) dx-$$

$$- 2\tau \int \Delta\varphi(|\nabla v|^2 - \tau^2|\nabla\varphi|^2 v^2)dx.$$

this identity and (14.60) give (14.59). ■

Theorem 14.9 *Let α be a positive number. Set*

$$H_\alpha = \left\{x \in \mathbb{R}^n : x_n < \frac{1}{2\alpha}\right\}.$$

Let

$$P = \Delta + \sum_{j=1}^n a_j(x)\partial_{x_j} + a_0(x), \quad in\ H_\alpha, \tag{14.61}$$

where a_j, for $j = 0, 1, \cdots, n$, satisfy (14.52), *for $\Omega = H_\alpha$. Let*

$$h(x_n) = -x_n + \frac{\alpha x_n^2}{2}. \tag{14.62}$$

Then there exist $C > 0$ and $\tau_0 \geq 1$ (C, τ_0 depend on α and M, the latter is defined in (14.53)*) such that the following inequality holds*

$$\begin{aligned} \tau^3 \int_{H_\alpha} |u|^2 e^{2\tau h(x_n)} dx + \tau \int_{H_\alpha} |\nabla u|^2 e^{2\tau h(x_n)} dx \leq \\ \leq C \int_{H_\alpha} |Pu|^2 e^{2\tau h(x_n)} dx, \end{aligned} \tag{14.63}$$

for every $u \in C_0^\infty(H_\alpha)$ and for every $\tau \geq \tau_0$.

Proof We prove (14.63) for the real and immaginary part of u, hence it is not restrictive to assume that u is a real-valued functions. Moreover, a slight modification of Proposition 14.1 yields that (14.63) is equivalent to

$$\begin{aligned} \tau^3 \int_{H_\alpha} |u|^2 e^{2\tau h(x_n)} dx + \tau \int_{H_\alpha} |\nabla u|^2 e^{2\tau h(x_n)} dx \leq \\ \leq C \int_{H_\alpha} |\Delta u|^2 e^{2\tau h(x_n)} dx, \end{aligned} \tag{14.64}$$

for every $u \in C_0^\infty(H_\alpha)$ and for every $\tau \geq \tau_0$ (with, maybe, different C and τ_0).
Let us choose in (14.59),

$$\varphi(x) = h(x_n)$$

and we have, for $\Omega = H_\alpha$,

$$\begin{aligned} \int_{H_\alpha} \left|\widetilde{L}_\tau v\right|^2 dx &= \int_{H_\alpha} |S_\tau v|^2 dx + \int_{H_\alpha} |A_\tau v|^2 dx + \\ &+ 4\tau\alpha \int_{H_\alpha} \left((\partial_n v)^2 + \tau^2 (1 - \alpha x_n)^2 v^2\right) dx - \\ &- 2\tau\alpha \int_{H_\alpha} \left(|\nabla v|^2 - \tau^2 (1 - \alpha x_n)^2 v^2\right) dx \geq \\ &\geq \int_{H_\alpha} |S_\tau v|^2 dx + 6\tau^3\alpha \int_{H_\alpha} (1 - \alpha x_n)^2 v^2 dx - \\ &- 2\tau\alpha \int_{H_\alpha} |\nabla v|^2 dx, \end{aligned} \tag{14.65}$$

for every $v \in C_0^\infty(H_\alpha)$ and for every $\tau > 0$.

Now, let γ be a positive number that we will choose. We have

$$\begin{aligned}
\int_{H_\alpha} |S_\tau v|^2 dx &= \int_{H_\alpha} \left|\left(\Delta v + \tau^2 (1-\alpha x_n)^2 v + \gamma \tau v\right) - \gamma \tau v\right|^2 dx \geq \\
&\geq -2\tau\gamma \int_{H_\alpha} v \left(\Delta v + \tau^2 (1-\alpha x_n)^2 v + \gamma \tau v\right) = \\
&= 2\tau\gamma \int_{H_\alpha} |\nabla v|^2 dx - 2\tau^3 \gamma \int_{H_\alpha} \left((1-\alpha x_n)^2 + \gamma \tau^{-1}\right) v^2 dx,
\end{aligned} \tag{14.66}$$

for every $v \in C_0^\infty(\Omega)$ and for every $\tau > 0$.

By (14.65) and (14.66) we have

$$\begin{aligned}
\int_{H_\alpha} \left|\widetilde{L}_\tau v\right|^2 dx \geq 2\tau^3 \int_{H_\alpha} \left((3\alpha - \gamma)(1-\alpha x_n)^2 - \gamma^2 \tau^{-1}\right) v^2 dx + \\
+ 2\tau(\gamma - \alpha) \int_{H_\alpha} |\nabla v|^2 dx,
\end{aligned} \tag{14.67}$$

for every $v \in C_0^\infty(H_\alpha)$ and for every $\tau > 0$.

Now let us choose

$$\gamma = 2\alpha,$$

and we have

$$\begin{aligned}
(3\alpha - \gamma)(1-\alpha x_n)^2 - \gamma^2 \tau^{-1} &\geq \frac{\alpha}{4} - 4\alpha^2 \tau^{-1} \geq \\
&\geq \frac{\alpha}{8}, \quad \forall x \in H_\alpha, \ \forall \tau \geq 32\alpha.
\end{aligned}$$

The just obtained inequality and (14.67) imply

$$\int_{H_\alpha} \left|\widetilde{L}_\tau v\right|^2 dx \geq \frac{\tau^3}{4} \int_{H_\alpha} v^2 dx + 2\tau\alpha \int_{H_\alpha} |\nabla v|^2 dx, \tag{14.68}$$

for every $v \in C_0^\infty(H_\alpha)$ and for every $\tau \geq 32\alpha$.

By (14.68) and by (14.58) we have

$$\begin{aligned}
\int_{H_\alpha} |L_\tau v|^2 dx &\geq \frac{1}{2} \int_{H_\alpha} \left|\widetilde{L}_\tau v\right|^2 dx - \tau^2 \alpha^2 \int_{H_\alpha} v^2 dx \geq \\
&\geq \frac{\tau^3}{16} \int_{H_\alpha} v^2 dx + \tau\alpha \int_{H_\alpha} |\nabla v|^2 dx,
\end{aligned} \tag{14.69}$$

for every $v \in C_0^\infty(H_\alpha)$ and for every $\tau \geq \max\{32\alpha, 16\alpha^2\}$.

Now, recalling that

$$v = e^{\tau h(x_n)} u,$$

by (14.69) we have for $\varepsilon \in (0, 1)$

$$\begin{aligned}
&\int_{H_\alpha} |\Delta u|^2 \, e^{2\tau h(x_n)} dx = \int_{H_\alpha} |L_\tau v|^2 \, dx \geq \\
&\geq \frac{\tau^3}{16} \int_{H_\alpha} v^2 dx + \varepsilon\tau\alpha \int_{H_\alpha} |\nabla v|^2 dx = \\
&= \frac{\tau^3}{16} \int_{H_\alpha} u^2 e^{2\tau h(x_n)} dx + \varepsilon\tau\alpha \int_{H_\alpha} |\nabla u + \tau(1 - \alpha x_n) u|^2 e^{2\tau h(x_n)} dx \geq \\
&\geq \tau^3 \left(\frac{1}{16} - \frac{\varepsilon\alpha}{4} \right) \int_{H_\alpha} u^2 e^{2\tau h(x_n)} dx + \varepsilon\tau\alpha \int_{H_\alpha} |\nabla u|^2 e^{2\tau h(x_n)} dx,
\end{aligned} \tag{14.70}$$

for every $v \in C_0^\infty(H_\alpha)$ and for every $\tau \geq \max\{32\alpha, 16\alpha^2\}$.

Let us choose $\varepsilon = \min\left\{\frac{1}{8\alpha}, \frac{1}{2}\right\}$ and by (14.70) we obtain (14.64), hence (14.63) is proved. ■

Let us introduce some notation that we will use in Theorem 14.10 below.

Let

$$g : \overline{B'_1} \to \mathbb{R},$$

be a continuous and strictly convex function such that

$$g(0) = 0$$

and let us denote by

$$\ell = \min_{\partial B'_1} g > 0.$$

Theorem 14.10 *Let $Q = B'_2 \times (0, \ell)$ and let $U \in H^2(Q)$ be a solution of the Cauchy problem*

$$\begin{cases} \Delta U + \sum_{j=1}^n a_j(x)\partial_j U + a_0(x) U = 0, & \text{in } Q \\ \\ U = 0, & \text{for } x \in Q \setminus \{x \in Q : \, g(x') < x_n\} \end{cases} . \tag{14.71}$$

Then

$$U = 0, \quad \text{in } Q. \tag{14.72}$$

Proof Let $\delta \in \left(0, \frac{1}{3}\right)$ and let $\zeta \in C^\infty(\mathbb{R})$ satisfy $0 \leq \zeta \leq 1$,

$$\zeta(x_n) = 1, \text{ for } x_n \leq \ell(1-2\delta); \quad \zeta(x_n) = 0, \text{ for } \ell(1-\delta) \leq x_n$$

and

$$\left|\zeta'(x_n)\right| \leq c(\delta\ell)^{-1} \text{ and } \left|\zeta''(x_n)\right| \leq c(\delta\ell)^{-2}, \text{ for } \ell(1-2\delta) \leq x_n \leq \ell(1-\delta),$$

where c is a constant.

Let us apply estimate (14.63) to

$$u = \zeta(x_n)U(x).$$

We have

$$P(\zeta U) = \zeta(x_n)PU + 2\zeta'(x_n)\partial_n U + \left(a_n(x)\zeta'(x_n) + \zeta''(x_n)\right) U,$$

where P is defined by (14.61), from which we have, taking into account (14.53),

$$|\Delta(\zeta U)| \leq c(\ell\delta)^{-2}\chi_I(x_n)\left(M\ell\delta\,|\partial_n U| + |U|\right), \tag{14.73}$$

where we set $I = (\ell(1-2\delta), \ell(1-\delta))$.

Let us choose in (14.63)

$$\alpha = \frac{1}{2\ell},$$

and let us denote by

$$J = \int_{Q\cap\{\ell(1-2\delta)<x_n<\ell(1-\delta)\}} \left(|U|^2 + (M\ell\delta)^2|\nabla U|^2\right) e^{2\tau h(x_n)} dx.$$

Hence by (14.63) and (14.73) we have (recall that C denotes a constant that may change from line to line)

$$\begin{aligned}
&\int_Q \left(\tau^3|\zeta U|^2 + \tau|\nabla(\zeta U)|^2\right) e^{2\tau h(x_n)} dx \leq \\
&\leq C \int_Q |P(\zeta U)|^2 e^{2\tau h(x_n)} dx \leq \\
&\leq C\,(\ell\delta)^{-4}\, J,
\end{aligned} \tag{14.74}$$

for every $\tau \geq \tau_0$, τ_0 depends on ℓ only.

Let us estimate from above J. Taking into account that

$$h(x_n) = -x_n + \frac{x_n^2}{4\ell}$$

and, consequently, h is a strictly decreasing function in $[0, 2\ell]$, we have

$$\begin{aligned} J &= \int_{Q\cap\{\ell(1-2\delta)<x_n<\ell(1-\delta)\}} \left(|U|^2 + (M\ell\delta)^2|\nabla U|^2\right) e^{2\tau h(x_n)} dx \leq \\ &\leq C e^{2\tau h(\ell(1-2\delta))} \int_{Q\cap\{\ell(1-2\delta)<x_n<\ell(1-\delta)\}} \left(|U|^2 + |\nabla U|^2\right) dx \leq \\ &\leq C e^{2\tau h(\ell(1-2\delta))} \|U\|^2_{H^1(Q)}, \end{aligned} \tag{14.75}$$

for every $\tau \geq \tau_0$, where C depends on M, ℓ and δ only.

Hence, by (14.74)–(14.75), we get

$$\begin{aligned} &\int_Q \left(\tau^3|\zeta U|^2 + \tau|\nabla(\zeta U)|^2\right) e^{2\tau h(x_n)} dx \leq \\ &\leq C\,(\ell\delta)^{-4}\, e^{2\tau h(\ell(1-2\delta))}\, \|U\|^2_{H^1(Q)}, \end{aligned} \tag{14.76}$$

for every $\tau \geq \tau_0$.

By the manner in which ζ is defined, we have trivially

$$\begin{aligned} &\int_Q \left(\tau^3|\zeta U|^2 + \tau|\nabla(\zeta U)|^2\right) e^{2\tau h(x_n)} dx \geq \\ &\geq \int_{Q\cap\{x_n<\ell(1-3\delta)\}} \left(|U|^2 + |\nabla U|^2\right) e^{2\tau h(x_n)} dx \geq \\ &\geq e^{2\tau h(\ell(1-3\delta))} \int_{Q\cap\{x_n<\ell(1-3\delta)\}} \left(|U|^2 + |\nabla U|^2\right) dx, \end{aligned} \tag{14.77}$$

for every $\tau \geq \tau_0$.

Now by (14.76) and (14.77) we have

$$\begin{aligned} &\int_{Q\cap\{x_n<\ell(1-3\delta)\}} |U|^2 dx \leq \\ &\leq C\,(\ell\delta)^{-4}\, e^{-2\tau(h(\ell(1-3\delta))-h(\ell(1-2\delta)))}\, \|U\|^2_{H^1(Q)}, \end{aligned} \tag{14.78}$$

for every $\tau \geq \tau_0$. Passing to the limit as $\tau \to \infty$, by (14.78) we have

$$U = 0, \quad \text{in } Q \cap \{x_n < \ell(1-3\delta)\}.$$

Finally, since δ is arbitrary we have

$$U = 0, \quad \text{in } Q.$$

■

Remark 14.6 Let Ω be an open set of $\mathbb{R}^n$. Let us observe that to require that

$$|\Delta U| \leq C\left(|\nabla U| + |U|\right), \qquad \text{in } \Omega, \tag{14.79}$$

is equivalent to require that there exist $a_j \in L^\infty(\Omega)$, $j = 0, \cdots, n$ such that U is a solution of the equation

$$\Delta U = -\sum_{j=1}^{n} a_j(x)\partial_j U - a_0(x)U, \quad \text{in } \Omega. \tag{14.80}$$

Indeed, it is obvious that if U satisfies (14.80) then U satisfies (14.79) with

$$C = \max_{0\leq j\leq n} \left\|a_j\right\|_{L^\infty(\Omega)}.$$

Conversely, if U satisfies (14.79), then we define, for $j = 1, \cdots, n$,

$$a_j(x) = \begin{cases} -\frac{\overline{\partial_j U(x)}(\Delta U(x))}{|\nabla U(x)|^2+|U(x)|^2} & \text{for } |\nabla U(x)|^2 + |U(x)|^2 > 0 \\ \\ 0, & \text{for } |\nabla U(x)|^2 + |U(x)|^2 = 0 \end{cases}$$

and, for $j = 0$,

$$a_0(x) = \begin{cases} -\frac{\overline{U(x)}(\Delta U(x))}{|\nabla U(x)|^2+|U(x)|^2}, & \text{for } |\nabla U(x)|^2 + |U(x)|^2 > 0 \\ \\ 0, & \text{for } |\nabla U(x)|^2 + |U(x)|^2 = 0 \end{cases}.$$

By (14.79) we have

$$\left\|a_j\right\|_{L^\infty(\Omega)} \leq C, \quad \text{for } j = 0, 1, \cdots, n.$$

In addition, if $|\nabla U(x)|^2 + |U(x)|^2 = 0$ then (14.79) implies

$$\Delta U(x) = 0 = -\sum_{j=1}^{n} a_j(x)\partial_j U(x) - a_0(x)U(x),$$

and, if $|\nabla U(x)|^2 + |U(x)|^2 > 0$ then

$$\begin{aligned} &-\sum_{j=1}^{n} a_j(x)\partial_j U(x) - a_0(x)U(x) = \\ &= \frac{\overline{\nabla U(x)}(\Delta U(x))}{|\nabla U(x)|^2 + |U(x)|^2} \cdot \nabla U(x) + \frac{\overline{U(x)}(\Delta U(x))}{|\nabla U(x)|^2 + |U(x)|^2} U(x) = \\ &= \Delta U(x). \end{aligned}$$

♦

Up to this point, we have established a local uniqueness result for the case where the initial surface is a continuous strictly convex function (corresponding to the graph of function g). Following Nirenberg, [101], we can exploit the specific properties of the Δ operator to achieve the global uniqueness for the Cauchy problem for quite general initial surface. To this purpose we derive some intermediate results that will serve as a stepping stone towards the more general uniqueness result.

The first intermediate result is the following

Proposition 14.11 *Let $M, R_2, R_1 > 0$, $R_1 < R_2$. Suppose that $U \in H^2(B_{R_2} \setminus \overline{B_{R_1}})$ satisfies what follows*

$$|\Delta U| \leq M\left(|\nabla U| + |U|\right), \quad in\ B_{R_2} \setminus \overline{B_{R_1}}, \tag{14.81}$$

$$U = 0, \quad in\ \mathbb{R}^n \setminus B_{R_2}. \tag{14.82}$$

Then

$$U = 0 \quad in\ B_{R_2} \setminus \overline{B_{R_1}}. \tag{14.83}$$

Proof By the invariance of Δ with respect to the rotations, by Remark 14.6 and by Theorem 14.10 we immediately obtain (14.83). ■

In the next Proposition we will use **the Kelvin transform** which is defined as follows. Let u be a sufficiently smooth function. Set

$$v(y) = |y|^{2-n}\, u\left(y|y|^{-2}\right). \tag{14.84}$$

We have

$$\Delta v(y) = |y|^{-2-n}\, (\Delta_x u)\left(y|y|^{-2}\right). \tag{14.85}$$

Theorem 14.12 (Weak Unique Continuation Property) *Let $M, R, \rho > 0$, $\rho < R$. Assume that $U \in H^2(B_R)$ satisfies the inequality*

$$|\Delta U| \leq M\left(|\nabla U| + |U|\right), \quad in\ B_R. \tag{14.86}$$

We have that, if

$$U = 0, \quad in\ B_\rho, \tag{14.87}$$

then

$$U = 0 \quad in\ B_R. \tag{14.88}$$

Proof It is not restrictive to assume $R = 1$ and, consequently, $\rho < 1$. Let us apply the Kelvin transform. Set

$$V(y) = |y|^{2-n}\, U\left(y|y|^{-2}\right), \quad \text{in } \mathbb{R}^n \setminus \overline{B_1}. \tag{14.89}$$

By (14.87) we have

$$V = 0, \quad \text{in } \mathbb{R}^n \setminus \overline{B_{1/\rho}}. \tag{14.90}$$

Moreover $V \in H^2\left(B_{1/\rho} \setminus \overline{B_1}\right)$ and

$$\left|(\nabla_x U)\left(y|y|^{-2}\right)\right| \leq C\left(|y|^{n-1}|V(y)| + |y|^n|\nabla_y V(y)|\right),$$

hence, by this inequality, by (14.85) and by (14.86) we obtain

$$|\Delta V| \leq CM\left(|\nabla V| + |V|\right), \quad \text{in } \mathbb{R}^n \setminus \overline{B_1}. \tag{14.91}$$

Now, by applying Proposition 14.11 we have

$$V = 0 \quad \text{in } B_{1/\rho} \setminus \overline{B_1}. \tag{14.92}$$

From which, taking into account (14.84), we immediately obtain the thesis. ■

We now prove the following

Theorem 14.13 *Let Ω be a connected open set of $\mathbb{R}^n$, let $\Sigma \subset \partial\Omega$ be a local graph of a function of class C^2. Assume $a_j \in L^\infty(\Omega)$, $j = 0, 1, \cdots, n$. Moreover, let $U \in H^2(\Omega, \mathbb{C})$ satisfy*

$$\begin{cases} \Delta U + \sum_{j=1}^n a_j(x)\partial_j U + a_0(x)U = 0, & in\ \Omega \\ U = 0, \quad on\ \Sigma & \\ \frac{\partial U}{\partial \nu} = 0, \quad on\ \Sigma & \end{cases}. \tag{14.93}$$

Then we have

$$U = 0 \quad in\ \Omega. \tag{14.94}$$

Remark 14.7 The conditions $U = \frac{\partial U}{\partial \nu} = 0$ on Σ have to be understood in the sense of traces or, equivalently, it should be understood as: $U\Psi \in H_0^2(\Omega)$, for every $\Psi \in C^\infty(\mathbb{R}^n)$ such that $\Psi = \frac{\partial \Psi}{\partial \nu} = 0$ on $\partial\Omega \setminus \Sigma$. ♦

Proof of Theorem 14.13 For any $K, r > 0$ let us denote

$$Q_{r,K} = B'_r \times (-Kr, Kr)\,.$$

Up to a rigid transformation of $\mathbb{R}^n$, we may assume there exist r_0 and $\psi \in C^2\left(\overline{B'_{r_0}}\right)$ such that

$$\psi(0) = |\nabla_{x'}\psi(0)| = 0$$

and

$$\text{graph}(\psi) \subset \Sigma,$$

where graph(ψ) is the graph of ψ. While eventually reducing r_0 we may assume that there exists $K_0 > 0$ such that

$$\|\psi\|_{C^2\left(\overline{B'_{r_0}}\right)} \leq K_0 r_0$$

and

$$\Omega \cap Q_{r_0,2K_0} = \left\{x \in B'_{r_0} \times \mathbb{R} : \quad \psi(x') < x_n < 2K_0 r_0\right\}.$$

Now, let us denote

$$Q^-_{r_0 2K_0} = \left\{x \in B'_{r_0} \times \mathbb{R} : \quad -2K_0 r_0 < x_n \leq \psi(x')\right\}$$

and let

$$\widetilde{\Omega} = \Omega \cup Q^-_{r_0 2K_0}.$$

Moreover, let $\widetilde{U}$ be the extension of U to 0 in $Q^-_{r_0,2K_0}$. We have $\widetilde{U} \in H^2\left(\widetilde{\Omega}\right)$ and

$$\Delta\widetilde{U} + \sum_{j=1}^{n} \widetilde{a}_j(x)\partial_j\widetilde{U} + \widetilde{a}_0(x)\widetilde{U} = 0,$$

where $\widetilde{a}_j$, are the extensions of a_j, $j = 0, 1, \cdots, n$, to 0 in $Q^-_{r_0,2K_0}$. We have trivially $\widetilde{a}_j \in L^\infty\left(\widetilde{\Omega}\right)$, for $j = 0, 1, \cdots, n$.

From now on, we argue as we did in the proof of Theorem 9.10. For the convenience of the reader here we repeat the main steps of this proof.

First of all, we note that $\widetilde{\Omega}$ is connected.

Then, set

$$A = \left\{x \in \widetilde{\Omega} : \exists \rho_x > 0 \text{ such that } \widetilde{U} = 0, \text{ in } B_{\rho_x}(x)\right\}. \tag{14.95}$$

By the definition of $\widetilde{U}$ we have $\mathrm{Int}\left(Q^-_{r_0,2K_0}\right) \subset A$, hence $A \neq \emptyset$ and it turns out, trivially, that A is an open set of $\widetilde{\Omega}$. To prove that A is also closed in $\widetilde{\Omega}$ (from which, since $\widetilde{\Omega}$ is connected, we get $A = \widetilde{\Omega}$ and, consequently, $\widetilde{U} \equiv 0$) we can follow exactly the same argument followed in the proof of Theorem 9.10, using, in this case, Proposition 14.12 instead of Theorem 9.9. ■

Remark 14.8 An important observation to make is that the weak unique continuation property (as stated in Theorem 9.9) allows us to establish the uniqueness of the Cauchy problem (14.93). On the flip side, if we assume the uniqueness of Cauchy problem (14.93), we can deduce the weak unique continuation property for the equation

$$\Delta U + \sum_{j=1}^{n} a_j(x)\partial_j U + a_0(x)U = 0, \qquad \text{in } \Omega,$$

where Ω is a connected open set of $\mathbb{R}^n$. Indeed, let ω a subset of Ω and let us assume that

$$U = 0, \quad \text{in } \omega.$$

Let $B_r(x_0) \Subset \omega$.

We have that U is a solution to the Cauchy problem

$$\begin{cases} \Delta U + \sum_{j=1}^{n} a_j(x)\partial_j U + a_0(x)U = 0, & \text{in } \Omega \\ U = 0, & \text{on } \partial B_r(x_0) \\ \frac{\partial U}{\partial \nu} = 0, & \text{on } \partial B_r(x_0) \end{cases}.$$

Hence, by the uniqueness for the Cauchy problem, we have

$$U = 0, \quad \text{in } \Omega.$$

♦

Exercise Take care of the details of the proof of Proposition 14.11. ♠

14.4 Derivation of Unique Continuation Properties and Quantitative Estimates

In this section, we discuss some uniqueness results for the Cauchy problem and quantitative estimates that can be derived by a Carleman estimate like (14.4).

Let Ω be an open set of $\mathbb{R}^n$, $P(x, D)$ operator (14.2), $x_0 \in \Omega$ and let $\phi \in C^2\left(\overline{\Omega}\right)$ be a real-valued function which satisfies $\phi\left(x_0\right) = 0$. Let us assume

$$\nabla\phi\left(x\right) \neq 0, \quad \forall x \in \Omega. \tag{14.96}$$

Let $R > 0$ be such that $\overline{B_R(x_0)} \subset \Omega$, $f \in L^2\left(B_R(x_0)\right)$ and let $U \in H^m\left(B_R(x_0)\right)$ satisfy

$$\begin{cases} P(x, D)U = f, & \text{in } B_R(x_0) \\ \\ U = 0, & \text{in } \{x \in B_R(x_0) : \phi(x) \geq 0\} \end{cases} . \tag{14.97}$$

Note that an equivalent formulation of the second condition in (14.97) is

$$\operatorname{supp} U \subset \{x \in B_R(x_0) : \phi(x) \leq 0\} .$$

Definition 14.14 Let

$$\Gamma = \{x \in \Omega : \phi(x) = 0\} . \tag{14.98}$$

We will say that a real-valued function $\varphi \in C^2\left(\overline{\Omega}\right)$ enjoys the **property of convexification with respect to Γ in** x_0 if $\varphi(x_0) = 0$, $\nabla\varphi(x_0) \neq 0$ and there exists $r > 0$ such that (Fig. 14.1)

$$\begin{aligned} &\left\{x \in \overline{B_r(x_0)} : \varphi(x) \geq 0\right\} \setminus \{x_0\} \subset \\ &\subset \left\{x \in \overline{B_r(x_0)} : \phi(x) > 0\right\} . \end{aligned} \tag{14.99}$$

We also say that the function φ is a convexification of Γ in x_0.

Let us note that (14.99) is equivalent to

$$\begin{aligned} &\left\{x \in \overline{B_r(x_0)} : \phi(x) \leq 0\right\} \subset \\ &\subset \left\{x \in \overline{B_r(x_0)} : \varphi(x) < 0\right\} \cup \{x_0\}. \end{aligned} \tag{14.100}$$

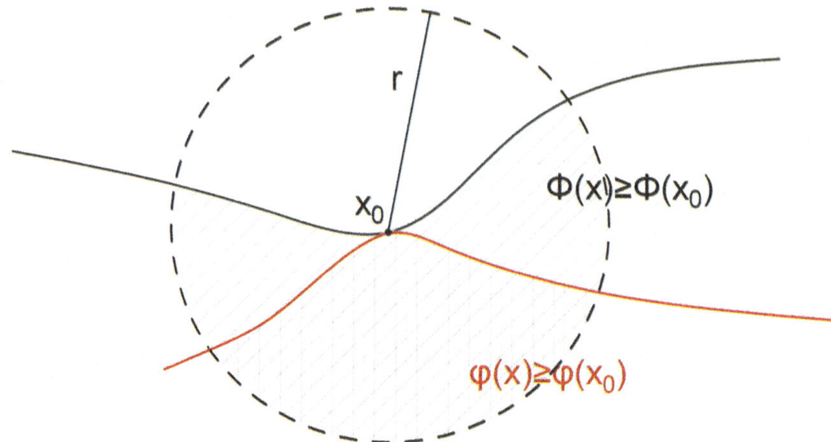

Fig. 14.1 The geometric property of the convexification of φ with respect to $\Gamma = \{x \in \Omega : \phi(x) = 0\}$ at x_0 is illustrated. Here $\phi(x_0) = 0$ and $\varphi(x_0) = 0$. Note that the region $\{x \in \overline{B_r(x_0)} : \varphi(x) \geq 0\} \setminus \{x_0\}$ (double dashed line) is contained within the region $\{x \in \overline{B_r(x_0)} : \phi(x) > 0\}$ (single dashed line)

Example 14.9 Let $\phi \in C^2(\Omega)$ be a real-valued function which satisfies $\phi(x_0) = 0$. Let us assume that ϕ satisfies (14.96). Set

$$\nu(x_0) = -\frac{\nabla\phi(x_0)}{|\nabla\phi(x_0)|}. \tag{14.101}$$

We have that there exists $r > 0$ such that (see the proof of Proposition 3.75)

$$\overline{B_r(x_0 - r\nu(x_0))} \cap \left\{x \in \overline{\Omega} : \phi(x) \leq 0\right\} = \{x_0\}.$$

Hence, for every $\delta \in (0, r]$ the function

$$\varphi(x) = \delta^2 - |x - (x_0 - \delta\nu(x_0))|^2, \tag{14.102}$$

is a convexification of $\Gamma = \{x \in \Omega : \quad \phi(x) = 0\}$ in x_0. ♣

Example 14.10 Let $\phi(x) = -x_n$. Let

$$\varphi_\delta(x) = -x_n + \frac{1}{2\delta}x_n^2 - \frac{\delta}{2}|x|^2.$$

We prove that φ_δ is a convexification of $\Gamma = \{x \in \Omega : \quad \phi(x) = 0\}$ in 0. More precisely we prove

$$\begin{aligned} &\left\{x \in \overline{B_\delta} : \varphi_\delta(x) \geq 0\right\} \setminus \{0\} \subset \\ &\subset \left\{x \in \overline{B_\delta} : \phi(x) > 0\right\}. \end{aligned} \tag{14.103}$$

Let $x \in \overline{B_\delta} \setminus \{0\}$ be such that $\varphi_\delta(x) \geq 0$. We have

$$-\frac{\delta}{2}|x|^2 \geq x_n \left(1 - \frac{1}{2\delta}x_n\right). \tag{14.104}$$

Now, since $x \in \overline{B_\delta}$ we have $|x_n| \leq \delta$, hence

$$1 - \frac{1}{2\delta}x_n \geq \frac{1}{2}.$$

Therefore, by (14.104) we have

$$x \in \overline{B_\delta} \setminus \{0\} \text{ and } \varphi_\delta(x) \geq 0 \Longrightarrow \phi(x) = -x_n > 0.$$

♣

Example 14.11 Let $\phi \in C^2(\Omega)$ be a real-valued function which satisfies $\phi(x_0) = 0$, where $x_0 \in \Omega$. Let us assume that ϕ satisfies (14.96). Let $\varphi \in C^2(\Omega)$ be a convexification of $\Gamma = \{x \in \Omega : \phi(x) = 0\}$ in x_0. If $h \in C^2(\mathbb{R})$ is an increasing real-valued function such that

$$h'(0) > 0,$$

then $h \circ \varphi$ is a convexification of Γ in x_0. As a matter of fact we have $(h \circ \varphi)(x_0) = 0$, $\nabla((h \circ \varphi))(x_0) \neq 0$ and

$$(h \circ \varphi)(x) \geq 0 \Longleftrightarrow \varphi(x) \geq 0.$$

♣

In the Proposition that follows, we prove a more general version of Example 14.10.

Proposition 14.15 *Let $\phi \in C^2\left(\overline{\Omega}\right)$ be a real-valued function which satisfies $\phi(x_0) = 0$. Assume*

$$\nabla\phi(x) \neq 0, \quad \forall x \in \Omega.$$

For any $\delta > 0$, denote by

$$\begin{aligned}\varphi_\delta(x) = &\nabla\phi(x_0) \cdot (x - x_0) + \frac{1}{2}\partial^2\phi(x_0)(x - x_0) \cdot (x - x_0) + \\ &+ \frac{1}{2\delta}\left(\nabla\phi(x_0) \cdot (x - x_0)\right)^2 - \frac{\delta}{2}|x - x_0|^2.\end{aligned} \tag{14.105}$$

Then $\varphi_\delta(x)$ is a convexification of $\Gamma = \{x \in \Omega : \phi(x) = 0\}$ in x_0.

Proof For the sake of brevity, let us assume that $x_0 = 0$ and let us prove that there exist $\delta_1 > 0$ such that

$$\begin{aligned} &\left\{x \in \overline{B_{\delta_1}} : \varphi_\delta(x) \geq 0\right\} \setminus \{0\} \subset \\ &\subset \left\{x \in \overline{B_{\delta_1}} : \phi(x) > 0\right\}. \end{aligned} \tag{14.106}$$

By the Taylor formula we have that there exists a function σ such that

$$\phi(x) = \nabla\phi(0) \cdot x + \frac{1}{2}\partial^2\phi(0)x \cdot x + |x|^2\sigma(x) \tag{14.107}$$

and

$$\lim_{x \to 0} \sigma(x) = 0. \tag{14.108}$$

We have

$$\begin{aligned} &\phi(x) - |x|^2\sigma(x) + \frac{1}{2\delta}\left(\phi(x) - \frac{1}{2}\partial^2\phi(0)x \cdot x - |x|^2\sigma(x)\right)^2 = \\ &= \nabla\phi(0) \cdot x + \frac{1}{2}\partial^2\phi(0)x \cdot x + \frac{1}{2\delta}\left(\nabla\phi(0) \cdot x\right)^2 = \\ &= \varphi_\delta(x) + \frac{\delta}{2}|x|^2. \end{aligned} \tag{14.109}$$

Hence, if

$$\varphi_\delta(x) \geq 0, \tag{14.110}$$

then we have

$$\begin{aligned} \phi(x) &\geq \frac{\delta}{2}|x|^2 - \frac{1}{2\delta}\left(\phi(x) - \frac{1}{2}\partial^2\phi(0)x \cdot x - |x|^2\sigma(x)\right)^2 + |x|^2\sigma(x) \geq \\ &\geq |x|^2\left(\frac{\delta}{2} + \sigma(x)\right) - \frac{3}{2\delta}\left[\phi^2(x) + \frac{1}{4}\left(\partial^2\phi(0)x \cdot x\right)^2 + |x|^4\sigma^2(x)\right]. \end{aligned}$$

Now, for $r > 0$, let us denote by

$$M = \left|\partial^2\phi(0)\right|, \quad M_{\sigma,r} = \sup_{x \in B_r} |\sigma(x)|.$$

By (14.109) we get that if $x \in \overline{B_r}$ and (14.110) holds, then

$$\phi(x)\left(1+\frac{3}{2\delta}\phi(x)\right) \geq |x|^2\left\{\frac{\delta}{2} - M_{\sigma,r} - \frac{3|x|^2}{2\delta}\left(\frac{M^2}{4} + M^2_{\sigma,r}\right)\right\}. \quad (14.111)$$

Now, (14.108) implies that there exists $\delta_0 > 0$ such that

$$M_{\sigma,r} \leq \frac{\delta}{6}, \quad \forall r \in (0, \delta_0].$$

Therefore by (14.111) we have: if

$$x \in \overline{B_{\delta_0}}, \quad |x|^2 \leq \frac{4\delta^2}{9M^2 + \delta^2} \quad \text{and} \quad \phi_\delta(x) \geq 0,$$

then

$$\phi(x)\left(1+\frac{3}{2\delta}\phi(x)\right) \geq \frac{\delta}{6}|x|^2. \quad (14.112)$$

Now, since $\psi(0) = 0$ there exists $\delta_1 > 0$,

$$\delta_1 \leq \min\left\{\delta_0, \sqrt{\frac{4\delta^2}{9M^2 + \delta^2}}\right\}$$

such that

$$|\phi(x)| \leq \frac{\delta}{3}, \quad \forall x \in B_{\delta_1}. \quad (14.113)$$

By (14.112) and (14.113) we have $\phi(x) \geq 0$ for all $x \in \left\{y \in \overline{B_{\delta_1}} : \varphi_\delta(y) \geq 0\right\}$. Hence

$$1 + \frac{3}{2\delta}\phi(x) \leq 1 + \frac{3}{2\delta}|\phi(x)| \leq \frac{3}{2}$$

and

$$\frac{3}{2}\phi(x) \geq \phi(x)\left(1+\frac{3}{2\delta}\phi(x)\right) \geq \frac{\delta}{6}|x|^2.$$

Therefore, if $x \in \left\{y \in \overline{B_{\delta_1}} : \varphi_\delta(y) \geq 0\right\}$ then

$$\phi(x) \geq \frac{\delta}{9}|x|^2.$$

In particular we obtain (14.106). ∎

Proposition 14.16 *With the notation of Definition 14.14, we have:*

(a) for every $s > 0$ *the interior part of the set*

$$K_s := \left\{x \in \overline{B_r(x_0)} : \varphi(x) \geq -s, \ \phi(x) \leq 0\right\}, \tag{14.114}$$

is nonempty;

(b) for every $\rho \in (0, r)$ *there exists* $\overline{s} > 0$ *such that*

$$K_{\overline{s}} := \left\{x \in \overline{B_r(x_0)} : \varphi(x) \geq -\overline{s}, \ \phi(x) \leq 0 \right\} \subset B_\rho(x_0). \tag{14.115}$$

Proof

(a) Let $s > 0$. Set

$$\nu(x_0) = -\frac{\nabla\phi(x_0)}{|\nabla\phi(x_0)|} = -\frac{\nabla\varphi(x_0)}{|\nabla\varphi(x_0)|}.$$

Since $\phi(x_0) = \varphi(x_0) = 0$ we have that there exists $\delta \in \left(0, \frac{r}{2}\right)$ such that

$$\varphi(x_0 + \delta\nu(x_0)) > -s, \ \ \phi(x_0 + \delta\nu(x_0)) < 0.$$

Hence, by the continuity of φ and ϕ, there exists $\delta_1 \leq \delta$ such that

$$\varphi(x) > -s, \ \ \phi(x) < 0, \ \ \forall x \in B_{\delta_1}(x_0 + \delta\nu(x_0)).$$

Therefore $x_0 + \delta\nu(x_0)$ is an interior point of K_s.

(b) We argue by contradiction. Assume that there exist $\rho \in (0, r)$ such that for every $j \in \mathbb{N}$ there exists $x_j \in \overline{B_r(x_0)}$ wich satisfies

$$\varphi\left(x_j\right) \geq -\frac{1}{j}, \ \ \phi\left(x_j\right) \leq 0, \ \ \left|x_j - x_0\right| \geq \rho. \tag{14.116}$$

By compactness, there exists a subsequence $\left\{x_j^*\right\}$ of $\left\{x_j\right\}$, and $x^* \in \overline{B_r(x_0)}$ such that

$$\left\{x_j^*\right\} \to x^*.$$

Hence, by (14.116), we get

$$\varphi\left(x^*\right) \geq 0, \ \ \phi\left(x^*\right) \leq 0, \ \ \left|x^* - x_0\right| \geq \rho. \tag{14.117}$$

Now by (14.100) we have that

$$\varphi\left(x^*\right) \geq 0 \text{ and } \phi\left(x^*\right) \leq 0 \Longrightarrow x^* = x_0.$$

which contradicts

$$\left|x^* - x_0\right| \geq \rho.$$

Therefore (b) holds. ∎

Remark 14.12 Let us note that sentence (b) of Proposition 14.16 implies

$$\partial\left(B_r(x_0)\right) \cap \partial K_{\overline{s}} = \emptyset.$$

♦

Theorem 14.17 *Let R be a positive number. Let Ω be an open set of $\mathbb{R}^n$, $P(x, D)$ the operator* (14.2), *$x_0 \in \Omega$ such that $\overline{B_R(x_0)} \subset \Omega$. Let $\phi \in C^2(\Omega)$ be a real-valued function such that $\phi(x_0) = 0$ and assume that* (14.96) *holds. Let $\varphi \in C^\infty(\Omega)$ be a real-valued function which satisfies the following property:*

(i) φ is a convexification of Γ in x_0, where Γ is defined by (14.98), *more precisely we assume*

$$\begin{aligned}&\left\{x \in \overline{B_R(x_0)} : \varphi(x) \geq 0\right\} \setminus \{x_0\} \subset \\ &\subset \left\{x \in \overline{B_R(x_0)} : \phi(x) > 0\right\}.\end{aligned} \tag{14.118}$$

(ii) The following Carleman estimate holds true ($\mu > 0$)

$$\tau^\mu \sum_{|\alpha| \leq m-1} \int_{B_R(x_0)} \left|D^\alpha u\right|^2 e^{2\tau\varphi} dx \leq C \int_{B_R(x_0)} \left|P(x, D)u\right|^2 e^{2\tau\varphi} dx, \tag{14.119}$$

for every $u \in C_0^\infty(B_R(x_0))$ and for every $\tau \geq \tau_0$.

Assume that $f \in L^2(B_R(x_0))$ and let $U \in H^m(B_R(x_0))$ satisfy (14.97). *Then there exists $t_0 > 0$ such that for every $0 < t_2 < t_1 < t_0$ we have*

$$\|U\|_{H^{m-1}(K_{t_2})} \leq C \left(\|U\|_{H^{m-1}(K_{t_0})}\right)^{\frac{t_2}{t_1}} \left(\|f\|_{L^2(K_{t_0})}\right)^{1-\frac{t_2}{t_1}}, \tag{14.120}$$

where C is a constant which depends on τ_0, $t_0 - t_1$, $\|\varphi\|_{C^m\left(\overline{B_R(x_0)}\right)}$ and $\max_{|\alpha| \leq m-1} \|a_\alpha\|_{L^\infty(B_R(x_0))}$ only. Here, for $t \in (0, t_0]$,

$$\|U\|_{H^{m-1}(K_t)} = (\sum_{|\alpha| \leq m-1} \int_{K_t} \left|D^\alpha U\right|^2 dx)^{1/2}.$$

In particular, if

$$f \equiv 0,$$

then there exists a neighborhood of x_0, $\mathcal{U}_{x_0}$, *such that*

$$U = 0, \quad in \ \mathcal{U}_{x_0}. \tag{14.121}$$

Proof By Proposition 14.16 we have there exists $t_0 > 0$ such that

$$K_{t_0} = \left\{ x \in \overline{B_R(x_0)} : \ \varphi(x) \geq -t_0, \ \phi(x) \leq 0 \right\} \subset B_{R/2}(x_0). \tag{14.122}$$

Let $t_1 \in (0, t_0)$. Let $h \in C^{\infty}(\mathbb{R})$ be a function such that

$$0 \leq h(t) \leq 1, \quad \forall t \in \mathbb{R},$$

$$h(t) = 0, \quad \forall t \in (-\infty, -t_0] \text{ and } h(t) = 1, \quad \forall t \in [-t_1, +\infty).$$

Moreover, let us assume that

$$\sup_{[-t_0, -t_1]} \left| \frac{d^k h}{dt^k} \right| \leq C \left(t_0 - t_1\right)^{-k}, \quad \text{for } k = 1, \cdots, m,$$

where C is a constant.

Set

$$\eta(x) = h\left(\varphi(x)\right), \quad x \in B_R(x_0). \tag{14.123}$$

Let us note that $\eta \in C^{\infty}\left(\overline{\Omega}\right)$ and

$$\eta(x) = 0, \quad \text{for } \varphi(x) \leq -t_0 \text{ and } \eta(x) = 1, \quad \text{for } \varphi(x) \geq -t_1.$$

Moreover, by (14.122), we have

$$\eta U \in H_0^m\left(B_R(x_0)\right).$$

By density, (14.119) holds for every $u \in H_0^m(B_R(x_0))$. Let us apply it to

$$u = \eta U.$$

Let us denote, for the sake of brevity, $P = P(x, D)$ and let us observe that

$$P(\eta U) = \eta P U + [P, \eta] U = \eta f + [P, \eta] U, \tag{14.124}$$

where $[P, \eta]U = P(\eta U) - \eta PU$ is the commutator of operators P and η (see also Sect. 15.2) so that it is a differential operators of order $m - 1$ in which the principal part of P does not occur and the function η only appears with its derivatives of order greater than or equal to 1 and less than or equal to m. Hence

$$|[P, \eta]U| \leq C\Big(\sum_{|\alpha|\leq m-1} \left|D^\alpha U\right|\Big)\chi_{K_{t_0}\setminus K_{t_1}}, \tag{14.125}$$

where $\chi_{K_{t_0}\setminus K_{t_1}}$ is the characteristic function of the set $K_{t_0} \setminus K_{t_1}$, C is a constant which depends on $t_0 - t_1$, $\|\varphi\|_{C^m\left(\overline{B_R(x_0)}\right)}$ and $\max_{|\alpha|\leq m-1} \|a_\alpha\|_{L^\infty(B_R(x_0))}$ only.

Now by (14.119), (14.124), (14.125), recalling that

$$PU = f, \quad \text{in } B_R(x_0)$$

and $U = 0$ in $\{x \in B_R(x_0) : \phi(x) \geq 0\}$, we have, for every $\tau \geq \tau_0$,

$$\begin{aligned}
&\tau^\mu \sum_{|\alpha|\leq m-1} \int_{K_{t_0}} \left|D^\alpha(\eta U)\right|^2 e^{2\tau\varphi} dx \leq C \int_{K_{t_0}} |f|^2 e^{2\tau\varphi} dx + \\
&+ C \sum_{|\alpha|\leq m-1} \int_{K_{t_0}\setminus K_{t_1}} \left|D^\alpha U\right|^2 e^{2\tau\varphi} dx \leq \\
&\leq C \|f\|^2_{L^2(K_{t_0})} + Ce^{-2\tau t_1} \|U\|^2_{H^{m-1}(K_{t_0})}.
\end{aligned} \tag{14.126}$$

Now, for every $\tau \geq \tau_1 := \max\{\tau_0, 1\}$, we have trivially

$$\begin{aligned}
\tau^\mu \sum_{|\alpha|\leq m-1} \int_{K_{t_0}} \left|D^\alpha(\eta U)\right|^2 e^{2\tau\varphi} dx &\geq \sum_{|\alpha|\leq m-1} \int_{K_{t_2}} \left|D^\alpha U\right|^2 e^{2\tau\varphi} dx \geq \\
&\geq e^{-2\tau t_2} \|U\|^2_{H^{m-1}(K_{t_2})},
\end{aligned}$$

and plugging what has been just obtained in (14.126) we get, for every $\tau \geq \tau_1$

$$\|U\|^2_{H^{m-1}(K_{t_2})} \leq Ce^{2\tau t_2} \|f\|^2_{L^2(K_{t_0})} + Ce^{-2\tau(t_1-t_2)} \|U\|^2_{H^{m-1}(K_{t_0})}. \tag{14.127}$$

If

$$\overline{\tau} \geq \tau_1,$$

then we choose $\tau = \overline{\tau}$ and by (14.127) we get

$$\|U\|_{H^{m-1}(K_{t_2})} \leq \sqrt{2C} \left(\|U\|_{H^{m-1}(K_{t_0})}\right)^{\frac{t_2}{t_1}} \left(\|f\|_{L^2(K_{t_0})}\right)^{1-\frac{t_2}{t_1}}. \tag{14.128}$$

On the other hand, if

$$\overline{\tau} < \tau_1,$$

then we have

$$\|U\|_{H^{m-1}(K_{t_0})} < e^{\tau_1 t_1} \|f\|_{L^2(K_{t_0})}$$

and we have trivially

$$\begin{aligned}\|U\|_{H^{m-1}(K_{t_2})} \leq \|U\|_{H^{m-1}(K_{t_0})} = \\ &= \left(\|U\|_{H^{m-1}(K_{t_0})}\right)^{\frac{t_2}{t_1}} \left(\|U\|_{H^{m-1}(K_{t_0})}\right)^{1-\frac{t_2}{t_1}} \leq \\ &\leq e^{\tau_1(t_1-t_2)} \left(\|U\|_{H^{m-1}(K_{t_0})}\right)^{\frac{t_2}{t_1}} \left(\|f\|_{L^2(K_{t_0})}\right)^{1-\frac{t_2}{t_1}}.\end{aligned} \tag{14.129}$$

Finally, if $f \equiv 0$ then (14.127) becomes

$$\|U\|^2_{H^{m-1}(K_{t_2})} \leq Ce^{-2\tau(t_1-t_2)} \|U\|^2_{H^{m-1}(K_{t_0})}, \quad \forall \tau \geq \tau_1 \tag{14.130}$$

and passing to the limit as τ goes to $+\infty$ (taking into account that $t_2 - t_1 > 0$) we get $U = 0$ in K_{t_2} and as t_2, t_1 are arbitrary in $(0, t_0)$ we get

$$U = 0, \quad \text{in } K_{t_0}$$

and since $U = 0$ in $\{x \in B_R(x_0) : \phi(x) \geq 0\}$, (14.121) follows. ■

14.5 Operators Whose Principal Part has Constant Coefficients

In the present section we consider the linear differential operators whose **principal part has constant coefficients**. In other words, we will consider the operators

$$Lu = P(D)u + M(x, D)u, \tag{14.131}$$

where $P(D)$ is a differential operator of order m whose coefficients are constant (real or complex) and $M(x, D)$ is a differential operator of order (less or equal to) $m - 1$ whose coefficients belong to L^∞.

The Main Theorem which we will prove here is due to **Nirenberg**, (see Theorem 14.18), [101]. Subsequently, we will briefly outline some applications of this theorem with short proofs. The estimates that can be derived from the Nirenberg theorem are not, properly speaking, Carleman estimates like (14.4), but with simple transformations, they provide some unique continuation properties.

14.5.1 The Nirenberg Theorem

Let us introduce and recall some notations.

Let $P(D)$ be the operator

$$P(D) = \sum_{|\alpha|\leq m} a_\alpha D^\alpha, \tag{14.132}$$

where $a_\alpha \in \mathbb{C}$, for every $\alpha \in \mathbb{N}_0^n$ satisfying $|\alpha| \leq m$. Let

$$p(\xi) = \sum_{|\alpha|\leq m} a_\alpha \xi^\alpha, \quad \forall \xi \in \mathbb{R}^n$$

be the symbol of $P(D)$. For each multi-index α, let us denote by

$$p^{(\alpha)}(\xi) = \partial_\xi^\alpha p(\xi)$$

and set

$$Q_1 = \left\{x \in \mathbb{R}^n : \left|x_j\right| < 1, j = 1, \cdots, n\right\}.$$

We prove the following

Theorem 14.18 (Nirenberg) *Let $N \in \mathbb{R}^n$, $|N| = 1$. Then there exists a constant C, depending on n and m only, such that for every $\alpha \in \mathbb{N}_0^n$ we have*

$$\int_{Q_1} e^{2\tau N\cdot x} \left|P^{(\alpha)}(D)u\right|^2 dx \leq C \int_{Q_1} e^{2\tau N\cdot x} \left|P(D)u\right|^2 dx, \tag{14.133}$$

for every $u \in C_0^\infty(Q_1)$ and for every $\tau \in \mathbb{R}$.

In order to prove Theorem 14.18 we need some preliminary results.

First of all, let us recall the following one-dimensional Poincaré inequality

$$\int_{-1}^{1} |u|^2 dt \leq \frac{4}{\pi^2} \int_{-1}^{1} \left|u'\right|^2 dt, \quad \forall u \in C_0^\infty((-1, 1)). \tag{14.134}$$

Now let us prove

Lemma 14.19 *There exists $C_0 > 0$ such that for each $\gamma \in \mathbb{C}$ we have*

$$\int_{-1}^{1} |u|^2 dt \leq C_0 \int_{-1}^{1} \left|u' - \gamma u\right|^2 dt, \quad \forall u \in C_0^\infty((-1,1)). \tag{14.135}$$

Proof Let $\gamma = \alpha + i\beta$, $\alpha, \beta \in \mathbb{R}$. We have

$$\begin{aligned}\left|u' - \gamma u\right|^2 &= \alpha^2 |u|^2 + \left|u' - i\beta u\right|^2 - 2\alpha \Re\left(\left(u' - i\beta u\right)\overline{u}\right) = \\ &= \alpha^2 |u|^2 + \left|\left(ue^{-i\beta t}\right)'\right|^2 - \alpha\left(|u|^2\right)'.\end{aligned}$$

Hence, as $u \in C_0^\infty((-1,1))$, taking into account (14.134), we get

$$\begin{aligned}\int_{-1}^{1} \left|u' - \gamma u\right|^2 dt &= \int_{-1}^{1} \left(\alpha^2 |u|^2 + \left|\left(ue^{-i\beta t}\right)'\right|^2\right) dt \geq \\ &\geq \int_{-1}^{1} \left|\left(ue^{-i\beta t}\right)'\right|^2 dt \geq \frac{\pi^2}{4} \int_{-1}^{1} \left|ue^{-i\beta t}\right|^2 dt = \\ &= \frac{\pi^2}{4} \int_{-1}^{1} |u|^2 dt.\end{aligned}$$

Therefore inequality (14.135) is proved with $C_0 = \frac{4}{\pi^2}$. ■

Let $a_1, \cdots, a_k \in \mathbb{C}$, $a_k \neq 0$. Let

$$q(\eta) = \sum_{j=0}^{k} a_j \eta^j, \quad \eta \in \mathbb{C}. \tag{14.136}$$

Let us consider the differential operator

$$Q(D_t) = \sum_{j=0}^{k} a_j D_t^j, \tag{14.137}$$

where $D_t = \frac{1}{i}\frac{d}{dt}$. Set

$$Q'(D_t) = \sum_{j=1}^{k} j a_j D_t^{j-1}.$$

We have the following

Lemma 14.20 *Let $k \in \mathbb{N}$. Then there exists $C_1 > 0$ depending on k only, such that we have*

$$\int_{-1}^{1} \left|Q'(D_t)u\right|^2 dt \leq C_1 \int_{-1}^{1} |Q(D_t)u|^2 dt, \quad \forall u \in C_0^\infty((-1,1)). \tag{14.138}$$

Proof It is not restrictive to assume that $a_k = 1$. Let $\gamma_1, \cdots, \gamma_k \in \mathbb{C}$ be the roots of the polynomial p, we have

$$q(\eta) = \prod_{1 \leq j \leq k} \left(\eta - \gamma_j\right).$$

Set

$$q_l(\eta) = \frac{1}{(\eta - \gamma_l)} \prod_{1 \leq j \leq k} \left(\eta - \gamma_j\right), \text{ for } l = 1, \cdots, k.$$

We have

$$q'(\eta) = \sum_{l=1}^{k} p_l(\eta), \quad \forall \eta \in \mathbb{C}. \tag{14.139}$$

Let $l \in \{1, \cdots, k\}$ be fixed, $u \in C_0^\infty((-1,1))$ and let us denote

$$v_l = Q_l(D_t)u.$$

Let us observe that

$$(D_t - \gamma_l)\, v_l = Q(D_t)u.$$

Now we apply Lemma 14.19 to v_l (where $\gamma = i^{-1}\gamma_l$) and we get

$$\begin{aligned} \int_{-1}^{1} |Q_l(D_t)u|^2 dt = \int_{-1}^{1} |v_l|^2 dt \leq \\ \leq C_0 \int_{-1}^{1} |(D_t - \gamma_l)\, v_l|^2 dt = \\ = C_0 \int_{-1}^{1} |Q(D_t)u|^2 dt. \end{aligned} \tag{14.140}$$

By (14.139) and (14.140) we have

$$\begin{aligned}\int_{-1}^{1}\left|Q'(D_t)u\right|^2 dt &\leq k\sum_{l=1}^{k}\int_{-1}^{1}|Q_l(D_t)u|^2\,dt \leq\\ &\leq C_0k^2\int_{-1}^{1}|Q(D_t)u|^2\,dt.\end{aligned} \tag{14.141}$$

Hence, inequality (14.138) is proved with $C_1 = k^2C_0$. ■

Theorem 14.21 (Hörmander) *Let $P(D)$ be a differential operator of order m with constant coefficients. Then there exists a constant C_2 which depends on m and on n only, such that we have, for any $\alpha \in \mathbb{N}_0^n$,*

$$\int_{Q_1}\left|P^{(\alpha)}(D)u\right|^2 dx \leq C_2\int_{Q_1}|P(D)u|^2\,dx, \quad \forall u \in C_0^\infty(Q_1). \tag{14.142}$$

Proof First of all we prove (14.142) for $\alpha = e_j$, $j = 1, \cdots, n$. It is not restrictive to assume $j = n$. for any $f \in L^2(\mathbb{R}^n)$ we set

$$\widehat{f}(\xi', x_n) = \mathcal{F}_{\xi'}(f(\cdot, x_n) = \int_{\mathbb{R}^{n-1}} f(x', x_n)e^{-ix'\xi'}dx', \quad \forall\xi' \in \mathbb{R}^{n-1}.$$

Let $u \in C_0^\infty(Q_1)$. We have

$$\mathcal{F}_{\xi'}(P(D)u) = P\left(\xi', D_n\right)\widehat{u}(\xi', x_n) \tag{14.143}$$

and by the Parseval identity, we have

$$\begin{aligned}&\int_{Q_1}\left|P^{(e_n)}(D)u\right|^2 dx =\\ &= \frac{1}{(2\pi)^{n-1}}\int_{\mathbb{R}^{n-1}}d\xi'\int_{-1}^{1}\left|P^{(e_n)}\left(\xi', D_n\right)\widehat{u}(\xi', x_n)\right|^2 dx_n.\end{aligned} \tag{14.144}$$

Now we apply Lemma 14.140 to the operator $Q(D_n) = P\left(\xi', D_n\right)$, where $\xi' \in \mathbb{R}^{n-1}$ is fixed. We obtain

$$\begin{aligned}\int_{-1}^{1}\left|P^{(e_n)}\left(\xi', D_n\right)\widehat{u}(\xi', x_n)\right|^2 dx_n &\leq\\ \leq C_1\int_{-1}^{1}\left|P\left(\xi', D_n\right)\widehat{u}(\xi', x_n)\right|^2 dx_n.\end{aligned} \tag{14.145}$$

By (14.144) and (14.145) we have

$$\int_{Q_1} \left|P^{(e_n)}(D)u\right|^2 dx \leq \frac{C_1}{(2\pi)^{n-1}} \int_{\mathbb{R}^{n-1}} d\xi' \int_{-1}^{1} \left|P\left(\xi', D_n\right)\widehat{u}(\xi', x_n)\right|^2 dx_n =$$

$$= C_1 \int_{Q_1} |P(D)u|^2 \, dx.$$

Since the previous proof can be repeated for any indices. We have that for each multi-indices α such that $|\alpha| = 1$ the estimate following holds

$$\int_{Q_1} \left|P^{(\alpha)}(D)u\right|^2 dx \leq C_1 \int_{Q_1} |P(D)u|^2 \, dx, \quad \forall u \in C_0^\infty(Q_1). \tag{14.146}$$

By iteration of (14.146) we get, for any $\alpha \in \mathbb{N}_0^n$,

$$\int_{Q_1} \left|P^{(\alpha)}(D)u\right|^2 dx \leq C_1^{|\alpha|} \int_{Q_1} |P(D)u|^2 \, dx, \quad \forall u \in C_0^\infty(Q_1).$$

Hence inequality (14.142) is now proved with $C_2 = C_1^{|\alpha|}$. ∎

Proof of Theorem 14.18 Let $u \in C_0^\infty(Q_1)$. Setting $v = e^{\tau N \cdot x} u$, we obtain

$$D_j u = e^{-\tau N \cdot x} \left(D_j + i\tau N_j\right) v.$$

For each multi-index α, we have

$$D^\alpha u = e^{-\tau N \cdot x} \left(D + i\tau N\right)^\alpha v.$$

Hence

$$e^{\tau N \cdot x} P(D)u = P(D + i\tau N)v, \quad e^{\tau N \cdot x} P^{(\alpha)}(D)u = P^{(\alpha)}(D + i\tau N)v.$$

Therefore by (14.142) we get

$$\int_{Q_1} e^{2\tau N \cdot x} \left|P^{(\alpha)}(D)u\right|^2 dx = \int_{Q_1} \left|P^{(\alpha)}(D + i\tau N)v\right|^2 dx \leq$$

$$\leq C_2 \int_{Q_1} |P(D + i\tau N)v|^2 \, dx =$$

$$= C_2 \int_{Q_1} e^{2\tau N \cdot x} \, |P(D)u|^2 \, dx,$$

where C_2 is the same constant of (14.142). ∎

Remark 14.13 Since $P(D)$ is an operator of order m, we have that there exists $\alpha \in \mathbb{N}_0^n$ such that $P^{(\alpha)}(\xi) = \alpha! a_\alpha \neq 0$. Therefore (14.133) gives, in particular,

$$\int_{Q_1} |u|^2\,dx \leq C_3 \int_{Q_1} |P(D)u|^2\,dx, \quad \forall u \in C_0^\infty(Q_1), \tag{14.147}$$

where

$$C_3 = C_2(\frac{1}{\alpha!}\min\{|a_\alpha| : a_\alpha \neq 0,\, |\alpha| = m\})^2.$$

By the proof of Theorem 14.18 we observe that, if $M(\xi)$ is a polynomial for which there exists a constant $C_4 > 0$ such that

$$\frac{|M(\xi + i\tau N)|^2}{\sum_{|\alpha|\leq m} \left|p^{(\alpha)}(\xi + i\tau N)\right|^2} \leq C_4, \quad \forall \xi \in \mathbb{R}^n, \quad \forall \tau \in \mathbb{R}, \tag{14.148}$$

then there exists a constant C such that

$$\int_{Q_1} e^{2\tau N\cdot x}\,|M(D)u|^2\,dx \leq C \int_{Q_1} e^{2\tau N\cdot x}\,|P(D)u|^2\,dx, \tag{14.149}$$

for every $u \in C_0^\infty(Q_1)$ and for every $\tau \in \mathbb{R}$.

Likewise, if $M_\tau(\xi)$ is a polynomial in the variable ξ depending by the parameter τ and if

$$\sup\left\{\frac{|M_\tau(\xi + i\tau N)|^2}{\sum_{|\alpha|\leq m} \left|p^{(\alpha)}(\xi + i\tau N)\right|^2} : (\xi, \tau) \in \mathbb{R}^{n+1}\right\} < +\infty, \tag{14.150}$$

then, for a constant C (independent of τ and u), we have

$$\int_{Q_1} e^{2\tau N\cdot x}\,|M_\tau(D)u|^2\,dx \leq C \int_{Q_1} e^{2\tau N\cdot x}\,|P(D)u|^2\,dx, \tag{14.151}$$

for any $u \in C_0^\infty(Q_1)$ and any $\tau \in \mathbb{R}$.

For instance, in the case

$$P(D) = -\left(D_1^2 + \cdots + D_n^2\right) = \Delta,$$

we have

$$p(\xi) = -\left(\xi_1^2 + \cdots + \xi_n^2\right), \quad \sum_{|\alpha|\leq 2} \left|p^{(\alpha)}(\xi)\right|^2 = |\xi|^4 + 4\left(n^2 + |\xi|^2\right),$$

from which we easily obtain that (14.148) is satisfied for all $N \in \mathbb{R}^n$, $|N| = 1$, provided $M(\xi) = \xi_j$ (as well as, of course, for $M(\xi) = 1$). Hence, we have

$$\int_{Q_1} \left(|u|^2 + |\nabla u|^2 + |D^2 u|^2\right) dx \leq C \int_{Q_1} |\Delta u|^2 \, dx, \tag{14.152}$$

for every $u \in C_0^\infty(Q_1)$, where

$$|D^2 u|^2 = \sum_{j,k=1}^{n} |D_{jk}^2 u|^2 = \sum_{j,k=1}^{n} |\partial_{jk}^2 u|^2.$$

Moreover, we have

$$\sum_{|\alpha|\leq 2} \left|p^{(\alpha)}(\xi + i\tau N)\right|^2 = \left(|\xi|^2 - \tau^2\right)^2 + 4\tau^2(\xi \cdot N)^2 + 4\left(|\xi|^2 + \tau^2\right) + 4n^2$$

and, setting

$$M_{0,\tau}(\xi) = \tau, \quad M_{j,\tau}(\xi) = \xi_j, \quad j = 1, \cdots, n,$$

we have (reader check), for $k = 1, \cdots, n$,

$$\sup \left\{ \frac{\left|M_{k,\tau}(\xi + i\tau N)\right|^2}{\sum_{|\alpha|\leq 2} \left|p^{(\alpha)}(\xi + i\tau N)\right|^2} : \quad (\xi, \tau) \in \mathbb{R}^{n+1} \right\} < +\infty.$$

Hence,

$$\int_{Q_1} e^{2\tau N\cdot x} \left(\tau^2 |u|^2 + |\nabla u|^2\right) dx \leq C \int_{Q_1} e^{2\tau N\cdot x} \, |\Delta u|^2 \, dx, \tag{14.153}$$

for every $u \in C_0^\infty(Q_1, \mathbb{C})$ and for every $\tau \in \mathbb{R}$. ♦

Remark 14.14 By what observed in Remark 1 we have that, if $p_m(\xi)$, is a homogeneous polynomial of degree m, then we have

$$\int_{Q_1} e^{2\tau N\cdot x} \, |u|^2 \, dx \leq C \int_{Q_1} e^{2\tau N\cdot x} \, |P_m(D)u|^2 \, dx, \tag{14.154}$$

for every $u \in C_0^\infty(Q_1)$ and for every $\tau \in \mathbb{R}$. ♦

Now, let us consider the following Cauchy problem. Let

$$h(x') = 1 - \sqrt{1 - |x'|^2},$$

$$\Lambda = \left\{(x', x_n) \in B_1' \times \mathbb{R} : h(x') < x_n < 1\right\}$$

and

$$\Gamma = \left\{(x', h(x')) : x' \in B_1'\right\}.$$

Let $a \in L^\infty(Q_1)$ and let $U \in H^m(Q_1)$ be a solution to the following Cauchy problem

$$\begin{cases} P(D)U + a(x)U = 0, & \text{in } Q_1 \\ \\ \operatorname{supp} U \subset \overline{\Lambda} \end{cases} . \tag{14.155}$$

In what follows we give a sketch of the proof of uniqueness for Cauchy problem (14.155). In particular we consider the case where $P(D) = P_m(D)$, $m \geq 1$. However for more details on this issue we refer to [101, Theorem 1] in which the uniqueness for problem (14.155) is proven under the sole condition that h is strictly convex.

Set

$$M = \|a\|_{L^\infty(Q_1)} . \tag{14.156}$$

Let us start by considering the simple case in which M is zero. In such a case we have

$$\begin{cases} P_m(D)U = 0, & \text{in } Q_1 \\ \\ \operatorname{supp} U \subset \overline{\Lambda} \end{cases} . \tag{14.157}$$

Let $\delta \in (0, \frac{1}{3})$. Let $\zeta \in C^\infty(\mathbb{R})$ satisfy $0 \leq \zeta \leq 1$, (Fig. 14.2)

$$\zeta(x_n) = 1, \text{ for } x_n \leq 1 - 2\delta, \quad \zeta(x_n) = 0 \quad \text{for } 1 - \delta \leq x_n < 1$$

and

$$\left|\zeta^{(k)}(x_n)\right| \leq c\delta^{-k}, \text{ for } 1 - 2\delta \leq x_n \leq 1 - \delta, \; k = 1, \cdots, m,$$

where c is a constant. Let us apply (14.154) to

$$u(x) = \zeta(x_n)U(x),$$

where

$$N = -e_n, \text{ and } \tau > 0.$$

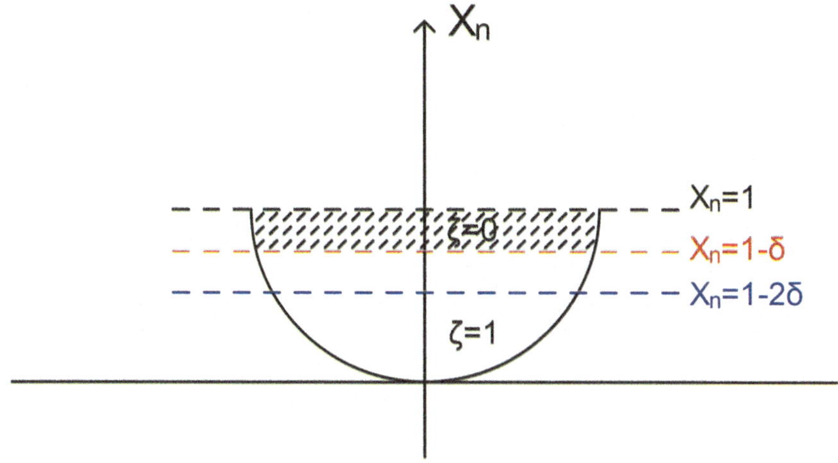

Fig. 14.2 $\zeta = 1$ for $x_n \leq 1 - 2\delta$, $\zeta = 0$ for $1 - \delta \leq x_n \leq 1$

We have

$$P_m(D)(\zeta U) = \zeta(x_n) P_m(D)U + [P_m(D), \zeta]U$$

as $P_m(D)U = 0$, we get

$$|P_m(D)(\zeta U)| \leq C\delta^{-m} \chi_{(1-2\delta,1-\delta)}(x_n) \sum_{k=0}^{m-1} \left|\partial_n^k U\right|. \tag{14.158}$$

Hence by (14.154) and (14.158) we get

$$\begin{aligned}
&\int_{Q_1} e^{-2\tau x_n} |\zeta U|^2 dx \leq C \int_{Q_1} e^{-2\tau x_n} \,|P_m(D)(\zeta U)|^2 \, dx \leq \\
&\leq C'\delta^{-2m} \sum_{k=0}^{m-1} \int_{Q_1 \cap \{1-2\delta < x_n < 1-\delta\}} e^{-2\tau x_n} \left|\partial_n^k U\right|^2 dx \leq \\
&\leq C' e^{-2\tau(1-2\delta)} \delta^{-2m} \, \|U\|^2_{H^{m-1}(Q_1)},
\end{aligned} \tag{14.159}$$

for every $\tau > 0$.

Now we have, trivially, for every $\tau > 0$

$$\begin{aligned}
&\int_{Q_1} e^{-2\tau x_n} \|\zeta U|^2 dx \geq \\
&\geq e^{-2\tau(1-3\delta)} \int_{\Lambda \cap \{x_n < 1-3\delta\}} |\zeta U|^2 dx.
\end{aligned} \tag{14.160}$$

By (14.159) and (14.160) we obtain

$$\int_{\Lambda\cap\{x_n<1-3\delta\}} |U|^2 dx \leq C' e^{-2\tau\delta} \delta^{-2m} \|U\|^2_{H^{m-1}(Q_1)} \to 0, \quad \text{as } \tau \to +\infty.$$

From which we have $U = 0$ in $\Lambda \cap \{x_n < 1 - 3\delta\}$ and, as δ is arbitrary, we get $U = 0$ in Λ.

In the general case, when a is not zero, one would like to argue in a similar manner, but one encounters an obstacle due to the fact that in expanding $|P_m(D)(\zeta U)|$, besides the terms present in (14.158) a new term, depending on U arises. To overcome this difficulty, we first apply a rescaling of estimate (14.154).

For any $r > 0$, set

$$Q_r = \left\{x \in \mathbb{R}^n : \quad |x_j| < r, \quad j = 1, \cdots, n\right\}.$$

Proposition 14.22 *Let $N \in \mathbb{R}^n$ ($|N| = 1$). There exists a constant C so that*

$$\int_{Q_r} e^{2\tau N\cdot x}|u|^2 dx \leq C r^{2m} \int_{Q_r} e^{2\tau N\cdot x} |P_m(D)u|^2 dx, \tag{14.161}$$

for every $r > 0$, for every $u \in H_0^m(Q_r)$ and for every $\tau \in \mathbb{R}$.

Proof Let $u \in H_0^m(Q_r)$ and let us denote by

$$\widetilde{u}(y) = u(ry), \quad \forall y \in Q_1.$$

We apply (14.154) to $\widetilde{u}$, replacing there τ by τr. We get

$$\int_{Q_1} e^{2\tau r N\cdot y}|u(ry)|^2 dy \leq C r^{2m} \int_{Q_1} e^{2\tau r N\cdot y} |(P_m(D))(ry)|^2 dy.$$

Now, by performing the change of variables $y = r^{-1}x$, we have

$$\int_{Q_r} e^{2\tau N\cdot x}|u(x)|^2 r^{-n} dx \leq C r^{2m} \int_{Q_r} e^{2\tau N\cdot x} |P_m(D)u(x)|^2 r^{-n} dx.$$

By last inequality we immediately obtain (14.161). ■

Now let us go back to Cauchy problem (14.155). Let $r \in (0, 1)$ be a number to be chosen later, set

$$\rho = \frac{r^2}{2}, \qquad R = \sqrt{2\rho - \rho^2}.$$

It is easy to check that $\rho, R < r$ hence

$$\Lambda_\rho := \left\{(x', x_n) \in B'_R \times \mathbb{R} : h(x') < x_n < \rho\right\} \subset Q_r.$$

Proposition 14.23 *Let $U \in H^m(Q_1)$ be a solution to Cauchy problem* (14.155). *Then there exists r_0 such that if $r \in (0, r_0]$ then*

$$U = 0, \quad in\ \Lambda_\rho. \tag{14.162}$$

Proof Let $\delta \in (0, \frac{1}{3})$ and $\zeta \in C^\infty(\mathbb{R})$ satisfy $0 \leq \zeta \leq 1$,

$$\zeta(x_n) = 1, \text{ for } x_n \leq \rho(1 - 2\delta), \quad \zeta(x_n) = 0 \quad \text{for } \rho(1 - \delta) \leq x_n < \rho$$

and

$$\left|\zeta^{(k)}(x_n)\right| \leq c(\delta\rho)^{-k}, \text{ for } \rho(1 - 2\delta) \leq x_n \leq \rho(1 - \delta),\ k = 1, \cdots, m,$$

where c is a constant.

Let us prepare to apply estimate (14.161) to

$$u = \zeta(x_n)U(x),$$

where $N = -e_n$ and $\tau > 0$.

Taking into account (14.156),

$$|P_m(D)(\zeta U)| \leq M\zeta\, |U| + c(\rho\delta)^{-m}\chi_I(x_n) \sum_{k=0}^{m-1} \left|\partial_n^k U\right|, \tag{14.163}$$

where we set $I = (\rho(1 - 2\delta), \rho(1 - \delta))$.

Let us denote by

$$J_1 = \int_{Q_r} e^{-2\tau x_n}\zeta^2(|U|^2\, dx,$$

and

$$J_2 = \sum_{k=0}^{m-1} \int_{Q_r \cap \rho\{(1-2\delta) < x_n < \rho(1-\delta)\}} e^{-2\tau x_n} \left|\partial_n^k U\right|^2 dx.$$

Hence by (14.161) and (14.163) we have (recall that C denotes a constant that may change from line to line)

$$\begin{aligned}
&\int_{Q_r} e^{-2\tau x_n}\left(|\zeta U|^2+r^2|\nabla(\zeta U)|^2\right)dx \leq \\
&\leq Cr^{2m}\int_{Q_r} e^{-2\tau x_n}\,|P_m(D)(\zeta U)|^2\,dx \leq \\
&\leq CM^2r^{2m}J_1+Cr^{2m}\,(\rho\delta)^{-2m}\,J_2,
\end{aligned} \tag{14.164}$$

for every $\tau>0$.

Proceeding in the same way as done in (14.159). We have

$$J_2 \leq Ce^{-2\tau\rho(1-2\delta)}\,\|U\|^2_{H^{m-1}(Q_r)}, \tag{14.165}$$

for every $\tau>0$.

Hence, by (14.164)–(14.165), we get

$$\begin{aligned}
&\int_{Q_r} e^{-2\tau x_n}|\zeta U|^2dx \leq CM^2r^{2m}J_1+ \\
&+C\left(r\rho^{-1}\delta^{-1}\right)^{2m}e^{-2\tau\rho(1-2\delta)}\,\|U\|^2_{H^{m-1}(Q_r)},
\end{aligned} \tag{14.166}$$

for every $\tau>0$.

By the manner in which ζ is defined, we have trivially

$$\int_{Q_r} e^{-2\tau x_n}|\zeta U|^2dx \geq \int_{Q_r\cap\{x_n<\rho(1-2\delta)\}} e^{-2\tau x_n}|U|^2dx, \tag{14.167}$$

for every $\tau>0$.

Concerning J_1, let us observe

$$\begin{aligned}
J_1 &= \int_{Q_r\cap\{x_n<\rho(1-2\delta)\}} e^{-2\tau x_n}\zeta^2\,|U|^2\,dx+ \\
&+\int_{Q_r\cap\{\rho(1-2\delta)<x_n<\rho(1-\delta)\}} e^{-2\tau x_n}\zeta^2\,|U|^2\,dx \leq \\
&\leq \int_{Q_r\cap\{x_n<\rho(1-2\delta)\}} e^{-2\tau x_n}\,|U|^2\,dx+e^{-2\tau\rho(1-2\delta)}\,\|U\|^2_{L^2(Q_r)},
\end{aligned} \tag{14.168}$$

for every $\tau>0$.

Now by (14.166)–(14.168) we have

$$\begin{aligned}
&\left(1-CM^2r^{2m}\right)\int_{Q_r\cap\{x_n<\rho(1-2\delta)\}} e^{-2\tau x_n}|U|^2dx \leq \\
&\leq Cr^{2m}\left(1+(\rho\delta)^{-2m}\right)e^{-2\tau\rho(1-2\delta)}\,\|U\|^2_{H^{m-1}(Q_r)},
\end{aligned}$$

for every $\tau>0$.

Now, let us choose $r = r_0 < 1$ satisfying $1 - CM^2 r_0^{2m} \geq \frac{1}{2}$ (here, recall that C does not depends by r) and denoting by ρ_0 and R_0 the values of ρ and R correspondingly to this choice of r we get

$$\int_{Q_{r_0} \cap \{x_n < \rho_0(1-2\delta)\}} e^{-2\tau x_n} |U|^2 dx \leq C e^{-2\tau\rho_0(1-2\delta)} \|U\|^2_{H^{m-1}(\Lambda_{\rho_0})}, \tag{14.169}$$

for every $\tau > 0$, where C depends on M and δ only.

On the other hand, we have, trivially

$$\int_{Q_{r_0} \cap \{x_n < \rho_0(1-2\delta)\}} e^{-2\tau x_n} |U|^2 dx \geq e^{-2\tau\rho_0(1-3\delta)} \|U\|^2_{L^2(\Lambda_{\rho_0(1-3\delta)})},$$

for every $\tau > 0$.

By the last obtained estimate and by (14.169) we get

$$\|U\|^2_{L^2(\Lambda_{\rho_0(1-3\delta)})} \leq C e^{-2\tau\rho_0\delta} \|U\|^2_{H^{m-1}(\Lambda_{\rho_0})}, \tag{14.170}$$

for every $\tau > 0$. Passing to the limit as $\tau \to +\infty$ we have

$$U = 0, \quad \text{in } \Lambda_{\rho_0(1-3\delta)}.$$

Since δ is arbitrary, we get (14.23). ■

Exercise 14.1 Prove Theorem 14.12 using estimate (14.153) instead of estimate (14.63). ♠

14.5.2 Brief Comments on the Heat Operator

The heat operator is given by (for $n > 1$)

$$P(D) = -\sum_{j=1}^{n-1} D_j^2 - iD_n = \Delta_{x'} - \partial_n.$$

So the principal part of the heat operator is

$$P_2(D) = -\sum_{j=1}^{n-1} D_j^2,$$

whose symbol is $-\sum_{j=1}^{n-1} \xi_j^2$. Let us notice that in the case of the heat operator the principal part $P_2(D)$ excludes the term $iD_n u$. Actually, if we employ directly estimate (14.18) we have (reader check)

$$\begin{aligned} &\int_{Q_1} e^{2\tau N\cdot x}\,|u|^2\,dx + \int_{Q_1} e^{2\tau N\cdot x}\,|\nabla u|^2\,dx \leq \\ &\leq C\int_{Q_1} e^{2\tau N\cdot x}\,|\Delta_{x'}u - \partial_n|^2\,dx, \end{aligned} \tag{14.171}$$

for every $u \in C_0^\infty(Q_1, \mathbb{C})$ and for every $\tau \in \mathbb{R}$.

As in previous situations, (14.171) implies the uniqueness for the Cauchy problem with **strictly convex initial surfaces** for the differential inequalities

$$|\Delta_{x'}U - \partial_n U| \leq M\left(|\nabla_{x'}U| + |U|\right), \tag{14.172}$$

where M is a positive number and U is enough regular.

When $n = 2$, one can exploit the particularity of the dimension two to prove the following unique continuation property, we refer to [101, Theorem 9] for the proof.

Let ω be an open of $\mathbb{R}^2 = \mathbb{R}_x \times \mathbb{R}_t$ contained in a strip $\mathcal{R}$. For $t_0 \in \mathbb{R}$, we denote by s_{t_0} straight line of equation $t = t_0$ and set (Fig. 14.3)

$$\mathcal{R}_\omega = \left\{(x, t_0) \in \mathcal{R} : s_{t_0} \cap \omega \neq \emptyset\right\}.$$

Let us assume

$$\left|\partial_x^2 U - \partial_t U\right| \leq M\left(|\partial_x U| + |U|\right), \quad \text{in } \mathcal{R}, \tag{14.173}$$

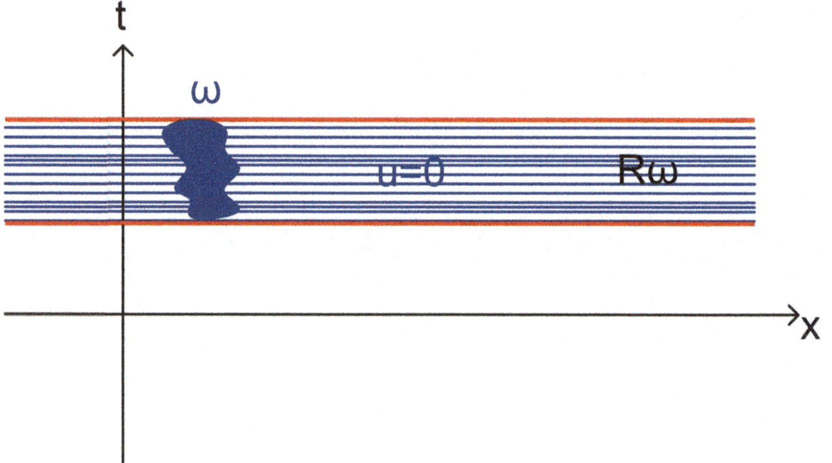

Fig. 14.3 If $u = 0$ in the open set ω, then u vanishes on R_ω, which is formed by the union of the characteristic lines $t = t_0$ that have a non-empty intersection with ω

then

$$U(x,t)=0, \quad \forall (x,t)\in \mathcal{R}_{\omega}. \tag{14.174}$$

From the previous result it follows, in particular, that if U is regular enough (it is sufficient that U, $\partial_x U$, $\partial_t U$, $\partial_x^2 U$ are continuous in $(0,1)\times(0,1)$) and if U is a solution of the Cauchy problem

$$\begin{cases} \partial_x^2 U-\partial_t U=a(x,t)\partial_x U+b(x,t)U, & \text{in } (0,1)\times(0,1) \\ U(0,t)=0, & \text{for } t\in(\alpha,\beta) \\ \partial_x U(0,t)=0, & \text{for } t\in(\alpha,\beta) \end{cases}, \tag{14.175}$$

where α, β are given numbers which satisfy $0<\alpha<\beta<1$ and

$$a,b\in L^{\infty}((0,1)\times(0,1)),$$

then

$$U=0, \quad \text{in } (0,1)\times(\alpha,\beta).$$

We do not enter into the details and refer the interested reader directly to [101, Theorem 9].

Remark 14.15 Since this book does not delve deeply into parabolic equations, of which the heat equation is a significant example, this seems to be the appropriate place to inform the reader that, over the years, there have been substantial advances in the study of unique continuation properties for equations of the form

$$\partial_t u-\sum_{j,k=1}^{n}\partial_{x_j}\left(a^{jk}(x,t)\partial_{x_k}u\right)+\sum_{j=1}^{n}b_j(x,t)\partial_{x_j}u+b_0(x,t)u=0, \tag{14.176}$$

where $\{a^{jk}(x,t)\}$ is a symmetric matrix whose entries are real-valued functions, sufficiently regular, and satisfying the uniform ellipticity condition

$$\lambda^{-1}|\xi|^2\leq\sum_{j,k=1}^{n}a^{jk}(x,t)\xi_j\xi_k\leq\lambda|\xi|^2, \quad \forall\xi\in\mathbb{R}^n, \quad \forall(x,t)\in\mathbb{R}^{n+1},$$

where $\lambda\geq 1$, and b_j, for $j=0,1,\ldots,n$, belong to $L^p_{loc}(\mathbb{R}^n)$ for certain $p\geq 1$. Here, we limit ourselves to mentioning the papers [2, 35–37, 62, 68, 117, 120, 137] which prove unique continuation properties on the characteristic planes $\{(x,t_0): x\in\mathbb{R}^n\}$, $t_0\in\mathbb{R}$, for equations of type (14.176). These results extend to the case of variable coefficients and arbitrary dimension $n\geq 1$ the results we briefly discussed above for the one-dimensional case. For a more comprehensive presentation of these results and a more extensive references, we refer the reader to the survey papers [138, 139]. ♦

14.6 Comments on the Necessary Conditions for the Carleman Estimates

In this section, we aim to briefly discuss the necessary conditions that the exponential weight φ must satisfy for a Carleman estimate to hold. We will restrict our discussion to the most elementary of operators, namely the operator

$$\frac{d}{dx}$$

on an interval of $\mathbb{R}$. At the end of this section, we will present a theorem on analogous necessary conditions for arbitrary linear differential operators. We will not provide the proof of this general theorem, but will instead offer a brief illustration of it.

We begin by the following elementary Lemma that will be useful here and in the next chapters.

Lemma 14.24 *Let d be a positive integer number. Let $\mathcal{X}$ be a compact set of $\mathbb{R}^d$ and let $f : \mathcal{X} \to \mathbb{R}$ and $g : \mathcal{X} \to \mathbb{R}$ be two continuous functions. The following conditions are equivalent:*

(i) $x \in \mathcal{X}$, $f(x) = 0 \Rightarrow g(x) > 0$;
(ii) There exists $C > 0$ such that $C\,(f(x))^2 + g(x) > 0$ for every $x \in \mathcal{X}$.

Proof If $\mathcal{X} = \emptyset$, the equivalence between (i) and (ii) is trivial. Let us suppose, accordingly, that $\mathcal{X} \neq \emptyset$ and that (i) apply. Set

$$\mathcal{X}_0 = \{x \in \mathcal{X} : f(x) = 0\}.$$

By (i), by the continuity of g, and since $\mathcal{X}$ is compact, there exists an open subset $\mathcal{V}_0$ of $\mathbb{R}^d$, which satisfies $\mathcal{X}_0 \subset V_0$ and

$$g(x) > 0, \quad \forall x \in \mathcal{V}_0 \cap \mathcal{X}.$$

If $\mathcal{X} \setminus \mathcal{V}_0 = \emptyset$, then (ii) is trivially satisfied. If $\mathcal{X} \setminus \mathcal{V}_0 \neq \emptyset$, we set

$$M_1 = \min_{\mathcal{X}\setminus\mathcal{V}_0} f^2 > 0, \qquad M_2 = \min_{\mathcal{X}\setminus\mathcal{V}_0} g$$

and let C be a positive number such that $CM_1 + M_2 > 0$. We have

$$C\,(f(x))^2 + g(x) \geq g(x) > 0, \quad \forall x \in \mathcal{V}_0 \cap \mathcal{X}$$

and

$$C\,(f(x))^2 + g(x) \geq CM_1 + M_2 > 0, \quad \forall x \in \mathcal{X} \setminus \mathcal{V}_0,$$

from which (ii) follows.

Let us suppose that (ii) holds, we have trivially that if $f(x) = 0$ then

$$g(x) = C\,(f(x))^2 + g(x) > 0.$$

■

Let $I = (-1, 1)$ and $\varphi \in C^2\left(\overline{I}\right)$; let us consider the following elementary Carleman estimate

$$\tau^{\mu} \int_I |u(x)|^2 e^{2\tau\varphi(x)} dx \leq C \int_I \left|u'(x)\right|^2 e^{2\tau\varphi(x)} dx, \tag{14.177}$$

for every $u \in C_0^{\infty}(I)$ and for every $\tau \geq \tau_0$. We will show how to derive an estimate like (14.177) and what conditions on μ and φ are necessary for them to hold.

Let us start by the following

Proposition 14.25 *Let*

$$A = \left\{x \in \overline{I} : \varphi'(x) = 0\right\}. \tag{14.178}$$

If

$$x \in A \Rightarrow \varphi''(x) > 0, \tag{14.179}$$

then there exist τ_0 and C such that

$$\tau \int_I |u(x)|^2 e^{2\tau\varphi(x)} dx \leq C \int_I \left|u'(x)\right|^2 e^{2\tau\varphi(x)} dx, \tag{14.180}$$

for every $u \in C_0^{\infty}(I)$ and for every $\tau \geq \tau_0$.

Remark 14.16 Of course if $A = \emptyset$ (14.179) is trivially satisfied. ♦

Proof Set

$$v = e^{\tau\varphi} u, \tag{14.181}$$

we have

$$e^{\tau\varphi} u' = e^{\tau\varphi} \left(e^{-\tau\varphi} v\right)' = v' - \tau\varphi' v. \tag{14.182}$$

Hence estimate (14.25) is equivalent to (we omit for brevity the integration set)

$$\frac{\tau}{C}\int |v|^2\,dx \le \int \left|v' - \tau\varphi' v\right|^2 dx$$

for every $v \in C_0^\infty(I)$ and for every $\tau \ge \tau_0$.

Now we estimate from below the integral on the right hand side. Spreading the square and integrating by parts we have

$$\begin{aligned}\int \left|v' - \tau\varphi' v\right|^2 dx &= \int \left(\left|v'\right|^2 - 2\tau\varphi'\Re\left(v'\overline{v}\right) + \tau^2\varphi'^2\,|v|^2\right)dx = \\ &= \int \left(\left|v'\right|^2 - \tau\varphi'\left(|v|^2\right)' + \tau^2\varphi'^2\,|v|^2\right)dx = \\ &= \int \left(\left|v'\right|^2 + \tau\left(\varphi'' + \tau\varphi'^2\right)|v|^2\right)dx.\end{aligned} \tag{14.183}$$

Now, as (14.179) holds, by applying Lemma 14.24 (with $f = \varphi'$ and $g = \varphi''$) we have that there exists τ_0 such that

$$\tau_0\varphi'^2(x) + \varphi''(x) > 0, \quad \forall x \in \overline{I}$$

and, setting

$$C^{-1} = \min_{\overline{I}}\left(\tau_0\varphi'^2 + \varphi''\right) > 0,$$

we have, by (14.183)

$$\begin{aligned}\int \left|v' - \tau\varphi' v\right|^2 dx &\ge \int \tau\left(\varphi'' + \tau\varphi'^2\right)|v|^2\,dx \ge \\ &\ge C^{-1}\tau\int |v|^2\,dx.\end{aligned} \tag{14.184}$$

From which, taking into account (14.181) and (14.182), (14.180) follows. ■

Remark 14.17 If $A = \emptyset$ then a stronger version of (14.25) holds true. More precisely, we have

$$\tau^2\int_I |u(x)|^2\,e^{2\tau\varphi(x)}dx \le C\int_I \left|u'(x)\right|^2 e^{2\tau\varphi(x)}dx, \tag{14.185}$$

for every $u \in C_0^\infty(I)$ and for every $\tau \ge \tau_0$. In this case, indeed, we have

$$m := \min_{\overline{I}}\left|\varphi'\right| > 0$$

as $A = \emptyset$. On the other hand, setting

$$m_1 = \|\varphi''\|_{L^\infty(I)},$$

and taking into account (14.183), we get

$$\begin{aligned}\int \left|v' - \tau\varphi' v\right|^2 dx &= \int \left(\left|v'\right|^2 + \tau^2\left(\tau^{-1}\varphi'' + \varphi'^2\right)|v|^2\right) dx \geq \\ &\geq \int \left(\left|v'\right|^2 + \tau^2\left(-\tau^{-1}m_1 + m^2\right)|v|^2\right) dx \geq \\ &\geq \frac{\tau^2 m^2}{2}\int |v|^2\, dx,\end{aligned} \tag{14.186}$$

for every $v \in C_0^\infty(I)$ and for every $\tau \geq \tau_0$, where $\tau_0 = \frac{2m_1^2}{m^2}$ e $C = \frac{2}{m^2}$. From which (14.185) follows. ♦

Remark 14.18 It is also evident that estimate (14.177) cannot be true for $\mu > 2$. As a matter of fact, we have

$$\int \left|v' - \tau\varphi' v\right|^2 dx \leq 2\int \left(\left|v'\right|^2 + \tau^2 \left\|\varphi'\right\|^2_{L^\infty(I)} |v|^2\right) dx \tag{14.187}$$

and so, if (14.177), and $\mu > 2$, would imply

$$\begin{aligned}&\int |v|^2\, dx \leq \\ &\leq 2\tau^{2-\mu}\int \left(\tau^{-2}\left|v'\right|^2 + \left\|\varphi'\right\|^2_{L^\infty(I)}|v|^2\right) dx \to 0, \quad \text{as } \tau \to +\infty,\end{aligned}$$

which is evidently absurd since v is an arbitrary function of $C_0^\infty(I)$. ♦

We establish some necessary conditions for estimate (14.177).

Proposition 14.26 *Let*

$$A_0 = \left\{x \in I : \varphi'(x) = 0\right\}. \tag{14.188}$$

If there exist C and τ_0 such that

$$\int_I |u(x)|^2 e^{2\tau\varphi(x)} dx \leq C \int_I \left|u'(x)\right|^2 e^{2\tau\varphi(x)} dx, \tag{14.189}$$

for every $u \in C_0^\infty(I)$ and for every $\tau \geq \tau_0$, then

$$x \in A_0 \Rightarrow \varphi''(x) \geq 0. \tag{14.190}$$

Proof We first notice that, by density, (14.189) is satisfied for each $u \in H_0^1(I)$.

In order to prove the Proposition we argue by contradiction and we assume that (14.190) does not hold. Let $x_0 \in I$ satisfy $\varphi'(x_0) = 0$ and $\varphi''(x_0) < 0$. For the purpose of simplifying the notations, since x_0 is an interior point of I, we assume that $x_0 = 0$. Hence we have

$$\varphi'(0) = 0 \quad \text{and} \quad \varphi''(0) < 0. \tag{14.191}$$

Since (14.189) is trivially equivalent to

$$\int_I |u(x)|^2 e^{2\tau(\varphi(x)-\varphi(0))} dx \le C \int_I \left|u'(x)\right|^2 e^{2\tau(\varphi(x)-\varphi(0))} dx, \tag{14.192}$$

for every $u \in H_0^1(I)$ and for every $\tau \ge \tau_0$, we may assume

$$\varphi(0) = 0.$$

Set

$$a = -\varphi''(0) > 0$$

and let $\psi \in H_0^1(I)$ be a function that we will choose later.

If $0 < \varepsilon \le \tau_0^{-1/2}$, we have

$$x \to \psi\left(\varepsilon\sqrt{\tau}x\right) \in H_0^1(I), \quad \forall \tau \ge \varepsilon^{-2}.$$

Now, introducing the following notation in (14.192)

$$u(x) = \psi\left(\varepsilon\sqrt{\tau}x\right),$$

we have

$$\int_I \left|\psi\left(\varepsilon\sqrt{\tau}x\right)\right|^2 e^{2\tau\varphi(x)} dx \le C\varepsilon^2\tau \int_I \left|\psi'\left(\varepsilon\sqrt{\tau}x\right)\right|^2 e^{2\tau\varphi(x)} dx.$$

By performing the change of variables $t = \sqrt{\tau}x$, we get

$$\int_{-\sqrt{\tau}}^{\sqrt{\tau}} |\psi(\varepsilon t)|^2 e^{2\tau\varphi(t/\sqrt{\tau})} dt \le C\varepsilon^2\tau \int_{-\sqrt{\tau}}^{\sqrt{\tau}} \left|\psi'(\varepsilon t)\right|^2 e^{2\tau\varphi(t/\sqrt{\tau})} dt. \tag{14.193}$$

We notice that the Taylor formula gives

$$\varphi(x) = -\frac{a}{2}x^2 + \frac{x^2}{2}\omega(x), \quad \forall x \in I,$$

where

$$\lim_{x\to 0}\omega(x) = 0. \tag{14.194}$$

Now we choose

$$\tau = \varepsilon^{-2},$$

and by (14.193) we get

$$\int_{-1/\varepsilon}^{1/\varepsilon} |\psi(\varepsilon t)|^2 e^{-t^2(a-\omega(\varepsilon t))} dt \le C\int_{-1/\varepsilon}^{1/\varepsilon} \left|\psi'(\varepsilon t)\right|^2 e^{-t^2(a-\omega(\varepsilon t))} dt. \tag{14.195}$$

Let us choose ψ such that

$$\psi(x) = \begin{cases} 2(x+1), & \text{for } x \in \left[-1, -\frac{1}{2}\right) \\ 1, & \text{for } x \in \left[-\frac{1}{2}, \frac{1}{2}\right) \\ 2(-x+1), & \text{for } x \in \left[\frac{1}{2}, 1\right] \end{cases}, \tag{14.196}$$

By (14.195) and (14.196) we have

$$\begin{aligned} \int_{-1/2\varepsilon}^{1/2\varepsilon} e^{-t^2(a-\omega(\varepsilon t)} dt &\le \int_{-1/\varepsilon}^{1/\varepsilon} |\psi(\varepsilon t)|^2 e^{-t^2(a-\omega(\varepsilon t)} dt \le \\ &\le C\int_{-1/\varepsilon}^{1/\varepsilon} \left|\psi'(\varepsilon t)\right|^2 e^{-t^2(a-\omega(\varepsilon t)} dt = \\ &= 8C\int_{1/2\varepsilon}^{1/\varepsilon} e^{-t^2(a-\omega(\varepsilon t)} dt. \end{aligned} \tag{14.197}$$

Passing to the limit as $\varepsilon \to 0$ and recalling (14.194), we get (by the Dominated Convergence Theorem)

$$0 < \int_{-\infty}^{+\infty} e^{-at^2} dt = \lim_{\varepsilon\to 0}\int_{-1/2\varepsilon}^{1/2\varepsilon} e^{-t^2(a-\omega(\varepsilon t)} dt \le \lim_{\varepsilon\to 0} 8C\int_{1/2\varepsilon}^{1/\varepsilon} e^{-t^2(a-\omega(\varepsilon t)} dt = 0$$

Which is, evidently, absurd. ■

Exercise 14.2

(a) Let A_0 be as in Proposition 14.26. If there exist C_0 and τ_0 such that

$$\tau\int_I |u(x)|^2 e^{2\tau\varphi(x)} dx \le C_0\int_I \left|u'(x)\right|^2 e^{2\tau\varphi(x)} dx, \tag{14.198}$$

for every $u \in C_0^\infty(I)$ and for every $\tau \geq \tau_0$, then

$$x \in A_0 \Rightarrow \varphi''(x) \geq \frac{1}{2C_0} > 0. \tag{14.199}$$

(b) If $1 < \mu \leq 2$, in (14.177), then $A_0 = \emptyset$ i.e. $\varphi'(x) \neq 0$ for every $x \in (-1, 1)$. [Hint: concerning (a) proceed as in the proof of Proposition 14.26; concerning (b), apply (a)].

We now state, without proof, a general theorem ensuring the validity of a Carleman estimate, referring to [56, Theorem 8.1.1] and [58, Theorem 28.2.1] for the full proof. Proposition 14.26 (and Exercise 14.2) can be regarded as a "miniature" version of this theorem.

Theorem 14.27 (Necessary Condition for the Carleman Estimate) *Let Ω be a bounded open set of $\mathbb{R}^n$. Let $\varphi \in C^\infty\left(\bar{\Omega}\right)$ satisfy $\nabla\varphi \neq 0$ in $\bar{\Omega}$. Let*

$$P(x, D) = \sum_{|\alpha| \leq m} a_\alpha(x) D^\alpha \tag{14.200}$$

be a linear differential operator of order m, such that $a_\alpha \in L^\infty(\Omega)$, $|\alpha| \leq m$ and $a_\alpha \in C^1\left(\bar{\Omega}\right)$, $|\alpha| = m$.

Suppose that there exist $C > 0$ and τ_0 such that for every $u \in C_0^\infty(\Omega)$ and for every $\tau \geq \tau_0$ we have

$$\tau \sum_{|\alpha| \leq m-1} \int_\Omega |D^\alpha u|^2 e^{2\tau\varphi} dx \leq C \int_\Omega |P(x, D)u|^2 e^{2\tau\varphi} dx. \tag{14.201}$$

Then, setting $\zeta = \xi + i\sigma\nabla\varphi(x)$ where $x \in \Omega$, $\xi \in \mathbb{R}^n$, $\sigma \in \mathbb{R}$, we have

$$\left.\begin{aligned} p_m(x, \zeta) = 0 \\ \sigma \neq 0 \end{aligned}\right\} \Longrightarrow \sum_{j,k=1}^n \partial^2_{jk}\varphi(x) p_m^{(j)}(x, \zeta) \overline{p_m^{(k)}(x, \zeta)} + \\ + \sigma^{-1} \Im \sum_{k=1}^n p_{m,k}(x, \zeta) \overline{p_m^{(k)}(x, \zeta)} > 0. \tag{14.202}$$

Here $p_m(x, \xi)$ is the principal symbol of $P(x, D)$. Namely

$$p_m(x, \xi) = \sum_{|\alpha|=m} a_\alpha(x) \xi^\alpha, \quad x \in \Omega, \ \xi \in \mathbb{R}^n$$

and

$$p^{(j)}(x, \xi) = \partial_{\xi_j} p(x, \xi), \qquad p_{,j}(x, \xi) = \partial_{x_j} p(x, \xi), \quad j = 1, \cdots, n.$$

We note that the expression on the right-hand side of the implication (14.202) can be written by Poisson brackets (see (15.3)) and it is equal to

$$\frac{i}{2\sigma}\left\{p_m(x,\xi+i\sigma\nabla\varphi),\overline{p_m(x,\xi+i\sigma\nabla\varphi)}\right\}.$$

We conclude this section with some examples illustrating Theorem 14.27.

Example 14.19 Let us consider the Cauchy–Riemann operator

$$P(D)=\frac{i}{2}\left(D_1+iD_2\right).$$

Let $\varphi\in C^\infty(\Omega)$ satisfy $\nabla\varphi\neq 0$ in Ω.

A necessary condition for the Carleman estimate to hold

$$\tau\int_{\mathbb{R}^2}|u|^2e^{2\tau\varphi}dx\leq C\int_{\mathbb{R}^2}|Pu|^2e^{2\tau\varphi}dx, \tag{14.203}$$

for every $u\in C_0^\infty\left(\mathbb{R}^2\right)$ and for every $\tau>\tau_0$ is that

$$\Delta\varphi(x)>0,\quad \forall x\in\mathbb{R}^2. \tag{14.204}$$

In fact, we apply condition (14.202) that in this case can be written

$$\left.\begin{array}{l}\xi_1=\sigma\varphi_{x_2}(x)\\ \xi_2=-\sigma\varphi_{x_1}(x)\\ \sigma\neq 0\end{array}\right\}\Longrightarrow\Delta\varphi(x)>0,$$

that is (14.204). ♣

Example 14.20 Let $\varphi\in C^\infty\left(\mathbb{R}^2\right)$ satisfy $\nabla\varphi\neq 0$ in Ω. We wish to prove that a necessary condition for the Carleman estimate to hold

$$\tau\int_{\mathbb{R}^2}\left(|u|^2+|\nabla u|^2\right)e^{2\tau\varphi}dx\leq C\int_{\mathbb{R}^2}|\Delta u|^2e^{2\tau\varphi}dx, \tag{14.205}$$

for every $u\in C_0^\infty\left(\mathbb{R}^2\right)$ and for every $\tau>\tau_0$ is that

$$\Delta\varphi(x)>0,\quad \forall x\in\mathbb{R}^2. \tag{14.206}$$

First, we note that (14.202) for the Laplace operator in $\mathbb{R}^n$ ($n \geq 1$) can be written

$$\left.\begin{array}{l} |\xi|^2 = \sigma^2 |\nabla\varphi(x)|^2 \\ \xi \cdot \nabla\varphi(x) = 0 \\ \sigma \neq 0 \end{array}\right\} \Longrightarrow Q(x,\xi,\sigma) = \sum_{j,k=1}^{n} \partial^2_{jk}\varphi(x)\xi_j\xi_k + \tag{14.207}$$

$$+\sigma^2 \sum_{j,k=1}^{n} \partial^2_{jk}\varphi(x)\varphi\partial_j\varphi(x)\partial_k\varphi(x) > 0.$$

Let us consider the case

$$n = 2.$$

Let us denote by

$$N(x) = \frac{\nabla\varphi(x)}{|\nabla\varphi(x)|}$$

and

$$\widetilde{Q}(x,\xi) = \sum_{j,k=1}^{2} \partial^2_{jk}\varphi(x)(\xi_j\xi_k + N_j(x)N_k(x)).$$

It is clear that (14.207) is equivalent to the following condition:
(★) if

$$\begin{cases} |\xi|^2 = 1, \\ N(x) \cdot \xi = 0, \end{cases} \tag{14.208}$$

then

$$\widetilde{Q}(x,\xi) > 0.$$

Let us suppose, then, (14.208) holds.
Since $\nabla\varphi(x) \neq 0$, we may assume $N_1 \neq 0$. Recalling that $|N| = 1$, by (14.208) we obtain

$$\begin{cases} \xi_1^2 + \xi_2^2 = 1 \\ \xi_1 N_1 + \xi_2 N_2 = 0 \end{cases} \Longleftrightarrow \begin{cases} \xi_2^2\left(\frac{N_2^2}{N_1^2} + 1\right) = 1 \\ \xi_1 = -\frac{\xi_2 N_2}{N_1} \end{cases} \Longleftrightarrow \begin{cases} |\xi_2| = |N_1| \\ \xi_1 = -\frac{\xi_2 N_2}{N_1} \end{cases}.$$

Let us suppose that $\xi_2 = |N_1|$, which implies $\xi_1 = -\frac{N_2}{N_1}|N_1|$. We have

$$\begin{aligned} Q &= \partial_1^2\varphi\xi_1^2 + 2\partial_{12}^2\varphi\xi_1\xi_2 + \partial_2^2\varphi\xi_2^2 + \partial_1^2\varphi N_1^2 + 2\partial_{12}^2\varphi N_1N_2 + \partial_2^2\varphi N_2^2 = \\ &= \partial_1^2\varphi N_2^2 + 2\partial_{12}^2\varphi\left(-\frac{N_2}{N_1}|N_1|\right)|N_1| + \partial_2^2\varphi N_1^2 \\ &\quad + \partial_1^2\varphi N_1^2 + 2\partial_{12}^2\varphi N_1N_2 + \partial_2^2\varphi N_2^2 = \\ &= \partial_1^2\varphi - 2\partial_{12}^2\varphi N_1N_2 + \partial_2^2\varphi + 2\partial_{12}^2\varphi N_1N_2 = \Delta\varphi. \end{aligned}$$

If $\xi_2 = -|N_1|$, we get a similar result. Therefore condition (★) is equivalent to (14.206). ♣

Example 14.19 Let us consider a radial function

$$\varphi(x) = f(|x|), \tag{14.209}$$

in $\Omega = B_1 \setminus \overline{B_r}$, $r \in (0, 1)$.

We would like to find some functions f such that φ satisfies condition (14.207). We will see that in some cases this is not possible. Let us proceed in a similar manner to the previous example. Let us calculate

$$\partial_j\varphi(x) = \frac{x_j}{|x|}f'(|x|), \quad j = 1, \cdots, n$$

$$\partial_{jk}^2\varphi(x) = \left(\frac{\delta_{jk}}{|x|} - \frac{x_jx_k}{|x|^3}\right)f'(|x|) + \frac{x_jx_k}{|x|^2}f''(|x|), \quad j,k = 1, \cdots, n$$

$$\sum_{j,k=1}^{n} \partial_{jk}^2\varphi(x)\xi_j\xi_k = \left(\frac{|\xi|^2}{|x|} - \frac{(\xi\cdot x)^2}{|x|^3}\right)f'(|x|) + \frac{(\xi\cdot x)^2}{|x|^2}f''(|x|),$$

$$\begin{aligned} \sum_{j,k=1}^{n} \partial_{jk}^2\varphi(x)\partial_j\varphi(x)\partial_k\varphi(x) &= \left(\frac{|\nabla\varphi(x)|^2}{|x|} - \frac{(\nabla\varphi(x)\cdot x)^2}{|x|^3}\right)f'(|x|) + \\ &\quad + \frac{(\nabla\varphi(x)\cdot x)^2}{|x|^2}f''(|x|). \end{aligned}$$

Since

$$|\nabla\varphi(x)|^2 = f'^2(|x|)$$

and

$$\nabla\varphi(x)\cdot x = |x|f'(|x|)$$

we have

$$\sum_{j,k=1}^{n} \partial_{jk}^2 \partial_{jk}^2 \varphi(x) \partial_j \varphi(x) \partial_k \varphi(x) = (f'(|x|))^2 f''(|x|).$$

Then by (14.207) we can write $Q(x, \xi, \sigma)$ as follows

$$Q(x, \xi, \sigma) = \left(\frac{|\xi|^2}{|x|} - \frac{(\xi \cdot x)^2}{|x|^3} \right) f'(|x|) + \\ + \frac{(\xi \cdot x)^2}{|x|^2} f''(|x|) + \sigma^2 (f'(|x|))^2 f''(|x|).$$

Let us suppose that the antecedent of condition (14.207) holds, i.e., let us suppose that

$$\begin{cases} |\xi|^2 = \sigma^2 |\nabla \varphi(x)|^2 = \sigma^2 (f'(|x|))^2 \\ \sigma \nabla \varphi(x) \cdot \xi = \sigma \frac{(\xi \cdot x)^2}{|x|^2} f'(|x|) = 0 \\ \sigma \neq 0 \end{cases},$$

namely

$$\begin{cases} |\xi|^2 = \sigma^2 (f'(|x|))^2 \\ \xi \cdot x = 0 \\ \sigma \neq 0 \end{cases}.$$

Hence $Q(x, \xi, \sigma)$ can be written as follows

$$Q(x, \xi, \sigma) = \frac{\sigma^2 (f'(|x|))^2}{|x|} f'(|x|) + \sigma^2 (f'(|x|))^2 f''(|x|).$$

We get

$$\sigma^{-2} Q(x, \xi, \sigma) = \frac{(f'(|x|))^2}{|x|} f'(|x|) + (f'(|x|))^2 f''(|x|).$$

Now we wish to characterize the functions

$$f : (0, 1) \to (0, +\infty)$$

which satisfy (17.4) and for which we have

$$\lim_{t \to 0} f(t) = +\infty$$

and

$$f' < 0, \quad \text{in } (0, 1).$$

The condition $Q > 0$ becomes

$$\frac{f'(t)}{t} + f''(t) > 0, \quad \forall t \in (0, 1).$$

To solve this differential inequality we set

$$f(t) = \psi(\log t),$$

from which we have

$$f'(t) = \psi'(\log t)\frac{1}{t},$$

$$f''(t) = \psi''(\log t)\frac{1}{t^2} - \psi'(\log t)\frac{1}{t^2}.$$

Hence the differential inequality can be written

$$\frac{f'(t)}{t} + f''(t) = \frac{1}{t^2}\psi''(\log t) > 0 \Longleftrightarrow \psi''(\log t) > 0.$$

Let $s = \log t$. Then, as $t \in (0, 1)$, $s \in (-\infty, 0)$, condition $Q > 0$ becomes:

$$\psi''(s) > 0 \Longleftrightarrow \frac{d^2}{ds^2}(f(e^s)) > 0 \quad \forall s \in (-\infty, 0).$$

Let us observe that there are functions that **do not satisfy** this condition. For $\alpha > 0$, we consider functions $f(t)$ of the type

$$f(t) = \left(\log \frac{1}{t}\right)^{\alpha}.$$

Let us calculate

$$f'(t) = \alpha\left(-\frac{1}{t}\right)\left(\log \frac{1}{t}\right)^{\alpha-1},$$

$$f'(t) = \frac{\alpha}{t^2}\left(\log \frac{1}{t}\right)^{\alpha-1} + \frac{\alpha(\alpha-1)}{t^2}\left(\log \frac{1}{t}\right)^{\alpha-2},$$

from which it follows that the condition on Q can be written as

$$Q = \frac{f'(t)}{t} + f''(t) = \frac{\alpha(\alpha-1)}{t^2}\left(\log\frac{1}{t}\right)^{\alpha-2} > 0.$$

Consequently if $\alpha < 1$, we have

$$\frac{f'(t)}{t} + f''(t) < 0.$$

If $\alpha = 1$, we have

$$\varphi = \log\frac{1}{|x|},$$

therefore

$$Q = 0.$$

♣

Chapter 15
Carleman Estimates and the Cauchy Problem for Operators with C^∞ Coefficients in the Principal Part

Abstract In this chapter, we focus on Carleman estimates for operators with smooth coefficients for the principal part and bounded coefficients for lower-order terms. We examine elliptic operators with complex-valued coefficients and operators with real-valued principal parts. We take an accessible approach to Carleman estimates, laying the groundwork for the later use of pseudodifferential methods. The focus is on differential operator symbols and their role in Carleman estimates with the large parameter introduced in the previous chapter. We introduce essential concepts such as the Poisson bracket, Sobolev spaces with large parameters, and the commutator of the self-adjoint and skew-adjoint parts of an operator. These lead to the proof of the localized Gårding inequality, a key tool for Carleman estimates. In subsequent sections, we develop a Carleman estimate with weight exponent and define a conjugate operator. We also provide approximation formulas for differential operators depending on a large parameter and explore commutators. These results are applied to establish Carleman estimates for elliptic operators and operators with real-valued principal parts. The chapter also covers the application of Carleman estimates to uniqueness in the Cauchy problem, introduces pseudo-convex surfaces, and briefly discusses pseudodifferential operators and the Sharp Gårding Inequality, illustrating their usefulness in proving Carleman estimates.

15.1 Introduction

In this chapter, we address Carleman estimates for operators

$$P(x, D) = \sum_{|\alpha| \le m} a_\alpha(x) D^\alpha$$

with $m \ge 1$. Let us recall that the principal part of $P(x, D)$ is given by

$$P_m(x, D) = \sum_{|\alpha| = m} a_\alpha(x) D^\alpha$$

S. Vessella, *Unique Continuation Properties for Partial Differential Equations*,
Birkhäuser Advanced Texts Basler Lehrbücher,
https://doi.org/10.1007/978-3-031-86366-0_15

and the symbol of $P_m(x, D)$ is the homogeneous polynomial in the variable ξ defined by

$$p_m(x, \xi) = \sum_{|\alpha|=m} a_\alpha(x)\xi^\alpha .$$

We consider cases where the coefficients of $P_m(x, D)$ are of class C^∞, while those of terms of lower order than m are of class L^∞. In this chapter, we focus on elliptic operators with complex-valued coefficients and operators whose principal part has real-valued coefficients.

While advanced texts such as [58, Vol. IV], [79], and [83], typically treat Carleman estimates through the formal machinery of pseudodifferential operators, here we establish an approach that prepares the reader to appreciate the pseudodifferential method without yet engaging with it fully. Through the C^∞ regularity assumptions on the coefficients of $P_m(x, D)$, we set the stage for the eventual application of pseudodifferential calculus, staying within a classical framework to emphasize the efficacy of symbol-based techniques. Here, our focus is on understanding differential operator symbols, which are polynomials in ξ, and, specifically for Carleman estimates, polynomials in ξ and the large parameter τ introduced in the previous chapter.

In addition to a careful analysis of symbols, we introduce certain key definitions in Sect. 15.2, including the Poisson bracket and Sobolev spaces with large parameter. We also introduce the commutator of the self-adjoint and skew-adjoint parts of an operator (Proposition 15.2) and we prove the localized Gårding inequality (Theorem 15.10), which serve as central tools in our study of Carleman estimates.

In Sect. 15.3, we develop a Carleman estimate with weight exponent φ and introduce the notion of a conjugate operator defined by

$$P_\tau v = e^{\tau\varphi} P_m(x, D) \left(e^{-\tau\varphi} v\right)$$

together with useful technical lemmas and preliminary theorems for Carleman estimates. In Sects. 15.4 and 15.5, we present approximation formulas for differential operators depending on a large parameter and commutators, applied in Sects. 15.6 and 15.7 to establish Carleman estimates for elliptic operators and for operators with a principal part having real-valued coefficients, respectively.

Section 15.8 introduces pseudo-convex and strongly pseudo-convex surfaces and applies the Carleman estimates to uniqueness in the Cauchy problem. Finally, in Sect. 15.9, we provide a concise look at pseudodifferential operators and the Sharp Gårding Inequality, illustrating their utility in proving Carleman estimates.

15.2 Some Definitions and Lemmas

In this section, we will introduce some notations, definitions, and a few lemmas that will facilitate the understanding of the rest of this chapter. In the sequel of this section, Ω is an open set of $\mathbb{R}^n$.

1. Let P and Q be two linear differential operators with $C^{\infty}(\Omega)$ coefficients, we call the **commutator operator** of P and Q the linear operator $[P, Q]$ defined as

$$[P, Q] = PQ - QP. \tag{15.1}$$

Let us note that, for every fixed P (for every fixed Q), $[P, \cdot]$ ($[\cdot, Q]$) is linear. Moreover we have $[P, Q] = -[Q, P]$ for every differential operators P and Q, and if P and Q have order 0, then $[P, Q] = 0$. More generally, if P and Q have order m_1 and m_2 respectively (with m_1 or m_2 greater than or equal to 1) then $[P, Q]$ is an operator of order $m_1 + m_2 - 1$ or less.
2. Let $m \in \mathbb{N}_0$. Let $a_\alpha \in C^0(\Omega)$ be a complex-valued function for $|\alpha| \leq m$. Let us consider the polynomial with respect to the variable ξ

$$p(x, \xi) = \sum_{|\alpha| \leq m} a_\alpha(x)\xi^\alpha, \quad x \in \Omega, \ \xi \in \mathbb{R}^n.$$

We denote by $\operatorname{Op}(p)$ the differential operator whose symbol is $p(x, \xi)$. Namely

$$(\operatorname{Op}(p)u)(x) = \sum_{|\alpha| \leq m} a_\alpha(x)\partial^\alpha u, \quad \forall x \in \Omega. \tag{15.2}$$

Let $p(x, \xi)$ be a polynomial with respect to the variable ξ, let us suppose that the coefficients of $p(x, \xi)$ are differentiable. Let us set

$$p^{(j)}(x, \xi) = \partial_{\xi_j} p(x, \xi), \qquad p_{,j}(x, \xi) = \partial_{x_j} p(x, \xi), \quad j = 1, \cdots, n.$$

Let us denote by

$$\overline{p}(x, \xi),$$

the polynomial in ξ whose coefficients are the complex conjugate of the coefficients of $p(x, \xi)$. Keep in mind that if $\zeta \in \mathbb{C}^n$, then

$$\overline{p(x, \zeta)} = \overline{p}\left(x, \overline{\zeta}\right).$$

If $a(x, \xi)$ and $b(x, \xi)$ are two $C^1(\Omega \times \mathbb{R}^n)$ functions in the variable (x, ξ), we define their **Poisson brackets**

$$\begin{aligned} \{a, b\}(x, \xi) &= \{a(x, \xi), b(x, \xi)\} = \\ &= \sum_{j=1}^{n} \left(\partial_{\xi_j} a(x, \xi)\partial_{x_j} b(x, \xi) - \partial_{\xi_j} b(x, \xi)\partial_{x_j} a(x, \xi)\right). \end{aligned} \tag{15.3}$$

Let us note that, for every fixed $a(x, \xi)$ (for every fixed $b(x, \xi)$), $\{a(x, \xi), \cdot\}$ ($\{\cdot, b(x, \xi)\}$) is linear. Moreover, for every $a(x, \xi)$ and $b(x, \xi)$, we have

$$\{a(x,\xi), b(x,\xi)\} = -\{b(x,\xi), a(x,\xi)\}$$

and if $a(x,\xi)$ and $b(x,\xi)$ are constant w.r.t. the variable ξ or w.r.t. the variable x, then $\{a(x,\xi), b(x,\xi)\} = 0$. More generally, if $a(x,\xi)$ and $b(x,\xi)$ are polynomials in the variable ξ and they have degree m_1 and m_2 respectively, w.r.t. the variable ξ (with m_1 or m_2 greater than or equal to 1), then $[\{a(x,\xi), b(x,\xi)\}$ is a polynomial in ξ of degree $m_1 + m_2 - 1$ or less.

Let $\widetilde{\Omega}$ and Ω be two open sets of $\mathbb{R}^n$ and let

$$\chi : \widetilde{\Omega} \times \mathbb{R}^n \to \Omega \times \mathbb{R}^n, \quad \chi(y,\eta) = (x(y,\eta), \xi(y,\eta)) \tag{15.4}$$

be a C^1 map. We say that χ is a **canonical map** provided the following relations holds, for every $h, k = 1, \cdots, n$

$$\begin{aligned}
\{x_k, x_h\} &= \sum_{j=1} \left(\partial_{\eta_j} x_k \partial_{y_j} x_h - \partial_{\eta_j} x_h \partial_{y_j} x_k\right) = 0, \\
\{\xi_k, \xi_h\} &= \sum_{j=1} \left(\partial_{\eta_j} \xi_k \partial_{y_j} \xi_h - \partial_{\eta_j} \xi_h \partial_{y_j} \xi_k\right) = 0, \\
\{\xi_k, x_h\} &= \sum_{j=1} \left(\partial_{\eta_j} \xi_k \partial_{y_j} x_h - \partial_{\eta_j} x_h \partial_{y_j} \xi_k\right) = \delta_{hk},
\end{aligned} \tag{15.5}$$

where δ_{hk} is the Kronecker's delta.

The following Propositions will be useful later.

Proposition 15.1

(i) Let $a(x,\xi)$, $b(x,\xi)$ and $c(x,\xi)$ be $C^1(\Omega \times \mathbb{R}^n)$ functions in the variable (x,ξ), then we have

$$\{a, bc\}(x,\xi) = c(x,\xi)\{a,b\}(x,\xi) + b(x,\xi)\{a,c\}(x,\xi). \tag{15.6}$$

(ii) Let $\widetilde{\Omega}$ and Ω be two open set of $\mathbb{R}^N$. Let $a(x,\xi)$ and $b(x,\xi)$ be as above and let χ be a C^1 map like (15.4). *Then we have*

$$\begin{aligned}
&\{a \circ \chi, b \circ \chi\}(y,\eta) = \\
&= \sum_{h,k=1}^{n} \left(\{\xi_k, x_h\}\left(\partial_{\xi_k} a \partial_{x_h} b - \partial_{\xi_k} b \partial_{x_h} a\right) \circ \chi\right)(y,\eta) + \\
&+ \sum_{h,k=1}^{n} \left(\{x_k, x_h\}\left(\partial_{x_k} a \partial_{x_h} b\right) \circ \chi\right)(y,\eta) + \\
&+ \sum_{h,k=1}^{n} \left(\{\xi_k, \xi_h\}\left(\partial_{\xi_k} a \partial_{\xi_h} b\right) \circ \chi\right)(y,\eta), \quad \forall (y,\eta) \in \widetilde{\Omega} \times \mathbb{R}^n.
\end{aligned} \tag{15.7}$$

(iii) If χ is a canonical map then

$$\{a\circ\chi, b\circ\chi\}(y,\eta) = (\{a,b\}\circ\chi)(y,\eta), \quad \forall(y,\eta)\in\widetilde{\Omega}\times\mathbb{R}^n. \tag{15.8}$$

Proof (i) We have

$$\begin{aligned}
\{a, bc\} &= \sum_{j=1}^{n}\left(\partial_{\xi_j} a\partial_{x_j}(bc) - \partial_{\xi_j}(bc)\,\partial_{x_j} a\right) = \\
&= \sum_{j=1}^{n}\partial_{\xi_j} a\left(c\partial_{x_j} b + b\partial_{x_j} c\right) - \sum_{j=1}^{n}\left(c\partial_{\xi_j} b + b\partial_{\xi_j} c\right)\partial_{x_j} a = \\
&= c\sum_{j=1}^{n}\left(\partial_{\xi_j} a\partial_{x_j} b - \partial_{\xi_j} b\partial_{x_j} a\right) + b\sum_{j=1}^{n}\left(\partial_{\xi_j} a\partial_{x_j} c - \partial_{\xi_j} c\partial_{x_j} a\right) = \\
&= c\{a,b\} + b\{a,c\}.
\end{aligned}$$

(ii) To check (15.7) we use the Einstein convention of repeated indices and, for the sake of brevity, we omit "$\circ\,\chi$" from the second equality onwards. We have by chain rule

$$\begin{aligned}
&\{a\circ\chi, b\circ\chi\} = \partial_{\eta_j}(a\circ\chi)\partial_{y_j}(b\circ\chi) - \partial_{y_j}(a\circ\chi)\partial_{\eta_j}(b\circ\chi) = \\
&= \left(\partial_{x_k} a\partial_{\eta_j} x_k + \partial_{\xi_k} a\partial_{\eta_j}\xi_k\right)\left(\partial_{x_h} b\partial_{y_j} x_h + \partial_{\xi_h} b\partial_{\eta_j}\xi_h\right) - \\
&- \left(\partial_{x_k} a\partial_{y_j} x_k + \partial_{\xi_k} a\partial_{y_j}\xi_k\right)\left(\partial_{x_h} b\partial_{\eta_j} x_h + \partial_{\xi_h} b\partial_{\eta_j}\xi_h\right) = \\
&= \partial_{x_k} a\partial_{x_h} b\left(\partial_{\eta_j} x_k\partial_{y_j} x_h - \partial_{y_j} x_k\partial_{\eta_j} x_h\right) + \\
&+ \partial_{x_k} a\partial_{\xi_h} b\left(\partial_{\eta_j} x_k\partial_{y_j}\xi_h - \partial_{y_j} x_k\partial_{\eta_j}\xi_h\right) + \\
&+ \partial_{\xi_k} a\partial_{x_h} b\left(\partial_{\eta_j}\xi_k\partial_{y_j} x_h - \partial_{y_j}\xi_k\partial_{\eta_j} x_h\right) + \\
&+ \partial_{\xi_k} a\partial_{\xi_h} b\left(\partial_{\eta_j}\xi_k\partial_{y_j}\xi_h - \partial_{y_j}\xi_k\partial_{\eta_j}\xi_h\right) = \\
&= \partial_{x_k} a\partial_{x_h} b\{x_h, x_k\} + \partial_{x_k} a\partial_{\xi_h} b\{x_k, \xi_h\} + \\
&+ \partial_{\xi_k} a\partial_{x_h} b\{\xi_k, x_h\} + \partial_{\xi_h} a\partial_{\xi_h} b\{\xi_k, \xi_h\} = \\
&= \partial_{x_k} a\partial_{x_h} b\{x_h, x_k\} + \left(\partial_{\xi_k} a\partial_{x_h} b - \partial_{\xi_k} b\partial_{x_h} a\right)\{\xi_k, x_h\} + \partial_{\xi_h} a\partial_{\xi_h} b\{\xi_k, \xi_h\}.
\end{aligned}$$

Finally, (iii) is an immediate consequence of (ii) and of (15.5). ■

Proposition 15.2 *Let $\widetilde{\Omega}$ and Ω be two open sets of $\mathbb{R}^n$. Let $g\in C^2\left(\widetilde{\Omega},\Omega\right)$ be a diffeomorphism. Let*

$$\chi:\widetilde{\Omega}\times\mathbb{R}^n\to\Omega\times\mathbb{R}^n, \quad \chi(y,\eta) = (x(y), \xi(y,\eta)),$$

where

$$\begin{cases} x(y) = g(y) \\ \xi(y, \eta) = \left(((\partial_y g(y)))^{-1} \right)^t \eta \end{cases}, \tag{15.9}$$

where $\partial_y g(y)$ *is the jacobian matrix of* g *in* y. *Then* χ *is a canonical map.*

Proof We set

$$h_{jm}(y) = \left((\partial_y g(y))^{-1} \right)_{jm}, \quad \text{for } j, m = 1, \cdots, n.$$

Map (15.9) can be written as

$$\begin{cases} x_j(y) = g_j(y), \quad j = 1, \cdots, n \\ \xi_i(y, \eta) = h_{mi}(y)\eta_m, \quad i = 1, \cdots, n \end{cases}. \tag{15.10}$$

We have to check (15.5). Since x_k, $k = 1, \cdots, n$, do not depend on η we have

$$\{x_k, x_h\} = 0, \quad \text{for } h, k = 1, \cdots, n. \tag{15.11}$$

Concerning $\{\xi_k, x_h\}$ we have (we use the Einstein convention of repeated indices)

$$\{\xi_k, x_h\} = \partial_{\eta_j} \xi_k \partial_{y_j} g_h = (\partial_y g(y))_{hj} \left((\partial_y g(y))^{-1} \right)_{jk} = \delta_{hk}. \tag{15.12}$$

Now we have to prove

$$\{\xi_i, \xi_j\} = 0, \quad \text{for } i, j = 1, \cdots, n. \tag{15.13}$$

First, let us recall that if $A(y)$ is a non singular matrix-valued function and $A^{-1}(y)$ is its inverse, then

$$\partial_{y_k} \left(A^{-1}(y) \right) = -A^{-1}(y) \left(\partial_{y_k} A(y) \right) A^{-1}(y), \quad k = 1, \cdots, n. \tag{15.14}$$

hence, choosing $A(y) = \partial_y g(y)$, (15.14) gives

$$\partial_{y_k} h_{mj} = -h_{ml} \partial_{y_k} \left(\partial_{y_s} g_l \right) h_{sj} = -h_{ml} \left(\partial^2_{y_k y_s} g_l \right) h_{sj}, \tag{15.15}$$

for $j, k, m = 1, \cdots, n$.

Now we have, for $i, j = 1, \cdots, n$,

$$\begin{aligned} \{\xi_i, \xi_j\} &= \partial_{\eta_k}\xi_i \partial_{y_k}\xi_j - \partial_{\eta_k}\xi_j \partial_{y_k}\xi_i = \\ &= \left(h_{ki}\partial_{y_k}h_{mj} - h_{kj}\partial_{y_k}h_{mi}\right)\eta_m. \end{aligned} \tag{15.16}$$

On the other hand, applying (15.15) we have

$$\begin{aligned} &-\left(h_{ki}\partial_{y_k}h_{mj} - h_{kj}\partial_{y_k}h_{mi}\right) = \\ &= h_{ki}h_{ml}\left(\partial^2_{y_k y_s}g_l\right)h_{sj} - h_{kj}h_{ml}\left(\partial^2_{y_k y_s}g_l\right)h_{si} = \\ &= h_{ml}\left(\left(\partial^2_{y_k y_s}g_l\right)h_{ki}h_{sj} - \left(\partial^2_{y_k y_s}g_l\right)h_{kj}h_{si}\right) = 0, \end{aligned} \tag{15.17}$$

for every $i, j, m = 1, \cdots, n$.

By (15.16) and (15.17) we obtain (15.13), which, with (15.11) and (15.12), concludes the proof. ∎

3. Let

$$P(x, D) = \sum_{|\alpha|\leq m} a_\alpha(x)D^\alpha$$

be a differential operator with $C^\infty(\Omega)$ coefficients. We recall that the formal adjoint operator of $P(x, D)$ is the operator $P^\star(x, D)$ such that

$$\left\langle v, P^\star(x, D)w\right\rangle_{L^2(\Omega)} = \langle P(x, D)v, w\rangle_{L^2(\Omega)}, \quad \forall v, w \in C_0^\infty(\Omega), \tag{15.18}$$

where ($L^2(\Omega) = L^2(\Omega; \mathbb{C})$)

$$\langle v, w\rangle_{L^2(\Omega)} = \int_\Omega v\overline{w}dx, \quad \forall v, w \in L^2(\Omega). \tag{15.19}$$

Let us note that

$$P^\star(x, D)w = \sum_{|\alpha|\leq m} D^\alpha\left(\overline{a}_\alpha w\right), \quad \forall w \in C_0^\infty(\Omega). \tag{15.20}$$

As a matter of fact we have, for every $v, w \in C_0^\infty(\Omega)$,

$$\begin{aligned} \langle P(x, D)v, w\rangle_{L^2(\Omega)} &= \sum_{|\alpha|\leq m}\int_\Omega a_\alpha(x)D^\alpha v\,\overline{w}\,dx = \\ &= \sum_{|\alpha|\leq m}\int_\Omega v\,(-1)^{|\alpha|}D^\alpha\left(a_\alpha\overline{w}\right)dx = \\ &= \int_\Omega v\,\overline{\sum_{|\alpha|\leq m} D^\alpha\left(\overline{a_\alpha}w\right)}\,dx. \end{aligned}$$

For a given differential operator $P(x, D)$, we call the **self-adjoint part** and the **skew-adjoint part** of $P(x, D)$, respectively, the operators

$$P^{(s)}(x, D) = \frac{1}{2}\left(P(x, D) + P^\star(x, D)\right), \tag{15.21}$$

$$P^{(a)}(x, D) = \frac{1}{2}\left(P(x, D) - P^\star(x, D)\right). \tag{15.22}$$

Let us notice that

$$\left(P^{(s)}(x, D)\right)^\star v = P^{(s)}(x, D)v, \quad \forall v \in C_0^\infty(\Omega), \tag{15.23}$$

$$\left(P^{(a)}(x, D)\right)^\star v = -P^{(a)}(x, D)v, \quad \forall v \in C_0^\infty(\Omega). \tag{15.24}$$

If the coefficients of $P(x, D)$ are **real-valued function** we also call $P^{(s)}(x, D)$ and $P^{(a)}(x, D)$ the **symmetric part** of $P(x, D)$ and the **anti-symmetric part** of $P(x, D)$ respectively.

The following simple Proposition will be very useful in the treatment of Carleman estimates.

Proposition 15.3 *Let $P(x, D)$ be a linear differential operators with $C^\infty(\Omega)$ coefficients, then we have, for every $v \in C_0^\infty(\Omega)$,*

$$\left[P^{(s)}(x, D), P^{(a)}(x, D)\right]^\star v = \left[P^{(s)}(x, D), P^{(a)}(x, D)\right] v, \tag{15.25}$$

$$\left\langle \left[P^{(s)}(x, D), P^{(a)}(x, D)\right] v, v \right\rangle_{L^2(\Omega)} \in \mathbb{R}, \tag{15.26}$$

and

$$\begin{aligned} &\|P(x, D)v\|^2_{L^2(\Omega)} = \\ &= \left\|P^{(s)}(x, D)v\right\|^2_{L^2(\Omega)} + \left\|P^{(a)}(x, D)v\right\|^2_{L^2(\Omega)} + \\ &+ \left\langle \left[P^{(s)}(x, D), P^{(a)}(x, D)\right] v, v \right\rangle_{L^2(\Omega)}, \quad \forall v \in C_0^\infty(\Omega). \end{aligned} \tag{15.27}$$

Proof Let us denote by P, $P^{(s)}$ and $P^{(a)}$ the operators $P(x, D)$, $P^{(s)}(x, D)$, $P^{(a)}(x, D)$ respectively.

Concerning (15.25), by (15.23) and (15.24) we get

$$
\begin{aligned}
\left[P^{(s)}, P^{(a)}\right]^{\star} &= \left(P^{(s)}P^{(a)} - P^{(a)}P^{(s)}\right)^{\star} = \\
&= \left(P^{(s)}P^{(a)}\right)^{\star} - \left(P^{(a)}P^{(s)}\right)^{\star} = \\
&= \left(P^{(a)}\right)^{\star}\left(P^{(s)}\right)^{\star} - \left(P^{(s)}\right)^{\star}\left(P^{(a)}\right)^{\star} = \\
&= -P^{(a)}P^{(s)} + P^{(s)}P^{(a)} = \\
&= \left[P^{(s)}, P^{(a)}\right].
\end{aligned}
\tag{15.28}
$$

Moreover, (15.26) is a trivial consequence of (15.25).

Concerning (15.27), we have

$$
\begin{aligned}
&\|Pv\|^2_{L^2(\Omega)} = \left\|P^{(s)}v + P^{(a)}v\right\|^2_{L^2(\Omega)} = \\
&= \left\|P^{(s)}v\right\|^2_{L^2(\Omega)} + \left\|P^{(a)}v\right\|^2_{L^2(\Omega)} + 2\Re\left(\left\langle P^{(s)}v, P^{(a)}v\right\rangle_{L^2(\Omega)}\right),
\end{aligned}
\tag{15.29}
$$

for all $v \in C^{\infty}(\Omega)$.

Now, taking into account that

$$
\left(P^{(s)}\right)^{\star} = P^{(s)} \quad \text{and} \quad \left(P^{(a)}\right)^{\star} = -P^{(a)},
$$

we have

$$
\begin{aligned}
&2\Re\left(\left\langle P^{(s)}v, P^{(a)}v\right\rangle_{L^2(\Omega)}\right) = \\
&= \left\langle P^{(s)}v, P^{(a)}v\right\rangle_{L^2(\Omega)} + \left\langle P^{(a)}v, P^{(s)}v\right\rangle_{L^2(\Omega)} = \\
&= -\left\langle P^{(a)}P^{(s)}v, v\right\rangle_{L^2(\Omega)} + \left\langle P^{(s)}P^{(a)}v, v\right\rangle_{L^2(\Omega)} = \\
&= \left\langle\left[P^{(s)}, P^{(a)}\right]v, v\right\rangle_{L^2(\Omega)},
\end{aligned}
\tag{15.30}
$$

for all $v \in C_0^{\infty}(\Omega)$.

By (15.29) and (15.30) we obtain (15.27). ■

15.2.1 Sobolev Spaces with Large Parameter

Definition 15.4 Let $n \in \mathbb{N}$ and $s \in \mathbb{R}$, $\tau \in [1, +\infty)$. The Sobolev space $H^s_\tau(\mathbb{R}^n)$ is the space of $f \in \mathcal{S}'(\mathbb{R}^n)$ which satisfy

$$\left(|\xi|^2 + \tau^2\right)^{s/2} \widehat{f}(\xi) \in L^2\left(\mathbb{R}^n\right), \quad \forall \tau \in [1, +\infty). \tag{15.31}$$

Equipped with the norm

$$f \to \|f\|_{H^s_\tau(\mathbb{R}^n)} = \left(\int_{\mathbb{R}^n} \left(|\xi|^2 + \tau^2\right)^s \left|\widehat{f}(\xi)\right|^2 d\xi\right)^{1/2}, \tag{15.32}$$

the space $H^s_\tau(\mathbb{R}^n)$ is a complex Hilbert space.

We denote $H^s_1(\mathbb{R}^n)$ simply by $H^s(\mathbb{R}^n)$.

Proposition 15.5 *Let $s = k \in \mathbb{N}_0$. Then norm* (15.32) *is equivalent to the norm*

$$f \to \left(\sum_{|\alpha| \le k} \tau^{2(k-|\alpha|)} \left\|D^\alpha f\right\|^2_{L^2(\mathbb{R}^n)}\right)^{1/2}. \tag{15.33}$$

More precisely, there exists $C \ge 1$ independent of τ such that, for every $f \in H^s_\tau(\mathbb{R}^n)$ we have

$$C^{-1} \|f\|^2_{H^s_\tau(\mathbb{R}^n)} \le \sum_{|\alpha| \le k} \tau^{2(k-|\alpha|)} \left\|D^\alpha f\right\|^2_{L^2(\mathbb{R}^n)} \le C \|f\|^2_{H^s_\tau(\mathbb{R}^n)}. \tag{15.34}$$

Proof The proof follows the same line as that of Theorem 5.40, but for the convenience of the reader we repeat it here.

Let $f \in H^k_\tau(\mathbb{R}^n)$. First we prove the second inequality from the left in (15.34). Since (see Sect. 4.8)

$$\widehat{D^\alpha f}(\xi) = \xi^\alpha \widehat{f}(\xi), \quad \text{for } |\alpha| \le k, \tag{15.35}$$

we have

$$\begin{aligned} &\sum_{|\alpha| \le k} \tau^{2(k-|\alpha|)} \left\|D^\alpha f\right\|^2_{L^2(\mathbb{R}^n)} = \\ &= (2\pi)^{-n} \sum_{|\alpha| \le k} \tau^{2(k-|\alpha|)} \int_{\mathbb{R}^n} \left|\xi^\alpha\right|^2 \left|\widehat{f}(\xi)\right|^2 d\xi. \end{aligned} \tag{15.36}$$

Now, since

$$\left|\xi^{\alpha}\right| \leq |\xi|^{|\alpha|},$$

we have

$$\begin{aligned} &\sum_{|\alpha|\leq k} \tau^{2(k-|\alpha|)} \left|\xi^{\alpha}\right|^2 \leq \sum_{|\alpha|\leq k} \tau^{2(k-|\alpha|)} |\xi|^{2|\alpha|} \leq \\ &\leq C_0 \sum_{j=0}^{k} \binom{k}{j} \tau^{2(k-j)} |\xi|^{2j} = C_0 \left(\tau^2 + |\xi|^2\right)^k, \end{aligned} \tag{15.37}$$

where C_0 is a constant. This inequality and (15.36) implies

$$\sum_{|\alpha|\leq k} \tau^{2(k-|\alpha|)} \left\| D^{\alpha} f \right\|^2_{L^2(\mathbb{R}^n)} \leq C_0 (2\pi)^{-n} \int_{\mathbb{R}^n} \left(\tau^2 + |\xi|^2\right)^k \left|\widehat{f}(\xi)\right|^2 d\xi. \tag{15.38}$$

This concludes the proof of the second inequality from the left in (15.34).

Now we prove the first inequality from the left in (15.34). Set

$$c_0 = \min\left\{ \sum_{j=1}^{n} \left|\eta_j^k\right|^2 \ : \ |\eta| = 1 \right\}.$$

We have $c_0 > 0$ and by homogeneity we obtain

$$c_0^{-1} \sum_{j=1}^{n} \left|\eta_j^k\right| \geq |\eta|^{2k}, \quad \forall \eta \in \mathbb{R}^n.$$

Hence, for every $\eta \in \mathbb{R}^n$ we have

$$\begin{aligned} \left(1 + |\eta|^2\right)^k &\leq 2^{k-1} \left(1 + |\eta|^{2k}\right) \leq \\ &\leq 2^{k-1} \max\left\{1, c_0^{-1}\right\} \left(1 + \sum_{j=1}^{n} \left|\eta_j^k\right|^2\right) \leq \\ &\leq 2^{k-1} \max\left\{1, c_0^{-1}\right\} \sum_{|\alpha|\leq k} |\eta^{\alpha}|^2. \end{aligned}$$

All in all, setting

$$C_1 = 2^{k-1} \max\left\{1, c_0^{-1}\right\},$$

we get

$$\left(1+|\eta|^2\right)^k \leq C_1 \sum_{|\alpha|\leq k} |\eta^\alpha|^2, \quad \forall \eta \in \mathbb{R}^n. \tag{15.39}$$

Now let $\eta = \frac{\xi}{\tau}$, for $\tau \geq 1$. By multiplying both the sides of (15.39) for τ^{2k}, we obtain

$$\left(\tau^2+|\xi|^2\right)^k \leq C_1 \sum_{|\alpha|\leq k} \tau^{2(k-|\alpha|)} |\xi^\alpha|^2, \quad \forall (\xi,\tau) \in \mathbb{R}^n \times [1,+\infty) \tag{15.40}$$

and, taking into account (15.36), we get

$$\begin{aligned}
&\sum_{|\alpha|\leq k} \tau^{2(k-|\alpha|)} \left\|D^\alpha f\right\|^2_{L^2(\mathbb{R}^n)} = \\
&= (2\pi)^{-n} \sum_{|\alpha|\leq k} \tau^{2(k-|\alpha|)} \int_{\mathbb{R}^n} |\xi^\alpha|^2 \left|\widehat{f}(\xi)\right|^2 d\xi \geq \\
&\geq C_1^{-1}(2\pi)^{-n} \int_{\mathbb{R}^n} \left(\tau^2+|\xi|^2\right)^k \left|\widehat{f}(\xi)\right|^2 d\xi.
\end{aligned} \tag{15.41}$$

By (15.38) and (15.41), (15.34) follows with $C = (2\pi)^{n/2} \max\left\{C_0^{1/2}, C_1^{1/2}\right\}$. ■

Proposition 15.6 *Let $s \in \mathbb{R}$, $\tau \in [1,+\infty)$, $k \in \mathbb{N}_0$. Then we have:*

(a) There exists $C \geq 1$ such that for every $f \in H^s_\tau(\mathbb{R}^n)$ we have

$$C^{-1} \|f\|^2_{H^s_\tau(\mathbb{R}^n)} \leq \sum_{|\alpha|\leq k} \tau^{2(k-|\alpha|)} \left\|D^\alpha f\right\|^2_{H^{s-k}_\tau(\mathbb{R}^n)} \leq C \|f\|^2_{H^s_\tau(\mathbb{R}^n)}. \tag{15.42}$$

(b) There exists $C > 0$ such that for every $f \in H^s_\tau(\mathbb{R}^n)$ we have

$$\left\|D^\alpha f\right\|^2_{H^{s-k}_\tau(\mathbb{R}^n)} \leq C\tau^{2(|\alpha|-k)} \|f\|^2_{H^s_\tau(\mathbb{R}^n)}, \quad |\alpha| \leq k, \tag{15.43}$$

that is, for $|\alpha| \leq k$

$$D^\alpha : H^s_\tau\left(\mathbb{R}^n\right) \to H^{s-k}_\tau\left(\mathbb{R}^n\right), \tag{15.44}$$

is a continuous map.

Proof Let us prove (a). Let $f \in H^s_\tau(\mathbb{R}^n)$. By (15.37) we have

$$\begin{aligned}
\|f\|^2_{H^s_\tau(\mathbb{R}^n)} &= \int_{\mathbb{R}^n} \left(\tau^2 + |\xi|^2\right)^s \left|\widehat{f}(\xi)\right|^2 d\xi = \\
&= \int_{\mathbb{R}^n} \left(\tau^2 + |\xi|^2\right)^k \left(\tau^2 + |\xi|^2\right)^{s-k} \left|\widehat{f}(\xi)\right|^2 d\xi \geq \\
&\geq \sum_{|\alpha|\leq k} \int_{\mathbb{R}^n} \tau^{2(k-|\alpha|)} \left|\xi^\alpha\right|^2 \left(\tau^2 + |\xi|^2\right)^{s-k} \left|\widehat{f}(\xi)\right|^2 d\xi = \\
&= \sum_{|\alpha|\leq k} \int_{\mathbb{R}^n} \tau^{2(k-|\alpha|)} \left(\tau^2 + |\xi|^2\right)^{s-k} \left|\widehat{D^\alpha f}(\xi)\right|^2 d\xi = \\
&= \sum_{|\alpha|\leq k} \int_{\mathbb{R}^n} \tau^{2(k-|\alpha|)} \left\|D^\alpha f\right\|^2_{H^{s-k}_\tau(\mathbb{R}^n)}.
\end{aligned}$$

Hence

$$\sum_{|\alpha|\leq k} \int_{\mathbb{R}^n} \tau^{2(k-|\alpha|)} \left\|D^\alpha f\right\|^2_{H^{s-k}_\tau(\mathbb{R}^n)} \leq \|f\|^2_{H^s_\tau(\mathbb{R}^n)}. \tag{15.45}$$

Similarly, by (15.40) we have

$$\begin{aligned}
\|f\|^2_{H^s_\tau(\mathbb{R}^n)} &= \int_{\mathbb{R}^n} \left(\tau^2 + |\xi|^2\right)^{s-k} \left(\tau^2 + |\xi|^2\right)^k \left|\widehat{f}(\xi)\right|^2 d\xi \leq \\
&\leq C_1 \int_{\mathbb{R}^n} \left(\tau^2 + |\xi|^2\right)^{s-k} \sum_{|\alpha|\leq k} \tau^{2(k-|\alpha|)} \left|\xi^\alpha\right|^2 \left|\widehat{f}(\xi)\right|^2 d\xi = \\
&= C_1 \sum_{|\alpha|\leq k} \tau^{2(k-|\alpha|)} \int_{\mathbb{R}^n} \left(\tau^2 + |\xi|^2\right)^{s-k} \left|\widehat{D^\alpha f}(\xi)\right|^2 d\xi = \\
&= C_1 \sum_{|\alpha|\leq k} \tau^{2(k-|\alpha|)} \left\|D^\alpha f\right\|^2_{H^{s-k}_\tau(\mathbb{R}^n)}.
\end{aligned}$$

The just obtained inequality and (15.45) imply (a).

Proposition (b) is now an immediate consequence of the second inequality from the left of (15.43). ■

Proposition 15.7 *If $s \leq t$ then $H^t_\tau(\mathbb{R}^n) \subset H^s_\tau(\mathbb{R}^n)$ and*

$$\|f\|_{H^s_\tau(\mathbb{R}^n)} \leq \tau^{-(t-s)} \|f\|_{H^t_\tau(\mathbb{R}^n)}, \quad \forall f \in H^t_\tau\left(\mathbb{R}^n\right). \tag{15.46}$$

Proof Let $f \in H^t_\tau(\mathbb{R}^n)$. Then we have

$$\begin{aligned}\|f\|^2_{H^t_\tau(\mathbb{R}^n)} &= \int_{\mathbb{R}^n} \left(\tau^2+|\xi|^2\right)^{t-s}\left(\tau^2+|\xi|^2\right)^{s}\left|\widehat{f}(\xi)\right|^2 d\xi \geq \\ &\geq \tau^{2(t-s)}\int_{\mathbb{R}^n}\left(\tau^2+|\xi|^2\right)^{s}\left|\widehat{f}(\xi)\right|^2 d\xi = \\ &= \tau^{2(t-s)}\|f\|^2_{H^s_\tau(\mathbb{R}^n)}.\end{aligned}$$

The just obtained inequality immediately implies (15.46) and also $H^t_\tau(\mathbb{R}^n) \subset H^s_\tau(\mathbb{R}^n)$. ■

Lemma 15.8 *Let Ω be an open set of $\mathbb{R}^n$, $a \in C^\infty(\Omega)$, $\alpha \in \mathbb{N}_0$. Let*

$$p(x,\xi) = a(x)\xi^\alpha$$

and

$$P(x,D) = \mathrm{Op}(p)(x,D) = a_\alpha(x)D^\alpha. \tag{15.47}$$

Then for any compact set K, $K \subset \Omega$, and for every $s \in \mathbb{R}$, there exists C such that

$$\left|\langle P(x,D)u, v\rangle_{L^2(\mathbb{R}^n)}\right| \leq C\,\|u\|_{H^s(\mathbb{R}^n)}\|v\|_{H^{|\alpha|-s}(\mathbb{R}^n)}, \tag{15.48}$$

for every $u, v \in C^\infty_K(\mathbb{R}^n)$.

Moreover

$$\left|\langle P(x,D)u, u\rangle_{L^2(\mathbb{R}^n)}\right| \leq C\,\|u\|^2_{H^{\frac{|\alpha|}{2}}(\mathbb{R}^n)}, \quad \forall u \in C^\infty_K\left(\mathbb{R}^n\right). \tag{15.49}$$

Proof Let $a \in C^\infty(\Omega)$. Let K be a compact subset of Ω. Let (see Lemma 3.49) $\varphi \in C^\infty_0(\Omega)$ satisfy $\varphi(x) = 1$ for every $x \in K$ and $0 \leq \varphi \leq 1$ in K_1, where K_1 is a compact subset of Ω such that $K \subset \overset{\circ}{K}_1$.

Let us denote by

$$Q(x,D) = \varphi(x)P(x,D) = \varphi(x)a(x)D^\alpha, \tag{15.50}$$

$$q(x,\xi) = \varphi(x)a(x)\xi^\alpha, \quad \forall\xi \in \mathbb{R}^n. \tag{15.51}$$

We have trivially

$$Q(x,D)u = P(x,D)u = \varphi(x)a(x)D^\alpha u, \quad \forall u \in C^\infty_K\left(\mathbb{R}^n\right). \tag{15.52}$$

Let $u \in C_K^\infty(\mathbb{R}^n)$. By the Fourier inversion formula we have

$$Q(x,D)u(x) = \frac{1}{(2\pi)^n}\int_{\mathbb{R}^n} e^{ix\cdot\xi} q(x,\xi)\widehat{u}(\xi)d\xi.$$

Hence

$$\begin{aligned}(\widehat{Q(x,D)}u)(\eta) &= \frac{1}{(2\pi)^n}\int_{\mathbb{R}^n}\left(\int_{\mathbb{R}^n} e^{ix\cdot\xi} q(x,\xi)\widehat{u}(\xi)d\xi\right)e^{-ix\cdot\eta}dx = \\ &= \frac{1}{(2\pi)^n}\int_{\mathbb{R}^n}\left(\int_{\mathbb{R}^n} e^{ix\cdot(\xi-\eta)} q(x,\xi)\widehat{u}(\xi)d\xi\right)dx.\end{aligned} \tag{15.53}$$

Since $u \in C_K^\infty(\mathbb{R}^n)$ we have $\widehat{u} \in \mathcal{S}(\mathbb{R}^n)$, and taking into account that

$$\left|e^{ix\cdot(\xi-\eta)} q(x,\xi)\widehat{u}(\xi)\right| \leq \|a\|_{L^\infty(\Omega)}\, \chi_K |\xi|^k \, |\widehat{u}(\xi)|,$$

where

$$k = |\alpha|,$$

we have, for every $\eta \in \mathbb{R}^n$,

$$\mathbb{R}^{2n} \ni (x,\xi) \to e^{ix\cdot(\xi-\eta)} q(x,\xi)\widehat{u}(\xi) \in L^1\left(\mathbb{R}^{2n}\right). \tag{15.54}$$

Therefore, by the Fubini–Tonelli Theorem and by (15.53) we get

$$\begin{aligned}(\widehat{Q(x,D)}u)(\eta) &= \frac{1}{(2\pi)^n}\int_{\mathbb{R}^n}\left(\int_{\mathbb{R}^n} e^{-ix\cdot(\eta-\xi)} q(x,\xi)\widehat{u}(\xi)dx\right)d\xi = \\ &= \frac{1}{(2\pi)^n}\int_{\mathbb{R}^n} \widehat{q_1}(\eta-\xi,\xi)\widehat{u}(\xi)d\xi,\end{aligned} \tag{15.55}$$

where $\widehat{q_1}(\cdot,\xi)$ is the Fourier transform of $q(\cdot,\xi)$.

Let $u, v \in C_K^\infty(\mathbb{R}^n)$. Set

$$f(\xi) = \left(1+|\xi|^2\right)^{s/2}\widehat{u}(\xi), \quad g(\eta) = \left(1+|\eta|^2\right)^{(k-s)/2}\widehat{u}(\eta), \tag{15.56}$$

and

$$K(\eta,\xi) = \frac{1}{(2\pi)^{2n}}\widehat{q_1}(\eta-\xi,\xi)\left(1+|\xi|^2\right)^{-s/2}\left(1+|\eta|^2\right)^{-(k-s)/2}. \tag{15.57}$$

With these notation we have

$$\begin{aligned}
&\langle Q(x,D)u, v\rangle_{L^2(\mathbb{R}^n)} = \\
&= \frac{1}{(2\pi)^{2n}} \left\langle (\widehat{Q(x,D)u}), \widehat{v}\right\rangle_{L^2(\mathbb{R}^n)} = \\
&= \frac{1}{(2\pi)^{2n}} \int_{\mathbb{R}^n} \left(\int_{\mathbb{R}^n} \widehat{q_1}(\eta-\xi,\xi)\widehat{u}(\xi)\overline{\widehat{v}(\eta)} d\xi \right) d\eta = \\
&= \int_{\mathbb{R}^n} \left(\int_{\mathbb{R}^n} K(\eta,\xi) f(\xi) \overline{g(\eta)} d\xi \right) d\eta = \\
&= \int_{\mathbb{R}^n} (Tf)(\eta) \overline{g(\eta)} d\eta,
\end{aligned} \tag{15.58}$$

where

$$(Tf)(\eta) = \int_{\mathbb{R}^n} K(\eta,\xi) f(\xi) d\xi. \tag{15.59}$$

Claim There exists C_s such that

$$\sup_{\xi\in\mathbb{R}^n} \int_{\mathbb{R}^n} |K(\eta,\xi)| d\eta \leq C_s, \quad \sup_{\eta\in\mathbb{R}^n} \int_{\mathbb{R}^n} |K(\eta,\xi)| d\xi \leq C_s, \tag{15.60}$$

actually, C_s depends not only on s, but also on φ, $\|a\|_{L^\infty(K)}$ and k.

Proof of the Claim Let $\beta \in \mathbb{N}_0^n$. For every $\zeta \in \mathbb{R}^n$ we have

$$\begin{aligned}
\left|\zeta^\beta \widehat{q_1}(\zeta,\xi)\right| &= \left| \int_{\mathbb{R}^n} D_x^\beta \left(e^{-ix\cdot\zeta}\right) q(x,\xi) dx \right| = \\
&= \left| \int_{\mathbb{R}^n} e^{-ix\cdot\zeta} D_x^\beta q(x,\xi) dx \right| \leq C_\beta \left(1+|\xi|^2\right)^{k/2}.
\end{aligned}$$

Therefore, for every $N \geq 0$ there exists C_N such that

$$|\widehat{q_1}(\zeta,\xi)| \leq C_N \left(1+|\xi|^2\right)^{k/2} \left(1+|\zeta|^2\right)^{-N}, \quad \forall \xi,\zeta \in \mathbb{R}^n.$$

This inequality, (15.57) and Lemma 4.24 give, for every $\eta, \xi \in \mathbb{R}^n$,

$$\begin{aligned}
&|K(\eta, \xi)| = \\
&= \frac{1}{(2\pi)^{2n}} \left| \widehat{q_1}(\eta - \xi, \xi) \left(1 + |\xi|^2\right)^{-s/2} \left(1 + |\eta|^2\right)^{-(k-s)/2} \right| \leq \\
&\leq C_N \left(1 + |\xi|^2\right)^{k/2} \left(1 + |\eta - \xi|^2\right)^{-N} \left(1 + |\xi|^2\right)^{-s/2} \left(1 + |\eta|^2\right)^{-(k-s)/2} = \\
&= C_N \left(\frac{1 + |\xi|^2}{1 + |\eta|^2}\right)^{\frac{k-s}{2}} \left(1 + |\eta - \xi|^2\right)^{-N} \leq \\
&\leq C'_N \left(1 + |\eta - \xi|^2\right)^{-N + \frac{|k-s|}{2}}.
\end{aligned}$$

Now, let us fix N such that

$$N > \frac{1}{2}(n + |k - s|)$$

and we obtain (15.60). **This concludes the proof of the Claim.**

Now, by (15.52), (15.58), (15.59) and by Lemma 3.43 we have, for any $u, v \in C_K^\infty(\mathbb{R}^n)$

$$\begin{aligned}
&\left|\langle P(x, D)u, v\rangle_{L^2(\mathbb{R}^n)}\right| = \left|\langle Q(x, D)u, v\rangle_{L^2(\mathbb{R}^n)}\right| = \\
&= \left|\int_{\mathbb{R}^n} (Tf)(\eta)\overline{g(\eta)}d\eta\right| \leq \|Tf\|_{L^2(\mathbb{R}^n)} \|g\|_{L^2(\mathbb{R}^n)} \leq \\
&\leq C_s \|f\|_{L^2(\mathbb{R}^n)} \|g\|_{L^2(\mathbb{R}^n)} = C_s \|u\|_{H^s(\mathbb{R}^n)} \|v\|_{H^{k-s}(\mathbb{R}^n)}.
\end{aligned}$$

The above inequality concludes the proof of inequality (15.48). Concerning (15.49), it is a trivial consequence of (15.48) for $s = \frac{|\alpha|}{2}$ and $u = v$ (recalling that $k = |\alpha|$). ■

Proposition 15.9 *Let Ω be an open set of $\mathbb{R}^n$ and $k \in \mathbb{N}$. Let $p(x, \xi, \tau)$ be a polynomial of degree k w.r.t. the variable (ξ, τ) whose coefficients belong to $L^\infty(\Omega)$, and let*

$$P(x, D, \tau) = \mathrm{Op(p)}(x, D, \tau).$$

Then we have:

(a) there exists $C > 0$ such that

$$\|P(x, D, \tau)v\|_{L^2(\mathbb{R}^n)} \leq C \|v\|_{H_\tau^k(\mathbb{R}^n)}, \quad \forall v \in C_0^\infty(\Omega), \ \forall \tau \geq 1. \tag{15.61}$$

(b) If the coefficients of $p(x,\xi,\tau)$ belong to $C^{\infty}(\Omega)$, then for any compact set K such that $K \subset \Omega$, there exists $C>0$ such that

$$\left|\langle P(x,D,\tau)v,v\rangle_{L^2(\mathbb{R}^n)}\right| \leq C\,\|v\|^2_{H_\tau^{k/2}(\mathbb{R}^n)}, \quad \forall v \in C_K^{\infty}(\mathbb{R}^n),\ \forall \tau \geq 1. \tag{15.62}$$

Proof First we prove (a).

Let us consider

$$p(x,\xi,\tau)=\sum_{|\alpha|+j\leq k} a_\alpha(x)\tau^j\xi^\alpha,$$

where $a_\alpha \in L^{\infty}(\Omega)$.

Let us note that, if $|\alpha|+j \leq k$ then we have

$$\tau^{2j}\left|\xi^\alpha\right|^2 \leq \tau^{2j}|\xi|^{2|\alpha|} \leq$$
$$\leq \left(\tau^2+|\xi|^2\right)^{|\alpha|+j} \leq \left(\tau^2+|\xi|^2\right)^k, \quad \forall(\xi,\tau)\in\mathbb{R}^n\times[1,+\infty).$$

Hence, for every $v \in C_0^{\infty}(\Omega)$ and $\tau \geq 1$ we have

$$\begin{aligned}\|P(x,D,\tau)v\|^2_{L^2(\mathbb{R}^n)} &= C\sum_{|\alpha|+j\leq k}\tau^{2j}\left\|D^\alpha v\right\|^2_{L^2(\mathbb{R}^n)} =\\ &= (2\pi)^{-n}C\sum_{|\alpha|+j\leq k}\int_{\mathbb{R}^n}\tau^{2j}\left|\xi^\alpha\right|^2|\widehat{v}(\xi)|^2\,d\xi \leq\\ &\leq \widetilde{C}\int_{\mathbb{R}^n}\left(\tau^2+|\xi|^2\right)^k|\widehat{v}(\xi)|^2\,d\xi =\\ &= \widetilde{C}\,\|v\|^2_{H_\tau^k(\mathbb{R}^n)},\end{aligned} \tag{15.63}$$

where $C, \widetilde{C}$ depends on $\|a_\alpha\|_{L^\infty(\Omega)}$, for $|\alpha| \leq k$. By (15.63) immediately follows (15.61).

Now we prove (b).

Let

$$P(x,D,\tau)=\sum_{|\alpha|+j\leq k} P_{\alpha,j}(x,D,\tau),$$

where

$$P_{\alpha,j}(x,D,\tau)=a_\alpha(x)\tau^jD^\alpha,$$

and $a_\alpha \in C^\infty(\Omega)$. Hence it not restrictive to prove (15.62) for any $P_{\alpha,j}(x,D,\tau)$, provided

$$|\alpha| + j \leq k.$$

By (15.49) we have, for every $u \in C_K^\infty(\mathbb{R}^n)$ and for every $\tau \geq 1$,

$$\begin{aligned}
\left|\left\langle P_{\alpha,j}(x,D,\tau)v, v\right\rangle_{L^2(\mathbb{R}^n)}\right| &\leq C\tau^j \|v\|^2_{H^{\frac{|\alpha|}{2}}(\mathbb{R}^n)} = \\
&= C\tau^j \int_{\mathbb{R}^n} \left(\tau^2 + |\xi|^2\right)^{\frac{|\alpha|}{2}} |\widehat{v}(\xi)|^2 \, d\xi \leq \\
&\leq C \int_{\mathbb{R}^n} \left(\tau^2 + |\xi|^2\right)^{\frac{|\alpha|+j}{2}} |\widehat{v}(\xi)|^2 \, d\xi \leq \\
&\leq C \int_{\mathbb{R}^n} \left(\tau^2 + |\xi|^2\right)^{\frac{k}{2}} |\widehat{v}(\xi)|^2 \, d\xi = \\
&= C \|v\|^2_{H_\tau^{k/2}}.
\end{aligned} \tag{15.64}$$

■

Theorem 15.10 (Localized Garding Inequality) *Let Ω be an open set of $\mathbb{R}^n$, $x_0 \in \Omega$ and $m \in \mathbb{N}$. Let $q(x,\xi,\tau)$ be a polynomial in the variable (ξ,τ), homogeneous of degree $2m$, whose coefficients belong to $C^\infty(\Omega)$. Let*

$$Q(x,D,\tau) = \mathrm{Op(q)}(x,D,\tau).$$

Then we have:

(a) For every $\varepsilon > 0$ there exists $r > 0$ and $C > 0$ such that $B_r(x_0) \Subset \Omega$ and

$$\begin{aligned}
&\left|\langle (Q(x,D,\tau) - Q(x_0,D,\tau))\, v, v\rangle_{L^2(\mathbb{R}^n)}\right| \leq \\
&\leq \varepsilon \|v\|^2_{H_\tau^m(\mathbb{R}^n)} + C \|v\|^2_{H_\tau^{m-1}(\mathbb{R}^n)},
\end{aligned} \tag{15.65}$$

for every $\tau \geq 1$ and for every $v \in C_0^\infty(\Omega)$ such that $\operatorname{supp} v \subset B_r(x_0)$.

(b) Assume there exists $\lambda_0 > 0$ such that

$$\Re\left(q(x_0,\xi,\tau)\right) \geq \lambda_0 \left(\tau^2 + |\xi|^2\right)^m, \quad \forall \xi \in \mathbb{R}^n, \ \forall \tau \geq 1. \tag{15.66}$$

Then there exist $C > 0$ and $r > 0$ such that

$$\Re\,\langle Q(x,D,\tau)v, v\rangle_{L^2(\mathbb{R}^n)} \geq \frac{\lambda_0}{2(2\pi)^n} \|v\|^2_{H_\tau^m(\mathbb{R}^n)} - C \|v\|^2_{H_\tau^{m-1}(\mathbb{R}^n)}, \tag{15.67}$$

for every $\tau \geq 1$ and for every $v \in C_0^\infty(\Omega)$ such that $\operatorname{supp} v \subset B_r(x_0)$.

Remark Properly speaking the localized Garding inequality is sentence (b) of Theorem 15.10. ♦

Proof It is not restrictive to assume $x_0 = 0 \in \Omega$.

We begin to prove (a).

Let us denote by

$$t(x,\xi,\tau) = q(x,\xi,\tau) - q(0,\xi,\tau) = \sum_{|\alpha|+j=2m} a_{\alpha j}(x)\tau^j \xi^\alpha,$$

$$T(x,D,\tau) = \mathrm{Op}(t)(x,D,\tau) = \sum_{|\alpha|+j=2m} a_{\alpha j}(x)\tau^j D^\alpha,$$

where $a_{\alpha j} \in C^\infty(\Omega)$ and

$$a_{\alpha j}(0) = 0, \quad \text{for } |\alpha| + j = 2m. \tag{15.68}$$

Let us denote by

$$T_{\alpha j}(x,D,\tau) = a_{\alpha j}(x)\tau^j D^\alpha, \quad \text{for } |\alpha| + j = 2m.$$

Hence

$$T(x,D,\tau) = \sum_{|\alpha|+j\le 2m} T_{\alpha,j}(x,D,\tau).$$

Let us note that it not restrictive to prove (15.65) for any $T_{\alpha,j}(x,D,\tau)$, provided

$$|\alpha| + j = 2m.$$

Let $\varepsilon > 0$. By (15.68) there exists $r > 0$ such that $B_r \Subset \Omega$ and

$$\left|a_{\alpha j}(x)\right| < \varepsilon, \quad \forall x \in B_r, \quad \text{for } |\alpha| + j = 2m. \tag{15.69}$$

Let $\alpha \in \mathbb{N}_0^n$ and $j \in \mathbb{N}_0$ be such that $|\alpha| + j = 2m$.

Let us distinguish two cases:

(i) $|\alpha|$ is an even number,
(ii) $|\alpha|$ is an odd number.

Case (i) Let $h \in \mathbb{N}_0$ be such that $|\alpha| = 2h$. Let us note that there exist $\beta, \gamma \in \mathbb{N}_0^n$ such that

$$|\beta| = |\gamma| = h, \text{ and } \alpha = \beta + \gamma.$$

As a matter of fact, up to a permutation of the indices, there exists $s \in \mathbb{N}$ such that $2s \le n$ and α_j are odd numbers for $j = 1, \cdots 2s$ and α_j are even numbers for $j = 2s+1, \cdots n$. Hence we can choose

$$\beta = ([\alpha_1/2], \cdots [\alpha_s/2], [\alpha_{s+1}/2] + 1, \cdots, [\alpha_{2s}/2] + 1, \alpha_{2s+1}/2, \cdots, \alpha_n/2),$$

$$\gamma = ([\alpha_1/2] + 1, \cdots [\alpha_s/2] + 1, [\alpha_{s+1}/2], \cdots, [\alpha_{2s}/2], \alpha_{2s+1}/2, \cdots, \alpha_n/2),$$

so that β and γ have the desired property.

Now we have, for every $v \in C_0^\infty(\Omega)$ such that $\operatorname{supp} v \subset B_r$

$$\begin{aligned}
&\langle T_{\alpha,j}(x, D, \tau)v, v\rangle_{L^2(\mathbb{R}^n)} = \\
&= \tau^j \langle a_{\alpha j}(x) D^\alpha v, v\rangle_{L^2(\mathbb{R}^n)} = \\
&= \tau^j \int_{\mathbb{R}^n} a_{\alpha j}(x) D^\gamma \left(D^\beta v\right) \overline{v} dx = \\
&= (-1)^h \tau^j \int_{\mathbb{R}^n} D^\beta v D^\gamma \left(a_{\alpha j}(x)\overline{v}\right) dx = \\
&= I_1 + I_2,
\end{aligned} \tag{15.70}$$

where

$$I_1 = (-1)^h \tau^j \int_{\mathbb{R}^n} a_{\alpha j}(x) D^\beta v D^\gamma \overline{v} dx, \tag{15.71}$$

if $h \geq 1$,

$$I_2 = (-1)^h \tau^j \int_{\mathbb{R}^n} D^\beta v \sum_{\delta<\gamma} \binom{\gamma}{\delta} D^\delta \overline{v} D^{\gamma-\delta} \left(a_{\alpha j}(x)\right). \tag{15.72}$$

and if $h = 0$, $I_2 = 0$.

First we estimate from above I_1. By (15.69) and by the Cauchy–Schwarz inequality we have, for every $\tau \geq 1$ and for every $v \in C_0^\infty(\Omega)$ such that $\operatorname{supp} v \subset B_r$,

$$\begin{aligned}
|I_1| &\leq \tau^j \varepsilon \left\| D^\beta v \right\|_{L^2(\mathbb{R}^n)} \left\| D^\gamma v \right\|_{L^2(\mathbb{R}^n)} \leq \\
&\leq C_0 \varepsilon \tau^j \|v\|^2_{H^h_\tau(\mathbb{R}^n)} = \\
&= C\tau^j \int_{\mathbb{R}^n} \left(\tau^2 + |\xi|^2\right)^h |\widehat{v}(\xi)|^2 d\xi \leq \\
&\leq C_0 \varepsilon \int_{\mathbb{R}^n} \left(\tau^2 + |\xi|^2\right)^{\frac{|\alpha|+j}{2}} |\widehat{v}(\xi)|^2 d\xi \leq \\
&\leq C_0 \varepsilon \int_{\mathbb{R}^n} \left(\tau^2 + |\xi|^2\right)^m |\widehat{v}(\xi)|^2 d\xi = \\
&= C_0 \varepsilon \|v\|^2_{H^m_\tau(\mathbb{R}^n)},
\end{aligned}$$

where C_0 does not depend on ε and τ.

Hence

$$|I_1| \leq C_0\varepsilon \|v\|^2_{H^m_\tau(\mathbb{R}^n)}, \tag{15.73}$$

for every $\tau \geq 1$ and for every $v \in C^\infty_0(\Omega)$ such that $\operatorname{supp} v \subset B_r$.

Concerning I_2, integrating by parts, by (15.72) we have

$$I_2 = \langle P_{2m-1}(x, D, \tau)v, v\rangle_{L^2(\mathbb{R}^n)},$$

where

$$P_{2m-1}(x, D, \tau)v = (-1)^h \sum_{\delta<\gamma} (-1)^{|\delta|} \binom{\gamma}{\delta} \tau^j D^\delta \left(D^{\gamma-\delta}\left(a_{\alpha j}(x)\right) D^\beta v\right).$$

Let us note that the symbol of $P_{2m-1}(x, D, \tau)$ is a polynomial in the variable (ξ, τ) of degree $2m-1$ w.r.t. the variable (ξ, τ). Therefore, by (15.62) we have there exists $C > 0$ such that, for every $\tau \geq 1$,

$$|I_2| \leq C \|v\|^2_{H_\tau^{m-\frac{1}{2}}(\mathbb{R}^n)}, \quad \forall v \in C^\infty_0(\mathbb{R}^n), \text{ s.t. } \operatorname{supp} v \subset B_r, \tag{15.74}$$

where C depends on $\left\|a_{\alpha j}\right\|_{C^{2m}(\overline{B}_r)}$, but does not depend on τ. On the other hand we have (by the inequality $ab \leq \varepsilon a^2 + (4\varepsilon)^{-1}b^2$)

$$\begin{aligned} &\|v\|^2_{H_\tau^{m-\frac{1}{2}}(\mathbb{R}^n)} = \\ &= \int_{\mathbb{R}^n} \left(\tau^2 + |\xi|^2\right)^{m-\frac{1}{2}} |\widehat{v}(\xi)|^2 d\xi = \\ &= \int_{\mathbb{R}^n} \left(\tau^2 + |\xi|^2\right)^{\frac{m}{2}} |\widehat{v}(\xi)| \left(\tau^2 + |\xi|^2\right)^{\frac{m-1}{2}} |\widehat{v}(\xi)|\, d\xi \leq \\ &\leq \varepsilon \|v\|^2_{H^m_\tau(\mathbb{R}^n)} + (4\varepsilon)^{-1} \|v\|^2_{H^{m-1}_\tau(\mathbb{R}^n)}. \end{aligned} \tag{15.75}$$

By this inequality, (15.70), (15.73), and (15.75) we have

$$\begin{aligned} &\left|\langle T_{\alpha,j}(x, D, \tau)v, v\rangle_{L^2(\mathbb{R}^n)}\right| \leq \\ &\leq |I_1| + |I_2| \\ &(C_0 + C)\,\varepsilon \|v\|^2_{H^m_\tau(\mathbb{R}^n)} + (4\varepsilon)^{-1} C \|v\|^2_{H^{m-1}_\tau(\mathbb{R}^n)}, \end{aligned} \tag{15.76}$$

for every $\tau \geq 1$ and for every $v \in C^\infty_0(\Omega)$ such that $\operatorname{supp} v \subset B_r \Subset \Omega$. This concludes the proof of Case (i).

Case (ii) Let $h \in \mathbb{N}_0$ be such that $|\alpha| = 2h+1$. Let us note that there exist $\beta, \gamma \in \mathbb{N}_0^n$ such that

$$|\beta| = h+1, \; |\gamma| = h \text{ and } \alpha = \beta + \gamma.$$

As a matter of fact, let $\alpha_j \geq 1$ then $\alpha - e_j$ is a multiindex such that $|\alpha - e_j| = 2h$. For what we noticed at beginning of Case (i), there exist $\widetilde{\beta}, \gamma \in \mathbb{N}_0^n$ such that $|\widetilde{\beta}| = |\gamma| = h$ and $\alpha - e_j = \widetilde{\beta} + \gamma$. Therefore, $\beta = \widetilde{\beta} + e_j$ and γ satisfy the desired property.

In this case we have $2h + 1 + j = 2m$, hence $j \geq 1$.

Likewise Case (i) we have

$$\langle T_{\alpha,j}(x, D, \tau)v, v\rangle_{L^2(\mathbb{R}^n)} = I_1' + I_2', \tag{15.77}$$

where

$$I_1' = (-1)^h \tau^j \int_{\mathbb{R}^n} a_{\alpha j}(x) D^\beta v D^\gamma \overline{v} dx \tag{15.78}$$

and

$$I_2' = \langle P^\bullet_{2m-1}(x, D, \tau)v, v\rangle_{L^2(\mathbb{R}^n)},$$

where

$$P^\bullet_{2m-1}(x, D, \tau)v = (-1)^h \sum_{\delta<\gamma} (-1)^{|\delta|} \binom{\gamma}{\delta} \tau^j D^\delta \left(D^{\gamma-\delta}\left(a_{\alpha j}(x)\right) D^\beta v\right).$$

Since the symbol of $P^\bullet_{2m-1}(x, D, \tau)$ is a polynomial in the variable (ξ, τ) of degree $2m-1$ w.r.t. the variable (ξ, τ), by (15.62) we have there exists $C > 0$ such that, for every $\tau \geq 1$

$$|I_2'| \leq C \|v\|^2_{H_\tau^{m-\frac{1}{2}}(\mathbb{R}^n)}, \quad \forall v \in C_0^\infty(\mathbb{R}^n), \text{ s.t. } \operatorname{supp} v \subset B_r, \tag{15.79}$$

where C depends on $\|a_{\alpha j}\|_{C^{2m}(\overline{B}_r)}$, but does not depend on τ.

Concerning $|I_1'|$, by (15.69) and by the Cauchy–Schwarz inequality we have, for every $\tau \geq 1$ and for every $v \in C_0^\infty(\Omega)$ such that $\operatorname{supp} v \subset B_r$,

$$\begin{aligned} |I_1'| &\leq C_1 \varepsilon \tau^j \|v\|_{H_\tau^{h+1}(\mathbb{R}^n)} \|v\|_{H_\tau^h(\mathbb{R}^n)} = \\ &= C_1 \varepsilon \left(\tau^{\frac{j-1}{2}} \|v\|_{H_\tau^{h+1}(\mathbb{R}^n)}\right) \left(\tau^{\frac{j+1}{2}} \|v\|_{H_\tau^h(\mathbb{R}^n)}\right), \end{aligned} \tag{15.80}$$

where C_1 does not depend on ε and τ. Now we have

$$\begin{aligned}\left(\tau^{\frac{j-1}{2}}\|v\|_{H_\tau^{h+1}(\mathbb{R}^n)}\right)^2 &= \tau^{j-1}\int_{\mathbb{R}^n}\left(\tau^2+|\xi|^2\right)^{h+1}|\widehat{v}(\xi)|^2\,d\xi \leq \\ &\leq \int_{\mathbb{R}^n}\left(\tau^2+|\xi|^2\right)^{\frac{2h+1+j}{2}}|\widehat{v}(\xi)|^2\,d\xi = \\ &= \|v\|^2_{H_\tau^m(\mathbb{R}^n)}\,.\end{aligned}$$

Similarly

$$\left(\tau^{\frac{j+1}{2}}\|v\|_{H_\tau^{h+1}(\mathbb{R}^n)}\right)^2 \leq \|v\|^2_{H_\tau^m(\mathbb{R}^n)}\,.$$

Hence, by (15.80) we have

$$\left|I_1'\right| \leq C_1\varepsilon\,\|v\|^2_{H_\tau^m(\mathbb{R}^n)}\,, \tag{15.81}$$

for every $\tau \geq 1$ and for every $v \in C_0^\infty(\Omega)$ such that $\operatorname{supp} v \subset B_r$.

Now proceeding in a similar manner as done in Case (i), by (15.79) (15.81), we obtain (15.65) in Case (ii). This concludes the proof of sentence (a).

Now we prove (b).

Let us observe that by (15.66) and by the Parseval formula we have

$$\begin{aligned}\Re\,\langle Q(0,D,\tau)v,v\rangle_{L^2(\mathbb{R}^n)} &= \frac{1}{(2\pi)^n}\int_{\mathbb{R}^n} q(0,\xi,\tau)\,|\widehat{v}(\xi)|^2\,d\xi \geq \\ &\geq \frac{\lambda_0}{(2\pi)^n}\int_{\mathbb{R}^n}\left(\tau^2+|\xi|^2\right)^m|\widehat{v}(\xi)|^2\,d\xi = \\ &= \frac{\lambda_0}{(2\pi)^n}\,\|v\|_{H_\tau^m(\mathbb{R}^n)}\,,\end{aligned} \tag{15.82}$$

for every $\tau \geq 1$ and for every $v \in C_0^\infty(\Omega)$.

Now, by (15.65) and (15.82) we have

$$\begin{aligned}&\Re\,\langle Q(x,D,\tau)v,v\rangle_{L^2(\mathbb{R}^n)} = \Re\,\langle Q(0,D,\tau)v,v\rangle_{L^2(\mathbb{R}^n)} + \\ &+ \Re\,\langle (Q(x,D,\tau)-Q(0,D,\tau))\,v,v\rangle_{L^2(\mathbb{R}^n)} \geq \\ &\geq \left(\frac{\lambda_0}{(2\pi)^n}-C_2\varepsilon\right)\|v\|_{H_\tau^m(\mathbb{R}^n)} - C\,\|v\|^2_{H_\tau^{m-1}(\mathbb{R}^n)}\,,\end{aligned}$$

where C_2 does not depends on ε and τ and C may depends on ε, but does not depend on τ.

Finally, let us choose

$$\varepsilon = \frac{\lambda_0}{2C_2(2\pi)^n}$$

and we get (15.67). ∎

15.3 Set Up of a Carleman Estimate: The Conjugate of a Differential Operator

Let Ω be an open set of $\mathbb{R}^n$ and let

$$P(x, D) = \sum_{|\alpha|\leq m} a_\alpha(x)D^\alpha \tag{15.83}$$

be a linear differential operator, where $a_\alpha \in L^\infty(\Omega)$.

As we observed in Sect. 14.3 with simple examples, the initial steps in proving a Carleman estimate involve introducing a transformation for an arbitrary function $u \in C_0^\infty(\Omega)$:

$$v = e^{\tau\varphi}u.$$

In this way, denoting, for the sake of brevity, by P the differential operator $P_m(x, D)$ (principal part of the operator $P(x, D)$) we introduce **the conjugate operator of** P which is defined by

$$P_\tau v = e^{\tau\varphi}P\left(e^{-\tau\varphi}v\right) = e^{\tau\varphi}P_m(x, D)\left(e^{-\tau\varphi}v\right); \tag{15.84}$$

after that, since

$$\int_\Omega |P_m(x, D)u|^2\, e^{2\tau\varphi(x)}dx = \int_\Omega |P_\tau v|^2\, dx,$$

it is clear that proving a Carleman estimate with exponential weight φ reduces to prove a suitable estimate from below of

$$\int_\Omega |P_\tau v|^2\, dx,$$

provided τ is **large enough**. The step involving the estimate from below is the most challenging part of proving a Carleman estimate. In certain cases, spreading the square of the operator and integrating by parts can lead to difficulties, especially when dealing with general operators (because the expression can be very long). This can even be the case for operators slightly more general than those encountered in Sect. 14.3. These difficulties also arise in the case of the Laplace operator. As a matter of fact, let us consider

$$P = \Delta = -\sum_{j=1}^{n} D_j^2,$$

we have

$$P_\tau v = e^{\tau\varphi}\Delta\left(e^{-\tau\varphi}v\right) = \Delta v - \tau\nabla\varphi\cdot\nabla v - \tau\Delta\varphi + \tau^2\left|\nabla\varphi\right|^2 v, \tag{15.85}$$

and, spreading the square we obtain an expression like

$$\sum_{|\alpha|,|\beta|\le 2} \tau^{\gamma_{\alpha,\beta}} a_{\alpha\beta}(x)\partial^\alpha v\partial^\beta v.$$

In order to prevent the development of the calculations from becoming complicated it is therefore wise to carefully study the most important properties of the conjugate operator.

We first observe that

$$P_\tau v = P_m(x, D + i\tau\nabla\varphi(x))v. \tag{15.86}$$

As a matter of fact we have

$$e^{\tau\varphi}D_j\left(e^{-\tau\varphi}v\right) = D_j v - \tau(D_j\varphi)v = D_j v + i\tau(\partial_j\varphi)v,$$

$$\begin{aligned} e^{\tau\varphi}D_k D_j\left(e^{-\tau\varphi}v\right) &= e^{\tau\varphi}D_k\left(e^{-\tau\varphi}(D_j v + i\tau(\partial_j\varphi)v)\right) = \\ &= (D_k + i\tau(\partial_k\varphi))(D_j + i\tau(\partial_j\varphi))v, \end{aligned}$$

$$\vdots$$

$$e^{\tau\varphi}D^\alpha\left(e^{-\tau\varphi}v\right) = (D + i\tau\nabla\varphi(x))^\alpha\, v,$$

for every $\alpha \in \mathbb{N}_0^n$. Now, since

$$P_m(x, D) = \sum_{|\alpha|=m} a_\alpha(x)D^\alpha, \tag{15.87}$$

we have

$$e^{\tau\varphi}P_m(x,D)\left(e^{-\tau\varphi}v\right)=\sum_{|\alpha|=m}a_\alpha(x)e^{\tau\varphi}D^\alpha\left(e^{-\tau\varphi}v\right)=$$

$$=\sum_{|\alpha|=m}a_\alpha(x)\left(D+i\tau\nabla\varphi(x)\right)^\alpha v=$$

$$=P_m(x,D+i\tau\nabla\varphi(x))v.$$

Now, let us consider the polynomial

$$p_m(x,\xi+i\tau\nabla\varphi(x))=\sum_{|\alpha|=m}a_\alpha(x)\left(\xi+i\tau\nabla\varphi(x)\right)^\alpha, \tag{15.88}$$

in the variable ξ and let us denote by

$$\widetilde{P}_m(x,D+i\tau\nabla\varphi(x))=\mathrm{Op}(p_m(x,\xi+i\tau\nabla\varphi(x))), \tag{15.89}$$

in other words, $\widetilde{P}_m(x,D+i\tau\nabla\varphi(x))$ is **the operator whose symbol is** $p_m(x,\xi+i\tau\nabla\varphi(x))$. Let us note that, unless $m=0,1$, the operator $\widetilde{P}_m(x,D+i\tau\nabla\varphi(x))$ does not equal to the operator $P_m(x,D+i\tau\nabla\varphi(x))$. For instance, if

$$P_2(x,D)=D_1^2,$$

we have

$$P_2(x,D+i\tau\nabla\varphi(x))v=D_1^2v+2i\tau\partial_1\varphi D_1v-\tau^2(\partial_1\varphi)^2v+\tau\left(\partial_1^2\varphi\right)v,$$

hence

$$\widetilde{P}_2(x,D+i\tau\nabla\varphi(x))v=P_2(x,D+i\tau\nabla\varphi(x))v-\tau\partial_1^2\varphi(x)v.$$

In general we have

$$\widetilde{P}_m(x,D+i\tau\nabla\varphi(x))=\sum_{|\alpha|+j=m}\tau^jb_{\alpha j}(x)D^\alpha, \tag{15.90}$$

where the coefficients $b_{\alpha j}(x)$ depend on $\nabla\varphi$, on the coefficients of $P_m(x,D)$ (but not on their derivatives) and **do not depend** on τ. Let us note that

$$p_m(x,\xi+i\tau\nabla\varphi(x)),$$

is a homogeneous polynomial w.r.t. the variable (ξ,τ) and $\widetilde{P}_m(x,D+i\tau\nabla\varphi(x))$ is homogeneous w.r.t. the variable (D,τ). We also say that the operator $\widetilde{P}_m(x,D+i\tau\nabla\varphi(x))$ is the **conjugate principal part** of the operator $P_m(x,D)$. Actually, we have

$$P_m(x, D+i\tau\nabla\varphi(x)) = \widetilde{P}_m(x, D+i\tau\nabla\varphi(x)) + R_{m-1,\tau}(x, D, \tau), \tag{15.91}$$

where

$$R_{m-1,\tau}(x, D) = \sum_{|\alpha|+j\leq m-1} \tau^j c_{\alpha j}(x) D^\alpha, \tag{15.92}$$

the coefficients $c_{\alpha j}$ depend on the coefficients of $P_m(x, D)$ (but not on their derivatives), on $\nabla\varphi$ and on the higher-order derivatives of φ and **do not depend on** τ. The second term on the right-hand side in (15.91), as we will realize soon, may be regarded as a harmless perturbation of the operator $\widetilde{P}_m(x, D+i\tau\nabla\varphi(x))$.

Lemma 15.11 *Let K be a compact subset of Ω. Let $P(x, D)$ be the differential operator* (15.83) *and $\varphi \in C^\infty(\Omega)$, then there exists $C > 0$ such that*

$$\begin{aligned} &\left|P_m(x, D+i\tau\nabla\varphi(x))v - \widetilde{P}_m(x, D+i\tau\nabla\varphi(x))v\right| \leq \\ &\leq C \sum_{|\alpha|\leq m-1} \tau^{m-1-|\alpha|} \left|D^\alpha v\right|, \end{aligned} \tag{15.93}$$

for every $v \in C^\infty_K(\Omega)$ and for every $\tau \geq 1$.

Proof By (15.91) and (15.92) we have for every $\tau \geq 1$

$$\begin{aligned} &\left|P_m(x, D+i\tau\nabla\varphi(x))v - \widetilde{P}_m(x, D+i\tau\nabla\varphi(x))v\right| = \\ &= \left|R_{m-1,\tau}(x, D, \tau)v\right| \leq \\ &\leq \sum_{|\alpha|+j\leq m-1} \tau^j \left|c_{\alpha j}(x) D^\alpha v\right| \leq \\ &\leq C \sum_{|\alpha|\leq m-1} \tau^{m-1-|\alpha|} \left|D^\alpha v\right|, \end{aligned}$$

where C depends on $\max_{|\alpha|+j\leq m-1} \left\|c_{\alpha,j}\right\|_{L^\infty(K)}$ and m only. ■

Now we prove the following

Lemma 15.12 *Let K be a compact subset of Ω. Let $\varphi \in C^\infty(\Omega)$. Then for every $k \in \mathbb{N}_0$ there exists a constant $C > 1$ such that*

$$\begin{aligned} C^{-1} \sum_{|\alpha|\leq k} \tau^{2(k-|\alpha|)} \left|D^\alpha\left(e^{\tau\varphi}u\right)\right|^2 &\leq \sum_{|\alpha|\leq k} \tau^{2(k-|\alpha|)} \left|D^\alpha u\right|^2 e^{2\tau\varphi} \leq \\ &\leq C \sum_{|\alpha|\leq k} \tau^{2(k-|\alpha|)} \left|D^\alpha\left(e^{\tau\varphi}u\right)\right|^2, \end{aligned} \tag{15.94}$$

for every $u \in C^\infty_K(\Omega)$ and for every $\tau \geq 1$.

Proof Both the inequalities are proved easily by means of the Leibniz formula. Here we limit ourselves to prove

$$\sum_{|\alpha|\leq k} \tau^{2(k-|\alpha|)} \left|D^\alpha u\right|^2 e^{2\tau\varphi} \leq C \sum_{|\alpha|\leq k} \tau^{2(k-|\alpha|)} \left|D^\alpha \left(e^{\tau\varphi} u\right)\right|^2. \tag{15.95}$$

We use the induction principle. If $k = 0$, then (15.95) is trivial. Let us suppose that

$$\sum_{|\alpha|\leq k} \tau^{2(k-|\alpha|)} \left|D^\alpha u\right|^2 e^{2\tau\varphi} \leq C_k \sum_{|\alpha|\leq k} \tau^{2(k-|\alpha|)} \left|D^\alpha \left(e^{\tau\varphi} u\right)\right|^2,$$

where $C_k \geq 1$ and we have

$$\begin{aligned}
\sum_{|\alpha|\leq k+1} \tau^{2(k+1-|\alpha|)} \left|D^\alpha \left(e^{\tau\varphi} u\right)\right|^2 &= \sum_{|\alpha|=k+1} \left|D^\alpha \left(e^{\tau\varphi} u\right)\right|^2 + \\
&+ \tau^2 \sum_{|\alpha|\leq k} \tau^{2(k-|\alpha|)} \left|D^\alpha \left(e^{\tau\varphi} u\right)\right|^2 \geq \\
&\geq \sum_{|\alpha|=k+1} \left|D^\alpha \left(e^{\tau\varphi} u\right)\right|^2 + C_k^{-1} \sum_{|\alpha|\leq k} \tau^{2(k+1-|\alpha|)} \left|D^\alpha u\right|^2 e^{2\tau\varphi}.
\end{aligned} \tag{15.96}$$

Let $\delta \in (0, 1)$ be to choose. By using the Leibniz formula we have, for $\tau \geq 1$,

$$\begin{aligned}
\sum_{|\alpha|=k+1} \left|D^\alpha \left(e^{\tau\varphi} u\right)\right|^2 \geq \delta \sum_{|\alpha|=k+1} \left|D^\alpha \left(e^{\tau\varphi} u\right)\right|^2 &\geq \\
\geq \delta \sum_{|\alpha|=k+1} \left|D^\alpha u\right|^2 e^{2\tau\varphi} &- \\
- \delta \widetilde{C}_k \sum_{|\alpha|\leq k} \tau^{2(k+1-|\alpha|)} \left|D^\alpha u\right|^2 e^{2\tau\varphi}&,
\end{aligned} \tag{15.97}$$

where $\widetilde{C}_k \geq 1$ is a suitable constant depending on k. By (15.96) and (15.97) we get

$$\begin{aligned}
\sum_{|\alpha|\leq k+1} \tau^{2(k+1-|\alpha|)} \left|D^\alpha \left(e^{\tau\varphi} u\right)\right|^2 &\geq \delta \sum_{|\alpha|=k+1} \left|D^\alpha u\right|^2 e^{2\tau\varphi} + \\
&+ \left(C_k^{-1} - \delta \widetilde{C}_k\right) \sum_{|\alpha|\leq k} \tau^{2(k+1-|\alpha|)} \left|D^\alpha u\right|^2 e^{2\tau\varphi}.
\end{aligned}$$

Now, we choose $\delta = \frac{1}{2C_k\widetilde{C}_k}$ and we get

$$\sum_{|\alpha|\leq k+1} \tau^{2(k+1-|\alpha|)} \left|D^\alpha u\right|^2 e^{2\tau\varphi} \leq 2C_k \sum_{|\alpha|\leq k+1} \tau^{2(k+1-|\alpha|)} \left|D^\alpha \left(e^{\tau\varphi} u\right)\right|^2,$$

which concludes the proof. ■

For the sake of brevity, in the sequel we will use the following notation. Let $f(\tau)$ and $g(\tau)$ be real-valued functions defined in the interval $[a, +\infty)$, $a \in \mathbb{R}$, when we write

$$f(\tau) \lesssim g(\tau),$$

or simply

$$f \lesssim g,$$

we mean that there exist $C > 0$ and $\tau_0 > 0$ such that

$$f(\tau) \leq Cg(\tau), \quad \forall \tau \geq \tau_0.$$

When we write

$$f(\tau) \gtrsim g(\tau),$$

we mean

$$g(\tau) \lesssim f(\tau).$$

Proposition 15.13 *Let K be a compact subset of Ω. Let $P(x, D)$ be the differential operator* (15.83) *and $\varphi \in C^\infty(\Omega)$, then the following propositions are equivalent*

(a)

$$\sum_{|\alpha|\leq m-1} \tau^{2(m-|\alpha|)-1} \int_\Omega \left|D^\alpha u\right|^2 e^{2\tau\varphi} dx \lesssim \int_\Omega |P_m(x, D)u|^2 e^{2\tau\varphi} dx, \tag{15.98}$$

for every $u \in C_K^\infty(\Omega)$.

(b)

$$\tau^{1/2} \|v\|_{H_\tau^{m-1}(\mathbb{R}^n)} \lesssim \|P_m(x, D + i\tau\nabla\varphi(x))v\|_{L^2(\mathbb{R}^n)}, \tag{15.99}$$

for every $v \in C_K^\infty(\Omega)$.

(c)

$$\tau^{1/2} \|v\|_{H_\tau^{m-1}(\mathbb{R}^n)} \lesssim \left\|\widetilde{P}_m(x, D + i\tau\nabla\varphi(x))v\right\|_{L^2(\mathbb{R}^n)}, \tag{15.100}$$

for every $v \in C_K^\infty(\Omega)$.

Proof First we observe that, Proposition 15.5 implies that (b) and (c) are equivalent respectively to

(b')

$$\sum_{|\alpha|\le m-1} \tau^{2(m-|\alpha|)-1} \int_\Omega \left|D^\alpha v\right|^2 dx \lesssim \int_\Omega |P_m(x, D+i\tau\nabla\varphi(x))v|^2\, dx, \tag{15.101}$$

for every $v \in C_K^\infty(\Omega)$;

(c')

$$\sum_{|\alpha|\le m-1} \tau^{2(m-|\alpha|)-1} \int_\Omega \left|D^\alpha v\right|^2 dx \lesssim \int_\Omega \left|\widetilde{P}_m(x, D+i\tau\nabla\varphi(x))v\right|^2 dx, \tag{15.102}$$

for every $v \in C_K^\infty(\Omega)$.

Thus, to prove the Proposition it is enough to prove "$(a) \Leftrightarrow (b')$" and "$(b') \Leftrightarrow (c')$".

Let us assume that (a) holds. Let $v \in C_K^\infty(\Omega)$ and $u = e^{-\tau\varphi}v$. By (a) we have

$$\begin{aligned} &\sum_{|\alpha|\le m-1} \tau^{2(m-|\alpha|)-1} \int_\Omega \left|D^\alpha u\right|^2 e^{2\tau\varphi} dx \lesssim \\ &\lesssim \int_\Omega |P_m(x, D+i\tau\nabla\varphi)v|^2\, dx, \quad \forall v \in C_K^\infty(\Omega). \end{aligned} \tag{15.103}$$

By inequality (15.94) (the first from the left) of Lemma 15.12, for $k = m-1$ we have trivially

$$\begin{aligned} &\sum_{|\alpha|\le m-1} \tau^{2(m-|\alpha|)-1} \int_\Omega \left|D^\alpha u\right|^2 e^{2\tau\varphi} dx = \\ &= \tau \sum_{|\alpha|\le m-1} \tau^{2(m-1-|\alpha|)} \int_\Omega \left|D^\alpha u\right|^2 e^{2\tau\varphi} dx \ge \\ &\ge C^{-1}\tau \sum_{|\alpha|\le m-1} \tau^{2(m-1-|\alpha|)} \int_\Omega \left|D^\alpha v\right|^2 dx = \\ &= C^{-1} \sum_{|\alpha|\le m-1} \tau^{2(m-|\alpha|)-1} \int_\Omega \left|D^\alpha v\right|^2 dx. \end{aligned} \tag{15.104}$$

By (15.103) and (15.104) we get (b') and consequently (b).

The implication "$(b')\Rightarrow(a)$" can be proved similarly to the previous one using the second inequality from the left of (15.94). This concludes the proof of "$(a)\Leftrightarrow(b')$".

Now, let us assume that (b') holds. Hence there exist $C_0>0$ and $\tau_0\geq 1$ such that

$$\sum_{|\alpha|\leq m-1}\tau^{2(m-|\alpha|)-1}\int_\Omega\left|D^\alpha v\right|^2dx\leq C_0\int_\Omega|P_m(x,D+i\tau\nabla\varphi)v|^2\,dx, \tag{15.105}$$

for every $v\in C_K^\infty(\Omega)$ and for every $\tau\geq\tau_0$.

By Lemma 15.11 we have that there exists $C_1>0$ such that

$$\begin{aligned}\int_\Omega|P_m(x,D+i\tau\nabla\varphi)v|^2\,dx&\leq 2\int_\Omega\left|\widetilde{P}_m(x,D+i\tau\nabla\varphi)v\right|^2dx+\\&+C_1\tau^{-1}\sum_{|\alpha|\leq m-1}\int_\Omega\tau^{2(m-|\alpha|)-1}\left|D^\alpha v\right|^2dx.\end{aligned} \tag{15.106}$$

for every $v\in C_K^\infty(\Omega)$ and for every $\tau\geq 1$.

By (15.105) and (15.106) we have there exist C_0 and τ_0 such that

$$\begin{aligned}&\left(1-C_0C_1\tau^{-1}\right)\sum_{|\alpha|\leq m-1}\tau^{2(m-|\alpha|)-1}\int_\Omega\left|D^\alpha v\right|^2dx\leq\\&\leq 2C_0\int_\Omega\left|\widetilde{P}_m(x,D+i\tau\nabla\varphi(x))v\right|^2dx,\end{aligned}$$

for every $v\in C_K^\infty(\Omega)$ and for every $\tau\geq\tau_0$. So that we have

$$\begin{aligned}&\sum_{|\alpha|\leq m-1}\tau^{2(m-|\alpha|)-1}\int_\Omega\left|D^\alpha v\right|^2dx\leq\\&\leq 4C_0\int_\Omega\left|\widetilde{P}_m(x,D+i\tau\nabla\varphi(x))v\right|^2dx,\end{aligned}$$

for every $v\in C_K^\infty(\Omega)$ and for every $\tau\geq\max\{\tau_0,2C_0C_1\}$ and we get (c'). The implication "$(c')\Rightarrow(b')$" can be proved similarly. This concludes the proof of "$(b')\Leftrightarrow(c')$" and by the equivalence between (b) and (b'), and between (c) and (c') observed above, the Proposition is completely proven. ∎

Remark 15.1 Likewise to Proposition 15.13 the following equivalence can be proved. Here K, $P(x,D)$ and φ satisfy the same hypotheses of Proposition 15.13. The following inequalities are equivalent

(a)

$$\sum_{|\alpha|\leq m} \tau^{2(m-|\alpha|)-1} \int_{\Omega} \left|D^{\alpha} u\right|^2 e^{2\tau\varphi} dx \lesssim \int_{\Omega} |P_m(x, D)u|^2 e^{2\tau\varphi} dx, \tag{15.107}$$

for every $u \in C_K^{\infty}(\Omega)$.

(b)

$$\tau^{-1/2} \|v\|_{H_\tau^m(\mathbb{R}^n)} \lesssim \|P_m(x, D + i\tau\nabla\varphi(x))v\|_{L^2(\mathbb{R}^n)}, \tag{15.108}$$

for every $v \in C_K^{\infty}(\Omega)$.

(c)

$$\tau^{-1/2} \|v\|_{H_\tau^m(\mathbb{R}^n)} \lesssim \left\|\widetilde{P}_m(x, D + i\tau\nabla\varphi(x))v\right\|_{L^2(\mathbb{R}^n)}, \tag{15.109}$$

for every $v \in C_K^{\infty}(\Omega)$. ♦

Let us introduce now the **self-adjoint part** and the **skew-adjoint part** of $\widetilde{P}_m(x, D + i\tau\nabla\varphi(x))$, respectively, as (see Sect. 15.2)

$$\begin{aligned} &\widetilde{P}_m^{(s)}(x, D + i\tau\nabla\varphi(x)) = \\ &= \frac{1}{2}\left(\widetilde{P}_m(x, D + i\tau\nabla\varphi(x)) + \widetilde{P}_m^{\star}(x, D + i\tau\nabla\varphi(x))\right), \end{aligned} \tag{15.110}$$

$$\begin{aligned} &\widetilde{P}_m^{(a)}(x, D + i\tau\nabla\varphi(x)) = \\ &= \frac{1}{2}\left(\widetilde{P}_m(x, D + i\tau\nabla\varphi(x)) - \widetilde{P}_m^{\star}(x, D + i\tau\nabla\varphi(x))\right), \end{aligned} \tag{15.111}$$

where $\widetilde{P}_m^{\star}(x, D+i\tau\nabla\varphi(x))$ is the adjoint operator of $\widetilde{P}_m(x, D+i\tau\nabla\varphi(x))$, namely

$$\begin{aligned} &\int_{\Omega} \widetilde{P}_m^{\star}(x, D + i\tau\nabla\varphi(x))v\, \overline{w} dx = \\ &= \int_{\Omega} v\, \overline{\left(\widetilde{P}_m(x, D + i\tau\nabla\varphi(x))w\right)} dx, \quad \forall v, w \in C_0^{\infty}(\Omega). \end{aligned}$$

Theorem 15.14 *Let Ω be an open set of $\mathbb{R}^n$, let K be a compact subset of Ω. Let $P(x, D)$ be the differential operator* (15.83) *and $\varphi \in C^{\infty}(\Omega)$, then there exists*

$C > 0$ *such that*

$$\begin{aligned}
&C\,\|v\|^2_{H^{m-1}_\tau(\mathbb{R}^n)} + C\,\|P_m(x, D + i\tau\nabla\varphi)v\|^2_{L^2(\mathbb{R}^n)} \geq \\
&\geq \left\|\widetilde{P}^{(s)}_m(x, D + i\tau\nabla\varphi)v\right\|^2_{L^2(\mathbb{R}^n)} + \left\|\widetilde{P}^{(a)}_m(x, D + i\tau\nabla\varphi)v\right\|^2_{L^2(\mathbb{R}^n)} + \\
&+ \Re\left\langle\left[\widetilde{P}^{(s)}_m(x, D + i\tau\nabla\varphi), \widetilde{P}^{(a)}_m(x, D + i\tau\nabla\varphi)\right]v, v\right\rangle_{L^2(\mathbb{R}^n)},
\end{aligned} \tag{15.112}$$

for every $v \in C^\infty_K(\Omega)$ *and for every* $\tau \geq 1$.

Proof By Proposition 15.3 we have

$$\begin{aligned}
&\|\widetilde{P}_m(x, D + i\tau\nabla\varphi)v\|^2_{L^2(\mathbb{R}^n)} = \\
&= \left\|\widetilde{P}^{(s)}_m(x, D + i\tau\nabla\varphi)v\right\|^2_{L^2(\mathbb{R}^n)} + \left\|\widetilde{P}^{(a)}_m(x, D + i\tau\nabla\varphi)v\right\|^2_{L^2(\mathbb{R}^n)} + \\
&+ \Re\left\langle\left[\widetilde{P}^{(s)}_m(x, D + i\tau\nabla\varphi), \widetilde{P}^{(a)}_m(x, D + i\tau\nabla\varphi)\right]v, v\right\rangle_{L^2(\mathbb{R}^n)},
\end{aligned} \tag{15.113}$$

for every $v \in C^\infty_K(\Omega)$ and for every $\tau \geq 1$.

By Lemma 15.11 we get

$$\begin{aligned}
&\|\widetilde{P}_m(x, D + i\tau\nabla\varphi)v\|^2_{L^2(\mathbb{R}^n)} \leq \\
&\leq 2\,\|P_m(x, D + i\tau\nabla\varphi)v\|^2_{L^2(\mathbb{R}^n)} + C\sum_{|\alpha|\leq m-1}\int_\Omega \tau^{2(m-|\alpha|-1)}\left|D^\alpha v\right|^2 dx \leq \\
&\leq 2\,\|P_m(x, D + i\tau\nabla\varphi)v\|^2_{L^2(\mathbb{R}^n)} + C'\,\|v\|^2_{H^{m-1}_\tau(\mathbb{R}^n)},
\end{aligned}$$

for every $v \in C^\infty_K(\Omega)$ and for every $\tau \geq 1$. The just obtained inequality and (15.113) immediately give (15.112). ■

Remark 15.2 Observe that

$$\left\langle\left[\widetilde{P}^{(s)}_m(x, D + i\tau\nabla\varphi), \widetilde{P}^{(a)}_m(x, D + i\tau\nabla\varphi)\right]v, v\right\rangle_{L^2(\mathbb{R}^n)}$$

is indeed a real number, nonetheless, it is convenient for us to keep "$\Re$" in the formula (15.112) to remind us that the result of this operation must be a real number. ♦

Remark 15.3 Theorem 15.14 clearly indicates to us that the crucial step in deriving a Carleman estimate consists of finding a suitable lower bound for the third term on the right of inequality (15.112) in which the commutator operator

$$\left[\widetilde{P}^{(s)}_m(x, D + i\tau\nabla\varphi(x)), \widetilde{P}^{(a)}_m(x, D + i\tau\nabla\varphi(x))\right], \tag{15.114}$$

appears. In the following sections, we will provide sufficient conditions to derive such commutator estimates. The first thing we will do to achieve this goal is to express inequality (15.112) differently through a careful approximation of commutator (15.114). ♦

15.4 Some Approximation Formulas for Operators Depending on a Parameter

In this section, we will provide some approximation formulas for operators and their symbols that depend on the parameter τ. These formulas will allow us, in particular, to write Theorem 15.14 in a form more suitable for our main purpose of proving the Carleman estimates. Let us begin with the following

Definition 15.15 Let Ω be an open set of $\mathbb{R}^n$ and let $m \in \mathbb{N}_0$. We define $\Sigma^m(\Omega)$ as the set of polynomials of degree less than or equal to m w.r.t. the variable $(\xi, \tau) \in \mathbb{R}^{n+1}$

$$p(x, \xi, \tau) = \sum_{|\alpha+j|\leq m} a_\alpha(x)\tau^j\xi^\alpha,$$

where $a_\alpha \in C^\infty(\Omega)$, for $|\alpha| \leq m$.

Let $p, q \in \Sigma^m(\Omega)$ and $k \leq m-1$, $k \in \mathbb{N}_0$. When we write

$$p(x, \xi, \tau) = q(x, \xi, \tau), \quad \text{Mod } \Sigma^k(\Omega),$$

we mean that there exist $r \in \Sigma^k(\Omega)$ such that

$$p(x, \xi, \tau) = q(x, \xi, \tau) + r(x, \xi, \tau), \quad \forall x \in \Omega, \ \forall (\xi, \tau) \in \mathbb{R}^{n+1}.$$

By Proposition 15.9 we have immediately that, if $p \in \Sigma^m(\Omega)$

$$p(x, \xi, \tau) = \sum_{|\alpha+j|\leq m} a_\alpha(x)\tau^j\xi^\alpha$$

and

$$P(x, D, \tau) = \text{Op(p)}(x, D, \tau) = \sum_{|\alpha+j|\leq m} a_\alpha(x)\tau^j D^\alpha,$$

then, for any compact set K such that $K \subset \Omega$ there exists $C > 0$ such that

$$\|P(x, D, \tau)v\|_{L^2(\mathbb{R}^n)} \leq C\,\|v\|_{H^k_\tau(\mathbb{R}^n)}, \quad \forall v \in C^\infty_K(\Omega), \ \forall \tau \geq 1, \tag{15.115}$$

and

$$\left|\langle P(x,D,\tau)v,v\rangle_{L^2(\mathbb{R}^n)}\right| \leq C\,\|v\|^2_{H^{k/2}_\tau(\mathbb{R}^n)}, \quad \forall v\in C^\infty_K(\mathbb{R}^n),\ \forall\tau\geq 1. \tag{15.116}$$

Proposition 15.16 *Let Ω be an open set of $\mathbb{R}^n$, $m_1, m_2 \in \mathbb{N}_0$ and let $p_1 \in \Sigma^{m_1}(\Omega)$, $p_2 \in \Sigma^{m_2}(\Omega)$. Set*

$$P_1(x,D,\tau) = \operatorname{Op}(p_1)(x,D,\tau), \quad P_2(x,D,\tau) = \operatorname{Op}(p_2)(x,D,\tau),$$

and

$$Q(x,D,\tau) = P_1(x,D,\tau)P_2(x,D,\tau).$$

Denote by $q(x,\xi,\tau)$ the symbol of $Q(x,D,\tau)$. Then we have $q \in \Sigma^{m_1+m_2}(\Omega)$. Moreover we have

(a)

$$\begin{aligned} q(x,\xi,\tau) &= p_1(x,\xi,\tau)p_2(x,\xi,\tau)+ \\ &+\frac{1}{i}\sum_{j=1}^{n}\left(p_1^{(j)}p_{2,j}\right)(x,\xi,\tau), \quad \operatorname{Mod}\ \Sigma^{m_1+m_2-2}(\Omega), \end{aligned} \tag{15.117}$$

where (see Sect. 15.2) $p_1^{(j)} = \partial_{\xi_j}p_1$ and $p_{2,j} = \partial_{x_j}p_2$, $j = 1,\cdots,n$.

(b) Let

$$C = [P_1, P_2],$$

where $[P_1, P_2]$ is the commutator of P_1 and P_2 and let $c(x,\xi,\tau)$ be the symbol of C then we have

$$c(x,\xi,\tau) = \frac{1}{i}\{p_1,p_2\}(x,\xi,\tau), \quad \operatorname{Mod}\ \Sigma^{m_1+m_2-2}(\Omega), \tag{15.118}$$

where $\{p_1,p_2\}$ is the Poisson brackets of p_1 and p_2 (see Sect. 15.2).

(c) Let us denote by

$$p_1^\star(x,\xi,\tau), \tag{15.119}$$

the symbol of $P_1^\star(x,D,\tau)$. Then we have

$$p_1^\star(x,\xi,\tau) = \overline{p_1(x,\xi,\tau)} + \frac{1}{i}\sum_{j=1}^{n}\overline{p_{1,j}^{(j)}(x,\xi,\tau)}, \quad \operatorname{Mod}\ \Sigma^{m_1-2}(\Omega). \tag{15.120}$$

Proof Let us prove (a). Let us consider the case

$$p_1(x,\xi,\tau)=a(x)\tau^h\xi^\alpha,\quad p_2(x,\xi,\tau)=a(x)\tau^h\xi^\beta,$$

where $h,k\in\mathbb{N}_0$ and $\alpha,\beta\in\mathbb{N}_0^n$. In this case we have, for every $v\in C^\infty(\Omega)$,

$$Q(x,D,\tau)v=P_1(x,D,\tau)\left(P_2(x,D,\tau)v\right)=\tau^{h+k}a(x)D^\alpha\left(b(x)D^\beta v\right)$$

and relation (15.117) that we wish to prove becomes

$$\begin{aligned}&q(x,\xi,\tau)=\tau^{h+k}a(x)b(x)\xi^{\alpha+\beta}+\\&+\frac{1}{i}\tau^{h+k}\sum_{j=1}^{n}\alpha_j a(x)\partial_{x_j}b(x)\xi^{\alpha-e_j+\beta},\ \text{Mod }\Sigma^{(|\alpha|+h+|\beta|+k)-2}(\Omega).\end{aligned}\tag{15.121}$$

Now, by the Leibniz formula, we have

$$\begin{aligned}a(x)D^\alpha\left(b(x)D^\beta v\right)&=a(x)\sum_{\gamma\le\alpha}\binom{\alpha}{\gamma}D^\gamma b(x)D^{\alpha-\gamma+\beta}v=\\&=a(x)b(x)D^{\alpha+\beta}v+\sum_{j=1}^{n}a(x)\alpha_j D_{x_j}b(x)D^{\alpha-e_j+\beta}v+R(x,D)v,\end{aligned}$$

where $R(x,D)$ is an operator of order $|\alpha|+|\beta|-2$, if $|\alpha|+|\beta|\ge 2$ and of order 0, if $|\alpha|+|\beta|\le 1$. Denoting by $r(x,\xi)$ the symbol of $R(x,D)$, we have that $\tau^{h+k}r(x,\xi)\in\Sigma^{(|\alpha|+h+|\beta|+k)-2}(\Omega)$ and this concludes the proof of (15.121).

Concerning the general case, let

$$P_1(x,D,\tau)=\sum_{|\alpha|+h\le m_1}a_\alpha(x)\tau^h D^\alpha,\quad P_2(x,D,\tau)=\sum_{|\alpha|+h\le m_2}b_\alpha(x)\tau^h D^\alpha.$$

By (15.121) we have, for every $v\in C^\infty(\Omega)$,

$$\begin{aligned}&Q(x,D,\tau)v=\\&=P_1(x,D,\tau)\left(P_2(x,D,\tau)v\right)=\sum_{|\alpha|+h\le m_1}\ \sum_{|\beta|+k\le m_2}a_\alpha(x)\tau^{h+k}D^\alpha\left(b_\beta(x)D^\beta v\right)=\\&=\mathrm{Op}\left(p_1p_2\right)(x,D,\tau)v+\frac{1}{i}\,\mathrm{Op}\left(\sum_{j=1}^{n}(p_1^{(j)}p_{2,j})\right)(x,D,\tau)v+S(x,D,\tau)v,\end{aligned}$$

where $S(x,D,\tau)=\mathrm{Op}(s)(x,D,\tau)$ and $s\in\Sigma^{m_1+m_2-2}(\Omega)$. Therefore (15.117) follows.

Concerning (b), it is a trivial consequence of (15.117).
Now we prove (c). Let

$$P_1(x, D, \tau) = \sum_{|\alpha|+h\le m_1} a_\alpha(x)\tau^h D^\alpha.$$

By (15.20), we get

$$P_1^\star(x, D, \tau)v = \sum_{|\alpha|+h\le m_1} \tau^h D^\alpha\left(\overline{a_\alpha(x)}v\right). \tag{15.122}$$

By linearity, it is enough to prove (15.120) in the case

$$P_1(x, D, \tau) = a(x)\tau^h D^\alpha.$$

In this case we have

$$P_1(x, D, \tau)v = \tau^h D^\alpha\left(\overline{a_\alpha(x)}v\right), \quad \forall v \in C^\infty(\Omega)$$

and relation (15.120) becomes

$$\begin{aligned} p_1^\star(x, \xi, \tau) &= \tau^h\overline{a(x)}\xi^\alpha + \\ &+ \frac{1}{i}\tau^h\sum_{j=1}^n \alpha_j\partial_{x_j}\left(\overline{a(x)}\right)\xi^{\alpha-e_j}, \text{ Mod } \Sigma^{(|\alpha|+h)-2}(\Omega). \end{aligned} \tag{15.123}$$

By the Leibniz formula we get

$$\begin{aligned} D^\alpha\left(\overline{a(x)}v\right) &= \sum_{\gamma\le\alpha}\binom{\alpha}{\gamma}D^\gamma\left(\overline{a(x)}\right)D^{\alpha-\gamma}v = \\ &= \overline{a(x)}D^\alpha v + \sum_{j=1}^n \alpha_j D_{x_j}\left(\overline{a(x)}\right)D^{\alpha-e_j}v + R(x, D)v, \end{aligned}$$

where $R(x, D)$ is an operator of order $|\alpha| - 2$, if $|\alpha| \ge 2$ and of order 0, if $|\alpha| \le 1$. Denoting by $r(x, \xi)$ the symbol of $R(x, D)$, we have that $\tau^h r(x, \xi) \in \Sigma^{(|\alpha|+h)-2}(\Omega)$ and this concludes the proof of (15.123) and, therefore, also the proof of (15.120). ■

Proposition 15.17 *Let Ω be an open set of $\mathbb{R}^n$, $m \in \mathbb{N}_0$ and let $p \in \Sigma^m(\Omega)$. Let*

$$P(x, D, \tau) = \text{Op}(p)(x, D, \tau), \quad Q(x, D, \tau) = P^\star(x, D, \tau)P(x, D, \tau).$$

Let $q \in \Sigma^{2m}(\Omega)$ be the symbol of $Q(x, D, \tau)$. Then we have

(a)

$$q(x,\xi,\tau) = |p(x,\xi,\tau)|^2 + \\ + \frac{1}{i}\sum_{j=1}^{n}\left(\overline{p_{,j}^{(j)}}\,p + \overline{p^{(j)}}\,p_{,j}\right)(x,\xi,\tau), \text{ Mod } \Sigma^{2m-2}(\Omega); \tag{15.124}$$

(b) for every compact subset K of Ω there exists C such that for every $v \in C_K^\infty(\mathbb{R}^n)$ we have

$$\left| \|P(x,D,\tau)v\|^2_{L^2(\mathbb{R}^n)} - \Re\left\langle \mathrm{Op}(|p|^2)v, v\right\rangle_{L^2(\mathbb{R}^n)} \right| \leq C\,\|v\|^2_{H_\tau^{m-\frac{1}{2}}(\mathbb{R}^n)}; \tag{15.125}$$

(b') if p is ***real-valued****, then for every compact subset K of Ω there exists C such that for every $v \in C_K^\infty(\mathbb{R}^n)$ we have*

$$\left| \|P(x,D,\tau)v\|^2_{L^2(\mathbb{R}^n)} - \Re\left\langle \mathrm{Op}(p^2)v, v\right\rangle_{L^2(\mathbb{R}^n)} \right| \leq C\,\|v\|^2_{H_\tau^{m-1}(\mathbb{R}^n)}. \tag{15.126}$$

Proof Let us prove (a). By (a) and (c) of Proposition 15.16, there exist (for the sake of brevity we omit the variable (x, ξ, τ)) $s_{m-2} \in \Sigma^{m-2}(\Omega)$ and $r_{2m-2} \in \Sigma^{2m-2}(\Omega)$ (s_{m-2} vanishes if $m = 0, 1$; r_{2m-2} vanishes if $m = 0$) such that

$$\begin{aligned} q = &\left(\overline{p} + \frac{1}{i}\sum_{j=1}^{n}\overline{p_{,j}^{(j)}} + s_{m-2}\right)p + \\ &+ \frac{1}{i}\sum_{k=1}^{n}\partial_{\xi_k}\left(\overline{p} + \frac{1}{i}\sum_{j=1}^{n}\overline{p_{,j}^{(j)}} + s_{m-2}\right)\partial_{x_k}p + r_{2m-2} = \\ &= |p(x,\xi,\tau)|^2 + \frac{1}{i}\sum_{j=1}^{n}\left(\overline{p_{,j}^{(j)}}\,p + \overline{p^{(j)}}\,p_{,j}\right) + \widetilde{r}_{2m-2}, \end{aligned}$$

where

$$\widetilde{r}_{2m-2} = s_{m-2}p + r_{2m-2} + \frac{1}{i}\sum_{k=1}^{n} s_{m-2,k}^{(k)}\,p_{,k} - \sum_{j,k=1}^{n} p_{,j}^{(j,k)}\,p_{,k}.$$

Since $\widetilde{r}_{2m-2} \in \Sigma^{2m-2}(\Omega)$ and

$$\frac{1}{i}\sum_{j=1}^{n}\left(\overline{p_{,j}^{(j)}}p+\overline{p^{(j)}}p_{,j}\right) \in \Sigma^{2m-1}(\Omega),$$

by (b) of Proposition 15.9, we obtain (15.124).

Let us prove (b). We have

$$\begin{aligned}&\|P(x,D,\tau)v\|^2_{L^2(\mathbb{R}^n)} - \Re\left\langle \mathrm{Op}(|p|^2)v, v\right\rangle_{L^2(\mathbb{R}^n)} = \\ &= \Re\left\langle T(x;D,\tau)v, v\right\rangle_{L^2(\mathbb{R}^n)}, \ \forall v \in C^\infty_K(\Omega),\end{aligned} \tag{15.127}$$

where

$$T(x,D,\tau) = P^\star(x;D,\tau)P(x;D,\tau) - \mathrm{Op}(|p|^2)$$

and, denoting by $t(x,\xi,\tau)$ the symbol of $T(x,D,\tau)$, by (15.124) we have

$$t \in \Sigma^{2m-1}(\Omega).$$

Hence, by Proposition 15.9 we derive that for every compact subset K of Ω there exists C such that for every $v \in C^\infty_K(\mathbb{R}^n)$ the following inequality holds

$$\left|\Re\left\langle T(x;D,\tau)v, v\right\rangle_{L^2(\mathbb{R}^n)}\right| \leq \|v\|^2_{H^{m-\frac{1}{2}}_\tau}. \tag{15.128}$$

Therefore, by (15.127) and (15.128) we obtain (15.125).

Let us prove (b'). Since p is real valued, the term

$$\frac{1}{i}\sum_{j=1}^{n}\left(\overline{p_{,j}^{(j)}}p+\overline{p^{(j)}}p_{,j}\right)(x,\xi,\tau),$$

is purely imaginary, hence

$$\Re\left\langle T(x;D,\tau)v, v\right\rangle_{L^2(\mathbb{R}^n)} = \Re\left\langle S_{2m-2}(x;D,\tau)v, v\right\rangle_{L^2(\mathbb{R}^n)}, \tag{15.129}$$

where $S_{2m-2}(x;D,\tau)$ is an operator whose symbol belongs to $\Sigma^{2m-2}(\Omega)$. Therefore, applying again Proposition 15.9 we have that for every compact subset K of Ω there exists C such that for every $v \in C^\infty_K(\mathbb{R}^n)$ the following inequality holds

$$\left|\Re\left\langle T(x;D,\tau)v, v\right\rangle_{L^2(\mathbb{R}^n)}\right| \leq C\|v\|^2_{H^{m-1}_\tau(\mathbb{R}^n)}. \tag{15.130}$$

Hence, by (15.127) and (15.130) we obtain (15.126). ■

Corollary 15.18 below is useful when dealing with differential operators whose coefficients of the principal part are real-valued functions.

Let

$$p_m(x,\xi) = \sum_{|\alpha|=m} a_\alpha(x)\xi^\alpha, \tag{15.131}$$

$$P_m(x,D) = \sum_{|\alpha|=m} a_\alpha(x)D^\alpha, \tag{15.132}$$

where $m \in \mathbb{N}$ and $a_\alpha \in C^\infty(\Omega)$, $|\alpha| = m$ are real-valued functions. Let $\varphi \in C^\infty(\Omega)$ and let us denote by

$$\mathfrak{a}_m(x,\xi,\tau) = p_m(x,\xi + i\tau\nabla\varphi(x)). \tag{15.133}$$

Corollary 15.18 *Let Ω be an open set of $\mathbb{R}^n$, $m \in \mathbb{N}_0$ and let $p_m(x,\xi)$ and $P_m(x,D)$ be as* (15.131) *and* (15.132) *respectively. Let K be a compact subset of Ω. Then there exists C such that we have*

$$\begin{aligned} &\left| \| \widetilde{P}_m(x,D+i\tau\nabla\varphi(x))v\|^2_{L^2(\mathbb{R}^n)} - \Re\left\langle \mathrm{Op}(|\mathfrak{a}_m|^2)v, v\right\rangle_{L^2(\mathbb{R}^n)} \right| \leq \\ &\leq C\tau \|v\|^2_{H^{m-1}_\tau(\mathbb{R}^n)}, \end{aligned} \tag{15.134}$$

for every $v \in C^\infty_K(\mathbb{R}^n)$ and for every $\tau \geq 1$, where $\widetilde{P}_m(x,D+i\tau\nabla\varphi(x))$ is defined by (15.89).

Proof Similarly to the proof of part (b) of Proposition 15.17 we can apply (15.124) and to this aim we compute (Mod $\Sigma^{2m-2}(\Omega)$) the real part of

$$\frac{1}{i}\sum_{j=1}^{n}\left(\overline{\mathfrak{a}^{(j)}_{m,j}}\mathfrak{a}_m + \overline{\mathfrak{a}^{(j)}_m}\mathfrak{a}_{m,j}\right)(x,\xi,\tau). \tag{15.135}$$

We restrict ourselves to considering a significant particular case, as it shows us the essential steps to take in the general case. Therefore we consider the case

$$p_m(x,\xi) = a(x)\xi^\alpha, \tag{15.136}$$

where $|\alpha| = m$ and $a \in C^\infty(\Omega)$, is a real-valued function. We have easily

$$\begin{aligned} &\frac{1}{i}\sum_{j=1}^{n}\left(\overline{\mathfrak{a}^{(j)}_{m,j}}\mathfrak{a}_m + \overline{\mathfrak{a}^{(j)}_m}\mathfrak{a}_{m,j}\right)(x,\xi,\tau) = \\ &= \frac{2}{i}\sum_{j=1}^{n}\alpha_j a_{x_j} a\,(\xi - i\tau\nabla\varphi)^{\alpha-e_j}(\xi + i\tau\nabla\varphi)^\alpha, \quad \mathrm{Mod}\ \Sigma^{2m-2}(\Omega). \end{aligned} \tag{15.137}$$

by using the binomial formula to calculate the real part of the right-hand side of (15.137), we get

$$\begin{aligned} &\Re\left(\frac{1}{i}\sum_{j=1}^{n}\left(\overline{\mathfrak{a}_{m,j}^{(j)}}\mathfrak{a}_m+\overline{\mathfrak{a}_m^{(j)}}\mathfrak{a}_{m,j}\right)\right)=\\ &=(-1)^{m-1}\tau\sum_{k=0}^{m-1}(-1)^k\tau^{2k}q_{2(m-k-1)}(x,\xi),\quad \text{Mod } \Sigma^{2m-2}(\Omega), \end{aligned} \tag{15.138}$$

where $q_{2(m-k-1)}(x,\xi)$, for $k=0,\cdots,m-1$, are polynomials of degree $2(m-k-1)$ w.r.t. the variable ξ and they are independent of τ.

Now we have

$$\begin{aligned} &\left\|\widetilde{P}_m(x,D+i\tau\nabla\varphi(x))v\right\|^2_{L^2(\mathbb{R}^n)}-\Re\left\langle \text{Op}(|\mathfrak{a}_m|^2)v,v\right\rangle_{L^2(\mathbb{R}^n)}=\\ &=\Re\left\langle T(x;D,\tau)v,v\right\rangle_{L^2(\mathbb{R}^n)},\quad \forall v\in C^\infty_K(\Omega), \end{aligned} \tag{15.139}$$

where

$$T(x,D,\tau)=\widetilde{P}_m(x,D+i\tau\nabla\varphi(x))^\star\widetilde{P}_m(x,D+i\tau\nabla\varphi(x))-\text{Op}(|\mathfrak{a}_m|^2)$$

and, denoting by $t(x,\xi,\tau)$ the symbol of $T(x,D,\tau)$ we have by (15.124) and (15.138)

$$t(x,\xi,\tau)=(-1)^{m-1}\tau\sum_{k=0}^{m-1}(-1)^k\tau^{2k}q_{2(m-k-1)}(x,\xi),\quad \text{Mod } \Sigma^{2m-2}(\Omega).$$

Hence we have $t\in\tau\Sigma^{2m-2}(\Omega)$ and by Proposition 15.9 we derive that for every compact subset K of Ω there exists C such that for every $v\in C^\infty_K(\mathbb{R}^n)$ and for every $\tau\geq 1$ the following inequality holds

$$\left|\Re\left\langle T(x;D,\tau)v,v\right\rangle_{L^2(\mathbb{R}^n)}\right|\leq\tau\left\|v\right\|^2_{H^{m-1}_\tau}.$$

Therefore, by (15.139) and (15.4) we obtain (15.134). ■

We conclude this section with a Lemma that will be useful in the sequel

Lemma 15.19 *Let $p\in C^1(\Omega\times\mathbb{R}^n)$ be a complex-valued function. Then, for every $(x,\xi)\in\Omega\times\mathbb{R}^n$ we have*

$$\{\overline{p},p\}(x,\xi)=2i\{\Re p,\Im p\}(x,\xi), \tag{15.140}$$

$$\Im\{\overline{p},p\}(x,\xi)=\frac{1}{i}\{\overline{p},p\}(x,\xi)=2\{\Re p,\Im p\}(x,\xi), \tag{15.141}$$

$$\Im \sum_{j=1}^{n} \overline{p^{(j)}}(x,\xi)p_{,j}(x,\xi) = \{\Re p, \Im p\}(x,\xi). \tag{15.142}$$

Proof Let

$$a = \Re p, \quad b = \Im p.$$

We have

$$\{\overline{p}, p\} = \{a - ib, a + ib\} = \{a, a\} + \{a, ib\} - \{ib, a\} - \{ib, ib\} = 2i\,\{a, b\}\,.$$

The above equality proves (15.140).

Identities (15.141) are trivial consequences of (15.140).

Concerning (15.142), by (15.140) it follows by

$$\begin{aligned}
\Im \sum_{j=1}^{n} \overline{p^{(j)}} p_{,j} &= \frac{1}{2i} \sum_{j=1}^{n} \left(\overline{p^{(j)}} p_{,j} - \overline{\overline{p^{(j)}} p_{,j}} \right) = \\
&= \frac{1}{2i} \sum_{j=1}^{n} \left(\overline{p^{(j)}} p_{,j} - p^{(j)} \overline{p_{,j}} \right) = \\
&= \frac{1}{2i} \{\overline{p}, p\} = \\
&= \{\Re p, \Im p\}\,.
\end{aligned}$$

■

15.5 The Calculation of the Commutator and Its Symbol

In this section we provide a refinement of the Theorem (15.14); the central point of this refinement consists of the calculation of the commutator (15.114) and its symbol.

Let

$$p_m(x,\xi) = \sum_{|\alpha|=m} a_\alpha(x)\xi^\alpha \tag{15.143}$$

be a homogeneous polynomial in the variable ξ, where

$$a_\alpha \in C^\infty(\Omega)\,, \quad \text{for } \alpha \in \mathbb{N}_0^n,\ |\alpha| = m, \tag{15.144}$$

let

$$P_m(x,D) = \mathrm{Op}(p_m)(x,D) = \sum_{|\alpha|=m} a_\alpha(x) D^\alpha. \tag{15.145}$$

Let

$$\varphi \in C^\infty(\Omega)$$

be a real-valued function.

Here and in the sequel we denote by (see (15.88))

$$\mathfrak{a}_m(x,\xi,\tau) := p_m(x,\xi+i\tau\nabla\varphi(x)) = \sum_{|\alpha|=m} a_\alpha(x)\left(\xi+i\tau\nabla\varphi(x)\right)^\alpha. \tag{15.146}$$

Let us recall that (see (15.89))

$$\widetilde{P}_m(x,D+i\tau\nabla\varphi) = \mathrm{Op}(p_m(x,\xi+i\tau\nabla\varphi)). \tag{15.147}$$

Denoting by $\mathfrak{a}_m^{(s)}(x,\xi,\tau)$ and $\mathfrak{a}_m^{(a)}(x,\xi,\tau)$ the symbols of $\widetilde{P}_m^{(s)}(x,D+i\tau\nabla\varphi)$ and $\widetilde{P}_m^{(a)}(x,D+i\tau\nabla\varphi)$ respectively, by Proposition 15.17 we immediately derive

$$\mathfrak{a}_m^{(s)} = \frac{\mathfrak{a}_m+\overline{\mathfrak{a}}_m}{2} = \Re\mathfrak{a}_m + h_{m-1} \tag{15.148}$$

and

$$\mathfrak{a}_m^{(a)} = \frac{\mathfrak{a}_m-\overline{\mathfrak{a}}_m}{2} = i\left(\Im\mathfrak{a}_m + k_{m-1}\right), \tag{15.149}$$

where h_{m-1} and k_{m-1} are real-valued and $h_{m-1}, k_{m-1} \in \Sigma^{m-1}(\Omega)$.

Moreover we define

$$\begin{aligned} &c_{2m-1,\varphi}(x,\xi,\tau) = \\ &= \tau\sum_{j,k=1}^n \partial^2_{x_jx_k}\varphi(x)\, p_m^{(j)}(x,\zeta)\overline{p_m^{(k)}(x,\zeta)} + \\ &+ \Im\left(\sum_{k=1}^n p_{m,k}(x,\zeta)\overline{p_m^{(k)}(x,\zeta)}\right), \quad \forall (x,\xi,\tau)\in\Omega\times\mathbb{R}^n\times\mathbb{R}, \end{aligned} \tag{15.150}$$

where

$$\zeta = \xi + i\tau\nabla\varphi(x), \quad \forall (x,\xi,\tau)\in\Omega\times\mathbb{R}^n\times\mathbb{R}. \tag{15.151}$$

Let us note that

$$c_{2m-1,\varphi}(x,\xi,\tau)=\frac{i}{2}\left\{p_m(x,\xi+i\tau\nabla\varphi(x)),\overline{p_m(x,\xi+i\tau\nabla\varphi(x))}\right\} \tag{15.152}$$

and for any $x\in\Omega$, $c_{2m-1,\varphi}(x,\xi,\tau)$ is a homogeneous polynomial of degree $2m-1$ in the variable (ξ,τ). Moreover, since the hessian matrix

$$\partial^2\varphi(x)=\left\{\partial^2_{x_jx_k}\varphi(x)\right\}_{j,k=1}^n,$$

is real symmetric, the first term on the right-hand side is a real number.

Theorem 15.20 *Let Ω be an open set of $\mathbb{R}^n$. Let $p_m(x,\xi)$ and $P_m(x,D)$ be defined by* (15.143) *and* (15.145) *respectively. Let $\varphi\in C^\infty(\Omega)$ be a real-valued function. Then for every compact subset K of Ω there exists C such that we have*

$$\begin{aligned}&C\|v\|^2_{H^{m-1}_\tau(\mathbb{R}^n)}+C\|P_m(x,D+i\tau\nabla\varphi(x))v\|^2_{L^2(\mathbb{R}^n)}\geq\\&\geq\|\mathrm{Op}(\Re\mathfrak{a}_m)v\|^2_{L^2(\mathbb{R}^n)}+\|\mathrm{Op}(\Im\mathfrak{a}_m)v\|^2_{L^2(\mathbb{R}^n)}+\\&+\Re\left\langle\mathrm{Op}(c_{2m-1,\varphi})v,v\right\rangle_{L^2(\mathbb{R}^n)},\end{aligned} \tag{15.153}$$

for every $v\in C^\infty_K(\Omega)$ and for every $\tau\geq 1$.

Proof By Proposition 15.9, taking into account (15.148) and (15.149) there exists C such that

$$\begin{aligned}&\left\|\widetilde{P}^{(s)}_m(x,D+i\tau\nabla\varphi(x))v\right\|^2_{L^2(\mathbb{R}^n)}\geq\\&\geq\frac{1}{2}\|\mathrm{Op}(\Re\mathfrak{a}_m)v\|^2_{L^2(\mathbb{R}^n)}-C\|v\|^2_{H^{m-1}_\tau(\mathbb{R}^n)}\end{aligned} \tag{15.154}$$

and

$$\begin{aligned}&\left\|\widetilde{P}^{(a)}_m(x,D+i\tau\nabla\varphi(x))v\right\|^2_{L^2(\mathbb{R}^n)}\geq\\&\geq\frac{1}{2}\|\mathrm{Op}(\Im\mathfrak{a}_m)v\|^2_{L^2(\mathbb{R}^n)}-C\|v\|^2_{H^{m-1}_\tau(\mathbb{R}^n)},\end{aligned} \tag{15.155}$$

for every $v\in C^\infty_K(\Omega)$ and for every $\tau\geq 1$.

Now let us prove

Claim The following inequality holds

$$
\begin{aligned}
&\Re\left\langle\left[\widetilde{P}_m^{(s)}(x, D+i\tau\nabla\varphi(x)), \widetilde{P}_m^{(a)}(x, D+i\tau\nabla\varphi(x))\right] v, v\right\rangle_{L^2(\mathbb{R}^n)} \geq \\
&\geq \Re\left\langle\left[\mathrm{Op}(\Re\mathfrak{a}_m), i\,\mathrm{Op}(\Im\mathfrak{a}_m)\right] v, v\right\rangle_{L^2(\mathbb{R}^n)} - C\,\|v\|^2_{H^{m-1}_\tau(\mathbb{R}^n)}
\end{aligned}
\tag{15.156}
$$

for every $v \in C^\infty_K(\Omega)$ and for every $\tau \geq 1$.

Proof of the Claim By (15.148) and (15.149) we have

$$
\begin{aligned}
&\left\langle\left[\widetilde{P}_m^{(s)}(x, D+i\tau\nabla\varphi(x)), \widetilde{P}_m^{(a)}(x, D+i\tau\nabla\varphi(x))\right] v, v\right\rangle_{L^2(\mathbb{R}^n)} = \\
&= \left\langle\left[\mathrm{Op}\left(\Re\mathfrak{a}_m + h_{m-1}\right), i\,\mathrm{Op}\left(\Im\mathfrak{a}_m + k_{m-1}\right)\right] v, v\right\rangle_{L^2(\mathbb{R}^n)} = \\
&= \left\langle\left[\mathrm{Op}\left(\Re\mathfrak{a}_m\right), i\,\mathrm{Op}\left(\Im\mathfrak{a}_m\right)\right] v, v\right\rangle_{L^2(\mathbb{R}^n)} + \\
&+ \left\langle\left[\mathrm{Op}\left(\Re\mathfrak{a}_m\right), i\,\mathrm{Op}\left(\Im k_{m-1}\right)\right] v, v\right\rangle_{L^2(\mathbb{R}^n)} + \\
&+ \left\langle\left[\mathrm{Op}\left(h_{m-1}\right), i\,\mathrm{Op}\left(\Im\mathfrak{a}_m\right)\right] v, v\right\rangle_{L^2(\mathbb{R}^n)} + \\
&+ \left\langle\left[\mathrm{Op}\left(h_{m-1}\right), i\,\mathrm{Op}\left(k_{m-1}\right)\right] v, v\right\rangle_{L^2(\mathbb{R}^n)}.
\end{aligned}
\tag{15.157}
$$

Now by (15.118) we have that the symbol of operators

$$
\left[\mathrm{Op}\left(\Re\mathfrak{a}_m\right), i\,\mathrm{Op}\left(\Im k_{m-1}\right)\right] \quad \text{and} \quad \left[\mathrm{Op}\left(h_{m-1}\right), i\,\mathrm{Op}\left(\Im\mathfrak{a}_m\right)\right],
$$

belong to $\Sigma^{2m-2}(\Omega)$ and the symbol of

$$
\left[\mathrm{Op}\left(h_{m-1}\right), i\,\mathrm{Op}\left(k_{m-1}\right)\right],
$$

belongs to $\Sigma^{2m-3}(\Omega)$. Hence, by (15.62) and (15.157) we get (15.156). **This concludes the proof of the Claim.**

By Theorem 15.14 and by (15.154)–(15.156) we have that for every compact subset K of Ω there exists C such that

$$
\begin{aligned}
&C\,\|v\|^2_{H^{m-1}_\tau(\mathbb{R}^n)} + C\,\|P_m(x, D+i\tau\nabla\varphi)v\|^2_{L^2(\mathbb{R}^n)} \geq \\
&\geq \|\mathrm{Op}(\Re\mathfrak{a}_m)v\|^2_{L^2(\mathbb{R}^n)} + \|\mathrm{Op}(\Im\mathfrak{a}_m)v\|^2_{L^2(\mathbb{R}^n)} + \\
&+ \Re\left\langle\left[\mathrm{Op}(\Re\mathfrak{a}_m), i\,\mathrm{Op}(\Im\mathfrak{a}_m)\right] v, v\right\rangle_{L^2(\mathbb{R}^n)},
\end{aligned}
\tag{15.158}
$$

for every $v \in C^\infty_K(\Omega)$ and for every $\tau \geq 1$.

Now, let us denote by

$$T = \left[\mathrm{Op}(\Re \mathfrak{a}_m), i\, \mathrm{Op}(\Im \mathfrak{a}_m)\right]$$

and by t the symbol of T. Then by (15.118) we have

$$t = \{\Re \mathfrak{a}_m, \Im \mathfrak{a}_m\} + r_{2m-2},$$

where $r_{2m-2} \in \Sigma^{2m-2}(\Omega)$. Hence, by (15.62) we have that for every compact subset K of Ω there exists C such that

$$\begin{aligned}
&\Re \left\langle \left[\mathrm{Op}(\Re \mathfrak{a}_m), i\, \mathrm{Op}(\Im \mathfrak{a}_m)\right] v, v\right\rangle_{L^2(\mathbb{R}^n)} = \\
&= \Re \left\langle \mathrm{Op}\left(\{\Re \mathfrak{a}_m, \Im \mathfrak{a}_m\}\right) v, v\right\rangle_{L^2(\mathbb{R}^n)} + \\
&+ \Re \left\langle \mathrm{Op}\left(r_{2m-2}\right) v, v\right\rangle_{L^2(\mathbb{R}^n)} \geq \\
&\geq \Re \left\langle \mathrm{Op}\left(\{\Re \mathfrak{a}_m, \Im \mathfrak{a}_m\}\right) v, v\right\rangle_{L^2(\mathbb{R}^n)} - C \left\|v\right\|^2_{H^{m-1}_\tau(\mathbb{R}^n)},
\end{aligned} \tag{15.159}$$

for every $v \in C^\infty_K(\Omega)$ and for every $\tau \geq 1$.

By (15.141) and by (15.152) we have

$$\begin{aligned}
&\{\Re \mathfrak{a}_m, \Im \mathfrak{a}_m\}(x, \xi, \tau) = \\
&= \frac{i}{2} \{\mathfrak{a}_m, \overline{\mathfrak{a}}_m\}(x, \xi, \tau) = \\
&= \frac{i}{2} \left\{p_m(x, \xi + i\tau\nabla\varphi), \overline{p_m(x, \xi + i\tau\nabla\varphi)}\right\} = \\
&= c_{2m-1,\varphi}(x, \xi, \tau).
\end{aligned} \tag{15.160}$$

Finally, (15.158), (15.159) and (15.160) give (15.153). ∎

Note Now we provide an outline of the approach we will follow from this point onward to derive some Carleman estimate for elliptic operators and, with some variations, for operators with real coefficients in the principal part.

Roughly anticipating what we will rigorously implement, we will consider an exponential weight φ for which the following property holds

$$\begin{aligned}
&p_m(x, \xi + i\tau\nabla\varphi(x)) = 0 \Rightarrow \\
&\Rightarrow c_{2m-1,\varphi}(x, \xi, \tau) > 0, \quad (\xi, \tau) \in \mathbb{R}^n \times \mathbb{R} \setminus (0, 0),\ \tau > 0.
\end{aligned} \tag{15.161}$$

Therefore, since $c_{2m-1,\varphi}(x, \xi, \tau)$ is a homogeneous polynomial of degree $2m - 1$ w.r.t. (ξ, τ), K is a compact subset of Ω we get

$$\begin{aligned}
&p_m(x, \xi + i\tau\nabla\varphi(x)) = 0 \Rightarrow \\
&\Rightarrow c_{2m-1,\varphi}(x, \xi, \tau) \geq C \left(|\xi|^2 + \tau^2\right)^{m-\frac{1}{2}}, \quad \forall x \in K,
\end{aligned} \tag{15.162}$$

for every $(\xi, \tau) \in \mathbb{R}^{n+1}$, $\tau > 0$, where K a compact subset of Ω.

Next, **roughly speaking**, we use (15.161) when $p_m(x, \xi + i\tau\nabla\varphi(x)) = 0$, and in cases where $p_m(x, \xi + i\tau\nabla\varphi(x)) \neq 0$ (where we cannot use (16.69)), we take advantage of the specific nature of the symbol p_m. In this procedure Lemma 14.24 will play an useful role.

Although it may seem plausible and intuitive that the algebraic properties of the symbols, particularly their sign, can translate into properties of the corresponding differential operators, this step is far from trivial. As we have already mentioned, to handle this delicate step, we will follow the classical path based on localization in the variable x. This method, originally used by Hörmander [56], has been largely supplanted by the more powerful theory of pseudodifferential operators, through which the 'translation' of the algebraic properties of symbols into properties of differential operators is more direct and general. Naturally, there is a price to pay for following the second path, and that is precisely the study of pseudodifferential operators, which is beyond the scope of this book. However, we will provide a brief overview of such pseudodifferential operators at the end of this chapter. On the other hand, the proofs of the Carleman estimates for the operators we consider here (elliptic or with real coefficients in the principal part) can be carried out without any problems using localization in the variable x. Moreover, this procedure is more elementary, allowing the proofs to be fully self-contained without resorting to other theories.

15.6 Carleman Estimate for the Elliptic Operators

Let $m \in \mathbb{N}$. Let Ω be an open set of $\mathbb{R}^n$. Let a_α be complex-valued functions. We recall that the operator

$$P(x, D) = \sum_{|\alpha| \leq m} a_\alpha(x) D^\alpha, \tag{15.163}$$

is elliptic in a point x_0 if

$$p_m(x_0, \xi) = \sum_{|\alpha| = m} a_\alpha(x_0)\xi^\alpha \neq 0, \qquad \forall \xi \in \mathbb{R}^n \setminus \{0\}. \tag{15.164}$$

We also say that $P(x, D)$ is elliptic in Ω if (15.164) holds for every $x_0 \in \Omega$. Let us note that if $a_\alpha \in C^0(\Omega)$, for $|\alpha| = m$, the ellipticity condition for the operator $P(x, D)$ is equivalent to the condition: for every K, compact subset of Ω there exists $\lambda > 0$ such that

$$|p_m(x, \xi)| \geq \lambda |\xi|^m, \qquad \forall \xi \in \mathbb{R}^n, \quad \forall x \in K. \tag{15.165}$$

For the sake of brevity, in the proof of Theorem below, for an open $\omega \subset \Omega$ we will identify $C_0^\infty(\omega)$ with the function space

$$\left\{u \in C_0^\infty(\Omega) : \operatorname{supp} u \subset \omega\right\}.$$

Theorem 15.21 *Let $\varphi \in C^\infty(\Omega)$ be a real-valued function which satisfies*

$$\nabla\varphi(x) \neq 0, \qquad \forall x \in \Omega. \tag{15.166}$$

Let $P(x, D)$ be an operator of order m whose coefficients belong to $L^\infty(\Omega)$. Let K be a compact subset of Ω and assume that the coefficients of the principal part $P_m(x, D)$ belong to $C^\infty(\Omega)$. Suppose that $P(x, D)$ satisfies the ellipticity condition (15.165)*. Moreover we assume that the following condition is satisfied:*

$$\left.\begin{array}{l} p_m(x, \xi + i\sigma\nabla\varphi(x)) = 0 \\ (\xi, \sigma) \neq (0; 0) \\ x \in \Omega \end{array}\right\} \Longrightarrow \omega_{2m-2,\varphi}(x, \xi, \sigma) > 0; \tag{15.167}$$

where

$$\omega_{2m-2,\varphi}(x, \xi, \sigma) = \frac{1}{\sigma} c_{2m-1,\varphi}(x, \xi, \sigma) =$$
$$= \frac{i}{2\sigma}\left\{p_m(x, \xi + i\sigma\nabla\varphi(x)), \overline{p_m(x, \xi + i\sigma\nabla\varphi(x))}\right\},$$

and $\{\cdot, \cdot\}$ is the Poisson bracket defined in (15.3)*.*

Then there exist constants C and τ_0 such that

$$\sum_{|\alpha|\leq m} \tau^{2(m-|\alpha|)-1} \int_\Omega \left|D^\alpha u\right|^2 e^{2\tau\varphi} dx \leq C \int_\Omega |P(x, D)u|^2 e^{2\tau\varphi} dx, \tag{15.168}$$

for every $u \in C_K^\infty(\Omega)$ and for every $\tau \geq \tau_0$.

Remark 15.4 Let us observe that the requirement $(\xi, \sigma) \neq (0, 0)$ in the antecedent of (15.167) is equivalent to requiring that both ξ and σ are nonzero. In fact, if

$$\xi = 0,$$

then, by

$$p_m(x, \xi + i\sigma\nabla\varphi(x)) = 0,$$

we have

$$(i\sigma)^m p_m(x, \nabla\varphi(x)) = 0.$$

In addition, since

$$p_m(x, \nabla\varphi(x)) \neq 0,$$

as $P_m(x, D)$ is elliptic and $\nabla\varphi(x) \neq 0$, we have $\sigma = 0$. Similarly, if $\sigma = 0$ by the ellipticity of $P_m(x, D)$ we have $\xi = 0$. ♦

Remark 15.5 If the coefficients of $P_m(x, D)$ are constants (let us rename it $P_m(D)$), condition (15.167) becomes:

If

$$\begin{cases} p_m(\xi + i\sigma\nabla\varphi(x)) = 0 \\ \\ (\xi, \sigma) \in \mathbb{R}^{n+1} \setminus \{(0,0)\} \end{cases},$$

then

$$\sum_{j,k=1}^{n} \partial^2_{x_j x_k}\varphi(x) p_m^{(j)}(\xi + i\tau\nabla\varphi(x))\overline{p_m^{(k)}(\xi + i\tau\nabla\varphi(x))} > 0.$$

In particular, if the Hessian matrix of φ is positive definite then the condition above is satisfied. ♦

In order to prove Theorem 15.21 we need some Lemma and Proposition that we state and prove below.

Lemma 15.22 *Let Ω, φ and $P(x, D)$ be as in Theorem 15.21. Assume that* (15.165) *and* (15.167) *are satisfied. Then, for every compact subset K of Ω, there exist two positive constants C_1 and C_2 such that for every $(x, \xi, \sigma) \in K \times \mathbb{R}^n \times \mathbb{R}$ we have*

$$\begin{aligned} &C_1 \left|\xi + i\sigma\nabla\varphi(x)\right|^{2m} \leq \\ &\leq \sigma c_{2m-1,\varphi}(x, \xi, \sigma) + C_2 \left|p_m\left(x, \xi + i\sigma\nabla\varphi(x)\right)\right|^2. \end{aligned} \tag{15.169}$$

Proof Let K be a (nonempty) compact subset of Ω and let us fix x^* in K. Let us denote by

$$N_{x^*} = \nabla\varphi\left(x^*\right),$$

$$\mathbf{S}_{x^*} = \left\{(\xi, \sigma) \in \mathbb{R}^n \times \mathbb{R} : \left|\xi + i\sigma N_{x^*}\right| = 1\right\}.$$

First let us prove the following

There exist two positive constants C_{1,x^*} and C_{2,x^*} such that

$$C_{1,x^*} \leq \mu c_{2m-1,\varphi}\left(x^*, \eta, \mu\right) + C_{2,x^*}\left|p_m\left(x^*, \eta + i\mu N_{x^*}\right)\right|^2, \tag{15.170}$$

for every $(\eta, \mu) \in \mathbf{S}_{x^*}$.

Proof of the Claim By (15.165), we have

$$\left(\mu c_{2m-1,\varphi}(x^*,\eta,\mu)+\left|p_m\left(x^*,\eta+i\mu N_{x^*}\right)\right|^2\right)_{|\mu=0}\geq\lambda,\quad\forall\eta,\ |\eta|=1.$$

By the compactness of $\mathbf{S}_{x^*}$ there exists $\mu_0>0$ such that

$$\mu c_{2m-1,\varphi}(x^*,\eta,\mu)+\left|p_m\left(x^*,\eta+i\mu N_{x^*}\right)\right|^2\geq\frac{\lambda}{2},\tag{15.171}$$

for every $(\eta,\mu)\in\mathbf{S}_{x^*}\cap\{|\mu|\leq\mu_0\}$.

Let us denote by $\mathcal{X}$ the compact set

$$\mathcal{X}=\mathbf{S}_{x^*}\cap\{|\mu|\geq\mu_0\}$$

(of course, if $\mathcal{X}=\emptyset$ the proof would be concluded). Taking into account Remark 15.4, (15.167) gives

$$\left|p_m\left(x^*,\eta+i\mu N_{x^*}\right)\right|=0,\ (\eta,\mu)\in\mathcal{X}\Longrightarrow\mu c_{2m-1,\varphi}(x^*,\eta,\mu)>0,$$

by Lemma 14.24 there exists $C_{x^*}>0$ such that

$$\mu c_{2m-1,\varphi}(x^*,\eta,\mu)+C_{x^*}\left|p_m\left(x^*,\eta+i\mu N_{x^*}\right)\right|^2>0,\quad\forall(\eta,\mu)\in\mathcal{K}.\tag{15.172}$$

By (15.171) and (15.172) we have

$$\mu c_{2m-1,\varphi}(x^*,\eta,\mu)+(C_{x^*}+1)\left|p_m\left(x^*,\eta+i\mu N_{x^*}\right)\right|^2>0,\tag{15.173}$$

for every $(\eta,\mu)\in S_{x^*}$. Hence (15.170) follows with

$$C_{1,x^*}=\min_{(\eta,\mu)\in\mathbf{S}_{x^*}}\left(\mu c_{2m-1,\varphi}(x^*,\eta,\mu)+C\left|p_m\left(x^*,\eta+i\mu N_{x^*}\right)\right|^2\right)$$

and

$$C_{2,x^*}=C_{x^*}+1.$$

This concludes the proof of the Claim.

Let $\varepsilon\in(0,1)$. Now taking into account that $\nabla\varphi(x)\neq0$ implies

$$0<\min_{x\in K}|N_x|\leq\max_{x\in K}|N_x|<+\infty,$$

we have that for any x^* there exists $\delta_{x^*,\varepsilon}$ such that

$$|N_x-N_{x^*}|<\varepsilon|N_x|,\ \forall x\in B_{\delta_{x^*,\varepsilon}}\left(x^*\right)\cap K\tag{15.174}$$

and by (15.170), we have

$$\frac{1}{2}C_{1,x^*} \le \mu c_{2m-1,\varphi}(x,\eta,\mu) + C_{2,x^*}\left|p_m\left(x,\eta+i\mu N_{x^*}\right)\right|^2, \tag{15.175}$$

for every $x \in B_{\delta_{x^*,\varepsilon}}(x^*) \cap K$ and for every $(\eta,\mu) \in \mathbf{S}_{x^*}$.

Let us denote by

$$F_{x^*}(x,\xi,\tau) = \tau c_{2m-1,\varphi}(x,\xi,\tau) + C_{2,x^*}\left|p_m\left(x,\xi+i\tau N_x\right)\right|^2, \tag{15.176}$$

for $(\xi,\tau) \in \mathbb{R}^{n+1}$.

Taking into account the homogeneity of $F_{x^*}(x,\xi,\tau)$ w.r.t. (ξ,τ), by (15.175) we have

$$F_{x^*}(x,\xi,\tau) \ge \left(\frac{1}{2}C_{1,x^*} - C_{3,x^*}\varepsilon\right)\left(|\xi|^2+\tau^2|N_x|\right)^m, \tag{15.177}$$

for every $x \in B_{\delta_{x^*,\varepsilon}}(x^*) \cap K$ and for every $(\xi,\tau) \in \mathbb{R}^{n+1}$, where C_{3,x^*} is a constant that depends on x^*.

Let

$$\varepsilon = \varepsilon^* := \frac{C_{1,x^*}}{4C_{3,x^*}}$$

and set

$$\delta_{x^*} := \delta_{x^*,\varepsilon^*}.$$

By (15.177) we have

$$F_{x^*}(x,\xi,\tau) \ge \frac{1}{4}C_{1,x^*}\left(|\xi|^2+\tau^2|N_x|\right)^m, \tag{15.178}$$

for every $x \in B_{\delta_{x^*}}(x^*) \cap K$ and for every $(\xi,\tau) \in \mathbb{R}^{n+1}$.

Since K is a compact set of $\mathbb{R}^n$ and $\left\{B_{\delta_{x^*}}(x^*)\right\}_{x^*\in K}$ is an open covering of K, there exists a finite subcovering of K, $\left\{B_{\delta_{x_j}}(x_j)\right\}_{j=1}^N$. Hence denoting by

$$C_1 = \frac{1}{4}\min_{1\le j\le N} C_{1,x_j}, \quad C_2 = \max_{1\le j\le N} C_{2,x_j},$$

by (15.176) and (15.178) we get (15.169). ■

Remark 15.6 In the Claim of the proof of Lemma 15.22 we have proven that if condition (15.167) holds then (15.170) holds. Actually, the reverse is also true. More precisely, if x^* is a point in Ω where (15.170) holds, then (15.167) holds for $x = x^*$. Hence, for $x = x^*$ condition (15.167) and (15.170) are equivalent.

Moreover, if condition (15.167) holds for $x = x^* \in \Omega$, then there exists a neighborhood, $\mathcal{U}$, of x^* such that for every $x \in \mathcal{U}$ (15.167) holds. As a matter of fact, assuming condition (15.167) holds then (15.170) holds. Hence, proceeding as in the proof of Lemma 15.22 there exist $\widetilde{C}_{1,x^*}$, $\widetilde{C}_{2,x^*}$ and δ_* positive numbers such that

$$C_{1,x^*} \leq \mu c_{2m-1,\varphi}(x, \eta, \mu) + C_{2,x^*} \left|p_m(x, \eta + i\mu N_x)\right|^2, \tag{15.179}$$

for every $x \in B_{\delta_*}(x^*)$ and for every $(\eta, \mu) \in \mathbf{S}_x$. In turns, (15.179) implies (15.167) for every $x \in B_{\delta_*}(x^*)$. ♦

Proof of Theorem 15.21 By (15.153) we have trivially there exists C such that we have

$$\begin{aligned} &C\tau \|v\|^2_{H^{m-1}_\tau(\mathbb{R}^n)} + C\tau \|P_m(x, D + i\tau\nabla\varphi(x))v\|^2_{L^2(\mathbb{R}^n)} \geq \\ &\geq \tau \Re \left\langle \mathrm{Op}(c_{2m-1,\varphi})v, v\right\rangle_{L^2(\mathbb{R}^n)}, \end{aligned} \tag{15.180}$$

for every $v \in C^\infty_K(\Omega)$ and for every $\tau \geq 1$.

Let now $x_0 \in K$ be a fixed point. We may assume $x_0 = 0 \in K$. We apply Theorem 15.10 to

$$q(x, \xi, \tau) = \tau c_{2m-1,\varphi}(x, \xi, \tau) + C_2 \left|p_m(x, \xi + i\tau\nabla\varphi(x))\right|^2, \tag{15.181}$$

$$Q(x, D, \tau) = \mathrm{Op}(q)(x, D, \tau). \tag{15.182}$$

To this purpose we denote by

$$C_3 = \min\left\{\frac{C_1\gamma^2}{2(2\pi)^n}, 1\right\},$$

where

$$\gamma = \min_{x\in K} |\nabla\varphi(x)|$$

and C_1 is introduced in Lemma 15.22. By such a Lemma 15.22 and by the above mentioned Theorem 15.10, there exists $r > 0$ such that for every $\tau \geq 1$ and for every $v \in C^\infty_0(B_r)$, $B_r \Subset \Omega$, we have

$$\Re \left\langle Q(x, D, \tau)v, v\right\rangle_{L^2(\mathbb{R}^n)} \geq C_3 \|v\|^2_{H^m_\tau(\mathbb{R}^n)} - C \|v\|^2_{H^{m-1}_\tau(\mathbb{R}^n)}. \tag{15.183}$$

By (15.125) we have

$$\begin{aligned}&\left|\Re\left\langle \mathrm{Op}(|p_m(x,\xi+i\tau\nabla\varphi(x))|^2)v,v\right\rangle_{L^2(\mathbb{R}^n)}\right|\leq\\&\leq\|P_m(x,D+i\tau\nabla\varphi(x))v\|^2_{L^2(\mathbb{R}^n)}+C\|v\|^2_{H^{m-\frac{1}{2}}_\tau(\mathbb{R}^n)}.\end{aligned}\tag{15.184}$$

Hence by (15.181), (15.183), and (15.184) we get

$$\begin{aligned}&\tau\Re\left\langle \mathrm{Op}(c_{2m-1,\varphi})v,v\right\rangle_{L^2(\mathbb{R}^n)}=\Re\left\langle Q(x,D,\tau)v,v\right\rangle_{L^2(\mathbb{R}^n)}-\\&-C_2\Re\left\langle \mathrm{Op}(|p_m(x,\xi+i\tau\nabla\varphi(x))|^2)v,v\right\rangle_{L^2(\mathbb{R}^n)}\geq\\&\geq C_3\|v\|^2_{H^m_\tau(\mathbb{R}^n)}-C\|v\|^2_{H^{m-1}_\tau(\mathbb{R}^n)}-\\&-C_2\|P_m(x,D+i\tau\nabla\varphi(x))v\|^2_{L^2(\mathbb{R}^n)}-CC_2\|v\|^2_{H^{m-\frac{1}{2}}_\tau(\mathbb{R}^n)}\geq\\&\geq\frac{C_3}{2}\|v\|^2_{H^m_\tau(\mathbb{R}^n)}-C_2\|P_m(x,D+i\tau\nabla\varphi)v\|^2_{L^2(\mathbb{R}^n)}-\\&-\widetilde{C}\|v\|^2_{H^{m-1}_\tau(\mathbb{R}^n)},\end{aligned}\tag{15.185}$$

for every $\tau\geq 1$ and for every $v\in C^\infty_0(B_r)$, $B_r\Subset\Omega$ (concerning the estimate of $\|v\|^2_{H^{m-\frac{1}{2}}_\tau(\mathbb{R}^n)}$ we proceed as in (15.75)).

By (15.180) and (15.185) we have

$$\begin{aligned}&C_4\tau\|v\|^2_{H^{m-1}_\tau(\mathbb{R}^n)}+C_4\tau\|P_m(x,D+i\tau\nabla\varphi(x))v\|^2_{L^2(\mathbb{R}^n)}\geq\\&\geq\tau\Re\left\langle \mathrm{Op}(c_{2m-1,\varphi})v,v\right\rangle_{L^2(\mathbb{R}^n)}\geq\\&\geq\frac{C_3}{2}\|v\|^2_{H^m_\tau(\mathbb{R}^n)},\end{aligned}\tag{15.186}$$

where $C_4=\max\left\{C+\widetilde{C},C+C_2\right\}$, for every $\tau\geq 1$ and for every $v\in C^\infty_0(B_r)$, $B_r\Subset\Omega$. Now, let us recall that

$$\tau\|v\|^2_{H^{m-1}_\tau(\mathbb{R}^n)}\leq\tau^{-1}\|v\|^2_{H^m_\tau(\mathbb{R}^n)}.$$

By this inequality and by (15.186) we have trivially

$$C_4\tau\|P_m(x,D+i\tau\nabla\varphi(x))v\|^2_{L^2(\mathbb{R}^n)}\geq\left(\frac{C_3}{2}-\tau^{-1}C_4\right)\|v\|^2_{H^m_\tau(\mathbb{R}^n)}.\tag{15.187}$$

Now let

$$\tau_* = \frac{4C_4}{C_3}.$$

By (15.187) we obtain

$$\tau^{-1} \|v\|^2_{H^m_\tau(\mathbb{R}^n)} \leq 4C_4 \|P_m(x, D + i\tau\nabla\varphi(x))v\|^2_{L^2(\mathbb{R}^n)}, \tag{15.188}$$

for every $\tau \geq \tau_*$ and for every $v \in C^\infty_0(B_r)$, $B_r \Subset \Omega$.

Now, by Remark 15.1 following Proposition 15.13 we have that for every $x_0 \in K$ there exist C_{x_0}, r_{x_0} and τ_{x_0} such that

$$\sum_{|\alpha|\leq m} \tau^{2(m-|\alpha|)-1} \int_\Omega |D^\alpha u|^2 e^{2\tau\varphi} dx \leq C_{x_0} \int_\Omega |P(x, D)u|^2 e^{2\tau\varphi} dx, \tag{15.189}$$

for every $u \in C^\infty_0(\Omega)$, $\operatorname{supp} u \subset B_{r_{x_0}}(x_0) \Subset \Omega$ and for every $\tau \geq \tau_{x_0}$.

Finally, by Lemma 14.2 we obtain estimate (15.168). ■

Remark 15.7 Let $P(x, D)$ be an elliptic operator in Ω, open set of $\mathbb{R}^n$. We observe that condition (15.167) of Theorem 15.21 is equivalent to Carleman estimate (15.168). Indeed, in Theorem 15.21 we have proved that condition (15.167) implies that for every K, compact subset of Ω, there exists C and τ_0 such that

$$\sum_{|\alpha|\leq m} \tau^{2(m-|\alpha|)-1} \int_\Omega |D^\alpha u|^2 e^{2\tau\varphi} dx \leq C \int_\Omega |P(x, D)u|^2 e^{2\tau\varphi} dx, \tag{15.190}$$

for every $u \in C^\infty_K(\Omega)$ and for every $\tau \geq \tau_0$. On the other side, if (15.190) holds, then from the arbitrariness of K, by virtue of Theorem 14.27 we have that condition (15.167) holds (take into account Remark 15.4 of this section).

We also observe, due to the previously noted equivalence, that if for every K, compact subset of Ω, there exists C and τ_0 such that

$$\tau \sum_{|\alpha|\leq m-1} \int_\Omega |D^\alpha u|^2 e^{2\tau\varphi} dx \leq C \int_\Omega |P(x, D)u|^2 e^{2\tau\varphi} dx, \tag{15.191}$$

for every $u \in C^\infty_K(\Omega)$ and for every $\tau \geq \tau_0$, then (15.190) holds. ♦

15.7 Carleman Estimate for Operators with Real-Valued Coefficients in the Principal Part

In the present section we provide the proof of a Carleman estimate for operators with real-valued and C^∞ coefficients in the principal part. We will use the same notations as in Sects. 15.5 and 15.6, except that the coefficients of the homogeneous polynomial in the variable ξ

$$p_m(x,\xi) = \sum_{|\alpha|=m} a_\alpha(x)\xi^\alpha, \tag{15.192}$$

and, consequently, the coefficients of the differential operator

$$P_m(x,D) = \mathrm{Op}(p_m)(x,D) = \sum_{|\alpha|=m} a_\alpha(x)D^\alpha, \tag{15.193}$$

are real-valued functions.

As in Sect. 15.6 we assume that

$$\varphi \in C^\infty(\Omega),$$

is a real-valued function.

We begin by providing, in Theorem 15.23 below, a modified version of Theorem 15.20.

Theorem 15.23 *Let Ω be an open set of $\mathbb{R}^n$. Let K be a compact subset of Ω. Let $p_m(x,\xi)$ and $P_m(x,D)$ be defined by (15.192) and (15.193) respectively. Let $\psi \in C^\infty(\Omega)$ be a real-valued function. Then there exists C such that*

$$\begin{aligned} &C\|v\|^2_{H^{m-1}_\tau(\mathbb{R}^n)} + C\left\|e^\psi P_m(x,D+i\tau\nabla\varphi(x))v\right\|^2_{L^2(\mathbb{R}^n)} \geq \\ &\geq \left\|e^\psi \mathrm{Op}(\Re\mathfrak{a}_m)v\right\|^2_{L^2(\mathbb{R}^n)} + \left\|e^\psi \mathrm{Op}(\Im\mathfrak{a}_m)v\right\|^2_{L^2(\mathbb{R}^n)} + \\ &+ \Re\left\langle \mathrm{Op}(\widetilde{c}_{2m-1,\varphi,\psi})v, v\right\rangle_{L^2(\mathbb{R}^n)}, \end{aligned} \tag{15.194}$$

for every $v \in C^\infty_K(\Omega)$ and for every $\tau \geq 1$, where

$$\begin{aligned} &\widetilde{c}_{2m-1,\varphi,\psi}(x,\xi,\tau) = e^{2\psi}c_{2m-1,\varphi}(x,\xi,\tau) + \\ &+ 2e^{2\psi}\Im\sum_{j=1}^n p_m(x,\xi+i\tau\nabla\varphi(x))\overline{p_m^{(j)}(x,\xi+i\tau\nabla\varphi(x))} \end{aligned} \tag{15.195}$$

and $c_{2m-1,\varphi}(x,\xi,\tau)$ is defined by (15.152).

Moreover, we recall (see (15.146) *)*

$$\mathfrak{a}_m(x,\xi,\tau) = p_m(x,\xi+i\tau\nabla\varphi(x)) = \sum_{|\alpha|=m} a_\alpha(x)\,(\xi+i\tau\nabla\varphi(x))^\alpha\,. \tag{15.196}$$

Proof Similarly to the proof of Theorem 15.20 we have

$$\begin{aligned} &C\,\|v\|^2_{H^{m-1}_\tau(\mathbb{R}^n)} + C\,\left\|e^\psi P_m(x,D+i\tau\nabla\varphi(x))v\right\|^2_{L^2(\mathbb{R}^n)} \geq \\ &\geq \left\|e^\psi\,\mathrm{Op}(\mathfrak{a}_m)v\right\|^2_{L^2(\mathbb{R}^n)}, \end{aligned} \tag{15.197}$$

for every $v \in C^\infty_K(\Omega)$ and for every $\tau \geq 1$.

Now, we have

$$\begin{aligned} &\left\|e^\psi\,\mathrm{Op}(\mathfrak{a}_m)v\right\|^2_{L^2(\mathbb{R}^n)} = \\ &= \left\|e^\psi\,\mathrm{Op}(\mathfrak{a}^{(s)}_m)v\right\|^2_{L^2(\mathbb{R}^n)} + \left\|e^\psi\,\mathrm{Op}(\mathfrak{a}^{(a)}_m)v\right\|^2_{L^2(\mathbb{R}^n)} + \\ &+ 2\Re\left\langle e^{2\psi}\,\mathrm{Op}(\mathfrak{a}^{(s)}_m)v,\,\mathrm{Op}(\mathfrak{a}^{(a)}_m)v\right\rangle_{L^2(\mathbb{R}^n)} = \\ &= \left\langle e^{2\psi}\,\mathrm{Op}(\mathfrak{a}^{(s)}_m)v,\,\mathrm{Op}(\mathfrak{a}^{(a)}_m)v\right\rangle_{L^2(\mathbb{R}^n)} + \\ &+ \left\langle e^{2\psi}\,\mathrm{Op}(\mathfrak{a}^{(a)}_m)v,\,\mathrm{Op}(\mathfrak{a}^{(s)}_m)v\right\rangle_{L^2(\mathbb{R}^n)} = \\ &= \Re\,\langle (Q_1 - Q_2)\,v,\,v\rangle_{L^2(\mathbb{R}^n)}, \end{aligned} \tag{15.198}$$

where

$$Q_1 := Q_1(x,D,\tau) = \mathrm{Op}(\mathfrak{a}^{(s)}_m)\left(e^{2\psi}\,\mathrm{Op}(\mathfrak{a}^{(a)}_m)\right) \tag{15.199}$$

and

$$Q_2 := Q_2(x,D,\tau) = \mathrm{Op}(\mathfrak{a}^{(a)}_m)\left(e^{2\psi}\,\mathrm{Op}(\mathfrak{a}^{(s)}_m)\right). \tag{15.200}$$

Let us denote by $q_j(x,\xi,\tau)$ the symbol of $Q_j(x,D,\tau)$, for $j = 1, 2$.

Claim The following relation holds

$$(q_1 - q_2)\,(x,\xi,\tau) = \widetilde{c}_{2m-1,\varphi,\psi}(x,\xi,\tau), \quad \mathrm{Mod}\ \Sigma^{2m-2}(\Omega), \tag{15.201}$$

where $\widetilde{c}_{2m-1,\varphi,\psi}(x,\xi,\tau)$ is defined by (15.195).

Proof of the Claim By (15.117) we have (for the sake of brevity we omit the variable (x, ξ, τ))

$$q_1 = e^{2\psi} \mathfrak{a}_m^{(s)} \mathfrak{a}_m^{(a)} + \frac{1}{i} \sum_{j=1}^{n} \left(\partial_{\xi_j} \mathfrak{a}_m^{(s)} \right) \partial_{x_j} \left(e^{2\psi} \mathfrak{a}_m^{(a)} \right) =$$

$$= e^{2\psi} \left(\mathfrak{a}_m^{(s)} \mathfrak{a}_m^{(a)} + \frac{1}{i} \sum_{j=1}^{n} \left(\partial_{\xi_j} \mathfrak{a}_m^{(s)} \right) \partial_{x_j} \left(\mathfrak{a}_m^{(a)} \right) + \right.$$

$$\left. + \frac{2}{i} \mathfrak{a}_m^{(a)} \sum_{j=1}^{n} \partial_{\xi_j} \mathfrak{a}_m^{(s)} \psi_{x_j} \right), \qquad \text{Mod } \Sigma^{2m-2}(\Omega).$$

Similarly we have

$$q_2 = e^{2\psi} \left(\mathfrak{a}_m^{(a)} \mathfrak{a}_m^{(s)} + \frac{1}{i} \sum_{j=1}^{n} \left(\partial_{\xi_j} \mathfrak{a}_m^{(a)} \right) \partial_{x_j} \left(\mathfrak{a}_m^{(s)} \right) + \right.$$

$$\left. + \frac{2}{i} \mathfrak{a}_m^{(s)} \sum_{j=1}^{n} \partial_{\xi_j} \mathfrak{a}_m^{(a)} \psi_{x_j} \right), \qquad \text{Mod } \Sigma^{2m-2}(\Omega).$$

Hence

$$\begin{aligned} q_1 - q_2 = & \frac{e^{2\psi}}{i} \left\{ \mathfrak{a}_m^{(s)}, \mathfrak{a}_m^{(a)} \right\} + \\ & + \frac{2e^{2\psi}}{i} \sum_{j=1}^{n} \left(\mathfrak{a}_m^{(a)} \partial_{\xi_j} \mathfrak{a}_m^{(s)} - \mathfrak{a}_m^{(s)} \partial_{\xi_j} \mathfrak{a}_m^{(a)} \right) \psi_{x_j}, \quad \text{Mod } \Sigma^{2m-2}(\Omega). \end{aligned} \tag{15.202}$$

By (15.148) and (15.149) (proceeding as in the proof of Theorem 15.20) we have

$$\begin{aligned} & \frac{e^{2\psi}}{i} \left\{ \mathfrak{a}_m^{(s)}, \mathfrak{a}_m^{(a)} \right\} = e^{2\psi} \left\{ \Re \mathfrak{a}_m, \Im \mathfrak{a}_m \right\} = \\ & = \frac{ie^{2\psi}}{2} \left\{ \mathfrak{a}_m, \overline{\mathfrak{a}}_m \right\} = \\ & = \frac{ie^{2\psi}}{2} \left\{ p_m(x, \xi + i\tau\nabla\varphi(x)), \overline{p_m(x, \xi + i\tau\nabla\varphi(x))} \right\} = \\ & = e^{2\psi} c_{2m-1,\varphi}(x, \xi, \tau), \qquad \text{Mod } \Sigma^{2m-2}(\Omega). \end{aligned} \tag{15.203}$$

Again, by (15.148) and (15.149) we have easily

$$\frac{2e^{2\psi}}{i}\sum_{j=1}^{n}\left(\mathfrak{a}_m^{(a)}\partial_{\xi_j}\mathfrak{a}_m^{(s)}-\mathfrak{a}_m^{(s)}\partial_{\xi_j}\mathfrak{a}_m^{(a)}\right)\partial_{x_j}\psi=$$

$$=2e^{2\psi}\Im\sum_{j=1}^{n}\mathfrak{a}_m\left(\partial_{\xi_j}\overline{\mathfrak{a}}_m\right)\partial_{x_j}\psi= \tag{15.204}$$

$$=2e^{2\psi}\Im\sum_{j=1}^{n}p_m(x,\zeta)\overline{p_m^{(j)}(x,\zeta)}\partial_{x_j}\psi,\quad \text{Mod }\Sigma^{2m-2}(\Omega),$$

where

$$\zeta=\xi+i\tau\nabla\varphi(x).$$

By (15.203) and (15.204) we get (15.201). **This concludes the proof of the Claim.**

By (15.62), (15.197), (15.198), and (15.201) we obtain (15.194). ■

Theorem 15.24 *Let $\varphi\in C^{\infty}(\Omega)$ be a real-valued function which satisfies*

$$\nabla\varphi(x)\neq 0,\qquad \forall x\in\Omega. \tag{15.205}$$

Let $P(x,D)$ be an operator of order m whose coefficients belong to $L^{\infty}(\Omega)$. Assume that the coefficients of the principal part $P_m(x,D)$ belong to $C^{\infty}(\Omega)$ and they are real-valued functions. Suppose that the following conditions are satisfied:

$$\left.\begin{array}{ll} p_m(x,\xi) & =0\\ \{p_m,\varphi\}(x,\xi) & =0\\ \xi\neq 0 & \\ x\in\Omega & \end{array}\right\}\Longrightarrow \{p_m,\{p_m,\varphi\}\}(x,\xi)>0; \tag{15.206}$$

and

$$\left.\begin{array}{l} p_m(x,\xi+i\sigma\nabla\varphi(x))=0\\ \sigma\neq 0\\ x\in\Omega \end{array}\right\}\Longrightarrow \omega_{2m-2,\varphi}(x,\xi,\sigma)>0; \tag{15.207}$$

where

$$\omega_{2m-2,\varphi}(x,\xi,\sigma)=\frac{1}{\sigma}c_{2m-1,\varphi}(x,\xi,\sigma)=$$
$$=\frac{i}{2\sigma}\left\{p_m(x,\xi+i\sigma\nabla\varphi(x)),\overline{p_m(x,\xi+i\sigma\nabla\varphi(x))}\right\}. \tag{15.208}$$

Then for every compact subset K of Ω there exist constants C and τ_0 such that

$$\sum_{|\alpha|\leq m-1} \tau^{2(m-|\alpha|)-1} \int_\Omega \left|D^\alpha u\right|^2 e^{2\tau\varphi} dx \leq C \int_\Omega |P(x,D)u|^2 e^{2\tau\varphi} dx, \tag{15.209}$$

for every $u \in C_K^\infty(\Omega)$ and for every $\tau \geq \tau_0$.

In preparation for the proof of Theorem 15.24, we need some lemmas, particularly Lemma 15.25, in which the hypotheses of Theorem 15.24 will be appropriately clarified.

We denote by

$$\widetilde{\omega}_{2m-2,\varphi,\psi}(x,\xi,\sigma) = \frac{1}{\sigma}\widetilde{c}_{2m-1,\varphi,\psi}(x,\xi,\sigma), \tag{15.210}$$

where $\widetilde{c}_{2m-1,\varphi,\psi}(x,\xi,\sigma)$ is defined by (15.195)

Let us note that

$$\begin{aligned} &\{p_m, \{p_m, \varphi\}\}(x,\xi) = \\ &= \sum_{j,k=1}^n \partial^2_{x_j x_k}\varphi(x) p_m^{(j)}(x,\xi) p_m^{(k)}(x,\xi) + \\ &+ \sum_{j,k=1}^n \left(p_m^{(j)}(x,\xi) p_{m,j}^{(k)}(x,\xi) - p_{m,j}(x,\xi) p_m^{(jk)}(x,\xi)\right) \partial_{x_k}\varphi(x). \end{aligned} \tag{15.211}$$

Lemma 15.25 *Let Ω, φ and $P(x,D)$ be as in Theorem 15.24. Let $\omega_{2m-2,\varphi}(x,\xi,\sigma)$ and $\widetilde{\omega}_{2m-2,\varphi,\psi}(x,\xi,\sigma)$ be defined by* (15.208) *and* (15.210)*, respectively. Then we have*

$$\lim_{\sigma\to 0} \omega_{2m-2,\varphi}(x,\xi,\sigma) = \{p_m, \{p_m, \varphi\}\}(x,\xi), \quad \forall (x,\xi) \in \Omega \times \mathbb{R}^n, \tag{15.212}$$

and

$$\begin{aligned} &\lim_{\sigma\to 0} \widetilde{\omega}_{2m-2,\varphi,\psi}(x,\xi,\sigma) = e^{2\psi(x)} \{p_m, \{p_m, \varphi\}\}(x,\xi) + \\ &+ 2e^{2\psi(x)} \left(\{p_m, \varphi\}\{p_m, \psi\}\right)(x,\xi), \quad \forall (x,\xi) \in \Omega \times \mathbb{R}^n. \end{aligned} \tag{15.213}$$

Proof By (15.208), taking into account that the coefficients of $p_m(x,\xi)$ are real-valued functions and applying the Taylor formula w.r.t. variable σ, we have

$$\begin{aligned}
&\omega_{2m-2}(x,\xi,\sigma) = \\
&= \frac{i}{2\sigma}\left\{p_m(x,\xi+i\sigma\nabla\varphi(x)), \overline{p_m(x,\xi+i\sigma\nabla\varphi(x))}\right\} = \\
&= \frac{i}{2\sigma}\left\{\{p_m + i\sigma\{p_m,\varphi\} + o(\sigma), p_m - i\sigma\{p_m,\varphi\} + o(\sigma)\right\}(x,\xi) = \\
&= \{p_m,\{p_m,\varphi\}\}(x,\xi) + o(1), \qquad \text{as } \sigma \to 0.
\end{aligned} \tag{15.214}$$

From this relation (15.212) immediately follows.

Concerning the proof of (15.213), it is sufficient to observe what follows. By (15.210) we have

$$\begin{aligned}
&\widetilde{\omega}_{2m-2,\varphi,\psi}(x,\xi,\sigma) = e^{2\psi(x)}\omega_{2m-2,\varphi}(x,\xi,\sigma) + \\
&+ \frac{2}{\sigma}e^{2\psi(x)}\Im\sum_{j=1}^{n} p_m(x,\xi+i\tau\nabla\varphi(x))\overline{p_m^{(j)}(x,\xi+i\tau\nabla\varphi(x))\partial_{x_j}\psi}.
\end{aligned} \tag{15.215}$$

Moreover, since the coefficients of $p_m(x,\xi)$ are real-valued functions by the Taylor formula w.r.t. variable σ we have

$$\begin{aligned}
&\frac{1}{\sigma}\Im\sum_{j=1}^{n} p_m(x,\xi+i\sigma\nabla\varphi(x))\overline{p_m^{(j)}(x,\xi+i\sigma\nabla\varphi(x))\partial_{x_j}\psi} = \\
&= \left(\sum_{j=1}^{n} p_m^{(j)}(x,\xi)\partial_{x_j}\varphi\right)\left(\sum_{k=1}^{n} p_m^{(k)}(x,\xi)\partial_{x_k}\psi\right) + o(1) = \\
&= (\{p_m,\varphi\}\{p_m,\psi\})(x,\xi) + o(1), \quad \text{as } \sigma \to 0.
\end{aligned} \tag{15.216}$$

Now by (15.214), (15.215), and (15.216) we derive (15.213). ∎

Lemma 15.26 *Let Ω, φ and $P_m(x,D)$ be as in Theorem 15.24. Assume that (15.206) and (15.207) are satisfied. Let K be a compact subset of Ω. Then there exists $A_1 > 0$ such that for every $A \geq A_1$ there exist $C_1 > 0$ and $C_2 > 0$ such that, for every $x \in K$ and for every $(\xi,\sigma) \in \mathbb{R}^{n+1} \setminus \{(0,0)\}$, we have*

$$\begin{aligned}
&C_1\left|\xi + i\sigma\nabla\varphi(x)\right|^{2(m-1)} \leq \\
&\leq \widetilde{\omega}^{(A)}_{2m-2,\varphi}(x,\xi,\sigma) + C_2\frac{\left|p_m(x,\xi+i\sigma\nabla\varphi(x))\right|^2}{\left|\xi+i\sigma\nabla\varphi(x)\right|^2},
\end{aligned} \tag{15.217}$$

where

$$\widetilde{\omega}^{(A)}_{2m-2,\varphi} := \widetilde{\omega}_{2m-2,\varphi,A\varphi}. \tag{15.218}$$

Proof Let us fix $x_0 \in K$. It is not restrictive to assume that $x_0 = 0 \in K$.

Let

$$\psi = A\varphi,$$

where A is to be chosen.

Let us note that (15.213) and (15.218) give, for every $\xi \in \mathbb{R}^n$

$$\lim_{\sigma\to 0} \widetilde{\omega}^{(A)}_{2m-2,\varphi}(0,\xi,\sigma) = e^{2A\varphi(0)}\left\{p_m,\{p_m,\varphi\}\right\}(0,\xi)+$$
$$+\,2e^{2A\varphi(0)}\left(\{p_m,\varphi\}(0,\xi)\right)^2.$$

Motivated by this limit relation, we define $\widetilde{\omega}^{(A)}_{2m-2,\varphi}(0,\xi,\sigma)$ at 0 by continuity. That is, we set

$$\begin{aligned}\widetilde{\omega}^{(A)}_{2m-2,\varphi}(0,\xi,0) &:= \lim_{\sigma\to 0} \widetilde{\omega}^{(A)}_{2m-2,\varphi}(0,\xi,\sigma) = \\ &= e^{2A\varphi(0)}\left\{p_m,\{p_m,\varphi\}\right\}(0,\xi) + 2e^{2A\varphi(0)}\left(\{p_m,\varphi\}(0,\xi)\right)^2,\end{aligned} \tag{15.219}$$

for every $\xi \in \mathbb{R}^n$.

Let us denote by

$$N = \nabla\varphi(0).$$

First, let us consider the case in which $P_m(0, D)$ is elliptic operator (with real-valued coefficients). Let us note that since the antecedent of implication (15.206) is false for $x = 0$, implication (15.206) is true. Moreover, by (15.207) we have, for every $A \in \mathbb{R}$

$$\left.\begin{array}{l} p_m(0,\xi+i\sigma N) = 0 \\ \sigma \neq 0 \\ x \in \Omega \end{array}\right\} \Longrightarrow \widetilde{\omega}^{(A)}_{2m-2,\varphi}(0,\xi,\sigma) > 0. \tag{15.220}$$

Proceeding similarly to what was done in the proof of Lemma 15.22 we have that for every $A \in \mathbb{R}$ there exist positive constants C_1 and C_2 such that

$$C_1\,|\xi+i\sigma N|^{2(m-1)} \leq \widetilde{\omega}^{(A)}_{2m-2,\varphi}(0,\xi,\sigma) + C_2\frac{|p_m(x,\xi+i\sigma N)|^2}{|\xi+i\sigma N|^2}, \tag{15.221}$$

for every $(\xi,\sigma) \in \mathbb{R}^n \setminus \{(0,0)\}$.

Now, let us consider the case in which $P_m(0, D)$ is not elliptic. Let us denote by

$$\mathcal{X}_1 = \left\{\xi \in \mathbb{R}^n : \; p_m(0,\xi) = 0, \; |\xi| = 1\right\},$$

(note that $\mathcal{X}_1 \neq \emptyset$).

Claim I There exists $A_0 > 0$ such that for every $A \geq A_0$ we have

$$\widetilde{\omega}^{(A)}_{2m-2,\varphi}(0,\xi,0) > 0, \quad \forall \xi \in \mathcal{X}_1. \tag{15.222}$$

Proof of the Claim I Let us denote by

$$\mathcal{X}_2 = \{\xi \in \mathcal{X}_1 : \{p_m, \varphi\}(0,\xi) = 0\}.$$

By the continuity of the function $\xi \to \{p_m, \{p_m, \varphi\}\}(0,\xi)$, by the compactness of $\mathcal{X}_2$ and by (15.206) we have

$$\varepsilon_1 := \min_{\xi \in \mathcal{X}_2} \{p_m, \{p_m, \varphi\}\}(0,\xi) > 0.$$

From the positivity of ε_1 and, again by the continuity and the compactness properties, we have there exists δ_1 such that

$$\widetilde{\omega}^{(A)}_{2m-2,\varphi}(0,\xi,0) \geq \frac{\varepsilon_1 e^{2A\varphi(0)}}{2}, \quad \forall \xi \in \mathcal{X}_{2,\delta_1}, \tag{15.223}$$

where

$$\mathcal{X}_{2,\delta_1} := \{\xi \in \mathcal{X}_1 : |\{p_m, \varphi\}(0,\xi)| \leq \delta_1\}.$$

Let us denote by

$$M_1 = \max_{|\xi|=1} |\{p_m, \{p_m, \varphi\}\}(0,\xi)|.$$

For every $\xi \in \mathcal{X}_1 \setminus \mathcal{X}_{2,\delta_1}$ we have

$$\begin{aligned}\widetilde{\omega}^{(A)}_{2m-2,\varphi}(0,\xi,0) &\geq e^{2A\varphi(0)}\left(-M_1 + 2A\left(\{p_m, \varphi\}(0,\xi)\right)^2\right) \geq \\ &\geq e^{2A\varphi(0)}\left(-M_1 + 2A\delta_1^2\right).\end{aligned}$$

Hence for every

$$A \geq \frac{M_1 + \frac{\varepsilon_1}{2}}{2\delta_1^2},$$

we have

$$\widetilde{\omega}^{(A)}_{2m-2,\varphi}(0,\xi,0) \geq \frac{\varepsilon_1 e^{2A\varphi(0)}}{2}, \quad \forall \xi \in \mathcal{X}_1 \setminus \mathcal{X}_{2,\delta_1}. \tag{15.224}$$

By (15.223) and (15.224) we get (15.222). **This concludes the proof of the Claim I.**

Let us denote by

$$\mathbf{S}_0 = \left\{(\xi, \sigma) \in \mathbb{R}^n : |\xi + i\sigma N| = 1\right\}.$$

Claim II Assuming still $P_m(0, D)$ is not elliptic, for every $A \geq A_0$ (A_0 as in Clam I) there exists $B > 0$ such that

$$\begin{aligned}&\widetilde{\omega}^{(A)}_{2m-2,\varphi}(0, \xi, \sigma) + B\frac{|p_m(0, \xi + i\sigma N)|^2}{|\xi + i\sigma N|^2} \geq \\ &\geq C\,|\xi + i\sigma N|^{2(m-1)}, \qquad \forall (\xi, \sigma) \in \mathbb{R}^{n+1} \setminus \{(0, 0)\}.\end{aligned} \tag{15.225}$$

Proof of the Claim II Let $(\xi, \sigma) \in \mathbf{S}_0$. By (15.208) and Claim I, we have, for every $A \geq A_0$,

$$((\xi, \sigma) \in \mathbf{S}_0,\ p_m(0, \xi + \sigma N) = 0) \Longrightarrow \widetilde{\omega}^{(A)}_{2m-2,\varphi}(0, \xi, \sigma) > 0.$$

Hence, this relation and Lemma 14.24 imply that for every $A \geq A_0$ there exists $B > 0$ such that

$$\widetilde{\omega}^{(A)}_{2m-2,\varphi}(0, \xi, \sigma) + B\,|p_m\,(0, \xi + i\sigma N)|^2 > 0, \ \ \forall (\xi, \sigma) \in \mathbf{S}_0. \tag{15.226}$$

Let us fix $A \geq A_0$ and let us consider the function

$$f(\xi, \sigma) := \widetilde{\omega}^{(A)}_{2m-2,\varphi}(0, \xi, \sigma) + B\frac{|p_m\,(0, \xi + i\sigma N)|^2}{|\xi + i\sigma N|^2}, \quad \text{for } (\xi, \sigma) \in \mathbb{R}^{n+1} \setminus \{(0, 0)\}.$$

Such a function f turns out to be continuous in $\mathbb{R}^{n+1} \setminus \{(0, 0)\}$. Therefore (15.226) imply

$$C := \min_{(\xi,\sigma)\in\mathbf{S}_0} f(\xi, \sigma) > 0.$$

Since f is homogeneous of degree $2m - 2$ we get

$$f(\xi, \sigma) \geq C\,|\xi + i\sigma N|^{2(m-1)}, \ \ \forall (\xi, \sigma) \in \mathbb{R}^{n+1} \setminus \{(0, 0)\}.$$

This concludes the proof of the Claim II.

To obtain (15.26), one can proceed similarly to what was done in Lemma 15.22, utilizing continuity and compactness. We leave the details to the reader. ∎

Let $N \in \mathbb{R}^n \setminus \{0\}$. For any $f \in C_0^\infty(\Omega)$, let us denote by

$$\|f\|_{-1,\tau,N}^2 = \frac{1}{(2\pi)^n} \int_{\mathbb{R}^n} |\xi + i\tau N|^{-2} \left|\widehat{f}(\xi)\right|^2 d\xi, \tag{15.227}$$

for every $\tau \geq 1$

Remark 15.8 Let us note that, as $f \in C_0^\infty(\Omega)$, we have $\widehat{f} \in \mathcal{S}(\mathbb{R}^n)$, hence (15.227) is well-defined. Clearly, the norm $\|\cdot\|_{H_\tau^{-1}(\mathbb{R}^n)}$ is equivalent to the norm $\|\cdot\|_{-1,\tau,N}$. By (15.227) and by the Parseval identity we have

$$\|f\|_{-1,\tau,N} \leq |\tau N|^{-1} \|f\|_{L^2(\Omega)}, \quad \forall f \in C_0^\infty(\Omega). \tag{15.228}$$

♦

The following Lemma holds.

Lemma 15.27 *Let $\rho > 0$. Let h be a Lipschitz continuous function defined in $B_\rho(x_0) \Subset \Omega$. Let*

$$M = [h]_{0,1,\overline{B_\rho(x_0)}}$$

denote the Lipschitz constant of h. Suppose that

$$h(x_0) = 0.$$

We have, for every $w \in C_0^\infty(B_\rho(x_0))$

$$\left\|h(\partial_j w - \tau N_j w)\right\|_{-1,\tau,N} \leq M\left(\rho + |\tau N|^{-1}\right) \|w\|_{L^2(B_\rho(x_0))}. \tag{15.229}$$

Proof We have

$$h(x)(\partial_j w - \tau N_j w) = (\partial_j - \tau N_j)(hw) - w\partial_j h.$$

Hence

$$\begin{aligned} \left\|h(\partial_j - \tau N_j w)\right\|_{-1,\tau,N} \leq &\left\|(\partial_j - \tau N_j)(hw)\right\|_{-1,\tau,N} + \\ &+ \left\|w\partial_j h\right\|_{-1,\tau,N}. \end{aligned} \tag{15.230}$$

Let us notice that

$$|h(x)| = |h(x) - h(x_0)| \leq M\rho, \quad \forall x \in B_\rho(x_0). \tag{15.231}$$

Now, (15.231) gives

$$\begin{aligned}
&\left\|(\partial_j w - \tau N_j)(hw)\right\|^2_{-1,\tau,N} = \\
&= \frac{1}{(2\pi)^n}\int_{\mathbb{R}^n} \frac{\left|\xi_j + i\tau N_j\right|^2 \left|\widehat{hw}(\xi)\right|^2}{|\xi + i\tau N|^2} d\xi \leq \\
&\leq \frac{1}{(2\pi)^n}\int_{\mathbb{R}^n} \left|\widehat{hw}(\xi)\right|^2 d\xi = \\
&= \int_{\mathbb{R}^n} |h(x)w(x)|^2 dx \leq M^2\rho^2 \|w\|^2_{L^2(B_\rho(x_0))}.
\end{aligned} \tag{15.232}$$

On the other hand we have

$$\begin{aligned}
\|w\partial_j h\|^2_{-1,\tau,N} &= \frac{1}{(2\pi)^n}\int_{\mathbb{R}^n} \frac{\left|\widehat{w\partial_j h}(\xi)\right|^2}{|\xi + i\tau N|^2} d\xi \leq \\
&\leq \frac{|\tau N|^{-2}}{(2\pi)^n} \left\|\widehat{w\partial_j h}\right\|^2_{L^2(\mathbb{R}^n)} = \\
&= |\tau N|^{-2} \|w\partial_j h\|^2_{L^2(B_\rho(x_0))} \leq \\
&\leq M^2 |\tau N|^{-2} \|w\|^2_{L^2(B_\rho(x_0))}.
\end{aligned} \tag{15.233}$$

Therefore, by (15.230), (15.232), and (15.233) we get (15.229). ■

Proposition 15.28 *Let Ω, φ and $P_m(x, D)$ be as in Theorem 15.24. Let K be a compact subset of Ω. Let $R = \operatorname{dist}(K, \partial\Omega)$. Then there exists a constant $C > 0$ such that*

$$\begin{aligned}
&\left\|\widetilde{P}_m(x_0, D + i\tau\nabla\varphi(x_0))v\right\|_{-1,\tau,N} \leq \\
&\leq C\left(\rho + \tau^{-1}\right) \|v\|_{H^{m-1}_\tau(\mathbb{R}^n)} + \\
&+ |\tau\nabla\varphi(x_0)|^{-1} \left\|\widetilde{P}_m(x, D + i\tau\nabla\varphi(x))v\right\|_{L^2(B_\rho(x_0))},
\end{aligned} \tag{15.234}$$

for every $x_0 \in K$, for every $\rho \in (0, R)$, for every $v \in C^\infty_0(B_\rho(x_0))$ and for every $\tau \geq 1$. $\widetilde{P}_m(x, D + i\tau\nabla\varphi(x))$ is defined in (15.88).

Proof In carrying forward the proof, we will repeatedly make use the triangle inequality and Lemma 15.27.

Let us assume that $x_0 = 0 \in K$ and set

$$N = \nabla\varphi(0).$$

We have, for any $v \in C_0^\infty(\Omega)$,

$$\begin{aligned} &\left\| \widetilde{P}_m(0, D + i\tau N)v \right\|_{-1,\tau,N} \leq \\ &\leq \left\| \widetilde{P}_m(x, D + i\tau N)v - \widetilde{P}_m(0, D + i\tau N)v \right\|_{-1,\tau,N} + \\ &+ \left\| \widetilde{P}_m(x, D + i\tau N)v \right\|_{-1,\tau,N} . \end{aligned} \tag{15.235}$$

Let us estimate from above the first term on the right-hand side of (15.235). We have

$$\begin{aligned} &\widetilde{P}_m(x, D + i\tau N)v - \widetilde{P}_m(0, D + i\tau N)v = \\ &= \sum_{|\alpha|=m} (a_\alpha(x) - a_\alpha(0)) (D + i\tau N)^\alpha v. \end{aligned} \tag{15.236}$$

We set

$$M = \max_{|\alpha|=m} [a_\alpha]_{0,1,K} .$$

Let $\rho \in (0, R)$. Let $|\alpha| = m$ and let $1 \leq j \leq n$ satisfy $\alpha \geq e_j$. Set $w_j = (D + i\tau\nabla\varphi(0))^{\alpha - e_j} v$. By Lemma 15.27 we have

$$\begin{aligned} &\left\| (a_\alpha(x) - a_\alpha(0)) (D + i\tau\nabla\varphi(0))^\alpha v \right\|_{-1,\tau,N} = \\ &= \left\| (a_\alpha(x) - a_\alpha(0)) \left(D_j + i\tau N_j\right) w_j \right\|_{-1,\tau,N} \leq \\ &\leq M \left(\rho + |\tau N|^{-1}\right) \left\| w_j \right\|_{L^2(B_\rho)} \leq \\ &\leq CM \left(\rho + |\tau N|^{-1}\right) \|v\|_{H_\tau^{m-1}(\mathbb{R}^n)} , \end{aligned} \tag{15.237}$$

(C is a constant that depends on m and n) for every $v \in C_0^\infty(\Omega)$ such that $\operatorname{supp} v \subset B_\rho$ and for every $\tau \geq 1$.

By (15.236) and (15.237) we have

$$\begin{aligned} &\left\| \widetilde{P}_m(x, D + i\tau N)v - \widetilde{P}_m(0, D + i\tau N)v \right\|_{-1,\tau,N} \leq \\ &\leq CM \left(\rho + |\tau N|^{-1}\right) \|v\|_{H_\tau^{m-1}(\mathbb{R}^n)} , \end{aligned} \tag{15.238}$$

for every $v \in C_0^\infty(B_\rho)$ and for every $\tau \geq 1$.

Now we estimate from above the second term on the right-hand side of (15.235). By (15.228) we have

$$\begin{aligned}
&\left\| \widetilde{P}_m(x, D + i\tau N)v \right\|_{-1,\tau,N} \le \\
&\le |\tau N|^{-1} \left\| \widetilde{P}_m(x, D + i\tau N)v \right\|_{L^2(B_\rho)} \le \\
&\le |\tau N|^{-1} \left\| \widetilde{P}_m(x, D + i\tau \nabla\varphi(x))v - \widetilde{P}_m(x, D + i\tau N)v \right\|_{L^2(B_\rho)} + \\
&+ |\tau N|^{-1} \left\| \widetilde{P}_m(x, D + i\tau \nabla\varphi(x))v \right\|_{L^2(B_\rho)},
\end{aligned} \tag{15.239}$$

for every $v \in C_0^\infty(B_\rho)$ such that $\operatorname{supp} v \subset B_\rho$ and for every $\tau \ge 1$. Now, let us denote by

$$Q_\alpha(x, D, \tau) = \operatorname{Op}\left((\xi + i\tau\nabla\varphi(x))^\alpha\right)$$

and

$$M_1 = \max_K \left|\partial^2\varphi\right|.$$

We have

$$\widetilde{P}_m(x, D + i\tau\nabla\varphi(x)) = \sum_{|\alpha|=m} a_\alpha(x) Q_\alpha(x, D, \tau) \tag{15.240}$$

and

$$\begin{aligned}
&\left|Q_\alpha(x, D, \tau) - (D + i\tau\nabla\varphi(0))^\alpha\, v\right| = \\
&= \left|\sum_{\beta<\alpha} \binom{\alpha}{\beta} \left((i\tau\nabla\varphi(x))^{|\alpha|-|\beta|} - (i\tau\nabla\varphi(0))^{|\alpha|-|\beta|}\right) D^\beta v\right| \le \\
&\le CM_1\rho \sum_{\beta<\alpha} \tau^{|\alpha|-|\beta|} \left|D^\beta v\right|,
\end{aligned} \tag{15.241}$$

for every $v \in C_0^\infty(B_\rho)$ and for every $\tau \ge 1$.

By (15.240) and (15.241) we have

$$\begin{aligned}
&\left|\widetilde{P}_m(x, D + i\tau\nabla\varphi(x)) - \widetilde{P}_m(x, D + i\tau\nabla\varphi(0))v\right| \le \\
&\le CM_1\rho \sum_{|\alpha|\le m-1} \tau^{m-|\alpha|} \left|D^\alpha v\right|,
\end{aligned}$$

for every $v \in C_0^\infty(B_\rho)$ and for every $\tau \ge 1$.

Therefore

$$\left\| \widetilde{P}_m(x, D + i\tau\nabla\varphi(x))v - \widetilde{P}_m(x, D + i\tau N)v \right\|_{L^2(B_\rho)} \leq \\ \leq CM_1\rho\tau \|v\|_{H^{m-1}_\tau(\mathbb{R}^n)}, \tag{15.242}$$

for every $v \in C_0^\infty(B_\rho)$ and for every $\tau \geq 1$.

By (15.239) and (15.242) we have

$$\left\| \widetilde{P}_m(x, D + i\tau N)v \right\|_{-1,\tau,N} \leq \\ \leq CM_1\rho|N|^{-1} \|v\|_{H^{m-1}_\tau(\mathbb{R}^n)} + |\tau N|^{-1} \left\| \widetilde{P}_m(x, D + i\tau\nabla\varphi(x))v \right\|_{L^2(B_\rho)}.$$

By this inequality, by (15.235) and (15.238) we obtain

$$\left\| \widetilde{P}_m(0, D + i\tau N)v \right\|_{-1,\tau,N} \leq \\ \leq C\left(\rho + \tau^{-1}\right) \|v\|_{H^{m-1}_\tau(\mathbb{R}^n)} + |\tau N|^{-1} \left\| \widetilde{P}_m(x, D + i\tau\nabla\varphi(x))v \right\|_{L^2(B_\rho)},$$

for every $v \in C_0^\infty(B_\rho)$ and for every $\tau \geq 1$, where C depends on M and M_1. This concludes the proof of (15.234) ■

Remark 15.9 Let us note that estimate (15.234) holds even if $P_m(x; D)$ is not with real coefficients, but we are not interested in such generality. ♦

Proof of Theorem 15.24 Let us denote by

$$\gamma = \max\left\{1, \max_{\overline{\Omega}} \frac{1}{|\nabla\varphi|}\right\}.$$

Let K be a compact subset of Ω and let $x_0 \in K$. Let us fix $A = A_1$, where A_1 is defined in Lemma 15.26. Here and in the sequel we denote by

$$\widetilde{\omega}_{2m-2,\varphi}(x, \xi, \tau) = \widetilde{\omega}^{(A_1)}_{2m-2,\varphi}(x, \xi, \tau).$$

By Lemma 15.26 we have

$$\gamma^{-2}C_1 \|v\|^2_{H^{m-1}_\tau(\mathbb{R}^n)} \leq \\ \leq \int_{\mathbb{R}^n} \widetilde{\omega}_{2m-2,\varphi}(x_0, \xi, \tau) |\widehat{v}(\xi)|^2 d\xi + \\ + C_2 \int_{\mathbb{R}^n} \frac{|p_m(x_0, \xi + i\tau\nabla\varphi(x_0))|^2 |\widehat{v}(\xi)|^2}{|\xi + i\tau\nabla\varphi(x_0)|^2} d\xi, \tag{15.243}$$

for every $v \in C_0^\infty(\Omega)$ and for every $\tau \geq 1$.

Let $\rho > 0$ satisfy $B_\rho(x_0) \Subset \Omega$ and let us apply Proposition 15.28 to estimate from above the second term on the right-hand side of (15.243). We get

$$\begin{aligned}
&\gamma^{-2}C_1 \|v\|^2_{H^{m-1}_\tau(\mathbb{R}^n)} \leq \\
&\leq \int_{\mathbb{R}^n} \widetilde{\omega}_{2m-2,\varphi}(x_0,\xi,\tau)\,|\widehat{v}(\xi)|^2\,d\xi + C\left(\rho+\tau^{-1}\right)^2 \|v\|^2_{H^{m-1}_\tau(\mathbb{R}^n)} + \\
&+\gamma^2\tau^{-2}\left\|\widetilde{P}_m(x,D+i\tau\nabla\varphi(x))v\right\|^2_{L^2(\mathbb{R}^n)},
\end{aligned} \tag{15.244}$$

for every $v \in C_0^\infty(B_\rho(x_0))$ and for every $\tau \geq 1$.

Let ε be a positive number that we will choose later. By (a) of Theorem 15.10 there exists $r \in (0, \varepsilon]$ such that

$$\begin{aligned}
&\left|\left\langle\left(\widetilde{\omega}_{2m-2,\varphi}(x,D,\tau)-\widetilde{\omega}_{2m-2,\varphi}(x_0,D,\tau)\right)v,\,v\right\rangle_{L^2(\mathbb{R}^n)}\right| \leq \\
&\leq \varepsilon^2 \|v\|^2_{H^{m-1}_\tau(\mathbb{R}^n)} + C\|v\|^2_{H^{m-2}_\tau(\mathbb{R}^n)},
\end{aligned} \tag{15.245}$$

($C = 0$, if $m < 2$) for every $v \in C_0^\infty(B_r)$ and for every $\tau \geq 1$.

By (15.245) we have

$$\begin{aligned}
&\frac{1}{(2\pi)^n}\int_{\mathbb{R}^n} \widetilde{\omega}_{2m-2,\varphi}(x_0,\xi,\tau)\,|\widehat{v}(\xi)|^2\,d\xi = \\
&= \left\langle\left(\widetilde{\omega}_{2m-2,\varphi}(x_0,D,\tau)-\widetilde{\omega}_{2m-2,\varphi}(x,D,\tau)\right)v,\,v\right\rangle_{L^2(\mathbb{R}^n)} + \\
&+\left\langle\widetilde{\omega}_{2m-2,\varphi}(x,D,\tau)v,\,v\right\rangle_{L^2(\mathbb{R}^n)} \leq \\
&\leq \varepsilon^2\|v\|^2_{H^{m-1}_\tau(\mathbb{R}^n)} + \left\langle\widetilde{\omega}_{2m-2,\varphi}(x,D,\tau)v,\,v\right\rangle_{L^2(\mathbb{R}^n)},
\end{aligned} \tag{15.246}$$

for every $v \in C_0^\infty(B_r)$ and for every $\tau \geq 1$.

Now, we choose $\rho = r$, by (15.244) and (15.246) (taking into account (15.93)) we get

$$\begin{aligned}
&\gamma^{-2}C_1\|v\|^2_{H^{m-1}_\tau(\mathbb{R}^n)} \leq \\
&\leq (2\pi)^n\left\langle\widetilde{\omega}_{2m-2,\varphi}(x,D,\tau)v,\,v\right\rangle_{L^2(\mathbb{R}^n)} + \\
&+C\left(\varepsilon+\tau^{-1}\right)^2\|v\|^2_{H^{m-1}_\tau(\mathbb{R}^n)} + \\
&+C\tau^{-2}\|P_m(x,D+i\tau\nabla\varphi(x))v\|^2_{L^2(\mathbb{R}^n)},
\end{aligned} \tag{15.247}$$

for every $v \in C_0^\infty(B_r(x_0))$ for every $\tau \geq 1$.

Now we multiply both the sides of (15.247) by τ and by (15.218) we have

$$\begin{aligned}
&\gamma^{-2}C_1\tau \|v\|^2_{H^{m-1}_\tau(\mathbb{R}^n)} \leq \\
&\leq (2\pi)^n \left\langle \widetilde{c}_{2m-1,\varphi,A_1\varphi}(x,D,\tau)v, v\right\rangle_{L^2(\mathbb{R}^n)} + \\
&+ C\tau\left(\varepsilon+\tau^{-1}\right)^2 \|v\|^2_{H^{m-1}_\tau(\mathbb{R}^n)} + \\
&+ C\tau^{-1}\|P_m(x, D+i\tau\nabla\varphi(x))v\|^2_{L^2(\mathbb{R}^n)},
\end{aligned} \tag{15.248}$$

for every $v \in C^\infty_0(B_r(x_0))$ and for every $\tau \geq 1$.

Now by (15.194) and (15.248) (taking into account that $r \leq \varepsilon$) we get

$$\begin{aligned}
&\gamma^{-2}C_1\tau \|v\|^2_{H^{m-1}_\tau(\mathbb{R}^n)} \leq \\
&\leq \widetilde{C}\left(1+\tau\varepsilon^2+\tau^{-1}\right) \|v\|^2_{H^{m-1}_\tau(\mathbb{R}^n)} + \\
&+ C\|P_m(x, D+i\tau\nabla\varphi(x))v\|^2_{L^2(\mathbb{R}^n)},
\end{aligned} \tag{15.249}$$

for every $v \in C^\infty_0(\Omega)$ such that $\operatorname{supp} v \subset B_r(x_0)$ and for every $\tau \geq 1$.

Let us choose

$$\varepsilon = \left(\frac{C_1}{2\gamma\widetilde{C}}\right)^{1/2}$$

and let us denote by

$$\tau_* = \max\left\{1, \frac{8\widetilde{C}\gamma^2}{C_1}\right\},$$

we have

$$\frac{\gamma^{-2}C_1}{4}\tau \|v\|^2_{H^{m-1}_\tau(\mathbb{R}^n)} \leq C\|P_m(x, D+i\tau\nabla\varphi(x))v\|^2_{L^2(\mathbb{R}^n)}, \tag{15.250}$$

for every $v \in C^\infty_0(B_r(x_0))$ and for every $\tau \geq \tau_*$.

Similarly to the proof of Theorem 15.21, we have that for every $x_0 \in K$ there exist C_{x_0}, r_{x_0} and τ_{x_0} such that

$$\sum_{|\alpha|\leq m-1} \tau^{2(m-|\alpha|)-1}\int_\Omega |D^\alpha u|^2 e^{2\tau\varphi}dx \leq C_{x_0}\int_\Omega |P(x,D)u|^2 e^{2\tau\varphi}dx,$$

for every $u \in C^\infty_0(\Omega)$, $\operatorname{supp} v \subset B_{r_{x_0}}(x_0) \Subset \Omega$ and for every $\tau \geq \tau_{x_0}$.

Finally, by Lemma 14.2 we obtain estimate (15.168). ■

Exercise 15.1 If condition (15.167) ((15.206), (15.207)) holds in $x = x_0$, where $x_0 \in \Omega$ then there exists a neighborhood of x_0, $\mathcal{U}_{x_0}$ such that (15.167) ((15.206), (15.207)) holds for every $x \in \mathcal{U}_{x_0}$. [Hint: use continuity and Lemma 14.24]. ♠

15.8 The Pseudo-Convexity Properties and the Cauchy Problem

In Sects. 15.6 and 15.7, we established Carleman estimates for elliptic operators (Theorem 15.21) and for operators with real coefficients in the principal part (Theorem 15.24). These estimates were derived under specific conditions on the exponential weight function φ: (15.167) for elliptic operators, and (15.206)–(15.207) for operators with real coefficients.

In this section, we apply these Carleman estimates to prove uniqueness for the Cauchy problem, focusing on a particular class of initial surfaces. The central concept here is that of a strongly pseudo-convex surface. Informally, these surfaces possess the property that certain appropriate convexifications (see Definition 14.14) meet the aforementioned conditions. By leveraging Theorem 14.17, we then prove the uniqueness for the Cauchy problem.

Definition 15.29 Let Ω be an open set of $\mathbb{R}^n$ and let $P(x, D)$ be a linear differential operator of order $m \in \mathbb{N}$. We assume:

(a) the coefficients of $P(x, D)$ belong to $L^\infty(\Omega)$, while the coefficients of the principal part belong to $C^1(\Omega)$;
(b) $P(x, D)$ is elliptic or has real coefficients in the principal part.

Denote by $p_m(x, \xi)$ the symbol of the principal part of $P(x, D)$. Let Γ be an oriented surface of class C^2,

$$\Gamma = \{x \in \Omega : \phi(x) = 0\}, \tag{15.251}$$

where $\phi \in C^2(\Omega)$, ϕ real-valued function, satisfies

$$\nabla\phi(x) \neq 0, \quad \forall x \in \Gamma. \tag{15.252}$$

Let $x_0 \in \Gamma$. If the principal part of $P(x, D)$ has real coefficient, we say that Γ is a **pseudo-convex surface** in x_0 w.r.t. $P(x, D)$, whenever

$$\left.\begin{aligned} p_m(x_0, \xi) &= 0 \\ \{p_m, \phi\}(x_0, \xi) &= 0 \\ \xi &\neq 0 \end{aligned}\right\} \Longrightarrow \{p_m, \{p_m, \phi\}\}(x_0, \xi) > 0. \tag{15.253}$$

If $P(x, D)$ is elliptic or has real coefficients in the principal part, we say that Γ is a **strongly pseudo-convex surface** in x_0 w.r.t. $P(x, D)$, whenever (15.253) is satisfied and

$$\left.\begin{array}{ll} p_m(x_0, \xi + i\sigma\nabla\phi(x_0)) & = 0 \\ \{p_m(x, \xi + i\sigma\nabla\phi(x)), \phi(x)\}_{|x=x_0} & = 0 \\ \sigma \neq 0 & \end{array}\right\} \Longrightarrow \omega_{2m-2,\phi}(x_0, \xi, \sigma) > 0, \tag{15.254}$$

where

$$\begin{aligned} \omega_{2m-2,\phi}(x, \xi, \sigma) &= \frac{1}{\sigma} c_{2m-1,\phi}(x, \xi, \sigma) = \\ &= \frac{i}{2\sigma}\left\{p_m(x, \xi + i\sigma\nabla\phi(x)), \overline{p_m(x, \xi + i\sigma\nabla\phi(x))}\right\}. \end{aligned} \tag{15.255}$$

Remark 15.10 If $P(x, D)$ is an elliptic operator, then condition (15.253) is immediately satisfied, because, due to the ellipticity of $P(x, D)$ the antecedent of the implication does not hold. Condition (15.253), for operators with real coefficients in the principal part, is nothing other than condition (15.206). Instead, condition (15.254) is weaker than condition (15.167) (in $x = x_0$), if $P(x, D)$ is elliptic, and it is weaker than condition (15.207), if $P(x, D)$ has real coefficients in the principal part. Indeed the condition

$$\{p_m(x, \xi + i\sigma\nabla\phi(x)), \phi(x)\}_{|x=x_0} = 0,$$

is present in the antecedent of condition (15.254), but does not appear in the antecedent of conditions (15.167) and (15.207). Also, let us note that Lemma 15.25 allows simplifying the condition of strong pseudo-convexity for operators with a principal part having real coefficients as follows:

$$\left.\begin{array}{l} p_m(x_0, \xi + i\sigma\nabla\phi(x_0)) = 0 \\ \{p_m(x, \xi + i\sigma\nabla\phi(x)), \phi(x)\}_{|x=x_0} = 0 \\ (\xi, \sigma) \neq (0, 0) \end{array}\right\} \Longrightarrow \tag{15.256}$$

$$\Longrightarrow \lim_{\varepsilon\to 0^+} \omega_{2m-2,\phi}(x_0, \xi, \sigma + \varepsilon) > 0.$$

As a matter of fact, if $\sigma \neq 0$ we have

$$\lim_{\varepsilon\to 0^+} \omega_{2m-2,\phi}(x_0, \xi, \sigma + \varepsilon) = \omega_{2m-2,\phi}(x_0, \xi, \sigma)$$

and (by Lemma 15.25) if $\sigma = 0$ then

$$\lim_{\varepsilon\to 0^+} \omega_{2m-2,\phi}(x_0, \xi, \varepsilon) = \{p_m, \{p_m, \phi\}\}(x_0, \xi).$$

◆

Remark 15.11 Let Ω, $P(x, D)$, ϕ and Γ be as in Definition 15.29. Let us suppose that x_0 satisfies the following condition

$$\left.\begin{array}{l} p_m(x_0, \xi + i\sigma\nabla\phi(x_0)) = 0 \\ (\xi, \sigma) \neq 0 \end{array}\right\} \Longrightarrow$$
$$\Longrightarrow \sum_{j=1}^{n} p_m^{(j)}(x_0, \xi + i\sigma\nabla\phi(x_0))\partial_{x_j}\phi(x_0) \neq 0. \tag{15.257}$$

Let us recall that

$$\sum_{j=1}^{n} p_m^{(j)}(x_0, \xi + i\sigma\nabla\phi(x_0))\partial_{x_j}\phi(x_0) = \{p_m(x + i\sigma\nabla\phi), \phi\}_{|x=x_0}. \tag{15.258}$$

Let us note that condition (15.257) is equivalent to

$$\begin{array}{l} \text{the polynomial } \sigma \to p_m(x_0, \xi + i\sigma\nabla\phi(x_0)) \\ \text{has no real multiple roots } \forall\xi \in \mathbb{R}^n \setminus \{0\}. \end{array} \tag{15.259}$$

Also note that a *necessary condition* in order that Γ satisfies (15.257) is that it is not characteristic in x_0 (see Sect. 15.8.2 below "The example of the wave operator" in which is proved that the condition is *not* sufficient). As a matter of fact, if

$$p_m(x_0, \nabla\phi(x_0)) = 0,$$

then

$$p_m(x_0, 0 + i\sigma\nabla\phi(x_0)) = (i\sigma)^m p_m(x_0, \nabla\phi(x_0)) = 0, \ \forall\sigma \in \mathbb{R} \setminus \{0\} \tag{15.260}$$

and

$$\begin{aligned} &\sum_{j=1}^{n} p_m^{(j)}(x_0, 0 + i\sigma\nabla\phi(x_0))\partial_j\phi(x_0) = \\ &= \sum_{j=1}^{n} (i\sigma)^{m-1} p_m^{(j)}(x_0, \nabla\phi(x_0))\partial_j\phi(x_0) = \\ &= m(i\sigma)^{m-1} p_m(x_0, \nabla\phi(x_0)) = 0, \ \forall\sigma \in \mathbb{R} \setminus \{0\}. \end{aligned} \tag{15.261}$$

From (15.260) and (15.261) it is clear that (15.257) is not satisfied if Γ is characteristic in x_0.

Now we observe that if condition (15.257) is satisfied then Γ is a strongly pseudo-convex surface in x_0 w.r.t. $P(x, D)$. As a matter of fact, if the coefficients of the principal part of $P(x, D)$ are real, then the antecedent of implication (15.253) is false as, for $\sigma = 0$ and $\xi \neq 0$, (15.257) becomes

$$p_m(x_0, \xi) = 0 \Longrightarrow \{p_m, \phi\}(x_0, \xi) \neq 0.$$

Likewise, taking into account (15.258) the antecedent of condition (15.254) is false.

Similarly we proceed if $P(x, D)$ is an elliptic operator.

It is also important to note that condition (15.257) **does not require** Γ to be orientable since condition (15.257) is trivially equivalent to

$$\left.\begin{array}{l} p_m(x_0, \xi + i\sigma\nabla(-\phi)(x_0)) = 0 \\ (\xi, \sigma) \neq 0 \end{array}\right\} \Longrightarrow$$

$$\Longrightarrow \sum_{j=1}^{n} p_m^{(j)}(x_0, \xi + i\sigma\nabla(-\phi)(x_0))\partial_{x_j}(-\phi)(x_0) \neq 0.$$

Condition (15.257), at least from a historical perspective, is important in its own right (see Theorem 15.34 and Remark 15.13 following that Theorem). ♦

Example 15.1 It is simple to check that for the elliptic operator in $x_0 \in \mathbb{R}^2$

$$P(x, D) = D_1 + ia(x)D_2,$$

where $a(x) = a_1(x)+ia_2(x)$, $a_2(x_0) \neq 0$, every C^2 oriented surface (15.251) which satisfies (15.252) is strongly pseudo-convex, as (15.257) is satisfied. As a matter of fact, the negation of the consequent of (15.257) is equivalent to

$$\begin{cases} \phi_{x_1}(x_0) + a_1(x_0)\phi_{x_2}(x_0) = 0 \\ \\ a_2(x_0)\phi_{x_2}(x_0) = 0 \end{cases}, \tag{15.262}$$

which gives $\nabla\phi(x_0) = 0$ hence, tacking into account condition (15.252), we have that condition (15.257) is satisfied. Hence, by Remark 15.11, Γ is strongly pseudo-convex in x_0 w.r.t. $P(x, D)$. ♣

Example 15.2 Let us consider the first-order operator with real coefficients $a_j \in C^1(\Omega)$, $j = 1, \cdots, n$,

$$P_1(x, D) = \sum_{j=1}^{n} a_j(x)D_j = \frac{1}{i}a(x) \cdot \nabla, \tag{15.263}$$

where, $a(x) = (a_1(x), \cdots, a_n(x))$. There is no difference between pseudo-convexity and strong pseudo-convexity and, as is easily checked, both the conditions are equivalent to the following one ($x_0 \in \Gamma$)

$$a(x_0) \cdot \nabla\phi(x_0) = 0 \quad \Longrightarrow$$
$$\Longrightarrow \sum_{j,k=1}^{n} \partial^2_{jk}\phi(x_0)a_j(x_0)a_k(x_0) + \sum_{j,k=1}^{n} \partial_k a_j(x_0)a_k(x_0)\partial_j\phi(x_0) > 0. \tag{15.264}$$

where $\phi \in C^2(\Omega)$ satisfies (15.252). ♣

Example 15.3 Here we give an example of an operator for which the pseudo-convexity and the strong pseudo-convexity does not coincide. Let us consider the fifth-order operator in $\mathbb{R}^2$

$$P_5(D) = \left(D_1^2 + D_2^2\right)^2 (D_1 - D_2), \tag{15.265}$$

whose symbol is

$$p_5(\xi) = \left(\xi_1^2 + \xi_2^2\right)^2 (\xi_1 - \xi_2). \tag{15.266}$$

Let

$$\phi(x) = x_1 + x_2 + \frac{\alpha}{2}x_1^2 - \frac{1}{2}x_2^2,$$

where

$$\alpha > 1.$$

Let

$$\Gamma = \left\{x \in \mathbb{R}^2 : \phi(x) = 0\right\}$$

and

$$x_0 = 0 \in \Gamma.$$

We have

$$\nabla\phi(x) \neq 0, \quad \forall x \in \Gamma.$$

Now let us examine the antecedent of implication (15.253). We have that

$$\begin{cases} p_5(\xi) = 0 \\ p_5^{(1)}(\xi)\phi_{x_1}(0) + p_5^{(2)}(\xi)\phi_{x_2}(0) = 0 \\ \xi \neq 0 \end{cases} \tag{15.267}$$

is solved by $\xi = (\xi_1, \xi_1)$, $\xi_1 \neq 0$.

Now we have

$$\partial^2\phi(0) = \begin{pmatrix} \alpha & 0 \\ 0 & -1 \end{pmatrix},$$

hence, for $\xi = (\xi_1, \xi_1)$, $\xi_1 \neq 0$, we have

$$\{p_5, \{p_5, \phi\}\}(0, \xi) = 16(\alpha - 1)\xi_1^2 > 0.$$

Therefore condition (15.253) is satisfied, consequently Γ is a pseudo-convex surface in 0 w.r.t. $P_5(D)$.

Now we check that Γ is a not strongly pseudo-convex surface in 0 w.r.t. $P_5(D)$. To this purpose, let us consider the equation

$$p_5\left(\xi + i\sigma\nabla\phi(0)\right) = 0,$$

which is equivalent to

$$\left(-2\sigma^2 + 2i\sigma\left(\xi_1 + \xi_2\right) + \xi_1^2 + \xi_2^2\right)^2 \left(\xi_1 - \xi_2\right) = 0.$$

We note that $\xi_0 = (1, -1)$, $\sigma_0 = \pm 1$ satisfy the last equation and furthermore, we have

$$p_5^{(j)}\left(\xi_0 + i\sigma_0\nabla\phi(0)\right) = 0, \quad \text{for } j = 1, 2.$$

Hence

$$\omega_{2m-2,\phi}(0, \xi_0, \sigma_0) =$$

$$= \sum_{j=1}^{2} \partial_{x_j x_k}\phi(0) p_5^{(j)}(\xi_0 + i\sigma_0\nabla\phi(0))\overline{p_5^{(k)}(\xi_0 + i\sigma_0\nabla\phi(0))} = 0.$$

Therefore Γ is not a strongly pseudo-convex surface in 0 w.r.t. $P_5(D)$.

15.8.1 Invariance of Pseudo-Convexity and Strong Pseudo-Convexity

Let Γ be an oriented surface of class C^2,

$$\Gamma = \{x \in \Omega : \phi(x) = 0\}, \tag{15.268}$$

where $\phi \in C^2(\Omega)$, ϕ real-valued function, satisfies

$$\nabla \phi(x) \neq 0, \quad \forall x \in \Gamma. \tag{15.269}$$

In the present subsection we will prove that the properties of pseudo-convexity and strong pseudo-convexity of Γ in a point x_0 (w.r.t. a given operator) are independent with respect to the choice of the function which define the oriented surface Γ and with respect to local smooth diffeomorphisms.

1. Let us suppose that the oriented surface Γ is defined not only by ϕ but also by ψ, where $\psi(x) = a(x)\phi(x)$, $a \in C^2(\Omega)$, $a > 0$, on Γ, then the properties of pseudo-convexity strong convexity in $x_0 \in \Gamma$ are independent of the choice of the function that defines Γ. First of all we note that the orientation of Γ remains unchanged, as

$$\nabla \psi(x) = a(x)\nabla \phi(x), \quad \forall x \in \Gamma$$

and $a > 0$ on Γ.

Let us begin to **examine the invariance of the pseudo-convexity property** (for operators whose principal part has real coefficients). Hence, let us assume that

$$\left.\begin{array}{ll} p_m(x_0,\xi) & = 0 \\ \{p_m,\phi\}(x_0,\xi) = 0 & \\ \xi \neq 0 & \end{array}\right\} \Longrightarrow \{p_m,\{p_m,\phi\}\}(x,\xi) > 0. \tag{15.270}$$

We have by Proposition 15.1

$$\begin{aligned} \{p_m,\psi\}(x,\xi) &= a(x)\{p_m,\phi\}(x,\xi) + \phi(x)\{p_m,a\}(x,\xi) = \\ &= a(x)\{p_m,\phi\}(x,\xi), \quad \forall x \in \Gamma \end{aligned} \tag{15.271}$$

and, similarly,

$$\begin{aligned} &\{p_m,\{p_m,\psi\}\}(x,\xi) = a(x)\{p_m,\{p_m,\phi\}\}(x,\xi)+ \\ &+2(\{p_m,a\}\{p_m,\phi\})(x,\xi) + \phi(x)\{p_m,\{p_m,a\}\}(x,\xi). \end{aligned} \tag{15.272}$$

Now, let us assume

$$\begin{cases} p_m(x_0, \xi) = 0 \\ \{p_m, \psi\}(x_0, \xi) = 0 \\ \xi \neq 0 \end{cases} . \tag{15.273}$$

By (15.271) and by (15.273) we have

$$\{p_m, \phi\}(x_0, \xi) = \frac{1}{a(x_0)} \{p_m, \psi\}(x_0, \xi) = 0. \tag{15.274}$$

On the other hand we have $\phi(x_0) = 0$, so that (15.272) gives

$$\{p_m, \{p_m, \psi\}\}(x_0, \xi) = a(x_0)\{p_m, \{p_m, \phi\}\}(x_0, \xi). \tag{15.275}$$

Now, (15.274) implies

$$\begin{cases} p_m(x_0, \xi) = 0 \\ \{p_m, \phi\}(x_0, \xi) = 0 \\ \xi \neq 0 \end{cases} , \tag{15.276}$$

which, in turns, as (15.270) holds, this implies (taking into account (15.275) and $a(x_0) > 0$)

$$\{p_m, \{p_m, \psi\}\}(x_0, \xi) > 0.$$

All in all we have proved that if we replace ϕ with ψ in implication (15.270), it still holds.

Now let us **examine the invariance of the strong pseudo-convexity** (for elliptic operators or for operators whose principal part has real coefficients). Let us assume that

$$\left.\begin{array}{ll} p_m(x_0, \xi + i\sigma\nabla\phi(x_0)) & = 0 \\ \{p_m(x, \xi + i\sigma\nabla\phi(x)), \phi(x)\}_{|x=x_0} & = 0 \\ \sigma \neq 0 & \end{array}\right\} \Longrightarrow \omega_{2m-2,\phi}(x_0, \xi, \sigma) > 0. \tag{15.277}$$

Let us recall that

$$\begin{aligned}
&\omega_{2m-2,\phi}(x_0,\xi,\sigma)=\\
&=\sum_{j,k=1}^{n}\phi_{x_jx_k}(x_0)p_m^{(j)}(x_0,\xi+i\sigma\nabla\phi(x_0))\,\overline{p_m^{(k)}(x_0,\xi+i\sigma\nabla\phi(x_0))}+\\
&+\frac{1}{\sigma}\Im\left(\sum_{k=1}^{n}p_{m,k}(x_0,\xi+i\sigma\nabla\phi(x_0))\,\overline{p_m^{(k)}(x_0,\xi+i\sigma\nabla\phi(x_0))}\right).
\end{aligned} \tag{15.278}$$

By assuming that

$$\begin{cases} p_m(x_0,\xi+i\sigma\nabla\psi(x_0))=0\\ \{p_m(x,\xi+i\sigma\nabla\psi(x)),\psi(x)\}_{|x=x_0}=0\\ \sigma\neq 0\end{cases}, \tag{15.279}$$

we want to prove that

$$\omega_{2m-2,\psi}(x_0,\xi,\sigma)>0. \tag{15.280}$$

Since $\phi(x_0)=0$ we have

$$\psi_{x_j}(x_0)=a(x_0)\phi_{x_j}(x_0),\quad \text{for } j=1,\cdots,n$$

and by Proposition 15.1 we have

$$\{p_m(x,\xi+i\sigma\nabla\psi(x)),\psi(x)\}_{|x=x_0}=a(x_0)\,\{p_m(x,\xi+it\nabla\phi(x)),\phi(x)\}_{|x=x_0},$$

where

$$t=a(x_0)\sigma.$$

Hence (15.279) becomes

$$\begin{cases} p_m(x_0,\xi+it\nabla\phi(x_0))=0\\ \{p_m(x,\xi+it\nabla\phi(x)),\phi(x)\}_{|x=x_0}=0\\ t\neq 0\end{cases}, \tag{15.281}$$

and, as (15.277) holds we have

$$\omega_{2m-2,\phi}(x_0,\xi,t)>0. \tag{15.282}$$

Now taking into account that $\phi(x_0) = 0$ we have

$$\psi_{x_j x_k}(x_0) = a(x_0)\phi_{x_j x_k}(x_0) + a_{x_k}(x_0)\phi_{x_j}(x_0) + a_{x_j}(x_0)\phi_{x_k}(x_0)$$

and taking into account that (15.281) implies, trivially,

$$\sum_{k=1}^{n} p_m^{(k)}(x_0, \xi + it\nabla\phi(x_0))\phi_{x_k}(x_0) = 0,$$

we have

$$\sum_{j,k=1}^{n} \psi_{x_j x_k}(x_0) p_m^{(j)}(x_0, \xi + i\sigma\nabla\psi(x_0)) \overline{p_m^{(k)}(x_0, \xi + i\sigma\nabla\psi(x_0))} =$$

$$= a(x_0) \sum_{j,k=1}^{n} \phi_{x_j x_k}(x_0) p_m^{(j)}(x_0, \xi + it\nabla\phi(x_0)) \overline{p_m^{(k)}(x_0, \xi + it\nabla\phi(x_0))} +$$

$$+ 2a(x_0)\Re\left(\sum_{j,k=1}^{n} p_m^{(j)}(x_0, \xi + it\nabla\phi(x_0)) \right.$$

$$\left. \times\, a_{x_j}(x_0) \overline{p_m^{(k)}(x_0, \xi + it\nabla\phi(x_0))}\phi_{x_k}(x_0) \right) =$$

$$= a(x_0) \sum_{j,k=1}^{n} \phi_{x_j x_k}(x_0) p_m^{(j)}(x_0, \xi + it\nabla\phi(x_0)) \overline{p_m^{(k)}(x_0, \xi + it\nabla\phi(x_0))}.$$

On the other hand, we have

$$\frac{1}{\sigma}\Im\left(\sum_{k=1}^{n} p_{m,k}(x_0, \xi + i\sigma\nabla\psi(x_0)) \overline{p_m^{(k)}(x_0, \xi + i\sigma\nabla\psi(x_0))} \right) =$$

$$= \frac{a(x_0)}{t}\Im\left(\sum_{k=1}^{n} p_{m,k}(x_0, \xi + it\nabla\phi(x_0)) \overline{p_m^{(k)}(x_0, \xi + it\nabla\phi(x_0))} \right).$$

Therefore, by (15.282), we have

$$\omega_{2m-2,\psi}(x_0, \xi, \sigma) = a(x_0)\omega_{2m-2,\phi}(x_0, \xi, t) > 0. \tag{15.283}$$

In other words, we have proved that if we replace ϕ with ψ in implication (15.277), it still holds.

2. Let us examine the invariance of pseudo-convexity and strong pseudo-convexity property with respect to diffeomorphisms.

Let Ω, $\widetilde{\Omega}$ be two open sets of $\mathbb{R}^n$. Let

$$\widetilde{\Omega} \ni y \to g(y) \in \Omega \tag{15.284}$$

be a C^2 diffeomorphism. In Sect. 9.4 we have seen (with a slight different formulation) that the principal part of the transformed operator of $P_m(x, D)$ defined as

$$P^{\bullet}(y, D_y)v(y) = (P_m(x, D)u(x))_{|x=g(y)}, \tag{15.285}$$

where

$$v(y) = u(g(y)),$$

is given by

$$p_m^{\bullet}(y, \eta) = p_m\left(g(y), (h(y))^t\, \eta\right), \tag{15.286}$$

where

$$h(y) = \left(\partial_y g(y)\right)^{-1}, \tag{15.287}$$

$\partial_y g(y)$ is the jacobian matrix of g.

Let Γ be like (15.268) where $\phi \in C^2(\Omega)$ satisfies (15.269). Set

$$\widetilde{\phi}(y) = \phi(g(y)), \quad y \in \widetilde{\Omega} \tag{15.288}$$

and

$$\widetilde{\Gamma} = \left\{y \in \widetilde{\Omega} : \widetilde{\phi}(y) = 0\right\} = g^{-1}(\Gamma).$$

Let $x_0 \in \Gamma$ and $y_0 = g^{-1}(x_0) \in \widetilde{\Gamma}$. We begin to prove that if Γ is a pseudo-convex surface in x_0 w.r.t. $P(x, D)$ (whose coefficients of the principal part are real-valued functions) then $\widetilde{\Gamma}$ is a pseudo-convex surface in y_0 w.r.t. $P^{\bullet}(y, D_y)$.

Let χ be the map

$$\widetilde{\Omega} \times \mathbb{R}^n \ni (y, \eta) \to \chi(y, \eta) = \left(g(y), (h(y))^t \eta\right) \in \Omega \times \mathbb{R}^n.$$

By (15.286) We have

$$p_m^{\bullet}(y, \eta) = (p_m \circ \chi)(y, \eta), \quad \forall (y, \eta) \in \widetilde{\Omega} \times \mathbb{R}^n.$$

By Proposition 15.2 we have

$$\left\{p_m^{\bullet}, \widetilde{\phi}\right\} = \{p_m \circ \chi, \phi \circ \chi\} = \{p_m, \phi\} \circ \chi$$

(of course, here ϕ is considered as a function of (y, η) constant w.r.t. η) and

$$\left\{p_m^\bullet, \left\{p_m^\bullet, \widetilde{\phi}\right\}\right\} = \{p_m, \{p_m, \phi\}\} \circ \chi,$$

from which it is clear that if Γ is a pseudo-convex surface in x_0 w.r.t. $P_m(x, D_x)$, then $\widetilde{\Gamma}$ is also pseudo-convex surface in y_0 w.r.t. $P_m^\bullet(y, D_y)$.

Now let us consider the strong pseudo-convexity property.

Recalling that

$$\nabla_y \widetilde{\phi}(y) = \left(h^{-1}(y)\right)^t (\nabla_x \phi)(y),$$

and setting

$$a_{m,\sigma}(x, \xi) = p_m(x, \xi + i\sigma\nabla\phi(x)),$$

we have

$$\begin{aligned} &p_m^\bullet\left(y, \eta + i\sigma\nabla_y\widetilde{\phi}(y)\right) = \\ &= \left(a_{m,\sigma} \circ \chi\right)(y, \eta) = \\ &= p_m\left(x, \xi + i\sigma\nabla\phi(x)\right)_{|(x,\xi)=\chi(y,\eta)} . \end{aligned} \tag{15.289}$$

Hence, applying again Proposition 15.2 we have

$$\begin{aligned} &\left\{p_m^\bullet\left(y, \eta + i\sigma\nabla_y\widetilde{\phi}(y)\right), \widetilde{\phi}(y)\right\} = \\ &= \left\{\left(a_{m,\sigma} \circ \chi\right)(y, \eta), (\phi \circ \chi)(y, \eta)\right\} = \\ &= \left(\left\{a_{m,\sigma}, \phi\right\} \circ \chi\right)(y, \eta) = \\ &= \{p_m(x, \xi + i\sigma\nabla_x\phi(x)), \phi(x)\}_{|(x,\xi)=\chi(y,\eta)} \end{aligned} \tag{15.290}$$

and, for $\sigma \neq 0$,

$$\begin{aligned} &\widetilde{\omega}_{2m-2,\widetilde{\phi}}(y, \eta, \sigma) := \\ &:= \frac{i}{2\sigma}\left\{p_m^\bullet\left(y, \eta + i\sigma\nabla_y\widetilde{\phi}(y)\right), \overline{p_m^\bullet\left(y, \eta + i\sigma\nabla_y\widetilde{\phi}(y)\right)}\right\} = \\ &= \frac{i}{2\sigma}\left\{\left(a_{m,\sigma} \circ \chi\right)(y, \eta), \left(\overline{a_{m,\sigma}} \circ \chi\right)(y, \eta)\right\} = \\ &= \frac{i}{2\sigma}\left\{a_{m,\sigma}(x, \xi), \overline{a_{m,\sigma}(x, \xi)}\right\}_{|(x,\xi)=\chi(y,\eta)} = \\ &= \frac{i}{2\sigma}\left\{p_m\left(x, \xi + i\sigma\nabla_x\phi(x)\right), \overline{p_m\left(x, \xi + i\sigma\nabla_x\phi(x)\right)}\right\}_{|(x,\xi)=\chi(y,\eta)} = \\ &= \omega_{2m-2,\phi}\left(x, \xi, \sigma\right)_{|(x,\xi)=\chi(y,\eta)} . \end{aligned} \tag{15.291}$$

By (15.289)–(15.291) it is clear that if Γ is a strongly pseudo-convex surface in x_0 w.r.t. $P(x, D)$ then $\widetilde{\Gamma}$ is a pseudo-convex surface in y_0 w.r.t. $P^\bullet(y, D_y)$ (both in the case where $P(x, D)$ has the principal part with real coefficients and in the case where $P(x, D)$ is elliptic).

15.8.2 The Case of the Second-Order Operators with Real Coefficients in the Principal Part

In the present subsection we consider the second-order operators whose coefficients of the principal part are real-valued function and we will see that, for this types of operators the pseudo-convexity and the strong pseudo-convexity properties coincide.

Let $\left\{g^{jk}(x)\right\}_{j,k=1}^{n}$ be a symmetric matrix, whose entries are the real-valued functions $g^{jk} \in C^\infty(\Omega)$, $j, k = 1, \cdots, n$, where Ω is an open set of $\mathbb{R}^n$. In this subsection we adopt the Einstein convention of repeated indices.

Let us denote

$$P_2(x, D) = g^{jk}(x) D^2_{x_j x_k}, \tag{15.292}$$

whose symbol is the quadratic polynomial in the variable ξ

$$p_2(x, \xi) = g^{jk}(x)\xi_j \xi_k. \tag{15.293}$$

As above, we denote by $\phi \in C^2(\Omega)$, a real-valued function, such that

$$\nabla\phi(x) \neq 0, \quad \forall x \in \Gamma, \tag{15.294}$$

where Γ is the oriented surface

$$\Gamma = \{x \in \Omega : \phi(x) = 0\}. \tag{15.295}$$

The following Proposition holds.

Proposition 15.30 *Let $P_2(x, D)$ and Γ be as in* (15.292) *and* (15.295) *respectively. Let $x_0 \in \Gamma$. Then Γ is strongly pseudo-convex in x_0 w.r.t. $P_2(x, D)$ if and only if Γ is pseudo-convex in x_0 w.r.t. $P_2(x, D)$.*

Proof Of course we only need to prove that "pseudo-convex in x_0 implies strongly pseudo-convex in x_0".

We have

$$p_2^{(j)}(x, \xi) = 2g^{jl}(x)\xi_l, \quad \text{for } j = 1, \cdots, n, \tag{15.296}$$

$$
\begin{aligned}
&p_2(x, \xi + i\sigma\nabla\phi(x)) = \\
&= p_2(x,\xi) + i\sigma p_2^{(j)}(x,\xi)\partial_j\phi(x) - \frac{\sigma^2}{2} p_2^{(jk)}(x,\xi)\partial_j\phi(x)\partial_k\phi(x) = \\
&= p_2(x,\xi) + i\sigma\,\{p_2,\phi\}\,(x,\xi) - \sigma^2 p_2\,(x, \nabla\phi(x))\,.
\end{aligned} \tag{15.297}
$$

From which we have

$$
\begin{aligned}
&\omega_{2,\phi}\,(x,\xi,\sigma) = \\
&= \frac{i}{2\sigma}\left\{p_2\,(x,\xi+i\sigma\nabla\phi(x))\,,\,\overline{p_2\,(x,\xi+i\sigma\nabla\phi(x))}\right\} = \\
&= \{p_2,\{p_2,\phi\}\}\,(x,\xi) + \sigma^2\,\{\{p_2,\phi\}\,,\,p(x,\nabla\phi(x))\}\,(x,\xi).
\end{aligned} \tag{15.298}
$$

Claim The following equality holds

$$
\{\{p_2,\phi\}\,,\,p_2(x,\nabla\phi(x))\}\,(x,\xi) = \{p_2,\{p_2,\phi\}\}\,(x,\nabla\phi(x))\,. \tag{15.299}
$$

Proof of the Claim First we note that, for $k = 1,\cdots,n$, we have

$$
p_2^{(jk)}(x,\xi)\partial_j\phi(x) = 2g^{jk}(x)\partial_j\phi(x) = p_2^{(k)}\,(x,\nabla\phi(x))\,.
$$

Hence we have

$$
\begin{aligned}
&\{\{p_2,\phi\}\,,\,p_2(x,\nabla\phi(x))\} = \\
&= \partial_{\xi_k}\left(\{p_2,\phi\}\right)\partial_{x_k}\left(p_2\,(x,\nabla\phi(x))\right) = \\
&= \left(p_2^{(jk)}(x,\xi)\partial_j\phi(x)\right)\left(p_{2,k}\,(x,\nabla\phi(x)) + p_2^{(j)}\,(x,\nabla\phi(x))\,\partial^2_{jk}\phi(x)\right) = \\
&= \partial^2_{jk}\phi(x)p_2^{(j)}\,(x,\nabla\phi(x))\,p_2^{(k)}\,(x,\nabla\phi(x)) + p_2^{(k)}\,(x,\nabla\phi(x))\,p_{2,k}\,(x,\nabla\phi(x)) = \\
&= \{p_2,\{p_2,\phi\}\}\,(x,\nabla\phi(x))\,.
\end{aligned}
$$

This concludes the proof of the Claim.

By (15.298) and (15.299) we have

$$
\begin{aligned}
\omega_{2,\phi}\,(x,\xi,\sigma) = {}&\{p_2,\{p_2,\phi\}\}\,(x,\xi) + \\
&+\sigma^2\,\{p_2,\{p_2,\phi\}\}\,(x,\nabla\phi(x)).
\end{aligned} \tag{15.300}
$$

Now we assume that

$$
\begin{cases}
p_2(x_0,\xi+i\sigma\nabla\phi(x_0)) = 0 \\
\{p_2(x,\xi+i\sigma\nabla\phi(x)),\phi(x)\}_{|x=x_0} = 0 \\
\sigma \neq 0
\end{cases}, \tag{15.301}
$$

and proceed to prove that

$$\omega_{2,\phi}(x,\xi,\sigma) > 0. \tag{15.302}$$

By (15.297) we have that conditions (15.301) can be equivalently written as

$$\begin{cases} p_2(x_0,\xi) = \sigma^2 p_2(x_0,\nabla\phi(x_0)) \\ \{p_2,\phi\}(x_0,\xi) = 0 \\ p_2(x_0,\nabla\phi(x_0)) = 0 \\ \sigma \neq 0 \end{cases}. \tag{15.303}$$

Now, by the third equality in (15.303) we have

$$\{p_2,\phi\}(x_0,\nabla\phi(x_0)) = p_2^{(j)}(x_0,\nabla\phi(x_0))\partial_j\phi(x_0) = 2p_2(x_0,\nabla\phi(x_0)) = 0.$$

Hence the following conditions are satisfied

$$\begin{cases} p_2(x_0,\nabla\phi(x_0)) = 0 \\ \{p_2,\phi\}(x_0,\nabla\phi(x_0)) \\ \nabla\phi(x_0) \neq 0 \end{cases}. \tag{15.304}$$

So that, since Γ is pseudo-convex in x_0 w.r.t. $P_2(x,D)$, by (15.304) we obtain

$$\{p_2,\{p_2,\phi\}\}(x_0,\nabla\phi(x_0)) > 0. \tag{15.305}$$

Moreover, if $\xi = 0$, we have trivially

$$\{p_2,\{p_2,\phi\}\}(x_0,0) = 0, \tag{15.306}$$

but, if $\xi \neq 0$, then by the first and the third equality in (15.303) we have

$$\begin{cases} p_2(x_0,\xi) = 0 \\ \{p_2,\phi\}(x_0,\xi) = 0 \\ \xi \neq 0 \end{cases},$$

and, again by the assumption that Γ is pseudo-convex in x_0 w.r.t. $P_2(x,D)$, we have

$$\{p_2,\{p_2,\phi\}\}(x_0,\xi) > 0, \quad \text{for } \xi \neq 0.$$

Finally, by this inequality, by (15.300), by (15.305) and by (15.306) we have for every $\sigma \neq 0$ and every $\xi \in \mathbb{R}^n$

$$\omega_{2,\phi}(x_0,\xi,\sigma) =$$
$$= \{p_2,\{p_2,\phi\}\}(x_0,\xi) + \sigma^2\{p_2,\{p_2,\phi\}\}(x_0,\nabla\phi(x_0)) \geq \tag{15.307}$$
$$\geq \sigma^2\{p_2,\{p_2,\phi\}\}(x_0,\nabla\phi(x_0)) > 0.$$

■

Remark 15.12

(a) It should also be noticed that if operator (15.292) is elliptic, then any Γ like (15.295) is a strongly pseudo-convex surface w.r.t. $P_2(x,D)$ in each of its points, because (see Remark 15.10 of the present section), the ellipticity of $P_2(x,D)$ implies the pseudo-convexity w.r.t. $P_2(x,D)$ in each points of Γ and Proposition 15.30 implies that Γ is a strongly pseudo-convex surface w.r.t. $P_2(x,D)$ in each point of Γ.

(b) Let us consider the case where g^{jk} are **real constants**. Note that if the matrix $\left\{g^{jk}\right\}_{j,k=1}^n$ is singular do not exist any pseudo-convex surfaces w.r.t. operator (15.292). As a matter of fact, as the matrix $\left\{g^{jk}\right\}_{j,k=1}^n$ is singular then there exists $\xi^0 \in \mathbb{R}^n \setminus \{0\}$ such that

$$p_2^{(j)}\left(\xi^0\right) = g^{jk}\xi_k^0 = 0, \quad \text{for } j = 1,\cdots,n. \tag{15.308}$$

Hence we have, for any $\phi \in C^2(\Omega)$

$$\begin{cases} p_2\left(\xi^0\right) = g^{jk}\xi_k^0\xi_j^0 = 0 \\ \{p_2,\phi\}\left(x,\xi^0\right) = g^{jk}\xi_k^0\partial_{x_j}\phi(x) = 0 \\ \xi^0 \neq 0 \end{cases}.$$

On the other hand, by (15.308) we have

$$\{p_2,\{p_2,\phi\}\}\left(x,\xi^0\right) = \partial^2_{x_j,x_k}\phi(x)p_2^{(j)}\left(\xi^0\right)p_2^{(k)}\left(\xi^0\right) = 0. \tag{15.309}$$

Therefore (15.253) does not hold.

(c) Nevertheless, if the matrix $\left\{g^{jk}(x)\right\}_{j,k=1}^n$ is not constant, it can well happen that it is singular at a point x_0 while, for a certain $\phi \in C^2(\Omega)$ such that $x_0 \in \Gamma$ (defined by (15.295)), Γ is a pseudo-convex surface in x_0 w.r.t. the operator $P_2(x,D)$.

Indeed, let us consider

$$\left\{g^{jk}(x)\right\}_{j,k=1}^2 = \begin{pmatrix} 1-x_1 & 0 \\ 0 & -x_1 \end{pmatrix}, \tag{15.310}$$

$$\phi(x) = x_1,$$

$$\Gamma = \left\{x \in \mathbb{R}^2 : \phi(x) = 0\right\}$$

and

$$P_2(x, D) = (1 - x_1)\, D_1^2 - x_1 D_2^2. \tag{15.311}$$

The matrix $\left\{g^{jk}(x)\right\}_{j,k=1}^{n}$ is singular in 0.
Since

$$p_2^{(1)}(x, \xi) = 2(1 - x_1)\xi_1, \quad p_2^{(2)}(x, \xi) = -2x_1\xi_2,$$

the antecedent of (15.253) becomes in 0

$$\begin{cases} \xi_1^2 = 0 \\ 2\xi_1 = 0 \\ \xi \neq 0 \end{cases},$$

which is satisfied for every $\xi = (0, \xi_2)$, $\xi_2 \neq 0$. For these values of ξ we have easily

$$\{p_2, \{p_2, \phi\}\}\,(0, (0, \xi_2)) = 2\xi_2^2 > 0. \tag{15.312}$$

Therefore (15.253) is satisfied in 0 and consequently Γ is a pseudo-convex surface in 0 w.r.t. the operator $P_2(x, D)$, although the matrix (15.310) is singular in 0. ♦

We conclude this subsection with a more detailed examination of condition (15.257) for second-order operator with real coefficients in the principal part.

To simplify the notation, we assume that $x_0 = 0$ and we denote by

$$\mathbf{g}(\xi, \eta) = g^{jk}(0)\xi_j\eta_k.$$

$$N = \nabla\phi(0).$$

Let us note that

$$p_2(0, \xi) = \mathbf{g}(\xi, \xi), \quad \{p_2, \phi\}\,(0, \xi) = 2\mathbf{g}(\xi, N). \tag{15.313}$$

The following will be useful.

Proposition 15.31 *Assume that*

$$p_2(0, N) \neq 0. \tag{15.314}$$

Then condition (15.257) *is equivalent to*

$$\left.\begin{array}{l} \mathbf{g}(\xi,\xi)=0 \\ \\ \xi \nparallel N \end{array}\right\} \Longrightarrow \mathbf{g}(\xi,N)\neq 0, \tag{15.315}$$

$\xi \nparallel N$ *mean "ξ and N linearly independent"* .

Proof We begin by noticing that the (15.257) is equivalent to

$$\left.\begin{array}{l} p_2(0,\xi+i\sigma N)=0 \\ \xi \nparallel N \\ (\xi,\sigma)\neq(0,0) \end{array}\right\} \Longrightarrow p_2^{(j)}(0,\xi+i\sigma N)N_j\neq 0. \tag{15.316}$$

Indeed, if (15.257) holds then trivially (15.316) holds. Let us suppose now that (15.316) holds, we have that (15.257) is trivially satisfied, for any $\xi \nparallel N$. Instead, if $\xi=\lambda N$, $\lambda\in\mathbb{R}$, we have that

$$p_2(0,\xi+i\sigma N)=0,$$

implies

$$0=p_2(0,\lambda N+i\sigma N)=(\lambda+i\sigma)^2 p_2(0,N),$$

from which, taking into account that $p_2(0,N)\neq 0$, we have $\sigma=\lambda=0$, that is $\xi=0$ and $\sigma=0$, consequently the antecedent of (15.257) is false and thus condition (15.257) is satisfied.

Now we prove that (15.315) and (15.316) are equivalent. Let us begin assuming that (15.316) holds true. To prove (15.315) let us suppose that $\xi \nparallel N$ and that

$$\mathbf{g}(\xi,\xi)=0.$$

Now, if it were

$$\mathbf{g}(\xi,N)=0, \tag{15.317}$$

we would have at the same time

$$\xi \nparallel N,$$

$$p_2(0,\xi+i0N)=\mathbf{g}(\xi,\xi)=0$$

and

$$p_2^{(j)}(0,\xi+i0N)N_j=2\mathbf{g}(\xi,N)=0.$$

So there would be a contradiction with (15.316), therefore (15.317) does not hold. Thus, if (15.316) holds true then (15.315) holds.

Now let us suppose that (15.315) holds and let us suppose that

$$\begin{cases} p_2(0, \xi + i\sigma N) = 0 \\ \xi \nparallel N \\ (\xi, \sigma) \neq (0, 0) \end{cases} . \tag{15.318}$$

If $\sigma = 0$, then, (15.318) implies $\xi \nparallel N$ and $\mathbf{g}(\xi, \xi) = p_2(0, \xi + i0N) = 0$, moreover (15.315) implies $\mathbf{g}(\xi, N) \neq 0$. Hence

$$p_2^{(j)}(0, \xi + i0N) = 2\mathbf{g}(\xi, N) \neq 0.$$

If $\sigma \neq 0$, recalling that N is a noncharacteristic direction—so that $\mathbf{g}(N, N) = p_2(0, N) \neq 0$—we have

$$p_2^{(j)}(0, \xi + i\sigma N)N_j = 2\mathbf{g}(\xi, N) + i\sigma\mathbf{g}(N, N) \neq 0.$$

All in all, if (15.315) holds then (15.316) holds. The proof is complete. ■

The Example of the Wave Operator

1. Let us consider the wave operator

$$P_2(D) = -\left(\Delta_{x'} - \partial_{x_n}^2\right). \tag{15.319}$$

Let us check for which $N \neq 0$ condition (15.257) is satisfied. It is not restrictive to assume

$$|N| = 1. \tag{15.320}$$

We first need to assume that N is a noncharacteristic direction for $P_2(\partial)$, i.e.

$$p_2(N) = |N'|^2 - N_n^2 \neq 0. \tag{15.321}$$

Now, let us see when (15.315) is satisfied (which, we recall, is equivalent to (15.257)). This condition can be written

$$\begin{cases} |\xi'|^2 - \xi_n^2 = 0, \\ \\ \xi \nparallel N, \end{cases} \Longrightarrow \xi' \cdot N' - \xi_n N_n \neq 0. \tag{15.322}$$

Let us first examine the **case** $n = 2$. In this case condition (15.321) can be written

$$N_1^2 - N_2^2 \neq 0. \tag{15.323}$$

The first condition of the antecedent of (15.322) can be written as $\xi_1^2 - \xi_2^2 = 0$ and it is equivalent to

$$\xi_2 = \pm\xi_1.$$

From which, by (15.321) and $\xi \neq 0$, we have

$$\xi_1 N_1 - \xi_2 N_2 = \xi_1 (N_1 \mp N_2) \neq 0.$$

Therefore, if $n = 2$, (15.257) is satisfied for all $N \in \mathbb{R}^2$ that satisfy (15.323), i.e. that are not a characteristic direction.

Now, let us consider the **case** $n \geq 3$. Since the scalar product in $\mathbb{R}^{n-1}$ is invariant w.r.t. the rotations, we may assume that

$$N = N_1 e_1 + N_n e_n.$$

In this way, conditions (15.320) and (15.321), can be written, respectively;

$$N_1^2 + N_n^2 = 1$$

and

$$N_1^2 - N_n^2 \neq 0.$$

Let us distinguish two cases

(a) $N_1^2 - N_n^2 < 0$, i.e. $|N'| < |N_n|$;
(b) $N_1^2 - N_n^2 > 0$, i.e. $|N'| > |N_n|$;

Case (a). Let $\xi \nparallel N$ be such that

$$|\xi'|^2 - \xi_n^2 = 0.$$

In particular, we have $\xi_n \neq 0$ and $\xi' \neq 0$ (as a matter of fact, if one of them is zero the other is also zero) moreover

$$|\xi'| = |\xi_n|.$$

Hence

$$|\xi' \cdot N'| \leq |\xi'||N'| = |\xi'||N_1| < |\xi'||N_n| = |\xi_n||N_n|,$$

from which we have

$$\xi' \cdot N' - \xi_n N_n \neq 0.$$

Therefore, in case (a), (15.315) is satisfied.

In case (b) it is simple to check that (15.315) is not satisfied. To check this, let

$$\xi_0 = e_1 N_n + e_2 \sqrt{N_1^2 - N_n^2} + e_n N_1.$$

We have

$$\begin{cases} |\xi_0'|^2 - \xi_{0,n}^2 = 0 \\ \\ \xi_0 \nparallel N \end{cases},$$

but

$$\xi' \cdot N' - \xi_n N_n = N_n N_1 - N_1 N_n = 0.$$

Hence, the condition that the direction $N = \nabla\phi(0)$ is non-characteristic for an operator $P(x, D)$, while necessary, is not, in general, sufficient for (15.257) *to hold (see Remark 15.11).* ♣

2. Let

$$\phi(x) = \frac{1}{2}\left(|x'|^2 - \theta^2 x_n^2\right). \tag{15.324}$$

For $x_0 \in \mathbb{R}^n \setminus \{0\}$, let

$$\Gamma = \left\{x \in \mathbb{R}^n : \phi(x) = \phi(x_0)\right\}. \tag{15.325}$$

We check that if $\theta \in (0, 1)$, then Γ is a pseudo-convex surface in x_0 w.r.t. wave operator (15.319).

Let us calculate

$$\nabla\phi(x) = \left(x_0', -\theta^2 x_n\right) \neq 0, \quad \text{in } \mathbb{R}^n \setminus \{0\},$$

$$\partial^2\phi(x) = \text{diag}\left(1, \cdots, 1, -\theta^2\right).$$

Now let us suppose

$$\begin{cases} |\xi'|^2 - \xi_n^2 = 0 \\ \xi' \cdot x' + \theta^2 \xi_n x_{0,n} = 0 \\ \xi \neq 0 \end{cases}, \tag{15.326}$$

and let us check that

$$\partial^2\phi(x_0)\widetilde{\xi}\cdot\widetilde{\xi} > 0, \tag{15.327}$$

where

$$\widetilde{\xi} = \left(\xi', -\theta^2\xi_n\right).$$

We have

$$\partial^2\phi(x_0)\widetilde{\xi}\cdot\widetilde{\xi} = |\xi'|^2 - \theta^2\xi_n^2.$$

On the other hand, by the first condition of (15.326) we have $|\xi'|^2 = \xi_n^2$ hence, taking into account that if ξ satisfies at the same time $\xi \neq 0$ and $|\xi'|^2 - \xi_n^2 = 0$, then $\xi' \neq 0$, we have

$$\partial^2\phi(x_0)\widetilde{\xi}\cdot\widetilde{\xi} = |\xi'|^2\left(1-\theta^2\right) > 0.$$

Therefore, we have proved that (15.326) implies (15.327). Hence, for any $x_0 \in \mathbb{R}^n \setminus \{0\}$, Γ is pseudo-convex surface in x_0 w.r.t. wave operator (15.319). ♣

15.8.3 *Strong Pseudo-Convexity and the Cauchy Problem*

As before, here too we denote by Ω an open set of $\mathbb{R}^n$ and by $\phi \in C^2(\Omega)$, a real-valued function,

$$\nabla\phi(x) \neq 0, \quad \forall x \in \Gamma, \tag{15.328}$$

where Γ is the oriented surface

$$\Gamma = \{x \in \Omega :\ \phi(x) = 0\}. \tag{15.329}$$

Let $x_0 \in \Gamma$ and for any $\delta > 0$, let us denote by

$$\begin{aligned}\varphi_\delta(x) = \nabla\phi(x_0)\cdot(x-x_0) + \frac{1}{2}\partial^2\phi(x_0)(x-x_0)\cdot(x-x_0)+\\ + \frac{1}{2\delta}\left(\nabla\phi(x_0)\cdot(x-x_0)\right)^2 - \frac{\delta}{2}|x-x_0|^2.\end{aligned} \tag{15.330}$$

By Proposition 14.15 we know that $\varphi_\delta(x)$ is a convexification of Γ in x_0 (see Definition 14.14).

We have

Proposition 15.32 *Let $P(x, D)$ be an operator either with principal part having real coefficients or elliptic in Ω. Let $x_0 \in \Gamma$. Assume that Γ is strongly pseudo-convex in x_0 w.r.t. $P(x, D)$. Then there exists $\delta_0 > 0$ such that for every $\delta \in (0, \delta_0]$, φ_δ satisfies* (15.206) *and* (15.207) *in x_0, provided the principal part of $P(x, D)$ has real coefficients, and φ_δ satisfies* (15.167) *in x_0, provided $P(x, D)$ is elliptic.*

Proof It is not restrictive to assume that $x_0 = 0 \in \Gamma$. We consider the case where the principal part of $P(x, D)$ has real coefficients (the case where $P(x, D)$ is elliptic can be proven similarly). Let us note that, as a consequence of Lemma 15.25, the conjunction of the two conditions (15.206) and (15.207) is equivalent to

$$\left.\begin{array}{l} p_m(0, \xi + i\sigma\nabla_x\varphi_\delta(0)) = 0 \\ \\ (\xi, \sigma) \neq (0, 0) \end{array}\right\} \Longrightarrow \omega^*_{2m-2,\varphi_\delta}(0, \xi, \sigma) > 0; \tag{15.331}$$

where

$$\omega^*_{2m-2,\varphi_\delta}(0, \xi, \sigma) = \begin{cases} \omega_{2m-2,\varphi_\delta}(0, \xi, \sigma), & \text{for } \sigma \neq 0 \\ \\ \{p_m, \{p_m, \varphi_\delta\}\}(0, \xi), & \text{for } \sigma = 0 \end{cases}. \tag{15.332}$$

Similarly (see also Remark 15.10 in this section) the condition of strong pseudo-convexity of Γ in 0 w.r.t. $P(x, D)$ can be written as

$$\begin{aligned} &\left.\begin{array}{l} p_m(0, \xi + i\sigma\nabla_x\phi(0)) = 0 \\ \{p_m(x, \xi + i\sigma\nabla_x\phi(x)), \phi(x)\}_{|x=0} = 0 \\ (\xi, \sigma) \neq (0, 0) \end{array}\right\} \Longrightarrow \\ &\Longrightarrow \omega^*_{2m-2,\phi}(0, \xi, \sigma) > 0. \end{aligned} \tag{15.333}$$

Thus, let us assume that (15.333) holds and let us prove (15.332). We have (we adopt the convention of repeated indices)

$$\begin{aligned} \partial_{x_j}\varphi_\delta(x) = \partial_{x_j}\phi(0) + \partial^2_{x_j x_k}\phi(0)x_k + \\ + \frac{1}{\delta}\left(\nabla_x\phi(0) \cdot x\right)\partial_{x_j}\phi(0) - \delta x_j, \end{aligned} \tag{15.334}$$

for $j = 1, \cdots, n$, and

$$\partial^2_{x_j x_k}\varphi_\delta(x) = \partial^2_{x_j x_k}\phi(0) + \frac{1}{\delta}\partial_{x_j}\phi(0)\partial_{x_k}\phi(0) - \delta\delta_{jk}, \tag{15.335}$$

for $j, k = 1, \cdots, n$.

By (15.334) we have $\nabla_x \varphi_\delta(0) = \nabla_x \phi(0)$, hence the antecedent of implication (15.331) is equivalent to

$$\begin{cases} p_m(0, \xi + i\sigma\nabla_x\phi(0)) = 0 \\ \\ (\xi, \sigma) \neq (0, 0) \end{cases} . \tag{15.336}$$

Moreover we have

$$\begin{aligned} &\omega^*_{2m-2,\varphi_\delta}(0, \xi, \sigma) = \omega^*_{2m-2,\phi}(0, \xi, \sigma) + \\ &+ \frac{1}{\delta} \left|\nabla_\xi p_m\left(0, \xi + i\sigma\nabla_x\phi(0)\right) \cdot \nabla_x\phi(0)\right|^2 - \\ &- \delta \left|\nabla_\xi p_m\left(0, \xi + i\sigma\nabla_x\phi(0)\right)\right|^2 . \end{aligned} \tag{15.337}$$

Let us denote

$$\mathbb{S}^n = \left\{(\xi, \sigma) \in \mathbb{R}^{n+1} : \ |\xi|^2 + \sigma^2 = 1\right\},$$

$$\mathcal{X} = \left\{(\xi, \sigma) \in \mathbb{S}^n : \ p_m\left(0, \xi + i\sigma\nabla_x\phi(0) = 0\right)\right\},$$

$$C_0 = \max_{(\xi,\sigma)\in\mathbb{S}^n} \left|\nabla_\xi p_m\left(0, \xi + i\sigma\nabla_x\phi(0)\right)\right|^2 .$$

By (15.337) we have

$$\begin{aligned} &\omega^*_{2m-2,\varphi_\delta}(0, \xi, \sigma) \geq \omega^*_{2m-2,\phi}(0, \xi, \sigma) + \\ &+ \frac{1}{\delta} \left|\nabla_\xi p_m\left(0, \xi + i\sigma\nabla_x\phi(0)\right) \cdot \nabla_x\phi(0)\right|^2 - \delta C_0, \end{aligned} \tag{15.338}$$

for every $(\xi, \sigma) \in \mathbb{S}^n$.

Now, by (15.333) we have

$$\left.\begin{array}{l} (\xi, \sigma) \in \mathcal{X} \\ \\ \nabla_\xi p_m\left(0, \xi + i\sigma\nabla_x\phi(0)\right) \cdot \nabla_x\phi(0) = 0 \end{array}\right\} \Longrightarrow \omega^*_{2m-2,\phi}(0, \xi, \sigma) > 0,$$

which implies, by Lemma 14.24, that there exist $C_1 > 0$ and $C_2 > 0$ such that

$$\omega^*_{2m-2,\phi}(0, \xi, \sigma) + C_1 \left|\nabla_\xi p_m\left(0, \xi + i\sigma\nabla_x\phi(0)\right) \cdot \nabla_x\phi(0)\right|^2 \geq C_2, \tag{15.339}$$

for every $(\xi, \sigma) \in \mathbb{S}^n$.

Now, let

$$\delta \le \delta_0 := \min\left\{\frac{1}{C_1}, \frac{C_2}{2C_0}\right\}.$$

By (15.338) and (15.339) we have, for every $\delta \in (0, \delta_0]$,

$$\omega^*_{2m-2,\varphi_\delta}(0, \xi, \sigma) \ge \frac{C_2}{2}, \quad \forall(\xi, \sigma) \in \mathbb{S}^n \tag{15.340}$$

and taking into account that $\omega^*_{2m-2,\varphi_\delta}(0, \xi, \sigma)$ is homogeneous of degree $2m - 2$, (15.331) is proved. ■

Theorem 15.33 *Let Ω be an open set of $\mathbb{R}^n$. Let $\phi \in C^2(\Omega)$ be a real-valued function and let Γ be the oriented surface defined in* (15.329). *Let $x_0 \in \Gamma$.*

Assume that

$$\nabla\phi(x) \neq 0, \quad \forall x \in \Gamma.$$

Let $P(x, D)$ be an operator of order $m \in \mathbb{N}$.

Assume that

(a) the coefficients of $P(x, D)$ belong to $L^\infty(\Omega; \mathbb{C})$ and they belong to $C^\infty(\Omega)$ in the principal part;
(b) $P(x, D)$ is elliptic or has real coefficients in the principal part.
(c) Γ is strongly pseudo-convex in x_0 w.r.t. $P(x, D)$.

Then there exist a neighborhood $\mathcal{U}_{x_0}$ and $R > 0$, $\overline{B_R(x_0)} \subset \Omega$, such that if $U \in H^m(B_R(x_0))$ satisfies

$$\begin{cases} P(x, D)U = 0, & \text{in } B_R(x_0) \\ U = 0, & \text{in } \{x \in B_R(x_0) : \phi(x) \ge 0\} \end{cases}, \tag{15.341}$$

then we have

$$U(x) = 0, \quad \forall x \in \mathcal{U}_{x_0}. \tag{15.342}$$

Proof Let us assume $x_0 = 0$. By Proposition 14.15 there exists $\delta_1 > 0$ such that

$$\begin{aligned} &\left\{x \in \overline{B_{\delta_1}} : \varphi_\delta(x) \ge 0\right\} \setminus \{0\} \subset \\ &\subset \left\{x \in \overline{B_{\delta_1}} : \phi(x) > 0\right\}. \end{aligned} \tag{15.343}$$

We may assume $\delta_1 \le \delta_0$, where δ_0 is defined in Proposition 15.32. Let us consider the case where $P(x, D)$ has real coefficients in the principal part. By

Proposition 15.32 we have that φ_{δ_1} satisfies (15.206) and (15.207) in 0. Hence we have

$$\left.\begin{array}{l} p_m(0, \xi + i\sigma\nabla_x\varphi_{\delta_1}(0)) = 0 \\ \\ (\xi, \sigma) \neq (0, 0) \end{array}\right\} \Longrightarrow \omega^*_{2m-2,\varphi_{\delta_1}}(0, \xi, \sigma) > 0; \tag{15.344}$$

where

$$\omega^*_{2m-2,\varphi_{\delta_1}}(0, \xi, \sigma) = \begin{cases} \omega_{2m-2,\varphi_{\delta_1}}(0, \xi, \sigma), & \text{for } \sigma \neq 0 \\ \\ \left\{p_m, \left\{p_m, \varphi_{\delta_1}\right\}\right\}(0, \xi), & \text{for } \sigma = 0 \end{cases}.$$

By Lemma 14.24, (15.344) implies there exists $C > 0$

$$\omega^*_{2m-2,\varphi_{\delta_1}}(0, \xi, \sigma) + C\frac{\left|p_m\left(0, \xi + i\sigma\nabla_x\varphi_{\delta_1}(0)\right)\right|^2}{|\xi|^2 + \sigma^2} > 0,$$

for every $(\xi, \sigma) \in \mathbb{S}^n = \left\{(\xi, \sigma) \in \mathbb{R}^{n+1} : |\xi|^2 + \sigma^2 = 1\right\}$. By continuity, there exists $R > 0$ such that

$$\omega^*_{2m-2,\varphi_{\delta_1}}(x, \xi, \sigma) + C\frac{\left|p_m\left(x, \xi + i\sigma\nabla_x\varphi_{\delta_1}(x)\right)\right|^2}{|\xi|^2 + \sigma^2} > 0, \tag{15.345}$$

for every $(\xi, \sigma) \in \mathbb{S}^n$ and for every $x \in B_R$. Now, (15.345) implies

$$\left.\begin{array}{l} p_m(x, \xi + i\sigma\nabla_x\varphi_{\delta_1}(x)) = 0 \\ (\xi, \sigma) \neq (0, 0) \\ x \in B_R \end{array}\right\} \Longrightarrow \omega^*_{2m-2,\varphi_{\delta_1}}(x, \xi, \sigma) > 0. \tag{15.346}$$

Consequently, Theorem 15.24 yields that there exist constants C and τ_0 such that

$$\sum_{|\alpha|\leq m-1} \tau^{2(m-|\alpha|)-1}\int_{B_R}\left|D^\alpha u\right|^2 e^{2\tau\varphi_{\delta_1}}dx \leq C\int_{B_R}|P(x, D)u|^2 e^{2\tau\varphi_{\delta_1}}dx,$$

for every $u \in C^\infty_0(B_R)$ and for every $\tau \geq \tau_0$. Therefore, Theorem 14.17 implies there exists a neighborhood $\mathcal{U}$ of 0, such that

$$U(x) = 0, \quad \forall x \in \mathcal{U}.$$

The proof in the elliptic case is quite the same to the previous one and we leave it to the reader. ■

Theorem 15.34 *Let Ω be an open set of $\mathbb{R}^n$. Let $\phi \in C^2(\Omega)$ be a real-valued function and let Γ be the oriented surface defined in* (15.329). *Let $x_0 \in \Gamma$.*
Assume that

$$\nabla\phi(x) \neq 0, \quad \forall x \in \Gamma.$$

Let $P(x, D)$ be an operator of order $m \in \mathbb{N}$. Assume that

(a) the coefficients of $P(x, D)$ belong to $L^\infty(\Omega)$ and they belong to $C^\infty(\Omega)$ in the principal part;
(b) $P(x, D)$ is elliptic or has real coefficients in the principal part.
(c)

$$\left.\begin{array}{l} p_m(x_0, \xi + i\sigma\nabla\phi(x_0)) = 0 \\ \\ (\xi, \sigma) \neq 0 \end{array}\right\} \Longrightarrow$$
$$\Longrightarrow \sum_{j=1}^{n} p_m^{(j)}(x_0, \xi + i\sigma\nabla\phi(x_0))\partial_{x_j}\phi(x_0) \neq 0. \tag{15.347}$$

Then there exist a neighborhood $\mathcal{U}_{x_0}$ and $R > 0$, $\overline{B_R(x_0)} \subset \Omega$, such that if $U \in H^m(B_R(x_0))$ satisfies

$$\begin{cases} P(x, D)U = 0, \quad in\ B_R(x_0) \\ \\ \frac{\partial^j U}{\partial \nu^j} = 0, \ on\ \Gamma, \ for\ j = 0, 1, \cdots, m-1 \end{cases}, \tag{15.348}$$

where $\Gamma = \{x \in B_R(x_0) : \ \phi(x) = 0\}$, $\nu = -\frac{\nabla\phi(x_0)}{|\nabla\phi(x_0)|}$ we have

$$U = 0, \quad in\ \mathcal{U}_{x_0}.$$

Proof The theorem is an immediate consequence of (15.344) and the fact that (15.348) implies the strong pseudo-convexity of the oriented surfaces

$$\Gamma_\pm = \{x \in B_R(x_0) : \ \pm\phi(x) = 0\}$$

in x_0 w.r.t. $P(x, D)$ (see Remark 15.11 of the present section). ■

Remark 15.13 It is important to point out at this stage that Theorem 15.34 was proven by Calderón, [20], before Theorem 15.33, which was instead proven by Hörmander in [56] and extends Calderón's Theorem to strongly pseudo-convex surfaces. It should also be mentioned that Calderón's original proof is based on pseudodifferential operators, which made their first appearance precisely in [20]. Among the various books that present the proof of Calderón's Theorem, following the author's original strategy but utilizing the more recent theory of pseudodifferential operators, we will limit ourselves here to mentioning [58, Ch. 28] and [142]. ♦

Remark 15.14 We have observed in "The example of the wave operator" of Sect. 15.8.2 that the condition that the direction $\nabla\phi(x_0)$ is non-characteristic for the wave operator (in three variables or more), while necessary, is not, sufficient for (15.347) be true. Now it is interesting to point out that it has been proved by Alinhac and Baouendi in [6] (also we refer to [83, Ch. 6]) that there exist $u, q \in C^\infty(\mathbb{R}^3, \mathbb{C})$ such that

$$\begin{cases} \partial_t^2 u - \partial_{x_1}^2 u - \partial_{x_2}^2 u + q(x,t)u = 0 \\ \\ \operatorname{supp} u = \{x_2 \geq 0\} \end{cases}, \tag{15.349}$$

or, in other words, **does not hold uniqueness** for the Cauchy problem with initial surface $\Gamma := \{x_2 = 0\}$, for the equation

$$\partial_t^2 u - \partial_{x_1}^2 u - \partial_{x_2}^2 u + q(x,t)u = 0,$$

where q is complex-valued function that belongs to $C^\infty(\mathbb{R}^3)$. Keep in mind that, when q is **analytic**, the Holmgren Theorem provides uniqueness for the Cauchy problem for the equation

$$\partial_t^2 u - \partial_{x_1}^2 u - \partial_{x_2}^2 u + q(x,t)u = 0,$$

with initial surface Γ (as it is a noncharacteristic surface). In this regard, it is important to recall that an in-depth study initiated by Robbiano in [111] and culminating in the works [59, 112, 127] has made it possible to extend the uniqueness theorems for the Cauchy problem—originally due to Hörmander, for operators with non-analytic coefficients, and to Holmgren, for operators with analytic coefficients—to operators with partially analytic coefficients. This class of operators, of course, includes wave operators like

$$\partial_t^2 - \partial_{x_1}^2 - \partial_{x_2}^2 + q(x,t),$$

with q analytic with respect to t and C^∞ with respect to the variable x. ♦

Let us note that, due to Proposition 15.31, for second-order operator with real-valued coefficients in the principal part, Theorem 15.34 is restated as follows

Theorem 15.35 *Let Ω be an open set of $\mathbb{R}^n$. Let $\phi \in C^2(\Omega)$ be a real-valued function and let Γ be the oriented surface defined in* (15.329). *Let $x_0 \in \Gamma$. Assume*

$$\nabla\phi(x) \neq 0, \quad \forall x \in \Gamma.$$

Let $P(x, D)$ be an operator of order 2. Assume that

(a) the coefficients of $P(x, D)$ belong to $L^\infty(\Omega; \mathbb{C})$ and that they belong to $C^\infty(\Omega)$ in the principal part;

(b) $P(x, D)$ has real-valued coefficients in the principal part.

(c)

$$p_2(x_0, \nabla\phi(x_0)) \neq 0 \tag{15.350}$$

and

$$\left.\begin{array}{l} p_2(x_0,\xi) = 0 \\ \\ \xi \not\parallel \nabla\phi(x_0) \end{array}\right\} \Longrightarrow \sum_{j=1}^{n} p_2^{(j)}(x_0,\xi)\partial_{x_j}\phi(x_0) \neq 0. \tag{15.351}$$

Then there exist a neighborhood $\mathcal{U}_{x_0}$ and $R > 0$, $\overline{B_R(x_0)} \subset \Omega$, such that if $U \in H^m(B_R(x_0))$ satisfies

$$\begin{cases} P(x, D)U = 0, & \text{in } B_R(x_0) \\ \\ U = \frac{\partial U}{\partial \nu} = 0, & \text{on } \Gamma \end{cases}, \tag{15.352}$$

where $\Gamma = \{x \in B_R(x_0) : \phi(x) = 0\}$, $\nu = -\frac{\nabla\phi(x_0)}{|\nabla\phi(x_0)|}$, we have

$$U = 0, \quad \text{in } \mathcal{U}_{x_0}. \tag{15.353}$$

15.8.4 Geometric Meaning of the Pseudo-Convexity Property

Let Ω be an open set of $\mathbb{R}^n$. Let

$$P(x, D) = \sum_{|\alpha|\leq m} a_\alpha(x) D^\alpha. \tag{15.354}$$

and let us assume that $P(x, D)$ has real and $C^1(\Omega)$ coefficients in the principal part.

We call a **bicharacteristic line**, or simply a **bicharacteristic**, a solution $(x(t), \xi(t))$, $t \in J$, where J is an interval of $\mathbb{R}$, of the Hamiltonian system associated to $p_m(x, \xi)$

$$\begin{cases} \dot{x}_j(t) = p_m^{(j)}(x(t), \xi(t)), & j = 1, \cdots, n \\ \\ \dot{\xi}_j(t) = -p_{m,j}(x(t), \xi(t)), & j = 1, \cdots, n \end{cases}. \tag{15.355}$$

Let us note that by (15.355) we have (we use the Einstein convention of repeated indices)

$$\begin{aligned}&\frac{d}{dt}p_m(x(t),\xi(t)) =\\&= \dot{x}_j(t)p_{m,j}(x(t),\xi(t)) + \dot{\xi}_j(t)p_m^{(j)}(x(t),\xi(t)) = 0,\end{aligned} \tag{15.356}$$

for every $t \in J$.

If

$$p_m(x(t),\xi(t)) = 0, \quad \forall t \in J,$$

we say that $(x(t),\xi(t))$ is a **null bicharacteristic line**. In particular, we have that if $(x(t),\xi(t))$ is a null bicharacteristic line in a point t_0 then it is a null bicharacteristic line in the whole interval J. As a matter of fact, if

$$p_m(x(t_0),\xi(t_0)) = 0,$$

then (15.356) implies $p_m(x(t),\xi(t)) = 0$, for every $t \in J$. Let us notice that in each point $(x(t),\xi(t))$ of a null bicharacteristic line, $\xi(t)$ is a characteristic direction for operator $P(x,D)$ in the point $x(t)$. The projection $x(\cdot)$, on $\mathbb{R}^n_x$ of a null bicharacteristic line is called **ray** of $P(x,D)$ (see Sect. 7.6).

The following simple Proposition will be useful to understand the geometrical meaning of pseudo-convexity condition.

Proposition 15.36 *Let $\phi \in C^2(\Omega)$. If $x(t)$ is a ray of $P(x,D)$, then*

$$\frac{d}{dt}\phi(x(t)) = \{p_m,\phi\}(x(t),\xi(t)), \quad \forall t \in J \tag{15.357}$$

and

$$\frac{d^2}{dt^2}\phi(x(t)) = \{p_m,\{p_m,\phi\}\}(x(t),\xi(t)). \tag{15.358}$$

Proof By (15.355) we have

$$\begin{aligned}\frac{d}{dt}\phi(x(t)) &= \partial_j\phi(x(t))\dot{x}_j(t) =\\&= p_m^{(j)}(x(t),\xi(t))\partial_j\phi(x(t)) =\\&= \{p_m,\phi\}(x(t),\xi(t)), \quad \forall t \in J\end{aligned}$$

and (see (15.211))

$$\begin{aligned}
&\frac{d^2}{dt^2}\phi(x(t)) = \frac{d}{dt}\left(p_m^{(j)}(x(t),\xi(t))\partial_j\phi(x(t))\right) = \\
&= \partial^2_{jk}\phi(x(t))\dot{x}_k(t)p_m^{(j)}(x(t),\xi(t)) + \\
&+\left(p_{m,k}^{(j)}(x(t),\xi(t))\dot{x}_k(t) + p_m^{(jk)}(x(t),\xi(t))\dot{\xi}_k(t)\right)\partial_j\phi(x(t)) = \\
&= \partial^2_{jk}\phi(x(t))p_m^{(j)}(x(t),\xi(t))p_m^{(k)}(x(t),\xi(t)) + \\
&+\left(p_{m,k}^{(j)}p_m^{(k)} - p_{m,k}p_m^{(jk)}\right)(x(t),\xi(t))\partial_j\phi(x(t)) = \\
&= \{p_m,\{p_m,\phi\}\}(x(t),\xi(t)).
\end{aligned}$$

■

Remark 15.15 Let us briefly focus on the case $m = 2$ to illustrate the geometric meaning of condition (15.351). Let $x_0 \in \Omega$ and let us suppose that $\xi_0 \in \mathbb{R}^n \setminus \{0\}$ is a characteristic direction in x_0, that is

$$p_2(x_0,\xi_0) = 0. \tag{15.359}$$

Let us consider the solution $(x(t),\xi(t))$ of system (15.355) satisfying the initial condition in $t_0 \in J$

$$x(t_0) = x_0, \quad \xi(t_0) = \xi_0. \tag{15.360}$$

We quickly notice that (15.357) allows us to write the consequent of condition (15.351) of Theorem 15.35, in $x = x_0$ and $\xi = \xi_0$ as follows

$$\frac{d}{dt}\phi(x(t))_{|t=t_0} \neq 0. \tag{15.361}$$

Taking into account that (15.359) holds, we can interpret condition (15.351) geometrically by stating that a bicharacteristic line $(x(t),\xi(t))$ passing through (x_0,ξ_0) has the property that the ray $x(t)$ crosses transversally through the level surface

$$\Gamma_\phi = \{x \in \Omega : \quad \phi(x) = \phi(x_0)\}.$$

♦

Remark 15.16 Let x_0, ξ_0 be as in Remark 15.15, but now we assume that $P(x,D)$ is operator (15.354) with $m \geq 1$ (and whose coefficients of the principal part are real-valued functions). Moreover let us assume

$$p_m(x_0, \xi_0) = 0. \tag{15.362}$$

Condition (15.357) and (15.358) allow us to write condition (15.253) (in (x_0, ξ_0)) in the form

$$\frac{d}{dt}\phi(x(t))_{|t=t_0} = 0 \Longrightarrow \frac{d^2}{dt^2}\phi(x(t))_{|t=t_0} > 0 \tag{15.363}$$

which is equivalent to fact that if t_0 is a critical point of $\phi(x(t))$ then it is a proper minimum point of the function $\phi(x(t))$.

Condition (15.363) can also be formulated in the following manner: let us consider the above defined level surface Γ_ϕ and the level set

$$\Omega_\phi^+ = \{x \in \Omega : \quad \phi(x) > \phi(x_0)\},$$

then (15.363) says that if the ray $x(t)$ is tangent in x_0 to the level surface Γ_ϕ then for every t in a neighborhood of t_0, we have, for $t \neq t_0$, $x(t) \in \Omega_\phi^+$. In other words, the ray $x(t)$ cannot "cross" the level surface Γ_ϕ at points where $x(t)$ is tangent to Γ_ϕ. ♦

15.8.5 Further Properties of Pseudo-Convex Surfaces

In this final subsection we give some further properties of pseudo-convex (and strongly pseudo-convex) surfaces w.r.t. the operator

$$P(x, D) = \sum_{|\alpha| \leq m} a_\alpha(x) D^\alpha, \tag{15.364}$$

where $P(x, D)$ is elliptic or has real-valued coefficients in the principal part. Again, let Ω be an open set of $\mathbb{R}^n$, $x_0 \in \Omega$, $\phi \in C^2(\Omega)$ be a real-valued function such that

$$\nabla\phi(x) \neq 0, \quad \forall x \in \Gamma, \tag{15.365}$$

where Γ is the oriented surface

$$\Gamma = \{x \in \Omega : \phi(x) = \phi(x_0)\}. \tag{15.366}$$

We want to study the relationships between the conditions of pseudo-convexity and strong pseudo-convexity, respectively, and the conditions

$$\left.\begin{array}{ll} p_m(x, \xi) & = 0 \\ \{p_m, \varphi\}(x, \xi) & = 0 \\ \xi \neq 0 & \\ x \in \Omega & \end{array}\right\} \Longrightarrow \{p_m, \{p_m, \varphi\}\}(x, \xi) > 0; \tag{15.367}$$

and

$$\left.\begin{array}{l} p_m(x, \xi + i\sigma\nabla\varphi(x)) = 0 \\ \sigma \neq 0 \\ x \in \Omega \end{array}\right\} \Longrightarrow \omega_{2m-2,\varphi}(x, \xi, \sigma) > 0; \tag{15.368}$$

where

$$\begin{aligned} &\omega_{2m-2,\varphi}(x, \xi, \sigma) = \frac{1}{\sigma} c_{2m-1,\varphi}(x, \xi, \sigma) = \\ &= \frac{i}{2\sigma}\left\{p_m(x, \xi + i\sigma\nabla\varphi(x)), \overline{p_m(x, \xi + i\sigma\nabla\varphi(x))}\right\}. \end{aligned} \tag{15.369}$$

We recall that conditions (15.367) and (15.368) are sufficient conditions for the Carleman estimates (15.168), (15.209) to hold (concerning (15.168), only (15.368) is required)

Proposition 15.37 *Let $\phi \in C^2(\Omega)$ be a real-valued function such that* (15.365) *where Γ is the oriented surface* (15.366). *Denote by*

$$\varphi_\lambda(x) = e^{\lambda\phi(x)}. \tag{15.370}$$

(a) If Γ is a pseudo-convex surface in x_0 w.r.t. to $P(x, D)$ then φ_λ satisfies (15.367), *in $x = x_0$, for λ large enough.*
(b) If Γ is a strongly pseudo-convex surface in x_0 w.r.t. to $P(x, D)$ then φ_λ satisfies (15.368), *in $x = x_0$, for λ large enough.*

Proof

(a) First, let us notice that

$$\{p_m, \varphi_\lambda\}(x, \xi) = \lambda e^{\lambda\phi(x)} \{p_m, \phi\}(x, \xi) \tag{15.371}$$

and

$$\begin{aligned} \{p_m, \{p_m, \varphi_\lambda\}\}(x, \xi) = \lambda e^{\lambda\phi(x)} \big(&\{p_m, \{p_m, \phi\}\}(x, \xi) + \\ &+ \lambda \left|\{p_m, \phi\}(x, \xi)\right|^2\big). \end{aligned} \tag{15.372}$$

Since Γ is a pseudo-convex surface in x_0 w.r.t. to $P(x, D)$ we have

$$\left.\begin{array}{ll} p_m(x_0, \xi) & = 0 \\ \{p_m, \phi\}(x_0, \xi) & = 0 \\ |\xi| = 1 & \end{array}\right\} \Longrightarrow \{p_m, \{p_m, \phi\}\}(x_0, \xi) > 0. \tag{15.373}$$

By Lemma 14.24 we derive that (15.373) is equivalent to the existence of $\lambda_0 > 0$ such that if $\lambda \geq \lambda_0$ we have

$$\left.\begin{array}{l} p_m(x_0, \xi) = 0 \\ \\ |\xi| = 1 \end{array}\right\} \Longrightarrow \{p_m, \{p_m, \phi\}\}(x_0, \xi) + \lambda \left|\{p_m, \phi\}(x_0, \xi)\right|^2 > 0. \tag{15.374}$$

By (15.372) and by homogeneity, (15.374) implies

$$\left.\begin{array}{l} p_m(x_0, \xi) = 0 \\ \\ \xi \neq 0 \end{array}\right\} \Longrightarrow \{p_m, \{p_m, \varphi_\lambda\}\}(x, \xi) > 0.$$

This concludes the proof of (a).

(b) It is simple to check that, if $\sigma \neq 0$,

$$\begin{aligned} \omega_{2m-2,\varphi_\lambda}(x, \xi, \sigma) &= e^{\lambda\phi(x)} \Big(\omega_{2m-2,\phi}(x, \xi, \tau\lambda e^{\lambda\phi(x)}) + \\ &+ \lambda \left|\nabla_\xi p_m \left(x, \xi + \sigma\lambda e^{\lambda\phi(x)} \nabla\phi(x)\right) \cdot \nabla_x \phi(x)\right|^2\Big) \end{aligned} \tag{15.375}$$

and, we recall, if $\tau = 0$, then (the continuous extension in $\tau = 0$) $\omega_{2m-2,\varphi_\lambda}(x, \xi, \tau) = \{p_m, \{p_m, \varphi_\lambda\}\}(x, \xi)$.

Since (a) has been proved, it is enough to prove that

$$\left.\begin{array}{l} p_m(x_0, \xi + i\sigma\nabla\varphi_\lambda(x_0)) = 0 \\ \\ \sigma \neq 0 \end{array}\right\} \Longrightarrow \omega_{2m-2,\varphi_\lambda}(x_0, \xi, \sigma) > 0 \tag{15.376}$$

Now, since Γ is a strongly pseudo-convex surface in x_0 w.r.t. to $P(x, D)$ we have

$$\left.\begin{array}{l} p_m(x_0, \xi + it\nabla\phi(x_0)) = 0 \\ \\ \nabla_\xi p_m(x_0, \xi + it\nabla\phi(x_0)) \cdot \nabla\phi(x_0) = 0 \end{array}\right\} \Longrightarrow \omega_{2m-2,\phi}(x_0, \xi, t) > 0,$$

(as usual $\omega_{2m-2,\phi}(x_0, \xi, t)_{|t=0} = \{p_m, \{p_m, \varphi_\lambda\}\}(x_0, \xi)$) that, by homogeneity is equivalent to the existence of $\lambda_0 > 0$ such that if $\lambda \geq \lambda_0$ we have

$$\left.\begin{array}{l} p_m(x_0, \xi + it\nabla\phi(x_0)) = 0 \\ \\ |\xi + it\nabla\phi(x_0)| = 1 \end{array}\right\} \Longrightarrow$$
$$\Longrightarrow \omega_{2m-2,\phi}(x_0, \xi, t)+$$
$$+ \lambda \left|\nabla_\xi p_m (x_0, \xi + t\nabla\phi(x_0)) \cdot \nabla_x\phi(x_0)\right|^2 > 0. \tag{15.377}$$

Taking into account (15.375), and setting $t = \sigma\lambda e^{\lambda\phi(x_0)}$ in (15.377) (for $\lambda \geq \lambda_0$) the latter implies (15.376).

■

Exercise 15.2 Let $P(x, D)$ be an elliptic operator or an operator with real-valued coefficients in the principal part. Let $\phi \in C^2(\Omega)$ satisfy (15.365). Let Ω' be a bounded open subset of Ω such that $\overline{\Omega'} \subset \Omega$, $x_0 \in \Omega'$. Prove that: if $\Gamma = \{x \in \Omega : \phi(x) = \phi(x_0)\}$ is a pseudo-convex (strongly pseudo-convex) surface in x_0 w.r.t. $P(x, D)$ then there exists $\varepsilon > 0$ such that for every $\psi \in C^2(\Omega)$ such that

$$\|\phi - \psi\|_{C^2\left(\overline{\Omega'}\right)} < \varepsilon$$

the oriented surface $\Gamma_\psi = \{x \in \Omega : \psi(x) = \psi(x_0)\}$ is a pseudo-convex (strongly pseudo-convex) surface in x_0 w.r.t. $P(x, D)$. [Hint: use continuity, homogeneity and Lemma 14.24].

♠

15.9 Brief Comments on the Use of Pseudodifferential Operators in the Proof of the Carleman Estimates

Trying to summarize the key points of the proofs of Theorems 15.21 (elliptic case) and 15.24 (operators with a principal part having real coefficients), we can certainly say that the Lemmas 15.22 and 15.26 play a decisive role, respectively. These two Lemmas establish very useful lower bounds for appropriate approximations of the commutator (see Sect. 15.5).

$$\left[\widetilde{P}_m^{(s)}(x, D + i\tau\nabla\varphi(x)), \widetilde{P}_m^{(a)}(x, D + i\tau\nabla\varphi(x))\right].$$

From these lower bounds, through the localized Gårding inequality Theorem 15.10, and a localization process in the variable x, the desired Carleman estimates are obtained.

We have followed the above-outlined procedure (in particular, localization in the variable x alone) to provide the reader with a completely self-contained (and, in a certain sense, elementary) framework for obtaining Carleman estimates for the operators considered so far. Historically, the localization procedure we followed is the one used by Hörmander in [56, Ch. VIII]. We will return to Hörmander's procedure in the next chapters, as it differs from the one we used to establish Theorems 15.14, 15.20, and 15.23, allowing for more accurate integration by parts and thus relaxing the regularity assumptions on the coefficients of the principal part of the operators.

However, as we already mentioned in Remark 15.13 of Sect. 15.8, both before and after the aforementioned book [56, Ch. VIII], the development of the theory of Carleman estimates has progressed through the theory of pseudodifferential operators, a theory developed by Hörmander himself, as well as by numerous other prominent authors. For an extensive bibliography on this topic, we refer the reader to [59, Vol. III, IV], [82, 128]. This development has allowed, in [59, Ch. XXVIII, Vol. IV], not only for the devising of more direct proofs of Carleman estimates for the operators we have considered (elliptic or with a principal part having real coefficients), but also for a significant broadening of the assumptions concerning the principal part of the operators. These assumptions pertain to *normally principal operators*, which were first introduced in [56, Ch. VIII], but were considerably generalized in [59, Ch. XXVIII, Vol. IV]. The principal part of a normally principal operator, $p_m(x,\xi)$, $(x,\xi) \in \Omega \times \mathbb{R}^n$, must satisfy the condition that for every compact subset K of Ω, there exists $C > 0$ such that

$$|\{p_m, \overline{p_m}\}(x,\xi)| \leq C|\xi|^{m-1}\,|p_m(x,\xi)|, \quad \forall (x,\xi) \in K \times \mathbb{R}^n.$$

Thus, as it is easy to check, these operators include both elliptic operators and those with a principal part having real coefficients. We do not delve into the exposition of normally principal operators and instead refer directly to [83, Ch. 5], where the main connections between Carleman estimates, local solvability of differential equations, and normally principal operators are discussed and Carleman estimates are proved.

A detailed study of the pseudodifferential operators goes beyond the limited scope we have set for this book. Here, we only intend to provide a brief overview of these operators in relation to their applications to the proofs of Carleman estimates, referring to [82], [83], and volumes III and IV of [59] for an extensive and in-depth treatment of the subject. In particular, we want to focus a bit on the so-called Sharp Gårding Inequality, which immediately leads, for example, from Lemma 15.22 to Theorem 15.21 (and similarly, from Lemma 15.26 to Theorem 15.24). This Sharp Gårding Inequality and the much more powerful Fefferman–Phong inequality, [40], which allows extending Carleman estimates to normally principal operators (see [83] and [59]), broadly speaking, allow us to translate an inequality of the type (15.169) into inequalities for operators.

To introduce pseudodifferential operators, we begin by noting that if $P(x, D)$ is a differential operator,

$$P(x, D) = \sum_{|\alpha| \leq m} a_\alpha(x) D^\alpha,$$

then, by the Fourier Transform Inversion Theorem, we have

$$a(x, D)u = \frac{1}{(2\pi)^n} \int_{\mathbb{R}^n} e^{i\xi\cdot x} a(x, \xi)\widehat{u}(\xi)d\xi, \quad \forall u \in C_0^\infty(\Omega), \tag{15.378}$$

where $p(x, \xi)$ is the symbol of $P(x, D)$. In the theory of pseudodifferential operators, generalizations of (15.378) are studied, considering instead of $p(x, \xi)$, a polynomial in ξ, functions $a(x, \xi)$ that are C^∞ in the variable (x, ξ) and whose derivatives have controlled growth, for example of the type, for every $\alpha, \beta \in \mathbb{N}_0^n$

$$\sup_{(x,\xi)\in K\times\mathbb{R}^n} \left|\partial_x^\alpha \partial_\xi^\beta a(x, \xi)\right| \leq C_{K,\alpha,\beta} \left(1 + |\xi|\right)^{m-|\beta|}, \tag{15.379}$$

where $m \in \mathbb{R}$, K is an arbitrary compact subset of $\mathbb{R}^n$, and C is a positive constant depending on K, α, β. Starting from $a(x, \xi)$, the operator $a(x, D)$ is defined by

$$a(x, D)u = \frac{1}{(2\pi)^n} \int_{\mathbb{R}^n} e^{i\xi\cdot x} a(x, \xi)\widehat{u}(\xi)d\xi, \quad \forall u \in C_0^\infty(\mathbb{R}^n), \tag{15.380}$$

$a(x, \xi)$ is called the symbol of the operator $a(x, D)$.

For instance if we have a simple differential operator, such as the Laplacian in $\mathbb{R}^n$:

$$P(x, D) = -\Delta = -\sum_{i=1}^{n} \frac{\partial^2}{\partial x_i^2}.$$

Here, the symbol is a polynomial function in ξ, specifically $p(x, \xi) = |\xi|^2$.

Unlike the Laplacian, which has a simple polynomial symbol, pseudodifferential operators allow for much greater flexibility by considering smooth symbols with more complex growth behavior, for general pseudodifferential operator, the symbol is no longer necessarily a polynomial, but a more general smooth function that can depend on both x and ξ, such as in the case of a symbol like:

$$a(x, \xi) = (1 + |\xi|^2)^{m/2},$$

where m is the order of the operator. This shows that the symbol is no longer restricted to a polynomial form as with the Laplacian, but can be more flexible and include controlled growth functions. This flexibility is very useful, in some cases essential, in the development of advanced estimates such as Carleman estimates, where operators of varying complexity play a fundamental role.

The study of these operators forms what is today considered the classical theory of pseudodifferential operators, on which the reader can find excellent and extensive introductions ([41, Ch. 8], [116, 121, 134], in addition to the previously mentioned books).

For the study of Carleman estimates, an extension is necessary, which involves considering the function a depending not only on x and ξ, but also on a parameter $\tau \geq 1$. An extensive treatment of these operators can be found in [79, Vol. I, Ch. 2], [84] (see also [83, Appendix A]).

The theory of these pseudodifferential operators develops through the following steps.

(i) A class of symbols, in our case S^m_τ (see Definition 15.38 below), is defined, and the algebraic properties of this class with respect to basic operations (such as addition, multiplication, reciprocal, etc.) and asymptotic expansions are studied.
(ii) Based on the class of symbols, the class of pseudodifferential operators $\mathrm{Op}(S^m_\tau)$ is defined, and the relationships between operator operations and algebraic operations in S^m_τ are investigated.
(iii) The continuity properties of the operators in $\mathrm{Op}(S^m_\tau)$ on various functional spaces are studied, particularly on Sobolev spaces (referred to as "continuity on $L^2(\mathbb{R}^n)$"). Subsequently, deeper properties are analyzed, with particular emphasis on the Sharp Gårding Inequality in our case.

The transition from (i) to (ii) can be made through oscillatory integrals, which were briefly introduced in Sect. 4.10.

To explore the topics for which we provide a brief summary below in more detail, we refer in particular to [84] and [83, Appendix A].

Definition 15.38

(a) Let $m \in \mathbb{R}$. We say that a function

$$a : \mathbb{R}^n \times \mathbb{R}^n \times [1, +\infty) \to \mathbb{C},$$

belongs to the **symbol class** S^m_τ if
(a) for every $\tau \geq 1$, $a(\cdot, \cdot, \tau) \in C^\infty(\mathbb{R}^n \times \mathbb{R}^n)$,
(b) for every $\alpha, \beta \in \mathbb{N}^n_0$, we have

$$q_{\alpha,\beta}(a) := \sup_{(x,\xi,\tau)\in\mathbb{R}^n\times\mathbb{R}^n\times[1,+\infty)} \left|\partial^\alpha_x \partial^\beta_\xi a(x, \xi, \tau)\right| < +\infty. \tag{15.381}$$

(b) Let Ω be an open set of $\mathbb{R}^n$. We say that a function

$$a : \Omega \times \mathbb{R}^n \times [1, +\infty) \to \mathbb{C}$$

belongs to the **symbol class** $S^m_\tau(\Omega)$ if
(a) for every $\tau \geq 1$, $a(\cdot, \cdot, \tau) \in C^\infty(\mathbb{R}^n \times \mathbb{R}^n)$,
(b) for every, K, compact subset of Ω and for every $\alpha, \beta \in \mathbb{N}^n_0$, we have

$$q_{K,\alpha,\beta}(a) := \sup_{(x,\xi,\tau)\in K\times\mathbb{R}^n\times[1,+\infty)} \left|\partial^\alpha_x \partial^\beta_\xi a(x, \xi, \tau)\right| < +\infty. \tag{15.382}$$

The space S^m_τ turns out to be a Fréchet space, equipped with a countable family of seminorms defined by

$$p_j(a) = \max_{|\alpha|,|\beta|\le j} q_{\alpha,\beta}(a).$$

Let $a \in S^m_\tau(\Omega)$. We define

$$a(x, D, \tau)u = \frac{1}{(2\pi)^n}\int_{\mathbb{R}^n} e^{i\xi\cdot x} a(x, \xi, \tau)\widehat{u}(\xi)d\xi, \quad \forall u \in C^\infty_0(\Omega). \tag{15.383}$$

It is useful (and simple) to note that $a(x, D, \tau)$ is a continuous operator from $C^\infty_0(\Omega)$ into $C^\infty(\Omega)$, for every $\tau \ge 1$. However (this will be useful shortly), if $v \in C^\infty_0(\Omega)$, then if we denote by $\chi \in C^\infty_0(\Omega)$ a function that equals 1 in a neighborhood of supp v, and set

$$b(x, \xi, \tau) = \chi(x)a(x, \xi, \tau),$$

we have $b \in S^m_\tau$, and we can write

$$\langle a(x, D, \tau)v, v\rangle_{L^2(\mathbb{R}^n)} = \langle b(x, D, \tau)v, v\rangle_{L^2(\mathbb{R}^n)}, \tag{15.384}$$

(keeping in mind the convention of considering $C^\infty_0(\Omega)$ as a subspace of $C^\infty(\Omega)$).

We say that a is the symbol of the operator $a(x, D, \tau)$. We also write $\mathrm{Op}(a)$ to denote $a(x, D, \tau)$. If, in addition, $a \in S^m_\tau$ then the following continuity property in Sobolev spaces holds: for every $s \in \mathbb{R}$, there exist constants $C_{m,s}$ (independent of τ) such that

$$\|\mathrm{Op}(a)u\|_{H^{s+m}_\tau(\mathbb{R}^n)} \le C_{m,s}\|u\|_{H^s_\tau(\mathbb{R}^n)}, \quad \forall u \in H^s_\tau(\mathbb{R}^n).$$

Theorem 15.39 *Let $m \in \mathbb{R}$. If $a \in S^m_\tau$, then the adjoint operator $(\mathrm{Op}(a))^\star$ has a symbol, denoted by $J\overline{a}$, given by the oscillatory integral*

$$J\overline{a}(x, \xi, \tau) = \frac{1}{(2\pi)^n}\iint_{\mathbb{R}^{2n}} e^{-iy\cdot\eta}\overline{a}(x + y, \xi + \eta, \tau)\,dyd\eta.$$

In addition, the following asymptotic expansion holds: for every $N \in \mathbb{N}$, there exists $r_N \in S^{m-N}_\tau$ such that

$$J\overline{a} = \sum_{|\alpha|<N}\frac{1}{\alpha!}D^\alpha_\xi \partial^\alpha_x \overline{a} + r_N. \tag{15.385}$$

Theorem 15.40 *Let $m_j \in \mathbb{R}$, $j = 1, 2$. If $a_j \in S_\tau^{m_j}$, $j = 1, 2$, then the composition* $\operatorname{Op}(a_1)\operatorname{Op}(a_2)$ *has a symbol, denoted by $a_1 \diamond a_2$, given by the oscillatory integral*

$$(a_1 \diamond a_2)(x, \xi, \tau) = \frac{1}{(2\pi)^n} \iint_{\mathbb{R}^{2n}} e^{-iy\cdot\eta} a_1(x, \xi + \eta, \tau) a_2(x + y, \xi, \tau) dy d\eta.$$

In addition, the following asymptotic expansion holds:

$$D_\xi^\alpha a_1 \partial_x^\alpha a_2 \in S_\tau^{m_1+m_2-|\alpha|}, \quad \forall \alpha \in \mathbb{N}_0^n, \tag{15.386}$$

and for every $N \in \mathbb{N}$, there exists $r_N \in S_\tau^{m_1+m_2-N}$ such that

$$a_1 \diamond a_2 = \sum_{|\alpha|<N} \frac{1}{\alpha!} D_\xi^\alpha a_1 \partial_x^\alpha a_2 + r_N. \tag{15.387}$$

Theorems 15.39 and 15.40 thus allow extending Propositions such as 15.9 and the results of Sect. 15.4 to the case of operators in $\operatorname{Op}\left(S_\tau^m\right)$, suitably formulated. In this regard, we have the following:

Corollary 15.41

(a) Let $m \in \mathbb{R}$. If $a \in S_\tau^m$ then we have

$$a^\star(x, D, \tau) = \overline{a}(x, D, \tau) + r_{m-1}(x, D, \tau), \tag{15.388}$$

where $r_{m-1} \in S_\tau^{m-1}$.

(b) Let $m_j \in \mathbb{R}$, $j = 1, 2$. If $a_j \in S_\tau^{m_j}$, $j = 1, 2$, then we have

$$a_1(x, D, \tau) a_2(x, D, \tau) = (a_1 a_2)(x, D, \tau) + s_{m_1+m_2-1}(x, D, \tau), \tag{15.389}$$

where $s_{m_1+m_2-1} \in S_\tau^{m_1+m_2-1}$;

(c) Let $m_j \in \mathbb{R}$, $a_j \in S_\tau^{m_j}$, $j = 1, 2$ be as above

$$[a_1(x, D, \tau), a_2(x, D, \tau)] = \{a_1, a_2\}(x, D, \tau) + t_{m_1+m_2-2}(x, D, \tau), \tag{15.390}$$

where $s_{m_1+m_2-2} \in S_\tau^{m_1+m_2-2}$.

Now we state the form of the Sharp Gårding Inequality that we will use in the brief illustration of the application of pseudodifferential operator theory to Carleman estimates (see the references mentioned above).

Theorem 15.42 (Sharp Gårding Inequality) *Let Ω be an open set of $\mathbb{R}^n$. Let $m \in \mathbb{R}$. Assume that $a \in S_\tau^m(\Omega)$ satisfies*

$$a(x, \xi, \tau) \geq 0, \quad \forall (x, \xi, \tau) \in \Omega \times \mathbb{R}^n \times [1, +\infty). \tag{15.391}$$

Then for every compact subset K of Ω there exists a constant C such that

$$\Re \langle a(x, D, \tau)v, v\rangle_{L^2(\mathbb{R}^n)} + C \|v\|^2_{H_\tau^{\frac{m-1}{2}}(\mathbb{R}^n)} \geq 0, \quad \forall v \in C_K^\infty(\Omega). \tag{15.392}$$

(take into account (15.384)*).*

At this point, we apply Theorem 15.42 to prove Theorem 15.21.

We recall that in Theorem 15.21, $P(x, D)$ is an elliptic operator, whose principal part, $P_m(x, D)$, has the symbol $p_m(x, \xi)$. Furthermore, $\varphi \in C^\infty(\Omega)$ satisfies (15.166) and (15.167).

Let us denote by

$$b(x, \xi, \tau) = |p_m(x, \xi + i\tau\nabla\varphi(x))|^2 \tag{15.393}$$

and

$$\begin{aligned} a(x, \xi, \tau) = \tau c_{2m-1,\varphi}(x, \xi, \tau) + C_2 b(x, \xi, \tau) - \\ - C_1 |\xi + i\tau\nabla\varphi(x)|^{2m}, \end{aligned} \tag{15.394}$$

where C_1 and C_2 are the constants given by Lemma 15.22 and (see (15.152))

$$c_{2m-1,\varphi}(x, \xi, \tau) = \frac{i}{2} \left\{p_m(x, \xi + i\tau\nabla\varphi(x)), \overline{p_m(x, \xi + i\tau\nabla\varphi(x))}\right\}.$$

We have $a, b \in S_\tau^{2m}(\Omega)$ and by Lemma 15.22 we have

$$a(x, \xi, \tau) \geq 0, \quad \forall (x, \xi, \tau) \in \Omega \times \mathbb{R}^n \times [1, +\infty). \tag{15.395}$$

Let K be a compact subset of Ω. By Theorem 15.42 we have, for every $v \in C_K^\infty(\Omega)$ and every $\tau \geq 1$

$$\begin{aligned} 0 &\leq \Re \langle a(x, D, \tau)v, v\rangle_{L^2(\mathbb{R}^n)} + C \|v\|^2_{H_\tau^{m-\frac{1}{2}}(\mathbb{R}^n)} = \\ &= \Re \left\langle \mathrm{Op}(\tau c_{2m-1,\varphi})v, v\right\rangle_{L^2(\mathbb{R}^n)} + C\Re \langle \mathrm{Op}(b)v, v\rangle_{L^2(\mathbb{R}^n)} + \\ &+ C \|v\|^2_{H_\tau^{m-\frac{1}{2}}(\mathbb{R}^n)} - \widetilde{C}_1 \|v\|^2_{H_\tau^m(\mathbb{R}^n)} \end{aligned} \tag{15.396}$$

and by (15.125) (taking into account (15.93)) we have

$$\begin{aligned} \Re \langle \mathrm{Op}(b)v, v\rangle_{L^2(\mathbb{R}^n)} \leq C \|P_m(x, D + i\tau\nabla\varphi(x))v\|^2_{L^2(\mathbb{R}^n)} + \\ + C \|v\|^2_{H_\tau^{m-\frac{1}{2}}(\mathbb{R}^n)} \end{aligned} \tag{15.397}$$

On the other hand by Theorem 15.20, we have

$$\begin{aligned} &\Re\left\langle \mathrm{Op}(\tau c_{2m-1,\varphi})v, v\right\rangle_{L^2(\mathbb{R}^n)} \leq \\ &\leq C\tau \left\|v\right\|^2_{H^{m-1}_\tau(\mathbb{R}^n)} + C\tau \left\|P_m(x, D + i\tau\nabla\varphi(x))v\right\|^2_{L^2(\mathbb{R}^n)}, \end{aligned} \tag{15.398}$$

for every $v \in C^\infty_K(\Omega)$ and every $\tau \geq 1$.

Now, by (15.396)–(15.398) we that there exists $\tau_0 \geq 1$ such that

$$\left\|v\right\|^2_{H^m_\tau(\mathbb{R}^n)} \leq C\tau \left\|P_m(x, D + i\tau\nabla\varphi(x))v\right\|^2_{L^2(\mathbb{R}^n)}, \tag{15.399}$$

for every $v \in C^\infty_K(\Omega)$ and every $\tau \geq \tau_0$.

By (15.399) and taking into account (15.108), we conclude the proof of Theorem 15.21.

Chapter 16
Carleman Estimates for Reduced Regularity Coefficients

Abstract In this chapter we prove Carleman estimates for operators with continuously differentiable or Lipschitz coefficients. The classical approach by Hörmander is applied to elliptic operators.

16.1 Introduction

In Chaps. 14 and 15, we provided an initial understanding of Carleman estimates by proving their application in investigating the uniqueness of the Cauchy problem and the unique continuation property for operators with C^∞ coefficients in their principal parts. In the present chapter and in Chap. 17 we will provide proofs of Carleman estimates for operators whose principal part has coefficients that are either of class C^1 or merely Lipschitz continuous. In this chapter we will consider elliptic operators of order m and follow the classical approach developed in [56] by Hörmander, which is based on careful integration by parts of the so-called differential quadratic forms. In Chap. 17, we will address second-order operators with real and Lipschitz coefficients in the principal part.

16.2 Quadratic Differential Forms and Their Integration by Parts

16.2.1 Definition and Motivations for the Study of Differential Quadratic Forms

Let Ω be a bounded open set of $\mathbb{R}^n$. Let $a_{\alpha\beta}$ be a finite set of complex-valued functions that we will assume at least of class $C^0\left(\overline{\Omega}\right)$, we call a **differential**

S. Vessella, *Unique Continuation Properties for Partial Differential Equations*, Birkhäuser Advanced Texts Basler Lehrbücher,
https://doi.org/10.1007/978-3-031-86366-0_16

quadratic form, the quadratic form

$$C^{\infty}\left(\overline{\Omega}\right) \ni u \to \sum_{\alpha,\beta} a_{\alpha\beta}(x) D^{\alpha} u \overline{D^{\beta} u} \in \mathbb{C}. \tag{16.1}$$

We adopt the notation

$$F\left(x, D, \overline{D}\right)[u, \overline{v}] = \sum_{\alpha,\beta} a_{\alpha\beta}(x) D^{\alpha} u \overline{D^{\beta} v}, \quad \forall u, v \in C^{\infty}\left(\overline{\Omega}\right). \tag{16.2}$$

Here, $F\left(x, D, \overline{D}\right)[u, \overline{v}]$ is a **sesquilinear form** (see Sect. 16.2.2) and the differential quadratic form (16.1) is simply $F\left(x, D, \overline{D}\right)[u, \overline{u}]$. If $a_{\alpha\beta}$ are constants, we denote the corresponding sesquilinear and differential quadratic forms by $F\left(D, \overline{D}\right)[u, \overline{v}]$ and $F\left(D, \overline{D}\right)[u, \overline{u}]$, respectively. In what follows we are particularly interested in the integration by parts of the forms (16.1) if $u \in C_0^{\infty}(\Omega)$.

Let us now attempt to explain the reason for our interest in the quadratic forms (16.1) in relation to the proof of the Carleman estimates for the operator

$$P(x, D) = \sum_{|\alpha| \leq m} a_{\alpha}(x) D^{\alpha}, \tag{16.3}$$

where $m \in \mathbb{N}$ and $a_{\alpha} \in L^{\infty}(\Omega)$. Let K be a compact subset of Ω. In Proposition 15.13 we have proved, in particular, the equivalence between the Carleman estimate

$$\sum_{|\alpha| \leq m-1} \tau^{2(m-|\alpha|)-1} \int_{\Omega} \left|D^{\alpha} u\right|^2 e^{2\tau\varphi} dx \lesssim \int_{\Omega} \left|P_m(x, D) u\right|^2 e^{2\tau\varphi} dx, \tag{16.4}$$

for every $u \in C_K^{\infty}(\Omega)$, and the estimate

$$\tau^{1/2} \|v\|_{H_{\tau}^{m-1}(\mathbb{R}^n)} \lesssim \left\|\widetilde{P}_m(x, D + i\tau\nabla\varphi(x)) v\right\|_{L^2(\mathbb{R}^n)}, \tag{16.5}$$

for every $v \in C_K^{\infty}(\Omega)$, where $\widetilde{P}_m(x, D + i\tau\nabla\varphi(x))$ is defined by (15.89).

The first step to obtain (16.5) is provided by Theorem 15.14, where the operator $\widetilde{P}_m(x, D+i\tau\nabla\varphi(x))$ is written as the sum of $\widetilde{P}_m^{(s)}(x, D+i\tau\nabla\varphi(x))$ and $\widetilde{P}_m^{(a)}(x, D+i\tau\nabla\varphi(x))$ (respectively, the self-adjoint part and the skew-adjoint part of $\widetilde{P}_m(x, D+i\tau\nabla\varphi(x))$. The equality (15.27) from Proposition 15.3 is then applied to estimate from below the L^2-norm of

$$\widetilde{P}_m(x, D + i\tau\nabla\varphi(x)).$$

The essential point of this process lies in writing the adjoint operator of $\widetilde{P}_m(x, D + i\tau\nabla\varphi(x))$ and in the integration by parts of (see (15.113))

$$\Re \int_{\mathbb{R}^n} \widetilde{P}_m^{(s)}(x, D + i\tau\nabla\varphi(x))v \left(\overline{\widetilde{P}_m^{(a)}(x, D + i\tau\nabla\varphi(x))v}\right) dx.$$

These procedures, which require the assumption that the coefficients of $P_m(x, D)$ are highly regular (we assumed $C^\infty(\Omega)$), lead to the identity

$$\Re \int_{\mathbb{R}^n} \widetilde{P}_m^{(s)}(x, D + i\tau\nabla\varphi(x))v \left(\overline{\widetilde{P}_m^{(a)}(x, D + i\tau\nabla\varphi(x))v}\right) dx =$$
$$= \Re \left\langle \left[\widetilde{P}_m^{(s)}(x, D + i\tau\nabla\varphi(x)), \widetilde{P}_m^{(a)}(x, D + i\tau\nabla\varphi(x))\right] v, v \right\rangle_{L^2(\mathbb{R}^n)}.$$

The method we aim to develop in this chapter involves considering, suitable approximations of $\widetilde{P}_m^{(s)}(x, D+i\tau\nabla\varphi)$ and $\widetilde{P}_m^{(a)}(x, D+i\tau\nabla\varphi)$ instead of the original operators.

Let us denote

$$S(x, D, \tau) = \frac{1}{2}\left(\widetilde{P}_m(x, D + i\tau\nabla\varphi(x)) + \overline{\widetilde{P}_m}(x, D - i\tau\nabla\varphi(x))\right), \tag{16.6a}$$

$$A(x, D, \tau) = \frac{1}{2}\left(\widetilde{P}_m(x, D + i\tau\nabla\varphi(x)) - \overline{\widetilde{P}_m}(x, D - i\tau\nabla\varphi(x))\right), \tag{16.6b}$$

where

$$\overline{\widetilde{P}_m}(x, D - i\tau\nabla\varphi(x)) = \mathrm{Op}\left(\overline{p_m(x, \xi + i\tau\nabla\varphi(x))}\right). \tag{16.7}$$

We have trivially

$$\widetilde{P}_m(x, D + i\tau\nabla\varphi) = S(x, D, \tau) + A(x, D, \tau). \tag{16.8}$$

Hence

$$\begin{aligned} &\int \left|\widetilde{P}_m(x, D + i\tau\nabla\varphi)v\right|^2 dx = \\ &= \int |S(x, D, \tau)v|^2 dx + \int |A(x, D, \tau)v|^2 dx + \\ &+ 2\int \Re\left(S(x, D, \tau)v\overline{A(x, D, \tau)v}\right) dx. \end{aligned} \tag{16.9}$$

Very similar to what was done in Theorem 15.14, the **crucial point in the proof of a Carleman estimate** lies in finding an appropriate estimate from below of the third integral on the right-hand side of (16.9), while trying to use as little regularity of the coefficients as possible. It is precisely for this purpose that we are interested in studying the quadratic forms (16.1) and, in particular, the form

$$2\Re\left(S(x,D,\tau)v\overline{A(x,D,\tau)v}\right),\quad v\in C_0^\infty(\Omega). \tag{16.10}$$

Remark 16.17 By (15.120) we have that, if $P_m(x, D)$ has C^∞ coefficients, then operators (16.6a), (16.6b) are the first-order approximations respectively, of the symmetric and the skew-symmetric parts of the operator $\widetilde{P}_m(x, D + i\tau\nabla\varphi(x))$, we refer to Remark 16.18 of Sect. 16.3 for more details. ♦

16.2.2 Brief Review of Sesquilinear Forms

We recall that a **sesquilinear form** on a complex vector space V is a function

$$\Phi : V \times V \to V$$

such that $f(\cdot, v)$ is linear for every $v \in V$ and $\Phi(u, \cdot)$ is antilinear for every $u \in V$. We say that a sesquilinear form on V is **hermitian**, if

$$\Phi(u,v)=\overline{\Phi(v,u)},\qquad \forall u,v\in V. \tag{16.11}$$

In the sequel of this section we will denote by $\mathcal{V}$ the space of sesquilinear forms on $C^\infty(\mathbb{R}^n)$.

Let us consider the sesquilinear forms

$$\Phi_{\alpha\beta} : C^\infty\left(\mathbb{R}^n\right)\times C^\infty\left(\mathbb{R}^n\right)\to C^\infty\left(\mathbb{R}^n\right), \tag{16.12}$$

$$\Phi_{\alpha\beta}(u,v)=D^\alpha u\overline{D^\beta v},\qquad \forall u,v\in C^\infty\left(\mathbb{R}^n\right). \tag{16.13}$$

In what follows, for any $\zeta \in \mathbb{C}^n$, $\zeta = (\zeta_1, \cdots, \zeta_n)$ and $x \in \mathbb{R}^n$, $x = (x_1, \cdots, x_n)$, we will denote by

$$\zeta\cdot x=\sum_{j=1}^n \zeta_j x_j.$$

The following Proposition holds true.

Proposition 16.1 *The family of sesquilinear forms* $\left\{\Phi_{\alpha\beta}\right\}_{\alpha\beta\in\mathbb{N}_0^n}$ *is linearly independent in* $\mathcal{V}$.

Proof Let us consider a finite linear combination of the forms $\Phi_{\alpha\beta}$, $\alpha, \beta \in \Lambda$, where Λ is a finite subset of $\mathbb{N}_0^n \times \mathbb{N}_0^n$ and let us assume that it vanishes identically. We have, for some $c_{\alpha\beta} \in \mathbb{C}$,

$$\sum_{\alpha,\beta\in\Lambda} c_{\alpha\beta}\Phi_{\alpha\beta}=0. \tag{16.14}$$

Now we prove that

$$c_{\alpha\beta}=0, \qquad \forall\alpha,\beta\in\Lambda. \tag{16.15}$$

Let us notice that (16.14) is equivalent to

$$\sum_{\alpha,\beta\in\Lambda} c_{\alpha\beta}D^{\alpha}u\overline{D^{\beta}v}=0, \quad \forall u,v\in C^{\infty}\left(\mathbb{R}^n\right). \tag{16.16}$$

Now, let $\zeta,\eta\in\mathbb{C}^n$ be arbitrary and put $u=e^{i\zeta\cdot x}$, $v=e^{i\eta\cdot x}$ in (16.16). We get

$$0=\sum_{\alpha,\beta\in\Lambda} c_{\alpha\beta}D^{\alpha}\left(e^{i\zeta\cdot x}\right)\overline{D^{\beta}\left(e^{i\eta\cdot x}\right)}=e^{i\zeta\cdot x}\overline{e^{i\eta\cdot x}}\sum_{\alpha,\beta\in\Lambda} c_{\alpha\beta}\zeta^{\alpha}\overline{\eta^{\beta}}.$$

Therefore

$$\sum_{\alpha,\beta\in\Lambda} c_{\alpha\beta}\zeta^{\alpha}\overline{\eta^{\beta}}=0, \quad \forall\zeta,\eta\in\mathbb{C}^n.$$

From which we obtain (16.15). ■

Let us denote by $\mathcal{W}$ the subspace of $\mathcal{V}$ generated by $\left\{\Phi_{\alpha\beta}\right\}_{\alpha\beta\in\mathbb{N}_0^n}$. We call **sesquilinear differential form** any element of the space $\mathcal{W}$. Thus an arbitrary element of $\mathcal{W}$ is

$$\Phi(u,v)=\sum_{\alpha,\beta} a_{\alpha\beta}D^{\alpha}u\overline{D^{\beta}v}, \tag{16.17}$$

where the sum is finite, and $a_{\alpha\beta}\in\mathbb{C}$.

The following Proposition holds true.

Proposition 16.2 *Let $\Phi\in\mathcal{W}$ be given by* (16.17) *the following conditions are equivalent.*

$$\Phi \text{ is an hermitian form}, \tag{16.18a}$$

$$\Phi(u,u)\in\mathbb{R}, \qquad \forall u\in C^{\infty}\left(\mathbb{R}^n\right), \tag{16.18b}$$

$$a_{\alpha\beta}=\overline{a_{\beta\alpha}}, \qquad \forall\alpha,\beta\in\mathbb{N}_0^n, \tag{16.18c}$$

$$\sum_{\alpha,\beta} a_{\alpha\beta}\zeta^{\alpha}\overline{\zeta^{\beta}}\in\mathbb{R}, \quad \forall\zeta\in\mathbb{C}^n. \tag{16.18d}$$

Proof We follow the pattern

$$(16.18a) \Longleftrightarrow (16.18b) \Longleftrightarrow (16.18c) \Longleftrightarrow (16.18d).$$

implication (16.18a) $\Longrightarrow$ (16.18b) is trivial.
Let us prove that (16.18b) $\Longrightarrow$ (16.18a).

Let $u, v \in C^\infty(\mathbb{R}^n)$ be arbitraries. By (16.18b) we have

$$\Phi(u+v, u+v) \in \mathbb{R}, \quad \Phi(u+iv, u+iv) \in \mathbb{R}. \tag{16.19}$$

Setting $z = \Phi(u, v)$, $w = \Phi(v, u)$, we get, by (16.19),

$$z + w = \Phi(u+v, u+v) - \Phi(u, u) - \Phi(v, v) \in \mathbb{R}$$

and

$$-iz + iw = \Phi(u+iv, u+iv) - \Phi(u, u) - \Phi(v, v) \in \mathbb{R}.$$

From which we have $\Im(z+w) = 0$ and $\Re(z-w) = 0$; that is

$$\begin{cases} z + w - \overline{(z+w)} = 0 \\ \\ z - w + \overline{(z-w)} = 0 \end{cases}$$

and adding member to member we have $2z - 2\overline{w} = 0$. Hence

$$\Phi(u, v) = \overline{\Phi(v, u)}.$$

Let us prove that (16.18b) $\Longrightarrow$ (16.18c).

Let us assume that (16.18b) holds true. By the equivalence proved previously we have

$$\Phi(u, v) = \overline{\Phi(v, u)}, \qquad \forall u, v \in C^\infty\left(\mathbb{R}^n\right).$$

Hence, for any $u, v \in C^\infty(\mathbb{R}^n)$ we get

$$\begin{aligned} \sum_{\alpha,\beta} a_{\alpha\beta} D^\alpha u \overline{D^\beta v} &= \overline{\sum_{\alpha,\beta} a_{\alpha\beta} D^\alpha v \overline{D^\beta u}} = \\ &= \sum_{\alpha,\beta} \overline{a}_{\alpha\beta} \overline{D^\alpha v} D^\beta u = \sum_{\alpha,\beta} \overline{a}_{\beta\alpha} \overline{D^\beta v} D^\alpha u \end{aligned}$$

(the last step is a mere change of indices). From what we have just obtained and setting

$$c_{\alpha,\beta} = a_{\alpha\beta} - \overline{a}_{\beta\alpha},$$

we have

$$\sum_{\alpha,\beta} c_{\alpha\beta} D^\alpha u \overline{D^\beta v} = 0 \tag{16.20}$$

and Proposition 16.1 gives $a_{\alpha\beta} - \overline{a}_{\beta\alpha} = c_{\alpha,\beta} = 0$, for any $\alpha, \beta \in \mathbb{N}_0^n$. Therefore (16.18c) holds.

Let us prove that (16.18c) $\Longrightarrow$ (16.18b).

Let us suppose

$$a_{\alpha\beta} = \overline{a}_{\beta\alpha}.$$

Let $u \in C^\infty(\mathbb{R}^n)$ be arbitrary. We get

$$\sum_{\alpha,\beta} a_{\alpha\beta} D^\alpha u \overline{D^\beta u} = \sum_{\alpha,\beta} \overline{a}_{\beta\alpha} D^\alpha u \overline{D^\beta u} =$$
$$= \overline{\sum_{\alpha,\beta} a_{\beta\alpha} \overline{D^\alpha u} D^\beta u} = \overline{\sum_{\alpha,\beta} a_{\alpha\beta} \overline{D^\beta u} D^\alpha u} = \overline{\Phi(u,u)}.$$

The implication (16.18c) $\Longrightarrow$ (16.18d) can be proved in a similar way to the previous one (just replace ζ to D).

Let us prove that (16.18d) $\Longrightarrow$(16.18c).

Let us assume that (16.18d) holds. Then

$$\sum_{\alpha,\beta} a_{\alpha\beta} \zeta^\alpha \overline{\zeta^\beta} = \overline{\sum_{\alpha,\beta} a_{\alpha\beta} \zeta^\alpha \overline{\zeta^\beta}} =$$
$$= \sum_{\alpha,\beta} \overline{a}_{\alpha\beta} \overline{\zeta^\alpha} \zeta^\beta = \sum_{\alpha,\beta} \overline{a}_{\beta\alpha} \overline{\zeta^\beta} \zeta^\alpha.$$

From what obtained above and setting $c_{\alpha,\beta} = a_{\alpha\beta} - \overline{a}_{\beta\alpha}$, we get

$$\sum_{\alpha,\beta} c_{\alpha\beta} \zeta^\alpha \overline{\zeta^\beta} = 0, \qquad \forall \zeta \in \mathbb{C}^n.$$

From which we have, for any $\gamma, \delta \in \mathbb{N}_0^n$,

$$\gamma!\delta! c_{\gamma\delta} = \partial_\zeta^\gamma \partial_{\overline{\zeta}}^\delta \left(\sum_{\alpha,\beta} c_{\alpha\beta} \zeta^\alpha \overline{\zeta^\beta} \right)_{|\zeta=0} = 0, \tag{16.21}$$

where, $\zeta = (\zeta_1, \cdots, z_n)$, $z_j = \xi_j + i\eta_j$,

$$\partial_{\zeta_j} = \frac{1}{2}\left(\partial_{\xi_j} - i\partial_{\eta_j}\right), \qquad \partial_{\overline{\zeta}_j} = \frac{1}{2}\left(\partial_{\xi_j} + i\partial_{\eta_j}\right), \quad j = 1, \cdots, n.$$

Finally, (16.21) gives (16.18c). ■

16.2.3 Integration by Parts of Differential Quadratic Forms

First, let us consider a differential quadratic form with constant coefficients

$$F(D, \overline{D})[u, \overline{u}] = \sum_{\alpha,\beta} a_{\alpha\beta} D^{\alpha} u \overline{D^{\beta} u}, \tag{16.22}$$

where $a_{\alpha\beta} \in \mathbb{C}$. In what follows we will do the convention of the "regrouped terms" according to which the terms with the same indices α and β occur only one time. With this convention, the following polynomial in ζ and $\overline{\zeta}$

$$F(\zeta, \overline{\zeta}) = \sum_{\alpha,\beta} a_{\alpha\beta} \zeta^{\alpha} \overline{\zeta^{\beta}}, \tag{16.23}$$

is uniquely associated to the form F. As a matter of fact, it turns out

$$F(\zeta, \overline{\zeta}) = e^{-2(\Im\zeta)\cdot x} F(D, \overline{D}) \left[e^{i\zeta\cdot x}, \overline{e^{i\zeta\cdot x}}\right].$$

Hence, if

$$F(D, \overline{D})[u, \overline{u}] = 0, \quad \forall u \in C^{\infty}\left(\mathbb{R}^n\right), \tag{16.24}$$

then

$$F(\zeta, \overline{\zeta}) = 0, \quad \forall \zeta \in \mathbb{C}^n. \tag{16.25}$$

Conversely, if (16.25) holds, then we have

$$\gamma!\delta! a_{\gamma\delta} = \partial_{\zeta}^{\gamma} \partial_{\overline{\zeta}}^{\delta} \left(\sum_{\alpha,\beta} a_{\alpha\beta} \zeta^{\alpha} \overline{\zeta^{\beta}} \right)_{|\zeta=0} = \partial_{\zeta}^{\gamma} \partial_{\overline{\zeta}}^{\delta} F(\zeta, \overline{\zeta})_{|\zeta=0} = 0,$$

from which we get (16.24). All in all (16.24) and (16.25) are equivalent.

We call the polynomial $F(\zeta, \overline{\zeta})$ the **symbol of the differential quadratic form** $F(D, \overline{D})$.

The following Proposition will be useful later on.

Proposition 16.3 *Let $F(D,\overline{D})[u,\overline{u}]$ be a differential quadratic form with constant coefficients. We have*

$$\begin{aligned}&\int_{\mathbb{R}^n} F(D,\overline{D})[u,\overline{u}]dx = \\ &= \frac{1}{(2\pi)^n}\int_{\mathbb{R}^n} F(\xi,\xi)\,|\widehat{u}(\xi)|^2\,d\xi, \quad \forall u\in C_0^\infty\left(\mathbb{R}^n\right).\end{aligned} \tag{16.26}$$

Proof Let $u \in C_0^\infty(\mathbb{R}^n)$ and

$$F(D,\overline{D})[u,\overline{u}] = \sum_{\alpha,\beta} a_{\alpha\beta} D^\alpha u \overline{D^\beta u}.$$

From the Parseval identity we have

$$\begin{aligned}\int_{\mathbb{R}^n} F(D,\overline{D})[u,\overline{u}]dx &= \sum_{\alpha,\beta} a_{\alpha\beta}\int_{\mathbb{R}^n} D^\alpha u\overline{D^\beta u}dx = \\ &= \frac{1}{(2\pi)^n}\sum_{\alpha,\beta} a_{\alpha\beta}\int_{\mathbb{R}^n}\widehat{D^\alpha u}\overline{\widehat{D^\beta u}}d\xi = \\ &= \frac{1}{(2\pi)^n}\sum_{\alpha,\beta} a_{\alpha\beta}\int_{\mathbb{R}^n}\xi^{\alpha+\beta}\,|\widehat{u}(\xi)|^2\,d\xi = \\ &= \frac{1}{(2\pi)^n}\int_{\mathbb{R}^n} F(\xi,\xi)\,|\widehat{u}(\xi)|^2\,d\xi.\end{aligned}$$

■

We are interested to examine **under what conditions** $F(D,\overline{D})[u,\overline{u}]$ **can be "written as a divergence" of some vector field.**

More precisely, we are interested in examining under which conditions there exist some differential quadratic forms $G_k(D,\overline{D})$, $k = 1,\cdots,n$ with constant coefficients, such that

$$F(D,\overline{D})[u,\overline{u}] = \sum_{k=1}^{n} \partial_k\left(G_k(D,\overline{D})[u,\overline{u}]\right). \tag{16.27}$$

For this purpose we consider the differential quadratic form with constant coefficients

$$G(D,\overline{D})[u,\overline{u}] = \sum_{\alpha,\beta} c_{\alpha\beta} D^\alpha u\overline{D^\beta u}$$

and we wish to express the symbol of $\partial_k\left(G(D,\overline{D})[u,\overline{u}]\right)$ through the symbol of $G(D,\overline{D})[u,\overline{u}]$. We have

$$\begin{aligned}\partial_k\left(G(D,\overline{D})[u,\overline{u}]\right) &= \sum_{\alpha,\beta} c_{\alpha\beta}\partial_k\left(D^\alpha u\overline{D^\beta u}\right) = \\ &= \sum_{\alpha,\beta} c_{\alpha\beta}\left(\partial_k D^\alpha u\overline{D^\beta u} + D^\alpha u\partial_k\overline{D^\beta u}\right) = \\ &= \sum_{\alpha,\beta} c_{\alpha\beta}\left(iD_k D^\alpha u\overline{D^\beta u} - iD^\alpha u\overline{D_k D^\beta u}\right).\end{aligned}$$

Hence, the symbol associated to $\partial_k\left(G(D,\overline{D})[u,\overline{u}]\right)$ is

$$\begin{aligned}\sum_{\alpha,\beta} c_{\alpha\beta}\left(i\zeta_k\zeta^\alpha\overline{\zeta^\beta} - i\zeta^\alpha\overline{\zeta_k\zeta^\beta}\right) &= i\left(\zeta_k - \overline{\zeta}_k\right)\sum_{\alpha,\beta} c_{\alpha\beta}\zeta^\alpha\overline{\zeta^\beta} = \\ &= i\left(\zeta_k - \overline{\zeta}_k\right)G(\zeta,\overline{\zeta}).\end{aligned}$$

Therefore, in order to have (16.27) it is necessary that

$$F(\zeta,\overline{\zeta}) = i\sum_{k=1}^{n}\left(\zeta_k - \overline{\zeta}_k\right)G_k(\zeta,\overline{\zeta}). \tag{16.28}$$

Now, setting $\zeta = \xi + i\eta$, where $\xi, \eta \in \mathbb{R}^n$, we have

$$F(\xi + i\eta, \xi - i\eta) = -2\sum_{k=1}^{n}\eta_k G_k(\xi + i\eta, \xi - i\eta). \tag{16.29}$$

In particular, we have

$$F(\xi,\xi) = 0, \qquad \forall\xi \in \mathbb{R}^n \tag{16.30}$$

and

$$G_k(\xi,\xi) = -\frac{1}{2}\frac{\partial}{\partial\eta_k}F(\xi + i\eta, \xi - i\eta)_{|\eta=0}, \qquad \forall\xi \in \mathbb{R}^n. \tag{16.31}$$

Therefore, a **necessary condition for** (16.27) **to hold is that** (16.30) **is satisfied**. We will see later that this condition is also **sufficient**. But first, let us provide the definition of the double, total and separated order of a differential quadratic form.

Definition 16.4 Let

$$F(D,\overline{D})[u,\overline{u}] = \sum_{\alpha,\beta} a_{\alpha\beta} D^{\alpha} u \overline{D^{\beta} u} \tag{16.32}$$

be a differential quadratic form with constant coefficients. We say that F has **double order** $(\mu; m)$ provided

$$a_{\alpha\beta} \neq 0 \quad \Longrightarrow \quad |\alpha| + |\beta| \leq \mu; \quad |\alpha|, |\beta| \leq m. \tag{16.33}$$

μ is called the **total order** and m is called the **separated order** of the differential quadratic form F.

It is evident that

$$\mu \leq 2m.$$

Moreover, when adopting the convention of the grouped terms, the previous definition uniquely determines the order of the differential quadratic form. Here and in the sequel we will naturally extend the notions of the double order and the total order to the symbol of a differential quadratic form.

Lemma 16.5 *Let $F(D,\overline{D})[u,\overline{u}]$ be a differential quadratic form with constant coefficients. Suppose that*

$$F(\xi,\xi) = 0, \qquad \forall \xi \in \mathbb{R}^n, \tag{16.34}$$

then there exist n differential quadratic forms $G_k(D,\overline{D})[u,\overline{u}]$, $k = 1, \cdots, n$, such that

$$F(D,\overline{D})[u,\overline{u}] = \sum_{k=1}^{n} \partial_k \left(G_k(D,\overline{D})[u,\overline{u}]\right) \tag{16.35}$$

and we have

$$G_k(\xi,\xi) = -\frac{1}{2}\frac{\partial}{\partial \eta_k} F(\xi + i\eta, \xi - i\eta)_{|\eta=0}, \qquad \forall \xi \in \mathbb{R}^n. \tag{16.36}$$

In addition, assume that $F(D,\overline{D})[u,\overline{u}]$ has a double order $(\mu; m)$, $m > 0$ then

(a) if $\mu < 2m$, the forms G_k, $k = 1, \cdots, n$, can be choosen of double order $(\mu - 1; m - 1)$;
(b) if $\mu = 2m$, the forms G_k, $k = 1, \cdots, n$, can be choosen of double order $(\mu - 1; m)$.

Proof Let us assume that (16.34) holds. Let us consider the polynomial

$$\eta \to F(\xi + i\eta, \xi - i\eta)$$

and let us apply the Taylor formula at $\eta = 0$. We have, for suitable polynomials f_k, $k = 1, \cdots, n$,

$$F(\xi + i\eta, \xi - i\eta) = \sum_{k=1}^{n} \eta_k f_k(\xi, \eta) = -\frac{i}{2}\sum_{k=1}^{n} \left(\zeta_k - \overline{\zeta}_k\right) f_k\left(\frac{\zeta + \overline{\zeta}}{2}, \frac{\zeta - \overline{\zeta}}{2i}\right).$$

Set

$$G_k\left(\zeta, \overline{\zeta}\right) = -\frac{1}{2} f_k\left(\frac{\zeta + \overline{\zeta}}{2}, \frac{\zeta - \overline{\zeta}}{2i}\right),$$

we get

$$F(\zeta, \overline{\zeta}) = i\sum_{k=1}^{n} \left(\zeta_k - \overline{\zeta}_k\right) G_k(\zeta, \overline{\zeta}) = -2\sum_{k=1}^{n} \eta_k G_k(\xi + i\eta, \xi - i\eta), \tag{16.37}$$

from which we have

$$\frac{\partial}{\partial \eta_k} F(\xi + i\eta, \xi - i\eta) = -2G_k(\xi + i\eta, \xi - i\eta), \qquad \forall \xi \in \mathbb{R}^n. \tag{16.38}$$

Hence

$$G_k(\xi, \xi) = -\frac{1}{2}\frac{\partial}{\partial \eta_k} F(\xi + i\eta, \xi - i\eta)_{|\eta=0}, \qquad \forall \xi \in \mathbb{R}^n.$$

Moreover, from the first equality in (16.37) (retracing to backward the calculations that led to (16.28)) we have

$$F(D, \overline{D})[u, \overline{u}] = i\sum_{k=1}^{n} \left(D_k - \overline{D}_k\right)\left(G_k(D, \overline{D})[u, \overline{u}]\right) = \sum_{k=1}^{n} \partial_k \left(G_k(D, \overline{D})[u, \overline{u}]\right).$$

Proof of (a) and (b)

Case (a), $\mu < 2m$ Let us show that if α', α'', β', β'' are multi-indices such that

$$\begin{cases} \alpha' + \beta' = \alpha'' + \beta'' \\ |\alpha'| + |\beta'| = |\alpha''| + |\beta''| \leq \mu \leq 2m - 1 \\ |\alpha'|, \ |\beta'|, \ |\alpha''|, \ |\beta''| \leq m \end{cases}, \tag{16.39}$$

then

$$\zeta^{\alpha'}\overline{\zeta^{\beta'}} = \zeta^{\alpha''}\overline{\zeta^{\beta''}} + i\sum_{j=1}^{n}\left(\zeta_j - \overline{\zeta}_j\right) h_j(\zeta,\overline{\zeta}), \tag{16.40}$$

where $h_j(\zeta,\overline{\zeta})$, $j = 1, \cdots, n$, have the total order less or equal than $\mu - 1$ and the separated order less or equal than $m - 1$.

Notice that by (16.39) we have either $\left|\alpha'\right| < m$ or $\left|\beta'\right| < m$ (likewise for $\left|\alpha''\right|$ and $\left|\beta''\right|$).

The proof consists of repeatedly applying both simple identities.

$$\overline{\zeta}_j = \zeta_j - \left(\zeta_j - \overline{\zeta}_j\right), \tag{16.41a}$$

$$\zeta_j = \overline{\zeta}_j + \left(\zeta_j - \overline{\zeta}_j\right). \tag{16.41b}$$

Let us consider $\zeta^{\alpha'}\overline{\zeta^{\beta'}}$ and let us suppose $\left|\alpha'\right| < m$. Identity (16.41a) allows us to move the factors from $\overline{\zeta^{\beta'}}$ to $\zeta^{\alpha'}$ as long as the exponent of ζ does not have modulus m, when this occurs identity (16.41b) is used. Let us see more in detail. If $\left|\alpha'\right| < m$ and $\beta' \neq 0$, for instance let $\beta'_j > 0$, then

$$\begin{aligned}\zeta^{\alpha'}\overline{\zeta^{\beta'}} &= \zeta^{\alpha'}\overline{\zeta^{\beta'-e_j}}\overline{\zeta}_j = \zeta^{\alpha'}\overline{\zeta^{\beta'-e_j}}\left(\zeta_j - \left(\zeta_j - \overline{\zeta}_j\right)\right) = \\ &= \zeta^{\alpha'+e_j}\overline{\zeta^{\beta'-e_j}} - \left(\zeta_j - \overline{\zeta}_j\right)\zeta^{\alpha'}\overline{\zeta^{\beta'-e_j}}.\end{aligned}$$

Let us notice that $\zeta^{\alpha'}\overline{\zeta^{\beta'-e_j}}$ has total order $\left|\alpha'\right| + \left|\beta' - e_j\right| \leq \mu - 1$ and separated order less or equal than $m - 1$ (recall $\left|\alpha'\right| < m$ and $\left|\beta'\right| \leq m$). If, on the other hand, $\left|\beta'\right| < m$ (and this includes the case $\beta' = 0$ which was neglected previously) we use identity (16.41b) and proceeding as above we reach (assuming $\alpha'_j > 0$, for some j) to

$$\zeta^{\alpha'}\overline{\zeta^{\beta'}} = \zeta^{\alpha'-e_j}\overline{\zeta^{\beta'-e_j}} + \left(\zeta_j - \overline{\zeta}_j\right)\zeta^{\alpha'-e_j}\overline{\zeta^{\beta'}}. \tag{16.42}$$

Similarly to the case $\left|\alpha'\right| < m$, we have that $\zeta^{\alpha'-e_j}\overline{\zeta^{\beta'}}$ has total order less or equal than $\mu - 1$ and separated order less or equal than $m - 1$. Repeatedly applying the procedure used above we arrive to (16.40), which in turn implies that there exists $\widetilde{F}$ such that

$$F(\zeta,\overline{\zeta}) = \widetilde{F}(\zeta,\overline{\zeta}) + i\sum_{j=1}^{n}\left(\zeta_j - \overline{\zeta}_j\right) G_j(\zeta,\overline{\zeta}), \tag{16.43}$$

where $G_j(\zeta, \overline{\zeta})$, $j = 1, \cdots, n$, has double order $(\mu - 1; m - 1)$ and

$$\widetilde{F}(\zeta, \overline{\zeta}) = \sum_{(\alpha,\gamma)\in\Lambda} c_\gamma \zeta^\alpha \overline{\zeta^{\gamma-\alpha}}, \tag{16.44}$$

where $c_\gamma \in \mathbb{C}$

$$\Lambda = \left\{(\alpha, \gamma) \in \mathbb{N}_0^n \times \mathbb{N}_0^n : |\gamma| \leq m, \ \alpha \leq \gamma\right\}$$

and also $\widetilde{F}$ has double order $(\mu; m)$. Notice that, thanks to (16.40) and (16.42), the summation in (16.44) is written in such a way that for a given sum of the multi-indices only one addend occurs.

By (16.34) and (16.43) we have

$$0 = \widetilde{F}(\xi, \xi) = \sum_{|\gamma|\leq m} N_\gamma c_\gamma \xi^\gamma, \quad \forall \xi \in \mathbb{R}^n, \tag{16.45}$$

where N_γ is the cardinality of the set $\left\{\alpha \in \mathbb{N}_0^n : \alpha \leq \gamma\right\}$, hence, $c_\gamma = 0$ from which we have $\widetilde{F} = 0$. Therefore

$$F(\zeta, \overline{\zeta}) = i \sum_{j=1}^{n} \left(\zeta_j - \overline{\zeta}_j\right) G_j(\zeta, \overline{\zeta})$$

and (16.35) is proved in case (a).

Case (b), $\mu = 2m$ Obviously, we can handle the terms of

$$F(\zeta, \overline{\zeta}) = \sum_{\alpha,\beta} a_{\alpha\beta} \zeta^\alpha \overline{\zeta^\beta}.$$

satisfying $|\alpha| + |\beta| < 2m$ (and $|\alpha|, |\beta| \leq m$) in the same way of case (a). Let us examine in which a way we can handle the terms such that $|\alpha| + |\beta| = 2m$. Since $|\alpha|, |\beta| \leq m$ we have $|\alpha| = |\beta| = m$.

By identities (16.41a) and (16.41b) we have

$$\zeta_j \overline{\zeta}_k = \zeta_k \overline{\zeta}_j + \left(\zeta_j - \overline{\zeta}_j\right) \overline{\zeta}_k - \left(\zeta_k - \overline{\zeta}_k\right) \overline{\zeta}_j,$$

for $j, k = 1, \cdots, n$.

Now, let us suppose that $\alpha', \alpha'', \beta', \beta''$ satisfy

$$\begin{cases} \alpha' + \beta' = \alpha'' + \beta'' \\ \\ |\alpha'| = |\beta'| = |\alpha''| = |\beta''| = m \end{cases},$$

then we have (if $\alpha'_j > 0$ and $\beta'_k > 0$)

$$\zeta^{\alpha'}\overline{\zeta^{\beta'}} = \zeta^{\alpha'-e_j}\overline{\zeta^{\beta'-e_k}}\zeta_j\overline{\zeta}_k = \zeta^{\alpha'-e_j+e_k}\overline{\zeta^{\beta'-e_k+e_j}}+$$
$$+\left(\zeta_j - \overline{\zeta}_j\right)h_1(\zeta,\overline{\zeta}) + \left(\zeta_k - \overline{\zeta}_k\right)h_2(\zeta,\overline{\zeta}),$$

where

$$h_1(\zeta,\overline{\zeta}) = \zeta^{\alpha'-e_j}\overline{\zeta^{\beta'}}, \qquad h_2(\zeta,\overline{\zeta}) = -\zeta^{\alpha'-e_j}\overline{\zeta^{\beta'-e_k+e_j}},$$

have double order $(\mu-1,m)$. From now on, one may proceed as in case (a) and we reach the conclusion.

■

In the case of a differential quadratic forms with variable coefficients we have the following

Lemma 16.6 *Let*

$$F(x,D,\overline{D})[u,\overline{u}] = \sum_{\alpha,\beta} a_{\alpha\beta}(x)D^\alpha u\overline{D^\beta u} \tag{16.46}$$

be a differential quadratic form with variable coefficients $a_{\alpha\beta} \in C^s(\Omega)$, where $s \in \mathbb{N}$. Suppose that F has double order $(\mu; m)$, $m > 0$, and that

$$F(x,\xi,\xi) = 0, \qquad \forall x \in \Omega,\ \forall \xi \in \mathbb{R}^n. \tag{16.47}$$

Then there exists a differential quadratic form $G(x,D,\overline{D})[u,\overline{u}]$ whose coefficients belong to $C^{s-1}(\Omega)$ such that

$$\int_\Omega F(x,D,\overline{D})[u,\overline{u}]dx = \int_\Omega G(x,D,\overline{D})[u,\overline{u}]dx, \quad \forall u \in C_0^\infty(\Omega) \tag{16.48}$$

and such that:

(a) if $\mu < 2m$, then $G(x,D,\overline{D})$ can be chosen of double order $(\mu-1; m-1)$;
(b) if $\mu = 2m$, then $G(x,D,\overline{D})$ can be chosen of double order $(\mu-1; m)$. Moreover

$$G(x,\xi,\xi) = \frac{1}{2}\sum_{k=1}^n \partial^2_{x_k\eta_k}F(x,\xi+i\eta,\xi-i\eta)_{|\eta=0}. \tag{16.49}$$

Proof Let $F_1,\cdots,F_N$ be a basis of the vector space (of finite dimension) of all quadratic forms H of double order $(\mu; m)$ with constant coefficients and satisfying

$$H(\xi,\xi) = 0, \qquad \forall \xi \in \mathbb{R}^n.$$

By Lemma 16.5 there exist differential quadratic forms with constant coefficients G_j^k, $j = 1, \cdots, N$, $k = 1, \cdots, n$, of double order $(\mu - 1; m')$, with $m' = m - 1$, provided $\mu < 2m$, and $m' = m$ provided $\mu = 2m$, such that

$$F_j(D, \overline{D})[u, \overline{u}] = \sum_{k=1}^{n} \partial_{x_k} \left(G_j^k(D, \overline{D})[u, \overline{u}] \right), \quad j = 1, \cdots, N.$$

Now, (16.47) implies that there exist $c_j \in C^s(\Omega)$, $j = 1, \cdots, N$, such that

$$F(x, D, \overline{D})[u, \overline{u}] = \sum_{j=1}^{N} c_j(x) F_j(D, \overline{D})[u, \overline{u}].$$

Hence, if $u \in C_0^\infty(\Omega)$, then integration by parts yields

$$\begin{aligned} \int_\Omega F(x, D, \overline{D})[u, \overline{u}] dx &= \sum_{j=1}^{N} \sum_{k=1}^{n} \int_\Omega c_j(x) \partial_{x_k} \left(G_j^k(D, \overline{D})[u, \overline{u}] \right) dx = \\ &= -\sum_{j=1}^{N} \sum_{k=1}^{n} \int_\Omega \partial_{x_k} c_j(x) G_j^k(D, \overline{D})[u, \overline{u}] dx. \end{aligned}$$

Thus, we can choose

$$G(x, D, \overline{D})[u, \overline{u}] = \sum_{j=1}^{N} \sum_{k=1}^{n} -\partial_{x_k} c_j(x) G_j^k(D, \overline{D})[u, \overline{u}].$$

From which we have

$$G(x, \xi, \xi) = \sum_{j=1}^{N} \sum_{k=1}^{n} -\partial_{x_k} c_j(x) G_j^k(\xi, \xi). \tag{16.50}$$

On the other hand by (16.36) we have

$$G_j^k(\xi, \xi) = -\frac{1}{2} \frac{\partial}{\partial \eta_k} F_j(\xi + i\eta, \xi - i\eta)_{|\eta=0}, \qquad \forall \xi \in \mathbb{R}^n.$$

Now, by the last equality we get

$$G(x,\xi,\xi)=\sum_{k=1}^{n}\sum_{j=1}^{N}-\partial_{x_k}c_j(x)G_j^k(\xi,\xi)=$$

$$=\frac{1}{2}\sum_{k=1}^{n}\left(\partial^2_{x_k\eta_k}\sum_{j=1}^{N}c_j(x)F_j(\xi+i\eta,\xi-i\eta)\right)_{|\eta=0}=$$

$$=\frac{1}{2}\sum_{k=1}^{n}\partial^2_{x_k\eta_k}F(\xi+i\eta,\xi-i\eta)_{|\eta=0}.$$

■

16.3 Carleman Estimates by Means Differential Quadratic Forms

In this section, we will see how we can apply differential quadratic forms to carry out proofs of Carleman estimates while reducing the regularity requirements on the coefficients of the principal part of differential operator (16.3).

We recall that (see (16.9))

$$\begin{aligned}&\int\left|\widetilde{P}_m(x,D+i\tau\nabla\varphi)v\right|^2dx=\\&=\int|S(x,D,\tau)v|^2\,dx+\int|A(x,D,\tau)v|^2\,dx+\\&+2\int\Re\left(S(x,D,\tau)v\overline{A(x,D,\tau)v}\right)dx,\end{aligned}\tag{16.51}$$

where $\widetilde{P}_m(x,D+i\tau\nabla\varphi(x))$ is defined by (15.89), $S(x,D,\tau)$ and $A(x,D,\tau)$ are defined by (16.6a) and (16.6b) respectively.

Now, let us consider the differential quadratic form

$$F(x,D,\overline{D},\tau)\,[v,\overline{v}]=2\Re\left(S(x,D,\tau)v\overline{A(x,D,\tau)v}\right),\tag{16.52}$$

whose symbol is

$$F(x,\zeta,\overline{\zeta},\tau)=2\Re\left(S(x,\zeta,\tau)\overline{A(x,\zeta,\tau)}\right),\quad\forall\zeta\in\mathbb{C}^n.$$

By the definition of $\widetilde{P}_m(x, D + i\tau\nabla\varphi(x))$ we have, for $\zeta = \xi + i\eta \in \mathbb{C}^n$,

$$\begin{aligned}
F(x, \zeta, \overline{\zeta}, \tau) &= 2\Re\left(S(x, \zeta, \tau)\overline{A(x, \zeta, \tau)}\right) = \\
&= \frac{1}{2}\Re\left((p_m(x, \zeta + i\tau\nabla\varphi(x)) + \overline{p}_m(x, \zeta - i\tau\nabla\varphi(x))\right) \times \\
&\times \left(\overline{p_m(x, \zeta + i\tau\nabla\varphi(x)) - \overline{p}_m(x, \zeta - i\tau\nabla\varphi(x))}\right)\Big) = \\
&= \frac{1}{2}\left(|p_m(x, \zeta + i\tau\nabla\varphi(x))|^2 - \left|\overline{p}_m(x, \zeta - i\tau\nabla\varphi(x))\right|^2\right).
\end{aligned}$$

Hence

$$\begin{aligned}
&F(x, \zeta, \overline{\zeta}, \tau) = \\
&= \frac{1}{2}\left(|p_m(x, \zeta + i\tau\nabla\varphi(x))|^2 - \left|\overline{p}_m(x, \zeta - i\tau\nabla\varphi(x))\right|^2\right).
\end{aligned} \tag{16.53}$$

By this equality we get, for every $\xi \in \mathbb{R}^n$,

$$\begin{aligned}
F(x, \xi, \xi, \tau) &= \frac{1}{2}\left(|p_m(x, \xi + i\tau\nabla\varphi(x))|^2 - \left|\overline{p}_m(x, \xi - i\tau\nabla\varphi(x))\right|^2\right) = \\
&= \frac{1}{2}\left(|p_m(x, \xi + i\tau\nabla\varphi(x))|^2 - \left|\overline{p_m(x, \xi + i\tau\nabla\varphi(x))}\right|^2\right) = 0.
\end{aligned}$$

Hence, **assuming that the coefficients of $P_m(x, D)$ belong to $C^1\left(\overline{\Omega}\right)$** we can apply Lemma 16.6. Using formula (16.49) and denoting by

$$G(x, \xi, \xi, \tau) := \frac{1}{2}\sum_{k=1}^{n} \partial^2_{x_k\eta_k} F(x, \xi + i\eta, \xi - i\eta, \tau)_{|\eta=0}, \tag{16.54}$$

we have

$$\begin{aligned}
2\int \Re\left(S(x, D, \tau)v\overline{A(x, D, \tau)v}\right) dx &= \int F(x, D, \overline{D}, \tau)[v, \overline{v}]dx = \\
&= \int G(x, D, \overline{D}, \tau)[v, \overline{v}]dx,
\end{aligned} \tag{16.55}$$

for every $v \in C_0^\infty(\Omega)$.

Now we calculate the expression on the right hand side in (16.54). Although the calculation is elementary, let us perform it in detail. We have

$$\begin{aligned}
&\partial_{\eta_k}\frac{1}{2}\left(|p_m(x,\xi+i\eta+i\tau\nabla\varphi(x))|^2-\right.\\
&\left.-\left|\overline{p}_m(x,\xi+i\eta-i\tau\nabla\varphi(x))\right|^2\right)_{|\eta=0}=\\
&=\partial_{\eta_k}\frac{1}{2}\left(|p_m(x,\xi+i\eta+i\tau\nabla\varphi(x))|^2-\right.\\
&\left.-|p_m(x,\xi-i\eta+i\tau\nabla\varphi(x))|^2\right)_{|\eta=0}=\\
&=2\Re\left(-ip_m(x,\xi+i\tau\nabla\varphi(x))\overline{p_m^{(k)}(x,\xi+i\tau\nabla\varphi(x))}\right).
\end{aligned}\tag{16.56}$$

Set for short

$$\zeta=\xi+i\tau\nabla\varphi(x)$$

and let us differentiate what obtained in (16.56) w.r.t. x_k. By (16.54) we get

$$\begin{aligned}
&G(x,\xi,\xi,\tau)=\\
&=\tau\sum_{j,k=1}^{n}\partial^2_{x_jx_k}\varphi(x)p_m^{(j)}(x,\zeta)\overline{p_m^{(k)}(x,\zeta)}+\\
&+\Im\left(\sum_{k=1}^{n}p_{m,k}(x,\zeta)\overline{p_m^{(k)}(x,\zeta)}\right)+\\
&+\Im\left[p_m(x,\zeta)\left(\sum_{k=1}^{n}\overline{p_{m,k}^{(k)}(x,\zeta)}-i\tau\sum_{j,k=1}^{n}\overline{p_m^{(k,j)}(x,\zeta)}\partial^2_{x_jx_k}\varphi(x)\right)\right].
\end{aligned}\tag{16.57}$$

Let us observe that we have

$$\begin{aligned}
&p_m(x,\xi+i\tau\nabla\varphi)=0\Rightarrow\\
&\Rightarrow\ G(x,\xi,\xi,\tau)=\frac{i}{2}\left\{p_m(x,\xi+i\tau\nabla\varphi),\overline{p_m(x,\xi+i\tau\nabla\varphi)}\right\}=\\
&=c_{2m-1,\varphi}(x,\xi,\tau),
\end{aligned}\tag{16.58}$$

where $\{\cdot,\cdot\}$ is the Poisson bracket defined in (15.3) and $c_{2m-1,\varphi}(x,\xi,\tau)$ is defined by (15.152). In order to check (16.58) it suffices to develop the Poisson bracket in (16.58), and to notice that the third term on the right hand side in (16.57) vanishes when $P_m(x,\xi+i\tau\nabla\varphi(x))=0$.

We already know that $G(x,\xi,\xi,\tau)$ is a homogeneous polynomial of degree $2m-1$ in the variables (ξ,τ). However, we will be interested in more precise information about the differential quadratic form $G(x, D, \overline{D}, \tau)$ or, equivalently on its symbol $G(x, \zeta, \overline{\zeta}, \tau)$, to this purpose we prove

Proposition 16.7 *Let $P_m(x, D)$ be the differential operator*

$$P_m(x, D) = \sum_{|\alpha|=m} a_\alpha(x) D^\alpha,$$

where $a_\alpha \in C^1\left(\overline{\Omega}\right)$, for $|\alpha| = m$. Let $F(x, \zeta, \overline{\zeta}, \tau)$ be defined by (16.53)*. Then there exists a differential quadratic form $G\left(x, D, \overline{D}, \tau\right)$ such that*

$$\int F\left(x, D, \overline{D}, \tau\right)[v, \overline{v}]dx = \int G\left(x, D, \overline{D}, \tau\right)[v, \overline{v}]dx. \tag{16.59}$$

Moreover

$$G\left(x, D, \overline{D}, \tau\right) = \sum_{h=0}^{2m-1} \tau^h G^{(h)}\left(x, D, \overline{D}\right), \tag{16.60}$$

where $G^{(h)}\left(x, D, \overline{D}\right)$ is a differential quadratic form which has double order $(2m - h - 1; m)$, for $h = 0, 1 \cdots, 2m - 1$.

If the ***coefficients*** a_α, ***for*** $|\alpha| = m$, ***are real valued functions*** *then* (16.59) *continues to hold true, but instead of* (16.60) *we have*

$$G\left(x, D, \overline{D}, \tau\right) = \tau \sum_{h=0}^{2m-2} \tau^h G^{(h)}\left(x, D, \overline{D}\right), \tag{16.61}$$

where $G^{(h)}\left(x, D, \overline{D}\right)$ is a differential quadratic form which has double order $(2m - h - 2; m)$, $h = 0, \cdots, 2m - 2$.

In any case $G(x, \xi, \xi, \tau)$ is given by (16.57)*.*

Proof By the Taylor formula we get

$$p_m(x, \zeta + i\tau\nabla\varphi(x)) = \sum_{k=0}^{m} \tau^k q_{m-k}(x, \zeta), \quad \forall \zeta \in \mathbb{C}^n, \tag{16.62}$$

where, for $k = 0, 1, \cdots, m$, $q_{m-k}(x, \zeta)$ are polynomials in the variable ζ of degree $m - k$. Moreover the coefficients of $q_{m-k}(x, \zeta)$ are of class $C^1\left(\overline{\Omega}\right)$. We have

$$|p_m(x, \zeta + i\tau\nabla\varphi(x))|^2 = \sum_{k,j=0}^{m} \tau^{k+j} q_{m-k}(x, \zeta)\overline{q}_{m-j}(x, \overline{\zeta}).$$

Hence, by (16.53) we get

$$F\left(x,\zeta,\overline{\zeta},\tau\right)=\sum_{k,j=0}^{m}\tau^{k+j}F_{kj}\left(x,\zeta,\overline{\zeta}\right),$$

where, for $j,k=1,\cdots,m$

$$F_{kj}\left(x,\zeta,\overline{\zeta}\right)=q_{m-k}(x,\zeta)\overline{q}_{m-j}(x,\overline{\zeta})-q_{m-k}(x,\overline{\zeta})\overline{q}_{m-j}(x,\zeta). \tag{16.63}$$

Each of the forms F_{kj} has double order $(2m-(k+j);m)$, furthermore by (16.63), since q_{00} has degree 0, we have

$$F_{mm}\left(x,\zeta,\overline{\zeta}\right)=0,\quad \forall\zeta\in\mathbb{C}^n. \tag{16.64}$$

Moreover

$$F_{kj}\left(x,\xi,\xi\right)=0,\qquad \forall\xi\in\mathbb{R}^n$$

and by Lemma 16.6—case (b)—we have that, for $j,k=1,\cdots,m$, where either j or k are different from m, there exist a differential quadratic form G_{kj} which have double order $(2m-(k+j)-1;m)$ and satisfying

$$\int F_{kj}\left(x,D,\overline{D}\right)[v,\overline{v}]dx=\int G_{kj}\left(x,D,\overline{D}\right)[v,\overline{v}]dx,\quad \forall v\in C_0^\infty(\Omega),$$

of course, since (16.64) holds, we may choose

$$G_{mm}\equiv 0.$$

Therefore, by the last obtained equality and setting

$$G^{(h)}\left(x,D,\overline{D}\right)=\sum_{k+j=h}G_{kj}\left(x,D,\overline{D}\right),\quad h=1,\cdots,2m-1,$$

$$G\left(x,D,\overline{D},\tau\right)[v,\overline{v}]=\sum_{h=0}^{2m-1}\tau^h G^{(h)}\left(x,D,\overline{D}\right)[v,\overline{v}],$$

we have that $G^{(h)}$ is a differential quadratic form which has double order $(2m-h-1;m)$ and

$$\begin{aligned}\int F\left(x,D,\overline{D},\tau\right)[v,\overline{v}]dx &= \sum_{k,j=0}^{m}\tau^{k+j}\int F_{kj}\left(x,D,\overline{D}\right)[v,\overline{v}]dx = \\ &= \sum_{k,j=0}^{m}\tau^{k+j}\int G_{jk}\left(x,D,\overline{D}\right)[v,\overline{v}]dx = \\ &= \sum_{h=0}^{2m-1}\tau^{h}\int G^{(h)}\left(x,D,\overline{D}\right)[v,\overline{v}]dx = \\ &= \int G\left(x,D,\overline{D},\tau\right)[v,\overline{v}]dx.\end{aligned} \tag{16.65}$$

If **the coefficients of** $P_m(x,D)$ **are real-valued** then also the coefficients of the polynomials $q_j(x,\zeta)$ in (16.62) are real-valued and (16.63) can be written as

$$F_{kj}\left(x,\zeta,\overline{\zeta}\right) = q_{m-k}(x,\zeta)q_{m-j}(x,\overline{\zeta}) - q_{m-k}(x,\overline{\zeta})q_{m-j}(x,\zeta). \tag{16.66}$$

Hence, besides (16.64), we have

$$F_{00}\left(x,\zeta,\overline{\zeta}\right) = 0, \quad \forall\zeta\in\mathbb{C}^n. \tag{16.67}$$

Therefore we have

$$F\left(x,\zeta,\overline{\zeta},\tau\right) = \tau\sum_{h=0}^{2m-2}\tau^{h}F^{(h)}\left(x,\zeta,\overline{\zeta}\right),$$

where

$$F^{(h)}\left(x,\zeta,\overline{\zeta}\right) = \sum_{k+j=h+1}F_{kj}\left(x,\zeta,\overline{\zeta}\right).$$

Hence $F^{(h)}$, for $h=0,\cdots,2m-2$, has double order $(2m-h-1;m)$. Therefore by applying Lemma 16.6—case (a)—there exist $G^{(h)}\left(x,D,\overline{D}\right)$, differential quadratic forms which have double order $(2m-h-2;m-1)$, such that

$$\int F^{(h)}\left(x,D,\overline{D}\right)[v,\overline{v}]dx = \int G^{(h)}\left(x,D,\overline{D}\right)[v,\overline{v}]dx, \quad \forall v\in C_0^\infty(\Omega),$$

and setting

$$G\left(x,D,\overline{D},\tau\right)[v,\overline{v}] = \tau\sum_{h=0}^{2m-2}\tau^{h}G^{(h)}\left(x,D,\overline{D}\right)[v,\overline{v}]$$

we get

$$\int F\left(x, D, \overline{D}, \tau\right)[v, \overline{v}]dx = \int G\left(x, D, \overline{D}, \tau\right)[v, \overline{v}]dx.$$

■

Now, we **broadly outline** the main ideas that are involved in proving a Carleman estimate. We come back, then, to the third integral in (16.51). Let $x_0 \in \overline{\Omega}$. We have from (16.26) and (16.55),

$$\begin{aligned}
2\int \Re\left(S(x, D, \tau)v\overline{A(x, D, \tau)v}\right) dx &= \int G(x, D, \overline{D}, \tau)[v, \overline{v}]dx = \\
&= \int G(x_0, D, \overline{D}, \tau)[v, \overline{v}]dx + \\
&+ \int \left(G(x, D, \overline{D}, \tau) - G(x_0, D, \overline{D}, \tau)\right)[v, \overline{v}]dx = \\
&= \frac{1}{(2\pi)^n}\int G(x_0, \xi, \xi, \tau)\left|\widehat{v}(\xi)\right|^2 d\xi + \\
&+ \underbrace{\int \left(G(x, D, \overline{D}, \tau) - G(x_0, D, \overline{D}, \tau)\right)[v, \overline{v}]dx}_{\mathcal{R}}.
\end{aligned} \tag{16.68}$$

The main idea that we will follow consist essentially in what follows:

(a) Choice of φ The choice of φ will be made so that we have

$$\begin{aligned}
&P_m(x, \xi + i\tau\nabla\varphi(x)) = 0 \Rightarrow \\
&\Rightarrow G(x, \xi, \xi, \tau) > 0, \quad (\xi, \tau) \in \mathbb{R}^n \times \mathbb{R} \setminus (0, 0), \tau > 0,
\end{aligned} \tag{16.69}$$

hence, by the homogeneity of G w.r.t. (ξ, τ) we get

$$\begin{aligned}
&P_m(x, \xi + i\tau\nabla\varphi(x)) = 0 \Rightarrow \\
&\Rightarrow G(x, \xi, \xi, \tau) \geq C\left(|\xi|^2 + \tau^2\right)^{m-\frac{1}{2}}, \quad \forall x \in \overline{\Omega},
\end{aligned} \tag{16.70}$$

for every $(\xi, \tau) \in \mathbb{R}^{n+1}$, $\tau > 0$.

(b) Next Steps Keeping in mind Lemma 14.2, we will exploit the local character of a Carleman estimate to focus on the case where **the support of** v **(and hence of** u**) is sufficiently small**. With this expedient, the term $\mathcal{R}$ on the right-hand side in (16.68) can be treated as a kind of rest and can be efficiently estimated from below by simultaneously exploiting the continuity of the coefficients of the quadratic form $G(x, D, \overline{D}, \tau)$ and Proposition 16.7.

Remark 16.18 Now, assuming that the coefficients of the principal of $P(x, D)$ are C^{∞}, we briefly compare the integral (see (16.55))

$$\int G(x, D, \overline{D}, \tau)[v, \overline{v}]dx,$$

with the integral of the commutator, used in Chap. 15,

$$\int \left[\widetilde{P}_m^{(s)}(x, D + i\tau\nabla\varphi(x)), \widetilde{P}_m^{(a)}(x, D + i\tau\nabla\varphi(x))\right] v\overline{v}dx. \tag{16.71}$$

We denote for brevity (see (15.110) and (15.111))

$$\begin{aligned}
&\mathcal{S}(x, D, \tau) = \widetilde{P}_m^{(s)}(x, D + i\tau\nabla\varphi(x)) = \\
&= \frac{1}{2}\left(\widetilde{P}_m(x, D + i\tau\nabla\varphi(x)) + \widetilde{P}_m^{\star}(x, D + i\tau\nabla\varphi(x))\right), \\
&\mathcal{A}(x, D, \tau) = \widetilde{P}_m^{(a)}(x, D + i\tau\nabla\varphi(x)) = \\
&= \frac{1}{2}\left(\widetilde{P}_m(x, D + i\tau\nabla\varphi(x)) - \widetilde{P}_m^{\star}(x, D + i\tau\nabla\varphi(x))\right).
\end{aligned}$$

We have

$$\begin{aligned}
&\int \left|\widetilde{P}_m(x, D + i\tau\nabla\varphi(x))v\right|^2 dx = \\
&= \int |\mathcal{S}(x, D, \tau)v|^2 dx + \int |\mathcal{A}(x, D, \tau)v|^2 dx + \\
&+ 2\int \Re\left(\mathcal{S}(x, D, \tau)v\overline{\mathcal{A}(x, D, \tau)v}\right) dx,
\end{aligned} \tag{16.72}$$

and (see Proposition 15.3)

$$\begin{aligned}
&2\int \Re\left(\mathcal{S}(x, D, \tau)v\overline{\mathcal{A}(x, D, \tau)v}\right) dx = \\
&= \int \left([\mathcal{S}(x, D, \tau), \mathcal{A}(x, D, \tau)]\, v\right) \overline{v}dx.
\end{aligned} \tag{16.73}$$

Now let us compare the integrals

$$2\int \Re\left(S(x, D, \tau)vA(x, D, \tau)v\right) dx, \qquad 2\int \Re\left(\mathcal{S}(x, D, \tau)v\overline{\mathcal{A}(x, D, \tau)v}\right) dx.$$

Set

$$R(x, D, \tau) = \frac{1}{2}\left(\frac{1}{i}\sum_{k=1}^{n} \mathrm{Op}\left(p_{m,j}^{(j)}\right)(x, D, \tau) + \mathrm{Op}\left(r_{m-2}\right)(x, D, \tau)\right),$$

where $r_{m-2} \in \Sigma^{m-2}(\Omega)$ (see (15.120)). We have

$$\mathcal{S}(x, D, \tau) = S(x, D, \tau) + R(x, D, \tau) \text{ and } \mathcal{A}(x, D, \tau) = S(x, D, \tau) - R(x, D, \tau)$$

then

$$2\Re\left(\mathcal{S}(x, D, \tau)v\overline{\mathcal{A}(x, D, \tau)v}\right) = 2\Re\left(S(x, D, \tau)v\overline{A(x, D, \tau)v}\right) - \\ - Q(x, D, \overline{D}, \tau)[v, \overline{v}] - |R(x, D, \tau)v|^2,$$

where

$$Q(x, D, \overline{D}, \tau) = 2\Re\left(S(x, D, \tau)v\overline{R(x, D, \tau)v} - R(x, D, \tau)v\overline{A(x, D, \tau)v}\right).$$

Recalling (16.6a) and (16.6b) we get

$$Q(x, D, \overline{D}, \tau)[v, \overline{v}] = Q_1(x, D, \overline{D}, \tau)[v, \overline{v}] + Q_2(x, D, \overline{D}, \tau)[v, \overline{v}],$$

where

$$Q_1(x, D, \overline{D}, \tau)[v, \overline{v}] = 2\Re\left(\widetilde{P}_m(x, D + i\tau\nabla\varphi(x))v\overline{R(x, D, \tau)v}\right)$$

and

$$Q_2(x, D, \overline{D}, \tau)[v, \overline{v}] = \\ = \Re\left(\overline{\widetilde{P}}_m(x, D - i\tau\nabla\varphi)v\overline{R(x, D, \tau)v} - \\ - R(x, D, \tau)v\left(\overline{\overline{\widetilde{P}}_m(x, D - i\tau\nabla\varphi)v}\right)\right) = 0.$$

Therefore

$$2\Re\left(\mathcal{S}(x, D, \tau)v\overline{\mathcal{A}(x, D, \tau)v}\right) = \\ = 2\Re\left(S(x, D, \tau)v\overline{A(x, D, \tau)v}\right) - \\ - 2\Re\left(\widetilde{P}_m(x, D + i\tau\nabla\varphi(x))v\overline{R(x, D, \tau)v}\right) - |R(x, D, \tau)v|^2$$

and by (16.55) and (16.73) we have

$$\begin{aligned}
&\int \left([\mathcal{S}(x,D,\tau),\mathcal{A}(x,D,\tau)]\,v\right)\overline{v}dx = \\
&= \int G(x,D,\overline{D},\tau)[v,\overline{v}]dx - \\
&- 2\Re \int \left(\widetilde{P}_m(x,D+i\tau\nabla\varphi(x))v\overline{R(x,D,\tau)v}\right)dx - \\
&- \int |R(x,D,\tau)v|^2\,dx.
\end{aligned} \tag{16.74}$$

♦

In what follows we need the following

Lemma 16.8 *Assume that the coefficients of operator*

$$P_m(x,D) = \sum_{|\alpha|=m} a_\alpha(x)D^\alpha, \tag{16.75}$$

belong to $C^0\left(\overline{\Omega}\right)$. *Let* $\varphi \in C^\infty\left(\overline{\Omega}\right)$. *Then for every* $\varepsilon > 0$ *there exists* $\delta > 0$ *such that*

$$\begin{aligned}
&\left|\frac{1}{(2\pi)^n}\int |p_m(x_0,\xi+i\tau\nabla\varphi(x_0))v|^2\,|\widehat{v}(\xi)|^2\,d\xi - \right.\\
&\left. - \int \left|\widetilde{P}_m(x,D+i\tau\nabla\varphi(x))v\right|^2 dx\right| \leq \varepsilon\,\|v\|^2_{H^m_\tau(\mathbb{R}^n)},
\end{aligned} \tag{16.76}$$

for every $v \in C^\infty_0(B_\delta(x_0)\cap\Omega)$, *for every* $\tau \in \mathbb{R}$ *and for every* $x_0 \in \overline{\Omega}$.

Proof Let $x_0 \in \overline{\Omega}$. Let recall that by (15.90) we have

$$\widetilde{P}_m(x,D+i\tau\nabla\varphi(x)) = \sum_{|\alpha|+j=m} \tau^j b_{\alpha j}(x)D^\alpha, \tag{16.77}$$

by the assumptions on φ and on the coefficients of $P_m(x,D)$, we have $b_{\alpha j} \in C^0\left(\overline{\Omega}\right)$, for any α and j such that $|\alpha|+j=m$.

Let $\varepsilon > 0$ and $\delta > 0$ be such that for any α and j satisfying $|\alpha|+j=m$ we have

$$\left|b_{\alpha j}(x) - b_{\alpha j}(x_0)\right| < \varepsilon, \quad \forall x \in B_\delta(x_0)\cap\overline{\Omega}$$

(δ independent of x_0). We obtain

$$\begin{aligned}&\left|\widetilde{P}_m(x, D+i\tau\nabla\varphi(x))v - \widetilde{P}_m(x_0, D+i\tau\nabla\varphi(x_0))v\right| \leq \\ &\leq \sum_{|\alpha|+j=m} |\tau|^j \left|b_{\alpha j}(x) - b_{\alpha j}(x_0)\right| \left|D^\alpha v\right| \leq \\ &\leq C\varepsilon \sum_{|\alpha|+j=m} |\tau|^j \left|D^\alpha v\right|, \quad \forall x \in B_\delta(x_0) \cap \overline{\Omega}.\end{aligned} \tag{16.78}$$

On the other hand

$$\left|\widetilde{P}_m(x, D+i\tau\nabla\varphi(x))v\right| \leq C \sum_{|\alpha|+j=m} |\tau|^j \left|D^\alpha v\right|, \quad \forall x \in \overline{\Omega}. \tag{16.79}$$

Now, taking into account the elementary inequality

$$\left||z|^2 - |w|^2\right| \leq (|z| + |w|)\,|z-w|, \quad \forall z, w \in \mathbb{C},$$

we have by (16.78) and (16.79), for every $x \in B_\delta(x_0) \cap \overline{\Omega}$,

$$\begin{aligned}&\left|\left|\widetilde{P}_m(x, D+i\tau\nabla\varphi(x))v\right|^2 - \left|\widetilde{P}_m(x_0, D+i\tau\nabla\varphi(x_0))v\right|^2\right| \leq \\ &\leq C\varepsilon \sum_{|\alpha|\leq m} \tau^{2(m-|\alpha|)} \left|D^\alpha v\right|^2.\end{aligned} \tag{16.80}$$

Therefore, for every $v \in C_0^\infty(B_\delta(x_0) \cap \Omega)$

$$\begin{aligned}&\frac{1}{(2\pi)^n}\int |p_m(x_0, \xi + i\tau\nabla\varphi(x_0))|^2 \, |\widehat{v}(\xi)|^2 \, d\xi - \\ &- \int \left|\widetilde{P}_m(x, D+i\tau\nabla\varphi(x))v\right|^2 dx = \\ &= \int \left(\left|\widetilde{P}_m(x_0, D+i\tau\nabla\varphi(x_0))v\right|^2 - \right. \\ &\left.\left|\widetilde{P}_m(x, D+i\tau\nabla\varphi(x))v\right|^2\right) dx \leq C\varepsilon \sum_{|\alpha|\leq m} \tau^{2(m-|\alpha|)} \int \left|D^\alpha v\right|^2 dx\end{aligned} \tag{16.81}$$

and similarly, for every $v \in C_0^\infty(B_\delta(x_0) \cap \Omega)$,

$$
\begin{aligned}
&\frac{1}{(2\pi)^n}\int |p_m(x_0,\xi+i\tau\nabla\varphi(x_0))v|^2\,|\widehat{v}(\xi)|^2\,d\xi- \\
&-\int \left|\widetilde{P}_m(x,D+i\tau\nabla\varphi(x))v\right|^2 dx \geq \qquad (16.82)\\
&\geq -C\varepsilon\sum_{|\alpha|\leq m}\tau^{2(m-|\alpha|)}\int \left|D^\alpha v\right|^2 dx.
\end{aligned}
$$

Finally (15.34), (16.81) and (16.82) implies (16.76). ■

16.4 Carleman Estimates for the Elliptic Operators with C^1 Coefficients in the Principal Part

Let $m\in\mathbb{N}$. Let Ω be a bounded open set of $\mathbb{R}^n$. Let a_α be complex-valued functions. We assume that the operator

$$
P(x,D)=\sum_{|\alpha|\leq m}a_\alpha(x)D^\alpha, \qquad (16.83)
$$

is elliptic in $\overline{\Omega}$. Namely, there exists a constant $\lambda>0$ such that

$$
|P_m(x,\xi)|\geq\lambda\,|\xi|^m, \qquad \forall\xi\in\mathbb{R}^n, \quad \forall x\in\overline{\Omega}. \qquad (16.84)
$$

Theorem 16.9 *Let $\varphi\in C^\infty\left(\overline{\Omega}\right)$ be a real-valued function which satisfies*

$$
\nabla\varphi(x)\neq 0, \qquad \forall x\in\overline{\Omega}. \qquad (16.85)
$$

Let $P(x,D)$ be an operator of order m whose coefficients belong to $L^\infty(\Omega,)$. Assume that the coefficients of the principal part $P_m(x,D)$ belong to $C^1\left(\overline{\Omega},\right)$. Let us suppose that $P(x,D)$ satisfies the ellipticity condition (16.84) *and that the following condition is satisfied:*

$$
\left.\begin{array}{l}
p_m(x,\xi+i\sigma\nabla\varphi(x))=0\\
(\xi,\sigma)\in\mathbb{R}^{n+1}\setminus\{(0,0)\}\\
x\in\overline{\Omega}
\end{array}\right\}\Longrightarrow \qquad (16.86)
$$

$$
\Longrightarrow\frac{i}{2\sigma}\left\{p_m(x,\xi+i\sigma\nabla\varphi(x)),\overline{p_m(x,\xi+i\sigma\nabla\varphi(x))}\right\}>0,
$$

where $\{\cdot,\cdot\}$ is the Poisson bracket defined in (15.3).

Then there exist constants C and τ_0 such that

$$\sum_{|\alpha|\leq m} \tau^{2(m-|\alpha|)-1} \int \left|D^{\alpha}u\right|^2 e^{2\tau\varphi} dx \leq C \int |P(x,D)u|^2 e^{2\tau\varphi} dx, \tag{16.87}$$

for every $u \in C_0^{\infty}(\Omega)$ and for every $\tau \geq \tau_0$.

Moreover C and τ_0 depend on λ, on the $L^{\infty}(\Omega)$ norms of a_{α}, $|\alpha| \leq m$, on the $L^{\infty}(\Omega)$ norms of ∇a_{α}, $|\alpha| = m$, and on the moduli of continuity of ∇a_{α}, for $|\alpha| = m$.

Proof Let $u \in C_0^{\infty}(\Omega)$. Set

$$v = e^{-\tau\varphi}u,$$

we have

$$e^{\tau\varphi} P_m(x,D)u = e^{\tau\varphi} P_m(x,D)\left(e^{-\tau\varphi}v\right) = P_m(x, D + i\tau\nabla\varphi(x))v \tag{16.88}$$

and, denoting by $\widetilde{P}_m(x, \xi + i\tau\nabla\varphi(x))$ the operator whose symbol is $p_m(x, \xi + i\tau\nabla\varphi(x))$, we get

$$\begin{aligned} &\int |P_m(x, D + i\tau\nabla\varphi(x))v|^2 dx \geq \\ &\geq \frac{1}{2}\int \left|\widetilde{P}_m(x, D + i\tau\nabla\varphi(x))v\right|^2 dx - \\ &- C_1 \|v\|^2_{H_{\tau}^{m-1}(\mathbb{R}^n)}, \end{aligned} \tag{16.89}$$

where C_1 depends by the L^{∞} norms of the coefficients of $P_m(x,D)$.

We now derive an appropriate estimate from below of the first term on the right-hand side in (16.89).

By (16.51) and (16.55) we get (by multiplying both equalities by τ)

$$\begin{aligned} &\tau \int \left|\widetilde{P}_m(x, D + i\tau\nabla\varphi(x))v\right|^2 dx \geq \\ &\geq 2\tau \int \Re\left(S(x,D,\tau)v\overline{A(x,D,\tau)v}\right) dx = \\ &= \tau \int G(x,D,\overline{D},\tau)[v,\overline{v}]dx, \end{aligned} \tag{16.90}$$

where $G(x,D,\overline{D},\tau)$ has been defined in Proposition 16.7.

Now let $x_0 \in \overline{\Omega}$ be a fixed point. We may assume $x_0 = 0 \in \overline{\Omega}$. We get

$$\begin{aligned} \tau \int G(x, D, \overline{D}, \tau)[v, \overline{v}]dx &= \tau \int G(0, D, \overline{D}, \tau)[v, \overline{v}]dx + \\ + \tau \int \left(G(x, D, \overline{D}, \tau) - G(0, D, \overline{D}, \tau)\right)[v, \overline{v}]dx &= \\ = \frac{\tau}{(2\pi)^n} \int G(0, \xi, \xi, \tau) \left|\widehat{v}(\xi)\right|^2 d\xi + \tau \mathcal{R}, \end{aligned} \tag{16.91}$$

where

$$\mathcal{R} = \int \left(G(x, D, \overline{D}, \tau) - G(0, D, \overline{D}, \tau)\right)[v, \overline{v}]dx.$$

By (16.60) we have

$$G\left(x, D, \overline{D}, \tau\right) = \sum_{h=0}^{2m-1} \tau^h G^{(h)}\left(x, D, \overline{D}\right), \tag{16.92}$$

where

$$G^{(h)}\left(x, D, \overline{D}\right)[v, \overline{v}] = \sum_{(\alpha,\beta)\in\Lambda_h} c_{\alpha\beta}^{(h)}(x) D^\alpha v \overline{D^\beta v}$$

and

$$\Lambda_h = \left\{(\alpha, \beta) \in \mathbb{N}_0^n : \quad |\alpha| \leq m, \ |\beta| \leq m, \ |\alpha| + |\beta| \leq 2m - h - 1\right\},$$

for $h = 0, 1 \cdots, 2m - 1$ and, further, $c_{\alpha\beta}^{(h)} \in C^0\left(\overline{\Omega}, \mathbb{C}\right)$ for $(\alpha, \beta) \in \Lambda_h$. Let ε be a positive number that we will choose later and let $\rho_1 > 0$ be such that

$$\left|c_{\alpha\beta}^{(h)}(x) - c_{\alpha\beta}^{(h)}(0)\right| < \varepsilon, \quad \forall x \in B_{\rho_1} \cap \overline{\Omega}, \ (\alpha, \beta) \in \Lambda_h, \ h = 0, 1 \cdots, 2m - 1.$$

We have, for every $\tau \geq 1$ and for every $x \in B_{\rho_1} \cap \overline{\Omega}$,

$$\begin{aligned}
&\left|\tau^{h+1}\left(G^{(h)}\left(x, D, \overline{D}\right) - G^{(h)}\left(0, D, \overline{D}\right)\right)[v, \overline{v}]\right| \leq \\
&\leq \sum_{(\alpha,\beta)\in\Lambda_h} \tau^{h+1}\left|c_{\alpha\beta}^{(h)}(x) - c_{\alpha\beta}^{(h)}(0)\right|\left|D^\alpha v\right|\left|\overline{D^\beta v}\right| \leq \\
&\leq \varepsilon \sum_{(\alpha,\beta)\in\Lambda_h} \tau^{2m-(|\alpha|+|\beta|)}\left|D^\alpha v\right|\left|D^\beta v\right| = \\
&= \varepsilon \sum_{(\alpha,\beta)\in\Lambda_h} \left(\tau^{m-|\alpha|}\left|D^\alpha v\right|\right)\left(\tau^{m-|\beta|}\left|D^\beta v\right|\right) \leq \\
&\leq C\varepsilon \sum_{|\alpha|\leq m} \tau^{2(m-|\alpha|)}\left|D^\alpha v\right|^2,
\end{aligned}$$

where C depends on m only. Let us notice that in the second inequality we have exploited that, for $\tau \geq 1$,

$$(\alpha, \beta) \in \Lambda_h \Rightarrow h + 1 \leq 2m - (|\alpha| + |\beta|) \Rightarrow \tau^{h+1} \leq \tau^{2m-(|\alpha|+|\beta|)}.$$

Therefore, for every $x \in B_{\rho_1} \cap \overline{\Omega}$,

$$\left|\tau\left(G\left(x, D, \overline{D}, \tau\right) - G\left(0, D, \overline{D}, \tau\right)\right)[v, \overline{v}]\right| \leq C\varepsilon \sum_{|\alpha|\leq m} \tau^{2(m-|\alpha|)}\left|D^\alpha v\right|^2.$$

Hence, by (15.34) we have, for any $\tau \geq 1$,

$$|\tau \mathcal{R}| \leq C\varepsilon \left\|v\right\|^2_{H^m_\tau(\mathbb{R}^n)}, \tag{16.93}$$

for every $v \in C_0^\infty(B_{\rho_1} \cap \Omega)$,

Now, by (16.90), (16.91) and (16.93) we get

$$\begin{aligned}
\tau \int \left|\widetilde{P}_m(x, D + i\tau\nabla\varphi)v\right|^2 dx \geq \frac{\tau}{(2\pi)^n} \int G(0, \xi, \xi, \tau)\left|\widehat{v}(\xi)\right|^2 d\xi - \\
- C\varepsilon \left\|v\right\|^2_{H^m_\tau(\mathbb{R}^n)},
\end{aligned} \tag{16.94}$$

for every $v \in C_0^\infty(B_{\rho_1} \cap \Omega)$ and for every $\tau \geq 1$.

Now, exactly as in (15.170) of Lemma 15.22 we have there exist two positive constants C_1 and C_2 such that

$$\begin{aligned}
&C_1 \left|\xi + i\sigma\nabla\varphi(0)\right|^{2m} \leq \\
&\leq \sigma G(0, \xi, \xi, \sigma) + C_2 \left|p_m\left(0, \xi + i\sigma\nabla\varphi(0)\right)\right|^2, \ \forall(\xi, \sigma) \in \mathbb{R}^{n+1}.
\end{aligned} \tag{16.95}$$

Now, we set

$$\gamma = \min\left\{1, \min_{\overline{\Omega}} |\nabla\varphi|\right\},$$

using (16.95) in (16.94) we get

$$\begin{aligned}
&\tau \int \left|\widetilde{P}_m(x, D + i\tau\nabla\varphi(x))v\right|^2 dx \geq \\
&\geq (2\pi)^{-n} C_1\gamma^2 \int \left(|\xi|^2 + \tau^2\right)^m |\widehat{v}(\xi)|^2 d\xi - \\
&- (2\pi)^{-n} C_2 \int |p_m(0, \xi + i\tau\nabla\varphi(0))|^2 |\widehat{v}(\xi)|^2 d\xi - \\
&- C\varepsilon \int \left(|\xi|^2 + \tau^2\right)^m |\widehat{v}(\xi)|^2 d\xi,
\end{aligned} \tag{16.96}$$

for every $v \in C_0^\infty(B_{\rho_1} \cap \Omega)$ and for every $\tau \geq 1$. By Lemma 16.8 there exists $\rho_2 \leq \rho_1$ such that for every $v \in C_0^\infty(B_{\rho_2} \cap \Omega)$ and for every $\tau \geq 1$ we have

$$\begin{aligned}
&(2\pi)^{-n} \int |p_m(0, \xi + i\tau\nabla\varphi(0))|^2 |\widehat{v}(\xi)|^2 d\xi \leq \\
&\int \left|\widetilde{P}_m(x, D + i\tau\nabla\varphi(x))v\right|^2 dx + \\
&+ C\varepsilon \|v\|^2_{H^m_\tau(\mathbb{R}^n)}.
\end{aligned} \tag{16.97}$$

By (16.96) and (16.97) we have

$$\begin{aligned}
&\tau \int \left|\widetilde{P}_m(x, D + i\tau\nabla\varphi(x))v\right|^2 dx \geq \\
&\geq \left((2\pi)^{-n} C_1\gamma^2 - C\varepsilon\right) \|v\|^2_{H^m_\tau(\mathbb{R}^n)} - \\
&- (2\pi)^{-n} C_2 \int \left|\widetilde{P}_m(x, D + i\tau\nabla\varphi(x))v\right|^2 dx,
\end{aligned} \tag{16.98}$$

for every $v \in C_0^\infty(B_{\rho_2} \cap \Omega)$. Now, let us choose

$$\varepsilon = \varepsilon_0 := \frac{(2\pi)^{-n} C_1\gamma^2}{2C}$$

and let us denote by $\overline{\rho}$ the value of ρ_2 when $\varepsilon = \varepsilon_0$. Moving the last integral of (16.98) to the left-hand side and we have

$$C_3\tau \int \left|\widetilde{P}_m(x, D + i\tau\nabla\varphi(x))v\right|^2 dx \geq$$
$$\geq \varepsilon_0 C^{-1} \|v\|^2_{H^m_\tau(\mathbb{R}^n)}, \tag{16.99}$$

($C_3 = 1 + (2\pi)^{-n} C_2$) for every $v \in C_0^\infty(B_{\overline{\rho}} \cap \Omega)$, for every $\tau \geq 1$.

At this point we use (16.89) and we have

$$C\tau \|v\|^2_{H^{m-1}_\tau(\mathbb{R}^n)} +$$
$$+ 2C_3\tau \int |P_m(x, D + i\tau\nabla\varphi(x))v|^2 dx \geq \tag{16.100}$$
$$\geq \varepsilon_0 C^{-1} \|v\|^2_{H^m_\tau(\mathbb{R}^n)},$$

for every $v \in C_0^\infty(B_{\overline{\rho}} \cap \Omega)$ and for every $\tau \geq 1$. Now we proceed in a similar way to the proof of Theorem 15.21 after (15.186) and we have that, for every $v \in C_0^\infty(B_{\overline{\rho}} \cap \Omega)$ and for every $\tau \geq \tau_0$, where $\tau_0 = \max\{2C^2\varepsilon_0^{-1}, 1\}$, we have

$$2C_3\tau \int |P_m(x, D + i\tau\nabla\varphi(x))v|^2 dx \geq \frac{\varepsilon_0 C^{-1}}{2} \|v\|^2_{H^m_\tau(\mathbb{R}^n)}.$$

Now, by Remark 1 following Proposition 15.13 (actually, a slight modification of it) we have that for every $x_0 \in \overline{\Omega}$ there exist C_{x_0}, r_{x_0} and $\tau_{x_0} \geq 1$ such that

$$\sum_{|\alpha| \leq m} \tau^{2(m-|\alpha|)-1} \int_\Omega \left|D^\alpha u\right|^2 e^{2\tau\varphi} dx \leq C_{x_0} \int_\Omega |P(x, D)u|^2 e^{2\tau\varphi} dx, \tag{16.101}$$

for every $u \in C_0^\infty(B_{r_{x_0}}(x_0) \cap \Omega)$ and for every $\tau \geq \tau_{x_0}$.

Finally, by using the partition of unity as in Lemma 14.2 we obtain estimate (15.168). ■

Remark In this context, it is certainly worth noting that the use of differential quadratic forms in [56, Ch. VIII] has proven effective, with appropriate modifications (see [62]), even for proving Carleman estimates for the so-called anisotropic operators, i.e., operators of the following form. Let Ω be an open, bounded set of $\mathbb{R}^n$, let $m = (m_1, \cdots, m_n)$ with $m_j \in \mathbb{N}$, and for $\alpha \in \mathbb{N}_0^n$, define

$$|\alpha : m| = \sum_{j=1}^n \frac{\alpha_j}{m_j}.$$

The anisotropic operators studied in [62] are of the type

$$P(x, D) = \sum_{|\alpha:m| \leq 1} a_\alpha(x) D^\alpha, \tag{16.102}$$

where the coefficients a_α are complex-valued, belong to $L^\infty(\Omega)$, and are of class $C^1\left(\overline{\Omega}\right)$ for $|\alpha : m| = 1$. We explicitly note that operators of type (16.102) include the heat operator

$$\sum_{j=1}^{n} D_j^2 + iD_n,$$

as well as the Schrödinger operator

$$\sum_{j=1}^{n} D_j^2 + D_n,$$

thus extending the results established in [56, Ch. VIII]. ♦

16.5 Elliptic Operators with Lipschitz Continuous Coefficients and the Cauchy Problem

In Theorem 16.9 we have assumed that the coefficients of the principal part are of class $C^1(\overline{\Omega})$ and it turns out that the constants, C and τ_0, in the estimate (16.87) depend on the **modulus of continuity** of the gradients of these coefficients. We will now see that with a relatively modest effort we can prove a Carleman estimate for elliptic operators with Lipschitz continuous coefficients in principal part. In this regard, it is useful to point out that this assumption cannot be substantially reduced as has been shown in the counterexamples of Mandache [90], Miller [94] and of Plĭs [105].

For any $x \in \mathbb{R}^n$ and $R > 0$ let us denote by

$$Q_R(x) = \left\{y \in \mathbb{R}^n : \quad |y_j - x_j| < R, \quad j = 1, \cdots, n\right\}.$$

Let us introduce a special **partition of unity** (see also [132]).

Let $\vartheta_0 \in C_0^\infty(\mathbb{R})$ satisfy

$$\vartheta_0(t) = \begin{cases} 1, & \text{for } |t| \leq 1 \\ \\ 0, & \text{for } |t| \geq 3/2 \end{cases}.$$

Let $0 \leq \vartheta \leq 1$ be such that

$$\vartheta(x) = \vartheta_0(x_1) \cdots \vartheta_0(x_n).$$

We have

$$\vartheta(x) = \begin{cases} 1, & \text{for } x \in Q_1(0) \\ \\ 0, & \text{for } x \in \mathbb{R}^n \setminus \overline{Q_{3/2}(0)} \end{cases}.$$

For any $\mu \geq 1$ and $g \in \mathbb{Z}^n$, let us denote by

$$x_g = g/\mu$$

and

$$\vartheta_{g,\mu}(x) = \vartheta(\mu(x - x_g)).$$

Hence, we have

$$\text{supp } \vartheta_{g,\mu} \subset \overline{Q_{3/2\mu}(x_g)} \subset Q_{2/\mu}(x_g)$$

and

$$|D^k \vartheta_{g,\mu}| \leq C_1 \mu^k (\chi_{Q_{3/2\mu}(x_g)} - \chi_{Q_{1/\mu}(x_g)}), \quad k = 0, 1, \cdots, m, \tag{16.103}$$

where $C_1 \geq 1$ depends on n only.

For any $g \in \mathbb{Z}^n$, set

$$A_g = \{g' \in \mathbb{Z}^n \ : \ \text{supp}\, \vartheta_{g',\mu} \cap \text{supp}\, \vartheta_{g,\mu} \neq \emptyset\},$$

then

$$\text{card}(A_g) \text{ depends only on } n. \tag{16.104}$$

Therefore we can define

$$\widetilde{\vartheta}_\mu(x) := \sum_{g \in \mathbb{Z}^n} \vartheta_{g,\mu}(x) \geq 1, \quad \forall x \in \mathbb{R}^n. \tag{16.105}$$

By (16.103), we get

$$|D^k \widetilde{\vartheta}_\mu| \leq C_2 \mu^k, \tag{16.106}$$

where $C_2 \geq 1$ depends on n only. Let us define

$$\eta_{g,\mu}(x) = \vartheta_{g,\mu}(x)/\widetilde{\vartheta}_\mu(x), \quad \forall x \in \mathbb{R}^n,$$

we have thus

$$\begin{cases} \eta_{g,\mu} \geq 0 \\ \sum_{g \in \mathbb{Z}^n} \eta_{g,\mu} = 1, \quad \text{in } \mathbb{R}^n \\ \text{supp } \eta_{g,\mu} \subset \overline{Q_{3/2\mu}(x_g)} \subset Q_{2/\mu}(x_g) \\ |D^\alpha \eta_{g,\mu}| \leq C_3 \mu^{|\alpha|} \chi_{\overline{Q_{3/2\mu}(x_g)}}, \quad \forall \alpha \in \mathbb{N}^n, \ 1 \leq |\alpha| \leq m \end{cases}, \tag{16.107}$$

where $C_3 \geq 1$ depends on n only.

Let $m \in \mathbb{N}$. Let

$$P(x, D) = \sum_{|\alpha| \leq m} a_\alpha(x) D^\alpha \tag{16.108}$$

be an elliptic operator. Let M_0, M_1, λ be positive constants and let us suppose that

$$\|a_\alpha\|_{L^\infty(Q_1)} \leq M_0, \quad \text{for } |\alpha| \leq m, \tag{16.109a}$$

$$|a_\alpha(x) - a_\alpha(y)| \leq M_1 |x - y|, \quad \forall x, y \in Q_1, \quad \text{for } |\alpha| = m, \tag{16.109b}$$

$$|P_m(x, \xi)| \geq \lambda \, |\xi|^m, \qquad \forall \xi \in \mathbb{R}^n, \quad \forall x \in \overline{Q}_1. \tag{16.109c}$$

Let $\varphi \in C^\infty\left(\overline{Q}_1\right)$. For $x, y \in \overline{Q}_1$, $(\xi, \sigma) \in \mathbb{R}^{n+1}$ set

$$\begin{aligned} &\mathcal{G}(x, y; \xi, \sigma) = \\ &= \sum_{j,k=1}^n \partial^2_{x_j x_k} \varphi(x) p_m^{(j)}(y, \xi + i\sigma \nabla\varphi(x)) \overline{p_m^{(j)}(y, \xi + i\sigma \nabla\varphi(x))}. \end{aligned} \tag{16.110}$$

Theorem 16.10 *Assume that operator* (16.108) *satisfies ellipticity condition* (16.109c) *and that its coefficients satisfy conditions* (16.109a) *and* (16.109b). *Moreover, assume that*

$$\left. \begin{array}{l} p_m(0, \xi + i\sigma \nabla\varphi(0)) = 0 \\ \\ (\xi, \sigma) \in \mathbb{R}^{n+1} \setminus \{(0, 0)\} \end{array} \right\} \Longrightarrow \mathcal{G}(0, 0; \xi, \sigma) > 0. \tag{16.111}$$

Then there exist $\overline{R} \in (0, 1]$, $\delta_0 \in (0, 1]$ $C_0 \geq 1$ *and* τ_0 *such that*

$$\sum_{|\alpha| \leq m} \tau^{2(m-|\alpha|)-1} \int \left|D^\alpha u\right|^2 e^{2\tau\varphi} dx \leq C_0 \int \left|P_m(\delta x, D) u\right|^2 e^{2\tau\varphi} dx, \tag{16.112}$$

for every $\delta \in (0, \delta_0]$, *for every* $u \in C_0^\infty\left(Q_{\overline{R}}\right)$ *and for every* $\tau \geq \tau_0$.

Proof Since $|p_m(0,\xi+i\sigma\nabla\varphi(0))|^2$ and $\left(|\xi|^2+\tau^2\right)\mathcal{G}(0,0;\xi,\sigma)$ are homogeneous polynomials of degree $2m$, (16.111) is equivalent to the following property: there exist positive constants C_1 and C_2 such that (Lemma 14.24)

$$C_2\,|p_m(0,\xi+i\sigma\nabla\varphi(0))|^2+\left(|\xi|^2+\sigma^2\right)\mathcal{G}(0,0;\xi,\sigma)\geq \geq C_1\left(|\xi|^2+\sigma^2\right)^m. \tag{16.113}$$

for every $(\xi,\sigma)\in\mathbb{R}^{n+1}$.

Let us denote by

$$H(x,y;\xi,\sigma)=C_2\,|p_m(y,\xi+i\sigma\nabla\varphi(x))|^2+\left(|\xi|^2+\sigma^2\right)\mathcal{G}(x,y;\xi,\sigma) \tag{16.114}$$

and let us notice that H is a continuous function. Moreover, by (16.113) we have trivially

$$H(0,0;\xi,\sigma)\geq C_1,\qquad \text{for all } (\xi,\sigma) \text{ such that } |\xi|^2+\sigma^2=1.$$

Hence, the continuity of H implies that there exists $\overline{R}_1\in(0,1]$ such that

$$H(x,y;\xi,\sigma)\geq\frac{C_1}{2},\qquad \text{for all } (\xi,\sigma) \text{ such that } |\xi|^2+\sigma^2=1,$$

for every $x,y\in\overline{Q}_{\overline{R}_1}$.

Therefore

$$C_2\,|p_m(y,\xi+i\sigma\nabla\varphi(x))|^2+\left(|\xi|^2+\sigma^2\right)\mathcal{G}(x,y;\xi,\sigma)\geq \geq\frac{C_1}{2}\left(|\xi|^2+\sigma^2\right)^m, \tag{16.115}$$

for every $x,y\in\overline{Q}_{\overline{R}_1}$. By the previous inequality we have that for every $\widetilde{y}\in Q_{\overline{R}_1}$ it occurs

$$\left.\begin{array}{l} p_m(\widetilde{y},\xi+i\sigma\nabla\varphi(x))=0 \\ \\ (\xi,\sigma)\in\mathbb{R}^{n+1}\setminus\{(0,0)\}\end{array}\right\}\Longrightarrow \mathcal{G}(x,\widetilde{y};\xi,\sigma)>0. \tag{16.116}$$

Now, for $y\in\overline{Q}_{\overline{R}_1}$ *fixed* and $\delta\in(0,1]$ to be chosen, let us consider the operator with constant coefficients w.r.t. the variable x

$$P_m(\delta y,D_x)=\sum_{|\alpha|=m}a_\alpha(\delta y)D_x^\alpha. \tag{16.117}$$

Of course $\delta y \in \overline{Q_{R_1}}$ (as $\delta \in (0, 1]$). Now, (16.116) (considered for $\tilde{y} = \delta y$) is nothing but condition (16.86) of Theorem 16.9) in the case where the coefficients of principal part of the operator are constants. Hence there exist $C_3 > 0$ and τ_1 such that

$$\begin{aligned}&\sum_{|\alpha|\leq m} \tau^{2(m-|\alpha|)-1} \int \left|D^\alpha u\right|^2 e^{2\tau\varphi(x)} dx \leq \\ &\leq C_3 \int |P_m(\delta y, D)u|^2 e^{2\tau\varphi(x)} dx,\end{aligned} \tag{16.118}$$

for every $u \in C_0^\infty\left(Q_{\overline{R}_1}\right)$ and for every $\tau \geq \tau_1$. Moreover $C_3 > 0$ and τ_1 **do not depend neither on** $y \in \overline{Q_{R_1}}$ **nor on** $\delta \in (0, 1]$.

We now use the partition of unity introduced above with

$$\mu = \sqrt{\varepsilon\tau}, \tag{16.119}$$

for $\tau \geq \tau^{(\varepsilon)} := \max\left\{\varepsilon^{-1}, \tau_1\right\}$ where $\varepsilon \in (0, 1]$ is to be chosen.

Let $u \in C_0^\infty\left(Q_{\overline{R}_1}\right)$. By the first relation of (16.107) we have

$$u = \sum_{g\in\mathbb{Z}^n} u\eta_{g,\mu}. \tag{16.120}$$

Now we apply (16.118) (for $y = x_g \in Q_{\overline{R}_1}$). We have, for every $\tau \geq \tau^{(\varepsilon)}$

$$\begin{aligned}&\sum_{|\alpha|\leq m} \tau^{2(m-|\alpha|)-1} \int \left|D^\alpha u\right|^2 e^{2\tau\varphi} dx \leq \\ &\leq c \sum_{g\in\mathbb{Z}^n} \sum_{|\alpha|\leq m} \tau^{2(m-|\alpha|)-1} \int \left|D^\alpha \left(u\eta_{g,\mu}\right)\right|^2 e^{2\tau\varphi} dx \leq \\ &\leq cC_3 \sum_{g\in\mathbb{Z}^n} \int \left|P_m\left(\delta x_g, D\right)\left(u\eta_{g,\mu}\right)\right|^2 e^{2\tau\varphi} dx,\end{aligned} \tag{16.121}$$

where the constant c appearing in the second inequality, by (16.104), depends on n only.

Now, let us estimate form above the last term on the right-hand side of (16.121). We have

$$\begin{aligned}\left|P_m\left(\delta x_g, D\right)\left(u\eta_{g,\mu}\right)\right|^2 &\leq 2\left|P_m\left(\delta x, D\right)\left(u\eta_{g,\mu}\right)\right|^2 + \\ &+ 2\left|\left(P_m\left(\delta x_g, D\right) - P_m\left(\delta x, D\right)\right)\left(u\eta_{g,\mu}\right)\right|^2.\end{aligned} \tag{16.122}$$

In order to estimate the **first term on the right-hand side in** (16.122) we notice that

$$\begin{aligned}P_m(\delta x, D)\left(u\eta_{g,\mu}\right) &= \eta_{g,\mu}\sum_{|\alpha|\leq m} a_\alpha(\delta x)D^\alpha u+\\&+\sum_{|\alpha|\leq m} a_\alpha(\delta x)\sum_{\beta<\alpha}\binom{\alpha}{\beta}D^\beta u D^{\alpha-\beta}\eta_{g,\mu} = \\&= \eta_{g,\mu}P_m(\delta x, D)\,u + P_\bullet(x, D, \mu)u,\end{aligned} \tag{16.123}$$

where we set

$$P_\bullet(x, D, \mu)u = \sum_{|\alpha|\leq m} a_\alpha(\delta x)\sum_{\beta<\alpha}\binom{\alpha}{\beta}D^\beta u D^{\alpha-\beta}\eta_{g,\mu}.$$

Let us note that this operator has order $m-1$. Moreover by (16.107) and (16.109a) we get

$$|P_\bullet(x, D, \mu)u| \leq CM_0\chi_{Q_{2/\mu}(x_g)}\sum_{|\beta|\leq m-1}\left|D^\beta u\right|\mu^{m-|\beta|}. \tag{16.124}$$

From (16.122), (16.124) and recalling (16.119) we have (for the first term on the right we use the trivial inequality $\eta_{g,\mu}^2 \leq \eta_{g,\mu}$)

$$\begin{aligned}\left|P_m(\delta x, D)\left(u\eta_{g,\mu}\right)\right|^2 &\leq \eta_{g,\mu}\left|P_m(\delta x, D)\,u\right|^2 +\\&+ CM_0^2\chi_{Q_{2/\mu}(x_g)}\sum_{|\alpha|\leq m-1}(\varepsilon\tau)^{m-|\alpha|}\left|D^\alpha u\right|^2.\end{aligned} \tag{16.125}$$

We now estimate the second term on the right-hand side in (16.122). Proceeding in a similar way as above, we have

$$\begin{aligned}&\left|\left(P_m\left(\delta x_g, D\right) - P_m(\delta x, D)\right)\left(u\eta_{g,\mu}\right)\right| \leq\\&\leq \sum_{|\alpha|=m}\left|\left(a_\alpha(\delta x) - a_\alpha(\delta x_g)\right)\right|\left|D^\alpha\left(u\eta_{g,\mu}\right)\right| =\\&= \eta_{g,\mu}\sum_{|\alpha|=m}\left|\left(a_\alpha(\delta x) - a_\alpha(\delta x_g)\right)\right|\left|D^\alpha u\right| +\\&+ CM_0\chi_{Q_{2/\mu}(x_g)}\sum_{|\alpha|\leq m-1}\left|D^\alpha u\right|\mu^{m-|\alpha|}.\end{aligned}$$

In order to estimate the second-to-last term, it must be taken into account that the estimate has to be done in the support of $\eta_{g,\mu}$. By (16.109b) we get, therefore,

$$\begin{aligned}&\left|\left(P_m\left(\delta x_g, D\right) - P_m\left(\delta x, D\right)\right)\left(u\eta_{g,\mu}\right)\right|^2 \leq \\ &\leq CM_1^2\eta_{g,\mu}\frac{\delta^2}{\varepsilon\tau}\sum_{|\alpha|=m}\left|D^\alpha u\right|^2 + \\ &+ CM_0^2\chi_{Q_{2/\mu}(x_g)}\sum_{|\alpha|\leq m-1}(\varepsilon\tau)^{m-|\alpha|}\left|D^\alpha u\right|^2.\end{aligned} \tag{16.126}$$

Now, we insert (16.125) and (16.126) into (16.122) and we get

$$\begin{aligned}\left|P_m\left(\delta x_g, D\right)\left(u\eta_{g,\mu}\right)\right|^2 &\leq 2\eta_{g,\mu}\left|P_m\left(\delta x, D\right)u\right|^2 + \\ &+ CM_1^2\eta_{g,\mu}\frac{\delta^2}{\varepsilon\tau}\sum_{|\alpha|=m}\left|D^\alpha u\right|^2 + \\ &+ CM_0^2\chi_{Q_{2/\mu}(x_g)}\sum_{|\alpha|\leq m-1}(\varepsilon\tau)^{m-|\alpha|}\left|D^\alpha u\right|^2.\end{aligned}$$

Inserting the latter into (16.121) we obtain

$$\begin{aligned}\sum_{|\alpha|\leq m-1}\tau^{2(m-|\alpha|)-1}\int\left|D^\alpha u\right|^2 e^{2\tau\varphi}dx + \frac{1}{\tau}\sum_{|\alpha|=m}\int\left|D^\alpha u\right|^2 e^{2\tau\varphi}dx \leq \\ \leq cC_3\int\left|P_m\left(\delta x, D\right)u\right|^2 e^{2\tau\varphi}dx + \\ + C_4\frac{\delta^2}{\varepsilon\tau}\sum_{|\alpha|=m}\int\left|D^\alpha u\right|^2 e^{2\tau\varphi}dx + \\ + C_5\sum_{|\alpha|\leq m-1}\int(\varepsilon\tau)^{m-|\alpha|}\left|D^\alpha u\right|^2 e^{2\tau\varphi}dx,\end{aligned}$$

where C_4 depends on M_1 only and C_5 depends by M_0 only. From which we have

$$\begin{aligned}&\sum_{|\alpha|\leq m-1}\tau^{(m-|\alpha|)}\left(\tau^{m-|\alpha|-1} - C_5\varepsilon^{m-|\alpha|}\right)\int\left|D^\alpha u\right|^2 e^{2\tau\varphi}dx + \\ &+ \frac{1}{\tau}\left(1 - C_4\frac{\delta^2}{\varepsilon}\right)\sum_{|\alpha|=m}\int\left|D^\alpha u\right|^2 e^{2\tau\varphi}dx \leq \\ &\leq cC_3\int\left|P_m\left(\delta x, D\right)u\right|^2 e^{2\tau\varphi}dx.\end{aligned} \tag{16.127}$$

Let us choose

$$\varepsilon = \varepsilon_0 := \frac{1}{2C_5},$$

$$\delta \leq \delta_0 := \sqrt{\frac{\varepsilon_0}{2C_4}}$$

and by (16.127) we get

$$\sum_{|\alpha|\leq m} \tau^{2(m-|\alpha|)-1} \int \left|D^{\alpha} u\right|^2 e^{2\tau\varphi} dx \leq 2cC_3 \int |P_m(\delta x, D)u|^2 e^{2\tau\varphi} dx, \tag{16.128}$$

for every $u \in C_0^{\infty}\left(Q_{\overline{R}_1}\right)$ and for every $\tau \geq \tau^{(\varepsilon_0)}$. Estimate (16.112) is proved. ■

Remark 16.19 Let us notice that (reader check) by the change of variables $X = \delta x$, (16.112) is become

$$\begin{aligned} \sum_{|\alpha|\leq m} \tau^{2(m-|\alpha|)-1} \delta^{m-|\alpha|} \int \left|D^{\alpha} u\right|^2 e^{2\tau\varphi(\delta^{-1}X)} dX \leq \\ \leq C_0 \int |P_m(X, D)u|^2 e^{2\tau\varphi(\delta^{-1}X)} dX, \end{aligned} \tag{16.129}$$

for every $u \in C_0^{\infty}\left(Q_{\delta\overline{R}}\right)$ and for every $\tau \geq \tau_0$. ♦

In the following Theorem we will apply estimate (16.129) to prove a uniqueness result for the Cauchy problem.

Theorem 16.11 *Let $\psi \in C^1\left(\overline{Q}_1\right)$ be real-valued function such that*

$$\nabla\psi(0) \neq 0. \tag{16.130}$$

Let $P(x, D)$ be operator (16.108) *and let us suppose that* (16.109) *holds true. Let $U \in H^m(Q_1)$ satisfy*

$$\begin{cases} P(x, D)U = 0, & \textit{in} \quad Q_1 \\ U(x) = 0 & \textit{in} \ \left\{x \in \overline{Q}_1 : \quad \psi(x) > \psi(0)\right\} \end{cases}. \tag{16.131}$$

Suppose that for every $\xi \in \mathbb{R}^n \setminus \{0\}$ we have

$$\sigma \to p_m(0, \xi + i\sigma\nabla\psi(0)) \ \textit{has no real multiple roots}\,. \tag{16.132}$$

Then there exist a neighborhood $\mathcal{U}_0$ of 0 *such that*

$$U = 0 \quad in \quad \mathcal{U}_0. \tag{16.133}$$

Proof It is not restrictive to assume $\psi(0) = 0$ and, since $\nabla\psi(0) \neq 0$, we may reduce to consider, up to isometries, the case where, for an appropriate $r_0 > 0$, we have

$$\left\{x \in Q_{r_0} : \quad \psi(x) = 0\right\} = \left\{(x', f(x')) : \quad x' \in Q'_{r_0}(0)\right\}, \tag{16.134}$$

($Q'_{r_0} = (-r_0, r_0)^{n-1}$) where $f \in C^1(Q'_{r_0})$, $f(0) = |\nabla f(0)| = 0$ and

$$\{x \in Q_1 : \quad \psi(x) > 0\} \cap Q_{r_0} = \left\{(x', x_n) \in Q_{r_0} : \quad x_n < f(x')\right\}.$$

Let us notice that in this way condition (16.132) becomes.

$$\sigma \to p_m(0, \xi - i\sigma e_n) \text{ has no real multiple roots }. \tag{16.135}$$

Now we use **Holmgren transformation** introduced in (9.149), that is we consider the transformation

$$\Lambda : \mathbb{R}^n_x \to \mathbb{R}^n_y, \quad x \to y = \Lambda(x', x_n) = \left(x', x_n + \frac{A}{2}|x'|^2\right), \tag{16.136}$$

(recall that Λ is a diffeomorphism) where $A > 0$ satisfies

$$A > \left\|\partial^2 f\right\|_{L^\infty(B'_{r_0})} \tag{16.137}$$

and where $\partial^2 f$ is the Hessian matrix of f. Let us fix A that satisfies (16.137) and we recall that, with this choice, the function

$$g(x') = f(x') + \frac{A}{2}|x'|^2, \tag{16.138}$$

is strictly convex and

$$g(0) = |\nabla g(0)| = 0. \tag{16.139}$$

Let us denote by $P^\bullet(y, D_y)$ the transformed operator of $P(x, D_x)$ by mean of Λ. Since

$$p_m(x, \xi) = \sum_{|\alpha|=m} a_\alpha(x)\xi^\alpha,$$

we have (compare with (9.26))

$$p_m^{\bullet}(y,\eta)=\sum_{|\alpha|=m} a_\alpha(\Lambda^{-1}(y))\left((\partial_x\Lambda(x))^t\,\eta\right)^\alpha_{|x=\Lambda^{-1}(y)}\,. \tag{16.140}$$

Thus condition (16.135) can be written (reader check)

$$\begin{aligned}&\text{for fixed } \xi\in\mathbb{R}^n\setminus\{0\},\\ &p_m^{\bullet}(0,\xi-i\sigma e_n)\ \text{ has no real multiple roots.}\end{aligned} \tag{16.141}$$

Moreover, (16.131) implies

$$\begin{cases} P^{\bullet}(y,D_y)\widetilde{U}=0, \quad \text{in} \quad Q_{r_0}\\ \\ \widetilde{U}(y)=0 \quad \text{in } \left\{(x',x_n)\in Q_{r_0}:\quad x_n<g(x')\right\}\end{cases}. \tag{16.142}$$

where $\widetilde{U}(y)=U\left(\Lambda^{-1}(y)\right)$. It turns out $\widetilde{U}\in H^m\left(Q_{r_0}\right)$.

We agree from here on to omit " $\bullet$" and "$^{\sim}$" from P and U, respectively, and to rename "x" the variable "y". Let

$$h(x_n)=-x_n+\frac{x_n^2}{2}$$

(let us notice that h is strictly decreasing in $0\le x_n\le 1$) and

$$\varphi(x)=h(\delta_0 x),$$

where δ_0 is defined in Theorem 16.10. We have

$$\nabla\varphi(0)=-\delta_0 e_n.$$

Moreover, (16.141) implies that if $\xi\in\mathbb{R}^n\setminus\{0\}$ and

$$p_m(0,\xi+i\sigma\nabla\varphi(0))=p_m(0,\xi-i\sigma\delta_0 e_n)=0,$$

then

$$\begin{aligned}&\mathcal{G}(0,0;\xi,\sigma)=\\ &=\sum_{j,k=1}^n\partial^2_{x_jx_k}\varphi(0)p_m^{(j)}(0,\xi+i\sigma\nabla\varphi(0))\overline{p_m^{(j)}(0,\xi+i\sigma\nabla\varphi(0))}=\\ &=\delta_0^2\left|p_m^{(n)}(0,\xi+i\sigma\nabla\varphi(0))\right|^2=\delta_0^2\left|p_m^{(n)}(0,\xi-i\sigma\delta_0 e_n)\right|^2>0.\end{aligned}$$

Therefore the assumptions of Theorem 16.10 are satisfied. Then Carleman estimate (16.129) holds. We may write such a Carleman estimate as (setting $\delta = \delta_0$)

$$\sum_{|\alpha|\leq m} \tau^{2(m-|\alpha|)-1} \int \left|D^\alpha u\right|^2 e^{2\tau h(x_n)} dx \leq \\ \leq C \int \left|P(x, D)u\right|^2 e^{2\tau h(x_n)} dx, \tag{16.143}$$

for every $u \in C_0^\infty\left(Q_{\delta_0 \overline{R}}\right)$ and for every $\tau \geq \overline{\tau}_0$ for a certain $\overline{\tau}_0 \geq \tau_0$. Set

$$r_1 = \min\left\{r_0, \delta_0 \overline{R}\right\}$$

and, for $\rho > 0$,

$$E_\rho = \left\{(x', x_n) \in Q_{r_1} : \quad g(x') < x_n < \rho\right\}.$$

For the strict convexity of g and by (16.139), we have that there exists $\rho_1 > 0$ such that

$$\overline{E}_{\rho_1} \subset Q_{r_1}.$$

Let $\rho_2 \in (0, \rho_1)$. Let $\eta \in C^\infty(\mathbb{R})$ be a function such that

$$0 \leq \eta(x_n) \leq 1, \quad \forall x \in \mathbb{R}; \quad \eta(x_n) = 1, \quad \forall x_n \leq \rho_2; \quad \eta(x) = 0, \quad \forall x_n \geq \rho_1.$$

Let us assume that

$$\eta^{(k)}(x_n) \leq C\left(\rho_1 - \rho_2\right)^{-k}.$$

By density, (16.129) holds for every $u \in H_0^m(Q_{r_1})$, hence, in particular, (16.129) holds for $u(x) = U(x)\eta(x_n)$. From now on, the proof is quite standard, we present it for completeness. Since $P(x, D)U = 0$ in Q_{r_1} we have.

$$|P(x, D)(U\eta)| \leq CM_0 \chi_{\mathbb{R}\setminus(\rho_2, \rho_1)} \sum_{|\alpha|\leq m-1} (\rho_1 - \rho_2)^{-|\alpha|} |D^\alpha U|$$

and, for any $0 < \rho < \rho_2$, by (16.129), we get

$$e^{2\tau h(\rho)} \sum_{|\alpha|\leq m} \tau^{2(m-|\alpha|)-1} \int_{E_\rho} \left|D^\alpha U\right|^2 dx \leq$$

$$\leq \sum_{|\alpha|\leq m} \tau^{2(m-|\alpha|)-1} \int_{E_{\rho_1}} \left|D^\alpha (U\eta)\right|^2 e^{2\tau h(x_n)} dx \leq$$

$$\leq C \int_{E_{\rho_1}} |P(x,D)(U\eta)|^2 e^{2\tau h(x_n)} dx \leq$$

$$\leq C M_0^2 e^{2\tau h(\rho_2)} \sum_{|\alpha|\leq m-1} \int_{E_{\rho_1}\setminus E_{\rho_2}} (\rho_1-\rho_2)^{-2|\alpha|} |D^\alpha U|^2 dx,$$

for every $\tau \geq \overline{\tau}_0$. Hence

$$\sum_{|\alpha|\leq m} \tau^{2(m-|\alpha|)-1} \int_{E_\rho} \left|D^\alpha U\right|^2 dx \leq$$

$$\leq C M_0^2 e^{-2\tau(h(\rho)-h(\rho_2))} \sum_{|\alpha|\leq m-1} \int_{E_{\rho_1}\setminus E_{\rho_2}} (\rho_1-\rho_2)^{-2|\alpha|} |D^\alpha U|^2 dx,$$

for every $\tau \geq \overline{\tau}_0$. Passing to the limit as $\tau \to +\infty$ which goes to infinity, and taking into account that $h(\rho) - h(\rho_2) > 0$ we have $U = 0$ in $Q_{r_1} \cap \{x_n \leq \rho\}$. ■

Remark 16.20 As stated at the beginning of this section, thanks to the examples developed in [90, 94], and [105], the Lipschitz continuity assumption on the coefficients of the principal part is essentially the minimal hypothesis for the validity of the unique continuation property for elliptic and parabolic equations. However, relatively recently, particularly in relation to control theory problems and quantitative error propagation estimates in transmission problems for differential equations, Carleman estimates have been established for elliptic and parabolic equations where the coefficients exhibit jump discontinuities. Among the numerous contributions on this matter, we mention [31, 42, 80, 81], for elliptic and parabolic second-order equations with real coefficients, [15, 44] for equations with complex coefficients, and [43] for second-order parabolic equations with real coefficients. In [80, 81], and [15], pseudodifferential operators are employed, and consequently, a C^∞ regularity is assumed outside the discontinuities in the coefficients of the principal part. Conversely, in [31, 43] and [44], using a partition of unity similar to that used in the proof of Theorem 16.10, the assumption on the principal part coefficients is reduced to Lipschitz continuity outside the discontinuities. In [43] and [42], Hölder estimates for error propagation across discontinuity surfaces are also provided. In [45], these error propagation estimates were applied to obtain size estimates (see Sect. 6.8) in the presence of inclusions in the case of complex coefficients. ♦

Chapter 17
Carleman Estimates for Second-order Operators with Real Coefficients in the Principal Part

Abstract In this chapter, we consider second-order operators with real coefficients in the principal part. The goal is to derive Carleman estimates in a more straightforward manner than in the previous chapters, assuming that the coefficients are Lipschitz continuous.

17.1 Introduction

In this chapter we will consider the second-order operators whose principal part (not necessarily elliptic) is given by (in this chapter we will adopt the Einstein convention of repeated indices)

$$P_2(x,\partial) = g^{jk}(x)\partial^2_{x_j x_k}, \tag{17.1}$$

where the matrix of coefficients $\left\{g^{jk}(x)\right\}_{j,k=1}^{n}$ is a symmetric real and invertible matrix, whose entries are the real-valued functions g^{jk} defined on a bounded open set $\Omega \subset \mathbb{R}^n$, on which we will make appropriate regularity assumptions. When we will refer to the symbol of the operator (17.1), here we will always refer to the polynomial in the variable ξ

$$p_2(x,\xi) = g^{jk}(x)\xi_j\xi_k. \tag{17.2}$$

We note that, with the notation for the derivatives used in the previous chapters, operator (17.1) can be written

$$P_2(x,\partial) = -g^{jk}(x)D^2_{x_j x_k} \tag{17.3}$$

and so polynomial (17.2) is simply the symbol of operator (17.3) with the sign changed. This abuse of notation will not create major problems, in particular, it will not create problems when we compare procedures and results found in this chapter with those in the previous chapters. The goal of this chapter is to obtain the

S. Vessella, *Unique Continuation Properties for Partial Differential Equations*, Birkhäuser Advanced Texts Basler Lehrbücher,
https://doi.org/10.1007/978-3-031-86366-0_17

Carleman estimates in a more straightforward manner compared to the previous two Chapters and assuming that the coefficients of (17.1) belong to $C^{0,1}$.

17.2 The Case of the Laplace Operator

The case of the Laplace operator will serve us somewhat as a model for more general operators of type (17.1).

Let us review some key steps of the proof of Theorem 16.9 (and Theorem 15.21) using Rellich identity (14.7) **to perform the integrations by parts.**

We begin by rewriting the statement of Theorem 16.9 in the case of the Laplace operator

Theorem 17.1 *Let Ω be a bounded open set of $\mathbb{R}^n$, and let $\varphi \in C^\infty(\bar{\Omega})$ be a real-valued function such that $\nabla\varphi \neq 0$ on $\bar{\Omega}$. Let us assume that the following implication holds true*

$$\left.\begin{array}{l} |\xi|^2 = \tau^2|\nabla\varphi(x)|^2 \\ \xi \cdot \nabla\varphi(x) = 0 \\ \tau \neq 0 \end{array}\right\} \Longrightarrow Q(x,\xi,\tau) = \partial^2_{jk}\varphi(x)\xi_j\xi_k + \tag{17.4}$$

$$+\, \tau^2\partial^2_{jk}(x)\varphi\partial_j\varphi(x)\partial_k\varphi(x) > 0.$$

Then there exist constants C and τ_0 such that

$$\tau^3\int_\Omega |u|^2 e^{2\tau\varphi}dx + \tau\int_\Omega |\nabla u|^2 e^{2\tau\varphi}dx +$$

$$+\,\tau^{-1}\int_\Omega |\partial^2 u|^2 e^{2\tau\varphi}dx \leq C\int_\Omega |\Delta u|^2 e^{2\tau\varphi}dx,$$

for every $u \in C_0^\infty(\Omega)$ and for every $\tau \geq \tau_0$.

Proof First, we observe that (17.4) is simply the rewriting of condition (16.86) of Theorem 16.9 in the case of the Laplace operator.

Let us denote by L the operator Δ. Let $u \in C_0^\infty(\Omega)$, it is not restrictive to assume that u is a real-valued function (nevertheless, we will continue to write $|v|^2$ instead of v^2). Set $v = e^{\tau\varphi}u$. We have

$$L_\tau v := e^{\tau\varphi}L(e^{-\tau\varphi}v) = \Delta v - \tau(\Delta\varphi)v - 2\tau\nabla\varphi\cdot\nabla v + \tau^2|\nabla\varphi|^2 v. \tag{17.5}$$

Now in the setting provided in Sect. 15.3, in the first line, (see in particular (15.91)) we have neglected the term $-\tau(\Delta\varphi)v$ and, in the middle part of the proof of Theorem 16.9, we have focused on the operator

$$\widetilde{P}_m(x, D + i\tau\nabla\varphi(x)),$$

which in this special case is given by

$$\widetilde{L}_\tau v = \Delta v - 2\tau \nabla\varphi \cdot \nabla v + \tau^2 |\nabla\varphi|^2 v. \tag{17.6}$$

Consequently, operators (16.6a) and (16.6b) in the case of the Laplace operator, are, respectively, given by

$$S_\tau v = \Delta v + \tau^2 |\nabla\varphi|^2 v$$

and

$$A_\tau v = -2\tau \nabla\varphi \cdot \nabla v.$$

Of course,

$$\widetilde{L}_\tau v = S_\tau v + A_\tau v.$$

By Proposition 14.8 we have

$$\int_\Omega \left|\widetilde{L}_\tau v\right|^2 dx = \int_\Omega |S_\tau v|^2 dx + \int_\Omega |A_\tau v|^2 dx + I, \tag{17.7}$$

for every $v \in C_0^\infty(\Omega)$ and for every $\tau \in \mathbb{R}$, where

$$\begin{aligned} I &= 2\int S_\tau v A_\tau v dx = \\ &= 4\tau \int_\Omega \left(\partial^2_{jk}\varphi \partial_j v \partial_k v + \tau^2 (\partial^2_{jk}\varphi \partial_j \varphi \partial_k \varphi)|v|^2\right) dx - \\ &- 2\tau \int_\Omega \Delta\varphi (|\nabla v|^2 - \tau^2 |\nabla\varphi|^2 |v|^2) dx, \end{aligned} \tag{17.8}$$

At this point one could continue without involving the Fourier transform, but for this approach we refer to [13] and [14].

Now, for any $x_0 \in \overline{\Omega}$ we write

$$I = I_{x_0} + R_{x_0},$$

where

$$\begin{aligned} I_{x_0} &= 4\tau \int \left(\partial^2_{jk}\varphi(x_0) \partial_j v \partial_k v + \tau^2 \partial^2_{jk}\varphi(x_0) \partial_j \varphi(x_0) \partial_k \varphi(x_0) |v|^2\right) dx - \\ &- 2\tau \int \Delta\varphi(x_0) (|\nabla v|^2 - \tau^2 |\nabla\varphi(x_0)|^2 |v|^2) dx, \end{aligned}$$

and, of course,

$$R_{x_0} = I - I_{x_0}.$$

Let ε be a positive number that we will choose later. We have there exists $\rho_1 > 0$ such that

$$\left|R_{x_0}\right| \leq \varepsilon \int \left(\tau|\nabla v|^2 + \tau^3|v|^2\right) dx, \quad \forall v \in C_0^\infty(B_{\rho_1}(x_0) \cap \overline{\Omega}).$$

By the Parseval identity we have

$$(2\pi)^n I_{x_0} = \int q(\xi, \tau) \, |\widehat{v}(\xi)|^2 \, d\xi,$$

where

$$q(\xi, \tau) = 4\tau \left(\partial^2_{jk}\varphi(x_0)\xi_j\xi_k + \tau^2\partial^2_{jk}\varphi(x_0)\partial_j\varphi(x_0)\partial_k\varphi(x_0)\right) -$$
$$- 2\tau\Delta\varphi(x_0)(|\xi|^2 - \tau^2|\nabla\varphi(x_0)|^2).$$

Now (17.1) implies there exist positive constants C_1 and C_2, such that

$$C_1 \, |\xi + i\tau\nabla\varphi(x_0)|^4 \leq \tau q(\xi, \tau) + C_2 \left|\sum_{j=1}^{n} \left(\xi + i\tau\partial_j\varphi(x_0)\right)^2\right|^2, \quad \forall(\xi, \tau) \in \mathbb{R}^{n+1}.$$

(as in (16.95)). At this stage, the most challenging part of the proof is complete. It should now be straightforward to assemble the various "pieces" as in the proof of Theorem 16.9. We encourage the reader to do so. ■

Exercise 17.1 Let $\varphi = e^{\lambda\psi}$, where $\lambda \in \mathbb{R}$ and

$$\nabla\psi(x) \neq 0, \quad \forall x \in \bar{\Omega}.$$

Prove that if $|\lambda|$ is large enough then (17.4) applies. ♠

Exercise 17.2 Prove that if $U \in H^2(B_1)$ is a solution to the equation

$$\Delta U = b(x) \cdot \nabla U + c(x)U = 0, \quad \text{in } B_1,$$

where $b \in L^\infty(B_1; \mathbb{R}^n)$ and $c \in L^\infty(B_1)$ then the following three sphere inequality holds true

$$\|U\|_{L^2(B_\varrho(0))} \leq C \, \|U\|^{\vartheta}_{L^2(B_r)} \, \|U\|^{1-\vartheta}_{L^2(B_1)}, \tag{17.9}$$

for $0 < r \leq \varrho \leq C^{-1}$, where $C \leq 1$ and $\vartheta \in (0,1)$ are constants depending on $\|b\|_{L^\infty(B_1;\mathbb{R}^n)}$, $\|c\|_{L^\infty(B_1)}$, on ϱ and on r.

[Hint: apply Theorem 17.1 where φ is a suitable radial function.] ♠

17.3 Second-order Operators with Lipschitz Coefficients in the Principal Part

We now consider a more general case related to the operator (17.1). Let M_0 and M_1 be given positive numbers and let us assume that

$$\left\|g^{jk}\right\|_{L^\infty(\Omega)} \leq M_0, \quad \text{for } j,k = 1,\cdots,n, \tag{17.10a}$$

$$\left|g^{jk}(x) - g^{jk}(y)\right| \leq M_1|x-y|, \quad \text{for } j,k = 1,\cdots,n,\ \forall x,y \in \Omega. \tag{17.10b}$$

Let us note that with conditions (17.10a) and (17.10b), operator (17.1) is not necessarily elliptic. As we noted in Sect. 14.2, to establish a Carleman estimate for operator (17.1), under assumption (17.10b), is equivalent to establish a Carleman estimate for the operator

$$L_g u = \partial_j\left(g^{jk}(x)\partial_k u\right). \tag{17.11}$$

As a matter of fact we have

$$(L_g - P_2)u = \partial_j\left(g^{jk}(x)\right)\partial_k u,$$

which is a first-order operator with bounded coefficients.

We begin by establishing an identity analogous to (14.7). To this purpose we introduce some notations. We set

$$\xi^{(g)} = \left\{g^{jk}(x)\xi_k\right\}_{j=1}^n$$

and for any function v, sufficiently regular, we set

$$\nabla^{(g)}v = \left\{g^{jk}(x)\partial_k v\right\}_{j=1}^n.$$

We further set,

$$\mathbf{g}(\xi,\eta) = g^{jk}(x)\xi_j\eta_k.$$

Using these notations we have

$$L_g u = \operatorname{div}\left(\nabla^{(g)} u\right). \tag{17.12}$$

Let us notice that if $g^{jk}(x) = \delta^{jk}$ then $\mathbf{g}(\xi, \eta) = \xi \cdot \eta$ and $L_g = \Delta$ and, if

$$\left\{g^{jk}(x)\right\}_{j,k=1}^{n} = \operatorname{diag}(1, \cdots, 1, -1),$$

then

$$\mathbf{g}(\xi, \eta) = \xi' \cdot \eta' - \xi_n \eta_n$$

and

$$L_g u = \Delta_{x'} u - \partial_n^2 u = \Box u.$$

Lemma 17.2 (Generalized Rellich Identity) *Let $\beta \in C^{0,1}(\Omega, \mathbb{R}^n)$, $\beta = \left(\beta^1, \cdots, \beta^n\right)$ and $v \in C^2(\Omega)$. Then we have*

$$\begin{aligned} 2\mathbf{g}(\beta, \nabla v) L_g v = {} & \operatorname{div}\left(2\mathbf{g}(\beta, \nabla v) \nabla v - \mathbf{g}(\nabla v, \nabla v) \beta\right) + \\ & + (\operatorname{div} \beta) \mathbf{g}(\nabla v, \nabla v) - 2\partial_l \beta^k g^{lj} \partial_j v \partial_k v + \\ & + \beta^k \left(\partial_k g^{lj}\right) \partial_l v \partial_j v, \quad a.e.\ x \in \Omega \end{aligned} \tag{17.13}$$

and

$$\begin{aligned} 2\int_\Omega \mathbf{g}(\beta, \nabla v) L_g v dx = {} & \int_\Omega (\operatorname{div} \beta) \mathbf{g}(\nabla v, \nabla v) - 2\partial_l \beta^k g^{lj} \partial_j v \partial_k v + \\ & + \int_\Omega \beta^k \left(\partial_k g^{lj}\right) \partial_l v \partial_j v dx, \quad \forall v \in C_0^\infty(\Omega). \end{aligned} \tag{17.14}$$

Proof First, we observe that identity (17.14) is an immediate consequence of (17.13) after its integration over Ω. Hence, it suffices to prove (17.13).

We have

$$\begin{aligned} 2\mathbf{g}(\beta, \nabla v) L_g v &= 2(\beta^k \partial_k v) \partial_l \left(g^{lj} \partial_j v\right) = \\ &= 2\partial_l \left(\beta^k \partial_k v g^{lj} \partial_j v\right) - 2\partial_l \left(\beta^k \partial_k v\right) g^{lj} \partial_j v = \\ &= \operatorname{div}\left(2\mathbf{g}(\beta, \nabla v) \nabla^{(g)} v\right) - 2\partial_l \left(\beta^k \partial_k v\right) g^{lj} \partial_j v = \\ &= \operatorname{div}\left(2\mathbf{g}(\beta, \nabla v) \nabla^{(g)} v\right) - 2\left(\partial_l \beta^k v\right) g^{lj} \partial_k v \partial_j v - \\ &\quad - 2\beta^k g^{lj} \partial_{lk}^2 v \partial_j v. \end{aligned} \tag{17.15}$$

Now, we notice that

$$\partial_k\left(g^{lj}\partial_l v\partial_j v\right)=g^{lj}\partial^2_{lk}v\partial_j v+g^{lj}\partial_l v\partial^2_{jk}v+\partial_k\left(g^{lj}\right)\partial_l v\partial_j v=$$
$$=2g^{lj}\partial^2_{lk}v\partial_j v+\partial_k\left(g^{lj}\right)\partial_l v\partial_j v,$$

from which we have

$$2g^{lj}\partial^2_{lk}v\partial_j v=\partial_k\left(g^{lj}\partial_l v\partial_j v\right)-\partial_k\left(g^{lj}\right)\partial_l v\partial_j v.$$

Using this identity, we transform the last term on the right-hand side of (17.15). We have

$$-2\beta^k g^{lj}\partial^2_{lk}v\partial_j v=-\beta^k\partial_k\left(g^{lj}\partial_l v\partial_j v\right)+\beta^k\partial_k g^{lj}\partial_l v\partial_j v=$$
$$=-\partial_k\left(\beta^k g^{lj}\partial_l v\partial_j v\right)+(\operatorname{div}\beta)g^{lj}\partial_l v\partial_j v+\beta^k\partial_k g^{lj}\partial_l v\partial_j v.$$

Hence

$$-2\beta^k g^{lj}\partial^2_{lk}v\partial_j v=-\operatorname{div}\left(\mathbf{g}\left(\nabla v,\nabla v\right)\right)+$$
$$+(\operatorname{div}\beta)\mathbf{g}\left(\nabla v,\nabla v\right)+\beta^k\partial_k g^{lj}\partial_l v\partial_j v$$

and using the just obtain equality in (17.15) we obtain (17.13). ■

17.3.1 *The Case of Constant Coefficients in the Principal Parts*

In the present subsection, similarly to what was done for the Laplace operator in Sect. 17.2, we will use the generalized Rellich identity (Lemma 17.2) to provide a more direct proof of Theorem 15.24 in the case of the operator whose principal part is given by

$$P_2(\partial)=g^{jk}\partial^2_{jk},\tag{17.16}$$

where $g^{jk}=g^{kj}$ are **real constants**. Let us start by reformulating the conditions (15.206) and (15.207) of Theorem 15.24 in the special case of operator (15.24). As the reader can easily check, such conditions (15.206) and (15.207) can be reformulated, respectively, as follows

$$\left.\begin{array}{l}\mathbf{g}(\xi,\xi)=0\\ \mathbf{g}(\xi,\nabla\varphi(x))=0\\ \xi\neq 0\\ x\in\Omega\end{array}\right\}\Longrightarrow\partial^2\varphi(x)\xi^{(g)}\cdot\xi^{(g)}>0\tag{17.17}$$

and

$$\left.\begin{array}{l}\mathbf{g}(\xi,\xi)=\tau^2\mathbf{g}(\nabla\varphi(x),\nabla\varphi(x))\\ \mathbf{g}(\xi,\nabla\varphi(x))=0\\ \tau\neq 0\end{array}\right\}\Longrightarrow \tag{17.18}$$

$$\Longrightarrow Q_\varphi := \partial^2\varphi(x)\xi^{(g)}\cdot\xi^{(g)} + \tau^2\partial^2\varphi(x)\nabla^{(g)}\varphi(x)\cdot\nabla^{(g)}\varphi(x) > 0.$$

Here $\varphi \in C^2(\overline{\Omega})$ satisfies

$$\nabla\varphi(x) \neq 0 \quad \forall x \in \overline{\Omega}. \tag{17.19}$$

Let us notice that if the matrix $\{g^{jk}\}_{j,k=1}^n$ **is singular then condition** (17.17) **does not hold** because there exists $\xi \in \mathbb{R}^n \setminus \{0\}$ such that

$$\xi^g = g \cdot \xi = 0.$$

It should also be noticed that if the operator $P_2(\partial)$ is elliptic, then condition (17.17) is trivially satisfied since the antecedent is false.

We now state and prove the following

Theorem 17.3 *Let Ω be a bounded open set of $\mathbb{R}^n$ and let*

$$P_2(\partial) = g^{jk}\partial^2_{jk},$$

where the matrix $\{g^{jk}\}_{j,k=1}^n$ is real, constant and symmetric. Suppose that $\varphi \in C^2(\bar{\Omega})$ satisfies conditions (17.17)–(17.19).

Then there exist constants C and τ_0 such that

$$\tau^3\int_\Omega |u|^2 e^{2\tau\varphi}dx + \tau\int_\Omega |\nabla u|^2 e^{2\tau\varphi}dx \leq C\int_\Omega |P_2(\partial)u|^2 e^{2\tau\varphi}dx, \tag{17.20}$$

for every $u \in C_0^\infty(\Omega)$ and for every $\tau \geq \tau_0$.

Proof Let us denote

$$L = P_2(\partial).$$

Let $u \in C_0^\infty(\Omega)$ (u real-valued function) and set $v = e^{\tau\varphi}u$. We have

$$\begin{aligned} P_{2,\tau}v &:= e^{\tau\varphi}L(e^{-\tau\varphi}v) = \\ &= L(v) + \tau^2\mathbf{g}(\nabla\varphi,\nabla\varphi)v - 2\tau\mathbf{g}(\nabla\varphi,\nabla v) - \tau L(\varphi)v. \end{aligned}$$

let us define

$$S_\tau v = L(v) + \tau^2 \mathbf{g}(\nabla\varphi, \nabla\varphi)v$$

and

$$A_\tau v = -2\tau \mathbf{g}(\nabla\varphi, \nabla v)$$

and set

$$\widetilde{P}_2(x, \partial, \tau)v = S_\tau v + A_\tau v, \tag{17.21}$$

Let us note that

$$P_{2,\tau} v - \widetilde{P}_2(x, \partial, \tau)v = -\tau L(\varphi)v. \tag{17.22}$$

In what follows, for the sake of brevity, we will omit the domain and the element of integration.

Let μ be a positive number which we will choose later, we have trivially

$$\begin{aligned} &\int \left|\widetilde{P}_2(x, \partial, \tau)v\right|^2 e^{2\mu\varphi} = \\ &= \int |S_\tau v|^2 e^{2\mu\varphi} + \int |A_\tau v|^2 e^{2\mu\varphi} + \\ &+ 2\int (S_\tau v A_\tau v)\, e^{2\mu\varphi} \geq \\ &\geq 2\int (S_\tau v A_\tau v)\, e^{2\mu\varphi} = \\ &= -2\tau \int 2\left(L(v) + \tau^2 \mathbf{g}(\nabla\varphi, \nabla\varphi)v\right) \mathbf{g}(\nabla\varphi, \nabla v) e^{2\mu\varphi} := \\ &:= -2\tau J_1 - 2\tau^3 J_2, \end{aligned} \tag{17.23}$$

where

$$J_1 = \int 2L(v)\mathbf{g}(\nabla\varphi, \nabla v)e^{2\mu\varphi} \tag{17.24}$$

and

$$J_2 = \int 2\mathbf{g}(\nabla\varphi, \nabla\varphi)\mathbf{g}(\nabla\varphi, \nabla v)v e^{2\mu\varphi}. \tag{17.25}$$

Let us Examine J_1
Let us apply (17.13) with

$$\beta = e^{2\mu\varphi}\nabla^{(g)}\varphi.$$

We get

$$\begin{aligned}
J_1 = &\int \left(\mathrm{div}(e^{2\mu\varphi}\nabla^{(g)}\varphi)\right)\mathbf{g}\left(\nabla v, \nabla v\right) - \\
&- 2\partial_l\left(e^{2\mu\varphi}g^{kj}\partial_j\varphi\right)g^{ls}\partial_s v\partial_k v = \\
&= \int \left(e^{2\mu\varphi}(L(\varphi))\mathbf{g}\left(\nabla v, \nabla v\right) + 2\mu e^{2\mu\varphi}\mathbf{g}\left(\nabla\varphi, \nabla\varphi\right)\mathbf{g}\left(\nabla v, \nabla v\right) - \right. \\
&\quad \left. -2e^{2\mu\varphi}g^{kj}\partial^2_{lj}\varphi g^{ls}\partial_s v\partial_k v - 4\mu e^{2\mu\varphi}\partial_l\varphi g^{kj}\partial_j\varphi g^{ls}\partial_s v\partial_k v\right) = \\
&= \int e^{2\mu\varphi}\left((L(\varphi))\mathbf{g}\left(\nabla v, \nabla v\right) - 2\partial^2\varphi\nabla^{(g)}v\cdot\nabla^{(g)}v + \right. \\
&\quad \left. +2\mu\mathbf{g}\left(\nabla\varphi, \nabla\varphi\right)\mathbf{g}\left(\nabla v, \nabla v\right) - 4\mu\left(\mathbf{g}\left(\nabla\varphi, \nabla v\right)\right)^2\right).
\end{aligned} \tag{17.26}$$

Let us Examine J_2

$$\begin{aligned}
J_2 &= \int 2e^{2\mu\varphi}\mathbf{g}(\nabla\varphi, \nabla\varphi)\mathbf{g}(\nabla\varphi, \nabla v)v = \\
&= \int e^{2\mu\varphi}\mathbf{g}(\nabla\varphi, \nabla\varphi)g^{ls}\partial_l\varphi\partial_s(v^2) = \\
&= -\int \partial_s\left(e^{2\mu\varphi}\mathbf{g}(\nabla\varphi, \nabla\varphi)g^{ls}\partial_l\varphi\right)|v|^2 = \\
&= -\int e^{2\mu\varphi}\left(2\partial^2\varphi\nabla^{(g)}\varphi\cdot\nabla^{(g)}\varphi + \mathbf{g}(\nabla\varphi, \nabla\varphi)L(\varphi) + 2\mu\left(\mathbf{g}(\nabla\varphi, \nabla\varphi)\right)^2\right)|v|^2.
\end{aligned}$$

By the above obtained equality and by (17.26) we have

$$\begin{aligned}
2\int\left(S_\tau v A_\tau v\right)e^{2\mu\varphi} &= -2\tau J_1 - 2\tau^3 J_2 = \\
&= 2\tau\int e^{2\mu\varphi}q_\mu(x, v, \nabla v, \tau),
\end{aligned} \tag{17.27}$$

where

$$q_\mu(x, v, \nabla v, \tau) = q_0(x, v, \nabla v, \tau) + \mu q_1(x, v, \nabla v, \tau), \tag{17.28}$$

$$\begin{aligned} &q_0(x, v, \nabla v, \tau) = \\ &= 2\left[\partial^2\varphi(x)\nabla^{(g)}v \cdot \nabla^{(g)}v + \tau^2\partial^2\varphi(x)\nabla^{(g)}\varphi(x) \cdot \nabla^{(g)}\varphi(x)|v|^2\right] + \\ &+ L(\varphi)\left[\tau^2\mathbf{g}(\nabla\varphi(x), \nabla\varphi(x))|v|^2 - \mathbf{g}(\nabla v, \nabla v)\right], \end{aligned} \tag{17.29}$$

and

$$\begin{aligned} &q_1(x, v, \nabla v, \tau) = \\ &= -2\mathbf{g}(\nabla\varphi(x), \nabla\varphi(x))\mathbf{g}(\nabla v, \nabla v) + 4\left(\mathbf{g}(\nabla\varphi(x), \nabla v)\right)^2 + \\ &+ 2\tau^2\left(\mathbf{g}(\nabla\varphi(x), \nabla\varphi(x))\right)^2|v|^2. \end{aligned} \tag{17.30}$$

Trivially, for any $x_0 \in \overline{\Omega}$, the right-hand side of (17.27) can be written

$$2\tau\int e^{2\mu\varphi}q_\mu(x, v, \nabla v, \tau) = I_{x_0} + R_{x_0},$$

where

$$I_{x_0} = 2\tau e^{2\mu\varphi(x_0)}\int q_\mu(x_0, v, \nabla v, \tau), \tag{17.31}$$

and

$$R_{x_0} = 2\tau\int\left(e^{2\mu\varphi(x)}q_\mu(x, v, \nabla v, \tau) - e^{2\mu\varphi(x_0)}q_\mu(x_0, v, \nabla v, \tau)\right). \tag{17.32}$$

Now, let $\rho \in (0, 1]$ be chosen. We get easily

$$\left|R_{x_0}\right| \leq C\rho(\mu+1)^2e^{2\mu\Phi_0}\int\left(\tau|\nabla v|^2 + \tau^3|v|^2\right)dx, \tag{17.33}$$

for every $v \in C_0^\infty(B_\rho(x_0) \cap \Omega)$, where

$$\Phi_0 = \max_{x\in\overline{\Omega}}|\varphi(x)|,$$

C does not depend by μ, but depends on the $C^2\left(\overline{\Omega}\right)$-norm of φ and by

$$M_0 = \max_{1\leq j,k\leq n}\left|g^{jk}\right|.$$

Now, the Parseval identity gives

$$I_{x_0} = \frac{2\tau e^{2\mu\varphi(x_0)}}{(2\pi)^n}\int Q_\mu(\xi, \tau)\,|\widehat{v}(\xi)|^2\,d\xi, \tag{17.34}$$

where

$$Q_\mu(\xi,\tau) = Q_0(\xi,\tau) + \mu Q_1(\xi,\tau), \tag{17.35}$$

$$\begin{aligned} Q_0(\xi,\tau) =& 2\left[\partial^2\varphi(x_0)\xi^{(g)}\cdot\xi^{(g)} + \tau^2\partial^2\varphi(x_0)\nabla^{(g)}\varphi(x_0)\cdot\nabla^{(g)}\varphi(x_0)\right] + \\ &+ L(\varphi)(x_0)\left[\tau^2\mathbf{g}(\nabla\varphi(x_0),\nabla\varphi(x_0)) - \mathbf{g}(\xi,\xi)\right] \end{aligned}$$

and

$$\begin{aligned} Q_1(\xi,\tau) =& -2\mathbf{g}(\nabla\varphi(x_0),\nabla\varphi(x_0))\left(\mathbf{g}(\xi,\xi) - \tau^2\mathbf{g}(\nabla\varphi(x_0),\nabla\varphi(x_0))\right) + \\ &+ 4\left(\mathbf{g}(\nabla\varphi(x_0),\xi)\right)^2. \end{aligned}$$

Now, we can prove, in a manner entirely analogous to the proof of Lemma 15.26, that there exist C_1, C_2 and μ_1, positive numbers, such that for every $\mu \geq \mu_1$ we have

$$|\xi + i\tau\nabla\varphi(x_0)|^2 \leq C_1 Q_\mu(\xi,\tau) + C_2\frac{|P_2(\xi + i\tau\nabla\varphi(x_0))|^2}{|\xi + i\tau\nabla\varphi(x_0)|^2}, \tag{17.36}$$

for every $(\xi,\tau) \in \mathbb{R}^{n+1}\setminus\{(0,0)\}$.

From now on, we **fix a value** μ for which (17.36) is satisfied (recalling that, however, μ depends on x_0). Set $N = \nabla\varphi(x_0)$. Using the notations introduced in (15.227), we have by (17.36)

$$\begin{aligned} \|v\|^2_{H^1_\tau(\mathbb{R}^n)} \leq& \frac{C_1}{(2\pi)^n}\int Q_\mu(\xi,\tau)\,|\widehat{v}(\xi)|^2\,d\xi + \\ &+ C_2\left\|\widetilde{P}_2(x_0,\partial,\tau)v\right\|^2_{-1,\tau,N}, \end{aligned} \tag{17.37}$$

where $\widetilde{P}_2(x,\partial,\tau)$ is defined in (17.21). We have easily

$$\left|\widetilde{P}_2(x,\partial,\tau)v - \widetilde{P}_2(x_0,\partial,\tau)v\right| \leq C\rho\left(\tau|\nabla v| + \tau^2|v|\right), \tag{17.38}$$

for every $v \in C^\infty_0(B_\rho(x_0)\cap\Omega)$. Now, by (15.228), (17.37) and by the triangle inequality, we get

$$\left\|\widetilde{P}_2(x_0,\partial,\tau)v\right\|_{-1,\tau}\leq\frac{1}{\tau|\nabla\varphi(x_0)|}\left\|\widetilde{P}_2(x_0,\partial,\tau)v\right\|_{L^2(\Omega)}\leq$$
$$\leq\frac{1}{\tau|\nabla\varphi(x_0)|}\left(\left\|\widetilde{P}_2(\cdot,\partial,\tau)v-\widetilde{P}_2(x_0,\partial,\tau)v\right\|_{L^2(\Omega)}+\left\|\widetilde{P}_2(\cdot,\partial,\tau)v\right\|_{L^2(\Omega)}\right)\leq$$
$$\leq\frac{1}{\tau|\nabla\varphi(x_0)|}\left[C\rho\tau\,\||\nabla v|+\tau|v|\|_{L^2(\Omega)}+\left\|\widetilde{P}_2(\cdot,\partial,\tau)v\right\|_{L^2(\Omega)}\right]\leq$$
$$\leq C\rho\,\|v\|_{H^1_\tau(\mathbb{R}^n)}+\frac{1}{\tau|\nabla\varphi(x_0)|}\left\|\widetilde{P}_2(\cdot,\partial,\tau)v\right\|_{L^2(\Omega)}.$$

Hence, we have

$$\left\|\widetilde{P}_2(x_0,\partial,\tau)v\right\|_{-1,\tau}\leq C\left(\rho\,\|v\|_{H^1_\tau(\mathbb{R}^n)}+\frac{1}{\tau}\left\|\widetilde{P}_2(\cdot,\partial,\tau)v\right\|_{L^2(\Omega)}\right),\tag{17.39}$$

for every $v\in C_0^\infty(B_\rho(x_0)\cap\Omega)$, where C depends on $C^1\left(\overline{\Omega}\right)$-norm of φ, on M_0 and on

$$m_2=\min_{\overline{\Omega}}|\nabla\varphi|.$$

By (17.39) and by (17.37) we obtain

$$\begin{aligned}2\tau\,\|v\|^2_{H^1_\tau(\mathbb{R}^n)}\leq&\frac{2\tau C_1}{(2\pi)^n}\int Q_\mu(\xi,\tau)\,|\widehat{v}(\xi)|^2\,d\xi+\\&+2\tau C\rho^2\,\|v\|^2_{H^1_\tau(\mathbb{R}^n)}+C\tau^{-1}\left\|\widetilde{P}_2(\cdot,\partial,\tau)v\right\|^2_{L^2(\Omega)},\end{aligned}\tag{17.40}$$

for every $v\in C_0^\infty(B_\rho(x_0)\cap\Omega)$.

Now we need to estimate the first term on the right-hand side in (17.40). From (17.31), (17.32) and (17.35) we have

$$\begin{aligned}&\frac{2\tau}{(2\pi)^n}\int Q_\mu(\xi,\tau)\,|\widehat{v}(\xi)|^2\,d\xi=2\tau\int q_\mu(x_0,v,\nabla v,\tau)dx=\\&=2\tau e^{-2\mu\varphi(x_0)}\int e^{2\mu\varphi(x_0)}q_\mu(x_0,v,\nabla v,\tau)dx=\\&=2\tau e^{-2\mu\varphi(x_0)}\int\Big(e^{2\mu\varphi(x_0)}q_\mu(x_0,v,\nabla v,\tau)-\\&-e^{2\mu\varphi(x)}q_\mu(x,v,\nabla v,\tau)\Big)\,dx+\\&+2\tau e^{-2\mu\varphi(x_0)}\int e^{2\mu\varphi(x)}q_\mu(x,v,\nabla v,\tau)dx=\\&=R_{x_0}+2\tau e^{-2\mu\varphi(x_0)}\int e^{2\mu\varphi(x)}q_\mu(x,v,\nabla v,\tau)dx.\end{aligned}\tag{17.41}$$

From what we have just obtained, from (17.27) and from (17.33) we have (recall that we have fixed μ)

$$\begin{aligned}
&\frac{2\tau}{(2\pi)^n}\int Q_\mu(\xi,\tau)\,|\widehat{v}(\xi)|^2\,d\xi \le \\
&\le C\rho(\mu+1)^2 e^{2\mu\Phi_0}\int\left(\tau|\nabla v|^2+\tau^3|v|^2\right)dx+ \\
&+C\int (S_\tau v A_\tau v)\,e^{2\mu\varphi}dx \le \\
&\le C\rho\tau\,\|v\|^2_{H^1_\tau(\mathbb{R}^n)}+C\int\left|\widetilde{P}_2(x,\partial,\tau)v\right|^2 e^{2\mu\varphi}dx,
\end{aligned} \tag{17.42}$$

for every $v\in C^\infty_0(B_\rho(x_0)\cap\Omega)$. In the last estimate from above we used that (see (17.23))

$$2\int (S_\tau v A_\tau v)\,e^{2\mu\varphi}dx \le \int\left|\widetilde{P}_2(x,\partial,\tau)v\right|^2 e^{2\mu\varphi}dx.$$

By (17.40) and (17.42) (recalling that $\tau\ge 1$) we get

$$2\tau\,\|v\|^2_{H^1_\tau(\mathbb{R}^n)} \le C\rho\tau\,\|v\|^2_{H^1_\tau(\mathbb{R}^n)}+C\left(1+\tau^{-1}\right)\int\left|\widetilde{P}_2(x,\partial,\tau)v\right|^2 dx,$$

for every $v\in C^\infty_0(B_\rho(x_0)\cap\Omega)$. Let us choose

$$\rho=\rho_0:=\min\left\{\frac{1}{C},1\right\}$$

and we get

$$\tau\,\|v\|^2_{H^1_\tau(\mathbb{R}^n)} \le C\int\left|\widetilde{P}_2(x,\partial,\tau)v\right|^2 dx,$$

for every $v\in C^\infty_0(B_{\rho_0}(x_0)\cap\Omega)$ and for every $\tau\ge 1$.

Now, recalling (17.22) and applying Proposition 15.13 we obtain

$$\tau^3\int_\Omega |u|^2e^{2\tau\varphi}dx+\tau\int_\Omega|\nabla u|^2e^{2\tau\varphi}dx\le C\int_\Omega|P_2(\partial)u|^2e^{2\tau\varphi}dx, \tag{17.43}$$

for every $u\in C^\infty_0(B_{\rho_0}(x_0)\cap\Omega)$ and for every $\tau\ge\tau^*$, where C and τ^* are suitable positive numbers depending by x_0. At this point we have only to apply Lemma 14.2 to conclude the proof. ∎

17.3.2 The Case of Lipschitz Continuous Coefficients in the Principal Parts

In the following Theorem we will consider the operator

$$P_2(x, \partial) = g^{jk}(x)\partial^2_{jk}, \quad \text{in } B_1, \tag{17.44}$$

where $\left\{g^{jk}(x)\right\}^n_{j,k=1}$ is a real symmetric matrix-valued function. Let $\varphi \in C^2\left(\overline{B_1}\right)$ satisfy

$$\nabla\varphi(x) \neq 0, \quad \forall x \in \overline{B_1}.$$

For the sake of brevity, in what follows we will say that φ satisfies condition (**S**) w.r.t. the operator $P_2(x, \partial)$ if we have

$$\left.\begin{array}{l} p_2(0,\xi) = 0 \\ p_2^{(j)}(0,\xi)\partial_j\varphi(0) = 0 \\ \xi \neq 0 \end{array}\right\} \Longrightarrow \partial^2_{jk}\varphi(0)p_2^{(j)}(0,\xi)p_2^{(k)}(0,\xi) > 0 \tag{17.45}$$

and

$$\begin{aligned} &\left.\begin{array}{l} p_2(0,\xi + i\tau\nabla\varphi(0)) = 0 \\ \\ \tau \neq 0 \end{array}\right\} \Longrightarrow \\ &\Longrightarrow \partial^2_{jk}\varphi(0)p_2^{(j)}(0,\xi + i\tau\nabla\varphi(0))\overline{p_2^{(k)}(0,\xi + i\tau\nabla\varphi(0))} > 0. \end{aligned} \tag{17.46}$$

Theorem 17.4 *Let $P_2(x, \partial)$ be operator (17.44). Assume that (17.10a) and (17.10b) are satisfied with $\Omega = B_1$. Suppose that $\varphi \in C^2\left(\overline{B_1}\right)$ satisfies condition* (**S**) *w.r.t. $P_2(x, \partial)$.*

Then there exist $\rho_0 \in (0, 1]$, $\delta_0 \in (0, 1]$, $C \geq 1$ and $\tau_0 \geq 1$ such that

$$\begin{aligned} &\tau^3\int_{B_1}|u|^2e^{2\tau\varphi}dx + \tau\int_{B_1}|\nabla u|^2e^{2\tau\varphi}dx \leq \\ &\leq C\int_{B_1}|P_2(\delta x, \partial)u|^2e^{2\tau\varphi}dx, \end{aligned} \tag{17.47}$$

for every $\delta \in (0, \delta_0]$, for every $u \in C^\infty_0(B_{\rho_0})$ and for every $\tau \geq \tau_0$.

Proof For the most part of the proof we repeat the steps we did in the proof of Theorem 17.3 by paying special attention to the additional terms that come up because now the coefficients of the operator are variables.

Let us denote

$$g_\delta^{jk}(x) = g^{jk}(\delta x), \quad j,k = 1,\cdots,n.$$

Let $u \in C_0^\infty(B_1)$ (u real-valued function) and set $v = e^{\tau\varphi}u$. We have

$$\begin{aligned} &P_2(\delta x, \partial, \tau)v := e^{\tau\varphi} P_2(\delta x, \partial)(e^{-\tau\varphi}v) = \\ &= P_2(\delta x, \partial)v - \tau(P_2(\delta x, \partial)\varphi)v + \tau^2 \mathbf{g}_\delta(\nabla\varphi, \nabla\varphi)v - 2\tau\mathbf{g}_\delta(\nabla\varphi, \nabla v) = \\ &= \partial_j\left(g_\delta^{jk}(x)\partial_k\right) - 2\tau\mathbf{g}_\delta(\nabla\varphi, \nabla v) - \\ &- \tau(P_2(\delta x, \partial)\varphi)v - \partial_j\left(g_\delta^{jk}(x)\right)\partial_k v. \end{aligned}$$

Let us denote

$$L_\delta(v) = \partial_j\left(g_\delta^{jk}(x)\partial_k v\right), \tag{17.48}$$

$$\widetilde{P}_2(x, \partial, \tau)v = S_\tau v + A_\tau v, \tag{17.49}$$

where

$$S_\tau v = L(v) + \tau^2 \mathbf{g}_\delta(\nabla\varphi, \nabla\varphi)v$$

and

$$A_\tau v = -2\tau\mathbf{g}_\delta(\nabla\varphi, \nabla v).$$

Let us notice that

$$P_2(\delta x, \partial, \tau)v - \widetilde{P}_2(x, \partial, \tau)v = -\tau(P_2(\delta x, \partial)\varphi)v - \partial_j\left(g_\delta^{jk}(x)\right)\partial_k v. \tag{17.50}$$

For the sake of brevity, we omit the domain and the element of integration. Let μ be a positive number that we will choose later. We have trivially

$$\begin{aligned} \int \left|\widetilde{P}_2(x, \partial, \tau)v\right|^2 e^{2\mu\varphi} &= \int |S_\tau v|^2 e^{2\mu\varphi} + \int |A_\tau v|^2 e^{2\mu\varphi} + \\ &+ 2\int (S_\tau v A_\tau v)\, e^{2\mu\varphi} \geq \\ &\geq 2\int (S_\tau v A_\tau v)\, e^{2\mu\varphi}. \end{aligned} \tag{17.51}$$

Set

$$J_1 = \int 2\,(L_\delta v)\,\mathbf{g}_\delta(\nabla\varphi, \nabla v)e^{2\mu\varphi} \tag{17.52}$$

and

$$J_2 = \int 2\mathbf{g}_\delta(\nabla\varphi, \nabla\varphi)\mathbf{g}_\delta(\nabla\varphi, \nabla v)ve^{2\mu\varphi}, \tag{17.53}$$

we have

$$2\int (S_\tau v A_\tau v)\,e^{2\mu\varphi} = -2\tau J_1 - 2\tau^3 J_2. \tag{17.54}$$

Now, to handle J_1, we apply (17.13) with

$$\beta_\delta = e^{2\mu\varphi}\nabla^{(g_\delta)}\varphi.$$

It should be kept in mind that now, with respect to proof of the Theorem 17.3, the coefficients of the operator $P_2(\delta x, \partial)$ depend on x, more precisely they are of the type $f(\delta x)$ where f is a Lipschitz continuous function. Hence we have

$$J_1 \le \int e^{2\mu\varphi}\Big((L_\delta(\varphi))\mathbf{g}_\delta\,(\nabla v, \nabla v) - 2\partial^2\varphi\nabla^{(g_\delta)}v \cdot \nabla^{(g_\delta)}v +$$

$$+2\mu\mathbf{g}_\delta\,(\nabla\varphi, \nabla\varphi)\,\mathbf{g}_\delta\,(\nabla v, \nabla v) - 4\mu\,(\mathbf{g}_\delta\,(\nabla\varphi, \nabla v))^2\Big) + C\delta\int e^{2\mu\varphi}|\nabla v|^2$$

and, similarly,

$$J_2 \le -\int e^{2\mu\varphi}\Big(2\partial^2\varphi\nabla^{(g_\delta)}\varphi \cdot \nabla^{(g_\delta)}\varphi + \mathbf{g}_\delta(\nabla\varphi, \nabla\varphi)L_\delta(\varphi) +$$

$$+2\mu\,(\mathbf{g}_\delta(\nabla\varphi, \nabla\varphi))^2\Big)\,|v|^2 + C\delta\int e^{2\mu\varphi}|v|^2,$$

in the last two inequalities, C depends on M_0, M_1 and $\|\varphi\|_{C^2(\overline{B_1})}$, but does not depend on μ. By these inequalities and by (17.54) we get

$$\begin{aligned} 2\int (S_\tau v A_\tau v)\,e^{2\mu\varphi} &= -2\tau J_1 - 2\tau^3 J_2 \ge \\ &\ge 2\tau\int e^{2\mu\varphi}q_\mu^{(\delta)}(x, v, \nabla v, \tau) - \\ &- C\delta\tau\int e^{2\mu\varphi}\left(\tau^2|v|^2 + |\nabla v|^2\right), \end{aligned} \tag{17.55}$$

where

$$q_\mu^{(\delta)}(x, v, \nabla v, \tau) = q_0^{(\delta)}(x, v, \nabla v, \tau) + \mu q_1^{(\delta)}(x, v, \nabla v, \tau), \tag{17.56}$$

$$\begin{aligned} & q_0^{(\delta)}(x, v, \nabla v, \tau) = \\ & = 2\left[\partial^2\varphi(x)\nabla^{(g_\delta)}v \cdot \nabla^{(g_\delta)}v + \right. \\ & \left. \tau^2(\partial^2\varphi(x)\nabla^{(g_\delta)}\varphi(x) \cdot \nabla^{(g_\delta)}\varphi(x))|v|^2\right] + \\ & + L_\delta(\varphi)\left[\tau^2\mathbf{g}_\delta(\nabla\varphi(x), \nabla\varphi(x))|v|^2 - \mathbf{g}_\delta(\nabla v, \nabla v)\right] \end{aligned} \tag{17.57}$$

and

$$\begin{aligned} & q_1^{(\delta)}(x, v, \nabla v, \tau) = \\ & = -2\mathbf{g}_\delta(\nabla\varphi(x), \nabla\varphi(x))\mathbf{g}_\delta(\nabla v, \nabla v) + 4\left(\mathbf{g}_\delta(\nabla\varphi(x), \nabla v)\right)^2 + \\ & + 2\tau^2\left(\mathbf{g}_\delta(\nabla\varphi(x), \nabla\varphi(x))\right)^2 |v|^2. \end{aligned} \tag{17.58}$$

Now we **examine the first addend on the right-hand side of** (17.55), namely

$$2\tau \int e^{2\mu\varphi} q_\mu^{(\delta)}(x, v, \nabla v, \tau).$$

Likewise to the proof of Theorem 17.3 we write

$$2\tau \int e^{2\mu\varphi} q_\mu^{(\delta)}(x, v, \nabla v, \tau) = I_0 + R_0,$$

where

$$I_0 = 2\tau e^{2\mu\varphi(0)} \int q_\mu^{(\delta)}(0, v, \nabla v, \tau) \tag{17.59}$$

and

$$R_0 = 2\tau \int \left(e^{2\mu\varphi(x)} q_\mu^{(\delta)}(x, v, \nabla v, \tau) - e^{2\mu\varphi(0)} q_\mu^{(\delta)}(0, v, \nabla v, \tau)\right). \tag{17.60}$$

Let $\rho \in (0, 1]$ be chosen; we get easily (recall $\delta \leq 1$)

$$|R_0| \leq C\rho(\mu + 1)^2 e^{2\mu\Phi_0} \int \left(\tau|\nabla v|^2 + \tau^3|v|^2\right) dx, \tag{17.61}$$

for every $v \in C_0^\infty(B_\rho)$, where

$$\Phi_0 = \max_{x \in \overline{B_1}} |\varphi(x)|,$$

C does not depend on μ, but depends on the $C^2\left(\overline{B_1}\right)$ norm of φ and on M_0.

Let us denote

$$g_0^{jk} = g^{jk}(0).$$

By the Parseval identity we have

$$I_0 = \frac{2\tau e^{2\mu\varphi(0)}}{(2\pi)^n} \int Q_\mu(\xi, \tau) \left|\widehat{v}(\xi)\right|^2 d\xi, \tag{17.62}$$

where

$$Q_\mu(\xi, \tau) = Q_0(\xi, \tau) + \mu Q_1(\xi, \tau), \tag{17.63}$$

$$\begin{aligned} Q_0(\xi, \tau) =& 2\left[\partial^2\varphi(0)\xi^{(g_0)} \cdot \xi^{(g_0)} + \tau^2 \partial^2\varphi(0)\nabla^{(g_0)}\varphi(0) \cdot \nabla^{(g_0)}\varphi(0)\right] + \\ & + L_0(\varphi)(0)\left[\tau^2 \mathbf{g}_0(\nabla\varphi(0), \nabla\varphi(0)) - \mathbf{g}_0(\xi, \xi)\right] \end{aligned}$$

and

$$\begin{aligned} Q_1(\xi, \tau) =& -2\mathbf{g}_0(\nabla\varphi(0), \nabla\varphi(0))\left(\mathbf{g}_0(\xi, \xi) - \tau^2 \mathbf{g}_0(\nabla\varphi(0), \nabla\varphi(0))\right) + \\ & + 4\left(\mathbf{g}_0(\nabla\varphi(0), \xi)\right)^2. \end{aligned}$$

Similarly to what was done in the proof of Theorem 17.3 can be proved the existence of constants C_1, C_2 and μ_1 such that for every $\mu \geq \mu_1$ we have

$$|\xi + i\tau\nabla\varphi(0)|^2 \leq C_1 Q_\mu(\xi, \tau) + C_2 \frac{|p_2(0, \xi + i\tau\nabla\varphi(0))|^2}{|\xi + i\tau\nabla\varphi(0)|^2}, \tag{17.64}$$

for every $(\xi, \tau) \in \mathbb{R}^{n+1} \setminus \{(0, 0)\}$.

From now on we fix a value μ for which (17.64) is satisfied Set $N = \nabla\varphi(0)$. By (17.64) we have

$$\begin{aligned} & 2\tau \left\|v\right\|^2_{H^1_\tau(\mathbb{R}^n)} \leq \\ & \leq \frac{2\tau C_1}{(2\pi)^n} \int Q_\mu(\xi, \tau) \left|\widehat{v}(\xi)\right|^2 d\xi + 2\tau C_2 \left\|P_2(0, \partial, \tau)v\right\|^2_{-1,\tau,N}. \end{aligned} \tag{17.65}$$

Now, **we estimate from above the last term on the right-hand side in** (17.65). For this purpose, we will repeatedly use the triangle inequality, (15.228) and Lemma 15.27.

First of all, set

$$P_{\bullet,2}(x,\partial,\tau)v = g_\delta^{jk}(x)\partial^2_{jk}v - 2\tau g_\delta^{jk}(x)\partial_j\varphi(0)\partial_k v + \tau^2 g_\delta^{jk}(x)\partial_j\varphi(0)\partial_k\varphi(0)v$$

and let us note that

$$P_{\bullet,2}(0,\partial,\tau) = p_2(0,\partial,\tau).$$

By this equality and by the triangle inequality we have

$$\begin{aligned} &\|P_2(0,\partial,\tau)v\|^2_{-1,\tau,N} = \left\|P_{\bullet,2}(0,\partial,\tau)v\right\|^2_{-1,\tau,N} \leq \\ &\leq 2\left\|P_{\bullet,2}(\cdot,\partial,\tau)v\right\|^2_{-1,\tau,N} + \\ &+2\left\|P_{\bullet,2}(\cdot,\partial,\tau)v - P_{\bullet,2}(0,\partial,\tau)v\right\|^2_{-1,\tau,N}. \end{aligned} \tag{17.66}$$

Let us estimate from above the first term on the right-hand side in (17.66). By the triangle inequality and by the definition of $\|\cdot\|_{-1,\tau,N}$ we obtain

$$\begin{aligned} &\left\|P_{\bullet,2}(\cdot,\partial,\tau)v\right\|^2_{-1,\tau,N} \leq C\tau^{-2}\left\|P_{\bullet,2}(\cdot,\partial,\tau)v\right\|^2_{L^2(B_1)} \leq \\ &\leq C\tau^{-2}\left\|P_{\bullet,2}(\cdot,\partial,\tau)v - P_2(\cdot,\partial,\tau)v\right\|^2_{L^2(B_1)} + \\ &+C\tau^{-2}\|P_2(\cdot,\partial,\tau)v\|^2_{L^2(B_1)}, \end{aligned} \tag{17.67}$$

on the other hand,

$$\begin{aligned} &P_{\bullet,2}(x,\partial,\tau)v - P_2(x,\partial,\tau)v = \\ &= -\partial_j\left(g_\delta^{jk}(x)\right)\partial_k v + 2\tau g_\delta^{jk}(x)\left(\partial_j\varphi(x) - \partial_j\varphi(0)\right)\partial_k v + \\ &+\tau^2 g_\delta^{jk}(x)\left(\partial_j\varphi(x)\partial_k\varphi(x) - \partial_j\varphi(0)\partial_k\varphi(0)\right)v. \end{aligned}$$

From the latter and the triangle inequality, we deduce easily that

$$\begin{aligned} &\tau^{-2}\left\|P_{\bullet,2}(\cdot,\partial,\tau)v - P_2(\cdot,\partial,\tau)v\right\|^2_{L^2(B_1)} \leq \\ &\leq C\delta^2\tau^{-2}\|\nabla v\|^2_{L^2(B_1)} + C\rho^2\|v\|^2_{H^1_\tau(\mathbb{R}^n)}, \end{aligned} \tag{17.68}$$

for every $v \in C_0^\infty(B_\rho)$.

By (17.67) and (17.68) we get

$$\begin{aligned}\left\|P_{\bullet,2}(\cdot,\partial,\tau)v\right\|^2_{-1,\tau,N} \leq C\tau^{-2}\left\|\nabla v\right\|^2_{L^2(B_1)} + C\rho^2\left\|v\right\|^2_{H^1_\tau(\mathbb{R}^n)} + \\ + C\tau^{-2}\left\|P_2(\cdot,\partial,\tau)v\right\|^2_{L^2(B_1)}.\end{aligned} \tag{17.69}$$

Now, let us estimate from above the second term on the right-hand side in (17.66). We notice

$$\begin{aligned}&P_{\bullet,2}(x,\partial,\tau)v - P_{\bullet,2}(0,\partial,\tau)v = \\ &= \left(g^{jk}_\delta(x) - g^{jk}_\delta(0)\right)\left(\partial_j - \tau\partial_j\varphi(0)\right)\left(\partial_k - \tau\partial_k\varphi(0)\right)v.\end{aligned}$$

and, for a fixed $k = 1, \cdots, n$, we set

$$w_k = \left(\partial_k - \tau\partial_k\varphi(0)\right)v.$$

By applying Lemma 15.27 we have, for every $v \in C^\infty_0(B_\rho)$,

$$\begin{aligned}&\left\|\left(g^{jk}_\delta(x) - g^{jk}_\delta(0)\right)\left(\partial_j - \tau\partial_j\varphi(0)\right)w_k\right\|^2_{-1,\tau,N} \leq \\ &\leq C\left(M_1^2\delta^2\rho^2 + |\tau\nabla\varphi(0)|^{-2}\right)\left\|w_k\right\|^2_{L^2(B_1)}.\end{aligned} \tag{17.70}$$

On the other hand, for $k = 1, \cdots, n$,

$$\left\|w_k\right\|^2_{L^2(B_1)} = \left\|\left(\partial_k - \tau\partial_k\varphi(0)\right)v\right\|^2_{L^2(B_1)} \leq C\left\|v\right\|^2_{H^1_\tau(\mathbb{R}^n)}.$$

By the latter and by (17.70) we get

$$\left\|P_{\bullet,2}(\cdot,\partial,\tau)v - P_{\bullet,2}(0,\partial,\tau)v\right\|^2_{-1,\tau,N} \leq C\left(M_1^2\rho^2 + \tau^{-2}\right)\left\|v\right\|^2_{H^1_\tau(\mathbb{R}^n)}. \tag{17.71}$$

By (17.66), (17.69) and (17.71) we have

$$\begin{aligned}&\left\|P_2(0,\partial,\tau)v\right\|^2_{-1,\tau,N} \leq \\ &\leq C\left(\rho^2 + \tau^{-2}\right)\left\|v\right\|^2_{H^1_\tau(\mathbb{R}^n)} + C\tau^{-2}\left\|P_2(\cdot,\partial,\tau)v\right\|^2_{L^2(B_1)},\end{aligned}$$

for every $v \in C^\infty_0(B_\rho)$, where C depends on M_0 and M_1.

By the latter and by (17.65) we have

$$\begin{aligned} 2\tau \left\|v\right\|^2_{H^1_\tau(\mathbb{R}^n)} \leq \frac{2\tau C_1}{(2\pi)^n}\int Q_\mu(\xi,\tau)\left|\widehat{v}(\xi)\right|^2 d\xi+ \\ +C\left(\rho^2\tau+\tau^{-1}\right)\left\|v\right\|^2_{H^1_\tau(\mathbb{R}^n)}+C\tau^{-1}\left\|P_2(\cdot,\partial,\tau)v\right\|^2_{L^2(B_1)}, \end{aligned} \tag{17.72}$$

for every $v \in C^\infty_0(B_\rho)$.

In order to estimate from above the first term on the right-hand side in (17.72) we proceed as in the proof of Theorem 17.3 (see (17.41)) and by (17.61), (17.62) we get

$$\frac{2\tau}{(2\pi)^n}\int Q_\mu(\xi,\tau)\left|\widehat{v}(\xi)\right|^2 d\xi = R_0+2\tau e^{-2\mu\varphi(0)}\int e^{2\mu\varphi(x)}q^{(\delta)}_\mu(x,v,\nabla v,\tau)dx \leq$$

$$\leq C\rho\tau\left\|v\right\|^2_{H^1_\tau(\mathbb{R}^n)}+2\tau e^{-2\mu\varphi(0)}\int e^{2\mu\varphi(x)}q^{(\delta)}_\mu(x,v,\nabla v,\tau)dx,$$

on the other hand, by (17.55) we have

$$2\tau\int e^{2\mu\varphi}q^{(\delta)}_\mu(x,v,\nabla v,\tau)dx \leq C\delta\tau\left\|v\right\|^2_{H^1_\tau(\mathbb{R}^n)}+2\int\left(S_\tau v A_\tau v\right)e^{2\mu\varphi},$$

hence, recalling (17.51),

$$\begin{aligned} &\frac{2\tau}{(2\pi)^n}\int Q_\mu(\xi,\tau)\left|\widehat{v}(\xi)\right|^2 d\xi \\ &\qquad \leq C(\delta+\rho)\tau\left\|v\right\|^2_{H^1_\tau(\mathbb{R}^n)}+C\int\left|P_2(x,\partial,\tau)v\right|^2 e^{2\mu\varphi}, \end{aligned}$$

for every $v \in C^\infty_0(B_\rho)$. By inserting the latter in (17.72) we have

$$\begin{aligned} 2\tau\left\|v\right\|^2_{H^1_\tau(\mathbb{R}^n)} &\leq C\left((\rho+\delta)\tau+\tau^{-1}\right)\left\|v\right\|^2_{H^1_\tau(\mathbb{R}^n)} \\ &\quad +C\left(1+\tau^{-1}\right)\left\|P_2(\cdot,\partial,\tau)v\right\|^2_{L^2(B_1)}, \end{aligned}$$

for every $v \in C^\infty_0(B_\rho)$. Let

$$\rho_0=\delta_0=\frac{1}{4C}.$$

Then

$$\tau\left\|v\right\|^2_{H^1_\tau(\mathbb{R}^n)} \leq 2C\left\|P_2(\cdot,\partial,\tau)v\right\|^2_{L^2(B_1)}, \tag{17.73}$$

for every $v \in C_0^\infty(B_{\rho_0})$, for every $\tau \geq 2$ and for every $\delta \leq \delta_0$. Taking into account (17.50), by (17.73) we easily deduce that there exists $\tau*$ such that

$$\tau \|v\|^2_{H^1_\tau(\mathbb{R}^n)} \leq 2C \|P_2(x, \partial, \tau)v\|^2_{L^2(B_1)},$$

for every $v \in C_0^\infty(B_{\rho_0})$ and for every $\tau \geq \tau_*$. Finally, applying Proposition 15.13 we obtain (17.47). ■

17.3.3 Application to the Cauchy Problem

By Carleman estimate (17.47), we can obtain an uniqueness result for the Cauchy problem for the operator

$$P(x, \partial)u = g^{jk}(x)\partial^2_{jk}u + b_k(x)\partial_k u + c(x)u, \tag{17.74}$$

where matrix $\{g^{jk}\}$ has real entries and satisfy conditions (17.10a) and (17.10b), $b_k \in L^\infty(B_1)$, $k = 1, \cdots, n$, $c \in L^\infty(B_1)$ (complex-valued functions). Below we state the theorem; the proof is only briefly mentioned as it is carried out in an analogous way to that of Theorem 16.11.

Theorem 17.5 *Let $\psi \in C^1\left(\overline{B_1}\right)$ be a real-valued function such that*

$$\nabla\psi(0) \neq 0.$$

Let $P(x, \partial)$ be operator (17.74). *Let $U \in H^2(B_1)$ be such that*

$$\begin{cases} P(x, \partial)U = 0, & in \quad B_1 \\ \\ U(x) = 0, & in \ \{x \in B_1 : \psi(x) > \psi(0)\} \end{cases}.$$

Suppose that

$$\left.\begin{array}{l} p_2(0, \xi + i\tau\nabla\psi(0)) = 0 \\ \\ (\xi, \tau) \neq (0, 0) \end{array}\right\} \Longrightarrow \tag{17.75}$$

$$\Longrightarrow p_2^{(j)}(0, \xi + i\tau\nabla\psi(0))\partial_j\psi(0) \neq 0,$$

($p_2(x, \xi) = g^{jk}(x)\xi_j\xi_k$).

Then there exists a neighborhood $\mathcal{U}_0$ of 0 such that

$$U = 0 \quad in \quad \mathcal{U}_0.$$

Proof The proof is carried out in a manner similar to that of Theorem 16.11. Therefore, first of all we write Carleman estimate (17.47) in the form

$$\tau^3\delta^4 \int |u|^2 e^{2\tau\varphi(\delta^{-1}X)} dX + \tau\delta^2 \int |\nabla u|^2 e^{2\tau\varphi(\delta^{-1}X)} dX \leq$$
$$\leq C_0 \int |P_2(X, \partial)u|^2 e^{2\tau\varphi(\delta^{-1}X)} dX,$$

for every $\delta \in (0, \delta_0]$, for every $u \in C_0^\infty(B_{\widetilde{\rho}_0}(0))$ and for every $\tau \geq \widetilde{\tau}_0$.

Next, by means of a diffeomorphism we reduce to the case in which $\left\{x \in B_{r_0} : \quad \psi(x) \leq 0\right\}$ is the epigraph of a function f strictly convex and such that $f(0) = |\nabla f(0)| = 0$. Like the proof of Theorem 16.11 we introduce the function

$$\varphi(x) = h(\delta_0 x),$$

where

$$h(x_n) = -x_n + \frac{x_n^2}{2}$$

and we check that φ satisfies condition (**S**). First we notice that by (17.75) we have that if $\tau = 0$ then $\xi \neq 0$. Hence

$$\left.\begin{aligned} & p_2(0, \xi) = 0 \\ & \xi \neq 0 \end{aligned}\right\} \Longrightarrow p_2^{(n)}(0, \xi) \neq 0,$$

the latter, in particular, implies that the antecedent of implication (17.45) is not satisfied by φ in 0 which, in turn, implies that (17.18) holds. Regarding (17.46), we have that if (recall $\nabla\varphi(0) = -\delta_0 e_n$)

$$\begin{cases} p_2(0, \xi - i\tau\delta_0 e_n) = 0 \\ \tau \neq 0 \end{cases},$$

then, (17.75) implies

$$p_2^{(n)}(0, \xi - i\tau\delta_0 e_n) \neq 0,$$

hence

$$\partial^2_{jk}\varphi(0) p_2^{(j)}(0, \xi - i\tau\delta_0 e_n)\overline{p_2^{(k)}(\xi - i\tau\delta_0 e_n)} = \delta_0^2 \left|p_2^{(n)}(0, \xi - i\tau\delta_0 e_n)\right|^2 > 0.$$

The remaining part of the proof is identical to that of Theorem 16.11. ■

17.4 Stability Estimate for the Wave Equation in a Cylinder

In this section we adopt the traditional notations: the "spatial coordinates" are denoted by $x_1, \cdots, x_n$, the "time coordinate" is denoted by t and, therefore, the wave operator is

$$\Box = \partial_t^2 - \Delta_x = \partial_t^2 - \left(\partial_{x_1}^2 + \cdots + \partial_{x_1}^2\right). \tag{17.76}$$

Let us denote by

$$\nabla_x = \left(\partial_{x_1}, \cdots, \partial_{x_1}\right), \qquad \nabla_{x,t} = (\nabla_x, \partial_t).$$

For any positive number T, we set

$$S_T = \mathbb{R}^n \times (-T, T).$$

The following theorem holds true (see also, [16], [108]).

Theorem 17.6 (Stability Estimate for the Wave Equation)
Let $T > 1$, $a \in L^\infty(S_T, \mathbb{R}^n)$, $b \in L^\infty(S_T)$ and $c \in L^\infty(S_T)$ be real-valued functions. Let $M \geq 1$. Assume that

$$\|a\|_{L^\infty(S_T,\mathbb{R}^n)} + \|b\|_{L^\infty(S_T)} + \|c\|_{L^\infty(S_T)} \leq M. \tag{17.77}$$

Let $F \in L^2(S_T)$ and $U \in C^\infty(S_T)$ satisfy

$$\begin{cases} \Box U + a \cdot \nabla_x U + b\partial_t U + cU = F, & \text{in } S_T \\ \\ U(x,t) = 0, & \text{for } |x| > 1,\ t \in (-T,T) \end{cases}. \tag{17.78}$$

Then

$$\int_{-T}^{T} \int_{B_1} \left(|U|^2 + |\nabla_{x,t} U|^2\right) dxdt \leq C \|F\|^2_{L^2(B_1 \times (-T,T))}, \tag{17.79}$$

where C depends on M and T.

The proof of Theorem 17.6 is based on what follows:

(a) an **energy estimate** for the equation in (17.78)
(b) Carleman estimate (17.20).

Lemma 17.7 (Energy Estimate) *Let $U \in C^2\left(\overline{B_1} \times (-T,T)\right)$ satisfy*

$$\begin{cases} \Box U + a \cdot \nabla_x U + b\partial_t U + cU = F, & \text{in } B_1 \times (-T,T) \\ \\ U(x,t) = 0, & \text{for } (x,t) \in \partial B_1 \times (-T,T) \end{cases}, \tag{17.80}$$

where a, b, c and F satisfy the same assumption of Theorem 17.6, then the following inequality holds true

$$\begin{aligned} &\int_{-T}^{T} \int_{B_1} \left(U^2(x,t) + U_t^2(x,t) + |\nabla_x U(x,t)|^2\right) dxdt \leq \\ &\leq CT\delta^{-1} \int_{-\delta}^{\delta} \int_{B_1} \left(U^2(x,t) + U_t^2(x,t) + |\nabla_x U(x,t)|^2\right) dxdt + \\ &+ C\, \|F\|^2_{L^2(B_1 \times (-T,T))}, \end{aligned} \tag{17.81}$$

where C depends on M and T only.

Proof of Lemma 17.7 By applying Lemma 17.2 with

$$\beta = (0, \cdots, 0, 1)$$

and

$$g = \operatorname{diag}(-1, \cdots, -1, 1),$$

we get

$$(\Box U)U_t = \frac{1}{2}\partial_t \left(U_t^2 + |\nabla_x U|^2\right) - \operatorname{div}_x (U_t \nabla_x U). \tag{17.82}$$

Moreover, by (17.80) and taking into account (17.77) we have easily

$$|(\Box U)U_t| \leq C(M+1)\left(U^2 + U_t^2 + |\nabla_x U|^2 + F^2\right). \tag{17.83}$$

Let $s, \sigma \in (-T,T)$, $s \leq \sigma$. By integrating both the sides of (17.82) over $B_1 \times [s,\sigma]$, we have from the divergence Theorem and from (17.83)

$$\begin{aligned}
&\int_{B_1}\left(U_t^2(x,\sigma)+|\nabla_x U(x,\sigma)|^2\right)dx-\\
&-\int_{B_1}\left(U_t^2(x,s)+|\nabla_x U(x,s)|^2\right)dx=\\
&=\int_s^\sigma\int_{B_1}\partial_t\left(U_t^2+|\nabla_x U|^2\right)dxdt= \qquad (17.84)\\
&=2\int_s^\sigma\int_{B_1}(\Box U)U_t dxdt\leq\\
&\leq C(M+1)\int_s^\sigma\int_{B_1}\left(U^2+U_t^2+|\nabla_x U|^2+F^2\right)dxdt.
\end{aligned}$$

Let us note that by the first Poincaré inequality (Theorem 5.14) we have

$$\begin{aligned}
C_*^{-1}\int_{B_1}|\nabla_x U(x,t)|^2\,dx&\leq\int_{B_1}\left(U^2(x,t)+|\nabla_x U(x,t)|^2\right)dx\leq\\
&\leq C_*\int_{B_1}|\nabla_x U(x,t)|^2\,dx,
\end{aligned} \qquad (17.85)$$

for every $t\in(-T,T)$, where $C\geq 1$ is a constant. Therefore, setting

$$E(t)=\int_{B_1}\left(U_t^2(x,t)+|\nabla_x U(x,t)|^2\right)dx,$$

by (17.84) and (17.85) we have

$$E(\sigma)\leq E(s)+C_1\,\|F\|^2_{L^2(B_1\times(-T,T))}+C_1\int_s^\sigma E(t)dt$$

($C_1=C_*C(M+1)$) and by the Gronwall inequality we obtain

$$E(\sigma)\leq\left(E(s)+C_1\,\|F\|^2_{L^2(B_1\times(-T,T))}\right)e^{2C_1T}. \qquad (17.86)$$

We notice that if $\sigma\leq s$ we obtain similarly the previous estimate provided we integrate both the sides of (17.82) over $B_1\times[\sigma,s]$ and we interchange σ and s. Therefore we have, for each $s,\sigma\in(-T,T)$,

$$E(\sigma)\leq\left(E(s)+C_1\,\|F\|^2_{L^2(B_1\times(-T,T))}\right)e^{2C_1T}. \qquad (17.87)$$

Now, by integrating with respect to s both members of (17.87) over $(-\delta,\delta)$, where $\delta\in(0,T)$, we have, for each $\sigma\in(-T,T)$,

$$2\delta E(\sigma)\leq 2C_2\int_{-\delta}^{\delta}E(s)ds+2\delta C_2\,\|F\|^2_{L^2(B_1\times(-T,T))} \qquad (17.88)$$

($C_2 = e^{2C_1T}$). Finally, integrating both the sides of (17.88) with respect to σ over $(-T, T)$ and taking into account (17.85), we obtain (17.81). ■

Remark 17.21 As can be seen immediately from the proof, it is not necessary for Lemma 17.81 that T be greater than 1. ♦

Proof of Theorem 17.6 In Sect. 15.8.2 we have proved that if $\theta \in (0, 1)$ the function (see (15.324))

$$\phi(x, t) = -\theta^2 t^2 + |x|^2,$$

has the property that, for any $(x_0, t_0) \in \mathbb{R}^{n+1} \setminus \{(0, 0)\}$,

$$\Gamma = \left\{(x, t) \in \mathbb{R}^{n+1} : \phi(x, t) = \phi(x_0, t_0)\right\}. \tag{17.89}$$

is a pseudo-convex surface in (x_0, t_0) w.r.t. wave operator (17.76). Let us fix θ in such a way that

$$\frac{1}{T} < \theta < 1. \tag{17.90}$$

(recall that $T > 1$) and let $\rho \in (0, 1/10)$ satisfy

$$\rho < \frac{\theta T - 1}{8}. \tag{17.91}$$

Proposition 15.37 implies that if λ is sufficiently large, then the functions

$$\varphi_0(x, t) = e^{\lambda\phi(x,t)}, \qquad \varphi_1(x, t) = e^{\lambda\phi(x-8\rho e_1,t)}, \tag{17.92}$$

satisfy condition (**S**) in $\left(B_R \setminus \overline{B_r}\right) \times (-T, T)$ for every $0 < r < 1 < R$.

Let us fix $\lambda > 0$ in such a way that φ_0 satisfy condition (**S**) (see Sect. 17.3.2) in $\left(B_2 \setminus \overline{B_\rho}\right) \times (-T, T)$ and, consequently φ_1 satisfies condition (**S**) in

$$\left(B_2(8\rho e_1) \setminus \overline{B_\rho(8\rho e_1)}\right) \times (-T, T).$$

For any $s > 0$ set

$$Z_{0,s} = \left\{(x, t) \in \overline{B_1} \times (-T, T) : \quad \varphi_0(x, t) \geq e^{\lambda s}\right\}$$

and

$$Z_{1,s} = \left\{(x, t) \in \overline{B_1} \times (-T, T) : \quad \varphi_1(x, t) \geq e^{\lambda s}\right\}.$$

Let us check that

$$\varphi_0(x, \pm T) \leq 1, \quad \varphi_1(x, \pm T) \leq 1, \quad \forall x \in B_1, \tag{17.93a}$$

$$\overline{B_1} \times [-2\rho, 2\rho] \subset Z_{0,s} \cup Z_{1,s}, \quad \forall s \in (0, 12\rho^2]. \tag{17.93b}$$

The first of (17.93a) is an immediate consequence of (17.90). Concerning the second of (17.93a), we observe that from (17.91) we have, for each $x \in B_1$

$$|x - 8\rho e_1| \leq 1 + 8\rho < \theta T,$$

which implies

$$\varphi_1(x, \pm T) = e^{\lambda(-\theta^2 T^2 + |x - 8\rho e_1|^2)} \leq 1,$$

for every $x \in B_1$.

Now let us check (17.93b). Set

$$\psi(x) = \max\left\{|x|^2, |x - 8\rho e_1|^2\right\},$$

we obtain easily

$$Z_{0,s} \cup Z_{1,s} = \left\{(x,t) \in \overline{B_1} \times (-T, T) : \quad e^{\lambda(-\theta^2 t^2 + \psi(x))} \geq e^{\lambda s}\right\}. \tag{17.94}$$

Let us note now that ψ can be written as

$$\psi(x) = \begin{cases} |x|^2, & \text{for } x_1 \geq 4\rho \\ (x_1 - 8\rho)^2 + x_2^2 + \cdots + x_n^2, & \text{for } x_1 < 4\rho \end{cases},$$

from which we have

$$\psi(x) \geq 16\rho^2, \quad \forall x \in \mathbb{R}^n.$$

Now, if $(x, t) \in \overline{B_1} \times [-2\rho, 2\rho]$, then

$$-\theta^2 t^2 + \psi(x) \geq -\theta^2 t^2 + 16\rho^2 > -4\rho^2 + 16\rho^2 = 12\rho^2 \geq s, \quad \forall s \in (0, 12\rho^2],$$

by the latter and by (17.94), we get $(x, t) \in Z_{0,s} \cup Z_{1,s}$ for $0 \leq s \leq 12\rho^2$. Hence (17.93b) is proved.

Let us apply Carleman estimate (17.20) to the operator $\square$ where $\varphi = \varphi_0$. Set

$$Q_T = \overline{B_1} \times (-T, T).$$

We get

$$\tau^3 \int_{Q_T} |u|^2 e^{2\tau\varphi_0} dxdt + \tau \int_{Q_T} |\nabla_{x,t} u|^2 e^{2\tau\varphi_0} dxdt \leq \\ \leq C \int_{Q_T} |\Box u|^2 e^{2\tau\varphi_0} dxdt, \tag{17.95}$$

for every $u \in C_0^\infty\left(\mathbb{R}^{n+1}\right)$, such that supp $u \subset Q_T = \overline{B_1} \times (-T, T)$ and for every $\tau \geq \tau_0$. Let $\widetilde{\eta} \in C^\infty(\mathbb{R})$ satisfy

$$\widetilde{\eta}(r) = 0,\ r \leq 9\rho^2; \quad 0 \leq \widetilde{\eta}(r) \leq 1,\ 9\rho^2 < r < 10\rho^2; \quad \widetilde{\eta}(r) = 1,\ r \geq 10\rho^2;$$

$$\left|\frac{d\widetilde{\eta}}{dr}\right| \leq C\rho^{-2}, \quad \left|\frac{d^2\widetilde{\eta}}{dr^2}\right| \leq C\rho^{-4}, \tag{17.96}$$

where C is a constant (independent by ρ). Set

$$\eta(x,t) = \widetilde{\eta}(-\theta^2 t^2 + |x|^2)$$

and let us apply estimate (17.95) to $U\eta$. By (17.77) we have

$$|\Box(\eta U)| = |\eta \Box U + 2\left(\partial_t \eta \partial_t U - \nabla_x \eta \cdot \nabla_x U\right) + U(\Box \eta)| \leq \\ \leq \eta M(|\nabla_{x,t} U| + |U|) + \eta |F| + C\rho^{-2} \chi_{Z_{0,9\rho^2} \setminus Z_{0,10\rho^2}} |\nabla_{x,t} U| + \\ + C\rho^{-4} \chi_{Z_{0,9\rho^2} \setminus Z_{0,10\rho^2}} |U|,$$

where C depends on T. Let us observe that (by inserting what was obtained in (17.95)), we have

$$\int_{Q_T} \left(\tau^3 |U\eta|^2 + \tau |\nabla_{x,t}(\eta U)|^2\right) e^{2\tau\varphi_0} dxdt \leq \\ \leq CM^2 \int_{Q_T} \eta^2 \left(|\nabla_{x,t} U|^2 + |U|^2\right) e^{2\tau\varphi_0} dxdt + \\ + C \int_{Q_T} |F|^2 e^{2\tau\varphi_0} dxdt + \\ + C\rho^{-8} \int_{Z_{0,9\rho^2} \setminus Z_{0,10\rho^2}} |U|^2 e^{2\tau\varphi_0} dxdt + \\ + C\rho^{-4} \int_{Z_{0,9\rho^2} \setminus Z_{0,10\rho^2}} |\nabla_{x,t} U|^2 e^{2\tau\varphi_0} dxdt, \tag{17.97}$$

for every $\tau \geq \tau_0$.

Now, let us estimate from below the left-hand side of (17.97)

$$\int_{Q_T} \left(\tau^3 |U\eta|^2 + \tau |\nabla_{x,t}(\eta U)|^2\right) e^{2\tau\varphi_0} dxdt \geq$$
$$\geq \tau \int_{Z_{0,10\rho^2}} \left(|U|^2 + |\nabla_{x,t} U|^2\right) e^{2\tau\varphi_0} dxdt \tag{17.98}$$

and let us estimate from above first integral on the right-hand side as follows

$$\int_{Q_T} \eta^2 \left(|\nabla_{x,t} U|^2 + |U|^2\right) e^{2\tau\varphi_0} dxdt =$$
$$= \int_{Z_{0,10\rho^2}} \left(|\nabla_{x,t} U|^2 + |U|^2\right) e^{2\tau\varphi_0} dxdt +$$
$$+ \int_{Z_{0,9\rho^2} \setminus Z_{0,10\rho^2}} \eta^2 \left(|\nabla_{x,t} U|^2 + |U|^2\right) e^{2\tau\varphi_0} dxdt \leq \tag{17.99}$$
$$\leq \int_{Z_{0,10\rho^2}} \left(|\nabla_{x,t} U|^2 + |U|^2\right) e^{2\tau\varphi_0} dxdt +$$
$$+ e^{2\tau e^{10\lambda\rho^2}} \int_{Z_{0,9\rho^2} \setminus Z_{0,10\rho^2}} \left(|\nabla_{x,t} U|^2 + |U|^2\right) dxdt.$$

By using (17.98) and (17.99) in (17.97), we obtain, by simple calculations (recall $\rho < 1$)

$$(\tau - CM^2) \int_{Z_{0,10\rho^2}} \left(|U|^2 + |\nabla_{x,t} U|^2\right) e^{2\tau\varphi_0} dxdt \leq$$
$$\leq C \int_{Q_T} |F|^2 e^{2\tau\varphi_0} dxdt + \tag{17.100}$$
$$+ C\rho^{-8} e^{2\tau e^{10\lambda\rho^2}} \int_{Z_{0,9\rho^2} \setminus Z_{0,10\rho^2}} \left(|U|^2 + |\nabla_{x,t} U|^2\right) dxdt,$$

for every $\tau \geq \tau_0$. We set $\tau_1 = \max\left\{\tau_0, (2CM^2)^{-1}\right\}$ and by (17.100) we obtain

$$\int_{Z_{0,10\rho^2}} \left(|U|^2 + |\nabla_{x,t} U|^2\right) e^{2\tau\varphi_0} dxdt \leq C \int_{Q_T} |F|^2 e^{2\tau\varphi_0} dxdt +$$
$$+ C\rho^{-8} e^{2\tau e^{10\lambda\rho^2}} \int_{Z_{0,9\rho^2} \setminus Z_{0,10\rho^2}} \left(|U|^2 + |\nabla_{x,t} U|^2\right) dxdt, \tag{17.101}$$

for every $\tau \geq \tau_1$. Now, in (17.101), we estimate trivially from below the integral on the left-hand side and we estimate trivially the integrals on the right-hand side. We get

$$\begin{aligned}
& e^{2\tau e^{12\lambda\rho^2}} \int_{Z_{0,12\rho^2}} \left(|U|^2 + |\nabla_{x,t}U|^2\right) dxdt \leq \\
& \leq \int_{Z_{0,10\rho^2}} \left(|U|^2 + |\nabla_{x,t}U|^2\right) e^{2\tau\varphi_0} dxdt \leq \\
& \leq Ce^{2\tau e^{\lambda}} \int_{Q_T} |F|^2 e^{2\tau\varphi_0} dxdt + \\
& + C\rho^{-8} e^{2\tau e^{10\lambda\rho^2}} \int_{Z_{0,9\rho^2} \setminus Z_{0,10\rho^2}} \left(|U|^2 + |\nabla_{x,t}U|^2\right) dxdt,
\end{aligned}$$

which implies

$$\begin{aligned}
& \int_{Z_{0,12\rho^2}} \left(|U|^2 + |\nabla_{x,t}U|^2\right) dxdt \leq \\
& \leq \int_{Z_{0,10\rho^2}} \left(|U|^2 + |\nabla_{x,t}U|^2\right) e^{2\tau\varphi_0} dxdt \leq \\
& \leq Ce^{2\tau\left(e^{\lambda} - e^{12\lambda\rho^2}\right)} \int_{Q_T} |F|^2 dxdt + \\
& + C\rho^{-8} e^{2\tau\left(e^{10\lambda\rho^2} - e^{12\lambda\rho^2}\right)} \int_{Z_{0,10\rho^2} \setminus Z_{0,9\rho^2}} \left(|U|^2 + |\nabla_{x,t}U|^2\right) dxdt \leq \\
& \leq Ce^{2\tau\left(e^{\lambda} - e^{12\lambda\rho^2}\right)} \int_{Q_T} |F|^2 dxdt + \\
& + C\rho^{-8} e^{2\tau\left(e^{10\lambda\rho^2} - e^{12\lambda\rho^2}\right)} \int_{Q_T} \left(|U|^2 + |\nabla_{x,t}U|^2\right) dxdt,
\end{aligned} \tag{17.102}$$

for every $\tau \geq \tau_1$. By Lemma 17.7 and by (17.102) we have

$$\begin{aligned}
& \int_{Z_{0,12\rho^2}} \left(|U|^2 + |\nabla_{x,t}U|^2\right) dxdt \leq Ce^{2\tau\left(e^{\lambda} - e^{12\lambda\rho^2}\right)} \|F\|^2_{L^2(B_1\times(-T,T))} + \\
& + C\rho^{-9} e^{2\tau\left(e^{10\lambda\rho^2} - e^{12\lambda\rho^2}\right)} \int_{-2\rho}^{2\rho} \int_{B_1} \left(|U|^2 + |\nabla_{x,t}U|^2\right) dxdt,
\end{aligned} \tag{17.103}$$

for every $\tau \geq \tau_1$, where C depends on M and T. At this point we note that, by using (17.20) for operator $\Box$ with $\varphi = \varphi_1$, we obtain an estimate similar to (17.103).

More precisely we have

$$\begin{aligned}&\int_{Z_{1,12\rho^2}}\left(|U|^2+|\nabla_{x,t}U|^2\right)dxdt\leq\\&\leq Ce^{2\tau\left(e^{\lambda}-e^{12\lambda\rho^2}\right)}\|F\|^2_{L^2(B_1\times(-T,T))}+\\&+C\rho^{-9}e^{2\tau\left(e^{10\lambda\rho^2}-e^{12\lambda\rho^2}\right)}\int_{-2\rho}^{2\rho}\int_{B_1}\left(|U|^2+|\nabla_{x,t}U|^2\right)dxdt,\end{aligned}\tag{17.104}$$

for every $\tau\geq\tau_1$. By (17.93b), (17.103) and (17.104) we have

$$\begin{aligned}&\int_{-2\rho}^{2\rho}\int_{B_1}\left(|U|^2+|\nabla_{x,t}U|^2\right)dxdt\leq\\&\leq Ce^{2\tau\left(e^{\lambda}-e^{12\lambda\rho^2}\right)}\|F\|^2_{L^2(B_1\times(-T,T))}+\\&+C\rho^{-9}e^{2\tau\left(e^{10\lambda\rho^2}-e^{12\lambda\rho^2}\right)}\int_{-2\rho}^{2\rho}\int_{B_1}\left(|U|^2+|\nabla_{x,t}U|^2\right)dxdt,\end{aligned}$$

For every $\tau\geq\tau_1$.

Let us choose $\tau=\tau_2\geq\tau_1$, where τ_2 satisfies

$$C\rho^{-9}e^{2\tau_2\left(e^{10\lambda\rho^2}-e^{12\lambda\rho^2}\right)}\leq\frac{1}{2}.$$

By this choice of τ the second term on the right-hand side of (17.103) is absorbed on the left-hand side, and we get (17.79). ■

Remark 17.22 Theorem 17.6 can be proved under less restrictive assumptions on U, but here we do not go into this question. Instead, we want to illustrate a simple and direct application of such a Theorem 17.6 to the proof **of the uniqueness of the following inverse problem**—which we discuss here only at the formal level—let F be a function depending only on the variable x, let U be the solution to the following direct problem

$$\begin{cases}\partial_t^2U-\Delta U=F(x), & \text{in } B_1\times(-T,T)\\ U(x,t)=h_0, & \text{on } \partial B_1\times(-T,T)\,,\\ U(x,0)=U_0(x),\quad U_t(x,0)=U_1(x), & \text{for } x\in B_1\end{cases}\tag{17.105}$$

where h_0, U_0, U_1 and F given functions. The proof of the uniqueness and the existence of solution to the direct problem (17.105), for a given $F\in L^2(B_1)$, can be found, for instance, in [38, Ch. 7]. Let us assume now to know also

$$\frac{\partial U}{\partial \nu}_{|\partial B_1\times(-T,T)} = h_1,$$

we wish to determine F. Note that F does not depend on t. Here we give a sketch of the proof of the uniqueness for the problem of determining F (inverse problem) by mean of h_0, h_1, U_0, U_1. Since the problem is linear, it is enough to prove that if

$$U_0 = U_1 = 0, \quad h_0 = h_1 = 0,$$

then

$$F = 0, \quad \text{in } B_1.$$

Let U satisfy

$$\begin{cases} \partial_t^2 U - \Delta U = F(x), & \text{in } B_1 \times (-T, T) \\ U = 0, \quad \frac{\partial U}{\partial \nu} = 0 & \text{on } \partial B_1 \times (-T, T) \\ U(x, 0) = 0, \quad U_t(x, 0) = 0, & \text{for } x \in B_1 \end{cases} . \tag{17.106}$$

Set

$$w = \partial_t U$$

and differentiate both the sides of the equation with respect to t. Since F does not depend on t we have

$$\begin{cases} \partial_t^2 w - \Delta w = 0, & \text{in } B_1 \times (-T, T) \\ \\ w = 0, \quad \frac{\partial w}{\partial \nu} = 0 & \text{on } \partial B_1 \times (-T, T) \end{cases} . \tag{17.107}$$

Applying to w estimate (17.79) we get

$$w = 0, \quad \text{in } B_1 \times (-T, T).$$

Hence

$$U(x, t) = U(x, 0) + \int_0^t w(x, s) ds = 0, \quad \text{in } B_1 \times (-T, T).$$

Therefore by (17.106) we have

$$F(x) = \partial_t^2 U - \Delta U = 0.$$

This establishes the uniqueness for the inverse problem. The approach we have outlined here is a simplified version of a more general method used to prove uniqueness and stability results for inverse problems involving evolution equations, where the goal is to determine time-independent coefficients of certain equations. For more in-depth study, we recommend referring to [16], [60] and [63]. ♦

Chapter 18
Optimal Three Sphere and Doubling Inequality for Second-Order Elliptic Equations

Abstract This chapter focuses on the strong unique continuation property for second-order elliptic equations with real coefficients in the principal part. The analysis begins with the essential concept of three-sphere inequalities, which play a crucial role in establishing this property. These inequalities are introduced alongside doubling inequalities, both derived using Carleman estimates applied to elliptic operators. The chapter explores the derivation of Carleman estimates for the Laplace operator and extends the approach to equations with variable coefficients, introducing technical challenges such as the use of geodesic polar coordinates.

18.1 Introduction

In this chapter we will prove the **strong unique continuation property** for the second-order elliptic equations with real coefficients (in the principal part). Now we recall briefly this property and provide an introduction to the chapter.

Let $\left\{a^{ij}(x)\right\}_{i,j=1}^{n}$ be a **symmetric matrix of real-valued functions**. We assume that the following uniform ellipticity condition is satisfied

$$\lambda^{-1}\left|\xi\right|^{2}\leq\sum_{i,j=1}^{n}a^{ij}(x)\xi_{i}\xi_{j}\leq\lambda\left|\xi\right|^{2},\quad\forall\xi\in\mathbb{R}^{n},\forall x\in B_{1},$$

where $\lambda\geq 1$. Let us assume that the function a^{ij}, $i,j=1,\cdots,n$ are Lipschitz continuous

$$\left|a^{ij}(x)-a^{ij}(y)\right|\leq\Lambda|x-y|,\quad\text{for } i,j\in\{1,\cdots,n\},\quad\forall x,y\in B_{1}.\tag{18.1}$$

Let $b^{i}\in L^{\infty}(B_{1})$, $i=1,\cdots,n$ and $c\in L^{\infty}(B_{1})$ (these coefficients can also be complex-valued) satisfy

S. Vessella, *Unique Continuation Properties for Partial Differential Equations*, Birkhäuser Advanced Texts Basler Lehrbücher,
https://doi.org/10.1007/978-3-031-86366-0_18

$$\left\|b^i\right\|_{L^\infty(B_1)} \leq M, \quad \text{for } i = 1, \cdots, n$$

and

$$\|c\|_{L^\infty(B_1)} \leq M. \tag{18.2}$$

We recall that the equation

$$Lu = \sum_{i,j=1}^{n} a^{ij}(x)\partial^2_{x^i x^j} u + \sum_{i=1}^{n} b^i(x)\partial_{x^i} u + c(x)u = 0, \quad \text{in } B_1, \tag{18.3}$$

enjoys the **strong unique continuation property** provided that any solution u to (18.3) satisfying the conditions

$$\int_{B_r} |u|^2 dx = O\left(r^m\right), \quad \text{as } r \to 0, \quad \forall m \in \mathbb{N}, \tag{18.4}$$

identically vanishes. We will establish the strong unique continuation property by deriving an optimal three-sphere inequality. A prototype of such an inequality is the Hadamard three-circle inequality for holomorphic functions, which we initially encountered in Sect. 12.5 and, concerning the equation $\partial_{\bar{z}} U + a(x, y)U = 0$, Theorem 14.5.

In a general sense, a three sphere inequality for solutions to Eq. (18.3) takes the form

$$\int_{B_\rho} |u|^2 dx \leq C\left(\int_{B_R} |u|^2 dx\right)^{1-\theta} \left(\int_{B_r} |u|^2 dx\right)^{\theta}, \tag{18.5}$$

where $0 < r < \rho < R \leq 1$, C and $\theta \in (0, 1)$ depend by λ, Λ, M and R, ρ (C and θ do not depend on u).

We say that (18.5) is an **optimal** three spheres inequality provided C **does not depend** on r and, for fixed R, ρ, we have

$$\theta \sim |\log r|^{-1}, \quad \text{as } r \to 0. \tag{18.6}$$

We recall that by $f(r) \sim g(r)$, as $r \to 0$ we means

$$0 < \lim_{r \to 0} \frac{f(r)}{g(r)} < +\infty.$$

First author that proved (18.5) was Landis in [72].

Using an argument similar to Remark 5 in Sect. 12.5, it becomes evident that if a function u satisfies an optimal three sphere inequality, then whenever u satisfies Eq. (18.4), it must be identically zero. Consequently, if an optimal three-sphere inequality is valid for Eq. (18.4), then the equation inherently possesses the strong unique continuation property.

Another type of inequality that implies the strong unique continuation property is the so-called **doubling inequality**. Such an inequality occurs in the form

$$\int_{B_{2r}} |u|^2 dx \leq K \int_{B_r} |u|^2 dx, \quad \forall r \in \left(0, \frac{1}{2}\right), \tag{18.7}$$

where K depends on u but **does not depend on** r. We will prove later in which a way (18.7) implies the strong unique continuation property. The main idea may be expressed as follows. Iterating inequality (18.7) we have, for every $j \in \mathbb{N}$

$$\int_{B_{1/2}} |u|^2 dx \leq K \int_{B_{1/4}} |u|^2 dx \leq \cdots \leq K^{j-1} \int_{B_{1/2^j}} |u|^2 dx,$$

which, together with (18.4), provides, for each $j, m \in \mathbb{N}$ (C_m depends on m only),

$$\int_{B_{1/2}} |u|^2 dx \leq K^{j-1} C_m \left(\frac{1}{2^j}\right)^m = C_m K^{-1} \left(\frac{K}{2^m}\right)^j. \tag{18.8}$$

Let now m satisfy

$$2^m > K$$

and passing to the limit as $j \to \infty$, we get by (18.8)

$$\int_{B_{1/2}(0)} |u|^2 dx = 0.$$

The optimal three sphere inequality and the doubling inequality will both be derived through an appropriate Carleman estimate applied to the elliptic operator L (or equivalently, to the principal part of that operator). We will approach this endeavor in two distinct phases: first, we will examine the case involving the Laplace operator Δ, and subsequently, we will delve into equations with variable coefficients. In both instances, the proofs of the Carleman estimates will be initiated by reexpressing the elliptic operators in polar coordinates. For the Laplace operator, this will be done using Euclidean polar coordinates, while for the case of variable coefficients, we will make use of geodesic polar coordinates with respect to the Riemannian structure induced by a metric that is conformal to

$$a_{ij} dx^i \otimes dx^j,$$

($\left\{a_{ij}(x)\right\}_{i,j=1}^n$ is the inverse of the matrix $\left\{a^{ij}(x)\right\}_{i,j=1}^n$). It is important to note that throughout this chapter, we will utilize the Einstein convention for repeated indices. Additionally, we will be more diligent in adhering to the notation regarding indices of tensor components. Although these notations are primarily required for Sects. 18.6 and 18.7, we will adopt them consistently in the preceding sections as well. The proof for the Laplace operator serves as the foundation for most of the primary challenges that will arise in the case of variable coefficients. Naturally, the latter case introduces additional technical complexities. Although the proofs in this chapter are essentially centered on the method and results of Aronszajn [8] and those by Aronszajn et al. in [9], it is worth noting that other proofs can also be found in the literature, such as those presented in [36] and [57].

The results presented in this chapter concern the strong unique continuation properties for equation (18.3) where the coefficients b^i, $i = 1, \cdots, n$ and c are L^∞ (even if only locally). However, we cannot omit mentioning certain notable results for cases where these coefficients are singular. In this regard, it is worth noting that Jerison and Kenig, in [64], established strong unique continuation for equation (18.3) with $a^{ij} = \delta^{ij}$, $b^i = 0$, $i, j = 1, \cdots, n$, and $c \in L^{\frac{n}{2}}$ for $n > 2$. Following a series of intermediate results, Koch and Tataru in [67] extended this to the case where a^{ij} is non-constant and Lipschitz, and b^i are also non-zero and singular, obtaining an essentially sharp result. The results of [64] and [67] are based on Carleman estimates but do not use integration by parts; rather, they rely on more sophisticated techniques beyond the scope of this treatment. For a more comprehensive historical and bibliographic overview, we refer the reader to [67].

18.2 Formulas for the Change of Variables of Second-Order Operators

We begin by deriving a formula to the change of variables of the operator

$$\operatorname{div}\left(A(x)\nabla u(x)\right), \tag{18.9}$$

where $A(x) = \left\{a^{ij}(x)\right\}_{i,j=1}^{n}$ is a symmetric matrix, a^{ij} are sufficiently smooth functions for $i, j = 1, \cdots, n$ (will suffice $a^{ij} \in C^{0,1}\left(\mathbb{R}^n\right)$).

Let us consider the case where $A(x)$ is the identity matrix. Let Λ be an open set of $\mathbb{R}^n$ and let $\Phi \in C^1\left(\overline{\Lambda}, \mathbb{R}^n\right)$ be a injective map such that

$$\det(J\Phi(x)) \neq 0, \quad \forall x \in \overline{\Lambda}, \tag{18.10}$$

where $J\Phi(x)$ is the jacobian matrix of Φ in x

$$J\Phi(x) = \begin{pmatrix} \partial_{x^1}\Phi^1((x) & \cdots & \partial_{x^n}\Phi^1((x) \\ \vdots & \ddots & \vdots \\ \partial_{x^1}\Phi^n((x) & \cdots & \partial_{x^n}\Phi^n((x) \end{pmatrix}.$$

Set

$$\Omega = \Phi(\Lambda)$$

and

$$\Psi = \Phi^{-1},$$

consequently $\Psi \in C^1\left(\overline{\Omega}, \mathbb{R}^n\right)$. Let us prove that, if $u \in C^2\left(\overline{\Lambda}\right)$, then

$$(\Delta u)(\Psi(y)) = \frac{1}{|\det(J\Psi(y))|}\operatorname{div}\left(B(y)v(y)\right), \tag{18.11}$$

where

$$v(y) = u(\Psi(y)) \tag{18.12}$$

and

$$B(y) = |\det(J\Psi(y))|\,(J\Psi(y))^{-1}\left((J\Psi(y))^{-1}\right)^T. \tag{18.13}$$

More generally, for operator (18.9) we have

$$\operatorname{div}\left(A(x)\nabla u(x)\right)_{|x=\Psi(y)} = \frac{1}{|\det(J\Psi(y))|}\operatorname{div}\left(\widetilde{A}(y)v(y)\right), \tag{18.14}$$

where v is given by (18.12) and

$$\widetilde{A}(y) = |\det(J\Psi(y))|\,(J\Psi(y))^{-1}A(\Psi(y))\left((J\Psi(y))^{-1}\right)^T. \tag{18.15}$$

Since (18.14) can be proved similarly, we limit ourselves to prove (18.11).

Proof of (18.11) Let $w \in C_0^\infty(\Omega)$. Let

$$\widetilde{w}(x) = w(\Phi(x)), \quad \forall x \in \Lambda.$$

Integrating by parts, we have

$$-\int_\Lambda \Delta u(x)\widetilde{w}(x)dx = \int_\Lambda \nabla u(x)\cdot\nabla\widetilde{w}(x)dx \tag{18.16}$$

and by the formula of change of variables for multiple integrals, we have

$$\begin{aligned}&\int_{\Lambda} \nabla_x u(x) \cdot \nabla_x \widetilde{w}(x) dx = \\&= \int_{\Omega} \left(\nabla_x u\right)\left(\Psi(y)\right) \cdot \left(\nabla_x \widetilde{w}\right)\left(\Psi(y)\right) \left|\det(J\Psi(y))\right| dy.\end{aligned} \tag{18.17}$$

Now, (18.12) gives

$$\nabla_y v(y) = \left(J\Psi(y)\right)^T \left(\nabla_x u\right)\left(\Psi(y)\right),$$

($\nabla_x u$ and $\nabla_y v$ are column vectors). Hence

$$\left(\nabla_x u\right)\left(\Psi(y)\right) = \left(\left(J\Psi(y)\right)^{-1}\right)^T \nabla_y v(y), \tag{18.18}$$

similarly

$$\left(\nabla_x \widetilde{w}\right)\left(\Psi(y)\right) = \left(\left(J\Psi(y)\right)^{-1}\right)^T \nabla_y w(y). \tag{18.19}$$

Substituting (18.18) and (18.19) into the integral on the right-hand side of (18.17) and recalling (18.16), we have

$$\begin{aligned}-\int_{\Lambda} \Delta u(x)\widetilde{w}(x) dx &= \int_{\Omega} B(y)\nabla_y v(y) \cdot \nabla_y w(y) dy = \\&= -\int_{\Omega} \operatorname{div}_y \left(B(y)\nabla_y v(y)\right) w(y) dy.\end{aligned} \tag{18.20}$$

Now, we change the variables in the integral on the left-hand side of (18.20) and we find

$$-\int_{\Omega} (\Delta u)(\Psi(y)) w(y) \left|\det(J\Psi(y))\right| dy = -\int_{\Omega} \operatorname{div}_y \left(B(y)\nabla_y v(y)\right) w(y) dy.$$

Since w is arbitrary in $C_0^\infty(\Omega)$, we get (18.11).

Exercise 18.1 Apply formula (18.11) to write the Laplace operator, in $\mathbb{R}^2$, in polar coordinates showing that, setting

$$(x^1, x^2) = \Psi(\varrho, \vartheta) = (\varrho \cos\vartheta, \varrho \sin\vartheta), \quad (\varrho, \vartheta) \in (0, +\infty) \times [0, 2\pi)$$

and

$$v(\varrho, \vartheta) = u(\Psi(\varrho, \vartheta)) = u(\varrho \cos\vartheta, \varrho \sin\vartheta),$$

we have

$$(\Delta u)(\varrho \cos\vartheta, \varrho \sin\vartheta) = \partial^2_\varrho v + \frac{1}{\varrho}\partial_\varrho v + \frac{1}{\varrho^2}\partial^2_\vartheta v. \tag{18.21}$$

♠

Exercise 18.2 Apply formula (18.11) to write the Laplace operator, in $\mathbb{R}^3$, in polar coordinates showing that, setting

$$(x^1, x^2, x^3) = \Psi(\varrho, \vartheta, \phi) = (\varrho \sin\vartheta \cos\phi, \varrho \sin\vartheta \sin\phi, \varrho \cos\vartheta), \tag{18.22}$$

where for $(\varrho, \vartheta, \phi) \in (0, +\infty)\times)(0, \pi) \times (0, 2\pi)$ and

$$v(\varrho, \vartheta, \phi) = u(\Psi(\varrho, \vartheta\phi)) = u(\varrho \sin\vartheta \cos\phi, \varrho \sin\vartheta \sin\phi, \varrho \cos\vartheta),$$

we have

$$\begin{aligned}(\Delta u)(\varrho \sin\vartheta \cos\phi, \varrho \sin\vartheta \sin\phi, \varrho \cos\vartheta) &= \\ = \partial^2_\varrho v + \frac{2}{\varrho}\partial_\varrho v + \frac{1}{\varrho^2 \sin\vartheta}\partial_\vartheta (\sin\vartheta\, \partial_\vartheta v) + \frac{1}{\varrho^2 \sin^2\vartheta}\partial^2_\phi v.\end{aligned} \tag{18.23}$$

♠

18.3 Polar Coordinates in $\mathbb{R}^n$. The Laplace–Beltrami Operator on the Sphere

In the n-dimensional case, it is feasible to represent the Laplace operator in polar coordinates using formulas analogous to (18.21) and (18.23). However, this approach often leads to cumbersome expressions. Therefore, we introduce a method here that begins by defining the Laplace operator on the sphere

$$\Sigma = \left\{x \in \mathbb{R}^n |\ |x| = 1\right\}.$$

This approach proves to be more elegant and efficient.

When working with polar coordinates in $\mathbb{R}^n$, if you intend to follow the approach outlined in Exercises 18.1 and 18.2 from the previous section, it would be advantageous to establish a transformation analogous to (18.22). To achieve this, you can utilize the following recursive formula

$$\begin{cases} r_n = |x| \\ x^n = r_n \cos \vartheta^n, & \vartheta^n \in [0, 2\pi) \\ r_{n-1} = r_n \sin \vartheta^n, & \vartheta^n \in [0, 2\pi) \\ x^{n-1} = r_{n-1} \cos \vartheta^{n-1}, & \vartheta^{n-1} \in [0, \pi) \\ \cdots \\ x^1 = r_1 \cos \vartheta^2, & \vartheta^2 \in [0, \pi) \end{cases}. \tag{18.24}$$

In this way, the "angular coordinates" are $\left(\vartheta^2, \vartheta^3, \cdots, \vartheta^n\right)$. By using (18.24) one reaches to the following formula of change of variables in polar coordinates (f integrable function)

$$\begin{aligned} \int_{B_R} f(x)dx &= \int_0^R \left(\int_\Sigma f(\rho\omega)\rho^{n-1} dS \right) d\rho = \\ &= \int_0^R \left(\int_{\partial B_\rho} f(y) dS_y \right) d\rho, \end{aligned} \tag{18.25}$$

where $R > 0$.

In order to write the Laplace operator in polar coordinates we proceed in two steps:

Step I. We will write the Laplace–Beltrami operator on the sphere Σ.
Step II. We will complete the transformation begun in Step I and we will prove Theorem 18.6.

Step I Let $u(\omega)$ be a real-valued function on Σ. We associate to u the homogeneous function of degree 0 defined by

$$\widetilde{u} : \mathbb{R}^n \setminus \{0\} \to \mathbb{R}.$$

$$\widetilde{u}(x) = u\left(\frac{x}{|x|}\right), \qquad x \in \mathbb{R}^n \setminus \{0\}. \tag{18.26}$$

If $\widetilde{u} \in C^k(\mathbb{R}^n \setminus \{0\})$, we say that u belongs to $C^k(\Sigma)$. We define the scalar product of two functions $u, v \in C^0(\Sigma)$ as

$$(u, v)_\Sigma = \int_\Sigma u(\omega)v(\omega)d\omega \tag{18.27}$$

($d\omega := dS$) and we set

$$\|u\|_{L^2(\Sigma)} = \int_\Sigma |u(\omega)|^2\, d\omega. \tag{18.28}$$

Hence $C^0(\Sigma)$, equipped with the scalar product (18.27), is a pre-Hilbertian space and $L^2(\Sigma)$ is the completion of that space.

For any $1 \le i \le n$ we define the operator d_i as follows. Let $u \in C^1(\Sigma)$. Let us denote

$$(d_i u)(\omega) = \left(\partial_{x^i}\widetilde{u}(x)\right)_{|\,x=\omega}, \quad \omega \in \Sigma. \tag{18.29}$$

Other symbols which are used in literature to denote $d_i u$ are: $\Omega_i u$, $\partial_{\omega^i} u$.

Remark 18.1 By (18.29) we have for $j, k = 1, \cdots, n$

$$\begin{aligned}(d_i u)(\omega) &= \left(\left(\partial_{x^i}\frac{x^k}{|x|}\right)\left(\partial_{x^k}\widetilde{u}\right)\left(\frac{x}{|x|}\right)\right)_{|\,x=\omega} = \\ &= \frac{1}{|x|}\left(\left(\partial_{x^i}\widetilde{u}\right)\left(\frac{x}{|x|}\right) - \frac{x^i}{|x|}\frac{x^k}{|x|}\left(\partial_{x^k}\widetilde{u}\right)\left(\frac{x}{|x|}\right)\right)_{|\,x=\omega}.\end{aligned} \tag{18.30}$$

Let us note that the function written in the brackets in the second line is the i-th component of

$$(\nabla\widetilde{u})\left(\frac{x}{|x|}\right) - \frac{x}{|x|}\left(\frac{x}{|x|}\cdot(\nabla\widetilde{u})\left(\frac{x}{|x|}\right)\right), \tag{18.31}$$

which, in turn, is the i-th tangential component on $\{|x| = 1\}$ of the gradient of $\widetilde{u}$. ♦

Remark 18.2 The operators d_i, $1 \le i \le n$, **are not independent**. As a matter of fact, by (18.29) and by the fact that $\widetilde{u}$ is a homogeneous function of degree 0, we have

$$\sum_{i=1}^n \omega^i\,(d_i u)(\omega) = \sum_{i=1}^n \left(\frac{x^i}{|x|}\left(\partial_{x^i}\widetilde{u}\right)\left(\frac{x}{|x|}\right)\right)_{|\,x=\omega} = 0. \tag{18.32}$$

Moreover, by (18.29) we have, for every $u, v \in C^1(\Sigma)$,

$$d_i(uv) = (d_i u)\, v + u\,(d_i v), \quad 1 \le i \le n. \tag{18.33}$$

Also let us note

$$\partial_{x^i}\widetilde{u}(x) = \frac{1}{|x|}(d_i u)\left(\frac{x}{|x|}\right), \quad x \in \mathbb{R}^n \setminus \{0\},\ 1 \le i \le n. \tag{18.34}$$

Indeed, by homogeneity we have

$$\widetilde{u}(x) = \widetilde{u}(\lambda x) \quad \forall \lambda \in \mathbb{R} \setminus \{0\}.$$

Hence

$$\partial_{x^i} \widetilde{u}(x) = \lambda \left(\partial_{x^i} \widetilde{u}\right)(\lambda x), \quad 1 \leq i \leq n,$$

from which, choosing $\lambda = \frac{1}{|x|}$, we get

$$\partial_{x^i} \widetilde{u}(x) = \frac{1}{|x|} \left(\partial_{x^i} \widetilde{u}\right) \left(\frac{x}{|x|}\right) = \frac{1}{|x|} (d_i u) \left(\frac{x}{|x|}\right), \quad 1 \leq i \leq n,$$

from which (18.34) follows. ♦

Proposition 18.1 *Let $n > 1$. We have*

$$(d_i u, v)_\Sigma = \left(u, d_i^* v\right)_\Sigma, \quad \forall u, v \in C^1(\Sigma), 1 \leq i \leq n, \tag{18.35}$$

where

$$d_i^* = (n-1)\omega^i - d_i, \quad 1 \leq i \leq n. \tag{18.36}$$

d_i^* is the formal adjoint of d_i with respect to the scalar product (18.27).

Proof First, let us notice that, for any $1 \leq i \leq n$, $\partial_{x_i} \widetilde{u}(x)$ is integrable over B_1 because

$$\left|\partial_{x_i} \widetilde{u}(x)\right| \leq \frac{1}{|x|} \max_\Sigma |\nabla \widetilde{u}|, \quad 1 \leq i \leq n$$

and $n > 1$.

Now, let us check that, for any $1 \leq i \leq n$, we have

$$\int_\Sigma d_i u(\omega) v(\omega) d\omega = (n-1) \int_{B_1} \partial_{x^i} \widetilde{u}(x) \widetilde{v}(x) dx. \tag{18.37}$$

We check (18.37). If $\varepsilon \in (0, 1)$ and $1 \leq i \leq n$, we have by (18.25) and by (18.34),

$$\begin{aligned}\int_{B_1 \setminus B_\varepsilon} \partial_{x^i} \widetilde{u}(x) \widetilde{v}(x) dx &= \int_{B_1 \setminus B_\varepsilon} \frac{1}{|x|} (d_i u) \left(\frac{x}{|x|}\right) v \left(\frac{x}{|x|}\right) dx = \\ &= \int_\varepsilon^1 \rho^{n-2} \int_\Sigma d_i u(\omega) v(\omega) d\omega = \\ &= \frac{1 - \varepsilon^{n-1}}{n-1} \int_\Sigma d_i u(\omega) v(\omega) d\omega.\end{aligned}$$

Passing to the limit as $\varepsilon \to 0$ we obtain (18.37).

Now, by (18.37) and integrating by parts we have

$$\begin{aligned}
\frac{1}{n-1}(d_i u, v)_\Sigma &= \int_{B_1} \partial_{x^i}\widetilde{u}(x)\widetilde{v}(x)dx = \\
&= \lim_{\varepsilon\to 0}\int_{B_1\setminus B_\varepsilon} \partial_{x^i}\widetilde{u}(x)\widetilde{v}(x)dx = \\
&= \lim_{\varepsilon\to 0}\int_{B_1\setminus B_\varepsilon} \left(\partial_{x^i}\left(\widetilde{u}\widetilde{v}\right) - \widetilde{u}\partial_{x^i}\widetilde{v}\right)dx = \\
&= \int_\Sigma u(\omega)v(\omega)\omega^i d\omega - \\
&- \lim_{\varepsilon\to 0}\left(\int_{\partial B_\varepsilon}\widetilde{u}(x)\widetilde{v}(x)\frac{x^i}{|x|}dS + \int_{B_1\setminus B_\varepsilon}\widetilde{u}(x)\partial_{x^i}\widetilde{v}(x)dx\right).
\end{aligned} \tag{18.38}$$

On the other hand

$$\int_{\partial B_\varepsilon}\widetilde{u}(x)\widetilde{v}(x)\frac{x^i}{|x|}dS = O\left(\varepsilon^{n-1}\right), \quad \text{as } \varepsilon \to 0,$$

hence we have

$$\lim_{\varepsilon\to 0}\int_{\partial B_\varepsilon}\widetilde{u}(x)\widetilde{v}(x)\frac{x^i}{|x|}dS = 0.$$

Therefore (18.38) gives

$$\frac{1}{n-1}(d_i u, v)_\Sigma = \int_\Sigma u(\omega)v(\omega)\omega^i d\omega - \int_{B_1}\widetilde{u}(x)\partial_{x^i}\widetilde{v}(x)dx$$

and employing (18.37) (interchange u with v in the latter) we get

$$\begin{aligned}
\frac{1}{n-1}(d_i u, v)_\Sigma &= \int_\Sigma u(\omega)\left(v(\omega)\omega^i - \frac{1}{n-1}d_i v(\omega)\right)d\omega = \\
&= \frac{1}{n-1}\left(u, (n-1)\omega^i v - d_i v\right)_\Sigma .
\end{aligned}$$

By the just obtained equality, (18.35) immediately follows. ■

Definition 18.2 (Laplace–Beltrami Operator on the Sphere) The Laplace–Beltrami operator on the sphere is defined as

$$\Delta_\Sigma = \sum_{i=1}^{n} d_i^2. \tag{18.39}$$

Proposition 18.3 *We have*

$$(\Delta_\Sigma u, v)_\Sigma = -\sum_{i=1}^{n} (d_i u, d_i v)_\Sigma\,, \quad \forall u, v \in C^2(\Sigma). \tag{18.40}$$

Proof By (18.32) and (18.35) we get

$$\begin{aligned}(\Delta_\Sigma u, v)_\Sigma &= \sum_{i=1}^{n}\left(d_i^2 u, v\right)_\Sigma = \sum_{i=1}^{n}(d_i u, d_i^* v)_\Sigma = \\ &= \sum_{i=1}^{n}\left(d_i u, (n-1)\omega^i v - d_i v\right)_\Sigma = \\ &= (n-1)\left(\sum_{i=1}^{n}\omega^i d_i u, v\right)_\Sigma - \sum_{i=1}^{n}(d_i u, d_i v)_\Sigma = \\ &= -\sum_{i=1}^{n}(d_i u, d_i v)_\Sigma\,.\end{aligned}$$

■

Remark 18.3 Proposition 18.3 implies

$$(\Delta_\Sigma u, u)_\Sigma = -\sum_{i=1}^{n}\|d_i u\|^2_{L^2(\Sigma)} \le 0, \quad \forall u \in C^2(\Sigma) \tag{18.41}$$

and

$$\int_\Sigma \Delta_\Sigma u(\omega)d\omega = (\Delta_\Sigma u, 1)_\Sigma = 0. \tag{18.42}$$

♦

Proposition 18.4 *We have, for* $i, j = 1, \cdots, n$,

$$d_j d_i u - d_i d_j u = \omega_j d_i u - \omega_i d_j u, \quad \forall u \in C^2(\Sigma). \tag{18.43}$$

Proof By (18.34) we have

$$(d_i u)\left(\frac{x}{|x|}\right) = |x|\partial_{x^i}\widetilde{u}(x), \quad 1 \le i \le n,\ \forall x \in \mathbb{R}^n \setminus \{0\}.$$

Hence

$$\begin{aligned}
d_j d_i u(\omega) &= \partial_{x^j}\left(|x|\partial_{x^i}\widetilde{u}(x)\right)_{|\,x=\omega} = \\
&= \left(\left(\partial_{x^j}|x|\right)\partial_{x^i}\widetilde{u}(x) + |x|\partial^2_{x^i x^j}\widetilde{u}(x)\right)_{|\,x=\omega} = \\
&= \left(\frac{x^j}{|x|}\partial_{x^i}\widetilde{u}(x)\right)_{|\,x=\omega} + \left(\partial^2_{x^i x^j}\widetilde{u}\right)(\omega) = \\
&= \omega^j d_i u(\omega) + \left(\partial^2_{x^i x^j}\widetilde{u}\right)(\omega)
\end{aligned} \tag{18.44}$$

and, similarly,

$$d_i d_j u(\omega) = \omega^i d_j u(\omega) + \left(\partial^2_{x^j x^i}\widetilde{u}\right)(\omega). \tag{18.45}$$

Subtracting (18.45) by (18.44) we obtain (18.43). ■

Step II We associate to $u \in C^2(B_R \setminus \{0\})$, where $R > 0$, the function $U \in C^2((0, R) \times (\mathbb{R}^n \setminus \{0\}))$

$$U(\rho, y) = u\left(\rho\frac{y}{|y|}\right), \quad \rho \in (0, R),\ y \in \mathbb{R}^n \setminus \{0\}. \tag{18.46}$$

It is obvious that if $\rho = |x|$ and $\omega = \frac{x}{|x|}$, then we have

$$U(\rho, \omega) = u\left(|x|\frac{x}{|x|}\right) = u(x). \tag{18.47}$$

Proposition 18.5 *Let U be defined by* (18.46)*. Then we have, for any $i = 1, \cdots, n$,*

$$\partial_{x^i} u(x) = \omega^i \partial_\rho U(\rho, \omega) + \frac{1}{\rho} d_i U(\rho, \omega). \tag{18.48}$$

where $\rho = |x|$ and $\omega = \frac{x}{|x|}$.

Proof By (18.47) we have

$$\begin{aligned}
\partial_{x^i} u(x) &= \left(\partial_{x^i}|x|\right)\partial_\rho U\left(|x|, \frac{x}{|x|}\right) + \\
&+ \sum_{k=1}^{n} \partial_{x^i}\left(\frac{x_k}{|x|}\right)\partial_{y_k} U\left(|x|, \frac{x}{|x|}\right) = \\
&= \frac{x^i}{|x|}\partial_\rho U\left(|x|, \frac{x}{|x|}\right) + \\
&+ \sum_{k=1}^{n}\left[\frac{\delta^{ik}}{|x|} - \frac{x^i x^k}{|x|^3}\right]\partial_{y^k} U\left(|x|, \frac{x}{|x|}\right).
\end{aligned} \tag{18.49}$$

Now, (18.30) implies

$$\sum_{k=1}^{n}\left[\frac{\delta^{ik}}{|x|}-\frac{x^i x^k}{|x|^3}\right]\partial_{y^k}U\left(|x|,\frac{x}{|x|}\right)=\frac{1}{|x|}d_iU(\rho,\omega).$$

By the latter and by (18.49) we get (18.48). ■

Theorem 18.6 *Let $u \in C^2\left(B_R \setminus \{0\}\right)$. Let U be defined by* (18.46). *Set*

$$U(\rho,\omega)=u(\rho\omega),\qquad \forall(\rho,\omega)\in(0,R)\times\Sigma. \tag{18.50}$$

We have $v \in C^2\left((0,R)\times\Sigma\right)$ and

$$(\Delta u)(\rho\omega)=\partial_\rho^2 v(\rho,\omega)+\frac{n-1}{\rho}\partial_\rho v(\rho,\omega)+\frac{1}{\rho^2}\Delta_\Sigma v(\rho,\omega). \tag{18.51}$$

Proof We denote, for the sake of brevity,

$$\rho(x)=|x|,\quad \omega(x)=\frac{x}{|x|}.$$

For any $i \leq i \leq n$, (18.48) implies

$$\begin{aligned}\partial_{x^i}^2u(x)&=\underbrace{\partial_{x^i}\left(\omega^i(x)\partial_\rho U(\rho(x),\omega(x))\right)}_{J_1}+\\&+\underbrace{\partial_{x^i}\left(\frac{1}{\rho(x)}d_iU(\rho(x),\omega(x))\right)}_{J_2}.\end{aligned} \tag{18.52}$$

Computation of J_1
We have

$$\begin{aligned}J_1&=\partial_{x^i}\left(\omega^i(x)\right)\partial_\rho U(\rho(x),\omega(x))+\omega^i(x)\partial_{x^i}\left(\partial_\rho U(\rho(x),\omega(x))\right)=\\&=\left(\frac{1}{|x|}-\frac{(x^i)^2}{|x|^3}\right)\partial_\rho U(\rho(x),\omega(x))+\\&+\omega^i(x)\left[\omega^i(x)\partial_\rho^2U(\rho(x),\omega(x))+\frac{1}{\rho(x)}\left(d_i\partial_\rho U\right)(\rho(x),\omega(x))\right]=\\&=\left(\frac{1}{\rho(x)}-\frac{\omega_i^2(x)}{\rho(x)}\right)\partial_\rho U(\rho(x),\omega(x))+\\&+(\omega^i(x))^2\partial_\rho^2U(\rho(x),\omega(x))+\frac{\omega^i(x)}{\rho(x)}\left(d_i\partial_\rho U\right)(\rho(x),\omega(x)).\end{aligned} \tag{18.53}$$

Computation of J_2
We have

$$\begin{aligned} J_2 &= \partial_{x^i}\left(\frac{1}{\rho(x)}\right) d_i U(\rho(x),\omega(x)) + \\ &+ \frac{1}{\rho(x)}\partial_{x^i}\left((d_i U)(\rho(x),\omega(x))\right) = \\ &= -\frac{\omega^i(x)}{\rho^2(x)} d_i U(\rho(x),\omega(x)) + \\ &+ \frac{\omega^i(x)}{\rho(x)}\partial_\rho\left(d_i U\right)(\rho(x),\omega(x)) + \frac{1}{\rho^2(x)}\left(d_i^2 U\right)(\rho(x),\omega(x)), \end{aligned} \tag{18.54}$$

in the last equality we have applied (18.48) to $(d_i U)(\rho(x),\omega(x))$.

Computation of Δ
Now, adding up (18.53) and (18.54) we obtain (we omit, the variables)

$$\begin{aligned} \Delta u &= \sum_{i=1}^n \partial_{x^i}^2 u = \\ &= \frac{n-1}{\rho}\partial_\rho U + \partial_\rho^2 U + \frac{1}{\rho}\sum_{i=1}^n \omega^i d_i\left(\partial_\rho U\right) - \\ &- \frac{1}{\rho^2}\sum_{i=1}^n \omega^i d_i U + \frac{1}{\rho}\sum_{i=1}^n \omega^i \partial_\rho\left(d_i U\right) + \frac{1}{\rho^2}\Delta_\Sigma U. \end{aligned} \tag{18.55}$$

Now, (18.32) implies

$$\sum_{i=1}^n \omega_i d_i U = 0, \quad \sum_{i=1}^n \omega^i d_i\left(\partial_\rho U\right) = 0.$$

By the first equality we get

$$\sum_{i=1}^n \omega_i \partial_\rho\left(d_i U\right) = \partial_\rho\left(\sum_{i=1}^n \omega^i d_i U\right) = 0.$$

Therefore, (18.55) gives

$$\Delta u(x) = \partial_\rho^2 U(\rho(x),\omega(x)) + \frac{n-1}{\rho(x)}\partial_\rho U(\rho(x),\omega(x)) + \frac{1}{\rho^2(x)}\left(\Delta_\Sigma U\right)(\rho(x),\omega(x))$$

and by (18.50) we get (18.51). ■

Remark 18.4 Generally speaking, even if $u \in C^2(B_R)$, $U(\rho,\omega)$ is not differentiable w.r.t. ρ in 0. However (18.32) and (18.48) give

$$\partial_\rho U(\rho(x),\omega(x)) = \nabla u(x)\cdot\omega(x), \tag{18.56}$$

from which we have

$$\left|\partial_\rho U(\rho,\omega)\right| \leq \max_{\overline{B_r}} |\nabla u|, \qquad \text{for } \rho \leq r \leq R. \tag{18.57}$$

♦

Remark 18.5 Let us observe that

$$\begin{aligned}\partial_\rho^2 U(\rho(x),\omega(x)) &= \sum_{ij=1}^{n} \partial^2_{x^i x^j} u(x)\frac{x^i x^j}{|x|^2} = \sum_{|\alpha=2|}\frac{2!}{\alpha!}\partial^\alpha u(x)\left(\frac{x}{|x|}\right)^\alpha = \\ &= \underbrace{\partial^2 u(x)}_{\text{Hessian matrix}}\ \frac{x}{|x|}\cdot\frac{x}{|x|}.\end{aligned} \tag{18.58}$$

Indeed we have

$$\begin{aligned}\partial_\rho^2 U(\rho(x),\omega(x)) &= \sum_{k=1}^{n}\omega^k(x)\partial_{x^k}\left(\sum_{j=1}^{n}\omega^j(x)\partial_{x^j}u\right) = \\ &= \sum_{k=1}^{n}\frac{x^k}{|x|}\sum_{j=1}^{n}\left[\left(\frac{\delta^{jk}}{|x|}-\frac{x^j x^k}{|x|^3}\right)\partial_{x^j}u + \frac{x^j}{|x|}\partial^2_{x^j x^k}u\right] = \\ &= \sum_{k=1}^{n}\frac{x^k}{|x|}\left[\frac{1}{|x|}\partial_{x^k}u - \frac{x^k}{|x|^2}\left(\frac{x}{|x|}\cdot\nabla u\right) + \sum_{j=1}^{n}\frac{x^j}{|x|}\partial^2_{x^j x^k}u\right] = \\ &= \sum_{ij=1}^{n}\partial^2_{x^i x^j}u(x)\frac{x^i x^j}{|x|^2}.\end{aligned}$$

Similarly can be checked that

$$\partial_\rho^m U(\rho(x),\omega(x)) = \frac{1}{|x|^m}\sum_{|\alpha|=m}\frac{m!}{\alpha!}\partial^\alpha u(x) x^\alpha. \tag{18.59}$$

♦

Remark 18.6 Of course, by (18.56) we have (for $r \leq R$)

$$\frac{\partial u}{\partial \nu}(x) = \partial_\rho U(r,\omega), \qquad \text{for } x \in \partial B_r, \tag{18.60}$$

where ν is the unit outward normal to ∂B_r. Similarly we have by (18.59),

$$\frac{\partial^m u}{\partial \nu^m}(x) := \frac{d^m}{dt^m} u(x+t\nu)_{|t=0} = \partial_\rho^m U(r,\omega).$$

♦

Exercise 18.3 Let $u \in C^2(B_R)$ and $U(\rho,\omega) = u(\rho\omega)$. Apply formula (18.51) to prove that for any $r < R$ we have

$$\int_\Sigma \partial_\rho U(\rho,\omega) d\omega = \frac{1}{r^{n-1}} \int_{B_r} \Delta u(x) dx.$$

♠

Exercise 18.4 Let u be a harmonic function in B_R. Prove that for any $r < R$ we have

$$\int_{\partial B_r} \frac{\partial^m u}{\partial \nu^m}(x) dS = 0.$$

♠

18.4 The Case of Laplace Leading Operator

In this section and in the sequel we will use the following **Hardy inequality**, [55], [125].

Lemma 18.7 (The Hardy Inequality) *If $f \in C_0^\infty((0,+\infty))$, then*

$$\int_0^{+\infty} \frac{f^2(s)}{s^2} ds \le 4 \int_0^{+\infty} (f'(s))^2 ds. \tag{18.61}$$

Note *The number* 4 *on the right-hand side of* (18.61) *is the best constant.*

Proof We set

$$g(s) = s^{-\frac{1}{2}} f(s). \tag{18.62}$$

and we have

$$f(s) = s^{\frac{1}{2}} g(s), \quad f'(s) = \frac{1}{2} s^{-\frac{1}{2}} g(s) + s^{\frac{1}{2}} g'(s).$$

Hence (18.61) is equivalent to

$$\int_0^{+\infty} s^{-1} g^2(s) ds \leq 4 \int_0^{+\infty} \left(\frac{1}{2} s^{-\frac{1}{2}} g(s) + s^{\frac{1}{2}} g'(s) \right)^2 ds. \tag{18.63}$$

Now, we have

$$\begin{aligned}
&4 \int_0^{+\infty} \left(\frac{1}{2} s^{-\frac{1}{2}} g(s) + s^{\frac{1}{2}} g'(s) \right)^2 ds = \\
&= 4 \int_0^{+\infty} \left(\frac{1}{4} s^{-1} g^2(s) + g(s) g'(s) + s \left(g'(s)\right)^2 \right) ds = \\
&= 4 \int_0^{+\infty} \left(\frac{1}{4} s^{-1} g^2(s) + \frac{1}{2} \left(g^2\right)'(s) + s \left(g'(s)\right)^2 \right) ds = \\
&= \int_0^{+\infty} \left(s^{-1} g^2(s) + 4s \left(g'(s)\right)^2 \right) ds \geq \\
&\geq \int_0^{+\infty} s^{-1} g^2(s) ds.
\end{aligned}$$

The proof is complete. ■

Theorem 18.8 (Carleman Estimate for Δ) *Let $\epsilon \in (0, 1]$. Let us define*

$$\rho(x) = \phi_\epsilon(|x|), \quad \textit{for } x \in B_1, \tag{18.64}$$

where

$$\phi_\epsilon(s) = \frac{s}{(1 + s^\epsilon)^{1/\epsilon}}. \tag{18.65}$$

Then there exist $\tau_1 > 1$ and $C > 1$, depending on ϵ only, such that

$$\begin{aligned}
\tau^3 \int \rho^{\epsilon - 2\tau} |u|^2 dx + \tau \int \rho^{2+\epsilon-2\tau} |\nabla u|^2 dx \leq \\
\leq C \int \rho^{4-2\tau} |\Delta u|^2 dx,
\end{aligned} \tag{18.66}$$

for every $u \in C_0^\infty(B_1 \setminus \{0\})$ and for every $\tau \geq \tau_1$.

Moreover, there exist $\tau_2 > 1$, $C > 1$, depending on ϵ only, such that

$$\begin{aligned}
\tau^3 \int \rho^{\epsilon - 2\tau} |u|^2 dx + \tau \int \rho^{2+\epsilon-2\tau} |\nabla u|^2 dx + \\
+ \tau^2 r \int \rho^{-1-2\tau} u^2 dx \leq C \int \rho^{4-2\tau} |\Delta u|^2 dx,
\end{aligned} \tag{18.67}$$

for every $r \in (0, 1)$, for every $u \in C_0^\infty\left(B_1 \setminus \overline{B}_{r/4}\right)$ and for every $\tau \geq \tau_2$.

Remark 18.7 We have

$$\frac{|x|}{2^{1/\epsilon}} \leq \rho(x) \leq |x|, \quad \forall x \in B_1. \tag{18.68}$$

♦

Proof It is not restrictive to assume that u is a real-valued function. First we prove (18.66), afterwards, with a few modifications we will prove (18.67).

Let u be an arbitrary function of $C_0^\infty(B_1 \setminus \{0\})$ and let us consider the n-dimensional Laplace operator in the polar coordinates (ϱ, ω), that is (recalling that $\Sigma = \partial B_1$)

$$\Delta u = u_{\varrho\varrho} + \frac{n-1}{\varrho} u_\varrho + \frac{1}{\varrho^2}\Delta_\Sigma u, \;\; \forall(\varrho, \omega) \in (0, \infty) \times \Sigma. \tag{18.69}$$

Let us perform the following change of variables

$$\varrho = e^t, \qquad \widetilde{u}(t, \omega) = u\left(e^t, \omega\right), \quad \forall(t, \omega) \in (-\infty, 0) \times \Sigma.$$

We have, for every $(t, \omega) \in (-\infty, 0) \times \Sigma$,

$$e^{2t}(\Delta u)(e^t, \omega) = \mathcal{L}\widetilde{u} := \widetilde{u}_{tt} + (n-2)\widetilde{u}_t + \Delta_\Sigma \widetilde{u}. \tag{18.70}$$

For the sake of brevity, for any function $h \in C_0^\infty(B_1 \setminus \{0\})$ we will write $h', h'', ...$ instead of $h_t, h_{tt}, ...$.

By (18.64) we have (we omit the subscript ϵ from now on)

$$\varphi(t) := \log(\phi(e^t)) = t - \epsilon^{-1}\log\left(1 + e^{\epsilon t}\right), \quad \forall t \in (-\infty, 0). \tag{18.71}$$

We get

$$\varphi'(t) = \frac{1}{1 + e^{\epsilon t}}, \quad \varphi''(t) = -\frac{\epsilon e^{\epsilon t}}{(1 + e^{\epsilon t})^2}, \quad \forall t \in (-\infty, 0). \tag{18.72}$$

Let

$$f(t, \omega) = e^{-\tau\varphi}\widetilde{u}(t, \omega), \quad \forall(t, \omega) \in (-\infty, 0) \times \Sigma.$$

We have

$$\mathcal{L}_\tau f := e^{-\tau\varphi}\mathcal{L}(e^{\tau\varphi} f) = \underbrace{b_0 f + b_1 f'}_{\mathcal{A}_\tau f} + \underbrace{a_0 f + f'' + \Delta_\Sigma f}_{\mathcal{S}_\tau f}, \tag{18.73}$$

where

$$a_0 = \tau^2 \varphi'^2 + \tau(n-2), \quad b_0 = \tau\varphi'', \quad b_1 = 2\tau\varphi' + (n-2). \tag{18.74}$$

Let us denote by $\int(\cdot)$ the integral $\int_{-\infty}^{0} \int_{\Sigma} (\cdot) d\omega dt$ and set

$$\gamma := \frac{1}{\varphi'} = 1 + e^{\epsilon t}. \tag{18.75}$$

We obtain

$$\int \gamma \left|\mathcal{L}_\tau f\right|^2 \geq 2 \int \gamma \mathcal{A}_\tau f \mathcal{S}_\tau f + \int \gamma \left|\mathcal{A}_\tau f\right|^2 \tag{18.76}$$

and

$$\begin{aligned} 2\int \gamma \mathcal{A}_\tau f \mathcal{S}_\tau f = & \underbrace{2\int \gamma \left(b_0 f + b_1 f'\right) \Delta_\Sigma f +}_{I_1} \\ & \underbrace{+2\int \gamma \left(b_0 f + b_1 f'\right)\left(a_0 f + f''\right).}_{I_2} \end{aligned} \tag{18.77}$$

We Examine I_1

For any function $f, g \in C^\infty(B_1 \setminus \{0\})$ and enough regular, let us denote by

$$\langle \nabla_\Sigma f, \nabla_\Sigma g\rangle = \sum_{i=1}^{n} d_i f d_i g, \qquad \left|\nabla_\Sigma f\right|^2 = \langle \nabla_\Sigma f, \nabla_\Sigma f\rangle.$$

Integrating by parts, using Proposition 18.3 and taking into account (18.75), we have

$$\begin{aligned} I_1 &= 2\int \left(\gamma b_0 f \Delta_\Sigma f + \gamma b_1 f' \Delta_\Sigma f\right) = \\ &= 2\int \left(-\gamma b_0 \left|\nabla_\Sigma f\right|^2 - \gamma b_1 \left\langle \nabla_\Sigma f, \nabla_\Sigma f'\right\rangle\right) = \\ &= 2\int \left(-\gamma b_0 \left|\nabla_\Sigma f\right|^2 - \frac{1}{2}\gamma b_1 \left(\left|\nabla_\Sigma f\right|^2\right)'\right) = \\ &= 2\int \left(-\gamma b_0 + \frac{1}{2}\left(\gamma b_1\right)'\right) \left|\nabla_\Sigma f\right|^2. \end{aligned}$$

By (18.72), (18.74), and (18.75) we obtain

$$\begin{aligned}-\gamma b_0+\frac{1}{2}(\gamma b_1)' &= -\gamma\tau\varphi''+\frac{1}{2}\gamma b_1'+\frac{1}{2}\gamma' b_1= \\ &= -\gamma\tau\varphi''+\gamma\tau\varphi''+\frac{1}{2}\left(2\tau\varphi'+n-2\right)\gamma'= \\ &= \frac{\epsilon e^{\epsilon t}}{2}\left(\frac{2\tau}{1+e^{\epsilon t}}+n-2\right)\geq\frac{\tau}{2}\epsilon e^{\epsilon t},\quad \forall\tau>0.\end{aligned} \tag{18.78}$$

Hence, we have

$$I_1\geq\int\tau\epsilon e^{\epsilon t}\left|\nabla_\Sigma f\right|^2,\quad \forall\tau>0. \tag{18.79}$$

Now We Examine I_2
Integration by parts gives

$$\begin{aligned}I_2 &= 2\int\gamma\left(b_0f+b_1f'\right)\left(a_0f+f''\right)= \\ &= 2\int\gamma\left(a_0b_0f^2+b_0ff''+b_1a_0f'f+b_1f'f''\right)= \\ &= 2\int\gamma a_0b_0f^2-(\gamma b_0f)'f'+\frac{1}{2}\gamma b_1a_0\left(f^2\right)'+\frac{1}{2}\gamma b_1\left(f'^2\right)'= \\ &= 2\int\left[\gamma a_0b_0-\frac{1}{2}(\gamma b_1a_0)'\right]f^2-(\gamma b_0)'ff'- \\ &\quad -\int\gamma b_0f'^2+\frac{1}{2}(\gamma b_1)'f'^2= \\ &= 2\int\underbrace{\left[\gamma a_0b_0-\frac{1}{2}(\gamma b_1a_0)'+\frac{1}{2}(\gamma b_0)''\right]}_{H_1}f^2- \\ &\quad -\int\underbrace{\left[\gamma b_0+\frac{1}{2}(\gamma b_1)'\right]}_{H_2}f'^2.\end{aligned} \tag{18.80}$$

Let now examine H_1.

Since by (18.74), H_1 is a polynomial of third degree w.r.t. τ, we begin by evaluating the coefficient of τ^3.

Let us notice that the terms of H_1 have the following behavior, as $\tau \to +\infty$

$$\gamma a_0 b_0 = O\left(\tau^3\right),$$

$$-\frac{1}{2}\left(\gamma b_1 a_0\right)' = O\left(\tau^3\right),$$

$$\frac{1}{2}\left(\gamma b_0\right)'' = O\left(\tau\right).$$

Hence, let us first examine the term

$$\widetilde{H}_1 := \gamma a_0 b_0 - \frac{1}{2}\left(\gamma b_1 a_0\right)'.$$

By (18.72), (18.74), and (18.75) we have

$$\begin{aligned}
\widetilde{H}_1 &= \gamma a_0 b_0 - \frac{1}{2}\left(\gamma b_1 a_0\right)' = \\
&= \gamma\left(\tau^2 \varphi'^2 + \tau(n-2)\varphi'\right)\tau\varphi'' - \frac{1}{2}b_1'\left(\gamma a_0\right) - \frac{1}{2}b_1\left(\gamma a_0\right)' = \\
&= \gamma\left(\tau^2 \varphi'^2 + \tau(n-2)\varphi'\right)\tau\varphi'' - \\
&- \frac{1}{2}\left(2\tau\varphi' + (n-2)\right)'\gamma\left(\tau^2\varphi'^2 + \tau(n-2)\varphi'\right) - \frac{1}{2}b_1\left(\gamma a_0\right)' = \\
&= -\frac{1}{2}b_1\left(\gamma a_0\right)' - \\
&= -\frac{1}{2}\left(\frac{2\tau}{1+e^{\epsilon t}} + n - 2\right)\left[\gamma\left(\tau^2\varphi'^2 + \tau(n-2)\varphi'\right)\right]' = \\
&= -\left(\frac{\tau}{1+e^{\epsilon t}} + \frac{n-2}{2}\right)\left(\tau^2\frac{1}{1+e^{\epsilon t}} + \tau(n-2)\right)' = \\
&= \left(\tau^3\frac{1}{1+e^{\epsilon t}} + \tau^2\frac{n-2}{2}\right)\frac{\epsilon e^{\epsilon t}}{(1+e^{\epsilon t})^2}.
\end{aligned} \tag{18.81}$$

Then, using the trivial inequality

$$\frac{1}{1+e^{\epsilon t}} \geq \frac{1}{2}, \qquad \forall t \in (-\infty, 0),$$

we have (for $t \in (-\infty, 0)$)

$$\widetilde{H}_1 \geq \frac{\tau^3}{8}\epsilon e^{\epsilon t}, \quad \forall \tau > 0. \tag{18.82}$$

and

$$\begin{aligned}\frac{1}{2}\left(\gamma b_0\right)'' &= \frac{\tau}{2}\left(\frac{-\epsilon e^{\epsilon t}}{1+e^{\epsilon t}}\right)'' = \\ &= -\frac{\tau}{2}\frac{\epsilon^3 e^{\epsilon t}\left(1-e^{\epsilon t}\right)}{(1+e^{\epsilon t})^3} \geq -\frac{\tau}{2}\epsilon^3 e^{\epsilon t}, \quad \forall \tau > 0.\end{aligned} \tag{18.83}$$

Inequalities (18.82) and (18.83) give

$$H_1 = \widetilde{H}_1 + \frac{1}{2}\left(\gamma b_0\right)'' \geq \frac{\tau^3}{8}\epsilon e^{\epsilon t} - \frac{\tau}{2}\epsilon^3 e^{\epsilon t} \geq \frac{\tau^3}{16}\epsilon e^{\epsilon t}, \quad \forall \tau > \sqrt{8}\epsilon. \tag{18.84}$$

Now, let us consider H_2.

$$\begin{aligned}H_2 &= \gamma b_0 + \frac{1}{2}\left(\gamma b_1\right)' = \\ &= \tau(1+e^{\epsilon t})\varphi'' + \frac{1}{2}\left[2\tau(1+e^{\epsilon t})\varphi' + (n-2)(1+e^{\epsilon t})\right]' = \\ &= -\tau\frac{\epsilon e^{\epsilon t}}{1+e^{\epsilon t}} + \frac{1}{2}\left[2\tau + (n-2)(1+e^{\epsilon t})\right]' = \\ &= -\tau\frac{\epsilon e^{\epsilon t}}{1+e^{\epsilon t}} + \frac{(n-2)\epsilon e^{\epsilon t}}{2}.\end{aligned} \tag{18.85}$$

Hence

$$-2H_2 \geq \frac{\tau}{2}\epsilon e^{\epsilon t}, \quad \forall \tau \geq 2(n-2). \tag{18.86}$$

By (18.80), (18.84), and (18.86) we have

$$I_2 \geq \int \frac{\tau^3}{8}\epsilon e^{\epsilon t} f^2 + \frac{\tau}{2}\epsilon e^{\epsilon t} f'^2, \quad \forall \tau \geq \tau_1, \tag{18.87}$$

where $\tau_1 = \max\{\sqrt{8}\epsilon, 2(n-2)\}$.

Now, (18.77), (18.79), and (18.87) give

$$2\int \gamma \mathcal{A}_\tau f \mathcal{S}_\tau f \geq \frac{\epsilon}{8}\int \left(\tau^3 f^2 + \tau\left(f'^2 + |\nabla_\Sigma f|^2\right)\right) e^{\epsilon t}, \quad \forall \tau \geq \tau_1. \tag{18.88}$$

By (18.76) and (18.88) we have

$$\int \gamma \left|\mathcal{L}_\tau f\right|^2 \geq \frac{\epsilon\tau^3}{8}\int f^2 + \frac{\epsilon\tau}{8}\int \left(f'^2 + |\nabla_\Sigma f|^2\right) e^{\epsilon t}, \tag{18.89}$$

for every $\tau \geq \tau_1$ and for every $f \in C_0^\infty((-\infty, 0) \times \Sigma)$.

Now, in order to obtain (18.66) we come back to u and to the original variables. Let us recall that $f(t,\omega) = e^{-\tau\varphi}u(e^t,\omega)$. By using (18.64), (18.70), and (18.73), we get

$$\begin{aligned}
&\int_{-\infty}^{0}\int_{\Sigma} |\mathcal{L}_\tau f|^2\, d\omega dt = \\
&= \int_{-\infty}^{0}\int_{\Sigma} e^{-2\tau\varphi(t)}e^{4t}|(\Delta u)(e^t,\omega)|^2 d\omega dt = \\
&= \int_{0}^{1}\int_{\Sigma} e^{-2\tau\varphi(\log\varrho)}\varrho^3|(\Delta u)(\varrho,\omega)|^2 d\omega d\varrho = \\
&= \int_{B_1} \rho^{-2\tau}|x|^{4-n}|\Delta u|^2 dx.
\end{aligned} \tag{18.90}$$

Similarly, we get

$$\int_{-\infty}^{0}\int_{\Sigma} f^2 e^{\epsilon t} d\omega dt = \int_{B_1} \rho^{-2\tau}|x|^{\epsilon-n}u^2 dx. \tag{18.91}$$

Concerning the second integral on the right-hand side of (18.89), let $\delta \in (0,1)$ be a number that we will choose later, we have

$$\begin{aligned}
&\int_{-\infty}^{0}\int_{\Sigma} e^{\epsilon t}\left(f'^2 + |\nabla_\Sigma f|^2\right) d\omega dt \geq \\
&\geq \delta \int_{-\infty}^{0}\int_{\Sigma} e^{\epsilon t}\left(f'^2 + |\nabla_\Sigma f|^2\right) d\omega dt \geq \\
&\geq \frac{\delta}{2}\int_{-\infty}^{0}\int_{\Sigma} e^{\epsilon t}e^{-2\tau\varphi(t)}\Big(|u_\varrho(e^t,\omega)|^2 e^{2t} + \\
&+ \left|\nabla_\Sigma u(e^t,\omega)\right|^2 - 2\tau^2|u(e^t,\omega)|^2\Big) d\omega dt = \\
&= \frac{\delta}{2}\int_{B_1} \rho^{-2\tau}|x|^{\epsilon-n}\left(|x|^2|\nabla u|^2 - 2\tau^2|u|^2\right) dx.
\end{aligned} \tag{18.92}$$

Now, let us choose $\delta = \frac{1}{2}$ so that, by (18.89) and (18.90)–(18.92), we have

$$\begin{aligned}
\int_{B_1} \rho^{-2\tau}|x|^{4-n}|\Delta u|^2 dx \geq &\frac{\varepsilon\tau}{16}\int_{B_1} \rho^{-2\tau}|x|^{\epsilon+2-n}|\nabla u|^2 dx + \\
&+ \frac{\epsilon\tau^3}{16}\int_{B_1} \rho^{-2\tau}|x|^{\epsilon-n}u^2 dx,
\end{aligned} \tag{18.93}$$

for every $u \in C_0^\infty(B_1 \setminus \{0\})$ and for every $\tau \geq \tau_1$. Finally, by (18.68), we substitute in (18.93) τ by $(\tau - \frac{n}{2})$ and we find inequality (18.66).

Let us now look at (18.67). Let $u \in C_0^\infty(B_1 \setminus \overline{B}_{r/4})$.

By (18.76) and (18.89) we have

$$\int |\mathcal{L}_\tau f|^2 \geq \frac{\epsilon}{8} \int \left(\tau^3 f^2 + \tau \left(f'^2 + |\nabla_\Sigma f|^2\right)\right) e^{\epsilon t} + \int \gamma \, |\mathcal{A}_\tau f|^2 . \tag{18.94}$$

To obtain the first term on the left-hand side of (18.67) we estimate from below the last term on the right-hand side of (18.94).

Let us note that by the trivial inequality $(a+b)^2 \geq \frac{1}{2}a^2 - b^2$ and by (18.72), (18.75) we have

$$\begin{aligned}\int \gamma \, |\mathcal{A}_\tau f|^2 &\geq \frac{1}{2} \int \gamma \left(2\tau\varphi' + n - 2\right)^2 f'^2 - \int \gamma \tau^2 \varphi''^2 f^2 \geq \\ &\geq \tau^2 \int f'^2 - \epsilon^2 \tau^2 \int e^{2\epsilon t} f^2, \quad \forall \tau > 0.\end{aligned} \tag{18.95}$$

By using inequality (18.95) in (18.94), we have

$$\begin{aligned}\int |\mathcal{L}_\tau f|^2 \geq &\tau^2 \int f'^2 + \epsilon\tau^3 \int \left(\frac{1}{8} - \epsilon\tau^{-1} e^{\epsilon t}\right) e^{\epsilon t} f^2 + \\ &+ \frac{\epsilon}{8}\tau \int \left(f'^2 + |\nabla_\Sigma f|^2\right) e^{\epsilon t}, \quad \forall \tau \geq \tau_1.\end{aligned} \tag{18.96}$$

Now, since $\left(\frac{1}{8} - \epsilon\tau^{-1} e^{\epsilon t}\right) \geq \frac{1}{16}$ for every $\tau \geq 4\epsilon$, by (18.96), we get

$$\int |\mathcal{L}_\tau f|^2 \geq \tau^2 \int f'^2 + \frac{\epsilon\tau^3}{16} \int e^{\epsilon t} f^2 + \frac{\epsilon}{8}\tau \int \left(f'^2 + |\nabla_\Sigma f|^2\right) e^{\epsilon t}, \tag{18.97}$$

for every $\tau \geq \tau_2$, where $\tau_2 = \max\{4\epsilon, \tau_1\}$.

Proposition 18.7 implies

$$\begin{aligned}\int_{-\infty}^0 f^2(t,\omega) e^{-t} dt &= \int_0^1 s^{-2} f^2(\log s, \omega) ds \leq \\ &\leq 4 \int_0^1 \left| \frac{\partial}{\partial s} f(\log s, \omega) \right|^2 ds = \\ &= 4 \int_{-\infty}^0 f'^2(t,\omega) e^{-t} dt, \quad \forall \omega \in \Sigma.\end{aligned} \tag{18.98}$$

On the other hand, since $u \in C_0^\infty(B_1 \setminus \overline{B}_{r/4})$, we have $f(t,\omega) = 0$ for every $t \leq \log(r/4)$ and for every $\omega \in \Sigma$, (18.98) gives

$$\int_{-\infty}^{0} f^2(t,\omega)e^{-t}dt \leq 4\int_{-\infty}^{\log\frac{r}{4}} f'^2(t,\omega)e^{-t}dt \leq \frac{16}{r}\int_{-\infty}^{0} f'^2(t,\omega)dt, \quad \forall\omega \in \Sigma.$$

Let us integrate over Σ both the sides of the just obtained inequality and let us use (18.97) to obtain

$$\int f^2e^{-t} \leq \frac{16}{r}\int f'^2 \leq \frac{16}{\tau^2 r}\int |\mathcal{L}_\tau f|^2, \quad \forall\tau \geq \tau_2. \tag{18.99}$$

By (18.99) and (18.97) we have

$$\begin{aligned} C\int |\mathcal{L}_\tau f|^2 \geq \epsilon\tau^3\int e^{\epsilon t}f^2 + \epsilon\tau\int \left(f'^2 + |\nabla_\Sigma f|^2\right)e^{\epsilon t} + \\ + \tau^2 r\int f^2e^{-t}, \qquad \forall\tau \geq \tau_2, \end{aligned} \tag{18.100}$$

where C is a constant.

Finally, by (18.93), (18.68) and by

$$\int_{-\infty}^{0}\int_{\Sigma} f^2e^{-t}d\omega dt = \int_{B_1}\rho^{-2\tau}|x|^{-1-n}u^2dx, \tag{18.101}$$

we obtain (18.67). ■

18.5 Proof of the Optimal Three Sphere and the Doubling Inequality

In the next theorem we will consider the equation

$$\Delta U = b(x)\cdot\nabla U + c(x)U, \quad \text{in } B_1, \tag{18.102}$$

where $b \in L^\infty(B_1;\mathbb{R}^n)$ and $c \in L^\infty(B_1)$. Moreover, set

$$M = \max\left\{\|b\|_{L^\infty(B_1;\mathbb{R}^n)}, \|c\|_{L^\infty(B_1)}\right\}. \tag{18.103}$$

Theorem 18.9 (Optimal Three Sphere and Doubling Inequality) *Assume that $U \in H^2(B_1)$ is a solution to Eq. (18.102). Let $x_0 \in B_1$ and $0 < R_0 \leq 1 - |x_0|$. Then there exists $C > 1$ depending on M only, such that, if $0 < 2r < R < \frac{R_0}{2}$ then*

$$\int_{B_R(x_0)} U^2 \leq C\left(\frac{R_0}{R}\right)^C \left(\int_{B_r(x_0)} U^2\right)^\theta \left(\int_{B_{R_0}(x_0)} U^2\right)^{1-\theta}, \tag{18.104}$$

where

$$\theta = \frac{\log \frac{R_0}{2R}}{\log \frac{2R_0}{r}}. \tag{18.105}$$

Moreover, if U does not vanish identically in $B_{R_0/4}(x_0)$ then the following doubling inequality holds

$$\int_{B_{2r}(x_0)} U^2 \leq CN_{x_0,R_0}^3 \int_{B_r(x_0)} U^2, \tag{18.106}$$

where

$$N_{x_0,R_0} = \frac{\int_{B_{R_0}(x_0)} U^2}{R_0 \int_{B_{R_0/4}(x_0)} U^2}. \tag{18.107}$$

In order to prove Theorem 18.9 we need the following

Lemma 18.10 *Under the same assumption of Theorem 18.9 we have, for every $x_0 \in B_1$, and for every R, r such that $0 < 2r < R < \frac{R_0}{2}$, where $R_0 \leq 1 - |x_0|$,*

$$\begin{aligned} &(2r)^{-2\tau} \int_{B_{2r}(x_0)} U^2 + R^{1-2\tau} \int_{B_R(x_0)} U^2 \leq \\ &\leq C\overline{M}^2 \left[\left(\frac{r}{4}\right)^{-2\tau} \int_{B_r(x_0)} U^2 + \left(\frac{R_0}{2}\right)^{-2\tau} \int_{B_{R_0}(x_0)} U^2\right], \end{aligned} \tag{18.108}$$

for every $\tau \geq \widetilde{\tau}_2$, where $\widetilde{\tau}_2$ and $C \geq 1$ depend on M only.

Proof of Lemma By a translation we may assume that $x_0 = 0$. Let r, R satisfy

$$0 < 2r < R < \frac{R_0}{2}. \tag{18.109}$$

Let $\eta \in C_0^\infty((0, R_0))$ satisfy

$$0 \leq \eta \leq 1, \tag{18.110}$$

$$\eta = 0, \quad \text{in } \left(0, \frac{r}{4}\right) \cup \left(\frac{2R_0}{3}, R_0\right); \quad \eta = 1, \quad \text{in } \left[\frac{r}{2}, \frac{R_0}{2}\right], \tag{18.111}$$

$$\left|\frac{d^k\eta}{dt^k}(t)\right| \leq Cr^{-k}, \quad \text{in } \left(\frac{r}{4}, \frac{r}{2}\right), \quad \text{for } 0 \leq k \leq 2, \tag{18.112}$$

$$\left|\frac{d^k\eta}{dt^k}(t)\right| \leq CR_0^{-k}, \quad \text{in } \left(\frac{R_0}{2}, \frac{2R_0}{3}\right), \quad \text{for } 0 \leq k \leq 2. \tag{18.113}$$

We define

$$\xi(x) = \eta(|x|). \tag{18.114}$$

By exploiting Carleman estimate (18.67) and fixing there $\epsilon = 1$, we get

$$\begin{aligned} \tau^3 \int_{B_{R_0}} \rho^{1-2\tau} u^2 dx + \tau \int_{B_{R_0}} \rho^{3-2\tau} |\nabla u|^2 + \\ + \tau^2 r \int_{B_{R_0}} \rho^{-1-2\tau} u^2 \leq C \int_{B_{R_0}} \rho^{4-2\tau} |\Delta u|^2, \end{aligned} \tag{18.115}$$

for every $u \in C_0^\infty\left(B_{R_0} \setminus \overline{B}_{r/4}\right)$ and for every $\tau \geq \tau_2$ (we recall that τ_2 and C depend neither on r nor on R_0 and that the value C may change from line to line).

Since $\xi U \in H_0^2(B_{R_0})$, by density we can apply Carleman estimate (18.115) to $u = \xi U$. Hence we find

$$\begin{aligned} \tau^3 \int_{B_{R_0}} \rho^{1-2\tau} \xi^2 U^2 + \tau \int_{B_{R_0}} \rho^{3-2\tau} |\nabla(\xi U)|^2 + \\ + \tau^2 r \int_{B_{R_0}} \rho^{-1-2\tau} \xi^2 U^2 \leq C \int_{B_{R_0}} \rho^{4-2\tau} |\Delta(\xi U)|^2, \end{aligned} \tag{18.116}$$

for every $\tau \geq \tau_2$.

Since we have

$$|\Delta(\xi U)|^2 \leq 2\xi^2 |\Delta U|^2 + C\left(\left|\partial^2 \xi\right|^2 U^2 + |\nabla \xi|^2 |\nabla U|^2\right), \tag{18.117}$$

setting

$$J_0 = \int_{B_{r/2} \setminus B_{r/4}} \rho^{4-2\tau} \left(r^{-4} U^2 + r^{-2} |\nabla U|^2\right), \tag{18.118}$$

$$J_1 = \int_{B_{2R_0/3} \setminus B_{R_0/2}} \rho^{4-2\tau} \left(R_0^{-4} U^2 + R_0^{-2} |\nabla U|^2\right), \tag{18.119}$$

we get

$$\begin{aligned}&\tau^3\int_{B_{R_0}}\rho^{1-2\tau}\xi^2U^2dx+\tau\int_{B_{R_0}}\rho^{3-2\tau}|\nabla(\xi U)|^2dx+\\&+\tau^2r\int_{B_{R_0}}\rho^{-1-2\tau}\xi^2U^2dx\leq\\&\leq C\int_{B_{R_0}}\rho^{4-2\tau}|\Delta U|^2+C\left(J_0+J_1\right),\end{aligned}\tag{18.120}$$

for every $\tau\geq\tau_2$.

Now we perform what follows: we use (18.109)–(18.114) and (18.117), we estimate trivially from below the left-hand side of (18.120) and we estimate trivially from above the right-hand side of (18.120), obtaining

$$\begin{aligned}&\tau^3\int_{B_{R_0/2}\setminus B_{r/2}}\rho^{1-2\tau}U^2+\tau\int_{B_{R_0/2}\setminus B_{r/2}}\rho^{3-2\tau}|\nabla U|^2+\\&+\tau^2r\int_{B_{R_0}}\rho^{-1-2\tau}\xi^2U^2dx\leq\\&\leq CM^2\int_{B_{R_0/2}\setminus B_{r/2}}\rho^{4-2\tau}\left(U^2+|\nabla U|^2\right)+\\&+C\overline{M}^2(J_0+J_1),\end{aligned}\tag{18.121}$$

for every $\tau\geq\tau_2$, where $\overline{M}=\sqrt{M^2+1}$.

By (18.121), we get

$$\begin{aligned}&\int_{B_{R_0/2}\setminus B_{r/2}}\left(\tau^3-CM^2\rho^3\right)\rho^{1-2\tau}U^2+\\&+\int_{B_{R_0/2}\setminus B_{r/2}}\left(\tau-CM^2\rho\right)\rho^{3-2\tau}|\nabla U|^2+\\&+\tau^2r\int_{B_{R_0}}\rho^{-1-2\tau}\xi^2U^2\leq C\overline{M}^2(J_0+J_1).\end{aligned}\tag{18.122}$$

By the latter, taking into account that $\rho\leq 1$ in B_{R_0}, we have

$$\begin{aligned}&\frac{\tau^3}{2}\int_{B_{R_0/2}\setminus B_{r/2}}\rho^{1-2\tau}U^2+\frac{\tau}{2}\int_{B_{R_0/2}\setminus B_{r/2}}\rho^{3-2\tau}|\nabla U|^2+\\&+\tau^2r\int_{B_{R_0}}\rho^{-1-2\tau}\xi^2U^2dx\leq C\overline{M}^2(J_0+J_1),\end{aligned}\tag{18.123}$$

for every $\tau \geq \widetilde{\tau}_2$, where

$$\widetilde{\tau}_2 = \max\left\{2CM^2, \tau_2\right\}.$$

Now, we estimate from above J_0 and J_1. By the Caccioppoli inequality (Theorem 6.7) and recalling (18.68), we have

$$\begin{aligned} J_0 = \int_{B_{r/2}\setminus B_{r/4}} \rho^{4-2\tau}\left(r^{-4}U^2 + r^{-2}|\nabla U|^2\right) \leq \\ \leq C\left(\frac{r}{4}\right)^{-2\tau} \int_{B_{r/2}} \left(U^2 + r^2|\nabla U|^2\right) \leq C\left(\frac{r}{4}\right)^{-2\tau} \int_{B_r} U^2, \end{aligned} \tag{18.124}$$

where C depends on M only.

Similarly we get

$$J_1 \leq C\left(\frac{R_0}{2}\right)^{-2\tau} \int_{B_{R_0}} U^2. \tag{18.125}$$

By (18.123)–(18.125) we have

$$\begin{aligned} \tau^2 r\int_{B_{R_0}} \rho^{-1-2\tau}\xi^2 U^2 + \tau^3 \int_{B_{R_0/2}\setminus B_{r/2}} \rho^{1-2\tau}U^2 dx \leq \\ \leq C\overline{M}^2\left(\left(\frac{r}{4}\right)^{-2\tau}\int_{B_r} U^2 + \left(\frac{R_0}{2}\right)^{-2\tau}\int_{B_{R_0}} U^2\right), \end{aligned} \tag{18.126}$$

for every $\tau \geq \widetilde{\tau}_2$.

Now, recalling that $2r < R < \frac{R_0}{2}$, by (18.111) we have

$$\tau^2 r\int_{B_{R_0}} \rho^{-1-2\tau}\xi^2 U^2 \geq (2r)^{-2\tau}\int_{B_{2r}\setminus B_{r/2}} U^2, \tag{18.127}$$

and

$$\tau^3 \int_{B_{R_0/2}\setminus B_{r/2}} \rho^{1-2\tau}U^2 \geq R^{1-2\tau}\int_{B_R\setminus B_{r/2}} U^2. \tag{18.128}$$

By (18.126), (18.127), and (18.128) we have

$$\begin{aligned}(2r)^{-2\tau} \int_{B_{2r}\setminus B_{r/2}} U^2 + R^{1-2\tau} \int_{B_R\setminus B_{r/2}} U^2 \le \\ \le C\overline{M}^2 \left[\left(\frac{r}{4}\right)^{-2\tau} \int_{B_r} U^2 + \left(\frac{R_0}{2}\right)^{-2\tau} \int_{B_{R_0}} U^2 \right],\end{aligned} \tag{18.129}$$

for every $\tau \ge \widetilde{\tau}_2$. Now, we add to both the sides of (18.129) the quantity

$$(2r)^{-2\tau} \int_{B_{r/2}} U^2 + R^{1-2\tau} \int_{B_{r/2}} U^2,$$

and we find (18.108) for $r < R/2$ and $R < R_0/2$. ■

Proof of Theorem 18.9 Let us suppose $x_0 = 0$, (18.108) gives, for $0 < 2r < R < \frac{R_0}{2}$,

$$R^{1-2\tau} \int_{B_R} U^2 \le C\overline{M}^2 \left[\left(\frac{r}{4}\right)^{-2\tau} \int_{B_r} U^2 + \left(\frac{R_0}{2}\right)^{-2\tau} \int_{B_{R_0}} U^2 \right], \tag{18.130}$$

for every $\tau \ge \widetilde{\tau}_2$.

Set

$$A(s) = \int_{B_s} U^2.$$

By (18.130) we have

$$A(R) \le CR^{-1}\overline{M}^2 \left[\left(\frac{4R}{r}\right)^{2\tau} A(r) + \left(\frac{2R}{R_0}\right)^{2\tau} A(R_0) \right], \tag{18.131}$$

for every $\tau \ge \widetilde{\tau}_2$. Let

$$\widehat{\tau} = \frac{\log \frac{A(R_0)}{A(r)}}{2 \log \frac{2R_0}{r}}. \tag{18.132}$$

If

$$\widehat{\tau} \ge \widetilde{\tau}_2, \tag{18.133}$$

we choose $\tau = \widehat{\tau}$ in (18.131) and since

$$\left(\frac{4R}{r}\right)^{2\widehat{\tau}} A(r) = \left(\frac{2R}{R_0}\right)^{2\widehat{\tau}} A(R_0),$$

we have

$$
\begin{aligned}
A(R) &\leq CR^{-1}\overline{M}^2\left[\left(\frac{4R}{r}\right)^{2\widehat{\tau}} A(r) + \left(\frac{2R}{R_0}\right)^{2\widehat{\tau}} A(R_0)\right] = \\
&= 2CR^{-1}\overline{M}^2\left(\frac{4R}{r}\right)^{2\widehat{\tau}} A(r) = R^{-1}\overline{M}^2 \left(A(r)\right)^{\theta}\left(A(R_0)\right)^{1-\theta},
\end{aligned}
\tag{18.134}
$$

where θ is given by (18.105). Whereas, if (18.133) does not hold, then

$$
\frac{\log \frac{A(R_0)}{A(r)}}{\log \frac{2R_0}{r}} \leq 2\widetilde{\tau}_2
$$

and multiplying both the sides of the last inequality by $\log \frac{R_0}{2R}$, we have

$$
\left(A(R_0)\right)^{\theta} \leq \left(\frac{R_0}{2R}\right)^{2\widetilde{\tau}_2}\left(A(r)\right)^{\theta}, \tag{18.135}
$$

from which we have trivially

$$
\begin{aligned}
A(R) &\leq A(R_0) = \left(A(R_0)\right)^{\theta}\left(A(R_0)\right)^{1-\theta} \leq \\
&\leq \left(\frac{R_0}{2R}\right)^{2\widetilde{\tau}_2}\left(A(r)\right)^{\theta}\left(A(R_0)\right)^{1-\theta},
\end{aligned}
\tag{18.136}
$$

which, together with (18.134), gives (18.104).

Now we prove (18.106).

Let us fix $R = \frac{R_0}{4}$ in (18.108). We have

$$
\begin{aligned}
&(2r)^{-2\tau}\int_{B_{2r}} U^2 + \left(\frac{R_0}{4}\right)^{1-2\tau}\int_{B_{R_0/4}} U^2 \leq \\
&\leq C\overline{M}^2\left[\left(\frac{r}{4}\right)^{-2\tau}\int_{B_r} U^2 + \left(\frac{R_0}{2}\right)^{-2\tau}\int_{B_{R_0}} U^2\right],
\end{aligned}
\tag{18.137}
$$

for every $\tau \geq \widetilde{\tau}_2$.

Now, by choosing $\tau = \tau_0$, where

$$
\tau_0 = \widetilde{\tau}_2 + \log_4\left(4C\overline{M}^2 N\right)
$$

and

$$N = \frac{\int_{B_{R_0}} U^2}{R_0 \int_{B_{R_0/4}} U^2}, \tag{18.138}$$

we have

$$\left(\frac{R_0}{4}\right)^{1-2\tau_0} \int_{B_{R_0/4}} U^2 \geq C\overline{M}^2 \left(\frac{R_0}{2}\right)^{-2\tau_0} \int_{B_{R_0}} U^2.$$

Hence, by (18.137), we obtain

$$(2r)^{-2\tau_0} \int_{B_{2r}} U^2 \leq C\overline{M}^2 \left(\frac{r}{4}\right)^{-2\tau_0} \int_{B_r} U^2. \tag{18.139}$$

By using (18.138) and (18.139), we have

$$\int_{B_{2r}} U^2 \leq CN^3 \int_{B_r} U^2, \tag{18.140}$$

where C depends on M only.

The proof is complete. ∎

Corollary 18.11 (Strong Unique Continuation for Laplace Operator) *Let $U \in H^2(B_1)$ be a solution to equation* (18.102). *Let $x_0 \in B_1$ and $0 < R_0 \leq 1 - |x_0|$.*

If U does not vanish identically in $B_{R_0/4}(x_0)$, then we have, for every $r < s \leq \frac{R_0}{16}$,

$$\int_{B_s(x_0)} U^2 \leq CN^3_{x_0,R_0} \left(\frac{s}{r}\right)^{\log_2\left(CN^3_{x_0,R_0}\right)} \int_{B_r(x_0)} U^2, \tag{18.141}$$

where N_{x_0,R_0} is defined by (18.107).

Moreover, if

$$\int_{B_r(x_0)} U^2 = O\left(r^m\right), \quad \text{as } r \to 0, \quad \forall m \in \mathbb{N}, \tag{18.142}$$

then

$$U = 0, \quad \text{in } B_1. \tag{18.143}$$

Proof We prove (18.141). Let us suppose that $x_0 = 0$ and let $r < s \leq \frac{R_0}{16}$. Set $j = \left[\log_2\left(sr^{-1}\right)\right]$ (we recall that $[a]$ is the integer part of a). We have

$$2^j r \leq s < 2^{j+1} r$$

and applying repeatedly (18.106) we obtain

$$\int_{B_s} U^2 \le \int_{B_{2^{j+1}r}} U^2 \le \left(CN^3\right)^{j+1} \int_{B_r} U^2 \le CN^3 \left(\frac{s}{r}\right)^{\log_2(CN^3)} \int_{B_r} U^2.$$

From which we get (18.141).

Now, let us suppose that (18.142) holds true. Hence let us suppose that there exists a sequence C_m such that

$$\int_{B_r} U^2 \le C_m r^m, \quad \text{for } r < 1, \quad \forall m \in \mathbb{N}. \tag{18.144}$$

Set

$$r_0 = \sup\left\{r \in [0,1] : \int_{B_r} U^2 = 0\right\} \tag{18.145}$$

(let us note that in (18.145) the "sup" is, actually, the maximum).

We distinguish two cases

(i) $r_0 = 0$,
(ii) $r_0 \in (0, 1]$.

In case (i) we have

$$\int_{B_r} U^2 > 0, \quad \forall r \in (0,1]. \tag{18.146}$$

Hence, setting

$$K = \log_2(CN_{0,1}^3)$$

by (18.141) and (18.144), we get, for r, s such that $r < s \le \frac{1}{4}$

$$\int_{B_s} U^2 \le CN_{0,1}^3 \left(\frac{s}{r}\right)^K \int_{B_r} U^2 \le CC_m N_{0,1}^3 s^K r^{m-K}. \tag{18.147}$$

Let $m > K$. Passing to the limit in (18.147) as r goes to 0. We obtain

$$\int_{B_s} U^2 = 0, \quad \text{for } s \le \frac{1}{4},$$

which contradicts (18.146).

Let us consider case (ii).

If $r_0 = 1$ there is nothing to prove. Let, therefore, $r_0 \in (0, 1)$. By the definition of r_0 we have

$$\int_{B_{r_0}} U^2 = 0. \tag{18.148}$$

Let

$$\delta < \min\left\{r_0, \frac{1-r_0}{15}\right\}$$

and let $\overline{x}$ be a point such that $|\overline{x}| = r_0 - \delta$. Setting $\overline{R} = 1 - |\overline{x}|$ we have

$$r_1 := r_0 - \delta + \frac{\overline{R}}{16} = r_0 + \frac{1 - r_0 - 15\delta}{16} > r_0.$$

Now, since (18.148) trivially implies

$$\int_{B_r(\overline{x})} U^2 = O\left(r^m\right), \quad \text{as } r \to 0, \ \forall m \in \mathbb{N},$$

repeating the argument of case (i) in the ball $B_{\overline{R}}(\overline{x})$, we reach

$$\int_{B_{\overline{R}/16}(\overline{x})} U^2 = 0.$$

Finally, since this equality holds for each $\overline{x}$ such that $|\overline{x}| = r_0 - \delta$, taking into account (18.148), we have

$$\int_{B_{r_1}} U^2 = 0,$$

which, as $r_1 > r_0$, contradicts the definition of r_0 given in (18.145). ■

Remarks and Comments

1. The optimal three sphere inequality can be obtained by the doubling inequality, (18.141), in a simple and direct way. As a matter of fact, by (18.141), using the elementary properties of the logarithmic function, we have, for $2r \leq s \leq \frac{R_0}{16}$,

$$\int_{B_s(x_0)} U^2 \leq \left(C N^3_{x_0,R_0}\right)^{2\log_2 \frac{s}{r}} \int_{B_r(x_0)} U^2. \tag{18.149}$$

Now, by (18.107) we have trivially

$$N_{x_0,R_0} = \frac{\int_{B_{R_0}(x_0)} U^2}{R_0 \int_{B_{R_0/4}(x_0)} U^2} \leq \frac{\int_{B_{R_0}(x_0)} U^2}{R_0 \int_{B_s(x_0)} U^2}.$$

By the latter and by (18.149) we get

$$\left(R_0 \int_{B_s(x_0)} U^2\right)^{1+6\log \frac{s}{r}} \leq \left(C \int_{B_{R_0}(x_0)} U^2\right)^{6\log_2 \frac{s}{r}} \int_{B_r(x_0)} U^2,$$

which gives

$$\int_{B_s(x_0)} U^2 \leq \frac{C}{R_0}\left(\int_{B_{R_0}(x_0)} U^2\right)^{1-\widetilde{\theta}} \left(\int_{B_r(x_0)} U^2\right)^{\widetilde{\theta}}$$

where

$$\widetilde{\theta} = \frac{1}{1+6\log_2 \frac{s}{r}}.$$

Let us notice that also $\widetilde{\theta}$ is an optimal exponent in the sense that, for fixed s, (18.6) holds.

2. Let us observe that three sphere inequality (18.104) has been proved using Carleman estimate (18.66), while to prove the doubling inequality we have used Carleman estimate (18.67), which differs from the estimate (18.66) for the occurrence of the term

$$\tau^2 r \int \rho^{-1-2\tau} u^2 dx.$$

The idea of including this term is indebted to Bakri [11, 12] and this idea simplifies the proof of the doubling inequality with respect to the previous proofs based on the Carleman estimates. It should also be pointed out that in the literature there are other methods to prove the doubling inequality, see for instance [49, 70]. We will briefly discuss the main ideas underlying such methods in Sect. 19.3.1. ♦

18.6 The Geodesic Polar Coordinates

In this section we give the definitions and the main properties of geodesic polar coordinates introduced by Aronszajn et al. in [9].

Let n be an integer number, $n \geq 2$. For any $r > 0$ we denote by $\widetilde{B}_r$ the set $B_r \setminus \{0\}$.

For any $A = \{a_{ij}\}_{i,j=1}^n$ real matrix, we denote by $|A|$ the norm of A, i.e.

$$|A|^2 = \sum_{i,j=1}^n a_{ij}^2.$$

In what follows we will use the Einstein convention of the repeated indices.

By I_n we denote the $n \times n$ identity matrix. Given two vectors $x, y \in \mathbb{R}^n$, $x = \left(x^1, \cdots, x^n\right)$, $y = \left(y^1, \cdots, y^n\right)$ we denote by

$$x \cdot y = \delta^{ij} x^i y^j = x^i y^i,$$

their Euclidean scalar product and by $|x| = \sqrt{x \cdot x}$ the Euclidean norm.

Let λ, $\lambda \geq 1$, Λ be positive numbers. Let $G(x) = \left\{g_{ij}(x)\right\}_{i,j=1}^n$ be a nonsingular symmetric real matrix whose entries are functions that belong to $C^\infty\left(\overline{B_2}\right)$. Let us denote by $G^{-1}(x) = \left\{g^{ij}(x)\right\}_{i,j=1}^n$ the inverse of $G(x)$. Let us suppose that

$$\lambda^{-1} |\xi|^2 \leq G(x)\xi \cdot \xi \leq \lambda |\xi|^2, \quad \forall \xi \in \mathbb{R}^n, \forall x \in B_2, \tag{18.150}$$

$$\left|\partial_{x^k} G\right| \leq \Lambda, \quad \text{for } k \in \{1, \cdots n\}, \quad \text{in } B_2. \tag{18.151}$$

Set

$$r(x) = |x|.$$

Let us denote by

$$\mu_0(x) = G^{-1}(x)\nabla r(x) \cdot \nabla r(x), \text{ for } x \in \widetilde{B}_2 \tag{18.152}$$

and

$$\widetilde{G}(x) = \mu_0(x)G(x), \text{ for } x \in \widetilde{B}_2. \tag{18.153}$$

Let us denote by $\widetilde{g}_{ij}(x)$, $i, j \in \{1, \cdots, n\}$ the entries of the matrix $\widetilde{G}(x)$.

We wish to introduce the geodesic polar coordinates with respect to the metric tensor

$$\widetilde{g}_{ij}(x) dx^i \otimes dx^j.$$

Then we will express in such geodesic polar coordinates the Laplace–Beltrami operator

$$\Delta_g(\cdot) = \frac{1}{\sqrt{g(x)}} \partial_{x^i}\left(\sqrt{g(x)} g^{ij}(x) \partial_{x^j} \cdot\right), \text{ in } B_2, \tag{18.154}$$

where

$$g(x) = \det G(x).$$

To perform the above mentioned transformation we assume also

$$G(0) = I_n, \tag{18.155}$$

For any $\overline{x} \in \overline{B}_1 \setminus \{0\}$ let us denote by $\Gamma(\sigma; \overline{x})$ the global solution of the following Cauchy problem

$$\begin{cases} \dot{\Gamma}(\sigma; \overline{x}) = \widetilde{G}^{-1}(\Gamma(\sigma; \overline{x}))\nabla r(\Gamma(\sigma; \overline{x})) \\ \\ \Gamma(r(\overline{x}), \overline{x}) = \overline{x} \end{cases}, \tag{18.156}$$

where $\dot{\Gamma}$ is the derivative of $\Gamma(\sigma; \overline{x})$ w.r.t. σ.

Remark 18.8 Let us observe that for any $\overline{x} \in \overline{B_1} \setminus \{0\}$, Γ is a geodesic line w.r.t. the Riemannian metric $\widetilde{g}_{ij}(x)\, dx^i \otimes dx^j$. Indeed, by (18.152) and (18.153) we have trivially that function $r(\cdot)$ is a solution to eikonal equation

$$\widetilde{g}^{ij}(x)\, \partial_{x^i} r \partial_{x^j} r = 1.$$

Hence, denoting by $p(\sigma) = \nabla r(\Gamma(\sigma; \overline{x}))$ we have that, see Sect. 7.6, (Γ, p) is a solution to Hamilton-Jacobi equations

$$\begin{cases} \dot{\Gamma} = \nabla_p H(\Gamma, p) \\ \\ \dot{p} = -\nabla_x H(\Gamma, p) \end{cases},$$

where

$$H(x, p) = \frac{1}{2}\widetilde{g}^{ij}(x)\, p_i p_j$$

is the Hamiltonian. Therefore Γ solves the Euler equation, see Sect. 7.7,

$$\frac{d}{d\sigma}\nabla_q L\left(\Gamma, \dot{\Gamma}\right) = L_x\left(\Gamma, \dot{\Gamma}\right),$$

where

$$L(x, q) = \frac{1}{2}\widetilde{g}_{ij}(x)\, q^i q^j.$$

Hence $\Gamma(\cdot, \overline{x})$ is a geodesic line w.r.t. the metric

$$\widetilde{g}_{ij}(x)\, dx^i \otimes dx^j.$$

♦

The following Proposition holds true.

Proposition 18.12 *Let $\Gamma(\cdot;\overline{x})$ be the global solution to Cauchy problem* (18.156). *Then*

$$\Gamma\left(\cdot;\overline{x}\right) \text{ is defined in the interval } (0,2), \tag{18.157}$$

and

$$r\left(\Gamma\left(\sigma;\overline{x}\right)\right)=\sigma, \text{ for every } \sigma\in(0,2)\,. \tag{18.158}$$

Proof The proof is the same of that given in the Remark after Theorem 7.13, we repeat with different notation for the reader's convenience. Let us denote by J the interval on which $\Gamma\left(\cdot;\overline{x}\right)$ is defined.

Claim We have

$$r\left(\Gamma\left(\sigma;\overline{x}\right)\right)=\sigma,\quad \forall\sigma\in J. \tag{18.159}$$

Proof of the Claim To prove (18.159) we note that Eq. (18.156) gives (we omit $\overline{x}$ in Γ)

$$\frac{d}{d\sigma}r\left(\Gamma(\sigma)\right)=\frac{d\Gamma^{i}(\sigma)}{d\sigma}\partial_{x^{i}}r(\Gamma(\sigma))=\widetilde{g}^{ij}(\Gamma(\sigma))\partial_{x^{i}}r(\Gamma(\sigma))r\partial_{x^{j}}r(\Gamma(\sigma))=1,$$

for every $\sigma\in J$. Therefore, there exists a constant c, such that

$$r\left(\Gamma\left(\sigma\right)\right)=\sigma+c,\quad \forall\sigma\in J.$$

By initial condition we have

$$\Gamma(r(\overline{x}),\overline{x})=\overline{x},$$

hence

$$r\left(\overline{x}\right)=r(\Gamma(r(\overline{x}),\overline{x}))=r\left(\overline{x}\right)+c,$$

consequently $c=0$ which implies (18.159). **This concludes the proof of the Claim.**

By (18.159) and by standard results of general theory of ordinary differential equations we have that Γ can be defined in the whole interval $(0,2)$ hence (18.157) is proved. ■

In order to introduce the geodesic polar coordinates we need some additional notations. Set $\Sigma=\partial B_1$. Let $\{U_\alpha,\varphi_\alpha\}_{\alpha\in\mathcal{J}}$ be a finite family of local maps which define an oriented C^∞ differentiable structure on Σ. For any $\alpha \in \mathcal{J}$, set $V_\alpha = \varphi_\alpha(U_\alpha)$. Let us denote by Φ the map

$$\Phi : B_1 \setminus \{0\} \to (0,1) \times \Sigma, \tag{18.160}$$

such that

$$\Phi(x) = (|x|, \Gamma(1;x)), \quad \forall x \in B_1 \setminus \{0\}. \tag{18.161}$$

By (18.156) we have easily that Φ is bijective, moreover

$$\Phi^{-1}(\varrho, p) = \Gamma(\varrho; p), \quad \forall (\varrho, p) \in (0,1) \times \Sigma. \tag{18.162}$$

For any $\alpha \in \mathcal{J}$, let us consider the following geodesic sector

$$\mathcal{I}(U_\alpha) = \{\Gamma(\varrho; p) : \varrho \in (0,1), p \in U_\alpha\}, \tag{18.163}$$

let us denote by Φ_α the map Φ expressed in the local coordinates, i.e.

$$\Phi_\alpha : \mathcal{I}(U_\alpha) \to (0,1) \times V_\alpha, \tag{18.164}$$

$$\Phi_\alpha(x) = (|x|, \varphi_\alpha(\Gamma(1;x))), \quad \forall x \in \mathcal{I}(U_\alpha). \tag{18.165}$$

For any $\alpha \in \mathcal{J}$ let us denote by

$$\Gamma_\alpha(\varrho, \theta) = \Gamma\left(\varrho, \varphi_\alpha^{-1}(\theta)\right), \quad \forall (\varrho, \theta) \in (0,1) \times V_\alpha. \tag{18.166}$$

We have

$$\Phi_\alpha^{-1}(\varrho, \theta) = \Gamma_\alpha(\varrho, \theta), \quad \forall (\varrho, \theta) \in (0,1) \times V_\alpha. \tag{18.167}$$

Let us note that, for any $\alpha, \alpha' \in \mathcal{J}$, we have

$$(\Phi_\alpha \circ \Phi_{\alpha'}^{-1})(\varrho, \theta) = \left(\varrho, \left(\varphi_\alpha \circ \varphi_{\alpha'}^{-1}\right)(\theta)\right), \quad \forall (\varrho, \theta) \in \Phi_\alpha(\mathcal{I}(U_\alpha \cap U_{\alpha'})).$$

Let us note that $\mathcal{I}(U_\alpha) \cap \mathcal{I}(U_{\alpha'}) = \mathcal{I}(U_\alpha \cap U_{\alpha'})$, hence $\{\mathcal{I}(U_\alpha), \Phi_\alpha\}_{\alpha \in \mathcal{J}}$ defines an oriented C^∞ differentiable structure on $B_1 \setminus \{0\}$.

Let us observe that by (18.156) and (18.166) we have

$$\begin{aligned} \partial_\varrho \Gamma_\alpha(\varrho, \theta) &= \widetilde{G}^{-1}(\Gamma_\alpha(\varrho, \theta)) \nabla r(\Gamma_\alpha(\varrho, \theta)) = \\ &= \frac{1}{\varrho} \widetilde{G}^{-1}(\Gamma_\alpha(\varrho, \theta)) \Gamma_\alpha(\varrho, \theta) \end{aligned} \tag{18.168}$$

and

$$\Gamma_\alpha(1,\theta) = \varphi_\alpha^{-1}(\theta). \tag{18.169}$$

Moreover, by (18.158) we have

$$|\Gamma_\alpha(\varrho,\theta)| = \varrho. \tag{18.170}$$

To save the sum index convention, in the next Proposition and in the sequel, we denote by θ^{j+1} the j-th component of θ, so $\theta = \left(\theta^2, \cdots, \theta^n\right)$.

Let $\alpha \in \mathcal{J}$ be fixed. Let $\eta = \left\{\eta^h\right\}_{h=2}^n$ be a vector. Set

$$y_\eta(\varrho,\theta) = \frac{1}{\varrho}\partial_\theta \Gamma_\alpha(\varrho,\theta)\eta, \tag{18.171}$$

where $\partial_\theta(\cdot)$ denotes the jacobian matrix w.r.t. the variables $\theta_2, \cdots, \theta_n$. Let us check that by (18.168) and (18.169) we obtain, respectively,

$$\partial_\varrho y_\eta = \frac{1}{\varrho}\left(\widetilde{G}^{-1}(\Gamma_\alpha) - I_n\right) y_\eta + \frac{1}{\varrho}(\partial_{x^k}\widetilde{G}^{-1}(\Gamma_\alpha)) y_\eta^k \Gamma_\alpha, \tag{18.172}$$

and

$$y_\eta(1,\theta) = \partial_\theta \varphi_\alpha^{-1}(\theta)\eta. \tag{18.173}$$

Equality (18.173) is an immediate consequence of (18.169). Concerning (18.172), first we set

$$\widetilde{y}_h = \frac{1}{\varrho}\partial_{\theta^h}\Gamma_\alpha(\varrho,\theta), \quad h = 2, \cdots, n,$$

and we have, for $h = 2, \cdots, n$,

$$\begin{aligned}
\partial_\varrho \widetilde{y}_h &= \partial_\varrho \left(\frac{1}{\varrho} \partial_{\theta^h} \Gamma_\alpha(\varrho, \theta) \right) = \\
&= -\frac{1}{\varrho^2} \partial_{\theta^h} \Gamma_\alpha(\varrho, \theta) + \frac{1}{\varrho} \partial^2_{\varrho\theta^h} \Gamma_\alpha(\varrho, \theta) = \\
&= -\frac{1}{\varrho} \widetilde{y}_h + \frac{1}{\varrho} \partial_{\theta^h} \left(\partial_\varrho \Gamma_\alpha(\varrho, \theta) \right) = \\
&= -\frac{1}{\varrho} \widetilde{y}_h + \frac{1}{\varrho^2} \partial_{\theta^h} \left(\widetilde{G}^{-1}(\Gamma_\alpha(\varrho, \theta)) \Gamma_\alpha(\varrho, \theta) \right) = \\
&= -\frac{1}{\varrho} \widetilde{y}_h + \frac{1}{\varrho^2} \partial_{\theta^h} \left(\widetilde{G}^{-1}(\Gamma_\alpha(\varrho, \theta)) \right) \Gamma_\alpha(\varrho, \theta) + \\
&+ \frac{1}{\varrho^2} \widetilde{G}^{-1}(\Gamma_\alpha(\varrho, \theta)) \partial_{\theta^h} \Gamma_\alpha(\varrho, \theta) = \\
&= -\frac{1}{\varrho} \widetilde{y}_h + \frac{1}{\varrho^2} \left(\partial_{x^k} \widetilde{G}^{-1}(\Gamma_\alpha(\varrho, \theta)) \partial_{\theta^h} \Gamma^k_\alpha(\varrho, \theta) \right) \Gamma_\alpha(\varrho, \theta) + \\
&+ \frac{1}{\varrho} \widetilde{G}^{-1}(\Gamma_\alpha(\varrho, \theta)) \widetilde{y}_h .
\end{aligned}$$

Hence, for $h = 2, \cdots, n$, we have

$$\begin{aligned}
\partial_\varrho \widetilde{y}_h = &\frac{1}{\varrho} \left(\widetilde{G}^{-1}(\Gamma_\alpha(\varrho, \theta)) - I_n \right) \widetilde{y}_h + \\
&+ \frac{1}{\varrho^2} \left(\partial_{x^k} \widetilde{G}^{-1}(\Gamma_\alpha(\varrho, \theta)) \partial_{\theta^h} \Gamma^k_\alpha(\varrho, \theta) \right) \Gamma_\alpha(\varrho, \theta).
\end{aligned}$$

By multiplying both the sides of the last equality by η^h and adding up the index h, we obtain (18.172). ■

Lemma 18.13 *Let us assume that* (18.150), (18.151), *and* (18.155) *hold true. Let* Γ_α, $\alpha \in \mathcal{J}$ *be defined by* (18.166). *We have*

$$C^{-1} \left| \partial_\theta \varphi_\alpha^{-1}(\theta) \eta \right| \leq \left| y_\eta(\varrho, \theta) \right| \leq C \left| \partial_\theta \varphi_\alpha^{-1}(\theta) \eta \right|, \tag{18.174}$$

for every $(\varrho, \theta) \in (0, 1) \times V_\alpha$ *and for every* $\eta \in \mathbb{R}^{n-1}$, *where* C *and* $C \geq 1$, *depend on* λ *and* Λ *only (here and in the sequel we omit the dependence on* n*).*

Proof Let us omit the index α. By (18.150), (18.153), (18.155), and (18.170) we have

$$\left| \partial_{x^k} \widetilde{G}^{-1}(\Gamma) y^k_\eta \right| \leq C \left| y_\eta \right|,$$

$$\left| \widetilde{G}^{-1}(\Gamma) - I_n \right| \leq C \varrho,$$

where C depends on λ and Λ only. Therefore, by (18.172) we have

$$\left|\partial_{\varrho} y_{\eta}\right| \leq C\left|y_{\eta}\right|, \tag{18.175}$$

where C depends on λ and Λ only.

By (18.173) and (18.175) we have

$$\left|y_{\eta}(\varrho, \theta)\right| \leq \left|\partial_{\theta} \varphi^{-1}(\eta\right| + C \int_{\varrho}^{1}\left|y_{\eta}(s, \theta)\right| ds, \quad \forall \varrho \in (0, 1], \tag{18.176}$$

By (18.176) and by the Gronwall inequality we get the second inequality of (18.174).

Now we prove the first inequality of (18.174). Let $\overline{\varrho}$ be fixed in (0, 1). We have

$$\left|y_{\eta}(\varrho, \theta)\right| \leq \left|y_{\eta}(\overline{\varrho}, \theta)\right| + \int_{\overline{\varrho}}^{\varrho}\left|\partial_{s} y_{\eta}(s, \theta)\right| ds, \quad \forall \varrho \in [\overline{\varrho}, 1],$$

hence by (18.175) we get

$$\left|y_{\eta}(\varrho, \theta)\right| \leq \left|y_{\eta}(\overline{\varrho}, \theta)\right| + C \int_{\overline{\varrho}}^{\varrho}\left|y_{\eta}(s, \theta)\right| ds, \quad \forall \varrho \in [\overline{\varrho}, 1].$$

By the Gronwall inequality we have

$$\left|y_{\eta}(\varrho, \theta)\right| \leq \left|y_{\eta}(\overline{\varrho}, \theta)\right| e^{C}, \quad \forall \varrho \in [\overline{\varrho}, 1], \tag{18.177}$$

where C depends on λ and Λ only. By (18.173) and (18.177) we have

$$\left|\partial_{\theta} \varphi^{-1}(\theta) \eta\right| e^{-C} \leq \left|y_{\eta}(\overline{\varrho}, \theta)\right|,$$

that gives the first inequality of (18.174). ■

The following Proposition holds true.

Proposition 18.14 *For any $\alpha \in \mathcal{J}$ let us denote by $\widetilde{b}_{\alpha,hk}$, $h, k \in \{1, \cdots, n\}$, the components of the metric tensor $\widetilde{g}_{ij}(x) dx^{i} \otimes dx^{j}$ with respect to the local coordinates $(\mathcal{I}(U_{\alpha}), \Phi_{\alpha})$. We have*

$$\widetilde{b}_{\alpha,hk}(\varrho, \theta) = \widetilde{g}_{ij}\left(\Gamma_{\alpha}(\varrho, \theta)\right) \partial_{\theta^{h}} \Gamma_{\alpha}^{i} \partial_{\theta^{k}} \Gamma_{\alpha}^{j}, \quad \textit{for } h, k \in \{2, \cdots, n\}, \tag{18.178}$$

$$\widetilde{b}_{\alpha,h1}(\varrho, \theta) = \widetilde{b}_{\alpha,1h}(\varrho, \theta) = 0, \quad \textit{for } h \in \{2, \cdots, n\} \tag{18.179}$$

and

$$\widetilde{b}_{\alpha,11}(\varrho, \theta) = 1. \tag{18.180}$$

Proof To simplify the notation, in what follows we omit the index α. Equality (18.178) are nothing but the rule of transformation of the components of the metric tensor.

Let us prove (18.179).

By (18.168) and (18.170) we get, for every $h \in \{2, \cdots n\}$,

$$\widetilde{b}_{h1} = \widetilde{g}_{ij}(\Gamma)\partial_{\theta^h}\Gamma^i\partial_\varrho\Gamma^j = \widetilde{g}_{ij}(\Gamma)\partial_{\theta^h}\Gamma^i\widetilde{g}^{jk}(\Gamma)\partial_{x^k}r(\Gamma) =$$
$$= \delta_i^k\partial_{\theta^h}\Gamma^i\partial_{x^k}r(\Gamma) = \partial_{\theta^k}(r(\Gamma)) = \partial_{\theta^k}\varrho = 0,$$

since $\widetilde{b}_{ij}$ is symmetric, we obtain (18.179).

Now, let us prove (18.180). We have

$$\widetilde{b}_{11} = \widetilde{g}_{ij}(\Gamma)\partial_\varrho\Gamma^i\partial_\varrho\Gamma^j = \widetilde{g}_{ij}(\Gamma)\widetilde{g}^{ik}(\Gamma)\partial_{x^k}r(\Gamma)\widetilde{g}^{jh}(\Gamma)\partial_{x^h}r(\Gamma) =$$
$$= \delta_j^k\widetilde{g}^{jh}(\Gamma)\partial_{x^k}r(\Gamma)\partial_{x^h}r(\Gamma) = \widetilde{g}^{kh}(\Gamma)\partial_{x^k}r(\Gamma)\partial_{x^h}r(\Gamma) = 1.$$

■

In formulas (18.181)–(18.186) below, we introduce some notations.

Set, for any $\alpha \in \mathcal{J}$,

$$\mu_\alpha = \mu_0 \circ \Phi_\alpha^{-1}, \tag{18.181}$$

$$\widetilde{b}_\alpha = \det\left\{\widetilde{b}_{\alpha,ij}\right\}_{i,j=1}^n, \tag{18.182}$$

$$\beta_{\alpha,hk} = \frac{1}{\varrho^2}\widetilde{b}_{\alpha,hk}, \text{ for } h,k \in \{2,\cdots,n\}, \tag{18.183}$$

$$\left\{\beta_\alpha^{hk}\right\}_{h,k=2}^n = \left(\left\{\beta_{\alpha,hk}\right\}_{h,k=2}^n\right)^{-1}, \tag{18.184}$$

$$\beta_\alpha = \det\left\{\beta_{\alpha,hk}\right\}_{h,k=2}^n. \tag{18.185}$$

In addition, set

$$\beta_\alpha = \varrho^{-2(n-1)}\widetilde{b}_\alpha. \tag{18.186}$$

The following Proposition holds true.

Proposition 18.15 *For every $\overline{\varrho} \in (0,1)$, $\beta_{\alpha,hk}(\overline{\varrho},\theta)$, $h,k \in \{2,\cdots,n\}$, are the components of a metric tensor on Σ with respect to the local maps $(U_\alpha,\varphi_\alpha)$.*

Proof Let $\overline{p} \in \Sigma$. Let $(U_\alpha,\varphi_\alpha)$, $(U_{\alpha'},\varphi_{\alpha'})$ be two coordinate neighborhoods such that $\overline{p} \in U_\alpha \cap U_{\alpha'}$. Let p be an arbitrary point of $U_\alpha \cap U_{\alpha'}$. Set

$$\theta^{(p)} = \varphi_\alpha(p), \qquad \widehat{\theta}^{(p)} = \varphi_{\alpha'}(p).$$

We have trivially,

$$p = \varphi_{\alpha}^{-1}\left(\theta^{(p)}\right) = \varphi_{\alpha'}^{-1}\left(\widehat{\theta}^{(p)}\right).$$

Moreover, since $\Gamma_{\alpha}\left(\cdot, \theta^{(p)}\right)$ and $\Gamma_{\alpha'}\left(\cdot, \widehat{\theta}^{(p)}\right)$ are solutions to Eq. (18.168) and

$$\Gamma_{\alpha}\left(1, \theta^{(p)}\right) = \Gamma_{\alpha'}\left(1, \widehat{\theta}^{(p)}\right) = p,$$

we have

$$\Gamma_{\alpha}\left(\cdot, \theta^{(p)}\right) = \Gamma_{\alpha'}\left(\cdot, \widehat{\theta}^{(p)}\right).$$

Therefore

$$\Gamma_{\alpha'}\left(\overline{\varrho}, \widehat{\theta}^{(p)}\right) = \Gamma_{\alpha}\left(\overline{\varrho}, \left(\varphi_{\alpha} \circ \varphi_{\alpha'}^{-1}\right)\left(\widehat{\theta}^{(p)}\right)\right).$$

Hence, if

$$\theta \in \varphi_{\alpha'}\left(U_{\alpha} \cap U_{\alpha'}\right),$$

then

$$\Gamma_{\alpha'}\left(\overline{\varrho}, \theta\right) = \Gamma_{\alpha}\left(\overline{\varrho}, \left(\varphi_{\alpha} \circ \varphi_{\alpha'}^{-1}\right)(\theta)\right).$$

Differentiating w.r.t. θ^{l} both the sides of the last equality, we obtain

$$\partial_{\theta^{l}}\Gamma_{\alpha'}\left(\overline{\varrho}, \theta\right) = \left(\partial_{\theta^{k}}\Gamma_{\alpha}\right)\left(\overline{\varrho}, \left(\varphi_{\alpha} \circ \varphi_{\alpha'}^{-1}\right)(\theta)\right)\partial_{\theta^{l}}\left(\varphi_{\alpha} \circ \varphi_{\alpha'}^{-1}\right)^{k}(\theta), \tag{18.187}$$

for every $l \in \{2, \cdots, n\}, \theta \in \varphi_{\alpha'}\left(U_{\alpha} \cap U_{\alpha'}\right)$.

Now

$$\beta_{\alpha', lm}\left(\overline{\varrho}, \theta\right) = \overline{\varrho}^{-2}\widetilde{g}_{ij}\left(\Gamma_{\alpha'}\left(\overline{\varrho}, \theta\right)\right)\partial_{\theta^{l}}\Gamma_{\alpha'}^{i}\partial_{\theta^{m}}\Gamma_{\alpha'}^{j}, \tag{18.188}$$

for $l, m \in \{2, \cdots, n\}, \theta \in \varphi_{\alpha'}\left(U_{\alpha} \cap U_{\alpha'}\right)$.

Therefore by (18.187) and (18.188) we obtain

$$\begin{aligned}
&\beta_{\alpha', lm}\left(\overline{\varrho}, \theta\right) = \\
&= \beta_{\alpha', ks}\left(\overline{\varrho}, \left(\varphi_{\alpha} \circ \varphi_{\alpha'}^{-1}\right)(\theta)\right)\partial_{\theta^{l}}\left(\varphi_{\alpha} \circ \varphi_{\alpha'}^{-1}\right)^{k}(\theta)\partial_{\theta^{m}}\left(\varphi_{\alpha} \circ \varphi_{\alpha'}^{-1}\right)^{s}(\theta),
\end{aligned} \tag{18.189}$$

for every $l, m \in \{2, \cdots, n\}$ and for every $\theta \in \varphi_{\alpha'}\left(U_{\alpha} \cap U_{\alpha'}\right)$.

Equality (18.189) proves that $\beta_{\alpha,hk}(\overline{\varrho},\theta)$, $h,k\in\{2,\cdots,n\}$, are the components of a tensor which is a metric tensor because the matrix $\left\{\beta_{\alpha,hk}(\overline{\varrho},\theta)\right\}_{h,k=2}^{n}$ is symmetric and positive. The proof is complete. ■

Now we begin to derive the expression of operator (18.154) in the polar coordinates introduced above.

Let $u\in C^{\infty}(B_2)$ and let us denote by w the function

$$(0,1)\times\Sigma\ni(\varrho,p)\to w(\varrho,p)=u(\Gamma(\varrho,p)). \tag{18.190}$$

Set

$$w_{\alpha}=u\circ\Phi_{\alpha}^{-1}, \tag{18.191}$$

by (18.167) we have

$$w\left(\varrho,\varphi_{\alpha}^{-1}(\theta)\right)=w_{\alpha}(\varrho,\theta),\quad \forall(\varrho,\theta)\in(0,1)\times V_{\alpha}, \tag{18.192}$$

hence, for any fixed $\varrho\in(0,1)$, $w_{\alpha}(\varrho,\cdot)$ is the expression of $w(\varrho,\cdot)$ in the local coordinates $(U_{\alpha},\varphi_{\alpha})$.

Now, for any fixed $\overline{\varrho}\in(0,1)$, Proposition 18.15 allows us to define on Σ the Riemannian structure, [18], induced by metric tensor whose components with respect coordinate neighborhood $(U_{\alpha},\varphi_{\alpha})$ are equals to $\beta_{\alpha,hk}(\overline{\varrho},\theta)$ for $h,k\in\{2,\cdots,n\}$.

Let us denote by $\langle\cdot,\cdot\rangle$ and $|.|_{\Sigma}$, respectively, the inner product and the associated norm on the above defined Riemannian structure. Let us denote by ∇_{Σ} and div_{Σ} the gradient and the divergence operators on Σ respectively.

We have

$$(\nabla_{\Sigma}w(\overline{\varrho},\cdot))\circ\varphi_{\alpha}^{-1}(\theta)=\left\{\beta_{\alpha}^{hk}(\overline{\varrho},\theta)\,\partial_{\theta^{k}}w_{\alpha}(\overline{\varrho},\theta)\right\}_{h=2}^{n},\quad \forall\theta\in V_{\alpha}. \tag{18.193}$$

Set

$$\mu(\varrho,p)=\mu_0(\Gamma(\varrho,p)),\quad \forall(\varrho,\theta)\in(0,1)\times\Sigma,$$

we have

$$\begin{aligned}&\mathrm{div}_{\Sigma}\left(\mu^{1-\frac{n}{2}}(\overline{\varrho},\cdot,)\,\nabla_{\Sigma}w(\overline{\varrho},\cdot)\right)\circ\varphi_{\alpha}^{-1}(\theta)=\\&=\frac{1}{\sqrt{\beta_{\alpha}(\overline{\varrho},\theta)}}\partial_{\theta^{h}}\left(\mu_{\alpha}^{1-\frac{n}{2}}(\overline{\varrho},\theta)\sqrt{\beta_{\alpha}(\overline{\varrho},\theta)}\beta_{\alpha}^{hk}(\overline{\varrho},\theta)\,\partial_{\theta^{k}}w_{\alpha}(\overline{\varrho},\theta)\right),\end{aligned} \tag{18.194}$$

for every $\theta\in V_{\alpha}$.

Let us note that Proposition 18.14 implies that the derivatives

$$\partial_\varrho \log \sqrt{\beta_\alpha(\overline{\varrho}, \theta)}, \quad \text{for } \alpha \in \mathcal{J},$$

are the expression in the coordinate neighborhoods of a C^∞ function on Σ.

Let us observe that by the equality

$$\Gamma_\alpha(\varrho, \theta) = \Gamma_{\alpha'}\left(\varrho, \left(\varphi_\alpha \circ \varphi_{\alpha'}^{-1}\right)(\theta)\right) \ \ \forall \theta \in \varphi_{\alpha'}\left(U_\alpha \cap U_{\alpha'}\right),$$

we get

$$\partial_{\theta^h}\Gamma_\alpha(\varrho, \theta) = \partial_{\theta^k}\Gamma_{\alpha'}\left(\varrho, \left(\varphi_\alpha \circ \varphi_{\alpha'}^{-1}\right)(\theta)\right)\partial_{\theta^h}\left(\varphi_\alpha \circ \varphi_{\alpha'}^{-1}\right)^k(\theta).$$

Let us denote, respectively, by $\mathcal{L}_1$, $\mathcal{L}_2$ and $\mathcal{L}$ the operators

$$\mathcal{L}_1 w = \mu\left(\frac{\partial^2 w}{\partial \varrho^2} + \frac{n-1}{\varrho}\frac{\partial w}{\partial \varrho} + \frac{1}{\varrho^2 \mu^{1-n/2}} div_\Sigma\left(\mu^{1-n/2}\nabla_\Sigma w\right)\right), \tag{18.195}$$

$$\mathcal{L}_2 w = \mu\frac{\partial}{\partial \varrho}\left(\log\left(\mu^{1-n/2}\sqrt{\beta}\right)\right)\partial_\varrho w \tag{18.196}$$

and

$$\mathcal{L} = \mathcal{L}_1 + \mathcal{L}_2. \tag{18.197}$$

We have the following

Proposition 18.16 (Transformation of the Operator Δ_g) *The following equality holds true*

$$\left(\left(\Delta_g u\right) \circ \Phi_\alpha^{-1}\right)(\varrho, \theta) = (\mathcal{L}w)\left(\varrho, \varphi_\alpha^{-1}(\theta)\right), \quad \forall(\varrho, \theta) \in (0, 1) \times V_\alpha. \tag{18.198}$$

Proof Let us denote by $\left\{b_\alpha^{ij}\right\}_{i,j=1}^n$ the inverse matrix of $\left\{\mu_\alpha^{-1}\widetilde{b}_{\alpha,ij}\right\}_{i,j=1}^n$ (recall that $\left\{\widetilde{b}_{\alpha,ij}\right\}_{i,j=1}^n$ is defined in Proposition (18.14)). Let us recall that $w_\alpha = u \circ \Phi_\alpha^{-1}$. By (18.153) and (18.179) we have

$$\begin{aligned}&\left(\Delta_g u\right)\left(\Phi_\alpha^{-1}(\varrho, \theta)\right) = \\ &= \frac{1}{\sqrt{b_\alpha}}\left(\partial_\varrho\left(\sqrt{b_\alpha}\mu_\alpha \partial_\varrho w_\alpha\right) + \partial_{\theta^h}\left(\sqrt{b_\alpha}b_\alpha^{hk}\partial_{\theta^k} w_\alpha\right)\right),\end{aligned} \tag{18.199}$$

where $b_\alpha = \det\left\{\mu_\alpha^{-1}\widetilde{b}_{\alpha,ij}\right\}_{i,j=1}^n$.

We have

$$b_\alpha = \mu_\alpha^{-n} \varrho^{2(n-1)} \beta_\alpha, \tag{18.200}$$

$$b_\alpha^{hk} = \varrho^{-2} \mu_\alpha \beta_\alpha^{hk}, \quad \text{for } h, k \in \{2, \dots, n\}. \tag{18.201}$$

By (18.194), (18.199), (18.200), and (18.201) we get

$$\begin{aligned}
&(\Delta_g u) \circ \Phi_\alpha^{-1} = \\
&= \left(\mu \left(\partial_\varrho^2 w + \frac{n-1}{\varrho} \partial_\varrho w + \frac{1}{\varrho^2 \mu^{1-n/2}} \operatorname{div}_\Sigma \mu^{1-n/2} \nabla_\Sigma w\right)\right) \circ \varphi_\alpha^{-1} \\
&+ \left(\mu \partial_\varrho \left(\log \mu^{1-n/2} \sqrt{\beta}\right) \partial_\varrho w\right) \circ \varphi_\alpha^{-1}.
\end{aligned}$$

■

In the next propositions we will estimate the tensors which occur in the transformed operator $\mathcal{L}$.

Proposition 18.17 *Let μ_α, $\{\beta_{\alpha,hk}\}_{h,k=2}^n$, β_α be defined by (18.181), (18.183), (18.186). For any $\alpha \in \Gamma$ and $(\varrho, \theta) \in (0,1) \times V_\alpha$, we have*

$$\lambda^{-1} \leq \mu_\alpha(\varrho, \theta) \leq \lambda, \tag{18.202}$$

$$\left|\partial_\varrho \beta_{\alpha,hk} \eta^h \eta^k\right| \leq C \beta_{\alpha,hk}(\varrho, \theta) \eta^h \eta^k, \quad \textit{for } \eta \in \mathbb{R}^{n-1}, \tag{18.203}$$

$$\left|\partial_\varrho \log \sqrt{\beta_\alpha}\right| \leq C, \tag{18.204}$$

$$\left|\partial_\varrho \mu_\alpha(\varrho, \theta)\right| \leq C, \tag{18.205}$$

where C depends on λ and Λ only.

Proof Let us omit the index α. By (18.151) and (18.152) we have

$$\lambda^{-1} \leq \mu_0(x) \leq \lambda,$$

by these inequalities and by (18.181) we obtain (18.202).

Now, we prove (18.203). For any vector $\{\eta^h\}_{h=2}^n$, let y_η be defined by (18.171). By (18.171), (18.178), and (18.183) we have

$$\beta_{hk}(\varrho, \theta)\, \eta^h \eta^k = \widetilde{G}(\Gamma(\varrho, \theta))\, y_\eta \cdot y_\eta. \tag{18.206}$$

Therefore

$$\partial_\varrho \beta_{hk}(\varrho,\theta)\eta^h\eta^k = 2\widetilde{G}(\Gamma)\partial_\varrho y_\eta \cdot y_\eta + \left(\partial_{x^k}\widetilde{G}(\Gamma)\partial_\varrho \Gamma^k y_\eta\right)\cdot y_\eta. \tag{18.207}$$

Let us recall that

$$\left|\partial_\varrho y_\eta\right| \leq C\left|y_\eta\right|.$$

By (18.150), (18.151) (18.153), (18.170), (18.207) and by the last inequality we get

$$\begin{aligned}\left|\partial_\varrho \beta_{hk}(\varrho,\theta)\eta^h\eta^k\right| &\leq C\left|y_\eta\right|^2 \leq \\ &\leq C\lambda^2\widetilde{G}(\Gamma)y_\eta\cdot y_\eta = \\ &= C\lambda^2\beta_{hk}(\varrho,\theta)\,\eta^h\eta^k,\end{aligned} \tag{18.208}$$

where C depends on λ and Λ only. Therefore (18.203) is proved.

In order to prove (18.204), recall that if $A(s)$ is a matrix-valued function of class C^1 such that $\det A(s) \neq 0$ then we have the following equality (we denote by $\mathrm{tr}(\cdot)$ the trace of the matrix in the brackets)

$$\frac{d}{ds}\log\left|\det A(s)\right| = \mathrm{tr}\left(\frac{dA(s)}{ds}A^{-1}(s)\right).$$

This equality and (18.203) give

$$\begin{aligned}\left|\partial_\varrho \log\sqrt{\beta}\right| &= \frac{1}{2}\left|(\partial_\varrho\beta_{ij})\beta^{ji}\right| = \frac{1}{2}\left|(\partial_\varrho\beta_{hk})\delta_i^h\delta_j^k\beta^{lm}\delta_l^i\delta_m^j\right| \leq \\ &\leq C\beta_{hk}\delta_i^h\delta_j^k\beta^{lm}\delta_l^i\delta_m^j = C\beta_{ij}\beta^{ji} = C(n-1),\end{aligned}$$

where C depends on λ and Λ only. Therefore (18.204) follows. ■

In order to prove Propositions 18.18 and 18.19 stated below we need a partition of unity $\{\zeta_\alpha\}_{\alpha\in\mathcal{J}}$ subordinate to the (finite) covering $\{(U_\alpha,\varphi_\alpha)\}_{\alpha\in\mathcal{J}}$. By this, we mean that for each $\alpha\in\mathcal{J}$, $\zeta_\alpha\in C^\infty(\Sigma)$, $\zeta_\alpha\geq 0$, $\mathrm{supp}\,\zeta_\alpha\subset U_\alpha$ and

$$\sum_{\alpha\in\mathcal{J}}\zeta_\alpha(p) = 1,\quad \forall p\in\Sigma.$$

Let us denote by $\widehat{\zeta_\alpha}$ the function $\zeta_\alpha\circ\varphi_\alpha^{-1}$ and set $\widetilde{\zeta}_\alpha(x) = \zeta_\alpha(\Gamma(1;x,))$. We have

$$\sum_{\alpha\in\mathcal{J}}\widetilde{\zeta}_\alpha(x) = 1,\quad \forall x\in\widetilde{B}_1.$$

Proposition 18.18 *For every $\varrho\in(0,1)$, let us denote by $d\Omega_\varrho$ the element of volume on Σ.*

Let f be a function belonging to $C^0(\overline{B_1})$. We have

$$\int_{B_1} f(x)\sqrt{\widetilde{g}(x)}dx = \int_0^{R_0} d\varrho \int_\Sigma f(\Gamma(\varrho, p))\varrho^{n-1}d\Omega_\varrho. \tag{18.209}$$

Proof For any $\sigma \in (0, 1)$ we have

$$\int_{B_1\setminus B_\sigma} f(x)\sqrt{\widetilde{g}(x)}dx = \sum_{\alpha\in\mathcal{J}} \int_{\mathcal{I}(U_\alpha)\setminus B_\sigma} f(x)\sqrt{\widetilde{g}(x)}\widetilde{\zeta}_\alpha(x)dx. \tag{18.210}$$

Now, in the integral on the right-hand side of (18.210), we perform the following change of variables:

$$x = \Phi_\alpha^{-1}(\varrho, \theta).$$

By such a change of variables and by (18.167), (18.182), (18.186) we get

$$\begin{aligned}&\int\limits_{\mathcal{I}(U_\alpha)\setminus B_\sigma} f(x)\sqrt{\widetilde{g}(x)}\,\widetilde{\zeta}_\alpha(x)\,dx = \\ &= \int_\sigma^1 \int\limits_{V_\alpha} f(\Gamma_\alpha(\varrho; \varphi_\alpha^{-1}(\theta))\varrho^{n-1}\sqrt{\beta_\alpha(\varrho,\theta)}\,\widehat{\zeta}_\alpha(\theta)d\theta d\varrho.\end{aligned} \tag{18.211}$$

By (18.210) and (18.211) we get

$$\begin{aligned}&\int_{B_1\setminus B_\sigma} f(x)\sqrt{\widetilde{g}(x)}dx = \\ &= \int_\sigma^1 d\varrho \sum_{\alpha\in\mathcal{J}} \int\limits_{V_\alpha} f\left(\Gamma\left(\varrho; \varphi_\alpha^{-1}(\theta)\right)\right)\varrho^{n-1}\sqrt{\beta_\alpha(\varrho,\theta)}\,\widehat{\zeta}_\alpha(\theta)\,d\theta = \\ &= \int_\sigma^1 d\varrho \int\limits_\Sigma f(\Gamma(\varrho, p))\,\varrho^{n-1}d\Omega_\varrho.\end{aligned}$$

Finally, by the latter we have

$$\begin{aligned}\int_{B_1} f(x)\sqrt{\widetilde{g}(x)}dx &= \lim_{\sigma\to 0}\int_{B_1\setminus B_\sigma} f(x)\sqrt{\widetilde{g}(x)}dx = \\ &= \int_0^1 d\varrho \int_\Sigma f(\Gamma(\varrho, p))\,\varrho^{n-1}d\Omega_\varrho.\end{aligned}$$

■

For any $\varrho \in (0,1)$ denotes by $\Xi^{(\varrho)}$ the covariant tensor of order 2 whose components with respect to coordinate neighborhoods $(U_\alpha, \varphi_\alpha)$ are equal to

$$\partial_\varrho \beta_\alpha^{lm}(\varrho,\cdot)\beta_{\alpha,lh}(\varrho,\cdot)\beta_{\alpha,mk}(\varrho,\cdot), \quad \text{for } h,k=2,\cdots,n.$$

Let us denote by $\ell(\varrho,\cdot)$ the $C^\infty(\Sigma)$ function whose expressions with respect to the coordinate neighborhoods $(U_\alpha, \varphi_\alpha)$ are equal to

$$\partial_\varrho \log\sqrt{\beta_\alpha(\varrho,\cdot)}.$$

In the following Proposition we will denote by $\int(\cdot)$ the integral $\int_0^{\varrho_0} d\varrho \int_\Sigma (\cdot) d\Omega_\varrho$.

Proposition 18.19 *Let $v_1, v_2 \in C^\infty((0,\varrho_0)\times\Sigma)$. Suppose either v_1 or v_2 of compact support. Let $h \in C^\infty((0,\varrho_0))$.*

Then we have

$$\int v_1 \partial_\varrho v_2 = -\int \left(v_1\ell + \partial_\varrho v_1\right) v_2. \tag{18.212}$$

If v_1 has compact support then

$$\begin{aligned}
&\int h(\varrho)\,\partial_\varrho v_1 div_\Sigma\left(\mu^{1-\frac{n}{2}}\nabla_\Sigma v_1\right) = \\
&= \frac{1}{2}\int h(\varrho)\,\mu^{1-\frac{n}{2}}\,\Xi^{(\varrho)}\left(\nabla_\Sigma v_1, \nabla_\Sigma v_1\right) + \\
&+ \frac{1}{2}\int \left(\partial_\varrho\left(h(\varrho)\mu^{1-\frac{n}{2}}\right) + h(\varrho)\mu^{1-\frac{n}{2}}\ell\right)|\nabla_\Sigma v_1|_\Sigma^2 .
\end{aligned} \tag{18.213}$$

Proof For every $\alpha \in \mathcal{J}$ we denote, respectively, by $v_{1,\alpha}(\varrho,\theta)$ and $v_{2,\alpha}(\varrho,\theta)$, the functions $v_1\left(\varrho, \varphi_\alpha^{-1}(\theta)\right)$ and $v_2\left(\varrho, \varphi_\alpha^{-1}(\theta)\right)$.

Let us prove (18.212).

We have

$$\begin{aligned}
&\int_0^{\varrho_0} d\varrho \int_\Sigma v_{1,\alpha}\partial_\varrho v_{2,\alpha} d\Omega_\varrho = \sum_{\alpha\in\mathcal{J}}\int_0^{\varrho_0} d\varrho \int_{V_\alpha} v_{1,\alpha}(\partial_\varrho v_{2,\alpha})\sqrt{\beta_\alpha}\widehat{\zeta_\alpha} d\theta = \\
&= -\sum_{\alpha\in\mathcal{J}}\int_0^{\varrho_0} d\varrho \int_{V_\alpha} \left(v_{1,\alpha}\partial_\varrho \log\sqrt{\beta_\alpha} + \partial_\varrho v_{1,\alpha}\right) v_{2,\alpha}\sqrt{\beta_\alpha}\widehat{\zeta_\alpha} d\theta = \\
&= -\int_0^{\varrho_0} d\varrho \int_\Sigma \left(v_1\ell + \partial_\varrho v_1\right) v_2 d\Omega_\varrho.
\end{aligned}$$

Let us prove (18.213). Let us suppose v_1 with compact support. By the divergence Theorem on the Riemannian manifold Σ we have

$$\begin{aligned}&\int_0^{\varrho_0} d\varrho \int_\Sigma f(\varrho)\partial_\varrho v_1 div_\Sigma \left(\mu^{1-\frac{n}{2}} \nabla_\Sigma v_1\right) d\Omega_\varrho = \\ &= - \int_0^{\varrho_0} d\varrho \int_\Sigma f(\varrho)\, \mu^{1-\frac{n}{2}} \left\langle \nabla_\Sigma v_1, \nabla_\Sigma \partial_\varrho v_1 \right\rangle d\Omega_\varrho .\end{aligned} \tag{18.214}$$

Now

$$\left\langle \nabla_\Sigma v_1, \nabla_\Sigma \partial_\varrho v_1 \right\rangle = \frac{1}{2}\partial_\varrho \left\langle \nabla_\Sigma v_1, \nabla_\Sigma v_1 \right\rangle - \frac{1}{2}\Xi^{(\varrho)} \left(\nabla_\Sigma v_1, \nabla_\Sigma v_1\right),$$

that, with (18.212) gives (18.213). ■

18.7 The Case of Variable Coefficients

In this section we will prove the Carleman estimate of Aronszajn et al. [9]. Basically we will proceed in a similar way to Sect. 18.4, however, instead of the Euclidean polar coordinates we will use the geodesic polar coordinates introduced in Sect. 18.6. Compared with the original proof of [9], the one we will prove here has some simplification.

Precisely we prove

Theorem 18.20 (Carleman Estimate for Δ_g) *Suppose that the matrix $G = \left\{g_{ij}(x)\right\}_{i,j=1}^n$ satisfies to (18.150), (18.151), (18.155) and $g_{ij} \in C^\infty(B_2)$, for $i, j = 1, \cdots, n$.*

Let $\epsilon \in (0, 1]$. We define

$$\rho(x) = \phi_\epsilon(|x|), \quad \forall x \in B_1 \setminus \{0\}, \tag{18.215}$$

where

$$\phi_\epsilon(s) = \frac{s}{(1+s^\epsilon)^{1/\epsilon}}. \tag{18.216}$$

Then there exist $r_0 \in (0, 1)$, $\overline{\tau} > 1$ and $C > 1$, which depend on ϵ, λ and Λ only, such that

$$\begin{aligned}&\tau^3 \int \rho^{\epsilon-2\tau}|u|^2 dx + \tau \int \rho^{2+\epsilon-2\tau}|\nabla u|^2 dx + \\ &\qquad + \tau^2 r \int \rho^{-1-2\tau} u^2 dx \le C \int \rho^{4-2\tau}|\Delta_g u|^2 dx,\end{aligned} \tag{18.217}$$

for every $r \in (0, r_0)$, for every $\tau \ge \overline{\tau}$ and for every $u \in C_0^\infty\left(B_{r_0} \setminus \overline{B}_{r/4}\right)$.

We start by the following simple estimation of the first-order operator $\mathcal{L}_2$ defined in (18.196).

Proposition 18.21 *The following estimate holds true*

$$|\mathcal{L}_2 w| \le C \left|\partial_\varrho w\right| \quad \forall w \in C^\infty\left((0,1)\times\Sigma\right), \tag{18.218}$$

where C depends on λ and Λ only.

Proof By (18.204) and (18.205) we have

$$\begin{aligned}
&\left|(\mathcal{L}_2 w)\left(\varrho, \varphi_\alpha^{-1}(\theta)\right)\right| = \\
&= \left|\mu_\alpha\left(\left(1-\frac{n}{2}\right)\partial_\varrho \log \mu_\alpha + \partial_\varrho \log\sqrt{\beta_\alpha}\right)\partial_\varrho w_\alpha\right| \le \\
&\le \left(\left|\mu_\alpha\left(1-\frac{n}{2}\right)\partial_\varrho \log \mu_\alpha\right| + \left|\partial_\varrho \log\sqrt{\beta_\alpha}\right|\right)\left|\partial_\varrho w_\alpha\right| \le \\
&\le C\left|\partial_\varrho w_\alpha\right|,
\end{aligned}$$

where C depends on λ and Λ only. ∎

Let u be an arbitrary function that belongs to $C_0^\infty\left(B_1 \setminus \{0\}\right)$ and let us denote by w the function (recall (18.190))

$$(0,1)\times\Sigma \ni (\varrho, p) \to w(\varrho, p) = u\left(\Gamma(\varrho, p)\right). \tag{18.219}$$

where Γ is defined in Proposition 18.12.

Now, we carry out the following change of variables

$$\varrho = e^t, \qquad \widetilde{w}(t,p) = w\left(e^t, p\right), \quad \forall (t,p) \in (-\infty, 0)\times\Sigma$$

and we adopt the following conventions: for any function $h(\varrho,\cdot)$ (or for every tensor) in which the variable ϱ occurs , we denote by $\widetilde{h}(t,\cdot)$ the function (or tensor) $h(e^t,\cdot)$. We will continue to denote by $\langle\cdot,\cdot\rangle$ and $|.|_\Sigma$, respectively, the inner product and the norm associated with it in the Riemannian structure induced by the metric tensor

$$\left\{\widetilde{\beta}_{\alpha,hk}\right\}_{h,k=2}^n.$$

We will still denote by ∇_Σ, div_Σ and $d\Omega_t$, respectively, the gradient, the divergence operators and the element of volume on Σ in the above mentioned structure. In particular, the local expression of $d\Omega_t$ is equal to $\sqrt{\beta_\alpha}d\theta$. Moreover we set

$$\mathcal{M}(\cdot) = \mathrm{div}_\Sigma\left(\widetilde{\mu}^{1-\frac{n}{2}}\nabla_\Sigma\cdot\right). \tag{18.220}$$

We have

$$e^{2t}\widetilde{\mu}^{-1}(\mathcal{L}_1 w)(e^t, p) = \mathcal{P}\widetilde{w}(t, p), \quad \forall (t, p) \in (-\infty, 0) \times \Sigma, \tag{18.221}$$

where

$$\mathcal{P}\widetilde{w} = \widetilde{w}_{tt} + (n-2)\widetilde{w}_t + \frac{1}{\widetilde{\mu}^{1-n/2}}\mathcal{M}\widetilde{w}. \tag{18.222}$$

For the reader's convenience, we reformulate Propositions 18.17 and 18.19 in the Riemannian structure induced by the metric tensor $\left\{\widetilde{\beta}_{\alpha,hk}\right\}_{h,k=2}^{n}$. Recall that $\{U_\alpha, \varphi_\alpha\}_{\alpha\in\mathcal{J}}$ is a (finite) family of coordinate neighborhoods defining on Σ a structure of C^∞ oriented differentiable manifold, where, for any $\alpha \in \mathcal{J}$, we set $V_\alpha = \varphi_\alpha(U_\alpha)$.

Proposition 18.22 *For any $\alpha \in \mathcal{J}$ and for any $(t, \theta) \in (-\infty, 0) \times V_\alpha$, we have*

$$\left|\partial_t \widetilde{\beta}_{\alpha,hk}\eta^h\eta^k\right| \le Ce^t\widetilde{\beta}_{\alpha,hk}(\varrho, \theta)\eta^h\eta^k, \quad \forall \eta \in \mathbb{R}^{n-1}, \tag{18.223}$$

$$\left|\partial_t \log\sqrt{\widetilde{\beta}_\alpha}\right| \le Ce^t, \tag{18.224}$$

$$|\partial_t\widetilde{\mu}_\alpha(t, \theta)| \le Ce^t, \tag{18.225}$$

where C depends on λ and Λ only.

By the convention introduced above, for any $t \in (-\infty, 0)$, let us denote by $\widetilde{\Xi}^{(t)}$ the covariant tensor satisfying

$$\widetilde{\Xi}^{(t)} = \Xi^{(e^t)}$$

and let us denote by

$$\widetilde{\ell}(t, \cdot) = \ell(e^t, \cdot).$$

In the next Proposition let us denote by $\int(\cdot)$ the integral $\int_{-\infty}^{t_0} dt \int_\Sigma(\cdot)d\Omega_t$.

Proposition 18.23 *Let $v_1, v_2 \in C^\infty((-\infty, t_0) \times \Sigma)$. Suppose either v_1 or v_2 of compact support. Let $h \in C^\infty((-\infty, t_0))$. Then we have*

$$\int v_1\partial_t v_2 = -\int \left(e^t v_1\widetilde{\ell} + \partial_t v_1\right) v_2. \tag{18.226}$$

If v_1 has compact support then

$$
\begin{aligned}
&\int h(t)\partial_t v_1 div_\Sigma\left(\widetilde{\mu}^{1-\frac{n}{2}}\nabla_\Sigma v_1\right) = \\
&= \frac{1}{2}\int e^t h(t)\widetilde{\mu}^{1-\frac{n}{2}}\,\widetilde{\Xi}^{(t)}\left(\nabla_\Sigma v_1, \nabla_\Sigma v_1\right) + \\
&+ \frac{1}{2}\int \left[\partial_t\left(h(t)\widetilde{\mu}^{1-\frac{n}{2}}\right) + e^t h(t)\widetilde{\mu}^{1-\frac{n}{2}}\widetilde{\ell}\right]|\nabla_\Sigma v_1|^2_\Sigma .
\end{aligned} \tag{18.227}
$$

Proposition 18.24 *We have*

$$
\left|\widetilde{\ell}(t,p)\right| \le C, \quad \forall (t,p) \in (-\infty,0)\times\Sigma, \tag{18.228}
$$

$$
\left|\widetilde{\Xi}^{(t)}(\nabla_\Sigma v, \nabla_\Sigma v)\right| \le C\,|\nabla_\Sigma v|^2_\Sigma, \quad \forall v \in C^\infty((-\infty,0)\times\Sigma), \tag{18.229}
$$

where C depends on λ and Λ only.

Proof Inequalities (18.228) and (18.229) are an immediate consequences of Proposition 18.17. ■

Proof of Theorem 18.20 For any smooth function v we write v', v'', ... instead of $\partial_t v, \partial_{tt} v, \ldots$.

By (18.71) we have, similarly to the proof of Theorem 18.20 (here and in the sequel we omit the subscript ϵ)

$$
\varphi(t) := \log(\phi(e^t)) = t - \epsilon^{-1}\log\left(1+e^{\epsilon t}\right), \quad \forall t \in (-\infty,0) \tag{18.230}
$$

and

$$
\varphi'(t) = \frac{1}{1+e^{\epsilon t}}, \quad \varphi''(t) = -\frac{\epsilon e^{\epsilon t}}{(1+e^{\epsilon t})^2}, \quad \forall t \in (-\infty,0). \tag{18.231}
$$

Let us define

$$
f(t,p) = e^{-\tau\varphi}w(t,p), \quad \forall (t,p) \in (-\infty,0)\times\Sigma, \tag{18.232}
$$

where w is defined in (18.219).

We have

$$
\mathcal{P}_\tau f := e^{-\tau\varphi}\mathcal{P}(e^{\tau\varphi}f) = \underbrace{b_0 f + b_1 f'}_{\mathcal{A}_\tau f} + \underbrace{a_0 f + f'' + \frac{1}{m}\mathcal{M}f}_{\mathcal{S}_\tau f}, \tag{18.233}
$$

where

$$
m = \widetilde{\mu}^{1-\frac{n}{2}}
$$

and

$$a_0 = \tau^2\varphi'^2 + \tau(n-2), \quad b_0 = \tau\varphi'', \quad b_1 = 2\tau\varphi' + (n-2). \tag{18.234}$$

Let us note that (18.202) gives

$$\lambda^{\frac{n}{2}-1} \leq m \leq \lambda^{1-\frac{n}{2}}. \tag{18.235}$$

Set

$$\gamma := \frac{1}{\varphi'} = 1 + e^{\epsilon t}. \tag{18.236}$$

We have

$$\int m\gamma \left|\mathcal{P}_\tau f\right|^2 \geq 2\int m\gamma \mathcal{A}_\tau f \mathcal{S}_\tau f + \int m\gamma \left|\mathcal{A}_\tau f\right|^2, \tag{18.237}$$

$$\begin{aligned} 2\int m\gamma \mathcal{A}_\tau f \mathcal{S}_\tau f &= \underbrace{2\int \gamma\left(b_0 f + b_1 f'\right)\mathcal{M}f +}_{I_1} \\ &+ \underbrace{2\int m\gamma\left(b_0 f + b_1 f'\right)\left(a_0 f + f''\right)}_{I_2}. \end{aligned} \tag{18.238}$$

We Examine I_1
We have

$$\begin{aligned} I_1 &= 2\int \left(\gamma b_0 f \mathcal{M}f + \gamma b_1 f'\mathcal{M}f\right) = \\ &= \underbrace{2\int \gamma b_0 f \mathcal{M}f}_{I_{11}} + \underbrace{2\int \gamma b_1 f'\mathcal{M}f}_{I_{12}}. \end{aligned} \tag{18.239}$$

By the divergence Theorem we obtain

$$I_{11} = 2\int \gamma b_0 f \operatorname{div}_\Sigma\left(m\nabla_\Sigma f\right) = -2\int m\gamma b_0 \left|\nabla_\Sigma f\right|^2_\Sigma. \tag{18.240}$$

By (18.227) we get

$$\begin{aligned} I_{12} = 2\int \gamma b_1 f' \mathcal{M} f = \int \left[(m\gamma b_1)' + e^t m\gamma b_1 \widetilde{\ell}\right] |\nabla_\Sigma f|^2_\Sigma + \\ + \int e^t m\gamma b_1 \widetilde{\Xi}^{(t)} (\nabla_\Sigma f, \nabla_\Sigma f) \,. \end{aligned} \tag{18.241}$$

By (18.239), (18.240), and (18.241) we have

$$\begin{aligned} I_1 = I_{11} + I_{12} = \int \left\{-2m\gamma b_0 + \left[(m\gamma b_1)' + e^t m\gamma b_1 \widetilde{\ell}\right]\right\} |\nabla_\Sigma f|^2_\Sigma + \\ + \int e^t m\gamma b_1 \widetilde{\Xi}^{(t)} (\nabla_\Sigma f, \nabla_\Sigma f) \,. \end{aligned} \tag{18.242}$$

We get (compare with (18.78))

$$\begin{aligned} & - 2m\gamma b_0 + \left[(m\gamma b_1)' + e^t m\gamma b_1 \widetilde{\ell}\right] = \\ & = 2m \left(-\gamma b_0 + \frac{1}{2} (\gamma b_1)'\right) + m'\gamma b_1 + e^t m\gamma b_1 \widetilde{\ell} \geq \\ & \geq \tau\epsilon e^{\epsilon t} + +m'\gamma b_1 + e^t m\gamma b_1 \widetilde{\ell}. \end{aligned} \tag{18.243}$$

Now, by (18.225) we have

$$\left|m'\right| \leq Ce^t, \tag{18.244}$$

where C depends on λ and Λ only. By this inequality, by (18.229) and by (18.242), taking into account (18.228), we have

$$I_1 \geq \tau \int e^{\epsilon t} \left(\epsilon - C_\star e^{(1-\epsilon)t}\right) |\nabla_\Sigma f|^2_\Sigma \,, \quad \forall \tau \geq 1, \tag{18.245}$$

where $C_\star$ depends on λ and Λ only.

Now, for every $t \geq T_1 := \frac{1}{1-\epsilon} \log \frac{\epsilon}{2C_\star}$ we have

$$\epsilon - C_\star e^{(1-\epsilon)t} \geq \frac{\epsilon}{2}$$

By this inequality and by (18.245) we have, for every $f \in C_0^\infty((-\infty, T_1) \times \Sigma)$,

$$I_1 \geq \frac{\epsilon\tau}{2} \int e^{\epsilon t} |\nabla_\Sigma f|^2_\Sigma \,, \quad \forall \tau \geq 1. \tag{18.246}$$

We Examine I_2
By (18.226) we have

$$
\begin{aligned}
I_2 &= 2\int m\gamma\left(b_0 f + b_1 f'\right)\left(a_0 f + f''\right) = \\
&= 2\int m\gamma\left(a_0 b_0 f^2 + b_0 f f'' + b_1 a_0 f' f + b_1 f' f''\right) = \\
&= 2\int m\gamma a_0 b_0 f^2 - \left(m\gamma b_0 f\right)' f' + \frac{1}{2} m\gamma b_1 a_0 \left(f^2\right)' + \frac{1}{2} m\gamma b_1 \left(f'^2\right)' - \\
&\quad - 2\int e^t m\gamma b_0 \widetilde{\ell} f f' = \\
&= 2\int m\gamma a_0 b_0 f^2 - \left(m\gamma b_0 f\right)' f' - \frac{1}{2}\left(m\gamma b_1 a_0\right)' f^2 - \frac{1}{2}\left(m\gamma b_1\right)' f'^2 - \\
&\quad - 2\int e^t m\gamma \widetilde{\ell}\left(b_0 f f' + \frac{1}{2} b_1 a_0 f^2 + \frac{1}{2} b_1 f'^2\right) \\
&== 2\int m\left[\left(\gamma a_0 b_0 + \frac{1}{2}\left(\gamma b_1 a_0\right)'\right) f^2 + \right. \\
&\qquad\qquad \left. + \left(\gamma b_0 + \frac{1}{2}\left(\gamma b_1\right)'\right) f'^2 - \left(\gamma b_0\right)' f f'\right] - \\
&\quad - 2\int e^t m\gamma \widetilde{\ell}\left(b_0 f f' - \frac{1}{2} b_1 a_0 f^2 - \frac{1}{2} b_1 f'^2\right) - \\
&\quad - 2\int m'\gamma\left(b_0 f f' + \frac{1}{2} b_1 a_0 f^2 + \frac{1}{2} b_1 f'^2\right) = \\
&= 2\int m\left[\left(\gamma a_0 b_0 + \frac{1}{2}\left(\gamma b_1 a_0\right)'\right) f^2 + \right. \\
&\quad \left. + \left(\gamma b_0 + \frac{1}{2}\left(\gamma b_1\right)'\right) f'^2 - \left(\gamma b_0\right)' f f'\right] - \\
&\quad - 2\int \gamma\left(m e^t \widetilde{\ell} + m'\right)\left(b_0 f f' + \frac{1}{2} b_1 a_0 f^2 + \frac{1}{2} b_1 f'^2\right) = I_{21} + I_{22}.
\end{aligned}
$$

Where we set

$$
\begin{aligned}
I_{21} = 2\int m &\left[\left(\gamma a_0 b_0 + \frac{1}{2}\left(\gamma b_1 a_0\right)'\right) f^2 + \right. \\
&\left. + \left(\gamma b_0 + \frac{1}{2}\left(\gamma b_1\right)'\right) f'^2 - \left(\gamma b_0\right)' f f'\right]
\end{aligned}
$$

and

$$I_{22} = -2 \int \gamma \left(me^t \widetilde{\ell} + m'\right) \left(b_0 f f' + \frac{1}{2} b_1 a_0 f^2 + \frac{1}{2} b_1 f'^2\right).$$

Now, let us estimate I_{21} and I_{22}.

We have (compare with (18.81) and (18.82)), for every $\tau > 0$

$$\gamma a_0 b_0 - \frac{1}{2} (\gamma b_1 a_0)' = \left(\tau^3 \frac{1}{1+e^{\epsilon t}} + \tau^2 \frac{n-2}{2}\right) \frac{\epsilon e^{\epsilon t}}{(1+e^{\epsilon t})^2} \geq \frac{\tau^3}{8} \epsilon e^{\epsilon t}, \tag{18.247}$$

and (compare with (18.85) and (18.86)) for every $\tau > 2(n-2)$

$$-\left(\gamma b_0 + \frac{1}{2} (\gamma b_1)'\right) = \tau \frac{\epsilon e^{\epsilon t}}{1+e^{\epsilon t}} - \frac{(n-2)\epsilon e^{\epsilon t}}{2} \geq \frac{\epsilon \tau}{4} e^{\epsilon t}. \tag{18.248}$$

Moreover

$$\left|(\gamma b_0)' f f'\right| = \tau \frac{\epsilon e^{\epsilon t}}{(1+e^{\epsilon t})^2} \left|f f'\right| \leq \epsilon e^{\epsilon t} \left(\tau^2 f^2 + f'^2\right).$$

By this inequality and by (18.247), (18.248) we have

$$I_{21} \geq \epsilon \int m e^{\epsilon t} \left(\frac{\tau^3}{8} f^2 + \frac{\tau}{4} f'^2\right), \quad \forall \tau \geq \overline{\tau}_1, \tag{18.249}$$

where $\overline{\tau}_1 = \max\{16, 2(n-2)\}$.

Now we estimate I_{22} and I_{23}.

By (18.228), (18.234), (18.235), and (18.244) we have

$$|I_{22}| \leq C_* \int e^t \left(\tau^3 f^2 + \tau f'^2\right), \quad \forall \tau \geq \overline{\tau}_1, \tag{18.250}$$

where C_* depends on λ and Λ only.

By (18.249) and (18.250) we have, for every $\tau \geq \overline{\tau}_1$

$$I_2 \geq I_{21} - |I_{22}| \geq \int e^{\epsilon t} \left(\frac{\epsilon \lambda^{\frac{n}{2}-1}}{8} - C_* e^{(1-\epsilon)t}\right) \left(\tau^3 f^2 + \tau f'^2\right). \tag{18.251}$$

Let us notice that for every

$$t \geq T_2 := \frac{1}{1-\epsilon} \log \left(\frac{\epsilon \lambda^{\frac{n}{2}-1}}{16 C_*}\right)$$

we have

$$\frac{\epsilon\lambda^{\frac{n}{2}-1}}{8}-C_*e^{(1-\epsilon)t}\geq\frac{\epsilon\lambda^{\frac{n}{2}-1}}{16},$$

this inequality and (18.251) imply that

$$I_2\geq\frac{\epsilon\lambda^{\frac{n}{2}-1}}{16}\int e^{\epsilon t}\left(\tau^3f^2+\tau f'^2\right),\tag{18.252}$$

for every $f\in C_0^\infty((-\infty,T_2)\times\Sigma)$ and for every $\tau\geq\overline{\tau}_1$. By (18.238), (18.246), and (18.252) we have,

$$\begin{aligned}2\int m\gamma\mathcal{A}_\tau fS_\tau f\geq\frac{\varepsilon\tau}{2}\int e^{\epsilon t}\left|\nabla_\Sigma f\right|_\Sigma^2+\\+\frac{\epsilon\lambda^{\frac{n}{2}-1}}{16}\int e^{\epsilon t}\left(\tau^3f^2+\tau f'^2\right),\end{aligned}\tag{18.253}$$

for every $\tau\geq\overline{\tau}_1$ and for every $f\in C_0^\infty((-\infty,T_3)\times\Sigma)$, where $T_3=\min\{T_1,T_2\}$.

Set

$$\epsilon_0=\epsilon\min\{\frac{1}{2},\frac{\lambda^{\frac{n}{2}-1}}{16}\}.$$

By (18.237) and (18.253) we get

$$\begin{aligned}&\int m\gamma\left|\mathcal{P}_\tau f\right|^2\geq\\&\geq\epsilon_0\int\left(\tau^3f^2+\tau\left(f'^2+\left|\nabla_\Sigma f\right|_\Sigma^2\right)\right)e^{\epsilon t}+\int m\gamma\left|\mathcal{A}_\tau f\right|^2,\end{aligned}\tag{18.254}$$

for every $\tau\geq\tau_1$ and for every $f\in C_0^\infty((-\infty,T_3)\times\Sigma)$.

In order to obtain the third term on the left-hand side of (18.217) we argue as in the proof of Theorem 18.8. For the sake of clarity, we repeat the most important steps.

By the trivial inequality $(a+b)^2\geq\frac{1}{2}a^2-b^2$ we have

$$\begin{aligned}\int m\gamma\left|\mathcal{A}_\tau f\right|^2&\geq\frac{\lambda^{\frac{n}{2}-1}}{2}\int\gamma\left(2\tau\varphi'+n-2\right)^2f'^2-\lambda^{\frac{n}{2}-1}\int\gamma\tau^2\varphi''^2f^2\geq\\&\geq\lambda^{\frac{n}{2}-1}\tau^2\int f'^2-\epsilon^2\lambda^{\frac{n}{2}-1}\tau^2\int e^{2\epsilon t}f^2,\quad\forall\tau>0.\end{aligned}$$

Plugging this inequality into (18.254) we have

$$\int m\gamma\, |\mathcal{P}_\tau f|^2 \geq \lambda^{\frac{n}{2}-1}\tau^2 \int f'^2 + \tau^3 \int \left(\epsilon_0 - \epsilon\lambda^{\frac{n}{2}-1}\tau^{-1}e^{\epsilon t}\right) e^{\epsilon t} f^2 + \\ + \epsilon_0 \tau \int \left(f'^2 + |\nabla_\Sigma f|^2\right) e^{\epsilon t}, \tag{18.255}$$

for every $\tau \geq \tau_1$ and for every $f \in C_0^\infty((-\infty, T_3) \times \Sigma)$.

By (18.255) we have

$$\int m\gamma\, |\mathcal{P}_\tau f|^2 \geq \\ \geq \lambda^{\frac{n}{2}-1}\tau^2 \int f'^2 + \frac{\epsilon_0 \tau^3}{2} \int e^{\epsilon t} f^2 + \epsilon_0 \tau \int \left(f'^2 + |\nabla_\Sigma f|_\Sigma^2\right) e^{\epsilon t}, \tag{18.256}$$

for every $\tau \geq \overline{\tau}_2$, where $\overline{\tau}_2 = \max\{2\epsilon\epsilon_0^{-1}\lambda^{\frac{n}{2}-1}, \overline{\tau}_1\}$ and for every $f \in C_0^\infty((-\infty, T_3) \times \Sigma)$.

Let $r \leq 4e^{T_3}$. Since $u = 0$ in $B_{r/4}$ we have $f(t, p) = 0$ for every $t \leq \log(r/4)$ and for every $p \in \Sigma$. Proceeding as in the proof of (18.100) we obtain

$$C \int |\mathcal{P}_\tau f|^2 \geq \tau^3 \int e^{\epsilon t} f^2 + \tau \int \left(f'^2 + |\nabla_\Sigma f|_\Sigma^2\right) e^{\epsilon t} + \\ + \tau^2 r \int f^2 e^{-t}, \quad \forall \tau \geq \overline{\tau}_2 \tag{18.257}$$

where C depends on ϵ, λ and Λ only.

Now we come back to the original coordinates. By (18.219), (18.221), (18.232), and (18.233) we have

$$\int_{-\infty}^0 dt \int_\Sigma |\mathcal{P}_\tau f|^2\, d\Omega_t = \\ = \int_{-\infty}^0 dt \int_\Sigma e^{-2\tau\varphi(t)} e^{4t} |\widetilde{\mu}^{-1}(\mathcal{L}_1 w)(e^t, p)|^2 d\Omega_t = \\ = \int_0^1 d\varrho \int_\Sigma e^{-2\tau\varphi(\log\varrho)} \varrho^3 |\mu^{-1}\mathcal{L}_1 w(\varrho, p)|^2 d\Omega_\varrho. \tag{18.258}$$

By Propositions 18.16 and 18.21 and by (18.258) we have

$$\int_{-\infty}^{0} dt \int_{\Sigma} |\mathcal{P}_\tau f|^2 \, d\Omega_t \leq C \int_0^1 d\varrho \int_\Sigma e^{-2\tau\varphi(\log\varrho)} \varrho^3 |\mathcal{L}w|^2 d\Omega_\varrho + \\ + C \int_0^1 d\varrho \int_\Sigma e^{-2\tau\varphi(\log\varrho)} \varrho^3 |\partial_\varrho w|^2 d\Omega_\varrho, \tag{18.259}$$

where C depends on λ and Λ only.

Moreover we have

$$\int_{-\infty}^{0} dt \int_{\Sigma} f^2 e^{-t} d\Omega_t = \int_0^1 d\varrho \int_\Sigma e^{-2\tau\varphi(\log\varrho)} \varrho^{-2} |w|^2 d\Omega_\varrho \tag{18.260}$$

and

$$\int_{-\infty}^{0} dt \int_{\Sigma} f^2 e^{\epsilon t} d\Omega_t = \int_0^1 d\varrho \int_\Sigma e^{-2\tau\varphi(\log\varrho)} \varrho^{\epsilon-1} |w|^2 d\Omega_\varrho. \tag{18.261}$$

Concerning the second integral on the right-hand side of (18.258), let $\delta \in (0, 1)$ be to choose later, we get

$$\begin{aligned}
&\int_{-\infty}^{0} dt \int_{\Sigma} e^{\epsilon t} \left(f'^2 + |\nabla_\Sigma f|_\Sigma^2 \right) d\Omega_t \geq \\
&\geq \delta \int_{-\infty}^{0} dt \int_{\Sigma} e^{\epsilon t} \left(f'^2 + |\nabla_\Sigma f|_\Sigma^2 \right) d\Omega_t \geq \\
&\geq \frac{\delta}{2} \int_{-\infty}^{0} dt \int_{\Sigma} e^{\epsilon t} e^{-2\tau\varphi(t)} \left(|w_\varrho(e^t, p)|^2 e^{2t} + \right. \\
&\left. + \left|\nabla_\Sigma w(e^t, p)\right|_\Sigma^2 - 2\tau^2 |w(e^t, p)|^2 \right) d\Omega_t = \\
&= \frac{\delta}{2} \int_0^1 d\varrho \int_\Sigma e^{-2\tau\varphi(\log\varrho)} \left(|w_\varrho(\varrho, p)|^2 + \right. \\
&\left. + \varrho^{-2} |\nabla_\Sigma w(\varrho, p)|_\Sigma^2 \right) \varrho^{\epsilon+1} d\Omega_\varrho - \\
&- \frac{\delta\tau^2}{2} \int_0^1 d\varrho \int_\Sigma e^{-2\tau\varphi(\log\varrho)} |w(\varrho, p)|^2 \varrho^{\epsilon-1} d\Omega_\varrho.
\end{aligned} \tag{18.262}$$

Now, plugging (18.259), (18.260), (18.261), and (18.262) into (18.257), we have

$$\tau^3\left(1-\frac{\delta}{2}\right)\int_0^1 d\varrho\int_\Sigma e^{-2\tau\varphi(\log\varrho)}|w|^2\varrho^{\epsilon-1}d\Omega_\varrho+$$

$$+\tau^2 r\int_0^1 d\varrho\int_\Sigma e^{-2\tau\varphi(\log\varrho)}|w|^2\varrho^{-2}d\Omega_\varrho+$$

$$+\frac{\delta}{2}\tau\int_0^1 d\varrho\int_\Sigma e^{-2\tau\varphi(\log\varrho)}\left(|w_\varrho|^2+\varrho^{-2}\left|\nabla_\Sigma w\right|_\Sigma^2\right)\varrho^{\epsilon+1}d\Omega_\varrho\leq \tag{18.263}$$

$$\leq C\int_0^1 d\varrho\int_\Sigma e^{-2\tau\varphi(\log\varrho)}\varrho^3|\mathcal{L}w|^2d\Omega_\varrho+$$

$$+C\int_0^1 d\varrho\int_\Sigma e^{-2\tau\varphi(\log\varrho)}\varrho^3|\partial_\varrho w|^2d\Omega_\varrho,$$

for every $\tau\geq\tau_2$ and for every $f\in C_0^\infty((0,r_0)\times\Sigma)$, where $r_0=e^{T_3}$. Now, let us choose $\delta=\frac{1}{2}$. It turns out that the last integral on the right-hand side of (18.263) can be absorbed by the third integral on the left-hand side. Finally, applying Proposition 18.18 and, taking into account that $2^{-1/\epsilon}|x|\leq\rho(x)\leq|x|$ and replacing τ by $(\tau-\frac{n}{2})$, we obtain inequality (18.217). ■

Corollary 18.25 *Assume that the entries of the matrix* $G=\left\{g_{ij}(x)\right\}_{i,j=1}^n$ *are of class* $C^{0,1}(B_2)$ *and that* G *satisfies* (18.150), (18.155) *and (instead of* (18.151)*) satisfies*

$$|G(x)-G(y)|\leq\Lambda\,|x-y|\,,\quad\forall x,y\in B_2, \tag{18.264}$$

then Carleman estimate (18.217) *continue to hold.*

More precisely, there exist C *and* $\widehat{\tau}_*$*, depending on* λ *and* Λ *only, such that*

$$\begin{aligned}\tau^3\int\rho^{\epsilon-2\tau}|u|^2dx+\tau\int\rho^{2+\epsilon-2\tau}|\nabla u|^2dx+&\\+\tau^2 r\int\rho^{-1-2\tau}u^2dx\leq C\int\rho^{4-2\tau}|\Delta_g u|^2dx,&\end{aligned} \tag{18.265}$$

for every $r\in(0,r_0)$*, for every* $\tau\geq\widehat{\tau}_*$ *and for every* $u\in C_0^\infty\left(B_{r_0}\setminus\overline{B}_{r/4}\right)$.

Proof Let $\psi\in C_0^\infty(\mathbb{R}^n)$ satisfy supp $\psi\subset B_1$, $\psi\geq 0$ and $\int_{\mathbb{R}^n}\psi dx=1$. Set

$$\psi_\nu(x)=\nu^n\psi(\nu x),\quad\nu\in\mathbb{N}$$

and

$$G_\nu(x)=\left(G\star\psi_\nu\right)(x)=\int_{\mathbb{R}^n}G(x-y)\psi_\nu(y)dy,\quad\nu\in\mathbb{N}.$$

We have that G_ν satisfies to (18.150), (18.151), (18.155). Moreover

$$G_\nu \in C^\infty(B_2)$$

and

$$\|G_\nu - G\|_{L^\infty(B_1)} \to 0, \quad \text{as } \nu \to \infty. \tag{18.266}$$

Let $r \in (0, r_0)$ and let u be an arbitrary function that belongs to $C_0^\infty\left(B_{r_0} \setminus \overline{B}_{r/4}\right)$. By (18.217) we have, for every $\nu \in \mathbb{N}$ and for every $\tau \geq \overline{\tau}$,

$$\begin{aligned} &\tau^3 \int \rho^{\epsilon-2\tau}|u|^2 dx + \tau \int \rho^{2+\epsilon-2\tau}|\nabla u|^2 dx + \\ &+ \tau^2 r \int \rho^{-1-2\tau} u^2 dx \leq C \int \rho^{4-2\tau}\left|\Delta_{g_\nu} u\right|^2 dx. \end{aligned} \tag{18.267}$$

On the other hand, by (18.151) we have (by using the convention of repeated index)

$$\int \rho^{4-2\tau}\left|\Delta_{g_\nu} u\right|^2 dx \leq 2\int \rho^{4-2\tau}\left|g_\nu^{ij}\partial^2_{x^i x^j} u\right|^2 dx + C\int \rho^{4-2\tau}|\nabla u|^2 dx,$$

where C **depends only on** λ and Λ. By the just obtained inequality and by (18.267) we have, for every $\nu \in \mathbb{N}$,

$$\begin{aligned} &\tau^3 \int \rho^{\epsilon-2\tau}|u|^2 dx + \int \rho^{2+\epsilon-2\tau}\left(\tau - C\rho^{2-\epsilon}\right)|\nabla u|^2 dx + \\ &+ \tau^2 r \int \rho^{-1-2\tau} u^2 dx \leq C \int \rho^{4-2\tau}\left|g_\nu^{ij}\partial^2_{x^i x^j} u\right|^2 dx, \end{aligned} \tag{18.268}$$

where C depends on λ and Λ only. Now let $\overline{\tau}_* \geq \overline{\tau}$ satisfy (recall that $\rho \leq 1$ in B_1) for every $\tau \geq \overline{\tau}_*$,

$$\tau - C\rho^{2-\epsilon} \geq \frac{\tau}{2}.$$

By (18.268) we have, for every $\nu \in \mathbb{N}$,

$$\begin{aligned} &\tau^3 \int \rho^{\epsilon-2\tau}|u|^2 dx + \frac{\tau}{2}\int \rho^{2+\epsilon-2\tau}|\nabla u|^2 dx + \\ &+ \tau^2 r \int \rho^{-1-2\tau} u^2 dx \leq C \int \rho^{4-2\tau}\left|g_\nu^{ij}\partial^2_{x^i x^j} u\right|^2 dx. \end{aligned} \tag{18.269}$$

Passing to the limit as $\nu \to \infty$ in (18.269) we obtain

$$\tau^3 \int \rho^{\epsilon-2\tau}|u|^2dx + \frac{\tau}{2}\int \rho^{2+\epsilon-2\tau}|\nabla u|^2dx+$$
$$+\tau^2 r\int \rho^{-1-2\tau}u^2dx \leq C\int \rho^{4-2\tau}\left|g^{ij}\partial^2_{x^i x^j}u\right|^2 dx.$$

By the just obtained inequality, employing

$$\int \rho^{4-2\tau}\left|g^{ij}\partial^2_{x_i x_j}u\right|^2 dx \leq 2\int \rho^{4-2\tau}\left|\Delta_g u\right|^2 dx + C\int \rho^{4-2\tau}\left|\nabla u\right|^2 dx$$

and repeating the arguments already used above, we have that there exists $\widehat{\tau}_* \geq \overline{\tau}_*$, where $\widehat{\tau}_*$ depends on λ and Λ only, such that

$$\tau^3 \int \rho^{\epsilon-2\tau}|u|^2dx + \tau\int \rho^{2+\epsilon-2\tau}|\nabla u|^2dx+$$
$$+\tau^2 r\int \rho^{-1-2\tau}u^2dx \leq C\int \rho^{4-2\tau}|\Delta_g u|^2dx,$$

for every $r \in (0, r_0)$, for every $\tau \geq \widehat{\tau}_*$ and for every $u \in C_0^\infty\left(B_{r_0} \setminus \overline{B}_{r/4}\right)$. ■

Now we can state the analog of Theorem 18.9 and of Corollary 18.11 for the solutions $U \in H^2(B_1)$ to the equation

$$LU = \sum_{i,j=1}^{n} a^{ij}(x)\partial^2_{x^i x^j}U + \sum_{i=1}^{n} b^i(x)\partial_{x^i}U + c(x)U = 0, \quad \text{in } B_1, \tag{18.270}$$

where $A(x) = \left\{a^{ij}(x)\right\}_{i,j=1}^n$ is a symmetric matrix whose entries are real-valued functions, such that

$$\lambda^{-1}|\xi|^2 \leq \sum_{i,j=1}^{n} a^{ij}(x)\xi_i\xi_j \leq \lambda|\xi|^2, \quad \forall\xi \in \mathbb{R}^n, \forall x \in B_1, \tag{18.271}$$

where $\lambda \geq 1$. Also we assume

$$\left|a^{ij}(x) - a^{ij}(y)\right| \leq \Lambda|x-y|, \quad \text{for } i, j \in \{1, \cdots, n\}, \quad \forall x, y \in B_1. \tag{18.272}$$

Moreover, $b^i \in L^\infty(B_1)$, $i = 1, \cdots, n$ and $c \in L^\infty(B_1)$ (these may be complex-valued coefficients) and set

$$M = \max\left\{\|b\|_{L^\infty(B_1,\mathbb{R}^n)}, \|c\|_{L^\infty(B_1)}\right\}, \tag{18.273}$$

where $b = \left(b^1, \cdots, b^n\right)$.

Theorem 18.26 *Assume that* $U \in H^2(B_1)$ *is a solution to Eq.* (18.271). *Let* $x_0 \in B_1$ *and* $0 < R_0 \leq 1 - |x_0|$. *Then there exist* $C_1 \geq C \geq 1$ *depending on* λ, Λ *and* M *only, such that, if* $0 < r < \frac{R}{C} < \frac{R_0}{C_1}$ *then*

$$\int_{B_R(x_0)} U^2 \leq C \left(\frac{R_0}{R}\right)^C \left(\int_{B_r(x_0)} U^2\right)^\theta \left(\int_{B_{R_0}(x_0)} U^2\right)^{1-\theta}, \tag{18.274}$$

where

$$\theta = \frac{\log \frac{R_0}{C_1 R}}{\log \frac{C_1 R_0}{r}}. \tag{18.275}$$

Moreover, if U *does not vanish identically in* $B_{R_0/C}(x_0)$ *the following doubling inequality holds true*

$$\int_{B_{2r}(x_0)} U^2 \leq C N_{x_0, R_0}^k \int_{B_r(x_0)} U^2, \tag{18.276}$$

where

$$N_{x_0, R_0} = \frac{\int_{B_{R_0}(x_0)} U^2}{R_0 \int_{B_{R_0/C}(x_0)} U^2} \tag{18.277}$$

and k *is a positive number* ($k \geq 3$).

Proof For fixed $x_0 \in B_1$, since $A(x_0)$ is a symmetric matrix, there exists a linear map S such that

$$SA(x_0)S^T = I_n.$$

Hence, we may perform the change of variables $y = S^{-1}(x - x_0)$ in Eq. (18.270) that allows to apply, after some simple modifications, the Carleman estimate (18.265) in a manner quite similar to what was done in the proof of Theorem 18.9. To obtain the analogon of Lemma 18.10, we may fix, for instance $\epsilon = \frac{1}{2}$. We leave the details to the reader. ■

Also, the following Corollary can be proved similarly to Corollary 18.11.

Corollary 18.27 (Strong Unique Continuation for Elliptic Equations) *Let* $U \in H^2(B_1)$ *be a solution to Eq.* (18.270). *Let* $x_0 \in B_1$ *and* $0 < R_0 \leq 1 - |x_0|$. *There exists* C *depending on* λ, Λ *and* M *only such that we have what follows. If* U *does* *not vanish identically in* $B_{R_0/C}(x_0)$ *then we have, for every* $r < s \leq \frac{R_0}{C}$,

$$\int_{B_s(x_0)} U^2 \leq CN^k_{x_0,R_0} \left(\frac{s}{r}\right)^{\log_2(CN^k_{x_0,R_0})} \int_{B_r(x_0)} U^2, \tag{18.278}$$

where N_{x_0,R_0} *is defined by* (18.277) *and* k *is the same number that occurs in* (18.276).

Moreover, if

$$\int_{B_r(x_0)} U^2 = O\left(r^m\right), \quad \text{as } r \to 0, \quad \forall m \in \mathbb{N}, \tag{18.279}$$

then

$$U = 0, \quad \text{in } B_1. \tag{18.280}$$

Chapter 19
Miscellanea

Abstract This final chapter begins with an overview of two alternative methods for addressing the unique continuation problem: log-convexity and frequency function approaches. We highlight their interconnections. Later, we introduce A_p weights and their relevance to inverse problems, followed by a discussion of the Runge property for the Laplace operator.

19.1 Introduction

In this final chapter, we will begin by providing a concise overview of two alternative approaches to address the unique continuation problem, which are distinct from the Carleman estimates. These methods are commonly known as the log-convexity method and the frequency function method. Interestingly, we will illustrate that these methods are closely interconnected. Subsequently, in Sect. 19.4, we will briefly introduce the concept of A_p weights and highlight their applications to inverse problems. Moving on to Sect. 19.5, we will delve into the Runge property associated with the Laplace operator.

19.2 The Backward Problem for the Heat Equation

Let us consider a rod of heat conducting material. Let π be the length of the rod, let us assume that its temperature is zero at its extremes and that the heat flows only in the direction of the axis of the rod. Let $u(x, t)$ be the temperature of the rod at the point x and the time t. If the initial temperature is $f(x)$, then u is a solution of the

S. Vessella, *Unique Continuation Properties for Partial Differential Equations*,
Birkhäuser Advanced Texts Basler Lehrbücher,
https://doi.org/10.1007/978-3-031-86366-0_19

following initial-boundary value problem for the heat equation

$$\begin{cases} u_t - u_{xx} = 0, & \text{for } (x,t) \in (0,\pi) \times (0,+\infty) \\ u(0,t) = u(\pi,t) = 0, & \text{for } t \in [0,+\infty) \\ u(x,0) = f(x), & \text{for } x \in [0,\pi] \end{cases} . \tag{19.1}$$

Also, we assume

$$f \in C^1([0,\pi]), \quad \text{and} \quad f(0) = f(\pi) = 0. \tag{19.2}$$

Let us recall that, [140], there exists an unique solution to problem (19.1) in the class $C^0([0,\pi] \times [0,+\infty)) \cap C^2((0,\pi) \times (0,+\infty))$ and it is given by

$$u(x,t) = \sum_{k=1}^{\infty} f_k \sin kx e^{-k^2 t}, \tag{19.3}$$

where

$$f_k = \frac{2}{\pi} \int_0^{\pi} f(x) \sin kx dx. \tag{19.4}$$

Moreover we have

$$\int_0^{\pi} u^2(x,t)dx \le \frac{\pi}{2} \int_0^{\pi} f^2(x)dx, \quad \forall t \ge 0$$

and, more generally

$$\int_0^{\pi} |\partial_x^m u(x,t)|dx \le C_{m,t} \int_0^{\pi} f^2(x)dx, \quad \forall t > 0.$$

These inequalities imply a continuous dependence of the solution of problem (19.1) by the initial datum f.

In the **backward problem** we are interested in determining the temperature u, if we know, instead of initial temperature, the temperature at an instant $t > 0$, say $t = 1$. Set

$$g(x) = u(x,1), \quad \text{in } [0,\pi]. \tag{19.5}$$

It is evident that, for $t > 1$ by the translation $t' = t - 1$ we reduce to problem (19.1). When $t < 1$ we will examine what happens for what concerns the uniqueness and continuous dependence of u by g.

Uniqueness By the linearity of the problem, it suffices to check that if $g \equiv 0$ then $u \equiv 0$. Now, since u is given by (19.3), we may consider the equation (of the unknown f)

$$\sum_{k=1}^{\infty} f_k \sin kx e^{-k^2} = 0, \quad \forall x \in [0, \pi], \tag{19.6}$$

from which, multiplying both the sides by $\sin mx$, for $m \in \mathbb{N}$ and integrating over $[0, \pi]$, we obtain

$$\frac{\pi}{2} f_m e^{-m^2} = 0, \quad \forall m \in \mathbb{N}.$$

Therefore $f_m = 0$ for every $m \in \mathbb{N}$. Hence

$$u \equiv 0.$$

Continuous Dependence and Conditional Stability Let us consider the sequence of functions

$$g_\nu(x) = e^{\nu} \sin \pi \nu x, \quad \nu \in \mathbb{N}.$$

It is easily checked that

$$u_\nu(x,t) = e^{\nu^2(1-t)} e^{\nu} \sin \pi \nu x, \quad \nu \in \mathbb{N}, \quad \nu \in \mathbb{N}.$$

Hence

$$\left\| g_\nu^{(n)} \right\|_{L^2([0,\pi])} \to 0, \quad \text{as } \nu \to \infty, \quad \forall n \in \mathbb{N},$$

($g_\nu^{(n)}$ is the n-th derivative of g), but

$$\|u_\nu(\cdot, t)\|_{L^2([0,\pi])} \to +\infty, \quad \text{as } \nu \to \infty, \quad \forall t \in (0,1).$$

In plain words, even if we had the estimates of the error of all derivatives of the datum g we could not control the error on $u(\cdot, t)$ when $t < 1$.

Let us denote by

$$\varepsilon := \|g\|_{L^2([0,\pi])} \tag{19.7}$$

and let us suppose the temperature at the initial time is bounded (in the $L^2([0, \pi])$ norm) by a known constant. More precisely, we suppose that

$$\|f\|_{L^2([0,\pi])} = \|u(\cdot, 0)\|_{L^2([0,\pi])} \leq E, \tag{19.8}$$

where $E > 0$ is known. By (19.3) and (19.5) we have

$$\sum_{k=1}^{\infty} f_k \sin kx e^{-k^2} = g(x), \quad \forall x \in [0, \pi],$$

from which we have

$$f_m = \frac{2}{\pi} g_m e^{-m^2}, \quad \forall m \in \mathbb{N},$$

where

$$g_m = \frac{2}{\pi} \int_0^{\pi} f(x) \sin mx dx, \quad \forall m \in \mathbb{N}.$$

Therefore (19.7) and condition (19.8) are expressed, respectively, by

$$\frac{\pi}{2} \sum_{k=1}^{\infty} f_k^2 e^{-2k^2} = \varepsilon^2$$

and

$$\frac{\pi}{2} \sum_{k=1}^{\infty} f_k^2 \leq E^2.$$

On the other hand, we are interested in estimating

$$\|u(\cdot, t)\|^2_{L^2([0,\pi])} = \frac{\pi}{2} \sum_{k=1}^{\infty} f_k^2 e^{-2k^2 t}$$

for $t \in (0, 1)$.

Applying the Hölder inequality we get

$$\begin{aligned}
\frac{\pi}{2} \sum_{k=1}^{\infty} f_k^2 e^{-2k^2 t} &= \frac{\pi}{2} \sum_{k=1}^{\infty} |f_k|^{2(1-t)} \left(|f_k|^2 e^{-2k^2}\right)^t \leq \\
&\leq \frac{\pi}{2} \left(\sum_{k=1}^{\infty} |f_k|^2\right)^{1-t} \left(\sum_{k=1}^{\infty} |f_k|^2 e^{-2k^2}\right)^t \leq \\
&\leq E^{2(1-t)} \varepsilon^{2t}.
\end{aligned}$$

Hence we have proved the following conditional stability estimate

$$\|u(\cdot, t)\|_{L^2([0,\pi])} \leq E^{1-t} \varepsilon^t, \quad \forall t \in (0, 1). \tag{19.9}$$

Remark It is easily checked that estimate (19.9) cannot be improved. Furthermore, (19.9) implies the log-convexity of the function

$$[0, \pi] \ni t \to \|u(\cdot, t)\|_{L^2([0,\pi])}$$

By this we mean that the function

$$F(t) = \log \|u(\cdot, t)\|_{L^2([0,\pi])} ,$$

is convex. ♦

19.3 The Log-convexity Method and the Frequency Function Method

19.3.1 The Log-convexity Method

The fundamental premise of the **log-convexity method** is rooted in the following straightforward observations.

Let us suppose that $F \in C^2([0, 1])$ is a nonnegative function, $F'' \geq 0$ in $[0, 1]$ and let us suppose

$$F''(t)F(t) - F'^2(t) \geq 0, \quad \forall t \in [0, 1]. \tag{19.10}$$

It is immediately checked that this inequality is equivalent to

$$F''(t)(F(t) + \gamma) - F'^2(t) \geq 0, \quad \forall t \in [0, 1], \ \forall \gamma > 0,$$

which, in turn, is equivalent to the log-convexity of $F + \gamma$ in $[0, 1]$. As a matter of fact we have,

$$\frac{d^2}{dt^2} \log(F(t) + \gamma) = \frac{F''(t)(F(t) + \gamma) - F'^2(t)}{F^2(t)} \geq 0.$$

Now, the log-convexity of $F + \gamma$ is equivalent to inequality

$$F(t) + \gamma \leq (F(0) + \gamma)^{1-t}(F(1) + \gamma)^t, \quad \forall t \in [0, 1], \ \forall \gamma > 0$$

hence

$$F(t) \leq (F(0))^{1-t}(F(1))^t, \quad \forall t \in [0, 1]. \tag{19.11}$$

By the latter inequality we derive that if one of values $F(0)$, $F(1)$ is zero then F vanishes identically.

Let us at once see an application of the aforementioned idea for proving the uniqueness and a conditional stability estimate for the following backward problem

$$\begin{cases} u_t - (a(x)u_x)_x = 0, & \text{for } (x,t) \in (0,1) \times (0,T) \\ u(0,t) = u(1,t) = 0, & \text{for } t \in [0,T] \\ u(x,T) = g(x), & \text{for } x \in [0,1] \end{cases}, \tag{19.12}$$

where $T > 0$. Let us suppose that $a \in C^1([0,1])$ and that there exists $u \in C^2([0,1] \times [0,T])$ solution to (19.12).

We define

$$F(t) = \int_0^1 u^2(x,t)dx, \quad t \in [0,T] \tag{19.13}$$

and we have

$$F'(t) = 2\int_0^1 u(x,t)u_t(x,t)dx. \tag{19.14}$$

Now, by the equation $u_t - (a(x)u_x)_x = 0$, taking into account that $u(0,t) = u(1,t) = 0$ and integrating by parts, we get

$$2\int_0^1 u(x,t)u_t(x,t)dx = 2\int_0^1 u(x,t)\left(a(x)u_x\right)_x dx = -2\int_0^1 a(x)u_x^2(x,t)dx.$$

Hence

$$F'(t) = -2\int_0^1 a(x)u_x^2(x,t)dx.$$

By using the just obtained equality, we calculate the second derivative of F

$$F''(t) = -4\int_0^1 a(x)u_x(x,t)u_{xt}(x,t)dx.$$

Now we integrate by parts, and we recall that $u_t(0,t) = u_t(1,t) = 0$ (which is obtained by differentiating $u(0,t) = u(1,t) = 0$ with respect to t), by using again the equation, we get

$$-4\int_0^1 a(x)u_x(x,t)u_{xt}(x,t)dx = 4\int_0^1 \left(a(x)u_x(x,t)\right)_x u_t(x,t)dx =$$

$$= 4\int_0^1 u_t^2(x,t)dx.$$

Hence

$$F''(t) = 4\int_0^1 u_t^2(x,t)dx \geq 0. \tag{19.15}$$

Now, by (19.13), (19.14) and (19.15) we have

$$F''(t)F(t) - F'^2(t) = 4\int_0^1 u_t^2(x,t)dx \int_0^1 u^2(x,t)dx - \\ -4\left(\int_0^1 u(x,t)u_t(x,t)dx\right)^2 \geq 0,$$

where the last inequality follows by the Cauchy–Schwarz inequality. Hence, by (19.11) we have

$$\int_0^1 u^2(x,t)dx \leq \left(\int_0^1 u^2(x,0)dx\right)^{1-\frac{t}{T}} \left(\int_0^1 u^2(x,T)dx\right)^{\frac{t}{T}}, \quad \forall t \in [0,T]$$

from which, recalling $u(x,T) = g(x)$ in $[0,1]$,

$$\int_0^1 u^2(x,t)dx \leq \left(\int_0^1 u^2(x,0)dx\right)^{1-\frac{t}{T}} \left(\int_0^1 g^2(x)dx\right)^{\frac{t}{T}}, \tag{19.16}$$

for every $t \in [0,T]$. By the last estimate we obtain the uniqueness for the backward problem (19.12). As a matter of fact, if $g \equiv 0$ then (19.16) implies $u \equiv 0$. Moreover, if we have the information

$$\int_0^1 g^2(x)dx \leq \varepsilon^2, \quad \text{(error)}$$

and

$$\int_0^1 u^2(x,0)dx \leq E^2, \quad \text{(a priori information)}$$

then we get the following conditional stability estimate

$$\left(\int_0^1 u^2(x,t)dx\right)^{1/2} \leq E^{1-\frac{t}{T}}\varepsilon^{\frac{t}{T}}, \quad \forall t \in [0,T]. \tag{19.17}$$

Remark In contrast to the Carleman estimates approach, the log-convexity method directly utilizes the equation as well as the initial and boundary data. Further elaboration and in-depth understanding of this method can be found in references such as [1, 102, 139]. It is important to note that despite the elegance and simplicity of the presented proof, the method itself may have its own challenges and intricacies. To loosely draw a comparison with the Carleman estimates approach, while the latter method relies on the choice of an appropriate weight, the log-convexity method's effectiveness hinges on the choice of the function F, which is similarly crucial and nontrivial. ♦

Let us revisit the backward problem with a dependence of the coefficient a on both x and t. In other words, we are looking at the following scenario

$$\begin{cases} u_t - (a(x,t)u_x)_x = 0, & \text{for } (x,t) \in (0,1) \times (0,T) \\ u(0,t) = u(1,t) = 0, & \text{for } t \in [0,T] \\ u(x,T) = g(x), & \text{for } x \in [0,1] \end{cases}, \tag{19.18}$$

where $a \in C^1([0,1] \times [0,+\infty))$. We denote

$$\lambda = \min_{(x,t)\in[0,1]\times[0,T]} a(x,t) > 0 \tag{19.19}$$

and

$$M = \max_{(x,t)\in[0,1]\times[0,T]} |a_t(x,t)|. \tag{19.20}$$

We are searching for a function μ, such that

$$\mu : [0,1] \to [0,T], \tag{19.21}$$

bijective, increasing, which satisfies $\mu \in C^2([0,1])$ and such that

$$[0,1] \ni s \to \Phi(s) := F(\mu(s)),$$

is log-convex, where F is given by

$$F(t) = \int_0^1 u^2(x,t)dx, \quad t \in [0,T].$$

We get

$$\dot{\Phi}(s) = F'(\mu(s))\dot{\mu}(s), \quad \forall s \in [0,1],$$

where $\dot{\Phi}$ denotes the derivative of Φ w.r.t. s. In addition

$$\ddot{\Phi}(s) = F''(\mu(s))\dot{\mu}^2(s) + F'(\mu(s))\ddot{\mu}(s), \quad \forall s \in [0,1]. \tag{19.22}$$

Hence

$$\ddot{\Phi}(s)\Phi(s) - \dot{\Phi}^2(s) = \left(F''(\mu(s))F(\mu(s)) - F'^2(\mu(s))\right)\dot{\mu}^2(s)(s) + \\ + F'(\mu(s))F(\mu(s))\ddot{\mu}(s), \qquad \forall s \in [0,1]. \tag{19.23}$$

Now, by the equation $u_t = (a(x,t)u_x)_x$, we get

$$\begin{aligned} F'(t) &= 2\int_0^1 u(x,t)u_t(x,t)dx = \\ &= 2\int_0^1 u(x,t)\left(a(x,t)u_x(x,t)\right)_x dx = \\ &= -2\int_0^1 a(x,t)u_x^2(x,t)dx \end{aligned}$$

and

$$\begin{aligned} F''(t) &= -4\int_0^1 a(x,t)u_x(x,t)u_{xt}(x,t)dx - 2\int_0^1 a_t(x,t)u_x^2(x,t)dx = \\ &= 4\int_0^1 u_t(x,t)\left(a(x,t)u_x(x,t)\right)_x dx - 2\int_0^1 a_t(x,t)u_x^2(x,t)dx = \\ &= 4\int_0^1 u_t^2(x,t)dx - 2\int_0^1 a_t(x,t)u_x^2(x,t)dx. \end{aligned}$$

Recalling (19.20), we have

$$F''(t) \geq 4\int_0^1 u_t^2(x,t)dx - 2M\int_0^1 u_x^2(x,t)dx, \qquad \forall t \in [0,T]. \tag{19.24}$$

By the last equality and by (19.23) we get (for the sake of brevity, we omit the variables)

$$\ddot{\Phi}(s)\Phi(s)-\dot{\Phi}^2(s)\geq\left[\left(4\int_0^1 u_t^2dx-2M\int_0^1 u_x^2dx\right)\left(\int_0^1 u^2dx\right)-\right.$$

$$\left.-4\left(\int_0^1 u_t u dx\right)^2\right]\dot{\mu}^2(s)+$$

$$+2\left(\int_0^1 u_t u dx\right)\left(\int_0^1 u^2dx\right)\ddot{\mu}(s)=$$

$$=4\left[\left(\int_0^1 u_t^2dx\right)\left(\int_0^1 u^2dx\right)-\left(\int_0^1 u_t u dx\right)^2\right]\dot{\mu}^2(s)+$$

$$+\left[-4M\dot{\mu}^2(s)\int_0^1 u_x^2dx+2\left(\int_0^1 u_t u dx\right)\ddot{\mu}(s)\right]\left(\int_0^1 u^2dx\right).$$

By applying the Cauchy–Schwarz inequality to the expression in the first square bracket, we obtain

$$\ddot{\Phi}(s)\Phi(s)-\dot{\Phi}^2(s)\geq\left[-4M\dot{\mu}^2(s)\int_0^1 u_x^2dx+\right.$$
$$\left.+2\left(\int_0^1 u_t u dx\right)\ddot{\mu}(s)\right]\left(\int_0^1 u^2dx\right). \tag{19.25}$$

On the other hand

$$\int_0^1 u_t u dx=\int_0^1 (a(x,t)u_x)_x\, u dx=-\int_0^1 a(x,t)u_x^2dx\geq-\lambda\int_0^1 u_x^2dx.$$

Now, proposing to find μ **concave**, by the last inequality and by (19.25) we have

$$\ddot{\Phi}(s)\Phi(s)-\dot{\Phi}^2(s)\geq-2\left[2M\dot{\mu}^2(s)+\lambda\ddot{\mu}(s)\right]\left(\int_0^1 u^2dx\right)\left(\int_0^1 u_x^2dx\right).$$

Hence, in order to

$$\ddot{\Phi}(s)\Phi(s)-\dot{\Phi}^2(s)\geq 0 \tag{19.26}$$

it suffices that μ satisfies the following conditions

$$\begin{cases} 2M\dot{\mu}^2(s)+\lambda\ddot{\mu}(s)\leq 0, & \forall s\in[0,T]\\ \dot{\mu}(s)\geq 0, & \forall s\in[0,T]\\ \ddot{\mu}(s)\leq 0, & \forall s\in[0,T]\\ \mu(0)=0,\ \mu(1)=T \end{cases}. \tag{19.27}$$

Let us notice that the condition $\ddot{\mu} \leq 0$ implies (see (19.22)) $\ddot{\Phi} \geq 0$.

Setting

$$\alpha = \frac{2M}{\lambda},$$

it is simple to check that

$$\mu(s) = T + \frac{1}{\alpha} \log \left[e^{-\alpha T} + \left(1 - e^{-\alpha T}\right) \right], \quad s \in [0, 1], \tag{19.28}$$

satisfies all conditions (19.27). In particular we get

$$2M\dot{\mu}^2(s) + \lambda \ddot{\mu}(s) = 0, \quad \forall s \in [0, T].$$

All in all, if we have

$$\int_0^1 g^2(x)dx \leq \varepsilon^2$$

and

$$\int_0^1 u^2(x, 0)dx \leq E^2,$$

we obtain the conditional stability estimate

$$\left(\int_0^1 u^2(x, t)dx \right)^{1/2} \leq E^{1-\mu^{-1}(t)} \varepsilon^{\mu^{-1}(t)}, \quad \forall t \in [0, T]. \tag{19.29}$$

where

$$\mu^{-1}(t) = \frac{e^{-\alpha(T-t)} - e^{-\alpha t}}{1 - e^{-\alpha T}}.$$

Let us note that as α goes to 0 (corresponding to the case in which a does not depend on t) $\mu^{-1}(t)$ goes to $\frac{t}{T}$, i.e. the exponent of the estimate (19.17).

The log-convexity method can also be applied to prove the uniqueness and conditional stability for the Cauchy problem. Perhaps, the first author to use it was M. M. Lavrent'ev in 1956, [75]. He applied the method to the Cauchy problem for the Laplace equation in a convex region. With some minor simplification, the situation considered is as follows (we consider only the uniqueness)

$$\begin{cases} u_{yy}(x,y) + \Delta_x u(x,y) = 0, & \forall (x,y) \in B_1 \times (0,1) \\ u(x,y) = 0 & \forall (x,y) \in \partial B_1 \times [0,1] \, , \\ u(x,0) = u_y(x,0) = 0, & \forall x \in \overline{B_1} \end{cases} \tag{19.30}$$

where

$$\Delta_x u(x,y) = \sum_{j=1}^{n} u_{x_j x_j}(x,y)$$

and we suppose that $u \in C^2\left(\overline{B_1} \times [0,1]\right)$.

Set

$$F(y) = \int_{B_1} u^2(x,y)dx. \tag{19.31}$$

We have

$$F'(y) = 2\int_{B_1} u(x,y)u_y(x,y)dx \tag{19.32}$$

and

$$F''(y) = 2\int_{B_1} \left(u_y^2(x,y) + u(x,y)u_{yy}(x,y)\right)dx. \tag{19.33}$$

Now, let us prove

$$\int_{B_1} u(x,y)u_{yy}(x,y)dx = \int_{B_1} u_y^2(x,y)dx. \tag{19.34}$$

First, we note that, by the equation $u_{yy} + \Delta_x u = 0$ and by the condition $u(x,y) = 0$ on $\partial B_1 \times [0,1]$ we have

$$\begin{aligned} \int_{B_1} u(x,y)u_{yy}(x,y)dx &= -\int_{B_1} \Delta_x u(x,y)u(x,y)dx = \\ &= \int_{B_1} |\nabla_x u(x,y)|^2 dx. \end{aligned} \tag{19.35}$$

On the other hand

$$\frac{d}{dy}\int_{B_1} u_y^2 dx = 2\int_{B_1} u_{yy}u_y dx =$$
$$= -2\int_{B_1} (\Delta_x u)u_y dx =$$
$$= -2\int_{B_1} \left[\mathrm{div}_x\left(\nabla_x u u_y\right) - \nabla_x u \cdot \nabla_x u_y\right] dx =$$
$$= 2\int_{B_1} \nabla_x u \cdot \nabla_x u_y dx = \frac{d}{dy}\int_{B_1} |\nabla_x u(x,y)|^2 dx.$$

By the just obtained equality and by (19.35) we get

$$\frac{d}{dy}\left(\int_{B_1} u_y^2(x,y)dx - \int_{B_1} |\nabla_x u(x,y)|^2 dx\right) = 0. \tag{19.36}$$

Now, since we have

$$u(x,0) = u_y(x,0) = 0,$$

we get

$$\int_{B_1} u_y^2(x,0)dx - \int_{B_1} |\nabla_x u(x,0)|^2 dx = 0$$

and by (19.36) we have

$$\int_{B_1} u_y^2(x,y)dx = \int_{B_1} |\nabla_x u(x,y)|^2 dx, \quad \forall y \in (0,1).$$

By the latter and by (19.35) we have (19.34) which, in turn (recall (19.33)), gives

$$F''(y) = 4\int_{B_1} u_y^2(x,y)dx.$$

By the just obtained equality, by (19.31), (19.32) and by the Cauchy–Schwarz inequality we have

$$F''(y)F(y) - F'^2(y) = 4\left(\int_{B_1} u_y^2 dx\right)\left(\int_{B_1} u^2 dx\right) - 4\left(\int_{B_1} uu_y dx\right)^2 \geq 0.$$

Hence $F(y)$ is a log-convex function, consequently we have

$$\int_{B_1} u^2(x,y)dx \leq \left(\int_{B_1} u^2(x,0)dx\right)^y \left(\int_{B_1} u^2(x,1)dx\right)^{1-y} = 0$$

for every $y \in [0, 1]$.

Therefore

$$u \equiv 0.$$

19.3.2 The Frequency Function Method

In Sect. 12.5 we saw various versions of the Hadamard three circle inequality for the holomorphic functions and for the harmonic functions in two variables (for the latter, see (12.114) and (12.115)). In Chap. 18, using the Carleman estimates, we have extended this inequality to the solutions of the second-order elliptic equations and we proved the doubling inequality. The frequency function method was used for the first time in [49] for the second-order elliptic equations with variable coefficients. Here, in order to present the main ideas of this method, we consider the case of the Laplace equation.

Let us propose to prove inequality (12.114) for the harmonic functions of n variables. Then, let u be a solution of the Laplace equation.

$$\Delta u = 0, \qquad \text{in } B_{R_0} \subset \mathbb{R}^n. \tag{19.37}$$

The inequality we are interested in can be written

$$H\left(r_2\right) \leq \left(H\left(r_1\right)\right)^{\theta_0} \left(H\left(r_3\right)\right)^{1-\theta_0}, \tag{19.38}$$

for $0 < r_1 < r_2 < r_3 < R_0$, where

$$H(r) = \int_{\partial B_r} u^2 dS, \tag{19.39}$$

and

$$\vartheta = \frac{\log\left(\frac{r_3}{r_2}\right)}{\log\left(\frac{r_3}{r_1}\right)}. \tag{19.40}$$

Now it is important to observe that (19.38) is equivalent to the log-convexity of

$$F(t) = H\left(e^t\right), \quad t \in \left(-\infty, \log R_0\right). \tag{19.41}$$

Therefore the derivative of $\log F(t)$ needs to be increasing. Let us calculate such a derivative

$$\frac{d}{dt}\log F(t) = \frac{F'(t)}{F(t)} = \frac{e^t H'\left(e^t\right)}{H\left(e^t\right)}, \quad t \in (-\infty, \log R_0).$$

This equality implies that (19.38) is equivalent to the fact that the function

$$\frac{rH'(r)}{H(r)}, \quad r \in (0, R_0), \tag{19.42}$$

is increasing. Now we calculate $H'(r)$. First we notice that

$$H(r) = \frac{1}{r}\int_{B_r} \operatorname{div}(xu^2)dx, \tag{19.43}$$

as a matter of fact, by the divergence Theorem we obtain

$$H(r) = \frac{1}{r}\int_{\partial B_r}\left(x \cdot \frac{x}{|x|}\right)u^2 dS = \frac{1}{r}\int_{\partial B_r}(x \cdot \nu)u^2 dS = \frac{1}{r}\int_{B_r}\operatorname{div}(xu^2)dx.$$

Now, set

$$I(r) = \int_{B_r} |\nabla u|^2 dx, \quad r \in (0, R_0) \tag{19.44}$$

and let us notice that

$$I(r) = \int_{B_r} \nabla u \cdot \nabla u dx = \int_{B_r} \operatorname{div}(u\nabla u)\, dx = \int_{\partial B_r} u\frac{\partial u}{\partial \nu}dS. \tag{19.45}$$

We have

$$\begin{aligned}
H'(r) &= \frac{1}{r}\int_{\partial B_r}\operatorname{div}(xu^2)dS - \frac{1}{r^2}\int_{B_r}\operatorname{div}(xu^2)dx \\
&= \frac{1}{r}\left\{n\int_{\partial B_r} u^2 dS + 2\int_{\partial B_r}(x\cdot\nabla u)u dS\right\} - \frac{1}{r}H(r) \\
&= \frac{n-1}{r}\int_{\partial B_r} u^2 dS + 2\int_{\partial B_r} u\frac{\partial u}{\partial \nu}dS = \frac{n-1}{r}H(r) + 2I(r).
\end{aligned}$$

Therefore

$$H'(r) = \frac{n-1}{r}H(r) + 2I(r), \quad r \in (0, R_0). \tag{19.46}$$

Hence, function (19.42) can be written

$$\frac{rH'(r)}{H(r)} = n - 1 + 2\frac{rI(r)}{H(r)}, \quad r \in (0, R_0)$$

In sum, the log-convexity of F is equivalent to the fact that the function

$$N(r) = \frac{rI(r)}{H(r)}, \quad r \in (0, R_0) \tag{19.47}$$

is increasing. $N(\cdot)$ is called the **frequency function** of u. The frequency function was introduced by Almgren, [7] and has taken this name because for a homogeneous harmonic polynomial of degree m, we have $N(r) = m$ for all r. For instance, in dimension 2, the homogeneous harmonic polynomials of degree m are (in polar coordinates) of the type

$$p_m(\varrho, \theta) = A\varrho^m \cos m\phi + B\varrho^m \sin m\phi$$

and it is easy to verify what has been asserted.

In the sequel, we will assume that $H(r) > 0$ for every $r \in (0, R_0)$, otherwise we can employ the device shown at the beginning of Sect. 19.3.1. We can also notice that if there exists $\overline{r} \in (0, R_0)$ such that $H(\overline{r}) = 0$ then $u = 0$ on ∂B_r from which, $u = 0$ in B_r and by the unique continuation property we have $u \equiv 0$, which make the (19.38) trivial. The second device (i.e., assuming a unique continuation property) is certainly legitimate, but it is somewhat reductive because by the method we are illustrating we can obtain independent proof of unique continuation property.

At this point, we state and prove the following.

Proposition 19.1 *If u is a non identically zero solution to* (19.37) *then $N(r)$ is an increasing function.*

Proof First we prove that

$$I'(r) = 2\int_{\partial B_r} \left(\frac{\partial u}{\partial \nu}\right)^2 dS + \frac{n-2}{r} I(r). \tag{19.48}$$

For this purpose we use the Rellich identity (Lemma 14.7) for $v \in C^2\left(B_{R_0}\right)$ and $\beta(x) = x$,

$$2(x \cdot \nabla v)\Delta v = \text{div}[2(x \cdot \nabla v)\nabla v - x|\nabla v|^2] + (n-2)|\nabla v|^2. \tag{19.49}$$

We get by (19.44)

$$I'(r) = \int_{\partial B_r} |\nabla u|^2 dS. \tag{19.50}$$

Now we apply identity (19.49) to the function u, we integrate both the sides of (19.49) over B_r and we recall that $\Delta u = 0$, obtaining

$$\begin{aligned} 0 &= \int_{\partial B_r} \left\{ 2(x \cdot \nabla u)\nabla u \cdot \frac{x}{|x|} - x \cdot \frac{x}{|x|}|\nabla u|^2 \right\} dS + (n-2) \int_{B_r} |\nabla u|^2 dx \\ &= \int_{\partial B_r} \left\{ 2r \left(\frac{\partial u}{\partial \nu} \right)^2 - r|\nabla u|^2 \right\} dS + (n-2) \int_{B_r} |\nabla u|^2 dx, \end{aligned}$$

from which, taking into account (19.50), we have

$$\begin{aligned} I'(r) &= \int_{\partial B_r} |\nabla u|^2 dS = 2 \int_{\partial B_r} \left(\frac{\partial u}{\partial \nu} \right)^2 dS + \frac{n-2}{r} \int_{B_r} |\nabla u|^2 dx \\ &= 2 \int_{\partial B_r} \left(\frac{\partial u}{\partial \nu} \right)^2 dS + \frac{n-2}{r} I(r). \end{aligned} \tag{19.51}$$

hence (19.48) is proved.

Differentiating both the sides of (19.47), we get

$$N'(r) = \frac{I(r)}{H(r)} + r\frac{I'(r)}{H(r)} - r\frac{I(r)H'(r)}{H^2(r)}. \tag{19.52}$$

Unless u is constant (and, therefore (19.38) becomes trivial) we have

$$N(r) > 0$$

and we divide both the sides of (19.52) by $N(r)$, obtaining

$$\frac{N'(r)}{N(r)} = \frac{1}{r} + \frac{I'(r)}{I(r)} - \frac{H'(r)}{H(r)},$$

from which, taking into account (19.46) and (19.48), we have

$$\begin{aligned} \frac{N'(r)}{N(r)} &= \frac{1}{r} + \frac{I'(r)}{I(r)} - \frac{\frac{n-1}{r}H(r) + 2I(r)}{H(r)} = \\ &= \frac{2-n}{r} + \frac{I'(r)}{I(r)} - 2\frac{I(r)}{H(r)} = \\ &= 2\left\{ \frac{\int_{\partial B_r} \left(\frac{\partial u}{\partial \nu}\right)^2 dS}{I(r)} - \frac{I(r)}{H(r)} \right\}. \end{aligned} \tag{19.53}$$

Now we express $I(r)$ by means of (19.45) and, by (19.53), we get

$$\frac{N'(r)}{N(r)} = 2\left\{\frac{\int_{\partial B_r}\left(\frac{\partial u}{\partial \nu}\right)^2 dS}{\int_{\partial B_r}\frac{\partial u}{\partial \nu}u dS} - \frac{\int_{\partial B_r}\frac{\partial u}{\partial \nu}u dS}{\int_{\partial B_r}u^2 dS}\right\} =$$

$$= \frac{2}{I(r)H(r)}\left\{\left(\int_{\partial B_r}\left(\frac{\partial u}{\partial \nu}\right)^2 dS\right)\left(\int_{\partial B_r}u^2 dS\right) - \left(\int_{\partial B_r}\frac{\partial u}{\partial \nu}u dS\right)^2\right\}.$$

Now, by the Cauchy–Schwarz inequality we have

$$\left(\int_{\partial B_r}\left(\frac{\partial u}{\partial \nu}\right)^2 dS\right)\left(\int_{\partial B_r}u^2 dS\right) - \left(\int_{\partial B_r}\frac{\partial u}{\partial \nu}u dS\right)^2 \geq 0$$

from which the thesis follows. ■

Let us summarize what has been obtained so far. By Proposition 19.1 we get the log-convexity of function (19.41), from which inequality (19.38) follows. By the latter, proceeding as done to prove (12.112) and using by (19.38), we obtain,

$$\int_{B_{r_2}} u^2 dx \leq \left(\int_{B_{r_1}} u^2 dx\right)^{\theta_0}\left(\int_{B_{r_3}} u^2 dx\right)^{1-\theta_0}, \tag{19.54}$$

for $0 < r_1 < r_2 < r_3 < R_0$, where θ is given by (19.40).

By the properties of the frequency function we derive a doubling inequality. Indeed the following holds true.

Proposition 19.2 *Let u be a non identically zero solution to* (19.37) *then*

$$\int_{B_{2r}} u^2 dx \leq 2^{2^n N(R_0)+1}\int_{B_r} u^2 dx, \quad \forall r \in \left(0, \frac{R_0}{2}\right] \tag{19.55}$$

and

$$\int_{B_{2r}} u^2 dx \leq \frac{2\int_{B_{R_0}} u^2 dx}{\int_{B_{R_0/4}} u^2 dx}\int_{B_r} u^2 dx, \quad \forall r \in \left[0, \frac{R_0}{4}\right], \tag{19.56}$$

In the proof of Proposition 19.2 we use

Lemma 19.3 *Let u be a solution to* (19.37) *then we have*

$$\frac{1}{r}\int_{B_r} u^2 dx \leq \int_{\partial B_r} u^2 dS \leq \frac{C}{r}\int_{B_{2r}} u^2 dx, \quad r \in \left(0, \frac{R_0}{2}\right], \tag{19.57}$$

where C depends on n only.

Proof of Lemma 19.57 Let us prove the first inequality of (19.57). By (19.46) we have $H'(r) \geq 0$. Hence

$$\begin{aligned}\int_{B_r} u^2 dx &= \int_0^r \left(\int_{\partial B_t} u^2 dS\right) dt \leq \\ &\leq \int_0^r \left(\int_{\partial B_r} u^2 dS\right) dt = \\ &= r\int_{\partial B_r} u^2 dS.\end{aligned} \tag{19.58}$$

Concerning the second inequality of (19.57), by (19.43) and by the Caccioppoli inequality, (6.7), we have

$$\begin{aligned}H(r) &= \frac{1}{r}\int_{B_r} \operatorname{div}(xu^2)dx = \\ &= \frac{n}{r}\int_{B_r} u^2 dx + \frac{1}{r}\int_{B_r} u\,(x\cdot\nabla u)\,dx \leq \\ &\leq \frac{n}{r}\int_{B_r} u^2 dx + \left(\int_{B_r} u^2 dx\right)^{1/2}\left(\int_{B_r} |\nabla u|^2 dx\right)^{1/2} \leq \\ &\leq \frac{n}{r}\int_{B_r} u^2 dx + \frac{\widetilde{C}}{r}\left(\int_{B_r} u^2 dx\right)^{1/2}\left(\int_{B_{2r}} u^2 dx\right)^{1/2} \leq \\ &\leq \frac{\widetilde{C}+n}{r}\int_{B_{2r}} u^2 dx,\end{aligned}$$

where $\widetilde{C}$ depends on n only. The second inequality in (19.57) is proved with $C = \widetilde{C} + n$. ∎

Proof of Proposition 19.2 By (19.46) and (19.47) we get

$$\frac{d}{dr}\left(\log\frac{H(r)}{r^{n-1}}\right) = \frac{2N(r)}{r}. \tag{19.59}$$

Let $\rho \in \left(0, \frac{R_0}{2}\right]$. Integrating both the sides of (19.59) over $[\rho, 2\rho]$ and recalling that N is increasing, we have

$$\log\left(\frac{H(2\rho)}{2^{n-1}H(\rho)}\right) = 2\int_\rho^{2\rho} \frac{2N(r)}{r} \leq (2\log 2)N(R_0).$$

Hence

$$H(2\rho) \le 2^{2^n N(R_0)} H(\rho), \quad \forall \rho \in \left(0, \frac{R_0}{2}\right],$$

Integrating over $[0, r]$, $r \in \left(0, \frac{R_0}{2}\right]$, we get (19.55).

Let now prove (19.56). Let $\rho \in \left(0, \frac{R_0}{4}\right]$ and $R \in \left[\frac{R_0}{4}, \frac{R_0}{2}\right]$. Integrating both the sides of (19.59) over $[\rho, 2\rho]$ and recalling that N is an increasing function, we have

$$\log\left(\frac{H(2\rho)}{2^{n-1}H(\rho)}\right) \le (2\log 2)N(R).$$

By the just obtained inequality, taking into account (19.59), we have

$$\frac{1}{R}\log\left(\frac{H(2\rho)}{2^{n-1}H(\rho)}\right) \le (2\log 2)\frac{N(R)}{R} = (\log 2)\frac{d}{dR}\left(\log\frac{H(R)}{R^{n-1}}\right).$$

Again we integrate both the sides of the last inequality w.r.t. R over $\left[\frac{R_0}{4}, \frac{R_0}{2}\right]$ so that we have

$$\log\left(\frac{H(2\rho)}{2^{n-1}H(\rho)}\right) \le \log\left(\frac{H\left(\frac{R_0}{2}\right)}{2^{n-1}H\left(\frac{R_0}{4}\right)}\right).$$

Now, Lemma 19.3 gives

$$H(2\rho) \le \frac{\int_{B_{R_0}} u^2 dx}{\int_{B_{R_0/4}} u^2 dx} H(\rho), \quad \forall \rho \in \left(0, \frac{R_0}{4}\right].$$

From which integration of both the sides over $[0, r]$, $r \in \left[0, \frac{R_0}{4}\right]$, of last inequality gives

$$\int_{B_{2r}} u^2 dx \le \frac{2\int_{B_{R_0}} u^2 dx}{\int_{B_{R_0/4}} u^2 dx} \int_{B_r} u^2 dx, \quad \forall r \in \left[0, \frac{R_0}{4}\right].$$

Therefore, we get (19.56). ■

Final Remark The proof of Proposition 19.1, based on the Rellich identity, differs from the proof given in [49] which is based on the transformation of the elliptic operator in polar coordinates. The proof based on the Rellich identity was given by [70] for the second-order elliptic operators. ♦

19.4 A Brief Review About the A_p Weights

In this section we will provide a brief summary of the A_p weights by referring to [26] and [47, Chapter 4] for further reading. In Eq. (3.47) we have introduced the maximal function related to a function $f \in L^1(\mathbb{R}^n)$. Basically equivalent to it is the following definition. In what follows we will denote by Q a closed cube whose sides are parallel to the axes. Let $f \in L^1(\mathbb{R}^n)$. We define the maximal Hardy-Littlewood function as

$$\mathcal{M}(f)(x) = \sup\left\{ \fint_Q |f(y)|dy : Q \ni x \right\}, \tag{19.60}$$

where, we recall,

$$\fint_Q |f(y)|dx = \frac{1}{|Q|}\int_Q |f(y)|dy.$$

It is simple to check that

$$\frac{\omega_n}{2^n n} M(f)(x) \leq \mathcal{M}(f)(x) \leq \omega_n n^{\frac{1}{2}n-1} M(f)(x), \quad \forall x \in \mathbb{R}^n, \tag{19.61}$$

where $M(f)$ is defined in (3.47). From inequalities (19.61) one can prove for $\mathcal{M}(f)$ properties similar to those of $M(f)$ some of which have been proved or presented in Sect. 3.6. In particular, the following apply (compare with Lemma 3.55 and with (3.52), respectively)

$$|E_t| \leq \frac{C_n}{t}\int_{\mathbb{R}^n} |f(x)|dx, \quad \forall f \in L^1\left(\mathbb{R}^n\right), \forall t > 0$$

(C_n depends on n only), where

$$E_t = \left\{x \in \mathbb{R}^n : M(f)(x) > t\right\}$$

and

$$\|M(f)\|_{L^p(\mathbb{R}^n)} \leq C_{n,p}\, \|f\|_{L^p(\mathbb{R}^n)}, \quad \forall f \in L^p\left(\mathbb{R}^n\right).$$

($C_{n,p}$ depends on n and p only).

The A_p weight were introduced by Muckenhoupt, [97], and by Coifman and Fefferman, [26] to answer to the following question:

Let $p \in (1, +\infty)$. Determine all measurable and nonnegative functions w such that

$$\int_{\mathbb{R}^n} |\mathcal{M}(f)(x)|^p w(x)dx \le C \int_{\mathbb{R}^n} |f(x)|^p w(x)dx, \quad \forall f \in L^p\left(\mathbb{R}^n\right), \tag{19.62}$$

where C depends on w only. The functions w which enjoys property (19.62) are called A_p **weight** . If $p = 1$, then we say that w is an A_1 weight if

$$\int_{\widetilde{E}_t} w(x)dx \le C \int_{\mathbb{R}^n} |f(x)| w(x)dx, \quad \forall f \in L^1\left(\mathbb{R}^n\right),$$

where

$$\widetilde{E}_t = \left\{x \in \mathbb{R}^n : \mathcal{M}(f)(x) > t\right\}.$$

The following Theorem can be proved ([47, Chap. 4, Sect. 2]).

Theorem 19.4 *Let* $w \in L^1_{loc}(\mathbb{R}^n)$, $w \ge 0$. *The following conditions are equivalent*

(a) there exists $p \in [1, +\infty)$ *such that* $w \in A_p$;
(b) if $p \in (1, +\infty)$, *then there exists* $C > 0$ *such that*

$$\left(\fint_Q w(x)dx\right)\left(\fint_Q w^{-\frac{1}{p-1}}(x)dx\right)^{p-1} \le C, \quad \textit{for every cube } Q \tag{19.63}$$

if $p = 1$, *then we have*

$$\left(\fint_Q w(x)dx\right) \operatorname{ess\,sup}\left(w^{-1}\right) \le C, \quad \textit{for every cube } Q; \tag{19.64}$$

(c) there exist $\delta > 0$ *and* $C > 0$ *such that*

$$\left(\fint_Q w^{1+\delta}(x)dx\right)^{\frac{1}{1+\delta}} \le C\fint_Q w(x)dx \quad \textit{for every cube } Q; \tag{19.65}$$

(d) there exist $s > 0$ *and* $C > 0$ *such that, for every cube* Q *and for every* $E \subset Q$, *E Lebesgue measurable set, we have*

$$\frac{|E|}{|Q|} \le C\left(\frac{\int_E w(x)dx}{\int_Q w(x)dx}\right)^s. \tag{19.66}$$

Comments and Remarks An inequality like (19.65) is called **"reverse Hölder inequality"**. It can be proved that Theorem 19.4 can be reformulated for weight functions w defined on an open set Ω, as long as we replace, in (b), (c) and (d) "for every cube Q" by "for every cube $Q \subset \Omega$. The following remark turns out to be very useful in proving quantitative estimates in inverse problems. Below we illustrate this idea in a simplified way. In Theorem 18.26 we proved that a nonzero if u is a solution of the elliptic equation

$$\sum_{i,j=1}^{n} a^{ij}(x)\partial^2_{x^i x^j}u + \sum_{i=1}^{n} b^i(x)\partial_{x^i}u + c(x)u = 0, \tag{19.67}$$

whose coefficients satisfy the hypotheses (18.271), (18.272) and (18.273) then u satisfies doubling inequality (18.276). Let us suppose that u satisfies (19.67) in $B_{\overline{R}}$, $\overline{R} > 1$ and let us write inequality (18.276) in the form

$$\int_{B_{2r}(x_0)} u^2 \leq CN_{x_0}^k \int_{B_r(x_0)} u^2, \quad x_0 \in B_1. \tag{19.68}$$

$$N_{x_0} = \frac{\int_{B_1(x_0)} u^2 dx}{\int_{B_{r_0}(x_0)} u^2 dx}. \tag{19.69}$$

where r_0 is a suitable point of $(0, 1)$ (here, $R_0 = 1$ and $r_0 = 1/C$, in (18.276)). Let us assume for simplicity

$$\overline{R} = \max\{1 + 2\sqrt{n}, 16\}.$$

Let us denote by

$$F = \frac{\int_{B_{\overline{R}}} u^2 dx}{\int_{B_{r_0/2}} u^2 dx}. \tag{19.70}$$

By the Caccioppoli inequality and the Sobolev Embedding Theorem, proceeding similarly to what was done in Lemma 6.21, we have

$$\left(\fint_{Q_r(x_0)} |u|^q dx\right)^{1/q} \leq \left(\fint_{Q_{2r}(x_0)} u^2 dx\right)^{1/2}, \quad \forall x_0 \in B_{r_0/2}, \tag{19.71}$$

where q is an arbitrary number of $(1, +\infty)$ when $n = 2$, and it is equal to $\frac{2n}{n-2}$ when $n \geq 3$. On the other hand (see Corollary 18.27),

$$\left(\fint_{Q_{2r}(x_0)} u^2 dx\right)^{1/2} \leq \left(\fint_{B_{2\sqrt{n}r}(x_0)} u^2 dx\right)^{1/2} \leq$$
$$\leq (CN_{x_0})^{\widetilde{k}} \left(\fint_{B_r(x_0)} u^2 dx\right)^{1/2} \leq$$
$$\leq (CN_{x_0})^{\widetilde{k}} \left(\fint_{Q_r(x_0)} u^2 dx\right)^{1/2}, \quad \forall x_0 \in B_{r_0/2},$$

where $\widetilde{k} = \frac{k}{2}\left(1 + \log_2(2\sqrt{2n})\right)$. By the just obtained inequality and by (19.71), we have

$$\left(\fint_{Q_r(x_0)} |u|^q dx\right)^{1/q} \leq (CN_{x_0})^{\widetilde{k}} \left(\fint_{Q_r(x_0)} u^2 dx\right)^{1/2}, \quad \forall x_0 \in B_{r_0/2}.$$

Moreover, by (19.70) we have trivially

$$N_{x_0} = \frac{\int_{B_1(x_0)} u^2 dx}{\int_{B_{r_0}(x_0)} u^2 dx} \leq \frac{\int_{B_{\overline{R}}} u^2 dx}{\int_{B_{r_0/2}} u^2 dx}.$$

Therefore we have, in particular, for any cube $Q \subset B_{r_0/2}$

$$\left(\fint_Q |u|^q dx\right)^{1/q} \leq F^{\widetilde{k}} \left(\fint_Q u^2 dx\right)^{1/2}. \tag{19.72}$$

Recalling that $q > 2$, we get by (19.72) that u^2 satisfies a reverse Hölder inequality, consequently u^2 is an A_p weight. In particular, Theorem 19.4 yields that for every $Q \subset B_{r_0/2}$ and for every $E \subset Q$, E Lebesgue measurable set, we have

$$\frac{|E|}{|Q|} \leq C \left(\frac{\int_E u^2 dx}{\int_Q u^2 dx}\right)^s, \tag{19.73}$$

where C depends by F. Let us suppose, now that the set E has positive measure, then, if we have some bounds on F (generally obtainable from values at the boundary of u), estimate (19.73) can be trivially rewritten

$$\int_Q u^2 dx \leq \left(\frac{C|Q|}{|E|}\right)^{1/s} \int_E u^2 dx \tag{19.74}$$

This estimate implies, in particular, that if u vanishes on a set E of positive measure then u vanishes identically in $B_{\overline{R}}$. Actually, (19.74) also allows us to control, in terms of the measure E only, the propagation of the error

$$\int_E u^2 dx \leq \varepsilon^2$$

on a "small" cube Q and from there on the whole $B_{\overline{R}}$.

Another trivial translation of (19.73) is

$$|E| \leq C|Q| \left(\frac{\int_E u^2 dx}{\int_Q u^2 dx} \right)^s,$$

which allows us to estimate the Lebesgue measure of E by the integral of u^2 on E itself. The latter observation is useful for finding **size estimates** of unknown inclusions in problems of the type considered in Sect. 6.8 (for details, see [4] and also [32], [17]). ♦

19.5 The Runge Property

As an introduction to the main topic of this section, we show by an example that it is not always possible to extend a harmonic function u from B_1 to an open set, Ω, such that $B_1 \Subset \Omega$. The example we present here is due to Hadamard, [53]. Let us consider the function whose expression in polar coordinates is given by

$$u(\rho, \phi) = \sum_{n=1}^{\infty} 2^{-n} \rho^{4^n} \sin(4^n \phi). \tag{19.75}$$

It is simple to check that $u \in C^0\left(\overline{B}_1\right) \cap C^2\left(B_1\right)$ and that u is harmonic in B_1. Now we check that

$$u \notin C^1\left(\overline{B}_1\right).$$

For this purpose we show that

$$\lim_{r \to 1^-} \int_{B_r} |\nabla u|^2 dx dy = \lim_{r \to 1^-} \int_0^r d\rho \int_0^{2\pi} \left(u_\rho^2 + \rho^{-2} u_\phi^2\right) \rho d\phi = +\infty. \tag{19.76}$$

We get

$$\int_0^{2\pi} \left(u_\rho^2 + \rho^{-2} u_\phi^2\right) \rho d\phi = 2\pi \sum_{n=1}^{\infty} 4^n \rho^{2 \cdot 4^n - 1}.$$

Hence

$$\int_0^r d\rho \int_0^{2\pi} \left(u_\rho^2 + \rho^{-2} u_\phi^2\right) \rho d\phi = \pi \sum_{n=1}^{\infty} r^{2\cdot 4^n},$$

from which (19.76) follows. It is therefore evident that u cannot be extended to a harmonic function in an open set containing $\overline{B}_1$.

The **Runge property** for the Laplace operator involves the question of whether a given function can be approximated by functions that are harmonic in a certain region. This concept has been studied not only for the Laplace operator but also for more general operators. However, in your current context, we are focusing specifically on the Laplace operator referring to the final comments for hints on further consideration.

We have

Theorem 19.5 *Let $\Omega_1 \Subset \Omega_2$ be two open sets of $\mathbb{R}^n$, where $\partial\Omega_2$ is of class $C^{1,1}$. Let us assume that $\Omega_2 \setminus \overline{\Omega_1}$ connected. Then for every u such that*

$$\Delta u = 0, \quad in \quad \Omega_1$$

and for every $\varepsilon > 0$ there exists $v \in H^1(\Omega_2)$ such that

$$\Delta v = 0, \quad in \quad \Omega_2$$

and

$$\|u - v\|_{L^2(\Omega_1)} < \varepsilon.$$

Proof Set

$$\mathcal{S}_1 = \left\{u \in H^1(\Omega_1) : \ \Delta u = 0 \text{ in } \Omega_1\right\}$$

and

$$\mathcal{S}_2 = \left\{v_{|\Omega_1} \ : \ \Delta v = 0 \text{ in } \Omega_2\right\}. \tag{19.77}$$

The property that we wish to prove is equivalent to the fact that $\mathcal{S}_2$ is dense in $\mathcal{S}_1$, with respect to the topology induced by $L^2(\Omega_1)$. We now prove this density property. To this aim, it suffices to prove that if $u \in \mathcal{S}_1$ and

$$\int_{\Omega_1} uv dx = 0, \quad \forall v \in \mathcal{S}_2, \tag{19.78}$$

then

$$u \equiv 0, \quad \text{in} \quad \Omega_1.$$

Let $\widetilde{u}$

$$\widetilde{u}(x) = \begin{cases} u, & \text{in } \Omega_1 \\ \\ 0, & \text{in } \Omega_2 \setminus \Omega_1 \end{cases}.$$

and let w satisfy

$$\begin{cases} \Delta w = \widetilde{u}, & \text{in} \quad \Omega_2 \\ \\ w \in H_0^1(\Omega_2) \end{cases}. \tag{19.79}$$

Since $\widetilde{u} \in L^2(\Omega_2)$ and $\partial\Omega_2$ is of class $C^{1,1}$, by Theorem 6.12 we have

$$w \in H^2(\Omega_2).$$

Now, for an arbitrary $v \in \mathcal{S}_2$ we have

$$\begin{aligned} 0 &= \int_{\Omega_1} uv dx = \int_{\Omega_2} \widetilde{u} v dx = \int_{\Omega_2} \Delta w v dx = \\ &= \int_{\partial\Omega_2} \frac{\partial w}{\partial \nu} v dS + \int_{\Omega_2} w \Delta v dx = \\ &= \int_{\partial\Omega_2} \frac{\partial w}{\partial \nu} v dS. \end{aligned} \tag{19.80}$$

Hence

$$\int_{\partial\Omega_2} \frac{\partial w}{\partial \nu} v dS = 0, \quad \forall v \in \mathcal{S}_2.$$

By the latter and by Theorem 6.3 we have

$$\int_{\partial\Omega_2} \frac{\partial w}{\partial \nu} \varphi dS = 0, \quad \forall \varphi \in H^{1/2}(\partial\Omega_2), \tag{19.81}$$

hence for $\varphi = \frac{\partial w}{\partial \nu}$ we have

$$\frac{\partial w}{\partial \nu} = 0, \quad \text{on} \quad \partial\Omega_2.$$

Therefore we have

$$\begin{cases} \Delta w = \widetilde{u}, & \text{in} \quad \Omega_2 \setminus \overline{\Omega}_1 \\ w = 0, & \text{on} \quad \partial\Omega_2 \\ \frac{\partial w}{\partial \nu} = 0, & \text{on} \quad \partial\Omega_2 \end{cases} . \tag{19.82}$$

Now, since $\Omega_2 \setminus \overline{\Omega}_1$ is connected, (19.82) implies, by the unique continuation property,

$$w = 0, \quad \text{in} \quad \Omega_2 \setminus \overline{\Omega}_1.$$

By this relationship, recalling that $w \in H^2(\Omega_2)$ we have that $w \in H_0^2(\Omega_1)$ from which we have

$$\int_{\Omega_1} u^2 dx = \int_{\Omega_1} u \Delta w dx = \int_{\Omega_1} \Delta u w dx = 0.$$

Hence

$$u = 0, \quad \text{in} \quad \Omega_1.$$

As we desired to prove. ■

Remarks

1. Theorem 19.5 continues to be true also if, instead of $\partial\Omega_2 \in C^{1,1}$, we assume only $\partial\Omega_2 \in C^{0,1}$ To prove this, one must first keep in mind that in (19.80) occurs

$$\langle \frac{\partial w}{\partial \nu}, v \rangle_{H^{-1/2}(\partial\Omega_2), H^{1/2}(\partial\Omega_2)}$$

instead of

$$\int_{\partial\Omega_2} \frac{\partial w}{\partial \nu} v dS$$

and, consequently, instead of (19.81), we have

$$\langle \frac{\partial w}{\partial \nu}, \varphi \rangle_{H^{-1/2}(\partial\Omega_2), H^{1/2}(\partial\Omega_2)} = 0, \quad \forall \varphi \in H^{1/2}(\partial\Omega_2).$$

Therefore, we likewise have that $\frac{\partial w}{\partial \nu} = 0$. However, since $\frac{\partial w}{\partial \nu} \in H^{-1/2}(\partial\Omega_2)$, it will be necessary to first reformulate Cauchy problem (19.82) in a weak form and then to prove the uniqueness for such a Cauchy problem, for both of which we refer the reader to the paper [3].

2. The assumption that $\Omega_2 \setminus \overline{\Omega}_1$ is connected cannot be dropped as the following simple **counterexample** shows. Let

$$\Omega_1 = \left(B_7 \setminus \overline{B}_5\right) \cup \left(B_3 \setminus \overline{B}_1\right), \qquad \Omega_2 = B_8,$$

let us notice that $\Omega_2 \setminus \overline{\Omega}_1$ is not connected.

Let u be the following function

$$u(x) = \begin{cases} 1, & \text{for } x \in B_3 \setminus \overline{B}_1 \\ & \\ 0, & \text{for } x \in B_7 \setminus \overline{B}_5 \end{cases}, \tag{19.83}$$

of course u is harmonic in Ω_1. Let ε be a given positive number which we will choose later and let $v \in H^1(\Omega_2)$ be a harmonic function satisfying

$$\|u - v\|_{L^2(\Omega_1)} < \varepsilon. \tag{19.84}$$

Let $x_0 \in \partial B_2$. Taking into account that $B_1(x_0) \subset \Omega_1$, by (19.84) we have

$$\begin{aligned} |v(x_0) - 1| &= \left| \frac{1}{|B_1(x_0)|} \int_{B_1(x_0)} (v - 1) dx \right| \leq \\ &\leq \frac{1}{|B_1(x_0)|^{1/2}} \left(\int_{B_1(x_0)} (v - 1)^2 dx \right)^{1/2} \leq \\ &\leq \frac{\varepsilon}{c_n}, \end{aligned}$$

where $c_n = |B_1|^{1/2}$. Hence we have

$$1 - \frac{\varepsilon}{c_n} < v(x) < 1 + \frac{\varepsilon}{c_n}, \quad \forall x \in \partial B_2. \tag{19.85}$$

Similarly we get

$$-\frac{\varepsilon}{c_n} < v(x) < \frac{\varepsilon}{c_n}, \quad \forall x \in \partial B_6. \tag{19.86}$$

Now, by (19.85), (19.86) and by using maximum principle we have

$$1 - \frac{\varepsilon}{c_n} < v(x) \leq \max_{x \in \partial B_6} v < \frac{\varepsilon}{c_n}, \quad \forall x \in \partial B_6,$$

that leads to a contradiction provided $\varepsilon < \frac{c_n}{2}$.

3. Theorem 19.5 can be extended to a large class of operators; in particular, it can be extended to the second-order elliptic operators with real coefficients whose formal adjoint enjoys the unique continuation property. Let us specify this a little. Let us consider, for instance, the operator

$$Lu = \operatorname{div}\left(A(x)\nabla u\right), \tag{19.87}$$

where $A \in L^{\infty}\left(\mathbb{R}^n; \mathbb{M}(n)\right)$ is a not necessarily symmetric matrix and such that $(\lambda \geq 1)$

$$\lambda^{-1}|\xi|^2 \leq A(x)\xi \cdot \xi, \quad \forall \xi \in \mathbb{R}^n, \ \forall x \in \mathbb{R}^n.$$

Then the formal adjoint of L is the operator

$$L^{\star}u = \operatorname{div}\left(A^T(x)\nabla u\right). \tag{19.88}$$

We say that L enjoys the **Runge property** provided it occurs what follows.

Let $\Omega_1 \Subset \Omega_2$ be two open sets of $\mathbb{R}^n$, like Theorem 19.5, then for every u such that

$$Lu = 0, \quad \text{in} \quad \Omega_1$$

and for every $\varepsilon > 0$ there exists $v \in H^1(\Omega_2)$ such that

$$Lv = 0, \quad \text{in} \quad \Omega_2$$

and

$$\|u - v\|_{L^2(\Omega_1)} < \varepsilon.$$

Lax in [77] has proved that the following conditions are equivalent

(a) L enjoys Runge property
(b) $L^{\star}$ enjoys the unique continuation property.

The proof of (b) $\Longrightarrow$ (a) is analogous to the proof of Theorem 19.5 and it s left to the reader as an exercise. Concerning the implication (a) $\Longrightarrow$ (b), we refer to [77].

4. The quantitative versions of the Runge property are also of interest (especially in the stability issue of inverse problems). That is, it is of interest to estimate appropriately from above in terms of ε, the quantity

$$\|v\|_{H^{1/2}(\partial\Omega_2)},$$

(by Hadamard example, illustrated at the beginning of this section, we should expect that, in general, as ε goes to 0 we should have $\|v\|_{H^{1/2}(\partial\Omega_2)}$ goes to infinity). A result in this regard is proven in [115]. ♦

References

1. S. Agmon, Unicité et Convexité dans les Problèmes Différentiels Séminaire de Mathématiques Supérieures, No. 13 (Été, 1965) (Les Presses de l'Université de Montréal, Montreal, 1966)
2. G. Alessandrini, S. Vessella, Remark on the strong unique continuation property for parabolic operators. Proc. Am. Math. Soc. **132**(2), 499–501 (2003)
3. G. Alessandrini, L. Rondi, E. Rosset, S. Vessella, The stability for the Cauchy problem for elliptic equations. Inverse Problems **25**, 1-47 (2009)
4. G. Alessandrini, A. Morassi, E. Rosset, Size estimates. Inverse problems: theory and applications. Contemp. Math. **333**, 1–33 (2003)
5. G. Alessandrini, A. Morassi, E. Rosset, The linear constraint in the Poincaré and Korn type inequalities. Forum Math. **20**, 557–569 (2008)
6. S. Alinhac, M.S. Baouendi, A non uniqueness result for operators of principal type. Math. Z. **220**(4), 635–653 (1995)
7. F.I. Almgren Jr., Dirichlet's problem for multiple valued functions and the regularity of mass minimizing integral currents, in *Minimal Submanifolds and Geodesic*, ed. by M. Obata (North–Holland, Amsterdam, 1979), pp. 1–6
8. N. Aronszajn, A unique continuation theorem for solutions of elliptic partial differential equations or inequalities of second order. J. Math. Pures Appl. **36**(9), 235–249 (1957)
9. N. Aronszajn, A. Krzywicki, J. Szarski, A unique continuation theorem for exterior differential forms on Riemannian manifolds. Ark. Matematik **4**(34), 417-453 (1962)
10. H. Bahouri, Dépendance non linéaire des données de Cauchy pour les solutions des équations aux dérivées partielles. J. Math. Pure Appl. **66**, 127–138 (1987)
11. L. Bakri, Quantitative uniqueness for Schrödinger operator. Indiana Univ. Math. J. **61**(4), 1565–1580 (2012)
12. L. Bakri, Carleman estimates for the Schrödinger operator. Applications to quantitative uniqueness. Commun. Part. Diff. Equ. **38**(1), 69–91 (2013)
13. L. Baldassari, Stime di Carleman per l'operatore di Laplace e applicazioni allo studio della continuazione unica e della stabilità, Tesi di Laurea Magistrale in Matematica , Marzo, 2017
14. L. Baldassari, S. Vessella, A straightforward proof of Carleman estimate for second–order elliptic operator and a three-sphere inequality. J. Inverse Ill-Posed Probl. **26**(3), 411–422 (2018)
15. M. Bellassoued, J. Le Rousseau, Carleman estimates for elliptic operators with complex coefficients. Part II: transmission problems. J. Math. Pures Appl. **115**(9),127–186 (2018)

S. Vessella, *Unique Continuation Properties for Partial Differential Equations*, Birkhäuser Advanced Texts Basler Lehrbücher,
https://doi.org/10.1007/978-3-031-86366-0

16. M. Bellassoued, M. Yamamoto, *Carleman Estimates and Applications to Inverse Problems for Hyperbolic Systems* (Springer, Tokyo, 2017)
17. E. Beretta, E. Francini, S. Vessella, Size estimates for EIT problem, with one measurement: the complex case. Rev. Mat. Iberoam. **30**(2), 551–580 (2014)
18. W.M. Boothby, *An Introduction to Differentiable Manifolds and Riemannian Geometry* (Academic Press, New York, 1975)
19. H. Brezis, *Analyse Fonctionelle*. Collectanea Mathematica Appl (Maîtres, Masson Paris, 1983). H. Brezis, Analisi funzionale, Ed. Liguori, Napoli, 1986. H. Brezis, Functional Analysis, Sobolev Spaces and Partial Differential Equations (Springer, New York, 2011)
20. A.P. Calderón, Uniqueness in the Cauchy problem for partial differential equations. Am. J. Math. **80**, 16–36 (1958)
21. T. Carleman, *Les Fonctions quasi-analitiques* (Gauthier-Villars, Paris 1926)
22. T. Carleman, Sur les systèmes linéaires aux dèrivèes partielles du primier ordre à deux variables. C. R. Acad. Sci. Paris **19**, 471–474 (1933)
23. T. Carleman, Sur un problème d'unicité pour les systèmes d'èquations aux dèrivèes partielles à deux variables indépendentes. Ark. Mat. Astr. Fys. **26B**, 1–9 (1939)
24. H. Cartan, *Elementary Theory of Analytic Functions of One or Several complex variable* (Dover Publications, Inc., New York, 1995)
25. V. Checcucci, A. Tognoli, E. Vesentini, *Lezioni di topologia generale* (Feltrinelli, Milano, 1968)
26. R. Coifman, C. Fefferman. Weghted norm inequalities for maximal functions and singular integrals. Studia Math. **5L**, 241–250 (1974)
27. R. Courant, D. Hilbert, *Methods of Mathematical Physics*, vol. II (John Wiley & Sons Inc., New York 1989)
28. R. Courant, F. John, *Introduction to Calculus and Analysis*, vol. I, II (Springer, New York, 1989)
29. D.G. de Figueredo, J.P. Gossez, Strict monotonicity of eigenvalues and unique continuation. Commun. Part. Differ. Equ. **17**(1, 2), 339–346 (1992)
30. E. DiBenedetto, *Real Analysis* (Birkhäuser/Springer, New York, 2016)
31. M. Di Cristo, E. Francini, C.-L. Lin, S. Vessella, J.-N. Wang, Carleman estimate for second order elliptic equations with Lipschitz leading coefficients and jump at an interface. J. Math. Pures Appl. (9) **108**(2), 163–206 (2017)
32. M. Di Cristo, E. Sincich, S. Vessella, Size estimates of unknown boundaries with Robin–type condition. Proc. R. Soc. Edinb. Sect. A **147**(4), 727–741 (2017)
33. J. Dgundji, *Topology* (Allyn and Bacon, Inc., Boston 1966)
34. P.L. Duren, Theory of H^p spaces (Academic Press, New York, 1970)
35. L. Escauriaza, J. Fernandez Unique continuation for parabolic operator. Ark. Mat. **41**, 35–60 (2003)
36. L. Escauriaza, S. Vessella, Optimal three cylinder inequalities for solutions to parabolic equations with lipschitz leading coefficients, in *Inverse Problems Theory and Applications Contemporary Mathematics*, vol. 333, ed. by G. Alessandrini, G. Uhlmann (American Mathematical Society, Providence, 2003), pp. 79–87
37. L. Escauriaza, F.J. Fernandez, S. Vessella, Doubling properties of caloric functions. Appl. Anal. **85** (2006). Special issue dedicated to the memory of Carlo Pucci ed R Magnanini and G Talenti
38. L.C. Evans, *Partial Differential Equations* (AMS, Providence, 1998)
39. L.C. Evans, R.E. Gariepy, *Measure Theory and Fine Properties of Functions* (CRC Press, Boca Raton, 1992)
40. C. Fefferman, D.H. Phong, On positivity of pseudo–differential operators. Proc. Natl. Acad. Sci. U.S.A. **75**(10), 4673–4674 (1978)
41. G. Folland, *Introduction to Partial Differential Equations* (Princeton University Press, Princeton, 1995)

42. E. Francini, C.-L. Lin, S. Vessella, J.-N. Wang, Three–region inequalities for the second order elliptic equation with discontinuous coefficients and size estimate. J. Differ. Equ. **261**(10), 5306–5323 (2016)
43. E. Francini, S. Vessella, Carleman estimates for the parabolic transmission problem and Hölder propagation of smallness across an interface. J. Differ. Equ. **265**(6), 2375–2430 (2018)
44. E. Francini, S. Vessella, J.-N. Wang, Carleman estimate for complex second order elliptic operators with discontinuous Lipschitz coefficients. J. Spectr. Theory **12**(2), 535–571 (2022)
45. E. Francini, S. Vessella, J.-N. Wang, Propagation of smallness and size estimate in the second order elliptic equation with discontinuous complex Lipschitz conductivity. J. Differ. Equ. **343**, 687–717 (2023)
46. A.V. Fursikov, O.Y. Imanuvilov, *Controllability of Parabolic Equations*. Lecture Notes, vol 34 (Seoul National University, Seoul, 1996)
47. J. Garcìa-Cuerva, J.L. Rubio De Francia, *Weighted Norm Inequalities and Related Topics* (North-Holland, Amsterdam, 1985)
48. L. Garding, *Some Points in Analysis ant Their History*. University Lectures Series, vol. 11 (American Mathematical Society, Providence, 1998)
49. N. Garofalo, F.H. Lin, Monotonicity properties of variational integrals, Ap weights and unique continuation. Indiana Univ. Math. J. **35**, 245–268 (1986)
50. D. Gilbarg, N. Trudinger, *Elliptic Partial Differential Equations of Second Order* (Grundlehren der Mathematischen Wissenschaften vol. 224), 2nd edn. (Springer, Berlin, 1983)
51. E. Giusti, Equazioni ellittiche del secondo ordine. Quaderni dell'Unione Matematica Italiana, Ed. Pitagora, Bologna, 1978
52. A. Gorny, Contribution a l'ètude des fonctions dérivables d'une variable réelle. Acta Math. **71**, 257–268 (1939)
53. J. Hadamard, Sur le principe de Dirichlet. Bull. S. M. F. **34**, 135–138 (1906)
54. J. Hadamard, *Le problème de Cauchy et les équations aux dérivées partielles linéaires hyperboliques* (Herman Paris, 1932)
55. G.H. Hardy, J.E. Littlewood, G. Pólya, *Inequalities* (Cambridge University Press, Cambridge, 1934)
56. L. Hörmander, *Linear Partial Differential Operators* (Springer, New York, 1963)
57. L. Hörmander, Uniqueness theorem for second order elliptic differential equations. Commun. Part. Differ. Equ. **8**(1), 21–64 (1983)
58. L. Hörmander, *The Analysis of Linear Partial Differential Operators*, vol. I-IV (Springer, New York, 1985)
59. L. Hörmander, On the uniqueness of the Cauchy problem under partial analyticity assumption. *Geometrical Optics and related Topics PNLDE (32)*, ed. by F. Colombini, N. Lerner (Birkhäuser, Boston, 1997)
60. M.V. Klibanov, A. Timonov, *Carleman Estimates for Coefficient Inverse Problems and Numerical Applications* (VSP, Utrecht–Boston, 2004)
61. V. Isakov, *Inverse Source Problems*. Mathematical Surveys and Monographs, vol. 34 (American Mathematical Society, Providence, 1990)
62. V. Isakov, Carleman type estimate in an anisotropic case and applications. J. Differ. Equ. **105**, 217–238 (1993)
63. V. Isakov, *Inverse Problems for Partial Differential Equations*. Applied Mathematical Sciences, vol. 127, 2nd edn. (Springer, New York, 2006)
64. D. Jerison, C.E. Kenig, Unique continuation and absence of positive eigenvalues for Schrödinger operators. Ann. Math. (2) **121**(3), 463–494 (1985)
65. F. John, Continuous dependence on data for solutions of partial differential equations with a prescribed bound. Commun. Pure Appl. Math. **XIII**, 551–586 (1960)
66. F. John, *Partial Differential Equations* (Springer-Verlag, New York, 1982)
67. H. Koch, D. Tataru, Carleman estimates and unique continuation for second-order elliptic equations with nonsmooth coefficients. Commun. Pure Appl. Math. **LIV**, 339–360 (2001)

68. H. Koch, D. Tataru, Carleman estimates and unique continuation for second order parabolic equations with nonsmooth coefficients. Commun. Part. Differ. Equ. **34**(4), 305–366 (2009)
69. A. Kufner, O. John, S. Fučik, *Function Spaces* (Noordhoff International Publishing/Academia, Leyden/Prague, 1977)
70. I. Kukavica, Quantitative uniqueness for second order elliptic operators. Duke Math. J. **91**, 225–240 (1998)
71. C. Kuratowski, *Topologie* (Polska Akademia Nauk, Warzawa, 1958)
72. E.M. Landis, A three sphere theorem. Dokl. Akad. Nauk SSSR **148**, 277–279 (1963). Engl. trans. Soviet Math. Dokl. **4**, 76–78 (1963)
73. S. Lang, *Real Analysis*, 2nd edn. (Addison–Wesley, London, 1983)
74. C. Laurent, M. Léautaud, Quantitative unique continuation for operators with partially analytic coefficients. Application to approximate control for waves. J. Eur. Math. Soc. **21**(4), 957–1069 (2019)
75. M.M. Lavrent'ev, On the Cauchy problem for Laplace equation. Izv. Akad. Nauk SSSR. Ser. Mat. **20**, 819–842 (1956)
76. M.M. Lavrent'ev, V.G. Romanov, S.P. Šišatskiĭ, *Nekorrektnye Zadachi Matematicheskoi Fiziki i Analiza* (Nauka, Moscow, 1980)
M.M. Lavrent'ev, V.G. Romanov, S.P. Šišatskiĭ, *Problemi non ben posti in fisica matematica e analisi, Pubblicazioni dell'Istituto di Analisi Globale e Applicazioni*, vol. 12 (IAGA, Firenze, 1983) (Italian Translation)
M.M. Lavrent'ev, V.G. Romanov, S.P. Šišatskiĭ, *Ill-Posed Problems of Mathematical Physics and Analysis*. Translations of Mathematical Monographs, vol. 64 (American Mathematical Society, Providence, 1986) (Engl. Transl.)
77. P.D. Lax, A stability theorem for solutions of abstract differential equations and its application to the study of local behavior of solutions of elliptic equations. Commun. Pure Appl. Math. **9**, 747–766 (1956)
78. P.D. Lax, *Functional Analysis* (Jonh Wiley & Sons, Inc., New York, 2002)
79. J. Le Rousseau, G. Lebeau, L. Robbiano, *Elliptic Carleman Estimates and Applications to Stabilization and Controllability*, vol. I, II (Birhäuser/Springer, Cham 2022)
80. J. Le Rousseau, N. Lerner, Carleman estimates for anisotropic elliptic operators with jumps at an interface. Anal. PDE **6**(7), 1601–1648 (2013)
81. J. Le Rousseau, L. Robbiano, Local and global Carleman estimates for parabolic operators with coefficients with jumps at interfaces. Invent. Math. **183**, 245–336 (2011)
82. N. Lerner, *Metric on the Phase Space and Nonselfadjoint Psuedo–differential Operators*. Theory and Applications, vol. 3 (Birkhäuser, Basel, 2010)
83. N. Lerner, *Carleman inequalities – An introduction and More* (Springer, Berlin 2019)
84. N. Lerner, L. Robbiano, Unicité de Cauchy pour des opérateurs de type principal. J. Anal. Math **44**, 32–66 (1984)
85. A. Logunov, E. Malinnikova Quantitative for solutions of elliptic equations. *Proceedings of the International Congress of Mathematicians–Rio de Janeiro 2018*. Invited Lectures, vol. III (World Silicotic Publication, Hackensack/New York, 2018), pp. 2391–2411
86. F. Maggi, *Sets of Finite Perimeter and Geometric Variational Problems – an Introduction to Geometric Measure Theory* (Cambridge University Press, Cambridge, 2012)
87. R. Magnanini, Dispense del Corso di Analisi III, Corso di Laurea Triennale in Matematica (2015). https://web.math.unifi.it/users/magnanin/Istit/a3gsm15.pdf
88. R. Magnanini, Dispense del Corso di Istituzioni di Analisi Superiore, Laurea Magistrale in Matematica (2016). https://web.math.unifi.it/users/magnanin/Istit/DispenseIAS2016.pdf
89. E. Malinnikova, S. Vessella, Quantitative uniqueness for elliptic equations with singular lower order terms. Math. Ann. **353**, 1157–1181 (2012)
90. N. Mandache, On a counterexample concerning unique continuation for elliptic equations in divergence form. Mat. Fiz. Anal. Geom. **3**, 308–331 (1996)
91. G. Métevier, Uniqueness and approximation of solutions of first order non linear equations. Invent. Math. **82**, 263–282 (1985)

92. G. Métevier, Counterexample to Holmgren's uniqueness theorem for analytic non–linear Cauchy problems. Inv. Math. **112**, 217–222 (1993)
93. K. Miller, Stabilized numerical prolongation with poles. SIAM J. Appl. Math. **18**, 346–363 (1970)
94. K. Miller, Nonunique continuation for uniformly parabolic and elliptic equations in selfadjoint divergence form with Hölder continuous coefficients. Arch. Ration. Mech. Anal. **54**, 105–117 (1974)
95. S. Mizohata, *Lectures on Cauchy Problem* (Tata Institute of Fundamental Research, Bombay, 1965)
96. C.B. Morrey, L. Nirenberg, On the analyticity of the solutions of linear elliptic systems of partial differential equations. Commun. Pure Appl. Math. **10**, 271–290 (1957)
97. B. Muckenhoupt, Weighted norm inequalities for the Hardy maximal function. Trans. Am. Math. Soc. **165**, 207–226 (1972)
98. N. Nadirashvili, Estimation of the solutions of elliptic equations with analytic coefficients which are bounded on some set. Vestnik Moskov. Univ. Ser. I Mat. Mekh. (2), 42–46 (1979) (Russian)
99. N. Nadirashvili, Uniqueness and stability of extension of solution of an elliptic equation from a set to a domain. Mat. Notes **40**(2), 623–627 (1986)
100. J. Neças, *Direct Methods in the Theory of Elliptic Equations* (Springer-Verlag, Berlin/Heidelberg, 2012)
101. L. Nirenberg, Uniqueness in cauchy problems for differential equations with constant leading coefficients. Commun. Pure Appl. Math. **10**, 89–105 (1957)
102. L.E. Payne *Improperly Posed Problems in Partial Differential Equations* (Society for Industrial and Applied Mathematics, Philadelphia, 1975)
103. I.G. Petrovsky, *Lectures on Partial Differential Equations* (Interscience Publishers, New York, 1954)
104. L.C. Piccinini, G. Stampacchia, G. Vidossich, Equazioni differenziali ordinarie in $\mathbb{R}^n$ (Liguori Editore, Napoli, 1978). L.C. Piccinini, G. Stampacchia, G. Vidossich, Ordinary differential equations in $\mathbb{R}^n$ (Springer-Verlag, New York, 1984)
105. A. Plǐs, On non-uniqueness in Cauchy problem for an elliptic second order differential equation. Bull. Acad. Pol. Sci. Sér. Sci. Math. Astron. Phys. **11**, 95–100 (1963)
106. C. Pucci, Discussione del problema di Cauchy per le equazioni di tipo ellittico. Ann. Mat. Pura Appl. **46**, 131–153 (1958)
107. C. Pucci, *Istituzioni di Analisi Superiore* (Unione Matematica Italiana, Bologna, 2013)
108. Rakesh, Carleman estimates for second order PDEs. http://www.bcamath.org/documentos_public/courses/3._Carleman_estimates.RAKESH.pdf
109. R. Regbaoui, Unique continuation from sets of positive measure, in *Carleman Estimates and Applications to Uniqueness and Control Theory (Cortona, 1999)*. Progress in Nonlinear Differential Equation and Application, vol. 46 (Birkhäuser, Boston, 2001), pp. 179–190
110. F. Rellich, Darstellung der eigenwerte von $\Delta u + \lambda u = 0$ durch ein Randitegral. Math. Zeit. **46**, 635–636 (1940)
111. L. Robbiano, Theorème d'unicité adapté au contrôle des solutions des problèmes hyperboliques. Commun. Part. Differ. Equ. **16**, 789–800 (1991)
112. L. Robbiano, C. Zuily, Uniqueness in the Cauchy problem for operators with partial holomorphic coefficients. Inventiones Math. **131**(3), 493–539 (1998)
113. H.L. Royden, *Real Analysis* (Macmillan P. C., New York, 1988)
114. W. Rudin, *Functional Analysis* (McGraw–Hill Publishing Company, New York, 1975)
115. A. Rüland, M. Salo, Quantitative Runge approximation and inverse problems. *International Mathematics Research Notices*, vol. 20 (Advance Access Publication, Oxford, 2018/2019), pp. 6216–6234
116. X. Saint Raymond, *Elementary Introduction of Pseudodifferential Operators* (CRC Press, Boston, 1991)
117. J.C. Saut, B. Scheurer, Unique continuation for some evolution equations J. Differ. Equ. **66**, 118–139 (1987)

118. P.E. Saylor, A priori estimates for continuation problems for elliptic and principally normal differential equations. Trans. Am. Math. Soc. **139**, 95–108 (1969)
119. L. Schwartz, *Théorie des Distributions* (Hermann & Cie, Paris, 1966)
120. C.D. Sogge, A unique continuation theorem for second order parabolic differential operators. Ark. Mat. **28**, 159–182 (1990)
121. J. Sjöstrand, *Microlocal Analysis for Differential Operators–An Introduction* (Cambridge University Press, Cambridge, 1994)
122. C.D. Sogge, *Fourier Integrals in Classical Analysis*. Cambridge Tracts in Mathematics, vol. 210, 2nd edn. (Cambridge University Press, Cambridge, 2017)
123. E.M. Stein, *Singular Integrals and Differentiability Properties of Functions* (Princeton University Press, Princeton, 1970)
124. E.M. Stein, *Harmonic Analysis: Real – Variable Methods, Orthogonality and Oscillatory Integrals* (Princeton University Press, Princeton, 1993)
125. G. Talenti, Sopra una disuguaglianza integrale. Annali Scuola Normale Superiore Pisa Classe di Scienze **21**(2), 167–188 (1967)
126. L. Tartar, *An Introduction to Sobolev Spaces and Interpolation Spaces* (Springer, New York, 2007)
127. D. Tataru, Unique continuation for solutions to PDE's; between Hörmander theorem and Holmgren theorem. Commun. Part. Differ. Equ. **20**, 855–884 (1995)
128. M.E. Taylor, *Pseudodifferential Operators* (Princeton University Press, Princeton, 1981)
129. A.N. Tikhonov, V.A. Arsenin, *Solutions of Ill–Posed Problems* (John Wiley and Sons, Toronto, 1977)
130. A. Tychonoff, Théoremes d'unicité pour l'équation de la chaleur. Mat. Sb. **42**, 199–216 (1935)
131. F. Treves, *Partial Differential Equations with Constants Coefficients* (Gordon and Breach, New York, 1966)
132. F. Treves, *Linear Partial Differential Equations* (Gordon and Breach Science Publishers, Inc., New York, 1970)
133. F. Treves, *Basic Linear Partial Differential Equations* (Academic Press, New York, 1975)
134. F. Treves, *Introduction to Pseudodifferential and Fourier Integral Operators*, vol. 2 (Plenum Press, New York, 1980)
135. S. Vessella, A continuous dependence result in the analytic continuation problem. Forum Math. **11**(6), 695–703 (1999)
136. S. Vessella, Quantitative continuation from a measurable set of solutions of elliptic equations. Proc R. Soc. Edinb. **130**(4), 909–923 (2000)
137. S. Vessella, Carleman estimates, optimal three cylinder inequality and unique continuation properties for solutions to parabolic equations Commun. Part. Differ. Equ. **28**, 637–676 (2003)
138. S. Vessella, Quantitative estimates of unique continuation for parabolic equations, determination of unknown time-varying boundaries and optimal stability estimates. Inverse Problems **24**, 023001 (81pp) (2008)
139. S. Vessella, *Unique Continuation Properties and Quantitative Estimates of Unique Continuation for Parabolic Equations*. Handbook of Differential Equations (Elsevier/North-Holland, Amsterdam, 2009), pp. 421–500
140. H.F. Weinberger, *A First Course in Partial Differential Equations* (Dover, New York, 1995)
141. K. Yosida, *Functional Analysis*, 6th edn. (Springer–Verlag, Berlin, 1995)
142. C. Zuily, *Uniqueness and Non–Uniqueness in the Cauchy Problem* (Birkhäuser, Boston, 1983)

Index

Symbols

S. Vessella, *Unique Continuation Properties for Partial Differential Equations*, Birkhäuser Advanced Texts Basler Lehrbücher,
https://doi.org/10.1007/978-3-031-86366-0

The manufacturer's authorised representative in the EU is Springer Nature Customer Service Centre GmbH, Europaplatz 3, 69115 Heidelberg, Germany. If you have any concerns regarding our products, please contact ProductSafety@springernature.com

Printed and bound by CPI Group (UK) Ltd, Croydon, CR0 4YY
07/07/2026
02160925-0001